Handbook of Sensory Physiology

Volume VII/5

The Visual System in Vertebrates

By

F. Crescitelli C. A. Dvorak D. J. Eder A. M. Granda
D. Hamasaki K. Holmberg A. Hughes N. A. Locket W. N. McFarland
D. B. Meyer W. R. A. Muntz F. W. Munz
E. C. Olson R. W. Reyer

Edited by

Frederick Crescitelli

With 284 Figures

Springer-Verlag Berlin Heidelberg New York 1977

Frederick Crescitelli

Department of Biology, University of California, Los Angeles, California 90024, USA

ISBN-13: 978-3-642-66470-0 e-ISBN-13: 978-3-642-66468-7
DOI: 10.1007/978-3-642-66468-7

Library of Congress Cataloging in Publication Data Main entry under title: The Visual system in evolution in vertebrates. (Handbook of sensory physiology; v. VII/5) 1. Vision. 2. Vetebrates-Evolution. I. Crescitelli, Frederick, 1909— II. Series. QP351.H34 vol. 7, no. 5 [QP475.5] 591.1'82'08s [596:01'823] 77—4371

Softcover reprint of the hardcover 1st edition 1977

Typesetting, printing and binding: Brühlsche Universitätsdruckerei Lahn-Gießen.

2122/3120-543210

Contents

List of Contributors

CRESCITELLI, F., Department of Biology, University of California, Los Angeles, California 90024, USA

DVORAK, C. A., Institute for Neuroscience and Behavior, University of Delaware, Newark, Delaware 19711, USA

EDER, D. J., Bascom Palmer Eye Institute, University of Miami, P. O. Box, 875, Biscayne Annex, Miami, Florida 33152, USA

GRANDA, A. M., Institute for Neuroscience and Behavoir, University of Delaware, Newark, Delaware 19711, USA

HAMASAKI, D., Bascom Palmer Eye Institute, School of Medicine, University of Miami, P. O. Box 875, Biscayne Annex, Miami, Florida 33152, USA

HUGHES, A., Department of Physilogy, The John Curtin School of Medical Research, P. O. Box 334, Canberra City, A. C. T. 2601, Australia

HOLMBERG, K., Zoologiska Institutionen, Universitetet Stockholm, Box 6801, Stockholm, Sweden

LOCKET, N.A., Department of Anatomy, Institute of Ophthalmology, University of London, Judd Street, London, WC1 H 9QS, Great Britain

MCFARLAND, W.N., Division of Biological Sciences, Cornell University, Ithaca, New York 14850, USA

MEYER, D. B., Department of Anatomy, School of Medicine, Wayne State University, Detroit, Michigan 48201, USA

MUNTZ, W. R. A., Department of Biology, University of Stirling, Stirling FK9 4LA, Scotland

MUNZ, F.W., Department of Biology, University of Oregon, Eugene, Oregon, 97403 USA

OLSON, E. C., Department of Biology, University of California, 405 Hilgard Avenue, Los Angeles 90024, California, USA

REYER, R. W., Department of Anatomy, School of Medicine, West Virginia University, Morgantown, West Virginia 26506, USA

Preface

The vertebrate eye has been, and continues to be, an object of interest and of inquiry for biologists, physicists, chemists, psychologists, and others. Quite apart from its important role in the development of ophthalmology and related medical disciplines, the vertebrate eye is an exemplar of the ingenuity of living systems in adapting to the diverse and changing environments in which vertebrates have evolved. The wonder is not so much that the visual system, like other body systems, has been able to adapt in this way, but rather that these adaptations have taken such a variety of forms.

In a previous volume in this series (VII/1) Eakin expressed admiration for the diversity of invertebrate photoreceptors. A comparable situation exists for the vertebrate eye as a whole and one object of this volume is to present to the reader the nature of this diversity. One result of this diversification of ocular structures and properties is that the experimental biologist has available a number of systems for study that are unique or especially favorable for the investigation of particular questions in visual science or neurobiology. This volume includes some examples of progress made by the use of such specially selected vertebrate systems. It is our hope that this comparative approach will continue to reveal new and useful preparations for the examination of important questions.

At the same time, investigators must be cognizant of the possible danger involved in the finding of a particularly useful animal. The squandering of species for experimental purposes is no less reprehensible than such squandering for nonscientific reasons. One useful role that the comparative approach might play is to guide investigators to species that are in no danger of extinction or to species that should, in fact, be brought under control. In this endeavor experimental scientists should cooperative with animal breeders, directors of zoological gardens and others who wish to preserve our freedom to utilize experimental animals, yet who also desire to maintain the balance of nature.

Undoubtedly, there are a number of ways in which knowledge of the vertebrate eye could be organized and presented. In this volume, I have selected the more classical approach, that of phylogeny. The account begins with an outline of vertebrate history pointing to the fact that several ancestral vertebrate lines are still represented by living species. Such is the case with the cyclostomes which are presently represented by both lines: the hagfishes and the lampreys.

Chapter 2 summarizes what is known of the morphologic features of the cyclostome eye. In addition, some information is available on the visual pigments and the electroretinography of the lamprey visual system. None of this information points to the presence, in these primitive vertebrates, of a more primitive or of an ancestral vertebrate system. Two types of visual cells—long and short—occur in the lamprey eye although the relation of these to rods and cones is still a matter of opinion. In addition, the two separate visual pigment systems, based on the two vitamins A, occur in lampreys.

Chapter 3 is concerned with the eye of teleosts, especially that of deep-sea forms. Here we learn of the multitudinous adaptations that have evolved to cope with the special environment of the ocean depths: how some species have "allowed" their eyes to degenerate while others have developed such extravaganzas as secondary globes, accessory retinas, tiered visual cells, aphakic apertures, extracorneal windows, etc. Most of the information is, of necessity, morphologic in nature although it should be noted that it was the examination of the photopigments of deep-sea fishes that gave an impetous to the concept of the evolution of visual pigments in relation to the quality of environmental light. This idea is discussed in Chapter 4, which, in turn, has a statement on the biologic meaning of color vision.

Continuing phylogenetically, Chapter 5 deals with amphibian vision and the several retinal systems that exist for eliciting specific behavior responses to specific visual stimuli. The next chapter by Reyer is of broad biologic interest for it treats the special properties of the amphibian eye in growth, differentiation, and regeneration. The amphibian eye is unique in respect to these properties, and it is unlikely that we have seen the full potential in this area of investigation. As Reyer points out, the future of this field lies in the direction of molecular biology.

The story proceeds next to the reptiles and to the terrestrial mode of life. Insufficient information is available on ophidian vision so this aspect of the subject is omitted. Underwood has recently written about the morphology of the snake eye and about the visual cells. Little is known about ophidian visual pigments except that the rattlesnake has rhodopsin. Much more is known about the eyes and retinas of lizards, and these reptiles are of special interest because most of them are diurnal and possess an assembly of cones in beautifully ordered array. Lizards have tempted investigators with the hope that cone visual pigments might be extracted out of the retina, but despite some unsubstantiated claims, we cannot state that a lacertilian cone pigment has been obtained in solution.

The situation is different with geckos from whose retinas a unique system of photopigments has been extracted. Attention was called to the gecko visual cells by Walls, who suggested the hypothesis that they represent a system undergoing active phylogenetic transmutation. It needs to be emphasized once again that the transmutation hypothesis in no way weakens or reduces the importance of the duplexity theory. The latter is concerned mainly with the functional attributes of fully evolved visual cells; the transmutation hypothesis is a statement of the biologic history of these cells in the vertebrate series. Following the discovery by Denton that the scotopic sensitivity of the Tokay gecko is at longer wavelengths than is expected from a rhodopsin-based system the retinal photopigments of geckos were shown to be easily extractable and were then found to have a set of unique properties unlike those of rhodopsin. This story is told in Chapter 7. The chelonian retina is presently under intense study, especially with respect to the electrophysiologic responses of the retinal cells to light. Chapter 8 gives an account of this as well as of the nature of the photoreceptors and the visual pigments contained therein. The interactions of photoreceptors with each other as well as with horizontal cells and bipolar cells is also included as well as how these interactions are related to the chromatic mechanisms of the turtle retina.

A volume of this character would be incomplete without some discussion of the pineal system. Accordingly, Chapter 9 gives an account of some present knowledge of this system. Far from being a degenerate or a vestigal structure, the pineal is now known to contain well-differentiated visual cells with photopigment, neurones which discharge impulses afferently and a color-coded excitatory-inhibitory set of neurones. In addition to its direct visual function as a third eye the pineal is also recognized as having light-mediated neuroendocrine relations to a number of body functions. The dual roles of the extracranial and intracranial segments of the pineal system and how these vary in the cold-blooded and warm-blooded vertebrates is also discussed in this chapter. The reader who is interested is advised to couple the reading of this chapter with a study of Dodt's chapter in Volume VII/3, Part B of this Handbook series.

The avian eye is the subject matter of Chapter 10, which considers such features of avian vision as visual acuity, the ability to discriminate colors, the judgment of distance, the analysis of movements, the methods of accommodation, the role of the retinal oil droplets, and the significance of the pecten. Much of the knowledge regarding the physiology of the bird eye is derived secondarily from deductions based on morphology; nevertheless it should be recalled that the only certain information we have about extracted cone pigments has come from the retinas of the chicken and turkey. The volume closes with a chapter dealing mainly with the mammalian eye in which certain anatomic, optical, and neurologic features of the eye and retina are related to such visual functions as image quality, accommodation, visual acuity, stereoscopic vision, central and peripheral vision, etc. These relations are discussed in the context of comparative ophthalmology taking into account phylogenetic relations as well as the environment and the habits of life of the animal within this environment. One of the lessons that is repeatedly brought out in this final chapter is how uncertain or questionable are many statements about vertebrate vision that have appeared in the literature. Clearly, much remains to be learned about the vertebrate eye and the editor hopes that this volume will serve to encourage investigators to examine some of the unanswered questions.

In the preparation of this volume discouraging delays and other problems were sometimes encountered. I wish to express my appreciation to the contributors of this volume for their patience and understanding and especially to the staff of Springer-Verlag for their forbearance during the periods of these difficulties.

Los Angeles, May, 1977 FREDERICK CRESCITELLI

Chapter 1

The History of the Vertebrates

E. C. OLSON, Los Angeles, Cal. (USA)

With 13 Figures

Contents

A. Introduction

The history of the vertebrates extends back in time nearly 500 millions of years. The first records are fragmentary and scattered in the rocks of the Ordovician period (Fig. 1). After a lengthy gap from the Middle Ordovician to the Late Silurian, the record is reasonably continuous to the present time. The remains of vertebrates, with relatively rare exceptions, consist of hard parts—bones, calcified cartilage, teeth and scales. The masticatory and locomotor systems are especially well documented and a great deal of the evolutionary history of vertebrates is written in the changes of structures related to these two systems. Less well documented, but still of primary importance, are the structures involved in reception and integration of information: the brain, cranial nerves, and sensory organs. Information available for interpretation of the history is much more extensive than that supplied directly by the hard parts of the animals. Many inferences about soft anatomy can be made on the basis of the bones and teeth, and by the use of living analogues much can be learned about function, physiology, and behavior. The sediments in which fossils are preserved offer another rich source of evidence. Directly, they tell something about conditions of burial and, to some extent, about living conditions. Relationships between sediment and deposits and the geological setting of the region in which they were formed increase the utility of the fossiliferous sediments. The associations of different animals in the sediments, their relative numbers and conditions of deposition, add the dimension of ecology to interpretations. All such bits of evidence, finally, can be placed in geographical and geological relationships with other fossil-producing sites, providing a vast spatial and temporal framework in which the evolution of vertebrates can be understood.

The history presented in this chapter has been developed through the efforts of many paleontologists working over the past centuries. With acceptance of evolutionary theory somewhat over 100 years ago, the history, which until then had

Period	First appearance	Principal vertebrate features	MY[a]
Quaternary		The age of *Homo*. Extinction of large vertebrates	2
	Hominids	Radiations of "advanced" mammals, with centers on several continents	
Tertiary		Last of "ancient" lines of placental mammals and initiation of "advanced" lines	
		Beginning of radiations of placental and marsupial mammals	63
		Massive extinctions on land and sea. Archosaurs dominant on land, Osteichthyes in the seas	
Cretaceous		Plesiosaurs, mosasaurs important marine reptiles. Marsupials increase near end of period. Placental mammals present	
		Flowering plants expand rapidly during the middle part of the period	
	Eutherian mammals		130

Period	First appearance	Principal vertebrate features	MY[a]
Jurassic	Anura Birds	Dinosaurs and other archosaurs dominant on land. Lizard radiations begin. Marine crocodiles, plesiosaurs, ichthyosaurs in seas, with holostean and some teleost osteichthyan fishes	
		Triconodont, symmetrodont, docodont, pantothere, and multituberculate mammals present	
			180
	Dinosaurs	Therapsids decrease, disappear by end of period	
	Pterosaurs	Last of labyrithondont amphibians	
	Plesiosaurs		
	Ichthyosaurs		
Triassic	Turtles	Turtles, nothosaurs, placodonts present in marine deposits	
	Mammals	Lepidosaurs prominent, rhynchosaurs and early lizard radiation	
	Archosaurs		230
		Reptiles increase, captorhinomorphs and synapsids	
		Therapsids dominant in Late Permian	
Permian		Amphibians prominent, some terrestrial, many aquatic. Osteichthyes abundant in fresh waters	
			280
		Rapid diversification of amphibians, reaching peak. Chondrichthyes dominant in marine waters	
Pennsylvanian		Paleoniscoids, crossopterygians and dipnoans dominant in fresh waters. Some expansion of reptiles	
	Reptiles		310
Mississippian		Labyrinthodont and lepospondylous amphibians present but uncommon. Chondrichthyes dominant in marine waters. Osteichthyes dominant in fresh waters	
	Lepospondyls		340
	Amphibians	Labyrinthodonts at very end of period. Agnatha at their peak in fresh waters. Placoderms highly varied and dominant in marine waters	
Devonian	Chondrichthyes		
	Osteichthyes	Actinopterygians, crossopterygians and actinopterygians in fresh (?marine) waters	
			400
	Placoderms		
Silurian	Acanthodians	Radiations of fish taking place, but except at end, little recorded	
			440
Ordovician	Heterostracans	Remains of heterostracan agnathans in near shore marine beds	
	Vertebrates		510
Cambrian	Chordates?	Uncertain remains of chordates	
			600

[a] Millions of years since the beginning of the period.

Fig. 1. Resume of the principal events in the history of the vertebrates based upon appearances of new groups, expansions and contractions, and extinctions during periods of geological time little integration, has taken on a meaning which has allowed it to contribute to a much clearer understanding of the nature and dynamics of the changes that have taken place in organisms since their appearance some 3.5 billion years ago. The light that this history has cast on origins of modern vertebrates is essential to an understanding of both what they are and why they are where they are.

B. Vertebrate Beginnings

The vertebrates form the major subphylum of the phylum Chordata, a group of organisms characterized by the presence of a notochord either in the adult or larval stages. This is one of some 25 or so phyla of metazoans, but, with the echinoderms, arthropods, and molluscs, it has become dominant among the macroorganisms of the earth. All of the major phyla of Metazoa for which there is an appreciable fossil record appeared during the early part of the Cambrian period (Fig. 1) or soon thereafter, dating back to a time some 600 million years ago. A few phyla appeared a little earlier and one or two later. The phylum Chordata appears to have been in existence since the Early Cambrian, but the tangible evidence of this is scant and equivocal. Relatively few early fossils can be assigned to the phylum and to this is added the problem that opinions upon the nature of the primitive chordates and ancestors of the vertebrates differ widely.

Some students consider that the homalozoan echinoderms, or carpoids, were ancestral to the chordates. These extinct organisms have been placed in a separate phylum, Calcichordata, by JEFFRIES (1972) and associated with the chordates rather than the echinoderms. If this view is accepted, then the early record of chordates is moderately good, but many students, especially specialists on echinoderms, dismiss the idea, believing rather that the initial chordates were soft-bodied and more like the living *Amphioxus*. Somewhat uncertain traces of soft-bodied creatures of this sort have been found in the Lower Cambrian rocks. Vertebrates proper appear during the Ordovician period and when unequivocal remains are encountered they are from well-developed, fishlike creatures. This leaves little doubt that the group was in existence during the Cambrian and in time a record may be uncovered.

Except for some problematical teeth from the Baltic and some toothlike remains from the Early Ordovician of North America, the first record of vertebrates comes from the Harding Sandstone, in the Middle Ordovician of Colorado. Most of the fossils from these beds are concentrations of fragments deposited in a nearshore, marine sand dating back about 500 million years in time. Enough larger pieces have been found to show the general structure of these fishlike animals and to place them with a particular group of jawless fishes known as Heterostraci (Fig. 2). Even the mere find of bone would have been sufficient to identify these animals as vertebrates, for this tissue is found in no other animals. Other features, in particular the neural crest, are more ubiquitous among the vertebrates, but their presence in fossils can only be inferred from the hard anatomy. It is assumed that these early creatures had a notochord and vertebrae, but this too is only inferential. Most of the early jawless vertebrates, however, did have heavy dermal skeletons which has made it possible to know a good deal about their external anatomy. The armor cover has given rise to the general name of ostracoderms, a term which cuts across taxonomic divisions.

The very earliest ostracoderms come from shallow water marine rocks, but most later forms are found in fresh-water deposits. This pattern of occurrence, plus physiologic and behavioral interpretations, has led to a long-standing and unresolved controversy as to whether vertebrates originated in fresh or marine waters. The fact that they are related to the strictly marine echinoderms and the very early appearance in marine rocks suggest marine origins and the majority of students currently appear to favor this interpretation.

C. The Jawless Vertebrates: Agnatha

This small but very important group of aquatic vertebrates is composed of fishlike animals which lack jaws formed from a visceral arch. Fortunately, remnants of the Agnatha, the lampreys and hagfishes or cyclostomes, have persisted to the present time. Without them interpretations of the fossil record of the Late Silurian and the Devonian, when agnathans were fairly abundant, would be much more difficult than it is, and even with this help much remains uncertain. After the record of vertebrates from the Middle Ordovician, almost nothing more appeared until the Late Silurian, some 75 million years later. Extensive evolution took place during the interval, but almost no record is found in the rocks which were formed largely under open seas where vertebrates seem not to have lived. Most agnathans of the Late Silurian and Devonian lived in fresh or brackish waters, and if this was the case during the later part of the Ordovician and Silurian, for which few such deposits are known, chances of finding remains are poor.

Unlike the lampreys and hagfishes, which totally lack bone, most of the early agnathans had heavy external skeletons, although one obscure group, the Coelolepida was made up of soft-bodied animals with only a shagreen of dermal denticles. Except for a fossil lamprey known from the Upper Pennsylvanian there is no certain record of agnathans from the Late Devonian to the present. The relationships of the jawless vertebrates to the jawed forms, the gnathostomes, is obscure. Although the gnathostomes certainly passed through an agnathous stage, no known agnathans appear to fill this role. Rather they represent a separate radiation and, because this was very extensive, they are best placed in a separate superclass Agnatha (see classification, Table 1). The two main branches of the agnathans are differentiated most readily by the nature of the nostril. One group, the Diplorhina, had paired narial organs, whereas the other, the Monorhina, had a single structure. Presumably the source of the gnathostomes lay with the former

Table 1. A classification of the fishes

Superclass Agnatha
Class Diplorhina
Subclass Heterostraci
Subclass Coelolepida
Class Monorhina
Subclass Osteostraci
Subclass Anaspida
Subclass Cyclostomata
Superclass Gnathostomata
Class Acanthodii
Class Placodermi
Order Arthrodira
Order Antiarcha
Class Chondrichthyes
Subclass Elasmobranchii
Order Selachii (sharks)
Order Batoidea (rays)
Subclass Holocephali (chimaeras)
Class Osteichthyes
Subclass Actinopterygii (ray-fins)
Superorder Chondrostei[a]
Superorder Holostei[a]
Superorder Teleostei[a]
Subclass Dipnoi (lungfishes)
Subclass Crossopterygii (lobe-fins)
Order Rhipidistia
Order Coelacanthini

[a] The Chondrostei, which include the paleoniscoids, the Holostei, and Teleostei are given a formal, taxonomic rank. They are, however, properly grade levels of development rather than true taxonomic units which had a single source. At least the holosteans and teleosteans are polyphyletic

rather than the latter. Both groups have basic vertebrate features, but they are widely separated and evolved in very different ways. The lampreys and hagfishes are placed with the monorhines by most persons and the soft-bodied coelolepids are generally considered diplorhines.

The evolution of the agnathans is best known from the Heterostraci, which are diplorhines, and the Osteostraci, which are the monorhines (Fig. 2). Associated with the latter are the Anaspida, a small group of fusiform fishes which have an obscure history. The heterostracans had no internal skeleton so that their evolution is known only from modifications of the exoskeleton and body form. The osteostracans, on the contrary, have well-developed internal ossifications and many of the details of their cranial anatomy have been deciphered (Fig. 3). The various pathways of nerves from the sensory organs have been identified and along with them a curious suite of heavy nerves which reach the brain in the vicinity of the facial nerve. These most probably are sensory nerves and the superficial organs from which they pass were probably important environmental sensors. The other sense organs, perhaps, were less developed. It should be noted, however, that these structures have also been interpreted as "electric organs" used not for sensing but for protection.

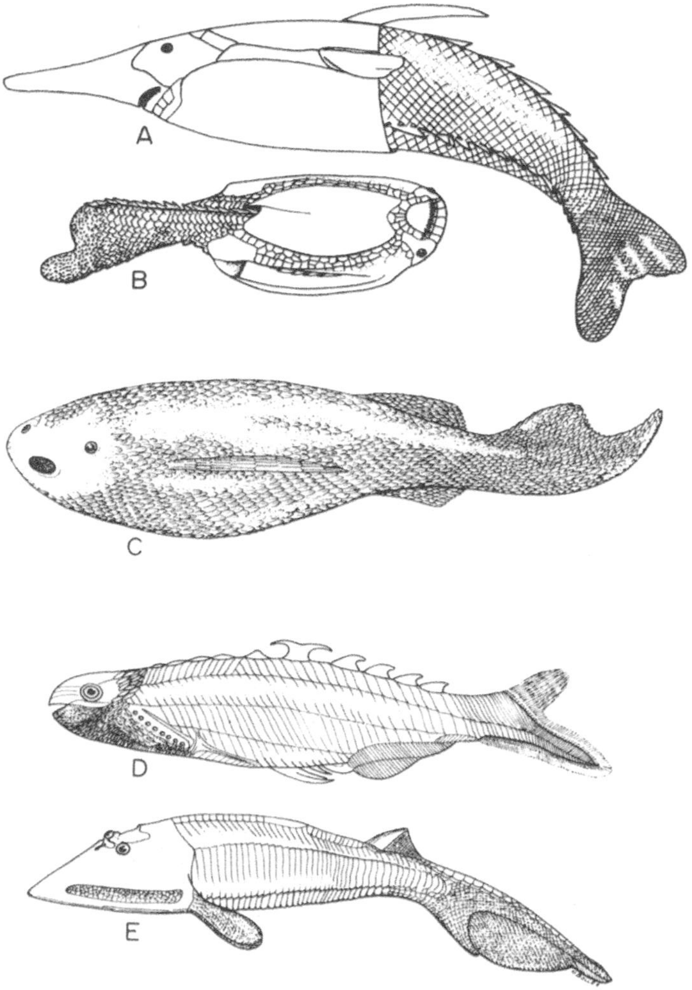

Fig. 2A—E. Representative agnathans. All are small, a few inches in length, but drawings are not to scale. (A) *Pteraspis* (after WHITE); (B) *Drepanaspis* (after HEINTZ); (C) *Phlebolepis* (after OBRUCHEV); (D) *Birkenia* (after STETSON); (E) *Hemiclaspis* (after STENSIO). (A—C) are Diplorhina and (D and E) are Monorhina

Most of the fishes, both heterostracans and osteostracans, were primarily bottom dwellers. They tended to be flat and the orbits, always small, were directed dorsally. Their lack of true jaws and the presence of various arrangements of bony plates in the oral region, suggest that some of them were bottom scavengers and "mud-grubbers." Others, such as the Anapsida, had circular mouths and filter feeding also may have been a common way of life. *Jaymoytius*, poorly known animal from the Middle Silurian (a rare record) also had this type of circular mouth and may well have been a filter feeder which lay near the ancestry of the modern lampreys.

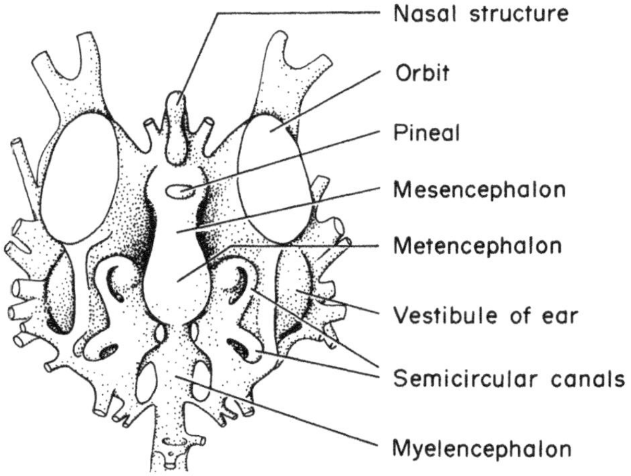

Fig. 3. Dorsal view of the reconstructed cranial cast of the monorhinan *Mimetaspis*, modified after STENSIO

After successful radiations during the Devonian, a time sometimes called the "Age of Fishes," the Agnatha no longer were an important part of the aquatic vertebrate faunas. Once the gnathostomes had come into existence and begun their radiations, the jawless fish seem to have been unable to meet the increasing competition.

D. Gnathostomes: The Fishes Proper

Transitional forms between the agnathans and gnathostomes are unknown, but embryology shows clearly that the gnathostomes did have an agnathan ancestry. Once they had come into existence, jaws became the key to all subsequent vertebrate evolution. As new potentials for feeding developed, complementary improvements in locomotion occurred. Agnathans depended largely upon median fins, the caudal fin in particular, for locomotion. No true paired fins were present. The median fins of gnathostomes were not much more efficient, but paired fins aided both in stabilization and in correcting for the downthrust of the caudal fin during forward motion. In more advanced gnathostomes the paired fins came to serve many other functions. Related to their primitive functions was the stabilization of the number of paired fins at two, one anterior and one posterior. One group of early, small gnathostomes, the acanthodians, included members with several fins, but this was exceptional.

After the origin of the gnathostomes, vertebrate evolution followed two broad evolutionary pathways, one in the water and one on the land. The aquatic radiation started first and, of course, was the source of the land radiations. The two courses show many parallels. In each efficiency in food procurement and locomotion was increased and as this took place new pulses of evolution occurred. Each repeated many of the adaptations of the radiation which it superseded, but always pushed beyond into new life zones. What is seen in the record is directly related to the evidence of skeletons and teeth, but it is possible to reconstruct behavioral, functional, and physiologic changes which lie at the heart of the evolutionary advancements.

I. Radiations of the Fishes

The first major expansion of the vertebrates took place during the Devonian period and involved both agnathans and gnathostomes. Among the latter were the acanthodians, placoderms, chondrichthyans, and all the major subdivisions of the osteichthyans (Table 1). Following the Devonian only two major groups remained prominent, Chondrichthyes and Osteichthyes, and among the latter, primarily the ray-finned fishes, the actinopterygians.

The earliest jawed fishes appeared during the Late Silurian and at that time only the placoderms and acanthodians were present. Some features of the acanthodians, especially their scales, suggest that they may lie close to the ancestry of the bony fishes, and it has been suggested that placoderms may have given rise to the sharks. In addition, some students have sought the origin of the Dipnoi, or lungfishes, among the placoderms, although they are usually placed with the osteichthyans. Finally, the chimaeras, or Holocephali, usually classified with the sharks, have been thought to have stemmed from a group of fishes, the extinct ptyctodonts, usually placed with the placoderms. Unless a good Late Silurian record is found, relationships are likely to remain uncertain.

1. Acanthodians and Placoderms

The acanthodians (Fig. 4) formed a conservative group of small fishes which persisted into the Permian and then became extinct. Their possible relationship to the osteichthyans has been noted as well as the fact that some had multiple pairs of fins, as many as seven in one genus. Some species lived in marine waters, others lived primarily in fresh water, and some acanthodians were euryhaline.

The placoderms, in contrast to the acanthodians, were a highly varied group which underwent a rapid adaptive radiation during the Devonian period. Some were but a few inches in length but others exceeded 30 feet and were dominant predators of the Devonian seas. Flat-bodied forms and mollusc feeders with crushing teeth developed and, although most were marine, some small placoderms, the antiarchs for example, lived in fresh water.

The acanthodians and the placoderms of the Late Silurian are considered by some authorities to represent a basic, early bifurcation of the two major groups of gnathostomes. One, the placoderms, thrived during the Devonian but became extinct soon after. According to this concept of two main lines, the placoderms gave rise to no other groups. The acanthodian line developed less rapidly but lay

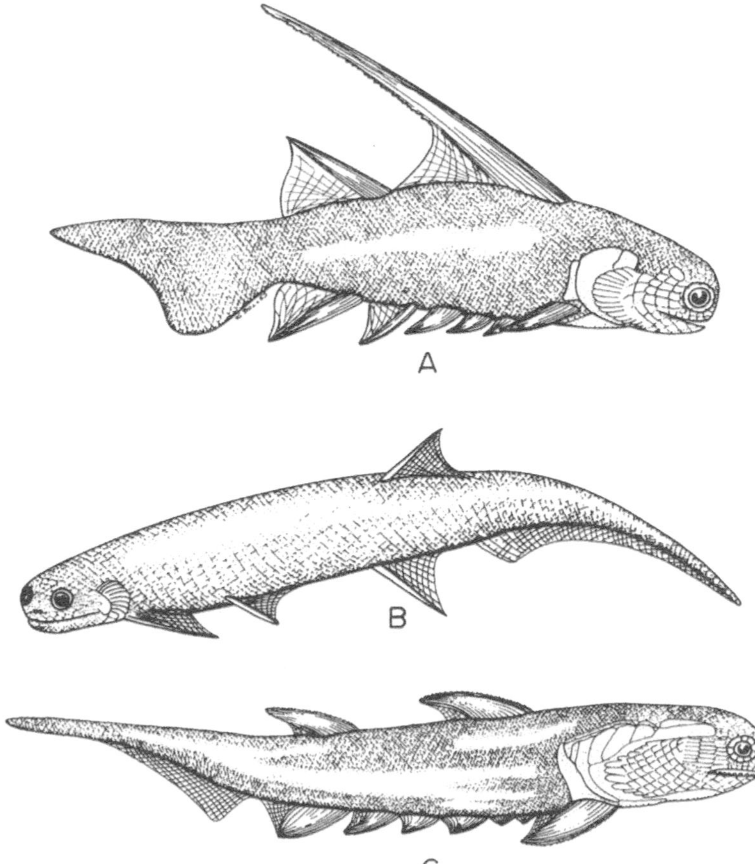

Fig. 4 A—C. Three acanthodians showing different patterns in this conservative, primitive group of gnathostomes. All small, under 6 in, but not to scale. Surface scale patterns of (A) and (B) restored. (A) *Parexus* (after WATSON); (B) *Homalacanthus* (after MILES); (C) *Climatius* (after WATSON)

at the base of all the other gnathostomes. If this concept is accepted, it places the chondrichthyans in the acanthodian line, but the position of the cartilaginous fishes is very uncertain and it is not unreasonable to suppose they may be closer to placoderms. It may well be that efforts to group and classify the early gnathostomes oversimplify the picture. Several lines may have diverged separately from a primitive Silurian stock, but if so, they did not leave any known record.

2. Chondrichthyans: Sharks, Rays, and Chimaeras

The elasmobranchs, sharks and rays, form a very coherent group which appeared unheralded in the Middle Devonian and has come through to the present with a fair degree of success. The chimaeras, on the other hand, have a rather

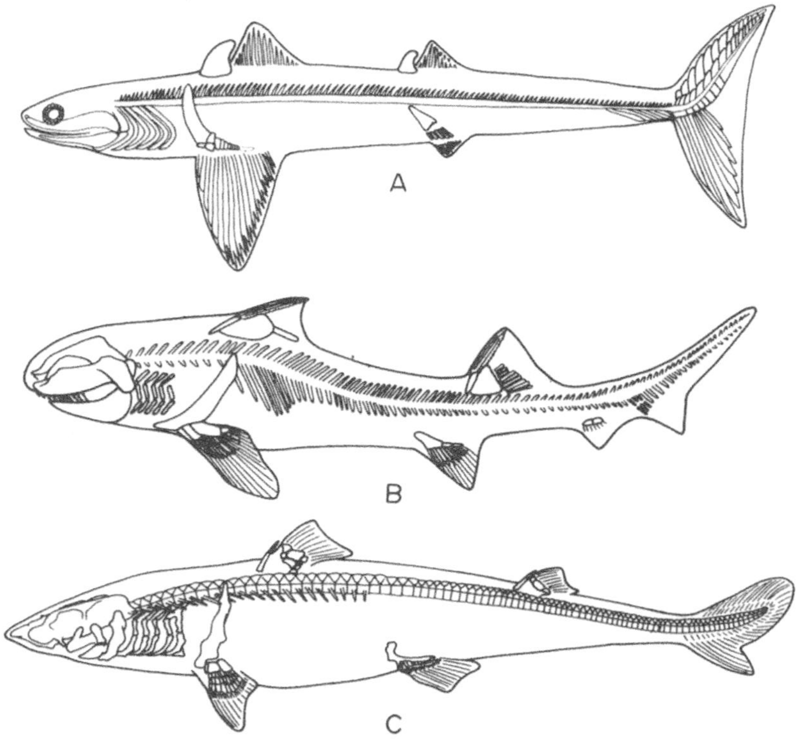

Fig. 5A—C. The three grades of chondrichthyan organization. (A) *Cladoselache*, a very primitive Paleozoic shark (after MILES); (B) *Hybodus*, representing an intermediate grade from the Mesozoic (after GREGORY); (C) *Squalus*, the dogfish shark of the present, representing the advanced grade (after WISCHNITZER). Not to scale

spotty fossil record and their evolution is not well understood. All chondrichthyans lack bone, except at the base of their dermal denticles. The fact that they are quite generalized in structure has suggested that they are very primitive gnathostomes, but the time of their appearance suggests the contrary. Although the extensive marine beds were deposited during the Early Devonian and Late Silurian, no traces of sharks have been found in them. Cartilage, of course, does not preserve well unless it is calcified, but shark teeth are easily recognized and very persistent as fossils. It has been suggested that sharks originated in fresh water and invaded the seas only during the Middle Devonian. This is possible, but many fresh water deposits are known from this time and they do not contain sharks. Whatever their source, once the chondrichthyans had appeared they left a fairly continuous record, but one which is quite spotty.

Sharks proper, the selachians, began their radiation during the Devonian, but because much of their record consists of teeth during the early and middle part of their history, the phylogeny cannot be followed in detail. Sharks progressed through a series of grades from the primitive one represented by *Cladoselache* (Fig. 5) to a late Paleozoic and early Mesozoic hybodont grade, represented by

Hybodus of the Triassic, and finally to a modern grade, traceable back to the Jurassic. Calcification of vertebrae was characteristic of the modern grade, so that its record is somewhat better. The changes represented by these grade levels are largely related to development of increasingly effective mastication and locomotor activities. The modern grade is typified by a hylostylic jaw suspension. Throughout their history the elasmobranchs showed tendencies to develop flattened, bottom-living types. During the Paleozoic this was expressed in a group of "pavement toothed" sharks, the bradydonts. During the Cretaceous, typical rays, in which the trend was carried to its fullest, developed.

The Holocephali, chimaeras or ratfish, probably arose from among the bradydonts. The record of these strange fishes, which are not abundant today, is obscure. They were present and fairly abundant and varied during the Jurassic. Some of the Paleozoic bradydonts show resemblances to them, but the links are tenuous.

3. Osteichthyes: The Bony Fishes

The bony fishes dominate the waters of the seas and continents today and have done so ever since the Devonian Period. Within the osteichthyans, the actinopterygians, or ray-fins, have undergone radiations which are crudely comparable to the adaptive radiations of the tetrapods. They have occupied almost every conceivable habitat in both fresh and salt waters, and among them the teleosts account for over 95 percent of all species of fishes living at the present time. The ray-fins appeared by the Middle Devonian, possibly in the Lower Devonian, along with the two other major groups of bony fishes, the lobefinned Crossopterygia and the lungfishes, or Dipnoi. By the end of the Devonian the latter two groups had passed their climax, but each persisted in reduced numbers and kinds.

Osteichthyes have well-ossified internal skeletons and primitively carried a heavy covering of scales. In general, scales tended to become lighter during evolution, losing their enameloid outer layers, especially as locomotor capacities improved. A second trend has been the loss of internal ossification, but the most successful osteichthyans, the teleosts, did not follow it but rather increased ossification of the vertebral column in relationship to development of rapid swimming and delicate maneuvering.

Primitively, all types of osteichthyans had lungs. These were retained in lungfishes and among rhipidistian crossopterygians provided a "preadaptation" to invasion of land and the origin of amphibians. In coelacanth crossopterygians and most actinopterygians the lungs became modified to form a hydrostatic organ, the gas bladder.

a) Dipnoi

There are the lungfish proper, represented by *Neoceratodus*, *Protopterus*, and *Lepidosiren*, which live today in Australia, Africa, and South America respectively. These genera are remnants of a once more extensive distribution, and they represent the principal types which have existed since the end of the Devonian period. After their appearance in the Early Devonian, the lungfishes underwent a modest, relatively rapid radiation. Soon specialized dentitions developed and these seem to have doomed the group to only moderate success. During the Permo-Carboniferous one group of lungfishes took up the habit of estivation, forming burrows somewhat like those of *Lepidosiren* and *Protopterus*. From this time on the lungfishes evolved but little, but because most of the fossils consist of teeth and ribs, it is difficult to judge the precise relationships between the ancient and living genera.

Lungfish, as air breathers, had a potential for invading the land and several students have argued that they were ancestral to the urodele amphibians. Neither the fossil evidence nor the evidence from embryology supports this contention, and the many resemblances of the anatomy of all amphibians argues strongly to the contrary.

b) Ray-Finned Fishes

The actinopterygians have a complex history, but one which is reasonably well known because of their well-ossified skeletons and scale-covered bodies. At the base of the radiation lay small, heavily scaled fishes, the palaeoniscoids (Fig. 6). They thrived during the Devonian and Carboniferous Periods. They were characterized by broad-based, relatively immobile paired fins, and a heterocercal tail, with the fin proper ventral to the upturned vertebral column. During propulsion by this organ, the tail tended to rise and the head to pitch downward. The anterior paired fins acted as lifts, counteracting this downward thrust, and locomotion was relatively inefficient. The paleoniscoids are grouped with a few living descendants such as sturgeons, *Acipenser*, and the Mississippi paddlefish, *Polyodon*, to form the Chondrostei, a grade rather than a valid taxonomic category.

The chondrosteans represent the first of three major waves of radiation which marked the history of the actinopterygians: the chondrosteans, holosteans, and teleosteans. Each grade shows advances in locomotor and masticatory structures, but in each there were similar adaptive radiations, producing many types of fishes adapted to many ways of life in different environments. With each new wave, the variety and number of habitats increased markedly. Each radiation, and some subradiations within each, was triggered by innovations on locomotion and feeding. Presumably, these were accompanied by changes in physiology as well, but this can only be indirectly inferred from differences in still living end members.

The teleosts, taking advantage of such modifications, underwent radiations much more extensive than those of any other fishes. Locomotor advances included ossification of the centra of the vertebral colum, lightening of scale cover, and development of a symmetrical, homocercal caudal fin, and narrow-based, highly mobile paired fins. Jaws have been highly modified, with the principal tooth-bearing elements tending to become free, mobile, and delicately controlled by intricate systems of muscles and ligaments. As a result the teleosts have become adapted to a multitude of different ways of feeding, utilizing virtually every food resource found in fresh and salt waters.

The holosteans were dominant during the first part of the Mesozoic era but were replaced largely by teleosts during the Cretaceous. They have left such descendants as *Amia* and the garpike, *Lepisosteus*. None were as progressive as the teleosts, but they included the roots of this more advanced group in two of

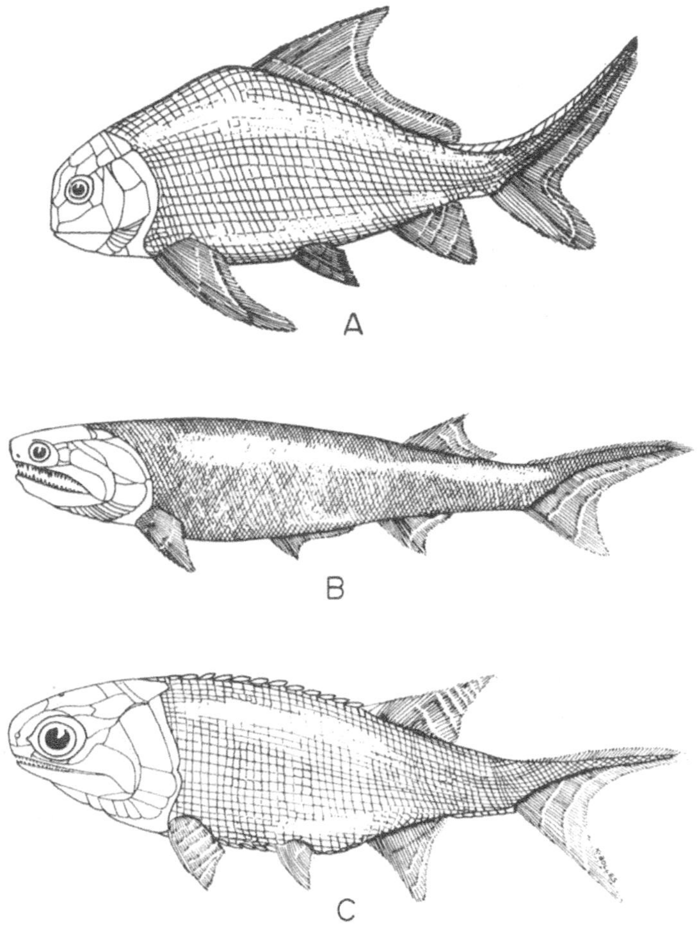

Fig. 6 A—C. Three paleoniscoid fishes showing some of the differences in patterns found in this highly varied group of primitive actinopterygians. (A) *Eurynothis;* (B) *Cheirolepis;* (C) *Stegorhachis* (all after LEHMAN). All small, but not to scale

their lineages. Among their many evolutionary accomplishments the teleosts developed air-breathing capacities in various ways. Several groups can and do go onto land for various purposes, including feeding and depositing eggs. They have a capacity for permanent invasion of the land, but in view of the competition of tetrapods, as well as various invertebrates, these probably will not be realized.

c) Lobe-Finned Fishes

The lobe-fins, or crossopterygians, had developed the ability to exploit the land at a time when there was little competition. They made the transition from water to land and gave rise to the first land animals, the amphibians. Crossopterygians were fairly "normal" fishes with fusiform bodies, heavy scales, strong jaws, and fleshy and skeletal elements in their fins. The fins were adapted to "walking" on the substrate under water and, by virtue of their lungs, these fishes appear to have been capable of living in shallow, stagnant waters where the oxygen content was low for part of the year. They were predators, probably feeding upon agnathans and paleoniscoids as well as upon invertebrate animals.

Many close resemblances between one group of crossopterygians, the rhipidistians, and the early amphibians show conclusively that the amphibian ancestry lay within this group. Both have a similar arrangement of skull bones, similar vertebrae, somewhat comparable ossifications in their paired fins or limbs, and teeth with deeply infolded enamel. The transition probably took place in the Late Devonian, for the first amphibians are found at the very end of the period. There have been many arguments about the precise source of amphibians and whether or not they originated from a single group of crossopterygians, or from two. The most compelling argument for a single origin is the consistency of the structure of the tetrapod limb throughout all vertebrates. It seems unlikely that such close resemblances in limb structure could have come from separate stocks among the fishes.

After the Devonian the crossopterygians played only a minor role in the world of vertebrates. The rhipidistians did not survive the Paleozoic and only few coelacanths, which had become marine, are known from the Mesozoic. Except for the living genus *Latimeria*, there is no record after the Cretaceous.

E. Amphibians

I. Origins and Principal Features

Much of what is known about the physiology and behavior of amphibians necessarily comes from studies of living representatives of the class, primarily from the Anura and Urodela. Naturally there is a tendency to carry this information to the fossil record and, more than we should, perhaps, to interpret fossils in the light of it. Within limits this is justified, but in some instances it may be misleading. The taxonomic distance between some of the living and fossil groups is large and the relationships are not fully understood.

The Paleozoic and Mesozoic amphibians are generally placed in two subclasses, the Lepospondyli and the Labyrinthodontia (Table 2). Some students recognize a third, Phyllospondyli, but for simplicity we will take the position that the members of this group can be distributed among the other two. The lepospondyls are mostly small forms, a few inches long, and few run over two to three feet in length. The group is characterized by the possession of spool-shaped intercentra of the vertebrae. All are Paleozoic and they reached the peak of development during the Pennsylvanian period. There is a wide variety of kinds and this subclass may in part be a "wastebasket" category into which various questionably related animals have been cast. The nature of the vertebrae gives some evidence of continuity and suggests a common ancestor.

Table 2. A classification of the Amphibia. The three living orders have been grouped as Lissamphibia, implying a common origin. It is by no means certain that this is the case. For simplicity the following assignments have been made. Ichthyostegalia are included as a suborder under Temnospondyli, although often they are ranked as an order. Anthracosaurs are divided often; they are ranked as an order. Anthracosaurs are divided into Embolomeri, which stands for several groups recognized by ROMER (1966), and the Batrachosauria including *Seymouria*, *Diadectes* and relatives, animals of somewhat uncertain relationships as discussed in the text

Class Amphibia
- Subclass Labyrinthodonia
  - Order Temnospondyli
    - Suborder Ichthyostegalia
    - Suborder Rhachitomi
    - Suborder Stereospondyli
  - Order Anthracosauria
    - Suborder Embolomeria
    - Suborder Batrachosauria
- Subclass Lepospondyli
  - Order Nectridea
  - Order Aïstopoda
  - Order Lysorophia
  - Order Microsauria
- Subclass Lissamphibia
  - Order Caudata (urodeles)
  - Order Anura (frogs and toads)
  - Order Caecilia (apodans)

The labyrinthodonts form a group which includes the dominant medium-to-large semiaquatic and terrestrial vertebrates of the Pennsylvanian or Early Permian. They continued in considerable numbers into the early part of the Jurassic period. Characteristic teeth with deeply infolded enamel are responsible for the name. This type of tooth was accompanied by vertebrae in which the centra were composed of an anterior element termed hypocentrum (intercentrum) and a posterior one, the pleurocentrum (true centrum). One or the other of these two tended to be emphasized in different groups and in some only one of the two may be present. Both within this subclass and between the two Paleozoic groups the structure of the vertebrae is important in classification, and this is true for "modern" amphibians as well.

Both lepospondyls and labyrinthodonts appeared early in the record of vertebrates, the former during the Mississippian period and the latter during the latest Devonian. When they first appeared the two groups were distinct and this is one of the reasons that amphibians have been considered to have arisen polyphyletically from the fishes. Regardless of the merit of this interpretation, which most students reject, the relationships between the two groups are not well understood. This is also true for the relationships between these amphibians and the moderns which are often placed in the subclass Lissamphibia. Recent evidence suggests that the anurans and urodeles may have arisen from the labyrinthodonts, and within this group from the family Dissorophidae, rhachitomous temnospondyls.

The amphibians reached their peak of diversity during the Pennsylvanian period, but during this period the reptiles came into prominence, severely limiting the ecological opportunities for the amphibians. The tendency among amphibians thereafter, through the Permian and Triassic, was to become increasingly aquatic and most of the late members were strictly aquatic animals. The development of frogs and urodeles during the middle and late Mesozoic began a new trend which led once again to successful terrestrial amphibians.

1. Lepospondyls

Some of the typical members of this group are shown in Fig. 7. Most of these animals were aquatic and even some of the earliest ones were limbless, presumably having lost limbs possessed by unknown ancestors. One group, called the

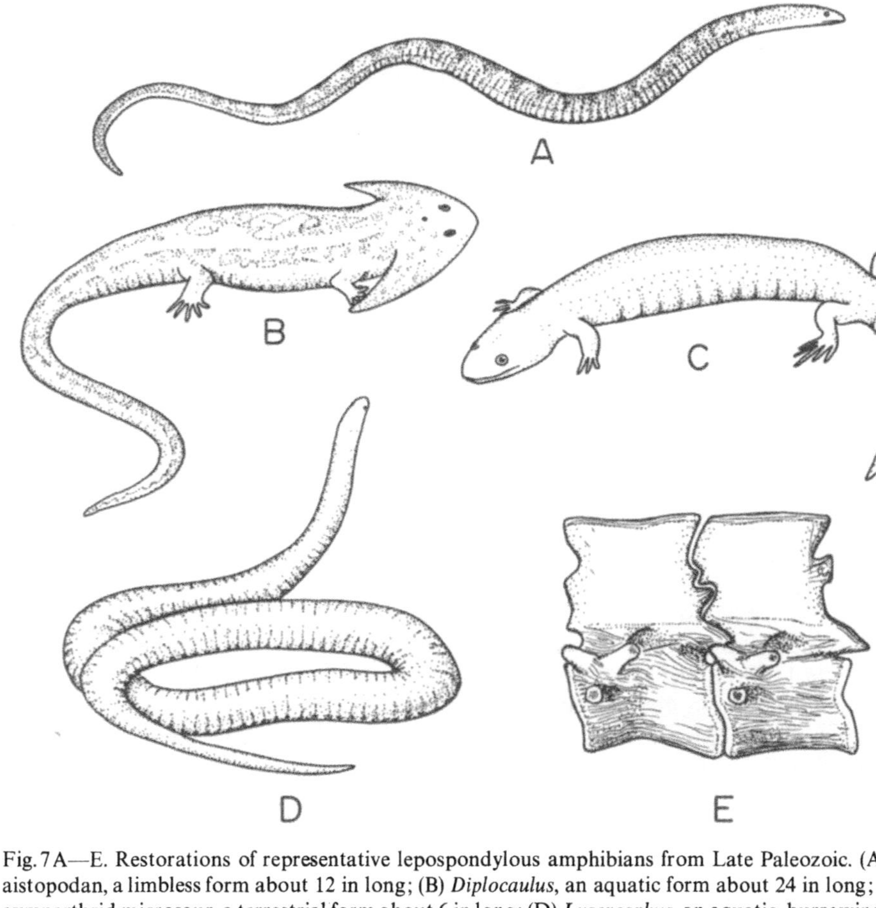

Fig. 7A—E. Restorations of representative lepospondylous amphibians from Late Paleozoic. (A) An aistopodan, a limbless form about 12 in long; (B) *Diplocaulus*, an aquatic form about 24 in long; (C) a gymnarthrid microsaur, a terrestrial form about 6 in long; (D) *Lysorosphus*, an aquatic, burrowing and aestivating animal, ranging up to about 24 in in length; (E) A lepospondylous vertebra showing spool-shaped intercentrum and also showing sutural connections between neural spines developed in this genus, *Crossotelos*

microsaurs, flourished during the Pennsylvanian and Early Permian, developing a number of salamanderlike terrestrial and semiterrestrial species. The microsaurs occupied an adaptive zone similar to that of some of the small urodeles today and they have been considered as ancestral to this group of lissamphibians.

The structure, however, is quite different, somewhat more like that of some of their contemporaries among the reptiles, with which they have been confused. The source of the caecilians has also been sought among some of the elongated microsaurs. One member of this group, *Lysorophus*, was an eellike animal which took up estivation in burrows, much like the contemporary Paleozoic lungfish, *Gnathorhiza*. Another group of lepospondyls, the Nectridea, included the odd-looking animal *Diplocaulus*, which was an immensely abundant bottom feeder during the Early Permian. Successful as many of these lepospondyls were, by the end of the Early Permian they were all but extinct and none survived far into the Late Permian.

2. Labyrinthodonts

This is a large group which includes the most familiar members of the Late Paleozoic amphibian radiation. It began at the end of the Devonian, but the major thrust of the radiation was during the Pennsylvanian and Permian. As the tabular classification shows (Table 2), the labyrinthodonts may be divided into two groups, the temnospondyls and the anthracosaurs. The former underwent the most successful amphibian adaptive radiation, but the latter gave rise to the reptiles.

a) Temnospondyls

These amphibians assumed a prominent role in the Pennsylvanian and Early Permian and they persisted into the early Jurassic. They represent the first truly successful radiation of vertebrates in the terrestrial environment. Few of them penetrated far beyond low, wet lands, suggesting that they were tied to moisture for reproduction. Most temnospondyls were carnivores and this too may have been a limiting factor, because except for insects and other small invertebrates, there were few food sources away from the water. Some temnospondyls were fairly large, for example *Eryops* which reached six to eight feet in length. Others were small, about the size of microsaurs. The most terrestrial forms are found among the Dissorophidae and some of these became quite reptilelike in many features. They represent one of several terrestrial radiations which took place during the late Paleozoic. For all their success, the terrestrial and semiterrestrial temnospondyls were unable to compete successfully with developing reptiles.

From the large rhachitimous eryopoid forms came highly aquatic stereospondyls which were successful throughout the Triassic. Many of these were large, flat-bodied forms which lived on the bottoms of lakes and streams. Their eyes were oriented dorsally and in some lines the skulls became very broad and short. Another branch of temnospondyls, the trematosaurs, became excellent swimmers and some of them appear to have invaded marine waters, something very unusual among the amphibians. These animals had long, tooth-bearing beaks and bodies well adapted to active swimming and maneuvering in open waters. Coincident with the evolution of marine reptiles, however, they became extinct, ending for once and for all this sort of amphibian adaptation.

b) Anthracosaurs

The anthracosaurs are particularly important, for they lie at the base of the reptilian radiations. Also they include a number of animals that have been variously classed as reptiles and as amphibians. By late Mississippian times the embolomerous anthracosaurs had developed many reptilian features. The vertebrae emphasized the pleurocentrum, as they do in reptiles, and many of the features of the dorsal skull roof and the brain case resembled those of reptiles. The well-known anthracosaurs are all fairly large, two or three feet in length. The anthracosaurs which we know are not themselves directly ancestral to reptiles and the real ancestors have so far escaped detection. Structurally the association is close, and certainly the beginnings of reptiles lay within this group.

A group of anthracosaurs called batrachosaurs includes animals which have found no comfortable home in any classification. Among the members are *Seymouria* and *Diadectes* of the Early Permian. Some students dispute their assignment to the amphibians. In early classifications of reptiles, after fossils had become fairly well known, the animals thought to be primitive reptiles were placed in the order Cotylosauria. This was an appropriate name, for it means stem-reptile, although not in the sense of its proposer, E.D. Cope. Within this order were three subgroups: Seymouriamorpha, Diadectomorpha, and Captorhinomorpha. Some seymouriamorphs are known to have had an aquatic phase in their development and were not amniotes. *Diadectes*, which is the heart of the Diadectomorpha, is an odd animal and neither it nor its relatives are typical reptiles. Nothing is known of its development, but some of its cranial features suggest that *Diadectes* is related to seymouriamorphs. There has been a recent tendency to class these two as anthracosaurs, that is, as amphibians, but this is not an entirely stable situation because it leaves other "diadectomorphs," procolophons and pareiasaurs which seem to be reptiles, hanging in the air. The amphibian allocation does, however, clear the decks for a more coherent consideration of the "true" reptiles.

Most interesting, aside from the always intriguing taxonomic tangle, is that *Diadectes* and its relatives and some of the seymouriamorphs are animals which invaded the land. They are thus linked ecologically with dissorophid amphibians and some of the microsaurs. At this time there were multiple colonizations of the land and presumably each of the colonizers solved the reproductive problems in its own way. The captorhinomorphs, all that now remains of the original cotylosaurs, surely developed an amniote egg, but what happened in other cases and particularly how the procolophons and pareiasaurs fit the picture remains uncertain.

F. Reptiles

I. Main Features of the Class

By definition, based on living animals, the reptiles are exothermic, amniote tetrapods. They are covered by corneal scales and lack fully developed four-chambered hearts. Many features of the skeletons and dentitions separate reptiles from amphibians, birds, and mammals, but at the transition zone between the classes, known only in the fossil record, the criteria become unclear and problems of assignment arise.

It appears that the captorhinomorphs lie at the base of all other reptiles and that this group came into existence during the early part of the Pennsylvanian period. The first amniote egg, however, is known only from the Early Permian, about 280 million years ago, long after the transition form anthracosaurs to captorhinomorphs had occurred. Presumably the amniote egg arose at the time of this transition, but exactly when and how remains to be discovered. The temporal region of the captorhinomorphs was fully roofed, a condition called anapsid. In descendant stocks several patterns of temporal fenestration developed and these were used as a basis for subdivision of major groups in early classifications. Four types were recognized: anapsid, synapsid, parapsid and diapsid (Fig. 8). Temporal fenestrae have proven only partially satisfactory criteria over the years, but remnants of the early subdivision still persist in current classifications as shown in Table 3.

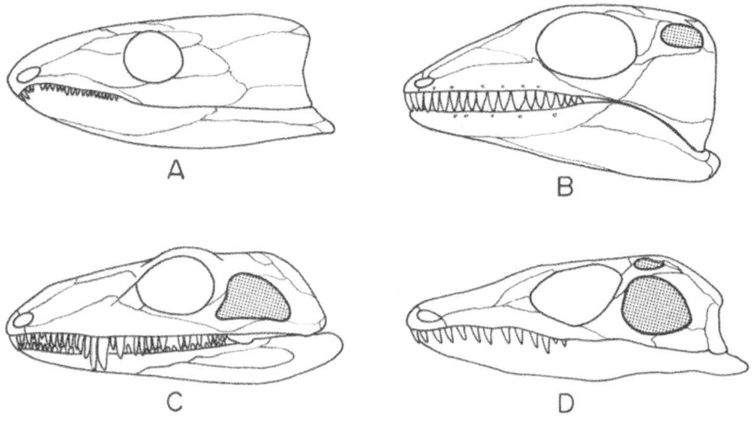

Fig. 8 A—D. Four types of temporal structure found among reptiles and initially used as a basis for classifying major groups. (A) *Anapsid* condition without a temporal fenestra, based on captorhinomorph *Captorhinus* but also found in other groups, including the Chelonia. (B) *Euryapsid* condition, with an "upper" temporal fenestra above the junction of postorbital and squamosal bones. Based on *Araeoscelis* (after VAUGHN) but also appearing in the Sauropterygia such as nothosaurs, plesiosaurs, and placodonts. (C) *Synapsid* condition, with a "lower" temporal fenestra primitive lying beneath the junction of postorbital and squamosal bones, but modified in descendant forms. Found in all of the mammallike reptiles (class Synapsida) and in the derived mammals. (D) *Diapsid* condition, with both "upper" and "lower" temporal fenestrae. Based on the primitive eosuchian *Youngina* from Late Permian and present in variously modified guises in all Lepidosauria and Archosauria

Table 3. Classification of the Reptilia. The composition of the subclasses Anapsida and Euryapsida is rather uncertain, as discussed in the text. Under Anapsida, Coltylosauria is not used, to avoid the confusion its many uses occasions. The Mesosauria are placed here, but their relationships are very uncertain. Procolophona includes procolophons, pareiasaurs, and related forms, all of uncertain taxonomic position. Chelonia, the turtles, appear in the record without any certain antecedants. They lack a temporal fenestra, and hence are best placed as Anapsida. The Euryapsida, as discussed in the text, include a group termed Araeosceloidea, a wastebasked for various reptiles which seem to have a temporal fenestra above the postorbital and squamosal and none below the junction of these bones. Relationships are unclear

- Class Reptilia
  - Subclass Anapsida
    - Order Captorhinomorpha
    - Order Mesosauria
    - Order Procolophona
    - Order Chelonia
  - Subclass Ichthyopterygia
    - Order Ichthyosauria
  - Subclass Lepidosauria
    - Order Eosuchia
    - Order Rhynchocephalia
    - Order Lacertilia
    - Order Ophidia
  - Subclass Archosauria
    - Order Thecodontia
    - Order Saurischia
    - Order Ornithischia
    - Order Pterosauria
  - Subclass Euryapsida
    - Order Araeosceloidea
    - Order Sauropterygia
      - Suborder Nothosauria
      - Suborder Plesiosauria
      - Suborder Placodontia
  - Subclass Synapsida
    - Order Pelycosauria
    - Order Therapsida

II. Patterns of Radiation

Once the amniote egg had been perfected, effective utilization of the relatively untouched resources of the continents became available to the tetrapods. A first wave of reptiles onto the continents took place during the Late Paleozoic and was coincident with the beginning of the decline of the major groups of amphibians. This first wave involved primarily the captorhinomorphs and synapsids, or, in the vernacular, the mammallike reptiles. It lasted until the end of the Triassic period. Vague roots of later radiations are found along with the two early groups, but it was not until the Triassic that their major deployments took place. The Mesozoic radiations on land were made by two diapsid groups, the lepidosaurs, lizards, snakes, and their relatives and the archosaurs, including the dinosaurs and flying reptiles. Matching these in the seas were several groups of marine reptiles. The lizards, snakes, and turtles were minor elements of the Mesozoic radiation, but beginning during the late Mesozoic they began a new pulse which flowered during the Cenozoic.

Interpretations of the course of the earliest terrestrial radiation and of the relationships of the animals involved is made difficult by the fact that the records come almost entirely from sediments formed in lowlands, mostly along the borders of the continents. What was occurring in the more upland areas is largely conjectural, indicated only by occasional animal remains washed into the lowlands and by retrospective inferences from what happened later. There is reason to suspect that most of the radiation of reptiles was taking place in the uplands and that what is found in the lowlands is a mixture of stocks which continued to exist in the ancestral habitat and migrants from the more progressive upland stocks which introgressed the lowland faunas.

Very early in their history the reptiles manifested a trend which was to be repeated over and over again, the tendency to return to aquatic habitats and to become well adapted to life in the water. Presumably the abundance of aquatic food and perhaps escape from terrestrial predators were responsible for this adaptive trend. The first striking case is the aquatic reptile *Mesosaurus* from the Late Pennsylvanian and Early Permian. This animal, which occurs in Africa and South America, was the first vertebrate to stimulate speculation about continental drift, now so fully accepted by most geologists and biogeographers.

The early synapsids, pelycosaurs, also showed aquatic trends and in turn members of almost every major subdivision followed this pattern. Not only semiaquatic reptiles such as the crocodiles evolved, but a host of highly specialized marine reptiles including some crocodiles, sea snakes, marine lizards, ichthyosaurs, the paddle-swimming plesiosaurs, and marine turtles, persistent anapsids.

Each aquatic radiation, of course, followed an initial terrestrial radiation, in which structures attained on land became basic to adaptations to aquatic life. The mammals arose from the synapsid terrestrial radiation. From another line, among the archosaurian diapsids, came the birds. The mammals and birds, like the reptiles before them, sent off branches which became successful in the seas.

1. Terrestrial Reptiles

The first captorhinomorphs, as we have noted, were a few inches in length and well adapted to terrestrial life. They became abundant and highly diversified during the Late Pennsylvanian, and produced several lines of large, anapsid reptiles, some about eight feet in length during the Permian. Initially the captorhinomorphs were invertebrates feeders, insectivores crudely, but later types were herbivores, representing one of the earliest developments of this mode of feeding among the reptiles. Probably derived from them were the large, heavy set anapsids of the Late Permian known as pareiasaurs and smaller active animals called procolophons.

Initially very close to captorhinomorphs, differing mainly in the presence of a lower temporal fossa, were the mammallike reptiles grouped as synapsids. Because of their importance in the origin of mammals we will defer consideration until the section on mammals.

The other terrestrial radiations took place primarily during the Mesozoic and resulted in the vast array of different animals which has given the name of "Age of Reptiles" to those times. The living snakes and lizards and the rhynchocephalian *Sphenodon*, plus many extinct reptiles form one large group, the diapsid subclass Lepidosauria. A major phase of the development of this subclass produced the present-day lacertilian and ophidian assemblages. The lizards began their modern radiation during the Jurassic and most modern families go well back into the record, some to Early Tertiary and some into the Mesozoic. During the Late Cretaceous the mosasaurs arose as a successful branch of marine lizards, stemming from varanidlike ancestors.

Snakes appeared during the Late Cretaceous and theirs has been a recent radiation. The source of snakes has been a matter of debate for many years. This has revolved around the distinctive nature of the eyes, which suggest derivation from a group in which there had been modification of the normal vertebrate eye. Ancestors have been sought among fossorial and nocturnal animals. On the other hand, skeletal structures have been used to support the position that the ancestry lay among the ancient varanids, in which eyes appear to have been normal.

The lepidosaurs can be traced far back in the fossil record. Animals very similar to *Sphenodon* occur in the mid-Jurassic and a peculiar side branch of the rhynchocephalians called rhynchosaurs (Fig. 9A) became adapted to mollusc feeding during the Triassic, when they were successful for a short time. At this time as well, there was an early radiation of very lizardlike animals which produced, among other types, a gliding lizard (Fig. 9B). A small suite of primitive lepidosaurs such as *Youngina* (Fig. 9C), shows the presence of this group in the middle Permian and far back in the Pennsylvanian has been found a lepidosaur diapsid called *Petrolacosaurus* (Fig. 9D).

The radiation of the archosaurian reptiles is classic, involving the dinosaurs, flying reptiles, and the crocodiles and alligators. Two general hypotheses have been advanced in explanation of the origins of the Archosauria. One places the origin among semiaquatic reptiles; the other seeks it among strictly terrestrial forebears. Both attempt to explain the predominantly bipedal tendencies within most lines of the archosaurs on the basis of the presumed early environment. The earliest representatives, from the latest Permian and earliest Triassic, have come from lowland deposits, supporting the first hypothesis. But, of course, these are the only sorts of deposits which have yielded any vertebrate fossils for this time. It has been suggested that the primitive archosaurs had roots in some of the semiaquatic, carnivorous synapsid pelycosaurs, specifically within the family Varanopsidae. The links at best are tenuous. The alternative interpretation is that the primitive, early archosaurs are side branches of a more general stock, which was basically more upland. In this event, there is little chance of finding actual ancestors except when they have been transported far from their habitats.

Although their origins are obscure, the archosaurs left an excellent record once they had become established. Best known, of course, are the two great stocks lumped popularly as dinosaurs. These include the two distinct radiations: the Saurischia, carnivores, and the giant sauropods such as *Diplodocus* and *Brontosaurus*; and the Ornithischia, including such animals as the duck-billed hadrosaurs, the horned ceratopsians, the armored ankylosaurs, and the strange stego-

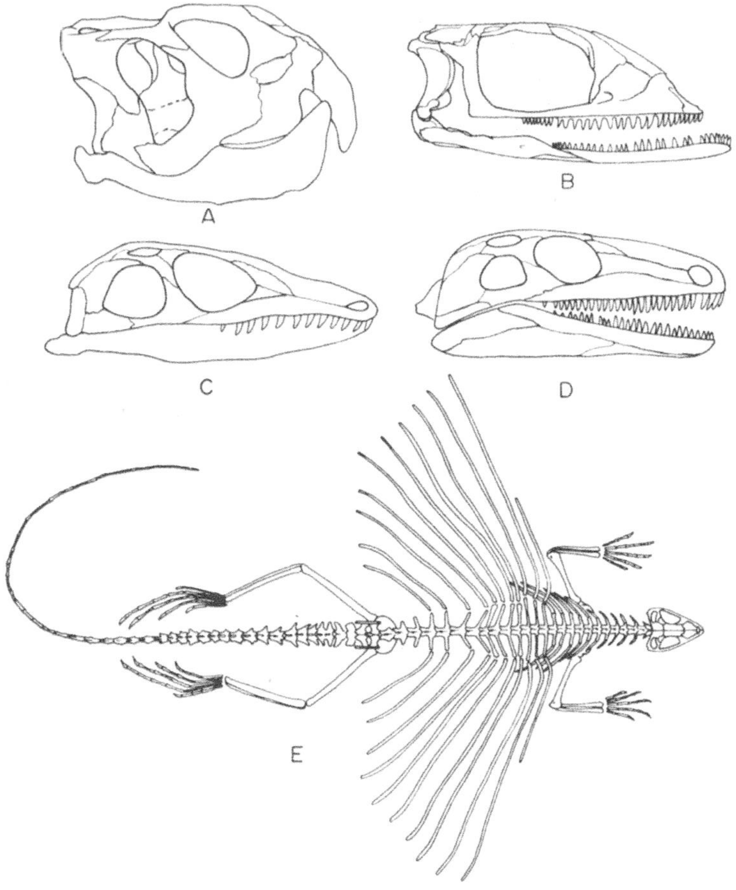

Fig. 9 A—E. Representative primitive lepidosaurs and some derived forms. (A) The extinct rhynochosaurid from the Triassic, a highly modified rhynochocephalian adapted to mollusc feeding. Skull about 12 in in length (after COLBERT). (B) Skull of extinct Triassic gliding lizard, *Kuehneosaurus*, with full skeleton shown in (E). Skull less than 1 in in length (after ROBINSON and COLBERT). (C) Skull of central, primitive lepidosaur *Youngina*, close to the base of the class, about 2 in long. (D) Skull of somewhat problematic Pennsylvania diapsid, *Petrolacosaurus*, skull about $1^1/_2$ in long (after PEABODY)

saurs with the great plates of bone down their backs and almost miniscule skulls and jaws. Along with these were many fast-running bipedal ornithischians grouped as ornithopods. Like the other ornithischians, they were basically herbivorous animals, but may have fed as well on eggs and small animals.

These two groups, as well as the crocodiles and pterosaurs, stemmed from a primitive, largely Upper Triassic assemblage known as Thecodontia. These were fast-running, small bipeds not more than a few feet in length. They spread rapidly into a wide range of environments, probably because of their ability to move rapidly and to utilize a wide range of food resources.

In recent years there has been renewed interest in the archosaurs and special attention has been directed to unravelling their behavioral and physiologic characteristics. Morphology, conditions of occurrence, and ecological considerations have supplied the main source of evidence, along, of course, with the use of modern analogues. Substantial agreement has been reached that the dinosaurs and pterosaurs were at least partially endothermic, although there are those who argue to the contrary. A fascinating problem of physiology and function involves the mechanisms of blood supply throughout the bodies of some of the giant terrestrial animals, especially supply of blood to the head and brain of the long-necked sauropods.

Debates continue on the life modes of many types. Were the giant saurpods aquatic, as was long thought to be the case? Were the duckbills aquatic and what use was made of the various crests on their highly modified skulls? Did the carnivores run or did they hop? Did some of the large dinosaurs hibernate? Certainly the pterosaurs flew and it seems that they must have been endotherms, but just how did they fly, with their large wings and their very light bodies, giving very low wing loading? Were they strictly gliders and, if so, how did they initiate flight? Most of the age-old solutions to these questions have been undergoing reexamination and so far have resulted in few definitive answers.

Of all the puzzles which surround the archosaurs, none is more fascinating than what appear to be mass extinctions at the end of the Cretaceous Period. Many kinds of terrestrial animals died out at this time and, except for the crocodiles, all the archosaurs disappeared. The birds survived, mammals survived, and lizards, snakes, and turtles continued on without appreciable loss in kinds or numbers. But in the seas none of the marine reptiles except the turtles passed into the Cenozoic. Some dropped out before the end of the Mesozoic, the ichthyosaurs and the marine crocodiles, for example. At about this time ammonite cephalopods, which had been extremely successful during the Upper Cretaceous, became extinct. Many explanations ranging from catastrophe to subtle ecology have been advanced to explain these events. Some make the assumption that all of the events are related; others treat them one by one. The ubiquitous nature of the extinctions has elicited cosmic hypotheses, involving great bursts of intense radiation, from solar explosions or reversals of magnetic fields, causing immediate extinction or gradual loss due to genetic disruption. Creeping ecological malaise, beginning with small events, perhaps climatic changes, disease, and floral changes (for various reasons) all have come in for considerations. It is a fascinating area for speculation, but so far it is not much more. With the techniques and insights developed to the present, the critical data for firm conclusions have not been found and it may well be that the critical events, or suites of events, were not recorded in such ways that we can ever more than guess. Our inability to explain the extinctions of many terrestrial animals that took place during the last 100,000 years does not auger well for success in understanding more remote events.

2. Mainly Aquatic Radiations

Two subclasses of reptiles which we have only mentioned to this point, the Euryapsida and Ichthyopterygia, include primarily marine animals. A rather puzzling fact about most marine groups, both reptiles and mammals, is that the transition from terrestrial to an aquatic way of life is not well recorded by fossils. The mesosaurs, turtles, ichthyosaurs, and mosasaurs all appear with very little ancestral background. The situation is somewhat better for plesiosaurs, where the ancestral nothosaurs are known, but we do not have well-defined ancestors for the nothosaurs. It appears that preservation along the strand line or in estuaries, where this shift presumably went on, tends to be poor. Marine reptiles tend to appear full-blown and somehow we are missing the actual and perhaps the conceptual keys as to how they originated.

The group called Euryapsida is perhaps composite, for a number of problematical, early reptiles have been placed in it along with the well-defined nothosaur-plesiosaur complex (Fig. 10 B—D). Among these problematical forms is *Araeoscelis* (Fig. 10 A), a small Permian reptile which has variously been placed with the captorhinomorphs, the lactertilians, and the synapsids, as well as the Euryapsida. It has an upper temporal fenestra, as do the better known euryapsids, the

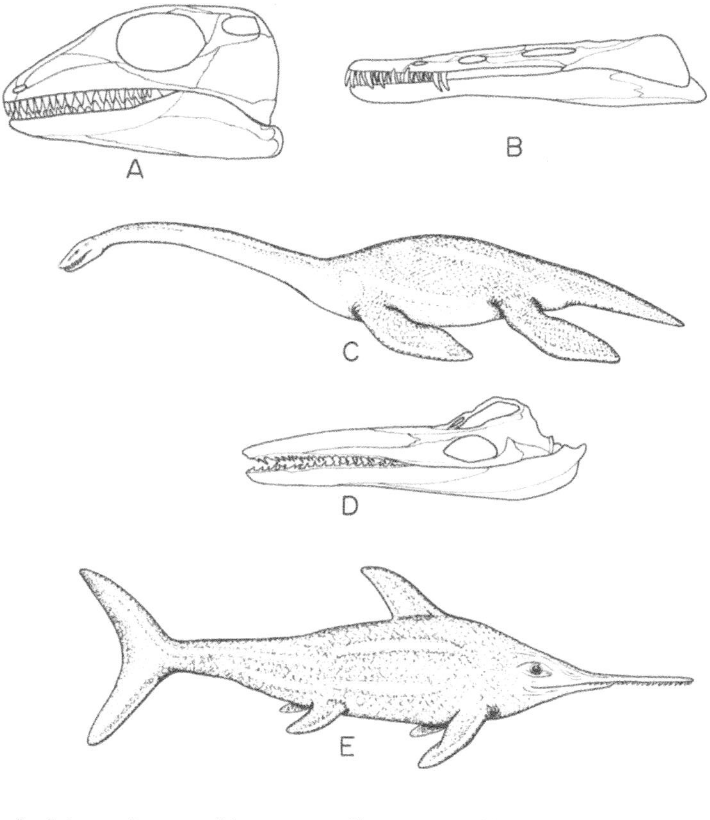

Fig. 10. (A) Skull of *Areoscelis*, a possible ancestor of later euryapsids such as nothosaurs and plesiosaurs (after VAUGHN). (B) Skull of partially aquatic nothosaur, about 10 in long (after COLBERT). (C) Reconstruction of Jurassic plesiosaur, total length up to 25 ft; (D) Skull of Triassic ichthyosaur, showing early conditions in this highly aquatic group (after GREGORY); (E) Reconstruction of highly developed, marine ichthyosaur from Jurassic, about 6 ft long

Sauropterygia. Along with some obscure genera from the Permian and Triassic, *Araeoscelis* is often placed in a "wastebasket" group termed Araeosceloidea, a suite of terrestrial and semiaquatic animals of uncertain relationships.

The euryapsid Sauropterygia fall into two well-known groups, the nothosaurs, which were partially marine, and the plesiosaurs, which were strictly so. These two represent a radiation in which swimming by the use of paddlelike limbs was carried to its highest among aquatic vertebrates, for in most lines the limbs are used more for guidance than for propulsion. The marine turtles, however, also employ their limbs for water locomotion, using the forelimbs almost as wings in the water.

The transition between the partially marine nothosaurs and the plesiosaurs is well documented, so much so that assignments of some animals to one or the other are essentially a matter of technical details of structure. The plesiosaurs were extremely successful during the Jurassic and Cretaceous after arising in the Triassic from the Middle Triassic nothosaurs. Before the mid-Triassic, there is a blank, and unless the source was among some of the araeosceloids, which is dubious, we have none. Also assigned to the Sauropterygia are the placodonts, bulbous-bodied marine reptiles with skulls and jaws highly adapted to mollusc feeding. The upper temporal opening suggests affinities with the nothosaurs and plesiosaurs, but the placodonts are different in many ways and may have been derived independently of the other euryapsid reptiles.

The most striking marine reptiles are the fishlike or porpoiselike ichthyosaurs. They were totally marine and even the most primitive, from the Upper Triassic, were little like their presumed terrestrial ancestors. This fact, of course, has spawned an immense amount of speculation about their source. Perhaps the most reasonable is that the ichthyosaurs arose from synapsids, or perhaps from captorhinomorphs as an independent line. Other suggested sources are: (1) the mesosaurs, which appears unlikely on structural grounds; (2) marine amphibians, the trematosaurs, implying a dual origin of reptiles, and, earlier, (3) even direct derivation from fishes. Ichthyosaurs, however, are definitely reptilian and derivation from anything other than reptiles stretches the concepts of parallelism or convergence beyond rational limits.

The principal structural features of a typical ichthyosaur are shown in Fig. 10E. In some ichthyosaurs the carbonized skin has been preserved indicating the nature of the body outline and the scaleless, soft body surface. The skin appears to have been more reptilian than mammalian, and probably did not have the specializations for rapid locomotion found among seals and whales. The orbits, and the eyes themselves, were remarkably large and the skull was strongly modified in accommodation to the eyes and by the presence of a long, porpoiselike beak. The paired fins were used for balance and guidance. Food was primarily fish and cephalopods, as revealed by stomach contents. The young were born alive and occasionally "embryos" have been found within the fossilized bodies of the females.

The ichthyosaurs thrived during the Late Triassic and Jurassic and continued in decreasing numbers during the Cretaceous period. Their decrease and final extinction coincided with the time of development of teleost fishes, including large, active predators, and likely was hastened by this great radiation.

The fossil record of reptiles during the Cenozoic is rather scant. The marine reptiles, except for the turtles, were gone. Strictly marine crocodiles, which existed during the Jurassic were absent, although during the early Cenozoic amphibious crocodilians were abundant and some approach dinosaurian dimensions. Lizards and snakes, being mostly small and delicate, tend not to preserve well as fossils. In addition to this, the interests of most paleontologists working on Cenozoic vertebrates have been centered on mammals, and reptilian and avian materials have not had the attention they have warranted.

Recently, increasing interest in paleoherpetology has helped to alleviate the situation and much better samples of terrestrial and fresh water reptiles of Cretaceous and Cenozoic ages are slowly being obtained. Still, both in the extinct and living faunas, the total contribution of reptiles to the full scope of tetrapod life is minute, compared to that of the birds and mammals. The glories of the Mesozoic were past, never, it would seem, to return.

G. Aves: The Birds

The broad patterns of avian evolution are fairly clear, but the details and precise times and courses of major events have yet to be worked out. The birds arose during the Jurassic period and had undergone a considerable radiation by the Late Cretaceous. Very primitive birds had by this time given way to birds not different in major features from those of today. During this radiation prior to the Late Cretaceous many, if not all, of the living orders were established, but many of the orders did undergo extensive evolution during the Cenozoic. This much is fairly clear. Problems of delving deeper into bird evolution on the basis of the fossil record stem from two sources. The first is simply the relative paucity of fossil remains. The usually small and delicate skeletons of birds are not conducive to preservation of more than fragmentary remains and life habits are such that fossilization is unlikely. The second factor is that the skeletons of most birds are generally similar and differentiation below the level of order is difficult when only single and incomplete bones are available. Also, I.e.: Because of the relative ease of finding and studying bones of reptiles and mammals, most vertebrate paleontologists working on materials from the Middle and Late Mesozoic and the Cenozoic have become paleoherpetologists or paleomammologists, while few have become paleoornithologists. Fossil bird remains often have been discarded or left to accumulate dust in some basement drawers.

In spite of these trends and difficulties, an interest in avian paleontology has been maintained from the very beginnings of vertebrate paleontology, a time when living birds already had attracted the attentions of many zoologists. Increased interest in recent years has resulted in the development of a fair fossil record and a steady increase in information about the evolution of birds. A generally accepted scheme of classification with some variants now exists. Two examples of it are shown in Table 4.

Table 4. Classification of the Aves. Two classifications are given. The structure is basically similar, but the terminology is different. BRODKORB's classification uses terms developed and used mostly during the nineteenth century, on the basis of priority. ROMER uses terms more familiar during the current century

BRODKORB (1963)

- Class Aves
  - Subclass Sauriurae
    - Order Archaeptorygiformes
  - Subclass Odontoholcae
    - Order Hesperornithiformes
  - Subclass Ornithurae
    - Infraclass Dromaeognathae
    - Infraclass Ratitae
    - Infraclass Carinatae

ROMER (1966)

- Class Aves
  - Subclass Archaeornithes (Sauriurae)
    - Order Archaeopterygiformes
  - Subclass Neornithes
    - Superorder Odontognathae (Odontoholcae)
      - Order Hesperoniformes
    - Superorder Palaeognathae
      - Including *Tinamus* and ratite orders
    - Superorder Neognathae

I. Origin of Birds

Few subjects have had more speculation than this one and few have had so little tangible evidence. Some students have argued for a polyphyletic origin of the birds, with several sources among the archosaurian reptiles. The running birds, or ratites, the penguins and the flying birds have been seen as coming from independent lines. Such hypotheses can neither be supported nor disproved by the use of fossil evidence, but the homogeneity of structure and physiology of living birds argues against a polyphyletic origin and the concept has few adherents today.

Entering into all hypotheses on the origin of birds are the Jurassic fossils placed in the genus *Archaeopteryx*. These have all come from the lithographic limestones of Solenhofen in Bavaria. Of the five specimens known, two are moderately complete, showing, among other things, the skull, including a brain case, the skeleton, and feathers. It has been argued on the basis of these remains that *Archaeopteryx* had nothing whatsoever to do with the origin of the birds, but represents an independent, feathered side branch of the archosaurs. Although *Archaeopterxy* is remote from all other birds in skeletal structures and no connection to later birds can be conclusively demonstrated, nothing bars it from an ancestral position and most students are willing to proceed on the hypothesis that this early birdlike creature does represent the early stock of the true birds.

If this is accepted, there still remains the interesting matter of whether birds originated from arboreal ancestors, which has been the popular idea for many years, or whether they originated directly from terrestrial ancestors. The usual version of terrestrial origin is one in which scales on the forelimbs became transformed into feathers as aids to balance and propulsion during rapid running, then to gliding, and finally for flight. Lately this idea has received almost no support. JOHN OSTRUM (1974), however, has made a detailed study of the problem of bird origins and has found considerable evidence in the foot structure supporting a terrestrial origin. He considers, as do many others, that feathers were initially developed in relationship to regulation of body temperature, not as original planing devices. Furthermore, Ostrum sees the origin among the primitive therapod dinosaurs rather than within some less well-defined group of primitive thecodonts. The earliest birds were in effect, small, active theropods which developed some degree of endothermy and had feathers. They probably fed on insects or perhaps small vertebrates and possibly used the feathered forelimbs as aids in gathering the food into the mouth.

This is an appealing hypothesis, and well supported by several lines of evidence. The next record of partially complete avian skeletons comes from the Upper Cretaceous some 75 million years later. By this time the birds were full-fledged avians and even their reptilian heritage had become severely masked. A few bird bones from the Lower Cretaceous show only that even then there were birds not greatly different from those of today.

II. Radiations of the Birds

The living birds, with relatively minor exceptions, can be allocated to about twenty-five orders below the superorder Neognathae. By far the greatest number of these are the passerine or perching birds of the Order Passeriformes. Some of these orders may have arisen during the Cenozoic, but at least fifteen have representatives in the Eocene and seven or eight are known from the Late Cretaceous. Essentially no intermediates between the orders are alive today and almost none which can be so identified have been found in the fossil record. Clearly the radiation of the birds of today is at a mature stage and it was, it would seem, largely well underway by the Late Cretaceous as far as orders are concerned. Below the ordinal rank, difficulties arise in coordinating the fossil evidence. Furthermore, since most of the criteria used for differentiation of living groups are not applicable to fossils, evidence from modern birds is not indicative nor have criteria in general been correlated with the osteological data which fossils provide. Slowly, owing to the painstaking studies of a few specialists, this is being corrected, but as yet nothing but the broad outlines of the history of evolution of most major groups can be understood.

Several interesting problems exist beyond those of the major radiations of the typical flying birds of today. These involve the Upper Cretaceous birds, *Ichthyornis* and *Hesperornis*, the origin and radiation of the ratites, the running birds, and the position of the penguins.

1. Upper Cretaceous Birds

Two genera, *Ichthyornis* and *Hesperornis*,first described many years ago from reasonably complete specimens, furnish some insight into conditions during the Upper Cretaceous. Both were first described as toothed birds and for this reason placed far apart from living birds. More recent studies by JOSEPH GREGORY (1952) have shown that *Ichthyornis* was not toothed and that the associated jaw upon which the toothed condition was based is almost surely that of a mosasaur reptile. This bird was rather gull- or ternlike and stood about eight inches tall. It is now generally placed with the Neornithes in a separate order fairly close to the Chardiformes which includes the gulls and auks. It was primitive, with such features as amphicoelous vertebrae, those in which the centra are biconcave. It may lie fairly close to the ancestry of the charadriiforms and, if so, is one of the rare cases in which the fossil record has produced such an ancestral type of bird.

Hesperornis is very different. It was a large, toothed bird, essentially wingless, and clearly flightless. Like *Ichthyornis* it was preserved in rocks formed in marine waters. It made its living diving for fish and had completely adapted to the aquatic way of life. BRODKORB (1963) (Table 4) placed it in a separate subclass whereas ROMER (1966), more conservatively, included it in a separate superorder under the subclass Neornithes. In any event, although *Hesperornis* is primitive in fundamental characters, as compared with living birds, it had already undergone a vast amount of evolution, losing its power of flight, assuming it had flying ancestors, and becoming adapted to a very special way of life. It had no known descendants.

These two birds give a momentary glance into the history of the Pre-Cenozoic radiations of birds.

2. Ratites

The Palaeognathae include the running birds, ratites, and the small, flying South American tinamou. All have a characteristic structure of the palate, indicated in the name Palaeognathae, and this at once distinguishes them from other birds and indicates a close relationship between the widely scattered members of the group. Ratites occur where placental carnivores did not develop, in the Australian region and South America or where these carnivores were not effective against mobile animals, as in the desert regions of North Africa and the Arabian Penninsula. Because ratites are large and do not fly, their fossil record is somewhat better than that of most groups of birds. Two extinct orders are known. One, the Aepyornithiformes, comes mainly from Madagascar, but is known from the Oligocene of Africa; the other, the Dinornithiformes, includes the great moa, *Dinornis*, which provided a food supply for man on the island of New Zealand. Most of the records go back only into the Pleistocene and Pliocene, so give little indication of the ratites.

Wherever mammalian placental carnivores are absent or ineffective, birds tend to become ground feeders and to reduce or lose their power of flight. Some of the neognathous birds have followed this course in New Zealand and in the

Americas, where *Diatryma* is known from the Eocene of North America and the giant, seven-foot-tall *Phororhachus* from the Miocene of South America. It can be logically argued that the ratites in different areas merely represent parallel adaptations to this particular mode of life. This may be the case, but the possession of the common palatal structure indicates clearly that they all must have stemmed from a common ancestral stock which had this feature. No such palate has developed in the neognathus running birds. At one time this stock may have been widespread, probably developing during the Cretaceous, perhaps before the major radiations of the neognathans began. Its remnants could have persisted as ground-living birds where adaptations to terrestrial life provided a special role which assured survival in spite of the new avian radiations.

This reconstruction of the history assumes that the ratites developed from flying ancestors, which their structure certainly suggests. If this is the case, the tinamou of South America may be a persistent remnant of the early ancestry. The tinamou is a small, somewhat chickenlike bird, which can fly, although awkwardly. Like the ratites these birds have a paleognathous palate, but unlike them the sternum is keeled, as in flying birds. If the ratites did come from such a stock, it must be assumed that features connected with flight were lost in their evolution, quite possibly in parallel in different areas.

3. Penguins

The penguins, or Sphenisciformes, have been the center of many debates. Like *Hesperornis* and the ratites the penguins have been advanced as evidence that the birds are polyphyletic and that not all had a flight stage in their history. Their fossil record is rather good for birds, but nevertheless scant. All finds have been in the Southern Hemisphere and they extend back to the Eocene in New Zealand, showing the group to be quite ancient.

Penguins have departed far from the usual bird morphology, even lacking the characteristic air sacs. Such modifications can be interpreted as adaptations to the aquatic way of life and this is usually done. They have also been used to argue that the penguins never went through a flight stage, but came directly from terrestrial ancestors. Currently, few if any serious students adhere to this concept, but the evolution of penguins does pose some biogeographic problems. They are present in areas that are widely separated by open seas. The Eocene penguins are like those living today in their basic organization and suggest that the group was in existence long before that time. Origins may have been sufficiently remote that the separated areas were then much closer together than they now are, allowing spread of the penguins from a common ancestor across lands that were not widely separated by the seas. Until the recent demonstration of the magnitude and recentness of continental drift, such an idea was unthinkable, but now it looms as a distinct possibility.

H. The Mammals

I. General Features of Mammalian History

The mammals are hairy, milk-giving endothermic vertebrates. They have much the best fossil record of any of the classes of vertebrates, being approached most closely by the Osteichthyes among the fishes. Their major record is from the Cenozoic era, a time during which there were partially separate radiations in North America, Eurasia, and Africa, where placentals dominated, and isolated radiations in Australia, by marsupials, and in South America, by a combination of placentals and marsupials. The placental mammals have been much the most successful and have largely replaced marsupials and mixed faunas wherever they have had access to their centers of development. The most recent product of placental evolution, *Homo*, has assumed a dominant role in the evolution of all groups of animals and plants.

In many respects the evolution of placental mammals repeated that of the reptiles, penetrating most of the life zones on land, in the water, and in the air. In each of these, however, the placentals carried the radiations farther, making use of a wider range of resources. Successful reptiles arose from several divergent sources and show a wide range of structural differences. The placentals, and marsupials as well, came from a single, well-stabilized ancestral source, and their

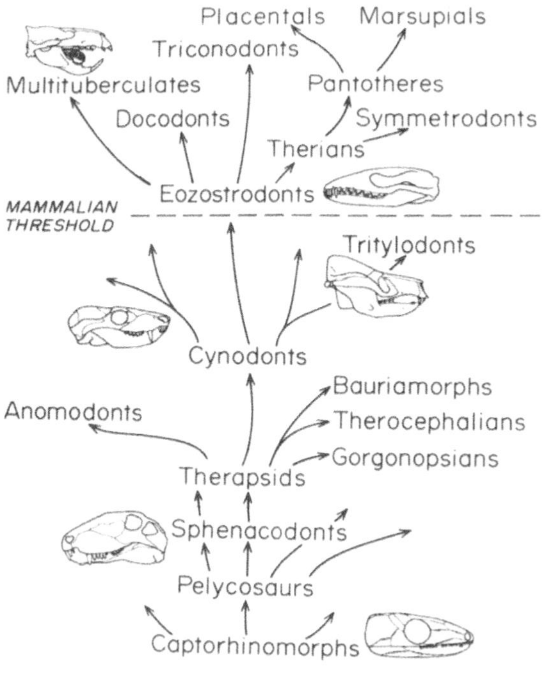

Fig. 11. General phylogentic chart of passage from primitive reptiles through synapsid reptiles to mammals

structure and physiology are similar throughout. There is great variation, but it is all based on a common ground plan. Studies of mammalian evolution have been centered much more upon dentitions than has been true for the reptiles. Teeth are not only the most universally preserved parts of mammals but in general they are excellent estimators of taxonomic levels. In addition dentitions are good indicators of dietary habits of mammals and from these other patterns of behavior can be inferred. To a major degree the paleontology of mammals has been written in the story of dental evolution.

Many features of the skulls and skeletons do, of course, differ greatly between major stocks of mammals and in different adaptive types. The skulls have offered intriguing information on the evolution of the brain; locomotor systems have been carefully analyzed, supplying, for example, critical information about the evolution of horses and the evolution of primate apes and hominids. Skeletal materials are available for most groups of Cenozoic mammals, although they are scant from very early Tertiary deposits, but they are very rare prior to the Cenozoic. During this early 100 million years or so of mammalian evolution, even skulls and jaws are rarities and studies must depend largely on the minute teeth of the primitive mammals. This time was largely before the placentals and marsupials began their radiations and it has become a somewhat special area of study of what are commonly called "mesozoic mammals." These mammals began during the Late Triassic period and only after about half of the Cretaceous period had passed did the beginnings of the true therians, the placentals, and marsupials appear in the record.

II. Origins and Early Radiations of Mammals

The beginnings of the class Mammalia are fairly well known. An outline of the course from primitive reptiles to mammals is shown in Fig. 11 and will form a guide for our discussions. Just when the mammals began is somewhat obscure, less because of the lack of data than because the transition from reptiles to mammals is well enough documented that differentiation of mammals from very mammallike reptiles poses problems. Definitions of mammals are based on living animals and involve morphology, functions, and physiology which are not preserved in the fossil record. But even were these criteria available nothing would be solved, for it is quite evident that "mammalian" characteristics did not appear all at once, full-blown, but rather arose in a mosaic pattern so that many animals, most conveniently grouped with the reptiles, had some "mammalian" features and others, which we choose to call mammals, had retained some "reptilian" features.

As there should be in evolutionary series, there are transitional forms between major groups and some of the fossils in the reptile-mammal series are of this type. Many taxonomic solutions have been proposed to take care of them. One is to call the "mammals" which have some reptilian characters "quasi-mammals." Another is to drop the boundary between reptiles and mammals deep into the synapsids, to the beginnings of the therapsids. Still another is to consider only the

Table 5. Classification of the Mammalia. Emphasis is placed on the subdivisions of the primitive, nontherian mammals, less familiar than the therians. There are many possible classifications. The one used groups the primitive forms and their derivatives under a single subclass, Prototheria, and makes the distinction between therian and nontherian mammals. Eozostrodonts and very primitive mammals with triconodont teeth may lie close to the ancestry of the therians, but the triconodonts proper, do not fall into this position. See Fig. 11

- Class Mammalia
  - Subclass Prototheria
    - Infraclass Eotheria
      - Order Eozostrodonta (Morganucodonta)
      - Order Docodonta
      - Order Triconodonta
    - Infraclass Ornithodelphia
      - Order Monotremata
    - Infraclass Allotheria
      - Order Multituberculata
  - Subclass Theria
    - Infraclass Pantotheria
      - Order Symmetrodonta
      - Order Pantotheria
    - Infraclass Metatheria (Marsupialia)
    - Infraclass Eutheria (Placentalia)

therian mammals as mammals and either put the others with the reptiles or set up a new class. No matter what is done, the transition merely is shifted from one level to another. A simple way of treating groups near this boundary is to consider that the group called Eozostrondonta (Morganucodonta) (Fig. 12) marks the beginning of mammals and that all other forms which are similar and largely at least derived from this type are also mammals, recognizing that they do have some

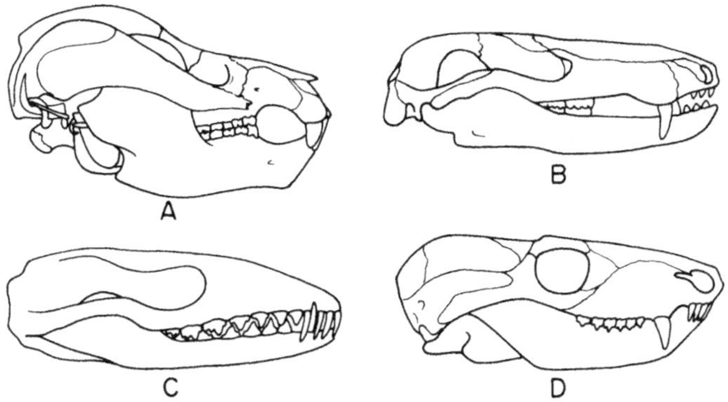

Fig. 12A—D. Skulls of advanced mammallike reptiles and very primitive mammals, all from Triassic. (A) *Oligokyphus*, a reptile with many mammalian features, about 2 in in length (after KUEHNE). (B) *Sinoconodon*, a very primitive mammal with shearing teeth and a nonmammalian type of tooth replacement. About $1^1/_2$ in long (after PATTERSON and OLSON). (C) *Morganucodon* (Eozostrodon), a very primitive mammal probably close to the ancestry of various later groups; about $^1/_2$ inch long (after RIGNEY). (D) *Thrinaxodon*, a cynodont reptile probably on the general line from which mammals arose, about 2 in long (after BRINK)

osteologic features of reptiles. This makes a fairly utilitarian separation and leaves only a few loose ends. This practice will be followed from here on.

Mammals arose directly from the synapsid, or the mammallike, reptiles. The synapsids in turn stemmed directly from primitive captorhinomorphs, so that the evolution of the synapsids and derived mammals was distinct from that of most other reptiles almost from the beginning, a time well over 300 million years ago. Possibly some other reptilian groups came from synapsids. The ichthyosaurs and archosaurs were mentioned in this regard earlier. At best the connections are tenuous. The primitive synapsids underwent a moderate radiation during the Late Paleozoic and along with the captorhinomorphs represent the first major reptilian radiation. Among them the carnivorous sphenacodonts lay close to the ancestry of some, perhaps all, of the more advanced synapsids, the therapsids.

During the Upper Permian and Triassic, the therapsids underwent a massive radiation which produced several lines of carnivores and herbivores. These lines are best known from deposits in Russia, South Africa, and South America. Therapsids persisted until well into the Triassic, at which time the "ruling reptiles" became dominant and the mammallike reptiles waned rapidly. Most of the groups of therapsids developed features that we now associate with mammals and many of these arose in parallel in diverging stocks. Some key innovation appears to be back of this extensive parallelism, and it has been suggested that incipient endothermy associated with increased activity and more efficient food procurement and preparation played this role.

Several lines, especially the cynodonts and bauriamorphs, came to resemble what we think of as mammals in many particulars. Within the cynodonts, which themselves underwent an extensive radiation, a number of lines approached the mammalian "threshold." Some end members, such as the tritylodonts, were herbivores which were extremely mammalian in general appearances (Fig. 12). Other herbivores, or omnivores, the gomphodont cynodonts, trended in somewhat different directions, but again were very mammallike.

As far as is known, the animals included in these lines did not lie directly on the course to mammals proper. This role seems to have been played by a group of insectivorous and carnivorous small cynodonts represented by genera such as *Thrinaxodon* (Fig. 12). The key to mammalian development appears to lie in the nature of the masticatory system—the jaws, the adductor musculature, and the dentitions. Only among the cynodonts did modifications of all of these structures combine to produce the complex that must have existed in the immediate ancestors of mammals. Other lines, with many mammalian features, but lacking the masticatory specializations, thrived for a period, became very mammallike, and then failed, presumably in competition with developing archosaurs and lepidosaurs.

The best evidence available today is that mammalian ancestry lay within the cynodonts. Whether just a single stock, or more than one, occupied this position has not been settled, but the tendency is to think that a very limited stock was involved.

It is now believed by many students that the animals which were ancestral to the mammals were adapted to nocturnal life, occupying niches in this habitat in which they found little competition from contemporary reptiles. The animals were extremely small, with skulls no more than one or two centimeters in length. They presumably fed on insects and other small invertebrates. Remote sensing most likely involved scent and sound, rather than a primary dependency on vision. These probably were haired animals, for if they were even partially endothermic the small size must have required very special physiologic adaptations. The fossil record of such small animals, particularly those living where there is heavy vegetation, is necessarily poor and, as far as the reptiles themselves are concerned, the predecessors of the first, very small mammals, are so far unknown.

The earliest known mammals, the eozostrodonts, were very small (Fig. 12). Some skull material is known as well as lower jaws with complete dentitions. Recently the postcranial anatomy has been described. The most primitive forms are, surprisingly, much the best known of the Mesozoic mammals, except for the odd group of multituberculates which appeared during the Jurassic period. Eozostrodonts are known from England, central Europe, China, and Africa. Just how they relate to each of the other groups of mammals is not certain, although it seems certain that they are not far from the central base of radiations, the position given in Fig. 11. On the other hand, it has been postulated that the multituberculates did not come from eozostrodonts and that the living monotremes may have had another origin. One concept is that the nontherian mammals, including such groups as triconodonts and docodonts, may have come from eozostrodonts and that the therians, at least the pantotheres, marsupials, and placentals, may have come from a different but closely related stock.

What we know of the triconodonts, symmetrodonts, and docodonts is derived largely from end members of their radiations, from times when they had departed far from whatever ancestral positions they may have occupied in the main stream of mammalian evolution. For each line, however, some more primitive members are known, if only poorly, and it is on these that the bifurcations in the diagram in Fig. 11 are based. In general a line through the eozostrodonts, primitive triconodonts, symmetrodonts, and panthotheres to the therians proper seems a reasonable one. Near the base lay the animal named *Khuneotherium* which shows trends toward the therians and can either be classed as a primitive symmetrodont or a basal pantothere. Any scheme that can be proposed today leaves a number of animals, such as *Docodon* and *Sinoconodon*, up in the air. Evolution at this level probably was much more complex than the current record indicates, but at least the general outlines are slowly emerging.

By the middle of the Cretaceous period true mammals had arisen from their pantothere ancestors. The story of their origin must be based almost entirely on teeth and jaws. Nothing is known of the skulls of panthotheres. Whether, for example, they had three ear ossicles or only one is uncertain, although it is assumed the typical mammalian condition had developed. A common ancestor to placentals and marsupials may have arisen during the Late Jurassic and by the end of the Lower Cretaceous there is fair evidence, from teeth and jaws only, that marsupials and placentals had differentiated. By the Late Cretaceous, the marsupials had undergone a modest radiation, although all of them were still small insectivorous or omnivorous animals.

Placentals, although relatively rare, seem to have diverged into insectivores, very primitive primates and condylarths, the ancestors of the great ferungulate stocks of the Cenozoic. In assessment of these assignments, it is essential to keep in mind that they can be made only on the basis of radiations that occurred long after the assignees lived. The primates, insectivores, and condylarths were all very similar, certainly related at no more than a familial level if looked at when they existed. By projecting lines of evolution of the three groups back to their beginnings, associations with different orders can be made, but this can be misleading with respect to the actual amount of divergence that the assigned animals actually has undergone.

III. The Cenozoic History of Mammals

The best known history of Cenozoic mammalian radiations is that of the placentals. The marsupials had their principal centers in the Australian region, where the record is extremely poor, and as carnivores in South America where, up to the mid-Pliocene, they left a small, but very interesting fossil record. The placental radiations were complex and well enough understood that the complexities are apparent, but not always decipherable. The continents were sufficiently separated by the time that major placental expansions began, that partially isolated evolution took place in North America, South America, Europe, central und eastern Asia. Migrations and countermigrations between these areas, alternating with times of isolation, resulted in introgressions of the established faunas. These often produced new complexes which in turn became centers of renewed expansions. Biogeography is an important aspect of the history of mammals of the Cenozoic, but biogeographic concepts are in a state of flux, especially with the implications of changing continental relationships related to the developing theory of plate tectonics.

1. The Marsupials

The principal locale of marsupials today is Australia. Small coenolestids occur in South America and opossums are spread over the New World continents. Only marsupials, plus some bats, were present in Australia until late in the Cenozoic. Just how the marsupials got to Australia, without placentals, has been a puzzle. With greater acceptance of the concept of continental drift, the idea of a southern continental deployment, from a southern center through South America, Antarctica, and Australia is receiving added attention. This contrasts sharply with the alternative hypothesis of a northern hemisphere radiation and passage to Australia from Asia by way of islands of the Malayan Straits. Under the second hypothesis, as well, marsupials gained entrance to South America from the north, again possibly by "island hopping."

Opossum-like marsupials were abundant in North America during the Late Cretaceous and present but less common in the Early Paleocene. They are hardly known from elsewhere during this time, but this could well be merely a matter, of preservation, for there are few mammal-producing beds of these ages. After the Paleocene, except for some remnants of these early forms, fossil marsupials are found in moderate abundance in South America and very sparingly in Australia. The history of Australia must be read largely from modern faunas and from the relatively few fossils that have been unearthed. The fossils indicate that, except perhaps for bats, the marsupials were the sole occupants of this region until late in the Pliocene. Then some rodents appeared, presumably from the mainland of Asia. The other placentals probably have all been introduced by man. Most of the modern stocks were differentiated by the Oligocene, and probably somewhat earlier. Gradually, intensive work is increasing the knowledge of Cenozoic mammals in Australia, but deposits are few and preservation is poor.

In South America, opossum-like didelphids and the coenolestids are found in the Paleocene and have persisted to the present. The borhyaenids are the marsupial carnivores of the mixed placental-marsupial faunal complexes of South America. This fauna dominated the continent until about the Middle Pliocene, when it was rapidly replaced by advanced placentals from the north. A variety of borhyaenids was present. Some of them were very similar to some of the carnivores of the present and recently extinct Australian faunas. Most striking is the saber-toothed marsupial, *Thylacosmilus*. As the North and South America faunas mingled, as first carnivores and then herbivores penetrated South America, the marsupial carnivores, along with most of the placental associates, became extinct.

2. The Placentals

a) North America

Placental radiations took place on each of the continents except Australia and Antarctica and each radiation has its distinctive features. Those of North America and Eurasia have many resemblances because there have been frequent interchanges. We will concentrate on the history of mammals of North America and then compare this history briefly with those of the other continents. The very early history of placentals is known almost exclusively from North America, where placentals were represented by a variety of insectivore-like mammals during the Late Cretaceous. At this time they were outnumbered many times over by the marsupials. Most of the placentals of this time can be placed in the order Insectivora, but also, as noted earlier, very primitive primates and condylarths have been found.

During the very early Paleocene much the same situation existed and then gradual differentiation produced about eight orders by the end of this epoch, the exact number depending upon taxonomic interpretations. Three or four more orders can be added to this number from other continents. In retrospect two general types of placentals can be recognized at this time. One included groups of animals which had evolved rapidly to become specialized for life in particular environments. Many were large herbivores, animals such as taeniodonts, tillodonts, and elephantine amblypods (Fig. 13), and others were creodont carnivores. Accompanying these, representing the other type, were smaller, less specialized creatures, insectivores, primates, the initial fissipede carnivores or miacids, and the basically herbivorous condylarths. As would be expected, because there had

Fig. 13 A—C. Restorations of three somewhat specialized, primitive mammals of early Tertiary radiation. (A) *Uintatherium*, from Eocene. Several feet high at shoulder. (B) *Phenacodus*, a condylarth, representing group from which various later lines of herbivorous mammals developed, but itself somewhat specialized. About 5 ft long. (C) *Ectogonus*, a taeniodont, an early side branch into herbovorous habits, specialized for feeding on harsh foods, such as roots and vines. About 4–5 ft in length. (All based on Scott)

been only moderate evolutionary divergence at this time, the taxonomy is confused and viewed quite differently by different students.

With the beginning of the Eocene two very important groups of herbivores, the perissodactyls and artiodactyls, were present and rodents appeared. Lagomorphs and rodents had arisen during the Paleocene but have not been found in North America. Perissodactyls already were diversified into five families by the

Early Eocene—Equidae, Tapiridae, Brontotheriidae, Rhinocerotidae, and Chalicotheriidae. Artiodactyls developed somewhat less rapidly, but several types were present early in the Eocene. True dogs, Canidae, came into existence during the Eocene but cats proper did not appear until the Oligocene.

The Oligocene epoch in North America represents a crossroads, with the old lines of specialized but basically primitive forms becoming extinct and the newer, more progressive lines becoming dominant. As is usual, the carnivores of the more archaic lines remained more prominent than their herbivorous counterparts, so that among the principal carnivores of the Oligocene the creodont hyaenodonts played an important role. The "progressive" lines were by no means "immune" to the phenomenon of early specialization and extinction. For example, among the perissodactyls, the large and temporarily very successful titanotheres (brontotheres), several groups of tapirs, and some of the aquatic rhinoceroses failed to survive. The later part of the Oligocene saw the beginnings of radiations that led to our present-day faunas.

The Miocene was marked by geological changes which initiated the development of grassland prairies, as the forested and semitropical conditions of the Eocene and Oligocene waned. With these changes came new radiations, producing the initial grazing animals among both the perissodactyls and artiodactyls. Carnivores followed these herbivores into the grasslands and became adapted to prairie life. Rodents, which had expanded greatly during the Eocene and Oligocene, were prominent members of these new faunas. Matching the plains radiation was continued evolution in the wooded areas, but the faunas are less well preserved and make up only a small part of the total known from the Miocene and Pliocene. By the middle of the Pliocene the North American plains radiations had reached their peaks and then the great diversification of horses, rhinoceroses, oreodonts, camels, prongbuck antelopes, and canoid carnivores slowed and then ceased. Representatives of only a few of the many lines lasted to form the faunas of the Late Pliocene and Pleistocene. During the Miocene and Pliocene new elements had been added from Europe and Asia, especially the proboscideans, and during the Late Pliocene bovids entered North America. These assumed prominent roles in the faunas and they remained important until the sharp decline in large mammals which occurred during the last part of the Pliocene. Man had arrived late, during the waning stages of the Pleistocene and how much his appearance had to do with this depletion of the faunas is still a matter of active debate.

b) Eurasia

For a time during the early part of the Tertiary, Europe and Asia appear to have been separated by a northern arm of the great Tethyan sea and this imposed a barrier to mammalian interchanges. Thus each developed a distinctive fauna. At the outset of the Tertiary, Europe was faunally similar to North America in some respects. The ancestral equid, *Hyracotherium*, for example, was present in both. European placental radiation, however, was marked by the development of semiaquatic animals and forest inhabitants. Large, amphibious rhinoceroses became characteristic of the Oligocene. Piglike artiodactyls were common, especially the anthracotheres and the equidlike palaeotheres were heavy, rhinoceroslike animals, in contrast to the small horselike equids such as *Mesohippus* of North America. Asiatic faunas had elements in part like those in North America in part European, and partly indigenous. Over that vast continent there were distinct subfaunal regions in India, southeastern and southwestern Asia, and central Asia.

After a persistent period of partial isolation, with only occasional intermingling of faunas, Asia and Europe witnessed the development of a cosmopolitan Early Pliocene assemblage known as the Pontain or *Hipparion* fauna. Each area in which this fauna was found had its own characteristics, reflecting the history of the region, but many cosmopolitan genera, such as the horse *Hipparion*, occurred throughout, from western Europe to the Pacific Ocean. During the Miocene and Pliocene, catarrhine primates, which were not to reach North America until the coming of man, appeared in deposits in Asia, Africa, and southern Europe. Monkeys were present and along with them apes of the *Dryopithecus* type which were close to the ancestry of modern apes and possibly to man as well. During this time, also, *Ramapithecus*, a presumed ancestral hominid, appeared in India and in Africa.

The dominant herbivores of the later Eurasian faunas were the bovidae, antelope, cattle, and forerunners of sheep and goats. They represent a Eurasian radiation which is a counterpart of the earlier equid-oreodont-cameloid complex of North America. One bovoid family, the Antilocapridae or pronghorned antelope, evolved in North America. Along with the bovoids, the deer and deerlike animals, cervoids, evolved. These, however, were present in both North America and the Old World and, because of the intermingling of the faunas of the two areas, the history of the parallel and intersecting cervoid faunas has been difficult to decipher. The Old World bovoid radiation was very rapid and so varied as to seem explosive. It produced many of the most successful of the large herbivores of the present world, all but replacing those with a North American heritage, most of which had evolved somewhat earlier and had begun to wane as the bovoids increased.

c) Africa

The Cenozoic of Africa has a spotty and unsatisfactory record, but it is sufficient to show that the continent was the center of a distinctive placental radiation. The Late Eocene and Oligocene phases of this evolution are recorded in beds of the Fayum region of Egypt. Thereafter scattered sites have produced isolated samples from the Miocene and fairly extensive ones from the Pliocene and Pleistocene. Recent studies related to man's evolution have enlarged the knowledge of mammals of the latest Miocene, Pliocene, and Pleistocene faunas immensely.

From the Late Eocene of the Fayum region have come remnants from faunas that lived in wet lowlands near the sea. The principal animals known belong to a suite of large herbivores often grouped together as subungulates. Proboscideans, sirenians, and hyracoids are the main constituents and some totally extinct genera accompany them. No traces of this assemblage are found elsewhere early in the Tertiary Period. Some of the subungulate types never left Africa and the hyracoids did not penetrate beyond Eurasia. The proboscidea and sirenians, however, spread rapidly and by Miocene the former had reached North America. They became prominent members of all faunas except those of Australia and South

America. As North and South America were joined during the Pliocene, proboscideans moved into South America where they enjoyed, success until their extinction during the late Pleistocene. Whales also may have had their origins in Africa, for there are many present in the Eocene beds. However, they were widespread at this time and the African concentration may be largely the result of the marine nature of many of the fossil-producing formations.

A second striking aspect of the Fayum fauna is the presence of catarrhine primates in beds of Lower Oligocene age. Finds in the Fayum region have produced the only anthropoid primates known from Early Oligocene times and from the Oligocene as a whole, except, curiously, from South America as discussed in the final section. The stocks that led to the Old World monkeys and to apes and man appear to have been separated by the time these primate-bearing rocks were deposited. The very well-known Lower Oligocene deposits from many other parts of the world have not produced anthropoids and it seems that at least the catarrhines had an African origin.

d) South America

The isolation of South America from the other continents of the world and its distinctive marsupialplacental fauna have already been mentioned. Several indigenous groups of ungulates developed on the continent from a condylarth base to produce a complex of animals similar but little related to the herbivores of other continents. Most prominent among these were the notoungulates, a large order which includes adaptive counterparts of many North American forms such as horses, deer, rhinoceroses, and even rodents. In addition there were the short-lived pyrotheres which were so like proboscideans, particularly the Old World Dinotheria, that it is hard to resist the thought that they were closely related. Other groups of South American ungulates developed somewhat camellike animals along with others which defy any efforts to find "look-alike" parallels in other faunas. These herbivores were associated with marsupial carnivores and also with a large array of distinctive placentals grouped as edentates. The last includes the armadillos, ground and tree sloths, glyptodonts, and anteaters.

The South American faunas reached a peak of diversity during the Oligocene and Miocene. Thereafter the ungulates decreased in diversity, although a few lines persisted, and the edentates became increasingly prominent in the faunas. Beginning in the Miocene and increasing during the Pliocene, invaders from North America entered the continent and came in contact with the South American faunas. The marsupials and South American ungulates, which had been waning prior to the invasion, failed to meet this competition. The migration between continents was not, however, all a one-way course. Some notoungulates penetrated well into Central America and edentates entered North America where, finding little competition, they thrived for most of the Pleistocene. The armadillos, of course, are a remnant of this invasion.

Two elements of the South American fauna, the placental rodents and primates, have excited interest and controversy. Both appeared during the Oligocene when it has been supposed that South America was well isolated from the rest of the world. Both have continued to the present day.

The rodents have the hystricomorph pattern of masseteric muscles, similar to that of the Old and New World porcupines and a number of African rodents. No rodents with this type of jaw musculature are known from the Early Tertiary of North America and this has posed some biogeographic problems. The possibility of transportation across the South Atlantic seemed so slight that it was concluded that the South American and African forms must have developed by parallelism and for this reason the two types of hystricomorphs have been separated into two suborders, the Hystricomorpha (Old World) and the Caviomorpha (New World). This became pretty well accepted. Now, however, with the new evidence on continental drift, the possibility that the two continents were much closer together during the Early Tertiary has renewed interest in the idea that these two may in fact have been derived from the common hystricomorph ancestry, probably of African origin.

The South American primates pose a somewhat different problem. They appeared first during the Upper Oligocene. This was somewhat later than the catarrhines had appeared in Africa. The South American primates are all platyrrhines, and include the cebids and the marmosets. The cebids, while very "monkeylike" have wide, flat nostrils which open forward, one more premolar tooth than the Old World monkeys, and prehensile tails. Marmosets fit the general pattern, but are less monkeylike. Once again the problem of the source of a group of animals with no indigenous predecessors arises. The idea that the Old and New World forms developed in parallel from lemuroid ancestors has been the most popular one. Again, however, the thought that the continents may have been closer together in the Early Tertiary has caused this to be questioned. But this still does not answer the questions of where the groups might have developed and what their common ancestor may have been. The earlier African primates are already somewhat more advanced than their South American counterparts, and the South American forms have specializations not found in the Old World anthropoids. Until additional discoveries have been made this problem must remain unsolved.

I. Summary Overview

We have gone over some 400 million years of vertebrate evolution briefly, bringing out the highlights of the record, the relationships of the groups of vertebrates, and the problems which the record has posed. This record shows a vast adaptive radiation, with many phases on land and in water. The major resources of the earth have been effectively utilized by the vertebrates, but during all of this history there has been little modification of the basic vertebrate structural pattern. The major organ systems, although modified, have remained intact. Each of the classes has its distinctive features and each had certain key specializations that made its radiation possible and at the same time limited the extent to which it could go. The two most distinctive features of vertebrates are the nervous system and the internal skeleton and success has depended in large part upon the utilization of these two systems by the processes of natural selection.

Major innovations led to the most recent and most successful radiations, those of the teleost fishes and the mammals. Teleosts seized primarily upon locomotor and masticatory modifications developed among the ancestral holosteans. The nervous system was relatively little modified. Mammals seized upon endothermy and brain modifications developed among their ancestors, the mammallike reptiles. The enlarged and complex brain of the placental mammals has been basic to their success. This has reached a peak in man who, during his 5 to 10 million years of hominid history, has developed properties almost nonexistent in any other group of animals. He has the ability to guide his own destiny and without question is the most influential species ever evolved with regard to affecting the course of organic evolution on the earth.

References

(Literature Cited)

Brodkorb, P.: Catalogue of fossil birds, Pt. 1. Archaeopterygiformes through Ardeiformes. Bull. Florida State Mus. Biol. Sci. **7**, 179—293 (1963).

Gregory, J.: The jaws of the Cretceous tooth birds, *Ichthyornis* and *Hesperornis*. Condor **54**, 73—88 (1952).

Jeffries, R. P. S.: The subphylum Calcichordata—primitive fossil chordates with echinoderm affinities. Bull. British Mus. (Nat. Hist.), Geology **16**, 243—339 (1968).

Ostrum, J. H.: *Archaeopteryx* and the origin of flight. Quart. Rev. Biol. **49**, 27—47 (1974).

Romer, A. S.: Vertebrate Paleontology, 3rd ed. Chicago: Univ. Chicago Press, 1966.

References for Additional Reading

Colbert, E. C.: Evolution of the Vertebrates, 2nd ed. New York: John Wiley & Sons, 1969.

Kurtén, B.: The Age of Mammals. New York: Columbia Univ. Press 1972.

Moy-Thomas, J. A.: Paleozoic fishes, 2nd ed., rev. by R. S. Miles. Philadelphia: W. B. Saunders Co. 1971.

Olson, E. C.: Vertebrate Paleozoology. New York: Wiley Interscience Series 1971.

Stahl, B. J.: Vertebrate History: Problems in Evolution. New York: McGraw-Hill Book Co. 1974.

Chapter 2

The Cyclostome Retina

KAJ HOLMBERG, Stockholm (Sweden)

With 11 Figures

Contents

A. Introduction

Hagfishes and lampreys are vertebrates classified systematically as Agnatha (jawless) or Cyclostomata (round-mouths). The rest of the vertebrates (fishes, amphibians, reptiles, birds, and mammals) belong to the Gnathostomata (jaw-possessing). Most research on retinas of vertebrates has been carried out on species which belong to the gnathostomes and this has in turn led to generalizations which are not always valid for the agnaths.

Since there is extensive anatomical and paleontologic evidence (e.g., JARVIK, 1964) to state, that agnaths and gnathostomes have evolved independently of each other, and that they represent two different lines of evolution within the vertebrates, one may ask if the retinas of hagfishes and lampreys have followed a different evolutionary line than those of gnathostomes. There are, indeed, observations which indicate that the retinas of recent agnaths, in some respects, differ from those of gnathostomes, and these differences may be due to the phylogenetic separation of agnaths from gnathostomes. Further research on the retinas of hagfishes and lampreys may reveal still more dissimilarities to the gnathostomes.

A characteristic feature of the hagfish eye, regardless of species so far examined, is the high degree of degeneration. The eyes are relatively small and invisible from the outside. Eye muscles, cornea, lens, and iris are always absent and, in certain species, the vitreous body as well. This degeneration of the eyes will be discussed with regard to the environmental conditions in which the animals live and in terms of degenerative evolution. Similarities to cave-living vertebrates are also pointed out.

An unsolved problem concerning the lamprey retina is the nature of the photoreceptor cells. Although the receptor cells in all lampreys so far studied are of two types which differ in the size of the inner segment, and thus in the position of the outer segment, it has hitherto not been possible to determine with certainty if the receptor cells are rods and/or cones. The conflicting conclusions reported in the literature may be due to the fact that criteria for distinguishing rods from cones in gnathostomes are not always valid for characterizing photoreceptor cells in agnaths. The problem is further discussed in this chapter on the basis of electron microscopic observations on the retina of the river lamprey (*Lampetra fluviatilis*) (ÖHMAN, 1971, 1974; HOLMBERG and ÖHMAN, 1976; ÖHMAN, 1976).

In this chapter emphasis will be on the retina. The reader who wishes detailed information on the morphology of the eyes of hagfishes is referred to the works of J. MÜLLER (1837), W. MÜLLER (1874), KRAUSE (1886), KOHL (1892), RETZIUS (1893), and DÜCKER (1924), and with regard to lampreys, the works of DÜCKER (1924), WALLS (1942), DUKE-ELDER (1958), and KLEEREKOPER (1972) should be consulted.

B. The Hagfish Retina

I. Eye Morphology

It was noted in the introduction that there are differences between hagfish genera with regard to the degree of degeneration of the eyes. The early studies demonstrated that the eyes in the two families of hagfishes, Eptatretidae and Myxinidae, can be classified into two types which coincide with the two families. The eyes of Eptatretidae (e.g., *Eptatretus stouti*, *E. burgeri*, *E. cirrhatus*, and *Paramyxine atami*) are larger than those of Myxinidae (e.g, *Myxine glutinosa* and *M. garmani*), ranging approximately between 0.8 and 1.5 mm in diameter depending on species, whereas the eyes of *Myxine* are about 0.5 mm in diameter. In the Eptatretidae, the eye is situated immediately beneath an unpigmented patch of skin which stands out distinctly against the surrounding pigmented skin. There is no orbit, but the eye lies in a simple connective tissue capsule which is embedded in adipose tissue except for the portion of the eye which comprises the opening of the eye cup. In the Myxinidae, on the other hand, there are no external signs as to the occurrence of eyes. The skin is uniformly pigmented and the eye is located deep under overlying muscle layers adjacent to the nasal capsule. In contrast to the eyes of Eptatretidae, those of Myxinidae are not embedded in adipose tissue. In both families the eyes lack a true cornea, lens, and iris. In the Eptatretidae there is a distinct vitreous body, whereas in the Myxinidae there is no vitreous body, but the retina fills almost the entire eyecup.

These morphologic characters indicate that the degeneration of the eyes in the Eptatretidae is less pronounced than in the Myxinidae. Now, is this difference also observable at the light- and electron microscopic level with regard to the organization of the retina and is the retina in the different species degenerated to such an extent that a function of the organ is out of question? Only a few electrophysiologic studies (KOBAYASHI, 1963, 1964) have been carried out on the retina of hagfishes, and with regard to the ultrastructure, only the receptor cells, the epithelial cells, and the optic nerve have been studied (HOLMBERG, 1969, 1970, 1971, 1972). The results do, however, suggest that the retina of hagfishes should have the capability to react on light stimulation.

II. The Organization of the Retina

It is important to emphasize that the conventional terms for the different layers in the gnathostome retina, e.g., the outer and inner nuclear layers, the outer and inner synaptic layers, will not be used in the description of hagfish retinas in order to avoid premature conclusions as to homologies between the hagfish retina and the gnathostome retina. Instead, the terms "layer of cell bodies" and "fibrous layer" will be used.

The retina in species belonging to the family Eptatretidae (e.g., *E. stouti* and *P. atami*) (Fig. 1) is differentiated into a receptor cell layer and two fibrous layers separated by a clearly defined inner layer of cell bodies. The nuclei of the receptor

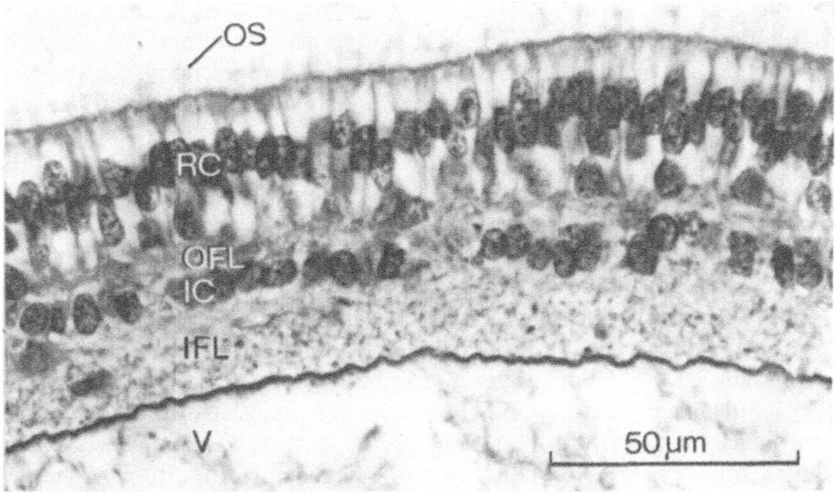

Fig. 1. Longitudinal section through retina of hagfish *E. stouti*. *OS*, outer segments of receptor cells; *RC*, receptor cell layer; *OFL*, outer fibrous layer; *IC*, inner layer of cell bodies; *IFL*, inner fibrous layer; *V*, vitreous body, ×660

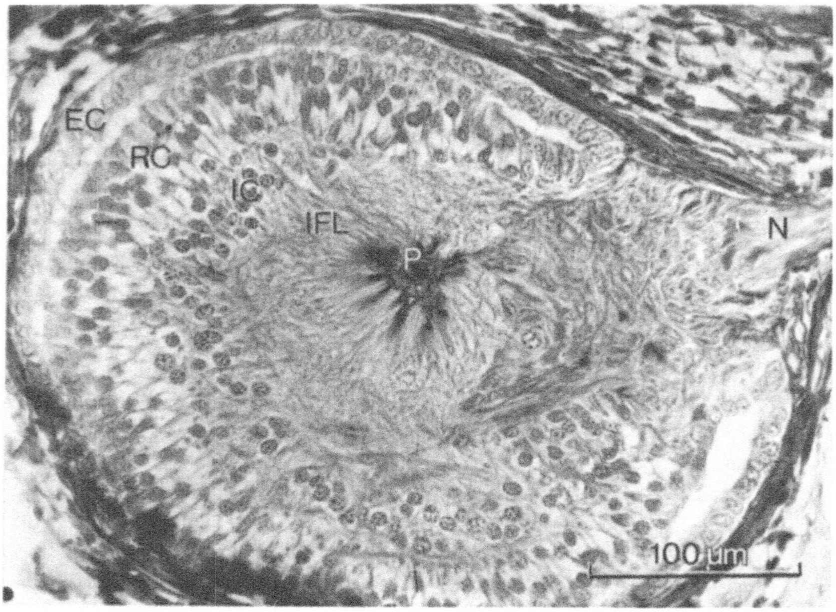

Fig. 2. Median section through eye of hagfish *M. glutinosa*. *EC*, epithelial cell layer; *RC*, receptor cell layer; *IC*, inner layer of cell bodies; *IFL*, inner fibrous layer; *P*, connective tissue plug, which enters eye laterally and extends to center of eyecup; *N*, optic nerve. Since there is no vitreous body, optic cup is practically filled by retina, ×280

cells are located approximately at the same level except in *E. cirrhatus*, in which the receptor cell layer seems to be differentiated into two nuclear layers (Fernholm and Holmberg, 1975). The outer segments are clearly recognized in the ventricular space between the epithelial cell layer and the retina. The inner fibrous layer, with few scattered cell bodies, extends to the vitreous body, where it ends in a distinct limiting membrane. The epithelial cells in hagfishes, regardless of species, do not contain pigment granules.

In the Myxinidae (e.g., *Myxine glutinosa*) the retina (Fig. 2) differs from that in the Eptatretidae in two respects. First, since there is no vitreous body, the cavity of the eye cup is practically filled by the retina with the exception of a connective tissue plug which enters the eye laterally and extends to the center of the eyecup. Thus, the inner fibrous layer is relatively thicker in *Myxine* than among the Eptatretidae. Second, the inner layer of cell bodies in *Myxine* is not clearly defined as is the case in the Eptatretidae. Therefore, the outer fibrous layer is sometimes difficult to delimit and may also be absent.

The connective tissue plug in *Myxine* has been suggested to be a rudiment of the vitreous body (e.g, Kohl, 1892) but the occurrence of this characteristic element is more likely a consequence of the degenerate condition of the eye (cf. p. 58). Within the connective tissue plug capillaries enter into the eyecup but otherwise the retina of *Myxine* is avascular. Among the Eptatretidae, only *E.*

stouti has a vascular retina, whereas in other species the choroidea seems to be the only blood supply.

A main question concerning the hagfish retina is whether the organ is capable of detecting light and of conveying nerve impulses to the brain through the optic nerve. The following account is based on earlier electron microscopic investigations on *E. stouti*, *E. burgeri*, *P. atami* and *M. glutinosa* (HOLMBERG, 1969, 1970, 1971, 1972; HOLMBERG and ÖHMAN, 1976).

III. The Receptor Cells

The receptor cells of *Eptatretus* and *Paramyxine* have cylindrical outer segments built up by numerous membranous discs arranged perpendicularly to the long axis of the outer segment (Fig. 3). The majority of the outer segments are in contact with the apical surface of the epithelial cells from which slender villi extend between the outer segments into the ventricular space. The outer segments are approximately 10 µm long and 3 µm in diameter. In *Myxine* (Fig. 4), the outer segments usually consist of whorls of membranes, 3–5 µm in diameter, which often give a disorganized impression. The arrays of membranes are frequently separated by wide spaces which may contain small vesicles. The outer segment of *Myxine* thus gives a clear impression of being degenerated to such an extent that the characteristic arrangement of outer segment membranes into well-ordered discs is no longer seen.

The outer segment is joined to the inner segment by a connecting piece which contains the usual arrangement of 9+0 microtubule doublets. The apical region of the inner segment is characterized by an aggregation of mitochondria (the "ellipsoid") which in *Eptatretus* (Fig. 3: E) and *Paramyxine* is more distinct and delimited than in *Myxine* (Fig. 4: E). At the level of the "ellipsoid", junctions between receptor cells and interposed glial cells are characteristically located where the receptor cells are slightly constricted. Basal to this constriction the receptor cells enlarge, reaching their widest extent in the nuclear region, beyond which the cell tapers off toward the receptor terminal.

From an ultrastructural point of view, the receptor terminal with its different components gives the most suggestive indications of a functional state (Figs. 5–7). Several invaginations of the receptor cell membrane are seen and these invaginations apparently contain the terminal portions of cell processes which most likely originate at the perikarya in the inner layer of cell bodies. The cytoplasm of the receptor terminal contains numerous synaptic vesicles but the synaptic organization is different in *Eptatretus* and *Paramyxine* on the one hand and *Myxine glutinosa* on the other. In *Eptatretus* (Figs. 5 and 6) and *Paramyxine* characteristic spherical organelles lie adjacent to the invaginated cell membrane which surrounds the entering processes. These organelles, commonly referred to as synaptic bodies, are 0.1–0.3 µm in diameter, and usually display an electron translucent core and an electron-dense rim (Fig. 6). They are always surrounded by synaptic vesicles regularly spaced around the organelle. Usually one or more synaptic

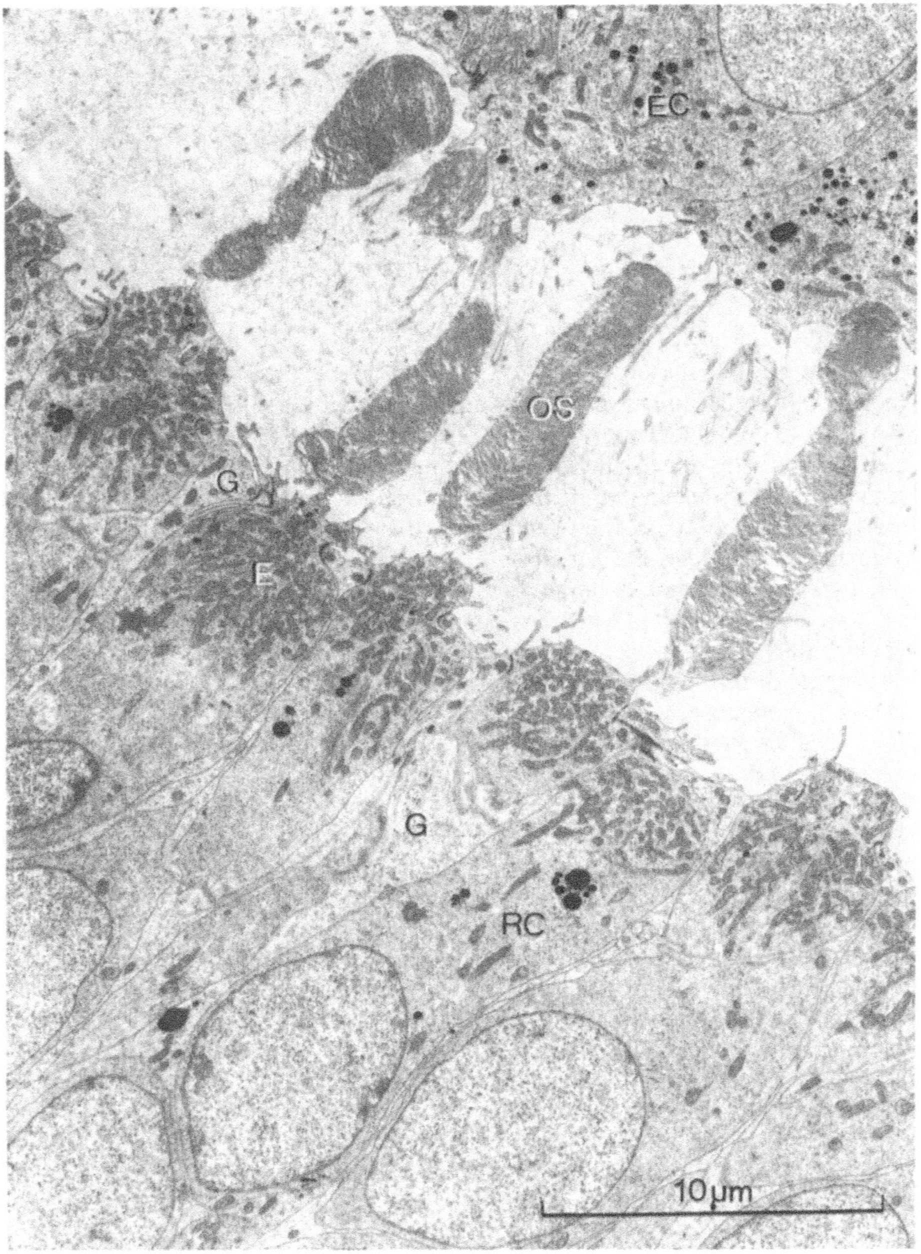

Fig. 3. Longitudinal section through part of receptor cell layer and epithelial cell layer in *E. stouti*. *E*, "ellipsoid"; *EC*, epithelial cells, which in hagfishes do not contain pigment granules; *G*, glial cells; *OS*, outer segment; *RC*, receptor cell, × 4500

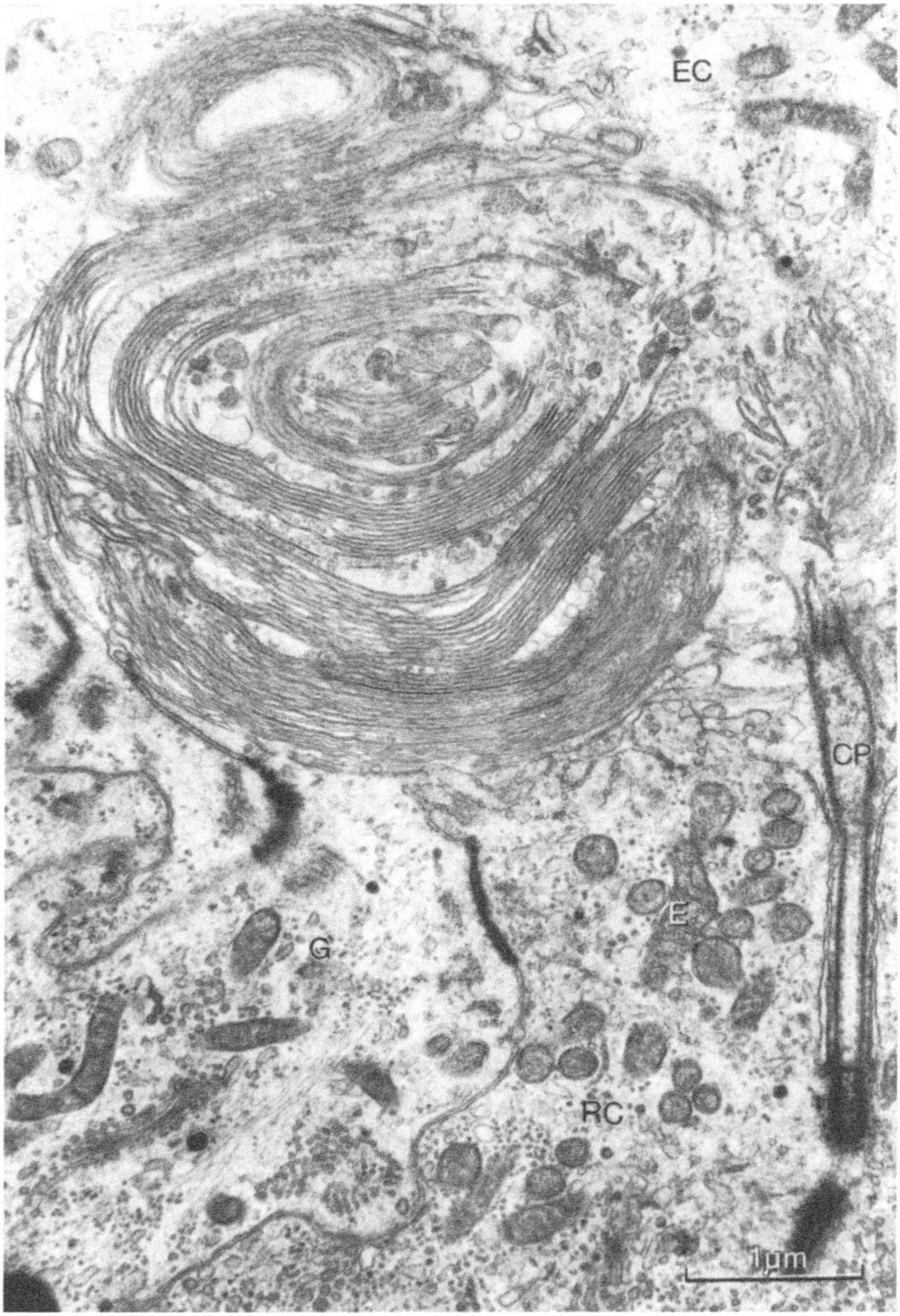

Fig. 4. Longitudinal section through apical part of receptor cell *(RC)* in *M. glutinosa*. Outer segment consists of whorls of membranes which give disorganized impression. *CP*, connecting piece; *E*, mitochondria of "ellipsoid"; *EC*, epithelial cell; *G*, glial cell, × 24000

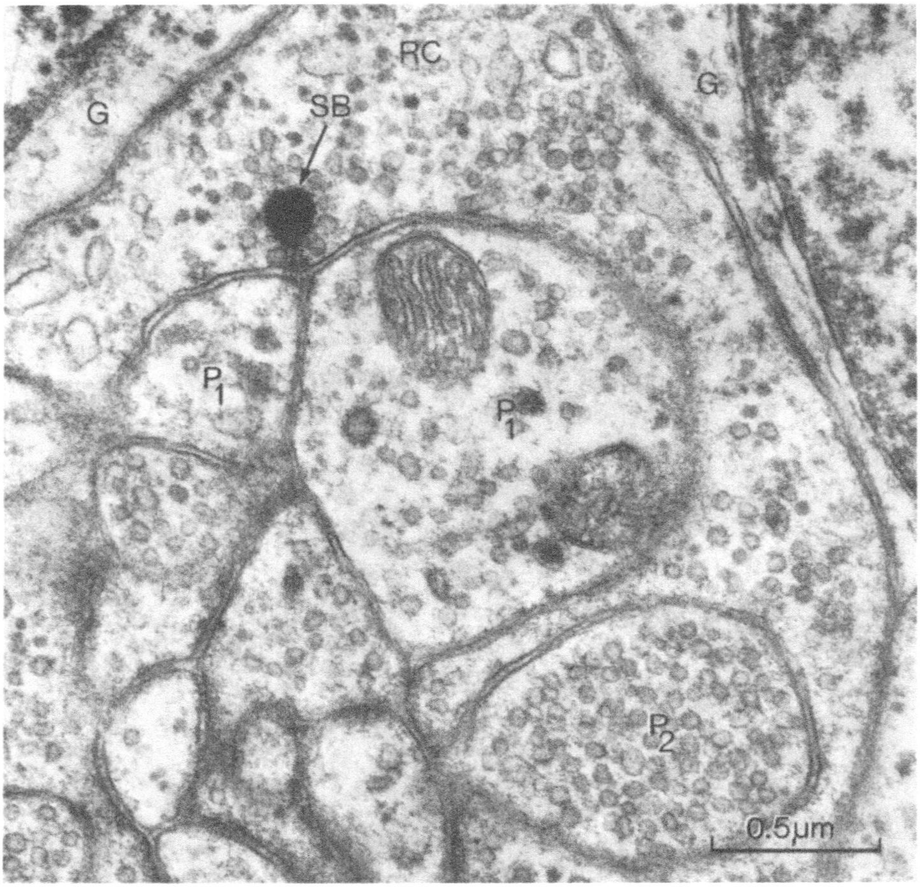

Fig. 5. Receptor cell terminal in *E. stouti*. A synaptic body *(SB)* is located at boundary between two entering cell processes (P_1) which contain few vesicles in contrast to process P_2. Notice synaptic vesicles in the receptor cell *(RC)* cytoplasm and surrounding synaptic body. *G*, glial cells, ×45000

bodies are located adjacent to one process but the occurrence of one synaptic body at the boundary between two processes is also seen (Fig. 5). At these points, presumably the synaptic sites, both the receptor cell membrane and the opposing membranes of the entering processes show increased electron density. In some cases the fusion of synaptic vesicles with the receptor cell membrane has been observed (Fig. 6, *arrow*), perhaps indicating the release of a transmitter substance into the intracellular cleft. Spherical synaptic bodies have not earlier been described in retinal photoreceptor cells of vertebrates. Rods and cones in gnathostome retinas so far examined (for references see, e.g., DOWLING, 1970; STELL, 1972) as well as the photoreceptor cells of lamprey (HOLMBERG and ÖHMAN, 1976; ÖHMAN, 1976) have the characteristic electron-dense synaptic ribbons or lamellae (Fig. 11). The presence of synaptic bodies in the receptor cells of *Eptatretus* and

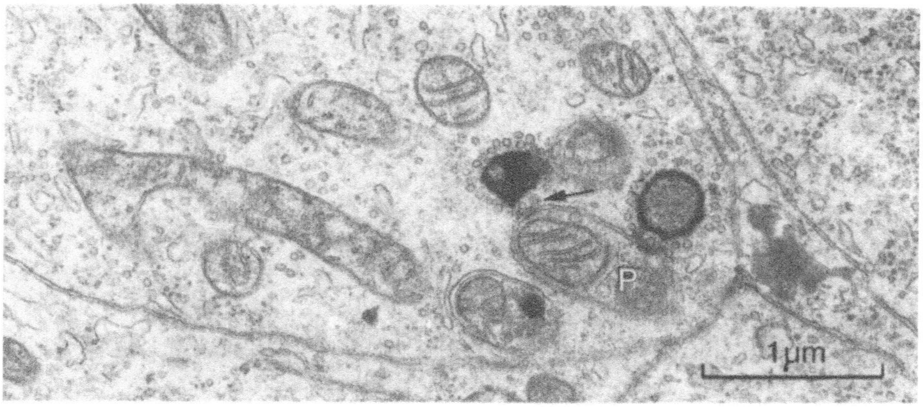

Fig. 6. Receptor cell terminal in *E. burgeri*. Two synaptic bodies, surrounded by synaptic vesicles, lie adjacent to entering cell process *(P)*. Fusion of vesicle with receptor cell membrane is seen at arrowhead, × 24000

Paramyxine is thus a unique feature in hagfishes, not found in cones and rods in gnathostomes. In *Myxine glutinosa* (Fig. 7) the synaptic sites at the receptor terminal are recognized only by the presynaptic aggregation of synaptic vesicles near membrane densities which occur both on the pre- and the postsynaptic membranes. Otherwise the receptor terminals of *Myxine* do not differ exceptionally from those of *Eptatretus* and *Paramyxine*.

Bipolar, horizontal, amacrine, and ganglion cells have not been identified with certainty in any hagfish retina. KOHL (1892) suggested that the inner layer of cell bodies and the inner fibrous layer in *Myxine glutinosa* contains different types of ganglion cells, but according to DÜCKER (1924), the inner layer of cell bodies in *Myxine* is composed of bipolar cells, and the ganglion cells are scattered throughout the inner fibrous layer. He estimated the number of ganglion cells to be between 150 and 180. Electron micrographs show, however, that the number of nerve fibers in the optic nerve is larger (HOLMBERG, 1972; cf. BRUESCH and AREY, 1942). About 800 and 1400 nerve fibers were counted in sections from two different nerves, and since the optic nerve consists mainly of ganglion cell axons, it is apparent that DÜCKER underestimated the number of ganglion cells in the retina.

The number and types of cell processes entering into each receptor terminal is also unknown and it has hitherto not been possible to correlate these processes with corresponding perikarya in the inner layer of cell bodies. On the other hand, the electron microscopic investigations on receptor terminals of *E. stouti*, *E. burgeri*, *P. atami* and *M. glutinosa* show that some of the processes entering the receptor terminals contain large numbers of vesicles which strongly resemble the synaptic vesicles found in the receptor cytoplasm (Fig. 5: P_2), while the majority of processes have no or few such vesicles (Fig. 5 P_1). Moreover, the synaptic bodies in *Eptatretus* and *Paramyxine* and the aggregation of synaptic vesicles near membrane densities in *Myxine* seem to be contiguous to processes without vesicles. Taken together, these observations suggest that the synaptic bodies and the mem-

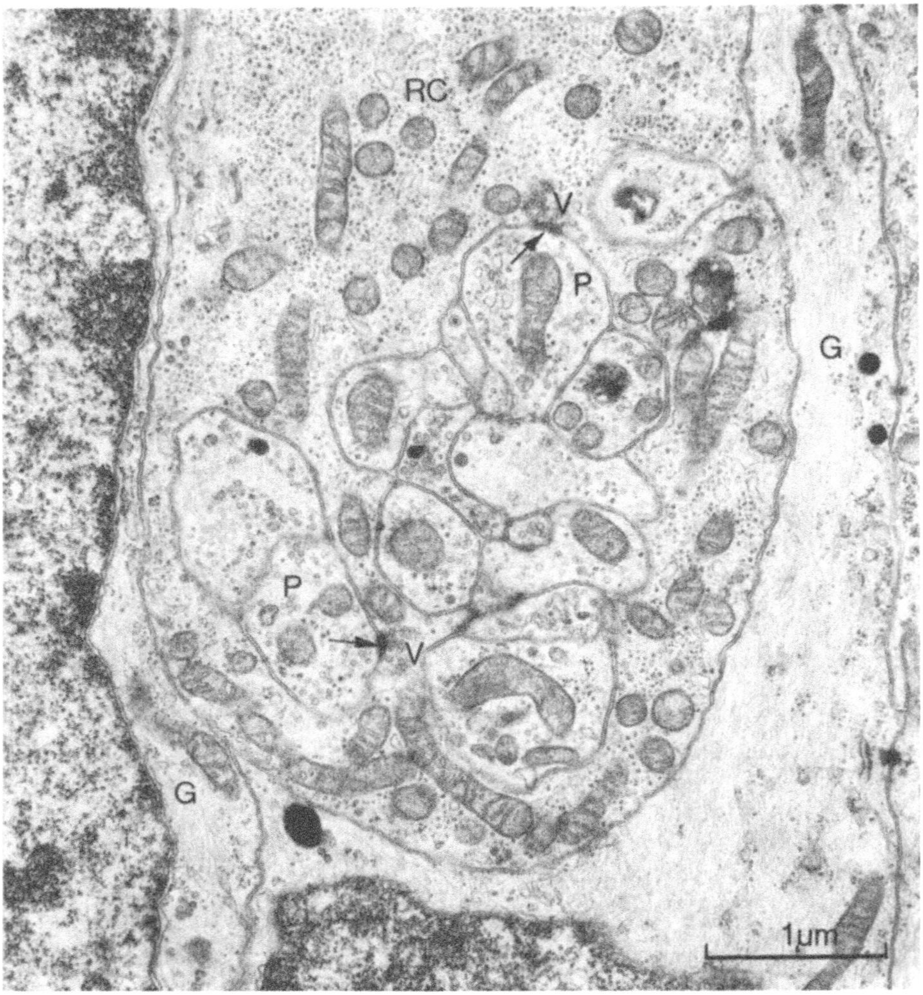

Fig. 7. Receptor cell terminal in *M. glutinosa*. Synaptic sites *(arrows)* between receptor cell *(RC)* and entering cell processes *(P)* are recognized by aggregations of synaptic vesicles *(V)* near membrane densities. Note absence of synaptic bodies in *Myxine*, *G*, glial cell, × 24000

brane densities with their associated vesicles are located on the presynaptic side of the invaginations, and that the adjacent processes would constitute postsynaptic elements, which may be interpreted as bipolar cell dendrites or horizontal cell processes. Those processes which contain large numbers of vesicles may also be horizontal cell processes in view of the fact that some horizontal cell processes in gnathostomes contain synaptic vesicles (cf. DUBIN, 1974). However, these possibilities are speculative because we must also take into consideration the phylogenetic position of the hagfishes and the degenerate state of the retina in different species.

The synaptic bodies in *Eptatretus* and *Paramyxine* have a striking resemblance to similar organelles in the receptor terminal of hair cells in the lateral-line organs of various species (e.g., YAMADA and HAMA, 1972; YAMADA, 1973; JØRGENSEN and FLOCK, 1973). In these sensory cells, the synaptic bodies are located presynaptically in the receptor terminal, and the postsynaptic processes are afferent nerve fibers with few or no vesicles. The efferent nerve fibers, which innervate the hair cells, contain large numbers of presynaptically located vesicles, whereas the postsynaptic region, in the receptor cytoplasm, is often recognized by the occurrence of subsynaptic cisterns.

The ultrastructural indications of a functional state described above in *Eptatretus* are supported by KOBAYASHI'S (1963, 1964) electrophysiologic investigations on hagfish eye. The animals used were most likely *E. burgeri* (not *M. garmani*, cf. FERNHOLM and HOLMBERG, 1975) and KOBAYASHI was able to record potentials induced in the dark-adapted eye. The responses to weak light (about 0.1 lux) were slow, positive, and monophasic potentials with long latency. In response to intense light (about 200 lux), the positive potentials were preceded by a small, negative potential.

IV. The Epithelial Cells

In all hagfish species so far examined, the epithelial cells do not contain pigment granules (W. MÜLLER, 1874; DÜCKER, 1924; HOLMBERG, 1970, 1971; FERNHOLM and HOLMBERG, 1975). The cells are arranged in a single layer which is separated from the endothelial cells of the choroidea by a collagenous layer. Slender villi, which do not contain any specific structures or organelles, extend from the apical surfaces of the cells between the outer segments (Fig. 3). Nearly all the cytoplasm is occupied by an elaborate smooth endoplasmic reticulum. Golgi membranes and lysosomelike spherical bodies are located in the apical region of the cells. The lateral cell membranes of neighboring cells interdigitate at irregular intervals, and cell junctions are located near the apical surface. In *E. stouti*, membranous inclusions have been found in the epithelial cells on a few occasions (HOLMBERG, 1971). These are enclosed within a membrane and they are usually seen near the apical surface of the cell or sometimes in the vicinity of the nucleus but never in the basal region. Inclusions farther away from the apical surface of the cell are more electron-dense and have the stacks of membranes more closely packed than those near the surface. These membranous inclusions resemble the outer segment fragments (phagosomes) found in the pigment epithelial cells of several gnathostome species (MARSHALL and ANSELL, 1971). It is tempting to conclude that the epithelial cells in *Eptatretus* phagocytize outer segment fragments. It is, however, necessary to investigate a larger sample and to use histochemical methods for the demonstration of enzyme activities in order to determine with certainty the nature of the membranous inclusions. This point is worth further study, because it could provide useful evidence regarding the presence of rods in the retina: in the gnathostome retina, the outer segments of rods but not cones are phagocytized by epithelial cells (Young and Bok, 1969; Ishikawa and Yamada, 1970; Spitznas and Hogan, 1970; Young, 1971; Ditto, 1975). Membranous inclusions do not occur in the epithelial cells of *M. glutinosa*.

V. Similarities to Cave-Living Vertebrates and Degenerative Evolution of the Eyes

It is evident from the preceding description that the retinas of *Eptatretus* and *Paramyxine* are less degenerated than the retina of *Myxine* and that the differences coincide with the location and the general morphology of the eye. In *Eptatretus* and *Paramyxine* the eyes are located immediately beneath the skin, they have a vitreous body, differentiated retinal layers, receptor cells with cylindrical outer segments with numerous membranous discs and synaptic bodies at the receptor terminals. The eyes of *Myxine* are situated deep under muscle layers, a vitreous body is absent, the retinal layers are less differentiated, and the receptor cells, with membranous whorls instead of stacks of discs in the outer segments, lack synaptic bodies.

These degenerative changes in hagfishes are similar to those found in the eyes of certain cave-living vertebrates. For instance, the eyes of *Haideotriton wallacei* (Brandon, 1968), a trogloditic salamander, are embedded deeply in tissues of the head, and each eye is surrounded by an extensive mass of adipose connective tissue. The average eye diameter in five species ranges from 170–200 μm. No extrinsic eye muscles are present. The lens, when present, is rudimentary in size and structure, and the vitreous cavity is completely supplanted by the relatively massive retina which lacks an outer synaptic layer (OSL). On the other hand, a conspicuous inner synaptic layer (ISL) is present, which agrees with the observation pointed out by Eigenmann (1909), that the ISL is relatively well developed in degenerate eyes among cave-living vertebrates, e.g., the salamanders *Typhlomolge rathbuni* and *Typhlotriton spelaeus* (cf. Fig. 2: *M. glutinosa*). Eigenmann and Denny (1900) suggested that loss of the OSL was the first step in retinal degeneration in *T. spelaeus*. However, Besharse and Brandon (1974) showed that reduction of the outer and inner segments of the photoreceptor cells in *T. spelaeus* occurs first, and this reduction is followed by loss of the OSL. These authors suggest that a reduction of the OSL may result from photoreceptor cell degeneration because the OSL is composed of synapses between neurons of the inner nuclear layer and the receptor cells. These observations agree with the conditions found in hagfishes. In *Eptatretus* and *Paramyxine* the receptor cells have well-developed outer segments and the outer fibrous layer is clearly recognized. In *Myxine* the receptor cells have disorganized outer segments and the outer fibrous layer is sometimes difficult to delimit and may also be absent.

It is evident that there is an evolutionary trend towards reduction of the visual capability and a degeneration of the eyes among hagfishes (cf. Fernholm and Holmberg, 1975). The eyes have been subjected to a degenerative evolution which in many respects is similar to that found in cave-living animals, and since the photic environment in which many hagfishes live resembles that found in deep caves, it is probable that this evolutionary trend is caused by the decreased selective pressure to retain a functional eye in a dark environment.

C. The Lamprey Retina

I. Eye Morphology

Although hagfishes and lampreys are classified systematically as agnaths, there are few organ characters common to both groups, and this applies especially well to the eye, which in the adult lampreys is a well-developed organ.

The eye is surrounded by large venous sinuses and together with the eye muscles and adipose tissue is protected within an orbital capsule which is formed of connective tissue. The cornea consists of two components: the dermal cornea, which is a transparent, thin layer of dermal epithelium without pigmentation, glands or blood vessels; and the scleral cornea, a proximal layer of thin tissue which is the transparent continuation of the sclera. The large and almost spherical lens is rather hard and rigid, held in place between the immobile pupil and the vitreous body.

II. The Organization of the Retina

In all lampreys so far studied, the retina (Fig. 8) is differentiated in a receptor cell layer, comprising an outer nuclear layer (ONL), and two synaptic layers separated by an inner nuclear layer (INL). The receptor cells are of two types, long and short, which differ in the size of the inner segment and hence in the position of the outer segment. The outer synaptic layer (OSL) is rather thin, whereas the inner synaptic layer (ISL), which extends from the INL to the vitreous body, is relatively thick. The INL comprises two layers of large horizontal cells and small bipolar cells. The occurrence of catecholamine and indoleamine containing neurons in the INL has been determined by fluorescence microscopy (EHINGER et al., 1977). The ganglion cells have been suggested by different authors to be located in the INL (cf. WALLS, 1942).

III. The Receptor Cells

One of the most controversial problems concerning the retina of lampreys is the nature of the long and short photoreceptor cells. Several authors (e.g., H. MÜLLER, 1856; LANGERHANS, 1876; KOHL, 1892; DÜCKER, 1924; FRANZ, 1932;

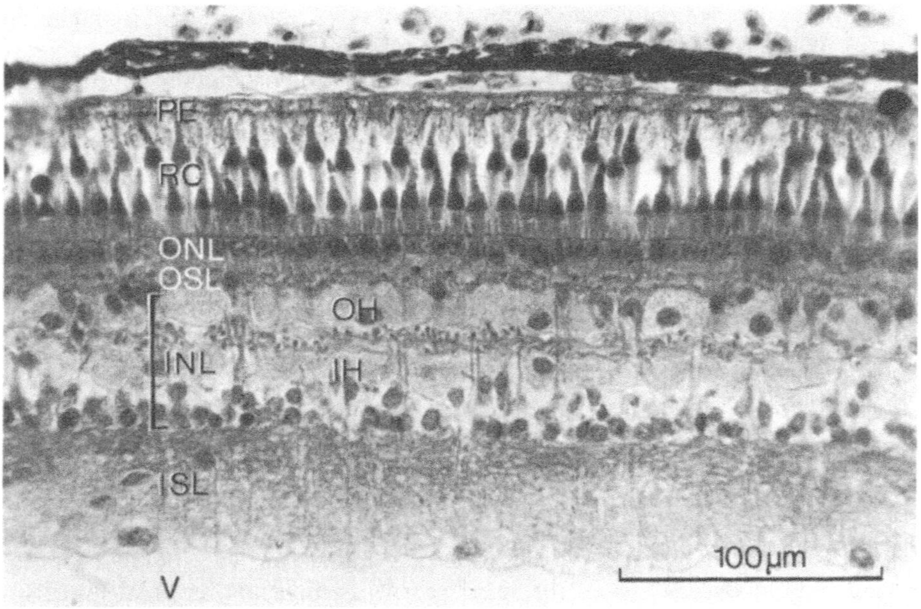

Fig. 8. Longitudinal section through retina of river lamprey, *L. fluviatilis*. *PE*, pigment epithelium; *RC*, receptor cells; *ONL*, outer nuclear layer; *OSL*, outer synaptic layer; *INL*, inner nuclear layer with outer *(OH)* and inner *(IH)* horizontal cells, *ISL*, inner synaptic layer; *V*, vitreous body, ×350 (Courtesy of P. ÖHMAN)

WALLS, 1935) have attempted to identify the two types of receptor cells with rods and cones in the retinas of gnathostomes. Their conclusions, based on light microscopic investigations, are conflicting since both long and short receptor cells have been suggested to be rods and cones (cf. WALLS, 1935; DUKE-ELDER, 1958). Rhodopsin and porphyropsin have been identified in *Petromyzon marinus* (WALD, 1957; CRESCITELLI, 1956, 1960), and KÜHNE (1878) observed that the dark-adapted retina of *Lampetra fluviatilis* had a faint pink colour which gradually disappeared in light. These observations indicate, at least with regard to *P. marinus*, that one or both of the receptor cell types may be rods.

In an attempt to find such differences between the long and short receptor cells which could help to identify these receptors with rods or cones of gnathostomes, ÖHMAN (1971, 1974, 1976) has applied light-, fluorescence- and electron microscopic methods on the retinas of adult *L. fluviatilis*. His results show that both receptor types resemble rods in many details. For instance, the outer segments of both long and short receptors (Fig. 9) have the double membrane discs enclosed by the cell membrane (a morphologic characteristic typical of rods), but at certain intervals the membrane makes infoldings between the discs (Fig. 10). These infoldings are as deep and wide as the individual discs of the outer segment. In a typical cone outer segment the discs are continuous with the cell membrane (cf. COHEN, 1972). The rodlike character of the outer segments is also indicated by the occurrence of membranous inclusions in the pigment epithelium (ÖHMAN,

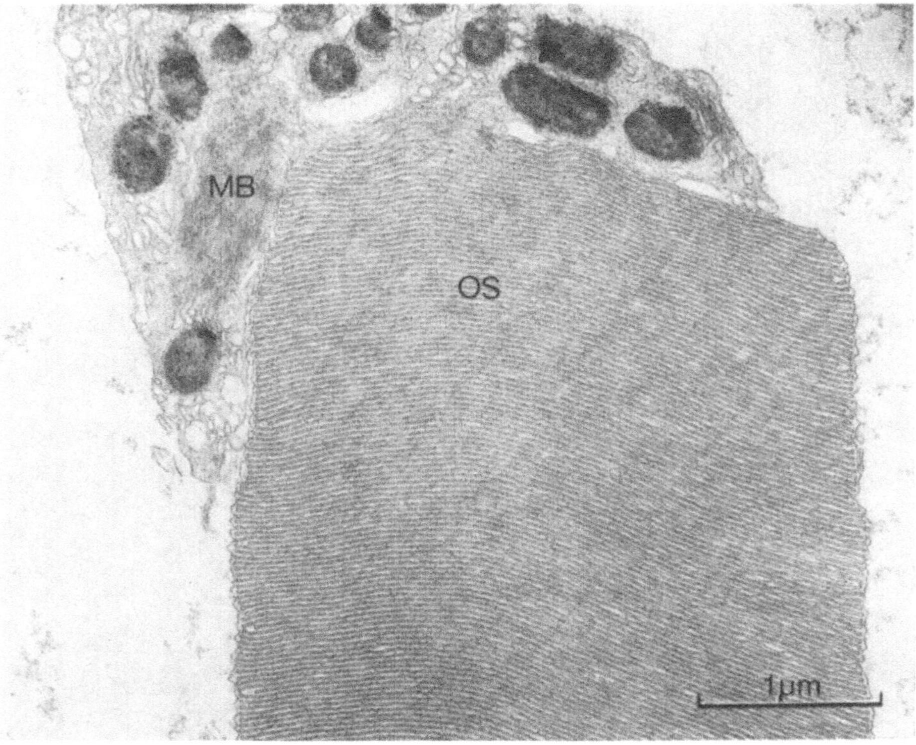

Fig. 9. Apex of long receptor outer segment *(OS)* in *L. fluviatilis*. Some pigment granules and one myeloid body *(MB)* are seen in pigment epithelial process, × 24000. (Courtesy of P. Öhman)

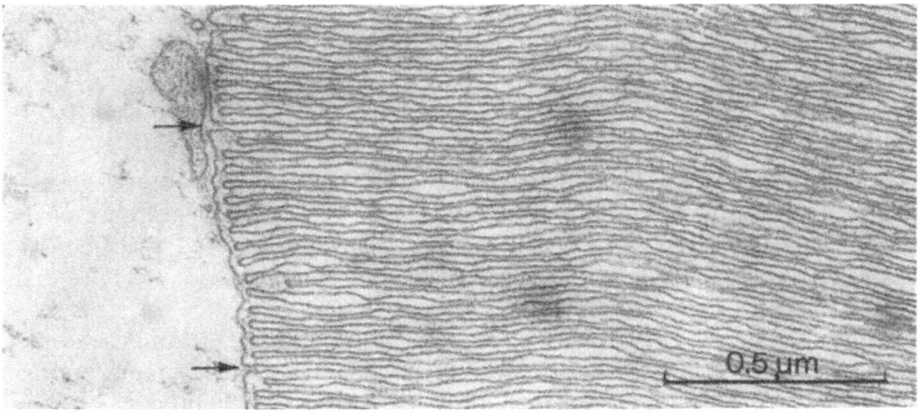

Fig. 10. Detail of short receptor outer segment in *L. fluviatilis*. Majority of discs are enclosed by cell membrane, but at certain intervals *(arrows)* membrane makes infoldings between discs, × 58000. (Courtesy of P. Öhman)

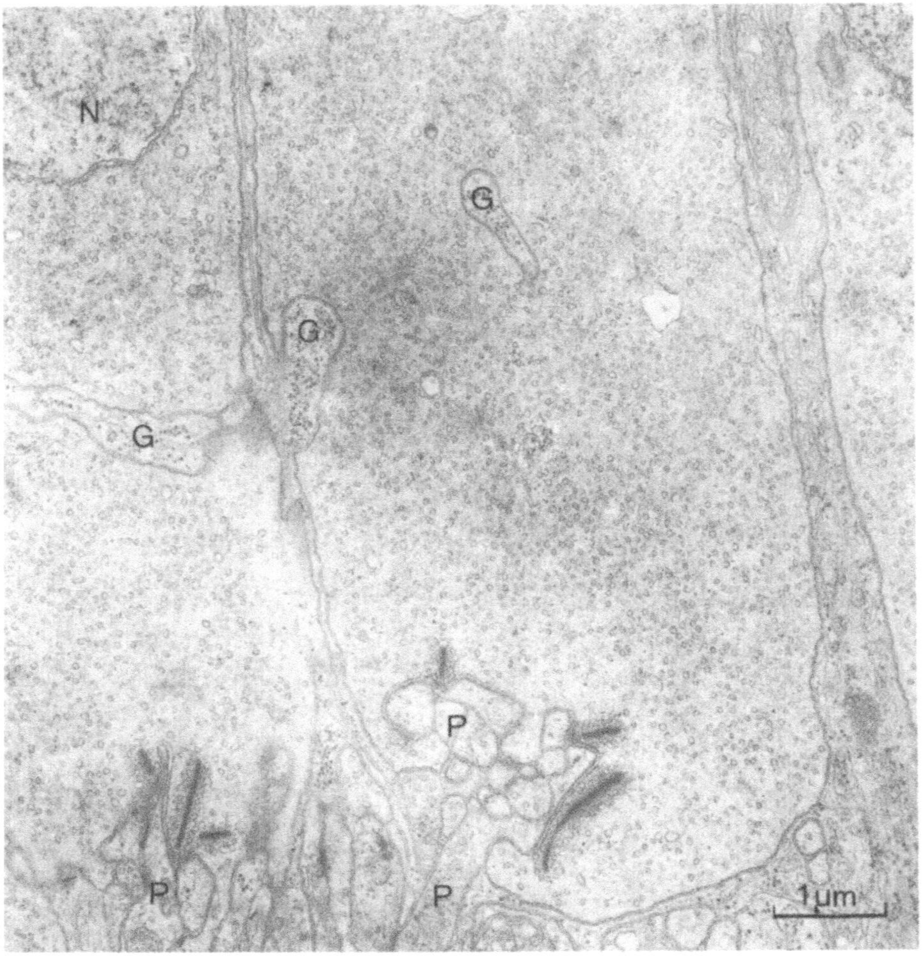

Fig. 11. Receptor cell terminal in *L. fluviatilis*. Synaptic ribbons lie adjacent to postsynaptic processes *(P)*. *G*, glial cell processes; *N*, nucleus of adjacent receptor cell, × 15000. (Courtesy of P. ÖHMAN)

1974). These inclusions may be phagosomes, i.e., rod outer segment fragments, phagocytized by the epithelial cells (YOUNG, 1971). Also, the appearance of the receptor terminals, which end bluntly or by a slight swelling, and which contain relatively few synaptic ribbons, coincides more with typical rod spherules than with the broader, polysynaptic pedicles of cones (Fig. 11).

Although both receptor types are mainly rodlike, they differ from each other in certain respects: a) after injection of Procion yellow into the vitreous body of living lampreys, the outer segments of long receptors emit an orange-yellow fluoresence (a characteristic of cones), while the short receptor outer segments show none or very faint autofluorescence (a characteristic of rods, cf. LATIES and LIEBMAN, 1970); b) after impregnation with osmium tetroxide (EAKIN and BRAN-

DENBURGER, 1970) large amounts of electron-dense material accumulate in the outer segments of long receptors and the disc membranes can hardly be distinguished. In short-receptor outer segments the faintly electron-dense material is evenly spread throughout the segment and most segments show characteristic transverse striations of dense material within the discs; c) in material fixed only with osmium tetroxide there occurs a disruption of discs into numerous vesicles in the long-receptor outer segments but not in the short ones. This phenomenon does not occur in retinas fixed with glutaraldehyde *and* osmium tetroxide.

Taken together, ÖHMAN'S results suggest that long and short receptors, although rodlike, apparently differ from each other, perhaps in the biochemical composition of the outer segment membranes. Furthermore, the occurrence of a c-wave and a cornea-positive off-response (d-wave) in the river lamprey ERG indicates that the retina is functionally similar to a mixed rod-cone retina (HOLMBERG et al., 1977).

IV. The Inner Nuclear Layer

Another controversial problem concerning the lamprey retina is the location of the ganglion cell bodies. In gnathostomes, these cells generally constitute a distinct layer separated from the INL by the ISL. The ganglion cell axons run in the optic nerve fiber layer, i.e., the innermost layer adjacent to the vitreous body.

In the lamprey retina, a distinct ganglion cell layer does not occur. According to FRANZ (1932), DUKE-ELDER (1958), and other authors, the ganglion cell bodies are located in the INL. If this is true, we may ask where the synaptic contacts between bipolar cell axons and ganglion cell dendrites are located: within the INL or in the ISL. In the ISL of gnathostomes, terminals of bipolar cell axons are always recognized by the occurrence of synaptic ribbons on the presynaptic side within the terminal. At these ribbon synapses generally two processes are simultaneously postsynaptic to the bipolar cell terminal: either two ganglion cell dendrites, a ganglion cell dendrite and an amacrine cell process, or two amacrine cell processes (cf. DOWLING, 1970; DUBIN, 1974). ÖHMAN (1976 and unpublished results) has described in silver-impregnated material from *L. fluviatilis* two morphologically different types of bipolar cells which send their axons into the ISL, and he has also observed synaptic ribbon contacts with postsynaptic elements. It is reasonable to assume that at least some of the postsynaptic processes at these ribbon synapses are ganglion cell dendrites. Hence, the opinion that the ganglion cell bodies in the lamprey are situated in the INL implies that these cells send their dendrites vitreally into the ISL to establish synaptic connections with bipolar cell axons, and also that they send their axons into this layer, which thus represents a combined ISL and optic nerve fiber layer.

The main glycogen store of the avascular lamprey retina seems to be the large horizontal cells (ÖHMAN, 1976) located in two distinct layers in the INL (Fig. 8: *OH* and *IH*). These cells, as well as the bipolar cells, make synaptic connections with the receptor cell terminals (HOLMBERG and ÖHMAN, 1976), but the relationship between long and short receptor cells, outer and inner horizontal cells, and the two types of bipolar cells remains to be determined.

V. The Pigment Epithelial Cells

The epithelial cells of lampreys have hitherto been investigated electron microscopically only in *Lampetra fluviatilis* (Öhman, 1974). These cells do not differ exceptionally from those of gnathostomes, but two characteristic features may be pointed out: first, the numerous and well-developed myeloid bodies which are distinct organelles occurring throughout the cell even in the larger cytoplasmic processes between the outer segments (Fig. 9: *MB*); second, the amount of pigment granules is relatively small compared with many other species having a dense pigmentation. The occurrence of membranous inclusions, which may be phagosomes, was pointed out earlier (p. 60).

Acknowledgments

The investigations on hagfish and lamprey retinas carried out by P. Öhman and myself were supported in part by grants 2124, 2127, and B 2935-003 from the Swedish Natural Science Research Council, and in part by grants from the Axel and Margaret Ax:hon Johnson Foundation, the Harald and Greta Jeansson Foundation, the Helge Ax:hon Johnson Foundation, the Henrik Granholm Foundation, and the Royal Swedish Academy of Science. This support is gratefully acknowledged.

References

Besharse, J. C., Brandon, R. A.: Postembryonic eye degeneration in the troglobitic salamander *Typhlotriton spelaeus*. J. Morph. **144**, 381—406 (1974).

Brandon, R. A.: Structure of the eye of *Haideotriton wallacei*, a North American troglobitic salamander, J. Morph. **124**, 345—352 (1968).

Bruesch, S. R., Arey, L. B.: The number of myelinated and unmyelinated fibres in the optic nerve of vertebrates. J. comp. Neurol. **77**, 631—655 (1942).

Cohen, A. I.: Rods and cones. Fuortes, M. G. F. (ed.). In: Handbook of Sensory Physiology, Vol. VII/2. Physiology of Photoreceptor Organs, Berlin-Heidelberg-New York: Springer-Verlag, 1972, pp. 63—110.

Crescitelli, F.: The nature of the lamprey visual pigment. J. gen. Physiol. **39**, 423—435 (1956).

Crescitelli, F.: Physiology of vision. Ann. Rev. Physiol. **22**, 525—528 (1960).

Ditto, M.: A difference between developing rods and cones in the formation of outer segment membranes. Vision Res. **15**, 535—536 (1975).

Dowling, J. E.: Organization of vertebrate retinas. Invest. Ophthalmol. **9**, 655—680 (1970).

Dubin, M. W.: Anatomy of the vertebrate retina. In: The Eye, Vol. VI. Davson, H., Graham, L. T. (eds.): New-York-London: Academic Press, 1974, pp. 227—256.

Duke-Elder, S.: The eye in evolution. In: System of ophthalmology, Vol. I. St. Louis: C. V. Mosby, 1958.

Dücker, M.: Über die Augen der Zyklostomen. Jena Z. Med. Naturw. **60**, 471—528 (1924).

Eakin, R., M., Brandenburger, J. L.: Osmic staining of amphibian and gastropod photoreceptors. J. Ultrastruct. Res. **30**, 619—641 (1970).

Ehinger, B., Holmberg, K., Öhman, P.: Aminergic and indoleamine accumulating neurons in the retina of the river lamprey (*Lampetra fluviatilis*). Acta zool. (Stockh.) **58**, 117—123 (1977).

Eigenmann, C. H.: Cave vertebrates of America. A study in degenerative evolution. The Carnegie Institution of Washington **104**, 1—241 (1909).

Eigenmann, C. H., Denny, W. A.: The eyes of the blind vertebrates of North America. III. The structure and ontogenetic degeneration of the eyes of the Missouri cave salamander, an account based on material collected with a grant from the Elizabeth Thompson Fund. Biol. Bull. **2**, 33—41 (1900).

Fernholm, B., Holmberg, K.: The eyes in three genera of hagfish (*Eptatretus*, *Paramyxine* and *Myxine*). A case of degenerative evolution. Vision Res. **15**, 253—259 (1975).

Franz, V.: Auge und Akkommodation von *Petromyzon (Lampetra) fluviatilis* L. Zool. Jb. (Zool.) **52**, 118—178 (1932).

Holmberg, K.: Hagfish eye: ultrastructure of retinal cells. Acta zool. (Stockh.) **50**, 179—183 (1969).

Holmberg, K.: The hagfish retina: fine structure of retinal cells in *Myxine glutinosa*, L., with special reference to receptor and epithelial cells. Z. Zellforsch. **111**, 519—538 (1970).

Holmberg, K.: The hagfish retina: electron microscopic study comparing receptor and epithelial cells in the Pacific hagfish, *Polistotrema stouti*, with those in the Atlantic hagfish, *Myxine glutinosa*. Z. Zellforsch. **121**, 249—269 (1971).

Holmberg, K.: Fine structure of the optic tract in the Atlantic hagfish, *Myxine glutinosa*. Acta zool. (Stockh.) **53**, 165—171 (1972).

Holmberg, K., Öhman, P.: Fine structure of retinal synaptic organelles in lamprey and hagfish photoreceptors. Vision Res. **16**, 237—239 (1976).

Holmberg, K., Öhman, P., Dreyfert, T.: ERG-recordings from the retina of the river lamprey *(Lampetra fluviatilis)*. Vision Res. **17**, 715—717 (1977) (in print).

Ishikawa, T., Yamada, E.: The degradation of photoreceptor outer segment within the pigment epithelial cell of rat retina. J. Electron. Microsc. (Tokyo) **19**, 85—91 (1970).

Jarvik, E.: Specializations in early vertebrates. Ann. Soc. roy. Zool. Belgique **94**, 11—95 (1964).

Jørgensen, J. M., Flock, Å.: The ultrastructure of lateral line sense organs in the adult salamander *Ambystoma mexicanum*. J. Neurocytol. **2**, 133—142 (1973).

Kleerekoper, H.: The sense organs. In: The Biology of Lampreys. Vol. II. Hardisty, H. W., Potter, I. C. (eds.): London-New York: Academic Press, 1972, pp. 373—404.

Kobayashi, H.: Preliminary report on the action potentials recorded from the rudimentary eye of hagfish, *Myxine garmani*. Zool. Mag. (Tokyo) **72**, 6—12 (1963).

Kobayashi, H.: On the photo-perceptive function in the eye of the hagfish, *Myxine garmani* Jordan et Snyder. J. Shimonoseki Univ. Fish. **13**, 67—83 (1964).

Kohl, C.: Rudimentäre Wirbelthieraugen. Zoologica (Stuttg.) **13**, 48—51 (1892).

Krause, C.: Die Retina. II. Die Retina der Fische. Cyclostomata. Int. Mschr. Anat. Histol. **3**, 8—21 (1886).

Kühne, W.: In: Untersuchungen aus dem physiologischen Institute der Universität Heidelberg, pp. 31, 455—456 (1878).

Langerhans, P.: Untersuchungen über *Petromyzon planeri*. Ber, Verh. dt. naturf. Ges. Freiburg **6**, 1—115 (1876).

Laties, A. M., Liebman, P.: Cones of living amphibian eye: selective staining. Science **168**, 1475—1477 (1970).

Marshall, J., Ansell, P. L.: Membranous inclusions in the retinal pigment epithelium: phagosomes and myeloid bodies. J. Anat. (Lond.) **110**, 91—104 (1971).

Müller, H.: Anatomisch-physiologische Untersuchungen über die Retina bei Menschen und Wirbelthieren. Z. wiss. Zool. λ, 1—122 (1956).

Müller, J.: Über den eigenthümlichen Bau des Gehörorgans bei den Cyclostomen, mit Bemerkungen über die ungleiche Ausbildung der Sinnesorgane bei den Myxinoiden. Abh. k. Akad. Wiss. zu Berlin (1837).

Müller, W.: Über die Stammesentwicklung des Sehorgans der Wirbelthiere. In: Beiträge zur Anatomie und Physiologie als Festgabe Carl Ludwig gewidmet. Leipzig: F. C. W. Vogel, 1874.

Öhman, P.: The photoreceptor outer segments of the river lamprey *(Lampetra fluviatilis)*. An electron-, fluorescence- and light microscopic study. Acta zool. (Stockh.) **52**, 287—297 (1971).

Öhman, P.: Fine structure of the retinal pigment epithelium of the river lamprey (*Lampetra fluviatilis*, Cyclostomi). Acta zool. (Stockh.) **55**, 245—253 (1974).

Öhman, P.: Fine structure of photoreceptors and associated neurons in the retina of *Lampetra fluviatilis* (Cyclostomi). Vision Res. **16**, 659—662 (1976).

Retzius, G.: Das Gehirn und das Auge von *Myxine*. Biol. Unters. **5**, 55—68 (1893).

Rochon-Duvigneaud, A.: L'oeil. In: Traité de Zoologie, Vol. XIII/1. Grassé, P.-P. (ed.). Paris: Masson et Cie Libraires de l'Académie de Médicine, 1958, pp. 58—65.

Spitznas, M., Hogan, M. J.: Outer segments of photoreceptors and the retinal pigment epithelium. Interrelationship in the human eye. Arch. Ophthal. (Chicago) **84**, 810—819 (1970).

Stell, W. K.: The morphological organization of the vertebrate retina. In: Physiology of Photoreceptor Organs, Fuortes, M. G. F. (ed.). Handbook of Sensory Physiology, Vol. VII/2, Berlin-Heidelberg-New York: Springer-Verlag 1972, pp. 111—213.

Wald, G.: The metamorphosis of visual systems in the sea lamprey. J. gen. Physiol. **40**, 901—914 (1957).

Walls, G. L.: The visual cells of lampreys. Br. J. Ophthal. **19**, 129—148 (1935).

Walls, G. L.: The vertebrate eye and its adaptive radiation. Bloomfield Hills: Cranbrook Institute of Science, 1942. Reprinted 1963 by Hafner, New York-London, 1963.

Yamada, Y.: Fine structure of the ordinary lateral line organ. I. The neuromast of lamprey, *Entosphenus japonicus*. J. Ultrastruct. Res. **43**, 1—17 (1973).

Yamada, Y., Hama, K.: Fine structure of the lateral-line organ of the common eel, *Anguilla japonica*. Z. Zellforsch. **124**, 454—464 (1972).

Young, R. W.: An hypothesis to account for a basic distinction between rods and cones. Vision Res. **11**, 1—5 (1971).

Young, R. W., Bok, D.: Participation of the retinal pigment epithelium in the rod outer segment renewal process. J. Cell Biol. **42**, 392—403 (1969).

Chapter 3

Adaptations to the Deep-Sea Environment

N. A. LOCKET, London (Great Britain)

With 66 Figures

Contents

A. Introduction

The Deep-Sea Environment

The world's oceans cover some 70% of the surface of the globe, and of this area 75% is deeper than 3000 m. Unaided, man can explore only the surface and edges of this mass of water; he requires diverse probes and samplers to investigate the habitat of deep-sea fishes.

The surface of the sea is subject to change, being exposed to the atmosphere. The temperature of surface waters differs widely in different regions, and winds cause the surface to be in constant motion, both on the small scale as waves and on the larger scale as currents. Below the surface the influence of the atmosphere is progressively reduced with depth, and a major feature of the depths is their constancy.

An important difference between the sea and land environments is concerned with the densities of water and air, respectively. Air is a gaseous fluid with a density much below that of the terrestrial animals which live immersed in it. It provides them with little support against gravity, and to overcome this force they must either live supported by a surface, or use aerodynamic forces to gain the lift necessary to support their weight in air. Not being dense, air offers relatively little resistance to movement through it.

Water is much more dense, and the density of aquatic animals is in many cases matched to the density of the water in which they live. If this equality is achieved, they need use no energy in counteracting gravity, and can consequently occupy any level in the sea, provided that other parameters are suitable. Such animals will, however, need to expend energy in moving through the dense medium to feed. Some relevant features of the deep-sea environment are now considered briefly; for fuller accounts see MARSHALL (1954, 1971) and HERRING and CLARKE (1971).

I. Physical Zonation

1. Temperature

Ecologists are familiar with the concept of a geographical range for a species, and how this may be determined by temperature. Such ranges also occur in the oceans, in which there are stretches of water unbroken from pole to equator. The surface temperatures in an ocean may range from $-2°$ C in arctic waters to $+25°$ C at the equator, and it is not surprising that different surface fauna characterize different geographical regions.

Land masses include regions of high ground and of low, and though these may be at the same latitude, the fauna is likely to be different due to the lower temperatures at altitude. Though the liquid surface of the ocean is not elevated like the land, the ocean floor lies at different depths below that surface, and temperatures in the sea vary, though not uniformly, with depth.

The temperature structure of the oceans is affected by the alteration of the density of sea-water with temperature. Warm sea-water is less dense than cool, with a density maximum at $+4°$ C. Due to this, warm water tends to lie above cooler, and in tropical latitudes a stable system is formed. A clearly defined warm surface layer builds up, separated from the underlying water by a more or less sharp temperature boundary, the thermocline. Such thermoclines also form in temperate latitudes in summer, but when the surface water cools down in winter the system loses stability, and mixing of the layers occurs (Fig. 1).

Temperature is an important parameter for many organisms; a species may live in shallow waters in polar seas, but in deeper waters, though of comparable temperature, nearer the tropics. This phenomenon, known as submergence, is demonstrated by bottom living echinoderms; the brittlestar *Ophiacantha bidentata* lives at 5 m in the Arctic, but only below 1000 m in mid-Atlantic (EKMAN,

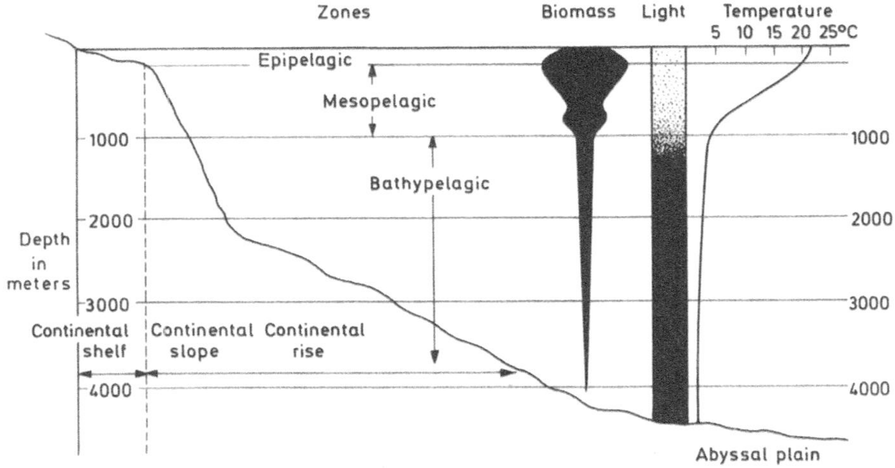

Fig. 1. Diagram to show relation of biomass, light penetration, and temperature to depth in oceans. Slope of bottom is much exaggerated. Benthic animals which live at bottom may be either benthopelagic, swimming above bottom, or benthal, living on bottom itself. Epipelagic, mesopelagic, and bathypelagic fauna discussed in text. (Redrawn from MARSHALL, 1971)

1935). The oceanic angler fishes *Himantolophus groenlandica* and *Ceratias holboelli* show the same tendency (BERTELSEN, 1951).

Below the thermocline the temperature falls slowly but more steadily with depth, from about +5° C at 1000 m to +2° C below 4000 m. Thus there are vast tracts of water with very even temperatures; one aspect of the constancy of the deep-sea environment.

2. Pressure

Water is a dense medium compared with air, and pressure in the sea rises one atmosphere for each 10 m depth, so that deep-sea animals are commonly exposed to pressures of some hundreds of atmospheres. Oceanographic equipment must be robust to withstand such extremes, but the tissues of the animals are permeated with water, and are neither more nor less compressible than the water around them. The examples of submergence show that a single species can tolerate a very wide range of pressures. Many animals, moreover, undertake diurnal vertical migrations which may range through hundreds of meters, thus involving changes of pressure of tens of atmospheres in a few hours. Vertical migrations are mainly concerned with feeding; these behaviour patterns, both of the fishes and of their prey, are closely connected with and influenced by the light regime in the ocean.

3. Light

All life in the oceans is ultimately dependent on light, and on the ability of the ocean flora to photosynthesize. This flora is not obvious to gross inspection, consisting of minute planktonic plants, but it is distributed through about 100 m of surface water. In this euphotic zone all the primary production is concentrated, and here too many of the small planktonic animals are found. The depth of the

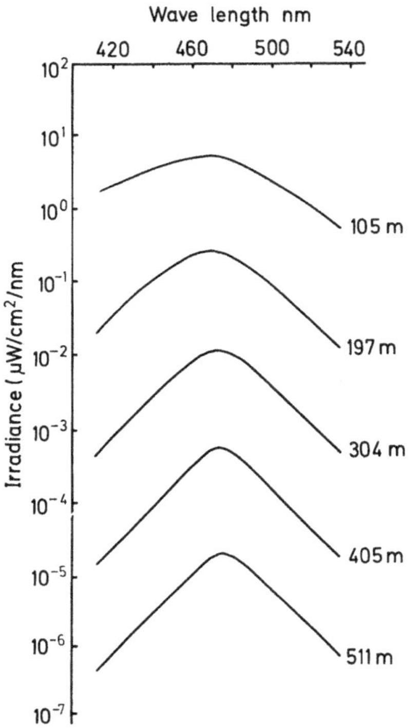

Fig. 2. Spectra of sunlight in various depths of sea water. These measurements, from near Canary Islands, show attenuation of light with depth. Note peaking of curves due to light of short wavelength being scattered, and of long wavelength absorbed, more than best-transmitted blue-green light. (From data in KAMPA, 1970)

euphotic zone varies with latitude and with season; 100 m is a figure for the tropics, it would be less away from the equator.

Not only is water much less transparent to visible light than is air but its spectral absorbance differs markedly. Short and long wave lengths are rapidly attenuated with depth, leaving only the blue-green light which penetrates to deeper levels (JERLOV, 1951, 1968) (Fig. 2).

Below the euphotic zone is a zone of twilight, in which the intensity falls progressively and the spectral band width narrows. This twilight zone may extend down to about 1200 m in the clearest ocean waters, and grades imperceptibly to the sunless waters below, which thus form the bulk of the ocean mass.

II. Faunal Zonation

1. Epipelagic Zone

The ocean fauna as well as the water column can be divided according to depth. Epipelagic animals are those which live in the sunlit surface waters. They include large predatory species such as tuna and sharks, besides numerous smaller forms such as flying fishes and oceanic squid (HARDY, 1959; MARSHALL, 1971). Most of the fishes in this zone are members of complex food chains leading back to the phytoplankton; though some of the larger ones feed directly on earlier members of the chain, thus the basking shark, *Cetorhinus maximus*, feeds on the zooplankton, as do the whalebone whales (HARDY, 1956).

2. Mesopelagic Zone

Below the euphotic zone with its epipelagic fauna is the mesopelagic zone. This is the twilight zone, characterized by the gradual falling off of sunlight with depth. The mesopelagic fauna depends on the zone above for its nutrition, and many mesopelagic animals make vertical migrations, ascending at night to feed on the rich plankton of the euphotic zone. The greatest biomass and faunal diversity of the ocean is concentrated in the mesopelagic zone, mostly in the shallower depths. This zone, extending from well-lit to dark regions, is one in which we might expect adaptations concerned with vision under conditions of dim illumination. It is also a zone in which concealment from the vision of predators will be of importance, and examples of adaptations to these conditions are numerous. The mesopelagic fauna includes a rich variety of fishes, many of which have silvery sides and dark backs, and it is the mesopelagic fauna which displays the greatest number and variety of luminous organs.

Descending through the mesopelagic zone, conditions range from variable near the surface to more constant at depth. Light falls off in intensity, the transmitted waveband becomes narrower, and its diurnal variation becomes less marked with increasing depth. Pressure increases steadily, as it does through the entire water column. Temperature falls gradually with depth, and below the thermocline is almost constant over large distances. As the water column is followed downward the available nutrients are less, and the biomass which depends on them also diminishes with depth. To some extent the mesopelagic fauna is independent of the level of nutrients at depth, since many of these animals migrate to the surface to feed.

The presence of mesopelagic organisms may sometimes be deduced from the presence of sonic scattering layers in midwater. Sonic scattering is caused by the presence of something less or more dense than water; in most cases it is due to bubbles of gas. These are found in siphonophores, and in the swim-bladders of many species of mesopelagic fish. In some cases it has been shown that a scattering layer is definitely due to fishes; BACKUS (1968) identified such a layer as due to shoals of a lantern fish, the myctophid *Ceratoscopelus maderensis*. Not infrequently, scattering layers are observed undergoing vertical migrations, and these may be related to the intensity of daylight (CLARKE and BACKUS, 1964). Such migrating layers may indeed follow an isolume, and BLAXTER and CURRIE (1967) were able to drive such a migration by means of artificial lights lowered into the sea.

Though the waters of the oceans are not still, at mesopelagic depths and below there are no sudden movements of water like those at the surface due to wave action. As depth increases the tendency is towards uniformity.

3. Bathypelagic zone

Except in the clearest tropical waters sunlight does not penetrate below the 1000 m level, so here there is constant darkness broken only by the flashes of bioluminescent animals. The temperature falls steadily but gradually with depth, and pressure continues to increase. This extensive zone contains the bathypelagic fauna, which is relatively sparse both in number of individuals and in variety of species. With light from above no longer a factor at these depths, countershading and photophores to help camouflage are not found. Even here light organs persist, but they are more concerned with feeding, as in deep-sea angler fishes, or perhaps with species and sex recognition, than with camouflage. Most bathypelagic fishes are dark brown or black; the crustaceans tend to be red, at least as seen when brought to the light at the surface, though as there is no red light at that depth, their red pigmentation is equivalent to the black of the fishes. (Malacosteid fishes have a special adaptation concerned with this situation, see section on visual pigments.)

The environment at these depths is characterized by its constancy, and this permits specialization to an extent which might be detrimental in more variable circumstances. Though currents flow at depth, they are on a large scale, and probably imperceptible to a small animal suspended in the water. Temperature is low, pressure high, and light virtually absent. An animal in these regions does not need to be robust, since it never encounters any water disturbance which it needs to counteract, but since food is scarce it must have an efficient economy. Bathypelagic fishes tend to have poorly mineralized bones and weak musculature, economical of metabolites and of energy to counteract gravity. This flimsiness makes it difficult to bring them to the surface undamaged in nets.

The low population density means that mates are scarce. This may be associated with the hermaphroditic condition in paralepidids and *Alepisaurus*, and with the evolution of parasitic males in the ceratioid angler fishes (MARSHALL, 1971; MEAD, 1960; BERTELSEN, 1951).

Besides their specialized reproductive arrangements, the ceratioids have a range of feeding specializations; relatively enormous jaws with teeth which fold inward but not out, and extremely distensible stomachs, enabling them to eat prey larger than themselves. Many have luminous lures which are assumed to attract prey, rather than the fish actively pursuing it. Bathypelagic fishes tend to have small and in some cases regressed eyes, since there is so little light by which to see; these fishes may find it more profitable to specialize in the use of other sensory systems, e.g. olfactory or acoustico-lateralis.

4. Benthal Zone

The bottom of the oceans is mainly smooth, and largely composed of soft ooze. The continental slopes range from about 200 m down to the abyssal plain at 3000 m, including both sunlit and sunless waters. The bottom itself is the habitat of two kinds of fishes; some, e.g. macrourids, morids, and brotulids are benthopelagic, swimming about clear of, but near, the bottom. Others, e.g. liparids, zoarcids, chlorophthalmids, and bathypteroids are benthic, resting on the bottom itself. The benthopelagic examples usually have swim-bladders, while the benthic forms do not (MARSHALL, 1971).

Thus, the oceans contain habitats which differ in terms of light, temperature, pressure, and constancy. The variations of these with depth are important factors in the lives of deep-sea fishes. Some of the problems and consequences of living in the deeps are examined next.

B. Problems and Consequences of Deep-sea Life

I. Adjustment of Buoyancy

Some of the problems met by deep-sea fishes are common to all marine fishes. Sea water is a relatively dense and viscous medium, and energy must be expended to move through it. Though the water is dense, the tissue components of a fish, and particularly its skeletal portions, are more so, and thus a fish will tend to sink unless it can overcome gravity. This may be achieved by the presence in the fish's body of substances with a lower specific gravity than sea water. Certain deep-sea sharks, e.g. *Centroscymnus*, have large livers which contain quantities of squalene, a hydrocarbon oil. This substance has a specific gravity of 0.86, compared to sea water at about 1.025, and enables the fish to attain neutral buoyancy. Oil is also present as droplets in many fish eggs found in the surface waters; some of these are the eggs of deep-sea fishes, which may be laid at depth.

Some pelagic sharks are negatively buoyant, and in these cases the fish is prevented from sinking only by its own swimming. These sharks have heterocercal tails, the upper lobe larger than the lower, which impart an element of downthrust at the tail of the fish. Their large pectoral fins are set at a positive angle of incidence, and have a hydrofoil section. These features provide lift sufficient to overcome gravity provided the fish swims constantly.

The swim-bladders of teleosts are familiar flotation devices, but depending on gases for their buoyancy, they are more sensitive to alteration of depth, and hence pressure, than are the oil floats of sharks. If a fish with a swim-bladder is to undertake vertical migrations, and many of them do, it must be able to alter the tension of gas in the swim-bladder to the appropriate level for the depth of the moment. The swim-bladders of these fishes have gas glands with which is associated a rete mirabile. This rete is composed of arterial and venous capillary limbs, and functions as a countercurrent exchange system. By means of the rete, the very large tensions necessary to secrete gas under the ambient conditions of pressure at depth are attained. The matter is discussed further by MARSHALL (1971), who has shown that the length of the retial capillaries can be correlated with the depth at which a species lives.

The poorly mineralized skeletons of many deep-sea species, and their flabby musculature, not only require less food for their maintenance, but are nearer the specific gravity of sea water than more robust tissues would be.

Many shallow living fishes are denser than sea water, and live resting on the bottom, swimming clear of it to feed, and sinking back after short forays. This life style is well-adapted to the shallows, where there is plenty of food, and where the bottom provides shelter from predators as well as support.

The bottom of the deep oceans also has a population of truly benthic fishes without swim-bladders; there is also a population of fishes which live at the same depths, but which swim just clear of the bottom. These benthopelagic fishes possess swim-bladders, and are neutrally buoyant (MARSHALL, 1971).

II. Nutrition

Marine animals which must feed frequently are usually confined to waters at or near the surface, where food is relatively abundant and easily found. Deeper living animals may be predators or scavengers; many migrate to the surface to feed, either by day among the surface organisms, or at night when many day-feeding animals spread downward. With increasing depth food becomes scarcer, and more extreme adaptations to secure its capture are found.

Consider the activities which a fish must pursue if it is to survive as an individual and a species. It must obtain sufficient food to support its own metabolism, and to provide the nutrients in its eggs. It must avoid attack from predators, and it must reproduce in sufficient numbers to maintain the species. These activities are now considered further.

1. Sensing the Environment

To capture its food, a predatory fish must first locate it, and then carry out a successful attack. Location may be achieved by any of several senses, of which vision is chiefly considered here. This sense is undoubtedly important in most fishes, but some bathypelagic and benthic examples from below 1000 m have degenerate eyes, and can scarcely gain much useful information about their food through sight.

The olfactory organs are well-developed in a number of species, including some of those, e.g. *Cyclothone* spp. which have small eyes of doubtful efficiency, though it may be that in these cases the developed olfactory sense is more concerned with reproductive than with feeding behaviour.

The lateral line system is well-developed in ceratioid angler fishes and in some other bathypelagic species (MARSHALL, 1967, 1971), and in these groups the neuromast organs are free on the surface or raised above the surface of the body on papillae. Such fishes tend to lie motionless in the water, and so "noise" due to the fishes' own movement through the water will not be a disturbing factor. Mesopelagic species, and certain more active benthopelagic examples such as macrourids, halosaurs, and brotulids, have neuromasts which are buried in canals filled with viscous fluid. This will tend to reduce noise while allowing the vibration-detecting neuromasts to pick up the signal of water disturbances from nearby animals.

Among benthic fishes, living right on the bottom, are some from great depths, and these generally have small or degenerate eyes. Examples are the brotulids *Leucicorus lusciosus*, *Typhlonus* spp., and *Bassogigas profundissimus*, the bathypteroids *Benthosaurus grallator* and *Bathypterois longipes;* and the liparid *Careproctus kermadecensis* (THINES, 1955; MUNK, 1966a). Life is concentrated in a single plane at the bottom, which will make it easier for fishes which live there to find food. Some are probably well-endowed with taste organs on their external surfaces, e.g. macrourids, morids, and brotulids (MARSHALL, 1971). Even without eyes they can still use their olfactory and lateralis systems as well as gustatory clues.

2. The Act of Predation

Having located its prey by whatever means, a predatory fish must make a successful attack on it. This is further discussed in connection with the arrangement and structure of the eyes, but note that a complete attack may be said to involve three stages. The first is that in which the fish becomes alerted to the presence of something in its vicinity. This something may be prey, or a predator, or perhaps a potential mate. Few things will be of neutral significance in so competitive a milieu as the deep sea.

Once alert, the fish must make the appropriate response to the object. If it is prey this second stage will be an approach, which may be stealthy or a sudden rush. If the object of attention is a potential predator the response should be escape, and if the object is a potential mate, reproductive behavior may be appropriate.

The third stage of the attack may be called the snap; it is the complex of movements, differing in various species, by which the prey is seized or engulfed.

3. Avoidance

The avoidance of predators may be achieved in various ways. Size itself is important, since a large fish is less likely to fall prey to others than a small one. But size carries the penalty of needing a larger share of the already scarce food. Some fishes are more or less armoured, with plates or spines of bone, e.g. melamphaeids and macrourids. Others are concealed by their colouration, dark in bathypelagic examples such as melamphaeids or searsids, largely transparent in scopelarchids, at least when young, and silvery-sided in many mesopelagic species. It is chiefly among the latter that bioluminescence plays a part in concealment, by completing the countershading achieved by silvery scales and dark pigment. Among bathypelagic fishes certain species, including the searsids, possess luminous organs of the kind which emit a luminous cloud into the water. In this they resemble the mysid crustacean *Gnathophausia* and the cephalopod mollusc *Heteroteuthis dispar*, both of which eject luminous secretion. At least in the latter case this is believed to function much as the ink of shallow-living squids and octopus does, to distract the predator.

4. Capacity to Eat Large Prey

Because food is scarce at depth, and the opportunity to feed may occur infrequently, it will be advantageous for a fish to be able to eat as large a meal as possible when the opportunity arises. Evermannellids, some ceratioid angler fishes, and chiasmodontids are among those best able to do this; they can eat a meal larger than themselves. Even fishes with less extreme adaptations have large and effective jaws bearing formidable teeth.

III. Reproduction

Olfaction is one sense in which certain deep-sea fishes show sexual dimorphism. The males of some ceratioid angler fishes have large olfactory organs, and some, though not all, have small eyes (MUNK, 1964b). Male *Cyclothone* spp. have larger olfactory organs than the females, though the females grow larger, but the difference is less marked than in the anglers. This predominance of the olfactory system in the males is almost certainly connected with the problem of finding a mate, rather than with feeding. Male ceratioids have a substantial liver and large reserves of fat, and are believed not to feed once they have metamorphosed. It is likely that the females secrete a pheromone, and that this enables the males to find the females in the dark waters in which they live. MARSHALL (1971) discusses this question, pointing out that according to BERTELSEN (1951) there may be no more than 30 m between individual ceratioids, although there will be more between individuals of a given species. *Cyclothone* spp. are commoner, and there may be as little as 3 m between individuals in their case. He also points out that individual members of a single species are more likely to be clumped than random in distribution.

C. Bioluminescence and its Relation to Vision

Among marine animals there are many which emit light, including some elasmobranch and numerous teleost fishes. Since the distribution of luminous animals extends from the surface to the deepest waters, their light is available for vision well below the depths at which sunlight no longer penetrates, and some consideration of the features of such animal light is relevant to studies on vision.

The literature on bioluminescence is extensive; reviews by NICOL (1969a) and BODEN and KAMPA (1974) form a good introduction, and ANCTIL (1971) gives an extensive bibliography on the bioluminescence of fishes. A further review will not be attempted here.

Among fishes the light organs are of several kinds; some depend on symbiotic bacteria for the light production, and others contain intrinsic luminous tissue.

The luminous portion of the system is commonly part of a complex photophore, and here, too, a wide variety of structures are found, including reflectors, diffusers, filters, occluders, and lenses.

The questions of particular interest concerning bioluminescent organisms in the present context concern the intensity of light they emit, and how its characteristics accord with what is known of the eyes of fishes. The luminescence of some marine animals is bright enough to be seen by man at distances of several meters when they are captured. Though such luminescence in captive animals may be brighter than that which occurs in more natural circumstances, and though water is less transparent than air, the light should be easily visible to deep-sea fishes at close range. Direct observations by BEEBE (1934) and RECHNITZER (quoted by BODEN and KAMPA, 1974) suggest that luminescence at depth is common, and certainly bright enough to be seen by man through a column of water and the window of the submersible.

Instrumental measurements give more objective information, and many have now been made. CLARKE and WERTHEIM (1956) used a bathyphotometer to measure luminous intensity at depth, and they recorded abundant flashes at depths below 200 m in the Atlantic. Between this depth and 300 m the flashes were 2–5 times greater than the downward irradiance, and below 360 m the light was as bright by night as it was by day. Some flashes were as much as 1000 times as bright as the background illumination.

The development of echo-sounding has led to the discovery of the sonic scattering layers in the oceans, and hence to investigation of their behaviour and constituent organisms. Some of these layers are due to populations of crustaceans, particularly euphausiids, which are known to be luminous. In other examples, a sonic scattering layer has been shown to be due to myctophid fishes; schools of *Tarletonbeania crenularis* were observed by BACKUS (1968) hanging motionless in the water: He saw no luminescence though the fishes have photophores.

BODEN and KAMPA (1974) point out that most records of bioluminescence show meso- and bathypelagic organisms emitting blue light around 470 nm, though those in coastal and shallow waters tend to give a greener light. Emission spectra, when obtained, approximate to the absorbance curves for visual pigments of animals from comparable depths. The match is not exact, but the overlap is enough to confirm that emitted light is spectrally suitable for vision. It is also close to the fraction of sunlight which penetrates ocean water best.

BODEN and KAMPA (1974) give a table of intensity measurements from their own work and that of others, notably CLARKE et al. (1962) and NICOL (1958). In these cases the luminescence was evoked by nonphysiological stimuli. KAY (1965), who used photic stimuli on the euphausiid *Meganyctiphanes norvegica*, however, obtained comparable results to KAMPA and BODEN (1957), who used fatal doses of ammonia. The values obtained for the radiant flux from myctophid fishes (*M. punctatum*) were 0.925×10^{-9}–52.345×10^{-9} μW cm^{-2} receptor surface at a distance of 1 m. Euphausiids, which occur at comparable depths and which are of suitable size to be eaten by many mesopelagic fishes, gave values 1.3×10^{-9}–3240×10^{-9} μW cm^{-2} at 1 m. Evidence that euphausiids can perceive the luminescence of others is given by MAUCHLINE (1960), who observed that if one captive example luminesced, others in nearby containers also flashed.

Measurements of threshold for vision in deep-sea fishes have not been carried out directly, but values based on human vision have been derived by DENTON and WARREN (1957), NICOL (1958, 1962), and CLARKE and DENTON (1962).

BODEN and KAMPA (1974) approach the problem indirectly by considering the populations of animals in the sonic scattering layers. These layers followed a certain light intensity from the sun, moon, stars and sky; the value was approximately 1×10^{-4} $\mu W\,cm^{-2}\,nm^{-1}$ at 470 nm (KAMPA and BODEN, 1957; BODEN and KAMPA, 1967; KAMPA, 1970). The scattering layer was about 60 m thick, and from attenuation curves the authors calculate that the intensity at the bottom of such a layer will by an order of magnitude less than the figure quoted for the top. They calculate that the constituent animals, including euphausiids and fishes, must be able to distinguish irradiance levels greater than 10^{-4} and less than 10^{-5} $\mu W\,cm^{-2}\,nm^{-1}$. They give the quantal equivalent of their figures, and from the diameter of the euphausiid corneal facets (about 60 μm) they show that at the bottom of the scattering layer each euphausiid ommatidium receives approximately 658 quanta per second at 470 nm. From their measurement of the luminescence of euphausiids they deduce that one euphausiid should be able to see the light from another at a distance of about 50 cm. These values and calculations are admittedly not for fishes, but they are from animals which live at the same depth, and which certainly fall prey to fishes.

NICOL (1969a) considers the question on the basis of figures obtained by himself (1958) and CLARKE et al. (1962) which place the light intensity of lantern fishes at about 1×10^{-6} $\mu W\,cm^{-2}$ at 1 m. From these figures, and from the values of sensitivity from DENTON and WARREN (1957) he estimates that the fishes should be able to see the lights of another at a distance of 10–16 m; this, he points out, would be an optimum distance, which in practice may well be considerably less.

The function of luminous organs has been the subject of much speculation, but in a number of cases there is evidence to confirm the suggested ideas. The use of light as a lure to attract prey is suggested by the presence of luminous barbels in stomiatoids, luminous fishing tentacles in ceratioid angler fishes, and the photophores in the mouths of, for example, *Neoscopelus* and *Chauliodus*.

Isolated lights flashing briefly must be a common indicator of the presence of some otherwise invisible animal at depth, and a number of organisms probably have a behaviour pattern which causes them to attack such an isolated light without identifying it further. Usually this will result in the animal catching a small crustacean or some other acceptable food, but if the light is the lure of a more powerful carnivore, the erstwhile predator may itself become the prey. NICOL (1969a) notes that lures which have been seen to luminesce have emitted feeble lights; this might be advantageous to their owners, since too bright a light might enable the prey to see the body of the predator and not merely the isolated light.

Ceratioid and searsid fishes, and a variety of crustaceans and cephalopods, are known to emit clouds of luminous secretion, probably in response to attack. Such discharges are believed to confuse or otherwise deter predators, so allowing the discharger to escape. The ink of shallow water cephalopods is used in this way, and the parallel between these and the deep-living squid *Heteroteuthis*, one of those which ejects a luminous cloud through its syphon, is close.

An obvious possible function of light organs is to illuminate potential prey. In some cases fishes have indeed been observed feeding on organisms which were illuminated by their photophores. BEEBE and VAN DER PYL (1944) observed *Gonichthys coccoi*, a myctophid, feeding on copepods which entered the light from its ventral photophores. The special case of the malacosteids, which were first observed by CHUN (1900) emitting green light from some photophores and red from another, is discussed in the section on visual pigments.

The patterns of photophores on the flanks of myctophids are used as recognition features by taxonomists, and it has been suggested that the same use may be made of them by the fishes themselves. PEARCY (1964) found that three species of Pacific myctophids occurred in strata at the same geographical location, though the layers overlapped. Such stratification might be due to schooling behaviour, but it could equally be due to the three species having different preferences for light levels.

Some fishes have different patterns of light organs in the two sexes, and in these cases it may be that they function as recognition signals between the sexes. Other sexual differences may also be present; BEEBE and CRANE (1939) quote the example of *Idiacanthus fasciola*, in which the male is much smaller than the female and so might well be eaten. Despite its small size, the male has a large cheek photophore, which is much less developed in the female.

A cursory look at a sample of deep-sea fishes is enough to show that the possession of ventral facing photophores is widespread. A number of examples, e.g. *Argyropelecus*, *Opisthoproctus*, also have dorsally directed tubular eyes. It appears likely from mere inspection that ventral photophores could function to camouflage the ventral silhouette of the fish against downwelling light. At depth the angular distribution of this light is constant, and though markedly unequal, in that there is much more downwelling than upwelling light, the distribution at mesopelagic depths is largely independent of the position of the sun in the sky and the presence or absence of clouds.

The work of DENTON and NICOL (1966 for references) has shown that the crystals of guanine present in the silvery surface tissues of fishes are often so arranged as to reflect light from the side of the fish in a way that matches the background light. This matching by reflection is efficient at angles of view to the side of the fish, but downwelling light is so much brighter than upwelling that reflection is not adequate to conceal the silhouette from below.

DENTON (1970) has shown that the photophores of *Argyropelecus* are so constructed that some of their light will undergo multiple reflections within the photophore before emerging, and that the angular distribution of the light from the ventral photophores matches that of downwelling daylight.

The hatchet fishes including *Argyropelecus*, and many other mesopelagic species have arrays of discrete photophores on their ventral surfaces, but *Opisthoproctus* has no externally obvious photophores. BERTELSEN and MUNK (1964) described the rectal light organ in these fishes, and the hyaloid body which lies beneath the skin of the curiously flat "sole" ventrally. They quoted the suggestion of CLARKE (1963) that ventral luminescence may serve to match downwelling light, but at that time there was no direct observation of the luminescence of *Opisthoproctus*. HERRING (1975) showed that BERTELSEN and MUNK's interpretation was correct, and that the sole of *Opisthoproctus* does indeed show luminescence all over.

Another case has been described in the myctophid *Tarletonbeania crenularis* by LAWRY (1974). This fish can be captured at the surface by night, and can be kept alive for some time. In total darkness the fish did not luminesce, but when illuminated from above by dim blue light its ventral photophores were luminous. LAWRY showed that the eye is so arranged as to see the light from a supraorbital photophore and from the surface simultaneously on adjacent regions of retina. If the ventral luminescence and the downwelling light could thus be matched, the fish would be camouflaged. Such matching might be achieved by altering the emitted light to match the natural, or vice versa by vertical migrations. From the observations of sonic scattering layers which follow isolumes, it appears that coarse matching may be done by vertical migration. Finer matching, as of transient changes in downwelling light due to clouds obscuring the sun, might well be carried out by adjusting the intensity of luminescence.

D. The Functional Anatomy of the Teleost Eye

The eyes of most vertebrates, including fishes, conform to a similar pattern. They consist of three layers, from without inward corneo-scleral, uveal, and retinal, and they have transparent media, the cornea, aqueous, lens, and vitreous, through which light passes before it reaches the sensitive elements of the retina.

The eyes of fishes show certain features which are connected with life in an aquatic environment; a brief account of the generalized teleost eye is now given (Fig. 3), and some terms which will be used below are defined (Fig. 5).

The outer coat of the eye includes the sclera, which in most cases forms a flattened hemisphere. The flattened portion abuts against the corresponding area of the other eye in the midline of the fish in many cases. The sclera is composed mainly of collagen, but is reinforced near the equator by a ring, sometimes incomplete, of hyaline cartilage. Some fishes, particularly large and fast-swimming ones, have plaques of bone in the sclera near the corneoscleral junction. The sclera is pierced by the optic nerve and by vessels, and receives the insertions of the extraocular muscles.

The cornea completes the outer coat of the eye; it is almost the full diameter of the eye in many cases, and its structure is more complex than it is in terrestrial animals. The inner (vitread) portion is part of the corneo-scleral coat, continuous with the sclera at its periphery. This portion consists of stroma, Descemet's membrane in those species in which it is present, and the endothelial lining. Overlying the deeper part is a dermal component; this may include more stroma, and the surface epithelium. Frequently there is a tissue space, crossed by tenuous fibres, between the superficial and deep parts, when movement may take place between the two. In many fishes the corneal epithelium is smoothly continuous with the surface epithelium of the head, and there is no conjunctival fold. In others, e.g.

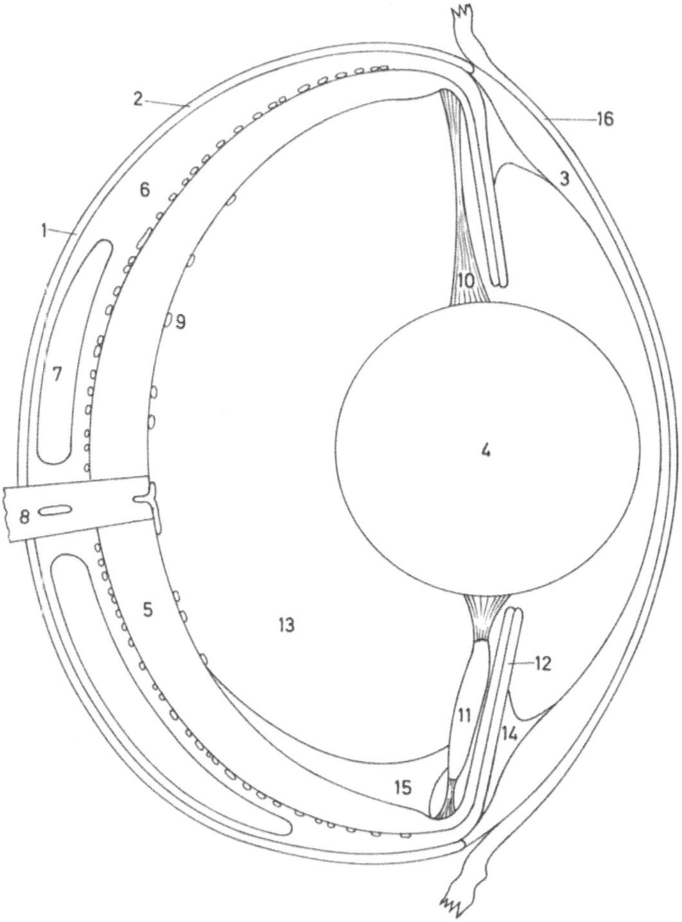

Fig. 3. Diagram of vertical section through generalized teleost eye. *1* Sclera. *2* Scleral cartilage. *3* Cornea. *4* Lens. *5* Retina. *6* Choroid. *7* Choroidal rete mirabile. *8* Optic nerve. *9* Hyaloid vessel on vitreal surface of retina. *10* Suspensory ligament of lens. *11* Lens muscle. *12* Iris. *13* Vitreous. *14* Annular ligament. *15* Falciform process. *16* Dermal component of cornea

some clupeids, salmonids, and carangids, so-called adipose eyelids are present. These are transparent, and streamline the otherwise protuberant eyes (MARSHALL, 1971). The corneal curvature is convex, so that the eye projects from the surface of the fish, but the very regular and symmetrical curves which characterize most terrestrial corneas are not always present in fishes. Such asymmetry is very marked in certain fast swimming fishes such as tuna. In these the corneal curvature hardly deviates from the surface contour of the streamlined head; the horizontal curvature is gentle and the vertical curve much more pronounced. Such a cylindrical cornea in a terrestrial animal would give rise to gross astigmatism, but living in water, fishes have liquids with almost identical refractive indices on each side of their cornea, so the optical effect of this corneal asymmetry is negligible.

The irido-corneal angle of the anterior chamber may be occupied by transparent tissue. In some cases this is an additional layer of cornea, the so-called autochthonous layer. In other examples the angle is occupied by cells which appear empty in sections. This cellular mass is known as the annular ligament, though it is not ligamentous in structure. Neither type of angle tissue is easy to see in life, suggesting that its refractive index is very close to that of the aqueous. The corneas of certain shallow-living fishes show marked iridescence, and others are yellow coloured (LYTHGOE, 1971, 1975, LOCKET, 1972). Those of deep-sea fishes are usually colourless and very transparent.

The uveal and retinal layers are closely associated, and together are referred to as the eye cup. The uveal coat consists of the choroid and the stromal tissue of the iris; the ciliary region so prominent in most terrestrial eyes is scarcely represented in fishes. The choroid is a vascular tissue, usually containing much dark pigment. Its deepest layer, the choriocapillaris, is closely applied to BRUCH's membrane and the retinal epithelium.

Present in the choroid of many teleosts is a structure called, in the older literature, the choroid gland, but which is better called the choroid rete mirabile. This rete is commonly horseshoe-shaped, the two limbs surrounding the optic nerve, but it may be disciform or divided into two separate lobes. Its structure and function are discussed further below (see the section on retinal nutrition and ocular vessels).

Covering the surface of the choroid there is commonly a silvery layer, the argentea, and in few teleosts but most elasmobranchs a reflecting tapetum is present in the deep layers of the choroid. These structures are also discussed further below (see section on the tapetum lucidum).

The iris in teleosts is generally immobile, and the pupil round, though a significant number have irregularly-shaped pupils. The commonest of these are those in which there is a gap in the rostral part of the pupil in a laterally situated eye. Such gaps uncover the margin of the lens, and so are called aphakic apertures. Such rostral and other types of aphakic apertures are particularly common in deep-sea fishes, and are discussed below. Animals which live in brightly lit environments often have mobile pupils, in some cases including lobes of the iris which cover all the pupil except for small but multiple gaps. Though such cases include shallow water elasmobranchs such pupils are not found in deep-sea fishes.

The retina is closely applied to the inner surface of the choroid, and in form is hemispherical or nearly so. Typically in shallow water fishes it is a well-developed and complex structure, containing many cells arranged in well-defined layers, and generally including two classes of visual cells, rods, and cones. The cones are generally larger than the rods; they may be of several types, including double (a conjoined pair in which the two elements are dissimilar) and twin (apparently identical elements). Many species have cones of several types arranged in rather precise geometrical patterns, with numerous small rods penetrating between the cones to lie more sclerad.

This at least is the situation in a light-adapted eye, but many teleosts show photomechanical movements of the visual cells and the pigment epithelium. In such a case the cones and rods have the positions described above in the light-adapted state, and the pigment granules are aggregated in the vitread tips of the pigment cell processes. In dark adaptation the rod myoids shorten, and those of the cones elongate, so that the rods now lie close to the outer limiting membrane and the cones are more sclerad. The pigment migrates to the sclerad ends of the pigment cells.

It is common to find different retinal structure in different regions of the fundus; the dorsal and ventral regions of the retina in for example *Callionymus lyra*, a bottom living shallow marine fish, are quite distinct (VILTER, 1947). In some cases an area temporalis is developed, and there may also be a fovea. Both may be present in deep-sea fishes, and are discussed later.

The retinal periphery shows a narrow region in which the structure changes from the highly organized functional retina to the undifferentiated two-layered posterior epithelium of the iris. In this region retinal growth and differentiation takes place during the growth of the fish (LYALL, 1957).

The lens of fishes is almost spherical. It is the main refracting element of the eye, since the cornea has so little optical effect, and is usually very transparent. The lens projects far through the pupil, and may project beyond the outline of the general body surface. The spherical shape of the lens means that it is optically symmetrical, forming an equally good image in any direction. This is in contrast to the situation in terrestrial animals, in which the cornea is a strong refracting element, and the lens flattened, so that there is a well-defined optic axis.

The combination of lens and retina in fishes is usually symmetrical; the lens being spherical and the retina hemispherical and concentric with the lens, the two have a constant relationship to each other. This was first observed by MATTHIESSEN (1880), who found that the focal distance of the fish lens, at which distance the retina is situated, is 2.55 times the lens radius, a relationship now known as Matthiessen's ratio.

The fish lens, though spherical, does not suffer the aberrations which a glass sphere of uniform refractive index does, indeed the image is surprisingly good across the whole diameter of the lens. PUMPHREY (1961) pointed out that in most cases the pupil is wide, exposing the full aperture, and that the lens is normally free from internal discontinuities and surface reflections. PUMPHREY quoted the observations of MAXWELL (1854) who considered a hypothetical universe, spherically symmetrical and boundless, in which all rays of light follow a circular course. Such a universe would have a finite refractive index at the centre, falling off in a prescribed way with increasing distance from the centre. MAXWELL was stimulated to consider such a system by the contemplation of the lens in fishes, and was aware that such lenses approached his ideal to a considerable extent.

MATTHIESSEN (1880) was aware that for a lens to possess the properties which he observed, the refractive index would have to fall progressively from the centre to the periphery.

Discussing the efforts made by himself, and by FLETCHER et al. (1954) at his instigation, to calculate the refractive index of such a fish lens, PUMPHREY stated that a homogeneous spherical lens with a focal length comparable to that observed in the fish lens would require a refractive index of 1.67, equal to that of a hard flint glass. In such a homogeneous lens the spherical aberration would be such that only 25% of the light from a distant point source would go anywhere near the focus, and only 5% near enough to fall on an element the size of a fish rod.

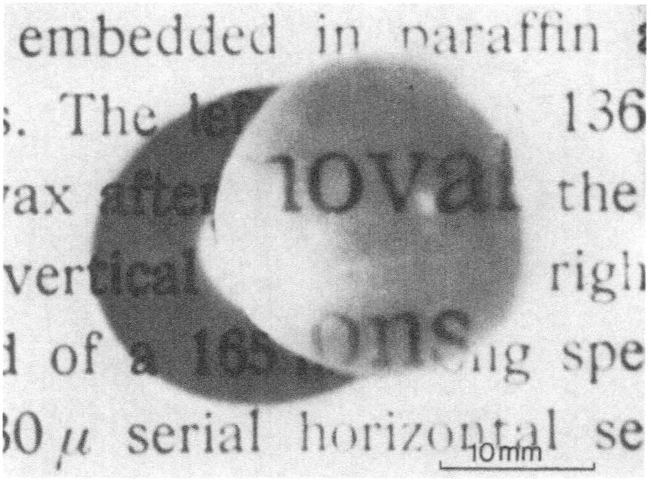

Fig. 4. Lens of *Mora moro*, deep-sea gadoid, immersed in sea water in acrylic box, which is resting on some print. Undistorted magnified image extends to edges of spherical lens, demonstrating that lens is free from spherical aberration. FLETCHER et al. (1954) computed how refractive index of spherical lens with these characteristics must decrease from centre to periphery

Direct measurement of the refractive index at points on a radius is not possible, since dissection in order to do so would destroy the optical properties of the lens. But if the lens can be shown to have undetectably little aberration then it must fit the computations of FLETCHER et al. PUMPHREY demonstrated this lack of aberration by placing the lens over a patterned surface and observing the resulting image. This experiment is repeated in Fig. 4, in which some print is photographed through a fresh lens of *Mora moro*, a deep-sea gadoid.

In most vertebrates the lens is flattened, and accommodation is brought about by contraction of the ciliary muscle. This relaxes the tension on the lens equator, and the elastic capsule then causes the lens to become more spherical. But the fish lens is already spherical, and cannot become more so to bring about accommodation. The teleost lens is held in position by a dorsal suspensory ligament, which arises from the inside of the iris near the ora terminalis. From the ventral aspect of the inside of the iris, also near the ora, and commonly associated with the choroid fissure, arises a small muscle, the retractor lentis (BEER, 1894) or campanula Halleri. This muscle inserts into the ventral aspect of the lens by a small tendon; the vessel and nerve to the lens muscle may be located in the falciform process, a thin plate or fold of connective tissue projecting into the vitreous from the choroidal fissure.

Views on the precise mechanism of accommodation in fishes have been diverse, and were reviewed by MUNK (1973). He stated that the lens is generally moved towards the zone of highest cone density, and that the movement is usually in a caudal direction. Such movements of the lens within the eye cup will bring one area of retina into focus, but at the same time will cause other areas to become more out of focus.

Fuller accounts of the teleost and other eyes are given by WALLS (1942), ROCHON-DUVIGNEAUD (1943), and DUKE-ELDER (1958).

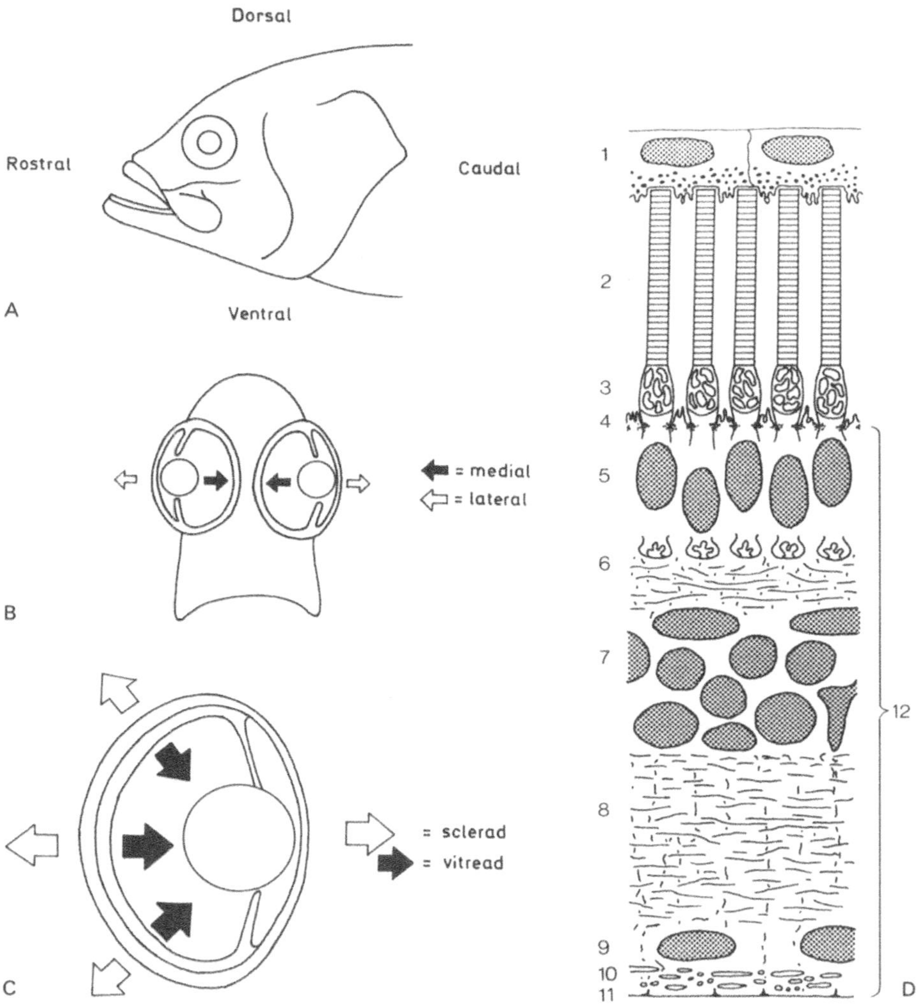

Fig. 5a–d. Diagrams to illustrate meanings of descriptive terms used. Terms in Fig. 5a and 5b are with respect to fish, and those in Fig. 5c with respect to eyeball. Terms in Fig. 5d are in common use except for inner retina, term useful in present context. 5d; *1* Pigment epithelium. *2* Outer segments. *3* Inner segments. *4* Outer limiting membrane. *5* Outer nuclear layer. *6* Outer plexifom layer. *7* Inner nuclear layer. *8* Inner plexiform layer. *9* Ganglion cell layer. *10* Nerve fibre layer. *11* Inner limiting membrane. *12* Inner retina

I. Terminology

Many terms commonly used in anatomical descriptions originated in human anatomy. Due to differences in posture, and to the relative proportions of the eye and other structures, some of these terms are ambiguous when applied to other animals. Describing the eyes, their position in the head, and their directions of gaze in fishes, I use terms related to the axes of the fish. Dorsal and ventral, rostral and caudal, lateral and medial are used as shown in Fig. 5a and 5b. For referring to the relative positions of structures within the eye, vitread and sclerad are used as defined in Fig. 5c. Though typical fish eyes face laterally, and not forward as in man, the terms anterior and posterior chambers are retained, since they are familiar.

The retina is conventionally described as being layered. Though these layers are in some cases not physiologically distinct, they are familiar and serve a descriptive purpose. They are used as shown in Fig. 5d. In the retina the extracellular space between the inner and outer segments of the visual cells and the radial fibre microvilli on the one hand, and the vitread faces of the pigment epithelium cells on the other, corresponds embryologically to the ventricles of the brain. This space in the retina is called the ventricular space.

In some cases it is useful to consider that part of the retina which lies between the outer and inner limiting membranes. This I refer to as the inner retina. The term neuro-retina is used to denote that part of the retina which is derived from the vitread layer of the optic cup: This includes all the retina except for the pigment epithelium.

In many deep-sea fishes the retinal and uveal components of the eye are closely related to each other, but easily separable from the lens and from the corneo-scleral coat. The uveo-retinal moiety, which I call the eye cup, should be distinguished from the optic cup. The latter is the cup-shaped structure formed in early embryonic life by the invagination of the primary optic vesicle. The optic cup consists of two layers of neuro-ectoderm: From the sclerad layer develops the pigment epithelium, and from the vitread layer the neuro-retina. The two layers extend over the vitread surface of the iris and are continuous with each other at the pupil margin.

Due to the marked asymmetry of certain deep-sea fish eyes, special terms are required. These are defined as they are introduced.

E. General Features of Deep-Sea Fish Eyes

Examination of the eyes of a number of deep-sea fishes shows that certain features, not commonly found in shallow water species, are rather often present in deep-sea forms.

The eyes may be of apparently normal proportions, but in many examples they appear large in relation to the size of the fish. In others however, the eyes are small, even to the point of degeneracy, a reduction found in species which live below the limit of penetration of sunlight. The relation of eye size to depth of habitat is discussed further below.

In other cases the whole eye is of normal size but the lens appears large. A disproportionately large lens would mean that Matthiessen's ratio is not obeyed, and in certain forms this is the case. Others have a large lens, but in an eye of unusual proportions, so that though Matthiessen's ratio is obeyed with respect to a particular region of retina, elsewhere in the retina it is not. The field of vision subserved by retina obeying Matthiessen's ratio is much more restricted in such asymmetrical eyes than in the more typical teleost eye.

The cornea and sclera are commonly thin, with the cornea very transparent. The shape of the corneo-scleral envelope varies markedly in many cases from the typical fish eye shape. Though many deep-sea fishes do have eyes of the normal shape, in others the eyes are more or less tubular. The proportions of corneal and scleral components in such eyes differ according to species; some only show a slight asymmetry, but in others the eye has become almost cylindrical. Tubular eyes and their features are discussed further below.

The uveal components are strikingly reduced in some deep-sea fishes, the choroid thin and containing few small blood vessels. Though there are certainly examples which have a well-developed choroid, including a choroidal rete, reduced choroids are perhaps more common. The iris in shallow living fishes is generally well-formed, and is often pigmented, or coated by the silvery argentea. Among deep-sea fishes, the argentea is found particularly in mesopelagic examples which have silvery scales on their bodies, such as myctophids. Bathypelagic and benthic fishes generally have dark irides.

Among shallow living fishes it is mainly those with laterally placed eyes and a fovea which show rostral aphakic apertures, (Fig. 6). Aphakic apertures are common in deep-sea fishes, in which they may be rostral as in foveate shallow water fishes, or circumlental, of approximately the same width all round. The occurrence and function of such apertures was discussed by Munk and Frederiksen (1974) and is further considered below.

The retina in most deep-sea fishes shows marked differences compared to shallow water fishes, in which the inner retina is preponderent. In deep-sea examples, on the other hand, a large proportion of the retinal thickness is commonly due to receptor outer segments. In almost all cases the visual cells are all rods: In some, the outer segments are particularly long; in others there is a great thickness of outer segment material, but this is due to the presence of more than one bank of rods. Such multibank retinae are considered further below. Most multibank retinae have well-defined layers of rod inner-outer segment complexes, but there are others in which the rods, though present in depth, are not arranged in discrete banks. In the majority, whatever the arrangement of rods in the radial direction, they are packed tightly together in the tangential direction. Not only are they tightly packed, but they are usually of very even diameter, when the packed array as seen in tangential sections shows marked regularity.

From the density and packing of the rods one might expect that such a retina would contain a high density of visual pigment, and this was shown to be the case by Denton and Warren (1957). Not only are the visual pigments present in large amounts, but their maximum absorbance is at a shorter wave length than are those of shallow water fishes.

The cones of shallow water fishes contain different pigments, with different wavelengths of maximum absorbance. Deep-sea fishes, with few recorded exceptions, have no cones, and only one visual pigment. An interesting exception, the malacosteid *Aristostomias scintillans*, studied by O'DAY and FERNANDEZ (1974) is discussed below.

The inner retina (as defined above), usually shows a marked paucity of cells compared to the same region in shallow water fishes. The outer nuclear layer, which contains the rod nuclei, may be thick and well-populated in those with many rods, or may contain fewer but closely packed nuclei in those with a single bank of long outer segments.

The inner nuclear layer typically contains the nuclei of the bipolar, amacrine, and horizontal cells, and those of the radial fibres. Those cell families probably occur in most deep-sea fishes, though in some retinae, showing degenerate characteristics, it is hard to identify them. Many deep-sea examples have apparently separate layers of bipolar and amacrine cells, separated by a layer of lamellated material derived from radial fibres. In certain other animals, e.g. gecko (PEDLER, 1963a), the radial fibres give rise to lamellae, but those in deep-sea fishes are particularly prominent and well-developed (LOCKET, 1971a, and below).

The horizontal cells of fishes are usually present in three layers, and make connections with particular families of visual cells. (STELL, 1975; STELL and LIGHTFOOT, 1975; and SCHOLES, 1975 for recent discussion of teleost horizontal cells.) In deep-sea fishes there may be one distinct layer of horizontal cells, as in *Sternoptyx diaphana* (LOCKET, 1970) and the scopelarchids (LOCKET, 1971b) but in other cases the horizontal cells are not recognizable as such by normal microscopic techniques. In these cases the cells may be absent, or may be of protoplasmic type as in mammals, and so not obvious morphologically.

The presence of large numbers of rods, but few bipolar cells and ganglion cells, implies that there is a high degree of convergence in deep-sea fish retinae, and the optic nerve fibre counts reported bear this out (MUNK, 1966a).

The pigment epithelium of the retina is prominently developed in most shallow water fishes, and in them commonly contains abundant granules of melanin located in processes which project vitreally between the outer segments of the visual cells. The processes, their contained pigment, and the outer segments all take part in the photomechanical movements with light and dark in such fishes. BRAUER (1908) commented on the poor development of the pigment epithelium in many deep-sea fishes. MUNK (1966a) did not find such a marked reduction of this epithelium as BRAUER had reported, and was inclined to think that the apparent reduction might have been due to artifacts. As seen in my material, fixed for electron microscopy and embedded in plastic, the pigment epithelium is in many cases thin, consisting of low cells without processes projecting vitreally, and making only small areas of contact with the outer segments. In this, my experience tends to confirm BRAUER's contention that the pigment epithelium is often poorly developed.

There are also cases, as BRAUER noted, in which the pigment epithelium is strikingly developed, but not in the same way as in shallow water fishes. In these cases the epithelium contains the reflecting or scattering material of a retinal tapetum lucidum, which may replace or coexist with the dark melanin granules. The types, structure, and function of tapeta are discussed in more detail below. (See section on tapetum lucidum.)

F. Adaptive Features of Deep-Sea Fish Eyes

I. Modifications of Otherwise Typical Lateral Eyes

Three environments were distinguished above on the grounds of their light regime. At the surface, light is abundant by day. In the mesopelagic zone light falls off, becomes more monochromatic, and the diurnal periodicity becomes less marked with depth. Below, at bathypelagic depths, sunlight does not penetrate, and what little light is present is due to bioluminescence.

In the mesopelagic zone the diurnal rhythm is apparent, and most animals in this zone perform vertical migrations, in some cases at least following an isolume. This is the zone in which ventral light organs are best developed, and in which luminescent camouflage against the downwelling daylight will be of most advantage.

In the lower reaches of the mesopelagic zone, and indeed in shallower water at dawn and dusk, the conditions of illumination range about threshold. Adaptations to increase absolute sensitivity may confer an advantage over a less sensitive animal at a given depth, or increase the depth range of a species.

1. Eye Size and Depth

Many mesopelagic fishes have large eyes, and the properties of such eyes, large in proportion to the fish, are now considered. In even diffuse illumination a large eye will intercept more quanta than a smaller one; from a near point source a large eye will receive a greater proportion of the emitted light than a smaller, because the large eye subtends a greater solid angle at the source. But the sensitivity will depend on the proportions of the eye, rather than its size, and in particular the optical properties of the eye and its components.

In a lens system such as that of the eye, or of a camera, the illumination of the image depends on the ratio of the focal length and the diameter of the pupil. This quantity, called the aperture, is adjustable in man by the mobile iris. In teleosts the iris is not mobile, and the lens projects through the pupil, the diameter of which is almost the same as that of the lens in many cases. Typically the teleost lens is spherical, and its radius and focal length are related by Matthiessen's ratio, regardless of the size of the eye.

Enlargement of the eye, as MARSHALL (1954) pointed out, does not result in greater retinal illumination, provided the enlargement is to scale. The area of a round pupil will be enlarged by a factor of four if its diameter is doubled, but so will that of the retinal image; the retinal illumination will be constant in eyes of different sizes providing they are of uniform proportions.

Though typical lateral eyes are of uniform proportions, some factors concerned with eye size may still be significant. A large eye necessarily has a longer light path through ocular tissues than a small one, and the light will be absorbed and scattered to some extent by these tissues, very transparent though they are in deep-sea fishes.

In near threshold conditions it will be advantageous to trap as high a proportion of the incident light as possible, and several features are found which will tend to increase this proportion. The light is absorbed in the outer segments by the visual pigment, first studied in deep-sea fishes by DENTON and WARREN (1957) and by MUNZ (1958). They reported high pigment densities, with maximum absorbance near the wave length of light best transmitted by ocean water. One way in which the quantity of visual pigment in a retina could be increased is by the presence of long outer segments; this adaptation is found in a number of species, including sternoptychids and searsids (LOCKET, 1970, 1971a). In *Diretmus argenteus* and *Scopelarchus sagax* there are areas in which the outer segments attain lengths of hundreds of micrometers (MUNK, 1966b; LOCKET, 1971b). The occurrence and implications of long outer segments are further considered below.

Though in shallow living fishes the cones in particular are often arranged in patterns, the visual cells do not give the impression of being packed very tightly. In deep-sea species they are; in many cases the rods are of very even thickness, and tangential sections show a close and uniform hexagonal array. In such a case there seems at first sight little opportunity of increasing the density of rod packing, but it is common in deep-sea fishes to find more than one bank of receptors. Such multiple bank retinae contain from two to six layers of rods, and if these are of uniform diameter, addition of a second bank would immediately double the number of rods in a given area. This feature is also discussed further below.

Even if the outer segments are long, or more than one bank present, it may be that some quanta escape absorption by the visual pigment. In diurnal animals the incident light not absorbed by visual pigment is absorbed instead by the melanin in the retinal epithelium and the choroid. If it were not, it would be scattered within the eye with consequent degradation of the image. In very dim conditions it may be advantageous for such scattering, or for ordered reflection, to occur, since the light may be absorbed on a second passage through the outer segments. Retinal or choroidal tapeta lucida produce such reflection or scattering; they occur in many deep-sea fishes, and are considered later.

2. Rostral and Circumlental Aphakic Apertures

The pupils of shallow water teleosts are in most cases round, and slightly smaller than the diameter of the lens. The lens projects through the pupil, and the iris is often somewhat bowed toward the cornea, forming a flattened cone rather than a plane. In some examples there is a considerable gap between the lens and the margin of the iris rostrally, e.g. *Serranus cabrilla* (BEER, 1894).

Rostral aphakic gaps of this type are also found in a number of deep-sea teleosts, and in these they are usually larger than in the shallow living species (Fig. 6). Other deep-sea species have an aphakic gap all round the lens, narrow in some, but in others so large as to leave scarcely any iris (Figs. 7a, b, c).

In deep-sea fishes with a rostral aphakic aperture this is often associated with the presence of an area temporalis, e.g. *Nansenia groenlandica* and *Bathylagus pacificus* (MUNK, 1966a). In other cases there is a fovea, of the shallow (concaviclivate) form, e.g. *Bathylagus benedicti* (= *B. euryops*) (VILTER, 1954a), and deeper (convexiclivate), in others, e.g. *Platytroctes procerus* (= *P. apus*) (BRAUER, 1908; LOCKET, 1971a), *Platytroctegen mirus* (MUNK, 1966a) and *Bathylaco nigricans* (MUNK, 1968). The area temporalis and both kinds of fovea are considered below.

MUNK and FREDERIKSEN (1974) considered the function of aphakic apertures, testing their ideas with mathematical models. They consider only eyes which conform to Matthiessen's ratio: In such an eye the lens subtends an angle of 47° at the retina, corresponding to an aperture of 0.8. This is true if the object is normal to the pupil plane, and light from it is not cut off in any way by the iris, but light striking the eye obliquely is partly so interrupted (Fig. 8). In such a case the retinal illumination can only be maximal over an area corresponding to the full aperture; objects in front of the fish, and thus oblique with respect to the laterally placed eyes, will not fall within this area.

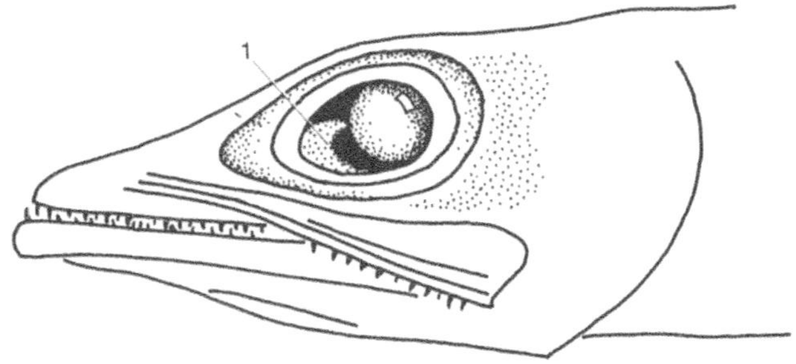

Fig. 6. Head of *Scopelosaurus* sp. This fish has a pronounced rostral aphakic gap, *1* frequently, as in this instance, associated with presence of temporally located fovea

MUNK and FREDERIKSEN showed that the rostral aphakic gap in shallow water fishes will admit a wide pencil of rays to the temporal area or fovea, so that the retinal image of a point source will have a bright central zone which is not large compared to the area of a single cone. They suggested that the rostral aphakic gap will enable full use to be made of the high cone density in that region of the retina.

They further suggested that the larger rostral aphakic apertures in deep-sea teleosts will increase the retinal illumination, and thus sensitivity, particularly in the caudal region, including the fovea where one is present. They did not regard the possible extension of the binocular field forward by the rostral gap as so important, since the location of the eyes, and the presence of sighting grooves along the snout is sufficient to give a large overlap of the two fields in any case, as it does in shallow water species.

Considering circumlental aphakic apertures, they showed that a comparatively small gap will considerably enlarge the area of retina which can receive an image at full aperture, and they suggested that this will facilitate recognition of laterally located objects under near-threshold conditions. In the extreme case of an object with the same radius as the lens, situated laterally and emitting light at just threshold intensity; if the pupil is of the same diameter as the lens, any movement of the object or the fish which caused the pupil plane to become oblique with respect to the object-lens-retina axis would cause the lens aperture to be partially occluded, so reducing the intensity of the image below threshold.

In those cases which have a large circumlental gap, the relative retinal illumination over the whole retina will be significantly increased, and in such cases the

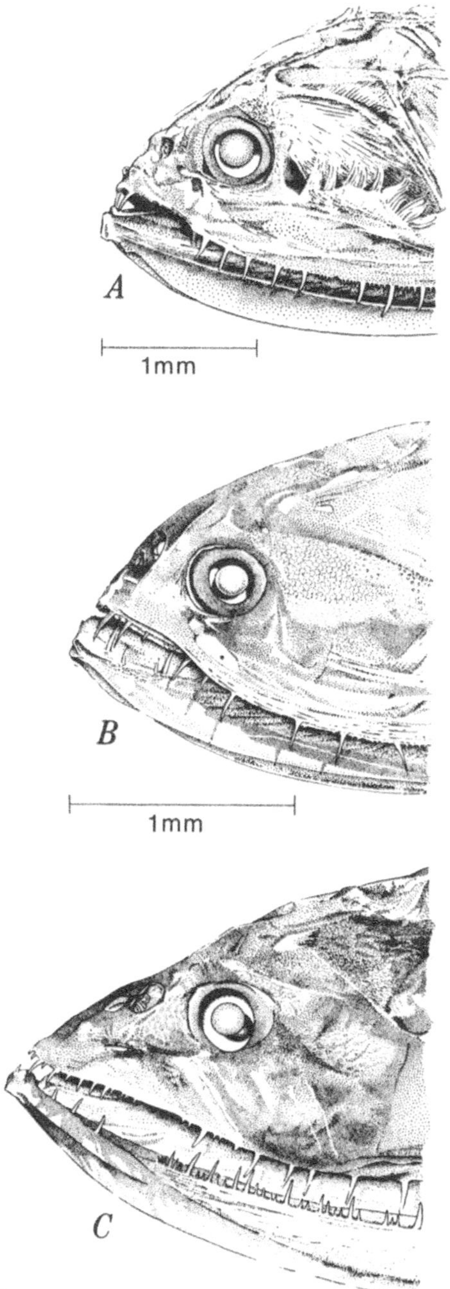

Fig. 7A–C. Heads of *Gonostoma denudatum*, A, *G. elongatum*, B, *G. bathyphilum*, C, to show circumlental aphakic apertures. (From MUNK and FREDERIKSEN, 1974)

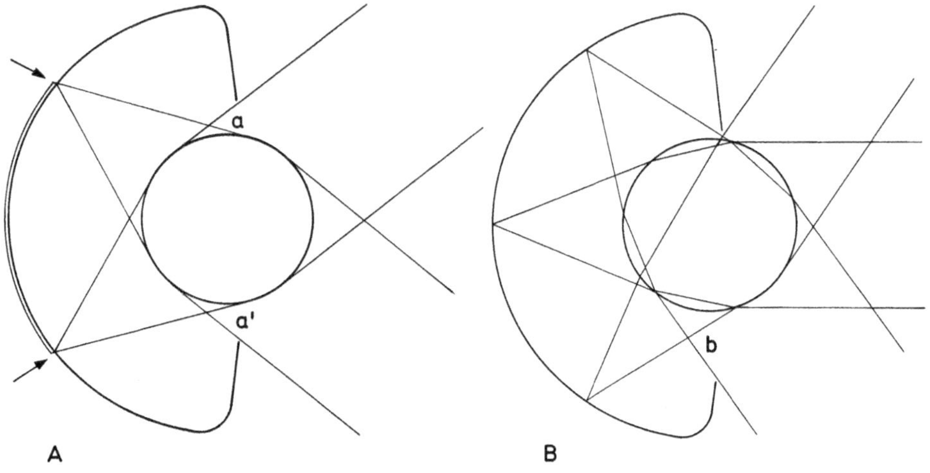

Fig. 8A and B. Diagrams to show function of aphakic apertures. In Fig. 8A small circumlental aphakic gap, a, a', allows area of retina between two arrows to be illuminated by full aperture of lens. In Fig. 8B rostral aphakic gap, b, allows full illumination of central and caudal parts of retina (after MUNK and FREDERIKSEN, 1974)

function of the gap is probably to increase the retinal sensitivity. They found that an increase in the pupil to lens ratio from 1 to 1.5 will double the retinal illumination in the peripheral retina.

In conditions of abundant light such a circumlental aphakic aperture might be a disadvantage; areas of retina would receive a focussed image through the lens, but this would fall on retina also diffusely illuminated through the aphakic gap. This dilution of the image would tend to reduce contrast and degrade the image. This may account for the relative absence of such apertures in shallow water fishes.

Though rostral aphakic apertures do occur in shallow water fishes, it is mainly in deep-sea species that they are found, and circumlental gaps are confined to this group. Where valid comparisons can be made, as in the case of the stomiatoids, the size of the circumlental gap tends to be larger in deeper living species. The aphakic gap of *Stomias boa* is small, whereas those of *Chauliodus sloani* and *C. danae* are larger; *Stomias* is a mesopelagic fish, but *Chauliodus* extends to bathypelagic depths. Within the genus *Gonostoma* there is a marked increase of pupil size with depth (MARSHALL, 1954). *Gonostoma denudatum*, caught at 100–500 m, has the largest eye and smallest aphakic gap; *G. elongatum* lives at 300–1700 m, and *G. bathyphilum* at 1000–3000 m, but mainly below 2000 m. The latter two species have smaller eyes than *G. denudatum*, and relative to eye size, *G. bathyphilum* has the largest aphakic gap.

MUNK and FREDERIKSEN did not discuss ventral aphakic gaps, but MARSHALL (1971), drew attention to such gaps in certain myctophids, in MUNK's account of *Omosudis lowei* (1965b) and ROFEN's (1966) drawings of barracudinas. Amongst the latter is *Notolepis rissoi*. Specimens of this fish which I have examined have a circumlental aphakic aperture, extending a little ventral in smaller specimens. Larger specimens, however, have a pronounced gap, extending almost to the retinal margin rostro-dorsally, and approaching it caudo-ventrally. The retina of this species is discussed below.

Among the myctophids listed by MARSHALL is *Tarletonbeania crenularis*. This species has been studied by LAWRY (1974), who was able to observe the luminescence of living specimens. He showed that the fish has a supraorbital photophore, and that light from this, and downwelling light from above, can pass through the ventral aphakic aperture and the lens to illuminate adjacent areas of retina. He also found that the fish exhibited luminescence when illuminated with dim blue light from above, but not in the dark. He concluded that this myctophid was able to monitor its own luminescence with respect to the ambient light.

II. Specializations in Binocular Vision

1. The Importance and Consequences of Binocular Vision

If the two eyes of an animal are so positioned that a single object in the environment is imaged simultaneously on the retinae of both eyes, the animal has a field of binocular vision. Some degree of binocular vision is common in vertebrates, but there are wide variations between and within natural groups in the extent of the binocular field.

Binocularity is best developed in active predators, on which it confers the ability to determine the location, and particularly the range, of prey with accuracy. This role in range finding is discussed below. Herbivorous animals do not need to hunt actively, and commonly have an extensive field of monocular vision giving them a good chance of detecting approaching predators.

Binocular vision is particularly well-developed in certain animals which hunt in conditions of dim illumination, e.g., owls; such conditions are present in the deeps, and many deep-sea fishes show adaptations for binocular vision. Apart from its function in predation these creatures' binocularity may confer an additional benefit upon them: PIRENNE (1943) found a difference of 10% between the monocular and binocular thresholds in man, and WEALE (1955) suggested that if the same is true for deep-sea fishes this might be of survival value to them.

The ability to see a single object simultaneously with both eyes, and to see it as one, despite the fact that there is an image in each eye, is a familiar experience for man, who has the additional ability to judge distances by means of binocular clues, though these are by no means the only ones available for the purpose. It is common experience that one can tell much about relative distances from a photograph, which shows that monocular clues are adequate for many of the distance judgments required by man.

WALLS (1942) discussed the monocular clues to distance and stated that they are also available to other vertebrates, though use of them will depend on the powers of learning and of observation, and on the animal's equipment of instincts. Closer examination of these clues, however, shows that they are not available to deep-sea fishes.

Most of the monocular clues depend on the presence of objects of reference in the visual field at the same time. Thus overlap and shadow casting, perspective, parallax, and the vertical nearness of the object to the horizon all depend on seeing such a reference object. Judgments based on the size of the retinal image imply a well-resolved image, recognition of the object seen and memory of the usual size of such objects. In deep-sea conditions the image may be a fleeting one at or near threshold intensity, or a pattern of luminous points rather than a clearly defined image. Even if the image is clear, a deep-sea fish is unlikely to have the mental equipment to recognize and remember an object.

WALLS' case of aerial perspective, in which distant objects seen through air look dim, bluish, or with hazy outlines, could not apply to a fish in deep water: In shallow and well-lit water this effect is certainly applicable, and over shorter distances than in air, due to the greater and more selective absorbance of water.

Deep-sea fishes are thus denied the necessary monocular clues to judge distances, but does this matter to the fish? Food is scarce at depth, so successful predation clearly does matter, and the important role of binocular vision in predation is now examined.

It was explained above that an act of predation may be divided into an attention stage, an approach, and a snap. The first is that in which the fish becomes aware of potential prey in the vicinity. This alerting must often be through vision, but olfaction and lateral line sensation are also modalities which could alert the fish. Visual alerting to an object to one side may well be monocular, and information on the whereabouts of the prey in the visual field can be derived from the location of the image on the single retina in terms of the horizontal and vertical meridians. This information may lead the fish to turn toward the prey, and begin the approach.

If the prey is itself immobile this simply involves swimming toward it, but if the prey is moving, the direction of the approach must be adjusted so as to keep the prey ahead; this could be adequately guided by a single eye, though in most cases there is a field of binocularity ahead of the fish, in which case both eyes will be used.

The final act, referred to above as the snap, though it may also be a sucking action, is that in which the prey is seized; in shallow water fishes, in which feeding may be watched, the snap is generally a sudden and rapid movement of the jaw apparatus. It may be the final abrupt element of a stealthy approach, as in the John Dory, *Zeus faber*, or the culmination of a rush at the prey, as in mackerel feeding on young fishes.

The jaws of many deep-sea fishes are relatively large, and may be specialized for the capture of large prey or small. The malacosteids, which have a long lower jaw bearing formidable teeth, are known to eat myctophid fishes almost, if not quite, as large as themselves. The skin of the floor of the mouth is deficient, and the tooth bearing bones and their associated muscles are thus relieved of the water resistance which would be caused by its presence. The feeding mechanism, described by GÜNTHER and DECKERT (1959), involves the head being cocked upwards as the lower jaw is first protruded, and then swept upwards to catch the prey. Another instance of a specialized feeding mechanism, in this case for small prey, is shown by *Stylephorus chordatus*, a mesopelagic fish with a long and fragile

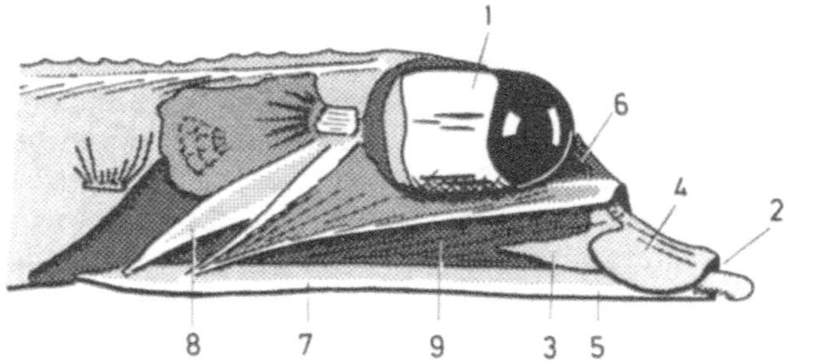

Fig. 9a. Head of *Stylephorus chordatus*, mouth closed. Tubular eyes, *1* lie parallel in head, and face rostrally. Small mouth, *2* is bounded above by maxilla, *3* and premaxilla, *4* and below by dentary, *5* Dorsal processes of maxilla and premaxilla are concealed beneath overhanging margin of cranium, just rostral to olfactory apparatus, *6*. Long lower jaw, *7* lies beneath cranium; its articulation with bony suspensorium, *8* is far posterior, beneath pectoral fin. Side of mouth between lower jaw and cranium is closed by lax membrane, *9*

Fig. 9b. Head of *Stylephorus chordatus*, mouth open. Anterior dorsal body muscles have contracted, and their tendon, *1* has rotated cranium dorsally through almost 90°. Eyes, almost immovable in orbits, now face dorsally, and are no longer directed to point where prey was located at commencement of snap. Suspensorium, rotating downward and forward, has brought articulation of lower jaw below eye. Opening lower jaw has drawn maxilla and premaxilla from beneath nasal bones, and mouth now opens far in front of cranium. Full extent of membrane closing sides of mouth is now revealed

ribbon-like body. The tail includes two rays nearly as long as the body again, and the fish does not appear well-adapted for active hunting. The head carries a pair of large tubular eyes which look rostrally, and a small mouth borne at the extremity of a remarkably protrusible jaw apparatus (Fig. 9). This jaw mechanism was described by REGAN (1924) who showed that mouth protrusion and opening was accompanied by marked elevation of the head, and so of the eyes, which are almost immobile in the head. In this case certainly, and probably in malacosteids too, the act of snapping includes a head movement which will cause the eyes to look away from the prey, but by the time this happens they have performed their function in locating and positioning it.

If an object is within the field of vision of the two eyes simultaneously, it will be imaged on both retinae. The position of the image will be different in the two eyes, however; the degree of disparity will depend on the range of the object, while the asymmetry of the two images will depend on the relation of the object to the sagittal plane of the fish. If the object and the fish are in the same plane, the retinal images will be symmetrical. We may suppose that when the object occupies the appropriate position, the pattern of retinal signals will trigger the snap. This does not imply that the images on the two retinae are clearly focussed, nor that they are well-resolved, and certainly not that the fish can "see" stereoscopically. The spatial sign of the disparate but symmetrical images on the two retinae is determined solely by the range of the object; its size will affect the size of the images, but not their relative positions.

Binocular vision results when the visual fields of the two eyes overlap, and the more they do so the greater the binocular field. The "normal" fish, with lateral facing eyes, has wide monocular fields but a fairly small binocular field ahead. Many deep-sea fishes have adaptations which increase the binocular field, presumably making for more effective predation, but which tend to reduce the monocular field, important in alerting the fish to possible prey and predators. Though some adaptations toward binocularity have this disadvantage, in others there are compensating adaptations which increase the monocular field while retaining a large binocular overlap.

2. Tubular Eyes

Among mesopelagic fishes are a number in which the eyes show a degree of asymmetry, resulting in a preferential axis of vision. This tendency is seen in connection with aphakic apertures, and reaches its most extreme form in tubular eyes. The term telescope eyes was applied to eyes of this type by CHUN (1893) who compared them to other eyes of exaggerated proportions in crustaceans and cephalopod molluscs. BRAUER (1901, 1902, 1908) was the first to study teleost tubular eyes in detail. He used CHUN's term, but without implying any telescopic function; he stated that he used it because the pair of eyes resemble an opera glass. BRAUER described such eyes in eight genera, including examples in which the eyes face dorsally and others in which they face rostrally. The eyes face dorsally in the species of *Argyropelecus*, *Rhynchohyalus*, *Dolichopteryx*, *Macropinna*, *Opisthoproctus*, scopelarchids, and some evermannellids. In various other species, e.g. *Gigantura*, *Winteria*, and *Stylephorus* (Fig. 9) the tubular eyes are directed rostrally. He regarded such eyes as an adaptation to the deep-sea environment, and his descriptions have provided the background against which more recent investigators have worked.

Among the examples studied by BRAUER, and later by CONTINO (1939) and MUNK (1966a) are fishes of the genus *Argyropelecus*: a description of such an eye follows (Fig. 10).

The lens is spherical, like that of other fish eyes, and is set in the oblique pupil at the dorsal extremity of the eye. It is covered by the thin, domed cornea, which follows the lens contours rather closely, so that the anterior chamber is shallow though extensive. The cornea extends far down the lateral wall of the eye, merging below with the sclera, which is thin and transparent over much of its extent. The eye cup, consisting of the retina and uveal tract, is more or less closely applied to the inner surface of the corneo-scleral coat.

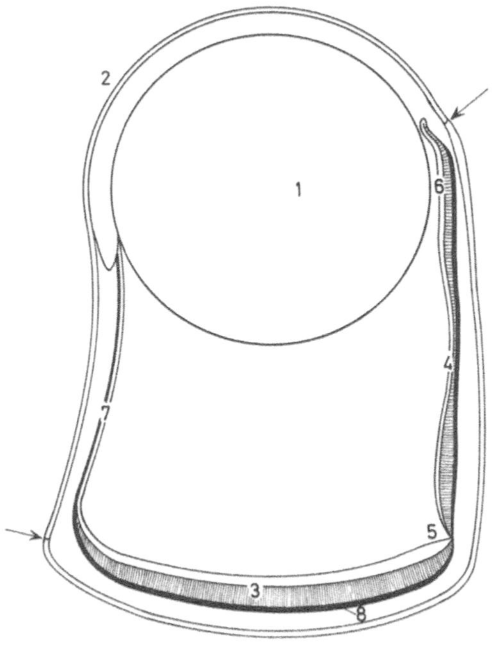

Fig. 10. Diagram of tubular eye in transverse section. Lens, *1* appears large for size of eye. It almost fills dorsal part of eye. Cornea, *2* is thin, and domed over lens dorsally, it extends far down lateral side of eye. Its limits are shown by arrow. Ventral part of eye contains main retina, *3* which conforms to Matthiessen's ratio. Accesory retina, *4* does not; it extends up medial wall of eye cup from choroid fissure, *5* where it is almost at focal distance of lens, to dorsal region of medial wall, *6* where it is almost in contact with lens. Lateral wall of eye cup is formed by iris, *7*. This extends from edge of main retina up to pupil margin. Choroid, *8* lies just sclerad to both main and accessory retinae

At the ventral extremity of the eye cup is the main retina, which lies at the focal distance from the lens, and corresponds to the retina of a normally proportioned eye in its curvature. Medially the main retina ends at a sharp boundary, marked by the choroidal fissure. Here the plane of the retina turns sharply upward, forming a flat medial wall to the eye cup, parallel to the corresponding wall of the other eye. The rostral, caudal, and lateral walls of the eye cup are curved, and their boundaries with the main retina are less sharp than the medial boundary. The lateral wall, and part of the rostral and caudal walls, are formed by two layered but nondifferentiated optic cup epithelium, overlaid by thin uveal tissue and the silvery argentea; these portions represent the iris.

The medial wall of the eye cup is lined by the accessory retina, which covers the full height of the medial wall, nearly touching the lens at its dorsal limit, and extends onto the rostral and caudal walls as well. The accessory retina, though a clearly differentiated retina, has a much less complex structure than the main retina. In such a tubular eye the main retina conforms to Matthiessen's ratio, but the accessory retina does not. Just dorsal to the choroid fissure it is almost at the focal distance, but its dorsal edge is almost in contact with the lens, and so well inside the focal distance.

BRAUER'S paper of 1902 discussed the eyes of deep-sea fishes including *Dissomma* (= *Scopelarchus*), and was originally presented verbally. The discussion which followed was reported, and included a remark by HESSE that the main retina might be in focus for near objects, and the accessory retina for more distant ones. As MUNK (1966a) pointed out, this idea was revived by WALLS (1942) who claimed that the tubular eye is bifocal, in the sense given above. MUNK made it clear that this notion is untenable; the accessory retina is situated so that it could not receive a focussed image through the lens. BRAUER was aware of this, and also

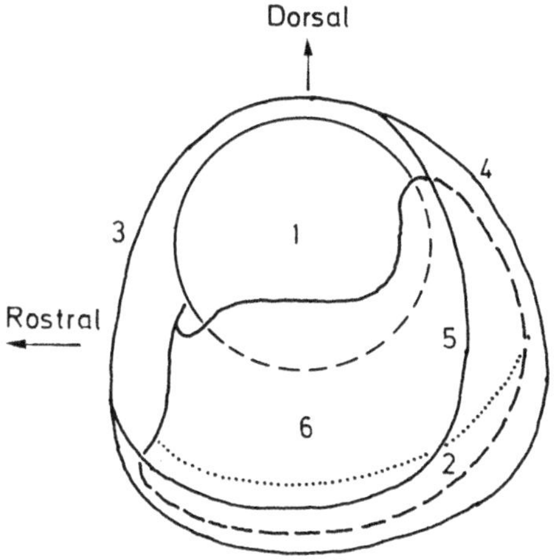

Fig. 11. Left eye of *Argyropelecus hemigymnus*, lateral view. Lens, *1* is situated far dorsally in eye, and projects above upper margin of eye cup. Floor of eye cup, *2* contains main retina, which conforms to Matthiessen's ratio, and extends up caudal wall of eye cup as shown. Cornea, *3* forms dome over lens, and extends over lateral side of eye. *4* Sclera. *5* Corneal margin. *6* Iris (lateral wall of eye cup)

that the accessory retina was unlikely to be brought into focus by movements of the lens. Such movements, produced by the lens muscle as in other fishes, would tend to move the lens toward the main retina, bringing more distant objects into focus.

Though the *Argyropelecus* eye is considered tubular, it is by no means a perfect cylinder. Seen in transverse sections through the head, as drawn by BRAUER, the lateral and medial walls are almost parallel, and the eye looks tubular. But an eye dissected from the fish and examined from the lateral side is almost round in shape. Unlike the normal round fish eye, however, the lens is markedly displaced dorso-rostrally, and the pupil is oblique (Fig. 11).

Seen from above in a horizontally bisected eye, the main retina is somewhat bean-shaped; about twice as long rostro-caudally as it is wide, with a straight medial border but with the lateral, rostral, and caudal borders curved. In such a view the choroid fissure is clearly visible, with the optic disc located near its caudal end. The structure of the main and accessory retinae are considered further below.

Among the early students of the tubular eye was FRANZ (1907). He observed that such eyes conformed to Matthiessen's ratio as far as the main retina was concerned, and suggested that the adaptation might owe its usefulness to the large lens, which would collect more light than a smaller one. He compared a tubular eye to one of normal shape but containing a lens of the same size, and suggested that such a tubular eye might be a compromise solution; as large an eye as possible in a fish of given size (Fig. 12). He was conscious that there would be no clear image on the accessory retina.

BRAUER (1908) accepted FRANZ's suggestion about the value of a large lens, regarding this as important for utilizing as much as possible of a light source and

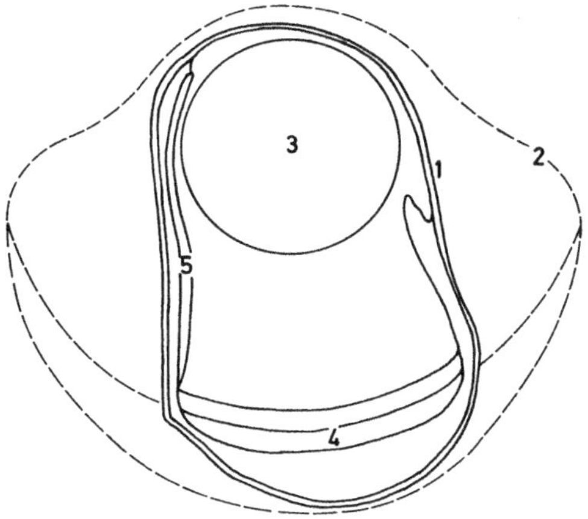

Fig. 12. Diagram of outline of tubular eye, *1* superimposed on outline of normal eye, *2* both containing lens, *3* of same size. Main retina of tubular eye, *4* corresponds to central part of retina of normal eye; both being at same distance from lens, they will receive images of same size. Visual field of tubular eye is reduced compared to that of normal eye. It is partly extended by accessory retina, *5* but this does not conform to Matthiessen's ratio, and so does not receive focussed image. (After FRANZ, 1907)

achieving maximal retinal illumination. He was also conscious that the binocular field would be enlarged, but at the expense of the all round field; this binocularity he thought would lead to better judgment of distances. He considered the main and accessory retinae as corresponding to the central and peripheral retina of a normal eye, regarding these as important for the perception of form and of movement, respectively. The latter he believed to be the more important attribute, since he thought there was insufficient light or transparency of the water at depth for the formation of good retinal images. The prevalence of dorsally directed eyes he regarded as associated with the predominantly downwelling light.

In an early paper (1901) BRAUER stated that the normal lens muscle of teleosts was absent in those with tubular eyes. Later (1902, 1908) he stated that the muscle is present in all but *Dissomma;* it is in fact present in that genus (now *Scopelarchus*) also (LOCKET, 1971 b).

The eyes of *A. hemigymnus* were the subject of CONTINO's monograph (1939). He had access to more than 100 specimens, including developmental stages, and was able to study the gross and microscopic anatomy in some detail. He described the two eyes in the orbits, noting their close contact in the midline, and consequently flat medial walls. He described the ocular muscles, though so closely do

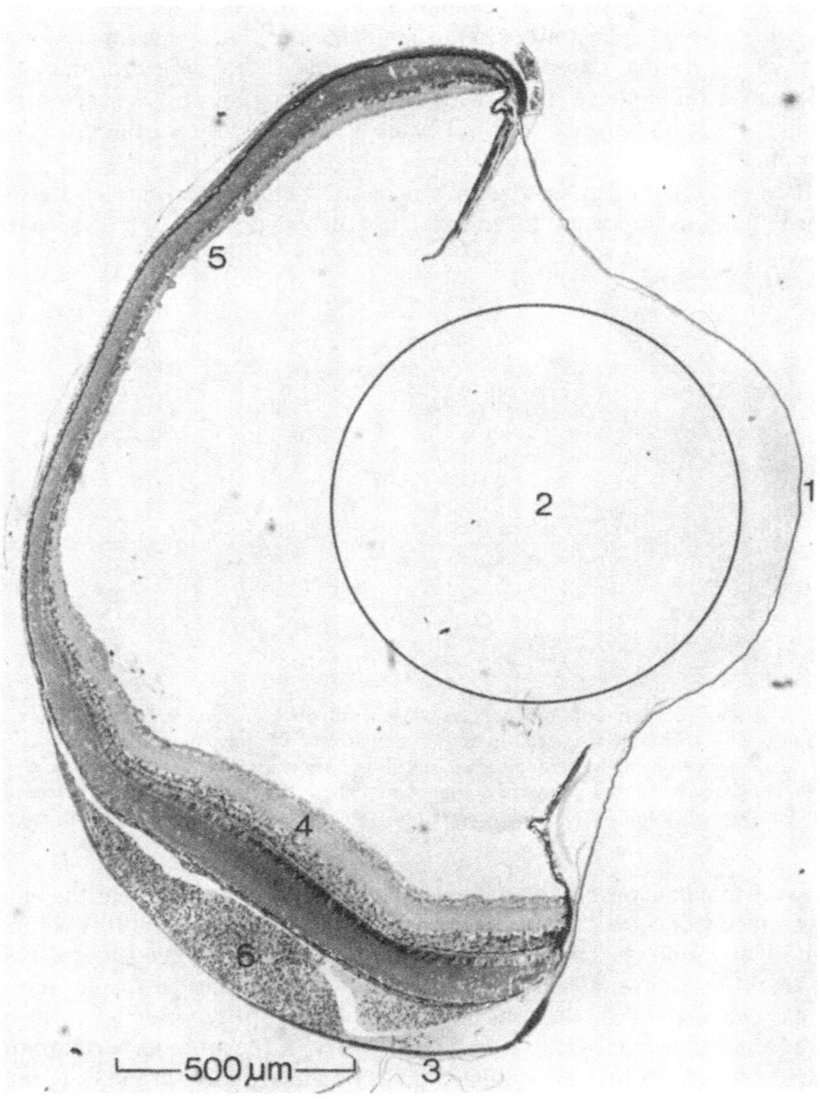

Fig. 13a. Vertical section of eye of *Valenciennellus tripunctulatus*. Seen laterally, eye appears of normal teleost configuration, except for slight dorsal displacement of lens. Section shows cornea, *1* bulging laterally over lens (removed in processing, but position shown), *2*. Sclera, *3* is almost hemispherical, but retina is notably thicker ventrally, *4* than dorsally, *5* This shows stage in differentiation into main and accessory retina. Choroid is thin over most of its extent, but substantial choroidal rete mirabile, *6* is present beneath ventral retina. Optical micrograph

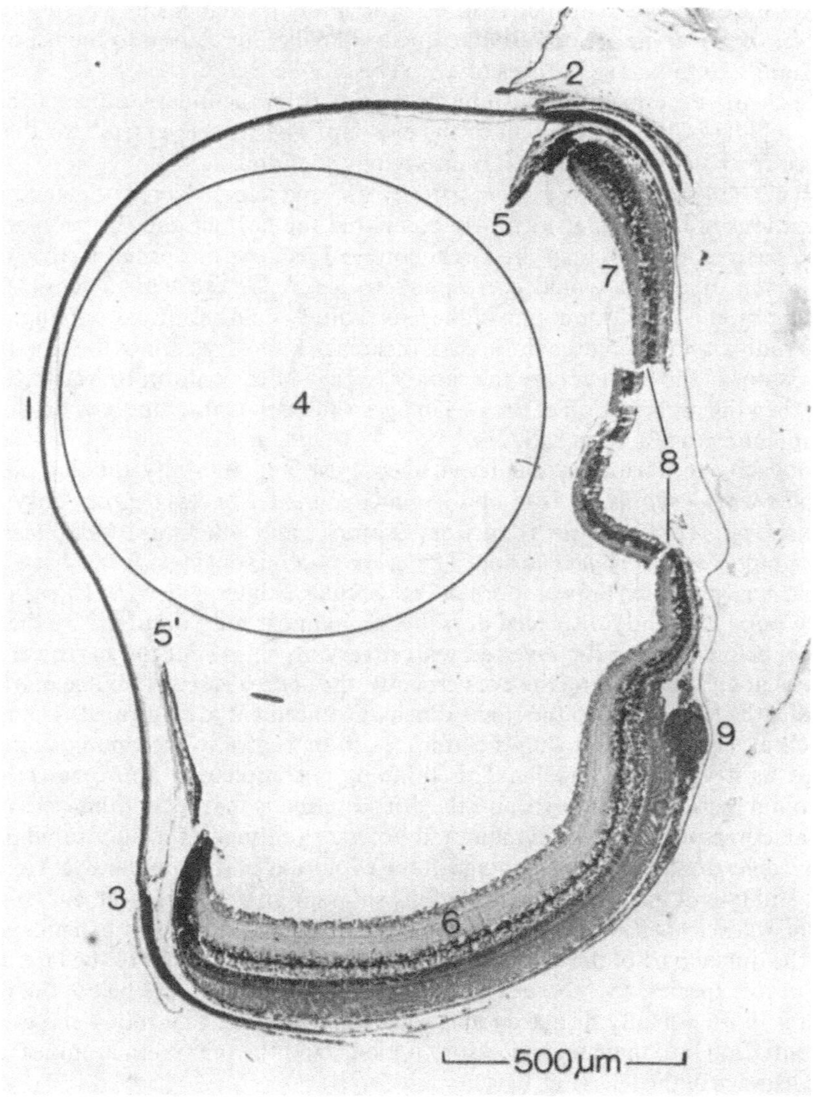

Fig. 13b. Vertical section of eye of *Vinciguerria poweriae*. Seen laterally this eye is almost round in outline, but there is definite dorsal displacement of lens. Cornea, *1* extends from medial wall of eye dorsally, *2* to most lateral extremity of floor of eye cup, *3* Lens, position shown, *4* is situated dorsally, and pupil, between *5* and *5'* is oblique. Retina is differentiated into main retina, *6* and accessory retina, *7* of which dorsalmost part is best developed. Accessory retina is somewhat damaged and displaced at *8*. There is small choroid rete mirabile, *9* Optical micrograph

the eyes fill the orbits that there can be little if any movement in the adult state. The eyes of the larvae are not tubular, and are smaller in relation to the head than in the adult; at this stage the eyes may well be mobile.

He clearly described the lens muscle, arising from the inner surface of the iris at the middle of the rostral wall of the eye cup. The muscle extends to the lens equator rostro-medially, where it is inserted by a tendon.

MUNK (1966a) reviewed the earlier work, and reconsidered the functional implications of the tubular form. He compared the normal and the tubular eyes on the basis of equal retinal area. A theoretical fish eye of normal shape with a lens 1.6 mm diameter would correspond to a tubular eye with a lens 3.0 mm diameter on this basis. Doubling of the lens radius would result in doubling of the image radius, and so enlarge the area of the image four times. But if the lens-retina proportions of the two eyes are the same, i.e., they both conform to Matthiessen's ratio, then the illumination of the two images will be the same, and not brighter in the tubular eye as BRAUER believed.

Between the normal symmetrical teleost eye and the fully tubular eye, for example of *Stylephorus*, lie examples in which a greater or less degree of asymmetry is shown. *Argyropelecus* is such an example, and others are *Valenciennellus tripunctulatus* and *Vinciguerria* spp. The latter two mesopelagic fishes have laterally placed eyes which appear more or less circular in lateral view. In *Valenciennellus* the pupil is slightly displaced dorsally, and almost fully occupied by the lens. The iris below the pupil is invested with silvery argentea, but the narrower strip above is not. On sectioning the eye vertically, the cornea is seen to bulge markedly opposite the lens, of which the slight dorsal displacement is confirmed. The retina in such a vertical section shows a thick portion ventrally, becoming gradually thinner as it is traced dorsally. This thinning is complete at approximately the horizontal meridian of the eye, and the dorsal retina is markedly thinner than the ventral, corresponding to the main and accessory retinae of a fully tubular eye.

Vinciguerria shows a further stage in the evolution of the tubular eye. Here the eye is still lateral in position, but the lens shows a greater displacement dorsally than in *Valenciennellus*. The cornea has a more regular curve, but extends nearly from the dorsal part of the medial wall over to the ventral border of the lateral eye wall. In this species, too, there is a patch of argentea on the iris below the pupil, which is more dorsally displaced and somewhat oblique. The retina shows clear differentiation into main and accessory regions, and the main retina only is at the focal distance of the lens (Fig. 13a, b.).

III. Devices Extending the Visual Fields

The advantages conferred by tubular eyes have an attendant drawback, in that specialization in vision in one direction limits the visual field in others: Binocular vision is advantageous for successful predation, but avoidance of predators requires a good all-around field. Certain deep-sea species have adaptations which extend the field of otherwise limited eyes.

The fields of tubular eyes usually overlap widely, but fishes with laterally set eyes may have a restriction on the field caused by their own body. Swimming movements will compensate for this to some extent; if the head moves from side to side, an object ahead will be within the field of the two eyes alternately. In some deep-sea and shallow species there is a rostral aphakic gap, and in some a sighting groove in the tissues of the snout. These features will help to extend the binocular field directly in front of the fish, perhaps favouring increased accuracy in the coordination of forward movement and mouth opening.

The tubular eyes of *Argyropelecus* spp. have a wide overlapping binocular field subserved by the main retinae of the two eyes. Their accessory retinae, lining the medial wall of the eye cup and extending onto the rostral and caudal walls, are not situated so that they will receive a focussed image. The neural organization of accessory retina appears much simpler than that of main retina (Figs. 49, 50), and even if the image were good it is unlikely that much information could be derived from it. But as a light sensitive area which will detect and signal the approximate direction of the stimulus, the accessory retina may well be adequate.

1. Retinal Diverticula

In certain other fishes with tubular eyes more complex adaptations to extend the field occur. Among these are retinal diverticula, which are found in several species. MUNK (1966a) lists *Opisthoproctus*, *Rhynchohyalus*, *Winteria*, *Dolichopteryx*, and *Gigantura*; FREDERIKSEN (1973) has added *Macropinna* and described *Dolichopteryx* further. These diverticula (Fig. 14) project from the eye cup, in most cases laterally. Their cavity is continuous with that of the eye cup through a slit-like opening, and their wall is lined with optic cup epithelium, here differentiated into retina. This portion of retina, called diverticulum retina by MUNK and by FREDERIKSEN (loc. cit.) is overlaid by choroid, the whole being contained in an outgrowth of the sclera (Figs. 14, 15).

At some point, caudo-laterally in *Macropinna* but facing medially in *Dolichopteryx*, the differentiated retina gives way to undifferentiated two-layered optic cup epithelium. The epithelium here is devoid of pigment, and the choroid is deficient over it, so that the area forms a window through which light can reach the interior of the diverticulum.

The opening from the diverticulum into the main eye cavity is narrow, and the retina at its margins contains pigment in its outer epithelial layer. Thus it is unlikely that light from the main cavity will affect the diverticulum retina significantly.

At first sight it seems improbable that a window, facing as it does onto the outside wall of the eye cup, could transmit light to any purpose. The outside wall of the eye cup in *Opisthoproctus grimaldii* and *O. soleatus*, however, is certainly silvery in life, due to the presence of the well-developed argentea. In *Macropinna*, FREDERIKSEN described several layers of elongated flat cells in this situation; they closely resembled the argentea in *Opisthoproctus*, though reflecting crystals were not found in the preserved specimens. (This absence of crystals is common in specimens which have been preserved for some time).

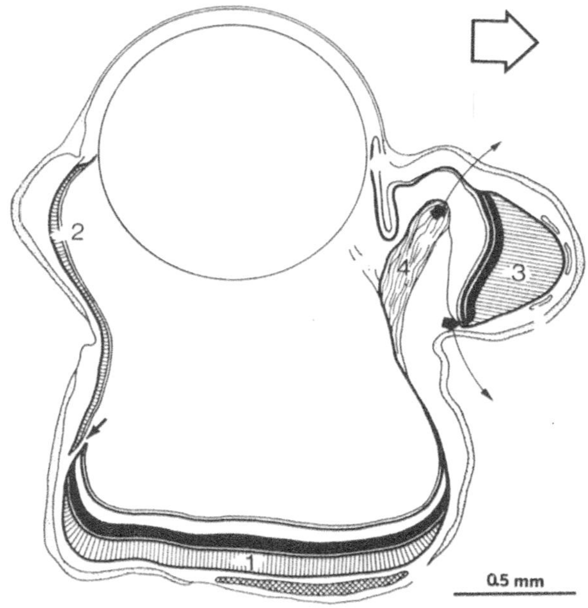

Fig. 14. Diagram of tubular eye of *Dolichopteryx longipes*. Main retina, *1* conforms to Matthiessen's ratio, and accessory retina, *2* does not. Lateral wall of eye is extended laterally, and houses diverticulum retina, *3*, which has longer rods than main retina. Choroid over upper wall of diverticulum is pigmented, and so is that over diverticulum retina, but between solid arrows, choroid is deficient, and retinal layer represented by nonpigmented two-layered epithelium. This transparent portion is window; light from lateral to fish is reflected from argentea, *4*, through window and onto diverticulum retina. (From FREDERIKSEN, 1973)

These diverticula are interpreted as receiving light from outside the visual field of the normal tubular eye. The light through the window reaches the diverticulum retina after being reflected by the argentea on the lateral eye wall. In *Macropinna*, *Dolichopteryx*, and *Rhynchohyalus* all the light so reflected will pass through the inner neural layers of the diverticulum retina before reaching the outer segments. In *Winteria telescopa* and *Opisthoproctus soleatus*, however, the transparent epithelium extends laterally to cover some of the differentiated retina, and not only undifferentiated window epithelium. In these cases the light can reach the outer segments of that part of the retina either through the window, traversing the neural layers, or directly through the transparent outer coats of the eye at that site. Thus some of the outer segments will receive light from both ends. It is possible that this might increase the light received by that part of the retina.

In the case of *Bathylychnops exilis* the specialization of the diverticulum is even more extreme (Fig. 16). PEARCY et al. (1965) described the eyes of this species; they noted large dorsally directed eyes, with a large field of binocular vision, and, projecting from each eye in a ventral and slightly caudal direction, a secondary globe.

This globe is lined by diverticulum retina, differentiated over most of its extent. The cavity of the secondary globe connects with that of the main globe

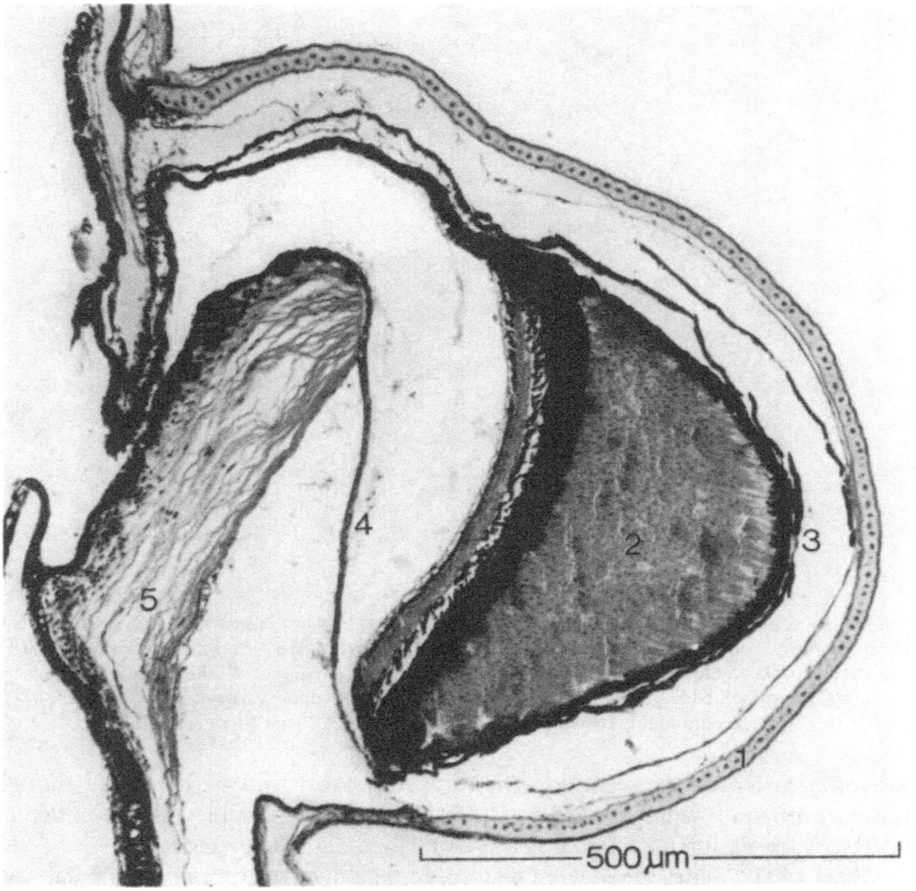

Fig. 15. Retinal diverticulum of *Dolichopteryx*. Wall of diverticulum includes scleral cartilage, *1*. Thick diverticulum retina, *2*, is backed by pigment epithelial cells and choroid, *3*. Window, *4*, is formed by two layers of low epithelial cells, and here choroid is deficient. Lateral wall of eye cup related to diverticulum contains well-developed argentea, *5*. This would probably have contained reflecting crystals in life. Cavity of diverticulum is continuous with that of main globe through slit medial to argentea. (From Frederiksen, 1973)

through a narrow slit, covered by a pigmented operculum so that light from the secondary globe cannot reach the main retina.

In the ventral part of the wall of the secondary globe there is a spherical lens-like structure, differentiated from and continuous with the scleral connective tissue. The deep surface of this lens is lined by two-layered epithelium continuous with the retina and pigment epithelium of the diverticulum retina, but devoid of pigment. The choroid is also deficient over the secondary lens, so light can pass directly through, being converged onto the diverticulum retina.

Bathylychnops shows another unusual feature. Caudal and dorsal to the secondary globe, and close to the corneal margin, are two structures differentiated from and continuous with the cornea. These corneal projections, which closely

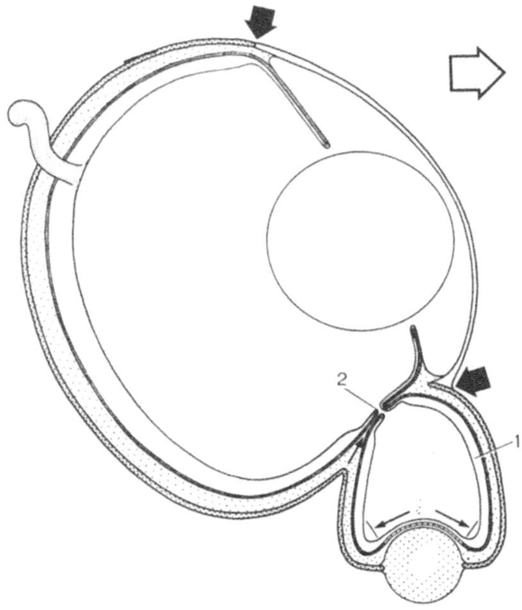

Fig. 16. Eye of *Bathylychnops exilis*. Cornea of main globe is between large solid arrows; Open arrow points laterally. Secondary globe protrudes ventrally from main globe. Its walls are formed from all normal layers of eye, but sclera is modified to form lens of secondary globe. Choroid and retina between thin arrows are modified to form window which admits light to secondary retina 1. Cavity of secondary globe communicates with that of main globe at 2. (From MUNK, 1966a)

resemble the lens of the secondary globe, are probably transparent in life, though they are opaque white in preserved specimens. It is possible that they act as accessory lenses, further extending the visual field of the main globe.

BRAUER (1908) described what he took to be a diverticulum in the medial wall of the retina of *Scopelarchus guntheri* (= *Dissomma anale*). MUNK (1966a) described the rostral lip of the choroid fissure in *Neoscopelarchoides* sp. as consisting of normally differentiated accessory retina overlaid by the temporal lip of the fissure, which consists of undifferentiated two-layered epithelium. I have not found diverticula in the scopelarchids I have examined, but they do possess another variety of field-extending device, now described.

2. Lens Pad in Scopelarchids

This structure in the eyes of scopelarchids was first described by BRAUER (1902, 1908), who called it the *Linsenpolster* or *Linsenkissen*. MUNK (1966a) translated these terms as lens pad, and this term was used by LOCKET (1971b). JOHNSON (1974), in his revision of the scopelarchids, called the lens pad the pearl organ, though he was aware of BRAUER'S and MUNK'S terminology. He stated that its function was obscure, suggesting that it may be simply glistening white connective tissue, though MERRETT et al. (1971) had used the term lens pad, and recorded my observation on its function.

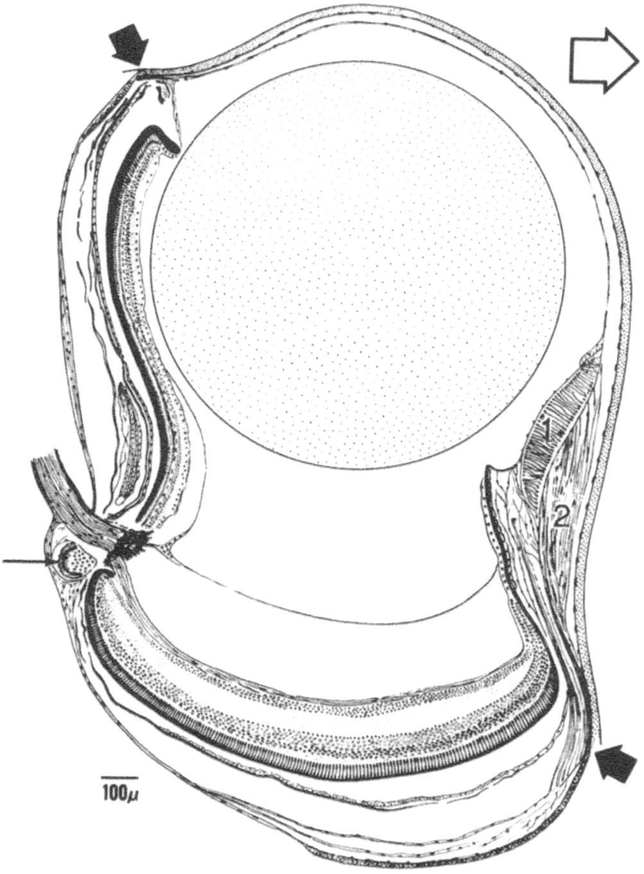

Fig. 17. Transverse section through eye of *Dissomma anale* Brauer, (= *Scopelarchus guntheri* Alcock). Thin cornea, between thick arrows, covers lens and lateral wall of eye cup, much as in *Argyropelecus*. Open arrow points laterally. Floor of eye cup is occupied by main retina and medial wall by accessory retina. Thin arrow indicates portion which BRAUER believed to be retinal diverticulum. Beneath cornea laterally, and related to iris, is lens pad, *1*. Ventral to pad is tissue which BRAUER believed to be unstriated muscle, *2*, but which is in fact modified cornea (see text). Lens looks disproportionately large; this is probably due to shrinkage of other components of globe during preparation. (After BRAUER, 1908; from MUNK, 1966a)

The scopelarchids have tubular eyes, similar in general form to those of *Argyropelecus* spp. Thus the floor of the eye cup contains the main retina, and the medial wall the accessory retina, which extends onto the rostral and caudal walls (LOCKET, 1971b). The pupil is oblique and slopes downward laterally. It is filled by the lens, and there is no aphakic gap. The cornea forms a dome concentric with the lens dorsally, and continues laterally over the lateral face of the eye cup. The eye cup itself is darkly pigmented on its flat medial face, which is apposed to the corresponding face of the other eye. Elsewhere the eye cup is covered by a silvery argentea, clearly visible through the cornea laterally, and through the almost transparent sclera.

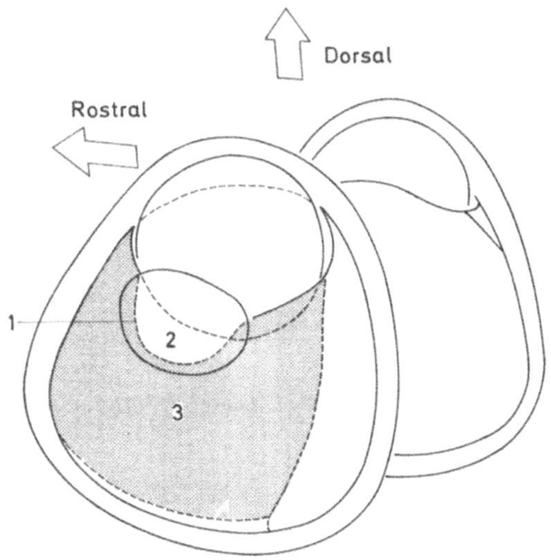

Fig. 18. Diagram to show position of lens pad in scopelarchid eye. Iris, shaded, forms lateral wall of eye cup; there is a dip in its border laterally, *1*, which is overlaid by lens pad, *2*. Space ventral to pad, *3*, is occupied by modified cornea, so that iris lies on deep plane, as shown in Fig. 17

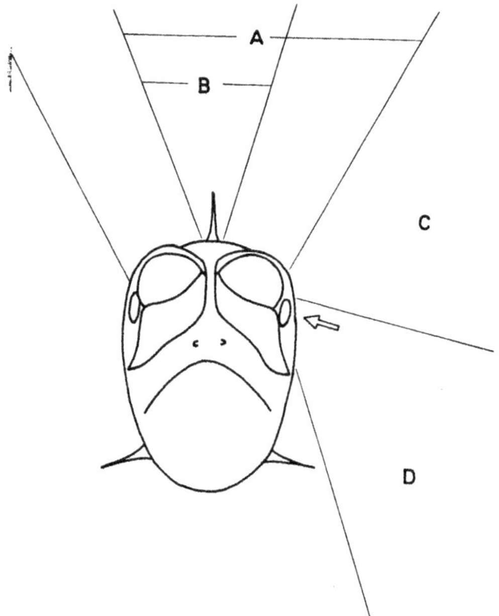

Fig. 19A–D. Diagram to show fields of view of *Scopelarchus* eyes, from rostrally. Main retina has field of view dorsal to fish, (A) which overlaps widely with that of other eye. Main retina fields of two eyes overlap to give dorsal field of binocular vision (B). More ventral part of accessory retina has monocular field laterally (C). Ventral field (D) is subserved by dorsal accessory retina, light passing through lens pad, arrowed

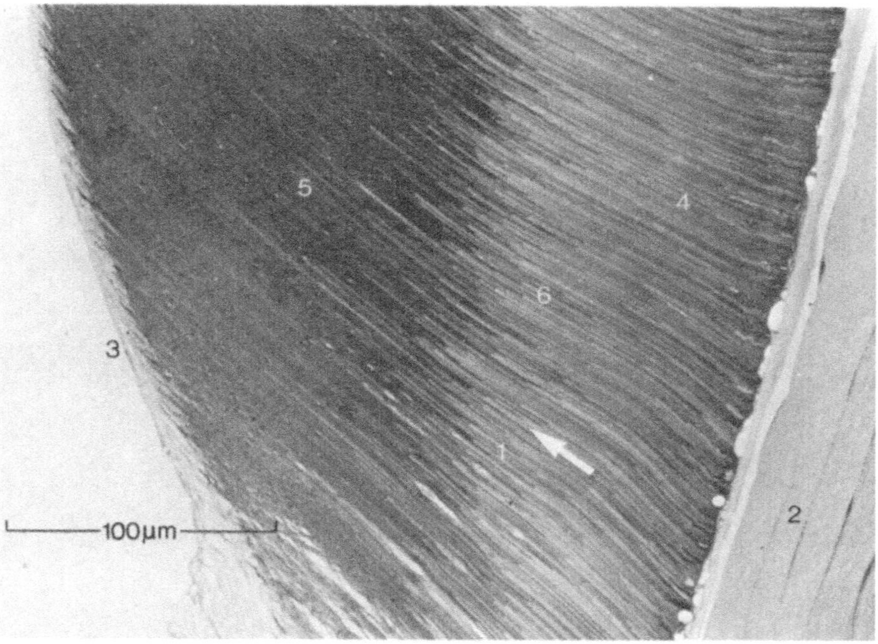

Fig. 20. Lens pad of *Benthalbella infans*. This micrograph is from near ventral border of pad. Here pad, *1*, is overlaid by diffuse cornea, *2*, on its external side. Deep face of pad, *3*, is lined by endothelium, continuous with that of cornea, but not visible in this micrograph. Outer lamellae, *4*, are less dense than inner ones, *5*. Zone of junction between two, *6*, shows regularity of spacing and alignment of lamellae. Light traverses lens pad along direction of lamellae, arrowed. Optical micrograph

In a fresh specimen viewed laterally, the wall of the eye cup appears to lie unusually deep, separated from the cornea by a space which looks optically empty. The lateral part of the pupil margin dips ventrally, and the lens pad is located just lateral to the lens and overlapping the pupil margin (Figs. 17, 18). The pad is elliptical, and in preserved specimens is pearly white and opaque, quite unlike the appearance in the living or freshly dead fish. The histological structure of the pad (*v. infra*) led me to think that it might be transparent in life, and this was confirmed by M. V. Angel (personal communication, 1969).

Later I was able to examine a specimen of the scopelarchid *Benthalbella infans* alive, and then to fix ocular tissues, including the lens pad, for optical and electron microscopy.

Ophthalmoscopy of the living specimen showed that the lens pad is not uniformly transparent. Seen from the ventro-lateral aspect the pad was freely transparent, and a blue to blue-green fundus reflex could be seen through it. As the point of view was moved toward the horizontal the reflex changed to blue, and was then extinguished (Fig. 19). The surface of the lens pad could now be seen with a lens of suitable power in the ophthalmoscope, and showed horizontal striations and iridescent colours. The tissue lying superficial and ventral to the pad was quite transparent, forming the optically empty space described above.

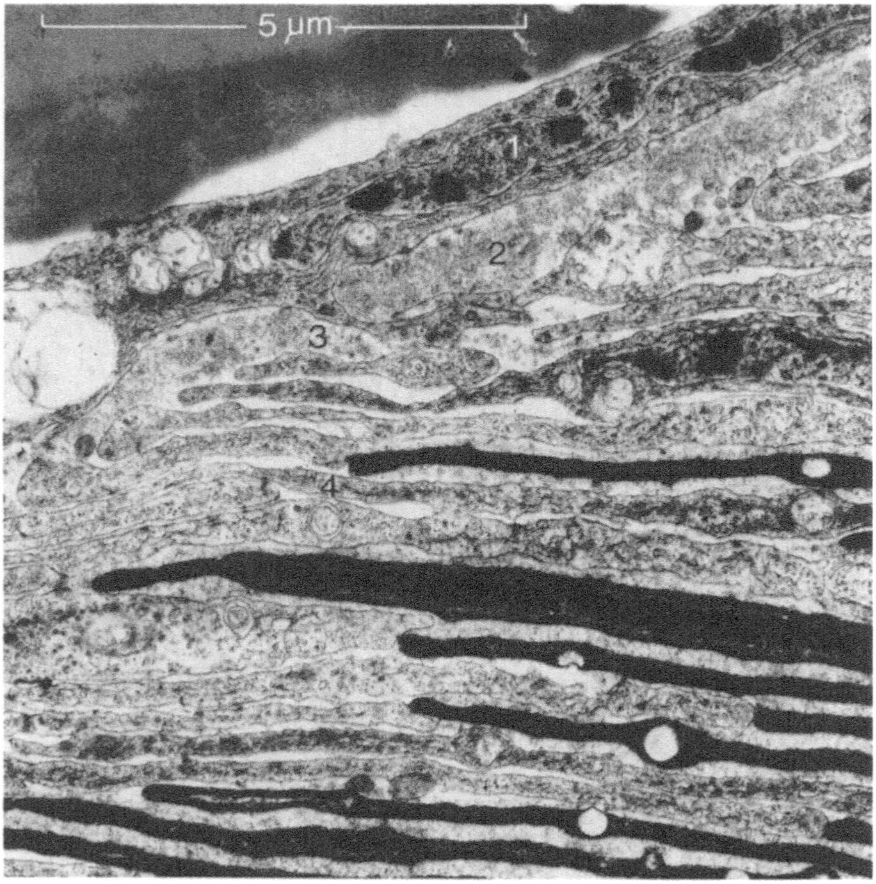

Fig. 21. Inner margin of lens pad of *Scopelarchus guntheri*, near its ventral border. Lens pad is bounded by endothelial cells, *1*, which are continuous with corneal endothelium dorsal to pad. Beneath these is layer, *2*, which is continuous with Descemet's membrane. At this level in pad there are gaps in this layer, and endothelial cells project through it, *3*. Deeper into pad cells become flat and regular. Those next to surface are pale, and abut end to end with darker lamella, *4*. Electron micrograph

BRAUER (1902, 1908), who examined transverse sections through the head of *S. guntheri*, described the lens pad as consisting of thin flat fibres, parallel to each other and perpendicular to the lens (Fig. 17). From sections in different planes it is clear that what BRAUER took to be fibres are in fact lamellae, and that many of these extend the full rostro-caudal width of the pad. Transverse sections (Fig. 20) show a dark staining inner portion and a paler outer portion, the junction having a brush-like appearance.

Electron microscopy of the lens pad shows further details of its structure, and suggests that it is of corneal origin. The structure may be best traced from within outward (Fig. 21). The pad is separated from the anterior chamber over most of its area by a single layered endothelium. This endothelium resembles the corneal endothelium, with which it is indeed continuous at the dorsal border of the pad.

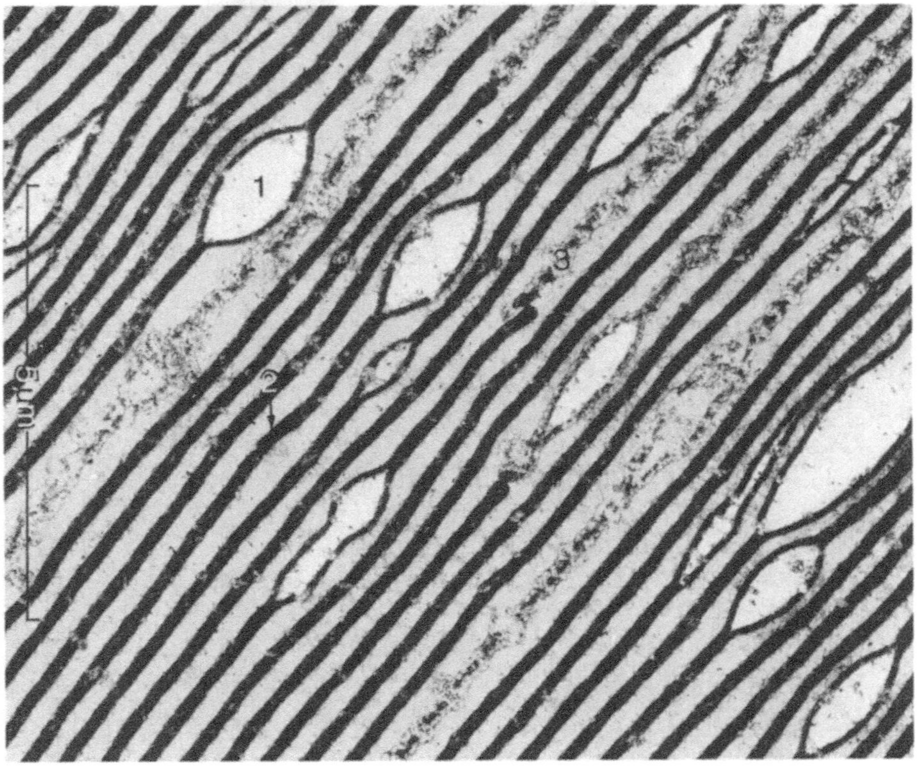

Fig. 22. Lamellae in substance of lens pad. Lamellae are mostly very even in thickness and contour; dilatations of central space, *1*, are probably artifacts. Dark staining lamellae, *2*, are separated by pale ground substance, and some of them abut against pale lamellae, *3*. Electron micrograph

Between the endothelium and the lamellated pad substance is a basement membrane, which has the same structure as Descemet's membrane of the cornea proper, with which it is in continuity. Uniform over most of the pad, this membrane shows irregularities toward the ventral border, and in places gaps, discussed below.

Lateral to the basement membrane, i.e. away from the anterior chamber, are the lamellae of the pad substance. Directly adjacent to the membrane a layer of irregular flattened cells slope ventrolaterally from it toward the pad substance, with the lamellae of which they become aligned. These cells resemble the endothelial cells where they lie against the basement membrane, but attain an even thickness and orientation where they come into relationship with the lamellae.

This layer of pale-staining cells gives way to a thick band of dark-staining lamellae, but the transition from pale cell to dark lamella is not at the same depth into the pad in every case, and the transition zone occupies some 6 μm. The lamellae themselves are modified cells, each having a flattened nucleus near its inner free border. In this region the lamellae are of slightly irregular thickness, but further into the pad substance they are very even in thickness and separation. The

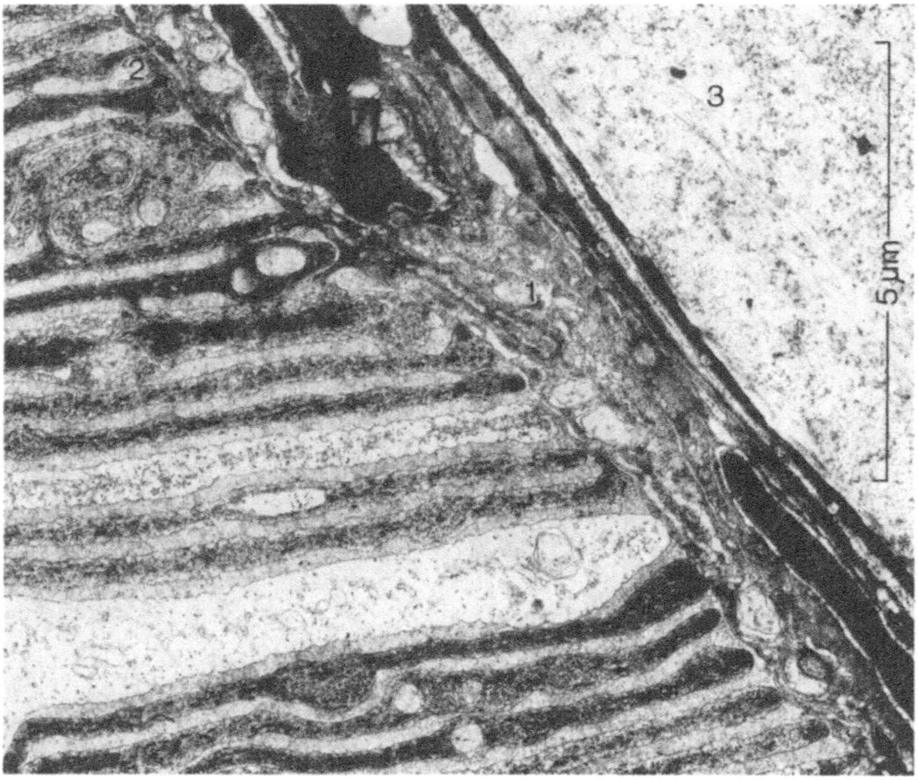

Fig. 23. Lateral boundary of lens pad is formed from layer of modified keratocytes, *1*. Lamellae, here rather irregular in outline, abut against keratocytes, which show protuberances where two come into contact, *2*. Lateral to keratocyte layer is part of diffuse cornea, *3*. Electron micrograph

main portion of the pad consists of an array of dark staining lamellae alternating with clear gaps; a measurement average from 70 lamellae and gaps showed a mean thickness of 203 nm (Fig. 22). The lamellae appear granular, and the centre of each is occupied by a discontinuous membrane-bounded space, probably representing endoplasmic reticulum, which shows dilatations in places, sometimes as a split in a lamella and sometimes of more bulbous shape. Between the dark lamellae, but not between the pale cells, is a lightly staining amorphous substance.

In the outer third of the pad the dark lamellae are replaced by pale lamellae, which abut edge to edge with the dark lamellae in the same way as the inner cells do, and here, too, is a transition zone, in this case 40 µm wide.

The regular external contour of the pad is marked by a layer of flat cells which resemble an epithelium, but there is no basement membrane associated with these cells, which lie from one to three deep, and there are no junctional complexes between them. These cells are in fact a rather well-defined layer of keratocytes which separate the lamellae of the pad from the substance of the cornea external to it (Fig. 23).

Dorsally the lens pad lies directly under the thin cornea, here called compact cornea to distinguish it from the tissue which lies between the pad and the compact cornea more ventrally (Fig. 23). This tissue, transparent and noticably less optically dense than the pad in life, is the portion which BRAUER believed to be composed of smooth muscle, but which, from its structure and affinities, I call diffuse cornea.

The upper border of the lens pad is clearly distinguishable in life, but the lower border is made out less easily, and here the most ventral lamellae of the pad appear to merge into the loose tissue of the iris stroma. At this level the Descemet-type membrane shows gaps, through which processes from the endothelial cells extend to become continuous with the most ventral lamellae. This suggests that the pad may be formed from the corneal epithelial cells during life; certainly it grows with the fish, and in small specimens the pad is absent (MERRETT et al., 1971).

At the rostral and caudal borders of the pad the lamellae appear to merge with the adjacent iris stroma. MUNK (1966a) described the lens pad in *Neoscopelarchoides* sp. as developed in the iris stroma, and the appearance in horizontal sections lends support to this suggestion. In vertical sections (Fig. 17), the pad is seen mainly above the iris margin, and here its appearance suggests a corneal origin. The presence of the Descemet-type membrane and the apparent migration of endothelial cells through this to form the pad lamellae tend to support a corneal derivation.

The function of the lens pad was first considered by BRAUER (1908), who believed that it was composed of fibres and that it supported the lens. He interpreted the diffuse cornea as smooth muscle, and suggested that by its contraction this muscle might alter the distance between the lens and the main retina, and thus function as an accommodatory device. Though BRAUER stated that the normal teleost lens muscle was absent in *Scopelarchus*, MUNK (1966a) noted its presence in *Neoscopelarchoides*, and it was certainly present in those scopelarchids I have examined.

BRAUER'S proposed function in accommodation depends on his erroneous identification of smooth muscle in the diffuse cornea, and his failure to find the normal lens muscle. From ophthalmoscopy of the fresh specimen it is clear that the lens pad has an optical function, transmitting and guiding light from below the fish to the ventro-lateral aspect of the lens. This light reaches the dorsal-most part of the accessory retina, and the visual field of the upward looking eye is thus extended to within about 20° of the vertical below (Fig. 19). Traversing the pad, the light passes approximately along the lamellae, which from their dimensions probably act by wave guiding rather than according to geometrical optics.

3. The Optical Fold in Evermannellids

a) Introduction

BRAUER (1908) described the eyes of three evermannellid species which he gave as *Evermannella atrata*, *E. indica*, and *E. balbo*. ROFEN (1966) divided the six species of evermannellids among three genera, and according to him, BRAUER'S *E.*

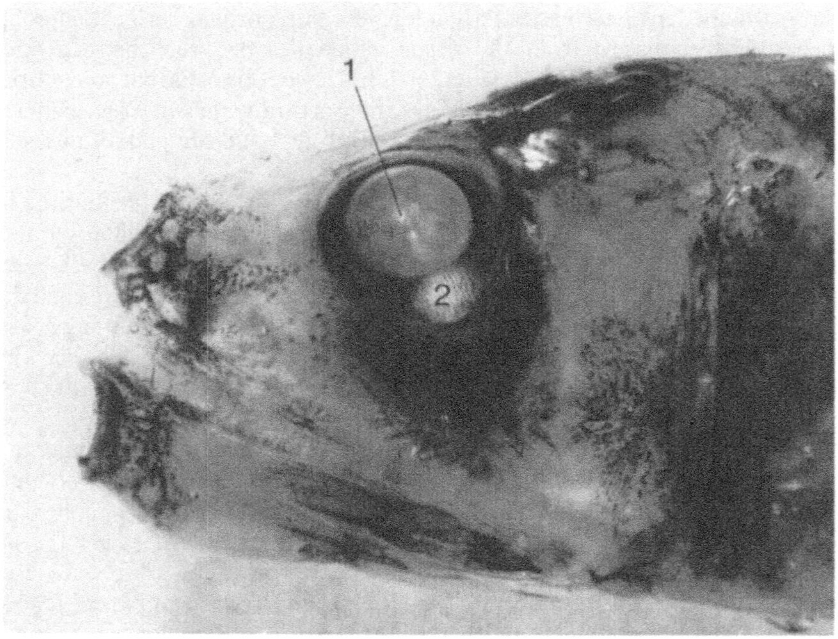

Fig. 24. Head and eye of *Evermannella indica* (preserved). Evermannelid fishes have large mouths with very long teeth, and can eat prey as large as themselves. In *E. indica* eyes are tubular, and face dorsally. Large lens, *1*, occupies oblique pupil. Overlying eye laterally is fold of skin, of which part is specialized, showing here as glistening patch, *2*. This is the part described as the optical fold, see text

atrata is *Coccorella atrata* (ALCOCK) 1893: *E. indica*, also studied by MUNK (1966a), and *E. balbo* remain in the genus *Evermannella*. This classification is followed here.

BRAUER (loc. cit) described the eyes of *E. indica* and *E. balbo* as situated laterally, somewhat tubular in form and looking dorsally, while those of *C. atrata* were of apparently normal form, though with a dorsal aphakic aperture. He noted that in all an unusual skin fold covered the lateral part of the eye, leaving the rostrodorsal part of the cornea free. The free margin of this fold was thin, and he stated that the epidermis which had covered it on the outer side was missing. The fold itself consisted of connective tissue, tightly stratified in the lateral part of the fold, but looser and wavy medially. The space between the two contained stacked plates of two kinds. Those of one type appeared pale, had no nuclei, and were not refractile; the other kind had nuclei, oriented with the plate, in their lateral ends and were strongly refractile. The only function BRAUER could suggest for this fold was protection of the eye, though why a pelagic fish should require such a thing was not clear to him.

MUNK (1966a), who described the retina of *E. indica*, made no comment on the fold, but ROFEN (1966) described a prominent white glistening spot on the side of the eye in both *Evermannella* (Fig. 24) and *Coccorella*, though not in *Odontostomops*, his third genus. He suggested that the "glistening spot" might be luminous in life, though it appeared more likely to him that it was a light reflector.

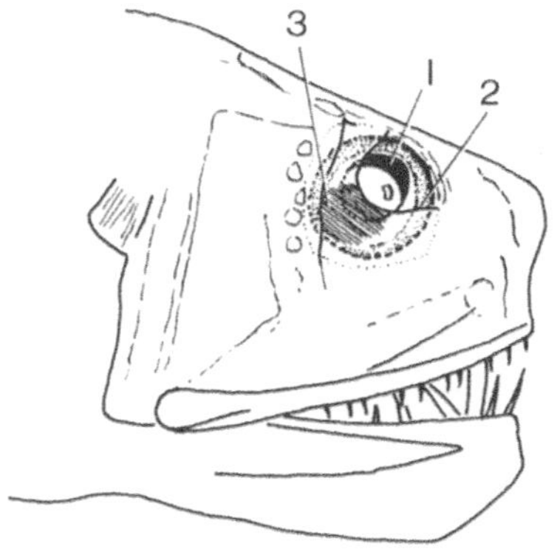

Fig. 25. Head of *Coccorella atrata*. Eye, of normal configuration but with dorsal aphakic aperture, *1*, is partially covered by fold or adipose eyelid, of which margin is marked *2*. Part of this fold is specialized for optical function of guiding light from below fish to lens, and eventually to dorsal-most part of accessory retina. This optical part of fold, *3*, is hatched in direction taken by lamellae

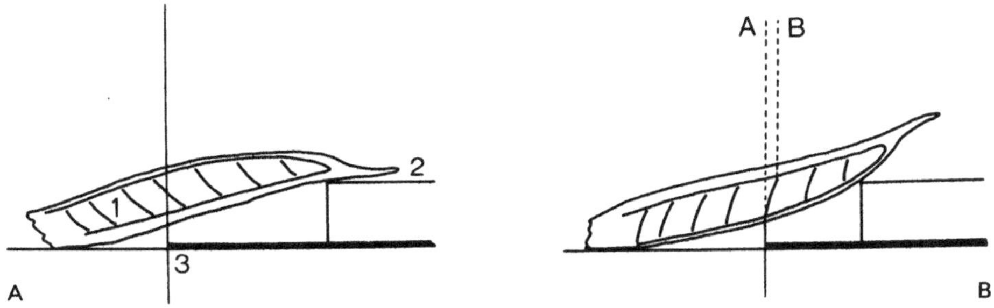

Fig. 26. Arrangement of optical fold used to obtain Figs. 27a and b. In Fig. 26a optical fold, *1*, is placed on microscope slide with its free dorsal border slightly elevated, *2*, and its external face upward. Opaque edge, *3*, is viewed through fold. In Fig. 26a light path is that which would be taken in life by light from above horizontal, but reversed. Figure 27a shows that image in this case is somewhat blurred, but not appreciably displaced. In Fig. 26b fold has been turned over, its external face downward. Light path is now that of light from below horizontal in life. Image is spread, Fig. 27b, due to light from edge being guided toward dorsal border of fold during its passage through lamellated part, from A to B

b) Observations on the Function of the Fold

BRAUER's figure of the fold in section recalled that of the scopelarchid lens pad, which suggested to me that they might have similar functions. In specimens preserved in glutaraldehyde or formaldehyde the fold is more or less transparent (Fig. 24) and it is very much so in the fresh condition (P. J. HERRING, personal communication). Figure 25 shows the relations of the eye and the fold in *C. atrata.*

The excised fold can be arranged on a microscope slide so that light passes through it in different directions, as shown in Figures 26a and b; the images obtained are shown in Figs. 27a and b.

Light from obliquely above the fish passes through the fold largely undeviated, but from obliquely below it is deviated along the plane of the laminae, and light from within 8° of the vertical below the fish can be directed into the lens of

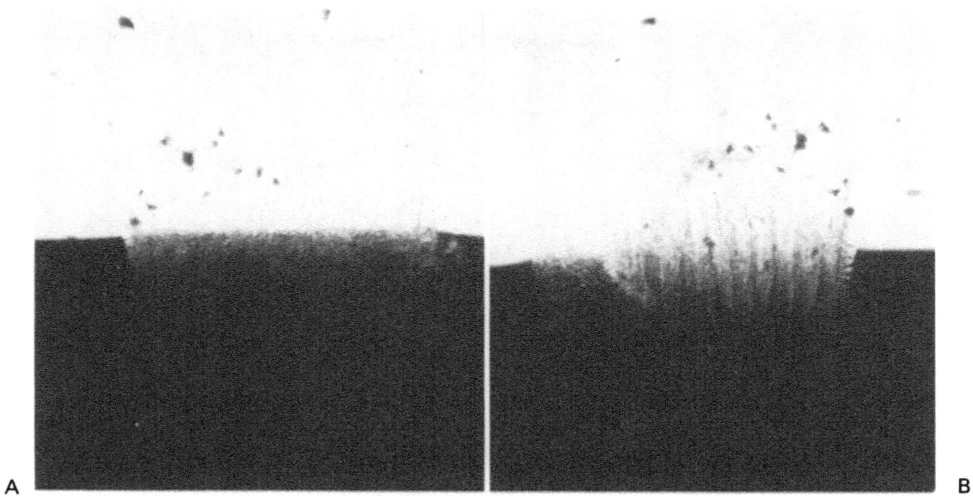

Fig. 27. In Fig. 27a optical fold of *Coccorella atrata* is arranged as in Fig. 26a, and edge seen through it. Image is somewhat blurred, but not much displaced toward dark side. In Fig. 27b fold is turned over as in Fig. 26b. Light from edge is now guided within fold, and appears to come from part of field which was dark in Fig. 27a

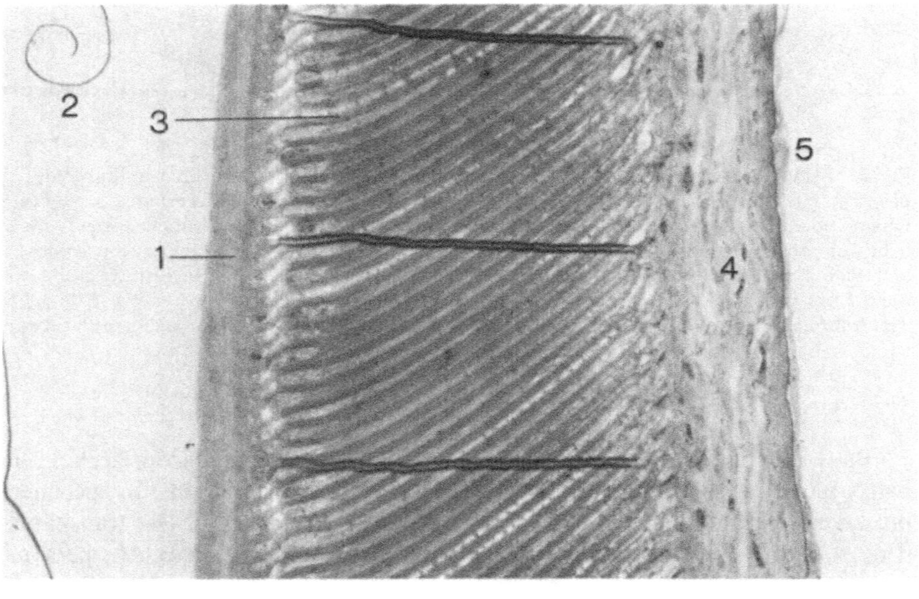

Fig. 28. Vertical section of optical fold of *Coccorella atrata*. Lateral side of fold is formed of uniform layer of fibrous material, *1*. Lying on this is single "cuticular" lamina, here detached and seen at *2*. Central part of fold is occupied by lamellated region, which contains raster of alternating cellular and fibrous plates, which slope dorsally, *3*. Medial third of fold consists of irregular fibrous tissue, with cells scattered in it, *4*. Bounding fold on medial side, which lies against eye, is single layer of epithelial cells, *5*, which have prominent basement membrane. Optical micrograph

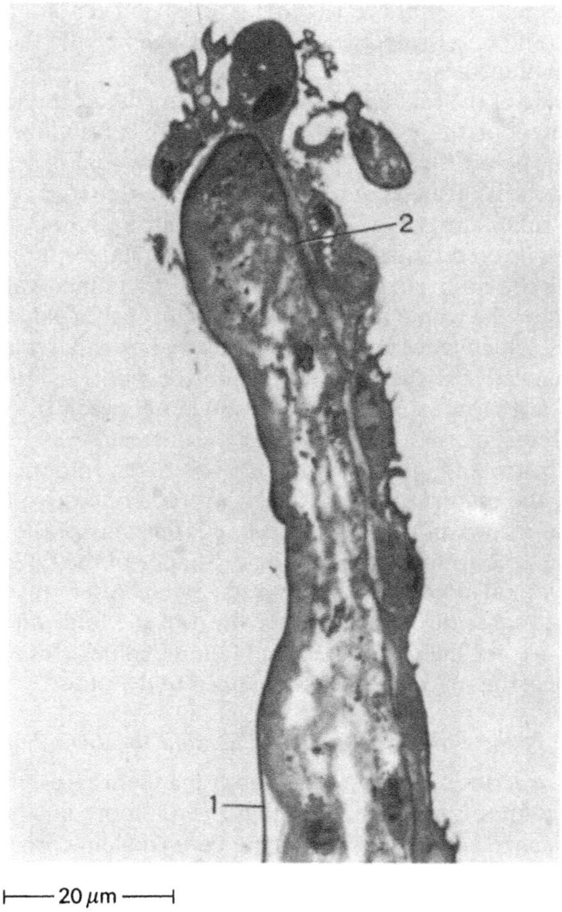

Fig. 29. Dorsal free border of optical fold of *Coccorella atrata*. "Cuticular" lamina on outside of fold, *1*, is continuous over border with basement membrane of epithelium, *2*, but cells of epithelium do not extend onto lateral face of fold. Optical micrograph

the eye. The function of the fold is thus comparable to that of the scopelarchid lens pad, serving as a light guide for light from below the fish, and so extending the field of upward facing tubular eyes.

c) Structure of the Fold

BRAUER's account of the structure of the fold is largely correct, but may now be extended in some respects. The presence of plates at an angle to the plane of the fold but parallel with each other is confirmed, and so is the presence of lateral and medial layers of connective tissue (Fig. 28). Though the inner side of the fold is covered by a single-layered epithelium, the outer surface lacks this covering. No specimen examined has shown any epithelium outside the fold, and microscopy of the free dorsal border suggests that there is no epithelial covering during life (Fig. 29). Between the epithelium on the inner side and the underlying connective tissue is a basement membrane which becomes thicker at the border, and is continued onto the outer side of the fold. It extends below the fold as the outer layer of the skin of the head.

The outer side of the fold consists of finely lamellated connective tissue, uninterrupted throughout the extent of the fold, and continuous with the dermal connective tissue below the fold. The inner side of the fold beneath the epithelium consists of irregularly layered connective tissue and scattered cells, approximately parallel to the inner surface. The region between the inner and outer layers is occupied by the layered tissue formed, as BRAUER stated, of lamellae which are continuous with the connective tissue of the outer and inner walls. These lamellae are continuous for the whole rostro-caudal extent of the fold, and between them are rows of cells, which taper from lateral to medial in transverse sections (Fig. 28). Though the connective tissue lamellae extend the whole length of the fold, individual cells do not. There are gaps of 10—30 μm between them, crossed by fine processes which make contact with those of the adjacent cells.

Light from below and laterally thus encounters the smooth and regular external membrane, the uniform layer of finely layered connective tissue, and then a raster of alternate rows of cells and lamellar extensions of the connective tissue. This alternation is continued through the substance of the fold, though it is most regular at the lateral border of the layered region. After traversing the layered region, the light passes through the inner third of the fold, where the connective tissue and cells are not highly ordered. In life the fold lies close against the lateral side of the cornea, the space between them open to the water.

d) Comparison of the Fold in Evermannellids and the Lens Pad in Scopelarchids

These two structures are interesting both for themselves, and as two similar solutions to a common problem, although the anatomy and optics differ in the two cases. The scopelarchid lens pad, being derived from corneal endothelium, is intraocular, but the evermannellid fold is developed in the extra-ocular "adipose eyelid". Both include modified cells in their system of alternate layers of more and less refractile material, but whereas the lens pad lamellae are about 200 nm in thickness, those of the fold are about 2.5 μm. The relation of these dimensions to the wave length of the light transmitted suggests that the fold may direct the light according to geometrical optics, but that the lens pad may function by wave guiding.

IV. Aberrant Eyes

1. Degenerate Eyes

Though the intensity of light falls off with depth in the oceans, and below 1000–1200 m there is no penetration of sunlight, yet fishes live at depths far below these, and so may live their whole lives in regions of darkness. Despite the absence of sunlight, the darkness is not absolute, since luminescent organisms occur at all depths, either among the plankton or living on the bottom.

It would be no surprise if some fishes from these regions had lost the use of their eyes, relying upon other senses to find their prey. As BRAUER (1908) and

MUNK (1965a, 1966a) show, the eyes of certain bathypelagic and benthal fishes are indeed small or regressed.

a) Benthal Fishes

Among BRAUER's examples was the elasmobranch *Benthobatis moresbyi*, from 820 m, of which a 44 cm specimen had an eye only 1.5 mm in diameter. This was merely a conical eye rudiment, containing traces of retina but neither iris, lens, nor pigment. In *Barathronus affinis*, a nonpigmented teleost trawled from 2919 m, the eye was represented merely by a small black point buried in the skin. Sections showed a cyst with crumpled walls and a minimal ring of cartilage. Some choroidal vessels could be identified, but there was no choroidal pigment. The iris was absent, but a lens rudiment was still present; BRAUER believed that even this might be totally absorbed later in the life of the fish. The retina was irregular, only the rods retaining the normal ordered pattern.

MUNK described the eyes of several other benthal species (1964a, 1965a), and reviewed his previous findings in his monograph (1966a) in which he also revised his naming of certain examples. MUNK noted that some fishes, e.g. *Bathymicrops regis*, *Sciadonus kullenbergi*, and *Gyronomimus* sp., have eyes which are clearly reduced and degenerate. Others have small eyes of aberrant structure (*Barathronus ericssoni*, *Ditropichthys storeri*) and still others have small eyes which appear normal on gross examination (*Echinomacrurus mollis*, *Careproctus kermadecensis*, *Bathypterois longipes*, *Bassogigas profundissimus*, and *Benthosaurus grallator*).

In the more extreme forms of ocular degeneration all that remains of the eye in the adult is a mass of pigmented cells enclosed in a fibro-cartilaginous capsule (*S. kullenbergi*), or remnants of a much degenerated retina among other pigmented cells. A folded PAS-positive membrane found by MUNK on the external side of the retina in *Bathymicrops* may represent the remnants of a lens.

MUNK described the eyes of *Barathronus ericssoni* as resembling those of *B. affinis*, which was among BRAUER'S examples. The retina proper was represented by folded and degenerate tissue, which included swollen and fragmented outer segments. Such swelling might be due to inadequate fixation, but MUNK described macrophages throughout the tissue, strongly suggesting that the disorganization was already present in life. In MUNK'S specimen there was no lens, though BRAUER found a structure which may have been a rudimentary lens in *B. affinis*. MUNK found a similar situation in *Ditropichthys*, as BRAUER had in a *Cetomimus* sp.

In other species, *B. profundissimus*, *B. longipes* and *B. grallator*, of those described by MUNK, the lens was disproportionately large, and the retina showed more or less degeneration. In *Careproctus* and *Echinomacrurus* the lens showed cataractous changes, and in both there was retinal degeneration and invasion by macrophages.

b) Ceratioid Angler Fishes

The eyes of ceratioid angler fishes have been studied by BRAUER (1908) and MUNK (1964b, 1966a). These fishes, of which the reproduction and development were described by BERTELSEN (1951), exhibit sexual dimorphism; the males, which in some species live a parasitic existence attached to the female, are in general much the smaller sex. Though the larval development takes place in the epipelagic zone, pairing and egg-laying take place in the unlit depths and the buoyant eggs rise to the surface. The young return to the depths on metamorphosis.

Most female ceratioids have a luminous organ on a chin barbel or a modified dorsal fin ray, but the males do not possess these organs. The eyes of the females grow little after metamorphosis, and according to BERTELSEN (1951), in the largest specimens they are so small and degenerate as to appear functionless. The eyes of the larvae, however, are well-developed in both sexes, while those of the adult males show marked differences with species. In the *Gigantactinidae* they are very small, in the *Melanocetidae* larger, in *Linophrynidae* tubular, and in the *Ceratiidae* large. Once the males have become parasitic, their eyes appear to degenerate, as do the olfactory organs, prominent in many free-living males. This prominence of eyes and olfactory organs in the free-living males are regarded by BERTELSEN as an adaptation of the males to the difficult task of finding a mate.

BRAUER observed that the eyes of the female ceratioids he studied were small, their diameter only 1/15—1/22 the length of the head. In *Aceratias*, *Melanocetus*, and *Gigantactis* the eye was located beneath a small opening in the black skin of the head, though in some specimens it was displaced and so not opposite the opening. This he realized might be due to shrinkage of the preserved specimens.

The sectioned eyes also appeared shrunken, and he commented that they appeared to be in an early stage of degeneration. Despite this, and their small size, the eyes contained all the normal components, though the lens-retina distance was less than Matthiessen's ratio. He concluded that the eyes were probably not very effective.

As MUNK (1964b) showed, the eye cups of the male *Cryptopsaras couesi* are shallow, very open rostrally, and asymmetrical in horizontal section. The lens is spherical and adequately developed, and a lens muscle is present. In front of the eye is a shallow sighting groove, and the impression of eyes adapted to binocular vision forward is reinforced by the increased density and length of the receptors in the caudal region of the retina. MUNK's figures confirm BRAUER's observation that Matthiessen's ratio is not obeyed in ceratioids, and in some, notably *Ceratias holboelli* and *Linophryne arborifera*, the lens-retina distance is by no means uniform. In *C. holboelli* the retinal cup is open anteriorly as in *Cryptopsaras*, and *Linophryne* males have tubular eyes in which the caudal retina may obey Matthiessen's ratio while the rostral periphery is close against the lens.

MUNK pointed out that the larvae of *Ceratias* and *Cryptopsaras* show a marked structural alteration of their retinae when they descend to bathypelagic depths at metamorphosis, the summation increasing considerably. The presence of sighting grooves and of the wide aphakic spaces in the males of those species will result in a wide binocular field rostrally.

c) *Cyclothone spp.*

The various species of *Cyclothone* are among the most successful in the deep-sea on grounds of their numbers and the extent of their living space (MARSHALL, 1971). These fishes, mostly diminutive, all have very small eyes, though these appear relatively normal on gross examination. *Cyclothone* species are mostly well-equipped with photophores; besides rows along the body, there is a prominent light organ rostro-ventral to the eye. Light from this photophore can probably not reach the eye directly, due to an investment of pigment.

BRAUER (1908) examined the eyes of several species of *Cyclothone* and noted the small size of the eyes. He also observed that the eye cup was shallow, and the lens large in proportion to the rest of the eye, almost filling the pupil, itself almost as wide as the whole eye in *C. microdon pallida* and *C. obscura*. The iris in these species is correspondingly reduced.

BRAUER found only rods in the retinae, and these were long, occupying one third of the retinal thickness in *C. signata;* in *C. microdon pallida* they accounted for 21 μm of the 38 μm retina. The lens was close to the retina in BRAUER's examples, and he gave figures of 1:1.3–1:2 for the lens-retina ratio. Of the retinal cells he found a single layer of rod nuclei over most of the retina, with two layers near the periphery. One layer of bipolar cells was also present, but he could recognize amacrines only near the periphery. There were no horizontal cells, and ganglion cells were few and far between; he counted 50 in a whole section. The whole retina appeared to him irregular, which he would have attributed to fixation artifacts except that the retinae of other fishes, treated in the same way, did not show the same appearance.

Species of *Cyclothone* were not among those studied by MUNK, but I have examined eyes of *C. braueri* and *C. pseudobscura*, fresh and sectioned. The shallow eye cup and large lens are prominent features, and the cornea arches across the whole width of the eye (Fig. 30). There is a mere frill of iris bordering the thick retina, so there is a wide aphakic gap. The lens is spherical and quite clear in life; both in the fresh state and in sections it appears close to the retina, but the small size of the whole eye, only some 650 μm diameter, makes the retina, which measures some 70–100 μm look disproportionately thick. The lens radius in *C. pseudobscura* measured 155 μm; the lens centre-retina distance was 227 μm measured to the inner limiting membrane, or 270 μm to the outer limiting membrane. If Matthiessen's ratio were obeyed this distance should be 395 μm, which is 65 μm outside the whole eye. Despite this disparity, light passing through the lens is strongly converged onto the retina, as can be seen through the aphakic gap of a fresh specimen. At the inner limiting membrane the blur circle would be 140 μm in diameter, and at the outer limiting membrane 96 μm. The rods are about 5.5 μm in diameter, and if they are tightly packed in hexagonal array about 140 would be included in such a blur circle. In some of the sections of *C. pseudobscura*, though not in others of the same species, there are areas in which two banks of rods are present, and in these regions the blur circle will contain twice as many rods: even in the most favourable cases there seems to be little chance of good resolution on optical grounds.

Closer examination of the eye showed that the retina is anomalous in some respects. Some of the rods in both species examined appear to lack nuclei, and the synaptic pedicles are hard to make out even by electron microscopy; they are distended and empty looking, containing few vesicles and only traces of synaptic ribbons. The inner nuclear layer contains few nuclei, which cannot be assigned to a particular cell type with any confidence. Horizontal cells are not recognizable, as BRAUER noted, and ganglion cells are indeed sparse. Some of the nuclei in the

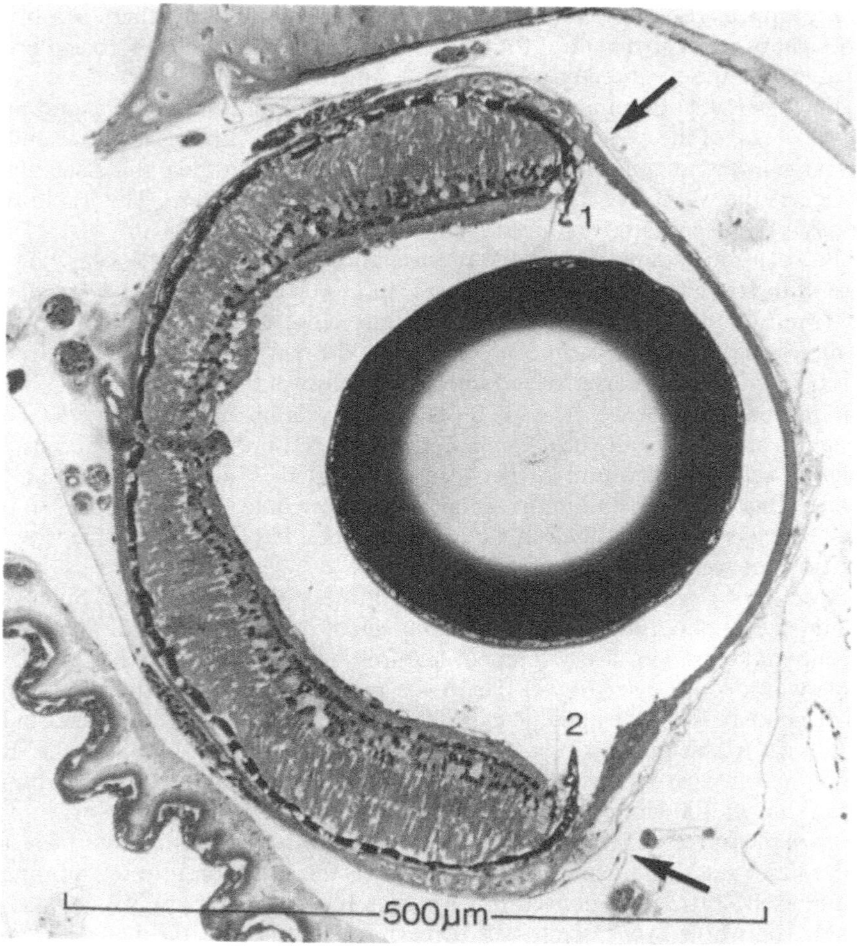

Fig. 30. Eye of *Cyclothone pseudobscura*. Very small eye appears dominated by lens. Cornea covers full diameter of eye, between arrows, and retina forms hemispherical cup, not conforming to Matthiessen's ratio. Iris, *1*, is small, and there is wide circumlental aphakic gap, *2*. Retina is anomalous; at periphery there are two banks of rods, but more central rods do not appear well-differentiated. Optical micrograph

ganglion cell layer are small and dark staining, and may well belong to glial cells. If this is so the convergence onto ganglion cells will be even greater than it appears at first sight.

It seems that the eyes of the *Cyclothone* spp. are not only physically small, but also poorly developed, suggesting that vision may be of minor importance in the lives of these fishes.

2. The Eyes of *Ipnops* Spp.

The fishes of the genus *Ipnops* were first known from the "Challenger" collection. Gunther (1878) described and named them, and in his first description expressed doubt about the function of the flattened areas on the head; he referred

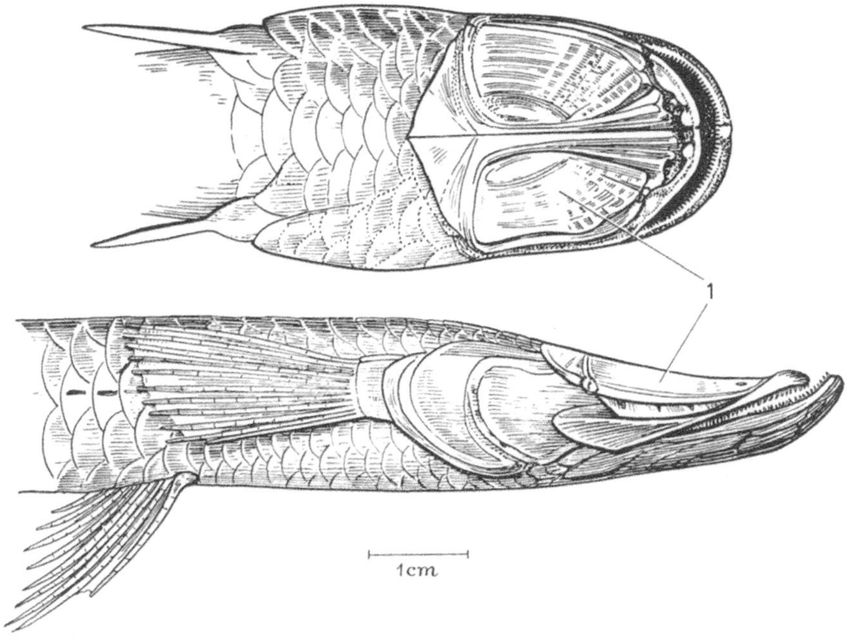

Fig. 31. Head of *Ipnops* sp. Flattened plates on dorsal surface of head, *1*, are greatly modified eyes, which were believed by early authors to have become luminous organs (From MUNK, 1959)

to "a most peculiar organ of vision (or luminosity), longitudinally divided into two symmetrical halves" (Fig. 31). In a later work (1880) he wrote; "The eye seems to have lost its function of vision and assumed that of producing light." In 1885 he referred to the organs as modified eyes, and gave a description of them by MOSELEY (1887). This author described a transparent membrane, probably representing the cornea, beneath which was a shallow fluid-filled chamber overlying a retina containing only rods, and a thin layer of nerve fibers. The choroid he described as formed of hexagonal areas, concave toward the surface, and associated with bundles of rods occupying the concavities. In this account MOSELEY suggested that the organs were devices for detecting the presence of small quantities of light at the expense of any apparatus for forming an image. In his appendix to GUNTHER (1887), MOSELEY gave the opinion that the organs must be regarded as luminous organs, a view which prevailed in the literature until MUNK (1959) published his observations on the organs in three specimens from the "Galathea" collections.

MUNK showed unequivocally that the organs are modified eyes, and that MOSELEY's account of 1885 was the nearest to the facts.

The eyes, which MEAD (1966) reported are brilliant yellow in life, consist of a flattened retina containing only rods and with very few other retinal cells (Fig. 32). The inner nuclear and ganglion cell layers are apparently absent, and probably the nuclei of second and third order neurons are among those which are scattered in the retina superficial to the layer of rods. The optic nerves are present, though not recognized by MOSELEY (1887); MUNK counted about 500 fibres in each, and traced the nerves, which cross entirely at the chiasma, to the tectum. MUNK found

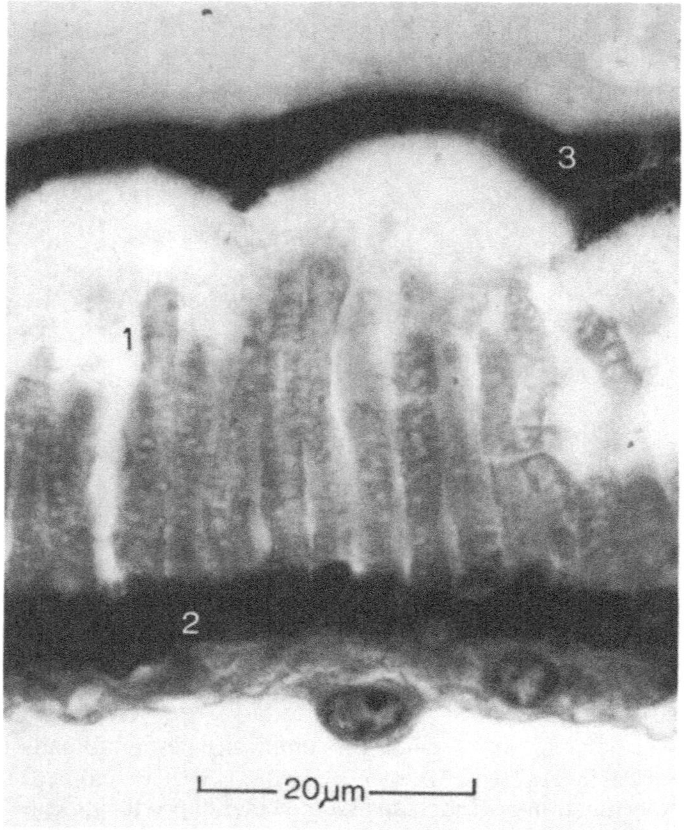

Fig. 32. Retina of *Ipnops*. Outer segments, *1*, and nuclei of rods can be distinguished, and other nuclei are present in more superficial part of retina, *2*. Retinal epithelium, *3*, contains dark pigment granules. (From MUNK, 1959. Optical micrograph)

that the pigment epithelium contained pigment in the sclerad portions of the cells only, and that this gave an appearance of cup-shaped cells such as MOSELEY had described. MUNK pointed out that the cells are in fact not cup shaped, and that there was no trace of bundles of rods in his preparations. It is likely that MOSELEY's material had shrunk on preservation, whereas MUNK was working with material fixed for histological examination, and that MOSELEY'S error was due to artifacts.

The surface of the head over the eyes is covered by a transparent membrane, which MUNK showed by alizarin staining to consist of thin bone; contained in this membrane are a pair of canals, one on each side of the midline, which contain lateral line organs and nerve fibres. MUNK suggested that these were infraorbital canals, and noted that supraorbital canals are also present on the transparent membrane.

Beneath the retina the choroid and sclera are represented. The choroid consists of rather few capillaries of the choriocapillaris among which are pigmented cells, and a large blood-filled sinus coextensive with the retina. The sclera is thin, and consists of fibrous tissue with a cartilaginous peripheral portion; in this it resembles the sclera of most teleost eyes. The peripheral rim is connected to the deep surface of the membranous surface bone by a thin membrane, probably representing the peripheral part of the scleral cornea.

In this strange eye there is no sign of any lens rudiment, iris, nor of the normal corneal structure.

Such an eye, consisting of a plate of poorly differentiated retina spread out beneath a transparent membrane, is unique. In most examples showing ocular degeneration there is greater disorganization of the eye tissues, though all are represented.

3. Stalked Eyes

Among the fishes collected by the "Valdivia" expedition were some small specimens which BRAUER named *Stylophthalmus paradoxus*. These fishes were distinguished by the presence of long stalks bearing the eyes at their tips. BRAUER recognized that these specimens resembled the larvae of stomiatoids, but he was unable to identify them further.

BEEBE (1934), who collected and examined large numbers of samples from a small area near Bermuda, was able to trace the growth pattern of *Stylophthalmus*, and to show that it was in fact the larval stage of the stomiatoid *Idiacanthus fasciola* (Fig. 33). His 129 specimens were all referable to this species, but he noted that the specimens on which BRAUER had worked belonged to more than one. Some were the young of *I. fasciola*, but others were not, though he could not positively identify them.

BRAUER'S description of the stalked eyes was confirmed by BEEBE, who also traced the disappearance of the stalks as the fish grew. The larval stage, when the fish is about 20 mm long, is the stage in which the stalks are most prominent, spanning a distance of almost three-quarters of the body length. At this stage the stalk itself contains a cartilaginous rod and the optic nerve, both enclosed in a transparent sheath, which extends to cover the eye itself. At this stage the eye consists of a shallow but thick eye cup, from which the relatively enormous lens projects. From BEEBE'S figures (Fig. 34) it is clear that Matthiessen's ratio does not apply at this stage. The cartilaginous rod is attached proximally to the frontal and parasphenoid bones, from which it extends the full length of the stalk to the globe. In the stalk it is anterior to the optic nerve, and at the eye it is inserted into the membranes covering the eye cup.

In the post-larval stage the stalks become reduced by the absorption of the optic nerve. The cartilaginous rod does not shrink in the same way; it first bows, tearing out through the membrane of the stalk (Fig. 34) and subsequently becomes curled up. The curled mass becomes buried in the orbital tissue in front of the globe, and is absorbed completely by the time the fish has reached adult size. The eye in the adult is a typical lateral eye of normal proportions, though it is elliptical in the post-larval stage when it first reaches the head on shortening of the stalk.

BEEBE concluded from catch analysis that the larvae, caught in August and September in those waters, form schools, and that their diet includes crustaceans.

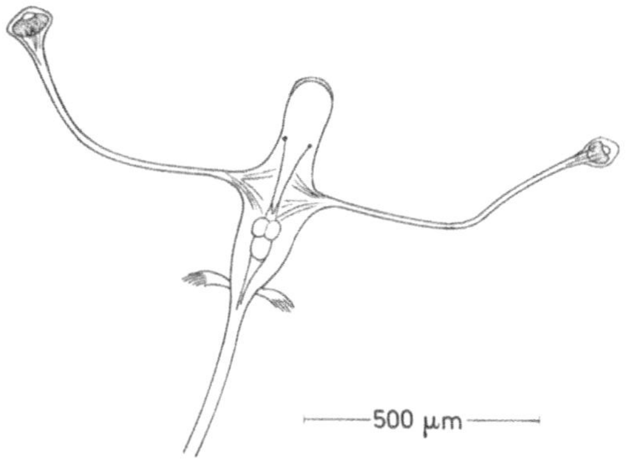

Fig. 33. Head of *Stylophthalmus*, a larval stomiatoid, perhaps *Idiacanthus* sp. Eyes are borne on stalks which project from sides of head. This stalk contains optic nerve, extraocular muscles and stiffening rod (see Fig. 34). (After BRAUER, 1908)

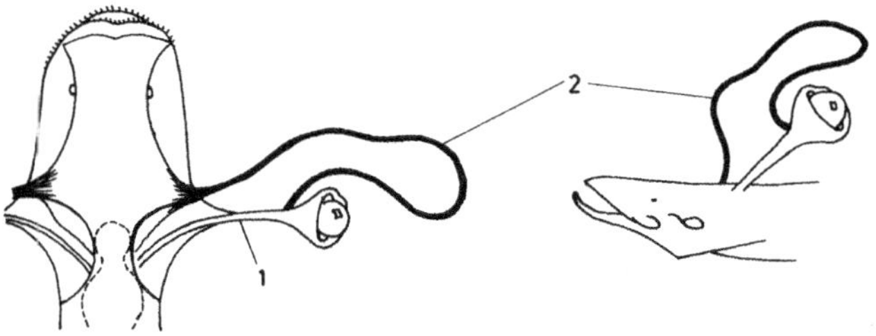

Fig. 34. Dorsal and lateral views of head of 40 mm male *Idiacanthus fasciola*. Optic nerve, *1*, has begun to retract, and stiffening rod, *2*, has torn free of eye stalk. It comes to be buried later in tissues rostral to orbit. (After BEEBE, 1934)

These findings, he suggested, indicate that the larval eyes are of use to the fish in behavioural terms.

Not only stomiatoids have stalk-eyed larvae; BEEBE also quotes bathylagids and myctophids. The development of the latter are described by MOSER and AHLSTROM (1970). The myctophid larval eye is vertically elongated in some species, though in others it is round. The adult myctophid always has a typical round lateral eye, to whichever group its larva belongs (Fig. 35). The stalked eyes occur in several of the elongated-eyed genera, and in some of these, notably *Gonichthys* and *Centrobranchus*, there is a ventral projection, apparently of solid choroidal tissue (Fig. 36). Both the stalks and the choroidal body disappear at metamorphosis, and even the youngest juveniles show neither. The stalk absorbs gradually in late larval development, but the choroidal body disappears more rapidly.

MOSER and AHLSTROM consider that a narrow, vertically elongated eye may be capable of rotation anteriorly and posteriorly, so increasing the field which may be scanned and increasing the possibility of binocular vision forward. As they point out, the mobility of the eyes could be further enhanced by the stalked condition. Though the function of the choroidal tissue mass remains obscure, they suggest that it may represent preformed choroidal tissue which could expand rapidly to supply the needs of the eye as it grows at metamorphosis.

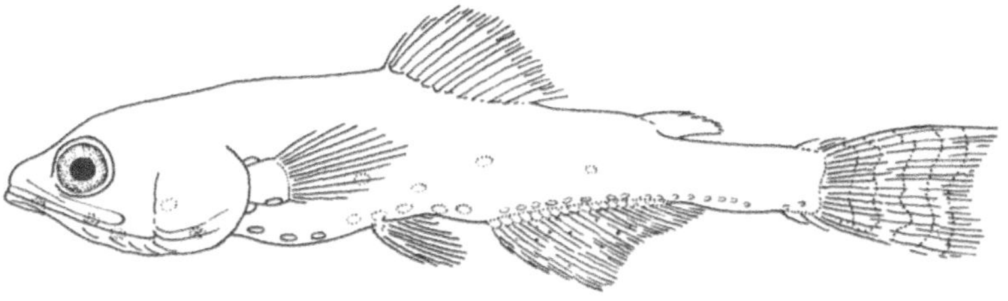

Fig. 35. Juvenile but metamorphosed specimen of myctophid, *Symbolophorus californiensis*. Eye is sessile, and of normal round shape. (After MOSER and AHLSTROM, 1970)

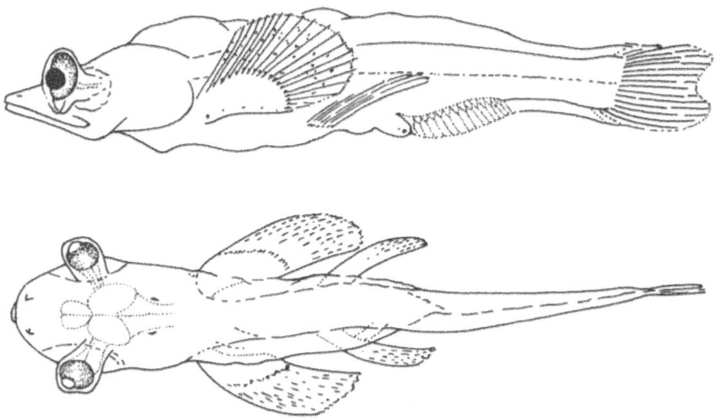

Fig. 36 A and B. A 9.6 mm larva of *Symbolophorus californiensis*, (A) lateral view; (B) dorsal view. Larval eyes are elongated dorso-ventrally, and are borne on short stalks. (After MOSER and AHLSTROM, 1970)

G. Retinal Adaptations in Deep-sea Fish Eyes

I. General Adaptive Features

1. Retinal Cell Types and their Proportions

Though the retina is commonly described as being composed of layers, easily recognized in thick sections of typical retina, the layering is but the visible sign of the ordering of different parts of the cells which make up the retina. When considering function, it is the cells which deserve attention, and in comparative studies it is the nature of the cells and the populations to which they belong that differ.

The literature on retinal morphology and function is large, and no attempt will be made to review it. The recent work of STELL (1972) and SCHOLES (1975) on teleost retinal structure give good accounts, and numerous references.

The various cell populations and their special features in deep-sea fishes are now considered.

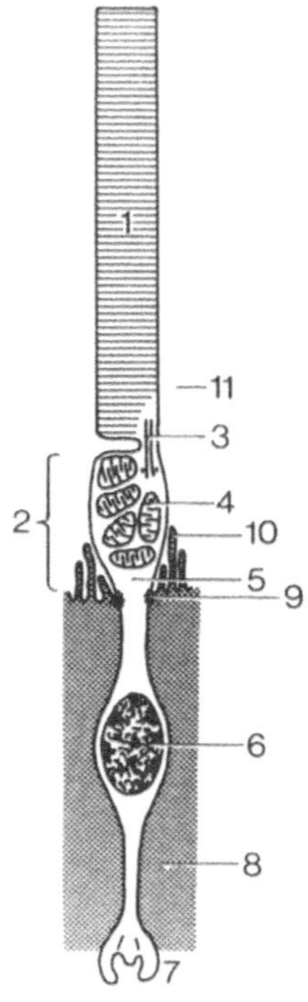

Fig. 37. Diagram of retinal rod. Outer segment, *1*, is cylindrical, and contains flattened saccules, or discs, which in most of outer segment are closed off from outer segment membrane. At base (vitread end) of outer segment discs may show continuity with membrane, see Fig. 39. Outer segment is connected to inner segment, *2*, by modified cilium, *3*, which in some cases appears to be sole connection between two. In other examples there is clear cytoplasmic continuity apart from cilium. Sclerad part of inner segment, *4*, called ellipsoid, contains closely packed mitochondria. Vitread to this is myoid region, *5*. Nucleus, *6*, is in outer nuclear layer. Synaptic spherule, *7*, is typically small and simple in organisation, with few synaptic ribbons. It is situated in outer plexiform layer. Vitread portions of rod are surrounded by radial fibre cytoplasm, *8*, which terminates at outer limiting membrane, *9*, as radial fibre microvilli, *10*. These project into ventricular space, *11*

a) Receptors

In shallow water teleosts receptors of two main types are found, the rods and cones familiar throughout the vertebrates. Deep-sea fishes, with few exceptions, lack cones. Their rods, however, show features that do not occur commonly in shallow living fishes. Rods in general have an outer segment that contains the visual pigment, an inner segment of which the sclerad portion is composed largely of mitochondria and called the ellipsoid, and a more vitread portion called the myoid. The myoid extends from the ellipsoid to the outer limiting membrane, and the cell is continued vitread to that "membrane" by the conducting fibre. The rod nucleus is usually situated vitread to the outer limiting membrane in the outer nuclear layer. Vitread to the nucleus the conducting fibre terminates in the synaptic pedicle, which is located in the outer plexiform layer (Fig. 37).

The outer segments in deep-sea fishes often account for more of the retinal thickness than they do in shallow living fishes. This preponderance may be due to small rods present in large numbers and closely packed, or to long outer segments. Since the visual pigment is the essential first link in the chain of events of photoreception, and is responsible for the capture of light quanta, it is not surprising to find large amounts of outer segment material with correspondingly high pigment densities in fishes which live in near-threshold conditions of light.

The inner segments and their contained mitochondria represent a site of metabolic activity in the receptor, and the work of YOUNG and his colleagues (YOUNG and BOK, 1969, for references) has shown that new outer segment material is added at the base of the outer segments in close association with the mitochondria of the ellipsoid. The ellipsoids of deep-sea fishes are similar in appearance to those of shallow-water fishes, and their important relationship with the outer segment is emphasized by the invariable association of the two. In retinae with multiple banks of rods the outer segment and ellipsoid are always found together; it is the myoid which is elongated, separating the ellipsoid-outer segment complex from the rest of the receptor cell.

Such myoids are in some cases very long and thin, traversing the ventricular space, and presumably conducting the excitation generated in the outer segment in response to light to the neuro-retina. In cases with long myoids, the materials destined for new outer segment discs must presumably travel along those myoids.

The rod nuclei are generally located in the outer nuclear layer, vitread to the outer limiting membrane, as they are in shallow living teleosts with duplex retinae. In some cases these nuclei are present in a layer several nuclei deep, and in others as a single layer of elongated nuclei arranged in pallisade formation. In certain other cases, notably the accessory retinae of tubular eyes (see below) the rod nuclei are sclerad to the outer limiting membrane, a situation typically occupied by cone nuclei in those fishes that have cones.

Typical rods and cones in duplex retinae also differ in the nature of their synaptic pedicles. Cones generally have a large synaptic pedicle with which processes from several horizontal cells and bipolar cells make contact, by invaginating and surface contacts. The typical rod pedicle is smaller, with few bipolar and horizontal cell contacts. These pedicles, called oligosynaptic by UNDERWOOD (1968) contain only one or two synaptic ribbons, whereas the complex (Underwood's polysynaptic) cone pedicles contain 5–12. Cone synaptic pedicles generally give rise to basal filaments, that ramify in the outer plexiform layer, and form interreceptor contacts (SCHOLES, 1975). The simple rod synapses typically lack basal filaments, though in the coelacanth *Latimeria*, each rod synapse gives rise to a single basal process, which in this case makes contact with a horizontal cell (LOCKET, 1973).

Deep-sea fishes generally have synapses of the rod type with correspondingly few ribbons, though in some cases more cone-like synapses are found. These special cases are considered later, in context.

The conducting fibre between the rod nucleus and synapse is generally thin, but in some cases, e.g. *Sternoptyx diaphana*, the synapses are directly applied to the vitread pole of the nucleus, in which case they are described as sessile synapses (LOCKET, 1970).

b) Inner Nuclear Layer

The inner nuclear layer typically contains the nuclei of bipolar, horizontal, and amacrine cells, and of radial (Müller's) fibres. Deep-sea fishes commonly have but few cells in this layer, but those which are present commonly form discrete layers, separated in some cases by lamellae derived from radial fibres, rather than the single thick layer of nuclei found in shallow water fishes.

An important component of the inner nuclear layer are the bipolar cells, whose processes contribute to the outer and inner plexiform layers. The bipolar processes in the outer plexiform layer make contacts with the synaptic pedicles of the visual cells, and in deep-sea fishes it is usual to find many more rods than bipolars, so that there is a large convergence of rods onto each bipolar. These rods may be spread out, so that the bipolar has a large "catchment area;" if the bipolar can be stimulated by less than the total number of rods with which it makes connection, then the potential resolving power may be limited by the size of the area sampled, a sufficient stimulus in any part of that area activating the bipolar, and being indistinguishable from a similar stimulus elsewhere in the area. If a bipolar is more easily stimulated by a low level of discharge from many receptors than from few, then such convergence will tend to increase sensitivity.

The bipolar cells are not directly responsible for the neural output from the retina; it is the ganglion cells, whose axons form the optic nerve fibres, which are. Further convergence usually occurs from the bipolars onto the ganglion cells, and the final convergence from rods to ganglion cells may be some hundreds to one. Exact measurements of convergence ratios depend on accurate identification of cell types, and accurate differential counting of elements of different shapes, sizes and distribution in three dimensions. Such accuracy is hard to achieve except in favourable cases, but the tendency to high convergence is easily recognised.

The sclerad part of the inner nuclear layer is occupied by horizontal cells, which in many fishes are large and stellate, with extensive areas of lateral junctions between adjacent cells. When they are of this kind they have the appearance of a perforated membrane, already familiar to KRAUSE (1886a,b). In shallow water teleosts there may be three or four layers of these distinctive horizontal cells; STELL and LIGHTFOOT (1975) and SCHOLES (1975) have given accounts of these and of their colour-specific connections. Some deep-sea fishes have stellate horizontal cells, e.g. *Sternoptyx* (LOCKET, 1970), but in others no such cells are visible, e.g. *Platytroctes* (LOCKET, 1971a). Still other examples have stellate horizontal cells in one region of the retina, but not in others; this is the case in *Scopelarchus*, in which the main retina contains a layer of stellate horizontal cells, but the accessory retina does not (LOCKET, 1971b).

The absence of stellate cells does not imply the absence of horizontal cells altogether; those of mammals are not of the stellate kind, and are difficult to distinguish morphologically in conventionally stained tissue. It is not possible to exclude the presence of such cells in deep-sea fishes on the present evidence.

The vitread part of the inner nuclear layer is usually occupied by amacrine cells, which may or may not be easy to distinguish from bipolars on morphological grounds.

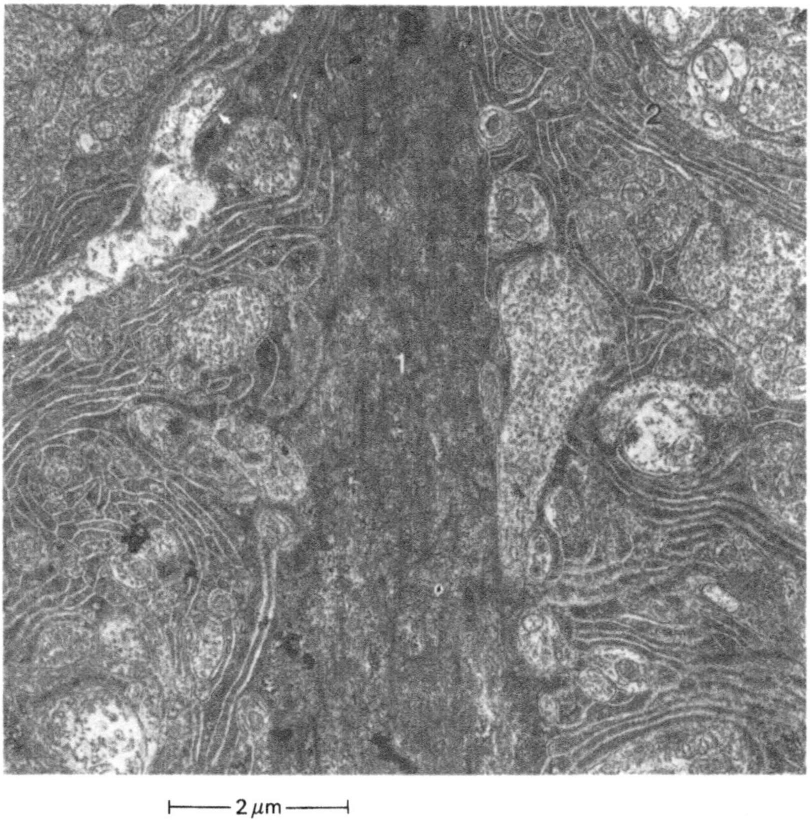

Fig. 38. Radial fibre, fovea of *Platytroctes apus*, Radial fibre trunk, *1*, easily recognized in optical micrographs, is seen traversing inner plexiform layer. Intricate lamellar systems, *2*, which permeate retina, are seen to be in continuity with radial fibre trunks. Electron micrograph

c) Radial Fibres

The radial fibres (of Müller) in deep-sea fishes are commonly developed into lamellar systems, sometimes of considerable complexity, which permeate the neuroretina between the inner and outer limiting membranes. These lamellae are convoluted among each other, and may be so extensive that their origin is hard to determine: sections from near the fovea of *Platytroctes apus* clearly showed the lamellar systems arising from typical radial fibre trunks, confirming their suspected radial fibre origin (LOCKET, 1971 a) (Fig. 38).

In many examples, including *Sternoptyx diaphana* and the main retina of *Scopelarchus guntheri*, the inner nuclear layer is divided by a layer composed of these radial fibre lamellae into two populations of nuclei, tentatively assigned to bipolar and amacrine cells. This subdivision of the inner nuclear layer appears to be peculiar to deep-sea fishes, and the lamellation of the radial fibres, though reported in other groups (e.g. geckos, PEDLER, 1963), is particularly prominent in the deep-sea fishes examined.

The contour of the inner limiting membrane at the fovea is of importance for Pumphrey's theory of foveal function (see below, G.V.); this contour in *Platytroctes* is determined by radial fibre elements. Radial fibres are present vitread to the nerve fibres elsewhere in the retina in many species (Pedler, loc.cit.). In this situation, at the fovea and indeed throughout the thickness of the neuroretina, the radial fibre appears to have an optical function, ensuring the optical homogeneity of the neuroretina through which light must pass to reach the sensitive elements.

Many fish retinae are avascular, in which case metabolites must diffuse from the choriocapillaris through the entire thickness of the retina. That portion of the retinal thickness accounted for by the space between the pigment epithelium and the outer limiting membrane contains the inner and outer segments of the receptors, which almost fill the ventricular space. Thus, the mitochondria contained in the inner segments are close to that space, and they will be exposed to the full concentration of nutrients diffusing from the choroidal blood.

In the neuroretina, between the inner and outer limiting membranes, the bodies and processes of the neural cells of the retina are surrounded by radial fibre cytoplasm, and the intercellular space is here represented by the 17 nm interval normally present between the membranes of adjacent cells. This distribution of radial fibre occupying the space between the retinal neurons is common to all vertebrate retinae, but the appearance of the radial fibres differs between species. In most retinae the radial fibres give rise to microvilli which project into the ventricular space between the visual cells, though in some retinae these are more developed than in others. The radial fibre microvilli are particularly prominent in the accessory retina of *Scopelarchus*, in which they fill almost all the space between the inner segments of the rods, which are arranged in groups in this retina (see below, G. III. 2).

The function of the microvilli has not been defined, but several possibilities exist. They may serve a mechanical function, maintaining the visual cells in a stable mutual orientation. Their function may be metabolic; the microvilli are the first part of the radial fiber cytoplasm which will be encountered by substances diffusing from the choroid, and the large surface area which they present might facilitate the passage of metabolites between the choriocapillaris and the retina. Mitochondria, which may be concentrated in sites of high metabolic activity, are plentiful in radial fibres, sometimes concentrated close to the outer limiting membrane.

It is also possible that the microvilli extend the optical function of the radial fibres in a sclerad direction. It is sometimes assumed that the image is brought to a focus at the level of the outer limiting membrane, and in many retinae at this level the homogeneous inner retina gives way to the rod and cone layer, in which the inner and outer segments of the visual cells lie free in the ventricular space. In such a case, if there are no microvilli, light which has entered a receptor will be confined to that cell, due to the difference of refractive index between the receptor and the ventricular fluid. If the space between the receptors is largely filled by radial fibre microvilli, as it certainly is in some cases, the difference of refractive index between these and the receptor elements may be small, in which case the receptors and microvilli will form an optically homogeneous mass, since the diameter and separation of the microvilli is less than the wave length of light. The light will not be constrained by refractive index differences until it reaches the level at which the microvilli end.

d) Relations of the Pigment Epithelium and the Receptors

In the retinae of higher animals, and indeed of many shallow water fishes, the pigment epithelium is a prominently developed layer of cells, closely adherent to Bruch's membrane and the choriocapillaris on its sclerad side, and intimately related to the outer segments of rods and cones on its vitread side. Among the activities of the pigment epithelium is its part in the metabolism of the visual pigment (BAUMANN, 1972, for full account). KÜHNE (1878) noted that regeneration of rhodopsin in the dark did not occur in the excised frog retina unless it was in contact with the pigment epithelium and WALD (1935) produced evidence for the interconversion of vitamin A and retinal during dark and light adaptation in the frog. DOWLING (1960) considered the same question in the albino rat. He found that in dark adapted rats 95% of the total vitamin A was bound in the form of retinal in the rhodopsin. The remaining 5% was in the form of vitamin A in the retina; none was in the pigment epithelium. On exposure to light, retinal was liberated and subsequently reduced to vitamin A. Some of this was transferred to the pigment epithelium, where 75–80% of the total ocular vitamin A, mainly esterified, was situated following one hour exposure to light. With dark adaptation, the level of retinal in rhodopsin rose, while the level of vitamin A in the epithelium fell, eventually becoming undetectable. This suggested that in the albino rat the pigment epithelium is essential for the changes in visual pigment which occur with light and dark adaptation.

The work of YOUNG and his colleagues (YOUNG and BOK, 1969, for references) established that in frogs, rats, and monkeys the rod outer segments undergo a process of replacement. By radioautography they showed that labeled amino acids are first concentrated in the inner segment, and then move to the base (vitread end) of the outer segment, where they form a dense labeled band. This band of label migrates along the outer segment as a discrete entity, and eventually disappears from the sclerad tip of the outer segment. At this time the label can be found in the pigment epithelium, in bodies called phagosomes by COHEN (1969). These bodies undergo a lytic process, and eventually the label disappears from the retina. These findings suggest that the amino acids are incorporated into the basal discs of the outer segment, that these are being constantly replaced, and eventually removed into the cells of the pigment epithelium, where they are broken down.

In light of these findings, i.e. retinal-vitamin A interconversion with light- and dark-adaptation, and turnover of the outer segments themselves, it is interesting to consider the situations found in deep-sea fishes. In many, the pigment epithelium is thin and poorly developed, sometimes without processes projecting into the ventricular space. In some cases the epithelium has acquired a function as a tapetum lucidum (see below, H. II. 2). In those cases the normal dark pigment may be absent, or confined to one part of the cells only. In the retinal tapetum of certain myctophids the cells are low and flat, containing layers of crystals, but apparently little else; even the nucleus is absent (Figs. 64, 65).

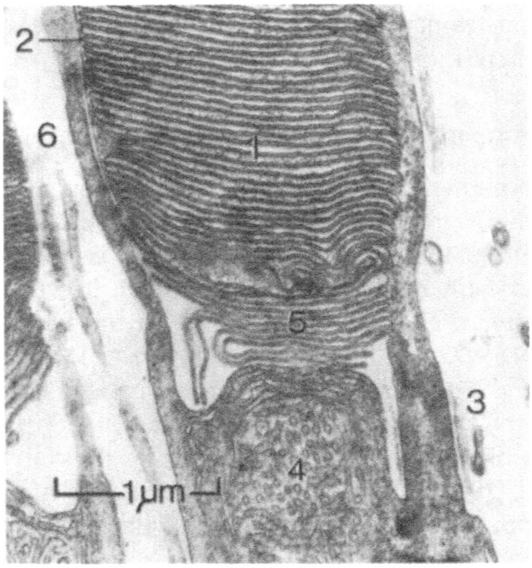

Fig. 39. Junction of inner and outer segments, accessory retina of *Scopelarchus guntheri*. Most of outer segment discs, *1*, are free of outer segment membrane, *2*. Cilium, *3*, is seen joining outer to inner segment, *4*. Few lamellae at base of outer segment appear to arise from cilium and lie free between inner and outer segments, *5*. Some of these are folded upon themselves. Calycal process, *6*, arises from inner segment, and lies free beside outer segment. Electron micrograph

In well-developed pigment epithelia, as in shallow water fishes, the processes and receptor outer segments have substantial areas of mutual contact, so that it is easy to visualize metabolic interchange taking place between them. In deep-sea fishes, however, the area may be small, as when only the tip of the outer segment is in contact with the surface of a flat epithelium cell. It may be that such a small area of contact is sufficient for the exchange required of it in the retina of a fish living in conditions of low light and high pressure, if the retina has a low metabolic rate.

DENTON and WARREN (1957) showed that the visual pigment density in some cases was such that most of the incident light would be absorbed in a single passage along the outer segment. If the pigment density is uniform throughout the outer segment, the vitread end will absorb most light, and so pigment bleaching will occur mostly at that end of the outer segment. How the products of such bleaching would reach the pigment epithelium, only in contact with the sclerad tips of the outer segments, in a retina with long and closely packed outer segments, is not clear. In such cases as *Diretmus*, in which the tips of the long outer segments are separated from the pigment epithelium by five banks of short outer segments, the problem is different. Here only the most sclerad bank of outer segments is in contact with the pigment epithelium. If visual pigment regeneration takes place in these retinae, it may do so by some process which does not involve the pigment epithelium (Fig. 46).

In an outer segment with a turnover process as in the cases studied by YOUNG, the length of the outer segment must be governed by the rate of formation and the rate of removal of outer segment material: If formation is faster than removal, the outer segment will grow longer. The factors which govern the rates of formation and removal are unknown, but it is possible that one such factor might be the rate at which the visual pigment is bleached. In an animal living in brightly illuminated conditions the bleaching and regeneration of the visual pigment takes place all the time, and if regeneration were to stop, the pigment would rapidly become bleached throughout, with consequent loss of function. Many deep-sea animals, however, live in conditions in which, for much of the time, there is no light, and so no light-induced breakdown of visual pigment. Some breakdown is likely to occur spontaneously, giving rise to retinal "noise," but such breakdown is probably minimized by the conditions of low temperature, high pressure, and the maximum absorbance being toward the shorter wave lengths of the spectrum (DENTON and NICOL, 1964).

If the bleaching of pigment, spontaneous and due to light, were sufficiently low, it might be that regeneration need not take place, the formation of new pigment being fast enough to meet the requirements. Such new pigment would presumably be formed at the outer segment base, where the new discs are added, and examination of this part of the rods in deep-sea fishes suggests that disc formation does occur in them, as in YOUNG'S animals. The appearances vary with species, and also between rods in a single retina, but in general it can be said that the basal discs are less regular in size, shape, and arrangement than those more sclerad in a rod. The surface membrane of the outer segment is often in continuity with that of several basal discs, and some of these may extend much further than others, being doubled on themselves, or projecting between the calyx processes and the outer segment. Examples are found in *Platytroctes* and in the scopelarchids (LOCKET, 1971 a, b) (Fig. 39).

2. Visual Pigments

The attenuation of light and the narrowing of the spectral band transmitted, with increasing depth, was mentioned in section A. The optical characteristics of different water masses vary (JERLOV, 1951, 1968), but most oceanic waters transmit blue-green light best; measurements by BODEN et al. (1960) in the Bay of Biscay showed a peak transmittance at 475–480 nm. Most of the light due to bioluminescence is also blue-green, and NICOL (1960) reported that the light of the lantern fish *Myctophum punctatum* was of wave lengths 410–600 nm, with the peak at about 470 nm. As CRESCITELLI (1972) pointed out, CLARKE (1936) had suggested that the maximum sensitivity of deep-sea fish eyes might be at a comparable wave length.

The matter was put to the test by DENTON and WARREN (1956, 1957) who examined the spectral absorbance of isolated retinae from species of deep-sea fishes. The retinae of all appeared golden to simple inspection, leading those authors to use the general term chrysopsin for the visual pigments they contained. Today it is usual to characterize such pigments by their absorbance maximum, and in deep-sea fishes this is about 480–485 nm. Similar results were obtained from deep-sea fishes by MUNZ (1957, 1958) who studied retinal extracts by partial bleaching and showed that in five cases there was only a single pigment, but in *Bathylagus wesethi* two pigments with λ_{max} 478 and 500 nm were present.

The presence of short-wavelength absorbing pigments has since been shown in deep-sea elasmobranchs by DENTON and SHAW (1963), and in a deep-sea holocephalian by DENTON and NICOL (1964). DARTNALL (1972) has reported a pigment from the coelacanth *Latimeria chalumnae* which absorbs maximally at 473 nm, and one with the same absorbance from the castor-oil fish *Ruvettus pretiosus*, a teleost which is caught at the same depth and localities as *Latimeria*.

DENTON and WARREN (1957) showed that the pigment density in the retinae they examined was high enough to absorb 95% of the light falling on the retina in some cases. This high pigment density appeared to be characteristic of deep-sea teleosts; elasmobranchs from comparable depths had about half the pigment density, but they had a brightly reflecting choroidal tapetum lucidum, which the teleosts did not have (DENTON and NICOL, 1964). The tapetum is discussed later, but DENTON and NICOL pointed out that by this means the elasmobranchs may attain the same quantum capture as the teleosts but with half the pigment, and hence with half the spontaneous pigment breakdown. This breakdown, not due to light, gives rise to retinal noise and its reduction will improve the signal to noise ratio of the retina.

The rule appeared to be for deep-sea fishes to have pigments with an absorbance more or less matched to the best penetrating light, but in a malacosteid, *Pachystomias*, DENTON et al. (1970) found a pigment which absorbs maximally in the red. These fishes have a suborbital photophore which emits red light, and a retinal tapetum (see below, H.II.2) which reflects red. Among the animals which they will encounter at depth are decapod crustaceans; many of these are red as seen in the laboratory, but would appear black if the light by which they are illuminated does not contain the long wave lengths. *Pachystomias* appears very well adapted to illuminate such prey and to see it: if the crustacean does not have a red-sensitive pigment itself it will be unable to see the light by which the malacosteid hunts it.

A related fish, *Aristostomias scintillans*, has been shown by O'DAY and FERNANDEZ (1974) to possess two visual pigments, of which the absorption maxima match the different light emitted by the suborbital (red) and postorbital (green) photophores.

3. Long Outer Segments

Among deep-sea fishes are a number in which the outer segments of the rods appear longer than they are in shallow water species. Such fishes include *S. diaphana* and *Platytroctes apus* (LOCKET, 1970, 1971a), in which the outer segments measure up to 150 μm. In certain other examples, notably *Diretmus argenteus*, this length is exceeded in part of the retina, though elsewhere the rods may not be outstandingly long (MUNK, 1966b).

The impression of great length in deep-sea fish rods might be due to unusual proportions of the various retinal elements; certainly the outer segments tend to account for more of the retinal thickness in deep-sea fishes. LOCKET (1971c) compared the retinae of two teleosts, *Sternoptyx* and a goldfish, in each of which the eye was of normal teleost configuration, and 6 mm in diameter. The *Sternoptyx* retina was 230 μm thick and the outer segments, 95 μm long, accounted for 41% of the retinal thickness. In the goldfish the retina was 260 μm thick, but the

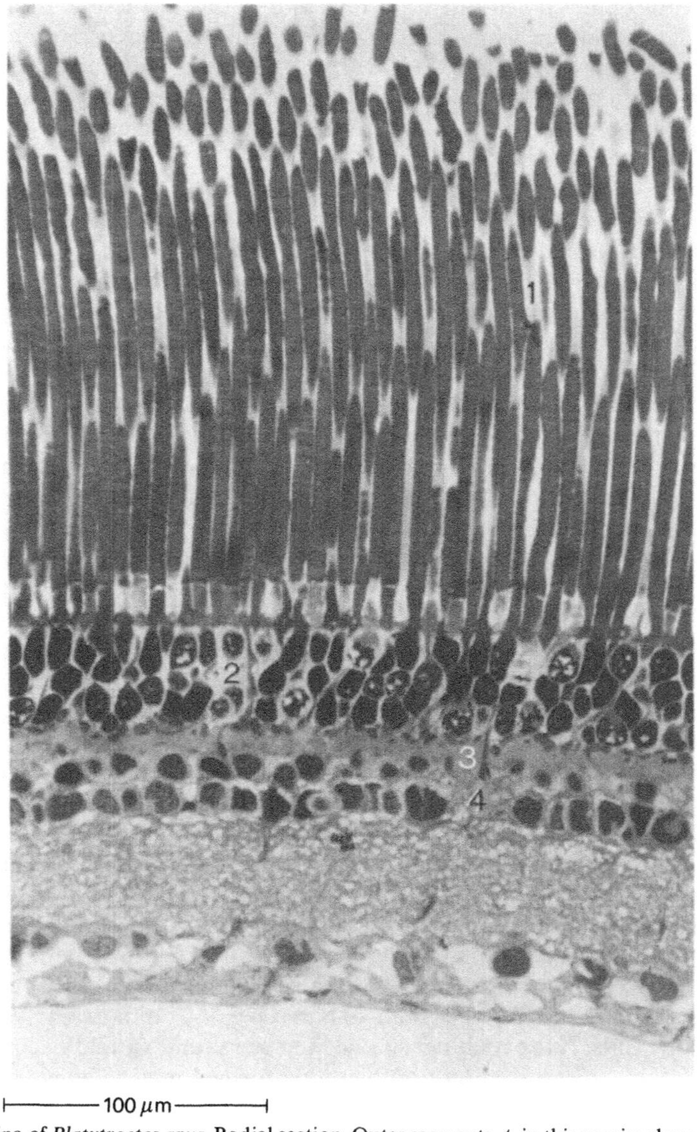

Fig. 40. Retina of *Platytroctes apus*. Radial section. Outer segments, *1*, in this species show great length often seen in deep-sea fishes. They are present in single bank, though their nuclei lie 3 deep in outer nuclear layer, *2*. Synaptic spherules can be seen as small dark bodies, often on curved stalks, in outer plexiform layer, *3*. Outer nuclear layer, *4*, contains rather few nuclei; it is not possible in this figure to distinguish between bipolar, amacrine, and horizontal cells and radial fibre nuclei. Optical micrograph

outer segments were only 40 µm long, representing 15% of the retinal thickness. In *Platytroctes* the outer segments accounted for more than half the retinal thickness, and were 150 µm long (Fig. 40).

Sternoptyx and *Platytroctes* are both species in which there is only one bank of outer segments. In certain other deep-sea species there is more than one such bank, a condition discussed further below, and in one of these, *D. argenteus*, the retina contains areas in which the outer segments reach extraordinary lengths. The eyes of this fish were described by MUNK (1966b), who found that a specimen 109 mm in length had a retina, part of which was 900 μm thick, of which 600 μm was due to the outer segments of one bank of long rods; five banks of short rods sclerad to the long rods were also present.

The work of DROZ, and YOUNG and his colleagues (YOUNG and BOK, 1969, for reference) has shown that rod outer segments are, at least in the animals studied, formed at their vitread end and eroded from the scleral end. The length of an outer segment in these circumstances will depend on the balance between the rates of formation and erosion, and if this balance is imperfect the outer segment will become longer or shorter with time, depending upon which process is the faster.

No experiments have been done with deep-sea fishes to determine whether outer segment turnover occurs in these species, but the appearances at the outer segment base suggest that *in vivo* disc formation may well take place, though they shed little light on a possible mechanism. In cases where the outer segment tips fail to reach the pigment epithelium, its cells might be unable to erode the sclerad ends of the outer segments; such lack of contact between outer segment tips and pigment epithelium cells might conceivably account for the great length of the long rods in *Diretmus*, but it cannot be the whole answer. Besides the long rods, *Diretmus* has five banks of short ones, and only the most sclerad of these makes contact with the pigment epithelium.

The case of *S. sagax* is different again (LOCKET, 1971b), (Fig. 41). In the main retina of this species there are three regions; the most caudal contains grouped rods in a single bank (discussed further below), and the middle region contains a single bank of rods with outer segments about 40 μm long in a 103 mm specimen. The rostral region is occupied by rods of which the outer segments are 230 μm in length, and the junction between the long and short rod regions is abrupt, extending across the retina as a distinct step, easily visible with the dissecting microscope. The other retinal components are almost identical in the two regions, and the reason for the change from the short to long outer segments is not known. A clue may be given by the fact that the long rods, (and the grouped retina), were only present in specimens longer than 40 mm. One of 28 mm examined had only uniform short rods. If the retina grows at the rostral and caudal edges only, and not at the lateral side, the pattern of the retina formed might change at a certain stage of development, but why it should do so, and in the way described, remains obscure. The closely related *S. guntheri* and *Benthalbella infans* do not show the region of long rods, although their retinae are otherwise very similar to that of *S. sagax*.

4. Multiple Bank Retinae

Among deep-sea fishes there are many which have multiple banks of receptors in the retina. The specialization affects those parts of the rods which lie between the outer limiting membrane and the pigment epithelium. An example with two banks is shown in Fig. 42, and the arrangement in a case with three banks is shown diagrammatically in Fig. 43.

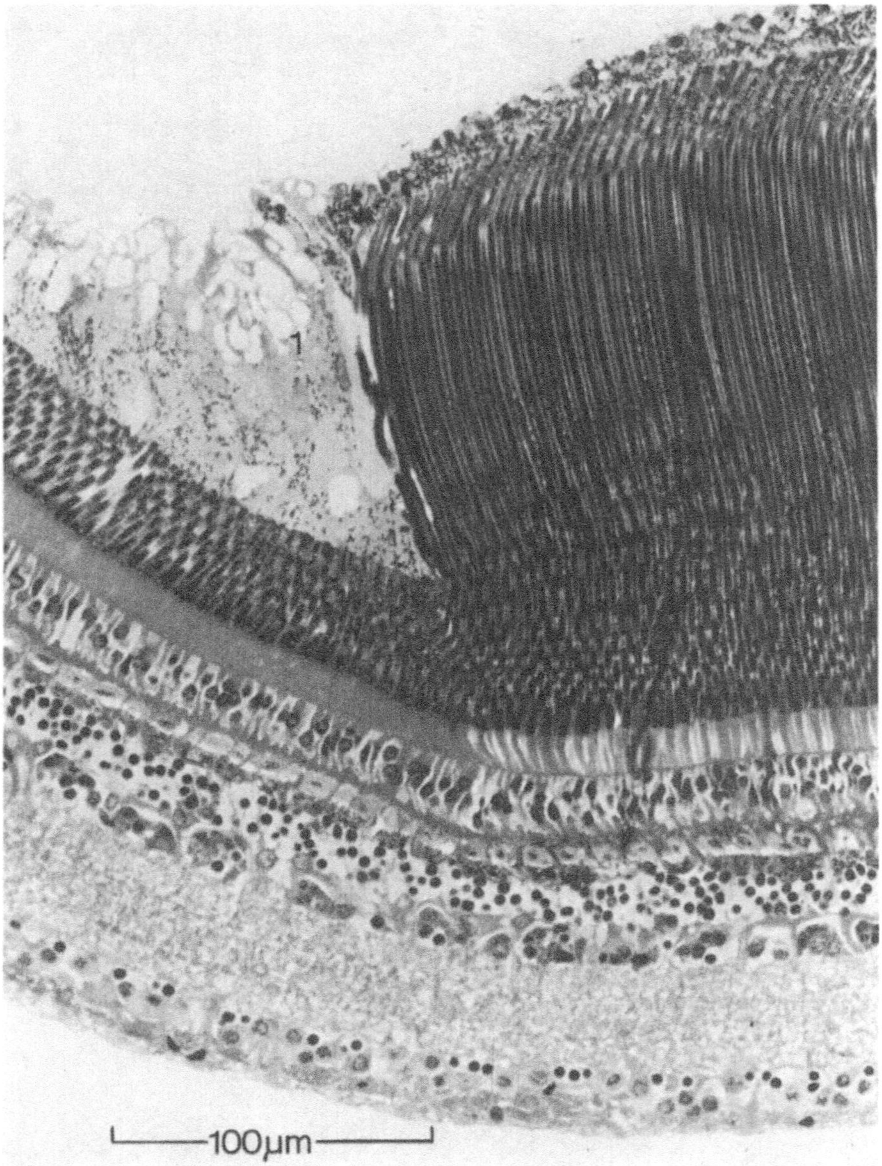

Fig. 41. Junction of short and long rod regions in main retina of *Scopelarchus sagax*. Short rod region corresponds closely to nongrouped main retina of *S. guntheri* shown in Fig. 51. Here short rods measure about 40 µm, but give way to an equally compact and regular array of rods 200 µm long. Junction is abrupt, and gives rise to step in retinal thickness. Step is to some extent filled out by elongated pigment cells, *1*. No difference appears in inner retina related to short and long rod regions. Optical micrograph

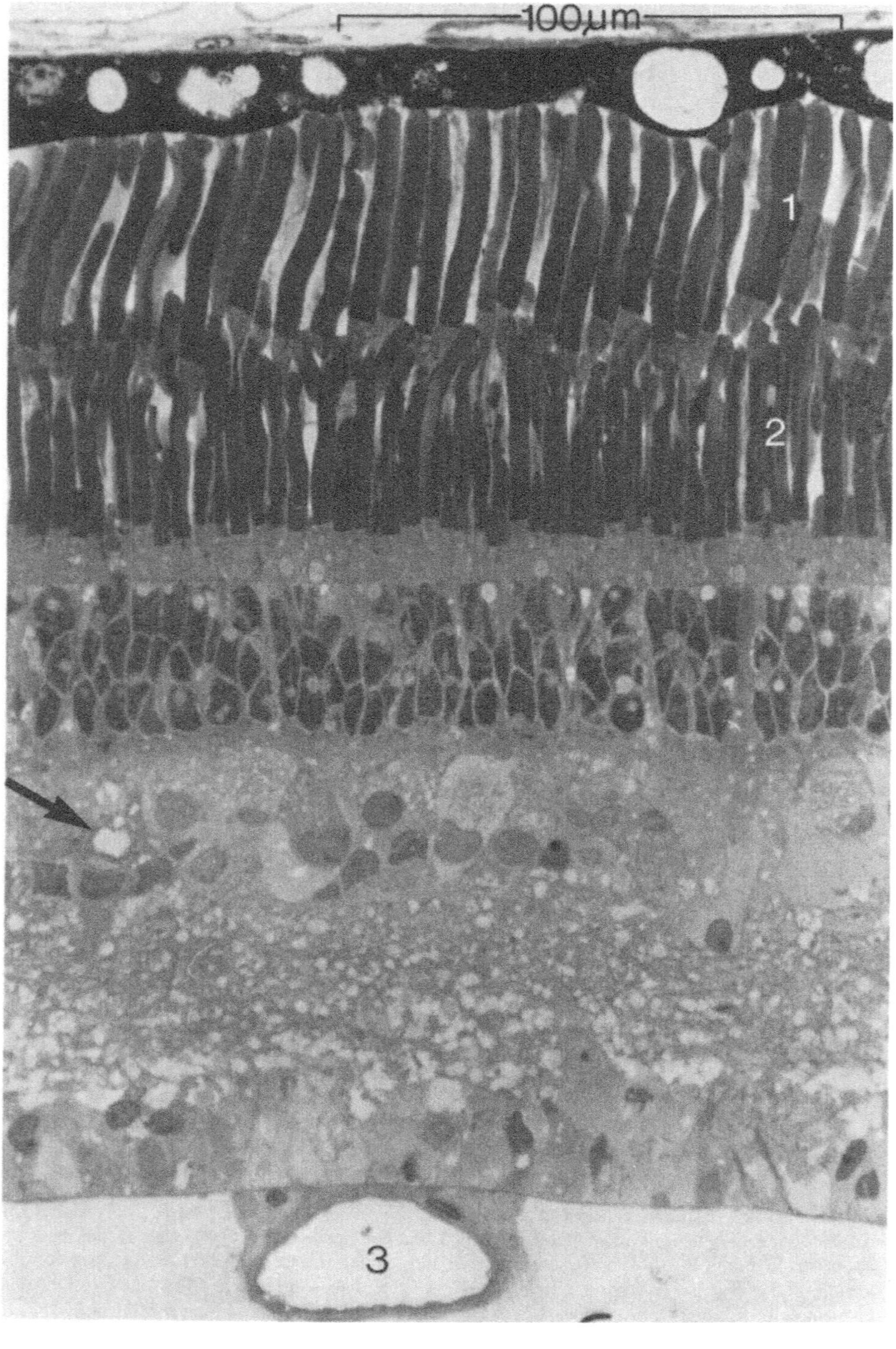

Fig. 42. Retina of *Serrivomer*, a deep-sea eel. There are two regular banks of rods, of which sclerad, *1*, are somewhat thicker than vitread, *2*. This retina, as in other eels, is vascularized; Hyaloid vessel is present on vitread surface, *3*, and intraretinal capillary is arrowed. Optical micrograph

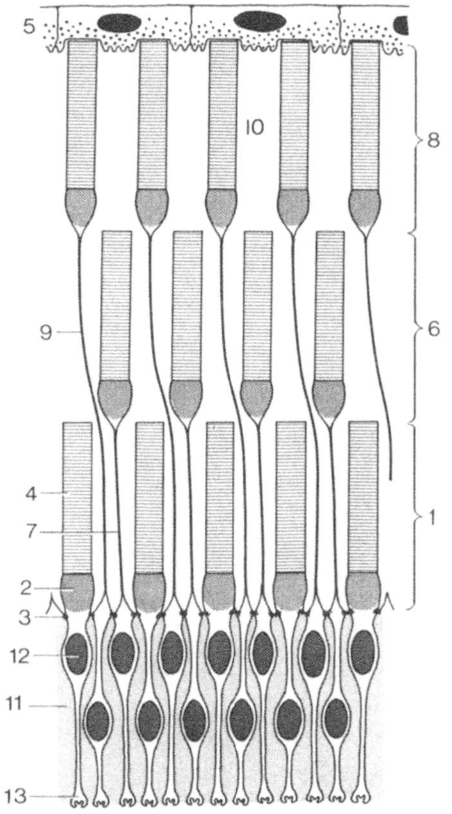

Fig. 43. Diagram of multibank retina. Vitread bank of rods, *1*, are like those of normal retina. Their ellipsoids, *2*, are just sclerad to outer limiting membrane, *3*. Their outer segments, *4*, fail to reach pigment epithelium, *5*, by twice their own length. Ellipsoid-outer segment complexes of middle bank of rods, *6*, are connected to perikaryon by slender filamentous myoid, *7*. They fall short of pigment epithelium by about their own length. Rods of sclerad bank, *8*, have long filamentous myoids, *9*, like those of middle bank, but longer. Tips of sclerad outer segments are only ones which come into contact with pigment epithelium. Outer segments, ellipsoids, and myoids lie in ventricular space. *10*. As in normal retinae, spaces between neurons in inner retina are occupied by radial fibre cytoplasm, *11*, rod nuclei, *12*, are situated in outer nuclear layer, and synaptic spherules, *13*, in outer plexiform layer

The most vitread bank closely resembles the rod array in a retina with only a single bank. The nuclei are located in the outer nuclear layer, vitread to the outer limiting membrane, and directly sclerad to that "membrane" is a short unspecialized myoid region. This leads, with little change of diameter, to the ellipsoid, from the sclerad end of which the cylindrical outer segment takes origin. As in the rods of a single bank retina, all the components of the vitread bank rods within the ventricular space are of approximately the same diameter, but unlike the single bank, the tips of the outer segments fail to meet the pigment epithelium, in this case by a distance of approximately twice the outer segment length.

The most sclerad bank of rods have outer segments and ellipsoids of approximately the same dimensions as those of the vitread bank, but the myoid shows a

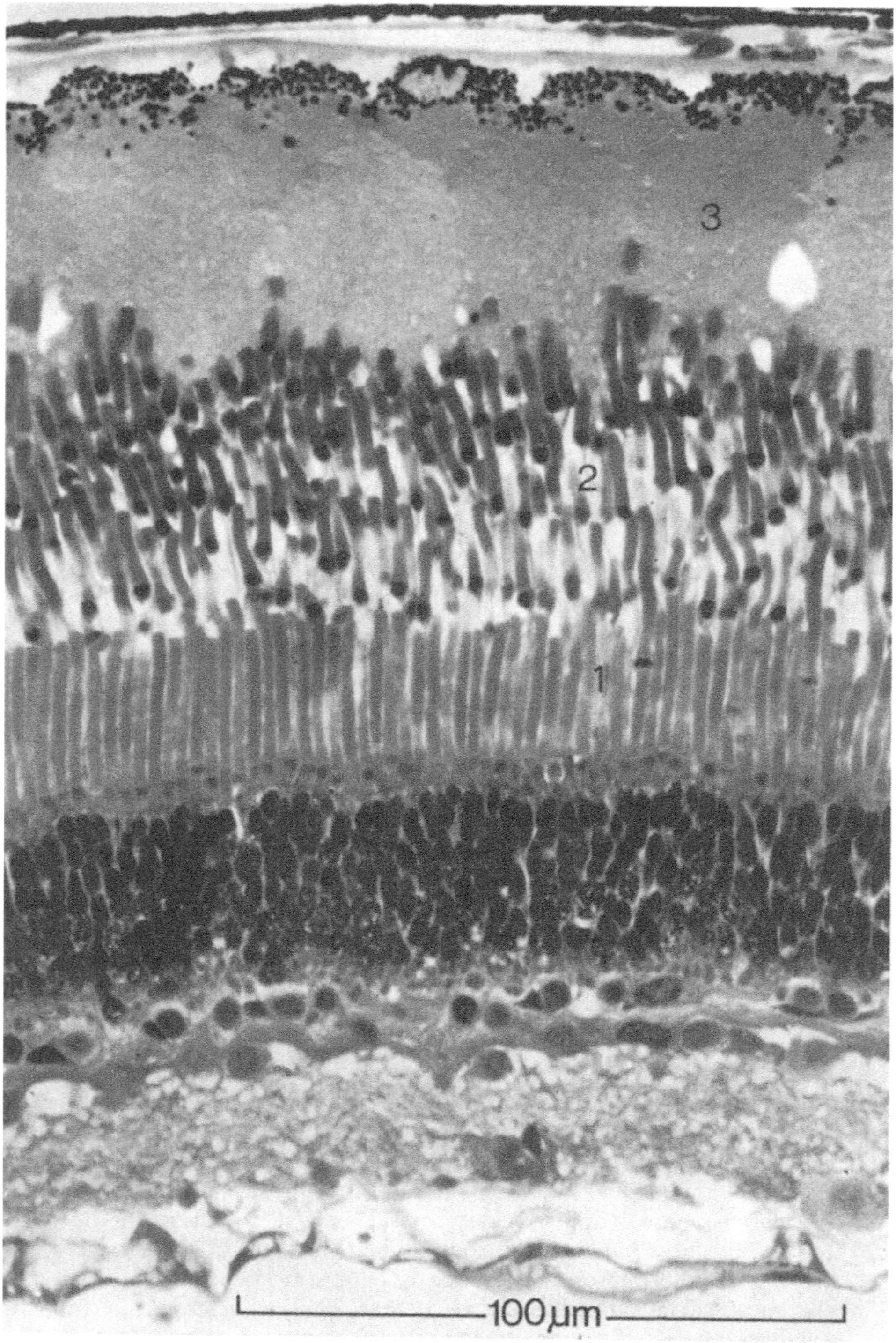
3
2
1
100 μm

striking difference; instead of being short and of similar diameter to the ellipsoid, it is narrowed to a filament, which extends from the outer limiting membrane to the ellipsoid, between the rods of the other banks. This sclerad bank shows normal areas of contact between the tips of the outer segments and the pigment epithelium.

The intermediate bank shows features of both the others. Its outer segments are of similar length, and its myoids are narrow and elongated. Like those of the vitread bank, the outer segments of the intermediate bank fail to reach the pigment epithelium.

The arrangement of rods in multiple banks was first described by VILTER (1953) in *B. benedicti*, and has since been observed in many other species. MUNK (1966a) lists 20 species in which it occurs, and other examples have been noted by LOCKET (1969) and MUNK (1966b). It is strange that BRAUER did not recognize multiple banks, though they are present in several of the fishes whose retinae he described.

In most multibank retinae the banks are arranged in a regular and closely packed formation (Fig. 42). Commonly, the vitread bank consists of narrow outer segments, the outer segment diameter being larger in the more sclerad banks. In some examples the more sclerad banks are less closely packed than the vitread; this is seen in *Malacosteus*, in which the sclerad of the two banks is irregularly packed (Fig. 44). This retina contains a well-marked retinal tapetum (q.v. below) with which the irregularity of the sclerad bank is associated. Many fishes of shallow water have a well-developed pigment epithelium, but in most it contains pigment granules rather than the reflecting particles which characterize tapeta. The rods of such shallow water species may be numerous and small, and may be distributed in depth like the layered rods of deep-sea species.

Most accounts of multiple bank retinae in the literature give a single figure, from two to six, for the number of banks in a given species. VILTER (1954a) in his account of the retina of *Bathylagus*, however, gave six banks in the shallow fovea, but only three in the extra-foveal retina, establishing that the number of banks may vary within a single retina.

Variation in the number of banks also occurs in *Stylephorus chordatus*, in which the main retina of the rostrally directed tubular eyes contains multiple banks. The main retina of this fish was described and figured by MUNK (1966a), who showed three banks of rods. I have examined seven specimens of different sizes among which there are retinae with three, four, or five banks (Fig. 45). In a given specimen the retina has the same number of banks almost all over; all the banks appear well-differentiated, and only at the periphery of the retina, where the main retina gives way to accessory retina or iris, does the number vary.

This variation of the number of banks with the size, and so probably the age, of the fish raises interesting points. First is the question of how the number of banks might be increased during life. The retinae of teleosts grow in area with the

---

Fig. 44. Retina of *Malacosteus niger*. This retina contains two banks of rods. Vitread bank, *1*, is regularly arranged, but sclerad bank, *2*, less so. Epithelium cells contain few pigment granules sclerally, but also diffuse tapetum, *3*. This is composed of spherules which contain red pigment astaxanthin; See also Fig. 66. Optical micrograph

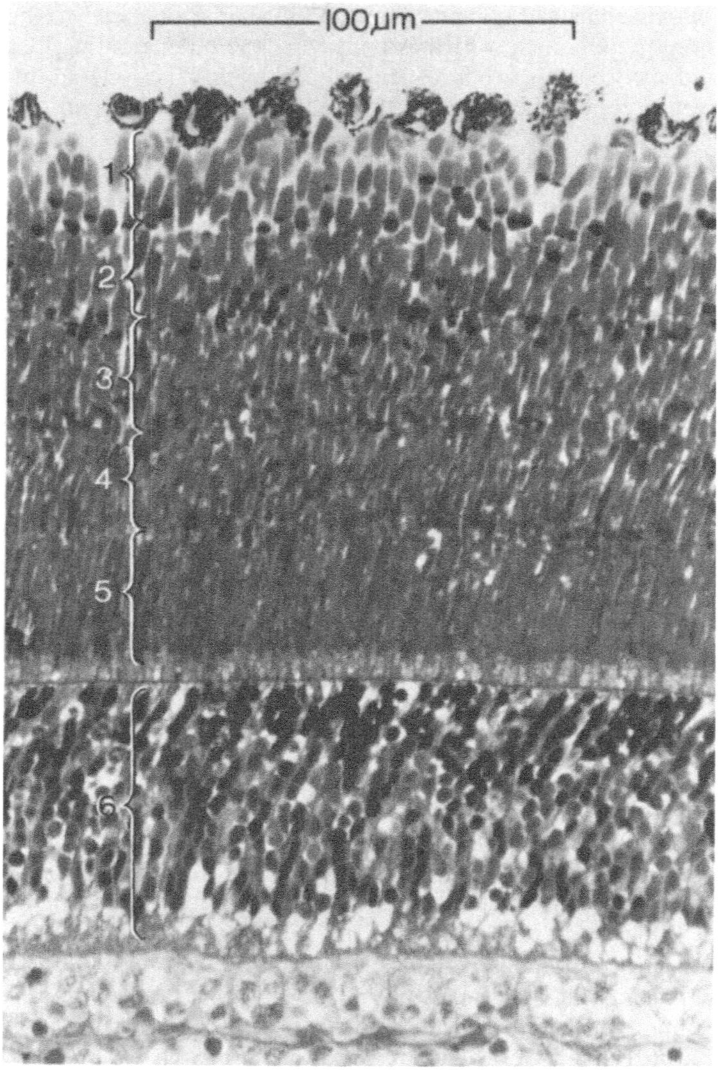

Fig. 45. Retina of *Stylephorus chordatus*. This retina is from specimen of S.L. 240 mm, and shows 5 banks of rods found in large specimens. Dimensions of outer segments are different in various banks; Those in sclerad banks, *1*, are short and fat, and those in vitread bank, *5*, are long and thin. Those in intervening banks, *2*, *3*, and *4*, show gradation between extremes. Note that tips of outer segments in bank 5 are about 100 µm from pigment epithelial cells, and light must traverse 4 banks of outer segments before it can reach sclerad bank. Nuclei of all rods are located in outer nuclear layer, *6*. Optical micrograph

increase in length of the fish, and in those which have been studied the growth has been found to occur at the retinal periphery, e.g. LYALL'S studies on trout (1957). The periphery of the retina in *Stylephorus* shows the region of change from simple two-layered epithelium to fully differentiated retina typically found at this situation. There is here a gradation from an apparently single bank right at the periphery to the number of banks characteristic of the retina in question. If the

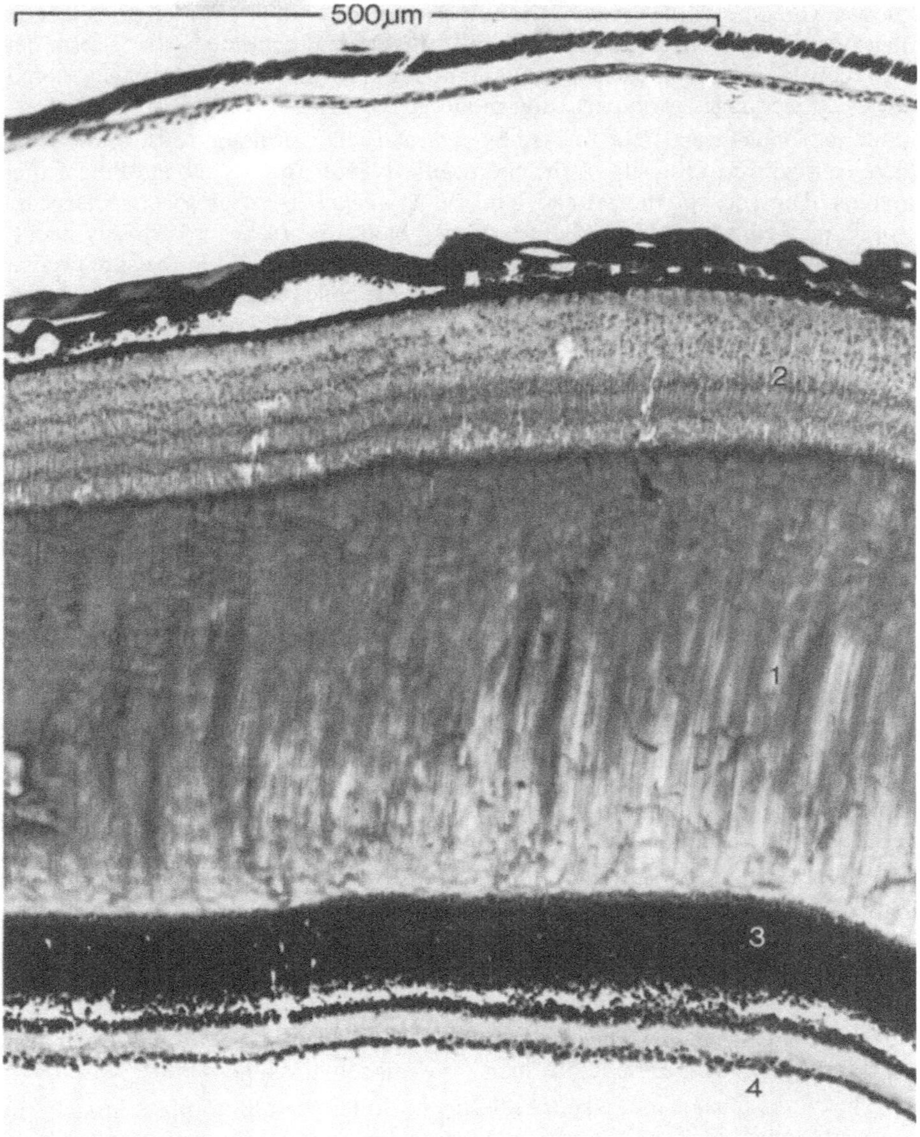

Fig. 46. Retina of *Diretmus argenteus*. This region contains bank of long rods, *1*, and 5 banks of short rods, *2*, sclerad to them. Outer nuclear layer, *3*, is thick and densely populated, and there are abundant ganglion cells, *4*. Optical micrograph

number of banks was dependent on their being laid down in depth as new retina was formed, one might expect more banks at the periphery than in the central retina. This is not the case; if anything, the tendency is for the central retina to contain more banks than the periphery.

If multiple banks are not laid down as such at the periphery, they must arise from already differentiated retina, but the mechanism by which they do so is not known. The possibilities seem to be either for already differentiated rods to alter their position to form an additional bank, or for undifferentiated cells to form new rods; consider the implications of these modes of formation, and whether there is any evidence suggesting a particular mode.

If new banks were to be formed from already differentiated rods, these would have to migrate to the site of the new bank, presumably by elongation of their myoids. The rods in the vitread bank in *Stylephorus* are uniform in size and closely packed, while those of more sclerad banks are rather less closely packed but are of greater diameter. If these outer segments are formed in the same way as those of the animals studied by YOUNG and his colleagues, that is to say by the addition of new discs at their vitread end, their diameter is presumably determined during the formative process. The vitread bank is so closely packed that the rods are in mutual contact, which would limit the diameter at which they could be formed. If the myoids of some became elongated, the outer segments would be displaced sclerally, when there would be room for larger diameter outer segments to be formed; this might account for the greater diameter observed in the more scleral banks.

The retina in teleosts differentiates at the periphery during growth, and it is commonly assumed that all the nuclei seen in the various layers of the mature retina belong to differentiated cells. The outer nuclear layer in *Stylephorus* is some 12–15 nuclei thick, and the population of nuclei appears morphologically heterogeneous. Some are pale staining, and the pale ones closest to the outer limiting membrane are larger than those elsewhere. The layer also contains small dark staining nuclei, and between the two a range of sizes and staining intensities. The ellipsoids also show a range, those closest to the outer limiting membrane containing pale staining mitochondria, while those in more sclerad banks are dark. The outer segments show the differences in thickness cited above, and in most of the material examined the more sclerad, larger, outer segments are less regular in form than those of the vitread bank. Whether this is a true indication of increasingly disordered structure in the more sclerad banks, or whether it is artifactual, is not clear from the evidence available.

To establish if there is a process of differentiation of stem cells into new rods during life, one would wish to see clear evidence of such a process, such as partially differentiated rods close to the outer limiting membrane. Such firm evidence is lacking at present, but such a process remains a possibility.

The banks of rods in the species discussed so far are all of approximately the same size, but this is not the case in *D. argenteus*, a laterally compressed silvery fish caught at mesopelagic depths. This fish has apparently typical lateral eyes, but the retina within is by no means typical, as shown by MUNK (1966b) (Fig. 46).

The eyes are not quite symmetrical, but slightly oval, the long axis directed rostro-dorsally, and there is a considerable binocular field in the direction of the long axes. MUNK found that the arrangement of visual cells differed in specimens of different sizes. A small specimen (26 mm) showed an area rostral to the choroid fissure in which there were two banks of short outer segments. This region separated two areas, one rostro-dorsal and the other caudo-ventral, in which there were both long and short outer segments (Fig. 47).

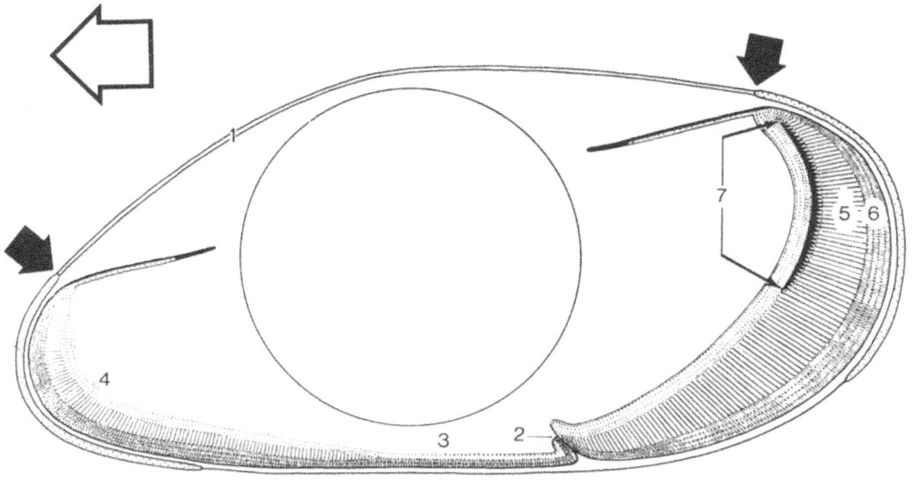

Fig. 47. Diagram of horizontal section through eye of *Diretmus argenteus*. Open arrow points rostrally, cornea, *1*, is between two large thick arrows. Rostral to choroid fissure, *2*, is area of retina with two banks of outer segments, *3*, and rostral to that the area containing long rods in addition, *4*, Caudal to choroid fissure is area containing very long rods, *5*, with several banks of short rods sclerad to them, *6*. Cones also are present in temporal area, between two small arrows, *7*. (From MUNK, 1966b)

In the rostro-dorsal area the outer segments of the vitread bank measured 65 μm, and there were five banks of short outer segments scleral to them. The long outer segments were shorter toward the periphery and toward the choroid fissure. The caudo-ventral region contained a vitread bank of 95 μm, and four banks of short outer segments. This region also contained, in the far temporal periphery, a region 0.8 mm in diameter, in which MUNK described cones situated vitread to the long outer segments.

In a 60 mm specimen the caudo-central region had a similar structure to the corresponding area in the smaller example, but there were six banks of short outer segments, and the long ones extended to 525 μm (Fig. 46). The cone area was somewhat larger, measuring 2 mm in diameter. The rostro-dorsal region contained only a small area, some 1.5 mm diameter, of long outer segments, here 95 μm long. The major portion of this region contained five banks of short outer segments, but just rostral to the choroid fissure there were only three banks.

Thus in *Diretmus* we find rods with long and short outer segments, the multi-bank short rods sclerad to the long ones; there is also an area of cones, situated in the region appropriate for binocular vision.

The function of multiple banks has been considered by several authors, and various theories have been advanced to explain the arrangement. The fact that multiple banks occur in a variety of deep-sea fishes, not all closely related, suggests that they may have arisen independently in the various groups. It also suggests that the possession of multiple banks confers some advantage in conditions of dim illumination.

VILTER (loc. cit.) made two suggestions on the multiple banks in *Bathylagus*; that they might be a static device obviating a need for dynamic accommodation, or that they might assist in gauging distances. The accommodatory function was supposed to depend on a focussed image being formed in one layer or another of the outer segments, depending on the distance of the object from the eye. This was closely linked to the proposed judgment of distances, which VILTER suggested might be achieved by an alteration of the quality of the image, depending upon which layer it was in, though he did not state what quality, nor in what way it might be altered.

MUNK (1966a) discussed these ideas, and pointed out that VILTER's theories were not tenable. MUNK stated that the total thickness of outer segment material is not increased over that found in deep-sea fishes with a single layer of long outer segments, nor that in shallow water species in their dark-adapted state. MUNK also stated (1963, 1966a) that the thickness of the outer segment layer, even in the fovea of *B. benedicti* where it measures 200 μm, is insufficient to make active accommodation unnecessary. He further pointed out that all the species with multiple banks which he had examined possessed a well-developed lens muscle, and presumably accommodated in the normal teleost manner.

DUKE-ELDER (1958) suggested that the multiple banks are a device to increase light sensitivity, by increasing the amount of visual pigment in the retina. The same effect, he suggested, was obtained by the long outer segments of other deep-sea fishes.

MUNK himself proposed that the multiple banks might increase sensitivity, but on different grounds to DUKE-ELDER. The fact that the total outer segment thickness is sometimes no more than in shallow water species makes DUKE-ELDER's hypothesis unlikely. MUNK proposed that the multiple banks would allow a greater degree of summation in a given area of retina than could be achieved by a single bank. If this suggestion is correct, counts of receptors and ganglion cells should support it, and MUNK examined such counts, both on his own material and from the literature. His comparison of deep-living with epipelagic species did not bear out his proposal, indeed he found greater summation in the figures for epipelagic fishes given by O'CONNELL (1963) than he found himself in *Bathylagus pacificus* and *Winteria telescopa*. He pointed out that the deeper living species might have a higher density of visual pigment in their outer segments than the shallow species, and so manage with a lower summation. Further doubt was cast on the summation theory by the higher summation in *Argyropelecus hemigymnus* than in *B. pacificus* and *Winteria*, though *Argyropelecus* has a relatively high pigment density (DENTON and WARREN 1957).

MUNK rightly expressed caution in making comparative cell counts. He pointed out that BRAUER's (1908) figures did not agree with his own, and attributed the discrepancies at least in part to the different methods of counting employed.

Comparison of the multiple bank retina with those having a single bank lends some support to MUNK's hypothesis that much summation might be achieved in a small area. If the eyes of deep-sea fishes obey Matthiessen's ratio, and most do, then the lens will subtend an angle of 47° at its focal point. With such an angle the depth of focus will be shallow, and so the level at which the image is in focus will be important. It is not known exactly where this plane is situated in deep-sea fishes, but let us assume that the light is focussed at the outer limiting membrane.

At that level the receptors are all of the same diameter, whichever bank they belong to, and thus the image will be "shared out" equally among them. The rod components situated sclerad to the outer limiting membrane are the myoids, ellipsoids, and outer segments. These three components are of the same diameter in the vitread bank, and light will traverse the inner segment with little disturbance.

MILLER and SNYDER (1972) showed that the myoids in the frog will act as light guides, and that their efficiency in this role is strongly dependent on their dimensions and on the wavelength of the light concerned. The myoids and ellipsoids of the vitread bank in a multibank retina may be about 3 μm in diameter, and the light predominating at depth has $\lambda = 480$ nm. These parameters are such that those inner segments are likely to behave as light guides, when they will conduct the light to the outer segment virtually without loss.

The outer segments, having appropriate dimensions and a high refractive index, will also behave as light guides, so that light entering them from their vitread end will not spread laterally during its passage along them. On leaving the tips of the vitread outer segments, the light will encounter the ellipsoids and outer segments of the next bank of rods, when it will again tend to become entrained due to light guiding (J. M. ENOCH, personal communication).

If a ganglion cell depends on a summated input originating in a critical number of rods to cause it to discharge, then, if they are arranged in multiple banks, the necessary number of rods might fall within the area of a small image, rather than the larger one which would be needed to illuminate the same number of rods arranged in a single bank.

YOUNG and his colleagues (YOUNG and BOK, 1969, for references) have shown that as well as being formed from their vitread end during life, outer segments are eroded at their sclerad ends, the portions removed being degraded in the cells of the pigment epithelium. The implications of this process, if it applies to rods in multiple banks, are also interesting. In many cases the pigment epithelium in deep-sea fishes is not prominently developed, and lacks the processes which project between the outer segments in shallow water fishes. In such a case the rods only make contact with the pigment epithelium where their tips abut against the flat surface of the pigment cell, and if the fish in question has a multibank retina it is only the tips of the sclerad bank of rods which make this contact. If the removal of outer segment fragments depends on the close proximity of the outer segment and pigment cell, it appears that the conditions for such removal would not be met for any bank of outer segments but the most sclerad. The pigment epithelium cells in multibank retinae do appear to contain phagosomes, so it is probable that phagocytosis of at least some outer segments takes place, but whether these include vitread outer segments, and if so how the fragments reach the pigment cells is not known.

5. Retinal Nutrition and Ocular Vessels

The tissue of the central nervous system is well-known to have a high metabolic rate, and the retina, a direct extension of central nervous tissue, shares in this activity. HOFFERT and FROMM (1972) showed that the retina of some shallow water

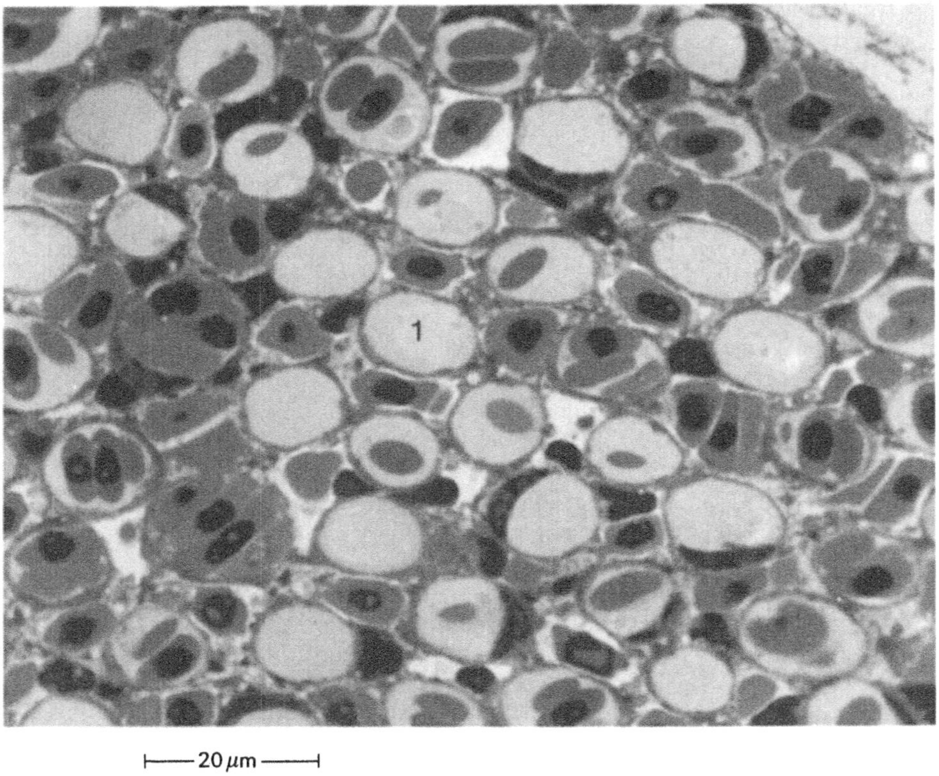

Fig. 48. Choroid rete mirabile, *Evermanella indica*. This section shows close packing of capillaries which make up rete; Alternation of arterial and venous limbs through tissue, seen particularly around *1*, is well-adapted for counter-current exchange of gases which occurs in rete. Optical micrograph

teleosts consumed oxygen vigorously, and although metabolic studies have not been carried out on the retinae of deep-sea fishes it is likely that these too have a relatively high metabolic rate.

In most teleosts the retina is avascular, although the members of one order, the Anguilliformes, do possess true retinal vessels. These are well-recognized in the common eels *Anguilla anguilla* and *A. rostrata*, and they are also present in the retinae of deep-sea eels, e.g. *Serrivomer* and *Synaphobranchus* (Fig. 42).

In some teleosts there is a system of hyaloid vessels, which ramify on the vitread surface of the retina, but do not penetrate the retinal substance itself. In most, however, there is no retinal vascular system, all the retinal metabolites coming from the choroidal circulation.

The choroidal vessels in many teleosts include a structure which was called the choroidal gland in the early literature, though ALBERS (1806) had shown that it is in fact an aggregation of small blood vessels. JONES (1838) introduced the term choroidal rete mirabile for this structure, and this is now the preferred term. The anatomy of the choroidal rete in teleosts was described by BARNETT (1951), who was the first to demonstrate that the rete contains an arterial and venous counter-current system (Fig. 48).

Consideration of the structure of the choroid rete, and of its similarity to the gas secreting gland of the swim-bladder, led WITTENBERG and WITTENBERG (1962) to believe that the rete might be responsible for gas secretion into the eye. Measurements of the oxygen tension in the vitreous near the retina showed that in those teleosts which possess a rete the tension was indeed high, from 250–800 mm Hg, whereas in teleosts without the rete the value was 10–30 mm Hg, figures comparable with those for other vertebrates. More recently WITTENBERG and WITTENBERG (1974) have reviewed the structure and function of the rete, and WITTENBERG and HAEDRICH (1974) have studied the distribution of the rete amongst teleosts, and its relation to the pseudobranch. The latter structure, which is not present in all teleosts, resembles a gill, and is so arranged that blood destined for the eye must pass through its vascular bed after leaving that in the gills. WITTENBERG and HAEDRICH found that almost all marine teleosts which lack a pseudobranch also lack the choroid rete, and they inferred that some function of the pseudobranch, which is required for the operation of the rete, may be dispensed with in fresh water. This function they concluded is to prevent the accumulation of carbon dioxide in the eye, which would otherwise occur due to the countercurrent concentration of gases brought about by the rete. They calculated that if the same degree of CO_2 concentration that occurs in the swim-bladder were to occur in the eye, the carbon dioxide concentration would be 13.5 mmolar. Such a concentration, ionized as it would be in the eye, might give rise to more acidity than the eye could compensate for. This conclusion is supported by experiments in which carbonic anhydrase activity was inhibited by the administration of acetazolamide (Diamox) (HOFFERT and FROMM, 1966; MAREN, 1967; MAETZ, 1956). Teleost retina has been shown by HOFFERT et al. (1973) to be resistant to the very high oxygen tensions to which it is exposed, and which would be sufficient to cause serious damage to mammalian retinae.

The silvery layers present in various situations in fishes, most of which depend on crystalline guanine for their reflectivity, have been shown by DENTON et al. (1970, 1972) to be markedly impermeable to gases. Such layers include the silvery layer investing the swim-bladder in some fishes and the argentea on the external surface of the choroid. The bright silvery tapetum of elasmobranchs owes its reflectivity to guanine crystals in the tapetal cells of the choroid, sclerad to the choriocapillaris, and so not located where it would hinder gas diffusion to and from the retina.

The tapeta of teleosts are generally located in the cells of the pigment epithelium, between the choroid and the neuroretina, a situation in which an impermeable layer might hinder the supply of oxygen to the retina from the choroidal blood. In this connection the situation in some myctophids is of interest: many of these fishes have a bright silvery tapetum formed of crystals, probably of guanine, densely packed in the pigment epithelial cells (Figs. 64, 65). From the findings of DENTON et al. one might expect that these crystal-packed cells would impede oxygen diffusion from the choriocapillaris to the retina. Myctophids, however, are among the fishes which WITTENBERG and HAEDRICH (1974) found to lack the choroidal rete, and several species do possess a well-developed system of hyaloid vessels on the vitread surface of the retina. Such a vascular arrangement would circumvent any barrier to diffusion due to the tapetum.

II. Main and Accessory Retinae of Tubular Eyes

1. Previous Work

The tubular eye of *Argyropelecus* spp. was described above as an example of the tubular form. The main and accessory retinae of that eye are now considered as examples of the two retinal types found in such tubular eyes.

BRAUER (1908) coined the terms *Haupt-* and *Nebenretina*, usually translated to main and accessory retina by authors writing in English, to denote the two types. He studied mainly *Argyropelecus hemigymnus*, but also a few examples of *A. affinis*, and *A. olfersi*, in which the structure was very similar. According to his description, the main retina of *A. hemigymnus* is about 150 μm thick, and occupies the ventral portion of the eye cup. The receptors are all rods, long and thin, and account for a third of the retinal thickness. There are two to three layers of nuclei in the outer nuclear layer; the inner nuclear layer is subdivided into two layers of horizontal cells, four of bipolars and two or three of amacrine cells. There are three layers of ganglion cells.

The accessory retina, lining the medial wall of the eye cup, is irregularly developed, but everywhere much less so than the main retina. The best developed region is at the dorsal aspect of the medial wall, near the lens, but even here the thickness of the accessory retina is only one-tenth of that of the main retina. The rods are thicker but shorter, and somewhat separated from each other, giving an impression of a rudimentary structure. A single layer of rod nuclei is present but horizontal cells are absent, and bipolar and amacrine cells together form only one layer. There are a few other nuclei scattered in the best developed part, but BRAUER could not be sure whether these belong to bipolars, ganglion cells or supporting cells.

CONTINO (1939) studied *A. hemigymnus* exclusively, but he had many examples, including developmental stages. He observed that the main retina became thinner toward the caudal end, where it slopes to merge with the caudal wall of the eye cup. The thickness is 150 μm, as in BRAUER's description, and the rods, 3 μm thick, account for 60 μm of this. He noted that the cells of the pigment epithelium are hexagonal, 8 μm thick, and contain brown pigment. Although he described them as hexagonal, his illustration shows pentagonal examples also, and gives an impression of irregularity.

The receptors consist of a single and uniform bank of rods, but the outer nuclear layer contains two layers of rounded nuclei, the whole appearing well-compacted. The outer plexiform layer is 8 μm thick, and immediately next to a layer of horizontal cells. These he described as long, with spindle shaped nuclei, and present in two or three rows. He described two populations of cells in the inner nuclear layer, and he noted that the retina often splits between the two during preparation. The inner plexiform layer is layered, and bounded vitreally by the ganglion cell layer, containing two kinds of nuclei.

Of the accessory retina, CONTINO wrote little; he described it as a two-layered membrane, 25 μm thick, and so one-sixth of the thickness of the main retina. The inner layer appeared as a fibrous web with two layers of nuclei, and the outer as a layer of lumpy cells full of pigment. In the dorsal part of the accessory retina some short outer segments were present.

MUNK (1966a) examined the eye of *A. olfersi*, and found that its structure accorded well with BRAUER's account. MUNK described the choroid fissure, extending from the retractor muscle on the rostral wall of the eye cup to the transition zone between main and accessory retinae. The fissure ends in the region of the optic disc, ventro-caudal in the medial wall of the cup. He gave no further account of the retinal structure.

I have examined material from *A. gigas*, *A. aculeatus* and *A. hemigymnus*, and can confirm BRAUER's and CONTINO's accounts in the main. The study has included tangential sections, which do not appear to have been examined by the earlier authors.

2. Main Retina (Fig. 49)

The pigment epithelium cells are polygonal, containing dark pigment granules, dark staining bodies believed to be phagosomes, and lobed nuclei. There are few, if any, processes extending between the outer segments, and no sign of a retinal tapetum. The outer segments are cylindrical, and closely packed. Being uniform in size, they form a regular lattice, best seen at the junction of inner and outer segments. The connecting cilia show well in such sections, and most are similarly oriented, as in the nongrouped main retina of *S. guntheri* (LOCKET, 1971b).

The rod nuclei are present as a single layer, but are somewhat elongated; although they retain the spacing of the inner and outer segments, it is less clear at this level due to variations of thickness of the nuclei. The vitread part of the outer nuclear layer contains a few nuclei which are larger than the majority, and which probably belong to displaced bipolar cells.

The rod synaptic pedicles as seen in tangential sections are arranged in a regular pattern like the outer segments, and they form a very even layer. Each pedicle appears to contain five or six synaptic ribbons, in which they are comparable to those of *S. guntheri* (LOCKET, loc. cit.). The outer plexiform layer is thin, as CONTINO stated, and vitread to it the layers of horizontal cells are present. Though CONTINO described these as spindle-shaped, tangential sections show that they are in fact stellate. In the present case there are at least two layers of horizontal cells: To the vitread side of each layer is a mass of pale staining stranded cytoplasm, without nuclei, which may represent arborizations of horizontal cell axons as described by STELL (1975).

The bipolar cells are present in a single layer. They are separated from the amacrine cells, also in a single layer, by a thin layer of tissue which appears amorphous by optical microscopy. This tissue contains scattered pyramidal nuclei, and by analogy with other deep-sea fishes, is probably composed of radial fibre lamellae.

The inner plexiform layer is bounded vitreally by a single layer of nuclei which are of two distinct types. Larger pale staining nuclei probably belong to ganglion cells, but there are approximately four times as many of a smaller, dark staining type. These are comparable in size to amacrine cell nuclei, and may be displaced amacrines or perhaps glial cells. In this retina the ganglion cell and nerve fiber layers are distinct.

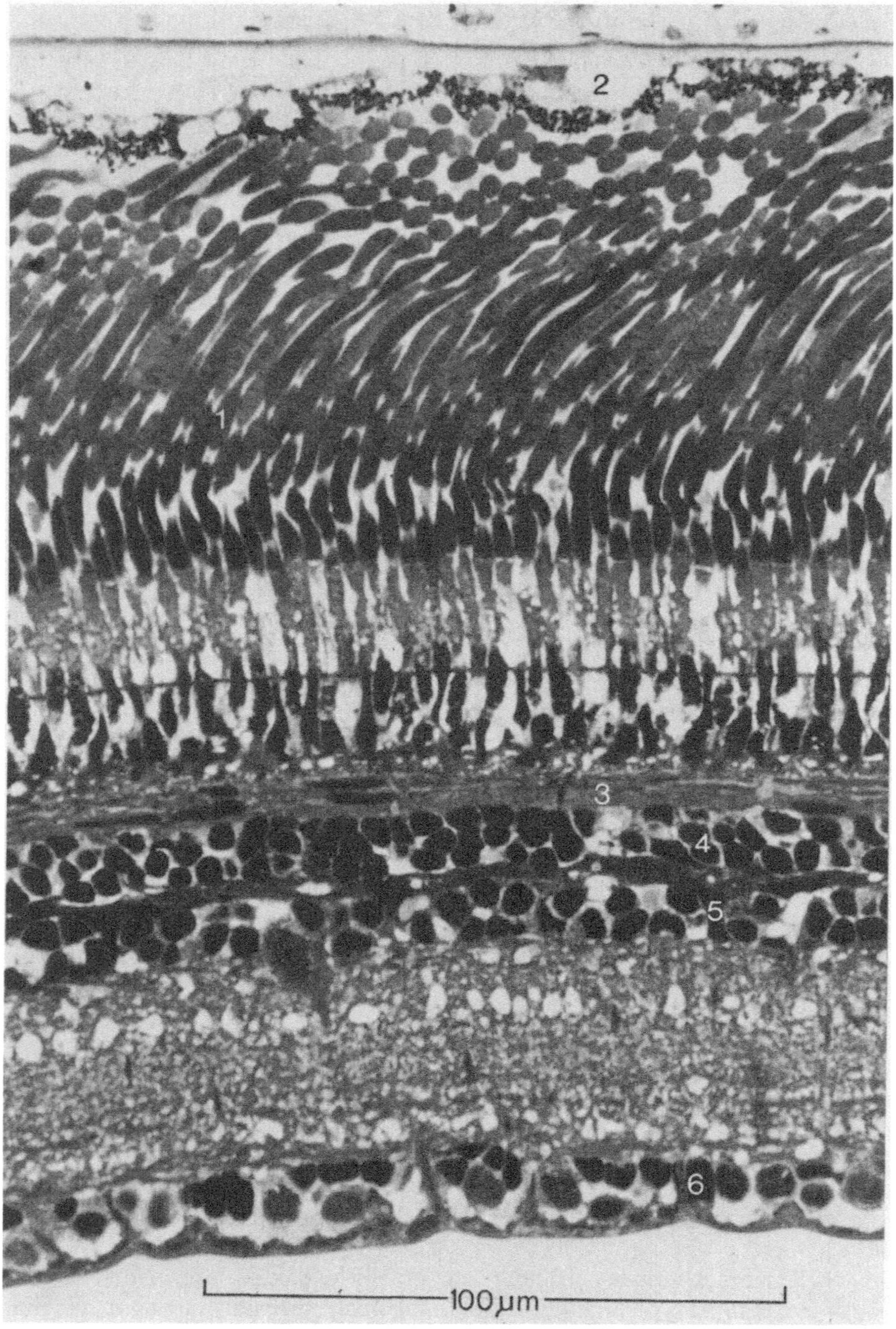

Fig. 49. Main retina of *Argyropelecus aculeatus*. This retina has a single bank of outer segments, *1*, which abut against low pigment epithelium cells, *2*. Inner retina contains layer of flat horizontal cells, *3*, and densely populated layers of bipolar, *4*, amacrine, *5*, and ganglion cells, *6*. Optical micrograph

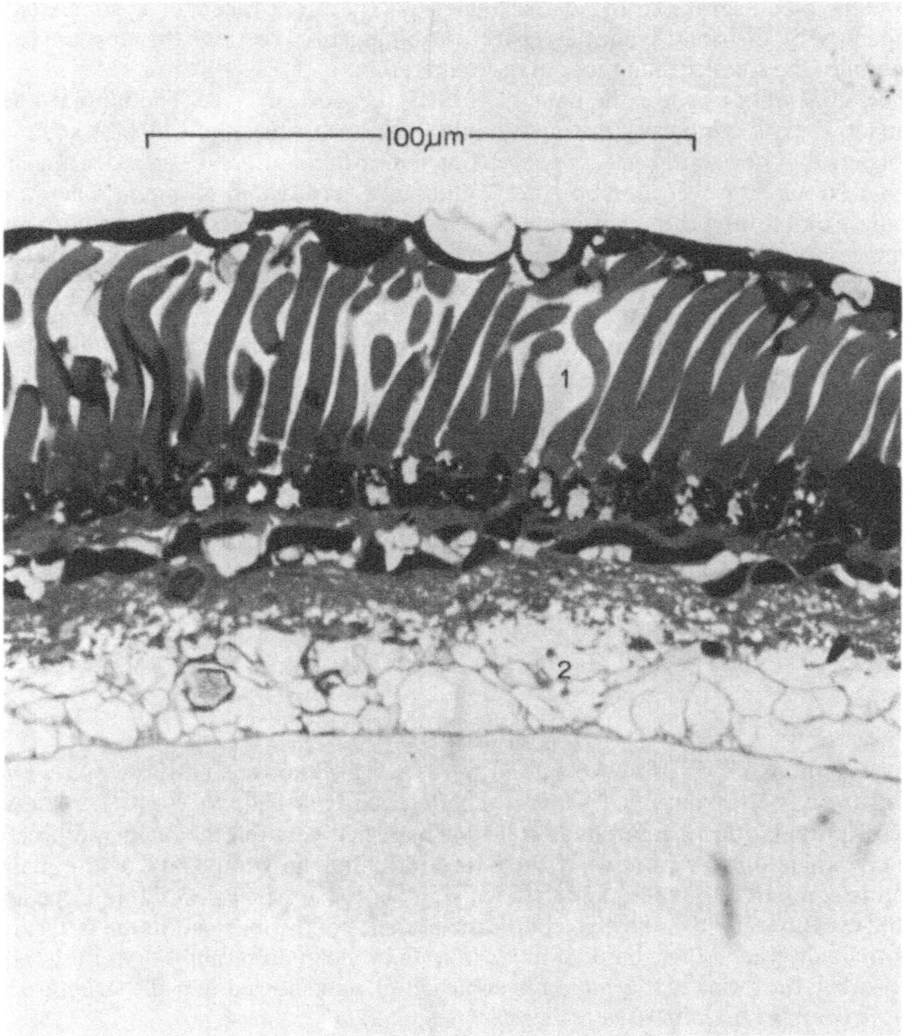

Fig. 50. Accessory retina of *Argyropelecus aculeatus*. This retina is thinner, and appears much less developed than main retina. Single bank of outer segments, *1*, are not so closely packed as those in main retina, but they are thicker. Inner retina, *2*, is sparsely populated, and cell types cannot be distinguished. Optical micrograph

3. Accessory Retina (Fig. 50)

The accessory retina does not conform to Matthiessen's ratio, and, as BRAUER and CONTINO stated, it is thinner and simpler than the main retina. The pigment epithelium consists of low cells without processes, but containing pigment granules. As in the main retina there is but one bank of outer segments; these are shorter and fatter than those of the main retina, as are the inner segments.

The rod nuclei are arranged in a single row; they are capped, or in some cases indented, by the small synaptic apparatus. In tangential sections this appears less complex and smaller than those in the main retina.

Radial sections show the paucity of the inner nuclear layer. The bipolar and amacrine cells are hardly distinguishable from each other, and at first sight it appears that horizontal cells are absent, as the earlier authors described. Tangential sections, however, do show horizontal cells, at least in *A. gigas*. They are stellate and thin; although it is impossible to distinguish discrete layers of them, elements of the horizontal cells interweave, suggesting that more than one layer is represented.

Radial fibre nuclei are scarce, and there is no layer of lamellar tissue such as separates the bipolar and amacrine nuclei in the main retina. Radial sections often fail to show ganglion cells at all, but tangential ones show that these cells are large, but sparsely distributed. Nerve fibres lie at the same level as the ganglion cells, and some dark staining nuclei among them may belong to glial cells.

Though certainly less complex than the main retina, the accessory retina appears to contain the normal retinal cell populations, and may be more than the ill-formed light receptor described by CONTINO.

III. Grouped Retinae

1. Introduction

The tubular eyes of *Argyropelecus* were described above, and those of *Scopelarchus* sp. were mentioned when discussing the lens pad. In BRAUER'S (1908) description of the scopelarchid eye he distinguished between the main and accessory retinae on the grounds of their location, and the complexity of the main retina compared to the accessory. His figures of the *Argyropelecus* and scopelarchid eyes are rather similar, except for the presence of the lens pad in the latter; in both he showed a single bank of uniform rods in the main retina. LOCKET (1971 b) described the retina of *S. guntheri* in some detail, and showed that the structure is more complex than BRAUER realized.

LOCKET showed that the rostral part of the main retina has a uniform structure similar to the corresponding region in *Argyropelecus;* the rods are present in a single bank and their tips are in contact with the low cells of the pigment epithelium, as BRAUER indicated.

Sections through the caudal half of the main retina show a very different arrangement of the rods. Here they are present in groups, each of which is set in a pit formed by the pigment epithelium cells. These cells are specialized in shape and by their contents of oriented crystals in this region, forming a complex retinal tapetum.

The differences between the grouped and nongrouped retinae are striking, and it seems strange that BRAUER did not observe the groups. He based his descriptions on transverse sections through the head of the intact fish, and it may be that he did not examine sections more caudal than those containing the maximum

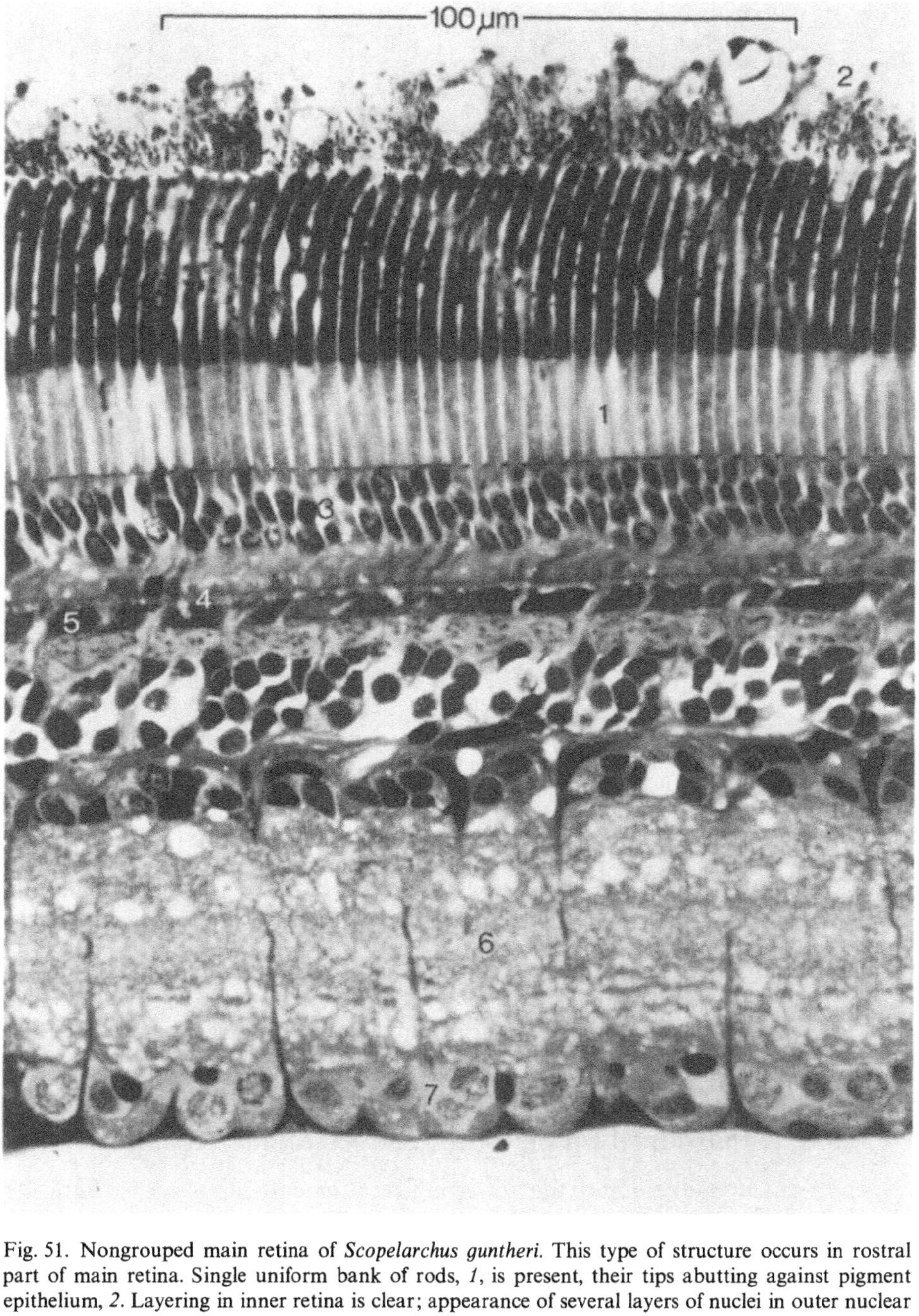

Fig. 51. Nongrouped main retina of *Scopelarchus guntheri*. This type of structure occurs in rostral part of main retina. Single uniform bank of rods, *1*, is present, their tips abutting against pigment epithelium, *2*. Layering in inner retina is clear; appearance of several layers of nuclei in outer nuclear layer, *3*, is artifact of plane of section. Outer plexiform layer, *4*, is thin; it is lined on sclerad side by synaptic pedicles, and on vitread by layer of horizontal cells, *5*. More detail of inner nuclear layer is shown in Fig. 52, but note here stratification within inner plexiform layer, *6*, and well-defined layer of ganglion cells, 7. Optical micrograph

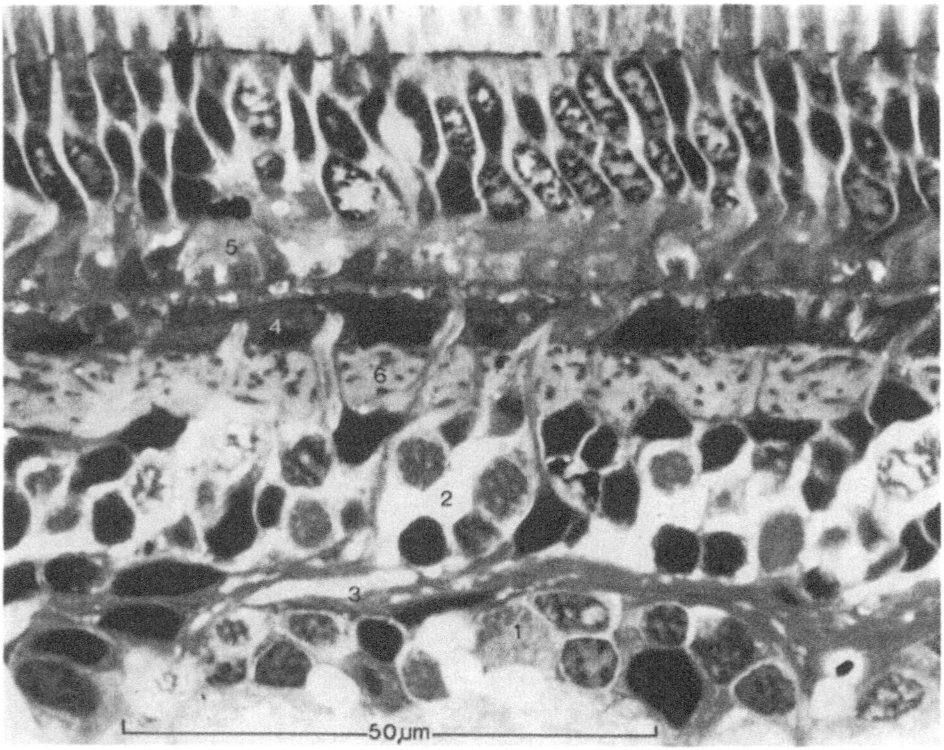

Fig. 52. Nongrouped main retina of *Scopelarchus guntheri*, from outer limiting membrane to inner nuclear layer. Amacrine cells, *1*, and bipolar cells, *2*, are separated from each other by layer of radial fibre lamellae, *3*. Single layer of horizontal cells, *4*, is present just vitread to outer plexiform layer, and short horizontal cell processes extend to synaptic pedicles, *5*. Layer vitread to horizontal cells, *6*, consists of horizontal cell processes, appearing here as dark spots, among swathes of extracellular microtubules, which are too small to be resolved by optical microscopy. Optical micrograph

diameter of the lens. Since the grouped retina is situated caudally, he might thus have missed it.

The grouped retina shares some features of the nongrouped, which has certain unusual features in the scopelarchids, and is therefore described first.

2. Nongrouped Main Retina, Scopelarchids (Figs. 51, 52)

The pigment epithelium of the nongrouped retina consists of rather flat cells, of irregular polygonal outline in tangential view. They do not give rise to processes extending between the outer segments, but contain both pigment granules, mostly in the sclerad part of the cell, and crystals, probably of guanine, located in the vitread region. These crystals are not oriented within the cell, and each crystal, which appears rectangular in radial sections, is surrounded by a membrane.

The rods are of uniform size and arrangement, their outer segments being about 55 µm long by 4 µm diameter. Their tips lie against the pigment epithelium, and phagosomes are present within the pigment cells. The inner segments are cylindrical, and of similar diameter to the outer segments.

The outer limiting membrane is clearly seen in radial sections, and the elongated rod nuclei lie in a single layer just vitread to it. The synaptic pedicles are situated all at the same level, and are more complex than the usual rod pedicles of deep-sea fishes. Those of *Scopelarchus* contain five or six synaptic ribbons, and give rise to basal filaments from their lateral borders. Such complex pedicles are usually associated with cones (PEDLER, 1969) but the outer segments of the scopelarchid receptor are cylindrical and rod-like.

The spaces between the stalks of the synapses is occupied by lamellae, each about 100 nm thick, derived from radial fibres; such lamellae are also present in the thin outer plexiform layer.

Vitread to the outer plexiform layer is a prominent layer of horizontal cells, somewhat rectangular in radial sections but stellate in tangential view. These cells have stout lateral processes which make contact with those of neighbouring cells, to form the sort of array which KRAUSE (1886, a, b) called the membrana fenestrata. From the sclerad aspect of the horizontal cells slender processes arise, which cross the outer plexiform layer to make contact with the synaptic pedicles of the receptors.

In many teleost retinae the horizontal, bipolar, and amacrine cells are closely associated in the inner nuclear layer, though the horizontal cells are usually easily recognized morphologically from the other two classes. In the scopelarchids there is a well-defined layer which separates the horizontal cells from the rest of the inner nuclear layer (Fig. 52). To optical microscopy this layer appears to consist of fine processes, interlacing in the horizontal plane, and embedded in an amorphous ground substance. The processes arise from the vitread surfaces of the horizontal cells; they extend for distances of 100 μm or more, tending gradually vitreally, and curving round radially directed bundles of processes, which are probably of bipolar origin. These horizontal cell processes form a web, set in the apparently amorphous ground substance; by electron microscopy it is seen to consist of microtubules each about 25 nm in diameter, which form swathes. The tubule diameter is less than the section thickness, so it is not possible to determine their origins unequivocally, but they appear to arise from the vitread horizontal cell processes. These tubules are extracellular, and extend in smaller swathes through the gaps between the horizontal cells in company with radially directed processes, probably of bipolar cells, some tubules reaching as far as the synaptic pedicles. The microtubule layer is up to 10 μm thick, and is bounded vitreally by a less definite layer of stellate horizontal cells. The lateral processes of these cells do not make broad junctions with neighbouring cells, but the cells do contribute processes to the web.

The presumed bipolar and amacrine cells, both abundant, are separated from each other by a layer of tissue which electron microscopy shows to be formed of lamellae derived from radial fibres.

The ganglion cell layer contains abundant nuclei and there are at least two populations of cells in this layer, which it is not possible to categorize with certainty. It is probable that small dark nuclei belong to glial cells, and that the true ganglion cells are those with large mottled nuclei.

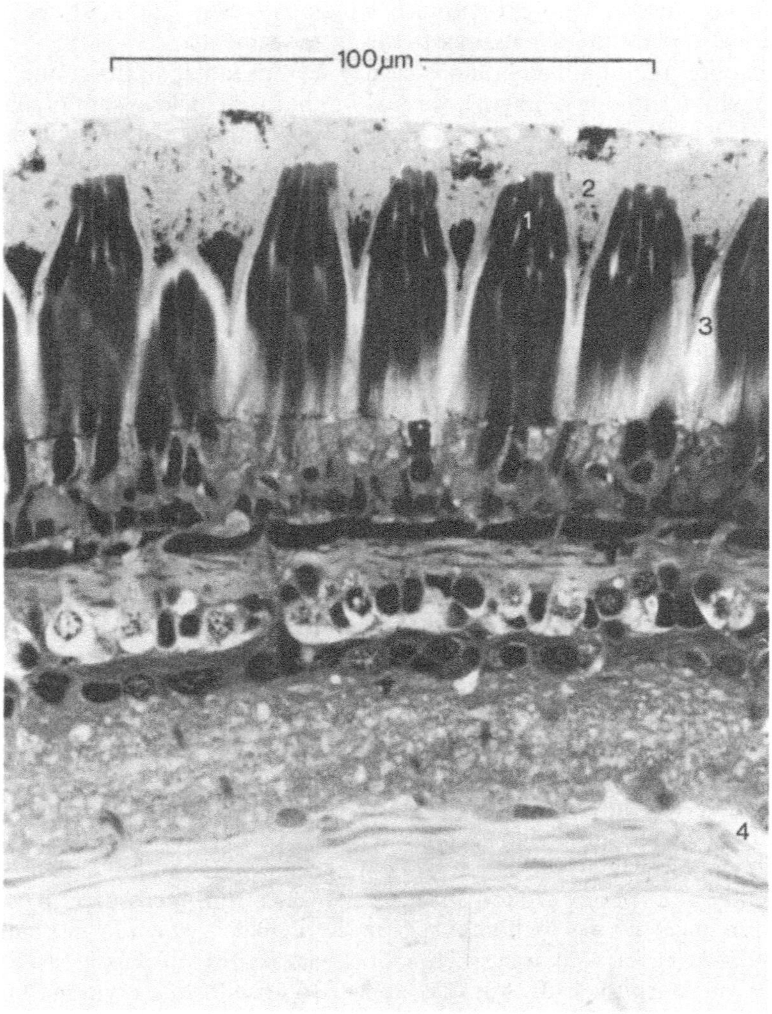

Fig. 53. Grouped main retina of *Scopelarchus guntheri*. Grouped retina is present in caudal part of main retina. Here rods are arranged in groups, *1*, each of which contains about 23 rods. Epithelial cells contain orientated crystals sclerally, *2*, and melanin granules in their vitread extremities, *3*. These cells are arranged to form pits, each of which houses group of rods (see Fig. 55). Inner retina contains same elements as in nongrouped region, but in smaller numbers; Ganglion cells, *4*, are particularly sparse. Optical micrograph

3. Grouped Main Retina, Scopelarchids (Figs. 53, 54, 55)

In the grouped retina the rods are present in bundles, each containing an average of 23 rods. The groups, which are regularly spaced in a 60° lattice, are only present sclerad to the outer limiting membrane, so include only the inner and outer segments of the rods. The inner segments are longer and the outer segments shorter than in the nongrouped retina (Fig. 53).

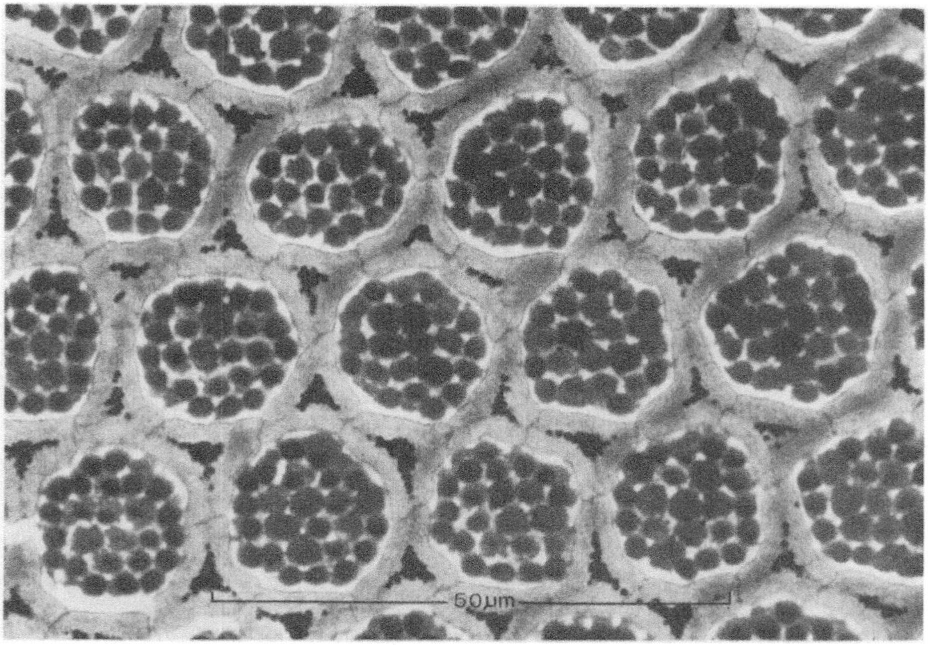

Fig. 54. Grouped retina of *Scopelarchus guntheri*. Tangential section at level of inner-outer segment junctions. Even size and arrangement of groups shows clearly. Pigment epithelium cells are so arranged that 6 contribute to each pit. Centre of each epithelium cell contains melanin granules, but rest of cytoplasm at this level is packed with crystals. Optical micrograph

The grouped region is also characterized by modification of the pigment epithelium to form a tapetum. At their bases the pigment cells are equilaterally triangular in tangential sections; they do not give rise to fine processes, but each cell projects vitreally for about 70 µm, ending between the inner segments close to the outer limiting membrane. The groups of rods are inserted into the tapetum in such a way that each group is located over the meeting angles of six adjacent pigment cells which form cup-shaped pits, an arrangement best shown by diagram (Fig. 55).

The pigment cells contain both dark pigment granules and refractile crystals. The pigment granules are present in a thin layer vitread to the nucleus, and as a core at the apex of the cell. The crystals, probably of guanine, are highly orientated here, unlike those in the nongrouped retina, and are present in layers, arranged parallel to the surface of the cell.

The dimensions and arrangement of the crystals suggest that they should function as a constructive interference reflector, (*v. infra*, tapetum lucidum), and my examination of a fresh specimen of the scopelarchid *Benthalbella infans* with the opthalmoscope and dissecting microscope confirmed that this is so. There was an orange fundus reflex, but no ophthalmoscopic difference could be detected between the grouped and nongrouped retina. With the dissecting microscope the difference could be seen easily: the grouped retina showed an even reticular pattern of the pigment when observed from the scleral side, and from the vitread

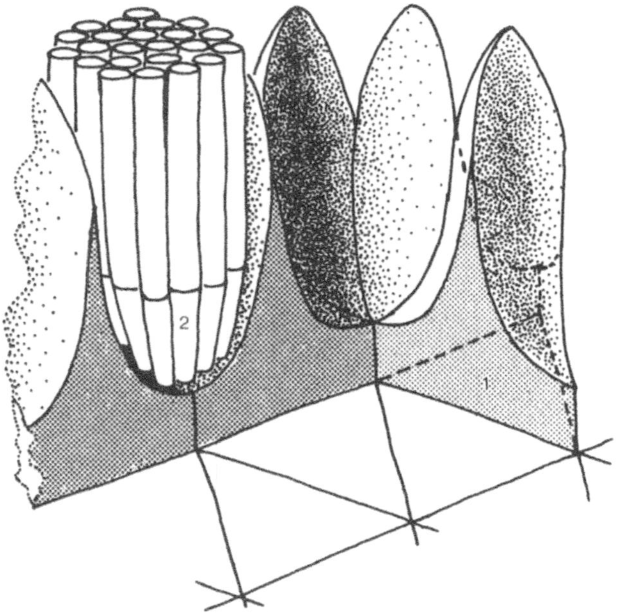

Fig. 55. Diagram to show arrangement of rods in groups and pigment epithelium cells in *Scopelarchus guntheri*. Pigment cells, *1*, are triangular in section at their bases, where they rest against BRUCH's membrane, and 6 contribute to pit surrounding each group. Outer segments, *2*, are located wholly within reflecting pit, and so are optically isolated from those of other groups. Light entering pit will be shared among all rods in group

side the quality of the reflection varied markedly, depending on the angle of view; if this angle was oblique there was little reflection, but if the retina was examined nearly at right angles to its surface, i.e., along the light path in life, the light was strongly reflected.

In the case of *Benthalbella* the bottom of the pits is lined with a flat formation of crystals, well-adapted to act as a small specular mirror. In *Scopelarchus*, however, the bottom of the pit is rounded.

Light which enters such a pit is likely to undergo multiple reflections within the pit and so be shared among the group of rods; reflection within the pit will also tend to prevent light from one pit reaching neighbouring groups.

Another feature suggests that the postulated sharing of information among the rods of a group does occur. This is the presence, in *Scopelarchus*,of dark staining plaques about 0.5 μm diameter at the sclerad end of the inner segments, which mutually connect most, if not all, of the rods within a group (Fig. 56). These junctions, which from their appearance may be structural junctions, or may be tight junctions associated with electrotonic synapses (BRIGHTMAN and REESE, 1969) do not occur in the accessory retina of *Scopelarchus* nor in either retina type in *Benthalbella*.

On the basis of the optical arrangement, and of the possible spread of activity at these plaques, I suggest that each group may function as a single "macroreceptor" (LOCKET, 1971b).

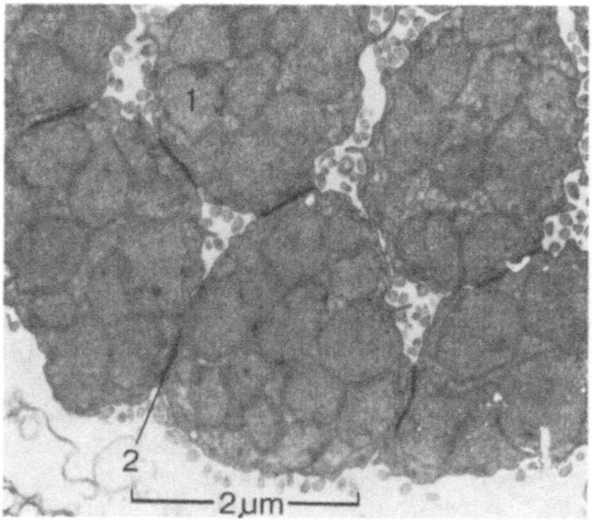

Fig. 56. Junctions between inner segments, grouped main retina of *Scopelarchus guntheri*. Inner segments are cut transversely just proximal to their junction with outer segments, and contain closely packed mitochondria, *1*, which have tubular cristae. Junctions, *2*, are present between inner segments; these plaques have same dimensions in radial sections, and are sites where external membranes of inner segments are in intimate contact. Whether these plaques are structural junctions of desmosome type, or junctions representing electrotonic synapses between rods within group is uncertain. Electron micrograph

4. Accessory Retina, Scopelarchids

The grouped retina so far considered is the grouped portion of the main retina; having tubular eyes the scopelarchids also have an accessory retina, lining the medial wall of the eye cup and not conforming to Matthiessen's ratio. This accessory retina also shows the grouped pattern, but here the groups are less regular in number of rods and in diameter and arrangement of the groups than in the main retina. The inner layers of the accessory retina, as in other tubular eyes, are thinner and less complex than those of the main retina.

5. Notolepis rissoi

A rather different situation occurs in *Notolepis rissoi*. This mesopelagic fish has a long snout, extending some three eye diameters in front of the orbit. The eyes are situated laterally, of normal shape, and with a marked, largely rostral, aphakic aperture.

Sections show that most of the retina is grouped, and that the groups are, as in the scopelarchids, isolated from each other by processes of pigment epithelium cells containing a reflecting material; the tapetal arrangement is less regular than in the scopelarchids, however. The tips of the pigment cells are the only parts of those cells to contain any black pigment; the rest of the cell appears empty in the preserved specimens examined, although they probably contained reflecting material in life. The pigmented tips extend as far as the inner segment region, optically isolating the groups from each other. As in *Benthalbella* the sclerad tips of the groups are square cut, and the tapetum beneath these tips probably behaves as a specular reflector (Fig. 57).

The synapses in the grouped retina are small (Fig 60), and electron microscopy shows them to be of the simple, rod type. The grouped arrangement extends to include the synapses in *Notolepis*, which is not the case, morphologically at least, in scopelarchids.

The eye of *Notolepis* is not tubular, and there is no accessory retina, but there is a region with a special structure. In the caudal part of the retina, behind the choroid fissure and extending almost, but not quite, to the iris root, is an area of single receptors rather than groups (Fig. 58). These receptors, which are substantially larger than the rods forming the groups, constitute an area temporalis, and are further discussed below in that context.

BRAUER'S (1908) illustration of the retina of *Coccorella atrata* (BRAUER'S *Evermannella*) shows the rods arranged in groups, with long processes from the pigment epithelium between them. He commented on the retinal tapetum and the sparkling granules it contains, but not on the grouped arrangement.

MUNK (1966a) found the same arrangements in *E. indica*, in the dorsal part of the accessory retina. Here the reticulation due to the grouped arrangement was easily seen in the dissected eye. Neither author found a tapetum in *E. indica*, though BRAUER stated that reflective material is present in the epithelium of *C. atrata*.

The grouped arrangement of visual cells is not confined to deep-sea fishes; FRANZ (1920) described the condition in mormyrids, MCEWAN (1938) also considered mormyrids, and compared them with *Elops* and *Megalops*, in both of which grouped retinae occur. MOORE (1944) found a similar appearance in *Hiodon*, and ENGSTRÖM (1963) recorded it additionally from *Gymnarchus* and *Gymnotus*. The knife-fish *Xenomystus* is another example, brought to my attention by MUNK (personal communication).

All these are fishes of shallow water; the mormyrids, *Gymnarchus*, and *Gymnotus* have electric organs, and probably locate their prey largely by changes in their electric field (MARSHALL, 1971). Although they live in shallow water, this is frequently turbid, in which vision will resemble that experienced by man in fog. Objects do not become visible until they are close, when they appear as a large image of low contrast. In such a case good acuity is less important than an ability to recognize small contrast differences, and it may be that the sharing of the image among a group of receptors is an adaptation to this end.

IV. Area Centralis

1. Characteristics of the Area

WALLS (1942) defined the area centralis as a circumscribed retinal area within which the retina is so constructed as to afford a marked local increase in resolving power. He regarded such an area as occurring mainly, if not exclusively, in diurnal animals, and noted that an area often contains a fovea.

Franz (1913) noted that many teleost fishes have regions of different structure in their retinae, and he cited Brauer's (1908) observations on deep-sea fishes as outstanding examples. These included such species as *Argyropelecus* and *Dissomma (Scopelarchus)* which have tubular eyes containing a main and accessory retina, and others such as *Platytroctes* and *Bathytroctes* which have a fovea.

It is certainly true that such fishes have retinal regions of widely diverse structure, but the area centralis is more than just an area of different structure. Duke-Elder (1958) agreed with Walls that an area is characteristic of diurnal forms and noted that it may be located centrally or temporally. He stated that the area may be rounded or band-shaped, and though usually single, may be double. The retinal structure in such an area was characterized by slender and closely packed receptors, associated with an increase in the number of bipolar and ganglion cells, and hence in the retinal thickness. Duke-Elder stated that most teleosts have an area without a fovea: although he did not mention deep-sea forms specifically, he suggested that in certain species, mainly nocturnal, the area would increase sensitivity rather than acuity, and expressed the opinion that this is the case in the pure rod fovea of *Bathytroctes*.

Munk (1970), describing the band-shaped areas in several shallow water and semiamphibious teleosts, suggested that they serve as movement detectors rather than as regions of high resolving power. Munk stated that an area is characterized by a maximum density of photoreceptors and a marked increase in the number of cells in the inner nuclear and ganglion cell layers.

Although most animals with areas are diurnal, there are some deep-sea fishes which may be regarded as having an area, in the sense that this is a region of retina with a specialized structure, but without at present attributing any special function to it. Other deep-sea species which certainly possess a well-marked fovea have already been mentioned and are discussed below.

2. The Area in Notolepis

N. rissoi, a mesopelagic fish, has a specialized retinal region which may be regarded as a true area in Walls' sense. The retina over most of the fundus shows a grouped structure (Fig. 57), similar to that in scopelarchids, and discussed above.

The retina caudal to the choroid fissure, and thus receiving images from in front of the fish, has a different structure. Here the receptors are not grouped but are single (Fig. 58). Their outer segments are cylindrical but at 3.5 μm diameter are considerably larger than those in the groups, which have a diameter of 1.4 μm. At the level of their inner segments these single receptors are tightly packed in a regular hexagonal array; although this regularity is also present at the outer segment level, the outer segments are narrower than the inner, and are separated from one another by a distance of the same order as their diameter. The space between the outer segments in the area is occupied by epithelial cell processes, which, like those in the grouped retina, contain pigment granules in their tips, and are probably packed with crystals in life.

The cylindrical outer segments of these area receptors are rod-like; their nuclei, however, are oval, show a tigroid pattern of chromatin, and are situated at

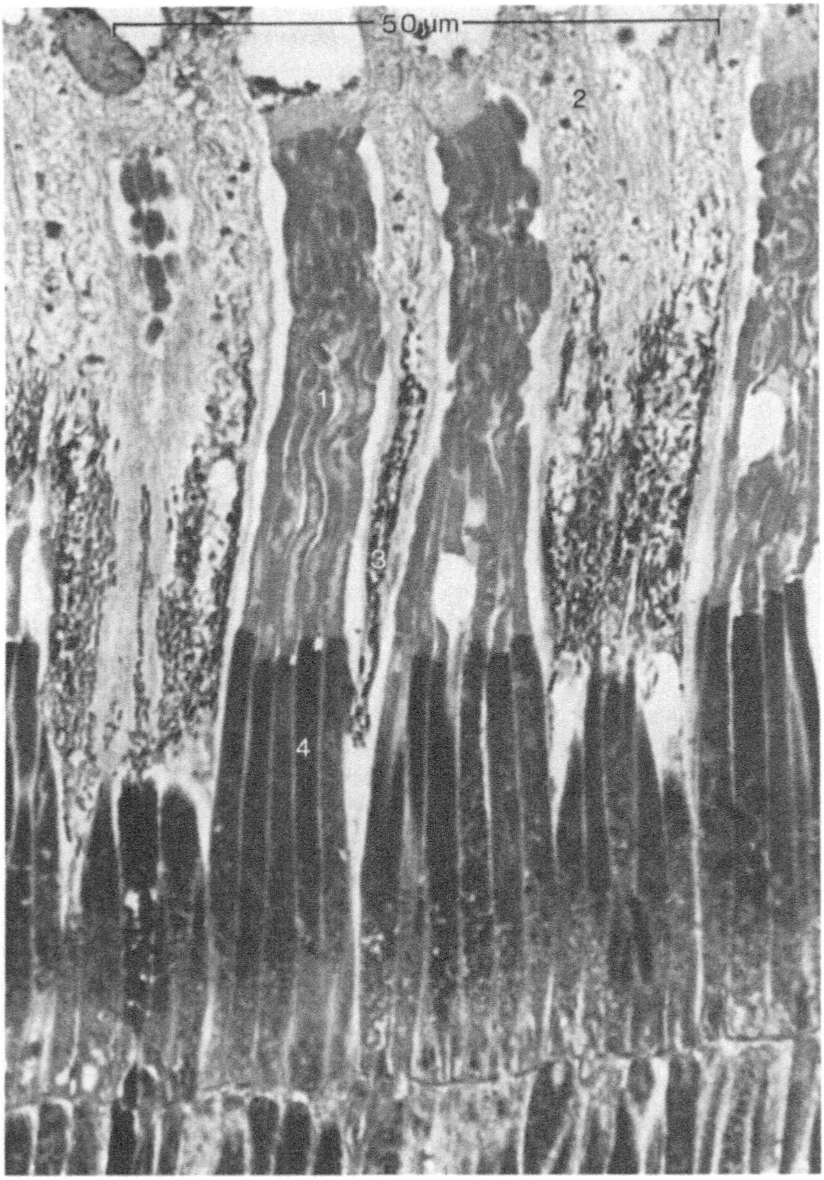

Fig. 57. Grouped retina of *Notolepis rissoi*. Rods in this species are arranged in groups over most of retina. Their slender outer segments, *1*, are inserted into pits among pigment epithelium cells, *2*. These cells contain melanin granules in their vitread processes, *3*, and probably have reflecting crystals in rest of cell during life. Epithelium cells invest rods to level of inner segments, *4*. Optical micrograph

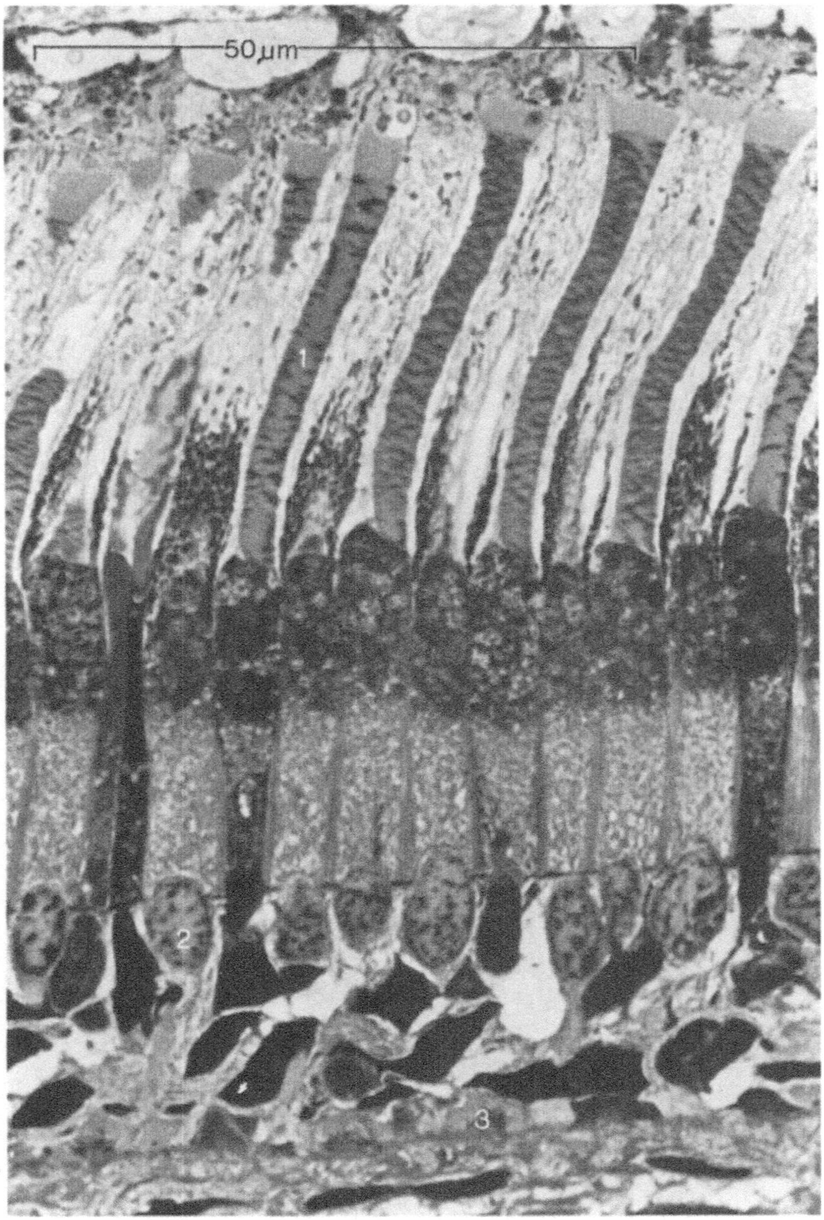

Fig. 58. Receptor cells from area of *Notolepis rissoi*. Area contains large single receptors; outer segments, *1*, of each is sheathed by pigment cell processes, as groups are elsewhere in retina. Nuclei of these receptors, *2*, resemble those of cones, and synaptic pedicles, *3*, are large and complex. Optical micrograph

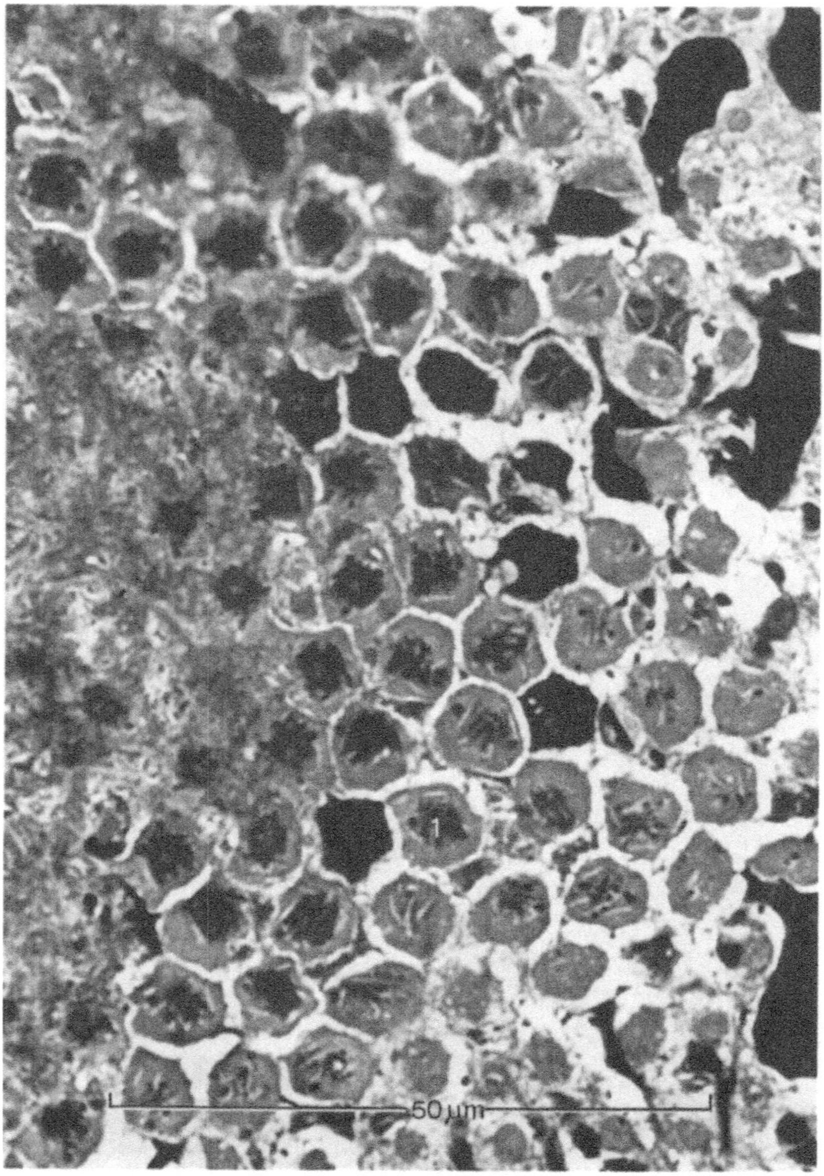

Fig. 59. Tangential section through synaptic pedicles, *1*, of receptors in area of *Notolepis*. This micrograph, to same scale as Fig. 60, shows even arrangement of large and complex synapses of single receptors in area. Optical micrograph

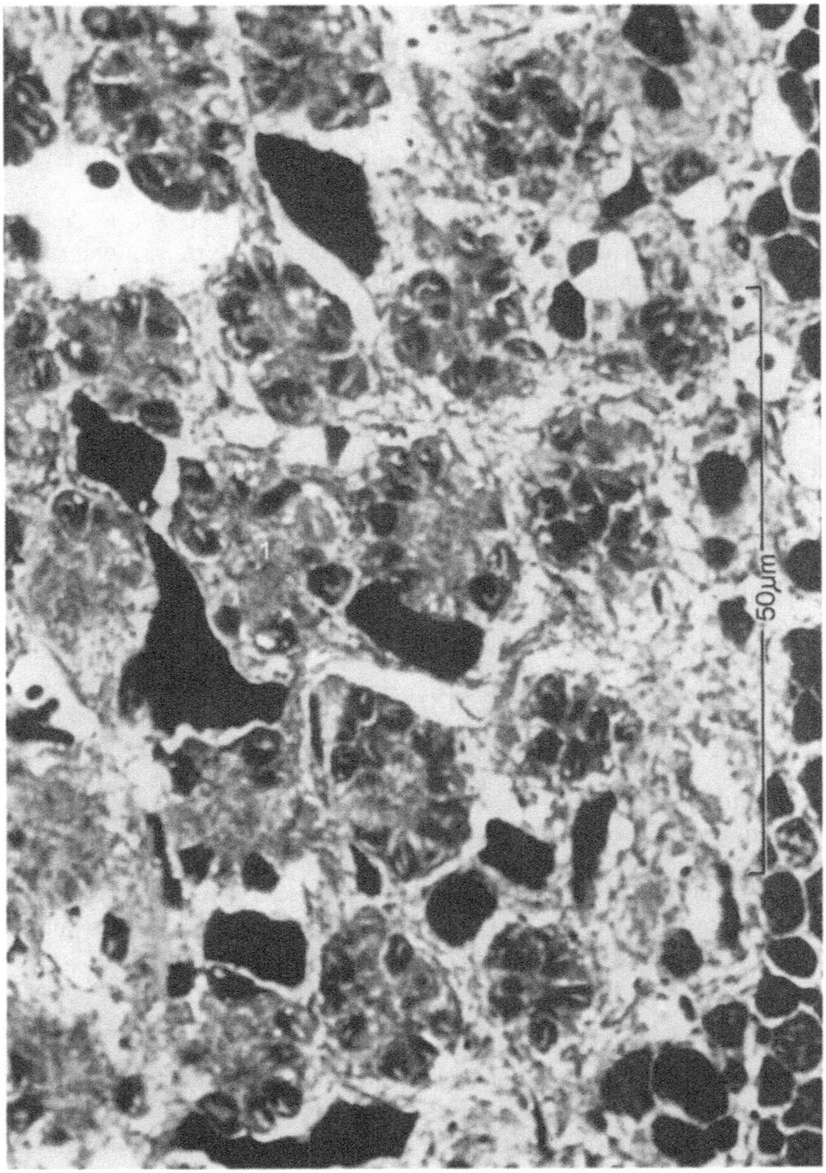

Fig. 60. Tangential section through synaptic pedicles of grouped rods in *Notolepis*. In this species, unlike scopelarchids, grouped arrangement is visible in synapses as well as in outer segment region. Each group, *1*, consists of cluster of small simple synapses. Optical micrograph

or just vitread to the outer limiting membrane, all characteristics usually associated with cones. The synaptic pedicles are twice the diameter of the grouped rod synapses, and are complex in structure like those of typical cones (Fig. 59). Teleost rod synapses are almost always of the simple, oligosynaptic type, like those of the grouped rods in the present case (Fig. 60). The large receptors in the temporal area resemble in some ways the rods of amphibia, though these animals have clearly separable rods and cones, all with complex synapses.

Comparing the retinae of scopelarchids with that of *Notolepis* we find regions of grouped and nongrouped retina in both. The grouped type is similar in the two, in that there are bundles of rods arranged so that they may function as macroreceptors, isolated from each other by reflecting layers of crystal-containing epithelium cells. The single receptors are arranged differently in the two, however: those of scopelarchids are closely packed together, and only their tips are related to the epithelial cells. In *Notolepis* the single outer segments are separated from each other, and are individually sheathed by tapetal cells. Thus, in their arrangement each *Notolepis* single receptor corresponds to a group of rods. The synaptic pedicles of both the grouped and nongrouped main retina of *Scopelarchus* are complex in type, and those of the accessory retina are simple in that species; in *Notolepis* it is only the single receptors of the area which have complex pedicles, those of the grouped rods are, as in scopelarchids, simple.

V. The Fovea

1. Types of Fovea and their Occurrence

In certain animals the retinae contain a small area of specialization where the retina is thinned. This depression, of which two types occur, is usually located in the area centralis (WALLS, 1942), and is called the fovea. The type which occurs in primates, including man, is shallow; a deeper type with steep walls is typified by the fovea of birds of prey. Though the two types appear as depressions in the retinal surface, their detailed structure and function is different.

The shallow type was called the concaviclivate fovea by WALLS (1937, 1940, 1942) and the steep-sided type the convexiclivate. Though WALLS differentiated between the two types structurally, he did not do so fully on functional grounds.

The fovea of man is of the concaviclivate type, and is the region of the retina giving the highest visual acuity. The foveal receptors are regarded as atypical cones, since their outer segments, longer than those of more peripheral cones, are cylindrical like rod outer segments, and thinner than those of typical cones. Characteristic of the primate fovea is the displacement of retinal tissue away from the foveal centre, which contains only elements of the cones and of radial fibres.

Some teleosts have foveae of the concaviclivate type; the fovea described by VILTER (1954a, c) in *B. benedicti* was of this form. This species, the first in which VILTER recognized the multiple bank arrangement of receptors, has a shallow fovea, in which the number of banks is increased from three to six, situated in the area temporalis. MUNK (1966a) found an area temporalis in *B. stilbius* and *B. pacificus*, in which there were six banks of rods rather than the three banks in the

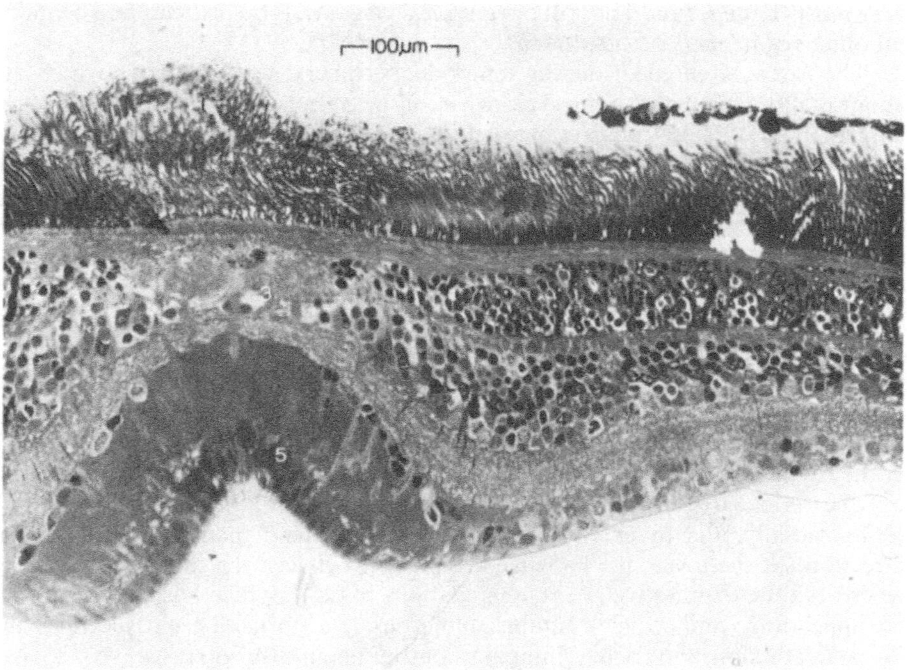

Fig. 61. Fovea of *Leptoderma macrops*. This fovea is similar to that of *Platytroctes*, and of other searsid fishes examined. Foveal rods, *1*, are thinner and more tightly packed than extrafoveal rods, *2*. The centre of fovea, *3*, is almost free of nuclei, but these are specially numerous in foveal shoulder, *4*. Vitread part of centre of fovea, *5*, is formed largely of lamellae and columns of radial fibre origin, which here determine inner contour of fovea. Optical micrograph

less differentiated retina. Only a 75 mm specimen of *B. pacificus* however had a fovea, which was very shallow; in a 134 mm and a 39 mm specimen of that species, and three specimens of *B. stilbius* he found no fovea, though the area was well-marked.

Convexiclivate foveae are known in several shallow water teleosts, e.g. *Syngnathus* (KRAUSE 1886a) and the sea horse *Hippocampus*. BRAUER (1908) described a pure rod fovea in a deep-sea fish which he named *Bathytroctes rostratus*. MUNK (1966a) pointed out that BRAUER'S specimens should, at least in part, be referred to *Searsia koefoedi*. MUNK himself examined *Platytroctegen mirus*, which, though not described in detail, apparently has a fovea similar to that described by BRAUER.

The searsid fishes, e.g. *Searsia*, *Platytroctes*, do indeed possess a fovea of convexiclivate form, so do the alepocephalid species *Leptoderma* and *Xenodermichthys* and the scopelosauridae (MARSHALL, 1966).

The eyes of *Platytroctes* are large and prominent, though situated laterally. There is a wide aphakic gap, particularly rostrally, which was noted by BRAUER in his description of the same species, called by him *P. procerus*. The retina of this species contains only a single bank of rods, but these are about 150 µm in length over most of the retina. They all have typical oligosynaptic pedicles and cylindrical outer segments (LOCKET, 1971a).

The fovea, situated in the far temporal periphery, contains rods which are about half the diameter of those elsewhere in the retina, but which are somewhat longer (Fig. 61). These rods are parallel to each other and to the extrafoveal rods in those parts which lie slerad to the outer limiting membrane. Vitread to this level, the foveal rods are angulated away from the centre, so that their conducting fibre is long compared to those of extrafoveal rods. Their nuclei are aligned with the conducting fibres, and also slope; they are largely displaced from the centre, and are present in a layer approximately eight nuclei thick at the edge of the foveal depression, compared to two or three loose layers extrafoveally. The rod synapses show no significant differences between foveal and extrafoveal receptors.

In the raised shoulders of the fovea are concentrated the nuclei, absent from the centre of the depression, of the inner nuclear and ganglion cell layers.

The contours of the foveal pit are determined by a layer, not present elsewhere in the retina, of radial fibre material. Radial fibres are of course present throughout the retina; extrafoveally they are represented by trunks which traverse the retina radially, and by extensive lamellae which extend mainly in a tangential direction. At the fovea, the lamellae are present between the other cells, as elsewhere, but the trunks give rise to aggregations of tissue which has a dense granular appearance and which contains numerous fine fibrils. This tissue lines the foveal depression, and so determines its contour against the vitreous.

Tangential sections of the fovea of *S. koefoedi* show that the contours are very regular, and though not so steep-sided as in birds or reptiles, there is a well-marked central pit.

2. Function of the Concaviclivate Fovea

The concaviclivate fovea is the type which occurs in man, and which is well-known to be the retinal region responsible for the most acute vision. This acuity depends partly on the optics of the eye; the terrestrial eye has a biconvex lens, situated behind the pupil and capable of deformation by the accommodatory mechanism. Much of the refractive power of such an eye depends on the cornea, steeply curved and with a large difference of refractive index between itself and the air. A consequence of the terrestrial arrangement is that an axial pencil of rays are refracted symmetrically, whereas oblique rays are refracted unequally. Since the axial rays suffer least from aberrations due to optical imperfections in the eye, the fovea receives the best image that the dioptric apparatus of the eye is capable of producing.

It is not sufficient merely to receive an optically good image. If this is to be appreciated the retina must be able to resolve the detail. The resolving power of the retina depends on several factors; the mosaic of sensitive elements must have the finest possible grain, and the retinal neurons must be able to encode the information provided by the receptors, so that it may be transmitted to the higher visual centres.

The primate fovea is generally stated to be avascular (MICHAELSON and CAMPBELL, 1940), and certainly it appears so in retinal digests or injected preparations.

Fluorescein angiography has shown that a few capillaries may traverse the fovea in man (BIRD and WEALE, 1974), but it remains true that the fovea is largely devoid of vessels. The interposition of moving blood between the incident light and the outer segments might degrade the image and so limit foveal acuity. If blood vessels are absent from a region of retina, the function of the retinal neurons might be embarrassed by the local ischaemia, and WEALE (1966) postulated that the neurons are absent in the foveal centre because they are displaced toward the surrounding blood vessels.

The deep-sea fishes that have a concaviclivate fovea do not have retinal vessels, and some have a greater thickness of cells at the fovea than away from it; examples with a multibank retina, e.g. *Bathylagus* (VILTER, 1954) have more banks at the fovea than they do elsewhere.

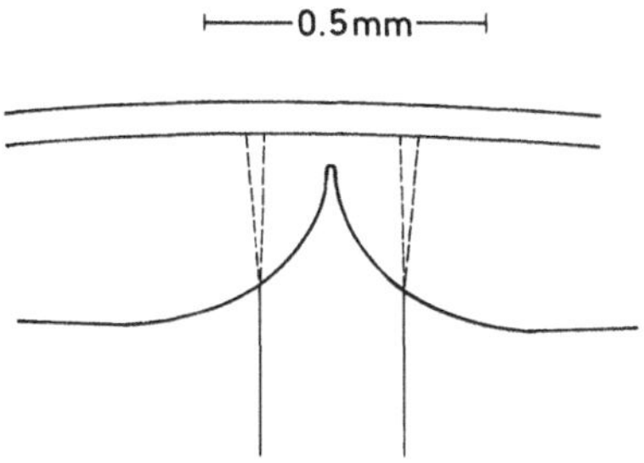

Fig. 62. Diagram to show how light refracted on passing from vitreous to retina at convexiclivate fovea will be deviated away from foveal centre. WALLS suggested this would magnify image, improving acuity, but PUMPHREY pointed out abrupt reversal of curvature in centre would distort image

3. Function of the Convexiclivate Fovea

The case of the convexiclivate fovea is different, and was first considered by WALLS (1937, 1940). He called attention to some figures for the refractive index of the retina and vitreous given by VALENTIN (1879), and on the basis of these and the curvature of the retino-vitreal boundary at the fovea, suggested that such a fovea would cause the image to be magnified. This magnification he believed would spread the image over a larger field of receptors, leading to improved visual acuity (Fig. 62).

The problem was reexamined by PUMPHREY (1948), who pointed out that though the rays through the foveal shoulders would be diverged in the sense that WALLS had shown, the abrupt reversal of curvature at the bottom of the foveal pit would give rise to distortion of the image, and so worsen resolution rather than the improvement postulated by WALLS.

PUMPHREY traced the behaviour of a ray of light moving across the fovea, and showed that the image would first lag, and then jump abruptly to a lead as it crossed the foveal centre (Fig. 63). Such a fovea has, as PUMPHREY showed, the property of making a symmetrical image asymmetrical unless it is centred exactly on the fovea. Such a fovea could signal small decentring movements of a centred image strongly, and so would be well suited to maintain fixation on an object once it had been seen. The possession of two such foveae within the binocular field is such a device as would enable a fish to approach a dimly seen prey with least chance of losing sight of it, and perhaps to respond with a snap when the prey is at the appropriate distance.

The best developed convexiclivate foveae are found in birds of prey, which have a feeding problem comparable to that of deep-sea fishes. Such birds must make an effective attack in midair, on an object of unknown size, with no nearby reference points. Consideration of a hawk catching a flying bird, or a swallow an insect, will show that the predator must strike or snap at exactly the right distance, if the attack is to be successful.

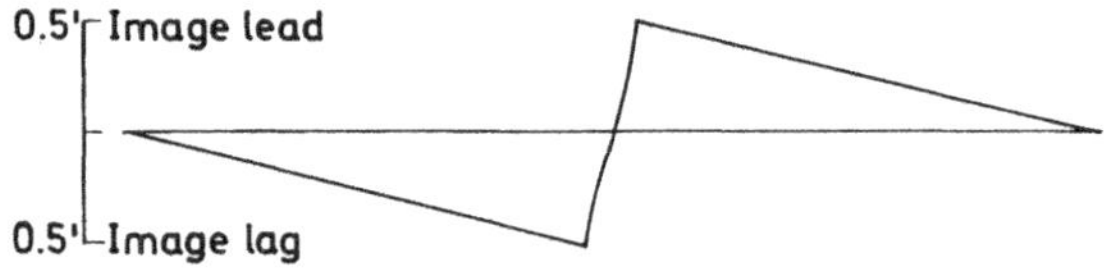

Fig. 63. If image passes across convexiclivate fovea it will first lag, and as it crosses centre jump abruptly to lead. PUMPHREY represented change by diagram such as this. Only when centred in fovea will symmetrical image still be symmetrical at level of receptors. Decentration of image will be exaggerated at receptor level, with opportunity for strong signalling of decentration

H. The Tapetum Lucidum

I. Tapetum and Argentea

A tapetum lucidum is a reflecting layer situated sclerad to the visual cells; light which has passed through the retina, instead of being absorbed by dark pigment in the retinal epithelium and choroid, is partially reflected, passing through the outer segments a second time, when a further fraction may be absorbed.

Though the tapetum and the argentea are quite distinct, both may be silvery, and the terms appear sometimes to have been confused in the literature. The argentea is a layer situated in the superficial part of the choroid, extending in many cases over the surface of the iris. The layer contains reflecting crystals of guanine, and, seen from the outside of the eye, appears bright and silvery. Covering the dark pigment of the choroid and retinal epithelium, and the blood in the choroidal vessels, but lying beneath the often transparent sclera and cornea, the argentea may well have a function in enabling the fish to match its background. The function of silvery surfaces in this context has been well-investigated, notably by DENTON and NICOL, and by LAND. (DENTON and LAND, 1971, for references.)

DENTON, et al. (1970, 1972) have postulated another function for such guanine containing layers in fishes. They found that the silvery layer which invests the swim-bladder in the conger eel conferred on that organ a marked impermeability to gases. The eye, through the medium of the choroidal rete, develops a high tension of oxygen, and those authors suggest that the argentea, present on both surfaces of the rete, may prevent loss of gas from the eye.

Fresh specimens of *Opisthoproctus soleatus* and *O. grimaldii* have silvery sides and a well-developed argentea, though neither are obvious in preserved specimens due to the disappearance of the guanine by the action of preserving fluids. These fishes have retinal diverticula (see section F.III.1.) in which the diverticulum retina faces a transparent "window" in the retina and choroid on the ventromedial side of the diverticulum. On the lateral wall of the eye, below this diverticulum, the argentea is a bright silver, and there can be little doubt that the light will be reflected by the argentea through the window, and so to the diverticulum retina.

II. Classification of Tapeta

1. Choroidal Tapeta

The tapetum proper occurs in a large variety of vertebrates, (WALLS, 1942), and may be of several kinds. Its location may be choroidal or retinal, and the nature of the reflecting material varies among the species. Choroidal tapeta were described by WALLS in three types: the tapetum fibrosum, tapetum cellulosum, and tapetum containing guanine.

a) Tapetum Fibrosum

The tapetum fibrosum occurs in mammals; mainly in ungulates and proboscideans, but also in whales. In these tapeta, a layer of collagenous tissue is present in the vitread part of the choroid, just sclerad to the choriocapillaris. This tissue resembles tendon, in which there are collagen fibres of similar size and arrangement, and the tapetum reflects light much as tendon does. WALLS stated that a tapetum fibrosum occurs in some pelagic marine teleosts, but he did not give the species.

b) Tapetum Cellulosum

A tapetum cellulosum occurs only in certain mammals: carnivores, seals and some lower primates. In a tapetum cellulosum the reflecting material, in the form of platelets or rods, is contained within cells in the choroid. The cells are usually flat and may be stacked in piles with their faces parallel to the plane of the retina. An example of this type of tapetum is found in the cat, in which rodlets 230 nm in diameter are arranged in regular arrays within the cells (PEDLER, 1963b). In the bush-baby the reflecting material is in the form of flat crystals (PIRIE, 1959, DARTNALL et al., 1965). The tapetum in both cases contains riboflavin, and in the bush-baby this may be the reflecting material (PIRIE, 1959). ELLIOTT and FUTTERMAN (1963) isolated the rodlets from the cat tapetum and found that they contain some, but not all, of the riboflavin present.

c) Tapetum Containing Guanine

The third type of choroidal tapetum considered by WALLS is that in which the reflecting material is guanine; such tapeta occur particularly in the elasmobranch fishes, many of which show brilliant eyeshine.

Crystals of guanine, a purine, occur very widely among fishes, and are responsible for the silvery surfaces in teleost scales. The architecture of such surfaces and the role of guanine in their reflectivity has been extensively studied in scales and in tapeta by DENTON and NICOL (1964, 1965a, b, 1966), BEST and NICOL (1967),

NICOL and VAN BAALEN (1968), and NICOL (1969b). LAND (1966) considered the optics of the eye of the scallop, which contains a reflector, also composed of guanine. DENTON and LAND (1971) compared the extent to which the reflecting surfaces in a variety of fishes and cephalopods conform to the dimensions which would give the "ideal" reflection, and found that the conformity is good in most cases.

The choroidal tapeta of elasmobranchs are formed by flat crystal-containing cells in which the crystal stacks are oriented approximately perpendicular to the direction at which light entering the eye will strike them. In some cases, e.g. *Squalus acanthias*, the tapetum is occlusible, pigment granules contained within cell processes moving in brighter conditions so that they cover and obscure the reflecting surfaces (DENTON and NICOL, 1964). The reflectivity of guanine reflecting surfaces can be very high, and in many cases, including shallow and deep-sea sharks, it is of specular character. DENTON and NICOL showed that the reflectivity of the dogfish tapetum, *(Scyliorhinus canicula)*,approached that of a sputtered aluminium mirror.

The high reflectivity of guanine tapeta depends on constructive interference of light, discussed by LAND (1972). A fraction of light will be reflected at a plane boundary between media of different refractive indices. If the media are present in lamellae of dimensions such that the optical thickness of each lamella is equal to a quarter of the wave length of light, then constructive interference will occur. The optical thickness is given by the actual thickness multiplied by the refractive index; so for light of 480 nm, and a system of lamellae of cytoplasm and guanine, with refractive index 1.36 and 1.85 respectively, the optical thickness will be 120 nm, and the actual thickness $\frac{120}{1.36} = 88.3$ nm for cytoplasm and $\frac{120}{1.85} = 64.9$ nm for guanine. Some constructive interference will occur if the reflective layers approximate to the ideal measurements, but the reflectivity will be less, and the band width greater in those circumstances.

Besides elasmobranchs, choroidal guanine tapeta also occur in chimaerids, bichirs *(Polypterus)*, and sturgeons (NICOL, et al., 1973, for references). The coelacanth *Latimeria* also has a specular choroidal tapetum of guanine, which is not occlusible (LENOBLE and LEGRAND, 1954, LOCKET, 1974), and the Australian lungfish *Neoceratodus* has a crystalline choroidal tapetum which is not specular (personal observation).

The tapetum in teleosts is generally retinal, but NICOL and ZYZNAR (1973) have described a choroidal tapetum, also specular and of guanine, in the big-eye *Priacanthus;* the gempylid *Ruvettus pretiosus*, which lives in the same waters as *Latimeria*, is another example of a teleost with a specular crystalline choroidal tapetum.

2. Retinal Tapeta

a) Specular Tapeta

Among deep-sea fishes are a number which display eyeshine. Few of these have been systematically studied to see whether they owe this to the presence of a tapetum, and if so what its nature may be. In some, the tapetum is so marked that

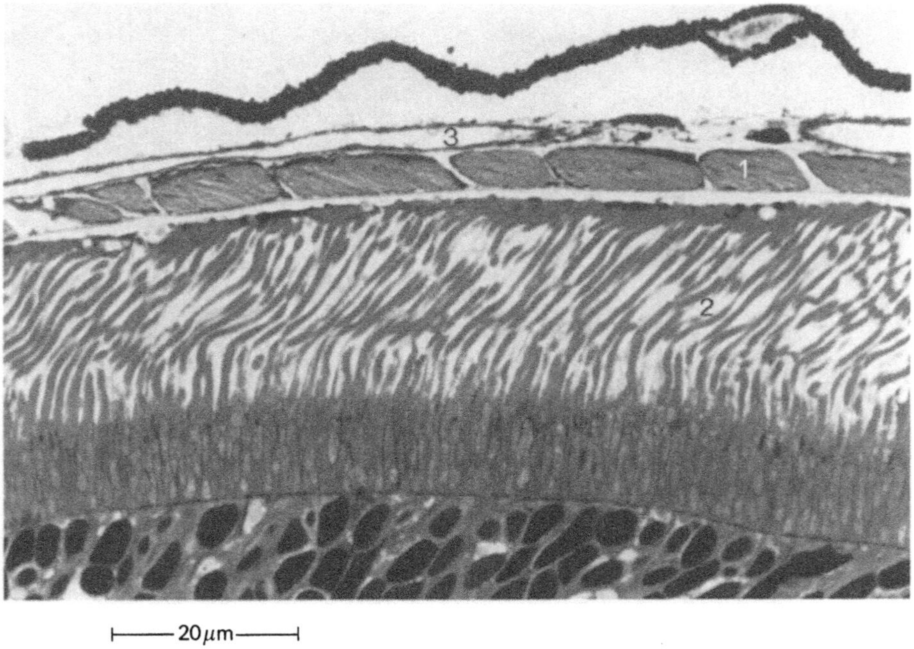

Fig. 64. Tapetum lucidum of myctophid, *Diaphus holti*. Cells of retinal pigment epithelium are modified to form reflecting elements. These cells, *1*, contain layers of guanine crystals, and in this micrograph, from periphery of retina, layers of crystals are not parallel to plane of retina, giving obliquely striated appearance. Thin rod outer segments, *2*, are also obliquely arranged, their long axes approximately normal to plane of reflecting crystals. Thin choroid, *3*, is sclerad to crystal-containing cells. Optical micrograph

it has been known since the earliest records of the fish, as in *Malacosteus* (CHUN, 1893). In other cases the eyeshine is well-known, but no record appears to have been made of the tapetum.

The eyeshine is sometimes easily visible without using special methods, particularly if there is an aphakic aperture or wide pupil. In other cases it is less obvious, but easily seen with an ophthalmoscope. Among the fishes which show eyeshine are the myctophids (lantern fishes). These fishes may sometimes be attracted to a light when they are at the surface at night, and bright reflections can then be seen from them. Some of these reflections are from their silvery flanks, but the eyeshine can also be seen under those conditions.

Not all myctophids possess a tapetum, but in those that do, e.g. *Diaphus*, *Ceratoscopelus*, the opthalmoscopic reflex is very bright. In examples I have observed the reflex was a clear blue-green colour, and NICOL (personal communication) reports that in others it is violet. On opening the eye and stripping off the retina, the tapetum is revealed lining most of the fundus. It has a tessellated appearance, each element measuring about 150 × 40 μm. Microscopy shows, as BRAUER (1908) described, that the retinal epithelium is devoid of pigment granules in the region of the tapetum (Fig. 64). The cells of the epithelium have become reflecting elements, and contain stacks of mutually oriented crystals. As in *Lati-*

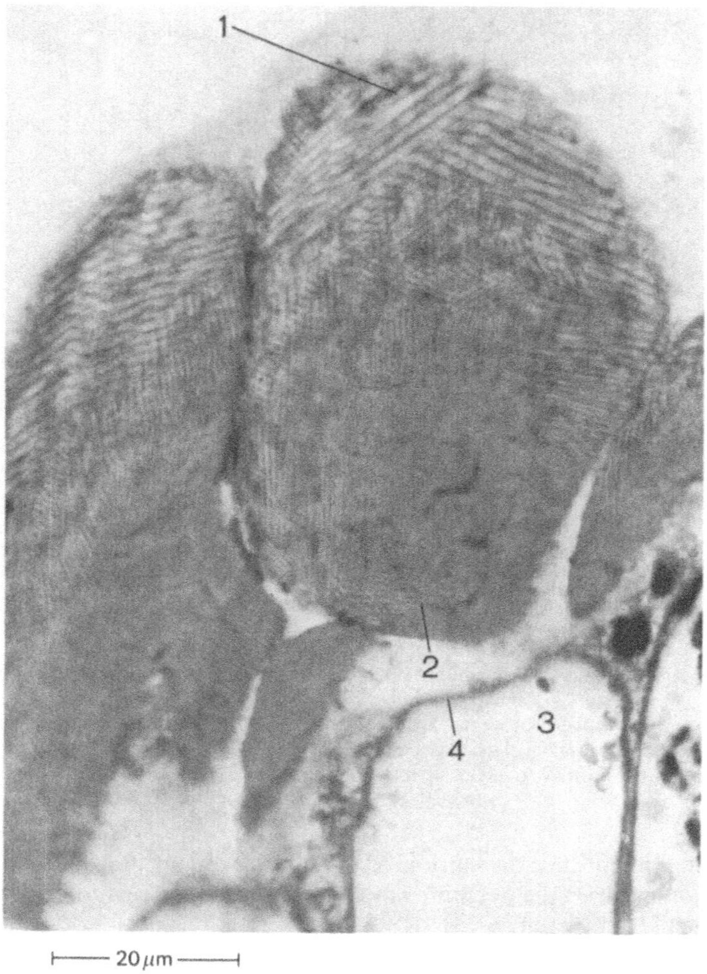

Fig. 65. Tapetal cells of *Diaphus holti*, tangential section. Cells are almost entirely filled by layers of guanine crystals. Their arrangement in layers, those of given layer being parallel to each other but at a large angle to those of adjacent layers is clearly suggested. More vitread layers, *1*, contain broader needle-shaped crystals than more sclerad layers, *2*. Choroidal vessels, *3*, are separated from epithelial cells by Bruch's membrane, *4*. Optical micrograph

meria the orientation is such that the plane of the crystal stacks is approximately normal to the incoming light, and electron microscopy of radial sections shows that the thickness and spacing of the crystals is comparable to that in *Latimeria* (Fig. 64).

In tangential sections a difference is apparent; the crystals in *Latimeria* are broad hexagonal plates, but in myctophids they are needle-shaped, about 0.5 μm across by 20 μm in length. These needles are arranged in layers, and within a layer the crystals are very regular in size and arrangement. Between the layers, however, there are minor differences in crystal width and large differences in orientation (Fig. 65).

Crystal-containing retinal tapeta are also present in the scopelarchids (Fig. 53, 54) (LOCKET 1971 b), evermannellids (BRAUER 1908), and probably in *Notolepis*. (The retinal structure of this fish, see section G. III. 5., suggests the presence of a tapetum, though no crystals remained in the specimens examined.) FREDERIKSEN (personal communication) has found a crystalline tapetum in *Omosudis lowei*, of which the retinal structure was described by MUNK (1965 b). Scopelarchids and *Notolepis* have regions in their retinae where the receptors are arranged in groups that are associated with a specialized tapetum. The crystal-containing processes of the pigment epithelial cells project between the groups, and the arrangement of the processes and the crystals within them is such as to optically isolate the groups from each other.

b) Diffuse Tapeta

In the tapeta discussed so far, the crystals are arranged in such a way as to give specular reflections. In the rostral region of the *Scopelarchus* main retina the rods are not grouped, but lie in a uniform layer, tightly packed and in contact with each other(Fig. 51). The epithelial cells in this nongrouped region are very different from those forming the tapetum in the grouped rod region. The former are flat, without processes, and the rod tips are in close apposition to these flat faces. The cells contain pigment granules, and also crystals. These are of comparable size and shape to those in the grouped retina, but are scattered through the cytoplasm and not oriented. These crystals, which measure about 0.16 μm in length, will scatter light randomly rather than reflect it in a specular manner, but they will still cause some light which would otherwise have been absorbed by melanin to reach the outer segments, and so perhaps contribute to sensitivity in conditions of low illumination.

Tapeta which do not involve specular reflection but scatter light instead, have been recognized in recent years in a wide variety of fishes as a result of the work of NICOL and his colleagues. (ARNOTT et al., 1970, NICOL et al., 1973, for further references). In some of these diffuse tapeta the reflecting substance is guanine, but in others it is some other substance; in many it is a lipid, and in others an unidentified substance which is neither guanine nor lipid.

In their survey of bony fish tapeta NICOL et al. found that such diffusely reflecting tapeta are the rule, and the only specular tapetum in their list is the choroidal tapetum of the big-eye, *Priacanthus;* those myctophids which have a specular tapetum may now be added.

Nonspecular tapeta have not been reported as such from deep-sea fishes, but they do in fact occur in several species. The fishes which live on the bottom in deep water are different in many ways from the meso- and bathypelagic species, and it is among these benthal fishes that diffuse tapeta seem to be most common. An example is *Mora moro*, a benthal gadoid, which has large eyes of normal teleost proportions. On dissecting a fresh eye of *Moro* the tapetal material is at once evident when the retina is cut. It appears as a dense white semifluid mass which exudes from the cut surface, becoming more mobile as it mixes with ocular fluids. Handling such a retina is difficult because retinal separation occurs so easily through the tapetal layer. By electron microscopy the material appears to consist of spheres of even size, about 450–500 nm in diameter. They are located in the retinal epithelium cells, which, though tall, have smooth vitread ends without villous processes, so the outer segments do not interdigitate with them. The epithelial cells also contain a few pigment granules basally, where the nucleus and some mitochondria are located. The rods in this species are numerous, and arranged in three or four ill-defined banks, of which only the most sclerad comes into relationship with the tapetum. The content of the spheres has not been chemically identified, but its cutting characteristics and affinity for osmium tetroxide suggest that it may be a lipid. The tapetum of the sea trouts *Cynoscion nebulosus* and *C. arenarius*, the first diffuse tapeta to be described by ARNOTT et al. (1970), contained spheres of almost identical appearance and size. Measurements on reflectance in those species showed that the tapetum reflected equally across the whole visible spectrum, and that the reflectivity was about 45%.

Since their first report, NICOL and his colleagues have studied numerous other diffuse tapeta, and found that the reflecting substance differs among species. Some certainly contain triglycerides, but in others the material is still unidentified.

In *Bathygadus melanobranchus*, a benthal gadoid from about 1500 m, the tapetal cells are not as tall as in *Mora*, and their apices are in rather close relationship with the rod outer segments; these, though present in depth, are not arranged in definite banks. The tapetal cells in *Bathygadus* contain dark pigment granules sclerally, and tapetal spheres vitreally. In the retinal periphery the tapetum is absent, and the epithelial cells here contain pigment but no reflecting spheres. This fish was examined with the ophthalmoscope while fresh, when the retinal reflex was yellow-orange in colour. The tapetum itself was white under the dissecting microscope: the yellow colour of the reflex may have been due to the visual pigment or to the epithelial pigment granules.

Of particular interest among diffuse tapeta is that of *Malacosteus* (Fig. 44). This fish has a wide aphakic aperture, and the retina is easily visible through this by simple inspection as well as through the lens, when this is clear, with the ophthalmoscope. The retinal reflex is a brilliant red, as observed on the "Valdivia" expedition by CHUN (1900) and described by BRAUER (1908). Simple dissection shows the extent of the red tapetum; it covers the fundus except for a rostral crescent, which appears grey. The red pigment tends to dissolve out of the tissues during processing, to some extent in ethanol, and strongly in epoxypropane, but despite this, the colour may persist in plastic-embedded sections. DENTON and HERRING (1971) reported that the red colour is due to astaxanthin, a pigment normally found in the tissues of deep-living crustaceans. Microscopy (Fig. 66) shows that the tapetum consists of spheres, similar to those found in *Mora*, and described in *Cynoscion* spp. by ARNOTT, et al. (1970).

3. Function of Tapeta

The function of tapeta is usually assumed to be the increase of sensitivity gained by reflecting light into the outer segments a second time. The specular tapeta of elasmobranchs and myctophids fulfil the requirements for such a tapetum well; any light which strikes a specular reflector at normal incidence will be

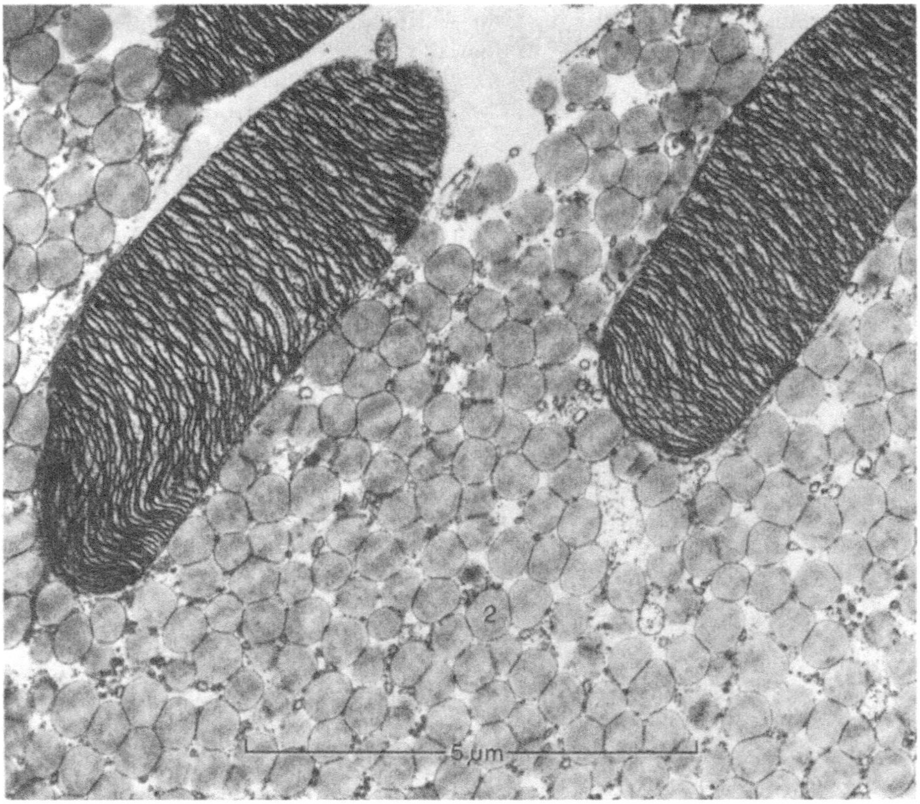

Fig. 66. Retinal tapetum lucidum of *Malacosteus niger*. Rod outer segments, *1*, are embedded among tapetal spheres, *2*. Spheres are of uniform size, about 0.5 µm in diameter, and contain red astaxanthin in life. They are confined to pigment epithelium cells, but cell boundaries are not shown well in this micrograph. Electron micrograph

reflected back along the same path, and this will occur to that fraction of the incident light which is not absorbed in either passage through the outer segments. As DENTON and NICOL (1964) point out, this light which leaves the eye, and thus contributes to eyeshine, has not been scattered within the eye in such a way as to degrade the image. The diffuse kind of tapetum is the reverse of specular, and here there may well be degradation of the image due to scatter. This will not matter if the retina is not neurally equipped to resolve fine detail, and the paucity of retinal neurons in benthal fishes examined suggest that this may well be the case. Such a retina is perhaps little more than a light detector, and as such the scattering of light will not be important, since the more light absorbed, by whichever outer segments, the greater the potential sensitivity of the retina as a whole.

The case of the scopelarchids and other fishes with grouped retinae is different. Here the potential resolution may be determined by the size of the groups and their separation from each other. Once light enters a tapetal pit containing a group it will be largely confined to that pit by the reflecting walls and any pigment they contain. In *Scopelarchus* the light will tend to undergo multiple reflection within the pit with the possibility of high absorption; this will also be the case for light entering the pit obliquely in *Benthalbella*. If the light enters axially however, it may well be reflected in a specular way from the flat bottom of the pit in that species.

The isolation of groups in tapetal pits seen in the scopelarchids is taken a stage further in *Notolepis;* in the area temporalis the reflecting pits contain single receptors rather than groups. Here we may suppose that the single receptor is receiving the maximum amount of light from its own domain, but the minimum stray light from the domains of neighbouring receptors. The presence of a "cap" of crystals arranged to give specular reflection through the length of the outer segment will enhance this discrete light catching.

The grouped arrangement coupled with a tapetum may have an effect peculiar to areas of retina which are not oriented along the main axis of the eye. Such areas occur in the accessory retina in scopelarchids, which is not only oblique to the incident light but is not at the focal point of the lens. Oblique and unfocussed light might be trapped into the pits, so contributing to an image which, though admittedly poor, may be better than one in an accessory retina without groups or in which there is no barrier to lateral spreading of light.

DENTON et al. (1970) observed a malacosteid fish, *Pachystomias*, alive, and confirmed CHUN's (1900) observation on *Malacosteus* that red light is emitted from the large photophore close below the eye. This light is produced by the luminescent tissue in the photophore proper, and traverses a layer of red tissue to reach the surface of the fish. Many deep-sea crustaceans are red in colour, and look very conspicuous when examined in a catch, but the red component of daylight is rapidly extinguished with depth, so such scarlet animals appear uniformly dark when seen under any appreciable depth of water. Many deep-living animals are luminescent, but the light they emit is in most cases blue or blue-green, and so is not reflected by red crustaceans. But malacosteid fishes emit red light, which will be well-reflected by the crustaceans, and the red tapetum is well-adapted to scatter this light once it has traversed the rods. DENTON et al. showed that *Pachystomias* rods contain a visual pigment with λ_{max} about 575 nm, which will absorb red light strongly.

Not only can such a fish thus illuminate and see a red crustacean, but it is possible that the crustacean will be unaware that it is being illuminated, since its visual pigment may be relatively insensitive to red light: The visual pigment of *Euphausia pacifica* absorbs maximally at 462 nm (KAMPA, 1955).

Another malacosteid, *Aristostomias scintillans*, has been studied by O'DAY and FERNANDEZ (1974). This fish has two large light organs close to the eye; the suborbital photophore was observed to emit red light, and the postorbital, green. Retinal extracts were prepared, and found to contain two pigments, with absorbance maxima at 551 and 526 nm and thus adapted for perception of the fishes' own luminescence.

DENTON and NICOL (1964) point out a further advantage which may result from the possession of a tapetum. In outer segments with a high density of visual pigment, as in deep-sea teleosts, there will be a tendency for pigment to break down spontaneously, giving rise to retinal noise. In threshold conditions this noise may be important if the signal is weak, when the signal to noise ratio becomes unfavourable. The noise increases with the pigment density, but the signal is increased by exposing the pigment to more photons. A tapetum will produce such an increase by redirecting the light through the visual pigment, and DENTON and NICOL suggested that this might account for the greater concentrations of pigment in teleosts without tapeta compared to elasmobranchs with tapeta from comparable depths.

I. Concluding Remarks

To review the salient points considered here, one might think first of the features of the deep-sea environment to which the fishes have adapted. Chief from the point of view of vision is the low light intensity at depth. Sunlight is attenuated to extinction with increasing depth, and adaptations to increase sensitivity will increase the depth at which a fish can use its visual system. Characteristic of the mesopelagic zone is the uneven distribution of light, downwelling light being much brighter than upwelling. Many mesopelagic fishes are adapted to cope with this situation; they may have downward shining photophores, of which the brightness must be matched against the downwelling light if the fish is to be concealed from predators below; the downwelling light is sampled visually, and as LAWRY (1974) showed in a myctophid, so sometimes is the fishes own luminescence. Below the limit of sunlight penetration the oceans are dark except for bioluminescence. This source, though weak by terrestrial standards, must be enough to enable fishes to see to a useful degree, if the development of their eyes is a criterion.

Besides the low light, the ocean depths are remarkable for the low density of biomass, and consequent scarcity of food for the individual. Adaptations which assist successful feeding are clearly of importance in this context, and include ocular adaptations as well as, for example, those of the jaws and belly which enable large prey to be eaten when the opportunity arises. Feeding in midwater, where there are no points of reference for judgments of size or range of an object seen, brings its own special problems, soluble only by binocular vision.

Among specializations tending to increase sensitivity are the possession of large eyes, or the segment of a large eye represented by the tubular eyes found mainly in mesopelagic fishes. That long outer segments are common suggests that it is advantageous for deep-sea fishes to have a high density of visual pigment, the essential medium for quantum capture. Such pigments are indeed present in high density, sometimes in long single outer segments, but also in shorter outer segments arranged in multiple banks. The majority of deep-sea visual pigments investigated have been found to absorb maximally near the wave length of light best transmitted through ocean water, which is also near the wave lengths maximally emitted by most luminous marine organisms. The transmitted light tends toward monochromacy as it becomes more attenuated, and most deep-sea fishes have a single visual pigment and a retina containing rods only, both features which will tend to favour sensitivity under those circumstances.

A further adaptation tending to make the best use of small amounts of light is the presence of a tapetum. In most teleosts this is of the diffuse kind, scattering light among the outer segments, and so promoting sensitivity at the expense of resolving power. Myctophids and many elasmobranchs have specular tapeta, which scatter light less than the diffuse kind, though that fraction of light which is reflected to emerge through the pupil as eyeshine is lost to the eye. The specialized tapeta associated with grouped retinae appear to represent a compromise solution, the receptor group probably behaving as a single "macro-receptor", isolated from its neighbours, but sharing the light reflected and scattered among its constituent elements.

Adaptations which may particularly help successful predation are those concerned with binocular vision. The essential role of binocularity in making a successful attack on an object seen dimly, and perhaps briefly, in a uniform environment, without any points of reference, was explained above, and adaptations toward binocularity are widespread. Tubular eyes are well-adapted to give a large field of binocular vision, in most cases upward, where silhouetting against the downwelling light is most likely to reveal prey. In some cases a portion of the retina subserving the binocular field is particularly developed, either as an area, or in some cases, a fovea. The remarkable property of accentuating eccenticity of the image possessed by the convexiclivate fovea was described.

The malacosteids are apparently unique in their adaptation concerned with red light, emitted by particular photophores in these fishes. Such light, well-reflected by their red tapetum, and absorbed maximally by their visual pigment, is reflected strongly by scarlet crustaceans among their potential prey.

Some of the features found in deep-sea fish eyes may be secondary adaptations to compensate for some disadvantage which might result from a primary adaptation. Thus, tubular eyes achieve large lenses and a good field of binocular vision, but at the expense of the visual field. The lens pad in scopelarchids and the optical fold in the evermannellids are examples of two convergently evolved adaptations using different structures to extend the otherwise limited field of all-around vision. The diverticula in certain other tubular-eyed species are further examples, and the adaptations reach their most extreme form in the secondary globe of *Bathylychnops*. The possession of a grouped accessory retina might suggest that this was an adaptation to compensate for the retina being too close to the lens to receive a focussed image, but this cannot be the explanation in all cases, since the scopelarchids have well-developed grouped regions in their main retinae, which do conform to Matthiessen's ratio.

In some deep-living species there appears to have been failure to adapt the eyes for better vision, and they appear as degenerate remains of doubtful functional value. The unique eyes of *Ipnops* show an extreme case in which the visual pigment-containing rods are almost the only apparently functional components of the eyes.

The deep ocean environment is so unlike our own terrestrial one that it is easy to make assumptions which are not supported by a closer examination of the true conditions. The statement by WALLS that the monocular clues to distance are available to all vertebrates is a case in point. It is only from two disparate images that an animal, be it bird or deep-sea fish, can obtain information about the distance of an object of unknown size seen isolated in space.

Some adaptations to the deep environment would be disadvantageous in a fish living in well-lit waters. The image in an eye with a large aphakic aperture would be much degraded in bright light, but in near threshold conditions the advantage of better retinal illumination may outweigh the disadvantage of image degradation in suprathreshold conditions.

Among the adaptations found are several in which the optics are somewhat removed from those of the "conventional" vertebrate eye. The widespread occurrence of multilayer reflectors in tapeta, as in the silvery scales of many fishes, is a case in point: Light guiding in receptor inner and outer segments is well-known, but the laminar structures in the optical fold of evermannellids and the lens pad of scopelarchids appear to have few parallels.

Finally, there remain certain features for which we have no explanation at present. Whether stalked eyes confer an advantage on their owner, and if they do so, how, is not clear. How such fishes as *Stylephorus* develop additional banks of rods in their formed and presumably functional multibank retina is not known, nor whether all the banks of such a retina are functional at one time.

A more extreme case is *Diretmus*, in which problems of long outer segments and those of multiple banks are compounded, and the various functions of different areas of retina with widely different structures are open to question. The processes which determine the length of outer segments, and their function once formed, are further highlighted by the abrupt alteration in length of rods at the step in the main retina of *Scopelarchus sagax*.

Though a number of problems have been solved since BRAUER first looked at deep-sea eyes, there are doubtless others still unrecognized, and certainly many which still await satisfactory solutions.

References

Albers, J. J. A.: Über das Auge des Kabeljau Gadus morhua und die Schwimmblase der Seeschwalbe, Trigla hirundo. Göttingen gelehrte Anzeiger **2**, 681—682 (1806).

Anctil, M.: A bibliography on bioluminescence in fishes. Sci. Rep. Yokosuka City Mus. **18**, 1—14 (1971).

Arnott, H. J., Maciolek, N. J., Nicol, J. A. C.: Retinal tapetum lucidum: A novel reflecting system in the eye of teleosts. Science **169**, 478—480 (1970).

Backus, R. H.: Solving the mystery of Alexander's acres. Oceanus **14**, 15—20 (1968).

Barnett, C. H.: The structure and function of the choroidal gland of teleostean fish. J. Anat. (Lond.). **85**, 113—119 (1951).

Baumann, C.: The regeneration and renewal of visual pigment in vertebrates. In: Handbook of Sensory Physiology. VII/1. Photochemistry of vision. Dartnall, H. J. A. (ed.). Berlin-Heidelberg-New York: Springer 1972.

Beebe, W.: Deep-sea fishes of the Bermuda oceanographic expeditions. Family Idiacanthidae. Zoologica, N. Y. **16**, 149—241 (1934).

Beebe, W., Crane, J.: Deep-sea fishes of the Bermuda oceanographic expeditions. Family Melanostomatidae. Zoologica, N. Y. **24**, 65—238 (1939).

Beebe, W., Pyl, M. van der: Eastern Pacific Expeditions of the New York Zoological Society. XXXIII. Pacific Myctophidae (Fishes). Zoologica, N. Y. **29**, 59—95 (1944).

Beer, T.: Die Accommodation des Fischauges. Pflügers Arch. ges. Physiol. **58**, 523—650 (1894).

Bertelsen, E.: The ceratioid fishes. Dana Rep. **39**, 1—276 (1951).
Bertelsen, E., Munk, O.: Rectal light organs in the argentinoid fishes *Opisthoproctus* and *Winteria*. Dana Rep. **62** (1964).
Best, A. C. G., Nicol, J. A. C.: Reflecting cells of the elasmobranch tapetum lucidum. Contr. Mar. Sci. **4**, 172—201 (1967).
Bird, A., Weale, R. A.: On the retinal vasculature of the human fovea. Exp. Eye Res. **19**, 409—417 (1974).
Blaxter, J. H. S., Currie, R. I.: The effect of artificial lights on acoustic scattering layers in the ocean. Symp. zool. Soc. Lond. **19**, 1—14 (1967).
Boden, B. P., Kampa, E. M.: The influence of natural light on the vertical migrations of an animal community in the sea. Symp. zool. Soc. Lond. **19**, 15—26 (1967).
Boden, B. P., Kampa, E. M.: Bioluminescence. In: Jerlov, N. G., Nielsen, E. S., (Ed.): Optical aspects of Oceanography. pp. 445—469. London-New York: Academic Press 1974.
Boden, B. P., Kampa, E. M., Snodgrass, J. M.: Underwater daylight measurements in the Bay of Biscay. J. mar. biol. Ass. U. K. **39**, 227—238 (1960).
Brauer, A.: Über einige von den Valdivia-Expedition gesammelte Tiefseefische und ihre Augen. Sitzungsberichte der Gesellschaft zur Beförderung des gesamten Naturwissenschaften zu Marburg, p. 115 (1901).
Brauer, A.: Über den Bau der Augen einiger Tiefseefische. Verh. dtsch. zool. Ges. **12**, 42—57 (1902).
Brauer, A.: Die Tiefseefische, 2. Anatomische Teil. Wiss. Ergebn. dtsch. Tiefsee-Exped. Valdivia **15**, 2 (1908).
Brightman, M. W., Reese, T. S.: Junctions between intimately apposed cell membranes in the vertebrate brain. J. Cell Biol. **40**, 648—677 (1969).
Chun, C.: Leuchtorgane und Facettenaugen. Ein Beitrag zur Theorie des Sehens in großen Meerestiefen. Biol. Zbl. **13**, 544 (1893).
Chun, C.: Aus den Tiefen des Weltmeeres. Schilderungen von der Deutschen Tiefsee-Expedition. pp. 549. Jena: Gustav Fischer 1900.
Clarke, G. L.: On the depth at which fish can see. Ecology **17**, 452—456 (1936).
Clarke, G. L., Backus, R. H.: Interrelations between vertical migrations of deep scattering layers, bioluminescence and changes in daylight in the sea. Bull. Inst. Oceanogr. Monaco **64**, 1—36 (1964).
Clarke, G. L., Conover, R. J., David, C. N., Nicol, J. A. C.: Comparative studies of luminescence in copepods and other marine animals. J. mar. biol. Ass. U. K. **42**, 541—564 (1962).
Clarke, G. L., Denton, E. J.: In: Hill, M. N. (Ed.): The Sea, Ideas and observations on progress in the study of the seas, Vol. 1, 456—458. New York: Interscience 1962.
Clarke, G. L., Wertheim, G. K.: Measurements of illumination at great depths and at night in the Atlantic Ocean by means of a new bathyphotometer. Deep Sea Res. **3**, 189—205 (1956).
Clarke, W. D.: Function of bioluminescence in mesopelagic organisms. Nature (Lond.) **198**, 1244—1246 (1963).
Cohen, A. I.: On the confusion between 'myeloid bodies' and 'phagosomes' in the pigment epithelium of the frog and other vertebrates. Vision Res. **9**, 1403—1404 (1969).
Contino, F.: Das Auge des Argyropelecus hemigymnus. Morphologie, Bau, Entwicklung und Refraktion. Albrecht v. Graefes Arch. Ophthal. **140**, 390—441 (1939).
Crescitelli, F.: The visual cells and visual pigments of the vertebrate eye. In: (Ed.): Dartnall, H. J. A. Handbook of Sensory Physiology, Vol. VII/1; pp. 245—363. Photochemistry of Vision. Berlin-Heidelberg-New York: Springer 1972.
Dartnall, H. J. A.: Visual pigment of the coelacanth. Nature (Lond.) **239**, 341—342 (1972).
Dartnall, H. J. A., Arden, G. B., Ikeda, Hisako, Luck, C. P., Rosenberg, M. E., Pedler, C. M. H., Tansley, Katherine: Anatomical electrophysiological and pigmentary aspects of vision in the bush baby: An interpretive study. Vision Res. **5**, 399—424 (1965).
Denton, E. J.: On the organization of reflecting surfaces in some marine animals. Phil. Trans. roy. Soc. B., **258**, 285—313 (1970).
Denton, E. J., Gilpin-Brown, J. B., Wright, P. G.: On the 'filters' in the photophores of mesopelagic fish and on a fish emitting red light and especially sensitive to red light. J. Physiol. (Lond.) **208**, 72P (1970).
Denton, E. J., Herring, P. J.: In: Report of the Council. J. mar. biol. Ass. U. K. **51**, 1035 (1971).
Denton, E. J., Land, M. F.: Mechanism of reflexion in silvery layers of fish and cephalopods. Proc. roy. Soc. B. **178**, 43—61 (1971).
Denton, E. J., Liddicoat, J. D., Taylor, D. W.: Impermeable "silvery" layers in fish. J. Physiol. (Lond.) **207**, 24P—25P 1970).

Denton, E. J., Liddicoat, J. D., Taylor, D. W.: The permeability to gases of the swimbladder of the conger eel (Conger conger). J. mar. biol. Ass. U. K. **52**, 727—746 (1972).

Denton, E. J., Nicol, J. A. C.: The chorioidal tapeta of some cartilaginous fishes (Chondrichthyes). J. mar. biol. Ass. U. K. **44**, 219—258 (1964).

Denton, E. J., Nicol, J. A. C.: Reflexion of light by external surfaces of the herring, *Clupea harengus*. J. mar. biol. Ass. U. K. **45**, 711—738 (1965 a).

Denton, E. J., Nicol, J. A. C.: Direct measurements on the orientation of the reflecting surfaces in the tapetum of *Squalus acanthias*, and some observations on the tapetum of *Acipenser sturio*. J. mar. biol. Ass. U. K. **45**, 739—742 (1965 b).

Denton, E. J., Nicol, J. A. C.: A survey of reflectivity in silvery teleosts. J. mar. biol. Ass. U. K. **46** 685—722 (1966).

Denton, E. J., Shaw, T. I.: The visual pigments of some deep sea elasmobranchs J. mar. biol. Ass. U. K. **43**, 65—70 (1963).

Denton, E. J., Warren, F. J.: Visual pigments of deep sea fish. Nature (Lond.) **178**, 1059 (1956).

Denton, E. J., Warren, F. J.: The photosensitive pigments in the retinae of deep-sea fish. J. mar. biol. Ass. U. K. **36**, 651—662 (1957).

Dowling, J.: Chemistry of visual adaptation in the rat. Nature (Lond.) **188**, 114—118 (1960).

Duke-Elder, S.: System of Ophthalmology, I. The Eye in Evolution. London: Kimpton 1958.

Ekman, S.: Zoogeography of the Sea. London: Sidgwick and Jackson 1953.

Elliott, J. H., Futterman, S.: Fluorescence in the tapetum of the cats eye. Identification, assay and localization of riboflavin in the tapetum and a proposed mechanism by which it may facilitate vision. Arch. Ophthal. (Chic.) **70**, 531—534 (1963).

Engström, K.: Cone types and cone arrangements in teleost retinae. Acta zool. (Stockh.) **44**, 179—342 (1963).

Fletcher, A., Murphy, T., Young, A.: Solutions of two optical problems. Proc. roy. Soc. Lond. A **223**, 216—225 (1954).

Franz, V.: Bau des Eulenauges und Theorie des Teleskopauges. Biol. Zbl. **27**, 271—278, 341—350 (1907).

Franz, V.: In: Oppel. (Ed.): Lehrbuch der Vergleichenden Mikroskopischen Anatomie der Wirbeltiere, Vol. 7, pp. 1—417, 1913.

Franz, V.: Zur mikroskopischen Anatomie der Mormyriden. Zool. Jahrb. (Jena) **42**, 91—148 (1920).

Frederiksen, R. D.: On the retinal diverticula in the tubular-eyed opisthoproctid deep-sea fishes *Macropinna microstoma* and *Dolichopteryx longipes*. Vidensk. Meddr. dansk naturh. Foren. **136**, 233—244 (1973).

Gunther, A. C. G. L.: Preliminary notices of deep-sea fishes collected during the voyage of H. M. S. Challenger. Ann. Mag. Nat. Hist. Ser. 5, 2, 17—28 (1878).

Gunther, A. C. G. L.: An introduction to the study of fishes. Edinburgh 1880.

Gunther, A. C. G. L.: Report on the scientific results of the voyage of H. M. S. Challenger during the years 1873—1876. Narrative. *1*, Pt. 1 (1885).

Gunther, A. C. G. L.: Report on the deep-sea fishes collected by H. M. S. Challenger during the years 1873—1876. Rep. Sci. Res. Challenger. **22**, pp 335 (1887).

Günther, K., Deckert, K.: Morphologie und Funktion des Kiefer- und Kiemen-Apparates von Tiefseefischen der Gattungen *Malacosteus* und *Photostomias* (Teleostei, Isospondyli, Stomiatoidea, Malacosteidae). Dana Rep. **49**, (1959).

Hardy, A. C.: The open sea: its natural history—the world of plankton. London: Collins 1956

Hardy, A. C.: The Open Sea: its natural history—fish and fisheries. London: Collins 1959.

Herring, P. J.: Bacterial luminescence in some argentinoid fishes. Barnes, H. (ed.). Aberdeen: Aberdeen University Press. Proc. 9th Europ. mar. biol. Symp. 563—572 (1975).

Herring, P. J., Clarke, M. R.: Deep Oceans. London: Arthur Barker 1971.

Hoffert, J. R., Fromm, P. O.: Effects of carbonic anhydrase inhibition on aqueous humour and blood bicarbonate ion in the teleost *(Salvelinus namaycush)*. Comp. Biochem. Physiol. **18**, 333—340 (1966).

Hoffert, J. R., Fromm, P. O.: Teleost retinal metabolism as affected by acetazolamide. Proc. Soc. exp. Biol. (N.Y.) **139**, 1060—1064 (1972).

Hoffert, J. R., Baeyens, D. A., Fromm, P. O.: The resistance of teleost ocular tissues to oxygen toxicity. Invest. Ophthal. (Chic.) **12**, 858—861 (1973).

Jerlov, N. G.: Optical studies of ocean water. Rept. Swed. Deep-Sea Exped. **3**, Physics, Chem. Fasc. 1, 1—59 (1951).

Jerlov, N. G.: Optical oceanography. Copenhagen: Elsevier 1968.

Johnson, R. K.: Revision of the Alepisauroid family Scopelarchidae, (Pisces, Myctophiformes). Fieldiana Zoology **66**, 249 pp (1974).
Jones, T. W.: On the so-called choroid gland or choroid muscle of the fish's eye. London Med. Gaz. **21**, 650—652 (1838).
Kampa, E. M.: Euphausiopsin, a new photosensitive pigment from the eyes of euphausiid crustaceans. Nature (Lond.) **175**, 996—997 (1955).
Kampa, E. M.: Underwater daylight and moonlight measurements in the eastern North Atlantic. J. mar. biol. Ass. U. K. **50**, 397—420 (1970).
Kampa, E. M., Boden, B. P.: Light generation in a sonic scattering layer. Deep Sea Res. **4**, 73—92 (1957).
Kay, R. H.: Light-stimulated and light-inhibited bioluminescence of the euphausiid *Meganyctiphanes norvegica* (G. O. Sars). Proc. roy. Soc. B. **162**, 365—386 (1965).
Krause, W.: Die Retina. II Die Retina der Fische (1). Int. Mschr. Anat. Physiol. **3**, 8—38 (1886a).
Krause, W.: Die Retina. II. Die Retina der Fische (2). Int. Mschr. Anat. Physiol. **3**, 41—73 (1886b).
Kuhne, W.: On the photochemistry of the retina and on visual purple. Foster, M. (ed.). London: Macmillan 1878.
Land, M. F.: A multilayer interference reflector in the eye of the scallop, *Pecten maximus*. J. exp. Biol. **45**, 433—447 (1966).
Land, M. F.: The physics and biology of animal reflectors. Progr. Biophys. molec. Biol. **24**, 75—106 (1972).
Lawry, J. V.: Jun. Lantern fish compare downwelling light and bioluminescence. Nature (Lond.) **247**, 155—157 (1974).
Lenoble, J., Le Grand, Y.: Le tapis de l'oeil du Coelacanthe (*Latimeria chalumnae* (Smith)). Bull. Mus. Nat. Hist. Nat. Paris, **2 Ser. 26**, 460—463 (1954).
Locket, N. A.: The retina of *Poromitra nigrofulvus* Garman, an optical and electron microscope study. Exp. Eye Res. **8**, 265—275 (1969).
Locket, N. A.: Retinal structure in a deep sea fish, *Sternoptyx diaphana* Hermann. Exp. Eye Res. **9**, 22—27 (1970).
Locket, N. A.: Retinal structure in *Platytroctes apus*, a deep sea fish with a pure rod fovea. J. mar. biol. Ass. U. K. **51**, 79—91 (1971a).
Locket, N. A.: Retinal anatomy in some scopelarchid deep-sea fishes. Proc. roy. Soc. Lond. B. **178**, 161—184 (1971b).
Locket, N. A.: Aspects of retinal structure in fishes. Ph. D. Thesis, University of London (1971c).
Locket, N. A.: The reflecting structure in the cornea of the serranid teleost *Nemanthias carberryi*. Proc. roy. Soc. B **182**, 249—254 (1972).
Locket, N. A.: Retinal structure in *Latimeria chalumnae*. Phil. Trans roy. Soc. B. **266**, 493—521 (1973).
Locket, N. A.: The choroidal tapetum lucidum of *Latimeria chalumnae*. Proc. roy. Soc. B. **186**, 281—290 (1974).
Lyall, A. H.: The growth of the trout retina. Quart. J. microsc. Sci. **98**, 101—110 (1957).
Lythgoe, J. N.: Iridescent corneas in fishes. Nature (Lond.) **233**, 205—207 (1971).
Lythgoe, J. N.: The structure and function of iridescent corneas in teleost fishes. Proc. roy. Soc. B. **188**, 437—457 (1975).
Maetz, J.: Le role biologique de l'anhydrase carbonique chez quelques teleosteens. Bull. Biol. France, Belgique (Paris) **Suppl 40**, 1—129 (1956).
Maren, T. H.: Carbonic anhydrase, chemistry, physiology and inhibition. Physiol. Rev. **47**, 595—781 (1967).
Marshall, N. B.: Aspects of deep sea biology. London: Hutchinson 1954.
Marshall, N. B.: Family *Scopelosauridae*. In: Fishes of the western North Atlantic. Mem. Sears Fdn. mar. Res. **I (5)**, 194—204 (1966).
Marshall, N. B.: The olfactory organs of bathypelagic deep-sea fishes. In: Marshall, N. B. (Ed.): Aspects of marine zoology. London: Symp. zool. Soc. (1967) **19**, 57—70.
Marshall, N. B.: Explorations in the life of fishes. Cambridge, Mass: Harvard University Press. 1971.
Matthiessen, L.: Untersuchungen über dem Aplanatismus und die Periscopie der Kristallinsen in den Augen der Fische. Pflügers Arch. ges. Physiol. **21**, 287—307 (1880).
Mauchline, J.: The biology of the euphausiid crustacean *Meganyctiphanes norvegica* (M. Sars). Proc. roy. Soc. Edinb. B **67**, 141—179 (1960).
Maxwell, J. C.: Some solutions of problems. Cambridge and Dublin Math. J. **8**, 188—195 (1854).
McEwan, M. R.: A comparison of the retina of the mormyrids with that of various other teleosts. Acta zool. (Stockh.) **19**, 427—465 (1938).

Mead, G. W.: Hermaphroditism in archibenthic and pelagic fishes of the order Iniomi. Deep Sea Res. **6**, 234—235 (1960).

Mead, G. W.: *Family Ipnopidae*. In: Fishes of the Western North Atlantic. Mem. Sears Fdn. mar. Res. **1(5)**, 147—161, 1966.

Merrett, N. R., Badcock, J., Herring, P. J.: Observations on bioluminescence in a scopelarchid fish, *Benthalbella*. Deep Sea Res. **18**, 1265—1267 (1971).

Michaelson, I. C., Campbell, A. C. P.: The anatomy of the finer retinal vessels, and some observations on their significance in certain retinal diseases. Trans. ophthal. Soc. U.K. **60**, 71—112 (1940).

Miller, W. H., Snyder, A. W.: Optical function of myoids. Vision Res. **12**, 1841—1848 (1972).

Moore, G. A.: The retinae of two North American teleosts with special reference to their tapeta lucida. J. comp. Neurol. **80**, 369—379 (1944).

Moseley, H. N.: Report on the structure of the peculiar organs on the head of *Ipnops*. Rep. Sci. Res. Challenger, Zool. **22**, 269—276 (1887).

Moser, H. G., Ahlstrom, E. H.: Development of lanternfishes (Family myctophidae) in the California current. Part. 1. Species with narrow-eyed larvae. Bull. Los Angeles County Mus. Nat. Hist. Sci. **7**, 1—145 (1970).

Munk, O.: The eyes of *Ipnops murrayi* Gunther 1887. Galathea Rep. **3**, 79—87 (1959).

Munk, O.: The eye of *Stomias boa ferox*. Vidensk. Meddr. dansk naturh. Foren. **125**, 353—359 (1963).

Munk, O.: The eyes of three benthic deep-sea fishes caught at great depths. Galathea Rep. **7**, 137—149 (1964a).

Munk, O.: The eyes of some ceratioid fishes. Dana Rep. No. **61**, 1—16 (1964b).

Munk, O.: Ocular degeneration in deep-sea fishes. Galathea Rep. **8**, 21—31 (1965a).

Munk, O.: *Omosudis lowei* Gunther 1887, A bathypelagic fish with an almost pure cone retina. Vidensk. Meddr. dansk naturh. Foren. **128**, 341—355 (1965b).

Munk, O.: Ocular anatomy of some deep-sea teleosts. Dana Rep. **70**, 1—62 (1966a).

Munk, O.: On the retina of *Diretmus argenteus* Johnson 1863 (Diretmidae, Pisces). Vidensk. Meddr. dansk naturh. Foren. **129**, 73—80 (1966b).

Munk, O.: On the eye and so-called preorbital light organ of the isospondylous deep-sea fish *Bathylaco nigricans* Goode and Bean, 1896. Galathea Rep. **9**, 211—218 (1968).

Munk, O.: On the occurence and significance of horizontal band-shaped retinal areas in teleosts. Vidensk. Meddr. dansk naturh. Foren. **133**, 85—120 (1970).

Munk, O.: Early notions of dynamic accommodatory devices in teleosts. Vidensk. Meddr. dansk naturh. Foren. **136**, 7—28 (1973).

Munk, O., Frederiksen, R. D.: On the function of aphakic apertures in teleosts. Vidensk. Meddr. dansk naturh. Foren. **137**, 65—94 (1974).

Munz, F. W.: Photosensitive pigments from retinas of deep-sea fishes. Science. **125**, 1142—1143 (1957).

Munz, F. W.: Photosensitive pigments from the retinae of certain deep-sea fishes. J. Physiol. (Lond.) **140**, 220—235 (1958).

Nicol, J. A. C.: Observations on luminescence in pelagic animals. J. mar. biol. Ass. U.K. **37**, 705—772 (1958).

Nicol, J. A. C.: Spectral composition of the light of the lantern fish *(Myctophum punctatum)*. J. mar. biol. Ass. U.K. **39**, 27—32 (1960).

Nicol, J. A. C.: Animal luminescence. In: Lowenstein, O. E. (Ed.): Advances in comparative physiology and biochemistry, Vol. 1, 217—273, 1962.

Nicol, J. A. C.: Bioluminescence. In. Hoar, W. S., Randall, D. J. (Ed.): Fish Physiology. Vol. 3, pp. 355—400, New York-London: Academic Press 1969a.

Nicol, J. A. C.: The tapetum lucidum of the sturgeon. Contr. mar. Sci. **14**, 7—18 (1969b).

Nicol, J. A. C., Arnott, H. J., Best, A. C. G.: Tapeta lucida in bony fishes (Actinopterygii): a survey. Can. J. Zool. **51**, 69—81 (1973).

Nicol, J. A. C., Baalen, C. van: Studies on the reflecting layers of fishes. Contr. mar. Sci. **13**, 65—88 (1968).

Nicol, J. A. C., Zyznar, E. S.: The tapetum lucidum in the eye of the big-eye *Priacanthus arenatus* Cuvier. J. Fish Biol. **5**, 519—522 (1973).

O'Connell, C. P.: The structure of the eye of *Sardinops coerulea*, *Engraulis mordax*, and four other pelagic marine teleosts. J. Morph. **113**, 287—329 (1963).

O'Day, W. T., Fernandez, H. R.: *Aristostomias scintillans* (Malacosteidae): A deep-sea fish with visual pigments apparently adapted to its own bioluminescence. Vision Res. **14**, 545—550 (1974).

Pearcy, W. G.: Some distributional features of mesopelagic fishes off Oregon. J. mar. Res. (Sears Fdn. Mar. Res.) **22**, 83—102 (1964).

Pearcy, W. G., Meyer, S. L., Munk, O.: A 'four-eyed' fish from the deep sea: *Bathylychnops exilis* Cohen, 1958. Nature (Lond.) **207**, 1260—1262 (1965).

Pedler, C. M. H.: The fine structure of the radial fibres in the reptile retina. Exp. Eye Res. **2**, 296—303 (1963a).

Pedler, C.: The fine structure of the tapetum cellulosum. Exp. Eye Res. **2**, 189—195 (1963b).

Pedler, C.: Rods and cones, a new approach. Int. Rev. gen. exp. Zool. **4**, 219—274 (1969).

Pirenne, M. H.: Binocular and uniocular threshold of vision. Nature (Lond.) **152**, 698—699 (1943).

Pirie, A.: Crystals of riboflavin making up the tapetum lucidum in the eye of a lemur. Nature (Lond.) **183**, 985—986 (1959).

Pumphrey, R. J.: The theory of the fovea. J. exp. Biol. **25**, 299—312 (1948).

Pumphrey, R. J.: Concerning vision. In: Ramsay, J. A., Wigglesworth, V. B. (Eds.): The cell and the organism. Essays presented to Sir James Gray. Cambridge: University Press 1961.

Regan, C. T.: The morphology of a rare oceanic fish, *Stylophorus chordatus* Shaw; based on specimens collected in the Atlantic by the Dana expeditions, 1920—1922. Proc. roy. Soc. B **96**, 193—207 (1924).

Rochon-Duvigneaud, A.: Les yeux et la vision des vertebrés. Masson, Paris (1943).

Rofen, R. R.: Fishes of the western North Atlantic. Family Paralepididae. Mem. Sears. Fdn. Mar. Res. **1 (5)**, 205—461 (1966).

Scholes, J. H.: Colour receptors, and their synaptic connexions in the retina of a cyprinid fish. Phil. Trans. roy. Soc. B **270**, 61—118 (1975).

Stell, W. K.: The morphological organisation of the vertebrate retina. In: Handbook of Sensory Physiology. Fuortes, M. G. F. Berlin: Springer VII/2 1972.

Stell, W. K.: Horizontal cell axons and axon terminals in goldfish retina. J. comp. Neurol. **159**, 503—520 (1975).

Stell, W. K., Lightfoot, D. O.: Color-specific interconnections of cones and horizontal cells in the retina of the goldfish. J. comp. Neurol. **159**, 473—502 (1975).

Thines, G.: Les poissons aveugles: (1) Origines, taxonomie, Repartition geographique, comportement. Ann. Soc. roy. Zool. Belg. **86**, 1—128 (1955).

Underwood, G.: Some suggestions concerning vertebrate visual cells. Vision Res. **8**, 483—488 (1968).

Valentin, G.: Ein Beitrag zür Kenntnis der Brechungsverhältnisse der Thiergewebe. Pflügers Arch. ges. Physiol. **19**, 78 (1879).

Vilter, V.: Dissociation spatiale des champs photo-sensoriels à cones et à bâtonnets chéz un poisson marin, le *Callionymus lyra*. C. R. Soc. Biol. (Paris) **141**, 344—346 (1947).

Vilter, V.: Existence d'une rétine a plusieur mosaïques photoreceptrice chéz un poisson abyssal bathypelagique, *Bathylagus benedicti*. C. R. Soc. Biol. (Paris). **147**, 1937—1939 (1953).

Vilter, V.: Différentiation fovéale dans l'appareil visuel d'un poisson abyssal, *Bathylagus benedicti*. C. R. Soc. Biol. (Paris). **148**, 59—63 (1954a).

Vilter, V.: Interprétation biologiques des trames photoréceptrices superposées de la rétine du *Bathylagus benedicti*. C. R. Soc. Biol. (Paris). **148**, 327—330 (1954b).

Vilter, V.: Relations neuronales dans la fovea à bâtonnets du *Bathylagus benedicti*. C. R. Soc. Biol. (Paris). **148**, 466—469 (1954c).

Wald, G.: Carotenoids and the visual cycle. J. gen. Physiol. **19**, 351—371 (1935-6).

Walls, G. L.: Significance of the foveal depression. Arch. Ophthal. (Chic.) **18**, 912—919 (1937).

Walls, G. L.: Postcript on image expansion by the foveal clivus. Arch. Ophthal. (Chic.) **23**, 831—832 (1940).

Walls, G. L.: The vertebrate eye and its adaptive radiation. Michigan: Cranbrook Institute of Science 1942.

Weale, R. A.: Binocular vision and deep-sea fish. Nature (Lond.) **175**, 996 (1955).

Weale, R. A.: Why does the human retina possess a fovea? Nature (Lond.) **212**, 255—256 (1966).

Wittenberg, J. B., Haedrich, R. L.: The choroid rete mirabile of the fish eye. II. Distribution and relation to the pseudobranch and to the swimbladder rete mirabile. Biol. Bull. **145**, 137—156 (1974).

Wittenberg, J. B., Wittenberg, B. A.: Active secretion of oxygen into the eye of fish. Nature (Lond.) **194**, 106—107 (1962).

Wittenberg, J. B., Wittenberg, B. A.: The choroid rete mirabile of the fish eye. I. Oxygen secretion and structure: Comparison with the swimbladder rete mirabile. Biol. Bull. **145**, 116—136 (1974).

Young, R. W., Bok, D.: Participation of the retinal pigment epithelium in the rod outer segment renewal process. J. Cell Biol. **42**, 392—403 (1969).

Chapter 4

Evolutionary Adaptations of Fishes to the Photic Environment

F. W. MUNZ*, Eugene, Oregon (USA), and W. N. MCFARLAND*, Ithaca, N.Y. (USA)

With 29 Figures

Contents

* The original work reported here was supported by grants (EY00323 and EY00324) from the Division of Research Grants and Fellowships, National Eye Institute, National Institutes of Health, U.S. Public Health Service.

A. Introduction

This chapter is concerned mainly with the evolutionary adaptation of visual systems in bony fishes living in aquatic environments. Some of the ideas, however, may be more generally applicable, both to other aquatic animals and to terrestrial vertebrates. Adaptations in all parts of the visual apparatus could be considered here. We shall concentrate on retinal mechanisms, especially on the receptors themselves. For more general reviews, see WALLS (1942), ROCHON-DUVIGNEAUD (1943), PRINCE (1956), and RODIECK (1973). Among more primitive fishes, a good deal is known about cyclostomes, and this has been recently reviewed by CRESCITELLI (1972) and BRIDGES (1972). Much less is known about the vision of elasmobranchs; this was also discussed by CRESCITELLI (1972).

Like most other vertebrates, bony fishes characteristically have duplex retinae, equipping them for vision at both low and high levels of light intensity. (Chapter 3 considers the interesting specializations of deep-sea fishes.) There is a high-sensitivity system, which is used when the light level is low. This is the *scotopic* system and is based on the photoreceptor cells that are called *rods*. In bright daylight, the high-acuity *photopic* system is functional. The *cones* are the photoreceptors of this system. In many bony fishes, the photopic system mediates the function of wavelength discrimination, which we shall consider to be synonymous with color vision.

In connection with vision, the word adaptation usually refers to the changes in sensitivity that follow an alteration in light intensity. These changes include bleaching or regeneration of visual pigments and altered patterns of neuronal interaction. But changes in pupil size, to us the most obvious part of the process, are not available to the majority of fishes, whose pupils are fixed in a wide-open position. In these species there is another process controlling the light intensity at the photoreceptor cells in the retina. Both the photoreceptors and the shielding pigment in the pigment epithelium are capable of "photomechanical" movements. The effect of these movements is to shield the highly sensitive rods from the dazzling effects of bright light and to return them to the focal plane of the eye at low light levels. It is not part of our purpose to describe the processes of light and dark adaptation, but we mention them to emphasize that in this chapter our interest is in adaptations in the evolutionary sense.

It should be self-evident that any understanding of visual functions must depend upon reliable measurements of the photic environments in which they are performed. It is also a truism that physiology usually makes sense only in relation to the relevant structures. Further, physiologic analyses (often at several levels of organization) must be considered in connection with the animal's behavior. We emphasize that the understanding of vision requires a multilevel, integrative approach. Light measurements, the morphology and histology of the visual system, biochemistry (visual pigments), electrophysiology, and visually significant behavior can all contribute to such understanding. This approach could be called "visual systems analysis;" to us it is the *environmental biology of vision*.

Before proceeding, it might be useful to explain why we have chosen to examine the visual systems of bony fishes. First, our entry into the field has been through the study of visual pigments—and here bony fishes have shown more diversity than all other vertebrates combined. But there is another important reason: water is an effective color filter, and there is great diversity among the photic environments in different water types. It has been suggested that, in evolution, the visual systems of fishes have become adapted to the characteristics of their particular photic environment (e.g., MUNZ, 1965). Such speculations were usually based, however, on only fragmentary knowledge of both the photic environment and the animal's behavior. We shall try here to integrate the data at many levels, from the environment to behavior, that are relevant to fish vision. Both scotopic and photopic systems will be considered.

We shall look closely at the relationship between the photic environment and the pigments of scotopic vision. In addition, the scotopic system is interesting because of its lability in many fishes. The most familiar case is the occurrence of both *rhodopsin* and *porphyropsin* in the same species. These visual pigments are based on vitamins A_1 and A_2, respectively. In a number of species, it has been shown that the proportion of these visual pigments is not constant within an individual. In nature the proportion may vary seasonally; in the laboratory, it can sometimes be altered experimentally. The visual significance of these changes (as well as that of porphyropsin itself) has not been explained completely. Less well recognized, but also of great potential interest, is the fact that there can be changes in the visual protein *(opsin)* manufactured during different stages in the life-cycle. It is not correct to assume that the larval or juvenile fish always has the same scotopic pigment as the adult.

Photopic vision has been more difficult to study, because most attempts to extract cone pigments have failed. Some information is available, however, through the method of microspectrophotometry (MSP). Because it can look at individual photoreceptor cells, MSP provides the best evidence on the retinal basis for color vision. Recently, however, we have succeeded in extracting cone pigments from certain fishes. These results will be described and related to the photic environment. From an analysis of the visual tasks performed by these fishes has come a better understanding of vision in the diverse photic environments faced by aquatic animals. Finally, from these ideas, we have developed a hypothesis about the evolution of color vision in the vertebrates.

B. Adaptations of the Scotopic System

I. Visual Pigments

1. Diversity and Phylogenetic Relationships

Visual pigments are identified by a procedure known as partial bleaching (DARTNALL, 1952). The retinae of dark-adapted animals are removed in dim red light and washed free of blood. Then a detergent such as digitonin is used to extract visual pigments. Spectrometric analysis exploits the fact that these pigments are photosensitive, but the light-absorbing impurities (e.g., hemoglobin) are not. Thus, the changes in spectral absorption that result from exposure to a bleaching light are more characteristic and reproducible than the directly measured absorbance spectra.

To illustrate partial bleaching, we have chosen an experiment with a retinal extract of a butterflyfish, *Chaetodon ephippium*. First, the absorbance spectrum of the unbleached extract is measured (Fig. 1, Curve 1). Then the extract is exposed to narrow-band, colored light, for a period long enough to bleach only a portion of the visual pigment. After this exposure the absorbance spectrum is recorded again (Curve 2). The bleaching process and measurement of the absorbance spectrum are repeated until all of the light-sensitive pigment is gone (Curves 3–5). The bleaching wavelength, light intensity, and exposure time are altered as necessary to control the amount of bleaching. The differences between successive absorbance spectra are calculated and plotted (Fig. 2A). Each difference spectrum is rescaled to 100% at its maximum (Fig. 2B). In the illustrated experiment, these curves are superimposable, indicating that only a single photosensitive pigment was present in the extract. When visual pigments are bleached, a product is formed. This absorbs light in the near ultraviolet region. In the illustrated experiment the spectral location of the bleaching product is characteristic of rhodopsins (pigments based on the aldehyde of vitamin A_1). Other partial bleaching experiments, on extracts containing mixtures of visual pigments, are illustrated in Section C. III.

It is now generally recognized that the scotopic pigments of fishes are remarkably diverse. Because the species-specific opsins affect the absorbance spectra of visual pigments, there is one series of rhodopsins and another series of porphyropsins (LYTHGOE, 1972b; ALI and WAGNER, 1975). The pigments in each series have spectra of the same shape when plotted against frequency, rather than wavelength (DARTNALL, 1953; MUNZ and SCHWANZARA, 1967; BRIDGES, 1972). They differ from each other in spectral location, which is expressed most conveniently as λ_{max}, the wavelength at which a visual pigment absorbs light maximally. The λ_{max} values of fish rhodopsins are known to range at least from 467 nm (in *Bathylagus microphthalmus*; DARTNALL and LYTHGOE, 1965, as corrected by LYTHGOE, 1972b) to 526 nm (in *Aristostomias scintillans*; O'DAY and FERNANDEZ, 1974). The λ_{max} values of porphyropsins range from 502 nm (in *Balistapus undulatus*; MUNZ and MCFARLAND, 1973) to 551 nm (in *A. scintillans*; O'DAY and FERNANDEZ, 1974). These visual pigments cover, therefore, two broadly overlapping spectral distributions.

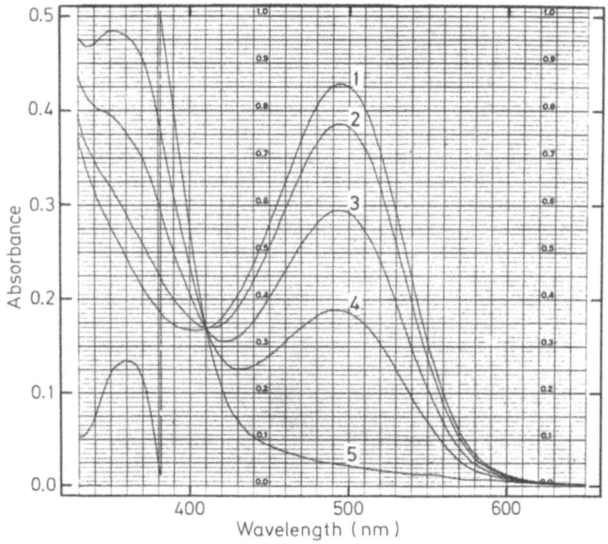

Fig. 1. A partial bleaching experiment (No. 4215) performed on a retinal extract of *Chaetodon ephippium*. Curve *1*, absorbance spectrum of the unbleached extract; Curve *2*, after 15 min exposure to deep-red light (λ675 nm); Curve *3*, after 20 s exposure to orange light (λ610 nm); Curve *4*, after 35 s further exposure to orange light; Curve *5*, after 6 min exposure to green light (λ560 nm). Note that the short-wave end of Curve *5* is offset by 0.5 absorbance units

This diversity of visual pigments is not closely correlated with phylogenetic relationships (MUNZ, 1971; CRESCITELLI, 1972; LYTHGOE, 1972a; MUNZ and MCFARLAND, 1973). Unrelated fishes often have the same visual pigment, as determined spectrometrically. Members of the same family may share the same pigment: the rhodopsins of 11 parrotfishes (Scaridae) range only from 483–485 nm in λ_{max}. In the Salmonidae and Characidae, many species have the same rhodopsin (P 503$_1$—the pigment is specified by λ_{max} and the subscript denotes the vitamin A) and/or the corresponding porphyropsin (P 527$_2$); but other species have different pigments. Species within the same genus may show substantial diversity (Table 1).

2. Habitat Depth

DENTON and WARREN (1956) first observed that the visual pigments of deep-sea fishes are more sensitive than other rhodopsins to blue light. In a sample of seven marine fishes from the Temperate Zone, it appeared to WALD et al. (1955) that the λ_{max} values of the visual pigments are correlated with depth. They suggested that the rhodopsins of surface species have λ_{max} values near 500 nm and those of deeper-living fishes are shifted to shorter wavelengths. This idea was later examined in a much larger sample of tropical marine fishes (MUNZ and MCFARLAND, 1973). Although judgments of depth distribution are imprecise, we recognized three overlapping categories: surface (0–5 m), intermediate (1–30 m) and

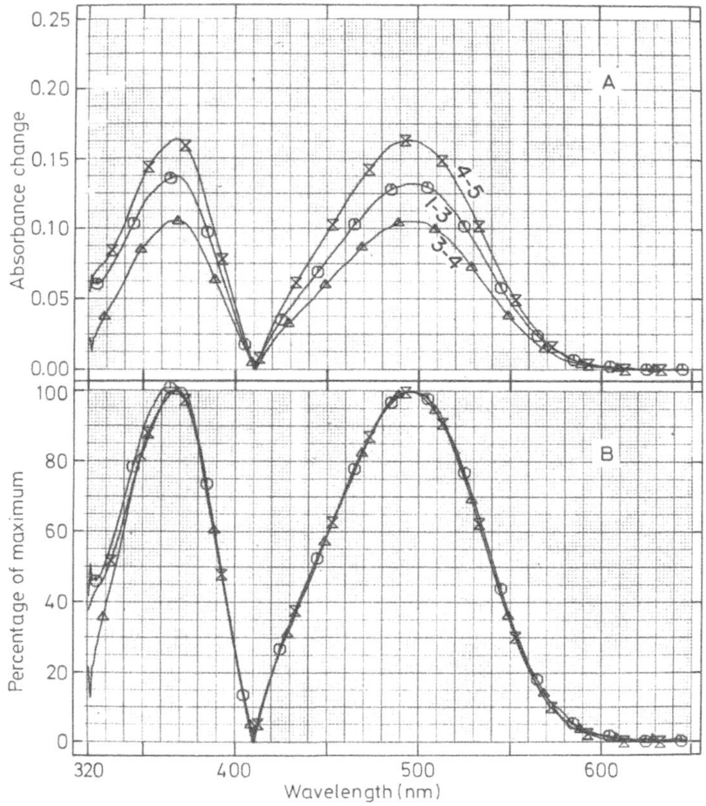

Fig. 2. (A) Difference spectra from experiment 4215. Curve 1–3, the cumulative result of the first two exposures to light: deep-red (λ675 nm) for 15 min and orange (λ610 nm) for 20 s; Curve 3–4, the result of 35 s further exposure to orange light; Curve *4–5*, the result of exposure to green light (λ560 nm) for 6 min. The symbols serve only to identify the curves; the data points occur at intervals of 1 nm. Absolute values are plotted; therefore, both the bleaching of the pigment and the bleaching product are positive. (B) The same difference spectra as in (A), each rescaled to 100% at the wavelength of maximal bleaching

deep-living species (20–200 m). When the λ_{max} values of their scotopic pigments are segregated in this way, there is a general correlation with depth (Fig. 3). The probable explanation for this relationship is discussed below in connection with the photic environment. Apparent exceptions are the parrotfishes (Scaridae), which live at intermediate depths but have a visual pigment expected in deeper-living species (see Section B. III. 4 for explanation) and a surface-limited flying fish (*Cypselurus simus*), which has the most blue-sensitive rhodopsin of all ($P475_1$). As described in Section C. V. 4, deep-swimming pelagic fishes are major predators on fishes at the surface, and they strike upward from a dimly lit and deep-blue background (Section C. IV). Perhaps its P475 helps *Cypselurus* to evade such predators. Recently MUNTZ (1976) has shown a similar correlation between the

Table 1. Spectral distribution of rhodopsins in genera of fishes in which five or more species have been analyzed. Data from LYTHGOE (1972b) and MUNZ and MCFARLAND (1973)

Family and genus	Number of species	Range in λ_{max} (nm)
Salmonidae		
Coregonus	6	509–520
Oncorhynchus	5	all 503
Salvelinus	6	503–512
Gadidae		
Gadus	5	493–500
Holocentridae		
Holocentrus[a]	9	481–502
Serranidae		
Epinephalus	9	493–497
Chaetodontidae		
Chaetodon	19	490–496
Cichlidae		
Tilapia	6	499–503
Pomacentridae		
Abudefduf	5	495–497
Pomacentrus	5	492–496
Scaridae		
Scarus	8	484–485
Acanthuridae		
Acanthurus	9	491—497

[a] Includes the subgenera *Adioryx* and *Flammeo*.

visual pigments of freshwater cichlid fishes and habitat depth in Lake Malawi. Some of these species have porphyropsins, and the extent of specific differences in the opsins has not been examined fully.

3. Patterns of Daily Activity

As with other animals, the feeding activity of most fishes is concentrated either during the day or at night. Components of the visual system have specialized in accord with these activity patterns. In fishes that feed diurnally, the retinal cone cells are more numerous and often smaller than they are in nocturnal species (MUNZ and MCFARLAND, 1973). In addition, certain species are best described as crepuscular predators, which feed most actively during the twilight periods at dawn and dusk (HOBSON, 1968, 1972). These are intermediate between diurnal and nocturnal fishes in the number of cones. Rods, the scotopic receptor cells, are present in all types, but form a greater proportion of the total in nocturnal species. The tropical marine fishes in our survey (MUNZ and MCFARLAND, 1973) were classified as diurnal (110 species), crepuscular (17), nocturnal (41), or arhythmic (11). There was no significant correlation of the scotopic pigments with these categories. In other words, the fishes with a given depth distribution have similar rhodopsins, regardless of their activity pattern.

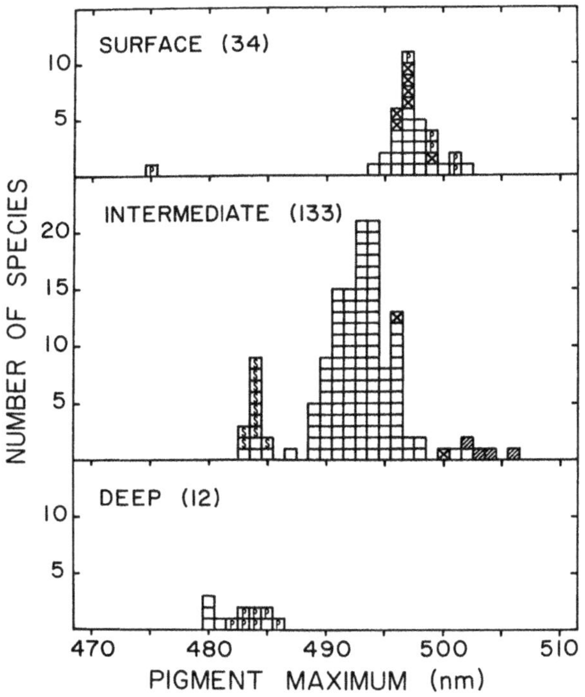

Fig. 3. Frequency distribution of the visual pigments of tropical marine fishes. The visual pigments are from species grouped by depth of habitat. Each square represents one species. Within a square, diagonal hatching signifies a porphyropsin (all others are rhodopsins); X, that the species is common in estuaries, rivers or along sandy beaches; P, an oceanic or pelagic species; and S, a scarid (parrotfish). (MUNZ and MCFARLAND, 1973)

4. Different Habitats

Several investigators have attempted to learn whether the spectral distribution of fish rhodopsins is correlated with the habitat. The impetus to this search was given by the discovery of blue-sensitive rhodopsins in deep-sea fishes, which apparently match the blue quality of transmitted sunlight and of bioluminescence (DENTON and WARREN, 1956, 1957; MUNZ, 1957, 1958a). The visual pigments of fishes from relatively clear, rocky-shore, and epipelagic habitats and from more turbid, sandy-shore, and estuarine environments were studied by MUNZ (1958c, 1964). He suggested that their rhodopsins can be correlated with habitat type and with the spectral characteristics of sunlight available in each habitat (MUNZ, 1965). His review was hampered by the relatively small size of his samples and by the inadequacy of data available on the spectral distribution of light under water. In fact, it was not really clear why, except in the deep-sea environment, the pigments of scotopic vision should be correlated with sunlight at all.

The first searching analysis of this problem was performed by LYTHGOE (1966, 1968). He considered not only the sensitivity to light reflected from an underwater target, but also the contrast between the target and the background against which it is viewed (these ideas are developed more fully in Section C.V). Using additional data on visual pigments (DARTNALL and LYTHGOE, 1965) and an equation

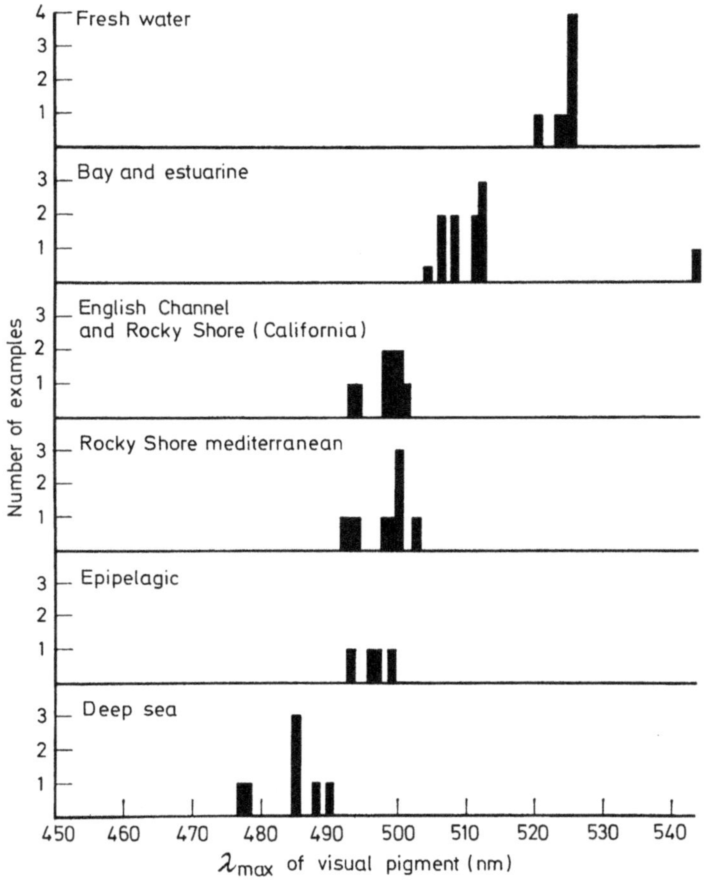

Fig. 4. Frequency distribution of the visual pigments of fishes from different habitats in the Temperate Zone. (From LYTHGOE, 1972a)

developed by DUNTLEY (1962) to quantify photocontrast, LYTHGOE concluded that, except in the deep-sea fishes, the rhodopsins have not been selected simply for maximal sensitivity to the background light available in the habitat. Rather, he suggested that there has been an evolutionary compromise between sensitivity and visual contrast. Although LYTHGOE'S exposition of the problem was penetrating, the application of his ideas was handicapped by a lack of reliable data on the spectral composition of underwater light, particularly at the times of day when scotopic vision is most important (Section B. III). In a recent review, LYTHGOE (1972a) has summarized the data indicating that the scotopic pigments of fishes are correlated with their habitats (Fig. 4). The trend appears to be as suggested earlier by MUNZ (1965): species living in clearer, bluer water generally have more blue-sensitive visual pigments. One would not, however, expect fishes from the turbid English Channel to have visual pigments as blue-sensitive as species from clearer, rocky-shore habitats. But, as already shown, deeper-dwelling fishes have pigments more blue-sensitive than surface species. Therefore, the inclusion of three cods (*Gadus* spp.) and *Capros aper* may have biased the sample from the English Channel toward shorter wavelengths.

One can accept, at least tentatively, the thesis that the scotopic pigments of Temperate Zone fishes are correlated with properties of the photic environment. Although the sample of tropical species has been much larger (MUNZ and MCFARLAND, 1973), too few have been examined from turbid water habitats to permit a similar analysis (Fig. 3). It is possible that the average λ_{max} of rhodopsins from fishes living in clear, blue, tropical waters is at a slightly shorter wavelength than that of Temperate Zone fishes from a comparable habitat; but the difference must be small (perhaps 6 nm). The visual significance of these differences will be examined in Section B. III.

Although the visual pigments of a substantial number of freshwater fishes have been identified, it is not possible to categorize most of the data by habitat. The fishes in SCHWANZARA'S (1967) extensive survey of tropical species were obtained from dealers, rather than being caught directly in their natural habitats. (In fact, the native habitats are scarcely known for some popular aquarium fishes.) This fact would not be expected to alter the λ_{max} values of either their rhodopsins or their porphyropsins, which are a function of species-specific proteins. But some species can have *paired pigments*, that is, mixtures of rhodopsin and the corresponding porphyropsin (same opsin). Only a series of wild fish, taken at different localities and seasons, could delineate these relationships (Section B. V). Nevertheless, several interesting conclusions were drawn by SCHWANZARA from her work. The average λ_{max} of porphyropsins from temperate fresh-water species was longer than the average λ_{max} of tropical species. This was also true within the large family Cyprinidae, which has both temperate and tropical members. Further, porphyropsins (either alone or mixed with rhodopsin) occurred in a greater proportion of the temperate than of the tropical species. Again, this same relationship also held within the Cyprinidae. It appeared, therefore, that fresh-water fishes from the North Temperate Zone tend to have a more red-sensitive scotopic visual system than those from the tropics. Some of the behavioral characteristics of fishes are probably not much altered by aquarium life. In SCHWANZARA'S sample, those species that scavenge at the bottom tended to have unmixed porphyropsins. Others feed at the water surface, and in these rhodopsin predominated. Thus, although there is clearly a need for data on the visual pigments of fresh-water fishes taken from different native habitats, SCHWANZARA'S work suggests that trends may exist, similar to those apparent in marine species. MUNTZ (1976) has recently found indications that the visual pigments of cichlid fishes from Malawi are correlated with spectral characteristics of the habitat. Species that dwell near the surface of Lake Malawi (clear water) are less red-sensitive than those from more turbid environments.

II. The Photic Environment

1. Twilight

Twilight is important in the scotopic vision of fishes (Section B. III). It has recently been described in detail (MCFARLAND and MUNZ, 1975a). The present treatment should give the background necessary for relating scotopic vision to

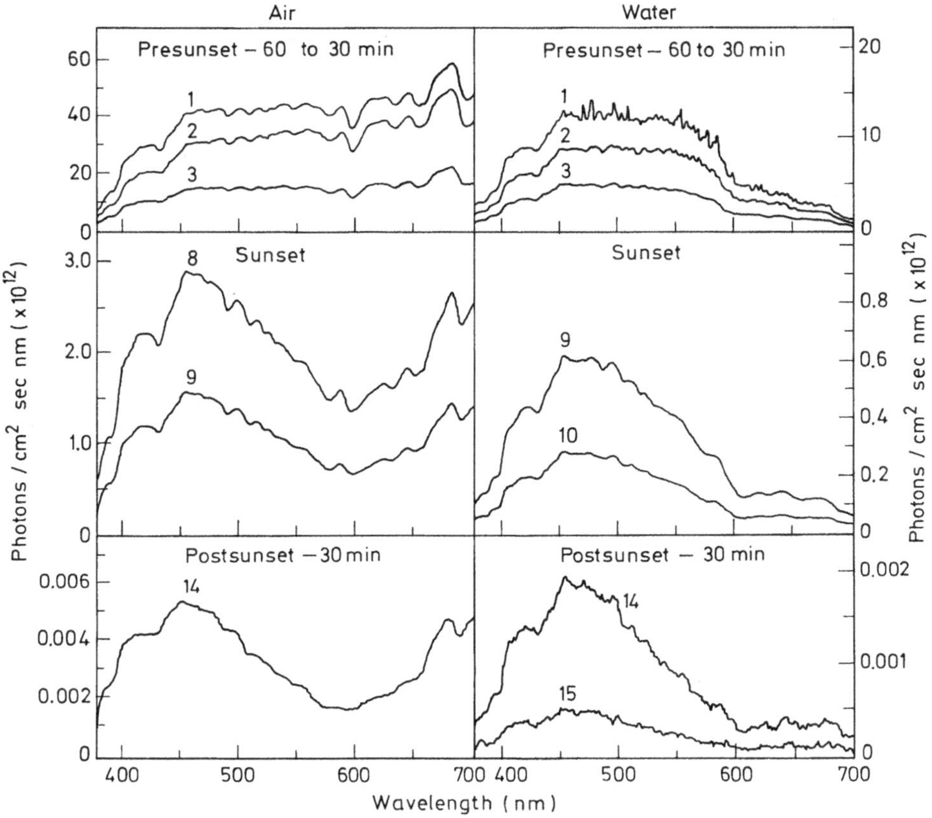

Fig. 5. Spectral changes in irradiance, as measured in air and for downwelling light at 3 m depth during sunset at Eniwetok Atoll. Actual times for each curve are: presunset, *1* = 57 min, *2* = 42 min, *3* = 26 min, *8* = 7 min, *9* = 2 min; postsunset, *10* = 3 min, *14* = 34 min, *15* = 39 min. Sequence recorded on 23 August 1970. Sunset occurred at 1828 h. (MUNZ and MCFARLAND, 1973)

twilight. Two essential features characterize twilight, but only one of them is obvious. There are, of course, rapid changes in the level of illumination (approximately 4 to 6 log units; see Fig. 10). Much less familiar is an alteration in the spectral distribution of light near sunrise and sunset.

The twilight spectrum is distinguished by a decrease in the yellow-to-orange region. This fact had been recognized earlier, at least to some extent (e.g., PITT and SELWYN, 1938; ROZENBERG, 1966). A particularly convincing demonstration was obtained at Eniwetok Atoll by MUNZ and MCFARLAND (1973), who recorded irradiance spectra in air and under water simultaneously, with identical spectroradiometers (Fig. 5). The qualitative differences between the two sets of curves are due to the attenuation of red light by water. As they were throughout the day, the presunset spectra were initially broad and relatively flat. [In order that they might be meaningful in a visual (photochemical) context, our measurements were made as numbers of photons, rather than in energy units.] Between 15 and 20 min before sunset, there was a greater decrease in the region from 550 to 650 nm than

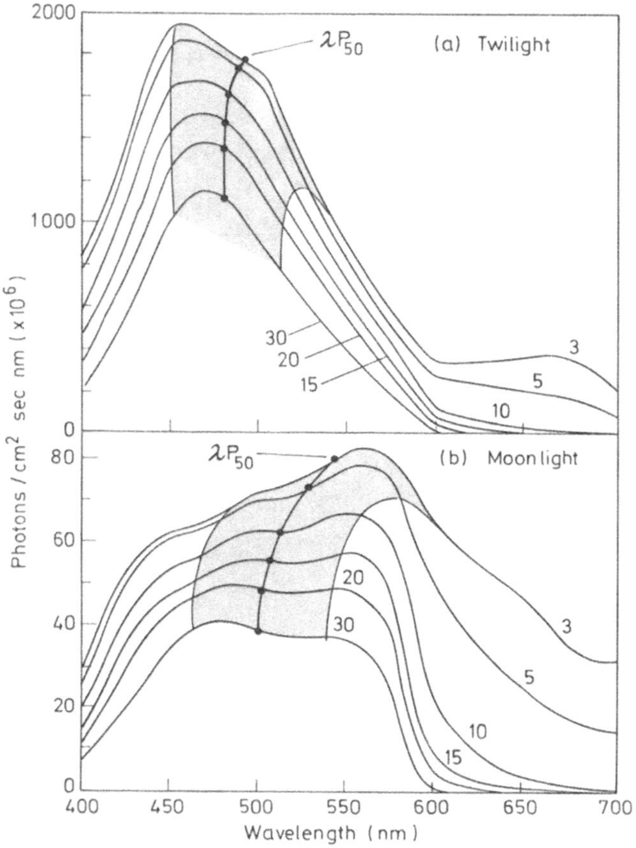

Fig. 6. Spectral irradiance of downwelling light during twilight and moonlight at various depths. Numbers on curves represent depths in m beneath the surface. The curves for 3 m were empirically derived (see Figs. 5 and 8). Curves for greater depths were calculated from our transmission values for water at Eniwetok. The closed circles represent the median wavelength (λP_{50}) in the spectral distribution (between 400 and 700 nm). The breadth of the semiquartile range is indicated by shading (MUNZ and MCFARLAND, 1973)

at other wavelengths. At 3 m depth under water, however, red light was so completely absorbed that the spectrum became narrower (more homochromatic) and appeared to shift to shorter wavelengths (Curve 9). This effect persisted throughout the twilight period, so long as there was enough light for the spectroradiometer to measure. To express this change numerically, we defined λP_{50}, the median wavelength in the spectral distribution of visible light (400–700 nm). (This is analogous to P_{50} in a cumulative frequency distribution. Since the measurements are in quantal units, λP_{50} is not directly related to energy.) Under water, λP_{50} shifted from 515 nm 1 h before sunset (Curve *1*), through 499 nm at 2 min before sunset (Curve 9), to 487 nm at 34 min after sunset (Curve *14*). The breadth of the spectrum was expressed by the semiquartile range (λP_{75}–λP_{25}); this decreased from 102 nm (Curve *1*) to 92 nm (Curve *14*) over the same time period. Generally similar changes in the reverse direction were measured at sunrise.

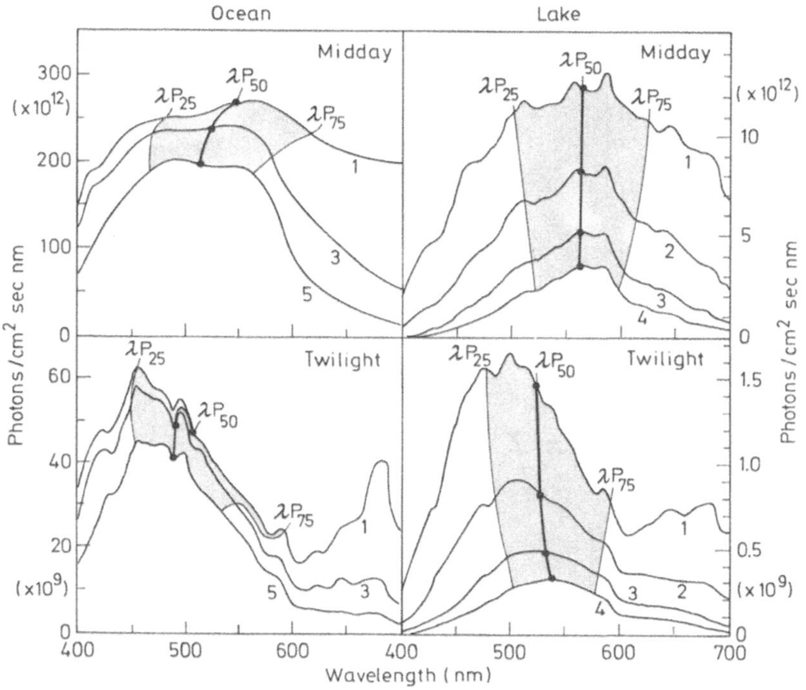

Fig. 7. A comparison between downwelling spectral irradiance in a clear, tropical sea (Eniwetok) and Silver Lake, Michigan, during midday and twilight. Numbers on curves represent depths in m beneath the surface. The closed circles and heavy line indicate λP_{50} and the shading shows the semiquartile range (λP_{75}–λP_{25}), as it changes with increasing depth (Mc Farland and Munz, 1975a)

These data were obtained in clear oceanic water. From the spectral transmission values that we observed, it was possible to calculate the effect of increasing depth on the twilight spectrum (at about the middle of the "Quiet Period;" see Section B. III. 1). Both the λP_{50} and semiquartile range are shown in Fig. 6A. At greater depths, the downwelling spectrum becomes more homochromatic and is shifted toward the blue. λP_{50} shifts from 490 nm at 3 m depth to 481 nm at 30 m.

These effects at twilight are of very general occurrence (Halldal, 1974; McFarland and Munz, 1975a) although their magnitude depends on water clarity, cloud cover, latitude, etc. One example of twilight measurements in fresh water is available. Irradiance spectra were measured in Silver Lake, Michigan (Fig. 7). The water of this moderately clear lake is "greenish" and the λP_{50} during twilight was 524 nm at a depth of 1 m. At greater depths, λP_{50} shifts to *longer* wavelengths (540 nm at 4 m). Much more data collection is needed in different habitats.

2. Moonlight

We were also able to measure the spectral distribution of moonlight at Eniwetok. Both in air and under water (3 m depth), the spectrum of downwelling irradi-

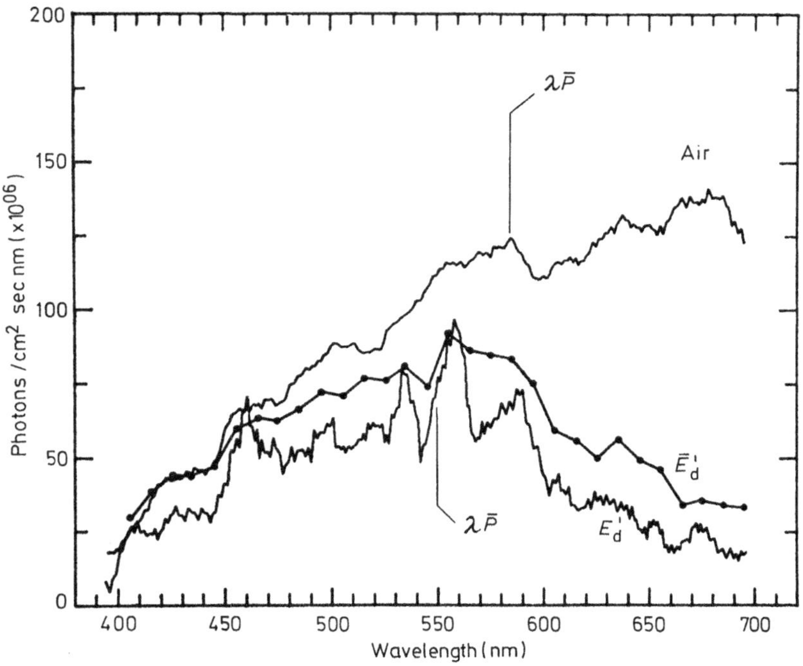

Fig. 8. The irradiance spectrum of moonlight in air and at 3 m depth. E_d' is a single curve, smoothed to minimize surface flicker; $\bar{E}_d'$ is the mean of three measured curves. $\lambda\bar{P}$ has the same meaning as λP_{50}. Peaks in the curves result from surface chop. Recorded at Eniwetok on 20 August 1970; moon at 15° from zenith; time 0200–0250 h (MUNZ and FARLAND, 1973)

ance was relatively broad (Fig. 8). In air, moonlight was somewhat "redder" (λP_{50} = 585 nm) than the midday sun (λP_{50} = 570 nm). Under water, the λP_{50} of moonlight was 548 nm, compared to 542 nm for sunlight. The total irradiance of moonlight was 3.0×10^{10} photons/cm$^2 \cdot$ s (400–700 nm) in air and 1.4×10^{10} photons at 3 m. From the spectral transmission values of the water, the effect of increasing depth was calculated (Fig. 6B). Under water, moonlight has a broader spectrum than twilight, and its λP_{50} values are at longer wavelengths. Partial spectra of moonlight have also been measured under water in the Atlantic Ocean by KAMPA (1970).

3. "Starlight"—the Night Sky

While studying scotopic vision, we tried to obtain the spectrum of the moonless night sky. Surprisingly little information is available on the irradiance of the night sky. "Starlight" is a misnomer for this light, because zodiacal light from the earth's atmosphere makes an important contribution to the total (ALLEN, 1973). From published sources (ALLEN, 1955; BABCOCK and JOHNSON, 1941), however, we compiled an approximate spectrum (Fig. 1 in MUNZ and MCFARLAND, 1973).

This was similar to the spectrum of RICHARDS (1969): both showed a predominance of red light.

Our confidence was not great in any of these spectra, and we eventually undertook to make our own measurements. These were made in the early morning of 23 October 1974, under a clear, moonless sky. The site was at Waldo Lake, Oregon, where the elevation is 1650 m. This is near the crest of the Cascade Range and is 90 km from any significant source of interfering light (Eugene-Springfield). The Gamma Model 3000R spectroradiometer was powered by a portable generator. This instrument normally measures irradiance, and its receptor is a circular piece of translucent quartz with a diameter of 3.25 cm. Behind this is a rectangular slit 11 × 0.75 mm, with two color filters attached. These help to flatten the spectral response curve of the instrument. First, the built-in lamp was used to calibrate the spectroradiometer in its normal configuration. The photomultiplier voltage was adjusted to give the greatest sensitivity that could be calibrated. When the night sky was scanned, however, only a noisy base line was obtained.

To increase sensitivity, we removed the translucent quartz receptor. Now the night sky showed a prominent, narrow band of light at 558 nm (the well-known auroral green line), superimposed on a less intense but noisy background spectrum. A further increase in sensitivity was obtained by removing the 0.75 mm wide entrance slit (with its attached filters) and replacing it with a 1.50 mm slit (without filters). Scanning at the slowest speed, a usable spectrum was then obtained (Fig. 9A). This spectrum was recorded at 0430 h, 3 h after moonset and $2^1/_2$ h before sunrise. A duplicate spectrum (not shown) was then recorded, which closely matched the first.

These measurements show both the auroral green line and a second, lower peak at 590 nm, which had been reported by BABCOCK and JOHNSON (1941). [The starlight spectrum of RICHARDS (1969) has less detail.] But the expected increase in red light was apparently not measured. Of course, removal of the quartz receptor and color filters and change in the slit altered the calibration. In particular, removal of the diffuser and exposure of the naked entry slit makes the instrument extremely directional in its response. The night sky (both the stars and zodiacal light) is, however, a greatly dispersed light source. It occurred to us that under a heavy cloud cover the daytime sky might share this property to an extent permitting calibration. This was confirmed by direct measurement on 30 October 1974, with a different receptor that collects light only from a 0.008 steradian cone (plane angle 6°). The calibration was performed by measuring six pairs of diffuse daylight spectra (on 29 and 30 October 1974), with the spectroradiometer in its normal configuration (3.25 cm irradiance receptor, narrow slit, and color filters) and with the naked, wider slit. The photomultiplier voltage was decreased as appropriate for the measurement of daylight, and the required difference in sensitivity for the two configurations (i.e., 100 ×) was within the scope of the instrument. The ratio of response in the two configurations was calculated at 5 nm intervals for each pair of spectra. Then the averages of these six sets of values were obtained. From these values, the average starlight spectrum was corrected to approximate spectral photon irradiance (Fig. 9B). Although the intensity of red light is greater than blue, the difference is not as large as previously observed; incidentally, the third peak at 630 nm had also been measured by BABCOCK and

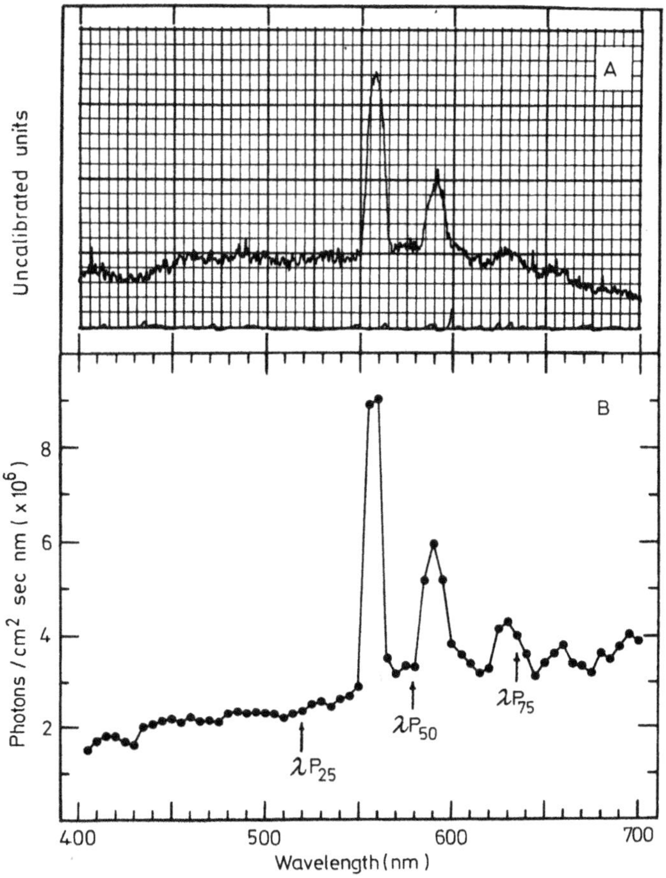

Fig. 9A and B. Spectral distribution of a moonless night sky. (A) Actual record, in uncalibrated units. This is not irradiance; for maximal sensitivity, the cosine receptor was removed from the spectroradiometer. (B) Same spectrum in calibrated units. λP_{50} and semiquartile range are indicated. Recorded at Waldo Lake, Oregon, on 23 October 1974, at 0430 h; elevation 1650 m

JOHNSON (1941). The spectrum is less regular but as broad as that of moonlight (Fig. 8), and its λP_{50} value (580 nm) is almost the same. The total irradiance was 9.3×10^8 photons/cm^2 · s (410–700 nm), about 1/30 the intensity of moonlight in air. We were disturbed, however, when we noticed that the spectral irradiance we measured was about 10 times greater than what we had calculated earlier from ALLEN's (1955) data (see Fig. 1 in MUNZ and MCFARLAND, 1973). Rechecking the old calculations revealed a decimal error in them, so that our new measurements are concordant with the old. Thus, we believe that the new spectrum of the night sky is an improvement over those from other sources; certainly, the sky itself must have been close to ideal. We realize, however, that to obtain this spectrum our spectroradiometer was pushed beyond the limits for which it was designed and that our compensatory calibration is approximate.

III. Function of the Scotopic System

1. Predation: The Quiet Period

There can be little doubt that predation is a potent selective force or that the evolution of scotopic vision has been profoundly affected by it. In a study of tropical marine fishes (MUNZ and MCFARLAND, 1973), our attention gradually became concentrated on the events that are associated with twilight. The impressive investigations of HOBSON (1965, 1968, 1972, 1974) have demonstrated that the short periods near sunrise and sunset are crucial to the survival of many fishes. In reef areas, predation by large piscivorous fishes is heavily concentrated during dawn and dusk. This fact was confirmed by examination of their gut contents and by direct observation. During the day, on the other hand, predators are often in close proximity to their potential prey, which are seemingly unconcerned (HOBSON, 1968). Unless a crippled individual attracts its attention, the predator may spend hours without making aggressive movements.

The most interesting feature of the activity of coral-reef fishes during twilight is what HOBSON (1972) called the Interim or "Quiet" Period. Like other fishes, most reef species have specialized patterns of feeding behavior and are either diurnal or nocturnal. At dusk the diurnal species take cover in the reef in a regular sequence, which has been carefully documented in Hawaii by HOBSON (1972). This orderly evacuation of the water column over the reef requires about 20 min, so that in shallow water few diurnal fishes are left by 10–15 min after sunset. Nocturnal fishes do not leave their daytime hiding places during the same interval, however, but emerge into the water column rather suddenly, about $^1/_2$ h after sunset. Thus, there is an astonishing decrease in the level of activity at dusk, culminating in a Quiet Period of about 20 min. During this time few fishes except stragglers and crepuscular predators are in the water column. And then, very abruptly, the water seems full of fishes once more; but now they are large-eyed, nocturnal forms. The same sequence occurs, in reverse order, at sunrise. COLLETTE and TALBOT (1972) have described very similar events in the Caribbean, as have DOMM and DOMM (1973) on the Great Barrier Reef. Fish predation on the reef reaches a maximum just before, and during, the Quiet Period (HOBSON, 1968, 1972; COLLETTE and TALBOT, 1972). Indeed, HOBSON (1972) suggested that "this well-ordered series of events is shaped by the threat of crepuscular predators."

Both HOBSON (1972) and COLLETTE and TALBOT (1972) related the changes in fish behavior at twilight to the light intensity. Thus, although the sequence of events was relatively constant, its timing with respect to sunset (or sunrise) was dependent on the cloud cover and water clarity, both of which affect light levels under water. Of course, the intensity changes rapidly during twilight; at Eniwetok Atoll we found (MUNZ and MCFARLAND, 1973) that it decreased 1000-fold in the 30 min after sunset (Fig. 10). During this time, the light level falls from what is adequate for photopic vision to values that require scotopic vision. It is no coincidence that the Quiet Period occurs then, because this transition must be even more awkward for bony fishes than it is for us. The processes of light and dark adaptation, particularly the photomechanical movements (Section A), are slow, needing up to more than 1 h for completion (ALI, 1971). Herein, we think,

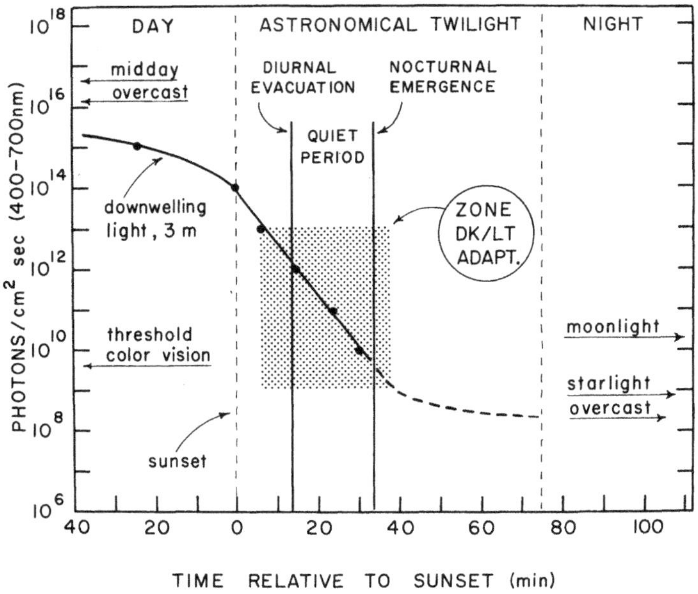

Fig. 10. Correlation between twilight light levels, retinal dark adaptation and the behavior of coral-reef fishes as night descends. Light intensities are the total photons between 400 and 700 nm, obtained from spectral irradiance measurements for downwelling light at 3 m beneath the surface. Values were obtained in the lagoon at Eniwetok Atoll on 23 August 1970, in very clear water. The other indicated values are also for downwelling light at 3 m beneath the surface. The midday (clear sky), overcast (day) and moonlight values were measured at Eniwetok. Starlight (clear sky) was measured in Oregon (Fig. 9), and the overcast value was calculated on the assumption that the percentage reduction in light would be similar to actual reductions measured during the daytime. The value for the threshold of color vision was recalculated from data given by CLARKE and DENTON (1962). The upper and lower thresholds for dark and light adaptation (stippled zone) were recalculated from data in BLAXTER (1970). Note the coincidence between actual light intensities, fish behavior and, retinal dark adaptation (MUNZ and MCFARLAND, 1973)

lies the explanation of the Quiet Period. Predators gain so great a visual advantage at this time that their prey, whether diurnal or nocturnal, dare not risk being away from cover even though planckters rapidly become more abundant in the water column during dusk (HOBSON, personal communication). In order to understand how the scotopic system functions, therefore, we must examine it in relation to the photic environment when it is being used.

2. Sensitivity or Contrast?

According to LYTHGOE (1966, 1968, 1972a), the rod pigments of fishes represent an evolutionary compromise between the requirements of sensitivity and visual contrast (Section B.I.4). Evaluation of his ideas requires reliable measurements of the photic environment when light levels are low enough for the scotopic system to function; unfortunately, this information was not then available, except for the deep-sea environment (KAMPA, 1961). It is generally agreed that the blue-sensitive pigments of deep-sea fishes maximize sensitivity to the blue sunlight that is transmitted down the water column. Marine bioluminescence is also generally blue or blue-green (NICOL, 1969). Whatever its source, therefore, background illumination in the deep sea is homochromatic. Targets seen as darker silhouettes against this background will be detected most readily by a blue-sensitive visual system. As LYTHGOE (1966) made clear, in this case both maximal sensitivity and greatest contrast are served by a visual pigment that matches blue light. This would also be true in viewing a luminescent blue light source.

Many aquatic predators, both bottom-dwelling and pelagic, gain a visual advantage over their prey by attacking from below. In this situation the prey is silhouetted against the brighter background of downwelling light, and the predator is difficult to see against the darker background of upwelling light. Countershading is an evolutionary countermeasure that has been adopted almost universally by diurnal pelagic fishes (Section C.IV.3). Its analogue for pelagic animals in the deep sea might be called "counterlighting." CLARKE (1963) pointed out that most of the photophores in these organisms are located beneath the pigmented parts of the body, with their light directed downward. This would tend to break up the animal's silhouette, as viewed from below. Other possible functions of bioluminescence have been discussed by NICOL (1969) and MARSHALL (1970).

A fascinating adaptation in stomiatoid dragonfishes has recently been described (in *Pachystomias* spp.) by DENTON et al. (1970) and (in *Aristostomias scintillans*) by O'DAY and FERNANDEZ (1974). Many species in this group possess large light organs near the orbit. The suborbital organ of both *Pachystomias* and *Aristostomias* has been observed to emit red light. This light is presumably directed forward and can be turned on and off (MARSHALL, 1970). Both genera have a red-sensitive visual pigment; that of *Aristostomias* has been identified as $P551_2$ (although this fish almost certainly has a pure-rod retina, O'DAY and FERNANDEZ also found a second, green-sensitive pigment, $P526_1$). Producing and sensing red light in a community specialized for existence in a blue environment could give these fishes private channels of intraspecific communication. Further, these stomiatoids could locate red prey (many deep-sea invertebrates are red; MARSHALL, 1954) without alerting them to danger. Here, then, seems to be a natural counterpart of the infrared "snooperscope!"

Relationships between the scotopic vision of fishes from other habitats and the photic environment have been more equivocal (LYTHGOE, 1966) until recently. In an extensive survey of visual pigments in tropical marine fishes (MUNZ and MCFARLAND, 1973), we showed that the rhodopsins of most shallow-water species are rather tightly grouped, with λ_{max} values at 489–499 nm (Section B.I). At Eniwetok we measured the intensity and spectral distribution of the downwelling sunlight and moonlight at a depth of 3 m (Figs. 7 and 8). Moonlight is only slightly more "reddish" than sunlight, and its spectrum is also relatively broad and flat (Section B.II.2). At shallow depths, therefore, moonlight would be "matched" by any visual pigment with its λ_{max} in the range from about 475 to 575 nm. The rhodopsins of reef fishes, however, are not scattered at random across the blue to yellow portions of the spectrum. On the other hand, we found that at twilight the underwater spectrum is more homochromatic than daylight, and it is shifted to shorter wavelengths (Fig. 5). There is a close match between the absorbance maxima of the scotopic pigments and the underwater spectrum during twilight periods. At greater depths, both the twilight spectrum and the scotopic pigments are shifted further to the blue (Figs. 6A and 3). This emphasis on the relationships between spectral sensitivity and the photic environment at twilight is consistent with HOBSON's (1968, 1972) demonstration that this is a critical period with respect to predation by fishes. The fact that the rod pigments match the underwater spectrum implies that the scotopic system is used primarily for the detection of dark objects against a lighter background (see Section C.V for a discussion of other possibilities). This fits our own visual impressions when diving at twilight. Just after sunset, when colors were fading rapidly, most reef fishes at 4–5 m depth were seen as dark silhouettes against a brighter, "greyish" background. In this situation, a rhodopsin matched to the available light would give both maximum sensitivity and the greatest visual contrast.

In summary, the short intervals of morning and evening twilight are critical in the survival of fishes living near coral reefs. Both their visual systems and their behavior help them to escape predation at these times. The spectral sensitivity of scotopic vision has been selected primarily to give maximal sensitivity to twilight. This is consistent with the nature of twilight vision under water, in that fishes are generally distinguished as darker than the background. Thus, it appears that the needs of both sensitivity and contrast have acted together to select rhodopsins that match the photic environment at twilight.

If our analysis of coral-reef fishes is attractive, can it also be applied in other shallow-water habitats? In no other case has the sampling of visual pigments been so extensive, but there are indications of correlation between pigment and habitat (Section B. I. 4). Still less reliable information is available on underwater photic environments, especially twilight and moonlight. It is probably safe, however, to conclude that a narrowing of the underwater spectrum and a shift to shorter wavelengths are general features of twilight in a wide variety of marine habitats (MCFARLAND and MUNZ, 1975a). The porphyropsins of many fresh-water fishes are generally more sensitive to green light than the rhodopsins of marine animals. Our measurements in a "greenish," fresh-water lake confirm that the twilight spectrum is not as blue as in clear, tropical seas and show that increasing depths can shift the twilight spectrum to longer wavelengths (Fig. 7). SCHWANZARA's (1967) conclusion that bottom-dwelling fresh-water fishes are more likely than surface-limited species to have porphyropsin may be a result of this shift. Although incomplete, this information and what is known about the visual pigments of fishes from other habitats (Section B. I. 4) are in harmony with our analysis.

We must also ask whether twilight periods are as critical for fishes in other habitats as they are near a coral reef. HOBSON's (1968) study of predator-prey relationships in a mixed rock-and-sand shore area of the Gulf of California showed that twilight is, indeed, critical in that habitat. Aggressive behavior of predators (especially the grouper, *Mycteroperca rosacea*) and schooling behavior of potential prey (the flatiron herring, *Harengula thrissina*) are both controlled by the light level. At twilight, the grouper rushes up from the bottom to strike its prey. The visual advantage in this encounter heavily favors the predator, but only when light levels are low. Similar intensive studies in other habitats have not yet been published. Although our experiences are not documented so convincingly as HOBSON'S, those of us who are fishermen can affirm that many fishes (and of those we seek many are predatory on other fishes) feed most actively in the early morning and the evening. And it is then that a dry fly, which must be seen as a dark object in a brighter background, is most effective.

3. Visual Advantages during Twilight

HOBSON (1968, 1972) showed that piscivorous fishes gain the advantage during twilight. We should consider the nature of this advantage and of the countermeasures taken by prey species. First, it should be stated that the scotopic pigments of the predators are similar to those of both diurnal and nocturnal prey (MUNZ and MCFARLAND, 1973). Many predators silhouette their prey and strike from below. The milling behavior of schooling prey fishes makes it more difficult for these (and other) predators to single out a target for attack (HOBSON, 1968). As evening twilight comes on, the diurnal prey fishes gradually approach the coral reef (HOBSON, 1972). Some, such as surgeonfishes (Acanthuridae), mill near the bottom in small schools. FRICKE (1973) has shown that a butterfly fish, *Chaetodon melanotus*, will flee from attack more readily at low light levels than during the day. Of the more solitary fishes, the small individuals are closest to the substrate—this is true both of the individuals within a species and of different species. This stratification by size is probably related to the generally greater swimming speed of larger fishes (FRY, 1971). A small fish is relatively safe for only a short distance from cover, and this distance would decrease as the light level falls. There may also be differences in visual ability, related to eye size; but these cannot be evaluated without more anatomic and physiologic information. It may seem odd that among diurnal fishes only the wrasses (Labridae) leave the water column as early as sunset (HOBSON, 1972). Why do the others expose themselves to such crepuscular predators as jacks (Carangidae) and some basses (Serranidae) as long as they do? Perhaps the diurnal fishes must first dark-adapt enough to see other predators, such as eels, which lurk within the caves of the reef. (This leaves unanswered the counterquestion of why the labrids are able to retire earlier than the others.)

Although the reef fishes have retinal rods with similar scotopic pigments, there are morphologic differences in their cones. Many diurnal fishes are specialized to feed on small objects such as coral polyps or other invertebrates (HOBSON, 1974). This demands good visual acuity, and the cone densities of these fishes are as high as $100000/mm^2$ (MUNZ and MCFARLAND, 1973). Even the diurnal herbivores (surgeonfishes) have densities of at least $40000/mm^2$. Their cones are much smaller than those few (3000–$9000/mm^2$) possessed by the nocturnal squirrelfishes (Holocentridae) and cardinalfishes (Apogonidae). Two crepuscular feeders, *Carangoides ajax* and *Epinephalus merra*, have intermediate cone densities (20000–$30000/mm^2$), but the individual cones are so large that together they may trap as many quanta as the more numerous, small cones of diurnal species. At light levels where the photopic vision of diurnal fishes begins to fail, the larger cones of crepuscular predators may still trap enough quanta to detect the location and motion of their targets. Thus, although their photopic acuity is never great, this system makes them dangerous at twilight. (As a partial compensation, diurnal species should enjoy a definite visual advantage during the day.) It would be interesting to know whether the photopic pigments of crepuscular predators are also specialized, but our information on cone pigments is limited to diurnal species (Section C).

4. Vision in Moonlight

However critical the twilight periods are for coral-reef fishes, most of them may also be active at night. This is true not only of nocturnal species but also, on bright moonlight nights, of many diurnal fishes (HOBSON, 1965 and personal communication). Butterflyfishes (Chaetodontidae), damselfishes (Pomacentridae), and surgeonfishes (Acanthuridae) reemerge into the water column, but stay closer to cover than during the day. On dark nights, they do not reappear over the reef. Since the moon provides light during part of more than half the nights in a year, it is an important component in the photic environment of both nocturnal and diurnal species.

Although the rhodopsins of reef fishes absorb twilight effectively, the match is not perfect. The λ_{max} values for the scotopic pigments of most species are in the range from 489 to 499 nm (Section B.I), but the underwater twilight spectrum (Fig. 6A) indicates that λ_{max} values from 470 to 485 nm would give the best match. Moonlight is less homochromatic under water and is centered at longer wavelengths than twilight (Fig. 6B). For most depths near the coral reef, a rhodopsin with its λ_{max} in the range between 490 and 515 nm would be most effective for moonlight vision. Thus, it seems likely that the rhodopsins of both diurnal and nocturnal reef fishes represent an evolutionary compromise between the somewhat different requirements of vision in twilight and moonlight. ("Starlight" is spectrally similar to moonlight, but less intense—Section B.II.3.) In other words, these pigments appear to give the best match overall to the major photic environments in which they are used. An apparent exception to this rule is instructive. As already noted, the parrotfishes (Scaridae) have a more blue-sensitive rhodopsin ($P484_1$) than most other species from intermediate depths (Fig. 3). Although they remain active well into the twilight period, parrotfishes do not venture out from cover at night. In fact, some individuals even secrete a mucoid "sleeping bag" in which they spend the night (WINN, 1955; WINN and BARDACH, 1960). Accordingly, there would be little reason for their scotopic pigment to match moonlight. The wrasses, in the related family Labridae, are still more exclusively diurnal, being seen seldom even during twilight (HOBSON, 1968, 1972). First to "bed" in the evening and last to "rise" in the morning, the functional significance of their rod pigments is in doubt. Perhaps it should not be surprising that wrasses possess a variety of visual pigments with no apparent relationship to either twilight or moonlight (MUNZ and MCFARLAND, 1973).

It appears, therefore, that the scotopic system of most reef fishes has been tuned, in an evolutionary sense, primarily to twilight. For most species moonlight is also important enough to have affected selection of the scotopic pigment. The photopic system of crepuscular predators has also become morphologically well-suited for vision in twilight. It would be interesting to know whether their cone pigments have specialized for crepuscular vision. If so, they would probably be the same as the rod pigments of these species.

IV. Significance of the Purkinje Shift

More than 150 years ago, PURKINJE (1825) discovered that the apparent brightness of red and blue objects shifted in accordance with the general level of light intensity. At dawn, blue objects appeared brighter than red although they might by equally bright during the day. This phenomenon was one of the fundamental facts incorporated into the "duplicity theory" of vertebrate vision as it was formulated by several authors in the late 1800's. Although its cause, a difference in the spectral sensitivity functions of the scotopic and photopic systems, has long been understood, the biologic significance of the Purkinje shift remained obscure (e.g., BLAXTER, 1970; LYTHGOE, 1972a). The problem was that moonlight is even "redder" than sunlight. Therefore, it seemed paradoxical that the maximal sensitivity of scotopic vision should be shifted toward the blue. This anomaly was clearly recognized by LE GRAND (1957). Largely overlooked has been LE GRAND'S suspicion that twilight might be a spectrally unique period, hinted at by his reference to PITT and SELWYN'S (1938) photographic demonstration of increased "blueness" during twilight. These authors, being interested in photography, dismissed the general significance of their results by stating, "This effect, occurring so near to the time of sunset, is a matter rather of interest than significance." The Purkinje shift is widely distributed among vertebrates with duplex retinae (in fishes, KOBAYASHI, 1962) and it has needed a convincing explanation. So far as we know, the tunas are the only vertebrates with substantial numbers of both rods and cones in which a Purkinje shift has not been shown to occur (see Section C.V.4 for a description and explanation).

It should be evident that if our interpretation of scotopic vision of coral-reef fishes is correct (Section B.III), it provides a logical and natural explanation for the Purkinje shift. Under water, twilight is, indeed, more homochromatic than sunlight and shifted to shorter wavelengths. If predatory activity at twilight has been a potent selective force as HOBSON (1968, 1972) believes, then it is reasonable that the scotopic system has evolved in relation to twilight. We have also shown that the twilight spectrum undergoes similar changes in temperate latitudes and in other water types, including fresh water (Section B.II; MUNZ and MCFARLAND, 1973; MCFARLAND and MUNZ, 1975a). If heavy predation at twilight is a common feature in these different habitats as we believe (Section B.III.2), this can account for the Purkinje shift more generally. And if allowance is made for the "greenish" nature of many fresh waters (Fig. 7), it also applies to the porphyropsins of fresh-water fishes. We have, in fact, extended very similar arguments to the scotopic vision of terrestrial vertebrates, including humans (MCFARLAND and MUNZ, 1975a). In air, the wavelengths containing yellow and orange light characteristically show a relative reduction in photons during twilight (Fig. 5). Thus, both blue and deep-red light are most available to terrestrial vertebrates. So far as we know, a scotopic visual pigment maximally sensitive to deep-red light has never been detected. Apparently shorter wavelengths have been preferred—at least, all mammals and birds have yielded rhodopsins with λ_{max} values clustered close to 500 nm. This would provide for maximum photosensitivity during twilight. The possibility of a study of fishes on a distant coral atoll in answering a basic visual question of long standing should perhaps be emphasized.

V. Lability in the Scotopic System

1. Changes in the Proportions of Rhodopsin and Porphyropsin

It has been recognized for about 20 years that, in fishes and amphibians possessing both rhodopsin and porphyropsin, the proportions of these pigments may change during the lifetime of the individual. This field has recently been reviewed in detail (BRIDGES, 1972; CRESCITELLI, 1972; BEATTY, 1975b). Therefore, we shall limit our summary of the evidence to bony fishes and then discuss possible functions of lability in the scotopic system.

A labile scotopic system was first discovered by accident in the rudd (*Scardinius erythrophthalmus*), a cyprinid fish (DARTNALL et al., 1961). Although this species is restricted to fresh water, both rhodopsin and porphyropsin are present in retinal extracts ($P507_1$ and $P535_2$ according to later determinations by BRIDGES and YOSHIKAMI, 1970). The proportions of these two pigments are not fixed within the individual, but vary seasonally and may be altered experimentally in the laboratory. Under natural conditions, rhodopsin attains its greatest percentage in summer and porphyropsin in winter. The proportion of porphyropsin can be increased by keeping the fish in darkness; in brightly lit outdoor tanks the proportion of rhodopsin increases. These changes are readily reversible and occur over a few days or weeks (Fig. 11). Other parameters, such as diet and photoperiod, were also implicated in the control of rhodopsin/porphyropsin balance. This pioneering work by DARTNALL et al. has been extended by other investigators to several additional species. Various extrinsic and intrinsic factors have been shown capable of affecting the regulation of the scotopic pigments. Unfortunately, it is still not possible to state how these factors interact in any species. Furthermore, it is now apparent that the regulatory processes are not the same in all the fishes that have been studied.

In other fishes with paired pigments, seasonal changes in the proportions of rhodopsin and porphyropsin are generally similar to those observed in the rudd (ALLEN et al., 1973). Another cyprinid (the golden shiner, *Notemigonus crysoleucas*, with $P503_1$ and $P529_2$) and a poeciliid (*Belonesox belizanus*, with $P498_1$ and $P522_2$) have more porphyropsin in winter than in summer (BRIDGES, 1964a, b, 1965b). Among individual *Notemigonus* collected from different habitats in the same region, those from clearer water consistently had less porphyropsin than those from a more turbid environment (BRIDGES, 1964a). This effect was superim-

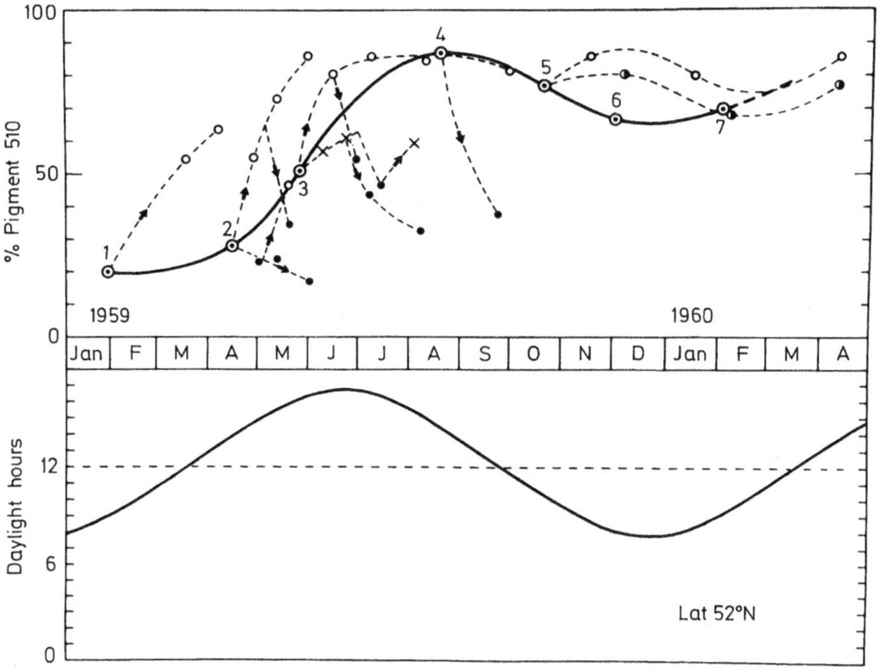

Fig. 11. (Upper) Changes in proportions of porphyropsin and rhodopsin (expressed as percentage P510, the rhodopsin) in retinal extracts of rudd. *Bull's-eye circles*, seven samples taken from a population living in a pond, during 1959–1960. Additional individuals from the same population were subjected to experimental treatment. *Open circles*, samples from rudd kept in brightly illuminated outdoor tanks under days of natural length. *Filled circles*, samples from rudd kept in total darkness. *X*, samples from an almost blacked-out tank outdoors. *Half-filled circles*, samples from an outdoor tank in which the light intensities were somewhat lower than the regular outdoor tanks. (Lower) The interval between sunrise and sunset, at the latitude of London. (From DARTNALL et al., 1961)

posed on the seasonal trends. The same seasonal changes were observed by ALLEN (1971) in a third cyprinid (the redside shiner, *Richardsonius balteatus*, with $P506_1$ and $P531_2$). The burbot (*Lota lota*, with $P503_1$ and $P527_2$), which is a fresh-water gadid, reaches a peak in A_2 pigment in the fall and minimum in the spring (BEATTY, 1969b). Among the salmonids (each species with $P503_1$ and $P527_2$, except for the brown trout), some have shown pronounced seasonal changes: juvenile coho salmon, *Oncorhynchus kisutch*, and king salmon, *O. tshawytscha* (BEATTY, 1966), and cutthroat trout, *Salmo clarki* (ALLEN et al., 1973). Others, however, have shown little or no seasonal change: landlocked sockeye salmon, *O. nerka* (BEATTY, 1969a), and a fluviatile stock of the brown trout, *S. trutta* (with $P508_1$ and $P534_2$; ALLEN et al., 1973). Although seasonal changes were first described in the visual pigments of rudd, they seem to be characteristic only of juveniles and young adults (BRIDGES and YOSHIKAMI, 1970). No evidence was found for such changes in older adults, even though these larger fish did respond in the typical way to experimental alteration of the light environment.

The effects of light and darkness on rhodopsin and porphyropsin have been tested by a number of investigators. The responses of the rudd were confirmed by

BRIDGES and YOSHIKAMI (1970) and MUNTZ and NORTHMORE (1971) and are as indicated:

$$\text{rhodopsin} \underset{\text{light}}{\overset{\text{darkness}}{\rightleftharpoons}} \text{porphyropsin} \qquad (\textit{Scardinius, Belonesox, Notemigonus})$$

This same pattern apparently occurs also in *Belonesox belizanus* (BRIDGES, 1965c). That is, in the laboratory, light favored the conversion of porphyropsin to rhodopsin while darkness prevented it. The opposite reaction has not been tested, however—fish initially with a high percentage of rhodopsin have not been kept in darkness to see whether the proportion of porphyropsin increases. More complete tests have been made with the golden shiner, *Notemigonus crysoleucas*; and this species also showed the "rudd response" (ALLEN and MCFARLAND, 1973). A similar increase in rhodopsin following exposure to bright light was also observed in juvenile coho salmon, but no changes were measured in an experiment in which the light intensity was lower (BEATTY, 1966). This generally self-consistent pattern was upset by ALLEN (1971), in carefully controlled experiments with *Richardsonius balteatus*. In the redside shiner, light favors porphyropsin and darkness (less effectively) favors rhodopsin:

$$\text{rhodopsin} \underset{\text{darkness}}{\overset{\text{light}}{\rightleftharpoons}} \text{porphyropsin} \qquad (\textit{Richardsonius}, \text{ some } \textit{Salmo})$$

Similar effects of light of this "anti-rudd" type have also been reported by ALLEN et al. (1973) in some trouts: rainbow (*Salmo gairdneri*, with $P503_1$ and $P527_2$), cutthroat, and brook (*Salvelinus fontinalis*, with $P503_1$ and $P527_2$). In brown trout, however, no clear effect of light was observed (ALLEN et al., 1973). These apparently contradictory responses to light under experimental conditions made the seasonal changes in rhodopsin/porphyropsin balance more interesting and more difficult to understand.

Experiments with light are complicated by the fact that intensity and photoperiod are necessarily involved, in addition to total quantum dosage. The spectral quality of the light used will also influence its effectiveness; indeed, the action spectrum may give insight into the underlying mechanisms. DARTNALL et al. (1961) suggested initially that photoperiod might be a factor in the action of light on rudd, because the very low light levels in an imperfectly darkened tank were nearly as effective as bright light in increasing the proportion of rhodopsin (Fig. 11). For several reasons, however, BEATTY (1966) doubted that photoperiod is an effective stimulus in salmon; but his experiments with juvenile coho salmon kept under different photoperiods were inconclusive. BRIDGES (1965c, 1972) has maintained that photoperiod is not effective in *Belonesox*, but his experimental design was not adequate to support this conclusion. BRIDGES and YOSHIKAMI (1970) stated that photoperiod is not involved in the regulation of scotopic pigments of the rudd, but performed no experiments to test this idea. It is somewhat surprising that no systematic investigation of the effects of light intensity, quantum dosage, photoperiod, or action spectrum has been undertaken on the rudd. It is clear, however, that long-day exposure to white light at moderate intensities is usually sufficient to increase the rhodopsin and that porphyropsin predominates in rudd kept under short days (BRIDGES and YOSHIKAMI, 1970; MUNTZ and

NORTHMORE, 1971; but see LOEW and DARTNALL, 1976). Only in an anti-rudd, *Richardsonius*, have the different parameters of photic stimulation been fully reported (ALLEN, 1971). At a given length of day, ALLEN found that the change to porphyropsin proceeded further under greater light intensities. In another experiment, in which the total quantum dosage was kept constant, he also showed that a 12-h light period was more effective in favoring porphyropsin than a short day (6-h), long day (18-h), or constant light. Thus, in the only experiments well-designed to test it, photoperiod was an effective stimulus—even more so than light intensity. Action spectra for the conversion of rhodopsin to porphyropsin in *Richardsonius* did not match the spectral absorption of any mixture of scotopic pigments, suggesting that cone vision might be involved (ALLEN, 1971). Additional data on rainbow trout and the common shiner, *Notropis cornutus*, also suggest that shifts in rhodopsin to porphyropsin may be under photopic and not scotopic control (Fig. 12). A recent investigation of rudd with MSP elegantly demonstrates that when the rods shift from rhodopsin to porphyropsin a similar change from A_1- to A_2-based visual pigments takes place in the cones (LOEW and DARTNALL, 1976). Clearly, further experiments with narrow-band light are needed to clarify whether the scotopic, the photopic, or both systems control light-induced changes in the visual pigments.

Water temperatures can also affect the proportions of scotopic pigments. So far, it appears that low temperatures favor porphyropsin and high temperatures, rhodopsin, in both "rudd response" fishes and the anti-rudd type. Juvenile golden shiners (*Notemigonus crysoleucas boscii*) with an initially high percentage of porphyropsin maintained this at 7° C, whatever the light treatment (ALLEN and MCFARLAND, 1973). At 21° C, the porphyropsin decreased slightly in continuous darkness and more rapidly (and completely?) in the light. Adults of another subspecies (*N. c. aureatus*) were also used in the same study. These fish initially had rhodopsin and porphyropsin in equal proportions. At 21° C, the porphyropsin decreased in an outdoor tank (but the rest of these fish at the higher temperature died before further sampling). At 7° C, the porphyropsin decreased slightly in fish kept outdoors; but it increased significantly in continuous darkness. MCFARLAND and ALLEN (1973) have reported similar temperature effects in another cyprinid, the common shiner (*Notropis cornutus*), and in rainbow trout. In this experiment, the fishes had high levels of porphyropsin initially; the porphyropsin remained higher in cold water than in warm, whatever the light conditions. A similar effect of water temperature has also been observed in rainbow trout by CRISTY (1976). Thus, it is clear that temperature affects the rate of change in all three species that have been examined. The final steady-state proportions are also affected in *Notemigonus*, but this is not known in *Notropis* or rainbow trout. ALLEN and MCFARLAND (1973) pointed out that these temperature effects could be responsible for seasonal changes in the proportions of rhodopsin and porphyropsin.

Endocrine regulation of the scotopic system was first indicated in fishes by NAITO and WILT (1962). In excised eyes of a sunfish (*Lepomis* sp.), thyroxine apparently decreased the conversion of vitamin A_1 to 3-dehydroretinal, the precursor of porphyropsin. More recent experiments with thyroxine have all used intact fishes (both salmonids and cyprinids). In every study, the proportion of

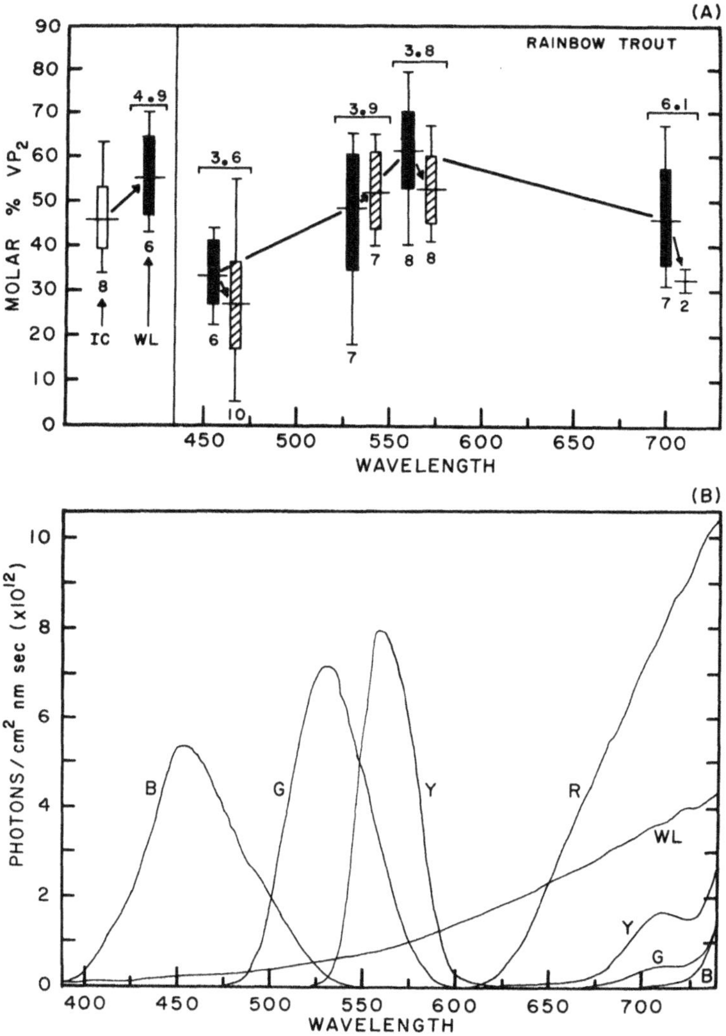

Fig. 12A and B. Action spectrum for the change of rhodopsin to porphyropsin in the rainbow trout. (A) Horizontal lines are the mean, vertical bars ±2 SE of the mean, and crossed vertical lines the range of per cent porphyropsin for each group of trout. Numbers below each range symbol are group size (*n*); values above each range symbol are the total number of photons/$cm^2 \cdot s \cdot 10^{-14}$ between 390 and 740 nm that were incident on the water surface. The *open bar* is the initial sample (IC); *black bars* are samples taken after 1 month in continuous light, either white (WL) or of wavelengths indicated by position of the graph; *hatched bars* are samples exposed for 2 months to the light treatment. (B) Quantal distribution of the lights used in this experiment. Data collected by W. N. MCFARLAND and D. M. ALLEN from March to May, 1972. A similar experiment on the common shiner, *Notropis cornutus*, was performed by J. CRAVER and W. N. MCFARLAND in 1974. The following mean values (±2 SE, *n*) for porphyropsin were obtained: initial sample = 17.1 (±11.2, 6); after 1 month, blue = 28.3 (±11.3, 6), green = 22.4 (±16.6, 6), yellow = 7.1 (±7.8, 10), red = 32.6 (±13.0, 9), and continuous darkness = 55.4 (±16.1, 9). This response is indicative of photopic control of rhodopsin-porphyropsin shifts, because the common shiner, like the rudd, increases porphyropsin in darkness and rhodopsin in the presence of white light (MCFARLAND and ALLEN, 1973)

porphyropsin was *increased* by thyroxine (i.e., opposite to the data of NAITO and WILT, 1962), whether it was administered in the aquarium water or injected intraperitoneally or intraocularly (MUNZ and SWANSON, 1965; BEATTY, 1969a, 1972; BRIDGES and YOSHIKAMI, 1970; ALLEN, 1971; JACQUEST and BEATTY, 1972; CRISTY, 1974). Intraperitoneal injections of the pituitary hormones, TSH (thyroid stimulating hormone; BEATTY, 1972) and prolactin (CRISTY, 1974), also increase the proportion of porphyropsin in salmonids. The normal physiologic roles of these hormones in visual pigment regulation are still not known, however.

In the first work on the variable visual pigments of the rudd (DARTNALL et al., 1961), a question was raised about the distribution of rhodopsin and porphyropsin within individual photoreceptor cells (rods) in the retina. The original guess, that they are mixed, has been confirmed in the rainbow trout by MSP (LIEBMAN, 1972). Under conditions that increased the rhodopsin, the two pigments were distributed homogeneously throughout the rod outer segments, indicating that conversion occurs independently of the basal replacement of discs, which had been shown by YOUNG (1967). Recently, however, LOEW and DARTNALL (1976) have found that rudd kept in continuous light for 4 weeks retained more porphyropsin at the apical ends of the rods, suggesting that conversion to rhodopsin paralleled disc replacement. But it seems odd that the apical ends should continue to retain more porphyropsin after 4 weeks in continuous light.

There might also be differences in the proportions of scotopic pigments between the two eyes or, more locally, in different retinal regions. BRIDGES (1964b) found only small differences between the two eyes of individual golden shiners. MUNZ and BEATTY (1965) were unable to find any differences between the two eyes, or between dorsal and ventral, anterior and posterior, or central and peripheral retinal regions in salmonids. In the rudd and in brown and rainbow trout, however, MUNTZ and NORTHMORE (1971) consistently found less porphyropsin in the dorsal half of the retina than in the ventral half. The intraretinal differences were smaller in the trouts than in rudd. This relationship held true in rudd, regardless of light-induced changes in the overall proportions of rhodopsin and porphyropsin. They found no consistent differences between the anterior and posterior retinal halves. Analyzing small portions of the intact retina, DENTON et al. (1971) confirmed these results in the rudd. The functional significance of intraretinal differences in the proportions of scotopic pigments remains obscure. BRIDGES (1964b) first emphasized that there can be substantial variation in the proportions of rhodopsin and porphyropsin among individual fish collected at the same location or held under the same experimental conditions. A possible function of such variation is discussed below.

An innovative approach toward understanding the regulation of the scotopic system was taken by BRIDGES and YOSHIKAMI (1970), who capped one or both eyes of the rudd (Table 2). When an eye was capped, its visual pigments behaved much as if the fish were being kept in darkness. Since the fish were actually kept under lighting conditions that favored rhodopsin, substantial differences developed between the capped and uncapped eyes, over 3 or 4 weeks. BRIDGES and YOSHIKAMI (1970; and see BRIDGES, 1972) concluded, therefore, that the control of rhodopsin/porphyropsin balance is localized within the eye, rather than systemic (e.g., hormonal). They noted, however, that in initially porphyropsin-rich fish the

Table 2. Analysis of visual pigment in rudd with neither eye, both eyes, or one eye capped. Data from BRIDGES and YOSHIKAMI (1970)

Type and treatment	Capping	No.	Days	% Porphyropsin		Difference: left eye minus right (%)
				Left eye	Right eye	
Porphyropsin-rich fish in the light	neither	2–5A	10	65	62	+3
		2–5B	10	75	75	0
		3–10A	28	7	1	+6
		3–10B	28	7	13	−6
Rhodopsin-rich fish in darkness	neither	2–6B	10	23	18	+5
		3–12A	28	62	59	+3
		3–12B	28	69	72	−3
Porphyropsin-rich fish in the light	both	2–3A	10	96	93	+3
		2–3B	10	96	96	0
Rhodopsin-rich fish kept in light	both	2–2A	10	0	1	−1
		2–2B	10	13	13	0
Rhodopsin-rich fish kept in light	both[a]	4–1A	21	51	32	+19
				Capped eye	Un-capped eye	Difference: capped minus uncapped eye
Porphyropsin-rich fish in the light	one	2–4A	10	88	78	+10
		2–4B	10	75	62	+13
		2–4C	10	83	69	+14
		4–12A	18	72	40	+32
		4–13A	22	55	28	+27
		4–14A	26	40	18	+22
		4–15A	26	59	36	+23
Rhodopsin-rich fish kept in light	one	4–3A	21	18	13	+ 5
		3–7B	28	51	18	+33
		6–2A	21	48	28	+20
		6–1A	28	51	18	+33

[a] Although both eyes were capped, only the left cap was opaque; the right was transparent.

percentage of porphyropsin was somewhat reduced in the capped eye (though less than in the uncapped eye; see Table 2) and suggested that some light must have leaked into the capped eye. Although this is a possible interpretation, careful examination of the data leads us to suggest that another is more likely: regulation of the scotopic pigments is neither entirely local nor entirely systemic, but the two levels of regulation are both involved. When two porphyropsin-rich fish were kept in the light for 10 days with neither eye capped (Table 2, fish Nos. 2-5A and 2-5B), their four eyes averaged 69.3% porphyropsin. When two others (Nos. 2-3A, 2-3B) were also kept in the light for 10 days, but with both eyes capped, their average porphyropsin percentage remained high (95.3%). In three fish (Nos. 2-4A, 2-4B, 2-4C) with only one eye capped, however, the mean percentage of porphyropsin was 82.0% after 10 days, indicating a small decrease. While one could imagine that there were no light leaks in either fish with double caps, but there were leaks in all three with single caps, our suggestion seems more probable. The experiments with rhodopsin-rich rudd also are informative. Unfortunately, we seem to lack controls for this group: rhodopsin-rich fish without eye caps, held in the light. But the initial percentage of porphyropsin was probably no more than 10% (see Table 2, fish Nos. 3-10A, 3-10B, 2-2A, 2-2B). When rhodopsin-rich fish were kept in the light for 21–28 days with one eye covered by an opaque cap and the other uncapped (Nos. 4-3A, 3-7B, 6-2A, 6-1A) or capped with a transparent spectacle (No. 4-1A), the average porphyropsin level in the unobscured eye was 21.8%. This suggests strongly that porphyropsin had been increasing in the uncovered eye, as well as in the eye with an opaque cap. (Whether the opaque caps leaked light would not affect this interpretation.) Thus, all these eye-capping experiments appear to support the interpretation that regulation of the scotopic pigments is partly localized within the eye and partly at a systemic level. It is unfortunate that these interesting experiments were successful with too small a number of fish to permit statistical analysis. Further work is badly needed, for this question is crucial for an understanding of vision in paired-pigment fishes.

To summarize this section briefly, it is now obvious that control of the scotopic system is complex in species with paired pigments; and it seems evident that the mechanisms are not the same in all species.

2. Possible Functions of Labile Systems

The summary of experimental evidence in the preceding section indicated the intensity of interest aroused by labile visual systems. (And the corresponding information on amphibians and lampreys was omitted entirely.) Probably every researcher in this area has pondered the possible functions served by labile systems. As already stated, porphyropsins appear to be well suited for vision in fresh water at twilight (Section B.III.2). Other suggestions of a special, nonvisual role of vitamin A_2 (WILLMER, 1956; WALD, 1957) have not been supported by more recent experiments. BRIDGES (1972) has clearly delineated the expected consequences of a variable rhodopsin/porphyropsin ratio: "Effectively, from the standpoint of spectral sensitivity, the paired-pigment system is equivalent to a single visual pigment with continuously variable λ_{max}" (p. 432). In the same volume, however, LYTHGOE (1972a) has taken an opposing position, because porphyropsins have a lower absorptivity (= molar extinction coefficient) than rhodopsins: "However, the improvement in sensitivity to be gained [at longer wavelengths] by having a porphyropsin instead of the homologous rhodopsin is small or indeed absent" (p. 594). (Fig. 13B indicates that this view is not correct.) Nearly every other author has concluded that the lability must have a visual function (e.g., DARTNALL et al., 1961, 1962; BRIDGES, 1965b; MUNZ, 1965; ALLEN, 1971; CRESCITELLI, 1972; MUNTZ, 1973).

At a given time, members of the same population may show substantial variation in the percentage of porphyropsin. This was first pointed out by BRIDGES (1964b), who repeatedly sampled the golden shiner *(Notemigonus crysoleucas)* in the same location. He suggested that in a schooling species like the shiner such variation could broaden the spectral sensitivity *of the school* and so have survival value for its members. This idea deserves testing, but that has not been attempted. Perhaps it should be mentioned, however, that similar variation can also occur among the individuals of other species which are not schooling: burbot (*Lota lota*; BEATTY, 1969b) and several trouts (ALLEN et al., 1973).

The most popular notion regarding the function of paired pigments is that they are adaptations to changes in the photic environment. Rather early, MUNZ (1965) emphasized that fresh waters are extremely variable photic environments. A given species is likely to inhabit rivers and lakes that differ greatly in color and clarity, both geographically and seasonally. Plankton levels, dissolved organic matter, and silt loads are all subject to local and seasonal fluctuations, profoundly influencing spectral transmission of the water. And there may also be seasonal or latitudinal differences in the spectrum of solar irradiance. One can readily imagine that, "mixtures of rhodopsin and porphyropsin, in changeable proportions, may be adapted to the highly variable nature of the photic environment in many freshwater habitats" (MUNZ, 1965). Similar thoughts have been expressed by BRIDGES (1965b, 1972). There is some support for this thesis, but all of it is indirect. Most persuasive are the results of seasonal sampling of golden shiners from two localities (BRIDGES, 1964b). Fish from a consistently clear habitat averaged about 20% less porphyropsin than those from a more turbid environment; this was true throughout the sampling period from October to March, superimposed on seasonal variations evident in both groups. It is a reasonable assumption that light in the clear habitat is more blue than in the turbid one. Therefore, these differences in the proportions of scotopic pigments are probably useful.

A somewhat similar situation in different Pacific salmons (*Oncorhynchus*) was revealed by BEATTY (1966). The species are all anadromous, but the juvenile sockeye (*O. nerka*) spends up to 3 years in glacial lakes before its seaward migration. Some populations, called kokanee, have become completely landlocked. At all stages of its life, the sockeye has less porphyropsin than the other salmons (BEATTY, 1966). And the kokanee, living its entire life in fresh water, has even less porphyropsin than migratory populations. If porphyropsin were viewed as an adaptation to existence in fresh water, these facts would make little sense. But the deep glacial lakes in which sockeye and kokanee lead a pelagic existence are generally blue in color. On the other hand, the streams and rivers inhabited by the other species are often green, or even brown during the rainy season. The lake trout (*Salvelinus namaycush*) is another salmonid which is adapted to a lacustrine habitat. This fish often lives near the bottom of large, deep lakes and is widely distributed across Canada and in the Great Lakes region. Sampling lake trout introduced into a very clear, blue lake (Odell Lake, Oregon), MUNZ (1965) found that the fish never had appreciable amounts of porphyropsin. Its rhodopsin was $P512_1$. Cayuga Lake, New York, is relatively clear, but more blue-green than Odell Lake. In Cayuga Lake this species has, throughout the year, mixtures of rhodopsin and porphyropsin (λ_{max} about 545 nm; MUNZ, 1965; a λ_{max} of 540 nm was reported by BRIDGES and DELISLE, 1974). Lake trout also occur in an atypical situation in Swan Lake, Alberta. This is a shallow (10 m), turbid, green lake. Fish taken during summer from Swan Lake had a great preponderance of porphyropsin (BEATTY and MUNZ, unpublished data, 1965). Thus, there is a striking parallel to the data on the golden shiner. It is tempting to conclude that rhodopsin/porphyropsin balance is related to light quality. Lacking direct light measurements, however, this is still only an assumption.

In his investigation of the redside shiner, ALLEN (1971) collected seasonal samples of fish from two different locations in Oregon. At both sites the fish had more porphyropsin during the winter. Throughout the sampling period, fish from a small tributary of the Willamette River had a higher percentage of porphyropsin than those living at the margin of a large reservoir (Hills Creek). The water was turbid at both locations during the winter. Measurements of spectral irradiance at a depth of 1 m indicated that daylight is richer in red light during the winter and in blue light during the summer. In winter, the stream appeared brown and the reservoir was green; these differences were not quantified spectroradiometrically, however. These results all appear to fit the same pattern. It seems probable that changes in the proportions of rhodopsin and porphyropsin can be related to differences in the photic environment. More measurements of light, especially twilight and moonlight, in fresh-water habitats are required before this interpretation can become accepted.

Discussion of the possible functions of labile scotopic systems rests upon a basic assumption. Changes in the proportions of rhodopsin and porphyropsin must alter the spectral sensitivity of scotopic vision. This assumption is amenable to behavioral testing and has been analyzed in a thorough investigation of the rudd by MUNTZ and NORTHMORE (1973). Four fish were used throughout the experiments and were kept under 12-h days for 5 weeks before training began. Over the next 2 months, they were kept at the same photoperiod and trained; during this time their scotopic sensitivity spectrum was tested. Then they were kept for 4 weeks with 18-h days (under conditions believed from prior experiments to produce about 90% rhodopsin, $P507_1$). After a further 2 months of testing, the fish were kept for 7 weeks with 4-h days (under conditions designed to change the visual pigment to 90% porphyropsin, $P535_2$). Finally, the scotopic spectral sensitivity was tested again. Their results did not show the expected spectral shifts in sensitivity, however (Fig. 13A). Instead, there was a vertical displacement, the fish being generally more sensitive to light in the long-day period than when they were kept under short days. This difference was greatest at 536 nm (0.7 log units, a five-fold difference). Thus, the "critical experiment" failed—to produce the expected result. In evaluating their own work, MUNTZ and NORTHMORE suggested that their conditioning lights may have been too dim to produce a high percentage of rhodopsin under long-day conditions. This seems unlikely, in view of their earlier success in controlling rhodopsin/porphyropsin balance in rudd (MUNTZ and NORTHMORE, 1970, 1971). They also noted that the fish must have used the central portion of the retinae in performing the visual tasks that were tested. But it is the dorsal retina that becomes rhodopsin-rich under long days (MUNTZ and NORTHMORE, 1971; DENTON et al., 1971). Although this appears to be a valid point, neither it nor the first argument could explain the five-fold difference in sensitivity that they measured at 536 nm. This difference is so large as to suggest some problem with their experimental procedure, for BRIDGES and YOSHIKAMI (1970) had found that the total amount of visual pigment was independent of the photic conditions in which rudd had been kept.

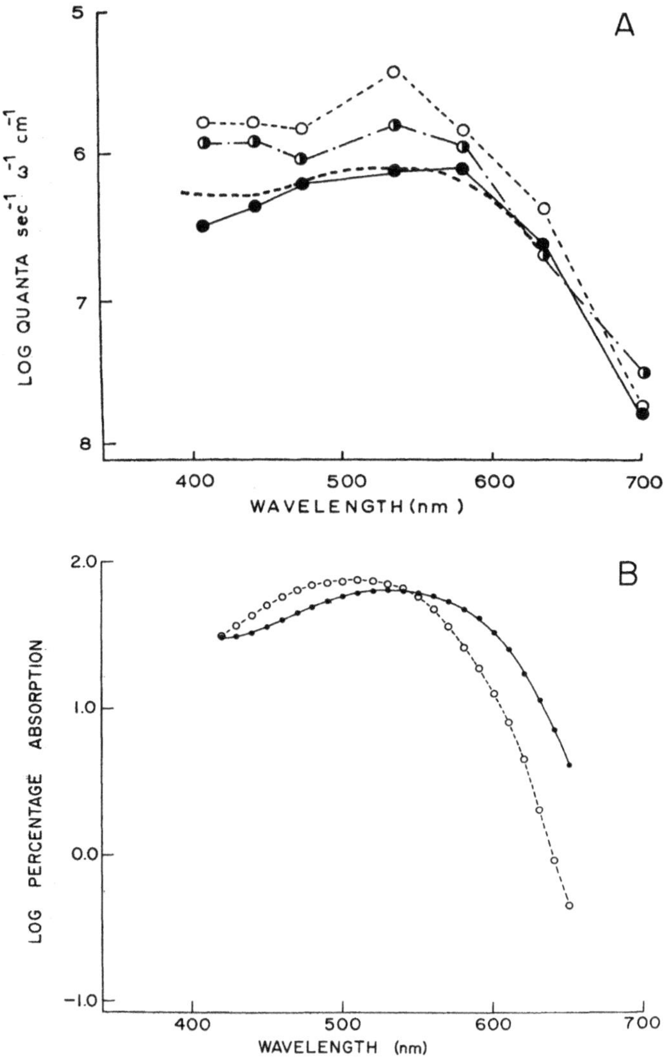

Fig. 13. (A) Mean scotopic thresholds for four rudd under days of different lengths. *Open circles* and *short-dashed line*, 18-h days; *half-filled circles* and *long-dashed line*, 12-h days; *filled circles* and *continuous line*, 4-h days. The *heavy dashed line* shows the spectral absorbance of a point on the retina of rudd C from DENTON et al. (1971). (From MUNTZ and NORTHMORE, 1973.) (B) Calculated absorption spectra of two different mixtures of the rudd rhodopsin and porphyropsin. *Open circles* and *dashed line*, 90% $P507_1$ and 10% $P535_2$; *filled circles* and *continuous line*, 10% $P507_1$ and 90% $P535_2$. Other assumptions as stated in text

To illustrate this problem, we have calculated the absorption spectra of the rudd pigments, mixed in different proportions. We assumed that the mixtures are equimolar and that the maximum absorbance of the pure rhodopsin is 0.6 (in accord with the calculations of 0.013 absorbance units per μm of outer segment length in several species by LIEBMAN, 1972). The maximum absorbance of porphyropsins is 0.70 that of rhodopsins (DARTNALL, 1968). The absorption spectra of a rhodopsin-rich (90% $P507_1$) and a porphyropsin-rich mixture (90% $P535_2$) were calculated, for these are the approximate proportions produced in the rudd by long-day and short-day conditions (MUNTZ and NORTHMORE, 1970, 1971). The calculated absorption spectra are shown in Fig. 13B, where they are scaled logarithmically to make them comparable to the behavioral data (Fig. 13A). As expected, there are substantial differences between the calculated absorption spectra, but only at longer wavelengths (eight-fold at 650 nm). The curves cross at about 544 nm, close to the wavelength at which MUNTZ and NORTHMORE found the greatest difference in scotopic sensitivity! We can conclude, therefore, that the behavioral data are not compatible with a simple photochemical interpretation. It must, of course, be kept in mind that the calculated absorption spectra would be comparable to the mixtures of scotopic pigments in the whole retina and not, necessarily, to those in the particular area actually used by the fish in these tests. We wonder, however, whether the training experiments were affected by the dim orange light that MUNTZ and NORTHMORE used in the fish tanks during the testing. At any rate, this crucial question must still be left unanswered: what are the visual consequences of changing the proportions of rhodopsin and porphyropsin?

Finally, it is possible that the shifts from rhodopsin to porphyropsin and vice versa, which tend to be seasonal, are as important in photopic as in scotopic vision. The circumstance that extractable visual pigments from fishes are mostly rod pigments has tended to obscure the possibility that the cone pigments also shift. LOEW and DARTNALL (1976) have shown in the rudd that rods and cones from any small patch of retina have about the same proportions of A_1 and A_2 pigments, apparently reflecting the proportions of vitamins A in the pigment epithelium. Indeed, they have suggested that changes in the proportions of rhodopsin and porphyropsin may be secondary to shifts in the photopic pigments. Although this idea is interesting, it cannot be applied to certain fishes, such as the parrotfishes (Section C. III. 1).

3. Changes in the Visual Protein

Migratory eels *(Anguilla)* have a fascinating scotopic system. Although the original investigations were performed on the European species *(A. anguilla)*, the results have never been fully published (BROWN and BROWN, reported by WALD, 1958, 1960) or the methods were imprecise (CARLISLE and DENTON, 1959). Generally similar results have recently been published on the North American species *(A. rostrata)* by BEATTY (1975a). Since his report is complete and supported by data, our discussion will be restricted to it. (Apparent differences between the two species may only be due to uncertainties about the results on *A. anguilla.*)

BEATTY (1975a) showed that "yellow eels" (preadult eels collected in fresh water before the downstream migration begins) had an ordinary rhodopsin/porphyropsin pair ($P501_1$/$P523_2$), with the porphyropsin predominating. "Silver eels" taken during the downstream migration toward the sea had small amounts of $P523_2$ and larger amounts of $P501_1$. In addition, however, they had a "deep-sea" rhodopsin, $P482_1$. After silver eels were transferred to sea water, the porphyropsin disappeared completely, leaving mixtures of the two rhodopsins. BEATTY states that small amounts of a second porphyropsin (based on the same opsin as $P482_1$) could also have been present in the silver eels, but that it was not detected in his partial bleaching experiments. In other words, *Anguilla* undergoes a double change in its scotopic system; both the prosthetic group and the opsin are altered. It is not known whether the conversion of opsins is complete (BEATTY thinks it probable) nor precisely how it occurs (for example, by the sequential replacement of discs in the rod outer segments). The net result is a change from a porphyropsin-rich condition typical of bottom-dwelling fresh-water fishes to a blue-sensitive rhodopsin typical of deep-sea fishes. As the oceanic migration and spawning of adult eels are believed to occur at considerable depths, this process offers further support for a *visual* interpretation of the meaning of labile scotopic systems. This story is still incomplete, for the visual pigments of eel larvae (leptocephalus) and elvers are so far unknown.

Intriguing as it is, the opsin change in *Anguilla* is evidently not unique. We found (MUNZ and MCFARLAND, 1973) that the larvae of a cardinalfish (*Apogon brachygrammus*) has a more blue-sensitive rhodopsin ($P482_1$) than do the adults ($P494_1$). This is a marine species from tropical coral reefs, and no change in the vitamin A occurs. Whether the larvae live at greater depths than the adults is not known. These scraps of information about ontogenetic changes in opsins leave much room for exploration. Especially interesting are the cases where there is a pronounced metamorphosis and the larvae are very different from the adults in their habits.

C. Adaptations of the Photopic System

I. Difficulty of Studying Cone Pigments

Function of the photopic visual system is based on its photoreceptors (cones) and the visual pigments they contain. Study of these pigments is therefore essential to an understanding of photopic vision. Unfortunately, it has been far more difficult to investigate cone pigments than their counterparts in the rods (DARTNALL, 1960). Attempts to extract cone pigments, even from the pure-cone retinae of certain reptiles, have usually failed or have produced ambiguous results. This poor success might be ascribed to any of several causes: 1) Cone pigments might not be solubilized by the extractant. 2) They might be destroyed by the extractant or by the procedure preliminary to extraction. 3) If extracted, the cone pigments might be too low in concentration to be detectable. 4) Even if their concentrations in the extract were not too low for analysis, the cone pigments might be masked by the much more abundant rod pigment.

Of course, mixtures of visual pigments have been extracted from many vertebrates, especially fishes and amphibians. As already described, these may be mixtures of scotopic visual pigments (Section B.V). Rod-cone pigment mixtures in extracts, of which the chicken and turkey system is best known, do occur. Relatively large amounts of iodopsin, a presumptive red-sensitive cone pigment, were extracted from the chicken retina by WALD et al. (1955). After extraction it is impossible to identify the cellular source of any visual pigment. It is only with MSP, therefore, that the red-sensitive pigment of certain birds has been definitely located in situ in the cones (LIEBMAN, 1972).

Recently MARC and SPERLING (1976) have used a cytochemical technique to locate visual pigments in particular classes of cones in the goldfish retina. They expose the retina to light of a given wavelength, and only those cones whose visual pigment absorbs it can then be stained by a redox dye. Although this method may not allow the precise identification of cone pigments, it will be valuable both in learning how many cone pigments are present in a given species and in associating known visual pigments with cone types in the retinal mosaic (Section C. II. 2). MSP remains the definitive method for identification of cone pigments, however.

II. Evidence from Microspectrophotometry

1. Identification of Cone Pigments

When HANAOKA and FUJIMOTO (1957) first applied MSP to visual pigments, they used a frog and the carp, *Cyprinus carpio*. Although they reported six different visual pigments in individual cones of the carp retina, these results are not reliable (see discussion by LIEBMAN, 1972). Their paper was important nevertheless, for it opened a new approach to the perplexing problems of photopic vision.

MARKS (1965), LIEBMAN and ENTINE (1964), and HÁROSI and MACNICHOL (1974) have examined the visual pigments in cones of the goldfish, *Carassius auratus*. In all studies, three classes of cones were found, each possessing a single visual pigment. MARKS (1965) reported that the three cone pigments had maxima at 455 nm (blue-sensitive), 530 nm (green-sensitive), and 625 nm (red-sensitive). LIEBMAN and ENTINE (1964) did not report λ_{max} values, except for a "537 µm pigment" mentioned in one figure legend. Neither paper indicated whether these pigments are based on vitamin A_1 or A_2. LIEBMAN (1972) has amended his earlier findings, giving λ_{max} values of 455, 535, and 620 nm, all in close agreement with the results of MARKS. HÁROSI and MACNICHOL (1974) obtained precisely the same λ_{max} values as MARKS for all three pigments. Although the product of bleaching is short-lived, its spectrum indicated that all three cone pigments are based on dehydroretinal (vitamin A_2).

Goldfish are known to possess color vision, and the results of training experiments (e.g., YAGER, 1969; YAGER and THORPE, 1970) confirm the implications of the MSP data: goldfish are, indeed, trichromatic. Their color vision is, therefore, generally similar to that of humans. It is not out of place here to indicate that TOMITA et al. (1967) concluded, from electrophysiologic evidence, that the retina of the closely related carp has three cone pigments. Clearly, the experimental evidence on cone pigments that is obtained with MSP can be related to neurophysiologic data and to psychophysical measurements.

Only a few other MSP results on the cone pigments of fishes have been published. An early paper by SVAETICHIN et al. (1965) attracted little attention (but see below). In the same laboratory, LAUFER and MILLÁN (1970) later found three cone pigments (P471, P568, P604) in *Eugerres plumieri*, but were unable to associate them with particular cell types. Recently, LIEBMAN (1973) has reported finding P620 in "principal cones" of a sunfish *(Lepomis microlophus)* and catfish *(Ictalurus punctatus)*, and (in collaboration with DARTNALL) in tench *(Tinca tinca)*. The sunfish also had P537 and the catfish, P543, in their "accessory cones." Another catfish *(I. melas)* had P543 in all its cones. Principal and accessory cones are the members of a pair of unequal double cones, such as occur in cyprinid fishes like the tench (see below). The meaning of these terms is not clear, however, in connection with the other species. ENGSTRÖM (1963b) found that *Lepomis macrochirus* has twin cones (Table 3), and a careful histological study by ARNOTT et al. (1974) has shown that two catfishes (*I. nebulosus* and *Arius felis*) have only single cones. Thus, the MSP data appear to be incomplete or ambiguous.

2. Double Cones and Twin Cones

In most vertebrate groups other than placental mammals, some of the retinal cones are joined through part of their length. Electron microscopy has shown that the members of these pairs of *double cones* are not fused into a syncytium, but are kept separate by their cell membranes (a special situation in certain deep-sea fishes is described by LOCKET in Chapter 4). The morphology of these cells is typically not the same, and the members of each pair are distinguishable as *chief* (= *principal*) and *accessory* cones. This topic has been reviewed recently by CRESCITELLI (1972).

In advanced bony fishes (teleosts), cone morphology is even more complex. In some families the double cone pairs are composed of dissimilar members, but in others the elements are so alike as to be called *identical twins* by some authors (e.g., WALLS, 1942). It is also typical of teleosts that the cones are laid out in a regular mosaic pattern. LYALL (1956, 1957a, b), ENGSTRÖM (1960, 1961, 1963a, b; ENGSTRÖM and AHLBERT, 1963), ANCTIL (1969), and WAGNER (1972, 1975) have made extensive histologic investigations of fish cones in the last 20 years. Differences in the staining properties of the conjugate elements led LYALL to avoid the term "twin cones." ENGSTRÖM also preferred the use of "double cones" for all paired elements. In most of the 39 families examined (Table 3), these cells are morphologically similar and of equal or nearly equal length. These he called "equal double cones," but we shall continue to use the term *twin* to emphasize their similarity (Fig. 14A). In a few teleost families, however, ENGSTRÖM found that the two components of each double cone are clearly dissimilar in diameter, length, organelles, etc. These he called "unequal double cones;" we shall restrict the term *double* to this group (Fig. 14B). LYALL (1956, 1957b) found that triple and quadruple cones can also occur; in the cyprinid fish *Phoxinus laevis* they predominate in most parts of the retina.

The single cones are not always all morphologically identical. In many groups of fishes there is only one type, which ENGSTRÖM (1960) called the "long single

Table 3. Retinal cone types and patterns. Data taken from ENGSTRÖM (1963b, Table 1), rearranged according to GREENWOOD et al. (1966); also from ANCTIL (1969) and from FINERAN and NICOL (1974). Parentheses around an entry mean that it is reported only in some of the species examined in that family

Order and family (No. of species studied)	Cone types						Pattern			
	(Long)[a] single	Short single	Twin	Double	Triple	Quadruple	Row	Square	Other	None
Anguilliformes										
Anguillidae (1)	×									
Clupeiformes										
Clupeidae (6)	×		×		(×)		×	×		
Mormyriformes										
Mormyridae (5)	×								×	
Gymnarchidae (1)	×								×	
Salmoniformes										
Salmonidae (7)	×	×	×		(×)			×		
Osmeridae (2)	(×)		×					×		
Esocidae (2)	×		×						×	
Cypriniformes										
Characidae (1)	×			×						×
Cyprinidae (15)	×	×		×	(×)	(×)	×			×
Lophiiformes										
Lophiidae (1)	×		×		×		×			×
Gadiformes										
Gadidae (10)	×		×		(×)		×			(×)
Merlucciidae (1)	×		×		×		×			
Zoarcidae (6)	×	(×)	×		(×)	(×)	(×)	(×)		(×)
Atheriniformes										
Cyprinodontidae (1)	×	×	×					×		
Poeciliidae (3)	×	×	×	×				×		
Gasterosteiformes										
Gasterosteidae (2)	×	×	×					×		
Syngnathidae (1)	×	×	×					×		
Scorpaeniformes										
Scorpaenidae (1)	×		×				×			
Triglidae (1)	×		×		×			×		
Cottidae (8)	×		×		(×)	(×)		×		
Agonidae (3)	×	(×)	×		×	×	(×)	(×)		
Cyclopteridae (2)	×		×		×			(×)		(×)
Perciformes										
Serranidae (3)	(×)		×							
Centrarchidae (1)	×		×					×		
Percidae (4)	×		×				(×)	(×)		
Sparidae (1)	×		×							
Toxotidae (1)	×		×							

[a] In some studies it is only implicit that the single cones are of the long type.

Table 3 (continued)

Order and family (No. of species studied)	Cones types						Pattern			
	(Long)[a] single	Short single	Twin	Double	Triple	Quadruple	Row	Square	Other	None
Perciformes										
Cichlidae (2)	×		×					×		
Labridae (9)	×	×		×				×		
Blenniidae (1)		×	×					×		
Anarhichadidae (1)	×	×	×					×		
Stichaeidae (1)	×		×					×		
Pholididae (3)	×		×		(×)		(×)	(×)		
Gobiidae (2)	×		×		(×)			×		
Scombridae (1)	×		×				×			
Anabantidae (3)	×		×					×		
Pleuronectiformes										
Pleuronectidae (7)	×	(×)	×		(×)	(×)		×		
Soleidae (1)	×		×		×					×
Tetraodontiformes										
Tetraodontidae (1)	×		×					×		

cone" from its typical shape; this is the "central single cone" of LYALL (1957a). Cones of this type are usually centered among four pairs of twin cones. In some species there is also a "short single cone" (= "additional single cone" of LYALL, 1957a).

Regular patterns of cones can be observed in the retinal mosaics of many fishes. Such patterns are often more prominent at the retinal periphery than at its center (LYALL, 1957b). The most commonly observed pattern is that already mentioned, in which each long single cone is at the center of a "square" of four twin cones. This arrangement occurs in adult salmonids (Fig. 15A; LYALL, 1957a) and in centrarchid sunfishes (ENGSTRÖM, 1963b). Young trout (LYALL, 1957a) and the stickleback (Gasterosteidae; ENGSTRÖM, 1963b) have basically the same pattern, but with a short single cone added at each corner of the twin-cone "square" (Fig. 15B). This developmental change in the retinal mosaic is characteristic of the brown trout, *Salmo trutta*; and LYALL (1957b) thought that it may result from the transmutation of short single cones into rods. The development of cone patterns has not been studied much in other species, except for *Lebistes reticulatus* (MÜLLER, 1951, 1952), in which it is always like that in young trout. Also of frequent occurrence is an organization into rows, without squares being evident (e.g., Gadidae). Several other basic cone patterns have also been observed; each of them is of limited occurrence. Unequal double cones are much less common among teleosts (Table 3). They are present, however, in cyprinids and labrids, two of the families of greatest interest to us. In both groups there are also long single cones and short single cones. Most cyprinids also have triple cones, and *Phoxinus*

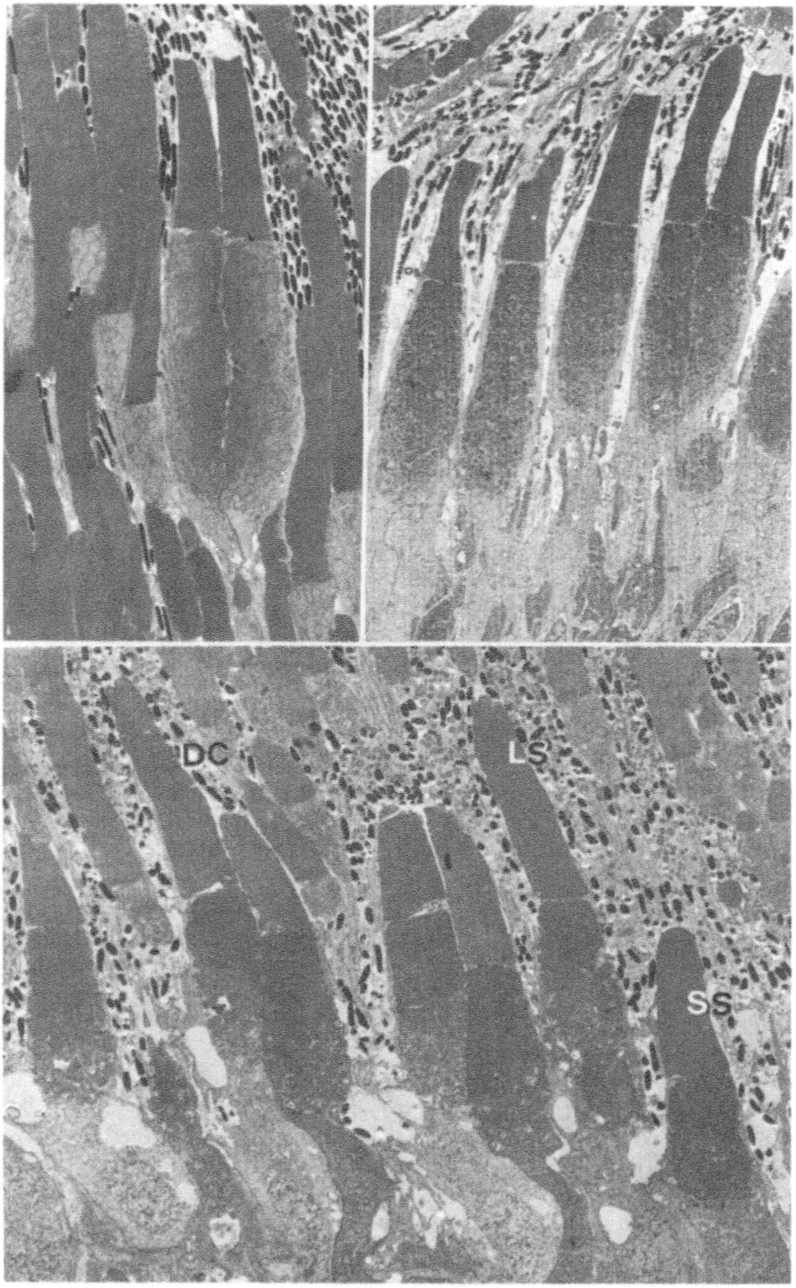

Fig. 14. (A – *upper panels*) Electron micrographs of retinal receptors from the rock bass, *Ambloplites rupestris*. *Left*, a pair of twin cones from partly dark-adapted retina (× 2600). *Right*, single cones and one almost equal twin cone pair from light-adapted retina (× 2250). (B – *lower panel*) Cones from light-adapted goldfish, *Carassius auratus*. DC, double cones; LS, long single cone; SS, short single cone (× 2250)

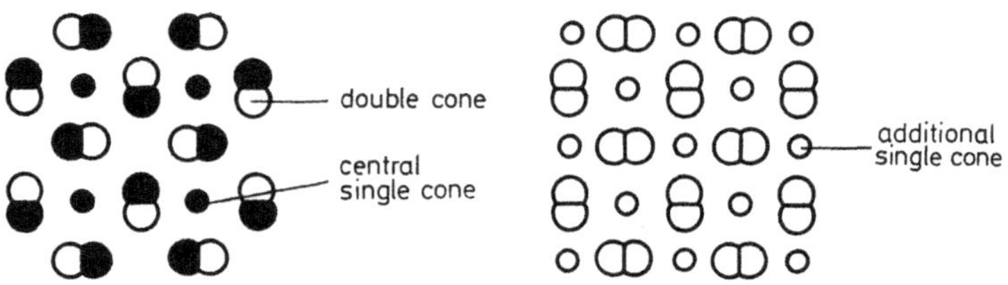

Fig. 15. Cone patterns in retinae of brown trout, *Salmo trutta*. *Left*, adult fish; *right*, young fish. Note: "double cone" is a true twin cone, "central single cone" is a long single cone, and "additional single cone" is a short single cone. (From LYALL, 1957b)

even has quadruple cones. And finally, the poeciliid fishes have both types of single cones, double cones, and twin cones. These fishes have certainly not been short-changed in their retinal morphology!

It is an attractive thought that the visual pigments in these different classes of cones may be related in a regular way to their morphology and patterns. One hypothetical arrangement was suggested by SVAETICHIN (1956), who recorded intracellularly from the retina of *Mugil*. He thought that the "double" cones might comprise mutually inhibitory opponent-pairs (red/green and yellow/blue), arranged around a central "luminosity" cone. But *Mugil* apparently has twin cones, rather than double. Later, SVAETICHIN et al. (1965) found that the conjugate elements in *Mugil* share the same visual pigment. In the goldfish, however, it has become clear that the double cones do have different visual pigments (STELL, 1975; MARC and SPERLING, 1976). The longer, principal member of each pair has the red-sensitive pigment (R); and the shorter member has the green-sensitive pigment (G). Thus, the double cones of goldfish are all R/G. About 80% of the long single cones are R and the rest, G. The shortest cones (short single and "miniature") all contain the blue-sensitive pigment (B). Thus, SVAETICHIN's ideas can be applied to the goldfish, but only in part. No yellow/blue pairs or "luminosity" cones have been found. Most important is the fact that the members of double-cone pairs have different visual pigments.

Another cyprinid fish, the rudd, has a very similar association between its visual pigments and the classes of cones (SCHOLES, 1975). These visual pigments have been identified by LOEW and DARTNALL (1976). Unlike the goldfish, which has visual pigments based on vitamin A_2, the rudd has mixtures of rhodopsin and porphyropsin in its rods (Section B.V). LOEW and DARTNALL found mixtures of A_1- and A_2-based visual pigments in its cones. Thus, the "red-sensitive" long single cone and principal member of double-cone pairs contain mixtures of $P562_1$ and $P620_2$. The accessory member of the double cones has mixtures of $P500_1$ and $P523_2$. The proportions of A_1 and A_2 pigments were about the same in cones and rods from the same patch of retina. After transfer from a 12-h day to continuous light, the proportion of A_1 pigments increased, in cones as well as rods. LOEW and DARTNALL also found a blue-sensitive pigment (with its λ_{max} between 450 and 460 nm) in short single cones from the retinal periphery.

Although the MSP literature on the cone pigments of fishes is limited, the results of two unpublished studies are available to us. The rock bass, *Ambloplites rupestris*, is a centrarchid fish. According to ENGSTRÖM (1963b), this family has long single cones and twin cones (Fig. 14A). LOEW and MCFARLAND (unpublished data, 1974) found only two cone pigments in *Ambloplites*; these pigments had their maxima at 532 and 604 nm. More important than the spectral maxima of these pigments are their cellular origins. P532 occurs only in single cones and is the only visual pigment that they contain. Both conjugate elements of each twin-cone pair share P604. Although this is, of course, a very different situation from that in the goldfish, it is not unique. The mahi mahi, *Coryphaena hippurus*, has both single and twin cones (TAMURA and WISBY, 1963); but the details of its cone morphology have not been reported. After successfully extracting three visual pigments from the retinae of this species (Section C.III.2), we enlisted Paul Liebman for an MSP study. MSP revealed two pigments in the cones: P463 in the single cones and P526/P526 in the twin cones (LIEBMAN, MCFARLAND, and MUNZ, unpublished data, 1974). Here again, therefore, the twin cones share the same visual pigment. Indeed, LIEBMAN (personal communication) has found this true in each species with twin cones that he has examined. We realize fully that it is early to be offering generalizations, but these findings certainly must alter the impressions left by experiments on cyprinid fishes, which have double cones.

SVAETICHIN et al. (1965) had also reported that the conjugate elements in twin cones share the same visual pigment. This early MSP study has been neglected by later reviewers (e.g., MUNZ, 1971; CRESCITELLI, 1972; LIEBMAN, 1972; DAW, 1973). The reasons may be that the MSP data were only a small part of a complex report, that the microspectrophotometer was evidently relatively crude (as these instruments go!), and that it and the experiments were inadequately described, both in the report and in the reference given therein (MITARAI et al., 1961). The goldfish and eight marine species (in the families Centropomidae, Lutjanidae, Gerridae, Pomadasyidae and Mugilidae) were studied by SVAETICHIN et al. (1965). Although their instrument was inadequate to identify visual pigments precisely, one feature of the results is significant. Unlike the goldfish, all the marine fishes had twin cones. In every species, these twins had the same visual pigment in both members. In a given species, there might be a single type (R/R, G/G, or C/C, i.e., cyan or blue-green sensitive), or a combination of two of these types. SVAETICHIN et al. indicated that in retinal mosaics with two types of twin cones, those on opposite sides of each "square" were alike. Thus, the arrangement of visual pigments in the mosaic was orderly. This was also true of the single cones. Every species had short single cones, which were always blue-sensitive. Six of the eight marine species also had long single cones; these were either C or G, but both types were never present in the same species. Only two cone pigments were present in two species, and the others all had three. Unfortunately, the uncertainties about this work demand that its conclusions be regarded with some skepticism. For example, if their MSP was side-on (which is only implicit), rather than axial, how were the twin-cone pigments located at particular positions in the retinal mosaic? Nevertheless, the corroborative evidence of more recent MSP shows that this report should not be ignored.

Most training experiments on color vision in fishes have used cyprinids as subjects. It is not clear whether this choice was dictated by extraneous factors, such as availability and familiarity, or whether these fishes were chosen because their color vision is unusually well developed. The German school of VON FRISCH used a minnow, *Phoxinus laevis*, as its favorite subject, but other species were also tested (reviewed by HERTER, 1953). The goldfish, *Carassius auratus*, has been studied by YAGER (1969) and YAGER and THORPE (1970). Still another cyprinid, the rudd *(Scardinius erythrophthalmus)*, was trained to discriminate between lights of different wavelengths by MUNTZ and NORTHMORE (1970). All of these species have long single, short single, and double cones. Only *Carassius* lacks triple cones, and only *Phoxinus* has quadruple cones. Although the experiments with *Phoxinus* are older, which means that colored papers were used rather than spectral lights, the behavioral evidence indicates that all three species probably have trichromatic color vision. The double cones, with their unlike visual pigments, are thought perhaps to enhance contrast by wavelength-dependent mutual inhibition.

It is our purpose to emphasize that cyprinid fishes are atypical teleosts in their cone morphology. And the cyprinids may have unusually well-developed color vision. Perhaps the first indication that other fishes are not all trichromatic was given by MACNICHOL and SVAETICHIN (1958), on the basis of intracellular retinal recordings (from horizontal cells). The S-potential recordings of TAMURA and NIWA (1967; also NIWA and TAMURA, 1969) led them to suggest that certain noncyprinids are totally, and some others partially, color-blind. Now the direct evidence of MSP indicates that *Ambloplites* and *Coryphaena* have only two classes of cones, rather than three; and this was also reported in two of the species studied by SVAETICHIN et al. (1965). (These thoughts are developed further in Section C.V, which relates photopic mechanisms to the photic environment.) Why cyprinids should need unusually good color vision is not clear. There are some similarities with labrids and poeciliids, which share a complex cone morphology, including double cones. All three families are large and widespread, with many sympatric species. Species recognition may partly depend on color vision. Some of these fishes also show pronounced sexual dimorphism in colors and patterns. But a convincing general explanation will require a great deal of further investigation.

In twin-cone fishes, the tests of color vision have been much less extensive. Without doubt, there is at least partial color vision in many (e.g., Gasterosteidae). Much of the experimental work was done with colored papers, rather than lights, and relied more or less on human color sense as reference. It is not possible to depend on this earlier work in detail. In no single case has it been established by psychophysical experiment that the photopic vision of a noncyprinid fish is monochromatic, dichromatic or trichromatic. Clearly, more behavioral testing is desirable, in combination with studies of cone morphology and of visual pigments by MSP. It should be possible to reveal the pattern of cone pigments in the retinal mosaic by means of the cytochemical method of MARC and SPERLING (1976).

Before abandoning this topic, we might ask how the twin cones function. They seem to be an evolutionary "invention" that is restricted to teleost fishes. Perhaps they arose from double cones, which are more widespread. According to WALLS (1942), twin cones are associated with bright light, being more common in fishes that live near the water surface. Twin cones are, however, particularly prominent in the cods (Gadidae), which often occur at greater depths. But if this correlation has any basis, it would fit an idea that has occurred to us. Later we shall emphasize the probable importance of bright flashes of sunlight reflected from the sides of many fishes, induced either by their own motion or by surface flicker (Section C. IV. 3). It may be that these flashes are crucial, both to predators and to their prey, in locating other fishes against the background. Since these flashes are typically of short duration (in the 10–100 ms range), the summation and interaction afforded by identical twins may help to identify bright targets without sacrificing acuity. These bright flashes are naturally most prominent near the surface and are confined to daytime. Are twin cones less common in families whose species are largely nocturnal? (Note: If this idea has any merit, one would expect that the visual pigment of the twin cones should be of the *offset* type—see Section C. V.)

III. Cone Pigments in Retinal Extracts

1. Green-Sensitive Pigments

Retinal extracts have sometimes contained visual pigments that are not easily referable to the retinal rods. Most of these "secondary" pigments have been more sensitive than rhodopsin (or porphyropsin) to the red lights used in bleaching the extracts. In conformity with the terminology used elsewhere in this chapter, their spectral maxima put them in the category of green-sensitive pigments. There is often too little of the additional visual pigment for it to be characterized successfully. But in other cases the method of partial bleaching (DARTNALL, 1962) has been used to study both photosensitive components. For example, MUNZ (1958a, 1964) found $P497_1$ and $P520_1$ in the labrid fish *Pimelometopon pulchrum*, $P500_1$ and $P520_1$ in the serranid *Paralabrax clathratus*, and $P486_1$ and a green-sensitive pigment with λ_{max} near 530 nm in the scombrids *Scomberomorus concolor* and *S. sierra*. More complex was the ternary mixture which DARTNALL (1955) described in the cyprinid *Alburnus lucidus*. Retinal extracts of this fish contained an apparent rhodopsin/porphyropsin pair ($P510_1/P533_2$), together with a more red-sensitive component (λ_{max} approximately 550 nm). In another cyprinid, *Tinca tinca*, DARTNALL (1952) found a blue-sensitive pigment ($P467_1$), in addition to the porphyropsin ($P533_2$). More recently, DARTNALL and LYTHGOE (1965) have reported mixtures of visual pigments in 18 species of marine fishes, but they did not provide any experimental data. In summarizing this work, we can say that rather frequently mixtures have been found in which only one photosensitive component can be ascribed to the scotopic system. Usually the additional or secondary visual pigment is more sensitive to red bleaching lights than the "main" rhodopsin (or porphyropsin). The possibility that some of these secondary pigments come from cones was recognized, but discussed only tentatively. BOWMAKER (1972) has also identified a possible cone pigment ($P544_2$) in sucrose suspensions of outer segments from a catfish (*Plecostomus plecostomus*). Unfortunately, the sources of visual pigments in outer-segment suspensions are just as ambiguous as in retinal extracts.

We have already described the rhodopsins of tropical marine fishes (Section B). In our survey of 179 species (MUNZ and MCFARLAND, 1973), we found

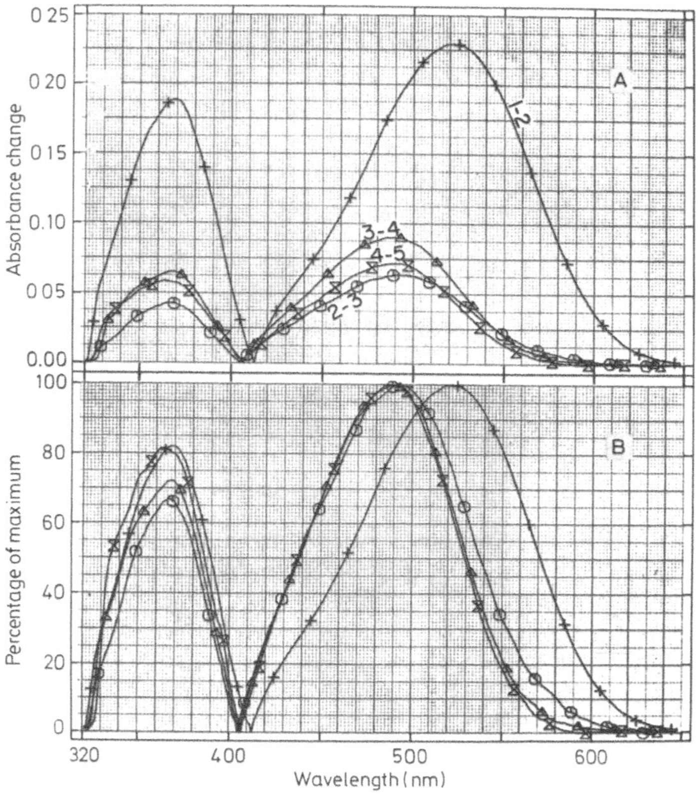

Fig. 16. (A) Difference spectra from a partial bleaching experiment performed on a retinal extract of *Scomberomorus commersoni*. Curve *1–2*, the result of exposure to deep-red light (λ675 nm) for 6 min; Curve *2–3*, the result of exposure to orange light (λ610 nm) for 20 s; Curve *3–4*, the result of 65 s further exposure to orange light; Curve *4–5*, the result of exposure to green light (λ560 nm) for 6 min. (B) The same difference spectra as in (A), each rescaled to 100% at the wavelength of maximal bleaching. (MUNZ and MCFARLAND, 1975)

mixtures of visual pigments in retinal extracts of 115 species. The additional pigments were extracted by routine methods and were stable, both thermally and in the presence of hydroxylamine. Their photosensitivities and products of bleaching appeared to be normal. These pigments were generally more sensitive to red light than the main, scotopic pigment. In 39 species enough of the additional pigment was extracted for it to be characterized by partial bleaching (MUNZ and MCFARLAND, 1975). In some species, such as the scombrid *Scomberomorus commersoni*, the green-sensitive component constituted more than half of the total photosensitive pigment, on an absorbance basis (Fig. 16). The initial difference spectrum, obtained after exposure to deep-red light, conformed to DARTNALL'S (1953) nomogram for visual pigments based on vitamin A_1. In three experiments, the mean estimate of λ_{max} was 520.5 nm (95% confidence limits ± 0.9 nm).

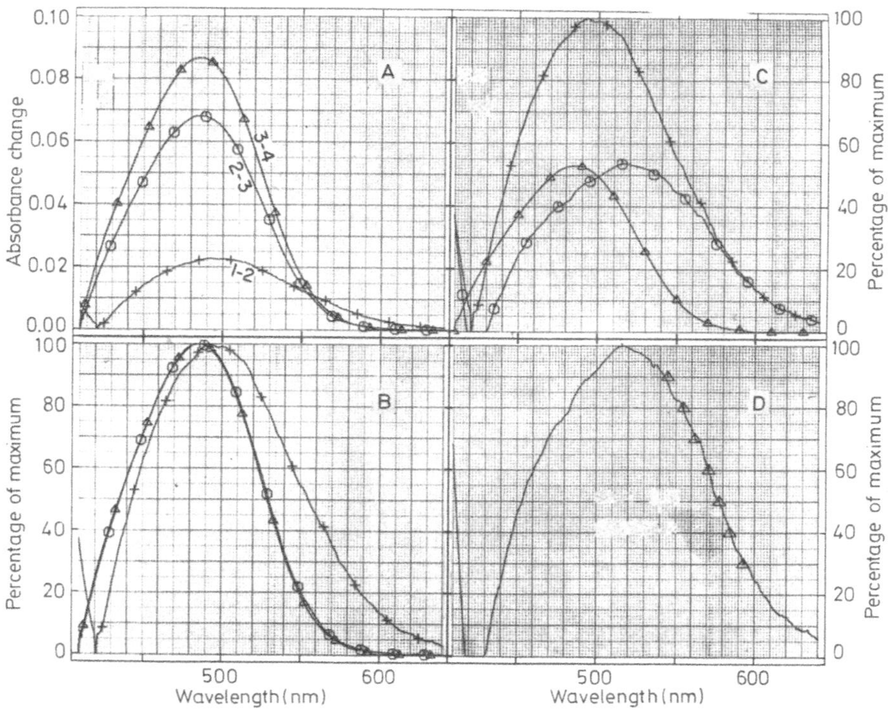

Fig. 17. (A) Difference spectra from a partial bleaching experiment performed on a retinal extract of *Scarus aeruginosus*. Curve *1–2*, after exposure to deep-red light (λ675 nm) for 16 min; Curve *2–3*, after exposure to orange light (λ610 nm) for 1 min; Curve *3–4*, after exposure to green light (λ560 nm) for 6 min. (B) The same difference spectra as in (A), each rescaled to 100 % at the wavelength of maximal bleaching. (C) Procedure used to correct the difference spectrum of the green-sensitive pigment. *Crosses* identify Curve *1–2*; *triangles*, Curve *3–4*, rescaled to 53% at its maximum; *circles*, the corrected difference. (D) The corrected difference spectrum of the green-sensitive pigment (λ_{max} 521.1 nm, SE $\pm$0.30 nm). *Triangles* are points calculated from the nomogram for this visual pigment (based on vitamin A_2). (From MUNZ and MCFARLAND, 1975)

In most of our extracts, the green-sensitive pigment was a smaller component of the total. In the parrotfish *Scarus aeruginosus*, for example, even the first exposure to deep-red light bleached both visual pigments to some extent. The first difference spectrum (Fig. 17A, B), therefore, represented a mixture of the two visual pigments. A corrective procedure was developed to help estimate λ_{max} of the green-sensitive pigment in such cases (MUNZ and MCFARLAND, 1975). It was assumed that this pigment should conform to the nomogram (in parrotfishes, the nomogram for vitamin A_2 pigments; MUNZ and SCHWANZARA, 1967); the presence of the scotopic pigment would make the difference spectrum too broad to match it. This was corrected by subtracting the scotopic pigment in the proportion (Fig. 17C) that leaves a remainder best fitting the nomogram (Fig. 17D). The last difference spectrum in the experiment was taken to represent the scotopic pigment. By an iterative, trial-and-error procedure, it was subtracted from the first difference spectrum in different proportions. In each iteration, nomogram fit

Table 4. Green-sensitive visual pigments[a]

Group	λ_{max}			Number of species
	Range (nm)	Mean (nm)	SD (±nm)	
A_1 Pigments				
Labrids and balistids	521–531	525.6	0.9	9
Other	518–526	521.7	0.7	15
Total	518–531	523.2	0.7	24
A_2 Pigments				
Labrids and balistids	531–541	536.9	2.1	4
Scarids	518–530	522.7	1.2	10
Total	518–541	526.8	2.1	14
Grand total	518–541	524.5	0.9	38

[a] Source: Table 3 in MUNZ and MCFARLAND (1975), + *Coryphaena hippurus* and *Chaetodontoplus mesoleucas*. *Mugil* excluded.

was indicated by the standard error of the estimated λ_{max}. In the illustrated experiment, the λ_{max} of the uncorrected initial difference spectrum was 501.9 nm (standard error ±1.58 nm). The corrective procedure reduced the standard error to a minimum of ±0.30 nm and indicated that the two visual pigments were present in about equal proportions in the difference spectrum. The λ_{max} of the green-sensitive pigment was estimated at 521.1 nm. We tested this procedure with a known rhodopsin/porphyropsin "pair" in retinal extracts of *Mugil cephalus* (BRIDGES, 1965a, 1967) and showed that it corrected for the presence of rhodopsin in the initial difference spectrum. Although *Scarus aeruginosus* also has a mixture of A_1/A_2 pigments like *Mugil*, these do not constitute a "pigment pair." The green-sensitive (A_2) pigment does not have its λ_{max} at the wavelength appropriate to its rhodopsin (MUNZ and SCHWANZARA, 1967; BRIDGES, 1972). This means that the two visual pigments of *Scarus* have different opsins as well as unlike prosthetic groups.

The green-sensitive pigments of the tropical marine fishes in our sample are tightly clustered with their λ_{max} values near 522 nm, except for members of the Labridae and Balistidae (Table 4). It is surprising that this is true both of the A_2 pigments of parrotfishes (Scaridae) and of the A_1 pigments (15 species in eight families). On the other hand, the green-sensitive pigments of labrids and balistids are shifted to longer wavelengths; this is especially true of the A_2 pigments in some members of these groups. Possible reasons for this are discussed in Section C.V.5.

The reasons which were advanced for believing that the green-sensitive pigments are cone pigments were circumstantial (MUNZ and MCFARLAND, 1975). It seemed no coincidence that these green-sensitive pigments occur in species that are diurnal, with large eyes and many retinal cones (MUNZ and MCFARLAND, 1973; TAMURA and WISBY, 1963). Moreover, they are particularly prominent

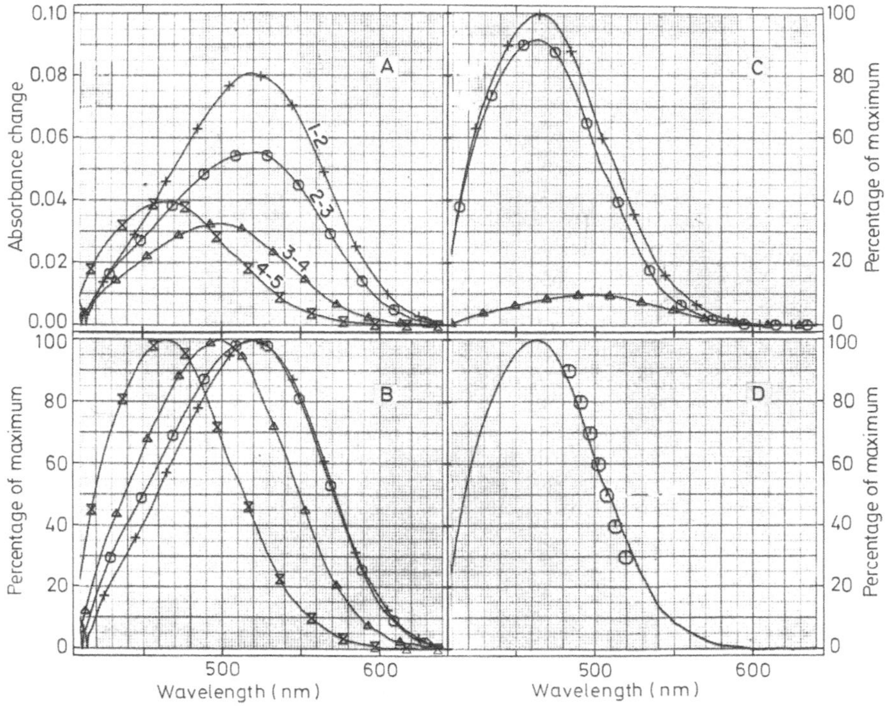

Fig. 18. (A) Difference spectra from a partial bleaching experiment performed on a retinal extract of *Coryphaena hippurus*. Curve *1–2*, after exposure to deep-red light (λ675 nm) for 150 s; Curve *2–3*, after further exposure to deep-red light for 405 s; Curve *3–4*, after exposure to orange light (λ610 nm) for 50 s; Curve *4–5*, after exposure to green light (λ560 nm) for 6 min. (B) The same difference spectra as in (A), each rescaled to 100% at the wavelength of maximal bleaching. (C) Procedure used to correct the difference spectrum of the blue-sensitive pigment. *Crosses* identify Curve *4–5*; *triangles*, Curve *3–4*, rescaled to 10% at its maximum; *circles*, the corrected difference. (D) The corrected difference spectrum of the blue-sensitive pigment (λ_{max} 467.1 nm, SE $\pm$1.01 nm). Circles are points calculated from the nomogram for this visual pigment (based on vitamin A_1). (From MUNZ and MCFARLAND, 1975)

(more than half of the total visual pigment extracted) in some pelagic species, which were caught in the light-adapted state and had their eyes removed immediately. Still more persuasive was the case of the mahi mahi (see below).

2. Visual Pigments of the Mahi Mahi

Initially, partial bleaching experiments with retinal extracts of the mahi mahi, *Coryphaena hippurus*, seemed like some others: there were mixtures of photosensitive pigments, much of the total being the green-sensitive component. But then it became clear that after the green-sensitive component had been bleached, the remaining visual pigment was itself a mixture. The retinal extracts were shown by partial bleaching to be ternary mixtures of three components (Fig. 18), all of them based on vitamin A_1: green-sensitive (P521), blue-green sensitive (P499), and

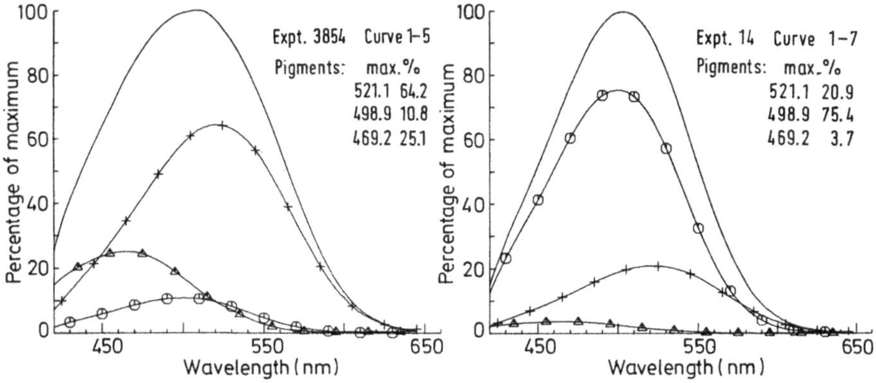

Fig. 19. Proportions of the three light-sensitive pigments of *Coryphaena hippurus* in two different partial bleaching experiments. Experiment 3854 was performed on the retinal extract of a light-adapted fish and Experiment 14 on a dark-adapted fish. In each panel the green-sensitive pigment is identified by *crosses*, the pigment sensitive to blue-green light by *circles*, and the blue-sensitive pigment by *triangles*. The percentages of the three components sum to 100% within rounding error. The total difference spectrum (*line without symbols*) is scaled to 100% at its maximum. (From Munz and McFarland, 1975)

blue-sensitive (P469). Identification of these pigments was complicated; special corrective procedures were required (Munz and McFarland, 1975). It was also possible to calculate the proportion of each component in the retinal extracts.

We believed that the middle pigment (P499) was the rhodopsin (i.e., rod pigment). *Coryphaena* also has many retinal cones (Tamura and Wisby, 1963) and we guessed that the green-sensitive and blue-sensitive components were cone pigments. This conclusion was supported by two facts: 1) Our initial experiments with light-adapted fish were relatively rich in the presumptive cone pigments. When we later obtained a dark-adapted individual, it was found to have much more of the middle component (Fig. 19; Table 5). All three pigments were present in extracts prepared from the dark-adapted retinae, but the total amount of pigment was much greater (more P499). 2) Tamura and Wisby (1963) found that the cone density was not constant over the whole retina, but the greatest concentration of cones occurred in the posterior-temporal region. By dividing retinae of light-adapted fish into quadrants, which were extracted separately, we showed

Table 5. Average proportions of the three photosensitive components in retinal extracts of *Coryphaena hippurus*

	Component		
	P469	P499	P521
Light adapted ($n=24$)	14.4%	31.6%	54.0%
Dark adapted ($n=12$)	6.8%	73.4%	19.8%

that the occurrences of the green-sensitive and blue-sensitive pigments were correlated with regions of greatest cone density (MUNZ and MCFARLAND, 1975).

More recently, these interpretations have been confirmed directly by MSP, in collaboration with LIEBMAN (Section C.II.2). This technique showed that the blue-green sensitive pigment is located within the rods. The twin cones have the green-sensitive pigment (both cells alike) and the single cones have the blue-sensitive pigment. Only our additional guess (MCFARLAND and MUNZ, 1975c) that the blue-green sensitive pigment might be a cone pigment (as well as the rod pigment) was not confirmed. There are also relatively minor differences in the estimates of λ_{max} made by MSP and ESP (extract spectrophotometry; BOWMAKER et al., 1975). Whatever the reasons for these discrepancies, they are not crucial to the argument. (This work will be reported in more detail elsewhere.)

It should also be mentioned that a retinal extract of the coral-reef fish *Chaetodontoplus mesoleucas*, gave evidence for three similar visual pigments (λ_{max} 448, 492, and 519 nm).

To summarize, evidence for visual pigments additional to the main or rod pigment has been found in many fishes, particularly in diurnal species with many retinal cones. In most cases, these additional pigments are more sensitive to red light than the rhodopsin of the rods. The correlation in occurrence with high retinal cone densities suggested that they are cone pigments. In the mahi mahi this interpretation has been confirmed directly.

IV. The Photic Environment

1. Spectral Distribution of Downwelling and Upwelling Light

The spectral distribution of daylight under water has been a subject of intense study for decades. The data document the general observation that open seas tend to be bluish, continental seas blue-green to greenish, and estuaries and fresh waters highly variable in hue (see Fig. 4 in MCFARLAND and MUNZ, 1975a). A description of the photic conditions applicable to the study of visual processes of fishes is so far available only for clear tropical seas (MCFARLAND and MUNZ, 1975b). Here we shall summarize the salient features that relate to the photopic vision of tropical pelagic fishes and of tropical coral-reef fishes.

Spectral irradiance of downwelling and upwelling light does not describe the backlighting along the varied lines of sight against which a fish might view a target. But it does provide an overview of the spectral range of photons that penetrate to various depths and, therefore, may be reflected from objects. Near the surface, downwelling spectral irradiance (H_d) is broad and flat, but it narrows and shifts to the blue at greater depth, i.e., the spectrum becomes more homochromatic with increased depth. This is familiar and results from the high transmission by clear oceanic water of light at shorter wavelengths. By calculating λP_{25}, λP_{50}, and λP_{75} (for explanation see Section B.II.1) a series of spectra can be indexed and compared. An example from the Kona coast of Hawaii is provided in Table 6. In considering photopic vision of fishes in clear tropical seas, two features of

Table 6. Changes in the spectral distribution of downwelling irradiance from 5 to 60 m depth in a clear tropical sea. Irradiance spectra were measured between 1135 and 1210 h on 5 May 1971 offshore from the Kona coast, Hawaii, where water depth was 500 m. Irradiance between 400 and 700 nm at 5 m was 87130×10^{12} photons/cm$^2 \cdot$ s; irradiance at greater depths is expressed as a percentage of that at 5 m. Spectral changes are indicated by the blue shift in the λP_{50} index with depth and by the decrease in the wavelength interval between λP_{75} and λP_{25} (see Section B.II.1 for explanation of the indices)

Depth (m)	Downwelling light index (nm)				Percentage irradiance
	λP_{25}	λP_{50}	λP_{75}	$(\lambda P_{75}-\lambda P_{25})$	
5	470	518	572	102	100
10	458	499	546	88	71
20	450	486	524	74	46
30	444	477	509	65	30
40	443	474	503	60	21
50	442	472	498	56	14
60	441	470	495	54	10

downwelling irradiance need emphasis: 1) The relatively greater absorption of light at the longer wavelengths produces homochromatic blue light at fairly shallow depths. Yet the total intensity of visible light does not diminish rapidly. At 50 m illumination is still intense, *albeit* blue (Table 6). 2) The spectrum of downwelling irradiance at a given depth beneath the surface is virtually independent of the total depth of the water column. This may seem trivial, but it is important when considering target detection in different locations. For example, a reflective object viewed horizontally at 3 m beneath the surface, whether in deep or shallow water, will be struck by downwelling light of the same intensity and spectral distribution (this assumes similar water masses in shallow and deeper situations, a typical condition around oceanic islands). The target, therefore, would have the same apparent photocontrast in both locations (Section C.V.2).

Upwelling irradiance (H_u) presents a very different picture. In the open ocean H_u is very blue and is spectrally constant, at least from the surface to depths of 60 m (Table 6). It is also far less intense than downwelling light (ratio $H_d/H_u = 100$ at 5 m, 63 below 30 m). These features are significant for target detection in the epipelagic zone. If one measures upwelling irradiance in shallower locations, where bottom reflections can occur, the photic conditions can be altered dramatically. This is especially true around coral reefs and beach areas where coral sands are highly reflective. A graphic example that compares H_u in the open ocean to shallow locations over reflective coral sands was reported by MCFARLAND and MUNZ (1975b, Fig. 6). It is immediately apparent that the upwelling reflections are spectrally broader than in deeper locations and would significantly alter the appearance of an object (fish, etc.) viewed horizontally.

In summary, downwelling light at any given depth in clear tropical seas is independent of total depth. Upwelling light in deep water is blue and homochromatic. In shallow water the H_u spectrum broadens and intensifies as the detector approaches a reflective substrate.

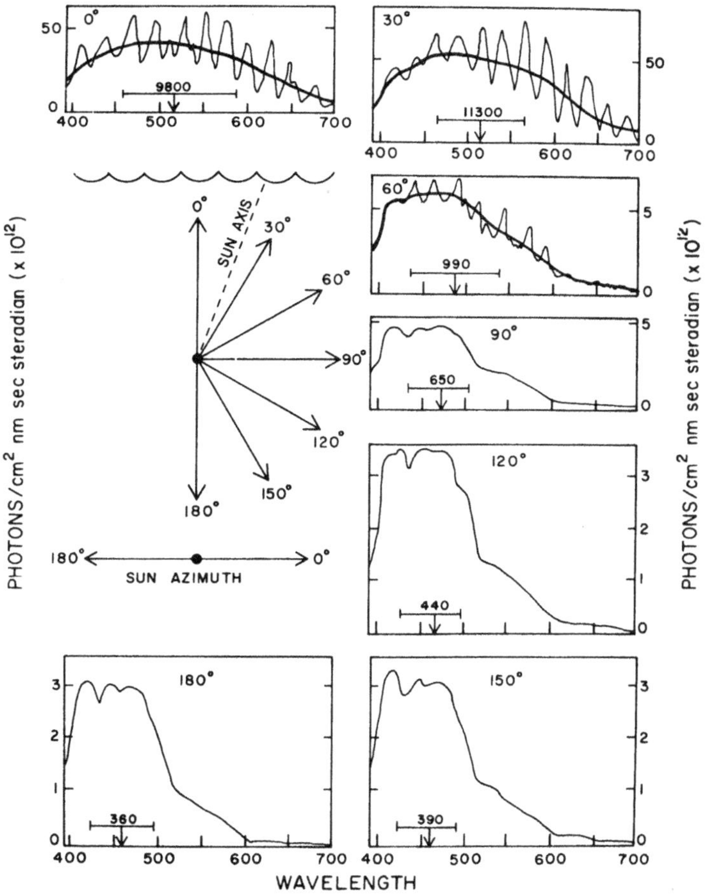

Fig. 20. The spectral distribution of light along different lines of sight 3 m beneath the surface in clear water of the open ocean. Each spectrum represents the photon-radiance measured along the indicated line of sight. Spectra were scanned with a receptor solid-angle window of 0.008 steradians. Beneath the water surface, the sun's axis was about 23°. All spectra were measured in the vertical plane of the solar azimith. Surface waves produced flicker in the three uppermost records; the heavy line in these records is the result of smoothing. The number under each spectrum, when multiplied by 10^{12}, is the total photon-radiance. The horizontal line just below this number is the interval of λP_{75}–λP_{25}. λP_{50} is indicated by the vertical line that crosses the horizontal line. These spectra were recorded on 5 August 1971, between 1107 and 1125 h, 3 km off the coast of Hawaii, near Kealakekua Bay; water depth was greater than 1000 m. The sky was clear

2. Angular Distribution of Light

To provide information pertinent to target detection, measurement of the backlighting immediately surrounding a target is required. This is accomplished by restricting the field of view of the receptor to measure radiance. In our investigations we use a cone that encompasses 0.008 steradians (cone with plane angle of ca. 6°). Spectral radiance can then be measured along varied lines of sight about a point under water to provide a three-dimensional view of the total radiance field. Several examples for clear seas were described by McFarland and Munz

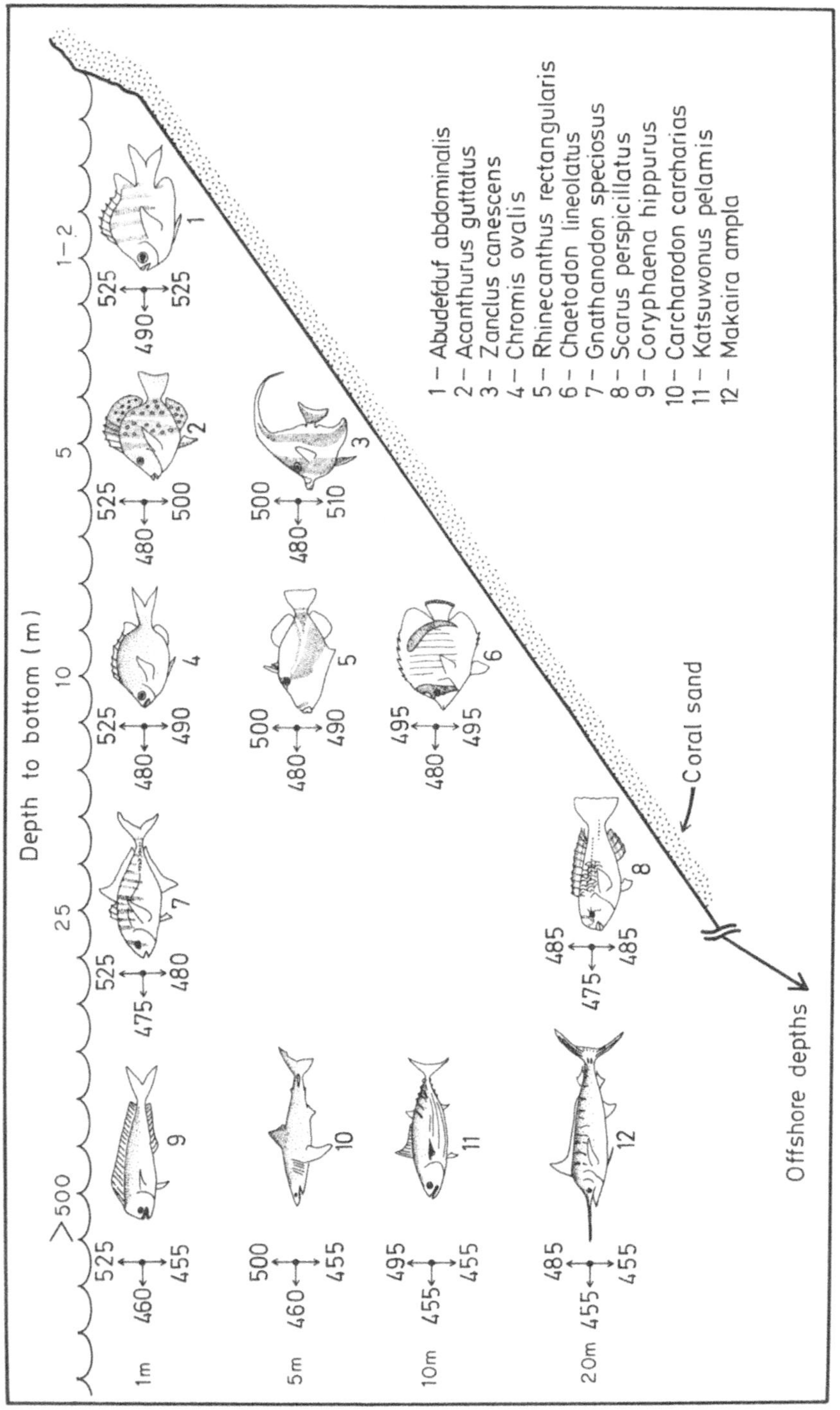
Depth to bottom (m)
1–2
5
10
25
>500
1m
5m
10m
20m
Coral sand
Offshore depths
1 – Abudefduf abdominalis
2 – Acanthurus guttatus
3 – Zanclus canescens
4 – Chromis ovalis
5 – Rhinecanthus rectangularis
6 – Chaetodon lineolatus
7 – Gnathanodon speciosus
8 – Scarus perspicillatus
9 – Coryphaena hippurus
10 – Carcharodon carcharias
11 – Katsuwonus pelamis
12 – Makaira ampla

(1975b; for a shallow-water example, see Fig. 3 in that paper). Similar measurements made in the open ocean are illustrated here as Fig. 20.

What do these spectral radiance measurements reveal about photic conditions in clear tropical seas? Near the surface in deep water the radiance spectrum toward the sun is relatively broad and intense (Fig. 20). Along other lines of sight the spectrum is narrowed (blue-shifted) and much less intense. As the receptor is lowered to greater depths the spectra become more symmetrical in the different directions, including the sun's axis. The depth of spectral symmetry over all lines of sight depends mostly on water clarity; in clear tropical seas it exceeds 100 m, but for most practical purposes the spectra are quite similar at depths greater than 30 m. Near the surface, therefore, the radiance spectrum is relatively uniform throughout the lower visual hemisphere. It becomes broad and intense only as the field of view approaches and includes the sun. At depths below 30 m the radiance spectra are more uniform along all lines of sight and become "brighter" but not spectrally "broader" when the line of sight includes the sun's axis.

As expected from the spectral irradiance measurements of downwelling and upwelling light, total water depth influences spectral radiance in the downward, but not the upward, field of view. At a given receptor depth, proximity to a reflective substrate (as in shallow water) tends to "broaden" and "brighten" spectral radiance. As with irradiance spectra, it is possible to index radiance spectra by calculating the λP_{50} values. A comparison of λP_{50} values for the radiance fields of the major coordinate lines of sight (zenith, horizontal, nadir) clearly specifies the spectral properties of the backlighting against which fish must detect objects (Fig. 21). Three features stand out: 1) As repeatedly stated, for the upward lines of sight the spectrum is dependent on depth beneath the surface, not total depth. 2) At any given depth beneath the surface, the spectrum along downward lines of sight is dependent on total depth and proximity to the bottom. 3) For the horizontal lines of sight, the spectrum is blue and independent of total depth or detector depth. The slight increase in λP_{50} for horizontal lines of sight just beneath the surface in shallow water (Fig. 20) is attributable to sand particles in the water and reflections from them.

Evaluation of these spectral differences reveals something special about the photic conditions of clear tropical seas. In general, pelagic habitats differ photically from shallow locations, such as coral reefs, even if the water masses have similar extinction properties. But the photic differences affect only the downward visual fields, not the upper and horizontal visual fields against which fishes must view objects.

Fig. 21. Spectral properties of the background light in clear tropical seas, both in the open ocean and near coral reefs. Total water depth (in m) is shown across the top. On the left are indicated the depths beneath the surface at which the light measurements were made. At each location, the median wavelength (λP_{50}) of the photon-radiance is shown for three lines of sight: upward to the zenith, horizontal, and downward to the nadir. In shallower water, a reflective coral sand bottom affects the horizontal and nadir light values. Fishes typical of these various locations are shown in outline (not to scale); the numbers beneath these outlines code to the species names in the lower right. This generalized figure is based on light measurements made at Eniwetok and Hawaii

3. Countershading

Organisms that inhabit the epipelagic regions of most seas are invariably countershaded or transparent and, therefore, are visually obscure. Most tropical, pelagic fishes appear blue dorsally. This hue fades laterally to a silvery flank and ventral surface. Diurnal coral-reef fishes have similar patterns, but dorsally their colors vary more and their ventral surfaces are as often white as silvery. In many species vertically arranged blotches are superimposed on the countershaded color patterns. These can function to obscure a fish viewed against the complex backdrops presented by coral. We reiterate these well-known facts of countershading only to emphasize that targets, be they predator or prey, have evolved patterns that minimize their apparent photocontrast when viewed against a given background. Since daylight is intense in clear tropical seas to considerable depths (Table 6), maximum photosensitivity is not a major visual problem for fishes trying to detect obscure targets—but the enhancement of photocontrast is.

4. Flicker

Flicker is a photic property often ignored in the measurement of underwater light. A rapid fluctuation in light intensity, produced by ripples and waves, can produce wild oscillations in spectral records. Calm conditions are preferred. But waves and ripples are common features of the oceans and their optical effects cannot be ignored, for they affect vision in both predator and prey.

Flicker is produced by ripples and waves and also by the movements of organisms. Thus, near the surface, intense flashes can be reflected from objects. Whether these flashes are detected, however, will depend on the line of sight from which the flashing object is viewed. Seen from below, flashes from a silvery fish moving near the surface would signal its presence best when viewed against a calm surface. If viewed against surface flicker, however, the fish might be rendered less visible, for it would flicker in synchrony with the flickering surface. Viewed from above, the dark blue dorsal surface of pelagic fishes tends to minimize reflections and flicker. When viewed horizontally, however, flashes from the silvery sides or bellies of fishes would be seen against a homochromatically blue background. Countershading functions to reduce this possibility, but movements or surface flicker can produce "intense" momentary flashes that give away the presence of another fish. It is our contention that success in locating prey by pelagic fishes may depend in part on detecting such flashes. We shall consider flicker again in connection with the possible role of visual pigments in pelagic fishes (Section C. V. 2).

V. Function of the Photopic System

1. The Contrast Hypothesis of Lythgoe

In order for an animal to see a target, at least two conditions must be met: 1) Enough light must come from the target to activate the visual receptors (the cones, in this case of photopic vision). In other words, the visual system must be sufficiently *sensitive.* 2) There must also be enough *visual contrast* so that the target can be distinguished from the background against which it is viewed. These conditions apply generally and are not limited to color vision. The two parameters, sensitivity and contrast, are not always mutually exclusive. In fact, the scotopic visual pigments of tropical marine fishes (Section B.III) offer an example. The rod pigments have undergone evolutionary selection so that their spectral locations permit maximal sensitivity to the background at twilight and also, therefore, maximal contrast between dark targets and the brighter background. The rod pigments can be said to be *matched* to the background light. This analysis appears to conform to the visual situation prevailing at twilight. When the sun is below the horizon, targets are generally silhouetted against the brighter background. We believe that this photomatching in the scotopic pigments applies to many vertebrates, including man (Section B.III.2; see also McFarland and Munz, 1975a).

Using an equation developed by Duntley (1962), Lythgoe (1966, 1968) examined the question of visual contrast under scotopic conditions. During the day, when light levels are much higher than at twilight, sensitivity of the visual system is somewhat less critical. It is also true that targets can be either "bright" (i.e., reflective) or "dark." We have applied Lythgoe's ideas to the photopic visual system (McFarland and Munz, 1975c). If the spectrum of background light is broad, then any visual pigment, with its much narrower absorption spectrum, will be "matched" to it. This would be true for fishes in shallow water, looking upward, or looking down at a reflective bottom. The horizontal spectrum is narrower, however, with its peak at short wavelengths. In the open ocean, the entire lower visual hemisphere is spectrally narrow and concentrated in the "blue" (Section C.IV). Under these conditions, the best visibility of a dark object requires a visual pigment matching the background (with its λ_{max} near the spectral maximum of the background).

A reflective target seen against such a blue background presents a different visual problem. A bright object may contrast poorly with a bright background. Now, if the light reflected from the target and that coming from the background were spectrally similar (e.g., blue fish viewed against a blue background), this problem would be insoluble. Near the surface of the sea, however, a bright target may reflect a relatively broad spectrum of light, similar to downwelling sunlight. As the light from the target travels greater distances, its longer wavelength components are gradually absorbed by the water. (Selective absorption of longer wavelengths and scattering of shorter wavelengths narrow the spectrum of the background light and make it "blue.") This means that at shorter ranges, at least, the spectrum of a "white" or "silver" target is broader than that of the background. If a visual pigment were spectrally *offset* from the background, rather than matched to it, the background would appear relatively dark. This would improve the visual contrast between a bright target and the darker background. The visual requirements of an animal could lead, therefore, to the evolution of photopic visual pigments which are either matched to the background spectrum or offset from it.

2. Surface-Limited Pelagic Fishes

The visual pigments of the mahi mahi, *Coryphaena hippurus*, have already been described (Sections C.II.2, C.III.2). This fish is also known, confusingly, as the "dolphin;" but it is not a cetacean. It is a large, diurnal predator with a

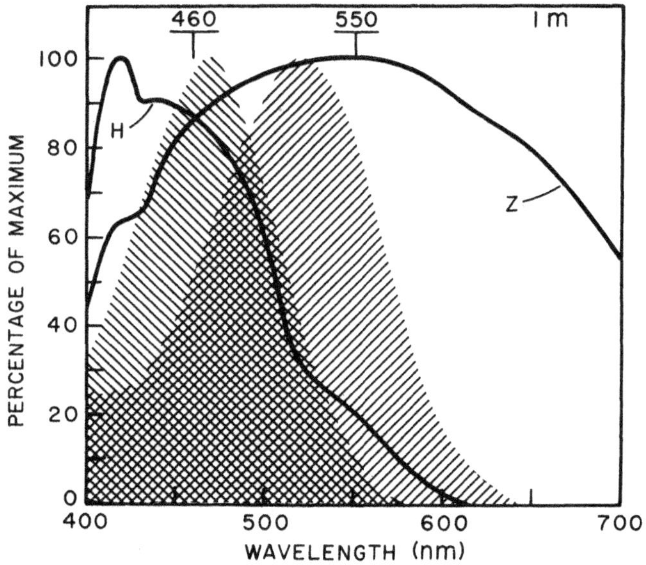

Fig. 22. Comparison of the background radiance spectra at 1 m depth (*heavy lines*: Z = zenith and H = horizontal line of sight) and the absorbance of the two cone pigments of the mahi mahi (*areas covered by diagonal lines*). The blue-sensitive pigment (P469) is to the left of the green-sensitive P521, and they overlap. The rod pigment (P499) has been omitted. Radiance and absorbance spectra have been scaled to the same maximum. The λP_{50} values of the horizontal (460 nm) and zenith (550 nm) radiance spectra are indicated. (MCFARLAND and MUNZ, 1975c)

circumtropical distribution in the open sea. It lives and feeds mainly in the uppermost 6 m of the water column, chasing fishes and crustaceans near the surface (ROSE and HASSLER, 1974). In its eyes, the visual axis is oriented forward along the horizontal line of sight, and there is well-developed binocularity (TAMURA and WISBY, 1963).

To understand how a visual system functions, the properties of the photoreceptors must be compared with the spectral distribution of light available. The sunlight spectrum is relatively invariant throughout the day. Cloud cover decreases total intensity and renders the light far less directional, but does not change its spectral distribution greatly. Near the surface of the open sea, the irradiance spectrum of downwelling light is relatively broad, but that of horizontal (and upwelling) light is narrow and shifted to shorter wavelengths (Fig. 22). Although these spectra have been scaled to the same maximum for comparison, at a depth of 1 m, the total irradiance of downwelling light is about 50 times greater than horizontal light (and 100 times greater than upwelling light). The absorbance spectra of the two cone pigments P469 and P521, are shown on the same figure. (Not knowing the concentrations of these visual pigments in the cones, we are unable to convert the absorbance values into absorption spectra, which would be a little broader.) Clearly, both visual pigments absorb downwelling light effectively. P469 is also well matched to the spectrum of horizontal background light, but the absorbance band of P521 is substantially offset from the spectrum of horizontal light.

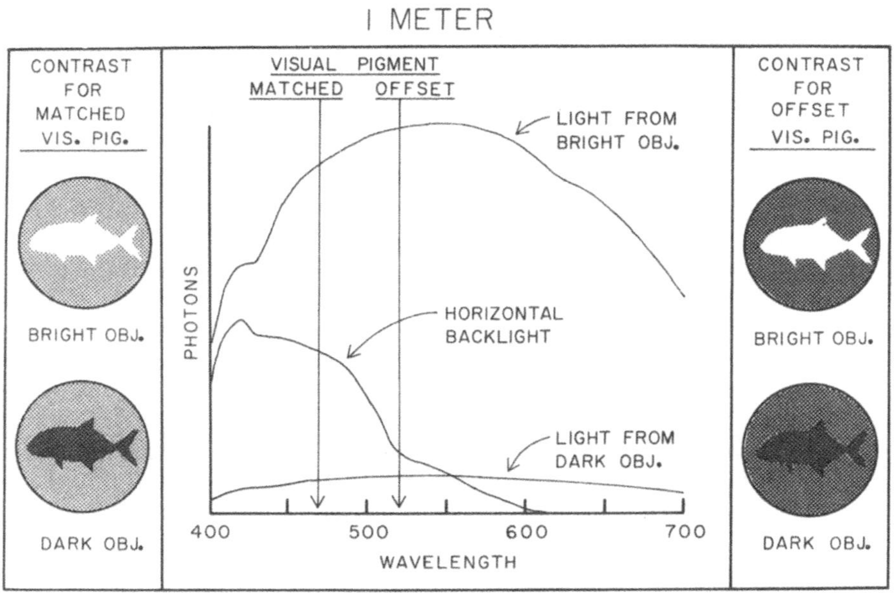

Fig. 23. Comparisons of the relative photocontrast of bright (reflective) and dark (nonreflective) objects viewed along the horizontal line of sight at 1 m beneath the surface by fishes with visual pigments matched to (P469), or (P521) offset from, the backlight. Both bright and dark targets are assumed to be grey. The insets portray the relative photocontrasts that would be produced by each visual pigment. The matched pigment enhances the contrast of a dark object and the offset pigment results in contrast enhancement of a bright target. (McFarland and Munz, 1975c)

The significance of the relationships between visual pigments and environmental light must depend on the properties of the targets that are viewed. Underwater measurements of the sunlight reflected from the side of a fish have not been made. Nor is much information available on the spectral reflectance of fishes (or other targets). The only measurements, those of Denton and Nicol (1966) on a smelt *(Osmerus eperlanus)*, showed that light of all wavelengths is reflected equally well. Assuming this in our own calculations, the light reflected from a bright object has the spectrum of downwelling sunlight, scaled down by a constant factor (Fig. 23). Similarly, a dark object has the same spectrum; we have assumed that its reflectivity is one-tenth that of the bright object. These two spectra are presented with that of the horizontal background, against which the mahi mahi views its prey and the λ_{max} values of the matched and offset cone pigments. In the panels of Fig. 23 we have attempted to portray the contrast or visibility of both bright and dark objects, as perceived by means of either cone pigment. If the mahi mahi views a dark target against the horizontal background, P469 should serve well to give maximal contrast *(lower left panel)*, but the offset pigment, P521, will work poorly *(lower right)*. On the other hand, a reflective target, such as a silver-sided fish, would appear to be brighter than the background *(upper panels)*. For this case, the offset P521 would give better contrast than the matched P469.

It has been stated (Section C. II. 2) that the conjugate elements in twin cones of the mahi mahi share the same visual pigment, P521. It should now be clear that this pigment is offset from the spectrum of the horizontal background light and that it would function best in perceiving bright targets against this background. We have also discussed (Section C. IV. 3) the reflection of flashes of light from the sides of fishes or other targets. These bright flashes are caused either by the target's own movement or by surface flicker. Under water, the human observer is often first alerted by such flashes to the presence of fishes in the water column. We suggest that these flashes are also important to fishes, particularly those, like the mahi mahi, which live in the pelagic environment, where most targets are subtly countershaded. Representing direct reflection of the sun's rays, these flashes are much brighter than the background; but they are also usually brief in duration (approximately 10–100 ms). It seems possible that the twin cones may function by summation to improve the perception of brief flashes without sacrifice of visual acuity. This interpretation is consistent with the exclusive possession of twin cones by teleosts and the fact that bright flashes are nearly restricted to underwater environments. (The reflections caused by surface flicker are not of vital importance to animals in air. Human technology has changed this, however, and recognizing the glint of metal or glass may be critical for survival, in real life as well as "Western" films.)

Although we have discussed only the mahi mahi, other pelagic fishes are also active mainly at the surface and are diurnal, visually oriented predators. The carangid fish *Elagatis bipinnulatus* is an example. In this species we extracted two visual pigments, $P500_1$ and $P520_1$, but have no definite information about their origins in rods and/or cones. The scombrid *Scomberomorus commersoni* also has two visual pigments, $P485_1$ and $P520_1$, which were identified in retinal extracts (MUNZ and MCFARLAND, 1975). In its visual behavior, this species is more or less intermediate between the mahi mahi and the tunas (see below). *S. cavalla*, the Caribbean counterpart of *S. commersoni*, has its visual axis directed upward and forward like the tunas (TAMURA and WISBY, 1963). It can take prey that it views horizontally while swimming near the surface, or it can strike upward at prey from below, like the tunas. As might be expected, the visual pigments of *Scomberomorus* also appear to be intermediate between those of the mahi mahi and the tunas (see below).

3. The Evaluation of Offsetting in Maximizing Photocontrast

All of the green-sensitive cone pigments that we extracted from tropical marine fishes have their λ_{max} values between 518 and 541 nm (Table 4). In four pelagic predators they are even more closely clustered at 520–521 nm. It is hard to dismiss the thought that natural selection underlies this narrow distribution. For example, the offset cone pigment, P521, of the mahi mahi must increase the photocontrast of bright silvery targets over that attained with the matched cone pigment, P469 (see Figs. 22 and 23). But visual pigments with λ_{max} values between 550 and 600 nm should effect an even greater contrast. This would result mainly from the lower quantum capture of such pigments of the blue background light, as compared to P521. Therefore, it is necessary to inquire why the offset cone pigments of tropical marine fishes should be green-sensitive, rather than yellow-sensitive or red-sensitive.

Several restraints suggest why offset pigments may be limited to the green region of the spectrum. 1) Although light reflected from a silver-sided fish near the surface would be rich in many wavelengths, it must suffer a greater attenuation at the longer wavelengths as target distance increases. Thus, reflected light will shift toward the blue, as well as become dimmer, at greater ranges. If photocontrast is to be enhanced at considerable distances from a target, an offset pigment must not be positioned too far toward longer wavelengths. Visual sensitivity at greater ranges, therefore, might limit the amount of offset. 2) If predators are to detect prey at greater depths as well as near the surface, an offset visual pigment must be positioned within the bandwidth of light available at these depths. Here the blue shift in downwelling light as it penetrates to increased depths would limit the degree of offset. 3) Targets in the epipelagic region of the seas are most often countershaded or transparent (Section C. IV. 3). Since blue is the most common color (in conjunction with silver) in countershaded pelagic organisms, the color range of targets is limited. Possession of P570 would represent a poor compromise, for few pelagic organisms are yellow. This visual pigment would be most sensitive to target reflections that contain longer wavelengths, a condition that prevails only when flicker produces a flash from the silvery region of a fish.

Whether any of these reasons do account for the natural selection of green-sensitive offset pigments is a question that requires a more quantitative approach. In an attempt to provide an initial answer, we have evaluated LYTHGOE's (1968) equation for visual contrast under water. Adapting the equations of DUNTLEY (1962, 1963), LYTHGOE included the properties of preretinal media and of visual pigments in his analysis. The resulting expressions are so complex that he treated only the simplest situation, which prevails along the horizontal line of sight. The essential optical parameters are shown in Fig. 24. For any visual pigment, it is necessary to integrate over the spectrum the absorption of light emanating from a target, $\int {}_tN_r \cdot V_p \cdot P \cdot d\lambda$, and the absorption of background light, $\int {}_bN \cdot V_p \cdot P \cdot d\lambda$. ($V_p$ is the absorption by the visual pigment, and P the absorption by the preretinal media.) Visual contrast, ${}_{\text{vis}}C_r$, is calculated by LYTHGOE's (1968) equation:

$$ {}_{vis}C_r = \frac{\int {}_tN_r \cdot V_p \cdot P \cdot d\lambda - \int {}_bN \cdot V_p \cdot P \cdot d\lambda}{\int {}_bN \cdot V_p \cdot P \cdot d\lambda} . $$

This does not take into account neural processing, such as lateral inhibition, in the retina or higher neural centers, that might further enhance visual response. Rather, it measures only the potential visual capability of an animal on the basis of the light, from the target and the surroundings, that is absorbed by similar photoreceptors. This distinction should be kept clearly in mind.

To calculate the visual contrast attainable with any visual pigment, several types of measurements are required: 1) the spectral radiance of the background, 2) the spectral radiance reflected from the target along the horizontal visual path, 3) the extinction properties of the water between target and viewer, and 4) the spectral absorption of the visual pigment and preretinal media. We have sufficient

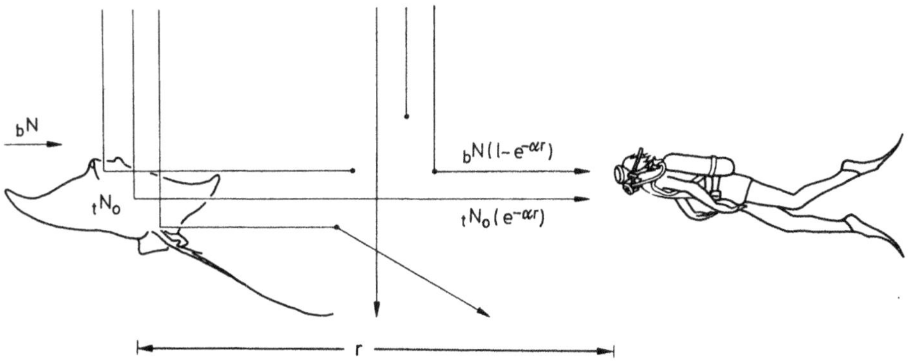

Fig. 24. Diagram to show the important features of horizontal radiance transfer through water. Daylight penetrating into the sea is the primary source of all the light that reaches the eye. The radiance, ${}_tN_0$, that is reflected by the object in the direction of the eye is partly absorbed, partly scattered out of the direction of the eye, and partly reaches the eye. The expression for the radiance reaching the eye is ${}_tN_0(e^{-\alpha r})$, where α is the beam attenuation coefficient and r is the length of the path of sight. Simultaneously there is a gain in brightness due to the light scattered from other directions into the eye. The amount of this "veiling brightness" reaching the eye is described by the expression ${}_bN(1-e^{-\alpha r})$, where ${}_bN$ is the radiance of the water background. The radiance ${}_tN_r$ reaching the eye from the object is thus: ${}_tN_r = {}_tN_0(e^{-\alpha r}) + {}_bN(1-e^{-\alpha r})$. (From LYTHGOE, 1972a)

data on items 1, 3, and 4 to evaluate the ${}_{\text{vis}}C_r$ over increasing visual ranges in clear tropical seas. To our knowledge, however, there are no data on spectral radiance from natural targets in the sea. This made necessary certain assumptions, which are explained below.

The spectral radiance along horizontal lines of sight is virtually identical from 1 to 10 m depth in clear tropical seas (Section C.IV.2). It is a relatively simple matter, therefore, to compare background absorption by different visual pigments, if we assume equal concentrations for each. Since the background radiance is very blue, blue-sensitive pigments will be more effective in catching photons than green-sensitive or yellow-sensitive pigments. The function $\int {}_bN \cdot V_p \cdot P \cdot d\lambda$ was evaluated separately for P469 and P521 of the mahi mahi, and for a hypothetical yellow-sensitive pigment, P570 (see Fig. 25D). (DARTNALL'S [1953] nomogram was used to generate the absorbance spectra, and a maximum absorbance of 0.4 [= 60% absorption] was assumed in each case [LIEBMAN, 1972]. We have measured the spectral transmission of the preretinal media of the mahi mahi [unpublished data, 1974].) The background absorptions for P469, P521, and P570 are in the ratio 100:90:60.

a) Grey Targets

To calculate ${}_{vis}C_r$ one must know the number of photons reflected from each target over the spectral range considered. Since we did not measure this in situ, it was necessary to assume a set of values for the spectral radiance of light striking the target that would, in turn, be reflected along the horizontal optical path. Our basic assumption was that a portion of the sun's radiance would be included as well as diffuse underwater radiation. The radiance level itself should fall, therefore, between the most intense radiance field (toward the sun) and the least intense

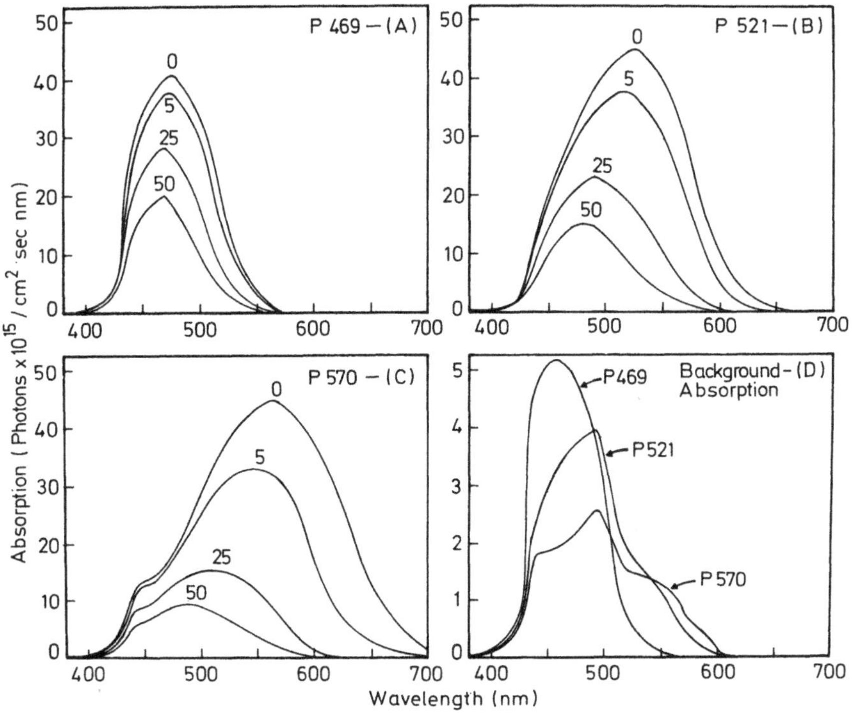

Fig. 25A–D. The calculated effectiveness of three visual pigments in absorbing the horizontal background light and that reflected from a bright target. All curves are for a horizontal line of sight 1 m beneath the surface in a clear tropical sea. (A) Absorption by the blue-sensitive cone pigment of the mahi mahi of light reflected from a target (reflectance assumed to be 0.25 at all wavelengths) at distances of 0, 5, 25, and 50 m. Additional assumptions are explained in the text. The total areas under these curves represent the photon capture by P469. (B) Absorption by the green-sensitive cone pigment (P521) of the mahi mahi of light reflected from a targed. Explanation as in (A). (C) Absoprtion of reflected light by a hypothetical yellow-sensitive pigment (P570). (D) The relative effectiveness of the three visual pigments in absorbing the horizontal background light. Note that the values on the vertical scale are 10 times smaller than in the other panels

field (nadir in the open sea). To simplify this complex optical problem further, we assumed that the spectrum reflected from the target would, on average, be most similar to the spectrum of downwelling irradiance at each depth. To provide for an intermediate intensity of this light, we chose that the spectral peak (550 nm at 1 m depth) of this spectrum would contain 30 times as many photons/$cm^2 \cdot s \cdot nm$ as the spectral peak of the background radiance (420 nm at 1 m depth). The number of photons at each wavelength is then easily calculated. If neutral grey targets are spectrally flat reflectors (the only set of measurements, which was made by DENTON and NICOL [1966] on the white belly of a fresh-water smelt, is consistent with this assumption), the spectral radiance emanating from targets of different reflectivities can be calculated. We chose a reflectance of 0.25 to represent a bright target and a reflectance of 0.005 for a dark target. Thus armed with a set of values for the spectral radiance emanating from bright and dark targets, we

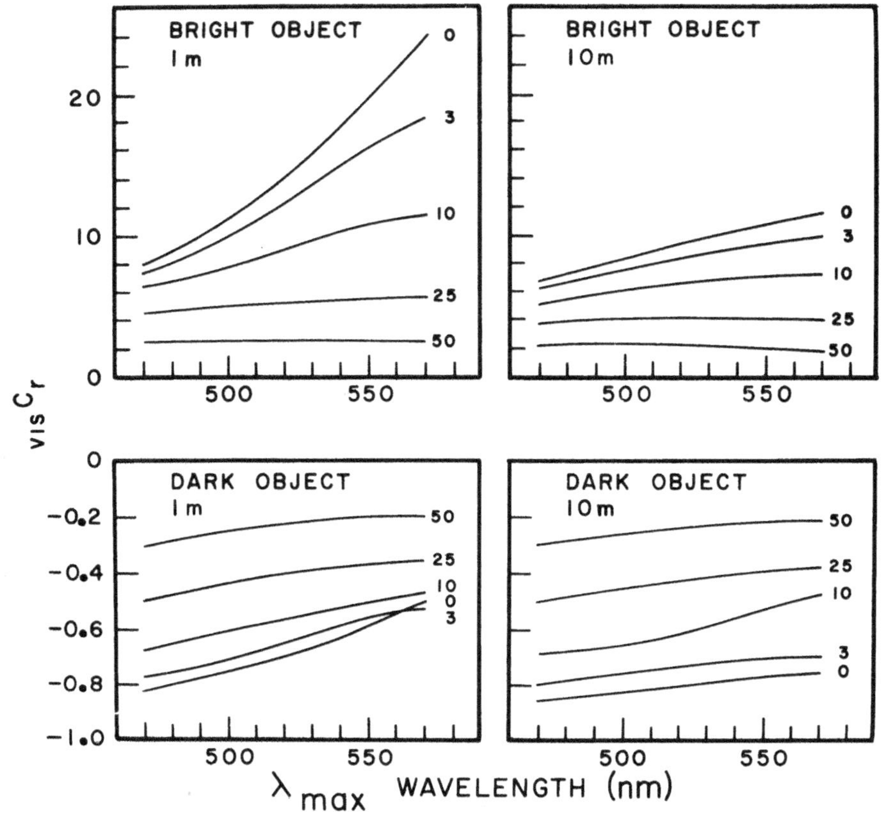

Fig. 26. The calculated visual contrast ($_{vis}C_r$) of bright and dark targets as perceived by various visual pigments (λ_{max} given on the abscissa). The *upper panels* are for bright targets (reflectance 0.25 at all wavelengths) and *lower panels* for dark targets (reflectance 0.005). The *two left-hand panels* show contrast at a depth 1 m beneath the surface; the *right-hand panels* show contrast at 10 m below the surface. Within a panel, each line represents the contrast achieved by different visual pigments at a particular distance (in m) between the object and the eye. For objects brighter than background, larger numbers indicate greater contrast; thus, in the *upper left panel*, at a range of 0 m, the contrast between a bright object and the background is much greater for P570 than for P469. At a range of 50 m, the contrast is poor for any visual pigment. For objects darker than background *(lower panels)*, the greatest possible contrast is −1.0

could calculate the number of photons absorbed by different visual pigments at increasing target distances (0 to 50 m). The function $\int {}_tN_r \cdot V_p \cdot P \cdot d\lambda$ was evaluated for P469, P521, and P570, at various ranges. Results of these calculations for a bright target are illustrated in Fig. 25A–C. With this information, we then calculated $_{vis}C_r$ for each pigment at every range. The contrast values clearly indicate that offset pigments do enhance visual contrast of bright targets and matched pigments enhance contrast of dark targets (Fig. 26). This simply confirms the conclusions of LYTHGOE.

Let us return to the question of why tropical marine fishes, like the mahi mahi, have only green-sensitive and not yellow-sensitive offset pigments. The evaluation of $_{vis}C_r$ for bright and dark grey targets provides no simple answer. Certainly, over short distances not exceeding 15 m, P570 would produce a greater photocontrast of bright targets than a green-sensitive pigment (Fig. 26). With increased range or depth, however, contrast enhancement is diminished and there is little difference between green- and yellow-sensitive pigments. Indeed, because of the enhancement at short ranges, one might predict that yellow-sensitive pigments would be favored in the surface fishes.

For dark grey targets a graded series exists: the more blue-sensitive the pigment, the greater the negative visual contrast (Fig. 26). With increased range and depth, the differences between the pigments decrease, but the overall relation is retained. The ${}_{vis}C_r$ attained by different visual pigments, therefore, presents a dilemma. Neither at greater ranges nor at increased depths (to 10 m) do the requirements of vision make clear why only green-sensitive pigments should be used for offsetting. So far, of course, this analysis has been restricted to spectrally neutral, grey or white targets.

b) Colored Targets

In the open sea, the need for countershading limits the range of colors that fishes may assume to the short end of the visible spectrum. For a target to blend into the background, the light reflected from it must match the background spectral radiance along the line of sight (DENTON and NICOL, 1966). The closer the match of spectral radiance, the less contrast will be apparent between target and background. As a result, target visibility will be reduced. Scrutiny of the spectral radiance field about a point in a clear tropical sea vividly demonstrates why countershading colors fall predominantly at shorter wavelengths (Fig. 20). Only toward the sun's axis does the background contain a significant proportion of light at wavelengths above 550 nm, and this is true only in the uppermost 10 m of the water column (Section C. IV).

In considering why green-sensitive cone pigments are dominant in tropical marine fishes, therefore, we should ask whether these pigments enhance the contrast of countershading colors more effectively than would blue-sensitive or yellow-sensitive pigments. Since no data are available for the spectral reflectance of natural targets from the sea, it is only possible to set up a series of spectral reflectances for "ideally" colored targets and then to recalculate the visual contrast for different visual pigments. In this analysis we chose to make the hypothetical colored targets very "homochromatic," with narrow and symmetrical spectral bandwidths. Three conditions were set up: a blue, a green, and a yellow target with peak reflectances at 480, 520, and 560 nm; half-bandwidths of 40 nm; and a total bandwidth of 80 nm (Fig. 27). Using the same data base and assumptions that were used to calculate ${}_{vis}C_r$ for the grey targets, the photocontrast was obtained again for P469, P521, and P570 at depths of 1 and 10 m with the three colored targets (Fig. 27). At all ranges, the ${}_{vis}C_r$ values show that a blue-sensitive pigment would be best for blue objects and poorest for yellow objects. The reverse is true for yellow objects, where P570 is best. But for objects of intermediate color, the green-sensitive pigment gives the greatest contrast; yet it also produces reasonable contrast with blue and yellow objects. Thus, the selection of green-sensitive cone pigments may well represent an evolutionary compromise. Their degree of offsetting from background assures that contrast enhancement to bluish and to greenish targets is high over usual visual ranges and at depths near the surface.

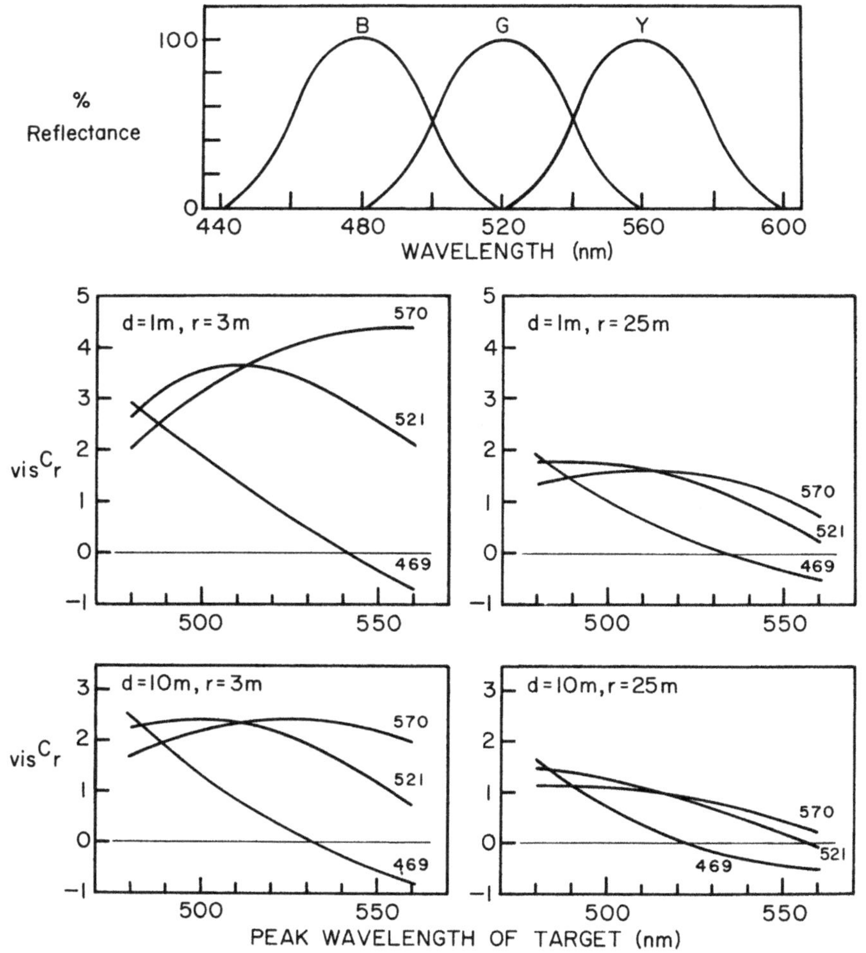

Fig. 27. The calculated visual contrast ($_{vis}C_r$) of hypothetical colored targets as perceived by three visual pigments. *(Top)* The spectral reflectance assumed for three such targets: blue, with a peak reflectance at 480 nm; green, at 520 nm; and yellow, at 560 nm. *(Four lower panels)* The two upper panels in this series show contrast at a depth of 1 m beneath the surface, the two lower panels show contrast 10 m below the surface. The two left-hand panels show contrast at a distance of 3 m between the target and eye, the right-hand panels at a distance of 25 m. Within the panels, each curve shows how the contrast achieved by one of the three visual pigments (P469, P521, P570) is affected by changing the spectral reflectance of the colored target

So far we have compared only $_{vis}C_r$, which is a ratio representing the light obsorbed from the target and the background, not the actual absorption of light from a target by each visual pigment. Thus, a given cone pigment may absorb less total light from target and background and yet yield a higher $_{vis}C_r$ than a pigment that absorbs more light. To demonstrate this, the relative absorption by each visual pigment of light from the blue, green, and yellow targets and of the background light are presented in Table 7. A specific example makes the point. The

Table 7. Relative quantal absorption by different visual pigments for a blue, green, and yellow target. Line of sight horizontal at 3 m depth in a clear tropical sea; calculated for visual pigments with absorbances of 0.4; B, G, and Y refer to blue, green, and yellow targets (see text for further description). Under the photic conditions described, P521 absorbed the maximum amount of light from the green target at 0 m range (assigned a value of 100)

Range (m)	P469			P521			P570		
	B	G	Y	B	G	Y	B	G	Y
0	90	44	6	80	100	72	50	83	95
3	86	41	6	76	91	62	48	75	79
10	78	36	8	69	74	44	44	61	53
25	64	29	11	57	50	26	37	40	27
Background		23			21			16	

${}_{vis}C_r$ at 10 m range for a green target is 2.5 for P521: (74–21)/21. For P570, ${}_{vis}C_r$ = 2.8: (61–16)/16. Yet P521 absorbs 1.2 to 1.3 times as many photons from the target and background as P570 (absorption ratio from target is 74/61 = 1.2; from background 21/16 = 1.3). The meaning of such analysis is uncertain, for we are not sure how the photoreceptors process and respond to this difference in absolute absorption. A complete answer would require neurophysiologic knowledge. Nevertheless, the analysis of photocontrast suggests that the answer may revolve about contrast enhancement of countershaded targets in the sea. How much blue-green light, or even green light, is contained in the light reflected from the blue dorsal surface of a fish we do not know. Measurements of the spectral radiance emanating from natural targets are mandatory before this problem can be pursued further.

4. Deep-Swimming Pelagic Fishes

Tunas, marlins, and the wahoo (*Acanthocybium solandri*) are active, diurnal predators with large eyes, but they live deeper beneath the surface than the species already discussed. Swimming depth varies in different species. The skipjack tuna (*Katsuwonus pelamis*) goes at least to 70 m, and the yellowfin tuna (*Neothunnus macropterus*) to 150 m (TALBOT and PENRITH, 1963; BLACKBURN, 1965). Although these fishes may feed at their swimming depth, they characteristically take prey located nearer the surface by striking up from below. In line with this behavior, the visual axis of the skipjack is oriented forward and upward; there are many cones in the retina (TAMURA and WISBY, 1963). We found only a single visual pigment in retinal extracts of these fishes (MUNZ and MCFARLAND, 1973, 1975). In the skipjack this was $P483_1$. The total range of λ_{max} in the visual pigments of wahoo, yellowfin, another tuna (*Euthunnus yaito*), and two marlins (*Istiompax marlina* and *Makaira ampla*) was only 482–486 nm.

As with the mahi mahi, we can compare the properties of the visual pigment with the light available for vision. We have plotted the spectra of downwelling and horizontal daylight at a depth of 25 m, where the skipjack swims (Fig. 28). The absorbance spectrum of its P483 is rather well matched to the light at this depth

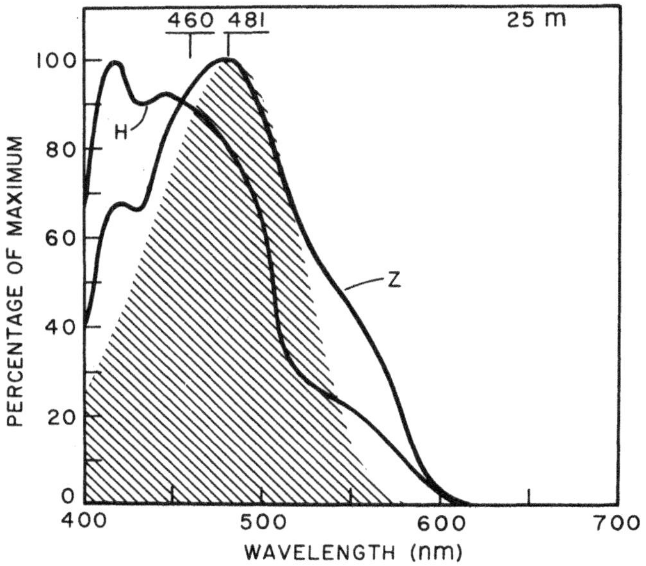

Fig. 28. Comparison of the background radiance spectra at 25 m depth (*heavy lines*: Z=zenith and H=horizontal line of sight) and the absorbance spectrum of visual pigment P483 of the skipjack (*area covered by diagonal lines*). Radiance and absorbance spectra have been scaled to the same maximum. The λP_{50} values of the horizontal (460 nm) and zenith (481 nm) radiance spectra are indicated. (MCFARLAND and MUNZ, 1975c)

in all directions of regard. Since the tunas have many retinal cones, we have assumed that the extract contains a cone pigment and a rod pigment, each with λ_{max} at 483 nm. The absorbance spectrum of the cone pigment is almost perfectly matched to the daylight downwelling from the surface (Fig. 28). This means that it would be excellent for silhouetting targets from below. In horizontal vision at 25 m depth, cone pigment P483 should also work quite well for both bright and dark targets (Fig. 29, *left panels*). An offset visual pigment, such as P521 of the mahi mahi, would not be as effective, however (*right panels*).

We have indicated that we found only a single visual pigment in retinal extracts of the skipjack and other tunas. The eyes of these diurnally feeding fishes are rich in cones, and we suggest that the rods and cones share the same visual pigment. There is support for such sharing in the MSP data of LIEBMAN (1972, 1973) on visual pigments of a frog, a salamander, two species of turtles, a sunfish (*Lepomis microlophus*), and two catfishes (*Ictalurus melas* and *I. punctatus*). DENTON et al. (1971) had already concluded that the rhodopsin of the dragonet (*Callionymus lyra*) is the pigment of some of its cones. Analysis of the photic environment and feeding behavior of tunas suggests that the light from both targets and backgrounds is so homochromatic that offset visual pigments would not be advantageous. These ideas received an independent confirmation from the electrophysiologic data of TAMURA et al. (1972). The spectral sensitivity based on the S-potential, retinal ganglion cell discharges, and the gross electroretinogram were recorded from the skipjack and two other species of tuna. In every case only

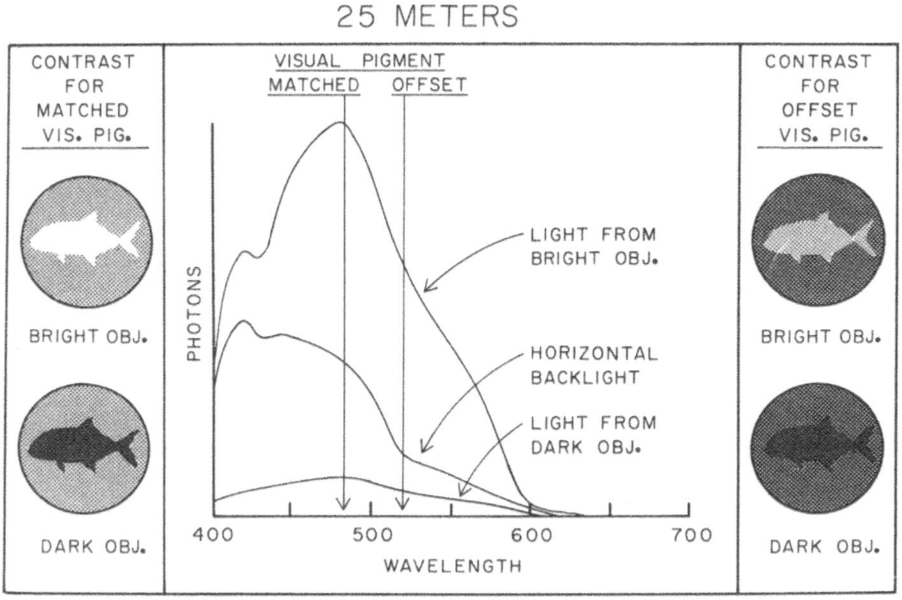

Fig. 29. Comparisons of the relative photocontrast of bright and dark objects viewed along the horizontal line of sight at 25 m beneath the surface by fishes with visual pigments matched to (P483), or (P521) offset from, the backlight. Assumptions as in Fig. 23. The matched pigment produces greater photocontrast than the offset pigment for both bright and dark objects. (McFarland and Munz, 1975c)

a spectral pattern of the luminosity type was obtained. There was never a response of the chromaticity type. The authors concluded that these tunas are probably color-blind. In their apparatus, narrow-band stimuli were obtained by interference filters. The filter at 497 nm produced a greater response than the adjacent 465- and 525-nm filters. Inspection of their records indicates that the maximum sensitivity is at a wavelength near 485 nm (rather than at 497 nm itself, as stated in their paper).

5. Coral-Reef Fishes

Reefs are more complex than pelagic environments, and the photic environment is complicated by varied backgrounds. Both the backgrounds (coral) and other reef animals are brightly and conspicuously colored. For coral-reef fishes, these color patterns may be cryptic or disruptive; or they can function in species recognition or as behavior releasers, etc. (Cott, 1940; Lorenz, 1962; Breder, 1972). The diurnal species typically have high densities of retinal cones. It seems probable that hue discrimination is of frequent occurrence in the photopic vision of these fishes.

Presumptive cone pigments were found in a high proportion of these fishes (84 of 99 species sampled; Munz and McFarland, 1973). These occurred in the retinal extracts as additional visual pigments, more red-sensitive than the scotopic rhodopsin (or rarely, porphyropsin). In more than 30 species, we were able to estimate the λ_{max} values of the presumptive cone pigment; the λ_{max} values ranged from 518 to 541 nm. We suspect, however, that in many species the "scotopic pigment" also occurs in one class of cones. Whether most of these species are trichromatic is even more uncertain in the absence of MSP data. Only in *Chaetodontoplus* have we found a mixture of three visual pigments (similar to *Coryphaena*). But they may, of course, occur more commonly because small quantities of a pigment may be undetectable in retinal extracts.

Lacking evidence of three pigments, we have tried to assess the possible roles of the binary mixtures in terms of contrast enhancement by matching or being offset from the background spectrum. In general, one cone pigment has its λ_{max} near 525 nm, and the other (also scotopic) pigment has its λ_{max} between 480 and 500 nm (MUNZ and MCFARLAND, 1973, 1975). In each species, the λ_{max} values and their spectral separation are like those of pelagic fishes that possess multiple pigments. We believe that these similarities result from the action of similar selective forces related to photic conditions and visual requirements. For example, at any given depth beneath the surface, the spectral distribution and total radiance of downwelling and horizontal light are similar for the reef and pelagic habitats (MCFARLAND and MUNZ, 1975b). Close to coral, the horizontal background would be different. The most dramatic difference, however, is in the downward visual field. In the coral-reef habitat, light upwelling after reflection from the substrate is often much broader in spectrum (like downwelling light) and more intense than upwelling light in the pelagic habitat (Section C.IV and MCFARLAND and MUNZ, 1975b). This is controlled by the reflective properties of the substrate and by the depth (optical path length).

In general, therefore, the more blue-sensitive pigment will *match* the spectrum of horizontal background light (except when looking directly at coral; in this case the green-sensitive pigment might provide the best match to the background). The green-sensitive pigment is typically *offset* from the horizontal background, enhancing contrast of reflective targets. Looking upward, any visual pigment matches the downwelling light, especially near the surface. These relationships are as already described for pelagic fishes. But looking downward, the situation is more complex. The spectrum of upwelling light reflected from white coral sand is broad, so that almost any visual pigment would match it. The light reflected from coral or coral rubble is even richer in longer wavelengths (Fig. 6 in MCFARLAND and MUNZ, 1975b). This means that, especially in shallow water, green-sensitive pigments may be used as either offset or matching pigments, depending on the background and direction of view.

We have already discussed the correlation between the spectral positions of scotopic pigments and habitat depth (Section B.I.2). There are also differences between the cone pigments of surface-limited and deep-swimming pelagic fishes, which are related to the photic environment. Is this true as well of the green-sensitive pigments of reef fishes? Most reef fishes are relatively shallow-water forms, being abundant only in the upper 20 m. A pomacentrid (*Chromis verater*) and a labrid (*Bodianus bilunulatus*) are characteristic of greater depths (approximately 20 to 100 m; BROCK and CHAMBERLAIN, 1968; HOBSON, 1974). In both species the more blue-sensitive pigment has its λ_{max} at 480 nm. As a cone pigment, $P480_1$ would be comparable to the visual pigment of tunas, matching both downwelling light and that from the horizontal background (away from coral). Unlike the tunas, however, these deeper-living diurnal reef species also have a green-sensitive visual pigment ($P518_1$ in *Chromis verater* and $P521_1$ in *Bodianus bilunulatus*). This suggests that, unlike the tunas, the reef fishes which inhabit a blue photic environment have maintained or evolved additional cone pigments to cope with the more complex backgrounds imposed by coral, rubble, and sand.

Another pomacentrid (*Abudefduf abdominalis*) is the only surface-limited reef species in our sample from which we extracted two visual pigments, $P496_1$ and $P527_1$ (MUNZ and MCFARLAND, 1973, 1975). We found no evidence for any more blue-sensitive pigment comparable to P469 of *Coryphaena*. (*Chaetodontoplus*, which has three visual pigments, is from intermediate depths.) Because of the reflections from coral and coral sand in its shallow-water habitat, P496 of *Abudefduf* should be generally well matched to the light in the horizontal field of view. The more blue-sensitive pigments of the many fishes that are intermediate in depth distribution have λ_{max} values in the range from 489 to 496 nm (Fig. 3), between those of surface and deeper-living species.

There may be a similar trend in the green-sensitive pigments of these fishes (Table 8). The deeper-dwelling *Chromis verater* and *Bodianus bilunulatus* have the lowest λ_{max} values, while the surface-limited *Abudefduf abdominalis* has the highest. This has been discussed in more detail elsewhere (MCFARLAND and MUNZ, 1975c), but cannot be supported statistically, because our sample included so few surface-limited and deep-water reef species. In general, however, it appears that the concepts developed in connection with pelagic fishes are also applicable to reef species. There is more variety of background in reef habitats, which may mean that a given visual pigment can function either as a matching or as an offset pigment, depending on the direction of view. Because of its effects on the photic environment, habitat depth may have influenced the selection of photopic pigments in reef fishes, as it has in pelagic species.

The A_2 pigments of scarids, labrids, and balistids merit special mention. All of the parrotfishes (Scaridae) that have been examined (one species by DARTNALL and LYTHGOE, 1965; 10 species by MUNZ and MCFARLAND, 1975) have an A_1 pigment with λ_{max} at about 484 nm and a green-sensitive A_2 pigment (Table 8). In no species is the spectral difference between the A_1 and A_2 pigments appropriate to the concept of "paired pigments" ($\lambda_1 = 0.60\,\lambda_2 + 186$; DARTNALL and LYTHGOE, 1965; see also BRIDGES, 1965a, 1972; MUNZ and SCHWANZARA, 1967). It is much more likely that the green-sensitive pigments of scarids are cone pigments. We suspect that the blue-sensitive pigment is derived from a second class of cones, in addition to being the scotopic pigment. The low λ_{max} and small variation of scarid blue-sensitive pigments have already been discussed (Section B.III.3). As parts of the photopic system, both the blue-sensitive and the green-sensitive pigments should function as in other reef fishes. Only the fact that the green-sensitive pigments are based on vitamin A_2 is unusual among marine fishes, but this is also characteristic of many of the closely related labrids. Their retinal histology has not been studied in enough detail to show whether scarids have double cones (like labrids) or the more typical twin cones.

It is less easy to characterize the visual pigments of wrasses (Labridae) or to understand their functions. Although it was reported relatively early by WALD

Table 8. Green-sensitive (presumptive cone) pigments extracted from the retinae of diurnal coral-reef fishes. Data from MUNZ and MCFARLAND (1975)

Visual pigment	Labrids and balistids	λ_{max} (nm)	Other	λ_{max} (nm)
			(nonscarids)	
A_1	*Bodianus bilunulatus*	521.4	*Heniochus acuminatus*	520.3
	Cheilinus diagrammus	524.2	*Abudefduf abdominalis*	527.1
	C. rhodochrous	526.0	*Chromis verater*	517.8
	C. trilobatus	530.8	*Dascyllus albisella*	524.0
	C. undulatus	525.1	*D. auranus*	525.2
	Choerodon zosteropherus	524.8	*D. reticulatus*	521.7
	Coris aygula	529.0	*Naso brevirostris*	522.6
	Rhinecanthus aculeatus[a]	523.6	*N. hexacanthus*	518.9
	R. verrucosus[a]	525.8	*Chaetodontoplus mesoleucas*	519.3
	Mean	525.6	Mean	521.9
	SD ±	2.8	SD ±	3.1
			(scarids)	
A_2	*Halichoeres centriquadrus*	538.5	*Calotomus spinidens*	517.8
	Thalassoma ballieui	541.0	*Scaridea zonarcha*	520.5
	T. duperrey	531.1	*Scarus aeruginosus*	525.6
	Balistapus undulatus[a]	536.9	*S. chlorodon*	524.4
			S. dubius	520.0
			S. fasciatus	520.7
			S. forsteri	519.2
			S. ghobban	526.9
			S. niger	530.2
			S. oviceps	522.1
	Mean	536.9	Mean	522.7
	SD ±	4.2	SD ±	3.9

[a] Balistid; others in this box are labrids.

(1941) that members of this exclusively marine family possess the "fresh-water pigment," porphyropsin, relatively little experimental work was done in the following 30 years. This cannot be due to unavailability, for this is a large and diverse family. Most species are tropical, and the family occurs in both the Indo-Pacific and the Atlantic Oceans. It is also represented in both the North Pacific and the Atlantic, even reaching the Baltic Sea. After an apparent, but somewhat uncertain, corroboration of WALD's report by KAMPA (1952), MUNZ (1958a) published the results of the first thorough investigation of visual pigments from a labrid fish, *Pimelometopon pulchrum*, from California. This species has two visual pigments, but both are based on vitamin A_1 ($P497_1$ and $P520_1$). A similar result was reported (without supporting data) by DARTNALL and LYTHGOE (1965; modified by LYTHGOE, 1972): *Labrus mixtus* has $P486_1$ and $P527_1$. On the other hand, WALD (1960) has confirmed (but also without publishing any data) the presence of A_2 pigments in three other wrasses: $P513_2$ in *Crenilabrus festivus* and *C. massa* and a mixture of $P492_1$ and $P510_2$ in *Labrus merula*. Now, according to the relationship worked out by MUNZ and SCHWANZARA (1967) and by BRIDGES (1972), $P492_1$ and $P510_2$ could be "paired pigments" based on the same opsin. Whether either or both could also be photopic pigments is, of course, unknown.

The paucity of spectrometric data on visual pigments of this interesting group is probably related to their extreme diurnality (HOBSON, 1972). It is typically difficult or impossible to free the retinae of labrids from the pigment epithelium, no matter how long they have been dark-adapted. Because of light-absorbing impurities and of instability, the retinal extracts are among the most difficult to study that we have encountered. In our survey of tropical marine fishes, we reported on 11 species of labrid fishes (MUNZ and MCFARLAND, 1973). The retinal extracts of all species had binary mixtures of visual pigments. In 10 of these, both visual pigments were identified (Table 8). In seven species both pigments are based on vitamin A_1; and in the other three, both are based on vitamin A_2. Two California labrids, each with two vitamin A_2-based visual pigments, can be added: *Oxyjulis californica* with $P512_2$ and $P538_2$, and *Halichoeres semicinctus* with $P513_2$ and $P537_2$. Thus, of the 17 labrids so far examined, nine species have only vitamin A_1, seven A_2 exclusively, and only *Labrus merula* has a mixture of A_1 and A_2 pigments which may be a rhodopsin-porphyropsin pair (WALD, 1960).

The triggerfishes (Balistidae) are not related to the wrasses and parrotfishes, but we found visual pigments based on vitamin A_2 in one species (Table 8). This was not altogether surprising, because rhodopsin-porphyropsin mixtures had been reported in two pufferfishes, which belong to the same order ($P500_1$ and $P522_2$ in *Sphaeroides annulatus*, MUNZ, 1958b; $P503_1$ and $P527_2$ in *Tetraodon schoutedeni*, SCHWANZARA, 1967). Of the six balistids that we studied (MUNZ and MCFARLAND, 1973), all had mixtures of visual pigments. But vitamin A_2 pigments occurred only in a single species. We were also able to identify both A_1 pigments in two of the other species. Because these balistid visual pigments are apparently so similar to those of labrids, we have included them in the same categories (Table 8).

Several interesting comparisons can be made between the visual pigments of labrids (and balistids) and those of other reef fishes. The A_2 pigments of labrids have a mean λ_{max} at a longer wavelength than the A_2 pigments of scarids, and the difference is highly significant ($P < 0.001$). This is also true of the comparison within the labrid-balistid grouping: the A_2 pigments have their mean λ_{max} at a longer wavelength than the A_1 pigments ($P < 0.001$). When the green-sensitive A_1 pigments of labrids and balistids are compared with those of other reef fishes, there is also a significant difference ($P = 0.016$ in a two-tailed t-test). If these same data are subjected to a two-way analysis of variance, both the vitamin A and the classification factors are themselves highly significant, but so is the interaction between factors ($P < 0.001$). To summarize, labrid and balistid fishes have green-sensitive, presumptive cone pigments based on both vitamins A_1 and A_2, with λ_{max} values at significantly longer wavelengths than those of other reef fishes. In contrast, the blue-sensitive A_1 pigments of the labrids and balistids are not located at longer wavelengths than those of other fishes ($P = 0.50$). The seven labrids and one balistid with vitamin A_2 systems, however, do have their "blue-sensitive" pigments with peaks at longer wavelengths (between $P502_2$ and $P513_2$). The functional meaning of this fact is not known.

Why should the green-sensitive pigments of labrids (and apparently balistids) have their maxima at longer wavelengths than those of other reef fishes? What is the functional significance of vitamin A_2 pigments in some of these fishes (and in parrotfishes)? The second question is perhaps a part of the first, because A_2

pigments are usually considered a device for shifting spectral sensitivity to longer wavelengths. We should ask why the parrotfishes have A_2 pigments at all, for these pigments are *not* shifted to longer wavelengths than the green-sensitive A_1 pigments of nonlabrids ($P = 0.30$). All three families are noteworthy for the richness and variety of their body coloration. Intricate patterns are of frequent occurrence. These groups are prominent among the diurnal fauna of coral reefs, and the labrids especially are abundant, with many species (HOBSON, 1974). Parrotfishes are herbivores, crushing dead coral for the algae it contains. But many labrids and balistids are omnivorous, often feeding on cryptically colored invertebrates that conceal themselves on the reef (HOBSON, 1974). Well-developed color vision might be particularly advantageous, therefore, both in feeding and in species recognition. For most of these species the food items are usually viewed at close range against the coral-reef background (rather than against blue water). And in many, the recognition marks and patterns are rich in red colors, as compared, for example, to the butterflyfishes (Chaetodontidae). These facts may be related to the significant shift of the green-sensitive pigments to longer wavelengths. Our conjectures are intended to help focus further studies, and not as a definitive interpretation.

The notion that labrid fishes may have unusually well-developed color vision receives some support from the histologic studies of ENGSTRÖM (1963a, b). Most teleost fishes have equal twin cones in the retina, together with one or two classes of single cones. Labrids, in contrast, have unequal double cones and two types of single cones. As already described (Section C. II. 2), it is the double cones in which the two members never have the same visual pigment, whereas twins apparently always share the same pigment. It would be interesting, indeed, to learn by MSP whether the double cones of labrids are like those of cyprinids in this respect. Our findings on the presumptive cone pigments of reef fishes merit one final caution, especially in connection with the labrids. Any truly red-sensitive visual pigments would probably have been bleached unintentionally by the red "safelights" that we used during retinal removal and extraction. It would be a mistake, therefore, to forget that any of these reef fishes, but particularly the labrids and balistids, may be true trichromats and have the ability to discriminate red light from other wavelengths. Once more, we can only say that MSP is needed.

VI. The Evolution of Color Vision

Color vision must have evolved independently in several different vertebrate groups, as well as in some invertebrates. The ability to discriminate between visual stimuli on the basis of wavelength requires two or more classes of photoreceptor cells, each with a different spectral sensitivity. The advantages of color vision seem so self-evident that it is difficult for us to imagine its gradual development in an evolutionary sequence. Several authors (e.g., PARSONS, 1915; EDRIDGE-GREEN, 1920; LADD-FRANKLIN, 1929; WILLMER, 1949; PICKFORD, 1951) have considered this problem, usually discussing possible modifications of the photoreceptors and neuronal layers in the retina, but without asking why these changes might have been favored by natural selection. WALLS (1942) not only asked why there should be color vision, but believed that the answer must be related to the requirements of contrast between objects and the backgrounds against which they are viewed. At that time, WALLS felt the lack of basic information so strongly that he urged "a moratorium on new genetic theories of color vision." How truly remarkable was the insight much earlier of Alfred Russel WALLACE (1891), who wrote:

"The primary necessity which led to the development of the sense of colour was probably the need of distinguishing objects much alike in form and size, but differing in important properties, such as ripe and unripe, or eatable and poisonous fruits, flowers with honey or without, the sexes of the same or of closely allied species. In most cases the strongest contrast would be the most useful, especially as the colours of the objects to be distinguished would form but minute spots or points when compared with the broad masses of tint of sky, earth, or foliage against which they would be set."

Clearly, WALLACE recognized not only the importance of photocontrast but also the relevance of a narrow-spectrum background. These two elements have been critical in our own thinking.

As a direct result of studying the visual mechanisms of tropical marine fishes, in the context of both their visual behavior patterns and the photic environments in which they live, we believe that the evolution of color vision has become clearer. We have analyzed photopic vision in terms of sensitivity and contrast (Section C.V). As LYTHGOE (1966, 1968) had argued earlier (but in connection with the scotopic system), the need for visual contrast can favor the selection of visual pigments *offset* from the spectrum of the background. But this reasoning applies only to relatively bright targets which have a broader spectrum than that of the background (or a different one). The recognition of dark targets, which are silhouetted against the background, is enhanced by visual pigments that *match* the background. Thus, the need to see different targets with maximal contrast can select for the multiple receptor mechanisms which are the basis of color vision. The mahi mahi and many other diurnal fishes have both offset and matched cone pigments; presumably they have at least some wavelength discrimination. Multiple cone types may not be universally advantageous, however; for the deeper-swimming tunas appear to be true monochromats. In their more homochromatic environment, both targets and the background are "bluish." Tunas apparently have both single and twin cones that contain the same visual pigment and which matches the available light. These ideas were developed in a study of pelagic fishes (McFARLAND and MUNZ, 1975c), primarily because light measurements in the open sea are simplest to interpret. But the photopic vision of fishes from coral reefs also conforms to these ideas; so far as we know, the data on fishes from other habitats are also compatible with them.

We have suggested that color vision may first have evolved among the vertebrates as long ago as the Silurian period (McFARLAND and MUNZ, 1975c). Some of the ostracoderms gave up their benthic habits at that time and became free-swimming. As this happened, their paired, dorsally located eyes moved away from the median eye, down onto the sides of the head. No doubt vision came to have increasing importance in these more pelagic animals for the location of their food.

So far as we can guess, the seas were blue then as now. For a benthic animal in shallow water, any visual pigment would match the broad spectrum of light downwelling from the surface. Away from the bottom, there could be selection of visual pigments matched to, or offset from, the blueness of the horizontal visual field. Multiple receptor mechanisms were developed for daytime use, when the light intensity permits the separate sensory pathways required for high visual acuity. These are the elements needed for color vision. Thus, we hypothesize that color vision first evolved in fishlike vertebrates to provide better photocontrast, and that a narrowing of the spectrum of background light gave impetus to the selective process.

Neither the need for visual contrast nor the narrowing of the spectrum of background light is restricted to fishes in aquatic habitats. One can easily imagine that color vision has evolved in terrestrial animals in response to the same factors. Do not birds, primates, ground squirrels, and others view targets against "the broad masses of tint of sky, earth, or foliage" that WALLACE (1891) described? But how far could the discussion be advanced beyond what he wrote 85 years ago? CRESCITELLI (1972) and DAW (1973) have recently reviewed aspects of this problem, and both of them were fully aware of the gaps in the available data. Rather than attempting to extend our theory, we should prefer to urge the need for information about the photic environments, visual behaviors, and photopic mechanisms of terrestrial animals. Only for a few species (fishes as well as other vertebrates), do we even know whether they are monochromats, dichromats, trichromats, or other. In some cases the scotopic system appears to be involved in color vision (see DAW, 1973), but this possibility has gone largely untested. We hope that the results of our investigation of photopic vision in fishes will encourage others to pursue these problems and that the ideas expressed in this chapter will offer them a basis for discussion.

D. Summary

The theme of this chapter is that multilevel approaches are essential for progress toward understanding vision. Photic environments must be measured with regard to their visual role: photoreceptors "count photons," rather than "reading out in microwatts;" and eyes are usually not irradiance detectors. The angular distribution of light is important, as is the specific reflectance of particular targets. This information about light must be integrated with anatomic, biochemical, and physiologic data. Consideration of visual behavior is also essential: when and how is a visual system used? Only in this way can we hope to learn how visual systems function and how they may have evolved.

We have taken an integrative approach to examine the scotopic systems of a large group of tropical marine fishes. Twilight periods are most precarious for both diurnal and nocturnal species, for predation is at a peak during these crucial sunrise and sunset periods. Light measurements, both in air and under water, have revealed that the spectrum changes during twilight, becoming less orange (more blue). This suggests an explanation for the otherwise paradoxical Purkinje shift: as moonlight and starlight are even somewhat more reddish than sunlight, why should scotopic sensitivities generally be shifted toward the blue? It is evident that the scotopic systems of both diurnal and nocturnal fishes have evolved to maximize sensitivity during twilight.

Much is now known about the distribution of visual pigments based on vitamin A_2 and about the various factors, such as temperature, light, and hormone treatment, that can affect the interconversion of vitamins A_1 and A_2 in some fishes. The biologic significance of porphyropsins remains enigmatic, however. Too little is known about photic environments in fresh water; more measurements are needed. Perhaps electrophysiologic or behavioral experiments will provide clues. Interactions between the photopic and scotopic systems must also be considered.

Morphologically, their retinal cones divide bony fishes into two broad categories: most species have twin cones, with nearly identical conjugate members in each pair; others have double cones, in which the conjugate members are dissimilar. Although very few species of either type have been studied, a basic dichotomy seems to exist with respect to the visual pigments of these cones. In twin cones, the members both possess the same visual pigment; in double cones, they do not. The advantage of dissimilar visual pigments in double cones seems obvious, in enhancing contrast; the special role of twin cones, however, is less clear. This dichotomy in cone types makes it unwise to generalize from the well-studied goldfish (a double-cone species) to any twin-cone fish. Trichromatic color vision has not yet been demonstrated in any bony fishes except a few cyprinids (e.g., goldfish). On the contrary, some twin-cone fishes seem to be monochromats. Further behavioral and electrophysiologic experiments are badly needed, and with a wider variety of fishes. MSP remains the technique of choice for the precise identification of cone pigments, but a new cytochemical method also promises to help localize these visual pigments in the retinal mosaic of cones.

Analysis of the photopic systems in tropical pelagic fishes suggests that different cone pigments have been selected to give maximal visual contrast. A visual pigment matched to the spectrum of background light will make a nonreflective (dark) target most readily visible. With reflective targets, on the other hand, a cone pigment that is offset from the background spectrum may enhance the contrast between the bright target and darker background. Thus, multiple cone pigments—and color vision—may have evolved to improve the contrast of varied targets in homochromatic environments.

References

Ali, M. A.: Les réponses rétinomotrices: caractères et mécanismes. Vision Res. **11**, 1225—1288 (1971).

Ali, M. A., Wagner, H.-J.: Visual pigments: phylogeny and ecology. In: Vision in fishes. New approaches in research. New York: Plenum 1975.

Allen, C. W.: Astrophysical quantities, 1st ed. London: Athlone, 1955.

Allen, C. W.: Astrophysical quantities, 3rd ed. London: Athlone, 1973.

Allen, D. M.: Photic control of the proportions of two visual pigments in a fish. Vision Res. **11**, 1077—1112 (1971).

Allen, D. M., McFarland, W. N.: The effect of temperature on rhodopsin-porphyropsin ratios in a fish. Vision Res. **13**, 1303—1309 (1973).
Allen, D. M., McFarland, W. N., Munz, F. W., Poston, H. A.: Changes in the visual pigments of trout. Canad. J. Zool. **51**, 901—914 (1973).
Anctil, M.: Structure de la rétine chez quelques téléostéens marins du plateau continental. J. Fish. Res. Bd. Can. **26**, 597—628 (1969).
Arnott, H. J., Best, A. C. G., Ito, S., Nicol, J. A. C.: Studies on the eyes of catfishes with special reference to the tapetum lucidum. Proc. roy. Soc. B **186**, 13—36 (1974).
Babcock, H. W., Johnson, J. J.: A spectrophotometric study of the light of the night sky. Astrophysical J. **94**, 271—275 (1941).
Beatty, D. D.: A study of the succession of visual pigments in Pacific salmon *(Oncorhynchus)*. Canad. J. Zool. **44**, 429—455 (1966).
Beatty, D. D.: Visual pigment changes in juvenile kokanee salmon in response to thyroid hormones. Vision Res. **9**, 855—864 (1969a).
Beatty, D. D.: Visual pigments of the burbot, *Lota lota*, and seasonal changes in their relative proportions. Vision Res. **9**, 1173—1183 (1969b).
Beatty, D. D.: Visual pigment changes in salmonid fishes in response to exogenous 1-thyroxine, bovine TSH and 3-dehydroretinol. Vision Res. **12**, 1947—1960 (1972).
Beatty, D. D.: Visual pigments of the American eel *Anguilla rostrata*. Vision Res. **15**, 771—776 (1975a).
Beatty, D. D.: Rhodopsin-porphyropsin changes in paired-pigment fishes. In: Vision in fishes. New approaches in research. New York: Plenum, 1975b.
Blackburn, M.: Oceanography and the ecology of tunas. Oceanogr. Mar. Biol. Ann. Rev. **3**, 299—322 (1965).
Blaxter, J. H. S.: Light. Fishes. In: Marine ecology. Vol. 1. Environmental factors. Part 1. New York: Wiley-Interscience, 1970.
Bowmaker, J. K.: Kundt's rule: the spectral absorbance of visual pigments *in situ* and in solution. Vision Res. **12**, 529—548 (1972).
Bowmaker, J. K., Loew, E. R., Liebman, P. A.: Variation in the λ_{max} of rhodopsin from individual frogs. Vision Res. **15**, 997—1003 (1975).
Breder, Jr., C. M.: On the relationship of teleost scales to pigment patterns. Contr. Mote Mar. Lab. **1**, 1—79 (1972).
Bridges, C. D. B.: Effect of season and environment on the retinal pigments of two fishes. Nature (Lond.) **203**, 191—192 (1964a).
Bridges, C. D. B.: Variation of visual pigment amongst individuals of an American minnow, *Notemigonus crysoleucas boscii*. Vision Res. **4**, 233—239 (1964b).
Bridges, C. D. B.: The grouping of fish visual pigments about preferred positions in the spectrum. Vision Res. **5**, 223—238 (1965a).
Bridges, C. D. B.: Variability and relationships of fish visual pigments. Vision Res. **5**, 239—251 (1965b).
Bridges, C. D. B.: Visual pigments in a fish exposed to different light-environments. Nature (Lond.) **206**, 1161—1162 (1965c).
Bridges, C. D. B.: Spectroscopic properties of porphyropsins. Vision Res. **7**, 349—369 (1967).
Bridges, C. D. B.: The rhodopsin-porphyropsin visual system. In: Handbook of sensory physiology. Vol. VII/1. Photochemistry of vision. Berlin-Heidelberg-New York: Springer 1972.
Bridges, C. D. B., Delisle, C. E.: Brief observations concerning the visual pigments of some selected fishes from Lake Heney, Quebec, a relict of glacial Lake Gatineau. Vision Res. **14**, 187—193 (1974).
Bridges, C. D. B., Yoshikami, S.: The rhodopsin-porphyropsin system in freshwater fishes—1. Effects of age and photic environment. Vision Res. **10**, 1315—1332 (1970).
Brock, V. E., Chamberlain, T. C.: A geological and ecological reconnaissance off western Oahu, Hawaii, principally by means of the research submarine "Asherah". Pacific Sci. **22**, 373—401 (1968).
Carlisle, D. B., Denton, E. J.: On the metamorphosis of the visual pigments of *Anguilla anguilla* (L.). J. mar. biol. Ass. U. K. **38**, 97—102 (1959).
Clarke, W. D.: Function of bioluminescence in mesopelagic organisms. Nature. (Lond.) **198**, 1244—1246 (1963).
Clarke, W. D., Denton, E. J.: Light and animal life. In: The sea, Vol. 1. New York: Interscience, 1962.
Collette, B. B., Talbot, F. H.: Activity patterns of coral reef fishes with emphasis on nocturnal-diurnal changeover. Bull. Nat. Hist. Mus. Los Angeles County **14**, 98—124 (1972).
Cott, H. B.: Adaptive coloration in animals. London: Methuen, 1940.

Crescitelli, F.: The visual cells and visual pigments of the vertebrate eye. In: Handbook of sensory physiology. Vol. VII/1, Photochemistry of vision. Berlin-Heidelberg-New York: Springer 1972.

Cristy, M.: Effects of prolactin and thyroxine on the visual pigments of trout, *Salmo gairdneri*. Gen. comp. Endocr. **23**, 58—62 (1974).

Cristy, M.: Effects of temperature and light intensity on the visual pigments of rainbow trout. Vision Res. **16**, 1225—1228 (1976).

Dartnall, H. J. A.: Visual pigment 467, a photosensitive pigment present in tench retinae. J. Physiol. (Lond.) **116**, 257—289 (1952).

Dartnall, H. J. A.: The interpretation of spectral sensitivity curves. Brit. med. Bull. **9**, 24—30 (1953).

Dartnall, H. J. A.: Visual pigments of the bleak *(Alburnus lucidus)*. J. Physiol. (Lond.) **128**, 131—156 (1955).

Dartnall, H. J. A.: Visual pigments of colour vision. In: Mechanisms of colour discrimination. New York-London-Oxford-Paris: Pergamon, 1960.

Dartnall, H. J. A.: The photobiology of visual processes. In: The eye. Vol. 2. The visual process. New York-London: Academic Press, 1962.

Dartnall, H. J. A.: The photosensitivities of visual pigments in the presence of hydroxylamine. Vision Res. **8**, 339—358 (1968).

Dartnall, H. J. A., Lander, M. R., Munz, F. W.: Periodic changes in the visual pigment of a fish. In: Progress in photobiology. Amsterdam: Elsevier, 1961.

Dartnall, H. J. A., Lythgoe, J. N.: The spectral clustering of visual pigments. Vision Res. **5**, 81—100 (1965).

Daw, N. W.: Neurophysiology of color vision. Physiol. Rev. **53**, 571—610 (1973).

Denton, E. J., Gilpin-Brown, J. B., Wright, P. G.: On the "filters" in the photophores of mesopelagic fish and on a fish emitting red light and especially sensitive to red light. J. Physiol. (Lond.) **208**, 72 P—73 P (1970).

Denton, E. J., Muntz, W. R. A., Northmore, D. P. M.: The distribution of visual pigment within the retina in two teleosts. J. mar. biol. Ass. U. K. **51**, 905—915 (1971).

Denton, E. J., Nicol, J. A. C.: A survey of reflectivity in silvery teleosts. J. mar. biol. Ass. U. K. **46**, 685—722 (1966).

Denton, E. J., Warren, F. J.: Visual pigments of deep-sea fish. Nature (Lond.) **178**, 1059 (1956).

Denton, E. J., Warren, F. J.: The photosensitive pigments in the retinae of deep-sea fish. J. mar. biol. Ass. U. K. **36**, 651—662 (1957).

Domm, S. B., Domm, A. J.: The sequence of appearance at dawn and disappearance at dusk of some coral reef fishes. Pacific Sci. **27**, 128—135 (1973).

Duntley, S. Q.: Underwater visibility. In: The sea. New York: Interscience, 1962.

Duntley, S. Q.: Light in the sea. J. opt. Soc. Am. **53**, 214—233 (1963).

Edridge-Green, F. W.: The physiology of vision, 2nd ed. London: G. Bell, 1920.

Engström, K.: Cone types and cone arrangement in the retina of some cyprinids. Acta Zool. **41**, 277—295 (1960).

Engström, K.: Cone types and cone arrangement in the retina of some gadids. Acta Zool. **42**, 227—243 (1961).

Engström, K.: Structure, organization and ultrastructure of the visual cells in the teleost family Labridae. Acta Zool. **44**, 1—41 (1963a).

Engström, K.: Cone types and cone arrangements in teleost retinae. Acta Zool. **44**, 179—243 (1963b).

Engström, K., Ahlbert, I. B.: Cone types and cone arrangement in the retina of some flatfishes. Acta Zool. **44**, 119—129 (1963).

Fineran, B. A., Nicol, J. A. C.: Studies on the eyes of New Zealand parrot-fishes (Labridae). Proc. roy. Soc. B **186**, 217—247 (1974).

Fricke, H. W.: Der Einfluß des Lichtes auf Körperfärbung und Dämmerungsverhalten des Korallenfisches *Chaetodon melanotus*. Marine Biol. **22**, 251—262 (1973).

Fry, F. E. J.: The effect of environmental factors on the physiology of fish. In: Fish physiology. Vol. VI. Environmental relations and behavior. New York-London: Academic Press, 1971.

Greenwood, H. P., Rosen, D. E., Weitzman, S. H., Meyers, G. S.: Phyletic studies of teleostean fishes, with a provisional classification of living forms. Bull. Am. Mus. Nat. Hist. **131**, 341—455 (1966).

Halldal, P.: Light and photosynthesis of different marine algal groups. In: Optical aspects of oceanography. New York: Academic Press, 1974.

Hanaoka, T., Fujimoto, K.: Absorption spectrum of a single cone in carp retina. Jap. J. Physiol. **7**, 276—285 (1957).
Hárosi, F. I., MacNichol, E. F., Jr.: Visual pigments of goldfish cones: spectral properties and dichroism. J. gen. Physiol. **63**, 279—304 (1974).
Herter, K.: Die Fischdressuren und ihre sinnesphysiologischen Grundlagen. Berlin: Akademie, 1953.
Hobson, E. S.: Diurnal-nocturnal activity of some inshore fishes in the Gulf of California. Copeia **1965**, 291—302.
Hobson, E. S.: Predatory behavior of some shore fishes in the Gulf of California. Bur. Sport Fish. and Wildlife, Res. Rep. **73**, 1—92 (1968).
Hobson, E. S.: Activity of Hawaiian reef fishes during the evening and morning transitions between daylight and darkness. Fish. Bull. Nat. Mar. Fish. Ser. **70**, 715—740 (1972).
Hobson, E. S.: Feeding relationships of teleostean fishes on coral reefs in Kona, Hawaii. Fish. Bull. Nat. Mar. Fish. Ser. **72**, 915—1031 (1974).
Jaquest, W. L., Beatty, D. D.: Visual pigment changes in the rainbow trout, *Salmo gairdneri*. Canad. J. Zool. **50**, 1117—1126 (1972).
Kampa, E. M.: [untitled] Year Book Am. Philos. Soc. 1952, 161—163 (1952).
Kampa, E. M.: Daylight penetration measurements in three oceans. Int. Un. Geod. Geophys. Monog. **10**, 91—95 (1961).
Kampa, E. M.: Underwater daylight and moonlight measurements in the eastern North Atlantic. J. mar. biol. Ass. U. K. **50**, 397—420 (1970).
Kobayashi, H.: A comparative study on electroretinogram in fish with special reference to ecological aspects. J. Shimonoseki Coll. Fish. **11**, 407—538 (1962).
Ladd-Franklin, C.: Colour and colour theories. New York: Harcourt, Brace and London: Kegan Paul, Trench, and Trubner, 1929.
Laufer, M., Millán, E.: Spectral analysis of L-type S-potentials and their relation to photopigment absorption in a fish *(Eugerres plumieri)* retina. Vision Res. **10**, 237—251 (1970).
Le Grand, Y.: Light, colour and vision. New York: Wiley, 1957, p. 123.
Liebman, P. A.: Microspectrophotometry of photoreceptors. In: Handbook of sensory physiology. Vol. VII/1, Photochemistry of vision. Berlin-Heidelberg-New York: Springer 1972.
Liebman, P. A.: Microspectrophotometry of visual receptors. In: Biochemistry and physiology of visual pigments. Berlin-Heidelberg-New York: Springer 1973.
Liebman, P. A., Entine, G.: Sensitive low-light-level microspectrophotometer: detection of photosensitive pigments of retinal cones. J. opt. Soc. Am. **54**, 1451—1459 (1964).
Loew, E. R., Dartnall, H. J. A.: Vitamin A_1/A_2-based visual pigment mixtures in cones of the rudd. Vision Res. **16**, 891—896 (1976).
Lorenz, K.: The function of colour in coral reef fishes. Proc. roy. Inst. Gr. Br. **39**, 282—296 (1962).
Lyall, A. H.: Occurrence of triple and quadruple cones in the retina of the minnow *(Phoxinus laevis)*. Nature (Lond.) **177**, 1086—1087 (1956).
Lyall, A. H.: The growth of the trout retina. Quart. J. microscop. Sci. **98**, 101—110 (1957a).
Lyall, A. H.: Cone arrangements in teleost retinae. Quart. J. microscop. Sci. **98**, 189—201 (1957b).
Lythgoe, J. N.: Visual pigments and underwater vision. In: Light as an ecological factor. Oxford: Blackwell, 1966.
Lythgoe, J. N.: Visual pigments and visual range underwater. Vision Res. **8**, 997—1012 (1968).
Lythgoe, J. N.: The adaptation of visual pigments to the photic environment. In: Handbook of sensory physiology, Vol. VII/1. Photochemistry of vision. Berlin-Heidelberg-New York: Springer 1972a.
Lythgoe, J. N.: List of vertebrate visual pigments. In: Handbook of sensory physiology, Vol. VII/1. Photochemistry of vision. Berlin-Heidelberg-New York: Springer 1972b.
MacNichol, Jr., E. F., Svaetichin, G.: Electric responses from isolated retinas of fishes. Amer. J. Ophthal. **46**, 26—40 (1958).
Marc, R. E., Sperling, H. G.: Color receptor identities of goldfish cones. Science **191**, 487—489 (1976).
Marks, W. B.: Visual pigments of single goldfish cones. J. Physiol. (Lond.) **178**, 14—32 (1965).
Marshall, N. B.: Aspects of deep sea biology. London: Hutchinson's, 1954.
Marshall, N. B.: The life of fishes. New York: Universe Books, 1970.
McFarland, W. N., Allen, D. M.: The effect of temperature on rhodopsin-porphyropsin ratios in paired pigment fishes. Am. Soc. Photobiol. Meeting Program, 102 [abstract] (1973).
McFarland, W. N., Munz, F. W.: The visible spectrum during twilight and its implications to vision. In: Light as an ecological factor, Vol. II. Oxford: Blackwell, 1975a.

McFarland, W. N., Munz, F. W.: Part II: The photic environment of clear tropical seas during the day. Vision Res. **15**, 1063—1070 (1975b).
McFarland, W. N., Munz, F. W.: Part III: The evolution of photopic visual pigments in fishes. Vision Res. **15**, 1071—1080 (1975c).
Mitarai, G., Svaetichin, G., Vallecalle, E., Fatehchand, R., Villegas, J., Laufer, M.: Glia-neuron interactions and adaptational mechanisms of the retina. In: Neurophysiologie und Psychophysik des visuellen Systems. Berlin-Göttingen-Heidelberg: Springer 1961.
Müller, H.: Über das Zapfenmosaik in der Netzhaut des „Guppy" *(Lebistes reticulatus)*. Naturwissenschaften **38**, 459—460 (1951).
Müller, H.: Bau und Wachstum der Netzhaut des Guppy *(Lebistes reticulatus)*. Zool. Jb. **63**, 275—324 (1952).
Muntz, W. R. A.: Yellow filters and the absorption of light by the visual pigments of some Amazonian fishes. Vision Res. **13**, 2235—2254 (1973).
Muntz, W. R. A.: Visual pigments of cichlid fishes from Malawi. Vision Res. **16**, 897—903 (1976).
Muntz, W. R. A., Northmore, D. P. M.: Vision and visual pigments in a fish, *Scardinius erythrophthalmus* (the rudd). Vision Res. **10**, 281—298 (1970).
Muntz, W. R. A., Northmore, D. P. M.: Visual pigments from different parts of the retina in rudd and trout. Vision Res. **11**, 551—561 (1971).
Muntz, W. R. A., Northmore, D. P. M.: Scotopic spectral sensitivity in a teleost fish *(Scardinius erythrophthalmus)* adapted to different daylengths. Vision Res. **13**, 245—252 (1973).
Munz, F. W.: Photosensitive pigments from the retinas of deep-sea fishes. Science **125**, 1142—1143 (1957).
Munz, F. W.: Photosensitive pigments from the retinae of certain deep-sea fishes. J. Physiol. (Lond.) **140**, 220—235 (1958a).
Munz, F. W.: Retinal pigments of a labrid fish. Nature (Lond.) **181**, 1012—1013 (1958b).
Munz, F. W.: The photosensitive retinal pigments of fishes from relatively turbid coastal waters. J. gen. Physiol. **42**, 445—459 (1958c).
Munz, F. W.: The visual pigments of epipelagic and rocky-shore fishes. Vision Res. **4**, 441—454 (1964).
Munz, F. W.: Adaptation of visual pigments to the photic environment. In: Ciba Foundation Symp. Colour vision: physiology and experimental psychology. London: J. & A. Churchill, 1965.
Munz, F. W.: Vision: visual pigments. In: Fish physiology, Vol. V. New York: Academic Press, 1971.
Munz, F. W., Beatty, D. D.: A critical analysis of the visual pigments of salmon and trout. Vision Res. **5**, 1—17 (1965).
Munz, F. W., McFarland, W. N.: The significance of spectral position in the rhodopsins of tropical marine fishes. Vision Res. **13**, 1829—1874 (1973).
Munz, F. W., McFarland, W. N.: Part I: Presumptive cone pigments extracted from tropical marine fishes. Vision Res. **15**, 1045—1062 (1975).
Munz, F. W., Schwanzara, S. A.: A nomogram for retinene$_2$-based visual pigments. Vision Res. **7**, 111—120 (1967).
Munz, F. W., Swanson, R. T.: Thyroxine-induced changes in the proportions of visual pigments. Amer. Zool. **5**, 683 [abstract] (1965).
Naito, K., Wilt, F. H.: The conversion of vitamin A_1 to retinene$_2$ in a fresh-water fish. J. biol. Chem. **237**, 3060—3064 (1962).
Nicol, J. A. C.: Bioluminescence. In: Fish physiology, Vol. III. New York: Academic Press, 1969.
Niwa, H., Tamura, T.: Investigations of fish vision by means of S-potential II. Spectral sensitivity and colour vision. Rev. Can. Biol. **28**, 79—88 (1969).
O'Day, W. T., Fernandez, H. R.: *Aristostomias scintillans* (Malacosteidae): a deep-sea fish with visual pigments apparently adapted to its own bioluminescence. Vision Res. **14**, 545—550 (1974).
Parsons, J. H.: An introduction to the study of colour vision. Cambridge: The University Press, 1915.
Pickford, R. W.: Individual differences in colour vision. London: Routledge and Kegan Paul, 1951.
Pitt, F. H. G., Selwyn, E. W. H.: Colour of outdoor photographic subjects. Photographic J. **78**, 115—121 (1938).
Prince, J. H.: Comparative anatomy of the eye. Springfield, Ill.: C. C. Thomas, 1956.
Purkinje, J. E.: Beobachtungen und Versuche zur Physiologie der Sinne, Vol. 2. Neue Beiträge zur Kenntnis des Sehens in subjektiver Hinsicht. Berlin: Tiskem a Nakladenn G. Reimera, 1825. (Reproduced in Opera Omnia, Vol. 1, Lhotaka ze Lhoty, Prague, 1919.)
Richards, E. A.: Contrast-enhancement in imaging devices by selection of input photosurface spectral response. Advan. Electronics and Electron Phys. **28** B, 661—675 (1969).

Rochon-Duvigneaud, A.: Les yeux et la vision des vertébrés. Paris: Masson, 1943.

Rodieck, R. W.: The vertebrate retina. Principles of structure and function. San Francisco: W. H. Freeman, 1973.

Rose, C. D., Hassler, W. W.: Food habits and sex ratios of dolphin *Coryphaena hippurus* captured in the western Atlantic Ocean off Hatteras, North Carolina. Trans. Am. Fish. Soc. **103**, 94—100 (1974).

Rozenberg, G. V.: Twilight—a study in atmospheric optics. New York: Plenum, 1966.

Scholes, J. H.: Colour receptors, and their synaptic connexions, in the retina of a cyprinid fish. Phil. Trans. roy. Soc. (Lond.) B **270**, 61—118 (1975).

Schwanzara, S. A.: The visual pigments of freshwater fishes. Vision Res. **7**, 121—148 (1967).

Stell, W. K.: Structural studies of functional pathways in goldfish retina. In: Vision in fishes. New approaches in research. New York: Plenum, 1975.

Svaetichin, G.: Spectral response curves from single cones. Acta Physiol. Scand. **39** (Suppl. 134), 17—46 (1956).

Svaetichin, G., Negishi, K., Fatehchand, R.: Cellular mechanisms of a Young-Hering visual system. In: Ciba Foundation Symp. Colour vision: physiology and experimental psychology. London: J. & A. Churchill, 1965.

Talbot, F. H., Penrith, M. J.: Synopsis of biological data on species of the genus *Thunnus (sensu lato)* (South Africa). F.A.O. Fish. Rep., No. 6 **2**, 608—646 (1963).

Tamura, T., Hanyu, I., Niwa, H.: Spectral sensitivity and color vision in skipjack tuna and related species. Bull. Jap. Soc. Sci. Fish. **38**, 799—802 (1972).

Tamura, T., Niwa, H.: Spectral sensitivity and color vision of fish as indicated by S-potential. Comp. Biochem. Physiol. **22**, 745—754 (1967).

Tamura, T., Wisby, W. J.: The visual sense of pelagic fishes, especially the visual axis and accommodation. Bull. mar. Sci. Gulf and Caribb. **13**, 433—448 (1963).

Tomita, T., Kaneko, A., Murakami, M., Pautler, E. L.: Spectral response curves of single cones in the carp. Vision Res. **7**, 519—531 (1967).

Wagner, H.-J.: Vergleichende Untersuchungen über das Muster der Sehzellen und Horizontalen in der Teleostier-Retina (Pisces). Z. Morphol. Tier. **72**, 77—130 (1972).

Wagner, H.-J.: Comparative analysis of the patterns of receptor and horizontal cells in teleost fishes. In: Vision in fishes. New approaches in research. New York: Plenum, 1975.

Wald, G.: The visual systems of euryhaline fishes. J. gen. Physiol. **25**, 235—245 (1941).

Wald, G.: The metamorphosis of visual systems in the sea lamprey. J. gen. Physiol. **40**, 901—914 (1957).

Wald, G.: The significance of vertebrate metamorphosis. Science **128**, 1481—1490 (1958).

Wald, G.: The distribution and evolution of visual systems. In: Comparative biochemistry, Vol. I. New York: Academic Press, 1960.

Wald, G., Brown, P. K., Brown, P. S.: Visual pigments and depths of habitat of marine fishes. Nature (Lond.) **180**, 969—971 (1957).

Wald, G., Brown, P. K., Smith, P. H.: Iodopsin. J. gen. Physiol. **38**, 623—681 (1955).

Wallace, A. R.: Natural selection and tropical nature, 2nd ed. London-New York: Macmillan, 1891.

Walls, G. L.: The vertebrate eye and its adaptive radiation. Bloomfield Hills, Mich.: Cranbrook Institute of Science Bull. **19**, 463—465 (1942).

Willmer, E. N.: Some evolutionary aspects of mammalian colour vision. Proc. Linnean Soc. (Lond.) **161**, 97—108 (1949).

Willmer, E. N.: [untitled comment, p. 101] In: The comparative endocrinology of vertebrates. Part II. The hormonal control of water and salt electrolyte metabolism in Vertebrates. London-New York: Cambridge University Press, 1956.

Winn, H. E.: Formation of a mucous envelope at night by parrot fishes. Zoologica **40**, 145—148 (1955).

Winn, H. E., Bardach, J. E.: Some aspects of the comparative biology of parrot fishes at Bermuda. Zoologica **45**, 29—34 (1960).

Yager, D.: Behavioral measures of spectral sensitivity in the goldfish following chromatic adaptation. Vision Res. **9**, 179—186 (1969).

Yager, D., Thorpe, S.: Investigations of goldfish color vision. In: Animal psychophysics: the design and conduct of sensory experiments. New York: Appleton-Century-Crofts, 1970.

Young, R. W.: The renewal of photoreceptor cell outer segments. J. Cell Biol. **33**, 61—72 (1967).

Chapter 5

The Visual World of the Amphibia

W. R. A. MUNTZ, Stirling (Great Britain)

With 6 Figures

Contents

A. Introduction

The function of all sensory systems is to receive and analyze information about the environment, which is then used to control behavior. The sense organs do not however pass on to the brain a faithful and unaltered representation of the outside world, but select and accentuate aspects that are important to the behavior and survival of the animal. The stimuli that are selected in this way vary widely with species. Thus in the case of vision, some invertebrates, such as bees, can respond to ultraviolet light, which cannot be seen by vertebrates such as man. The visual world of bees is consequently very different from ours, and some stimuli, such as the nectar guides of flowers, are clearly visible to them but cannot be seen by us.

The types of stimuli to which animals respond not only vary greatly between species, but also vary within species depending on the behavior pattern being studied. The visual system of the animal apparently analyzes incoming information in different ways according to the circumstances. Several examples of the dependence of sensory processes on behavior are discussed below; more detailed reviews of the topic are available in VON UEXKÜLL (1957) and MUNTZ (1971). In the present context the important point is that any assessment of the visual world of amphibia must depend on behavioral studies, as well as on electrophysiologic and anatomical data. Unfortunately, the amphibia have proved rather intractable behaviorally, and in particular it has proved impossible to train them satisfactorily. Consequently, although many behavioral experiments have been carried out, these have almost always involved innate behavior patterns, and the data in this field are on the whole rather imprecise. On the other hand, the amphibia have proved to be exceptionally convenient for physiologic experiments, and have been widely used for this purpose. Much of the earliest work on the electrophysiology of vision, as well as on the visual pigments, was carried out on these animals, and newer techniques, such as intracellular recordings from the various elements of the retina, have also frequently been developed with amphibian subjects. The purpose of this chapter is not, however, to review the physiologic and anatomical literature in detail, but rather to assess the functional significance of amphibian visual mechanisms in relation to such behavioral data as we possess.

One of the most striking features of the electrophysiologic data on the visual systems of amphibia is that a very high degree of visual coding occurs in the retina itself. This is particularly well documented for various species of frog *(Rana)*, and will be described in more detail in Section B I below. Such data have often led to the suggestion that the frog retina responds largely or entirely to stimuli that are of biological importance to the animal, and that there is little or no further processing by more central parts of the nervous system. BARLOW (1953b), for example, discussing the characteristics of the retinal ganglion cells as revealed by his microelectrode studies, concluded that the receptive fields of the "on-off" units would be nicely filled by the image of a fly at 2″ distance, and may act as "fly detectors". Similar views have been expressed by LETTVIN et al. (1959), who concluded that the operations of the retinal ganglion cells have more the flavor of perceptions than of sensations, and suggested that some of the cells are ideally suited for the detection of bugs and could therefore be considered as "bug perceivers". Such suggestions imply that the visual world of the frog is very largely determined at the retinal level, and that the retinal output may therefore be connected with little further processing into the appropriate motor systems. Section B II considers the degree to which such a view of the frog's visual world is valid.

Another characteristic of the amphibia that is particularly striking is of course the fact that they live both in the water and on the land, and most of them pass from the former to the latter at metamorphosis. The visual environment is very different in the two cases (this volume), and the habits and feeding behavior of the amphibia also usually change markedly when they leave the water. Section C endeavors to assess whether corresponding adaptive changes occur in their visual systems at this time.

B. Vision in Adult Amphibia

I. Electrophysiologic Data on the Perception of Contour and Form

The frog retina is a highly complex structure (Fig. 1), and on anatomical grounds alone we would expect it to perform quite complicated analytical operations on the incoming visual information. The most straightforward way to find out what these operations are (as opposed to how they are performed) is to record from the axons of the retinal ganglion cells, since these are the pathways that carry the final output of the retina up to the brain.

Successful recordings from single vertebrate retinal ganglion cells were first achieved by HARTLINE (1938), using bullfrogs *(Rana catesbiana)*. He developed a technique of dissecting out small bundles of nerve fibers from the surface of the retina, for a distance of 1–2 mm. The bundles were cut where they entered the optic nerve, and the cut end lifted on to a wick electrode. Recordings from any moderately large bundle revealed a massed discharge from many fibers, but it proved possible in many cases to dissect the bundle down further until only one fiber remained active. As a result of such experiments HARTLINE divided the ganglion cells into three classes, depending on whether they discharged at the onset or cessation of light, or at both, and he called these three classes "on" fibers, "off" fibers, and "on-off" fibers, respectively. In later experiments (1940a, 1940b) HARTLINE used a small exploring spot of light to map out the portion of the retinal surface over which stimulation was effective in producing action potentials. The receptive fields measured in this way were found to be roughly circular, and to approach a millimeter in diameter. The large size of such a receptive field is perhaps best appreciated if we remember that the diameter of the whole eye is only about 8 mm.

HARTLINE'S work was confirmed and extended by BARLOW (1953a, 1953b), using *Rana temporaria* and *R. esculenta*. He also substituted the use of microelectrodes for HARTLINE'S microdissection technique. The same three types of ganglion cell were found, and, as in HARTLINE'S experiments, the receptive fields were surprisingly large, being as much as 1 mm in diameter in the peripheral retina if strong stimuli were used. A central plateau of higher sensitivity was also apparent, which varied from about 0.6 mm in the peripheral retina to 0.2 mm in the central retina. BARLOW further found that the "on-off" fibers were surrounded by an inhibitory zone which, while it did not give responses when stimulated on its own, was able, when so stimulated, to decrease the responsiveness of the more central areas of the receptive field.

BARLOW'S experiments also showed a marked difference between the summation and inhibition properties of "off" and "on-off" fibers. In the former type, discharges occurred whenever there was a reduction in the total effective light reaching the receptive field, and in general the greater the reduction in the light the larger the response. It did not matter how the reduction in effective light was caused: dimming the light or moving the stimulus to a less sensitive part of the visual field, for example, resulted in similar discharges. It thus appears that this

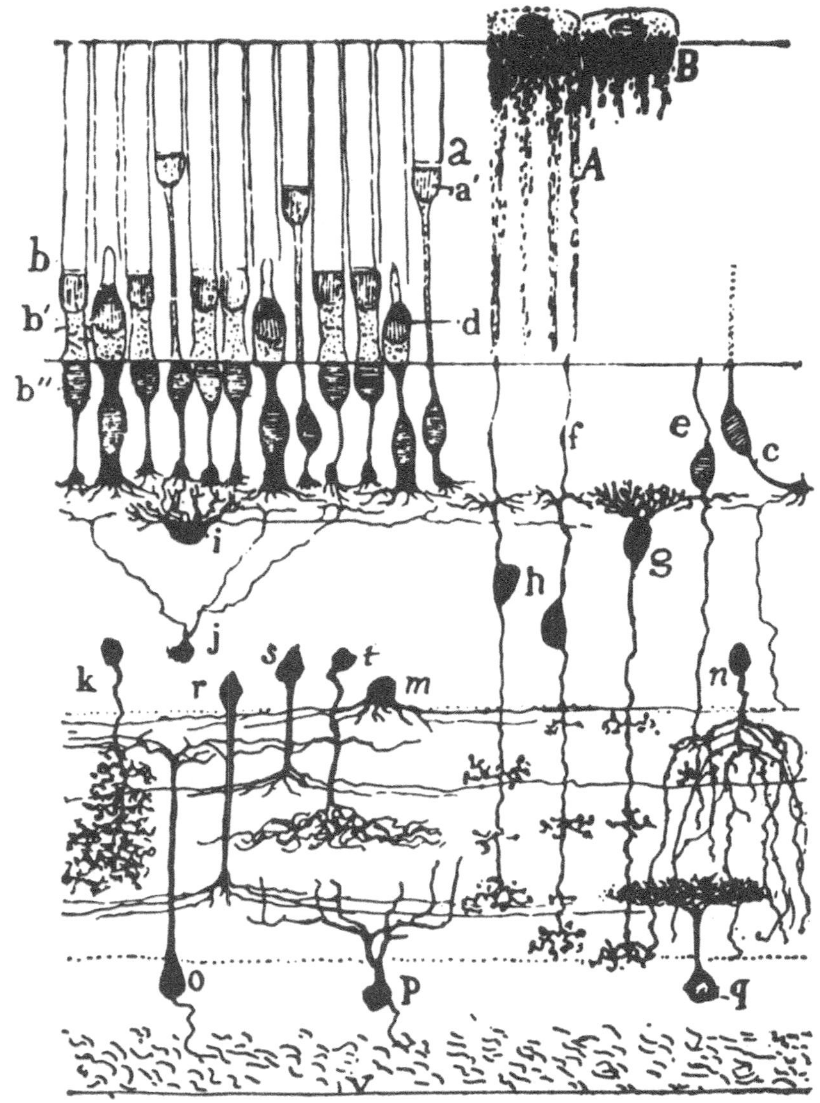

Fig. 1. General structure of retina of frog, from RAMÓN Y CAJAL (1911)

type of ganglion cell responds to the summated light reaching the receptive field, and cannot differentiate between the contributions from the different parts. On the other hand the "on-off" units, while they will respond to increases and decreases of light in a similar fashion, also respond strongly to changes in the spatial pattern of the light reaching the receptive field, even when this does not alter the total effective light. This shows that some form of spatial analysis is occurring within their receptive fields.

BARLOW's experiments thus demonstrated a considerable degree of sensory integration within the retina. He concluded that the retina acts as a filter, rejecting unwanted information and passing on useful information to the brain. The strongest example of this viewpoint is his conception of the "on-off" fibers as "fly detectors", which has already been mentioned above.

The analysis of the frog's visual system in functional terms was further developed by LETTVIN et al. (1959) (see also MATURANA et al., 1960; LETTVIN et al., 1961). They decided to use as wide a range of visual stimuli as possible, such as stationary and moving edges, geometrical figures, and objects that a frog might in nature be disposed to flee from or eat; and then to try and discover what common features were extracted by the different types of optic nerve fibre. This approach may be contrasted to previous work in this field, in which the stimuli were almost invariably restricted to spots of light. The experiments were carried out on *Rana pipiens*, and the type of microelectrode used allowed recordings to be made from unmyelinated as well as myelinated fibers. Four main types of fiber, each uniformly distributed across the retina, were described.

(I) *Sustained Contrast Detectors, or Class 1 Units.* These fibers are unmyelinated, and have receptive fields subtending approximately 2°. A sustained output occurs whenever there is an edge in the receptive field. The fibers show some memory: that is, they continue to fire for a time (up to 1 min) after all the lights are turned off. They are not sensitive to changes in the general level of the illumination.

(II) *Net Convexity Detectors, or Class 2 Units.* Fibers of this group are also unmyelinated and fail to respond to changes in the general level of illumination. They have receptive fields subtending about 7°, and respond optimally to small dark objects. If such an object is moved into the receptive field and left there the fiber continues to respond, but if the lights are now turned off for a moment the fiber ceases to respond and remains silent. Any object with a dark convexity or tongue extending into the field is effective in evoking responses, but a straight edge subtending more than 7° is ineffective. Jerky movements are more effective than smooth movements, and many of the fibers are directionally sensitive, only responding if a convex edge is moved into the field in a certain direction. Finally, a checkerboard pattern or a pattern of repeated dots is ineffective provided that the interval between the elements of the pattern is less than about half the receptive field, but if an individual member of the pattern is moved relative to the others a response will occur.

The distinction between the class 1 und class 2 units was not always clear-cut, and occasional units were found that had intermediate properties. The vast majority of examples, however, could be placed unequivocally in one or other of the classes.

(III) *Moving Edge Detectors, or Class 3 Units.* In contrast to the above two groups these are myelinated, and have receptive fields subtending about 12°. They respond to increases and decreases in the general illumination, and are equivalent to the "on-off" fibers of HARTLINE and BARLOW. However, as BARLOW noted, they respond as well or better to changes in the pattern of the illumination within their receptive fields, and are particularly responsive to moving edges.

(IV) *Net Dimming Detectors, or Class 4 Units.* These are also myelinated, and have receptive fields subtending about 15°. They respond to decreases in the illumination, whether these are caused by turning the lights off or moving a dark object into the receptive field, and are equivalent to the "off" fibers of HARTLINE and BARLOW.

Occasional units were also found which did not have distinct receptive fields, and which apparently responded to the absolute level of illumination over a wide area and with a long time constant.

In general the scheme of classification proposed by LETTVIN et al. has been followed by later workers, with occasional minor modifications (e.g., REUTER 1969). Most recent studies have concentrated on obtaining more detailed information on the stimulus properties of the ganglion cells, especially of the class 2 units. GAZE and JACOBSEN (1963), for example, have attempted to account for the properties of the class 2 units in terms of conventional concentric fields, and a detailed parametric study of the properties of this group of fibers has been carried out by GRÜSSER and his collaborators (GRÜSSER et al., 1968, 1967, 1964). These experiments have shown that class 2 units have an excitatory receptive field (ERF) of 2.5°–4° diameter in which small dark moving stimuli evoke responses, surrounded by an inhibitory receptive field (IRF) of at least 6°–8° in diameter. Within the ERF the relation between the average impulse frequency R and the velocity of the stimulus V is given by

$$R = k \cdot V^c \text{ impulses/sec}.$$

The value of k varies with many factors, such as the position of the stimulus in the receptive field, the level of the illumination, and the contrast of the stimulus, but the value of c is remarkably independent of these factors, with a value of about 0.7, and is also unaffected by whether the movement is real or stroboscopic. The same equation can also describe the responses of the class 1 and class 3 units if the exponent is given a value of 0.5 for the former class, and 0.95 for the latter.

One interesting feature that may be noted from the above summary is that in the absence of movement (AUTRUM, 1959) or changes in light intensity there will be very little activity in the retina, except in the class 1 fibers. This means that, since amphibians do not show eye movements, in a stationary world almost no visual information will reach the brain. It has been suggested (SHIPPERHEYN, 1965) that breathing movements might result in sufficient movement of the eyes for some ganglion cell activity to occur, but EWERT and BORCHERS (1974) have shown that, at least for the "on-off" fibers of toads, this is not so. In this experiment, a microelectrode holder was fixed to the head so that recordings could be made from conscious moving animals. No activity of the "on-off" fibers occurred in the absence of appropriate stimulation. Furthermore, no activity occurred when the animals spontaneously closed their eyes, suggesting the presence of efferent control.

The properties of the retinal ganglion cells may also vary with time. Thus EWERT and SIEFERT (1974) have reported seasonal changes in the properties of the class 2 units of toads *(Bufo bufo)*. During the winter large (4°–16°) black stimuli are optimal if they are in the lower visual field, and small (1°–4°) white stimuli if they are in the upper visual field. In the summer the situation is reversed: small white stimuli are optimal in the upper visual field and large black stimuli in the lower visual field. EWERT and SIEFERT have shown that corresponding behavioral changes occur as well.

Many of the recordings of LETTVIN et al. were made from the optic nerve just behind the eye, but records were also made from the surface layers of the tectum, where most of the fibers terminate as dense arborizations. These results indicate that the four groups of fibers they describe terminate in four separate layers of the tectum, each of which shows a continuous map of the retina. The four maps are in register with each other. The most superficial contains the endings of the sustained contrast detectors, and the deeper maps the terminals of the net convexity detectors, the moving edge detectors, and the net dimming detectors, in that order. The separation between the two most superficial maps is not as distinct as the separations between the other maps. WALLS (1953), discussing a similar layered arrangement that occurs in the visual projection to the lateral geniculate nucleus of primates, likens the different layers to a pile of maps of a country, which might for example show geodectic, climatologic, and agricultural information, respectively. In the case of the frog tectum there are similarly four maps of the visual world, each containing information about one of the four events to which the retinal ganglion cells are sensitive.

Apart from the terminal arborizations of the optic nerve fibers, the optic tectum contains a variety of different neurons (SZEKELY, 1971), which are presumably responsible for the further analysis of the visual information coming from the retina. A brief description of the properties of some of these units is given in LETTVIN et al. (1961) who distinguish between two types, "newness" neurones and "sameness" neurones. The former have visual fields of about 30° in diameter, and generally respond to a variety of moving objects, with different neurons having different optimal stimuli. The most striking property of the neurons, however, is the stimulus specific manner in which they adapted. If, for example, an object is moved along a given diameter of the receptive field, adaptation is rapid but specific to that particular diameter, so that a new stimulus that moves along a diameter at 90° to the first will evoke a full response. The properties of the "sameness" neurons are even more complex and difficult to specify. LETTVIN et al. state that they respond to the movements of small stimuli anywhere over a very wide range of the total visual field, except for a null patch, located differently for different cells, in which all stimuli are ineffective. These cells also exhibit a form of "attention," so that if they are responding to one stimulus, they fail to "notice" a second stimulus introduced into their visual field.

The presence of cells with the general properties described by LETTVIN et al. (1961) has been confirmed on several occasions, both in the optic tectum and in other areas of the brain. Thus MUNTZ (1962a) detected units of this type in the frog *(Rana pipiens)* diencephalon, EWERT (1971) and EWERT and BORCHERS (1971) reported them from the tectum and caudal thalamus of toads *(Bufo bufo)*, LIEGE and GALAND (1972) have found them in the telencephalon, diencephalon, mesencephalon, and metencephalon of *Rana esculenta*, and INGLE (1973a) and BROWN and INGLE (1973) in the tectum and thalamus of *Rana pipiens*. All these authors agreed that the properties of the units are very complex, and roughly as described by LETTVIN et al. (1961). It is clear that they often have very large visual fields, often as big as the total field of the eye, that they may be binocular or monocular, and are often responsive to several modalities. Some units have been shown to respond to stationary as well as moving objects, and their properties may be labile: the "null" area of "sameness" neurons, for example, can vary between 60° and 10° during an experiment, for no clear reason (EWERT, 1971).

While it is clear that the tectal and other central nervous units have many very interesting properties, the wide variety of units, and the difficulty of describing their characteristics precisely, make it difficult to relate their properties to the behavior of the whole animal. One interesting attempt in this direction has, however, been carried out by EWERT and his collaborators (EWERT and VON WIETERSHEIM, 1974a, b; EWERT et al., 1974), for the toad *(Bufo bufo)*. This will be returned to in Section B II below.

While the tectum is not the only visual center in the frog's brain, it does receive the majority of the optic nerve fibers, and the retinotectal system forms a very striking example of peripheral coding, with the different types of information separately coded by the retina and subsequently kept distinct in the central nervous system. The net convexity detectors are particularly noteworthy, and it is very tempting to suppose that they have the specific function of detecting prey objects. Since their output feeds into a separate tectal map, which could presumably itself be fairly directly connected to the motor system (electrical stimulation of the tectum is known to cause motor movements, see Section B5 below), one can imagine a fairly straightforward sensory-motor arc controlling feeding behavior. The other classes of fiber could control other simple types of behavior in a similarly direct fashion: net dimming detectors, for example, could mediate escape behavior. The properties of the tectal units obviously suggest that many behavior patterns will depend on stimulus parameters in quite complex ways, but simple types of behavior at least, such as feeding or flight, could depend directly on the output of the retinal ganglion cells.

II. Behavioral Evidence on the Perception of Contour and Form

A straightforward way to estimate the extent to which different behavior patterns depend on the outputs of the different types of retinal ganglion cell is to test the relative effectiveness of a variety of stimuli in releasing both behavioral and electrophysiologic responses. If it could be shown, for example, that the relative effectiveness of different stimuli in releasing feeding behavior correlated accurately with their relative effectiveness in causing class 2 unit activity, this would be reasonable evidence for the view that feeding behavior depends primarily on the outputs of these units.

A direct comparison of this type has been carried out by GRÜSSER and GRÜSSER-CORNEHLS (1968), using *Rana esculenta* and *R. temporaria*. A variety of stimuli were presented to the frogs under natural conditions and their responses to them noted, and the same stimuli were used in electrophysiologic experiments. The results are summarized in Table 1. It is clear that there is no one-to-one correspondence between the behavior and the electrophysiology, and GRÜSSER and GRÜSSER-CORNEHLS concluded that additional neural networks in the optic tectum must be involved. Nevertheless, there is some correlation between the two

Table 1. Comparison of electrophysiologic and behavioral data on the frog, translated from Table 4 of GRÜSSER and GRÜSSER-CORNEHLS (1968)

Stimulus	Representing	Responses of units of class 1	2	3	4	Behavior
2° white spot, moved on a dark background at $<0.5°\ sec^{-1}$	worm	++	+	0	0	Prey capture
3° black spot, moved on a light background at $5°\ sec^{-1}$	running fly	++	+++	(+)	0	Prey capture
6° black spot, moved on a light background at $20°\ sec^{-1}$	flying dragonfly	+	++	+++	+	Prey capture
20° black surface, moved on a light background at $10°\ sec^{-1}$	enemy	+ (Edge)	0	++ (Edge)	++	Flight
20° white surface, moved on a dark background at $10°\ sec^{-1}$	enemy	+ (Edge)	0	+ (Edge)	+ (Edge)	Weak flight reaction
Dark gray outlined surface, $>10°$, moved with the background. (= movement of the frog itself).	hiding place	0	0	(+)	++	Hiding
Many black spots moved on a light background at $0.5–10°\ sec^{-1}$	many worms or flies in one place	(+)	0	0	0	No prey capture
Dark—light stripes moved at $0.1–10°\ sec^{-1}$ (optokinetic stimulus)	movements of the visual world due to the frog's own movement	+	0	(+)	+	Turning, nystagmus.

+ + + indicates a strong neural response, (+) a weak response

sets of results: the class 2 units in particular, at any rate over the rather limited set of stimuli used in these experiments, were only activated by stimuli that also caused feeding behavior.

Although most of the available electrophysiologic data have been obtained with frogs, most of the detailed behavioral work has been done on toads, presumably because these animals are more amenable to such experiments. It is possible that the retinal ganglion cells of toads have different properties than those of frogs: it has, for example, been reported that the class 1 units are relatively scarcer in the former than the latter (INGLE, 1971).

The comparison of behavioral results obtained with toads and electrophysiologic data from frogs is not therefore strictly justified. Nevertheless some conclusions can be reached. In particular, many of the toad's behavioral responses are released by stimuli with complex properties that are quite different from the properties of any known amphibian retinal units. Thus the stimuli that produce feeding behavior in *Bufo bufo* (EWERT, 1968) would certainly not be predicted from the known properties of frog class 2 units. Here the shape, movement, and contrast of the stimulus were all found to be important parameters. The optimal rate of stimulus movement was between 30° and 60° per second. It was also found that, for rectangles moved horizontally, the effectiveness of the stimulus depended on the ratio of their horizontal and vertical extents. Thus horizontal rectangles were more effective than vertical rectangles, and the effectiveness of a horizontal rectangle of a given height increased as its length was increased. There was also an interaction between shape and direction of contrast; white stimuli depend much less on the ratio of their horizontal and vertical extents than black stimuli, and in fact were altogether more effective in releasing feeding behavior (black stimuli, in contrast, were more effective in releasing escape behavior). Seasonal variations in the effectiveness of different stimuli have also been shown (EWERT and SIEFERT, 1974).

EWERT and his collaborators (EWERT and VON WIETERSHEIM, 1974a, b; EWERT et al., 1974) attempted to account for some of these behavioral data in terms of the properties of tectal and thalamic/pretectal neurons. Behaviorally, for moving black stimuli, elongation of the stimuli in the direction of movement increases the probability that it will be treated as prey, whereas increasing the width of the stimulus in the direction perpendicular to the direction of movement increases the probability of flight responses. Microelectrode recordings from the optic tectum (stratum griseum and album centrale) revealed units with excitatory receptive fields (ERFs) of about 27°, some of which (tectum 1 units) responded mainly to stimuli elongated in their direction of movement. Other units (tectum 2 units) differed in that, while responding to elongated stimuli in a similar way to the tectum 1 units, their response rate also decreased if the width of the stimulus was increased. Finally, units were detected in the thalamic/pretectal region with large circular ERFs (about 46°), and which responded only to stimuli that were wide in the direction perpendicular to their direction of movement: that is, to stimuli that the animals would treat as predators in behavioral experiments. On the basis of these data EWERT and his collaborators proposed that the recognition of prey and predators depends on additive and subtractive processes between tectal and thalamic/pretectal units; and in particular that the responses of the tectal 2 units result from excitatory tectal 1 and inhibitory thalamic/pretectal inputs, and underlie feeding behavior.

Some further support for this idea was obtained from lesion experiments (EWERT and VON WIETERSHEIM, 1974b), since isolating the tectum from the thalamus and pretectum by a knife cut was found to disinhibit tectal 2 units, so that they now came to respond to wide stimuli that were previously ineffective. These lesions also, however, disinhibited the tectal 1 units, so that the wiring hypothesis summarized in the above paragraph cannot be entirely correct. The effect of the lesions on the tectal units was also strikingly similar to their effect on the behavior of the animals to the same sorts of stimuli, resulting in stimuli that had previously evoked flight responses now evoking feeding responses (see also Section BV below).

Other experiments that have been done on the feeding behavior of anurans have shown that larger species of frog (*Rana catesbiana*) will attempt to eat larger prey objects than smaller species (*R. pipiens*) (BIERSNER and MELZAK, 1966), and that larger individuals of a given species *(R. temporaria)* feed on larger prey than the smaller individuals (HOUSTON, 1973). Frogs *(Rana pipiens)* also tend to select smaller prey objects on those occasions when their response is slow (over 2s) and cautious than when their response is rapid, and alcohol intoxication results in their preference shifting to larger objects (INGLE, 1973b). Some frogs will attempt to eat larger food objects in the presence of blue light (BIERSNER and MELZAK, 1966), and the odor of crushed mealworms also results in toads responding to larger food objects (EWERT, 1970). In the reverse direction, HEATWOLE and AUDREY (1968) have shown that partial satiation shifts the preference of toads to smaller food objects. Finally, it has been shown that when stimuli are presented at different distances, frogs respond on the basis of real size, not the retinal angle subtended (INGLE, 1971). Data of this type suggest that if the class 2 units do indeed mediate the feeding response, the sizes of their receptive fields are not immutable. Variations in receptive field size have, of course, been shown for neurons at later stages in the central nervous system (see Section BI above).

Experiments carried out with natural stimuli also show that toads are capable of quite fine discriminations between food objects. Thus various species have been shown to avoid red ants (BUYTENDIJK, 1918), various caterpillars (SCHAEFFER, 1911), and bumble bees and their robber fly mimics (BROWER et al., 1960), while in each case continuing to eat other types of prey, and EIBL-EIBESFELDT (1952) has shown that *Bufo bufo* will come to prefer the size of prey that it has experienced most.

Finally, inhibitory and disinhibitory effects have been shown to affect the feeding behavior of anurans over large visual angles, showing that interactions must occur between the outputs of separate ganglion cells. Thus in toads, EWERT and HÄRTER (1969) have shown that feeding responses to a central moving stimulus are inhibited by simultaneously presenting peripheral stimuli, even though the stimuli are not all moving together, a phenomenon they ascribe to "central nervous inhibition," and EWERT and INGLE (1971) have shown, in both toads *(Bufo bufo)* and frogs *(Rana pipiens)* that habituating the feeding response to stimuli presented in the lateral part of the visual field actually increases the animals' responsiveness to stimuli presented in the frontal part of the visual field.

The stimuli that evoke flight have also been studied in detail by EWERT and REHN (1969), using *Bufo viridis.* They found that stimuli presented above the animals are much more effective than stimuli presented at the same level as the animals, and that the effectiveness increases with increasing contrast of the stimulus against the background. Stimuli are maximally effective when moved at 30° s^{-1}, and the optimal stimulus size is 50°, which is much bigger than the receptive field of any of the units described by LETTVIN et al. (1959) for the frog, though of the same order of size as the fields of some of the tectal neurons. The essential characteristic of effective stimuli is in fact the presence of detectable contours or parts ("retinal events") separated by 50° along the axis of movement (see Fig. 2). Here again, the results would not be predicted from the properties of any known amphibian ganglion cells.

It is clear from the electrophysiologic data that a great deal of visual analysis is performed at the retinal level in frogs, and this is presumably also true in many other amphibian species. The output of the retina is a highly coded representation of the pattern of light reaching it, and since this output constitutes the only visual information that reaches the brain, the characteristics of the retinal ganglion cells obviously determine the frog's visual world to a large degree. However, in many cases behavior patterns depend on the characteristics of the stimuli in ways that are much more complex than would be expected from our knowledge of the retinal ganglion cells. Further processing of the visual information is obviously occurring, which is presumably accomplished by neurons in the optic tectum and elsewhere. At present, however, our knowledge of the properties of these neurons is too fragmentary to allow any proper accounting of these behavior patterns to be made.

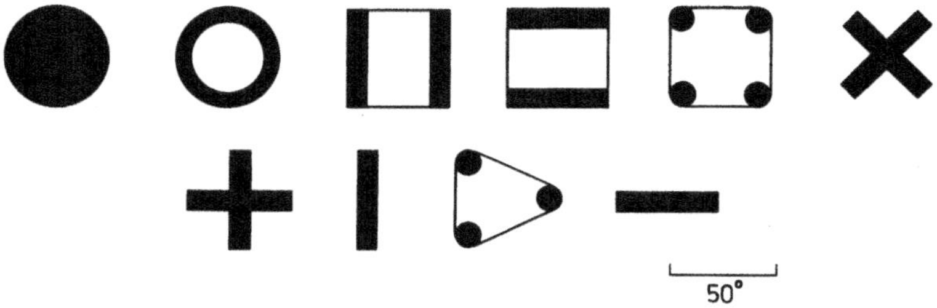

Fig. 2. Stimuli evoking escape responses in toad, *Bufo viridis*, from EWERT and REHN (1969). When moved horizontally to right, stimuli on top line are all effective in evoking escape responses, while those in bottom line are much less effective

III. Physiologic Data on the Perception of Wavelength

1. Visual Pigments

Vision ultimately depends on the absorption of light quanta by the visual pigments, and those aspects of vision that vary with wavelength, namely sensitivity and color vision, depend on the fact that the visual pigments absorb light more readily at some wavelengths than others. This wavelength dependence is largely determined by the absorptive properties of the pigments themselves, but it is also affected by other structures in the eye, such as the lens, cornea, oil droplets, and tapetum, that either selectively filter or reflect the light before it reaches the receptors (MUNTZ, 1972).

Tapeta have not been reported in amphibians, and there appear to be no data on the spectral properties of the cornea. In the frog *(Rana pipiens* and *R. esculenta)* the lens acts as a cut-off filter, transmitting completely at long wavelengths but with the absorbance increasing rapidly below 400 mm (DENTON, 1955; KENNEDY and MILKMAN, 1956). In *R. esculenta* the center of the lens proved to be less heavily pigmented than the periphery. A comparison of the dark adapted spectral sensitivity of intact and aphakic frogs, measured by means of the electroretinogram (ERG), has shown that the absorption of the lens can have a significant effect on vision at short wavelengths (KENNEDY and MILKMAN, 1956). An added complication is that the frog lens fluorescees strongly (MERKER, 1934), and lens fluorescense has been shown to affect vision in some animals (e.g., the cat; DODT and WALTHER, 1958).

Many of the cones of anurans contain pale yellow oil droplets, which lie between the outer segments and the incoming light. Microspectrophotometric measures of oil droplets from frog pigment epithelium have shown that they absorb negligibly above about 520 nm (LIEBMAN, 1972). Provided that the droplets within the cones have similar characteristics, their effects on spectral sensitivity should not be large since those cones that have droplets have visual pigments that absorb maximally at about 575 nm (LIEBMAN and ENTINE, 1968).

The limited data that are available therefore suggest that the effects of the inert filtering structures of the amphibian eye will not be large, and will be restricted to short wavelengths. This means that the spectral sensitivities of the different retinal receptors will depend almost entirely on the absorptive properties of their visual pigments. Since recent reviews of amphibian visual pigments are available (e.g., CRESCITELLI, 1972), only a brief summary need be given here.

Visual pigment extracts have been prepared and analyzed from about 15 anuran and 3 urodele species. In general, extracts from adult terrestrial amphibia are dominated by a single vitamin A_1-based pigment absorbing maximally between 502 and 504 nm, while aquatic forms yield a vitamin A_2-based pigment absorbing maximally at about 522 nm. As we shall see below, a similar relationship between aquatic and terrestrial environments and the type of pigment found in extracts also often occurs during the lifetime of an individual as it passes from the water to the land at metamorphosis. Comparison of the absorptive properties of the extracted pigments with microspectrophotometric and electrophysiologic data shows that these pigments come from the rods.

Although in general one or other of the two pigments described in the previous paragraph forms the major component in retinal extracts prepared from adult amphibia, in some cases a smaller amount of the second pigment can also be detected. For example, while extracts from *Xenopus laevis*, an anuran that is aquatic even in its adult stages, contain predominantly $VP523_2$, small amounts of $VP502_1$ can also be detected (DARTNALL, 1954, 1956; WALD, 1955). Microspectrophotometric measurements on *Xenopus* rods by DOBROWLSKI et al. (1955) raise the possibility that the rod pigment of this species shows seasonal variations, which would be of great interest in view of the well-known seasonal variations that occur in the visual pigments of fishes (see e.g., BRIDGES, 1972 for a review). The bullfrog *(Rana catesbiana)* is another species in which both A_1- and A_2-based pigments occur in the adult (REUTER et al., 1971). In this case, the dorsal part of the retina is dominated by the A_2-based pigment, and the ventral half by the A_1-based pigment. It would be interesting to know whether a similar distribution of pigments occurs within the retinas of *Xenopus*.

Apart from these two rod pigments, extracts sometimes yield small quantities of a third pigment that has its maximal absorbance at about 430 nm (DARTNALL, 1967). DENTON and WYLIE'S (1955) observations on intact retinas, and LIEBMAN and ENTINE'S (1968) microspectrophotometric data show that this pigment comes from the green rods.

The visual pigments of the red and green rods are thus well known from both extraction procedures and microspectrophotometry. Extraction procedures have yielded no information on the cone pigments, but these have been studied in a few species by microspectrophotometry (LIEBMAN and ENTINE, 1968; LIEBMAN, 1972).

The data for *Rana pipiens* show that the single oil-droplet bearing cones have a pigment with its maximal absorption at about 575 nm, as does the principle element of the double cones, while the accessory element of the double cones has the same pigment as the red rods. LIEBMAN (1972) has also stated that microspectrophotometric data show that the visual pigments of *Salamandra tigrinum* are identical to those of *Rana*, but that in *Necturus* only one cone pigment has been detected, absorbing maximally at about 575 nm.

2. Electrophysiologic Data

a) Spectral Sensitivity

Almost all the electrophysiologic data available on spectral sensitivity and wavelength discrimination in amphibia have been obtained with various species of *Rana*, and this was the case in all the experiments described below unless stated otherwise. There is at present no reason to suppose that the different species of *Rana* differ in the way in which spectral coding is carried out by their retinas.

Spectral sensitivity curves may readily be obtained from frog eyes using standard electrophysiologic techniques, such as the ERG or single unit recording with microelectrodes. In all cases the method involves finding the intensities of different monochromatic lights necessary to evoke some standard response, such as an ERG b-wave of a given size, or a given number of nerve impulses. Sensitivity is defined as the reciprocal of the number of quanta at a given wavelength necessary to evoke this standard response, and a plot of sensitivity so defined against wavelength constitutes a spectral sensitivity curve.

The usual result when dark adapted preparations are used is a smooth bell-shaped spectral sensitivity curve, maximal at about 500 nm, and when light adapted preparations are used a rather similar curve with its maximum shifted to about 560 nm (e.g., GRANIT, 1955; CHAPMAN, 1961 and many others). The scotopic results agree well with the absorption spectrum of the rod visual pigment, as we should expect. There is, however, no such straightforward agreement between the photopic results and the cone pigments, for while microspectrophotometry has shown two cone pigments absorbing maximally at 502 and 575 nm, the photopic spectral sensitivity curve is maximal at about 560 nm. It is possible that the photopic results are due to the combined output of the two cone types, though the fact that the sensitivity curve is usually rather narrower than the characteristic absorption spectrum of a single visual pigment, as given by DARTNALL'S nomogram (1953), weighs against this suggestion (e.g., GRANIT, 1955). Data are also available showing that if the two cone types do contribute together to the photopic sensitivity, they must cooperate to yield a remarkably stable and unvarying spectral sensitivity curve, and behave in many ways as a single photoreceptor. DONNER and RUSHTON (1959), for example, found that many ganglion cells that were maximally sensitive at 560 nm exhibited the following three characteristics, all of which suggest that only a single photoreceptor type was involved. (i) For any two wavelengths of light it was always possible to find an intensity of each such that one could be substituted for the other without causing any response (silent substitution). When this was achieved the relative intensities of the two stimuli agreed with expectations based on the 560 nm photopic sensitivity curve. (ii) The logarithmic range of intensities over which the substitution remained silent (the Fechner fraction) did not vary with wavelength. (iii) Various chromatic adaptations had no effect on the shape of the spectral sensitivity curve. REUTER and VIRTANEN (1972) have similarly concluded that the retina contains a system maximally sensitive at about 560 nm that behaves like a single mechanism.

Accepting the conclusion that the two cone types act together as a single unit means that while frogs possess three distinct classes of photopic receptor, functionally they only possess two; and it seems unlikely that the potential analyzing power of the visual system would be sacrificed in this way. An alternative, and perhaps more likely, possibility is that the 560 nm photopic sensitivity curve represents the output of a single receptor class, whose spectral sensitivity does not agree with the spectral absorbance of the underlying visual pigment as determined by microspectrophotometry. A situation of this sort might arise, for example, through the action of inert filtering pigments.

b) Wavelength Discrimination

Ganglion cells in frog retinas responsive to narrow bands of the spectrum were first reported by GRANIT and SVAETICHIN (1939) and GRANIT (1942). Cells of this type were assumed to underlie color vision and were called modulators. Granit and his collaborators recorded from groups of optic nerve fibers as well as from single units; they also used diffuse light and often did not distinguish between the "on"- and "off"-components of the response. It is likely for these reasons that the modulators represent quite complicated interactions between different receptor types. Nevertheless, the experiments represent the first physiologic evidence that frogs are capable of discriminating wavelenghts.

The technique of "silent substitution" has also been used to show that the frog retina is capable of color vision. The technique, which appears to have been used first by FORBES and BURLEIGH (1955) on turtles and frogs, involves finding whether, for two monochromatic lights, it is possible to find an intensity of each such that one can be substituted for the other without causing any electrical response. Both BONGARD and SMIRNOV (1957), using *Rana ridibunda*, and DONNER and RUSHTON (1959), using *Rana temporaria*, have found that with many (but not all) ganglion cells it is impossible to match two monochromatic lights in this way, showing that these cells are receiving inputs from more than one type of receptor. BONGARD and SMIRNOV further showed that any monochromatic light could be matched with a mixture of a red and a blue primary, if the primaries were in the appropriate proportions and the mixture had the correct intensity: that is, the cells were dichromatic.

It is now known that many of the ganglion cells of the frog are color-coded in much the same way as those of many other animals, with two different receptor types contributing to the "on"- and "off"-responses (REUTER, 1969; REUTER and VIRTANEN, 1972; BÄCKSTRÖM and REUTER, 1975). The two receptor types involved appear to be the green rods, absorbing maximally at 430 nm, and the 560 nm cone mechanism. Stimulation of the green rods causes responses that differ in many ways from those that result from cone stimulation, which is of particular interest

Table 2. Summary of frog ganglion cell types, from REUTER and VIRTANEN (1972)

Class of cell	Green rod "on", cone "on"	Green rod "on", cone "on-off"	Green rod "on", cone "off"	Green rod "on-off", cone "on-off"	Green rod "on-off", cone "off"	Cone "on-off"	Cone-"off"	Totals
1		15	3			1	1	20
2		16	1			5	2	24
1 or 2		10	2			7	2	21
3		9		27		3		39
4			17		1		3	21
5							2	2
pure "on"	2							2

in view of the data showing that the behavior of frogs is often dominated by blue light (see below). In the first place, there is a prolonged discharge on green rod stimulation, whereas cone stimulation results in a brief burst; the frequency of the response is, however, lower (LIBERMAN, 1957; ORLOV, 1961; MUNTZ, 1962a; REUTER, 1969; REUTER and VIRTANEN, 1972). Secondly, it is much more common for the green rods to contribute to the "on"-responses of ganglion cells than to the "off"-responses. Thus pure "on"-cells, which have been detected in isolated retinas (LIBERMAN, 1957; REUTER and VIRTANEN, 1972), and project strongly to the diencephalon (MUNTZ, 1962a), receive excitation from both green rods and cones. In other classes of ganglion cell the green rods usually contribute to "on"-responses but seldom to "off"-responses, whereas the cones contribute to both types of response. This is clearly shown in Table 2, which summarizes the data obtained by REUTER and VIRTANEN on 129 ganglion cells. The experiments were done on excised opened eyes from *Rana temporaria*. A yellow adapting light was also used in order to decrease the activity of the red rods and cones, and thus facilitate the study of the green rods. Out of the 129 ganglion cells studied, green rod contributions to the "off"-response were found in only 27 class 3 cells and one class 4 cell; and in the class 3 cells, although the green rods participated in both the "on"- and the "off"-responses, the "on"-response often dominated the "off"-response.

Ganglion cells with different receptors contributing to their "on"- and "off"-responses can obviously code wavelength information in the same way as the opponent ganglion cells of many other species. It is clear that the pure "on"-fibers are also capable of wavelength discrimination, even though they do not show opponent responses (MUNTZ, 1962a). Thus their high responsiveness to short wavelengths is apparent for highly unsaturated stimuli, such as colored papers, as well as for monochromatic stimuli (Fig. 3), and it was possible to show that they responded more to an unsaturated blue stimulus than to an unsaturated green stimulus even when the intensity of the former was reduced to such an extent that the green stimulus contained more light throughout the visible spectrum. This shows that inhibition must be involved, even though it never reaches sufficient strength to reverse the sign of the response. Secondly, the prolonged discharge in response to short wavelength stimulation is qualitatively quite different from the

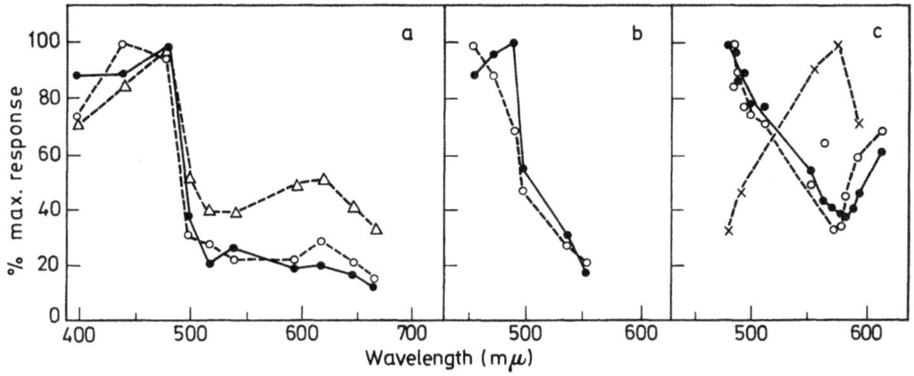

Fig. 3 a–c. Variation of response of individual "on"-fibers with changes in wavelength, expressed as percentage of maximum response (from MUNTZ, 1962a). (a) Responses of 3 fibers to highly saturated colored stimuli produced by interference filters. (b) Responses of 2 fibers to less saturated coloured stimuli produced with Wratten gelatine filters. (c) Responses of 3 fibers to highly unsaturated stimuli of light reflected from Munsell papers. Crosses in c show responses of a fiber when a white shutter, brighter than papers, was closed in front of them; Circles the responses of 2 fibers to the papers exposed by opening black shutter

brief burst of impulses occurring in response to light of longer wavelengths, and this is true irrespective of the intensities and saturations of the stimuli; in fact any stimulus that appeared blue to humans also released a prolonged response in the "on"-fibers.

IV. Behavioral Data on Spectral Sensitivity and Wavelength Discrimination

Three behavioral methods have been used to study spectral sensitivity and wavelength discrimination in amphibia, namely feeding, the optomotor response, and phototactic behavior. Surprisingly, in view of its convenience, the first of these methods appears only to have been used once, in an experiment by ZIPSE (1935) designed to determine the sensitivity of amphibia to ultraviolet light. This author reported that feeding could occur in light of wavelength down to 313 nm in *Bufo*, *Hyla*, and *Rana*. In view, however, of the rapid increase in the absorption of the lens below 400 nm (see above), this result should probably be viewed with caution.

The optomotor response has been more widely used to determine the visual capabilities of the amphibia, particularly by BIRUKOW (BIRUKOW, 1949; BIRUKOW and KNOLL, 1952) and his collaborators. The results show a clear Purkinje shift in adult frogs *(Rana temporaria)*, but the stimuli in the optomotor drum consisted of colored and gray papers so they cannot be used to determine accurate spectral sensitivity curves. BIRUKOW (1950) also used the technique to study color vision in this species. He found that when green and yellow papers were used in conjunction with gray, a gray could always be found such that no response was elicited. When red or blue papers were used, however, he was unable to find a matching gray, suggesting that these stimuli were being responded to on the basis of wavelength. This result has been criticized by WALLS (1942), who suggested that BIRUKOW could have found a gray to match the red and blue stimuli if he had used many more shades than he did. This criticism receives some support from an experiment by BURGERS (1952) on *Xenopus laevis*, in which red-gray, green-gray, and blue-gray matches were all obtained in an optomotor situation. Nevertheless, BIRUKOW's data are in some ways similar to data obtained in phototactic situations.

Many amphibia show a strong visually determined preference for the brighter of two stimuli when they are given the choice of moving toward stimuli of different intensities, or of turning toward or away from a light source. Since it is not clear whether this behavior should be called positive phototropism or positive phototaxis (see below), the term photopositive response will be used here.

It has been known for a long time that in many amphibia the photopositive response is much more strongly released by short wavelength (blue) stimuli than by stimuli of longer wavelengths (e.g., TORELLE, 1903; PEARSE, 1910), and the early work also strongly suggested that this is a genuine example of color vision, and not merely a question of the animals being more sensitive to short wavelengths. TORELLE, for example, used a variety of stimuli obtained with colored cloths and colored solutions, and on some occasions made serious attempts to ensure that the blue stimuli contained less energy than the red or green stimuli, but the blue preference remained marked in all cases. More recent data have amply confirmed this supposition in *Rana temporaria* (MUNTZ, 1962b), *R. catesbiana* (CHAPMAN, 1966), and *R. pipiens* (JAEGER and HAILMAN, 1971). Thus MUNTZ showed that adding green light to blue light reduced the latter's attractiveness, in spite of the fact that the animals were photopositive and the stimulus must now have appeared brighter to them, and CHAPMAN equated the intensities of blue and green stimuli according to the frog's spectral sensitivity curve, determined with the electroretinogram, and then showed that the blue preference was unaffected even when the intensity of the green stimulus was varied over 4 log units. JAEGER and HAILMAN were able to show that their animals preferred short wavelength stimuli to other longer wavelength broad-band stimuli that contained more energy at every wavelength over the spectrum. This means, of course, that regardless of the frog's spectral sensitivity the second stimulus must have appeared brighter than the first, and since they also showed that the frogs were monotonically photopositive with intensity, the fact that the short wavelength stimulus was preferred must mean that the animals were capable of discriminating wavelengths.

It is highly likely that the green rods contribute to the photopositive response, since these are known to be responsive in the appropriate part of the spectrum. Some indirect evidence also supports this view. Thus *Salamandra salamandra*, a urodele that is said to lack green rods (WALLS, 1942), is photonegative as opposed to photopositive and fails to show short wavelength sensitivity (MUNTZ, 1963a), and *Rana temporaria* tadpoles, which are photopositive, also fail to show short wavelength sensitivity, which again may be correlated with the late development of the green rods in this species (MUNTZ, 1963b). Clearly, however, other receptors apart from the green rods must contribute to the behavior, and in particular some receptors must contribute in an inhibitory fashion if wavelength discrimination is to be achieved. It has been suggested (MUNTZ, 1963b, 1966) that the green rods and the cones act synergistically in an excitatory fashion, and are opposed by an inhibitory influence from the red rods. This hypothesis was based on the following evidence.

(i) Young *Rana temporaria* tadpoles are photopositive, and most sensitive to green (about 550 nm) light (MUNTZ, 1963b; HAILMAN, 1975). During growth, the strength of the photopositive response decreases, and the characteristic blue responsiveness of the adults slowly develops. Histologic and electrophysiologic studies suggest that in this species the cones develop before the red rods, and that the green rods develop only just before metamorphosis (see below). According to the hypothesis the late development of the red rods, which are inhibitory, accounts for the decrease in the strength of the photopositive response during development, and the even later development of the green rods for the fact that the blue responsiveness only develops just before metamorphosis.

(ii) Adult *Rana pipiens* are more strongly photopositive when light-adapted than when dark-adapted (BOYCOTT et al., 1964; MUNTZ, 1966). Their blue responsiveness is, however, less accentuated under these conditions, and in fact light-adapted animals are only most responsive to short wavelengths when relatively intense stimuli are used: with low intensity stimuli they are most responsive in the green (MUNTZ, 1966). These data can again be accounted for by assuming that the red rods are inhibitory, for their contribution should increase on dark adaptation, both accentuating wavelength discrimination and reducing total sensitivity.

Although the hypothesis can account for the above data in detail, the evidence for it is indirect. Recent work has shown that the accessory elements of the double cones contain the same pigment as the red rods (LIEBMAN and ENTINE, 1968), and it is quite possible that these, not the red rods, are acting as the inhibitory receptors.

The characteristics of the photopositive behavior are similar in many respects to those of the "on"-fibers that project to the diencephalon (MUNTZ, 1962a, b), suggesting that these fibers may underlie the behavior. Thus in both cases the response is largely independent of the intensity and saturation of the stimuli, and they can both be shown to be genuine examples of color vision, and not merely cases of high short wavelength sensitivity. The form of the spectral response curve of the "on"-fibers is also very similar to that obtained in the behavioral situation (Fig. 4). Some lesion and stimulation studies have also indicated that the photopositive response may be more dependent on the diencephalon than other forms of behavior (see below).

Although this blue sensitive photopositive behavior is very common in amphibians, it is not shown by all species. For example, MUNTZ (1963) compared two urodeles, *Triturus cristatus* and *Salamandra salamandra*. The former is aquatic, photopositive, and most sensitive to short wavelengths, whereas the latter is terrestrial, photonegative, and most responsive in the green (about 570 nm). JAEGER and HAILMAN (1971) similarly compared two anurans, *Rana pipiens* and *Eleutherodactylus rhodopsis*. The former is semi-aquatic, photopositive, and most responsive to short wavelengths; and this last characteristic was unequivocally demonstrated to be a true example of color vision. The latter is terrestrial, photonegative, and most sensitive to green, and the behavior involved sensitivity, not

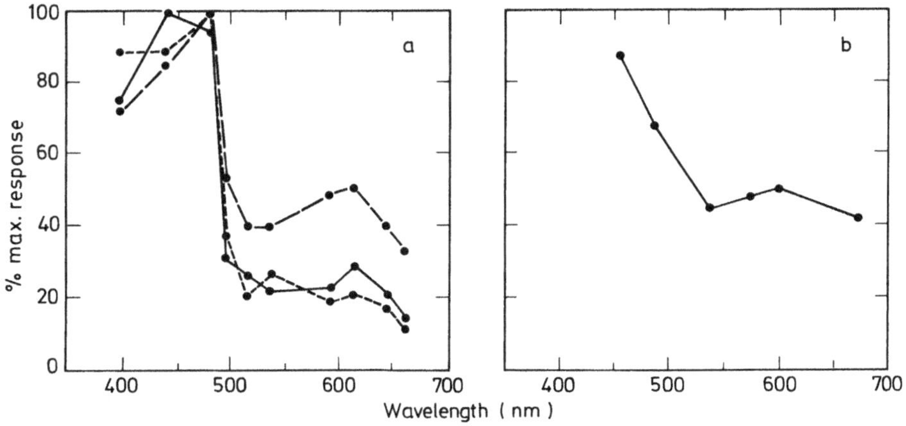

Fig. 4a and b. Comparison between the spectral properties of the "on"-fibers and the photopositive behavior of frogs (from MUNTZ, 1971). (a): Number of impulses as percentage of maximum for 3 "on"-fibers. (b): Percent choices of differently illuminated windows in the behavioral situation shown in Fig. 5

wavelength discrimination. JAEGER and HAILMAN (1973) continued this work in a comprehensive study of the responses of 121 species of Anura to white lights of different intensities. Eighty-seven percent of the species were found to be photopositive, 8% photonegative, while 5% preferred intermediate intensities. A later study (HAILMAN and JAEGER, 1974) extended the work to spectral stimuli. In all, 127 species of anurans were tested; 101 of them were clearly most responsive to blue stimuli, and a further 6 species were probably of this type. Specially constructed tests on 16 of the blue responsive species showed that the animals were using color vision in each case. Eleven species gave a U-shaped spectral response curve, responding mostly to violet and red and least to green, and finally 9 species were unclassifiable. The data from this and the previous study also showed clearly that the blue responsiveness occurred in photopositive species, and the U-shaped response curve in photonegative species. The U-shaped response curve is thus equivalent to the results for *Salamandra* and *Eleutherodactylus* described above: the animals responded least to green stimuli, but since they were photonegative this shows that they were in fact *most* sensitive to this part of the spectrum. It was also found that blue responsiveness was not restricted to aquatic species. The phylogenetic distribution of species showing the different types of behavior will, however, be considered below, when the adaptive significance of the behavior is discussed.

Some studies have suggested that the photopositive behavior is very stable. Thus JAEGER and HAILMAN (1973) compared the same species obtained from different geographic locations, males and females, behavior at 7° C and 25° C, aquatic against dry conditions, animals collected in spring and autumn, and the effects of repetitive tests. These comparisons were carried out on a variety of species, without finding any effects. Nevertheless, data are available showing that even those amphibia that show clear blue sensitive photopositive behavior may not do so all the time. Part of this variability depends on the form of the apparatus used to test the behavior. Most of the experiments described above used similar experimental situations, in which the animals were placed in a dark starting box facing two windows or screens, which were illuminated with different intensities and colors of light (Fig. 5). If the animals did not respond spontaneously they were usually forced to respond, either by mechanical stimulation or by a small electric shock. An experiment by BOYCOTT et al. (1964) on *Rana temporaria* showed that in such situations the level of illumination in the starting box is a relevant factor, for when a light, as opposed to a dark, starting box was used strong and consistent photonegative behavior was obtained. The animals in this experiment tended to be unexpectedly photonegative throughout, though the strength of the photone-

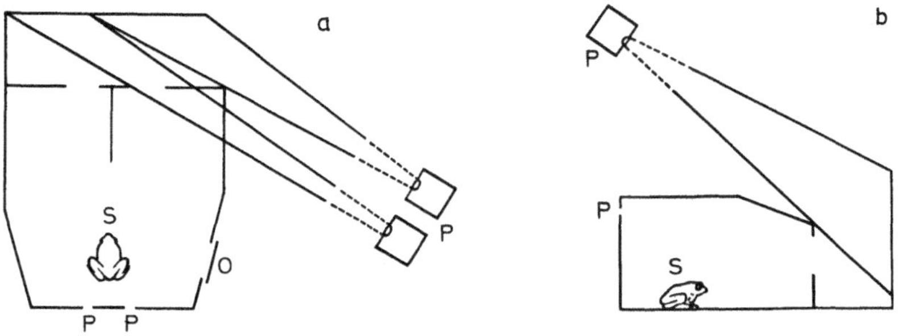

Fig. 5a and b. Typical apparatus for studying photopositive behavior of anurans (from MUNTZ, 1962b). (a) apparatus seen from above; (b) seen from side. *P*, projectors for providing stimuli; *O*, opening for inserting subject, *S*; *p*, peepholes. Animals respond by jumping through windows toward illuminated screen

gative behavior was decreased by having a dark starting box, by having the animals light-adapted as opposed to dark-adapted, and by forcing the animals to jump by means of a rod: it was, however, only when all these factors were combined that the usual photopositive behavior was obtained. The effect of forcing was very marked, and many more animals required forcing when the dark start box was used. These data agree with early work, which showed that frogs may often sit in the dark facing the light (TORELLE, 1903): that is, when undisturbed they seek the darkness, but take up such a position that if subsequently disturbed they will jump toward the light. As PEARSE (1910) put it, "mechanical stimulation furnishes the impulse to locomotion, while light is effective in determining the direction of movement." The relationship between the photopositive behavior and forcing was also clearly present in the data of MUNTZ (1962b), as the re-analysis presented in BOYCOTT et al. (1964) shows: both the relative responsiveness to blue and the degree to which the behavior was photopositive were markedly greater on those trials on which forcing was required.

Other factors may also affect the behavior. STIER (1926), for example, found a relationship between feeding and the sign of the behavior in *Diemyctylus viridescens*. He first tested the animals' behavior, and then subsequently dissected them, and found that 80% of the photopositive animals had full stomachs, while all the photonegative animals had empty stomachs. Feeding similarly caused previously photonegative animals to become photopositive. It is also possible that occluding the pineal eye increases the strength of the photopositive behavior in *Rana temporaria* (MROSOVSKY and TRESS, 1966), and as we have seen, dark adaptation can have a marked effect (BOYCOTT et al., 1964; MUNTZ, 1966; HAILMAN and JAEGER, 1976). Finally, "attention" may also come into play, as suggested by MAST (1911), who showed that if a toad *(Bufo americanus)*, moving towards a source of light, entered a second beam of light reaching it from the side, it ignored it and continued toward the first source even if the lateral beam was considerably brighter.

The adaptive significance of these orienting responses to light is not clear. MUNTZ (1962b) suggested that the photopositive response of many species will tend to take them to the water when frightened, since the open space over water will be less green, and contain more short wavelength light, than the vegetation on the bank. Such a hypothesis can, however, only account for a limited amount of the data: it cannot, for example, deal with the effects of altering the illumination in the start box, or of feeding. Furthermore, JAEGER and HAILMAN'S studies showed that many photopositive species are terrestrial and do not escape into the water. In such amphibia the photopositive behavior would tend to take them out into the open when frightened, and it seems probable that this would decrease their chances of escape.

As a result of their studies HAILMAN and JAEGER (1974, 1976) have proposed that each species has an optimum ambient illumination (OAI) at which its visual system operates most efficiently, and which it actively seeks by phototaxis. If the stimulus intensities used during an experiment lie below the OAI the animals will show photopositive behavior and a preference for short wavelength light, while if the stimulus intensities lie above the OAI they will show photonegative behavior and a U-shaped spectral response curve. They concluded that photonegative species are all nocturnal or inhabitants of dimly lit environments, and have low OAI's, whereas photopositive species are diurnal and have high OAI's. They also proposed that all species are potentially capable of either photopositive or photonegative behavior under appropriate circumstances. For example, dark adaptation should lower the OAI and consequently make the animals more photonegative in their behavior, and they should also be more photonegative when more intense stimuli are used, since these will be more likely to lie above the OAI. Both these predictions were confirmed on a number of species (HAILMAN and JAEGER, 1976). It seems possible that some of the other effects described above, such as those of feeding and the brightness of the start box, could also be interpreted as alterations in the OAI.

Although HAILMAN and JAEGER'S hypothesis is very plausible, it does present some minor difficulties. In the first place, according to their view, the function of phototaxis is to keep the animal at an illumination level at which its visual system operates optimally. If this is the case, however, it is difficult to see why wavelength discrimination, and not simply a brightness preference, is involved. Secondly, as we have seen above, amphibians commonly sit in the dark facing toward the light, and it is only when they are disturbed from such a position that they show the characteristic spectrally-dependent phototactic behavior. Most of their time is therefore not spent at the illumination levels which they would chose in a typical phototactic experiment, which would be required by HAILMAN and JAEGER'S hypothesis.

V. Lesion and Stimulation Studies

Several lesion and stimulation studies have shown that different visually guided behavior patterns involve different areas of the central nervous system. Thus EWERT (1968), working with the toad *(Bufo bufo)*, has shown that lesions to the dorsal caudal thalamus abolish flight reactions in this species, but potentiate feeding behavior, so that the animals now give prey-catching responses to stimuli that had previously evoked flight responses. Electrical point stimulation of the central nervous system gave compatible results: that is, stimulation of the dorsal caudal thalamus resulted in escape movements such as turning, crouching, and jumping, whereas prey-catching movements such as snapping were evoked by stimulation of the tectum (see also Section B II above). INGLE (1973c) obtained similar results with *Rana pipiens*. He found that following unilateral removal of the tectum, animals did not respond to moving food or threat objects in the contralateral visual field, but still successfully avoided barriers in this field; and in fact continued to do this even if the contralateral eye was sewn shut, so that no direct retinal input to the tectum remained. Furthermore, the optic fibers from the eye contralateral to the lesion eventually regenerated into the remaining ipsilateral tectum, following which responses to food or threat object in the field of this eye became directed to the mirror image position. Stationary barriers were, however, still responded to correctly. These results are compatible with the view that flight responses, and the mechanisms necessary for the animals to orient and move correctly in space, are dependent on the thalamus, whereas feeding mechanisms, and the recognition of food and other moving objects, depend on the integrity of the tectum. The results are also in reasonable agreement with electrophysiologic recordings from the diencephalon, which, as we have seen, suggest that the photopositive escape behavior is mediated by the "on"-fibers projecting to this part of the brain; and with the presence of various classes of optic nerve fiber in the tectum apparently specialized for the detection of various types of moving object. The "on"-fibers were, however, recorded from the anterior dorsal thalamus, whereas EWERT's lesions were to the caudal dorsal thalamus, so it is likely that some further processing occurs beyond the "on"-fibers before the final centers responsible for escape behavior are reached.

KICLITER (1973) also studied the effects of lesions on various types of visual behavior in *Rana pipiens*. He used three tasks in all: a wavelength discrimination task, a flux discrimination task, and a movement detection task. The first two tasks were tested in a two-choice apparatus typical of those used to study amphibian photopositive behavior. The wavelength discrimination task involved blue and green stimuli, with the latter made slightly brighter on the basis of electrophysiologic spectral sensitivity data. Intact animals, as usual, consistently chose the blue stimulus. The flux discrimination task measured the preference for different intensities of tungsten light in the same apparatus: intact animals chose the more intense stimulus. Movement detection was tested by observing the animal's responses to meal worms and to a small black moving dowel.

The results showed that all these behavior patterns depend on the lateral eyes. Lesions to the anterior dorsal thalamus impaired both the wavelength and the flux discrimination tasks, though they did not bring performance down to chance, and did not affect movement detection. Complete tectal removal abolished both movement detection and wavelength descrimination, and impaired but did not abolish flux discrimination. Removal of the retinal afferent input to the tectum by means of knife cuts through the optic tract, abolished movement detection but left both wavelength and flux discrimination unimpaired.

It seems possible that the flux and wavelength discrimination tasks used in KICLITER's experiments were essentially the same, except that the former task was rather easier and so survived the lesions better (LASHLEY, 1929). The results in general agree with those of EWERT and INGLE, and with electrophysiologic data, but the separation achieved was not complete, indicating that although different behavior patterns depend largely on certain areas of the central nervous system, other areas also contribute to them.

C. Vision Before and After Metamorphosis

The majority of the amphibia pass their larval stages in the water, and move to the land following metamorphosis. This change from an aquatic to a terrestrial mode of life is of course accompanied by large changes in the visual environment to which the animals are exposed, and we should expect corresponding changes to occur in their visual systems. Thus, the spectral characteristics of the light will alter, for there is almost always a greater predominance of long wavelength light in freshwater than on land. Underwater stimuli will also usually have much lower contrasts than those on land, due to light attenuation and the introduction of "veiling" light through scatter. Optical problems are also introduced through the loss of the cornea's focusing power underwater. Factors such as these all mean that the animals must solve very different visual problems before and after metamorphosis. The characteristics of different photic environments, including freshwater and terrestrial environments, have been dealt with in detail elsewhere.

The development of the amphibian retina has been studied in detail by a variety of techniques. Many of these studies have been concerned with developmental problems (e.g., REYER, this volume), and are outside the scope of this chapter. The large number of amphibia that are known, and the wide range of habitats that they occupy, make generalizations difficult. Nevertheless, some tentative conclusions are possible.

(i) The lateral eyes of some species are not functional at hatching. This conclusion has been reached on histologic grounds for *Rana temporaria*, *Xenopus laevis*, and *Rana pipiens* (SAXÉN, 1954; NILSSON, 1964). CRESCIELLI and NILSSON (1966) also investigated the development of vision in *Rana pipiens* by means of the electroretinogram (ERG), and found that electrical activity could first be detected when the tadpoles were 7–9 days old, just after the outer segment disks of the receptors first appear. The ERG at this stage was, however, purely cornea-negative, and quite unlike the ERG of adult animals. The adult pattern developed slowly, with cornea-positive potentials appearing between the 11th and 17th days, and the final form of the response being achieved at around the 20th day.

It should be remembered that tadpoles may be able to respond to light even in the absence of functional lateral eyes, using either the pineal eye or the dermal light sense (HAMASAKI, this volume; STEVENS, 1963).

(ii) In some species the cones develop before the rods. This conclusion was reached at the beginning of this century on the basis of light microscopy (BERNARD, 1900; CAMERON, 1905), and has been confirmed more recently by SAXÉN'S (1954) detailed histologic study on the retinal development of *Rana temporaria* and *Xenopus laevis*. In both these species SAXÉN showed that not only do the cones approach their adult form more quickly than the rods, but their photomechanical movements also develop earlier.

Behavioral and electrophysiologic data also confirm that the cones develop earlier than the rods in *Rana temporaria*. Thus BIRUKOW (1949) used the optomotor response to study the development of vision in this species. In his experiments the optomotor drum was lined with alternate strips of colored and gray paper, and the relative brightness of the colored strips estimated by varying the brightness of the gray strips until no response occurred. He found that in young tadpoles brightness estimated in this way was greatest for yellow stimuli, that the relative sensitivity to the different colors of strip was the same in light and dark adapted animals, and the same as in light-adapted adults. As development progressed the color of the stimuli to which the dark-adapted animals responded best gradually shifted towards shorter wavelengths. The adult condition, in which responsiveness was maximal in the blue-green, was only achieved after metamorphosis. BIRUKOW'S experiments were performed with colored paper for stimuli, and so cannot be used to derive accurate spectral sensitivity curves. Nevertheless, the results clearly indicate that the photopic spectral sensitivity of *Rana temporaria* tadpoles is the same as that of adults from the earliest stages (17–19 mm long) at which testing was carried out, whereas the adult scotopic spectral sensitivity only develops slowly. Similar results were obtained by MUNTZ (1964), using the ERG as a measure of visual function. Here again the photopic and scotopic spectral sensitivities of the youngest tadpoles (21–24 mm long) used were identical, and the same as the photopic spectral sensitivity of adults. With older animals, the scotopic spectral sensitivity gradually approached the adult form, and the absolute sensitivity also improved by about 1.5 log units.

The fact that the rod outer segments of some species are conical-shaped during the early stages of development led BERNARD (1900) and CAMERON (1905) to suggest that the rods develop from the cones. MUNTZ (1964) also suggested, on the basis of his ERG scotopic spectral sensitivity data, that the rods develop through a conelike stage. It is clear, however, that the cones and rods of many species are distinct from an early stage (e.g., SAXÉN, 1954; NILSSON, 1964). Furthermore, MUNTZ'S (1964) conclusion depended on the statement of COLLINS et al. (1953) that all of the rod pigment of *Rana temporaria* tadpoles is based on vitamin A_1, whereas later work has shown that in fact substantial amounts of vitamin A_2-based pigments are present as well (DARTNALL, 1962; MUNTZ and REUTER, 1966). It is more likely, therefore, that the changes in the ERG-determined scotopic spectral sensitivity of this species during development reflect the gradual replacement in the rods of A_2-based pigment by A_1-based pigment.

Histologic (CAMERON, 1905), behavioral (MUNTZ, 1963), and electrophysiologic (REUTER, 1969) data, as well as pigment extracts (MUNTZ and REUTER, 1966; REUTER, 1969), all suggest that the third major receptor type found in the retina of *Rana temporaria*, the green rod, probably develops last, only becoming functional at metamorphosis.

(iii) It is clear that there is a wide variation between species in the manner in which vision develops. For example, although the evidence suggests that in *Rana temporaria* the rods develop later than the cones, NILSSON (1964) reported that in *Rana pipiens* outer segment disk formation starts at the same time in both types of receptor. SAXÉN's histologic data (1954) show that the development of the retina in *Rana temporaria* proceeds slowly, and that the adult form is only achieved after metamorphosis, whereas in *Xenopus laevis* the adult structure is attained at an early stage. SAXÉN suggests that *Xenopus* tadpoles and adults have similar retinal structures because they occupy the same habitat, and therefore face the same visual problems.

The development of the visual pigments has been studied in a variety of species, and here also considerable differences occur. The most straightforward finding is that the larvae have A_2-based visual pigments and the adults A_1-based visual pigments, which agrees with the well-known general tendency for animals occupying freshwater environments to have the former type of pigment, while animals occupying terrestrial environments have the latter type of pigment. The best documented species of this type is *Rana pipiens*, for which microspectrophotometric data have shown that the pigments of the cones, as well as of the rods, shift from the A_2-based to the A_1-based form during development (LIEBMAN and ENTINE, 1968).

The situation exemplified by *Rana pipiens* is by no means universal. In *Xenopus laevis*, for example, over 90% of the adult visual pigment is based on vitamin A_2 (DARTNALL, 1956). The tadpoles, however, have considerably much greater amounts of A_1-based pigment (about 40%), so that the metamorphic shift in this species is in the reverse direction from that of *Rana pipiens* (CRESCITELLI, 1973). In other species no metamorphic change in visual pigments has yet been detected. Thus extracts from both the adults and tadpoles of *Bufo boreas halophilis* and *Bufo bufo* yield A_1-based pigments only (CRESCITELLI, 1958; MUNTZ and REUTER, 1966). Finally, indirect evidence suggests that in other species various intermediate situations may occur. Thus, as we have seen above, both behavioral and ERG data have shown that the photopic spectral sensitivity of *Rana temporaria* tadpoles is identical with that of adults, whereas the scotopic spectral sensitivity is displaced toward long wavelengths. This suggests that the tadpole cones already contain the adult A_1-based pigment whereas the rods do not: that is, the cones develop faster than the rods in both their structure and their visual pigments. If this interpretation is correct, this species may be contrasted with *Rana pipiens*, where as we have seen the anatomical evidence suggests that there is no difference in the rate of development of the rods and cones, and microspectrophotometric evidence similarly shows no difference in the time during development at which the pigments change.

Table 3. Frequency of different types of ganglion cell in *Rana temporaria* tadpoles and adults, expressed as percentages. In all, 55 units from tadpoles and 110 units from adults were studied. Data from REUTER (1969)

Type of unit	Tadpoles	Adults
"on"-units	10.9	9.1
"on"-units, small field	9.1	2.7
"on-off" units	14.6	26.4
"on-off" units, small field	38.2	13.7
"off"-units	25.4	24.5
convex edge detectors	1.8	23.6

The "on-off" units are equivalent to the class 3 units of MATURANA et al. (1960), the "off-units to their class 4 units, and the "on-off" small field units are probably equivalent to their class 1 units. It is not clear how the "on"-units with small fields relate to the scheme of MATURANA et al.

Even in species showing a clear metamorphic visual pigment shift the transition is not always complete. Adult *Xenopus*, as we have seen, retain a small percentage of the A_1-based pigment (DARTNALL, 1956), and, in the reverse direction, REUTER et al. (1971) have shown that adult *Rana catesbiana* retain considerable amounts of A_2-based pigment, concentrated in the dorsal part of the retina.

The general problems of correlating the absorptive properties of the visual pigments of tadpoles with the available light in their environment are the same as for the fishes (e.g., MUNZ and MCFARLAND, this volume). In the case of the amphibia, no satisfactory explanation has yet been put forward to account for the known variations in the pattern of visual pigment development. For example, *Bufo* tadpoles with A_1-based visual pigments often occur in the same bodies of water as other species of tadpole that have A_2-based pigments, and do not differ in their behavior in any obvious way. Some factors also affect the visual pigments of tadpoles and fishes in very different ways. Thus many species of fishes have two extractable visual pigments, based on vitamin A_1 and vitamin A_2 respectively, and in such species increasing daylength often causes the proportion of the A_1-based pigment to increase (e.g., BRIDGES, 1972). On the other hand, increasing daylength causes the proportion of the A_2-based visual pigment to increase in *Rana pipiens* tadpoles (BRIDGES, 1974), and yet the freshwater environment is presumably much the same for both tadpoles and fishes.

(iv) *Rana temporaria* tadpoles feed mostly on algae, and on microscopic and dead animals, and they can feed in the dark. Adults, on the other hand, feed on living prey and hunt by sight. Many other amphibians show a similar change in feeding behavior on metamorphosis. Form vision may therefore be more important for adults than for tadpoles, and the retinal mechanisms for form vision may be correspondingly reduced or absent during the larval stages. Some electrophysiologic data are available that support this view. Table 3 summarizes the results obtained by REUTER (1969) on single retinal units of *Rana temporaria* tadpoles and adults. The major difference found was that convex edge detectors were scarce in tadpoles, a finding that can obviously be correlated with the change in feeding behavior at metamorphosis if this type of unit does indeed underlie the detection of prey. Ganglion cell recordings have also been obtained from the tadpoles of *Rana catesbiana*, *R. pipiens*, and *R. clamitans* (POMERANZ, 1972). It was found that the tadpoles lacked the class I units of MATURANA et al. (1960), that the class 2 units were limited to an equatorial strip of retina, and that the class 4 units gave transient responses, as opposed to the continuous responses of similar cells in adults. Class 3 units were similar in tadpoles and adults.

POMERANZ's data differ in some respects from those of REUTER, particularly in the number of class 1 units found. Possible reasons for this include the different species that were used, and also the fact that rather different schemes for classifying the units were used in the two cases.

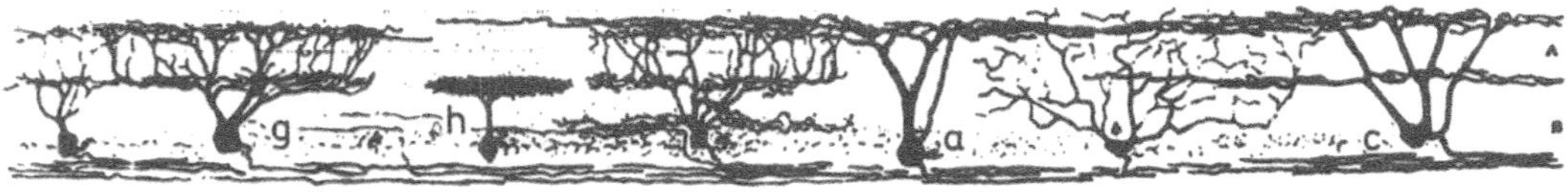

Fig. 6. Ganglion cell types in frog retina, from RAMÓN Y CAJAL (1973). In LETTVIN et al.'s terminology (1961), *h* is an example of one-level constricted field cell, *g* of many-level E distribution cell, *c* of many-level H distribution cell, and *a* of single-level broad field cell

In 1961, LETTVIN et al. hypothesised that the different classes of retinal unit that they described could be correlated with the different types of retinal ganglion cell that can be found in the frog retina. CAJAL's description (1973) of the retinal anatomy of the frog suggests that there are four main types of ganglion cell, which LETTVIN et al. refer to as one-level constricted field, many-level E distribution, many-level H distribution, and single-level broad field cells (see Fig. 6). On the basis of indirect evidence, such as comparisons between the size of the cells' dendritic fields and the receptive field sizes of the different classes of retinal unit, they suggested that these four types of ganglion cell underlie the class 1, class 2, class 3, and class 4 units, respectively. POMERANZ (1972) presented histologic data that support this view, for he reported that one-level constricted field cells are absent in tadpoles, and that many-level E distribution cells are restricted to an equatorial strip. This of course agrees with his data on the occurrence and distribution of class 1 and class 2 units.

D. Summary

It is clear that the visual systems of many anuran amphibians are characterized by a high degree of peripheral coding. This may well be a characteristic of their other sensory systems as well: it has been shown, for example, that the ear of the bullfrog *(Rana catesbiana)* acts as a peripheral filter tuned to certain components of the bullfrog's call (FRISHKOPF and GOLDSTEIN, 1963; CAPRANICA, 1965). Peripheral coding is, however, by no means confined to the amphibia, but occurs in the visual systems of all the other vertebrates that have been studied as well; and in some cases, such as the rabbit, it may rival the degree of coding found in the anuran eye (OYSTER, 1968). In general, peripheral coding appears to be less well developed in higher animals, such as the cat and the monkey, and it is tempting to suppose that as we ascend the phylogenetic scale the central nervous system becomes better developed, behavior more flexible, and the amount of peripheral coding decreases (MUNTZ, 1971). While there may be some truth in this view, it must be remembered that even in primates, the ganglion cells show specific opponent responses to stimuli of different wavelengths, and have receptive fields organized into concentric excitatory and inhibitory zones. These characteristics still represent a considerable degree of peripheral coding: the ubiquitous concentric fields of ganglion cells give them the properties of "discontinuity detectors," and even in man the greater part of color vision is achieved at a retinal level (e.g., DE VALOIS and WALRAVEN'S (1967) demonstration that color contrast is a retinal process).

The highly organized outputs of the different classes of frog retinal ganglion cell obviously determine this animal's visual world in the sense that they represent the only visual information that reaches the brain. Attempts to correlate the characteristics of the retinal ganglion cells with the behavior of the whole animal have, however, been only partly successful. Although there is, for example, some evidence that the class 2 units of the retina, together with the optic tectum, are especially concerned with feeding behavior; whereas blue-sensitive retinal units, and the pretectal and thalamic regions of the brain, are particularly relevant to escape behavior; it is also clear that the behavior in both cases is too flexible and plastic to be explained completely in peripheral terms.

Attempts to correlate laboratory studies on amphibian vision with the problems the animals face in their normal environments have also only met with limited success. Thus one of the most striking and consistent pieces of amphibian visual behavior in the laboratory is the high short wavelength preference many species show in their orienting responses to light, and yet in spite of the large number of studies that have been devoted to this topic the functional significance of the behavior is still uncertain. The significance of the distribution and ontogenetic changes in the visual pigments of amphibia is also still unknown. These difficulties probably arise in large part because of the limitations in our knowledge of how the amphibians behave in their natural environments. In spite of these limitations, however, the body of data that has now been accumulated on the vision of the amphibia is very considerable, and the relatively limited repertoire of visual behavior that they show, together with the strikingly different visual problems that they face at different times during their life cycle, makes the group very promising for future study.

References

Autrum, H.: Das Fehlen unwillkürlicher Augenbewegungen beim Frosch. Naturwissenschaften **46**, 435 (1959).

Bäckström, A.-C., Reuter, T.: Receptive field organization of ganglion cells in the frog retina: contribution from cones, green rods and red rods. J. Physiol. **246**, 79—107 (1975).

Barlow, H. B.: Action potentials from the frog's retina. J. Physiol. **119**, 58—68 (1953a).

Barlow, H. B.: Summation and inhibition in the frog's retina. J. Physiol. **119**, 69—88 (1953b).

Bernard, H. H.: Studies in the retina: rods and cones in the frog and in some other amphibia. Part I. Q. J. Microsc. Sci. **43**, 23—47 (1900).

Biersner, R., Melzak, R.: Approach-avoidance responses to visual stimuli in frogs. Exp. Neurol. **15**, 418—424 (1966).

Birukow, G.: Die Entwicklung des Tages- und des Dämmerungssehens im Auge des Grasfrosches (Rana temporaria L). Z. vergl. Physiol. **31**, 322—347 (1949).

Birukow, G.: Vergleichende Untersuchungen über das Helligkeits- und Farbensehen bei Amphibien. Z. vergl. Physiol. **32**, 348—382 (1950).

Birukow, G., Knoll, M.: Tages- und Dämmerungssehen von Froschlarven nach Aufzucht in verschiedenen Lichtbedingungen. Naturwiss. **21**, 494—495 (1952).

Bongard, M.M., Smirnov, M.S.: Spectral sensitivity curves for receptors connected to single optic nerve fibres in the frog. Biophysics **2**, 328—333 (1957).

Boycott, B.B., Mrosovsky, N., Muntz, W.R.A.: Black and white preferences in the frog (Rana temporaria), and other Anura. J. exp. Biol. **41**, 865—877 (1964).

Bridges, C.D.B.: The rhodopsin-porphyropsin visual system. In Handbook of Sensory Physiology, vol. VII/1, Photochemistry of Vision, ed. H.J.A. Dartnall, Berlin-Heidelberg-New York: Springer-Verlag 1972.

Bridges, C.D.B.: Effects of light and darkness on the visual pigments of amphibian tadpoles. Vision Res. **14**, 779—794 (1974).

Brower, L.P., Brower, J.V.Z., Westcott, P.W.: Experimental studies of mimicry. 5. The reactions of toads (Bufo terrestris) to bumblebees (Bombus americanorum) and their robberfly mimics (Mallophora bomboides), with a discussion of aggressive mimicry. Amer. Nat. **94**, 343—356 (1960).

Brown, W., Ingle, D.J.: Receptive field changes produced in frog thalamic units by lesions of the optic tectum. Brain Res. **59**, 405—409 (1973).

Burgers, A.C.J.: Optomotor reactions of Xenopus laevis. I. Physiol. comparata et oecol. **2**, 272—281 (1952).

Buytendijk, F.J.J.: L'instinct d'alimentation et l'expérience chez les crapauds. Néerl. Physiol. **2**, 217—228 (1918).

Cameron, J.: The development of the retina in amphibia: an embryological and cytological study. J. Anat. Physiol. **39**, 135—153, 332—361, 471—488 (1905).

Capranica, R.R.: The evoked vocal response of the bullfrog—a study of communication by sound. Research monograph 33. Cambr. Mass., M.I.T. (1965).

Chapman, R.M.: Spectral sensitivity of single neural units in the bullfrog retina. J. opt. Soc. Amer. **51**, 1102—1112 (1961).

Chapman, R.M.: Light wavelength and energy preferences of the bullfrog: evidence for colour vision. J. comp. physiol. Psychol. **61**, 429—435 (1966).

Collins, F.D., Love, R.M., Morton, R.A.: Studies on vitamin A. 25. Visual pigments in tadpoles and adult frogs. Biochem. J. **53**, 632—636 (1953).

Crescitelli, F.: The natural history of visual pigments. Proc. 19th Ann. Biol. Coll. Oregon State College, April 1958, 30—51 (1958).

Crescitelli, F.: The visual cells and visual pigments of the vertebrate eye. In Handbook of Sensory Physiology, vol. VII/1, Photochemistry of Vision, ed. H.J.A. Dartnall. Berlin-Heidelberg-New York: Springer-Verlag 1972.

Crescitelli, F.: The visual pigment system of Xenopus laevis: tadpoles and adults. Vision Res. **13**, 855—865 (1973).

Crescitelli, F., Nilsson, S.E.C.: Electroretinogram of the frog during embryonic development. Science **151**, 1545—1547 (1966).

Dartnall, H.J.A.: The interpretation of spectral sensitivity curves. Brit. Med. Bull. **9**, 24—30 (1953).

Dartnall, H.J.A.: A study of the visual pigments of the clawed toad. J. Physiol. **125**, 25—42 (1954).

Dartnall, H.J.A.: Further observations on the visual pigments of the clawed toad, Xenopus laevis. J. Physiol. **134**, 327—338 (1956).

Dartnall, H.J.A.: The photobiology of visual processes. In The Eye, vol. 2, Ed. H. Dawson, Academic Press, London, New York (1962).

Dartnall, H.J.A.: The visual pigment of the green rods. Vision Res. **7**, 1—16 (1967).

Denton, E.J.: Absorption du cristallin de Rana esculenta et d'Anguilla vulgaris. Bull. Mus. nat. Hist. nat., Paris 2e série, **27**, 418—425 (1955).

Denton, E.J., Wyllie, J.H.: Study of the photosensitive pigments in the pink and green rods of the frog. J. Physiol. **127**, 81—89 (1955).

Dobrowlski, J.A., Johnson, B.K., Tansley, K.: The spectral absorption of the photopigment of Xenopus laevis measured in single rods. J. Physiol. **130**, 533—542 (1955).

Dodt, E., Walther, J. B.: Fluorescence of the crystalline lens and electroretinographic sensitivity determinations. Nature (Lond.) **181**, 286—287 (1958).

Donner, K. O., Rushton, W. A. H.: Retinal stimulation by light substitution. J. Physiol. **149**, 288—302 (1959).

Eibl-Eibesfeldt, I.: Nahrungserwerb und Beuteschema der Erdkröte (Bufo bufo L.), Behaviour **4**, 1—35 (1952).

Ewert, J. P.: Der Einfluß von Zwischenhirndefekten auf die Visuomotorik im Beute- und Fluchtverhalten der Erdkröte (Bufo bufo L.). Z. vergl. Physiol. **61**, 41—70 (1968).

Ewert, J. P.: Neural mechanisms of prey-catching and avoidance behaviour in the toad (Bufo bufo L.), Brain Behav. Evol. **3**, 36—54 (1970).

Ewert, J. P.: Single unit response of the toad's (Bufo americanus) caudal thalamus to visual objects. Z. vergl. Physiol. **74**, 81—102 (1971).

Ewert, J. P., Borchers, H. W.: Reaktionscharakteristik von Neuronen aus dem Tectum opticum and Substectum der Erdkröte Bufo. Z. vergl. Physiol. **71**, 165—189 (1971).

Ewert, J. P., Borchers, H. W.: Antwort von retinalen Ganglienzellen bei freibeweglichen Kröten (Bufo bufo L.). J. comp. Physiol. **92**, 117—130 (1974).

Ewert, J. P., Härter, H. A.: Der hemmende Einfluß gleichzeitig bewegter Beuteattrappen auf Beutefangverhalten der Erdkröte (Bufo bufo L.). Z. vergl. Physiol. **64**, 135—153 (1969).

Ewert, J. P., Ingle, D.: Excitatory effects following habituation of prey-catching activity in frogs and toads. J. comp. physiol. Psychol. **77**, 369—374 (1971).

Ewert, J. P., Rehn, B.: Quantitative Analyse der Reiz-Reaktionsbeziehungen bei visuellem Auslösen des Fluchtverhaltens der Wechselkröte (Bufo viridis Laur.) Behaviour **35**, 212—234 (1969).

Ewert, J. P., Siefert, G.: Seasonal change of contrast detection in the toad's (Bufo bufo L.) visual system. J. comp. Physiol. **94**, 177—186 (1974).

Ewert, J. P., Hock, F. J., von Wietersheim, A.: Thalamus, Praetectum, Tectum: Retinale Topographie und physiologische Interaktionen bei der Kröte Bufo bufo (L.). J. comp. Physiol. **92**, 343—356 (1974).

Ewert, J. P., von Wietersheim, A.: Musterauswertung durch Tectum- und Thalamus/Praetectum-Neurone im visuellen System der Kröte Bufo bufo (L.). J. comp. Physiol. **92**, 131—148 (1974a).

Ewert, J. P., von Wietersheim, A.: Einfluß von Thalamus/Praetectum-Defekten auf die Antwort von Tectum-Neuronen gegenüber bewegten visuellen Mustern bei der Kröte Bufo bufo (L.). J. comp. Physiol. **92**, 149—160 (1974b).

Forbes, A., Burleigh, S., Neyland, M.: Electric responses to colour shift in frog and turtle retina. J. Neurophysiol. **18**, 517—535 (1955).

Frishkopf, L. S., Goldstein, M. H.: Responses to acoustic stimuli from single unites in the eight nerve of the bullfrog. J. acoust. Soc. Amer. **35**, 1219—1228 (1963).

Gaze, R. M., Jacobsen, M.: 'Convexity detectors' in the frog's visual system. J. Physiol. **169**, 1P—3P (1963).

Granit, R.: Colour receptors of the frog's retina. Acta physiol. Scand. **3**, 137—151 (1942).

Granit, R.: Receptors and sensory perception. New Haven: Yale Univ. Press 1955.

Granit, R., Svaetichin, G.: Principles and technique of the electrophysiological analysis of colour reception with the aid of microelectrodes. Upsala läkaref. förh. **65**, 161—177 (1939).

Grüsser, O. J., Grüsser-Cornehls, U.: Neurophysiologische Grundlagen visueller angeborener Auslösemechanismen beim Frosch. Z. f. vergl. Physiol. **59**, 1—24 (1968).

Grüsser, O. J., Grüsser-Cornehls, U., Bullock, T. H.: Functional organisation of receptive fields of movement-detecting neurones in the frog's retina. Pflügers Arch. ges. Physiol. **279**, 88—93 (1964).

Grüsser, O. J., Grüsser-Cornehls, U., Finkelstein, D., Henn, V., Patutschnick, M., Butenandt, E.: A quantitative analysis of movement-detecting neurones in the frog's retina. Pflügers Arch. ges. Physiol. **293**, 100—106 (1967).

Grüsser, O. J., Grüsser-Cornehls, U., Licker, M. D.: Further studies on the velocity function of movement detecting class-2 neurons in the frog retina. Vision Res. **8**, 1173—1186 (1968).

Hailman, J. P.: Phototaxis in amphibians and reptiles: a non-technical review. Wisc. Herpetol. Soc. Newsltr. **14**, 3—6 (1975).

Hailman, J. P., Jaeger, R. G.: Phototactic responses to spectrally dominant stimuli and use of colour vision by adult Anuran amphibians: a comparative survey. Animal Behav. **22**, 757—795 (1974).

Hailman, J. P., Jaeger, R. G.: A model of phototaxis and its evaluation with Anuran amphibians. Behaviour **56**, 215—249 (1976).

Hartline, H. K.: The response of single optic nerve fibres of the vertebrate eye to illumination of the retina. Amer. J. Physiol. **121**, 400—415 (1938).
Hartline, H. K.: The receptive fields of optic nerve fibres. Amer. J. Physiol. **130**, 690—699 (1940a).
Hartline, H. K.: The effects of spatial summation in the retina on the excitation of the fibres of the optic nerve. Amer. J. Physiol. **130**, 700—711 (1940b).
Heatwole, H., Audrey, H.: Motivational aspects of toad feeding. Copeia **4**, 692 (1968).
Houston, W. W. K.: The food of the common frog, Rana temporaria, on high moorland in northern England. Journ. Zool. **171**, 153—166 (1973).
Ingle, D.: Prey catching behaviour of Anurans toward moving and stationary objects. Vision Res. Suppl. **3**, 447—456 (1971).
Ingle, D.: Disinhibition of tectal neurons by pretectal lesions in the frog. Science **180**, 422—424 (1973a).
Ingle, D.: Size preferences for prey-catching in frogs: relationship to motivational state. Behav. Biol. **9**, 485—491 (1973b).
Ingle, D.: Two visual systems in the frog. Science **181**, 1053—1055 (1973c).
Jaeger, R. G., Hailman, J. P.: Two types of phototactic behaviour in Anuran amphibians. Nature **230**, 189—190 (1971).
Jaeger, R. G., Hailman, J. P.: Effects of intensity on the phototactic responses of adult Anuran amphibians: a comparative survey. Z. Tierpsychol. **33**, 352—407 (1973).
Kennedy, D., Milkman, R. D.: Selective light absorption by the lenses of lower vertebrates, and its influence on spectral sensitivity. Biol. Bull. **111**, 375—386 (1956).
Kicliter, E.: Flux, wavelength and movement discrimination in frogs: forebrain and midbrain contributions. Brain. Behav. Evol. **8**, 340—365 (1973).
Lashley, K. S.: Brain mechanisms and intelligence. Chicago: Univ. Chicago Press 1929.
Lettvin, J. Y., Maturana, H. R., McCulloch, W. S., Pitts, W. H.: What the frog's eye tells the frog's brain. Proc. Inst. Radio Engr. **47**, 1940—1951 (1959).
Lettvin, J. Y., Maturana, H. R., Pitts, W. H., McCulloch, W. S.: Two remarks on the visual system of the frog. In Sensory Communication, ed. W. A. Rosenblith. Cambridge: M.I.T. Press, and New York-London: Wiley & Sons Inc. 1961.
Liberman, E. A.: The nature of the information arriving at the brain by one nerve fibre from two retinal receptors in the frog. Biophysics **2**, 424—428 (1957).
Liebman, P. A.: Microspectrophotometry of photoreceptors. In Handbook of Sensory Physiology, Vol. VII/1, Photochemistry of Vision, ed. H. J. A. Dartnall. Berlin-Heidelberg-New York: Springer-Verlag 1972.
Liebman, P. A., Entine, G.: Visual pigments of frog and tadpole (Rana pipiens). Vision Res. **8**, 761—775 (1968).
Liege, B., Galand, G.: Single-unit responses in the frog's brain. Vision Res. **12**, 609—622 (1972).
Martof, B. S.: The behaviour of Fowler's toad under various conditions of light and temperature. Physiol. Zool. **35**, 38—46 (1962).
Mast, S. O.: Light and the behaviour of organisms. New York: Wiley & Sons Inc. 1911.
Maturana, H. R., Lettvin, J. Y., Pitts, W. H., McCulloch, W. S.: Physiology and anatomy of vision in the frog. J. gen. Physiol. 43 Suppl., 129—175 (1960).
Merker, E.: Die Sichtbarkeit ultravioletten Lichtes. Biol. Rev. **9**, 49—78 (1934).
Mrosovsky, N., Tress, K. H.: Plasticity of reactions to light in frogs and a possible role for the pineal eye. Nature **210**, 1174—1175 (1966).
Muntz, W. R. A.: Microelectrode recordings from the diencephalon of the frog (Rana pipiens), and a blue-sensitive system. J. Neurophysiol. **25**, 699—711 (1962a).
Muntz, W. R. A.: Effectiveness of different colours of light in releasing the positive phototactic behaviour of frogs, and a possible function of the retinal projection to the diencephalon. J. Neurophysiol. **25**, 712—720 (1962b).
Muntz, W. R. A.: Phototaxis and green rods in urodeles. Nature **199**, 620 (1963a).
Muntz, W. R. A.: The development of phototaxis in the frog (Rana temporaria). J. exp. Biol. **40**, 371—379 (1963b).
Muntz, W. R. A.: The development of photopic and scotopic vision in the frog (Rana temporaria). Vision Res. **4**, 241—250 (1964).
Muntz, W. R. A.: The photopositive response of the frog (Rana pipiens) under photopic and scotopic conditions. J. exp. Biol. **45**, 101—111 (1966).

Muntz, W. R. A.: Sensory processes and behaviour. In Psychobiology, ed. J. L. McGaugh. Academic Press 1971.

Muntz, W. R. A.: Inert absorbing and reflecting pigments. In Handbook of Sensory Physiology, vol. VII/1, Photochemistry of Vision, ed. H. J. A. Dartnall. Berlin-Heidelberg-New York: Springer-Verlag 1972.

Muntz, W. R. A., Reuter, T.: Visual pigments and spectral sensitivity in Rana temporaria and other European tadpoles. Vision Res. **6**, 601—618 (1966).

Nilsson, S. E. G.: Receptor cell outer segment development and ultrastructure of the disk membranes in the retina of the tadpole (Rana pipiens). J. ultrastruct. Res. **11**, 581—620 (1964).

Orlov, O. Yu.: Reactions of the optic nerve of the frog to various colour stimuli. Biophysics **6**, 48—52 (1961).

Oyster, C. W.: The analysis of image motion by the rabbit retina. J. Physiol. **199**, 613—636 (1968).

Pearse, A. S.: The reactions of amphibians to light. Proc. Amer. Acad. Arts. Sci. **45**, 159—208 (1910).

Pomeranz, B.: Metamorphosis of frog vision: changes in ganglion cell physiology and anatomy. Exp. Neurol. **34**, 187—199 (1972).

Ramón y Cajal, S.: Histologie du système nerveux, 2 vols. Maloine, Paris.

Ramón y Cajal, S.: The Vertebrate Retina. Translation by D. Maguire and R. W. Rodieck in The Vertebrate Retina by R. W. Rodieck. San Francisco: Freeman & Co. 1973.

Reuter, T.: Visual pigments and ganglion cell activity in the retinae of tadpoles and adult frogs (Rana temporaria L.) Acta Zool. Fenn. **122**, 1—64 (1969).

Reuter, T., Virtanen, K.: Border an colour coding in the retina of the frog. Nature **239**, 260—263 (1972).

Reuter, T., White, R. H., Wald, G.: Rhodopsin and porphyropsin fields in the adult bullfrog retina. J. gen. Physiol. **58**, 351—371 (1971).

Saxén, L.: The development of the visual cells. Ann. Acad. Sci. fenn. Series A. **23**, 1—93 (1954).

Schaeffer, J. J.: Habit formation in frogs J. Anim. Behav. **1**, 309—335 (1911).

Shipperheyn, J. J.: Contrast detection in frog retina. Acta physiol. pharmacol. neerl. **13**, 231—277 (1965).

Stevens, D. M.: The dermal light sense. Biol. Rev. **38**, 204—240 (1963).

Stier, T. J. B.: Reversal of phototropism in Diemyctylus viridescens. J. gen. Physiol. **9**, 521—523 (1926).

Székely, G.: The mesencephalic and diencephalic optic centres in the frog. Vision Res. **11** Suppl. 269—280 (1971).

Torelle, E.: The response of the frog to light. Amer. J. Physiol. **9**, 466—488 (1903).

De Valois, R. L., Walraven, J.: Monocular and binocular aftereffects of chromatic adaptation. Science **155**, 463—465 (1967).

von Uexküll, J.: A stroll through the worlds of animals and men. In C. H. Schiller (Ed.), Instinctive Behaviour. New York: International University Press 1957.

Wald, G.: Visual pigments and vitamins A of the clawed toad, Xenopus laevis. Nature **175**, 390—394 (1955).

Walls, G. L.: The vertebrate eye and its adaptive radiation. Michigan: Cranbrook Inst. Sci. 1942.

Walls, G. L.: The lateral geniculate nucleus and visual histophysiology. Univ. Calif. Publ. Physiol. 9, No. 1 (1953).

Zipse, W.: Können unsere heimischen Frösche und echten Kröten ultraviolettes Licht sehen? Zool. Jahrb. **55**, 487—524 (1935).

Chapter 6

The Amphibian Eye: Development and Regeneration

R. W. REYER, Morgantown, W.Va. (USA)

With 30 Figures

Contents

A. Introduction

The eyes of amphibians, both anurans and urodelans, have interested students of development since the beginning of the twentieth century. SPEMANN investigated the roles of the optic vesicle and surface ectoderm in lens formation (SPEMANN, 1903, 1905, 1907, 1912, 1938) before he began his better known studies on neural induction and the "organizer." In the United States, W. LEWIS (1904, 1907a, b) and HARRISON (1920, 1929) also employed this system of lens induction in their work on dependent and independent development. MANGOLD (1931) wrote a very complete review of the experiments on amphibian eyes up to that time. These early investigations consisted of the following types of microsurgical experiments: 1. isolating the prospective lens-forming ectoderm by removing the eye-forming region of the neural plate or the optic vesicle, 2. combining optic vesicle and ectoderm of different prospective significance located in other regions of the head or flank, and 3. transplanting the prospective lens ectoderm to other parts of the body. Several different species were employed and these were found to differ in the ability of the prospective lens ectoderm to differentiate into a lens without an eyecup and in the capacity of the optic vesicle to induce a lens from foreign ectoderm. Additional investigations seemed to indicate that these properties varied inversely in different species and led SPEMANN (1938) to propose the hypothesis of "double assurance" in lens formation. This concept postulates two processes, each of which can lead to lens development under experimental conditions: 1. self-differentiation of prospective lens ectoderm, and 2. lens induction by the optic vesicle. In normal development, these two mechanisms reinforce each other. In a later section of this account, it will be seen how recent investigations have led to refinements in our knowledge of the interactions leading to the development of a lens in its proper relation to the eyecup.

Another aspect of eye development that interested early investigators was the organization of the eye-forming region in the anterior neural plate. Together with adjacent parts of the brain, this seems to become determined as a general forebrain area under the inductive influence of the rostral part of the chordamesoderm, in particular, the median prechordal plate and flanking prechordal mesoderm. The experiments of ADELMANN (1929a, b, 1930, 1932, 1934, 1936, 1937) demonstrated that the parts of the two eyes are not present in the tissue of the neural plate as a mosaic but rather as an eye "field" with the highest eye-forming capacity in the midline which then declines laterally. This region can form from one to four eyes depending on experimental manipulation.

When an optic vesicle invaginates to form an eyecup, the outer layer thins out and develops into the pigmented retina while the inner layer thickens to form the neural retina. Quite early in studies on eye morphogenesis, it was found that each of these parts could form the other during head and tail-bud stages and that the pigmented retinal layer retained the ability to develop into neural retina for a considerable part of embryogenesis (DRAGOMIROV, 1932, 1933, 1934, 1936; IKEDA, 1937, 1938).

Interest in the capability of the amphibian eye for regeneration preceded the experimental studies on eye development. In 1781, BONNET found that a small normal-appearing eye had been reconstituted 1 year after partial removal of a salamander eye including the intact lens. This observation was made only on a living animal. The first histologic study of a similar case in adult *Triturus* was that of COLUCCI (1891) who found a new lens being regenerated from the dorsal iris epithelium of the reconstituting eye. Following lens extirpation through the cornea, leaving the remainder of the eye intact, the events of lens regeneration from the iris were studied histologically by WOLFF (1895, 1901, 1904), MÜLLER (1896) and BRACHET and BENOIT (1899) in larval and adult *Triturus* and by FISCHEL (1900, 1903) in *Salamandra*. Lens regeneration from the iris is frequently called "Wolffian" lens regeneration although WOLFF was not the first to observe it. In 1914, WACHS made the most thorough study to date of lens regeneration from the dorsal iris epithelium of larval *Triturus taeniatus* and *T. cristatus*. He also initiated the experimental analysis of these events using microsurgical techniques of lens removal and reimplantation, lens displacement, and the transplantation of pieces of dorsal iris into eyes with and without a lens or into extraocular sites. He reached several conclusions most of which have been confirmed by later investigators. Two of the most important of these are: 1. lens regeneration is initiated by lens removal and blocked by the presence of a lens, and 2. pieces of iris can regenerate a lens when lying within a lentectomized eye but not in extraocular locations unless neural retina is also present.

That regeneration in the newt eye is not limited to the formation of a new lens was also demonstrated first by GRIFFINIE and MARCCHIO (1889) and later, by WACHS (1920). The latter author removed the lens and neural retina from eyes of *Triturus* larvae and observed the regeneration of a new neural retina from both the proximal, inner layer of iris epithelium and the pigmented retinal epithelium (tapetum). After neural retina regeneration was well advanced, a lens started to develop from the pupillary margin of the dorsal iris so that a normal eye was finally reconstituted. The first observations on the regeneration of the optic nerve and return of vision were made by MATTHEY (1926, 1927). When the optic nerve was severed, it regenerated completely in both larval *Salamandra salamandra* and adult *Triturus cristatus*. In the adult *Triturus*, vision was found to return after both optic nerve section and eye transplantation.

In the account which follows, an attempt will be made to show how the experimental analysis of the differentiation of this complex organ and of its dramatic capability for self-repair has resulted in important contributions to our understanding of embryonic development and of regeneration.

B. Structure of the Amphibian Eye

Before discussing the experimental analysis of development and regeneration, it seemed desirable to briefly describe the anatomy of this typical vertebrate eye. WALLS (1942), ROCHON-DUVIGNEAUD (1943), and DUKE-ELDER (1958) give thorough accounts based on the older work. The frog eye (*Rana*, several species) and the newt eye (*Notophthalmus*, [*Triturus*]) are quite similar in structure with some minor differences. Most of the previous descriptions have used the frog eye as an example. We shall consider eyes of both anurans and urodeles but will use the newt eye for the general description because it has been the object of many of the recent experimental studies.

Figure 1 is a transverse section through the entire eye of an adult newt (*Notophthalmus viridescens*). This eye is almost spherical and is composed of the usual three tunics: *tunica fibrosa*, *tunica vasculosa*, and *tunica interna*. The outer tunic consists of a rather thin, fibrous sclera and the slightly thicker *substantia propria* of the cornea. The former is supported by a cup-shaped scleral cartilage in the frog but not in the adult newt. In anuran tadpoles, an outer cornea of epithelium and connective tissue is present which is continuous with the skin of the head. Beneath this is a very thin, inner cornea continuous with the outer surface of the ciliary body. At metamorphosis, the outer and inner corneas fuse to complete the

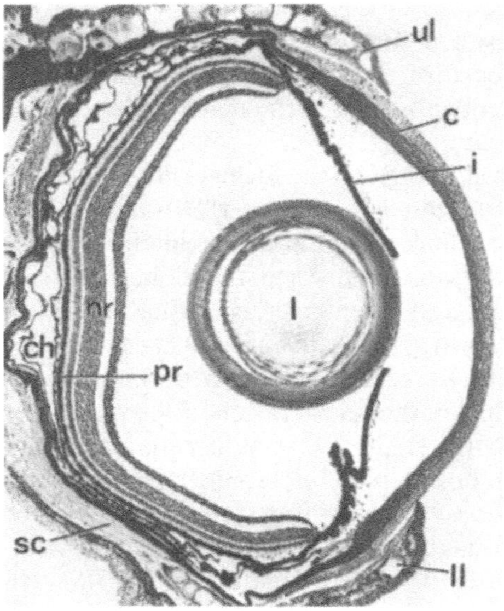

Fig. 1

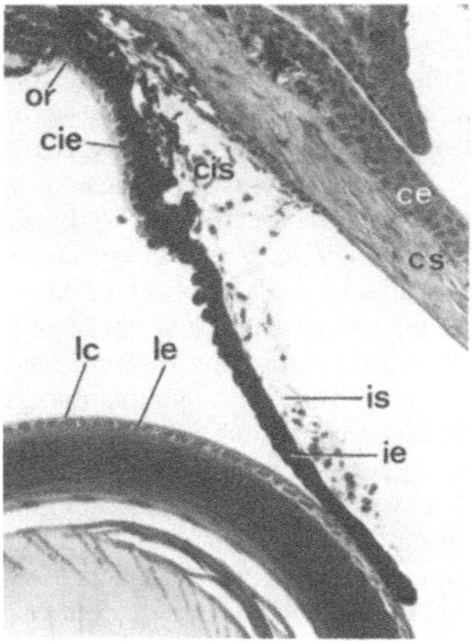

Fig. 2

Fig. 1. Vertical, meridional section of the eye of the newt, *Notophthalmus viridescens*. × 30. *c* = cornea, *l* = lens, *i* = iris, *ul* = upper lid, *ll* = lower lid, *nr* = neural retina, *pr* = pigmented retina, *ch* = choroid, *sc* = sclera

Fig. 2. Detail of cornea, lens, dorsal iris, and ciliary body. (Masson's trichrome stain) × 93. *lc* = lens capsule, *le* = lens epithelium, *ie* = iris epithelium, *is* = iris stroma, *cie* = ciliary epithelium, *cis* = stroma of ciliary body, *or* = *ora serrata*, *ce* = corneal epithelium, *cs* = corneal stroma (substantia propria)

substantia propria of the typical cornea. The vascular tunic or uvea consists of 1. a choroid which is made up of heavily pigmented lamellae, passing between numerous veins, and an internal choriocapillaris, 2. the stroma of the ciliary body, triangular in cross section, which is composed of a meshwork of vascular connective tissue containing the *tensor choroideae* and *protractor lentis* muscles, and 3. the stroma of the iris (Fig. 2). There are numerous ciliary processes in the frog eye but only a single, midventral ciliary fold in the newt eye. In the frog, both dorsal and ventral *protractor lentis* muscles pass from the corneoscleral junction through the ciliary body to insert on hypertrophied ciliary processes. When these muscles contract in accommodation, the lens is pulled forward by means of the zonula fibers. Only a ventral *protractor lentis* muscle, having a similar function, is present in the newt eye. This muscle originates from the corneoscleral junction, passes through the surface layers of the ciliary body and perforates the pigmented iris epithelium near its junction with the ciliary epithelium. The iris epithelium projects inward as a horizontal fold on either side of this cleft. The fold ventral to the muscle is continuous with the single ciliary process where the muscle terminates. Stone (1966 b) followed the development of this iris cleft from a remnant of the choroid (fetal) fissure. From the ciliary body, zonula fibers pass to the lens capsule. Dorsally and ventrally, in both frogs and newts, smaller bundles of smooth

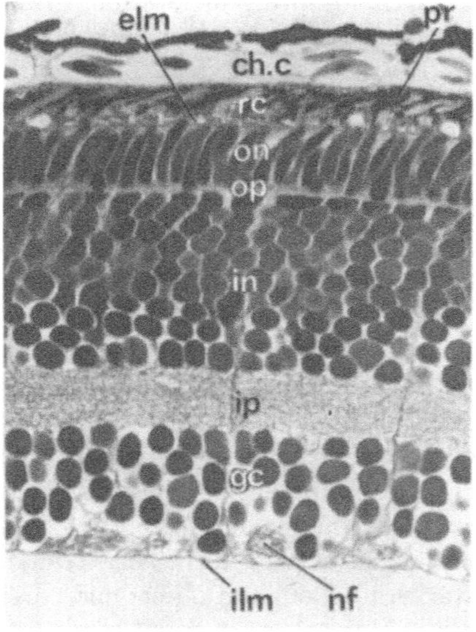

Fig. 3. Neural and pigmented retina (1.0 μ plastic section). × 273. *ilm* = internal limiting membrane, *nf* = nerve fiber layer, *gc* = ganglion cell layer, *ip* = inner plexiform layer, *in* = inner nuclear layer, *op* = outer plexiform layer, *on* = outer nuclear layer, *elm* = external limiting membrane, *rc* = layer of rods and cones, *pr* = pigmented retinal epithelium, *ch.c.* = choriocapillaris

muscle, the *tensor choroideae*, extend from the sclera in a meridional direction to insert on the choroid close to the *ora serrata*. In the stroma of the iris are located the iridial vessels and the *sphincter* and *dilator pupillae* muscles. There are also scattered melanophores and iridophores, giving the characteristic pattern of pigmentation.

The *tunica interna* consists of the neural retina *(pars optica retinae)* and pigmented retina *(stratum pigmenti retinae)* which extend to the *ora serrata*, the two layered ciliary epithelium *(pars ciliaris retinae* and *stratum pigmentum corporis ciliaris)* and the iris epithelium *(pars iridica retinae* and *stratum pigmenti iridis)*. The cells of both epithelial layers of the iris, the outer ciliary epithelium, and the pigmented retinal epithelium are packed with melanosomes. In the adult newt eye, the border between *pars iridica retinae* and *pars ciliaris retinae* is marked by a change from a pigmented to an unpigmented epithelium.

The lens of the amphibian eye is large in proportion to the size of the eyeball, especially in urodeles. It is nearly spherical in urodeles and slightly flattened in anurans. Otherwise, it has the typical structure of lens epithelium and concentric layers of slender, elongated lens fibers which meet in an anterior vertical suture and a posterior horizontal suture. The lens is suspended by the zonula fibers running from the equatorial lens capsule to the ciliary body. The space remaining within the eyeball between the zonula, the lens, and the neural retina is occupied by the vitreous body.

In amphibians, the pigmented retinal epithelium is characterized by long cytoplasmic processes which extend between the parts of the photoreceptors protruding beyond the external limiting membrane. In response to bright light, the pigment granules move into these processes, thus shielding the outer segments of the rods and cones. The reverse occurs in dim light. An extensive smooth endoplasmic reticulum is present in the pigmented retinal cells of the newt (DICKSON and HOLLENBERG, 1971; KEEFE, 1971). Fragments of rod outer segments are also observed embedded in the cytoplasm of these cells.

The layers of the amphibian neural retina are those typical of all vertebrate eyes (Fig. 3). Depending on the species, the outer nuclear layer comprises one to three layers of nuclei, the outer plexiform layer is narrow, the inner nuclear layer is five to eight nuclei thick, the inner plexiform layer is wider than the outer plexiform layer, and the ganglion cell layer has one to four layers of nuclei. Four kinds of photoreceptors have been described: red rods, green rods, cones, and double cones. A great deal of new information on their structure has been obtained by electron microscopy. In a series of papers, NILSSON (1964a–c, 1965) described the structure and development of the inner and outer segments of these four types of photoreceptors and the interreceptor contacts in the frog, *Rana pipiens*. Previously, DE ROBERTIS and LASANSKY (1961), FERNANDEZ-MORÁN (1961), MOODY and ROBERTSON (1960) and others had studied the organization of the double membrane discs in the outer segment of the frog rod. Electron-microscopic studies on urodele photoreceptors have recently been reported by DICKSON and HOLLENBERG (1971) and KEEFE (1971). The following summary is based mainly on the work of NILSSON and KEEFE.

The four different kinds of photoreceptors are shown in Fig. 4, adapted from a drawing by NILSSON. These will now be described briefly. The most common rod is the red rod. This highly specialized cell consists of an elongated, cylindrical, outer segment (27–45 μ long), an eccentric connecting stalk (present in all of the photoreceptors), an inner segment comprising an ellipsoid packed with elongated mitochondria and a myoid of irregular electron density, and a tapering rod fiber containing an electron-dense nucleus adjacent to the external limiting membrane and terminating in a synaptic body or rod spherule of uneven density and rather irregular shape. Slender extensions of the inner segment project along the proximal part of the outer segment. The green rod differs from the red rod in having a shorter outer segment (20–30 μ long) and an elongate inner segment within which lies an ellipsoid with smaller mitochondria and a slender myoid; the nucleus is located among the cone nuclei in the outer nuclear layer. The single cone has a much smaller, tapering, outer segment (7–13 μ long) and a plump inner segment containing both an ellipsoid of mitochondria, less densely packed than in the rod, and a darkly staining oil or lipid droplet; the cone fiber does not taper, contains a less electron-dense nucleus, and ends in a larger, more homogeneous synaptic body. The double cone consists of two, somewhat modified cones in close contact with each other except for their outer segments. The principal member has a tapering myoid and a denser fiber than the accessory member. A lipid droplet is present in the former and lacking in the latter which is, however, the only photoreceptor in the frog to have a paraboloid containing glycogen.

Much attention has been focused on the structure of the outer segment because this is the site where light is absorbed by the visual pigments resulting in the generation of an electrical potential. Electron-microscopic studies have shown

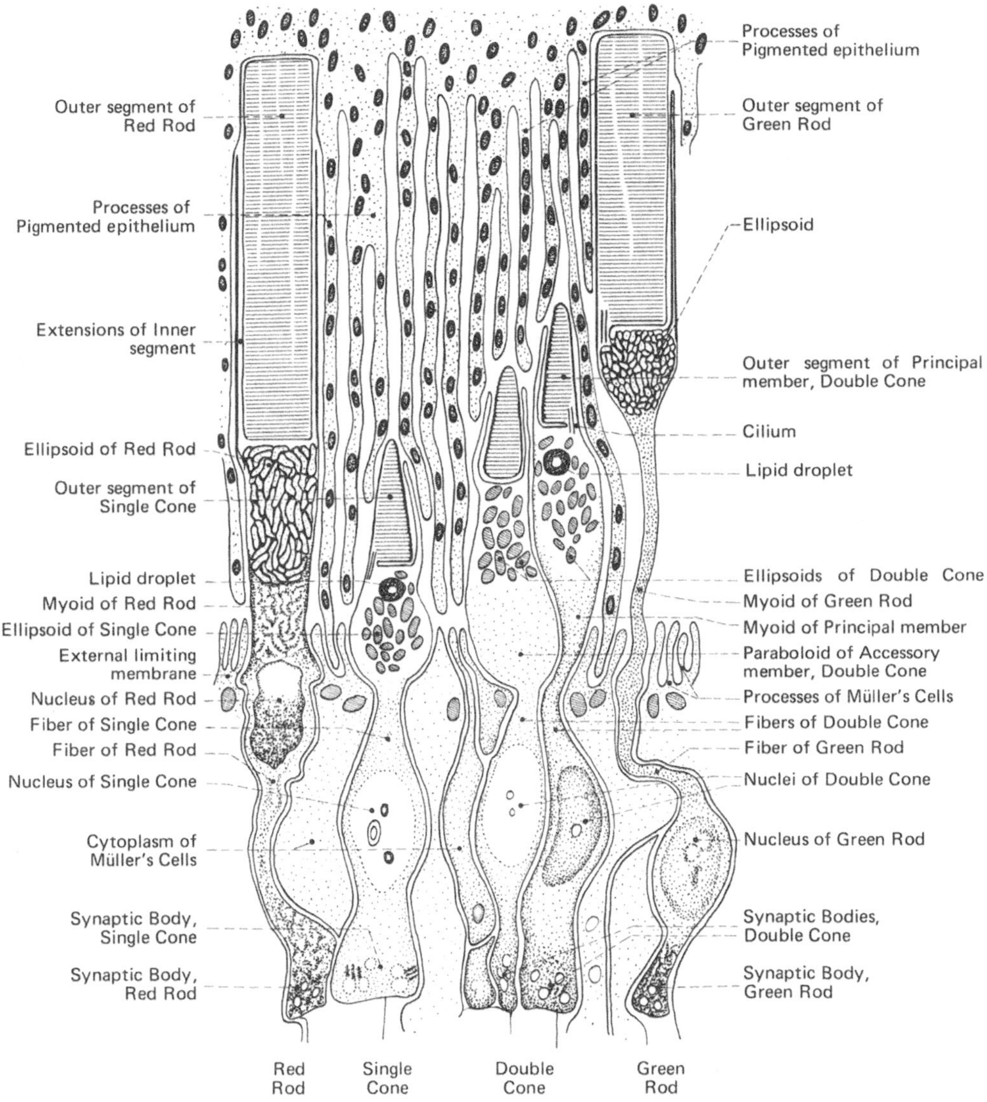

Fig. 4. The four types of photoreceptors found in frog neural retina. Diagram based on electron microscopic studies. (Modified from Fig. 1, NILSSON, 1964a, by permission of the author and Academic Press)

that the outer segments of both rods and cones are composed of stacks of flattened discs bounded by membranes. These disc membranes are continuous with the cell membrane all along one side of the cone, while in the rod this continuity is present only at the base of the outer segment (NILSSON, 1965; COHEN, 1968). Proceeding distally, the rod discs lose their connection with the cell membrane and become "free floating." Rod discs also have folds or incisures, extending radially from the periphery, which divide the disc into lobules. The presence or absence of a space between the membranes within a disc, the size of this space and that of the spaces between discs is sensitive to the tonicity of the solutions used in fixation and postfixation treatment of the tissue (COHEN, 1971; HELLER et al., 1971). In rod discs of the frog, NIR and PEASE (1973) recently found a consistent intradiscal space of 15–20 Å between two parallel trilaminar unit membranes with the total thickness of the discs varying between 165 and 210 Å, depending on the treatment. These probably represent the most reliable figures available at present.

Of considerable interest is the discovery by YOUNG that there is a continual formation of new rod discs at the base of the outer segment, displacement of these discs distally, and shedding of discs from the apical end of the rod outer segment. Fragments of shed discs are then phagocytized by cells of the pigmented retinal epithelium. These dynamic events were traced by studying protein synthesis in the photoreceptors, following injection of tritiated amino acids, using autoradiography at both light and electron-microscopic levels (YOUNG, 1967, 1969, 1970, 1971; YOUNG and BOK, 1969, 1970). When similar studies were performed on cones, there was an incorporation of the radioactive amino acids into insoluble proteins but there was no evidence of localization of radioactivity in cone discs or of a movement of radioactive discs along the cone outer segment. Thus, there appears to be a distinct difference in the physiology of maintenance in the rod and in the cone.

KEEFE'S findings on the newt photoreceptors (KEEFE, 1971) agree with those of NILSSON for the frog. However, some differences were observed but it is not certain whether these apply to all urodelan and anuran photoreceptors. These differences were the absence of a lipid droplet in the single cone and principal member of the double cone, the presence of a glycogen-rich paraboloid in all four kinds of rods and cones, cytoplasmic extensions from the accessory member which clasp the principal member of the double cone, resemblance of the inner segment of the accessory member of the double cone to that of a rod, and more extensive areas of cell-to-cell contact among the different photoreceptors.

The main supporting cells for the neural retina are Müller's cells. Their nuclei are usually located in the inner nuclear layer but their cytoplasm extends the entire width of the retina. Junctional associations of Müller's cells with each other and with the photoreceptors constitute the external (outer) limiting membrane (KEEFE, 1971) or external junctional zone (HENDRICKSON, 1966). Fine cytoplasmic processes of Müller's cells extend beyond this membrane between the inner segments of the photoreceptors as they pass through the external limiting membrane.

The conducting and association cells of the amphibian retina and their synaptic organization have been studied primarily in the frog *(Rana pipiens)* and mudpuppy *(Necturus maculosus)* by DOWLING and his co-workers (DOWLING, 1968, 1970; DOWLING and WERBLIN, 1969; WERBLIN, 1970, 1971, 1972; WERBLIN and DOWLING, 1969; MILLER and DOWLING, 1970). This work combined electron-microscopic studies on the pattern of synaptic contacts with intracellular, electrophysiologic recording techniques following stimulation of the retina by light. As in the retina of all vertebrates, there are five principal kinds of neurons in the amphibian retina. These are the photoreceptors with nuclei in the outer nuclear layer, the horizontal cells, the bipolar cells, and the amacrine cells with nuclei in the inner nuclear layer, and the ganglion cells with nuclei in the ganglion cell layer (Fig. 3).

The synaptic organization of the frog and mudpuppy retina is similar to that of mammals. Synapses are located in the outer and inner plexiform layers and are of two kinds: 1. ribbon synapses made by the synaptic bodies of the photoreceptors and by the axons of the bipolar cells, and 2. conventional synapses made by the processes of horizontal cells and amacrine cells. For details of these synaptic connections and the electrical responses of the different kinds of cells to visual stimuli, the reader is referred to DOWLING's and WERBLIN's reviews (DOWLING, 1970; WERBLIN, 1972).

C. Development of the Amphibian Eye

As can be seen from the discussion of its anatomy, the amphibian eye is a complex organ which must meet the criteria for a delicate optical instrument as well as those of a neural organ capable of generating, transmitting and, to some degree, integrating nerve impulses in response to stimulation by light. The purpose of this section is to discuss the morphogenesis and differentiation of the eye and how this is controlled so that the resulting organ can meet these stringencies for normal functioning.

I. Localization and Morphogenesis

The embryonic development of the amphibian eye has been reviewed by MANGOLD (1931), TWITTY (1955), LOPASCHOV (1961, 1963), LOPASCHOV and STROEVA (1961, 1963), and COULOMBRE (1965). The discussions of the earlier literature by the Russian authors are comprehensive. In the following account, space is insufficient for another detailed review. Instead, an attempt will be made to emphasize certain aspects concerning which recent investigations have brought increased understanding.

Maps showing the prospective developmental fate of different parts of the early embryo can be obtained by applying vital stains such as Nile blue sulfate or neutral red to small areas of the gastrula or neurula and then following the movement of these stained parts during subsequent stages of development. In this way, the locations of the anlagen of the eyes and other head organs in the ectoderm have been mapped at the early neural fold stage as shown in Fig. 5A (MANCHOT, 1929; WOERDEMAN, 1929; CARPENTER, 1937; TEN CATE, 1956; C.-O. JACOBSON, 1959; VON WOELLWARTH, 1960). Figure 5B, modified from Fig. 85 of HOLTFRETER and HAMBURGER (1955), shows the tissues lying beneath the ectoderm. The neural plate and folds are underlain by prechordal plate and mesoderm rostrally, followed caudally by notochord and dorsal mesoderm. The head ectoderm, destined to form nose and lens, is underlain by the entoderm of the rostral

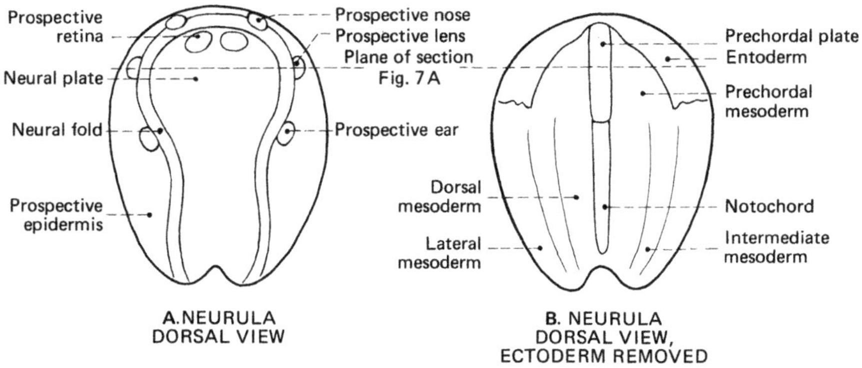

Fig. 5. Diagram showing dorsal views of early neurula of urodele (Harrison stage 15). A. The ectodermal areas which will develop into the organs of special sense. B. The regions of mesoderm and entoderm lying beneath the neural plate, folds, and adjacent ectoderm. Composite diagrams modified from CARPENTER (1937), C.-O. JACOBSON (1959), A. JACOBSON (1963a), VON WOELLWARTH (1960), ADELMANN (1932), HOLTFRETER and HAMBURGER (1955)

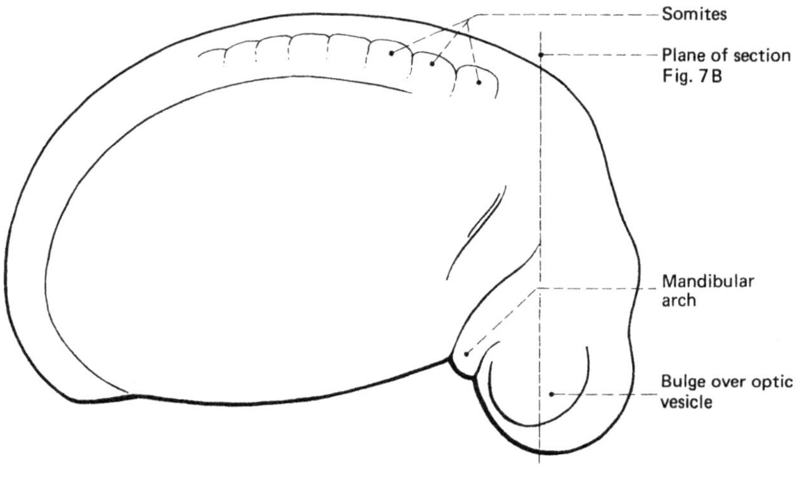

Fig. 6. Diagram of lateral view of early head and tail-bud embryo of urodele (Harrison stage 24)

archenteron while the prospective ear ectoderm rests on the most rostral part of the lateral mesoderm, the prospective heart region (WILENS, 1955). A cross section through the rostral neural plate of an early neurula is illustrated in Fig. 7A, p. 320. During closure of the neural folds, the retinal anlagen on each side evaginate laterally to form the optic vesicles which are clearly delineated at an early head and tail-bud stage (Fig. 6). The apex of the optic vesicle contacts the surface ectoderm (Fig. 7B, p. 320). Formation of the optic cup takes place by the invagination of the apical part of the optic vesicle into the proximal part which thins to become a simple squamous epithelium. During this invagination, there is a transitory gap in the midventral border of the eyecup (the fetal or choroid fissure) which is continuous with a groove in the ventral optic stalk. At the same time, the inner layer of epithelium, overlying the optic vesicle, thickens to form a lens placode, rounds up into a lens vesicle, and detaches from the surface ectoderm to occupy the cavity in the optic cup (Figs. 7C–E, p. 321). The inner layer of the optic cup has become thicker preparatory to differentiating into neural retina. Lens fibers develop by the elongation of the lens vesicle cells next to the retina followed by addition of secondary lens fibers in concentric layers at the equatorial zone. The corneal epithelium develops from the surface ectoderm remaining over the eyecup (Figs. 7E and F, p. 321). Later on, mesenchyme will aggregate around the eye and also become associated with the corneal epithelium and the margin of the eyecup to lay down the connective tissue coats of corneal and iris stroma, choroid and sclera.

II. Interactions in the Development of Pigmented and Neural Retina

As summarized in the Introduction, it has been known since the studies of ADELMANN that, at the early neurula stage, eyes can develop from any part of an "eye field" which includes a larger region than the prospective eye-forming area itself. For example, in *Triturus alpestris*, eyes differentiated in 85–100% of small grafts obtained from five regions in the midline of the presumptive diencephalon of stage $14^1/_2$ neurulae (stages from HARRISON, 1969). Only two of these actually contained prospective eye tissue (BOTERENBROOD, 1970). When the rostral neural plate and one-half of the transverse neural fold of *Rana temporaria* and *R. esculenta* was cultured as an explant, both brain and neural retina differentiated. In the interzone between these parts, incompletely developed iris and also lentoids of irregular lens fibers and lens epithelium were found even though no epidermal ectoderm had been included with the explant (HOPERSKAYA, 1968, 1972). Eyes are a characteristic component of the complex of parts which comprises the archencephalon and are induced by the neuralizing factor originating from the underlying prechordal plate and mesoderm (SAXÉN and TOIVONEN, 1962). Disturbances in the continued action of this inductor result in abnormalities in the paired nature of the eyes such as synophthalmia and cyclopia (ADELMANN, 1936; VON WOELLWARTH, 1952, 1971). In a study of the development of the head, MANGOLD and VON WOELLWARTH (1950) and MANGOLD (1957) combined the neural plate and prospective epidermis from the head region of early neurulae with different parts of the mesoderm and entoderm. The presence of the prechordal plate appeared to be the most important and the prechordal mesoderm the next most important element in ensuring the presence of paired eyes.

DRAGOMIROV (1932, 1933, 1934, 1936) demonstrated that the inner as well as the outer layers of the eyecup can form both neural retina and pigmented retina when isolated during early head and tail-bud stages. There appear to be several conditions operating during the early development of the optic vesicle and cup which contribute to the segregation of neural and pigmented retina. DRAGOMIROW (1936, 1937, 1940), IKEDA (1937, 1938), and DETWILER and VAN DYKE (1951, 1953, 1954) transplanted optic vesicles or cups and ear vesicles next to each other

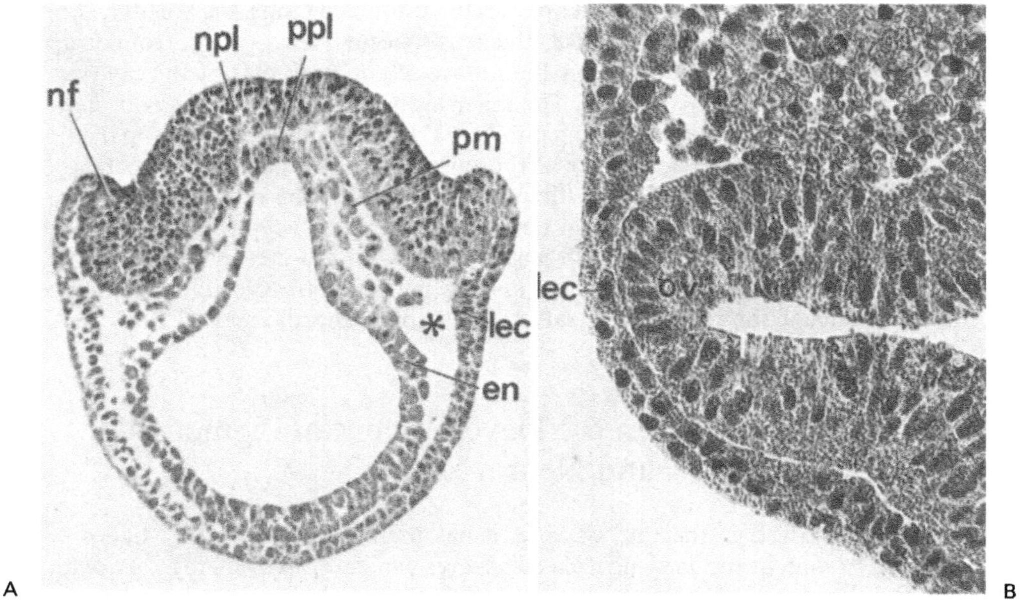

Fig. 7. Normal development of eye of urodele, *N. viridescens*. A. Cross section through prechordal region of neurula of stage 15. *npl* = neural plate, *nf* = neural fold, *ppl* = prechordal plate, *pm* = prechordal mesoderm, *en* = entoderm, *lec* = prospective lens ectoderm. Space (*) between entoderm and surface ectoderm is an artifact. × 63. B. Cross section through optic vesicle and prospective lens ectoderm at stage 22. *ov* = optic vesicle. × 240. C. Cross section through invaginating optic vesicle at stage 28. *ap* = apical portion, *pp* = proximal portion. × 240. D. Cross section through the eyecup and lens placode at stage 34. *lpl* = lens placode, *nr* = prospective neural retina, *pr* = prospective pigmented retina. × 240. E. Cross section through eyecup and lens vesicle with primary lens fibers at stage 36. *le* = lens epithelium. *lf* = lens fibers. × 240. F. Cross section through eyecup and lens at stage 41. The three nuclear layers are now established. *ce* = corneal epithelium, *gc* = ganglion cell layer, *in* = inner nuclear layer, *on* = outer nuclear layer. × 240

during head and tail-bud stages of several amphibians (*Triturus taeniatus, Ambystoma maculatum, A. tiginum, Hynobius unnangso, Pelobates fuscus*) and observed that accessory neural retina differentiated from the prospective pigmented epithelium when it was in close contact with the epithelium of the auditory labyrinth. Prolonged contact appeared to be necessary.

The presence of a lens affects the invagination of the eyecup. When lens formation was blocked in *Notophthalmus viridescens*, eyecup formation occurred in the absence of a lens, but the vitreous chamber was smaller, the anterior chamber absent, and the neural retina was thicker than normal and often folded (Reyer, 1950). Explants of optic vesicles without lens ectoderm usually failed to invaginate. When the lens-forming ectoderm was included, then the optic vesicle formed an eyecup which enclosed the lens (Lopashov, 1946, 1963; Lopashov and Stroeva, 1961). Eye rudiments, together with or without mesenchyme, were also transplanted or explanted, either exposed to the surroundings or within ectodermal coats (Lopashov, 1946, 1963; Lopashov and Hoperskaya, 1967). From the results of these different experiments, some of the tissue interactions that may be

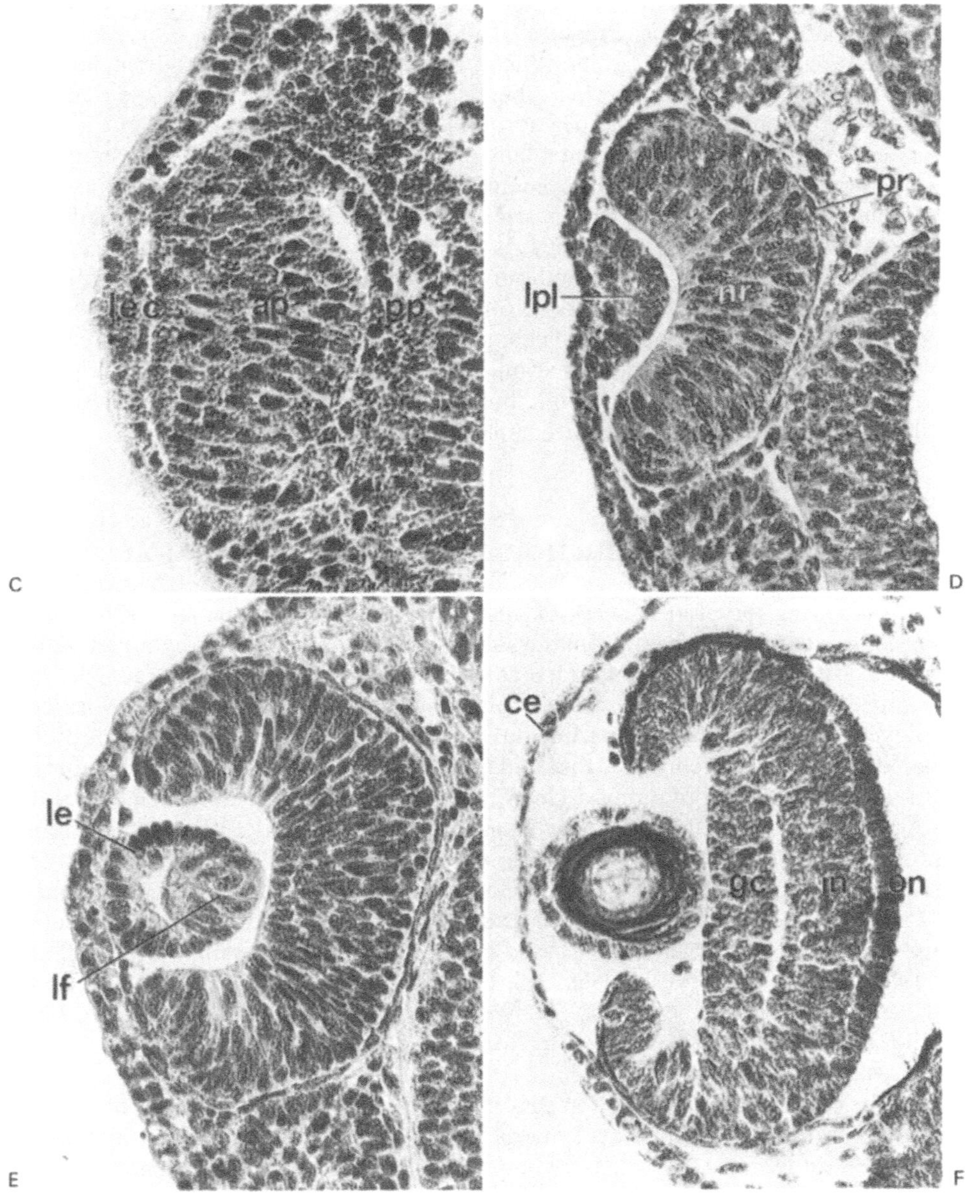

involved in the modeling of a normal eyecup can be summarized as follows. Adhesion of the outer part of the optic vesicle to the overlying ectoderm and crowding of the cells stimulates differentiation of neural retina from this part while contact of the remainder with the surrounding mesenchyme and close proximity to developing blood vessels promotes thinning and formation of pigmented epithelium. Contact of the margin of the eyecup with the lens stimulates differentiation of the iris. Pressure of fluids within the neural tube and optic vesicle are counteracted by its attachment to the overlying ectoderm and the restrictions of space within which the eye is developing. Thus, the eye does not swell and become vesicular under normal conditions. The elongation of the inner layer of ectodermal cells to form a lens placode and vesicle and the invagination of the optic vesicle to form an eyecup take place at the same time. The area of mutual adhesion is restricted in size. From their studies on chick embryos, ZWAAN and HENDRIX (1973) believe that resistance to expansion combined with continued cell division in the ectoderm result in the formation of the lens placode and its subsequent bulging and rounding up into a lens vesicle. Unfortunately, no similar studies of cell kinetics and changes in cell number, shape, surface area, and volume have been performed on amphibian eyes. Furthermore, the development of lens vesicles and lenses without the presence of an eyecup, to be described shortly, must involve different mechanisms.

III. Lens Induction and Cytodifferentiation

Following the pioneer work of SPEMANN and LEWIS on the role of the optic vesicle in lens development from the surface ectoderm, a great many investigators studied the extent to which lens formation is dependent on the optic vesicle. For many years their results remained confusing due to the variety of species employed, the variations in experimental procedures, most of which did not utilize adequate labeling techniques to distinguish host and donor tissues, and the technical difficulties in obtaining a clean removal of all lens-forming cells from the optic anlage, particularly in older stages. These results have been reviewed and analyzed by TEN CATE (1956), REYER (1954a, 1958a, b, 1962b), TWITTY (1955), A. JACOBSON (1958), and LOPASHOV and STROEVA (1961). Rather than repeat these discussions here, an attempt will be made to present a unified account of inductive interaction in lens development based mainly on the more recent work of LIEDKE (1951, 1955), TEN CATE (1956), A. JACOBSON (1958, 1963a–c, 1966), BALINSKY (1951, 1957), MANGOLD (1961), VON WOELLWARTH (1961), REYER (1958a, b, 1960, 1962b, 1966a), TAHARA (1962a, b), MICHAEL and NIEUWKOOP (1967), and MICHAEL (1968).

The concept of lens induction that is most widely accepted at the present time is that a series of inductors act synergically and sequentially in time on a population of ectodermal cells which respond by initiating the development of a lens. This hypothesis is depicted graphically in Fig. 8 from A. JACOBSON (1966). Toward the end of gastrulation (yolk plug stage), the invaginating entoderm and chordamesoderm has moved far enough rostrally so that the prospective lens ectoderm is now underlain by the entoderm of the rostral archenteron. This relationship continues through early neurula stages (Figs. 5A and B, 7A). At this time also, the border of the lateral (prospective heart) mesoderm lies just caudal to the prospective lens ectoderm. Inductive stimuli could therefore pass from these tissues into the adjacent, overlying ectoderm. Due to the tissue movements during elevation

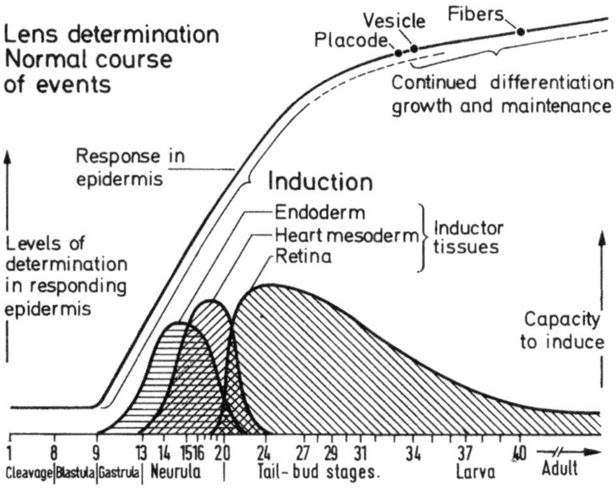

Fig. 8. A graphic model of lens induction in salamander. *Abscissa:* relative time of development represented by Harrison stages. *Left-hand ordinate:* logarithmic response curve for epidermis. *Right-hand ordinate:* linear scale indicating inducing capacity of inductors. (From Fig. 1, A. JACOBSON, 1966, by permission of author and A.A.A.S.)

of the neural folds, this relationship is ended and the prospective lens ectoderm soon comes into contact with the surface of the optic vesicle (Figs. 6, 7B). During these tissue movements, this ectoderm might also be briefly exposed to the lateral edge of the prechordal mesoderm. JACOBSON believes that the entoderm, lateral mesoderm, and optic vesicle act in sequence on the reacting ectoderm, each one raising the level of determination for lens until differentiation begins with the appearance of the lens placode.

Some of the more important evidence for these conclusions will now be presented. It is possible to prevent eyecup development on one or both sides of an embryo by removing one-half or all of the rostral neural plate and folds including the prospective retinal area. These embryos are allowed to develop until a lens has appeared in the eye on the opposite side or in a normal control. Using this procedure, lenses have been observed in the absence of the eyecup in *Rana esculenta* (SPEMANN, 1912, 1938; TEN CATE, 1956), *R. limnocharis* and *Rhacophorus schlegelii arboreus* (TAHARA, 1962a), *Xenopus laevis* (BALINSKY, 1951), *Triturus alpestris* (MANGOLD, 1961; VON WOELLWARTH, 1961) *T. pyrrhogaster* and *H. nebulosus* (TAHARA, 1962a), and *Taricha torosa* (A. JACOBSON, 1958). In *R. esculenta, T. alpestris*, and *T. torosa*, the formation of these lenses was enhanced by raising the embryos preoperatively at cool temperatures, 12° C, 14° C, or 16° C, respectively. Most of these "free" lenses were small cell masses or vesicles but some had lens fibers. The best developed lenses were observed in *Rhacophorus* and *T. alpestris*. Lens development was rare or absent in most other species tested in this way. Under the conditions of this experiment, the prospective lens ectoderm has been exposed to the influence of foregut entoderm, lateral and prechordal mesoderm, but not of the optic vesicle. These pre-optic inductors have been sufficient to initiate lens development. VON WOELLWARTH observed fewer free lenses when the neural folds were not extirpated with the neural plate. The presence of neural crest in these cases appeared to inhibit lens differentiation. Isolated lenses have also been observed in the head region of host embryos following implantation of various abnormal inductors, such as alcohol-fixed guinea pig thymus, into the blastocoel of early gastrulae (TOIVONEN, 1945; BECKER, 1959, 1960).

In an extensive investigation, A. JACOBSON (1958, 1963a) combined various possible lens-inducing tissues with the prospective lens ectoderm of an early neurula of *Taricha torosa* in explants to Niu-Twitty salt solution. When isolated alone, neurula, prospective lens ectoderm failed to form any lenses. When combined in the explant with foregut entoderm from the same stage, small lens vesicles developed from this ectoderm in 31% of the cases; when combined with the rostral part of the lateral plate mesoderm (prospective heart), then lens formation occurred in only 14% of the cases while lens vesicles were found 42% of the time when both the entoderm and lateral mesoderm were present with the prospective lens ectoderm in the explant. However, removal of the entoderm from head and tail-bud (stage 22), late neurula (stage 19), and early neurula (stage 14) embryos resulted in only a small reduction in the frequency of lens development which was still 75% of normal following operation at the youngest stage (A. JACOBSON, 1963b). Apparently, action of the entoderm for less than the normal period of time is usually sufficient preparation for the ensuing action of the optic vesicle. Similar evidence for the operation of lens inductors before the appearance of the optic vesicle has been obtained by other investigators in different species. TAHARA (1962b) observed the formation of small lens vesicles and lenses when prospective lens ectoderm from middle neural fold stages was explanted into salt solution and also when transplanted to the trunk or posterior head region in *Rhacophorus*. In *Ambystoma maculatum*, HARRISON (1920) obtained well-developed lenses when prospective lens ectoderm from a closing neural fold or early head and tail-bud stage was transplanted to the ventral caudal head region near the heart. REYER (1962b) implanted prospective lens and nose ectoderm from early neurula (stages 13–15), late neurula (stages 18–20), and head and tail-bud (stages 21–24) embryos into the dorsal fin or lentectomized eye of host larvae. In both locations, the implants from the younger donors rarely formed a few lens fibers and these were close to or in contact with a nose, also of donor origin; implants from the older donors occasionally formed tiny, isolated lenses and also larger lenses with fibers when associated with a nose. Similar implants from early neurulae, combined with the underlying foregut entoderm, were made into the dorsal fin. Well-developed lenses with fibers were found in 32% of the cases (Fig. 10). Some of these lenses were associated with nasal tissue but others were not (REYER, 1960). Probably due to the more favorable *in vivo* environment, lens differentiation progressed much further in these cases than in explants of similar tissues by A. JACOBSON.

Another approach to determining the role of pre-optic inductors in lens development is the transplantation of either prospective trunk or prospective lens ectoderm from donors of different ages to the eye region of hosts of different ages. In this way, the duration of exposure of the ectoderm to the foregut entoderm and to the optic vesicle can be manipulated. In *Ambystoma maculatum*, prospective ventral trunk ectoderm from gastrulae (stages $10^3/_4$–12) was induced to form lens

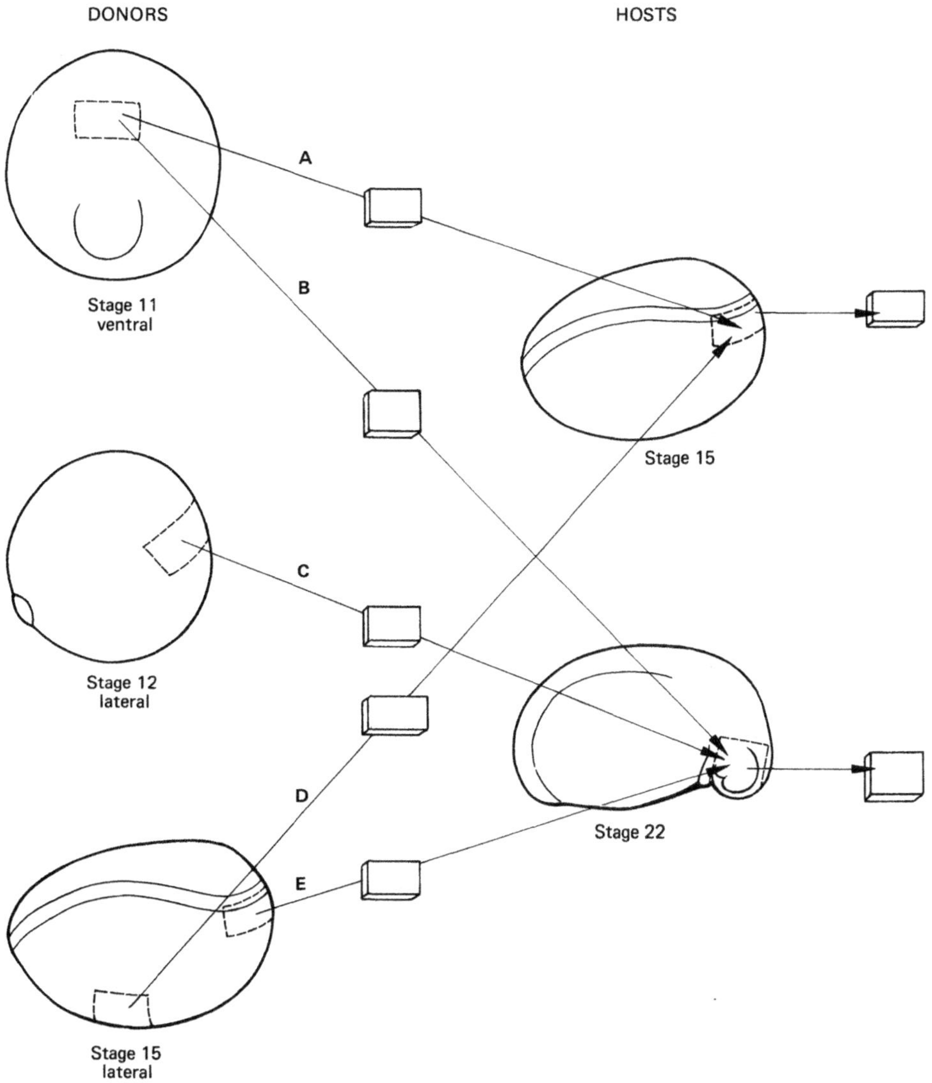

Fig. 9 A–E. Diagram illustrating transplantation of gastrula, prospective trunk ectoderm from a donor to replace the lens-forming ectoderm of (A) a neurula or (B) a head and tail-bud host; the transplantation of prospective lens ectoderm from a gastrula donor (C) or a neurula donor (E) over the optic vesicle of a head and tail-bud host; the transplantation of prospective trunk ectoderm from a neurula donor to replace the prospective lens ectoderm of a neurula host (D)

vesicles and lenses when transplanted over the anterior entoderm in place of the prospective lens ectoderm of early neurulae (stages 14–15; Figs. 9A, 11), but not when grafted over the optic vesicle of early head and tail-bud hosts (stages 20–23; Fig. 9B). With the onset of neurulation (stages 13–18), competence for lens formation was lost from the prospective trunk ectoderm when similarly transplanted (Fig. 9D). Grafts of *Ambystoma* ectoderm to *Notophthalmus* hosts or *Not-*

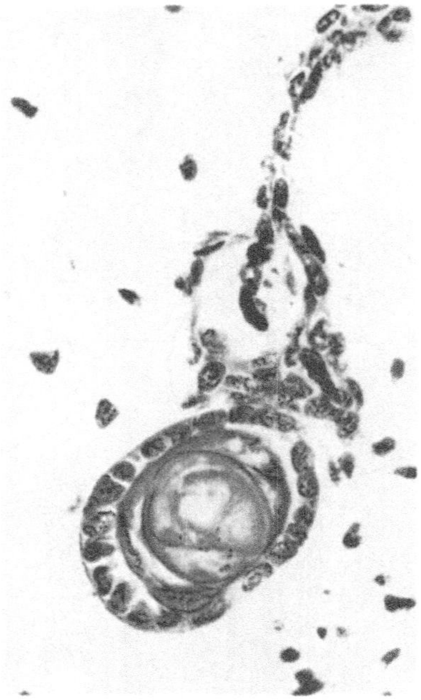

Fig. 10

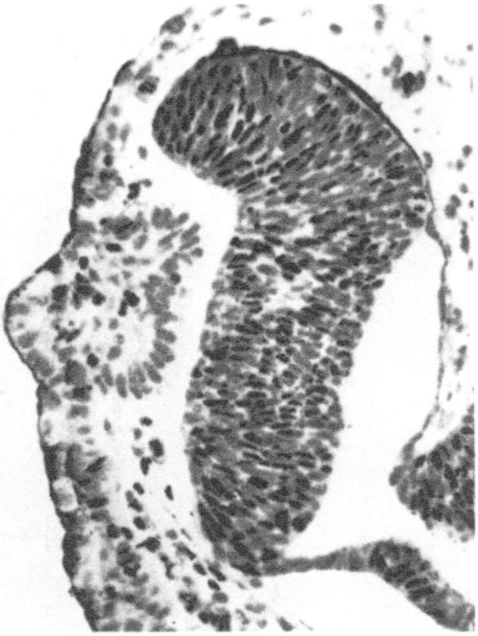

Fig. 11

Fig. 10. Lens with epithelium and fibers which has developed in the larval dorsal fin from an implant of prospective lens ectoderm combined with underlying entoderm of anterior archenteron wall derived from stage 15 neurula donor in *A. maculatum*, 25 days after implantation. × 230

Fig. 11. A partly doubled lens vesicle which has formed from prospective trunk ectoderm of stage 11 gastrula donor grafted in place of lens-forming ectoderm of stage 14 neurula host in *A. maculatum*, 5 days after operation. × 170. (From Fig. 7, REYER, 1958a, by permission of the Wistar Institute)

ophthalmus ectoderm to *Ambystoma* hosts behaved in the same manner. Host and donor tissues were identified by means of vital staining with Nile blue sulfate or by interspecific differences in pigmentation (STONE and DINNEAN, 1943; REYER, 1958a). LIEDKE (1951, 1955) obtained similar results with *Ambystoma* but also illustrated two cases where trunk ectoderm from stages 19 and 22 formed a lens vesicle after transplantation to a stage 14 host indicating a longer duration of lens competence than that found by REYER. See discussion in REYER (1958a) for an evaluation of this evidence.

In *Taricha torosa*, A. JACOBSON (1958) observed that lens induction occurred from 71% of the grafts when ventral trunk ectoderm from stage 11–14 donors was transplanted to stage 15–17 hosts, but was reduced to 28% when this ectoderm was transplanted to stage 18–20 hosts. BRAHMA (1959) reported lens induction from presumptive ectoderm of early and late gastrulae grafted over the optic vesicles of stage 24 *Xenopus laevis* embryos following removal of the prospective lens ectoderm (stages from NIEUWKOOP and FABER, 1967). However, no method of marking was used to ensure that the lenses observed did not come from incompletely removed host lens ectoderm and all lenses illustrated were already detached from the ectoderm. When *Xenopus* optic vesicles were combined in explants with axolotl early gastrula ectoderm, the one lens illustrated resembled *Xenopus* rather than axolotl tissue although the author believed that this lens was induced from the latter. SIRLIN and BRAHMA (1959) also explanted stage 26–30 optic vesicles, labeled with DL-3-phenyl alanine-2 C^{14}, in sandwiches of unlabeled *Xenopus* gastrula ectoderm. Autoradiographs were made after 6–48 hrs of culture. Label was observed in both the eyecups and the associated lens masses but not in the covering ectoderm. The authors interpreted these results as showing transfer of labeled molecules from optic vesicle to lens during lens induction from the gastrula ectoderm. However, here also, some prospective lens ectoderm might have adhered to the optic vesicles so that both were labeled together.

To determine when the head ectoderm, including the prospective lens cells, acquired the capacity to respond by lens formation to the action of the optic vesicle alone in *A. maculatum*, prospective head ectoderm from gastrulae and neurulae was transplanted over the optic vesicle of head and tail-bud hosts (stages 20–24; Figs. 9C and E). The invaginating entoderm first underlies the head ectoderm at stage 12. Lenses did not develop from young gastrula ectoderm, not yet exposed to entoderm (stages 10–$11^1/_2$), but formed consistently from ectoderm derived from yolk plug or older embryos (stages 12–17) (REYER, 1958b).

In summarizing these transplantation experiments, it can be assumed that early gastrula ectoderm is labile and its course of development is not yet specified. When this ectoderm from either prospective trunk or head is placed in the head region of an older neurula host, the underlying entoderm provides the first stimulus for lens development, to be followed by a second induction from the optic vesicle. If this first entodermal stimulus is eliminated, then the ectoderm does not readily respond to the action of the optic vesicle alone. Once gastrula ectoderm from the head region has been underlain by the invaginating entoderm at late gastrula and neurula stages, then it can be induced to form a lens by the optic vesicle of a head and tail-bud host. Variations among different species in the relative importance of the several lens-inducing stimuli and in the duration of lens competence in the reacting ectoderm could account for many of the differences in the results of experiments on lens induction reported in the literature.

There is now good evidence that the eyecup is not only important in lens induction but also in the further growth and differentiation of the lens. REYER (1962b, 1966a) implanted lens vesicles, removed from the eyes of *Ambystoma maculatum* or *Notophthalmus viridescens* embryos, into the larval dorsal fin or into the anterior chamber or vitreous chamber of the larval eye either with the host lens present or after lentectomy. In the dorsal fin or anterior chamber, the lens vesicles differentiated into lenses with fibers but remained quite small. When located within the pupil or vitreous chamber of lentectomized eyes, they not only differentiated into normal lenses but also had the same rate of growth as lenses in the normal eyes of the donors (Fig. 12). The presence of a host lens appeared to retard this growth rate somewhat. These results show that the older larval eye still has the capacity to stimulate the further differentiation and growth of the embryonic lens vesicle.

Compared to the number of investigations that have been carried out on avian and mammalian lenses, studies on the cytodifferentiation of the embryonic, am-

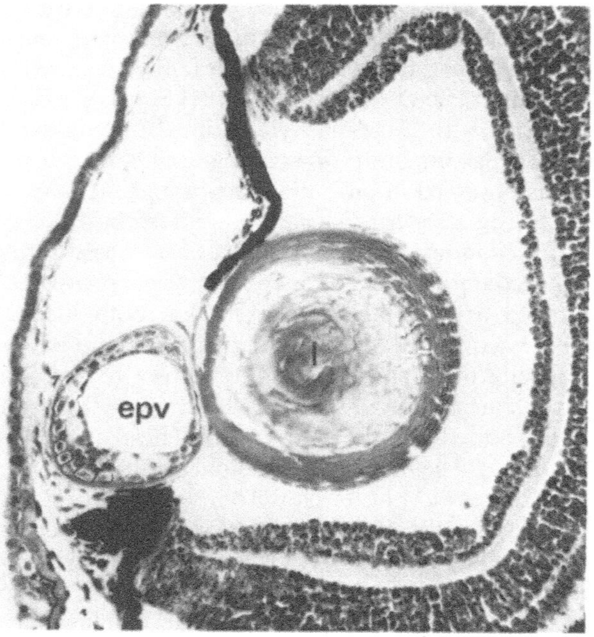

Fig. 12. A large lens which has developed from the lens vesicle of a stage 35 donor implanted into pupil of lentectomized eye of larval host in *A. maculatum*, fixed after 25 days. *l* = lens, *epv* = epidermal vesicle. × 122. (From Fig. 19, REYER, 1962b, and Fig. 9, REYER, 1965, by permission of the Wistar Institute and F. K. Schattauer-Verlag)

phibian lens are very limited. The specificity, sequence of appearance, and localization of the lens crystallins have been studied by McDEVITT et al. (1969) and McDEVITT (1972) in *Rana pipiens*. They prepared antisera to the total proteins of adult frog lenses and to a gamma crystallin fraction of these lenses obtained by chromatography on DEAE-cellulose and Sephadex G-100. Gamma globulin fractions were isolated from these antisera for immunologic assays. Antibodies to total lens proteins were absorbed with mouse tissue powder and frog liver powder before use. Homogenates of pooled embryonic lenses (Shumway stage 25, SHUMWAY, 1940) and of older tadpole lenses (Taylor-Kollros stages I, VI, XXIV, TAYLOR and KOLLROS, 1946) were tested by immunoelectrophoresis and Ouchterlony immunodiffusion against antibodies to total lens proteins. Gamma crystallins with very small amounts of beta crystallins were detected in the stage 25 embryonic lenses. Additional beta and, finally, alpha crystallins appeared in older tadpole lenses. Using direct as well as indirect immunofluorescence and antibodies to total lens proteins, the presence of crystallins was first observed throughout lens vesicles in which the prospective lens fibers had just begun to elongate. When gamma crystallin antibodies were employed, fluorescence was limited to these elongating cells (Fig. 13). With further development of the lens, the gamma crystallins were localized in the lens fibers only, while the lens epithelium reacted only with antibodies to total lens proteins and is, therefore, believed to

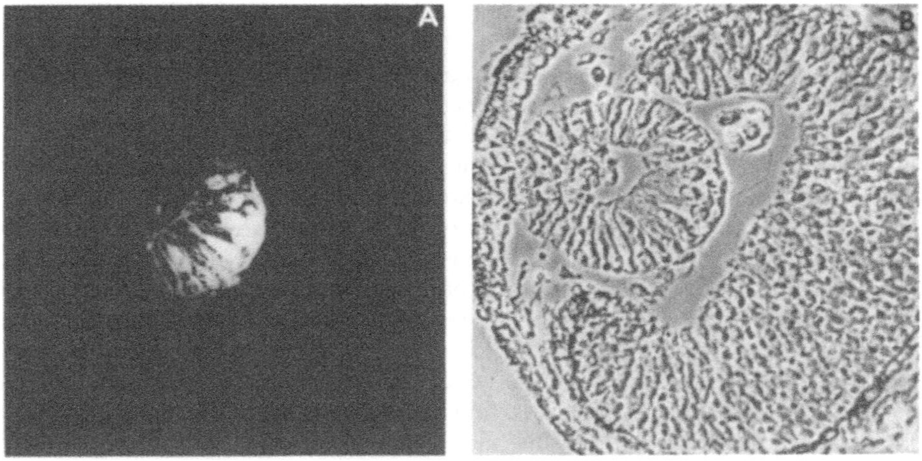

Fig. 13. Section through eyecup and lens vesicle of *R. pipiens* embryo which has been treated with anti-gamma-crystallins immunofluorescence reagent. A. Fluorescence photomicrograph showing a positive reaction only over the elongating prospective lens fiber cells in the internal layer of the lens vesicle. × 114. B. Phase-contrast photomicrograph of the same section. × 114. (From Fig. 10, McDEVITT et al., 1969, by permission of the authors and Academic Press)

contain first the beta and later the alpha crystallins observed in the immunodiffusion experiments. There was no reactivity of the antisera to extraocular tissues or to ocular tissues other than the lens with any of the techniques employed. Similar results were obtained using *Xenopus laevis*, *Rana temporaria*, *Ambystoma mexicanum*, *Pleurodeles waltlii*, or *Triturus pyrrhogaster* embryos (BRAHMA and McDEVITT, 1972a, 1974; McDEVITT and BRAHMA, 1973; TAKATA et al., 1964b). Isoelectric focusing of lens extracts from larval and adult *Xenopus laevis* also demonstrated a gradual decrease in gamma crystallins and an increase in beta crystallins in lenses from older animals and a large increase in alpha crystallins after metamorphosis (BRAHMA and BOURS, 1972). Synthesis of the crystallins, therefore, appears to be specific for lens and begins only as the lens fibers start to differentiate morphologically.

IV. Growth and Differentiation of the Neural Retina

The development of the amphibian neural retina has also been the subject of much investigation. As indicated in the discussion of ocular anatomy, this part of the eye is a neural tissue which is specialized for photoreception and connected by the optic nerve in a very precise manner with the optic tectum. The latter is the main center for central reception of visual stimuli in the amphibians. In investigating the ontogeny of the retina, there are not only the usual processes of cell proliferation and cytodifferentiation to consider, but also problems relevant to neuroembryology such as the formation of synaptic patterns within the retina and the mechanisms controlling the synaptic connections of ganglion cell axons onto neurons within the optic tectum. These must provide for the patterned topographic representation of retinal stimuli on the tectum.

Following optic cup formation, there is a period during which the cells of the inner layer continue to proliferate and the eyecup deepens (Fig. 7). In *Rana pipiens*, a separate ganglion cell layer, inner nuclear layer, and outer nuclear layer, as well as rod and cone cell processes were first observed at Shumway stages 22 or 23 (MACDONALD, 1969). Incorporation of ^{3}H-thymidine continued in all cells of the *Xenopus laevis* neural retina through NIEUWKOOP and FABER stage 28 (M. JACOBSON, 1968 b). Before differentiation of the nuclear layers, mitotic figures were observed in the outer part of the neural retina. Presumably, the cell kinetics of the embryonic *Xenopus* and *Rana* retina as well as that of other amphibians corresponds to that found in the mouse where DNA synthesis occurs in the middle layers of the neural retina, the nuclei move to the outer surface to divide, and then the daughter cells again move inward to repeat the cycle of interkinetic nuclear migration (SIDMAN, 1961). Beginning at stage 29 in *Xenopus*, some of the prospective ganglion cells stopped synthesizing DNA. Subsequently, there was a cessation of DNA synthesis throughout the fundic region, first in the prospective ganglion cells, next in the receptor cells, and last, by stage 42, in the inner nuclear layer. Axon outgrowth from ganglion cells began at stages 34 to 35 and the outer segments of receptor cells appeared at stage 36 (M. JACOBSON, 1968 b). During the early development of the optic vesicle and cup of *Xenopus*, DIXON and CRONLY-DILLON (1972) observed *zonulae adhaerentes* between cells at the outer margin of the prospective neural retina and *maculae adhaerentes diminutae* scattered between the other cells. "Gap" junctions (30–90 Å space, 1.0 μ in diameter) were found between these cells throughout the developing neural retina until stage 30 when they disappeared from the central region but remained at the periphery. Between stages 28 and 32, there was an increase in the number and length of the polyribosomes. This is the time when retinal polarity becomes specified as will be described later (GRILLO and ROSENBLUTH, 1972).

The growth in size of the neural retina during late embryonic and larval stages is due to continued cell proliferation in a marginal growth zone at the periphery which contributes cells to all layers of the retina. Following colchicine treatment in *Rana pipiens* tadpoles, arrested mitoses accumulated only in the outer layers of this marginal growth zone over a period of 72 hrs. Autoradiographs were prepared at intervals from 1 hr to 15 days after injection of ^{3}H-thymidine using tadpoles and metamorphic stages XIII-XXV of TAYLOR and KOLLROS. One to four hours after injection, labeled nuclei were found only in the marginal growth zone. With fixation after longer intervals of time, scattered labeled cells appeared well into the fundic regions of the inner nuclear layer within 72 hrs while addition of labeled cells to the outer nuclear and ganglion cell layers occurred much more slowly by a uniform apposition of labeled cells to the margin of the unlabeled fundic retina (HOLLYFIELD, 1968). Two processes appear to be involved here. First, there is tangential growth in size of all layers of the retina by the addition of new cells produced by mitoses of a precursor cell population in the marginal growth zone; second, there is also active migration of cells from this marginal zone far into the inner nuclear layer. This could account for the differential growth in thickness of the inner nuclear layer from two cells thick at stage III to 10–11 cells thick by the end of metamorphosis at stage XXV.

The first type of peripheral growth was also observed in *X. laevis* by both HOLLYFIELD (1971) and STRAZNICKY and GAZE (1971) using similar procedures of ^{3}H-thymidine injection at different ages and fixation after varying lengths of time. The former author also found a few scattered, labeled cells in the inner nuclear layer of the fundus. These cells appeared to divide in the outer nuclear layer and then return to the inner nuclear layer, presumably contributing to this part. The latter authors injected isotope into embryonic and larval eyes of several ages and allowed some of their animals to survive until 3 months after metamorphosis. They found that cells labeled at embryonic stages 30 and 35 were finally located in the neural retina around the optic nerve head. Cells labeled at later stages occupied concentric positions at successively larger distances from the optic nerve head indicating that the neural retina is laid down gradually in circumferential increments.

Most of the studies on cytodifferentiation have dealt with the photoreceptors. NILSSON (1964c) described the ultrastructural changes during the differentiation of the outer segments of the rods and cones in *Rana pipiens*. These structures begin development as modified cilia, containing microtubules, which arise from centrioles acting as basal bodies. This ciliary structure is retained in the connecting stalk beyond which there is an enlargement of the outgrowing process. Proximally, on the side opposite the connecting stalk, the cell membrane invaginates to form a pouch which then flattens to become a disc of closely apposed membranes arranged perpendicular to the longitudinal axis of the receptor cell. New discs of increasing diameter are added continuously at the proximal end of the outer segment and the older discs are transposed distally, thereby giving a tapering shape at first to both rod and cone outer segments. In cones, the membranes of the discs retain continuity with the cell membrane while the rod discs are quickly pinched off to become entirely separate. Thus, continuity of disc and cell membrane is only seen at the proximal end of rod outer segments. The formation of new discs at the base and shedding of discs at the apex brings about the uniform diameter of the rod outer segment and continues throughout life. Disc replacement is arrested in the cone outer segment at a time when the tapered shape is still present (YOUNG, 1971). While the outer segments are growing in length, processes of the inner segments and of the pigmented epithelium grow outward and inward respectively to surround each photoreceptor outer segment. When *Rana pipiens* eyecups were cultured in hanging drops of Niu-Twitty saline solution, normal differentiation of the rod outer segments occurred only when the pigmented retinal epithelium was in close proximity (HOLLYFIELD and WITKOVSKY, 1974).

NILSSON and CRESCITELLI (1969, 1970) correlated the first appearance of the cornea negative (a-wave) and the cornea positive (b-wave) potential in the electroretinogram of tadpole eyes (*Rana pipiens* and *Rana catesbiana*) with changes in the ultrastructure of the differentiating neural retina. The first electrical activity in response to a flash of light was a slow negative potential. This appeared in eyes where the photoreceptor outer segments had begun to form and some outer segment discs were present. At this time, the synaptic complexes in the outer plexiform layer were still immature with no synaptic vesicles and only an occasional synaptic ribbon. When a cornea-positive potential was elicited by photic stimulation, these eyes were found to have mature synapses, containing synaptic ribbons and synaptic vesicles, between the synaptic bodies of the photoreceptors and postsynaptic cell processes in the outer plexiform layer. Therefore, the negative response seems to be produced by stimulation of the photoreceptors alone, while presence of both a- and b-waves appears to depend on transmission of a graded potential from the photoreceptors across a synapse to additonal cells in the neural retina.

V. Ontogeny of Neuronal Specificity in the Retinotectal System

The developing amphibian eye has been an important subject for investigations on the role of neuronal specificity in neurogenesis. The concept of neuronal specificity was originally proposed by SPERRY (1943a, b, 1944, 1951a, b, 1963, 1965; SPERRY and HIBBARD, 1968) as an explanation for the return of oriented visuomotor responses following neural retina regeneration in newts or optic nerve regeneration in anurans. The results of the regeneration experiments will be discussed later. SPERRY suggested that the ganglion cells of the neural retina differ from each other in some cytochemical properties, probably distributed in gradients along the nasotemporal and dorsoventral axes. Within the optic tectum, similar cytochemical differences are present such that an optic nerve fiber originating from a specified ganglion cell synapses only with a tectal cell for which it has a cytochemical affinity. In optic nerve regeneration, the axons return to and make functional synapses with those tectal cells with which they share chemospecificity.

It soon became evident that neuronal specificity in the optic system must arise during the embryonic development of the retina and optic tectum. Therefore, the ontogeny of the retinotectal connections has been studied extensively. It is well known from the classical experiments of HARRISON (summarized in HARRISON, 1969) on the limb and ear that organs which exhibit asymmetry pass through a critical stage during embryonic development when they change from a nonpolarized organ anlage to one whose main axes of symmetry are fixed. The same situation holds for the functional polarity of the eye. This was first studied by rotating the optic vesicle and overlying ectoderm through 180° at different stages of embryonic development and later on, at larval stages, testing for normal or reversed vision by observing the visuomotor responses to a red lure or the optokinetic response to the movement of vertical stripes on a rotating drum. In *Ambystoma maculatum*, fixation of retinal polarity occurred at stage 36 (HARRISON) at a time when the lens vesicle was just detaching from the surface ectoderm and the neural retina was still undifferentiated (STONE, 1969). SZEKELY (1954, 1957, 1966) found an even earlier determination of the nasotemporal axis at sometime before stage 25 in *Triturus vulgaris* and *T. cristatus*. At this time, the dorsoventral axis was not yet fixed. STONE (1960) also observed that retinal polarity was retained in eyes that had been transplanted with normal orientation, at stage 36 or later, to the flank of a host embryo and then, after a month's residence on the flank, retransplanted to the orbit of a second host larva where the eye reestablished functional connections with the optic tectum.

Recently, these events have been much more thoroughly analyzed by M. JACOBSON and HUNT using *Xenopus laevis*. In their studies, various experimental procedures were performed on the eye in embryos and normality of vision was assayed in tadpoles. After metamorphosis, electrical recordings were made from the optic tectum during visual stimulation. This procedure for mapping the projection of the visual field onto the optic tectum was developed by GAZE (1958, 1959, 1960) and has been used extensively by many investigators to assay the effects of both experimentally altered eye development and the regeneration of the neural retina and/or optic nerve on the pattern of neural connectivity between retina and tectum. This work has been reviewed in detail by GAZE (1970, 1974), M. JACOBSON (1966, 1969, 1970), M. JACOBSON and HUNT (1973) and HUNT and M. JACOBSON (1974a). In brief, the recording procedure consists of inserting a microelectrode into the surface of the exposed optic tectum at a series of points determined by the intersecting lines of a grid. Spots of light or black discs are projected onto a perimeter covering the visual field of the eye contralateral to this tectum. For each electrode position in the tectum, the effective stimulus area in the visual field and the point within this area giving the largest evoked response is determined. From these data, a map of the visuotectal projection is drawn. It is assumed that this is an inverted representation of the retinotectal projection and that each numbered recording in the tectum is produced by the terminal arborizations of optic nerve fibers carrying the nerve impulses generated by the stimulation of this particular locus in the retina. Another assumption is that these arborizations are the site of synapses with tectal neurons. These assumptions have recently been criticized, especially for experimentally altered eyes, by HUNT and M. JACOBSON (1974a).

M. JACOBSON (1968a) rotated eyes 180° at stages 28 to 35 (NIEUWKOOP and FABER, 1967) in *Xenopus laevis* and found a critical period during which the nasotemporal axis was specified at stage 30 and the dorsoventral axis at stage 31 (Fig. 14), at a time before visible differentiation of the ganglion cells and outgrowth of their axons but just after the first of the prospective ganglion cells have ceased DNA synthesis at stage 29 (M. JACOBSON, 1968b). However, in *Rana pipiens*, specification of retinal polarity had already occurred at stage 17 (SHUMWAY) shortly before DNA synthesis was first terminated in some of the cells in the ganglion cell layer at stage 18 (SHARMA and HOLLYFIELD, 1974). HUNT and M. JACOBSON (1972a, b, 1973a, 1974b) continued this analysis with a large series of experiments in which they transplanted eyes between donor and host embryos of different ages. In some experiments, they also interposed a period of culture *in vitro* or transplantation to the flank, belly, or orbit of an intermediate host before grafting the eye to the orbit of the final host. At each transplantation, the eye could be normally orientated or rotated 180°. The final analysis of visuomotor responses and retinotectal projections was made for the experimental eye on the final host. Several important conclusions can be derived from their work. There is a labile determination of the axes of symmetry in the eye preceding the specification of polarity at stage 31. This was shown by culturing stage 22–23 eyes for 3–4

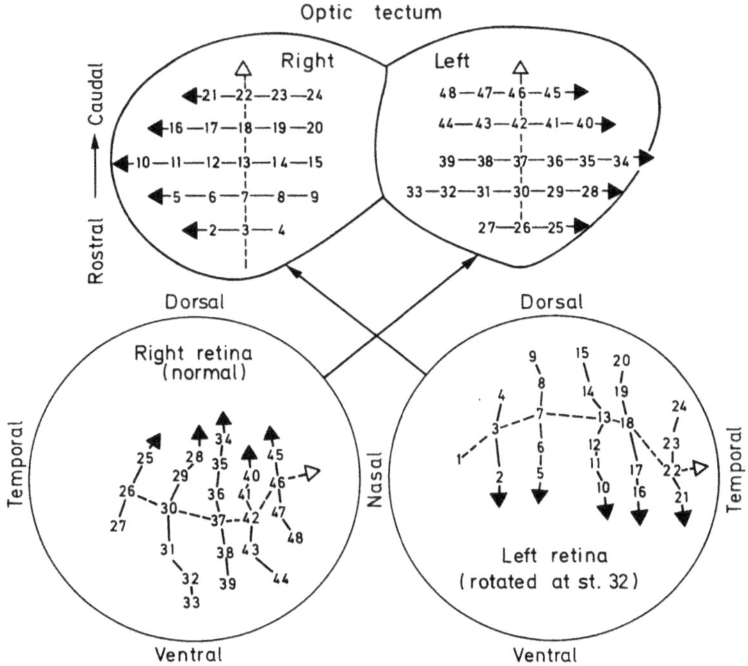

Fig. 14. Retinotectal projection in adult *Xenopus*. A normal right retina projects to left tectum (ventral retina to medial tectum; dorsal retina to lateral tectum; nasal retina to caudal tectum; temporal retina to rostral tectum). Left eye had been rotated 180° at stage 32 during embryonic development. Projection from the left retina is inverted for both the dorsoventral and nasotemporal axes showing that specification had occurred prior to stage 32 (Nieuwkoop and Faber, 1967). (From Fig. 6, M. Jacobson, 1968a, by permission of the author and Academic Press)

days *in vitro*, and then transplanting these now well-differentiated eyes to the orbit of stage 39–40 hosts. The resulting visuotectal projections were normal if the original axes of the transplant corresponded to those of the host and inverted if the eye was rotated 180°. Similar results were obtained when stage 25–26 eyes were grafted to the midventral belly of stage 26 hosts, allowed to remain through the critical period, and then transplanted to the orbit of stage 39–40 hosts. Orientation on the intermediate host had no effect. Although polar influences thus appeared to be lacking in the ventral midline, this was not true for grafts to the flank of an intermediate host. When stage 28 eyes were transplanted to the above location and allowed to pass through their critical period here before transfer to the orbit of a stage 32–34 host, then the polarity of the eye was determined by its orientation with respect to the body axes of the intermediate host. Specification was neither accelerated by transplanting eyes from younger donors to orbits of older hosts nor retarded by transplanting eyes from older donors to orbits of younger hosts. Once specification of the axes of symmetry had taken place at stage 31, these could no longer be modified under a number of experimental conditions. These conditions included 1. prolonged residence as a flank graft on an intermediate host followed by transplantation to the orbit of an older, stage 50 host so that a larger neural retina now came to innervate a larger optic tectum, 2. culture for 6–10 days *in vitro* before grafting to the orbit of a stage 38 host, or 3. transfer to the orbit of a younger, stage 28 host so that the specified eye was exposed to the influences present here when the host went through its own critical period. In all of these situations, there was no change in the polarity of the experimental eye and the final visuotectal projection depended on the relation of the original axes of the eye to the axes of the final host. HUNT and M. JACOBSON concluded that the polarity of the eye rudiment at stages 22–28 is in a stable but reversible condition. If the eye is rotated, there is an interaction with the axial cues of the embryo which changes the polarity of the eye rudiment to conform to its new position. When the development of the eye reaches the critical period, the positional information becomes fixed, first for the nasotemporal axis and, very soon, for the dorsoventral axis. This results in a stable and irreversible state of polarity. This polarization of the retina appears to be correlated with the terminal differentiation of the ganglion cells in the central retina. Treatment of embryos at stage 23–25 with BrUdR (5-bromodeoxyuridine) resulted in a delay in both retinal cytodifferentiation and in axial specification beyond stage 31–32 when tested by 180° eye rotation (HUNT, 1975).

Does this initial polarization fix the locus specificity of every ganglion cell present at this early stage as well as that of all new ganglion cells formed during the subsequent growth of the retina, or is this property more that of an embryonic field, applying only to the ganglion cell population as a whole and still capable of being modified at the cellular level? A number of lines of evidence favor the latter hypothesis. A great deal of work has been performed by GAZE and his collaborators in which the embryonic *Xenopus* eye is modified by surgical intervention at the time when its polarity is specified but the optic nerve fibers have not yet reached the optic tectum. Either half eyes (nasal or temporal) or compound eyes (double nasal, double temporal, or double ventral) were produced by removing half of the eye or by fusing half-eyes at stage 32. These abnormal eyes subsequently innervated the contralateral optic tectum. Visuotectal projections were determined after metamorphosis (GAZE et al., 1963, 1965; M. JACOBSON, 1966, 1969, 1970; GAZE, 1970; STRAZNICKY et al., 1971, 1974). If the positional specification of the ganglion cells and the matching specificities in the tectal neurons were both present at the time of operation, the half-eyes would be expected to project to the corresponding half of the tectum and double nasal and double temporal eyes would have duplicate projections to the corresponding half-tectum. The part of the tectum normally innervated by the missing part of the eye would lack an input of optic nerve fibers. Unexpectedly, however, the visuotectal projections from some of the half-eyes showed a normal pattern over the whole optic tectum. In the compound nasal eyes, the visual field for each nasal half-retina projected in correct order over the entire rostrocaudal extent of the optic tectum, each point on the tectum receiving stimuli from two points in the visual field (Fig. 15A). Each half-retina, therefore, filled the available tectal space. Corresponding results were obtained with double temporal eyes. This adjustment did not appear to involve hypoplasia of the denervated part of the tectum since reinnervation of this tectum after metamorphosis by directing to it regenerating, optic nerve fibers from the normal ipsilateral eye, resulted in an approximately normal visuotectal projec-

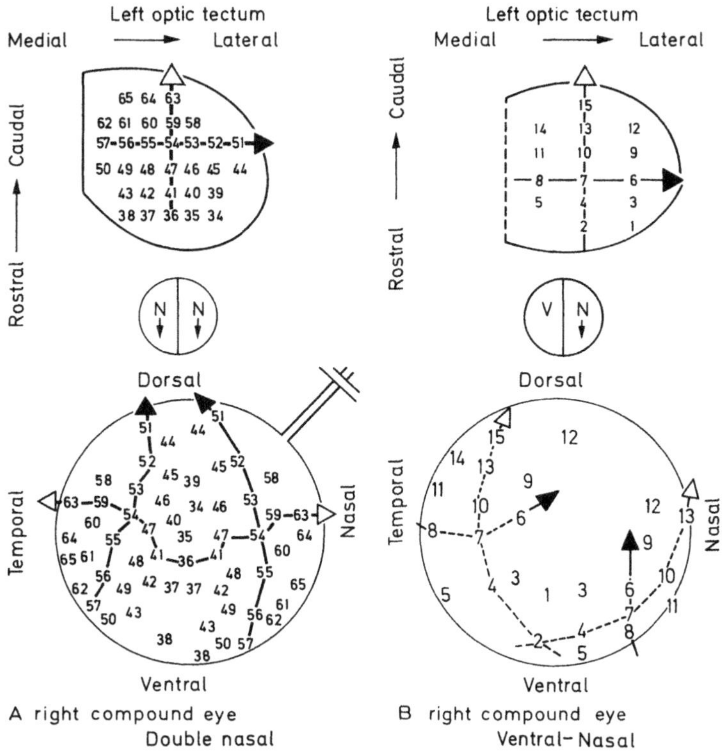

Fig. 15. A. Retinotectal projection in adult *Xenopus* from a compound right eye to left optic tectum. This compound eye has developed from two nasal half-eyes fused together at stages 30–32 of embryonic development. B. Retinotectal projection in adult *Xenopus* from compound eye to left optic tectum. This compound eye has developed from fusion of nasal half-eye with a ventral half-eye at stage 32. (Modified from Fig. 3, M. Jacobson, 1969, and Fig. 2a, Hunt and Jacobson, 1973b, by permission of the authors and A.A.A.S.)

tion. However, in these animals, the lines of field positions tended to run in slanted, sometimes zigzag lines instead of vertical straight lines. At the same time, a compound visuotectal projection was found when optic nerve fibers from the compound eye reinnervated the normal optic tectum (Straznicky et al., 1971). Autoradiographic studies by Feldman and Gaze (1972) demonstrated no DNA synthesis at stages 45–47 at the line of fusion in double nasal eyes so that no new cells with more temporal specificities were being produced at this site at this time. These results indicate a capacity for regulation of either the retinal or of the tectal specificities. Hunt and M. Jacobson (1973b, 1974a) reported experiments where right nasal half-eyes were combined with right ventral half-eyes or with left temporal half-eyes. In both of these cases, each half of the compound eye again projected to the entire optic tectum rather than to those parts corresponding to the retinal quadrants present (Fig. 15B). There was also some skewing of the pattern on each side of the dorsoventral axis in the retina. These last results demonstrate not merely a filling of available tectal space by a spreading out of the gradient of retinal locus specificities but, apparently, an actual modification of these specificities during the continued differentiation and growth of these compound eyes. After metamorphosis, this capacity for regulation appears to be lost. Following removal of the rostral, caudal, medial, or lateral half of the optic tectum and sectioning of the contralateral optic nerve, defects remained in the visual field after optic nerve regeneration which corresponded to the part of the optic tectum removed (STRAZNICKY, 1973).

Recently, the development of visuotectal projections from half-eyes has been analyzed further (FELDMAN and GAZE, 1975; BERMAN and HUNT, 1975; HUNT and BERMAN, 1975). Nasal, temporal, or ventral half-eyes were produced by ablation of the other half at stages 25–26, 31–32, and 37–39. After surgery, the eyes rounded up and the majority produced a normal visuotectal projection over the entire optic tectum. However, there were some cases at all stages where duplicated retinotectal projections developed. Nasal half-eyes gave double nasal, temporal half-eyes gave double temporal, and ventral half-eyes gave double ventral projections identical to those obtained with the compound double nasal, double temporal, and double ventral eyes. When the half-eye fragments, prepared at stage 31–32, were subjected to various experimental conditions of rotation and transplantation, the reconstituted polarity (either normal or double) remained stable and did not change. By grafting a normal eye and an experimental eye to the same orbit so that they both innervated the same optic tectum, it could be shown that the locus specificities in the tectum were normal since a normal eye and an eye fragment which had acquired normal polarity both had duplicate projections across the entire tectum. This result also demonstrated that the ganglion cells of the eye fragment had acquired the full range of locus specificities characteristic of the normal eye. These results have been reviewed in more detail by HUNT (1975).

STRAZNICKY and GAZE (1971, 1972) compared the pattern of growth in the retina and tectum of *Xenopus* using autoradiography. As described previously, the retina grows by the addition of concentric rings of new cells at its margin. The ganglion cells which first differentiate and send optic nerve fibers to the tectum lie in the center of the retina. On the other hand, the tectum was found to grow in a linear fashion from ventrorostral to mediocaudal. When ^{3}H-thymidine was administered early at stage 35 or 45 and the specimens fixed at metamorphosis, the labeled cells were found at the ventrorostral pole of the tectum and also, in the latter case, along the rostral part of the ventrolateral border. Cells labeled at stage 50 came to lie in a band, comprising the entire tectal thickness, and running at an angle from mediorostral to laterocaudal. When labeled at successively older stages, new bands of labeled cells formed on the medial and caudal side of the above. Connections from retina to tectum could first be demonstrated electrophysiologically at stages 43–44. However, the first recordings of an ordered visuotectal projection could only be obtained from the laterorostral part of the tectum at stage 47 when the receptive fields were still very large. In older tadpoles (stages 54–62), more and more of the caudal and medial tectum became involved. In these tadpoles, the nasal part of the visual field projected to a small region of the rostral tectum while the temporal visual field was expanded over a larger area of more caudal tectum. A pattern characteristic of the adult, with uniform spacing and smaller receptive fields, appeared in metamorphosing tadpoles at stages 63–64 (GAZE et al., 1972; GAZE et al., 1974). By means of a technique designed to record monophasic, postsynaptic potentials, following direct stimulation of optic nerve fibers, CHUNG et al. (1974a, b) found that functional synapses between optic nerve fibers and tectal neurons were located in the same areas of the developing tectum as the visuotectal projections recorded as presynaptic potentials. Quite similar evidence for a rostrocaudal sequence of innervation in the optic tectum was also obtained by tracing degenerating optic nerve fibers after eye extirpation and by following transport of ^{3}H-proline injected into the eye (SCOTT, 1974).

Because of the topologic problems presented by an organ developing concentrically (the retina) sending nerve fibers to establish connections in an organ growing curvilinearly (the tectum), GAZE and his co-workers have postulated a continual shifting of these connections in the optic tectum such that the fibers from the oldest, central part of the retina move in a caudal direction along the tectum and fibers from other parts of the nasotemporal axis of the retina shift in a corresponding manner. Some evidence for such a shift was obtained in the recordings from successively older tadpoles just described. Presumably this would involve continual changes in the tectal cell locus specificities. Another possible hypothesis to explain these events of normal development as well as the results of the different compound eye experiments would be regulation of the retinal cell locus specificities by cellular interactions up to the time of metamorphosis. There might also be a parallel regulation in both retina and tectum. Similar problems have also been found in studies on the adult goldfish retinotectal system where partial ablation or division of the optic tectum resulted in adjustments in the visuotectal projections, recorded electrophysiologically after optic nerve regeneration (reviewed by GAZE, 1974).

In a recent critical review of the evidence for neuronal specificity obtained from studies of the visual system, HUNT and M. JACOBSON (1974a) question many of the assumptions underlying these analyses. In particular, they feel that the long time lag between surgical interventions on embryonic eyes and the visuotectal recordings in metamorphosed frogs introduces the opportunity for many as yet undetected changes during the development of the eye and optic tectum. They are also sceptical as to whether the electrical changes recorded from the tectum in response to point stimulation of the retina can be interpreted as representing the site of functional optic nerve synapses with tectal neurons under different experimental conditions in the absence of postsynaptic recording. The results of postsynaptic recording reported by CHUNG et al. (1974b) answers the latter question for normally oriented eyes in *Xenopus* tadpoles. It can be concluded that studies of the ontogeny of function in the retinotectal system in amphibians has contributed to the hypothesis of neuronal specificity both important supporting data and also problems that yet await solution.

D. Regeneration in the Amphibian Eye

Although its structure and functional organization is as complex as that found in the higher vertebrates (birds and mammals), nevertheless two of the most important constituents of the adult newt eye can be replaced by metaplasia of pigmented epithelia within the eye. These parts are the neural retina and the crystalline lens. The literature on neural retina regeneration has been reviewed briefly by REYER (1962a), SQUIRES (1972), and KEEFE (1973a) while that on lens regeneration has been the subject of several thorough reviews (REYER, 1954a, 1962a; STONE, 1965; SCHEIB, 1965; YAMADA, 1966, 1967a, b, 1972).

I. Regeneration of the Neural Retina

Since the first detailed study by WACHS in 1920, investigations on neural retina regeneration have involved two principal experimental techniques: (1) surgical removal of the neural retina and lens through an incision in the eyeball, and (2) devascularization of the eye by transplantation or by cutting the blood vessels so that the old neural retina and lens completely degenerate. Formation of a new neural retina and regeneration of the optic nerve back to the optic tectum take place in quite similar fashion following both procedures so they will be discussed together.

There are two tissues involved in reconstituting the new neural retina: (1) the outer pigmented retinal epithelium *(stratum pigmenti retinae)*, and (2) the inner epithelial layer at the *ora serrata* and adjacent *pars ciliaris retinae*, termed the "anterior complex" by KEEFE. Depending on the exact site where the neural retina detaches during retinectomy, there are variations in the number of cells remaining at the *ora serrata*. STONE (1950a, b, 1959b) believed that the entire neural retina regenerated from the pigmented retina alone. This was questioned by HASEGAWA (1958) who observed a contribution from both pigmented and ciliary retina following extirpation of the lens and neural retina. When he removed the pigmented retinal epithelium in addition to the neural retina, both of these layers regenerated from the cut margin of the *inner pars ciliaris retinae* and the outer *stratum pigmentosum corporis ciliaris* (HASEGAWA, 1965). Based on autoradiographic studies with ^{3}H-thymidine in adult newts, the consensus of most investigators now is that the central part of the neural retina is regenerated from the pigmented retinal epithelium while the peripheral part is replaced by cells of the anterior complex (GAZE and WATSON, 1968; MITASHOV, 1968; REYER, 1971b; KEEFE, 1973a, d).

In some normal adult eyes of *N. viridescens*, REYER (unpublished observation) observed an occasional nucleus incorporating ^{3}H-thymidine, located at the edge of the neural retina adjacent to the *ora serrata*, although KEEFE (1971, 1973a) has not found this to occur in his normal animals. KEEFE studied neural retina regeneration in autoradiographs following i.p. injection of 40 μCi of ^{3}H-thymidine at 1–81 days after devascularization of the eye followed by fixation at 1 or 7 days after injection. REYER (1971b and unpublished observation) removed the neural retina through a dorsal incision in the eyeball and injected ^{3}H-thymidine (1–2 μCi/g body wt) into groups of animals daily from 1 to 10 days after operation and then at frequent intervals up to 35 days. Eyes were fixed for autoradiography at 3 hrs or 1 day after injection and subsequently at frequent intervals, from 2 to 25 days. Following surgical retinectomy, labeled nuclei were observed in the cells attached to the *ora serrata* 1 day after operation which then increased in number up to 4 days (Fig. 16A). In sections bleached to remove the melanin, cells in the

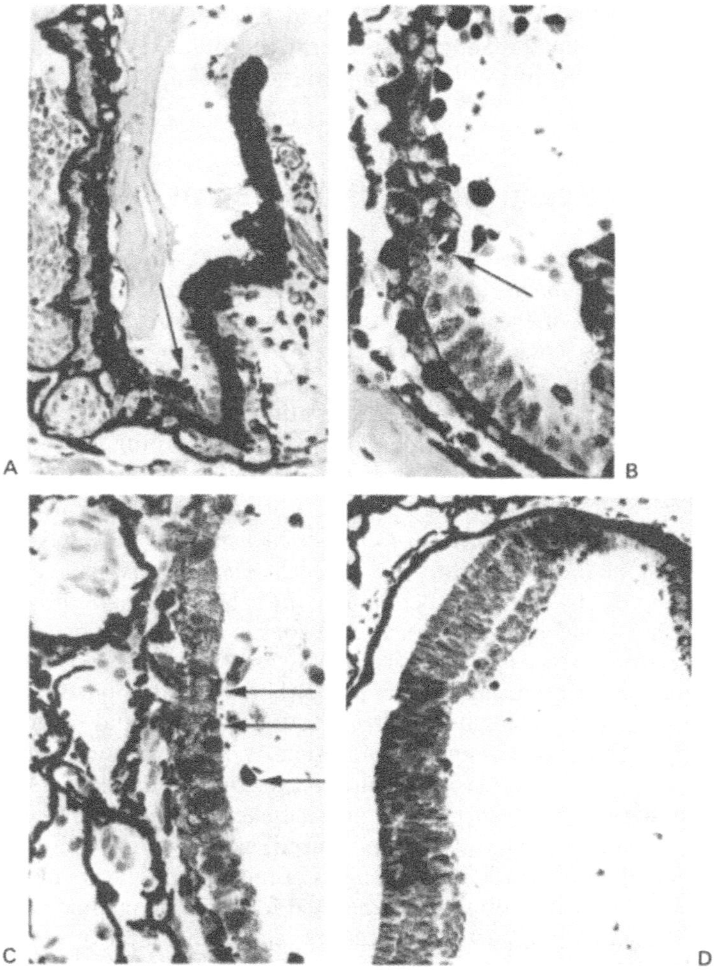

Fig. 16 A–D. Autoradiographs of neural retina regeneration following retinectomy in adult *N. viridescens*. (A) Six days after operation and 2 days after i.p. injection of ^{3}H-thymidine at 4 days after retinectomy. Arrow indicates a group of heavily labeled cells among the unpigmented cells remaining adjacent to the ventral *ora serrata*. Pigmented epithelium is thickened with beginning depigmentation over nuclei but is unlabeled. × 105. (B) Fifteen days after operation and 3 hrs after isotope injection. Label is present over many cells in both unpigmented ventral regenerate and depigmenting, pigmented epithelium. Arrow indicates junction between these two components. × 175. (C) Twenty-two days after operation and 2 days after isotope injection. Pigmented epithelium has formed layer of low columnar depigmented cells most of which are labeled. Scattered pigmented cells are still present *(arrows)*. × 175. (D) Differentiating neural retina 35 days after operation and 3 hrs after isotope injection. Labeled cells are still present in peripheral growth zone adjacent to *ora serrata* and in parts of retina where nuclear layers have not yet become separate. × 105

pigmented retinal epithelium or iris epithelium did not incorporate ^{3}H-thymidine until 4 days after retinectomy. The number of labeled cells in this layer increased rapidly between 5 and 9 days (REYER, unpublished observation). On the other hand, KEEFE (1973a) did not observe labeling of cells in the pigmented retinal epithelium or in the *ora serrata* region until 10 days after devascularization. This delay can be attributed to the gradual removal of neural retina by degeneration and the absence of a blood supply during most of the 1st week. In his studies, proliferation of the unpigmented cells in the anterior complex resulted in the formation of an annular band of labeled, unpigmented epithelium extending posteriorly from the *ora serrata*. In eyes where the neural retina had been removed surgically, either through a nasal incision (HASEGAWA, 1958, on *Triturus pyrrhogaster*) or a dorsal incision (REYER, 1971b, on *N. viridescens*), a similar regeneration from the anterior complex was observed in that part of the eyeball not disturbed by the incision. From 10 to 25 days after retina removal, a distinct zone of junction was distinguishable where the anterior, unpigmented regenerate met the regenerate developing from the depigmenting retinal epithelium (Fig. 16B). By injecting ^{3}H-thymidine 3 or 4 days after operation, it was possible, in some cases, to label the cells of the anterior complex and not the pigmented retinal epithelium. When fixed at different lengths of time up to 15 days after injection, these labeled cells could be followed in the regenerate from this region. After exposure to the isotope 6 or more days after operation, the cells of both the anterior complex and the pigmented epithelium labeled extensively (Fig. 16B). Regeneration from the anterior complex stayed ahead of that from the pigmented epithelium up to 25 days after retinectomy. After complete loss of pigment from the latter tissue, the extent of the contribution from these two sources could no longer be distinguished. With colchicine treatment, KEEFE (1973a) observed numerous mitotic figures in both regions. He also reported localized areas of increased regeneration around the optic nerve head and posterior vascular ring.

While actively engaged in DNA synthesis and mitosis, the cells of the pigmented retinal epithelium changed from squamous to cuboidal in shape, formed two layers, and gradually lost their melanin pigment (Fig. 16C). When studied with the electron microscope, these proliferating, depigmented cells were observed to contain abundant polysomes, a small amount of granular endoplasmic reticulum, small mitochondria, occasional lipid droplets, and a nucleus with a large nucleolus and some aggregated chromatin (J. WILLIAMS, 1964; KEEFE, 1973b). Specialized intercellular connections were lacking. Embedded in this epithelium were macrophages containing either abundant melanin granules or else some melanin and many lipid droplets (Fig. 17). KEEFE believes that these macrophages probably developed directly from the pigmented retinal epithelium. During the process of depigmentation, they had the appearance of removing melanin granules either by direct contact with pigmented cells or by phagocytizing discharged melanin (J. WILLIAMS, 1964; KEEFE, 1973b). Continued cell division resulted in the formation of a pseudostratified epithelium. The outer layer, lying on Bruch's membrane, now began to redifferentiate into pigmented epithelium. These cells became flattened, separated from the remaining layers, and began to resynthesize melanin granules while the nuclei of the developing neural retina became more elongated (Figs. 16C and D). Mitotic figures were now confined to

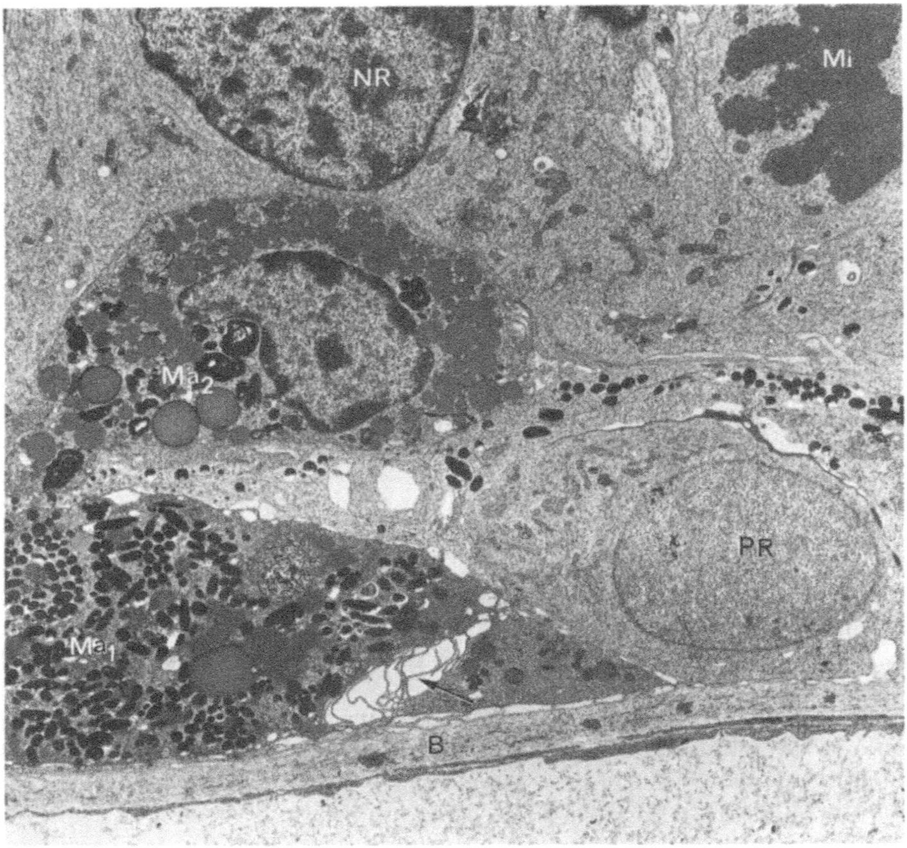

Fig. 17. A low-power electron micrograph of neural retina (NR) regenerating from pigmented retinal epithelium (PR) 26 days after retinal degeneration was initiated by severing blood supply. This view shows a squamous layer of cells resting on Bruch's membrane (B) and basal part of adjacent layer of low columnar cells. Depigmentation is complete except for scattered melanosomes in vitreal ends of basal cell layer. Two macrophages have invaded the epithelium, one containing a great deal of phagocytized melanin (Ma_1) and one with many lipid droplets (Ma_2). First macrophage has slender cytoplasmic processes along one border *(arrow)*. Depigmented cells are characterized by free ribosomes and granular endoplasmic reticulum as well as elongate mitochondria. One cell is in mitosis (Mi). × 3900. (From Fig. 4, KEEFE, 1973b, by permission of the author and Wistar Institute)

the external part of the neural retina where lateral intercellular attachments (*maculae* and *zonulae adhaerentes*) appeared as the first indication of a new external limiting membrane (KEEFE, 1973b).

As differentiation of the new neural retina proceeded, the ganglion cell layer separated from the other neural retinal cells by formation of the inner plexiform layer. This was followed shortly by the outer plexiform layer demarcating the external and internal nuclear layers. With the appearance of a defined nuclear layer, incorporation of ^{3}H-thymidine ceased in that layer (Fig. 16D). Following retinectomy, this was observed first in the more peripheral neural retina and, a little later on, in the fundus (REYER, 1971 b). After DNA synthesis and cell division had stopped throughout most of the regenerated neural retina, a marginal growth zone between the differentiated retina and the *pars ciliaris retinae* remained. These cells continued to incorporate ^{3}H-thymidine, to divide and differentiate thus providing for further growth of the retina around its ciliary margin. Differentiation of the photoreceptor cells was the same as that found during embryonic development and described earlier (KEEFE, 1973 b; HENDRICKSON, 1964).

Neural retina regeneration has also been studied in the adult European newt, *Triturus cristatus*. In this species, there was continual growth of the retina in the normal adult by means of cell division at the *ora serrata* where labeled cells were consistently observed after ^{3}H-thymidine injection (GAZE and WATSON, 1968; MITASCHOV, 1968, 1969c; KEEFE, 1973d). When fixed over a period of time, these labeled cells moved centrally into the differentiated neural retina. Following surgical retinectomy and injection of ^{3}H-thymidine after 4–44 days, MITASCHOV (1968) observed incorporation of the isotope, 4 days after operation, in both the ciliary region and in the pigmented epithelium, and found the same pattern of neural retina regeneration in *Triturus cristatus* as has been described already for *Triturus pyrrhogaster* and *Notophthalmus viridescens*. GAZE and WATSON (1968) followed regeneration and growth of the neural retina over an extensive period of time after devascularization and labeling with ^{3}H-thymidine. They concluded that there was an extensive contribution to the final regenerate, by cells labeled in the growth area at the ciliary margin. Contrary to the observations in *N. viridescens*, KEEFE (1973d) found that the pigmented retinal epithelium separated into two layers during degeneration of the neural retina after devascularization in *T. cristatus*. The inner layer depigmented and formed the anlage for a new neural retina while the outer phagocytized the remaining external portion of the old neural retina and then reconstituted the pigmented epithelium lying on Bruch's membrane. Scattered cells in the degenerating neural retina also incorporated ^{3}H-thymidine and divided before cytolyzing. By 14 days after either retinectomy or devascularization, the cell cycle was accelerated in the regenerating neural retina when compared to cells remaining in the pigmented epithelium (MITASCHOV, 1969a, c, 1970). By plotting the percentage of labeled mitoses in these tissues against time after injection with ^{3}H-thymidine, the time for a complete cell cycle was found to be 43 hrs in the pigmented epithelium. This was reduced to 21 or $23^1/_2$ hrs in the regenerating neural retina due to a decrease in the time for the S phase from 19 hrs to 13 or $16^1/_2$ hrs and for G_1+M from $20^1/_2$ hrs to $5^1/_2$ or $4^1/_2$ hrs (first figures for retinectomized eyes, second for devascularized eyes).

HASEGAWA (1958) and REYER (1971 b) also investigated neural retina regeneration in larval stages of *Triturus pyrrhogaster* and *Notophthalmus viridescens*. Normal larval eyes have a conspicuous growth zone at the ciliary margin where ^{3}H-thymidine is incorporated and mitoses are continually taking place. After retinectomy, some of this tissue still remained in the eye. As in the adult, neural retina regeneration was initiated by increased proliferation of these cells. In the fundus, cells of the pigmented epithelium also began to label with ^{3}H-thymidine, 3–5 days after retinectomy, and then depigmented where they were not already covered by the unpigmented tissue regenerating from the ciliary margin. Regeneration was more rapid in larvae than in adults and there was variation in the proportion of the new retina contributed by the two sources, apparently depending on the amount of tissue left behind at the *ora serrata* after retinectomy. When the entire iris and *pars ciliaris retinae* was removed together with the neural retina, regeneration was much delayed. By 20–25 days after operation, some of the remaining pigmented retinal epithelium became partly depigmented. These cells were actively dividing but no differentiated neural retina was observed by this time (REYER, unpublished observation). Unlike eyes of adults, larval eyes of urodeles are resistant to a temporary loss of blood supply and can be transplanted without degeneration of the neural retina (STONE, 1930, 1964a, b; STONE et al., 1937) so that this technique cannot be used to study regeneration. It can be concluded, therefore, that a new neural retina regenerates from two different cell populations in all larval and adult newts of the family Salamandridae that have been studied so far.

Do amphibians other than newts have an equal capacity to reconstitute a neural retina? Another urodele (*Ambystoma maculatum*) and several anurans have been investigated. After lentectomy and retinectomy in *A. maculatum* larvae, retina regeneration from the remaining cells of the marginal growth zone occurred as in larvae of *N. viridescens* but with little evidence of a contribution from the pigmented retinal epithelium although the latter did partially depigment and some incorporation of ^{3}H-thymidine was observed (REYER, 1971b). Neural retina regeneration from the pigmented retina failed up to 60 days after complete removal of lens, iris, *pars ciliaris retinae*, and neural retina. In adults of this species, the eye was severely injured by retinectomy. Large blood-filled cysts formed which displaced part of the pigmented retinal epithelium anteriorly toward the pupil so that it often fused with the iris. Nevertheless, a neural retina regenerated from the marginal cells of the anterior complex. Occasionally, partly depigmented cells from the pigmented retina fused with this regenerate and may have contributed some tissue to the final neural retina (REYER, unpublished observation). When adult *Ambystoma maculatum* eyes were transplanted, the neural retina degenerated but the lens remained intact and a new neural retina subsequently regenerated from the ciliary margin (STONE and COLE, 1943; STONE and ELLISON, 1945).

In anurans, removal of the retina and lens in adult *Bombinator* (STROEVA, 1956) or retina and iris in tadpoles of *Bufo viridis* (LOPASHOV, 1949, 1955, 1959), *Rana esculenta* or *R. temporaria* (SOLOGUB, 1968) failed to elicit neural retina regeneration from the pigmented retinal epithelium. If the lens and the inner, unpigmented layer of the iris remained, then a new neural retina regenerated from the latter source in tadpoles of the last three species (LOPASHOV, 1955, 1959; DABAGYAN and SHERESHEVA, 1966; DABAGYAN and TRET'YAKOVA, 1967). Secondary irises developed as the regenerating neural retina came into contact with the lens which had not been removed (DABAGYAN et al., 1966). Immediately after metamorphosis, neural retina regeneration still occurred but at a slower rate. Normal electroretinograms were obtained after differentiation of the regenerated neural retina was complete (DABAGYAN et al., 1969). When the lens was left in the eye during neural retina regeneration, it did not degenerate. However, there was a temporary cessation of secondary lens fiber formation and an overgrowth of lens epithelium to cover the posterior pole. As soon as part of the neural retina had differentiated, then the lens epithelium adjacent to it once more started to form lens fibers (SOKOLOVA et al., 1970).

SATO (1953) reported neural retina regeneration in *Rana temporaria* tadpoles from isolated pieces of pigmented epithelium which were obtained either from dorsal iris or from the region of the adjacent ciliary and dorsal neural retina. These were stripped of their mesodermal coats and then implanted into lentectomized host eyes. In the same species, LOPASHOV and SOLOGUB (1972) investigated the conditions necessary for neural retina regeneration from pieces of pigmented retina removed from more posterior regions of the eyes of either tadpoles or adult frogs. No retina formed when this tissue was transplanted alone into the enucleated orbit of host tadpoles. If a piece of neural retina was included in the transplant, then new neural retina differentiated from the pigmented epithelium. When exposed to host neural retina by implantation into a lentectomized tadpole eye, the isolated, pigmented retinal epithelium was also able to regenerate new neural retina. The authors list four necessary conditions for this metaplasia of pigmented epithelium into neural retina. These are separation from mesenchymal envelopes, an adequate mass of tissue, cell proliferation, and the presence of some other neural retina. This requirement for neural retina resembles that found for neural retina regeneration from pigmented retinal epithelium following removal of the neural retina from the eyes of 4-day-old chick embryos (COULOMBRE and COULOMBRE, 1965; COULOMBRE, 1970).

The optic nerve itself readily regenerates in all amphibians. In the newt, it contains mixed fascicles of myelinated and unmyelinated nerve fibers enclosed by processes of centrally located ependymoglial cells (TURNER and SINGER, 1974a). When the optic nerve was cut and the cut ends replaced in the dural sheath, Wallerian degeneration occurred very rapidly in the central stump and was complete after 10–14 days. The myelin was broken down by the hypertrophied ependymoglial cells as well as by invading neutrophils and mononuclear phagocytes. With the electron microscope, fine, regenerating nerve fibers were observed running in troughs between the glial cells by 4 days after nerve transection. These regenerating fibers then increased in number and size. There was extensive destruction of regenerating axons from 6–28 days so that the population of regenerating nerve fibers was labile for a considerable period of time. This axon degeneration might be due to failure of the regenerating fibers to make appropriate connections in the optic tectum (TURNER and SINGER, 1974b; 1975; TURNER, 1975).

Although the chemical constitution and metabolism of the neural retina has been investigated in a number of species, there has been only one study of biochemical events during the metaplasia of pigmented retina into neural retina. In the normal *Notophthalmus* neural retina, SQUIRES (1972) determined, by microchemical methods, the activities of acid and alkaline phosphatase, five respiratory enzymes, and acetylcholinesterase as well as their histochemical localization. The activities of these enzymes were then followed during the course of retinal degeneration and regeneration subsequent to eye removal and reimplantation. Each assay was made on a single neural retina or regenerate.

During the first 3 weeks of retinal degeneration, all enzyme activities studied fell precipitously except for acid phosphatase which increased for 2 weeks. Acid

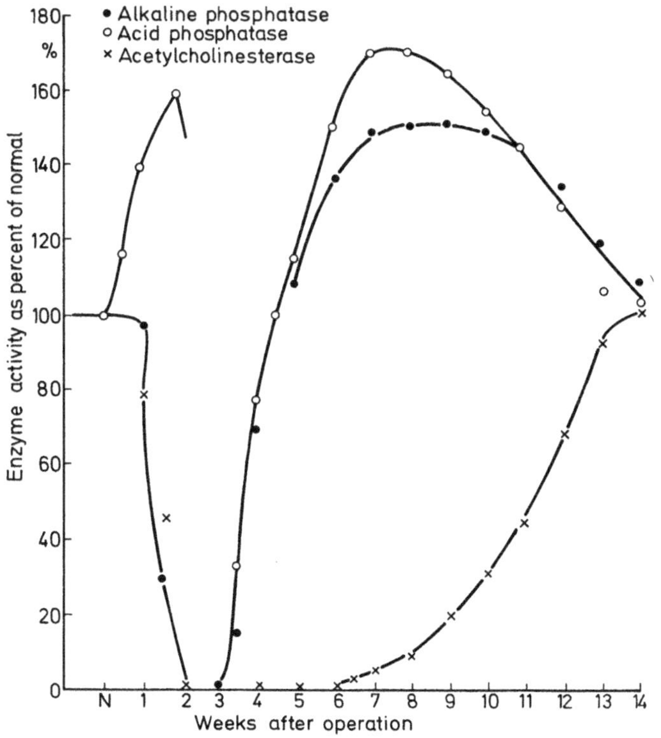

Fig. 18. Comparison of specific activities of alkaline phosphatase (●), acid phosphatase (○), and acetylcholinesterase (×) in degenerating and regenerating neural retina determined weekly after cutting blood supply to eye and expressed as a percent of activity of this enzyme in normal retina (From SQUIRES, 1972, by permission of the author)

and alkaline phosphatases were the first to show increased levels of activity as the new neural retina began to form after 3–4 weeks and rose quickly to above normal levels before falling again to normal (Fig. 18). Increased activities of the five respiratory enzymes appeared next, 4–5 weeks after operation. Of these, 6-P-GDH activity rose most rapidly, MDH, and COX activity advanced together at a slightly slower rate and LDH and aldolase activity increased at the slowest rate (Fig. 19). Acetylcholine esterase activity did not begin to reappear until 6–7 weeks after operation, probably at the onset of functional differentiation (Fig. 18). When the enzymes were first present in a regenerating retina, their activity was distributed diffusely over all parts of the regenerate. Normal localization and levels of activity were not attained until histologic differentiation had taken place, 14 or more weeks after devascularization of the eye. Following the initial fall during retinal degeneration, the rates of respiration and aerobic and anaerobic glycolysis, as determined manometrically per μg protein, all gradually increased during regeneration to that characteristic of the normal neural retina. There was also quite a close correlation between the rate of anaerobic glycolysis and the activities of aldolase and LDH and a closer correlation between the rate of respiration and the cytochrome oxidase and malate dehydrogenase activities. So far, it has not been possible to isolate pure pigmented retinal epithelium for enzyme assay.

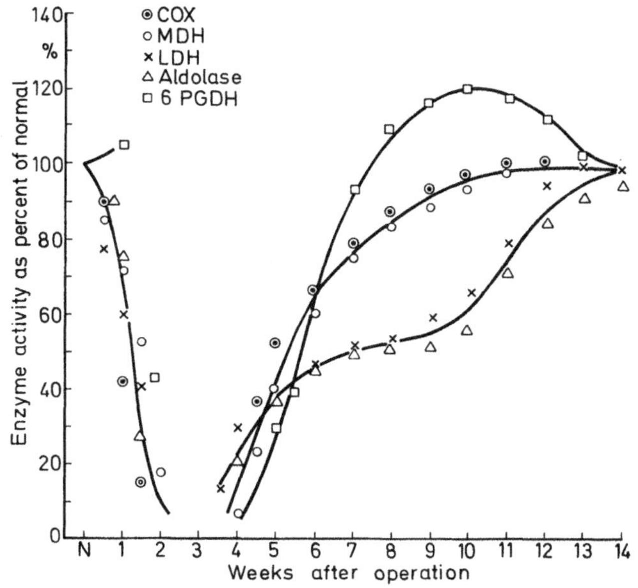

Fig. 19. Comparison of specific activities of five respiratory enzymes in degenerating and regenerating neural retina plotted in the same way as in Figure 18. Enzymes are cytochrome oxidase (COX ⊙), malate dehydrogenase (MDH ○), lactate dehydrogenase (LDH ×), aldolase △, and 6-phosphogluconate dehydrogenase (6 PGDH □). (From SQUIRES, 1972, by permission of the author)

Because the activities of these enzymes in pigmented retinal epithelium are therefore unknown, it cannot be determined whether the morphologic metaplasia of this tissue into neural retina also involves changes in enzyme populations or in the metabolic events catalyzed by these enzymes. Studies of RNA and protein synthesis have not yet been made on this system.

II. Restoration of Vision

Restoration of vision and the reappearance of a normal electroretinogram (DABAGYAN and OGANESYAN, 1971) follows optic nerve and neural retina regeneration in amphibians. The visual response is normal if the eye remains normally oriented but is disoriented to correspond with any rotation of the eye in adult *Notophthalmus viridescens* (SPERRY, 1943a, b; STONE, 1947, 1948, 1953, 1963a; STONE and ZAUR, 1940; STONE and FARTHING, 1942; STONE and ELLISON, 1945). SPERRY originally proposed the hypothesis of neuronal specificity to account for these results as has already been discussed. These early experiments have been summarized by STONE (1947), GAZE (1960, 1970), and REYER (1962a).

Electrical recording from the optic tectum following its reinnervation by optic nerve fibers from a regenerated neural retina in the newt has been employed by GRAFSTEIN (1964), GRAFSTEIN and BURGEN (1964), CRONLY-DILLON (1968), and CRONLY-DILLON and LEVINE (1971). With normal orientation of the eye, the pattern of tectal responses to stimuli in the visual field was essentially normal except for a temporary expansion of the receptive field sizes for the peripheral retina (CRONLY-DILLON, 1968). Nasal or temporal half-eyes regenerated projections that could have either a normal pattern extending over the entire optic tectum, or an abnormal overlapping pattern (CRONLY-DILLON and LEVINE, 1971). GRAFSTEIN (1964) and GRAFSTEIN and BURGEN (1964) removed the anterior part of the eye by an incision around the eyeball a short distance posterior to the *ora serrata*, extirpated the neural retina from both parts, and replaced the anterior part after 180° rotation, leaving the posterior part unrotated. Variable results were obtained with these procedures. In some cases, the tectal projections from the central and peripheral visual fields were both normal; in others both were rotated. The central field projection was rostral to its normal location on the optic tectum in the latter group. Occasionally, the central field had a double representation on the tectum. LEVINE and CRONLY-DILLON (1974) and LEVINÉ (1975) have performed the same operation on adult *T. cristatus* employing both a limbal and a postlimbal incision. With both procedures, these authors found either completely rotated visuotectal projections (corresponding to the polarity of the rotated anterior part of the eye) or a rotated peripheral eyefield combined with a normal central or dorsal eyefield. No completely normal, unrotated visuotectal projections were observed. After a postlimbal incision, these differences could be due in part to varying contributions to the final neural retina from the central, unrotated, and the peripheral, rotated pigmented epithelium and anterior complex with one or the other part assuming dominance in reestablishing retinal polarity in the cases with entirely normal or completely rotated visuotectal projections. However, following the limbal incision, nearly all of the regenerated retina was initially derived from the unrotated posterior part. In those cases where the final result was a completely rotated visuotectal projection, the rotated anterior part must have repolarized the retina regenerated from the posterior part.

To summarize the preceding discussion, we know that the pigmented retinal epithelium of amphibians has the capacity for metaplastic change into the neural retina, an entirely different, complex neural tissue, from which it separated during ontogeny when the outer and inner layers of the eyecup were established. These changes involve extensive loss of melanin pigment which is removed by macrophages, initiation of cell division, and formation of several layers of unpigmented cells which then stop dividing and differentiate into neural and supporting elements. In addition to this source of neural retina, there is also a contribution from cells remaining at the junction of *pars ciliaris retinae* and *ora serrata*. The relative amount of the contributions from these two sources depends on species, age, and the nature of the experimental intervention. This lability of pigmented retina seems to be greatest in newts but is present to some degree in other urodeles and in anurans. Initiation of neural retina regeneration by metaplasia of the pigmented retina can be elicited in newts merely by separation of the old neural retina from the pigmented epithelium and also from grafts of pigmented epithelium into the eye chamber (STONE, 1950b) as well as after retina degeneration or removal. In anurans, removal of the choroid coat and the presence of a piece of neural retina also seems to be necessary. Since neural retina regeneration from the anterior complex takes place together with that from the pigmented epithelium in newt eyes, a stimulus from developing neural retina is also acting on the pigmented epithelium under these conditions. When a new neural retina has been restored, new optic nerve fibers grow out from the ganglion cells, invade the stump of the old optic nerve, and follow it to reinnervate the optic tectum. In spite of the fact that the central area of the new neural retina is derived from a variable amount of pigmented epithelium and the peripheral part comes from the anterior complex, nevertheless, normal retinotectal projections are reestablished and vision is restored after a period of several months.

III. Regeneration of the Lens from the Cornea

Two parts of the eye have the capacity to form a new lens when the old one is removed. These are (1) the corneal epithelium in older embryos and young larvae of the urodele *Hynobius unnangso* (IKEDA, 1936b, 1939) and in tadpoles of the anuran *Xenopus laevis* (FREEMAN, 1963; OVERTON, 1965; WAGGONER, 1973), and (2) the iris epithelium at the pupillary margin of the dorsal iris in newts of the family Salamandridae (REYER, 1954a; STONE, 1967).

Before discussing the latter in more detail, lens regeneration in *Xenopus* tadpoles will be summarized. The new lens develops from the inner cell layer of the outer cornea by elongation of these cells, formation of a solid cellular aggregate, and subsequent differentiation of lens fibers from the central and inner part facing the vitreous. The basement membrane of the outer cornea encloses the regenerating lens and contributes to the lens capsule after separation of this lens from the cornea (FREEMAN, 1963; WAGGONER, 1973). There is an increase in granular endoplasmic reticulum during early stages followed by its breakdown into vesicles and the formation of increased numbers of free ribosomes within the lens cell aggregate. Microfilaments appear and the nucleoli enlarge when lens fibers begin to differentiate (OVERTON, 1965). Increased incorporation of ^{3}H-thymidine was first observed when the corneal cells started to form an aggregate and stopped completely when lens fiber differentiation began (WAGGONER, 1972; WAGGONER and REYER, 1975). Immunofluorescence studies with antisera to adult *Xenopus* lenses have given somewhat conflicting results. CAMPBELL (1965) observed fluorescence, indicating the presence of lens antigens, in the inner cells of the corneal epithelium by 1 day after lentectomy while BRAHMA and MCDEVITT (1972b) failed to find any reaction to the antiserum until the beginning of lens fiber differentiation. Lenses regenerated from corneas transplanted into lentectomized eyes or into regenerating limbs but did not regenerate from corneas implanted in the dorsal fin or under the skin (FREEMAN, 1963; WAGGONER, 1973). When pieces of cornea were cultured in clotted fowl plasma immersed in one-half strength medium 199, lenses developed which gave a positive immunofluorescent reaction for *Xenopus* lens antigens (CAMPBELL and JONES, 1968).

IV. Regeneration of the Lens from the Iris

Lens regeneration from the iris has been investigated more intensively and over a much longer period of time than has lens regeneration from the cornea. In 1962 when I last reviewed this subject for the twentieth Growth Symposium on

Regeneration, the main events and the tissue interactions involved had been clarified by the thorough investigations of WACHS, Sato, and STONE over a period of nearly 50 years. These data had been derived almost entirely by means of the microsurgical methods of experimental morphology and little was then known of the molecular biology of this process. However, even at that time, research into lens regeneration had already reached a turning point for, in the year 1963, YAMADA and his colleagues at the Oak Ridge National Laboratory published the first results of a continuing series of investigations into the repertoire of cytologic and molecular changes which underlie the formation of a transparent lens from the densely pigmented iris. YAMADA has summarized these studies in a series of reviews (YAMADA, 1966, 1967a, b, 1972). The earlier morphologic work has also been reviewed several times (REYER, 1954a, 1962a; STONE, 1959a, 1965; SCHEIB, 1965). In the following account, I shall attempt to describe briefly our present understanding of this most interesting example of the transformation of one differentiated tissue into another.

Lens regeneration was divided into 13 morphologic stages by SATO (1940). This staging has been described and illustrated further by REYER (1954a, 1962a) for larval stages and by STONE and STEINITZ (1953a) and YAMADA (1967b) for adult stages of *Notophthalmus viridescens*. Very briefly, the histologic changes, visible with the light microscope, are swelling of the pupillary margin of the dorsal iris (stage I, Fig. 20A, p. 352), separation of the inner and outer laminae of iris epithelium (stage II), depigmentation of cells in the inner lamina at the pupillary margin (stage III, Fig. 20B, p. 352), formation of a vesicle of basophilic, depigmented, low columnar epithelial cells (stage IV, Fig. 20C, p. 353), enlargement of the lens vesicle and thickening of the inner part (stage V), further elongation of cells in the inner layer (stage VI) and their differentiation into acidophilic primary lens fibers (stage VII, Fig. 20D, p. 353). Formation of additional primary lens fibers reduces the size of the optic vesicle cavity (stage VIII) and secondary lens fibers are then laid down on the outside of the primary fibers at their junction with the lens epithelium (stage IX, Fig. 20E, p. 353). The lens continues to grow by differentiation of new secondary fibers which completely surround the primary fibers (stage X) and finally enclose them symmetrically (stage XI, Fig. 20F, p. 353). In these last stages, the lens epithelium is reduced to a single layer of cells which detaches from the dorsal iris. In stages XII and XIII, the lens grows in size and the nuclei of the central lens fibers no longer stain. Lens regeneration can be inhibited or greatly retarded by 1500 R of X-irradiation (DONALDSON, 1972), and retarded proportionally by decreasing the temperature from 26° to 18° or 12° C (DONALDSON and CHAN, 1973) or by keeping the animals postoperatively in total darkness (BORSUK and POPOV, 1968).

1. Cytologic and Macromolecular Events

When observed with the electron microscope (EGUCHI, 1963; KARASAKI, 1964; DUMONT et al., 1970; DUMONT and YAMADA, 1972), the normal dorsal iris of the adult newt (*T. pyrrhogaster or N. viridescens*) is composed of a double layer of cuboidal epithelial cells, packed with melanosomes, and a stroma on the outer surface. The stroma is a loose connective tissue containing small blood vessels, fibroblasts, mast cells, iridophores, and fine collagenous fibers. A basal lamina covers the inner surface of the iris epithelium and extends between epithelium and stroma. The nuclei of the epithelial cells have an indented nuclear membrane, are of quite uniform electron density, and may have a small, mostly fibrous nucleolus. Except for the melanin granules, the cytoplasmic organelles are rather sparse, consisting of microfilaments near the inner plasma membrane, and a few scattered mitochondria, ribosomes, and elements of the endoplasmic reticulum. Adjacent epithelial cells adhere by means of desmosomes. Some authors describe larger melanosomes in the inner epithelial lamina and smaller, more oval ones in the outer lamina.

Following removal of the lens through an incision in the cornea, the first visible change is the enlargement of the nucleolus and an increase in its 100–200 Å, granular component. These changes of nucleolar activation take place by 2 days after lentectomy and become even more prominent after 4 days providing morphologic evidence for the production of ribosomal RNA in the nucleolus. Significant metabolic changes are occurring at the same time. Two days after lentectomy, REESE et al. (1969) reported increased incorporation of ^{14}C-uridine into the 18 S and 28 S ribosomal RNA and 4 S RNA isolated by sucrose density gradient centrifugation from the RNA obtained by phenol extraction of homogenates from dorsal iris. This incorporation continued to increase for 10 days after lens removal. Following pulse labeling of only 30 min with this isotope, the label appeared in the heavier, 40 S rRNA precursor fraction. During the first 4 days, YAMADA and KARASAKI (1963) also found that the nuclei of these cells labeled more heavily in autoradiographs after injection of ^{3}H-cytidine and EGUCHI and ISHIKAWA (1960) demonstrated an increased uptake of ^{32}P-orthophosphoric acid. Furthermore, EGUCHI (1964) and KARASAKI (1964) observed the beginning of an augmentation in the number of cytoplasmic ribosomes in electron micrographs. Following continuous exposure to ^{3}H-thymidine for either the first 24 hrs or the second 24 hrs after lens removal, no label was detected in autoradiographs of iris epithelial cells, indicating the absence of DNA synthesis for the first 2 days (YAMADA, 1971). YAMADA and his co-workers interpret these results as indicating that increased production of ribosomal RNA is the first event taking place in lens regeneration, occurring already during regeneration stage I. This could be due to more rapid processing of the 40 S ribosomal RNA precursor, amplification of the ribosomal RNA cistrons of DNA, increased transcription of these cistrons, or any combination of the above events.

There is now evidence for some selective amplification of the cistrons coding for ribosomal RNA during the first week after lentectomy. COLLINS (1972) used the method of RNA-DNA hybridization to compare the number of RNA cistrons in the DNA of normal dorsal iris and dorsal iris 7 days after lentectomy. He found that the DNA of the iris initiating lens regeneration contained 60% more rRNA cistrons than the DNA of normal iris (31 000 vs. 19 900). Therefore, the fourfold increase in ribosomal RNA synthesis discussed previously is apparently due to both amplification of the cistrons for rRNA and increased transcription of these cistrons. Transcription of DNA-dependent RNA would seem to be involved in the events preceding lens fiber formation because daily injections of actinomycin D (an inhibitor of RNA synthesis) into adult newts (1 ug/g body wt) over a period

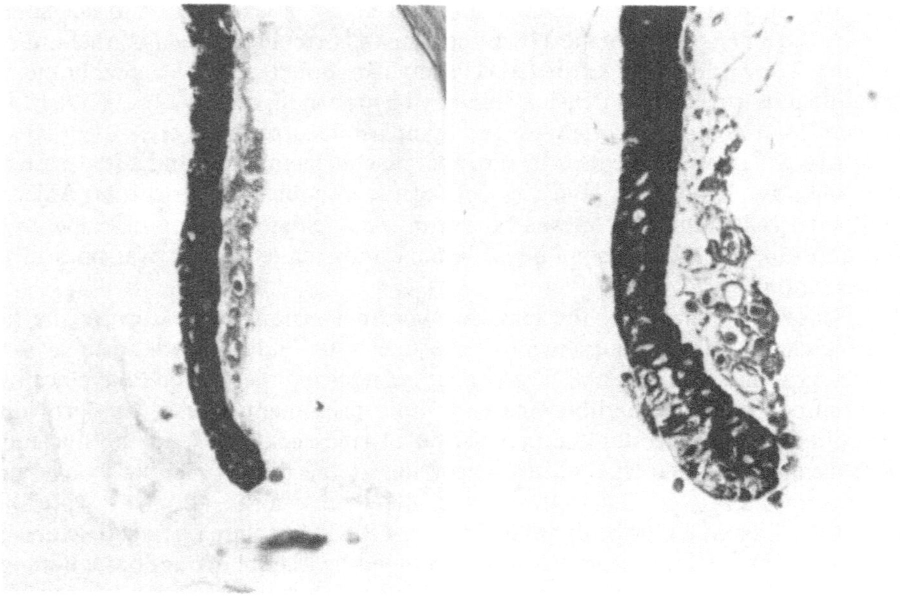
A B

of 10 days prevented lens regeneration beyond stage II when the eyes were fixed up to 15 days after lentectomy. Injections given for only 4 consecutive days retarded lens regeneration if administered before lens fibers had formed and resulted in the destruction of only the lens epithelium if given after lens fibers were already developing (YAMADA and ROESEL, 1964).

The cells of the normal iris epithelium neither synthesize DNA nor replicate (YAMADA and ROESEL, 1969, 1971). Following lentectomy, uptake of ^{3}H-thymidine and labeling of a significant number of these cells is first observed in the inner lamina after 4 days. Labeling spreads to the outer lamina as well and the total number of labeled cells increases for the next 10 days during the proliferative period (EISENBERG and YAMADA, 1966; YAMADA and ROESEL, 1969; REYER, 1971a; EGUCHI and SHINGAI, 1971). Mitotic figures are also first observed on the 4th day after lens removal and increase rapidly from 0 to 1.27% of the total iris epithelial cell population on day 7 (YAMADA and ROESEL, 1971). While most pronounced in the dorsal iris, DNA synthesis and cell division take place throughout the iris epithelium. By plotting the percent of labeled mitoses against time after injection of ^{3}H-thymidine (method of QUASTLER and SHERMAN, 1959), ZALIK and YAMADA (1967) determined the minimum duration of the DNA-S period in 15-day regenerates (stages VI–VIII) to be 19 hrs, the G-2 period, 2 hrs, and the G-1 period, 40 hrs. The DNA synthetic index was shown to increase from 0.1 at stage I to 0.29 at stage VIII. Therefore, assuming a constant time for the S period, the mean cell generation time (T) decreased from approximately 200 hrs at the onset of mitosis at stage I to 65 hrs at stage VIII. In cultures of iris epithelial cells, values of 36 hrs for S, 6 hrs for G_2, 25 hrs for G_1, and 1.8–2.1 hrs for M were obtained (HORSTMAN and ZALIK, 1974). Using adult *Triturus cristatus* in which both the lens (stages VI–

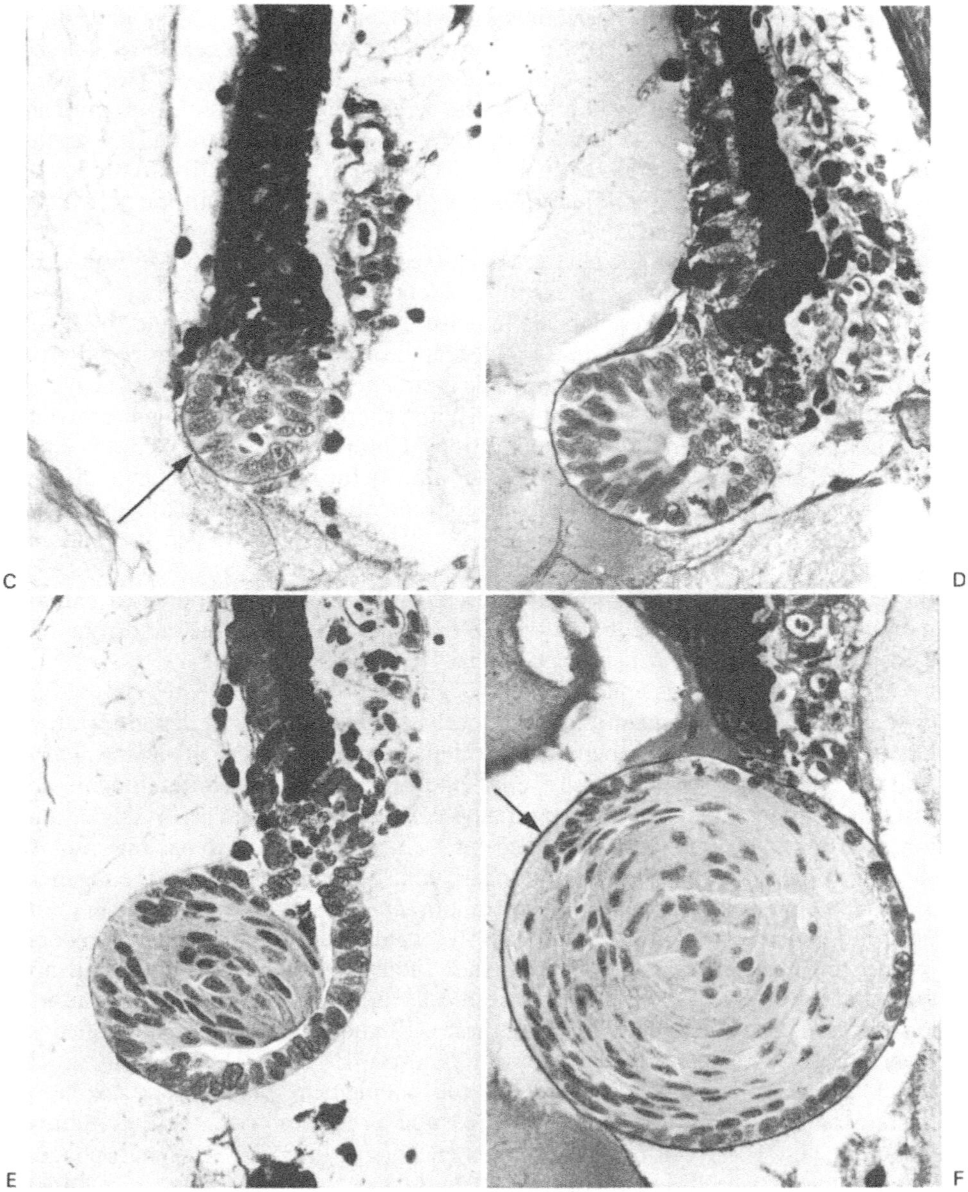

Fig. 20 A–F. Lens regeneration from dorsal iris in adult *N. viridescens*. (A) Stage I, 5 days after lentectomy, hematoxylin (H) and PAS stain. × 180. (B) Stage III, 10 days after lentectomy, H and PAS stain. × 180. (C) Stage IV, 15 days after lentectomy, H and PAS stain. Arrow shows first indication of lens capsule forming from basement membrane. × 180. (D) Stage VII, 15 days after lentectomy, Himes Moriber triple stain. × 180. (E) Stage IX, 20 days after lentectomy, H & E stain. × 180. (F) Stage XI, 25 days after lentectomy, Himes Moriber triple stain. Lens capsule indicated by arrow. × 180

IX) and neural retina were regenerating, MITASCHOV (1969b) confirmed these *in vivo* results for the S and G_2 periods. However, from a curve depicting percent of labeled mitoses for two successive cell cycles, he was able to calculate a much shorter period for the whole cell cycle and G_1. His data give $T=23.5$ hrs, $S=16$ hrs (after subtracting 3 hrs for the availability time of ^{3}H-thymidine), $G_2=2$ hrs and $G_1+M=5.5$ hrs. MICHEL and YAMADA (1974) give 45 hrs as the average cell cycle time for iris epithelial cells regenerating a lens. Irradiation with x-rays of 1500 R greatly inhibited uptake of ^{3}H-thymidine and mitotic activity of these cells both *in vivo* and *in vitro*.

In order to study possible changes in the secondary structure of DNA preceding its replication, adult newt irises in the G_0 and G_1 stages of the cell cycle (up to 5 days after lentectomy) were labeled in culture with ^{3}H-thymidine and the DNA was extracted and purified. These DNA preparations were characterized by hydroxyapatite chromatography, isopycnic gradient centrifugation in CsCl and Cs_2SO_4 (Hg) to determine buoyant density, and sucrose density gradient centrifugation under neutral and alkaline conditions. Compared to DNA of the normal iris, late G_1 stage DNA exhibited a greater affinity for nitrocellulose, eluted from the hydroxyapatite at a lower phosphate concentration, and formed a peak at 13S rather than 19S when denatured by 1N NaOH before sucrose density gradient centrifugation. These data indicate the presence of nicks or gaps in the DNA of the late G_1 stage. This "activated" prereplicative DNA was found to exhibit template activity when incubated with rat liver DNA polymerase and four nucleoside triphosphates (COLLINS, 1974a, b).

Elongation of the iris epithelial cells begins by 4 days after lentectomy. Four days later (8 days postlentectomy), these cells are columnar and depigmentation has been initiated. During the course of depigmentation, there are great alterations in the morphology of the iris epithelial cells. Intercellular attachments are lost, large intercellular spaces appear, the cells become irregular in shape, and long cytoplasmic processes are formed into which the melanosomes migrate. At the same time, a second type of cell appears in large numbers around the regenerate. These cells have a nucleus with chromatin condensations, considerable granular endoplasmic reticulum, numerous lysosomes, and many thin, elongated projections from their surface. They invade the iris epithelium, being found not only on the surface of the iris but between cells and in the space between the iris laminae. EGUCHI (1963) first described these cells and they were later identified as macrophages by DUMONT (DUMONT and YAMADA, 1972; YAMADA and DUMONT, 1972). These authors believe that iris cells depigment both by extrusion of melanosomes and by loss of the cytoplasmic processes with their contained melanosomes. These cell fragments are then phagocytized by the macrophages. There may also be direct transfer of melanin from an iris epithelial cell to a macrophage attached to this cell (EGUCHI, 1963). The entire process is termed "cytoplasmic shedding." Volume measurements with the light microscope on 0.25–0.5 μ-thick, serial Epon sections showed that both nuclear volume and total cell volume increases during the first 8 days of regeneration followed by a constant nuclear volume and a decrease in cytoplasmic volume over the next 7 days thus confirming the electron microscopic evidence for cytoplasmic shedding. When a cell culture, derived from iris epithelial cells, was treated with cyclic

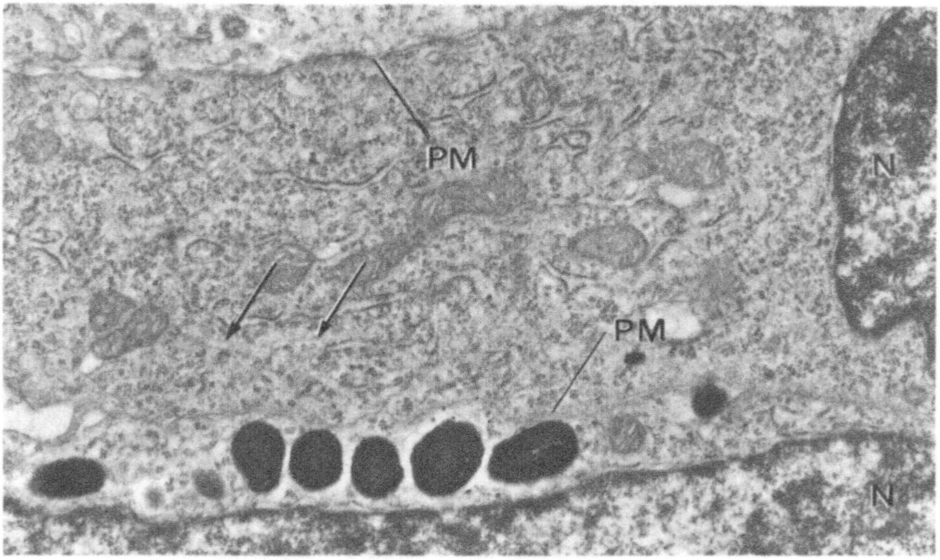

Fig. 21. Electron micrograph of the cytoplasm of a depigmented iris epithelial cell and narrow bands of two adjacent cells showing many polyribosomes, granular endoplasmic reticulum, mitochondria, parts of two nuclei, and a few melanosomes in one of the adjacent cells, 15 days after lentectomy. Arrows indicate a few scattered microtubules within cytoplasm. *PM* = plasma membranes between cells, *N* = nuclei. × 15800. (From Fig. 15, DUMONT and YAMADA, 1972, by permission of the authors and Academic Press)

adenosine monophosphate (cyclic AMP) or adenosine, the cytoplasm was temporarily redistributed from a flat, sheetlike configuration to a stellate array of long slender processes somewhat resembling the long processes formed during cytoplasmic shedding within the eye (ORTIZ et al., 1973).

Other evidence for activation of the cell surface has been provided by the cell electrophoresis studies of ZALIK and SCOTT (1972, 1973) which demonstrated a decrease in the electrophoretic mobilities of trypsin-dissociated iris cells during the first 10 days after lens removal which was then followed by an increase. Treatment of cells from normal iris with RNase, hyaluronidase, chondroitinase, or neuraminidase reduced their electrophoretic mobility. Sensitivity to the first three enzymes was lost in cells from 5–7-day regenerates but not until after 10 days for neuraminidase. According to ZALIK and SCOTT, these results point to a loss of negatively charged, RNA-containing acid mucopolysaccharides and glycoproteins from the cell surface during dedifferentiation of the iris epithelial cell. IDOYAGA-VARGAS and YAMADA (1974) demonstrated that the activity of the enzyme N-acetylglucosaminidase (releases N-acetylglucosamine from mucopolysaccharides) in the dorsal iris increases over a period of 17 days after lentectomy which includes the time period studied by ZALIK and SCOTT as well as that when cytoplasmic shedding occurs. These changes in the cell surface may be related to the loss of cell adhesion observed with the electron microscope. Further study is needed to clarify any role they might play in activating the nucleoprotein metabolism.

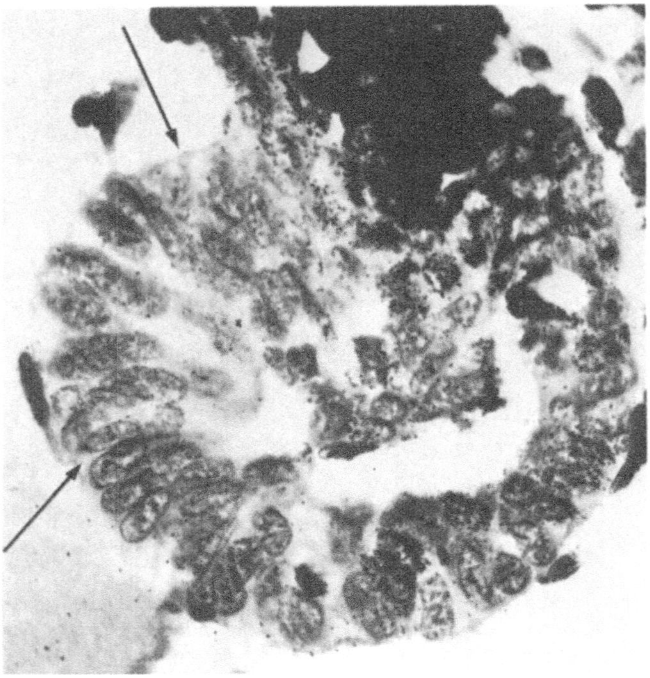

Fig. 22. Lens regeneration stage VI, 15 days after lentectomy and 3 hrs after injection of ^{3}H-thymidine. Label is present over cells of lens vesicle except for elongating primary lens fibers *(between arrows)* which are unlabeled. × 425. (From Fig. 18, REYER, 1971a, by permission of Academic Press)

Concomitant with cytoplasmic shedding and loss of pigment, there is an increase of cytoplasmic organelles in the cells of the regenerating lens. Mitochondria become more abundant and may be very elongate and complex in shape. There is some increase in granular endoplasmic reticulum and a great increase in free ribosomes which are organized as polysomal clusters, rosettes, or helices (KARASAKI, 1964; YAMADA, 1967b, DUMONT and YAMADA, 1972; REYER, unpublished observation). Microfilaments are aggregated just beneath the plasmalemma while microtubules are scattered throughout the cytoplasm (Fig. 21). Following this period of dedifferentiation and multiplication, some of the depigmented cells facing the vitreous body and neural retina enter their terminal cell cycle. Following this, at stages IV or V, DNA synthesis and cell division ceases (EISENBERG and YAMADA, 1966; YAMADA, 1966, 1967b; REYER, 1971a) (Fig. 22) and these cells start to redifferentiate in a new direction into lens fibers.

Employing the technique of immunofluorescence, using antibodies to whole newt lenses, TAKATA et al. (1964a) first detected the presence of lens antigens in a few prospective lens fibers at stage IV with a marked increase at stage V. When antibodies to the gamma crystallin fraction of newt lenses were used in the reagent (TAKATA et al., 1965, 1966), fluorescence was first observed in these same cells at stage V. No lens antigens were found in any other ocular tissues. By combining autoradiography after injection of ^{3}H-thymidine with immunofluorescence, YAMADA (1966) concluded that alpha and beta crystallins appeared no earlier than 3 hrs after the final DNA-S period while $1\frac{1}{2}$ days elapsed between this S period and the synthesis of gamma crystallins in the same cells. As the lenses grew in size, gamma crystallins were restricted to the lens fibers while alpha and beta crystallins were present in both lens fibers and lens epithelium. Therefore, differentiating primary or secondary lens fibers in the regenerating lens never synthesize gamma crystallins until they have stopped replicating, but cells in the secondary lens fiber line, while still part of the lens epithelium, can synthesize alpha and beta crystallins at the same time that DNA synthesis and cell division are taking place. As shown in autoradiographs by YAMADA and TAKATA (1963), the period of lens fiber development, extending 12–25 days after lentectomy, is also characterized by rapidly increasing incorporation of DL-leucine-4,5 ^{3}H (up to 40 times control levels) into the lens epithelium of the regenerating lens. In electron micrographs, fine filamentous material (<60 Å in thickness) can be seen throughout the cytoplasm of differentiating lens fibers so that there is a gradual increase in electron density (EGUCHI, 1964; KARASAKI, 1964) with a concomitant decrease in the other cytoplasmic organelles, especially ribosomes. As the new lens increases in size, the basal lamina of the iris epithelium becomes thicker and develops into the lens capsule (REYER, 1967, unpublished observation; KARASAKI, 1964).

The activities of several enzymes have been shown to increase in the dorsal iris following lentectomy. These enzymes are alkaline phosphatase (EGUCHI and ISHIKAWA, 1963), dipeptidase (DE VINCENTIIS et al., 1965), tyrosinase (ACHAZI and YAMADA, 1972), and galactosyl-transferase (IDOYAGA-VARGAS, personal communication). THORPE et al. (1974) also observed a decrease in the amount of cyclic AMP (adenosine 3′, 5′-monophosphate) present in the dorsal iris during the first 2 days followed by a return over the next 2 days. At the same time, the activity of glutamine synthetase increased sevenfold. The significance of these changes is not yet well understood. Lens regeneration always occurs from the mid-dorsal iris in the normal eye. Nevertheless, in the ventral half of the iris, the same general changes in RNA synthesis, DNA synthesis, beginning depigmentation, protein synthesis, and enzyme activation were observed as in the dorsal iris. However, these changes were temporary and less in amount. Furthermore, no crystallins were formed anywhere except in the regenerating lens.

Ever since the first investigations on lens regeneration, three principal questions have been asked about this tissue transformation and its controls: (1) Is the lens derived exclusively from cells of the iris epithelium? (2) What is the source and nature of the intraocular stimulus acting during lens regeneration? (3) What inhibits lens regeneration in the normal eye and how does lentectomy release the iris from this inhibition? Evidence for the metaplastic conversion of iris epithelial cells into lens fibers is very convincing but not yet established with absolute certainty. Data supporting this hypothesis are retention of the epithelial organization of the iris as it dedifferentiates and changes into a lens, autoradiographic tracing of pigmented iris cells, labeled with ^{3}H-thymidine, into lens fibers when eyes are fixed at different times after injection of the isotope (EISENBERG and YAMADA, 1966; REYER, 1971a; EGUCHI and SHINGAI, 1971), and lens regeneration from pieces of newt iris (epithelium and stroma) implanted into the optic cavity of a lentectomized eye of the same species (WACHS, 1914; SATO, 1930; MIKAMI, 1941; STONE, 1952; REYER, 1956, 1966b) or of a different, nonregenerating species (IKEDA, 1934; AMANO and SATO, 1940; AMATATU and FUJITA, 1941; REYER, 1956) or into the blastema of the regenerating limb (REYER et al., 1973). Recently, more critical evidence has been obtained by YAMADA using a system of organ culture for the iris and other tissues. In previous attempts to culture iris alone or iris and cornea together (STONE and GALLAGHER, 1958; EGUCHI, 1967; EISENBERG-ZALIK and SCOTT, 1969), the isolated iris did not form any lens fibers unless, before explantation, it had already initiated lens fiber formation by remaining *in vivo* for at least 15 days after lentectomy. Addition of neural retina to such a culture stimulated development of larger lenses from explanted 15-day regenerates but failed to elicit lens differentiation from normal dorsal iris (EGUCHI, 1967). Synthesis of rRNA and initiation of DNA synthesis were found to be the same in cultures of isolated dorsal iris, maintained for 4 days, as previously found for the beginning of lens regeneration *in vivo* (REESE, 1973). However, there was a decrease in these activities in cultures maintained for longer periods of time (JAUKER and YAMADA, 1973). After modifying the culture medium[1], YAMADA et al. (1973) and YAMADA and MCDEVITT (1974) cultivated newt dorsal iris, dorsal iris removed 8 days after lentectomy, and dorsal iris epithelium alone for 20–21 days. They also combined whole dorsal iris, dorsal iris epithelium, dorsal iris stroma, ventral iris, and ventral iris epithelium with *R. pipiens* neural retina in similar cultures. When the neural retina was absent, no lenses developed; however, lens differentiation was consistently observed from both the dorsal iris with stroma and the dorsal iris epithelium deprived of its stroma when tadpole neural retina was present. A few lenses also developed from the ventral iris epithelium lacking stroma. Immunofluorescence studies demonstrated the presence of gamma crystallins in these lenses (Fig. 23A and B). Dissociated iris epithelial cells from both dorsal and ventral iris have also been grown in monolayer cultures (in diluted Leibovitz medium L-15 plus 10% fetal calf serum) where they depigmented and proliferated in the absence of other tissues (HORSTMAN and ZALIK, 1974; EGUCHI et al., 1974). After 50 days in culture, the latter authors observed the formation of lentoids containing crystallins and exhibiting the fine structure of lens fibers. These experiments support the hypothesis that iris epithelial cells transform into lens cells since the epithelium was removed from other possible sources of cells in the bloodstream and stroma. This would seem to eliminate the possibility that macrophages invade the iris epithelium and then develop into lens fibers. As yet, however, no one has obtained lens fiber differentiation from clones of iris epi-

[1] The following culture medium was used for organ culture of iris and neural retina:

72% medium 199, containing Earle's salts and glutamine which was buffered with 1.25 mg/ml sodium bicarbonate.

8% fetal calf serum

20% distilled water

100 U/ml penicillin, 100 µg/ml streptomycin, and 0.25 µg/ml Fungizone.

All components were obtained from Grand Island Biological Co.

The iris or iris and neural retina was cultured on lens paper supported by a stainless steel grid at the surface of the medium in a Falcon plastic organ culture dish and incubated at 25° ± 1.0° C with the pH maintained at 7.5 ± 0.1 in an automatic CO_2 incubator. The culture medium was exchanged for fresh culture medium every 3–4 days.

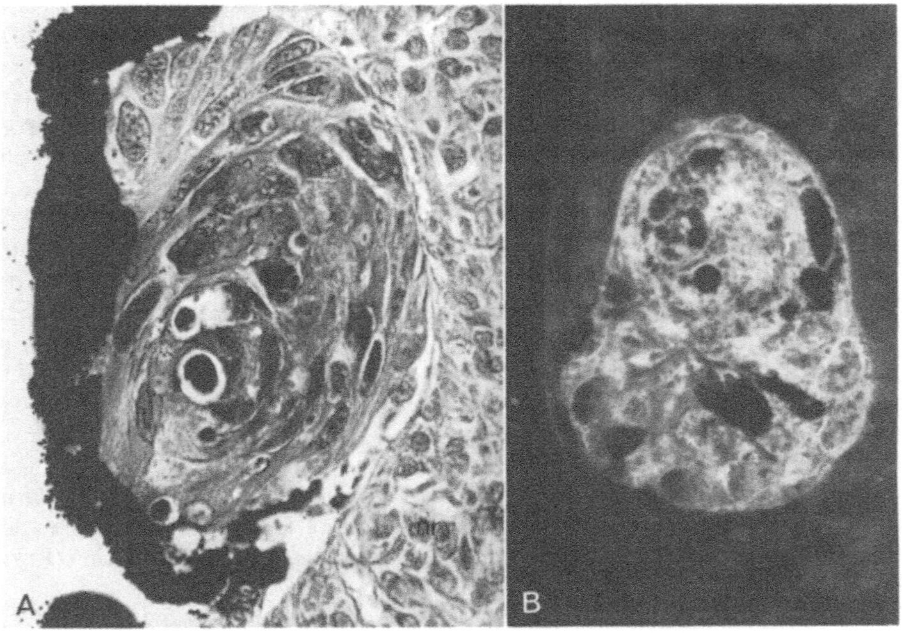

Fig. 23. A and B. Lens regeneration after organ culture of dorsal iris epithelium from adult *N. viridescens* on neural retina from a *R. pipiens* tadpole for 19–21 days at 25° C. (A) Iris and regenerating lens of stage IX in contact with neural retina *(nr)*. Bodian stain. × 330. (B) Fluorescence photomicrograph of a similar regenerating lens treated with anti-gamma crystallins immunofluorescence reagent. × 330. (From Figs. 2B and 7A, YAMADA and McDEVITT, 1974, by permission of the authors and Academic Press)

thelial cells, derived from a single cell, as has been done for lens formation from the pigmented retina in chick embryos by EGUCHI and OKADA (1973).

Early studies demonstrated that the competence to develop into a lens is highest in the mid-dorsal pupillary margin and then decreases circumferentially around the pupillary margin of the iris and meridionally toward the *ora serrata* and pigmented retina mid-dorsally (reviewed in REYER, 1954a, 1962a). Usually no lenses regenerate from the ventral iris. However, a few cases were observed in organ culture and EGUCHI and WATANABE (1973) occasionally elicited lens regeneration from ventral iris in the lentectomized eye by exposing the iris to the carcinogen, N-methyl-N'nitro-N-nitrosoguanidine for 60 minutes either *in situ* or in culture followed by implantation into a host, lentectomized eye. Therefore, ventral iris still retains the capacity for lens regeneration if suitably stimulated.

2. Role of the Neural Retina

A great deal of experimental evidence has been obtained which demonstrates that lens regeneration from the iris is dependent on a stimulus coming from the neural retina. The nature of this stimulus is still unknown. The earlier work has been reviewed most recently by REYER (1962a), STONE (1965), and SCHEIB (1965). To summarize, if dorsal iris was isolated from the neural retina by transplanting it to extraocular sites such as under the skin and into the peritoneal cavity, larval dorsal fin, or brain ventricle, it usually remained as pigmented iris or else partially depigmented without any lens fiber formation (IKEDA, 1935, 1936a; STONE, 1958a; REYER, 1966b; REYER et al., 1973). When an impermeable barrier of pliofilm separated the neural retina from the iris in the lentectomized eye or else the lens, pigmented retina, neural retina, and choroid were extirpated (STONE, 1958a, b), then there was no lens regeneration from the iris. Removal of neural retina together with the lens delayed but did not prevent lens regeneration which proceeded in parallel with neural retina regeneration (ZALOKAR, 1944; STONE and STEINITZ, 1953a; REYER, 1971a; POWELL and POWERS, 1973). The previously described organ culture experiments also provide a very clear demonstration of the action of the neural retina. The normal polarity of the regenerating lens depends on the spatial localization of this stimulus. Lens fibers always develop from the lens vesicle cells facing the neural retina. When the dorsal iris was turned inside out before lentectomy (STONE, 1954), a lens still regenerated with normal polarity; following the same operation during the course of lens regeneration, new lens fibers differentiated from the lens epithelium now facing the neural retina (REYER, 1974).

Although the role of the neural retina is well established, some recent investigations have shown that it is not the only source of stimuli for lens formation. STONE (1958a) observed two small fiber lentoids formed from dorsal iris implants into the body cavity and REYER et al. (1973) found that lens vesicles and small lenses occasionally developed from subcutaneous iris implants and rarely from implants into the brain or brain ventricle. Furthermore, when dorsal iris was inserted into the 8–20-day-old blastema of the regenerating forelimb and allowed to grow in this site for 20–70 days, then lens vesicles, lenses, and lentoids differentiated in 43–71% of the implants in different experiments (Fig. 24). When the limbs were denervated at the time of iris implantation, lens regeneration occurred in only 14% of the cases thereby indicating a probable role of regenerating nerve fibers. POWELL and POWERS (1973) found that a spinal ganglion, previously innervating a 10–14-day regenerating limb, stimulated the initial stages of lens regeneration (II–VI) when implanted for 10–18 days in the eye of an adult after extirpation of both lens and neural retina. Several tissues have been tested in organ culture for their ability to stimulate lens regeneration from the iris. Spleen or lung from adult newts induced depigmentation of the iris but no lens formation (YAMADA et al., 1973) while skeletal muscle also stimulated depigmentation and, rarely, the formation of a few lens fibers (CONNELLY et al., 1973). Lenses with fibers were obtained more frequently when the iris was cultured with newt spinal ganglion, newt pituitary or frog pituitary, reaching 60–90% in the last two situations. With no other tissue present, depigmentation was found more frequently when the iris and cornea were cultured together than in cultures of iris alone. In the former preparation, a few lens fibers differentiated in one out of 69 cases. Stimulation of lens regeneration by the pituitary gland *in vitro* would imply a regulatory role for pituitary hormones *in vivo*. There is some evidence in support of this. Following hypophysectomy, lens regeneration in lentectomized eyes was

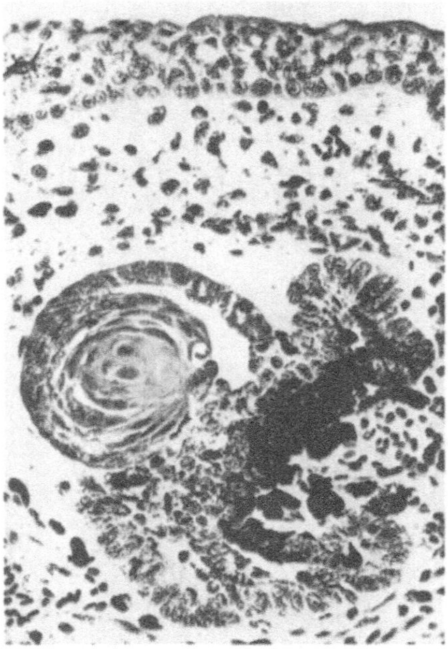

Fig. 24. A regenerating lens of stage IX, attached to pigmented and depigmented iris epithelium, which has developed 20 days after implantation of iris into a 9-day-old limb regenerate. × 150. (From Fig. 15, REYER et al., 1973, by permission of Academic Press)

retarded, both the rate of differentiation in terms of morphologic stages and the rate of growth in size (SCHOTTÉ and MURPHY, 1953; TAKANO, 1962; CONNELLY et al., 1973) or only the growth rate alone (STONE and STEINITZ, 1953b). Injections of ACTH counteracted the effect of hypophysectomy in TAKANO's experiments.

Therefore, the best answer to the second question is that the dorsal iris of newts of the family Salamandridae have an intrinsic capacity for lens formation which remains latent unless stimulated. This stimulus normally comes from the neural retina but the iris will also respond to factors present in the pituitary gland, regenerating limb, and spinal ganglion. The nature of the stimulus is unknown.

3. Role of the Lens

When embryonic lens development from the surface ectoderm was blocked, a lens differentiated later on, in older embryonic stages, from the dorsal iris of this lensless eye (REYER, 1950, 1954b). Following lentectomy, it also appears to be the absence of a lens that allows lens regeneration to begin. Several hypotheses have been proposed to explain how the presence of the lens blocks lens regeneration in the normal eye (discussed in REYER, 1954a, p. 13, and 1962a, pp. 236—237). The two most generally accepted are that the lens either produces a diffusible, inhibitory factor or else neutralizes the retinal factor. However, some recent work appears to indicate that the position of the lens with respect to the iris is also important.

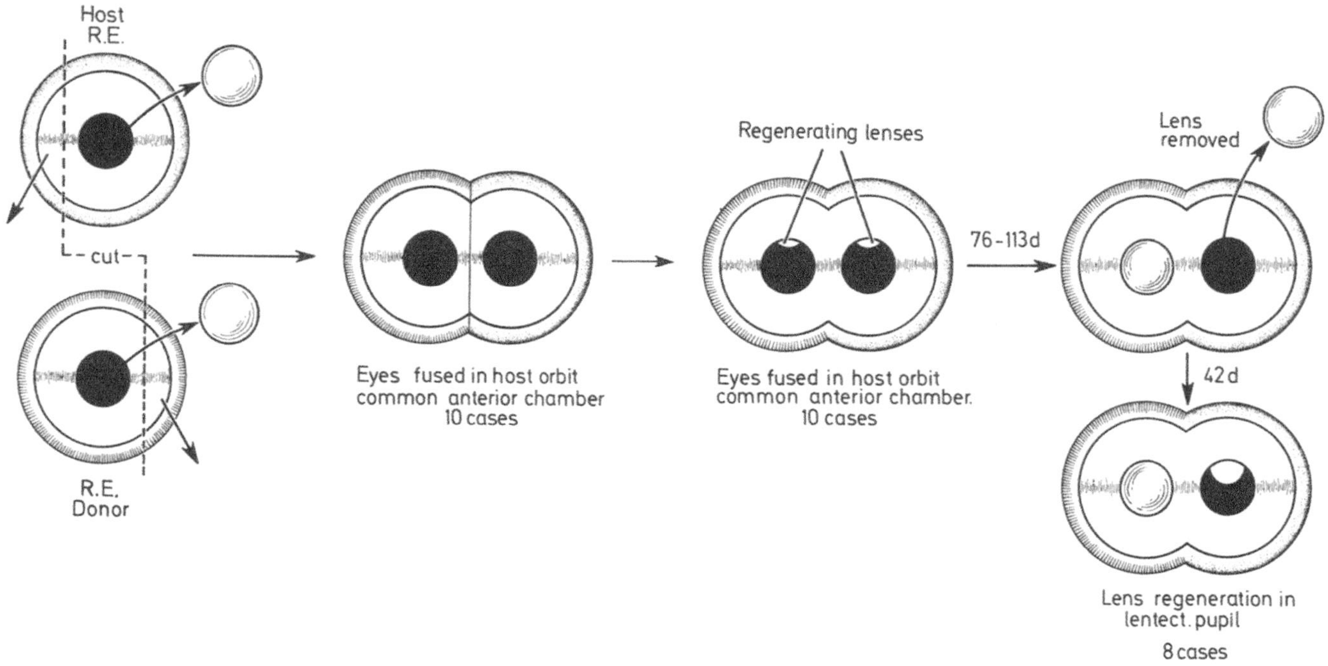

Fig. 25. Diagram illustrating operations on adult *N. viridescens* in which two eyes were fused to form a compound eye with extra neural retina, two pupils and a common anterior chamber. After two new lenses had regenerated from the two dorsal irises, one of these lenses was removed. A second new lens regenerated in this lentectomized pupil. (Drawn from description in Stone, 1963b)

If the lens is the source of a diffusible, inhibitory factor, this should be present in the aqueous humor bathing the lens. Attempts to demonstrate this have given conflicting results. STONE reported inhibited or retarded lens regeneration in some lentectomized eyes of adult *Notophthalmus viridescens* when small amounts of aqueous humor from eyes with intact lenses were injected daily with a glass micropipette into the anterior chamber of lentectomized eyes for 19–96 days. Before injection, an equal volume of aqueous humor was withdrawn from the host eye. The stages of regeneration were compared to those in eyes similarly injected with a saline solution (STONE and VULTEE, 1949; STONE, 1963b). However, GOSS (1961, 1964) repeated these experiments and failed to observe any slowing of lens regeneration. TAKANO et al. (1957) undertook similar experiments, using adult *Triturus pyrrhogaster*, and again failed to observe any retardation of regeneration for 29 days. Injection of lens extracts also had no effect.

In an attempt to remove the variables introduced by the trauma of continual injections while still exposing the iris to aqueous humor containing the postulated lens inhibitor, STONE (1963b, 1966a) produced experimental eyes with two pupils, two lenses, and a common anterior chamber or a single pupil and anterior chamber with two lenses. The procedure employed consisted in lentectomizing and fusing two eyes with large nasal and temporal wounds or else two dorsal one-half eyes in the orbit of a host animal. Following regeneration of the neural retina and two lenses in these eyes over a period of 76–154 days, one of the lenses was extirpated (Figs. 25, 26). In the first experiment, a lens regenerated from the dorsal iris in the lentectomized pupil in spite of aqueous humor bathing this iris and the lens in the adjacent pupil (8 out of 10 cases). In the second experiment, there was no lens regeneration in the single pupil in the presence of the remaining lens (5 cases). STONE (1966a) also produced eyes with two lenses by implanting a segment of dorsal iris into an eye at the time of lentectomy so that a lens would develop from both the host and graft iris. When one of these lenses was later removed, no lens regeneration was observed in the presence of the second lens. Conflicting results were obtained by L. WILLIAMS (1970a, b) and L. WILLIAMS and HIGGINBOTHAM (1975), in a similar experiment where part of the ventral iris of the host eye was removed before replacing it with a piece of dorsal iris. Only the dorsal lens regenerate, derived from the host, was extirpated 53–91 days after the original lens had been removed (Figs. 27, 28). In specimens fixed over a period of 9–40 days after the second lentectomy, a lens regenerated normally from the host dorsal iris in 35 out of 49 cases even though a second healthy lens was present in the same eye (Fig. 29). In those cases where lens regeneration was inhibited or retarded, the second lens was located close to the pupillary margin of the host dorsal iris.

The discrepancy in these results following removal of one of two lenses lying in a single pupil could have been due to the proximity of the remaining lens to the pupillary margin of the host dorsal iris (close in STONE's experiments, farther removed in WILLIAMS'). Evidence in support of this comes from experiments of EGUCHI (1961) where the lens was displaced in the pupil of a larval, cyclopean eye or, by means of a glass thread, in a normal adult eye. If a space was maintained between the reacting iris and displaced lens, then lens regeneration occurred from the iris even though the lens was present and in good condition. Lens regeneration from the dorsal iris in eyes where the lens was ventrally displaced by means of

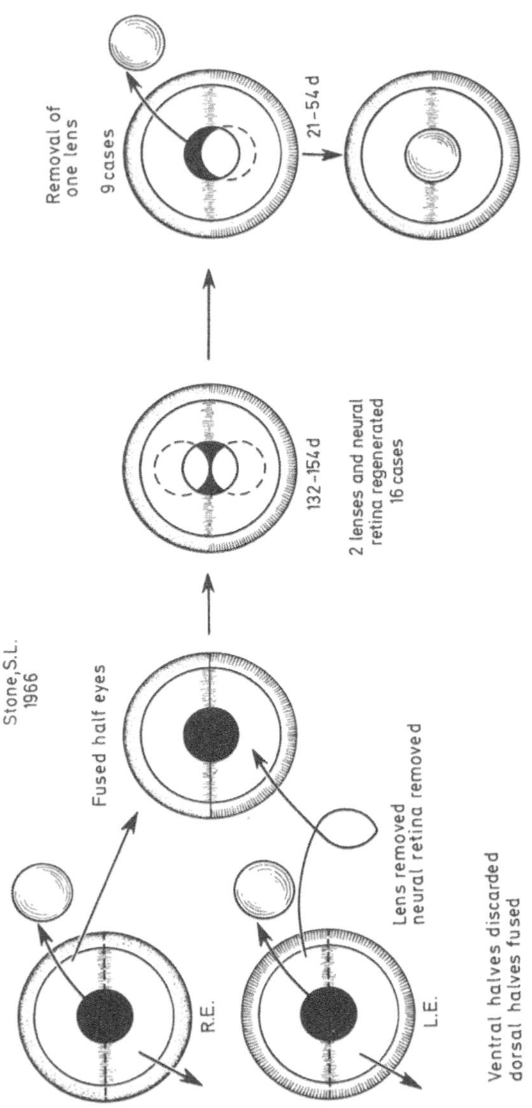

Fig. 26. Diagram illustrating operations in which two dorsal half-eyes were fused to form a compound eye with a normal amount of neural retina and a single pupil. Two lenses regenerated from two dorsal irises bounding the single pupil. When one of these lenses was removed, a new lens did not regenerate in the pupil already containing the other lens. (Drawn from description in STONE, 1966a)

a cellophane membrane dividing the eye cavity could also have been due to the spatial separation of lens from iris (STONE, 1966a). In a different experiment, lenses were observed to regenerate from pieces of iris lying within the vitreous body between the host lens and the neural retina (UNO, 1943; TAKANO, 1959; REYER, 1966b) (Fig. 30). However, the frequency of lens regeneration and the rate of growth in size of these lenses was less than that of lenses developing from iris implants into lentectomized eyes (REYER, 1966b).

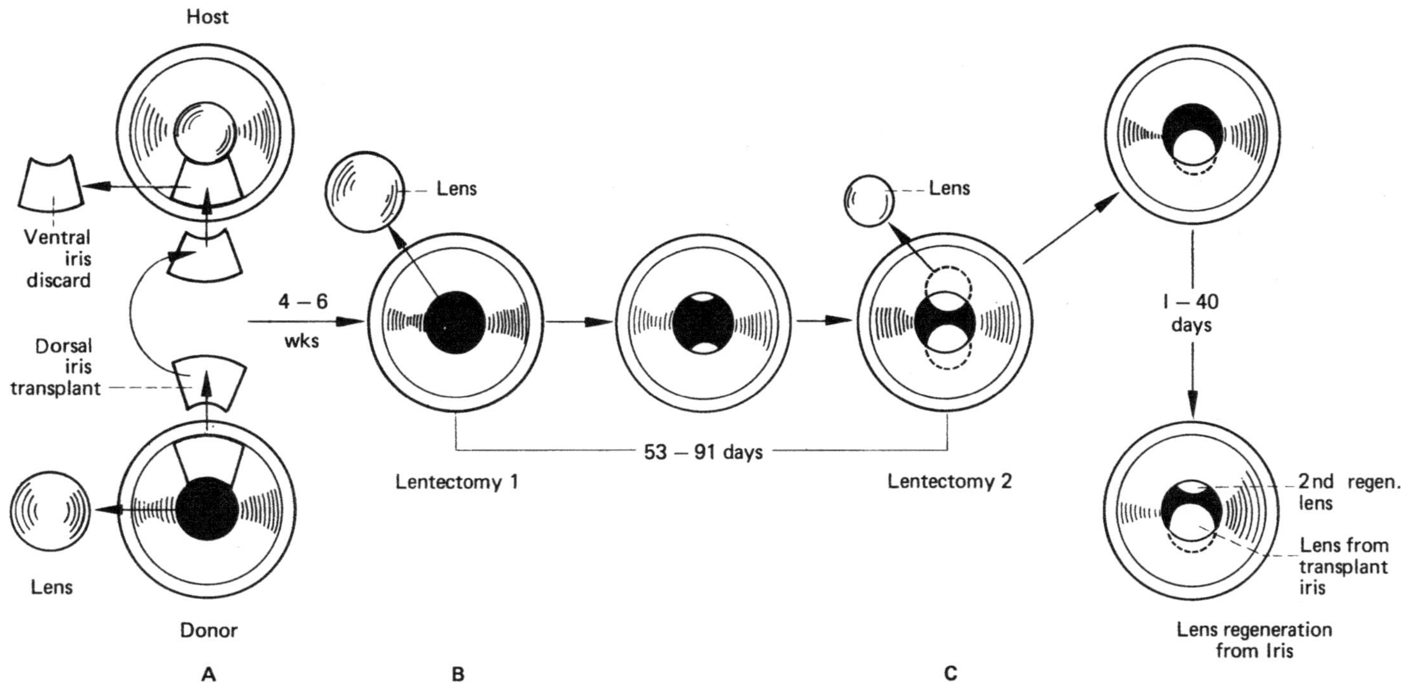

Fig. 27 A–C. Diagram illustrating operations in which (A) a sector of ventral iris was removed from a host eye and replaced by a sector of dorsal iris from a donor eye. After healing for 4–6 weeks, the host lens was removed (B). Two lenses regenerated from the two dorsal irises bounding the pupil. Lens derived from host dorsal iris was removed (C) 53–91 days after lentectomy 1, and the newts were fixed over a period of 40 days. (Modified from Fig. 1, L. WILLIAMS and HIGGINBOTHAM, 1975, by permission of authors and Wistar Institute)

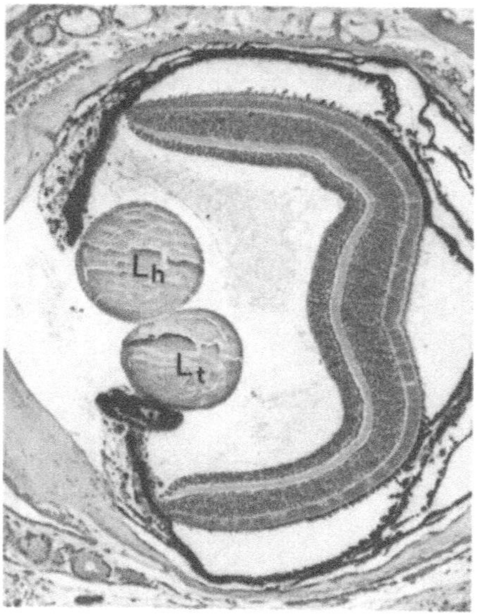

Fig. 28

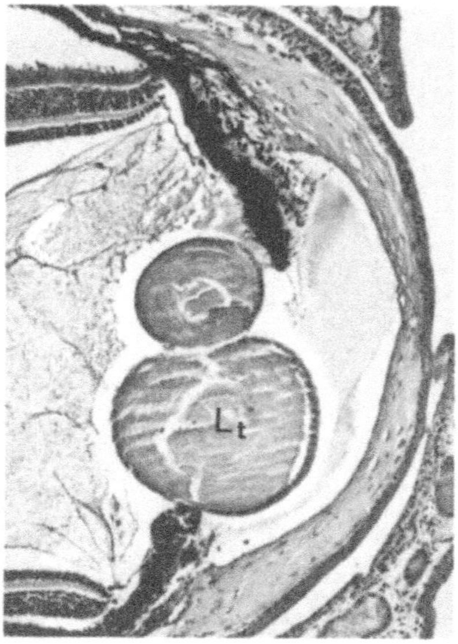

Fig. 29

Fig. 28. Section through an eye in which a sector of dorsal iris was transplanted in place of a sector of ventral iris, followed later by removal of the lens. Two lenses of stage XII and of approximately equal size have regenerated from the two dorsal irises, 70 days after lentectomy 1. L_h = lens from host dorsal iris, L_t = lens from transplanted dorsal iris. In experiments of this series, L_h would now be removed in lentectomy 2. × 45. (From Fig. 2, L. WILLIAMS and HIGGINBOTHAM, 1975, by permission of authors and Wistar Institute)

Fig. 29. A second lens of stage XII has regenerated from host dorsal iris, 40 days following removal of first lens regenerate (lentectomy 2), although a large lens, derived from the transplant iris, is present in the same pupil. × 107. (From Fig. 11, L. WILLIAMS and HIGGINBOTHAM, 1975, by permission of authors and Wistar Institute)

The only attempt to isolate an inhibitory substance from the lens was that of SMITH (1965) who fractionated macerated lenses by means of starch gel electrophoresis into seven bands of protein. Starch gel plugs, taken from each of these bands, were inserted into the pupils of lentectomized eyes. Some retardation of lens regeneration was observed 15 days after operation, with plugs from two of the seven bands and little or no retardation with the others or with plugs of the starch gel alone.

From the data reviewed here, it is clear that the manner in which the lens blocks lens regeneration is still not well understood. Additional work should be directed to studying the effect of the lens and lens constituents on lens regeneration in organ culture. Furthermore, a clarification of the fine structure of iris, suspensory ligament, lens capsule, and vitreous body in newts before and after lentectomy should be undertaken in order to determine the nature of any structural barriers intervening between neural retina and iris that might be disrupted by lentectomy.

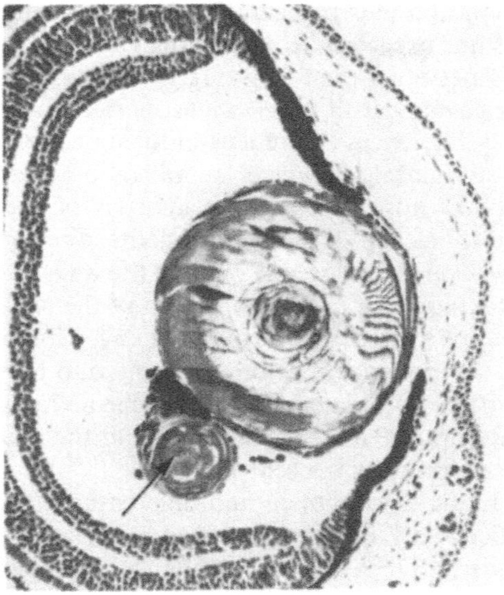

Fig. 30. A stage X lens *(arrow)* has regenerated from dorsal iris implanted into vitreous chamber following removal and replacement of host lens in larva of *N. viridescens*, 25 days after operation. × 88

Although the newt eye has a structural and functional complexity equal to that of the mammalian eye, it has an amazing capacity for the replacement of the neural retina, optic nerve, and lens with subsequent restoration of visual function. This shows that regeneration is not necessarily limited to the restoration of systems with a simple organization.

E. Conclusions and Discussion

As can be seen from the research surveyed in this chapter, the amphibian eye has proved to be an excellent subject for several different kinds of experimental studies in developmental biology. Its unique properties include easy accessibility to microsurgical intervention in embryonic stages, a highly developed capacity for regeneration, the presence of a simple morphology and very specific biochemical properties in the lens and a complicated morphology and intricate neuronal connections in the neural retina. With respect to structure and function, it has the same complexity and is subject to the same optical requirements as a human eye. Investigations on development and regeneration in amphibian eyes have contributed greatly to our understanding of embryonic induction, control of morphogenesis, cytodifferentiation, metaplasia, and neuronal specificity.

A foundation has now been laid on which future research can build and some likely paths of attack are discernible. At the level of tissue interactions, the mechanisms of lens induction and lens regeneration have been quite well established. There are, however, some aspects where additional study is needed. Among these are the capacity of tissues other than neural retina to stimulate lens regeneration from the iris, the precise manner in which a lens inhibits lens regeneration as well as the similarities and differences in the lens-inducing action of embryonic optic vesicle and the lens-stimulating action of neural retina. Nevertheless, the most promising approach now appears to be the analysis of neural retina and lens differentiation at the molecular level following the direction to which Tuneo Yamada and his co-workers have already pointed the way. The ultimate goal is to describe the events taking place in differentiation of the lens, of photoreceptors, and of neuronal synaptic patterns in terms of changes in specific molecules located within cells, as components of cell membranes and between cells. As in all analyses of differentiation, the role of changes in gene action and therefore, of the control of DNA transcription, RNA translation, and the final assembly of polypeptides to form specific proteins is a central theme.

Taking lens and retina development and differentiation as specific examples, a greatly oversimplified, working hypothesis can be suggested. It can be assumed that, in gastrula ectoderm, DNA cistrons coding for the lens crystallins (C) or for the enzymes and other specific proteins necessary for photoreception (P) or for the enzymes involved in melanogenesis (M) either are not being transcribed or are being transcribed at a very slow rate. After the optic vesicle has grown out to contact the surface ectoderm and then invaginated to form the optic cup, transcription of the (P) cistrons in the prospective photoreceptors increases and there is repression of the (C) and (M) cistrons in the developing neural retina. In a similar fashion, transcription of the (M) cistrons for melanogenesis is finally restricted to the prospective pigmented epithelium of the retina and iris and to the choroid while transcription of the (C) cistrons for the lens crystallins is limited to the lens placode and vesicle. This differential gene activity would become established gradually as the parts of the eye become determined and would continue to characterize these parts after development has been completed.

In those few species capable of regenerating neural retina or lens by metaplasia of the pigmented epithelium, the repression of the cistrons coding for photoreception and for lens crystallins in the pigmented epithelium is not permanent. Following an appropriate stimulus, the cistrons for melanogenesis are repressed, melanin synthesis stops, and the melanin present is shed and phagocytized. The cistrons coding either for lens crystallins or else for components characteristic of the neural retina such as the photoreceptor opsins and the enzyme, glutamine synthetase, are derepressed and one differentiated tissue (iris or pigmented retinal epithelium) changes into another (lens or neural retina). This is very unusual behavior for a vertebrate tissue where the differentiated state is usually stable. As an exception to the rule, these tissue metaplasias within the amphibian eye provide a unique experimental system for the investigation of gene action in differentiation. It is likely that controls also act later on in the path from gene to functional protein at the level of the translation of messenger RNA into polypeptides by such mechanisms as degradation, inactivation or stabilization of mRNA. The assembly of polypeptides into proteins and the activation or inactivation of enzymes already present could also be regulated.

There is some evidence supporting this hypothesis from work on the amphibian eye. In lens regeneration from the iris, there is amplification of the DNA cistrons coding for ribosomal RNA and a rapid increase in rRNA synthesis during the first few days after lentectomy. DNA synthesis and cell replication follow closely, the melanosomes are shed, polyribosomes appear in the cytoplasm, and then crystallins are synthesized in the differentiating lens fibers. All of these events point toward a renewed transcription of the cistrons coding for the RNA templates which direct the synthesis of the different subunits of the alpha, beta, and gamma crystallins as well as those coding for ribosomal RNA since both are needed in the production of these specific structural proteins. However, specific messenger RNAs for any of these crystallins have not yet been isolated from these small, regenerating lenses.

Therefore, for further evidence, it is necessary to consider the results of studies on other species. Investigations on calf and chick lenses have provided some information on gene action in lens differentiation. By zonal centrifugation through a sucrose density gradient, two RNAs which sedimented at 14S and 10S were isolated from SDS-dissociated calf lens polysomes. When these RNAs were added to a cell-free system from Krebs II ascites cells or rabbit reticulocytes containing all of the components necessary for protein synthesis, including amino acid(s) labeled with ^{35}S, ^{3}H, or ^{14}C, then following incubation, labeled alpha crystallins were isolated by SDS acrylamide and alkaline urea polyacrylamide gel electrophoresis from these cultures. Similarly, labeled alpha crystallin (αA_2) was synthesized by oocytes of *Xenopus laevis* injected with the 14S RNA and cultured in a medium containing ^{35}S-methionine. These RNAs were, therefore, considered to be messenger RNAs, the 14S mRNA acting as the template for the αA_2 crystallin subunit and the 10S mRNA as the template for the αB crystallin subunits (MATTHEWS et al., 1972; BERNS et al., 1972a, b; CHEN et al., 1973). Similar RNA fractions were also isolated by affinity chromatography on a poly(rU)-agarose or oligo(dT)-cellulose column (CHEN et al., 1974). These mRNAs were found to have reverse transcriptase activity so that complementary DNA was synthesized in a reaction mixture of the four nucleotides (dATP, dCTP, dGTP, ^{3}H-dTTP), oligo (dT) primer, and avian myeloblastosis virus (AMV) RNA-dependent DNA polymerase to which the mRNA templates were added. This cDNA hybridized with the mRNA template originally added to the culture (CHEN et al., 1973, 1974). By employing methods previously worked out for the isolation of mouse globin mRNA, including zonal centrifugation of dissociated lens polysomes and polyacrylamide gel electrophoresis, WILLIAMSON et al. (1972) isolated two RNA fractions with sedimentation coefficients of 9S and 15S from lenses of 1-day-old posthatched chicks. When these mRNAs were incubated with lysates of duck reticulocytes or Landschütz ascites cells and ^{3}H-leucine, labeled proteins were synthesized which gave the typical banding pattern of crystallin subunits after electrophoresis on polyacrylamide gel with 8M urea. ZELENKA and PIATIGORSKY (1974) also prepared a mRNA fraction from homogenates of lens fibers from 15-day-old chick embryos by chromatographing crude RNA, obtained by phenol-chloroform extraction, on an oligo (dT)-cellulose column. When this mRNA was added to cell-free systems derived from Krebs II ascites tumor cells or rabbit reticulocytes and containing ^{3}H-leucine, a labeled protein was synthesized which co-migrated with ^{35}S-labeled delta crystallin during electrophoresis on SDS-acrylamide-agarose gel. Digestion of this ^{3}H-protein and ^{14}C-delta crystallin with trypsin yielded similar peptides.

Very extensive studies have been made on the crystallins in the chick embryo and posthatched chick. These have shown that delta (FISC), beta, and alpha crystallins appear in that order in the developing lens of the chick embryo (from 50 hrs to $4^1/_2$ days of incubation) and are present in both lens epithelium and lens fibers (ZWANN, 1968, 1974; ZWANN and IKEDA, 1968; BRAHMA and VAN DOORENMAALEN, 1971; TRUMAN et al., 1972; KATOH and YOSHIDA, 1973). In older lenses (7–18 days of incubation), the lens fibers synthesize primarily the delta crystallins and do so at a constant rate during ontogeny while the lens epithelium synthesizes alpha and beta crystallins at a steadily increasing rate and delta crystallins at a decreasing rate as the embryo grows older (YOSHIDA and KATOH, 1971b). The subunit composition of these three classes of crystallins and their immunologic relationships have been analyzed in detail (CLAYTON, 1969, 1970, 1974; YOSHIDA and KATOH, 1971a; TRUMAN and CLAYTON, 1974; CLAYTON and TRUMAN, 1974). The beta crystallins are the most complex and have been shown to be a family of heteropolymers composed of different combinations of subunits. There is an increase in the number of different beta crystallin molecules during ontogeny (TRUMAN et al., 1972). Delta crystallin is believed to be a tetramer of four subunits of the same size each having a molecular weight of 45000–50000 daltons giving a total molecular weight of 200000 for native delta crystallin (PIATIGORSKY et al., 1974). The crystallins of the calf lens (α, β, γ) have also been thoroughly investigated and these results are reviewed by PAPACONSTANTINOU (1967), PALMER and PAPACONSTANTINOU (1969), DELCOUR and PAPACONSTANTINOU (1972), CLAYTON (1970, 1974), SPECTOR (1969), and WALEY (1969).

Some evidence has been obtained on the role of crystallin messenger RNAs in chick lens development. YOSHIDA and KATOH (1972) studied the incorporation of labeled amino acids into crystallins during the culture of lenses from 10–16-day-old embryos with and without actinomycin-D in the culture medium. The nonreplicating lens fibers were separated by dissection from the replicating lens epithelial cells and their crystallin content analyzed separately. Crystallin synthesis was inhibited by actinomycin-D in lens epithelium from embryos $11^2/_3$ days old or younger but occurred normally in lens epithelium from $12^2/_3$-day or older embryos and in lens fibers from embryos at all ages tested. The authors believe that this actinomycin-D-insensitive, crystallin synthesis occurs on stable mRNA templates and can therefore take place even though RNA synthesis has been blocked. CLAYTON et al. (1972, 1974) studied crystallin mRNA turnover without the use of inhibitors. This involved culturing lenses from 8-day-old embryos or 1-day-old posthatched chicks for 3 hrs with ^{14}C-amino acids or with ^{3}H-uridine followed by 5–13 hrs culture in cold medium, isolation of the labeled polysomes, and their fractionation by treatment with antisera to total crystallins, to specific crystallins, or to their subunits. This fractionation was based on the immunologic specificity of the nascent polypeptide chains. The half-life of mRNA for the different crystallins was estimated. For total crystallins, this was found to increase from approximately $3^1/_2$ hrs in 8-day-old embryonic lenses to 20 hrs in lenses from 1-day-old posthatched chicks. By using these methods together with saturation labeling of all RNA in the embryo following *in vivo* injection of ^{32}P-orthophosphate at 3 days of incubation, it was possible to estimate, by the ratio of recovered counts $^{3}H/^{32}P$ and $^{14}C/^{32}P$, the amount of RNA or protein synthesized, during culture of the lens, in a population of polysomes corresponding to a particular crystallin antibody. In 8- and 15-day-old embryos, the alpha crystallin mRNAs were turning over more rapidly (were less stable) than the anodal, beta crystallin mRNAs while turnover of total crystallin mRNA was occurring at an intermediate rate. As lens fibers age, there is a gradual decrease and final cessation of transcriptional activity with concurrent degradation of ribosomal RNA and loss of ribosomes (ZAPISEK and PAPACONSTANTINOU, 1973). The DNA content of the nuclei of lens fibers also gradually decreases, accompanied by single strand scissions in the DNA molecules (MODAK, 1972).

During the differentiation of the bovine lens from fetus to calf and adult, PAPACONSTANTINOU and his co-workers found that synthesis of gamma crystallins was restricted to the elongating lens fibers while synthesis of alpha and beta crystallins occurred in both lens epithelium and lens fibers. In the epithelial cells, incorporation of ^{14}C-amino acids into the crystallins was inhibited by actinomycin-D while it was stimulated in the lens fibers in the presence of this antibiotic. Therefore, there is a change from short-lived to long-lived messenger RNA templates during lens fiber differentiation (PAPACONSTANTINOU, 1967). Alpha crystallins were isolated by chromatography on DEAE-cellulose or by sucrose density gradient centrifugation, dissociated with 7M urea and then fractionated into subunits by further DEAE-cellulose chromatography or acrylamide gel electrophoresis. Four subunits were found in the calf and adult lens and only three in the fetal lens. During culture of the adult or calf lens with ^{14}C-amino acids or ^{3}H-leucine, isotope was incorporated only into the latter three subunits (αA_2, αB_2, and αB_1). It would appear that the αA_1 subunit is not synthesized on an mRNA template but is formed post-translationally from subunit αA_2 during the aging of the lens fibers (PALMER and PAPACONSTANTINOU, 1969; DELCOUR and PAPACONSTANTINOU, 1972).

Evidence for the regulation of gene action during differentiation of the neural retina comes from the studies of MOSCONA and his collaborators on the chick embryo (MOSCONA, 1971, 1972; MOSCONA et al., 1972; GARFIELD and MOSCONA, 1974). These deal with the enzyme, glutamine synthetase, which catalyzes the combination of NH_3 with glutamic acid to form glutamine. In the neural retina, the activity of this enzyme increases rapidly by a factor of 100 between the 16th day and hatching. This increased activity is due to enzyme synthesis and probably is induced *in vivo* by an adrenal, corticosteroid hormone. In 9–12-day-old embryos, precocious synthesis of this enzyme occurred both in cultures of neural retina and *in vivo*, when 11β-hydroxycorticosteroids such as corticosterone, aldosterone, and hydrocortisone were added to the culture medium or injected into the embryo. This increase in the specific activity of the enzyme during 24 hrs of culture required the continued presence of the inducer, protein synthesis, and RNA synthesis. It was blocked by cycloheximide or actinomycin-D if these were present throughout the culture period. Induction of glutamine synthetase was also found to be dependent on the histotypic association of the retinal cells as determined in cell dissociation and reaggregation experiments and was blocked by prior treatment with BrdU (5-bromodeoxyuridine) which interfered with the morphologic differentiation of the neural retina (MORRIS and MOSCONA, 1971; MORRIS, 1973). However, concurrent cell replication and DNA synthesis was not necessary during the time of enzyme induction.

Estimates of the actual amounts of enzyme protein synthesized during culture of retinas for 15 minutes with ^{14}C-amino acids were obtained by precipitating this protein with antiserum to glutamine synthetase and then measuring the radioactivity in the precipitate. In the absence of induction, there was a continuing low level of enzyme synthesis. When hydrocortisone was added, then both the amount and the activity of the enzyme increased rapidly during the entire 24 hrs of culture. Treatment of cultures with the inducer plus cycloheximide and actinomycin-D in both high and low doses in different combinations, for varying lengths of time, and in different sequences gave some insight into the control mechanisms involved in induction of this enzyme. Some of the more important results were: (1) Addition of cycloheximide to the induced culture at different times after the beginning of the culture period stopped the increase in specific activity of glutamine synthetase. (2) Similar addition of a high dose of actinomycin-D (10 μg/ml) at different times slowed the rate of increase less and less until it had essentially no effect when added 4 or more hrs after initiation of culture. (3) When a low dose of actinomycin-D (0.05 μg/ml) was added after 4 hrs of culture, the increase in enzyme activity was arrested. (4) If the culture was exposed to cycloheximide for the initial 4 hrs, followed by its removal and replacement by the high dose of actinomycin-D, then after a 4-hrs-delay, the rate of increase in specific activity paralleled the induced control without inhibitors. (5) Following withdrawal of the inducer after 6 hrs of culture, the normal increase in enzyme specific activity was maintained by the addition, at that time, of the high dose of actinomycin-D. (6) In the absence of the inductor, glutamine synthetase activity was increased above basal levels by 5 hrs culture with cycloheximide, removal of the cycloheximide, and subsequent culture with the high dose of actinomycin-D.

MOSCONA has proposed a complex model of controls at the levels of transcription, translation, and post-translation to explain these results. In this model, there are five genetic elements: a regulator gene (R) whose product (r) regulates transcription of the other genes, a structural gene for glutamine synthetase, suppressor, and desuppressor genes whose products act on RNA templates at a post-transcriptional level, and a gene whose product acts posttranslationally in enzyme degradation. Transcription of the glutamine synthetase cistron is hindered by the repressor (r) but there is some leakage of labile mRNA templates resulting in a low basal level of enzyme synthesis in uninduced cells. The steroid inducer binds the repressor (r) thereby allowing transcription of the glutamine synthetase cistron and both the desuppressor and the suppressor cistrons. This results in the formation, accumulation, and translation of mRNA templates for glutamine synthetase since the suppressor is now inactivated by the desuppressor. High doses of actinomycin-D block all transcription so that any previously accumulated templates for glutamine synthetase continue to be translated in the absence of the suppressor, and degradation of the enzyme stops. Low doses of actinomycin-D are assumed to block selectively transcription of the desuppressor gene so that suppressors accumulate and continuously inactivate those mRNA templates for the enzyme which are still being transcribed. Recently, MOSCONA and WIENS (1975) have found that proflavine inhibits the induction of glutamine synthetase and is believed to act directly on the DNA to prevent transcription of functional templates for synthesis of this enzyme.

Although limited by the much smaller amounts of tissue available, some of the techniques used in these studies on the control of differentiation in the chick lens and neural retina should be applicable to the investigation of control mechanisms in the development and regeneration of the lens and neural retina in the amphibian eye. Extension of the tissue and organ culture techniques now available for amphibia to include cloned, pigmented epithelial cells from iris and retina would also be valuable. The elucidation of these control systems for eye development and differentiation is the challenge for the future and here these studies join with the many others seeking to understand the origin and maintenance of the diverse tissues of the animal body during its life history from zygote to senescence.

F. Acknowledgments

The research reviewed in this account, which was performed by the author, was supported by Research Grant 5 RO1 EY00196 from the National Eye Institute of the National Institutes of Health.

The author wishes to thank Drs. JAMES DUMONT, R. KEVIN HUNT, ANTONE JACOBSON, MARCUS JACOBSON, J. RICHARD KEEFE, DAVID MCDEVITT, SVEN ERIK NILSSON, JERRY SQUIRES, LEAH WILLIAMS, and TUNEO YAMADA for permission to reproduce illustrations from their work; the author also wishes to acknowledge permissions from Academic Press Inc. to reproduce copyrighted material from the *Journal of Ultrastructure Research* and *Developmental Biology*, from the Wistar Institute of Anatomy and Biology to reproduce material from the *Journal of Experimental Zoology*, from the American Association for the Advancement of Science to reproduce material from *Science*, and from F. K. Schattauer-Verlag to reproduce material from *The Structure of the Eye II. Symposium.* The technical assistance of Mrs. ALICE BARNETT and Mrs. DOROTHY HERITAGE is also much appreciated.

G. Appendix Added in Proof

The review of the literature, included in this chapter, extends through July, 1975. Since submission of the manuscript, numerous papers have been published dealing with various aspects of the development and regeneration of the amphibian eye. Some of the more significant are summarized here and bibliographic references cited.

HAYES (1976) studied the distribution of gap junctions in the developing optic vesicle and eyecup of *Xenopus laevis*. These did not begin to disappear from the central region of the retina until stage 35, well after retinal polarization at stage 32. In developing *Rana pipiens* tadpoles, CURRIE and COWAN (1975) investigated the sequential innervation of the optic tectum by retinal axons from the contralateral eye, following intraocular injection of ^{3}H-proline at different stages. The first labeling was observed in the rostrolateral tectum at stage 22 and then spread sequentially in a caudodorsomedial direction paralleling the gradient of cell proliferation and cytodifferentiation. These results are similar to those previously described for *Xenopus*.

In the continuing studies on neuronal specificity in the development of the retino-tectal projections in *Xenopus laevis*, HUNT and FRANK (1975) combined left temporal half eyes (NT axis normal, DV axis inverted) with right nasal half eyes in the right orbit at stage 31—32 following axial specification. The two half eyes fused and either remained permanently combined or else the host, nasal half eye was removed later on. This fusion of half eyes resulted in a reprogramming of the left temporal half eye by its association with the right nasal half eye so that the NT axis became inverted and the DV axis normal, giving duplicated, mirror symmetrical, visuo-tectal projections from the compound eye. When the nasal half eye was removed after 16 or 32 hours fusion, it was found that the NT axis of the temporal half eye was modified first, followed by the DV axis.

Several investigators have studied this question of neuronal specificity by surgical modification of the optic tectum. In embryonic stages (21–24, 37) of *Xenopus*, before innervation of the tectum by retinal nerve fibers at stage 45, CHUNG and COOKE (1975) excised the prospective midbrain region, rotated it 180° and then replaced it. Immediately after metamorphosis, the visuo-tectal projections were studied. Due to small variations in the level of the incisions through the brain, animals developed single or double tecta associated with a single or duplicated diencephalon. In all cases, the functionally rostral margin(s) of the optic tectum or tecta (receiving a projection from nasal visual field) lay next to the diencephalon. These results show that the polarity of the optic tectum can be regulated up to stage 37, 24 hours before innervation, and that the diencephalon appears to play a role in fixing this polarity. LEVINE and JACOBSON (1974) tested the stability of the locus specificities in the optic tectum in young, postmetamorphic *Xenopus* by rotating the dorsal part of the optic tectum 180° or 90°. In the majority of cases, the visuo-tectal projections were correspondingly rotated when mapped after a recovery period of 1–8 months. However, a few normal projection maps were found. Two types of result were also obtained in adult *Rana catesbeiana*, (JACOBSON and LEVINE, 1975a, b) when part of one optic tectum was excised or non-corresponding parts of the right and left tecta were exchanged. After the former procedure, a permanent scotoma in the visual field resulted in some cases while, in others, there was reconstitution of normal visuo-tectal projections. Following the latter procedure, either a combination of scotoma plus reduplication of visuo-tectal projections was observed or normal visuo-tectal projections were restored. A topic not considered in this chapter was binocular vision and the intertectal neuronal connections involved (KEATING et al., 1975; KEATING and FELDMAN, 1975). Recent reviews of the whole field have been published by KEATING (1976) and MEYER and SPERRY (1976).

By means of scanning electron microscopy, CONNELLY (1977b) observed that one third of the dorsal half of the iris thickens to form the lens vesicle. HIGGINBOTHAM (1976, 1977) demonstrated that the normal zonula in the adult newt is a meshwork of fine fibrils closely associated with the anterior surface of the vitreous body. A stronger bundle extends as a fan-shaped band mid-ventrally from the *protractor lentis* muscle to the lens. After lentectomy, it was not until late stages of regeneration (40 to 80 days) that these fibrils were reconstituted to connect the ciliary retina with the new lens. By means of a statistical, iterative estimation technique, YAMADA et al. (1975) calculated the cell cycle parameters S, G_2, G_1, and T for beginning lens regenerates, 6 days after lentectomy, from data on labeled mitotic figures in eyes fixed at 4–8 hour intervals up to 112 hours after injection of ^{3}H-thymidine. Values of 27, $7^1/_2$, 11 and 46 hrs were calculated for the lens-forming cells and $40^1/_2$, 8, 30 and $78^1/_2$ hrs for non lens-forming iris epithelial cells. These are more reliable estimates of G_1 and T than those of ZALIK and YAMADA (1967). On the basis of values for S and G_2 alone, it would appear that cell-cycle parameters become shorter as differentiation proceeds from 6 to 15 days after lentectomy. The prospective primary lens fiber cells probably go through 4 cell cycles from the initiation of mitosis on day 4 to the beginning of lens fiber differentiation on day 10.

REESE et al. (1976), REESE (1977) have established a cell line, derived from the iris epithelium of adult *Notophthalmus viridescens*, which has been maintained in tissue culture for 10 years. Initially, there was loss of pigment which failed to reappear. The cells also became heteroploid. Lentoid bodies did not develop in these cultures. Studies on RNA synthesis using ^{14}C-uridine, ^{3}H-uridine and (^{3}H-methyl) methionine demonstrated incorporation of uridine into rRNA (28S, 18S) after long-term labeling. A larger rRNA precursor (40S) was isolated after 30–90 min incorporation of (^{3}H-methyl methionine). ORTIZ and YAMADA (1975) have continued their studies on induction of the stellate configuration in cultures of iris epithelial cells by adenosine and cAMP. They demonstrated increased activity when adenosine was combined with theophylline or with cAMP and suggested that intracellular cAMP may be the active molecule. Injection of prostaglandins (PGE_1 or $PGF_{1\alpha}$) or of the prostaglandin inhibitor (indomethacin) into lentectomized newts had no effect on lens regeneration from the iris (CONNELLY, 1977a). POWELL and SEGIL (1976) implanted whole newt pituitary glands into the anterior chamber next to the dorsal iris of adult newts with intact lens. In 16 out of 17 cases, lens regenerates of various stages, including some with lens fibers, had developed from the host dorsal iris 9–20 days after implantation even though the host lens was present. IDOYAGA-VARGAS et al. (1976) and REYER (1977a, b) have recently published the results of experiments already summarized in this chapter.

References

Achazi, R., Yamada, T.: Tyrosinase activity in the Wolffian lens regenerating system. Develop. Biol. **27**, 295—306 (1972).

Adelmann, H. B.: Experimental studies on the development of the eye. I. The effect of the removal of median and lateral areas of the anterior end of the urodelan neural plate on the development of the eyes. (Triton taeniatus and Amblystoma punctatum). J. exp. Zool. **54**, 249—290 (1929a).

Adelmann, H. B.: Experimental studies on the development of the eye. II. The eye-forming potencies of the median portions of the urodelan neural plate. (Triton taeniatus and Amblystoma punctatum). J. exp. Zool. **54**, 291—317 (1929b).

Adelmann, H. B.: Experimental studies on the development of the eye. III. The effect of the substrate (‚Unterlagerung') on the heterotopic development of median and lateral strips of the anterior end of the neural plate of Amblystoma. J. exp. Zool. **57**, 223—281 (1930).

Adelmann, H. B.: The development of the prechordal plate and mesoderm of Amblystoma punctatum. J. Morph. **54**, 1—67 (1932).

Adelmann, H. B.: A study of cyclopia in Amblystoma punctatum with special reference to the mesoderm. J. exp. Zool. **67**, 217—282 (1934).

Adelmann, H. B.: The problem of cyclopia. Part I. Quart. Rev. Biol. **11**, 161—182; Part II. Quart. Rev. Biol. **11**, 284—304 (1936).

Adelmann, H. B.: Experimental studies on the development of the eye. IV. The effect of the partial and complete excision of the prechordal substrate on the development of the eyes of Amblystoma punctatum. J. exp. Zool. **75**, 199—237 (1937).

Amano, U., Sato, J.: Über die xenoplastische Implantation der larvalen Iris des Triturus pyrrhogaster (Boie) in das entlinste Auge der Larven des Hynobius nebulosus Schlegel. Jap. J. med. Sci. (I. Anat.) **8**, 75—81 (1940).

Amatatu, H., Fujita, T.: Über die Beziehungen der Wirkung des Retinafaktors zur Wolffschen Linsenregeneration mit der Zeitdauer nach der Linsenentfernung. Acta med. nagasak. **2**, 143—149 (1941).

Balinsky, B. I.: On the eye cup-lens correlation in some South African amphibians. Experientia (Basel) **7**, 180—181 (1951).

Balinsky, B. I.: On the factors determining the size of the lens rudiment in amphibian embryos. J. exp. Zool. **135**, 255—299 (1957).

Becker, U.: Die Bedeutung des reagierenden Gewebes für die Ausbildung von freien Linsengebilden bei Triturus vulgaris. Wilhelm Roux' Arch. Entwickl.-Mech. Org. **151**, 188—217 (1959).

Becker, U.: Untersuchungen über die Abhängigkeit der Linsenbildung von der Wirtsregion bei Triturus vulgaris. Wilhelm Roux' Arch. Entwickl.-Mech. Org. **152**, 339—372 (1960).

Berman, N., Hunt, R. K.: Visual projections to the optic tecta in Xenopus after partial extirpation of the embryonic eye. J. comp. Neurol. **162**, 23—42 (1975).

Berns, A. J. M., Strous, G. J. A. M., Bloemendal, H.: Heterologous in vitro synthesis of lens α-crystallin polypeptide. Nature (New Biol.) **236**, 7—9 (1972a).

Berns, A. J. M., Van Kraaikamp, M., Bloemendal, H., Lane, C. D.: Calf crystallin synthesis in frog cells: the translation of lens-cell 14S RNA in oocytes. Proc. nat. Acad. Sci. (Wash.) **69**, 1606—1609 (1972b).

Bonnet, C.: Oeuvres d'histoire naturelle et de philosophie. Neuchatel: Chez S. Fauche, 1781, Vol. XI, pp. 175—179.

Borsuk, R. A., Popov, V. V.: Importance of light and darkness for lens regeneration in Triturus taeniatus. Dokl. Akad. Nauk SSSR **180**, 1500—1503 (281—284) (1968).

Boterenbrood, E. C.: Differentiation in small grafts of the median region of the presumptive prosencephalon. J. Embryol. exp. Morph. **23**, 751—759 (1970).

Brachet, A., Benoit, F.: Sur la régénération du cristallin chez les amphibiens urodèles. Bibliogr. anat. **7**, 277—295 (1899).

Brahma, S. K.: Studies on the process of lens induction in Xenopus laevis (Daudin). Wilhelm Roux' Arch. Entwickl.-Mech. Org. **151**, 181—187 (1959).

Brahma, S. K., Bours, J.: Thin layer isoelectric focusing of the soluble lens extracts from larval stages and adult Xenopus laevis. Exp. Eye Res. **13**, 309—314 (1972).

Brahma, S. K., van Doorenmaalen, W. J.: Immunofluorescence studies of chick lens FISC and α-crystallin antigens during lens morphogenesis and development. Ophthal. Res. **2**, 344—357 (1971).

Brahma, S. K., McDevitt, D. S.: Ontogeny and localization of gamma crystallins in Ambystoma mexicanum, Pleurodeles waltlii and Rana temporaria normal lens development. J. Cell Biol. **55** (2), 26A (1972a).

Brahma, S. K., McDevitt, D. S.: Ontogeny and localization of lens-specific proteins, the crystallins, in Xenopus laevis regenerating lens development. Ophthalmic Res. **3**, 29 (1972b).

Brahma, S. K., McDevitt, D. S.: Ontogeny and localization of gamma-crystallins in Rana temporaria, Ambystoma mexicanum and Pleurodeles waltlii normal lens development. Exp. Eye Res. **19**, 379—387 (1974).

Campbell, J. C.: An immuno-fluorescent study of lens regeneration in larval Xenopus laevis. J. Embryol. exp. Morph. **13**, 171—179 (1965).
Campbell, J. C., Jones, K. W.: The in vitro development of lens from cornea of larval Xenopus laevis. Develop. Biol. **17**, 1—15 (1968).
Carpenter, E.: The head pattern in Amblystoma studied by vital staining and transplantation methods. J. exp. Zool. **75**, 103—129 (1937).
Chen, J. H., Lavers, G. C., Spector, A.: Synthesis of complementary DNA from lens mRNA with RNA-dependent DNA polymerase. Biochem. Biophys. Res. Comm. **52**, 767—773 (1973).
Chen, J. H., Lavers, G. C., Spector, A., Schutz, G., Feigelson, P.: Translation and reverse transcription of lens mRNA isolated by chromatography. Exp. Eye Res. **18**, 189—199 (1974).
Chung, S. H., Bliss, T. V. P., Keating, M. J.: The synaptic organization of optic afferents in the amphibian tectum. Proc. roy. Soc. B **187**, 421—447 (1974a).
Chung, S. H., Cooke, J.: Polarity of structure and of ordered nerve connections in the developing amphibian brain. Nature **258**, 126—132 (1975).
Chung, S. H., Keating, M. J., Bliss, T. V. P.: Functional synaptic relations during the development of the retino-tectal projection in amphibians. Proc. roy. Soc. B **187**, 449—459 (1974b).
Clayton, R. M.: Properties of the crystallins of the chick in terms of their subunit composition. Exp. Eye Res. **8**, 326—339 (1969).
Clayton, R. M.: Problems of differentiation in the vertebrate lens. Curr. Top. Develop. Biol. **5**, 115—180 (1970).
Clayton, R. M.: Comparative aspects of lens proteins. In: The Eye, comparative physiology, H. Davson and L. T. Graham, Jr., ed. New York: Academic Press 1974, Vol. V, pp. 399—494.
Clayton, R. M., Truman, D. E. S.: The antigenic structure of chick β-crystallin subunits. Exp. Eye Res. **18**, 495—506 (1974).
Clayton, R. M., Truman, D. E. S., Campbell, J. C.: A method for direct assay of messenger RNA turnover for different crystallins in the chick lens. Cell Differentiation **1**, 25—35 (1972).
Clayton, R. M., Truman, D. E. S., Hannah, A. I.: RNA turnover and translational regulation of specific crystallin synthesis. Cell Differentiation **3**, 135—145 (1974).
Cohen, A. I.: New evidence supporting the linkage to extracellular space of outer segment saccules of frog cones but not rods. J. Cell Biol. **37**, 424—444 (1968).
Cohen, A. I.: Electron microscope observations on form changes in photoreceptor outer segments and their saccules in response to osmotic stress. J. Cell Biol. **48**, 547—565 (1971).
Collins, J. M.: Amplification of ribosomal ribonucleic acid cistrons in the regenerating lens of Triturus. Biochem. **11**, 1259—1263 (1972).
Collins, J. M.: Structural changes in deoxyribonucleic acid during early stages of lens regeneration in Triturus. J. biol. Chem. **249**, 1839—1847 (1974a).
Collins, J. M.: Template ability of activated DNA from the regenerating lens. Biochem. Biophys. Res. Comm. **57**, 359—364 (1974b).
Colucci, V. L.: Sulla rigenerazione parziale dell'occhio nei tritoni. Istogenesi e svilluppo. Studio sperimentale. Mem. R. Accad. Bologna Sez. Sci. nat. (Ser. 5), **1**, 167—203 (1891).
Connelly, T. G., Ortiz, J. R., Yamada, T.: Influence of the pituitary on Wolffian lens regeneration. Develop. Biol. **31**, 301—315 (1973).
Connelly, T. G.: Influence of indomethacin on lens regeneration in the newt Notophthalmus viridescens. Wilhelm Roux's Arch. Develop. Biol. **181**, 103—106 (1977a).
Connelly, T. G.: Scanning electron microscopy of lens regeneration in the newt. Anat. Rec. **187**, 555—556 (1977b).
Coulombre, A. J.: The Eye. In: Organogenesis, Dehaan, R. L., and Ursprung, H. (eds.) New York: Holt, Rinehart and Winston, 1965, pp. 219—251.
Coulombre, J. L.: Influence of mouse neural retina on regeneration of chick neural retina from chick embryonic pigmented epithelium. Nature (Lond.) **228**, 559—560 (1970).
Coulombre, J. L., Coulombre, A. J.: Regeneration of neural retina from the pigmented epithelium in the chick embryo. Develop. Biol. **12**, 79—92 (1965).
Cronly-Dillon, J. R.: Pattern of retinotectal connections after retinal regeneration. J. Neurophysiol. **31**, 410—418 (1968).
Cronly-Dillon, J. R., Levine, R.: Retinotectal projection in half eyes in the adult newt, Triturus vulgaris. J. Physiol. (Lond.) **216**, 62P—63P (1971).
Currie, J., Cowan, W. M.: The development of the retino-tectal projection in Rana pipiens. Develop. Biol. **46**, 103—119 (1975).

Dabagyan, N. V., Oganesyan, R. O.: An electrophysiological study of restoration of the function in the regenerating retina of adult newts. Ontogenesis, Akad. Nauk. SSSR **2**, 327—329 (1971) (In Russian)

Dabagyan, N. V., Sheresheva, E. L.: Regeneration of retina in larvae of anurous amphibia. Arh. Anat. Gistol. Embriol. **50**, 12—19 (1966) (In Russian, from Biol. Abstr.)

Dabagyan, N. V., Tret'yakova, L. I.: Regeneration of the retina in tailless amphibians during their metamorphosis. Dokl. Akad. Nauk SSSR **179**, 746—749 (160—163) (1967).

Dabagyan, N. V., Oganesyan, R. O., Popov, V. V.: Restoration of function in the regenerating retina of tailless amphibian larvae. Dokl. Akad. Nauk. SSSR **185**, 957—960 (257—261) (1969).

Dabagyan, N. V., Stroeva, O. G., Sheresheva, E. L.: Induction of the anlage of the iris by the lens in the regenerating retina of tadpoles of tailless amphibians. Dokl. Akad. Nauk. SSSR **167**, 933—955 (213—216) (1966).

Delcour, J., Papaconstantinou, J.: Biochemistry of bovine lens proteins IV. Synthesis and aggregation of α-crystallin subunits in differentiating lens cells. J. biol. Chem. **247**, 3289—3295 (1972).

De Robertis, E., Lasansky, A.: Ultrastructure and chemical organization of photoreceptors. In: The Structure of the Eye, Smelser, G. K. (ed.) pp. 29—49. New York: Academic Press 1961.

Detwiler, S. R., van Dyke, R. H.: Recent experiments on the differentiation of the labyrinth in Amblystoma. J. exp. Zool. **118**, 389—406 (1951).

Detwiler, S. R., van Dyke, R. H.: The induction of neural retina from the pigment epithelial layer of the eye. J. exp. Zool. **122**, 367—383 (1953).

Detwiler, S. R., van Dyke, R. H.: Further experimental observations on retinal inductions. J. exp. Zool. **126**, 135—156 (1954).

De Vincentiis, M., Magni, A., Materazzi, G.: Attività dipeptidasica delle metà dorsale e ventrale dell' iride nel corso della rigenerazione del cristallino in Triton cristatus. Atti Accad. Naz. Lincei Cl. Sci. Fis. Mat. Nat. Rend. (Ser. 8) **38**, 259—263 (1965).

Dickson, D. H., Hollenberg, M. J.: The fine structure of the pigment epithelium and photoreceptor cells of the newt, Triturus viridescens dorsalis (Rafinesque). J. Morph. **135**, 389—432 (1971).

Dixon, J. S., Cronly-Dillon, J. R.: The fine structure of the developing retina in Xenopus laevis. J. Embryol. exp. Morph. **28**, 659—666 (1972).

Donaldson, D. J.: Effects of X-irradiation on lens regeneration in adult Triturus viridescens. Anat. Rec. **172**, 45—56 (1972).

Donaldson, D. J., Chan, F.: The influence of temperature on lens regeneration in the adult newt, Triturus viridescens. Growth **37**, 69—76 (1973).

Dowling, J. E.: Synaptic organization of the frog retina: an electron microscopic analysis comparing the retinas of frogs and primates. Proc. roy. Soc. B **170**, 205—228 (1968).

Dowling, J. E.: Organization of vertebrate retinas. Invest. Ophthalmol. **9**, 655—680 (1970).

Dowling, J. E., Werblin, F. S.: Organization of retina of the mudpuppy, Necturus maculosus. I. Synaptic structure. J. Neurophysiol. **32**, 315—338 (1969).

Dragomirow, N.: Über Entwicklung von Augenbechern aus transplantierten Stückchen des embryonalen Tapetums. Wilhelm Roux' Arch. Entwickl.-Mech. Org. **126**, 636—662 (1932).

Dragomirow, N.: Über Koordination der Teilprozesse in der embryonalen Morphogenese des Augenbechers. Wilhelm Roux' Arch. Entwickl.-Mech. Org. **129**, 522—560 (1933).

Dragomirow, N.: Über die Determination der Augenbecherblätter bei Triton taeniatus. Wilhelm Roux' Arch. Entwickl.-Mech. Org. **131**, 540—542 (1934).

Dragomirow, N.: Über Induktion sekundäre Retina im transplantierten Augenbecher bei Triton and Pelobates. Wilhelm Roux' Arch. Entwickl.-Mech. Org. **134**, 716—737 (1936).

Dragomirow, N.: The influence of the neighboring ectoderm on the organization of the eye rudiment. Dokl. Akad. Nauk. SSSR **15**, 61—64 (1937) (From Lopaschov, 1963).

Dragomirow, N.: Duration of action necessary for the induction of the retina in amphibian embryos. Dokl. Akad. Nauk. SSSR **26**, 515—518 (1940) (From Lopaschov, 1963).

Duke-Elder, S.: System of ophthalmology. Vol. I. The eye in evolution, chap. XII, The eyes of amphibians, pp. 333—351. St. Louis: C. V. Mosby Co. 1958.

Dumont, J. N., Yamada, T.: Dedifferentiation of iris epithelial cells. Develop. Biol. **29**, 385—401 (1972).

Dumont, J. N., Yamada, T., Cone, M. V.: Alteration of nucleolar ultrastructure in iris epithelial cells during initiation of Wolffian lens regeneration. J. exp. Zool. **174**, 187—204 (1970).

Eguchi, G.: The inhibitory effect of the injured and the displaced lens on the lens-formation in Triturus larvae. Embryologia (Nagoya) **6**, 13—35 (1961).

Eguchi, G.: Electron microscopic studies on lens regeneration I. Mechanism of depigmentation of the iris. Embryologia (Nagoya) **8**, 45—62 (1963).

Eguchi, G.: Electron microscopic studies on lens regeneration II. Formation and growth of lens vesicle and differentiation of lens fibers. Embryologia (Nagoya) **8**, 247—287 (1964).

Eguchi, G.: In vitro analyses of Wolffian lens regeneration: Differentiation of the regenerating lens rudiment of the newt, Triturus pyrrhogaster. Embryologia (Nagoya) **9**, 246—266 (1967).

Eguchi, G., Ishikawa, M.: ^{32}P-uptake by dorsal and ventral halves of the iris at the beginning of lens regeneration in the newt. Embryologia (Nagoya) **5**, 219—226 (1960).

Eguchi, G., Ishikawa, M.: Alkaline phosphatase in the dorsal and the ventral halves of the iris during early stages of lens-regeneration in the newt. Embryologia (Nagoya) **7**, 295—305 (1963).

Eguchi, G., Okada, T. S.: Differentiation of lens tissue from the progeny of chick retinal pigment cells cultured in vitro: A demonstration of a switch of cell types in clonal cell culture. Proc. nat. Acad. Sci. (Wash.) **70**, 1495—1499 (1973).

Eguchi, G., Shingai, R.: Cellular analysis on localization of lens forming potency in the newt iris epithelium. Develop. Growth Differ. (Japan) **13**, 337—349 (1971).

Eguchi, G., Watanabe, K.: Elicitation of lens formation from the "ventral iris" epithelium of the newt by a carcinogen N-methyl-N'-nitro-N-nitrosoguanidine. J. Embryol. exp. Morph. **30**, 63—71 (1973).

Eguchi, G., Abe, S.-I., Watanabe, K.: Differentiation of lens-like structures from newt iris epithelial cells in vitro. Proc. nat. Acad. Sci. (Wash.) **71**, 5052—5056 (1974).

Eisenberg, S., Yamada, T.: A study of DNA synthesis during the transformation of the iris into lens in the lentectomized newt. J. exp. Zool. **162**, 353—368 (1966).

Eisenberg-Zalik, S., Scott, V.: In vitro development of the regenerating lens. Develop. Biol. **19**, 368—379 (1969).

Feldman, J. D., Gaze, R. M.: The growth of the retina in Xenopus laevis: an autoradiographic study. II. Retinal growth in compound eyes. J. Embryol. exp. Morph. **27**, 381—387 (1972).

Feldman, J. D., Gaze, R. M.: The development of half-eyes in Xenopus tadpoles. J. comp. Neurol. **162**, 13—22 (1975).

Fernández-Morán, H.: The fine structure of vertebrate and invertebrate photoreceptors as revealed by low-temperature electron microscopy. In: The structure of the eye, Smelser, G. K. (ed.), pp. 521—556. New York: Academic Press 1961.

Fischel, A.: Über die Regeneration der Linse. Anat. Hefte **14**, 1—256 (1900).

Fischel, A.: Weitere Mittheilungen über die Regeneration der Linse. Wilhelm Roux' Arch. Entwickl.-Mech. Org. **15**, 1—138 (1903).

Freeman, G.: Lens regeneration from the cornea in Xenopus laevis. J. exp. Zool. **154**, 39—66 (1963).

Garfield, S., Moscona, A. A.: Glutamine synthetase in the embryonic chick neural retina: the effect of cycloheximide on the conservation of labile templates for enzyme synthesis. Mech. Ageing and Develop. **3**, 253—269 (1974).

Gaze, R. M.: The representation of the retina on the optic lobe of the frog. Quart. J. exp. Physiol. **43**, 209—214 (1958).

Gaze, R. M.: Regeneration of the optic nerve in Xenopus laevis. Quart. J. exp. Physiol. **44**, 290—308 (1959).

Gaze, R. M.: Regeneration of the optic nerve in amphibia. Int. Rev. Neurobiol. **2**, 1—40 (1960).

Gaze, R. M.: The formation of Nerve Connections: A Consideration of Neural Specificity, Modulation and Comparable Phenomena. New York: Academic Press, 1970

Gaze, R. M.: Neuronal specificity. Brit. med. Bull. **30**, 116—121 (1974).

Gaze, R. M., Chung, S. H., Keating, M. J.: Development of the retinotectal projection in Xenopus. Nature (New Biol.) **236**, 133—135 (1972).

Gaze, R. M., Jacobson, M., Székely, G.: The retino-tectal projection in Xenopus with compound eyes. J. Physiol. **165**, 484—499 (1963).

Gaze, R. M., Jacobson, M., Székely, G.: On the formation of connexions by compound eyes in Xenopus. J. Physiol. **176**, 409—417 (1965).

Gaze, R. M., Keating, M. J., Chung, S. H.: The evolution of the retinotectal map during development in Xenopus. Proc. roy. Soc. B **185**, 301—330 (1974).

Gaze, R. M., Watson, W. E.: Cell division and migration in the brain after optic nerve lesions. In: Growth of the Nervous System, G. E. W. Wolstenholme, M. O'Connor, (eds.), pp. 53—76. Boston: Little Brown and Co. 1968.

Goss, R. J.: Factors affecting lens regeneration in the newt. Carnegie Inst. Wash. Year Book **60**, 430—431 (1961).

Goss, R. J.: Adaptive Growth, pp. 100—103. London: Logos Press 1964.

Grafstein, B.: Functional organization in regeneration of amphibian visual pathways. Bol. Inst. Estud. méd. biol. (Méx.) **22**, 217—230 (1964) (From Biol. Abstr.).

Grafstein, B., Burgen, A. S. V.: Pattern of optic nerve connections following retinal regeneration. Progr. Brain Res. **6**, 126—138 (1964).

Griffinie, L., Marcchio, G.: Sulla rigenerazione totale della retina nei tritoni. Rif. med. Jannar (1889) (From Keefe, 1973a).

Grillo, M. A., Rosenbluth, J.: Ultrastructure of developing Xenopus retina before and after ganglion cell specification. J. comp. Neurol. **145**, 131—140 (1972).

Harrison, R. G.: Experiments on the lens in Amblystoma. Proc. Soc. exp. Biol. (N.Y.) **17**, 199—200 (1920).

Harrison, R. G.: Correlation in the development and growth of the eye studied by means of heteroplastic transplantation. Wilhelm Roux' Arch. Entwickl.-Mech. Org. **120**, 1—55 (1929).

Harrison, R. G.: Organization and Development of the Embryo. New Haven: Yale Univ. Press 1969.

Hasegawa, M.: Restitution of the eye after removal of the retina and lens in the newt, Triturus pyrrhogaster. Embryologia (Nagoya) **4**, 1—32 (1958).

Hasegawa, M.: Restitution of the eye from the iris after removal of the retina and lens together with the eye-coats in the newt, Triturus pyrrhogaster. Embryologia (Nagoya) **8**, 362—386 (1965).

Hayes, B. P.: The distribution of intercellular gap junctions in the developing retina and pigment epithelium of Xenopus laevis. Anat. Embryol. **150**, 99—111 (1976).

Heller, J., Ostwald, T. J., Bok, D.: The osmotic behavior of rod photoreceptor outer segment discs. J. Cell Biol. **48**, 633—649 (1971).

Hendrickson, A. E.: Regeneration of the retina in the newt (Diemictylus v. viridescens): An electron microscope study. Diss. Abstr. **25** (5), 2702 (1964).

Hendrickson, A. E.: Landolt's club in the amphibian retina: A Golgi and electron microscope study. Invest. Ophthalmol. **5**, 484—496 (1966).

Higginbotham, L.: Morphology of the accomodation apparatus in the newt eye. J.S.C. med. Assoc. **72**, 69 (1976).

Higginbotham, L.: The redevelopment of the zonular apparatus during lens regeneration in the newt eye. Anat. Rec. **187**, 604 (1977).

Hollyfield, J. G.: Differential addition of cells to the retina in Rana pipiens tadpoles. Develop. Biol. **18**, 163—179 (1968).

Hollyfield, J. G.: Differential growth of the neural retina in Xenopus laevis larvae. Develop. Biol. **24**, 264—286 (1971).

Hollyfield, J. G., Witovsky, P.: Pigmented retinal epithelium involvement in photoreceptor development and function. J. exp. Zool. **189**, 357—378 (1974).

Holtfreter, J., Hamburger, V.: Amphibians. In: Analysis of Development, Willier, B. H., Weiss, P. A., Hamburger, V. (eds.), pp. 230—296. Philadelphia: W. B. Saunders 1955.

Hoperskaya, O. A.: Differentiation of the optic part of the neural plate in the early neurula of frog embryos in the absence of subsequent induction factors. Dokl. Akad. Nauk. **180**, 1012—1015 (267—270) (1968).

Hoperskaya, O. A.: Differentiation of lens and iris-like tissue in explants of the anterior part of the frog neural plate. Wilhelm Roux' Archiv. Entwickl.-Mech. Org. **171**, 1—16 (1972).

Horstman, L. P., Zalik, S. E.: Growth of newt iris epithelial cells in vitro. A study of the cell cycle. Exp. Cell Res. **84**, 1—14 (1974).

Hunt, R. K.: Developmental programming for retinotectal patterns. In: Cell Patterning, Ciba Foundation Symp. **29** (new series), pp. 131—159. Amsterdam: Associated Scientific Publishers 1975.

Hunt, R. K., Berman, N.: Patterning of neuronal locus specificities in retinal ganglion cells after partial extirpation of the embryonic eye. J. comp. Neurol. **162**, 43—70 (1975).

Hunt, R. K., Frank, E.: Neuronal locus specificity: trans-repolarization of Xenopus embryonic retina after the time of axial specification. Science **189**, 563—565 (1975).

Hunt, R. K., Jacobson, M.: Development and stability of positional information in Xenopus retinal ganglion cells. Proc. nat. Acad. Sci. (Wash.) **69**, 780—783 (1972a).

Hunt, R. K., Jacobson, M.: Specification of positional information in retinal ganglion cells of Xenopus: Stability of the specified state. Proc. nat. Acad. Sci. (Wash.) **69**, 2860—2864 (1972b).

Hunt, R. K., Jacobson, M.: Specification of positional information in retinal ganglion cells of Xenopus: Assays for analysis of the unspecified state. Proc. nat. Acad. Sci. (Wash.) **70**, 507—511 (1973a).

Hunt, R. K., Jacobson, M.: Neuronal locus specificity: Altered pattern of spatial deployment in fused fragments of embryonic Xenopus eyes. Science **180**, 509—511 (1973b).

Hunt, R. K., Jacobson, M.: Neuronal specificity revisited. Curr. Top. Develop. Biol. **8**, 203—259 (1974a).
Hunt, R. K., Jacobson, M.: Specification of positional information in retinal ganglion cells of Xenopus laevis: Intra-ocular control of the time of specification. Proc. nat. Acad. Sci. (Wash.) **71**, 3616—3620 (1974b).
Idoyaga-Vargas, V., Yamada, T.: Glucosaminidase and dedifferentiation of newt iris epithelium. Differentiation **2**, 91—98 (1974).
Idoyaga-Vargas, V., Yamada, T., Michel, C.: Cell type conversion and galactosyltransferase in lens regeneration. Develop. Biol. **49**, 563—568 (1976).
Ikeda, Y.: Beitrag zur Analyse der Wolffschen Linsenregeneration durch xenoplastische Implantation der Iris in das entlinste Auge bei Triton und Hynobius. Arb. anat. Inst. Sendai **16**, 69—82 (1934).
Ikeda, Y.: Über die Regeneration von Augenbechern an verschiedenen Körperstellen durch isolierte Irisstücke. Arb. anat. Inst. Sendai **17**, 11—54 (1935).
Ikeda, Y.: Neue Versuche zur Analyse der Wolffschen Linsenregeneration. Arb. anat. Inst. Sendai **18**, 1—16 (1936a).
Ikeda, Y.: Beiträge zur Frage der Fähigkeit zur Linsenregeneration bei einer Art von Hynobius (Hynobius unnangso Tago). Arb. anat. Inst. Sendai **18**, 17—50 (1963b).
Ikeda, Y.: Über die Bildung akzessorischer Retina aus dem Tapetum bei Hynobius. Wilhelm Roux' Arch. Entwickl.-Mech. Org. **136**, 676—680 (1937).
Ikeda, Y.: Über die wechselseitigen Beziehungen der Sinnesorgane untereinander in ihrer normalen und experimentell bedingten Entwicklung. Arb. anat. Inst. Sendai **21**, 1—44 (1938).
Ikeda, Y.: Zur Frage der Linsenpotenz der Hornhaut in spätembryonalen und larvalen Stadien bei einer Art von Hynobius (Hynobius unnangso Tago). Arb. anat. Inst. Sendai **22**, 27—52 (1939).
Jacobson, A. G.: The roles of neural and non-neural tissues in lens induction. J. exp. Zool. **139**, 525—557 (1958).
Jacobson, A. G.: The determination and positioning of the nose, lens and ear. I. Interactions within the ectoderm, and between the ectoderm and underlying tissues. J. exp. Zool. **154**, 273—284 (1963a).
Jacobson, A. G.: The determination and positioning of the nose, lens and ear. II. The role of the endoderm. J. exp. Zool. **154**, 285—292 (1963b).
Jacobson, A. G.: The determination and positioning of the nose, lens, and ear. III. Effects of reversing the antero-posterior axis of epidermis, neural plate and neural fold. J. exp. Zool. **154**, 293—304 (1963c).
Jacobson, A. G.: Inductive processes in embryonic development. Science **152**, 25—34 (1966).
Jacobson, C.-O.: The localization of the presumptive cerebral regions in the neural plate of the axolotl larva. J. Embryol. exp. Morph. **7**, 1—21 (1959).
Jacobson, M.: Starting points for research in the ontogeny of behavior. In: Major Problems in Developmental Biology, (ed.) M. Locke, pp. 339—383. New York: Academic Press 1966.
Jacobson, M.: Development of neuronal specificity in retinal ganglion cells of Xenopus. Develop. Biol. **17**, 202—218 (1968a).
Jacobson, M.: Cessation of DNA synthesis in retinal ganglion cells correlated with the time of specification of their central connections. Develop. Biol. **17**, 219—232 (1968b).
Jacobson, M.: Development of specific neuronal connections. Science **163**, 543—547 (1969).
Jacobson, M.: Developmental Neurobiology. New York: Holt, Rinehart, and Winston, 1970.
Jacobson, M., Hunt, R. K.: The origin of nerve-cell specificity. Sci. Amer. **228**, (No. 2) 26—35 (1973).
Jacobson, M., Levine, R. L.: Plasticity in the adult frog brain: filling the visual scotoma after excision or translocation of parts of the optic tectum. Brain Res. **88**, 339—345 (1975a).
Jacobson, M., Levine, R. L.: Stability of implanted duplicate tectal positional markers serving as targets for optic axons in adult frogs. Brain Res. **92**, 468—471 (1975b).
Jauker, F., Yamada, T.: Progressive alteration in the pattern of nucleic acid metabolism in the newt iris in cultivation. J. exp. Zool. **183**, 145—151 (1973).
Karasaki, S.: An electron microscopic study of Wolffian lens regeneration in the adult newt. J. Ultrastruct. Res. **11**, 246—273 (1964).
Katoh, A., Yoshida, K.: Delta crystallin synthesis during chick lens differentiation. Exp. Eye Res. **15**, 353—360 (1973).
Keating, M. J.: The formation of visual neuronal connections: an appraisal of the present status of the theory of "Neuronal Specificity". In: Studies on the Development of Behavior and the Nervous System, Vol. 3: Neural and Behavioral Specificity, Gottlieb, G. (ed.), pp. 59—110. New York: Academic Press 1976.

Keating, M.J., Feldman, J.D.: Visual deprivation and intertectal neuronal connexions in Xenopus laevis. Proc. Roy. Soc. Lond. B **191**, 467—474 (1975).

Keating, M.J., Beazley, L., Feldman, J.D., Gaze, R.M.: Binocular interaction and intertectal neuronal connexions: dependence upon developmental stage. Proc. Roy. Soc. Lond. B **191**, 445—466 (1975).

Keefe, J.R.: The fine structure of the retina in the newt, Triturus viridescens. J. exp. Zool. **177**, 263—294 (1971).

Keefe, J.R.: An analysis of urodelian retinal regeneration: I. Studies of the cellular source of retinal regeneration in Notophthalmus viridescens utilizing ^{3}H-thymidine and colchicine. J. exp. Zool. **184**, 185—206 (1973a).

Keefe, J.R.: An analysis of urodelian retinal regeneration: II. Ultrastructural features of retinal regeneration in Notophthalmus viridescens. J. exp. Zool. **184**, 207—232 (1973b).

Keefe, J.R.: An analysis of urodelian retinal regeneration: III. Degradation of extruded melanin granules in Notophthalmus viridescens. J. exp. Zool. **184**, 233—238 (1973c).

Keefe, J.R.: An analysis of urodelian retinal regeneration: IV. Studies of the cellular source of retinal regeneration in Triturus cristatus carnifex using ^{3}H-thymidine. J. exp. Zool. **184**, 239—258 (1973d).

Levine, R.: Regeneration of the retina in the adult newt, Triturus cristatus, following surgical division of the eye by a limbal incision. J. exp. Zool. **192**, 363—380 (1975).

Levine, R., Jacobson, M.: Deployment of optic nerve fibers is determined by positional markers in the frog's tectum. Exp. Neurol. **43**, 527—538 (1974).

Levine, R., Cronly-Dillon, J.R.: Specification of regenerating retinal ganglion cells in the adult newt, Triturus cristatus. Brain Res. **68**, 319—329 (1974).

Lewis, W.H.: Experimental studies on the development of the eye in amphibia. I. On the origin of the lens. Rana palustris Amer. J. Anat. **3**, 505—536 (1904).

Lewis, W.H.: Experimental studies on the development of the eye in amphibia. III. On the origin and differentiation of the lens. Amer. J. Anat. **6**, 473—509 (1907a).

Lewis, W.H.: Lens-formation from strange ectoderm in Rana sylvatica. Amer. J. Anat. **7**, 145—169 (1907b).

Liedke, K.B.: Lens competence in Amblystoma punctatum. J. exp. Zool. **117**, 573—592 (1951).

Liedke, K.B.: Studies on lens induction in Amblystoma punctatum. J. exp. Zool. **130**, 353—380 (1955).

Lopashov, G.V.: Ground processes of eye organization in amphibians. Dokl. Akad. Nauk. SSSR **53**, 177—180 (1946).

Lopashov, G.V.: On the significance of different processes in the restoration of amphibian eyes. Dokl. Akad. Nauk SSSR **61**, 581—584 (1949) (From Lopaschov and Sologub, 1972).

Lopashov, G.V.: On the quantitative regularities in the regeneration of the retina. Dokl. Akad. Nauk SSSR **105**, 599—602 (1955) (In Russian).

Lopashov, G.V.: Comparative studies of the transformation capacity of the eye layers at various stages of development in vertebrates. Proc. XV Int. Zool. Congr. (Lond.), 613—615 (1959).

Lopashov, G.V.: The development of the vertebrate eye in relation to the problem of embryonic induction. Ann. Zool. Soc. Zool. Bot. fenn. Vanamo **22**, (No. 6) 1—17 (1961).

Lopashov, G.V.: Developmental Mechanisms of Vertebrate Eye Rudiments (English translation by Jean Medawar). Oxford: Pergamon Press, 1963.

Lopashov, G.V., Hoperskaya, O.A.: Nature of cell interaction during differentiation of amphibian eye rudiments. Dokl. Akad. Nauk SSSR **175**, 962—965 (442—445) (1967).

Lopashov, G.V., Sologub, A.A.: Artificial metaplasia of pigmented epithelium into retina in tadpoles and adult frogs. J. Embryol. exp. Morph. **28**, 521—546 (1972).

Lopashov, G.V., Stroeva, O.G.: Morphogenesis of the vertebrate eye. Adv. Morphogen. **1**, 331—377 (1961).

Lopashov, G.V., Stroeva, O.G.: Development of eyes in light of experimental investigations (In Russian). Moscow: Akademia Nauk SSSR 1963.

MacDonald, E.L.: The development of the eye of Rana pipiens at two different temperatures. J. Morph. **129**, 445—472 (1969).

Manchot, E.: Abgrenzung des Augenmaterials und anderer Teilbezirke in der Medullarplatte; die Teilbewegungen während der Auffaltung (Farbmarkierungsversuche an Keimen von Urodelen). Wilhelm Roux' Arch. Entwickl.-Mech. Org. **116**, 689—708 (1929).

Mangold, O.: Das Determinationsproblem. III. Das Wirbeltierauge in der Entwicklung und Regeneration. Ergebn. Biol. **7**, 193—403 (1931).

Mangold, O.: Experimente zur Entwicklungsphysiologie des Urodelenkopfes. Verhandl. Anat. Gesellschaft **54** (Freiburg), 3—53 (1957).

Mangold, O.: Freie Linsen in augenlosen Köpfen und Isolaten von Triturus alpestris. Acta Morph. Acad. Sci. hung. **10**, 153—176 (1961).

Mangold, O., von Woellwarth, C.: Das Gehirn von Triton, ein experimenteller Beitrag zur Analyse seiner Determination. Naturwiss. **37**, 365—372, 390—395 (1950).

Mathews, M. B., Osborn, M., Berns, A. J. M., Bloemendal, H.: Translation of two messenger RNAs from lens in a cell free system from Krebs II ascites cells. Nature (New Biol.) **236**, 5—6 (1972).

Matthey, R.: La greffe de l'oeil. I. Étude histologique sur la greffe de l'oeil chez la larve de Salamandre (Salamandra maculosa). Rev. suisse Zool. **33**, 317—334 (1926).

Matthey, R.: La greffe de l'oeil. Étude expérimentale de la greffe de l'oeil chez le Triton (Triton cristatus). Wilhelm Roux' Arch. Entwickl.-Mech. Org. **109**, 326—341 (1927).

McDevitt, D. S.: Sequential appearance of α, β, and γ crystallins in embryonic and metamorphic Rana pipiens. Experientia (Basel) **28**, 1215—1217 (1972).

McDevitt, D. S., Brahma, S. K.: Ontogeny and localization of the crystallins during embryonic lens development in Xenopus laevis. J. exp. Zool. **186**, 127—140 (1973).

McDevitt, D. S., Meza, I., Yamada, T.: Immunofluorescence localization of the crystallins in amphibian lens development, with special reference to the gamma crystallins. Develop. Biol. **19**, 581—607 (1969).

Meyer, R. L., Sperry, R. W.: Retinotectal specificity: chemoaffinity theory. In: Studies on the Development of Behavior and the Nervous System, Vol. 3, Neural and Behavioral Specificity, Gottlieb, G. (ed.), pp. 111—149. New York: Academic Press 1976.

Michael, M. I.: The induction of the head sense placodes in the axolotl, (Amblystoma mexicanum). Effect of removal of the entire prospective brain area s.s. or including the inner half or the entire neural fold. Alexandria med. J. **14**, 53—64 (1968).

Michael, M. I., Nieuwkoop, P. D.: Experimental analysis of the development of the head sense placodes of Ambystoma mexicanum. Proc. kon. ned. Akad. Wet. (Ser. C) **70**, 272—279 (1967).

Michel, C., Yamada, T.: Cellular studies of X-ray induced inhibition of lens regeneration. Differentiation **2**, 193—202 (1974).

Mikami, Y.: Experimental analysis of the Wolffian lens-regeneration in adult newt, Triturus pyrrhogaster. Jap. J. Zool. **9**, 269—302 (1941).

Miller, R. F., Dowling, J. E.: Intracellular responses of the Müller (glial) cells of mudpuppy retina: their relation to b-wave of the electroretinogram. J. Neurophysiol. **33**, 323—341 (1970).

Mitashov, V. I.: Autoradiographic investigations into the regeneration of the retina in pectinate newts (Triturus cristatus). Dokl. Akad. Nauk SSSR **181**, 1510—1513 (411—415) (1968).

Mitashov, V. I.: Characterization of mitotic cycles of pigmented epithelium and retinal rudiment cells during regeneration of the retina in adult newts (Triturus cristatus and Triturus taeniatus). Dokl. Akad. Nauk SSSR **189**, 666—669 (831—833) (1969a).

Mitashov, V. I.: Characteristics of the mitotic cycle during lens regeneration in an experiment involving complete removal of the retina and lens in adult newts (Triturus cristatus). Dokl. Akad. Nauk SSSR **189**, 913—916 (834—836) (1969b).

Mitashov, V. I.: Dynamics of DNA synthesis in pigment epithelium cells throughout the eye restitution after a surgical removal of the retina in adult newt, Triturus cristatus. Tsitologiya, Akad. Nauk SSSR **11**, 434—446 (1969c) (In Russian).

Mitashov, V. I.: The dynamics of DNA synthesis in pigment epithelium cells of adult newts Triturus taeniatus during restitution of the eye after the cutting of the optic nerve and blood vessels. Tsitologiya, Akad. Nauk SSSR **12**, 1521—1529 (1970) (In Russian).

Modak, S. P.: A model for transcriptional control in terminally differentiating lens fiber cells. In: Cell Differentiation, R. Harris, P. Allin, and D. Viza, (eds.) pp. 339—342. Copenhagen: Munksgaard, 1972.

Moody, M. F., Robertson, J. D.: The fine structure of some retinal photoreceptors. J. biophys. biochem. Cytol. **7**, 87—92 (1960).

Morris, J. E.: Histogenesis and dependent glutamine synthetase inducibility in embryonic neural retina. Irreversible inhibition of differentiation by 5-bromodeoxyuridine. Develop. Biol. **35**, 125—142 (1973).

Morris, J. E., Moscona, A. A.: The induction of glutamine synthetase in cell aggregates of embryonic neural retina: correlations with differentiation and multicellular organization. Develop. Biol. **25**, 420—444 (1971).

Moscona, A. A.: Control mechanisms in hormonal induction of glutamine synthetase in the embryonic neural retina. In: Hormones in Development, M. Hamburgh, and E. J. W. Barrington (eds.) pp. 169—189. New York: Appleton-Century-Crofts 1971.

Moscona, A. A.: Induction of glutamine synthetase in embryonic neural retina: a model for the regulation of specific gene expression in embryonic cells. In: FEBS Symposium Vol. 24, Biochemistry of Cell Differentiation, A. Monroy and R. Tsanev, (ed.), pp. 1—23. London: Academic Press 1972.

Moscona, A. A., Wiens, A. W.: Proflavine as a differential probe of gene expression: inhibition of glutamine synthetase induction in embryonic retina. Develop. Biol. **44**, 33—45 (1975).

Moscona, M., Frenkel, N., Moscona, A. A.: Regulatory mechanisms in the induction of glutamine synthetase in the embryonic retina: Immunochemical studies. Develop. Biol. **28**, 229—241 (1972).

Müller, E.: Über die Regeneration der Augenlinse nach Exstirpation derselben bei Triton. Arch. mikr. Anat. **47**, 23—33 (1896).

Nieuwkoop, P. D., Faber, J.: Normal table of Xenopus laevis (Daudin). Amsterdam: North-Holland Publ. Co. 1967.

Nilsson, S. E. G.: An electron microscopic classification of the retinal receptors of the leopard frog (Rana pipiens). J. Ultrastruct. Res. **10**, 390—416 (1964a).

Nilsson, S. E. G.: Interreceptor contacts in the retina of the frog (Rana pipiens). J. Ultrastruct. Res. **11**, 147—165 (1964b).

Nilsson, S. E. G.: Receptor cell outer segment development and ultrastructure of the disc membranes in the retina of the tadpole (Rana pipiens). J. Ultrastruct. Res. **11**, 51—620 (1964c).

Nilsson, S. E. G.: The ultrastructure of the receptor outer segments in the retina of the leopard frog (Rana pipiens). J. Ultrastruct. Res. **12**, 207—231 (1965).

Nilsson, S. E. G., Crescitelli, F.: Changes in ultrastructure and electroretinogram of bullfrog retina during development. J. Ultrastruct. Res. **27**, 45—62 (1969).

Nilsson, S. E. G., Crescitelli, F.: A correlation of ultrastructure and function in the developing retina of the frog tadpole. J. Ultrastruct. Res. **30**, 87—102 (1970).

Nir, I., Pease, D. C.: Ultrastructural aspects of discs in rod outer segments. Exp. Eye Res. **16**, 173—182 (1973).

Ortiz, J. R., Yamada, T., Hsie, A. W.: Induction of the stellate configuration in cultured iris epithelial cells by adenosine and compounds related to adenosine 3′:5′-cyclic monophosphate. Proc. nat. Acad. Sci. (Wash.) **70**, 2286—2290 (1973).

Ortiz, J. R., Yamada, T.: Synergistic effects of adenosine and compounds related to adenosine 3′-5′-cyclic monophosphate on dedifferentiating iris epithelial cells in culture. Differentiation **4**, 135—142 (1975).

Overton, J.: Changes in cell fine structure during lens regeneration in Xenopus laevis. J. Cell. Biol. **24**, 211—222 (1965).

Palmer, W. G., Papaconstantinou, J.: Aging of α-crystallins during development of the lens. Proc. nat. Acad. Sci. (Wash.) **64**, 404—410 (1969).

Papaconstantinou, J.: Molecular aspects of lens cell differentiation. Science **156**, 338—346 (1967).

Piatigorsky, J., Zelenka, P., Simpson, R. T.: Molecular weight and subunit structure of delta-crystallin from embryonic chick lens fibers. Exp. Eye Res. **18**, 435—446 (1974).

Powell, J. A., Powers, C.: Effect on lens regeneration of implantation of spinal ganglia into the eyes of the newt, Notophthalmus. J. exp. Zool. **183**, 95—114 (1973).

Powell, J. A., Segil, N.: Secondary lens formation caused by implantation of pituitary into the eyes of the newt, Notophthalmus. Develop. Biol. **52**, 128—140 (1976).

Quastler, H., Sherman, F. G.: Cell population kinetics in the intestinal epithelium of the mouse. Exp. Cell Res. **17**, 420—438 (1959).

Reese, D. H.: In vitro initiation in the newt iris of some early molecular events of lens regeneration. Exp. Eye Res. **17**, 435—444 (1973).

Reese, D. H.: Ribosomal RNA synthesis in the eastern North-American newt, Notophthalmus viridescens. Differentiation **7**, 99—106 (1977).

Reese, D. H., Puccia, E., Yamada, T.: Activation of ribosomal RNA synthesis in initiation of Wolffian lens regeneration. J. exp. Zool. **170**, 259—268 (1969).

Reese, D. H., Yamada, T., Moret, R.: An established cell line from the newt Notophthalmus viridescens. Differentiation **6**, 75—82 (1976).

Reyer, R. W.: An experimental study of lens regeneration in Triturus viridescens viridescens. II. Lens development from the dorsal iris in the absence of the embryonic lens. J. exp. Zool. **113**, 317—353 (1950).

Reyer, R. W.: Regeneration of the lens in the amphibian eye. Quart. Rev. Biol. **29**, 1—46 (1954a).

Reyer, R. W.: Further studies on lens development from the dorsal iris of Triturus viridescens viridescens in the absence of the embryonic lens. J. exp. Zool. **125**, 1—15 (1954b).

Reyer, R. W.: Lens regeneration from homoplastic and heteroplastic implants of dorsal iris into the eye chamber of Triturus viridescens and Amblystoma punctatum. J. exp. Zool. **133**, 145—190 (1956).

Reyer, R. W.: Studies on lens induction in Amblystoma punctatum and Triturus viridescens viridescens. I. Transplants of prospective belly ectoderm. J. exp. Zool. **138**, 505—556 (1958a).

Reyer, R. W.: Studies on lens induction in Amblystoma punctatum and Triturus viridescens viridescens. II. Transplants of prospective lateral head ectoderm. J. exp. Zool. **139**, 137—180 (1958b).

Reyer, R. W.: The influence of the cephalic archenteron wall on lens and nose development in Amblystoma punctatum. Anat. Rec. **136**, 264—265 (1960).

Reyer, R. W.: Regeneration in the amphibian eye. In: Regeneration (20th Growth Symposium), D. Rudnick (ed.), pp. 211—265. New York: Ronald Press 1962a.

Reyer, R. W.: Differentiation and growth of the embryonic nose, lens, and corneal anlagen implanted into the larval eye or dorsal fin in Amblystoma punctatum. J. exp. Zool. **151**, 123—154 (1962b).

Reyer, R. W.: Stimulation and inhibition of the growth and differentiation of embryonic and regenerating lenses. In: The Structure of the Eye, II. Symposium, J. W. Rohen (ed.), pp. 495—513. Stuttgart: F. K. Schattauer-Verlag, 1965.

Reyer, R. W.: The influence of neural retina and lens on the development of embryonic lens vesicles in Amblystoma punctatum and Triturus viridescens. J. exp. Zool. **162**, 99—132 (1966a).

Reyer, R. W.: The influence of neural retina and lens on lens regeneration from dorsal iris implants in Triturus viridescens larvae. Develop. Biol. **14**, 214—245 (1966b).

Reyer, R. W.: Lens regeneration following removal of the lens alone or after removal of the lens and neural retina. DNA synthesis and lens capsule differentiation. Anat. Rec. **157**, 308 (1967).

Reyer, R. W.: DNA synthesis and incorporation of labeled iris cells into the lens during lens regeneration in adult newts. Develop. Biol. **24**, 533—558 (1971a).

Reyer, R. W.: The origins of the regenerating neural retina in two species of urodele. Anat. Rec. **169**, 410—411 (1971b).

Reyer, R. W.: Differentiation of lens fibers from lens epithelium in Ambystoma maculatum larvae and the repolarization of reversed, regenerating lenses in adult Notophthalmus viridescens. Amer. Zool. **14**, 1302 (1974).

Reyer, R. W.: Morphological evidence for lens differentiation from intraocular implants of lens epithelium in Ambystoma maculatum. Exp. Eye Res. **24**, 511—522 (1977a).

Reyer, R. W.: Repolarization of reversed, regenerating lenses in adult newts, Notophthalmus viridescens. Exp. Eye Res. **24**, 501—509 (1977b).

Reyer, R. W., Woolfitt, R. A., Withersty, L. T.: Stimulation of lens regeneration from the newt dorsal iris when implanted into the blastema of the regenerating limb. Develop. Biol. **32**, 258—281 (1973).

Rochon-Duvigneaud, A. J. F.: Les yeux et la vision des vertébrés. Paris: Masson et Cie. 1943.

Sato, T.: Beiträge zur Analyse der Wolff'schen Linsenregeneration. I. Wilhelm Roux' Arch. Entwickl.-Mech. Org. **122**, 451—493 (1930).

Sato, T.: Vergleichende Studien über die Geschwindigkeit der Wolffschen Linsenregeneration bei Triton taeniatus und bei Diemyctylus pyrrhogaster. Wilhelm Roux' Arch. Entwickl.-Mech. Org. **140**, 570—613 (1940).

Sato, T.: Über die Ursachen des Ausbleibens der Linsenregeneration und zugleich über die linsenbildende Fähigkeit des Pigmentepithels bei den Anuren. Wilhelm Roux' Arch. Entwickl.-Mech. Org. **146**, 487—514 (1953).

Saxén, L., Toivonen, S.: Primary Embryonic Induction. London: Logos Press, 1962

Scheib, D.: Recherches récentes sur la régénération du cristallin chez les vertébrés. Evolution du problème entre 1931 et 1963. Ergebn. Anat. Entwickl.-Gesch. **38**, 45—114 (1965).

Schotté, O. E., Murphy, G. W.: Regeneration of the lens in absence of the pituitary in the adult newt (Triturus viridescens). J. Morph. **93**, 447—464 (1953).

Scott, T. M.: The development of the retino-tectal projection in Xenopus laevis: an autoradiographic and degeneration study. J. Embryol. exp. Morph. **31**, 409—414 (1974).

Sharma, S. C., Hollyfield, J. G.: Specification of retinal central connections in Rana pipiens before the appearance of the first post-mitotic ganglion cells. J. comp. Neurol. **155**, 395—408 (1974).

Shumway, W.: Stages in the normal development of Rana pipiens. I. External form. Anat. Rec. **78**, 139—148 (1940).

Sidman, R. L.: Histogenesis of mouse retina studied with thymidine-H^3. In: The Structure of the Eye, Smelser, G. K. (ed.) pp. 487—506. New York: Academic Press 1961.
Sirlin, J. L., Brahma, S. K.: Studies on embryonic induction using radioactive tracers. II. The mobilization of protein components during induction of the lens. Develop. Biol. **1**, 234—246 (1959).
Smith, S. D.: The effects of electrophoretically separated lens proteins on lens regeneration in Diemyctylus viridescens. J. exp. Zool. **159**, 149—166 (1965).
Sokolova, Z. A., Farberov, A. I., Popov, V. V.: Disturbance and regulation of fiber formation in the lens of common frog larvae after removal of the retina. Dokl. Akad. Nauk SSSR **194**, 1229—1232 (513—516) (1970).
Sologub, A. A.: On the capacity of eye pigment epithelium for transformation into retina in anuran amphibia tadpoles. Tsitologiya, Akad. Nauk SSSR **10**, 1526—1532 (1968). (In Russian).
Spector, A.: Physiological chemistry of the eye. Arch. Ophthalmol. **81**, 127—143 (1969).
Spemann, H.: Über Linsenbildung bei defekter Augenblase. Anat. Anz. **23**, 457—464 (1903).
Spemann, H.: Über Linsenbildung nach experimenteller Entfernung der primären Linsenbildungszellen. Zool. Anz. **28**, 419—432 (1905).
Spemann, H.: Neue Tatsachen zum Linsenproblem. Zool. Anz. **31**, 379—386 (1907).
Spemann, H.: Zur Entwicklung des Wirbeltierauges. Zool. Jb. (Abt. 3) **32**, 1—98 (1912).
Spemann, H.: Embryonic Development and Induction. New Haven: Yale Univ. Press, 1938
Sperry, R. W.: Effect of 180 degree rotation of the retinal field on visuomotor coordination. J. exp. Zool. **92**, 263—279 (1943a).
Sperry, R. W.: Visuomotor coordination in the newt (Triturus viridescens) after regeneration of the optic nerve. J. comp. Neurol. **79**, 33—55 (1943b).
Sperry, R. W.: Optic nerve regeneration with return of vision in anurans. J. Neurophysiol. **7**, 57—70 (1944).
Sperry, R. W.: Mechanisms of neural maturation. In: Handbook of Experimental Psychology, S. S. Stevens (ed.) pp. 236—280. New York: Wiley 1951a.
Sperry, R. W.: Regulative factors in the orderly growth of neural circuits. Growth **15** (Suppl.) Growth Symp. **10**, 63—87 (1951b).
Sperry, R. W.: Chemoaffinity in the orderly growth of nerve fiber patterns and connections. Proc. nat. Acad. Sci. (Wash.) **50**, 703—710 (1963).
Sperry, R. W.: Embryogenesis of behavioral nerve nets. In: Organogenesis, R. L. Dehaan, H. Ursprung (eds.) pp. 161—186. New York: Holt, Rinehart, and Winston 1965.
Sperry, R. W., Hibbard, E.: Regulative factors in the orderly growth of retino-tectal connexions. In: Growth of the Nervous System, G. E. W. Wolstenholme, M. O'Conner (eds.) pp. 41—52. Boston: Little Brown and Co. 1968.
Squires, J. E.: Biochemical and metabolic changes during the degeneration and regeneration of the neural retina of Notophthalmus viridescens. Diss., Yale Univ. (1972).
Stone, L. S.: Heteroplastic transplantation of eyes between the larvae of two species of Amblystoma. J. exp. Zool. **55**, 193—261 (1930).
Stone, L. S.: Return of vision and functional polarization in the retinae of transplanted eyes. Trans. ophthal. Soc., U.K. **67**, 349—368 (1947).
Stone, L. S.: Functional polarization in developing and regenerating retinae of transplanted eyes. Ann. N.Y. Acad. Sci. **49**, 856—865 (1948).
Stone, L. S.: Neural retina degeneration followed by regeneration from surviving retinal pigment cells in grafted adult salamander eyes. Anat. Rec. **106**, 89—109 (1950a).
Stone, L. S.: The role of retinal pigment cells in regenerating neural retinae of adult salamander eyes. J. exp. Zool. **113**, 9—31 (1950b).
Stone, L. S.: An experimental study of the inhibition and release of lens regeneration in adult eyes of Triturus viridescens viridescens. J. exp. Zool. **121**, 181—223 (1952).
Stone, L. S.: Normal and reversed vision in transplanted eyes. Arch. Ophthal. (Chicago) **49**, 28—35 (1953).
Stone, L. S.: Further experiments on lens regeneration in eyes of the adult newt Triturus v. viridescens. Anat. Rec. **120**, 599—624 (1954).
Stone, L. S.: Inhibition of lens regeneration in newt eyes by isolating the dorsal iris from the neural retina. Anat. Rec. **131**, 151—172 (1958a).
Stone, L. S.: Lens regeneration in adult newt eyes related to retina pigment cells and the neural retina factor. J. exp. Zool. **139**, 69—84 (1958b).

Stone, L. S.: Regeneration of the retina, iris, and lens. In: Regeneration in Vertebrates, C. S. Thornton (ed.) pp. 3—14. Chicago: Univ. of Chicago Press 1959a.

Stone, L. S.: Experiments testing the capacity of iris to regenerate neural retina in eyes of adult newts. J. exp. Zool. **142**, 285—308 (1959b).

Stone, L. S.: Polarization of the retina and development of vision. J. exp. Zool. **145**, 85—95 (1960).

Stone, L. S.: Vision in eyes of several species of adult newts transplanted to adult Triturus v. viridescens. J. exp. Zool. **153**, 57—68 (1963a).

Stone, L. S.: Experiments dealing with the role played by the aqueous humor and retina in lens regeneration of adult newts. J. exp. Zool. **153**, 197—210 (1963b).

Stone, L. S.: Return of vision in transplanted larval eyes of cave salamanders. J. exp. Zool. **156**, 219—228 (1964a).

Stone, L. S.: Return of vision in larval eyes exchanged between Amblystoma punctatum and the cave salamander, Typhlotriton spelaeus Invest. Ophthalmol. **3**, 555—565 (1964b).

Stone, L. S.: The regeneration of the crystalline lens. Invest. Ophthalmol. **4**, 420—432 (1965).

Stone, L. S.: Experiments dealing with the inhibition and release of lens regeneration in eyes of adult newts. J. exp. Zool. **161**, 83—94 (1966a).

Stone, L. S.: Development, polarization and regeneration of the ventral iris cleft (remnant of choroid fissure) and protractor lentis muscle in urodele eyes. J. exp. Zool. **161**, 95—108 (1966b).

Stone, L. S.: An investigation recording all salamanders which can and cannot regenerate a lens from the dorsal iris. J. exp. Zool. **164**, 87—104 (1967).

Stone, L. S.; Cole, C. H.: Grafted eyes of young and old adult salamanders (Amblystoma punctatum) showing return of vision. Yale J. Biol. Med. **15**, 735—754 (1943).

Stone, L. S.: Dinnean, F. L.: Lens induction in the salamander (Amblystoma punctatum) with special reference to the conditions in experimentally produced cyclopia. Yale J. Biol. Med. **16**, 31—42 (1943).

Stone, L. S., Ellison, F. S.: Return of vision in eyes exchanged between adult salamanders of different species. J. exp. Zool. **100**, 217—227 (1945).

Stone, L. S., Farthing, T. E.: Return of vision four times in the same adult salamander eye (Triturus viridescens) repeatedly transplanted. J. exp. Zool. **91**, 265—286 (1942).

Stone, L. S., Gallagher, S. B.: Lens regeneration restored to iris membranes when grafted to neural retina environment after cultivation in vitro. J. exp. Zool. **139**, 247—262 (1958).

Stone, L. S., Steinitz, H.: The regeneration of lenses in eyes with intact and regenerating retina in adult Triturus v. viridescens. J. exp. Zool. **124**, 435—468 (1953a).

Stone, L. S., Steinitz, H.: Effects of hypophysectomy and thyroidectomy on lens and retina regeneration in the adult newt, Triturus v. viridescens. J. exp. Zool. **124**, 469—503 (1953b).

Stone, L. S., Vultee, J. H.: Inhibition and release of lens regeneration in the dorsal iris of Triturus v. viridescens. Anat. Rec. **103**, 560 (1949).

Stone, L. S., Zaur, I. S.: Reimplantation and transplantation of adult eyes in the salamander (Triturus viridescens) with return of vision. J. exp. Zool. **85**, 243—269 (1940).

Stone, L. S., Ussher, N. T., Beers, D. N.: Reimplantation and transplantation of larval eyes in the salamander (Amblystoma punctatum). J. exp. Zool. **77**, 13—48 (1937).

Straznicky, K.: The formation of the optic fibre projection after partial tectal removal in Xenopus. J. Embryol. exp. Morph. **29**, 397—409 (1973).

Straznicky, K., Gaze, R. M.: The growth of the retina in Xenopus laevis: an autoradiographic study. J. Embryol. exp. Morph. **26**, 67—79 (1971).

Straznicky, K., Gaze, R. M.: The development of the tectum in Xenopus laevis: an autoradiographic study. J. Embryol. exp. Morph. **28**, 87—115 (1972).

Straznicky, K., Gaze, R. M., Keating, M. J.: The retinotectal projections after uncrossing the optic chiasma in Xenopus with one compound eye. J. Embryol. exp. Morph. **26**, 523—542 (1971).

Straznicky, K., Gaze, R. M., Keating, M. J.: The retinotectal projection from a double-ventral compund eye in Xenopus laevis. J. Embryol. exp. Morph. **31**, 123—137 (1974).

Stroeva, O. G.: Study of retina regenerative ability in adult Bombinator and rats. Izv. Akad. Nauk SSSR (Ser. Biol.) **5**, 76—84 (1956) (In Russian).

Székely, G.: Zur Ausbildung der lokalen funtionellen Spezifität der Retina. Acta biol. Acad. Sci. hung. **5**, 157—167 (1954).

Székely, G.: Regulationstendenzen in der Ausbildung der „funktionellen Spezifität" der Retinaanlage bei Triturus vulgaris. Wilhelm Roux' Arch. Entwickl.-Mech. Org. **150**, 48—60 (1957).

Székely, G.: Embryonic determination of neural connections. Adv. Morphogen. **5**, 181—219 (1966).
Tahara, Y.: Formation of the independent lens in Japanese amphibians. Embryologia (Nagoya) **7**, 127—149 (1962a).
Tahara, Y.: Determination of the independent lens rudiment in Rhacophorus schlegelii arboreus Okada et Kawano. Embryologia (Nagoya) **7**, 150—168 (1962b).
Takano, K.: On the lens-effect in the Wolffian lens regeneration in Triturus pyrrhogaster. Mie med. J. **8**, 385—403 (1959).
Takano, K.: On the role of the hypophysis in lens regeneration in Triturus pyrrhogaster. Mie med. J. **11**, 499—508 (1962).
Takano, K., Yoshida, Y., Ohashi, T., Ogasawara, T., Takeuchi, A., Masaki, H., Miyazaki, A., Mikami, Y.: Experimental analysis of the effect of lens upon the Wolffian lens regeneration in adults of the newt, Triturus pyrrhogaster. Mie med. J. **7**, 257—272 (1957).
Takata, C., Albright, J. F., Yamada, T.: Lens antigens in a lens-regenerating system studied by the immunofluorescent technique. Develop. Biol. **9**, 385—397 (1964a).
Takata, C., Albright, J. F., Yamada, T.: Study of lens antigens in the developing newt lens with immunofluorescence. Exp. Cell Res. **34**, 207—210 (1964b).
Takata, C., Albright, J. F., Yamada, T.: Lens fiber differentiation and gamma crystallins: Immunofluorescent study of Wolffian regeneration. Science **147**, 1299—1301 (1965).
Takata, C., Albright, J. F., Yamada, T.: Gamma crystallins in Wolffian lens regeneration demonstrated by immunofluorescence. Develop. Biol. **14**, 382—400 (1966).
Taylor, A. C., Kollros, J. J.: Stages in the normal development of Rana pipiens larvae. Anat Rec. **94**, 7—24 (1946).
Ten Cate, G.: The intrinsic embryonic development. Verh. kon. ned. Akad. Wet. **51** (No. 2), 1—257 (1956).
Thorpe, C. W., Bond, J. S., Collins, J. M.: Early events in lens regeneration: changes in cyclic AMP concentrations during initiation of RNA and DNA synthesis. Biochem. biophys. Acta **340**, 413—418 (1974).
Toivonen, S.: Zur Frage der Induktion selbständiger Linsen durch abnorme Induktoren im Implantatversuch bei Triton. Ann. Zool. Soc. Zool. Bot. fenn. Vanamo **11** (No. 3), 1—28 (1945).
Truman, D. E. S., Clayton, R. M.: The subunit structure of chick β-crystallins. Exp. Eye Res. **18**, 485—494 (1974).
Truman, D. E. S., Brown, A. G., Campbell, J. C.: The relationship between the ontogeny of antigens and of the polypeptide chains of the crystallins during chick lens development. Exp. Eye Res. **13**, 58—69 (1972).
Turner, J. E.: Non-glial phagocytes within the degenerating optic nerve of the newt (Triturus viridescens). J. exp. Zool. **193**, 87—100 (1975).
Turner, J. E., Singer, M.: An electron microscopic study of the newt (Triturus viridescens) optic nerve. J. comp. Neurol. **156**, 1—18 (1974a).
Turner, J. E., Singer, M.: The ultrastructure of regeneration in the severed newt optic nerve. J. exp. Zool. **190**, 249—268 (1974b).
Twitty, V.: Eye. In: Analysis of Development, B. H. Willier, P. A. Weiss, V. Hamburger, eds., pp. 402—414. Philadelphia: W. B. Saunders 1955.
Uno, M.: Zur Frage des Mechanismus der Wolffschen Linsenregeneration. Jap. J. med. Sci. (I. Anat.) **11**, 75—100 (1943).
Wachs, H.: Neue Versuche zur Wolffschen Linsenregeneration. Wilhelm Roux' Arch. Entwickl.-Mech. Org. **39**, 384—451 (1914).
Wachs, H.: Restitution des Auges nach Extirpation von Retina und Linse bei Tritonen. (Neue Versuche zur Wolffschen Linsenregeneration.) (Zweiter Teil). Wilhelm Roux' Arch. Entwickl.-Mech. Org. **46**, 328—390 (1920).
Waggoner, P. R.: Lens regeneration from the cornea in larval Xenopus laevis. Diss., West Virginia Univ. (1972).
Waggoner, P. R.: Lens differentiation from the cornea following lens extirpation or cornea transplantation in Xenopus laevis. J. exp. Zool. **186**, 97—110 (1973).
Waggoner, P. R., Reyer, R. W.: DNA synthesis during lens regeneration in larval Xenopus laevis. J. exp. Zool. **192**, 65—72 (1975).
Waley, S. G.: The lens: function and macromolecular composition. In: The eye, vol. 1, vegetative physiology and biochemistry, 2nd. edit., H. Davson, ed., pp. 299—379. New York: Academic Press 1969.

Walls, G. L.: The Vertebrate Eye and Its Adaptive Radiation. Bloomfield Hills, Mich.; The Cranbrook Press 1942.
Werblin, F. S.: Response of retinal cells to moving spots; Intracellular recording in Necturus maculosus. J. Neurophysiol. **33**, 342—350 (1970).
Werblin, F. S.: Adaptation in a vertebrate retina: Intracellular recording in Necturus. J. Neurophysiol. **34**, 228—241 (1971).
Werblin, F. S.: Functional organization of a vertebrate retina: Sharpening up in space and intensity. Ann. N.Y. Acad. Sci. **193**, 75—85 (1972).
Werblin, F. S., Dowling, J. E.: Organization of the retina of the mudpuppy, Necturus maculosus. II. Intracellular recording. J. Neurophysiol. **32**, 339—355 (1969).
Wilens, S.: The migration of heart mesoderm and associated areas in Amblystoma punctatum. J. exp. Zool. **129**, 579—606 (1955).
Williams, J. A. Jr.: The regeneration of the neural retina from retina pigment cells in adult newt eyes: an electron microscopic study. Diss. Abstr. **25**, (8) 4368 (1964).
Williams, L. A.: Studies on lens and iris regeneration in Notophthalmus viridescens viridescens (Salamandridae). Diss., West Virginia Univ. (1970a).
Williams, L. A.: The effect of a normal lens on lens regeneration in Notophthalamus viridescens viridescens. Amer. Zool. **10**, 322—323 (1970b).
Williams, L. A., Higginbotham, L. T.: The role of a normal lens in Wolffian lens regeneration. J. exp. Zool. **191**, 233—252 (1975).
Williamson, R., Clayton, R., Truman, D. E. S.: Isolation and identification of chick lens crystallin messenger RNA. Biochem. Biophys. Res. Comm. **46**, 1936—1943 (1972).
Von Woellwarth, C.: Die Induktionsstufen des Gehirns. Wilhelm Roux' Arch. Entwickl.-Mech. Org. **145**, 582—668 (1952).
Von Woellwarth, C.: Über das Anlagenmuster und die Kinematik des Ektoderms, insbesondere der präsumptiven Epidermis, im Neurula- und Schwanzknospenstadium von Triturus alpestris. Wilhelm Roux' Arch. Entwickl.-Mech. Org. **152**, 602—631 (1960).
Von Woellwarth, C.: Die Rolle des Neuralleistenmaterials und der Temperatur bei der Determination der Augenlinse. Embryologia (Nagoya) **6**, 219—242 (1961).
Von Woellwarth, C.: Versuche zur Entstehung des Organmusters der vorderen Gehirnregion bei Amphibien. Wilhelm Roux' Arch. Entwickl.-Mech. Org. **167**, 199—208 (1971).
Woerdeman, M. W.: Experimentelle Untersuchungen über Lage und Bau der augenbildenden Bezirke in der Medullarplatte beim Axolotl. Wilhelm Roux' Arch. Entwickl.-Mech. Org. **116**, 220—241 (1929).
Wolff, G.: Entwicklungsphysiologische Studien. I. Die Regeneration der Urodelenlinse. Wilhelm Roux' Arch. Entwickl.-Mech. Org. **1**, 380—390 (1895).
Wolff, G.: Entwicklungsphysiologische Studien. II. Weitere Mittheilungen zur Regeneration der Urodelenlinse. Wilhelm Roux' Arch. Entwickl.-Mech. Org. **12**, 307—351 (1901).
Wolff, G.: Entwicklungsphysiologische Studien. III. Zur Analyse der Entwicklungspotenzen des Irisepithels bei Triton. Arch. mikr. Anat. **63**, 1—9 (1904).
Yamada, T.: Control of tissue specificity: The pattern of cellular synthetic activities in tissue transformation. Amer. Zool. **6**, 21—31 (1966).
Yamada, T.: Cellular synthetic activities in induction of tissue transformation. In: Cell Differentiation (Ciba Symp.), A.V.S. De Reuck, J. Knight, eds., pp. 116—130. Boston: Little, Brown and Co. 1967a.
Yamada, T.: Cellular and subcellular events in Wolffian lens regeneration. Curr. Top. Develop. Biol. 2, 247—283 (1967b).
Yamada, T.: A test for DNA synthesis associated with nucleolar activation in newt iris epithelial cells. Exp. Cell Res. **67**, 227—230 (1971).
Yamada, T.: Control mechanisms in cellular metaplasia. In: Cell Differentiation, R. Harris, P. Allin, D. Viza, eds., pp. 56—60. Copenhagen: Munksgaard 1972.
Yamada, T., Dumont, J. N.: Macrophage activity in Wolffian lens regeneration. J. Morph. **136**, 367—384 (1972).
Yamada, T., Karasaki, S.: Nuclear RNA synthesis in newt iris cells engaged in regenerative transformation into lens cells. Develop. Biol. **7**, 595—604 (1963).
Yamada, T., McDevitt, D. S.: Direct evidence for transformation of differentiated iris epithelial cells into lens cells. Develop. Biol. **38**, 104—118 (1974).
Yamada, T., Roesel, M. E.: Effects of actinomycin D on the lens regenerating system. J. Embryol. exp. Morph. **12**, 713—725 (1964).

Yamada, T., Roesel, M. E.: Activation of DNA replication in the iris epithelium by lens removal. J. exp. Zool. **171**, 425—432 (1969).
Yamada, T., Roesel, M. E.: Control of mitotic activity in Wolffian lens regeneration. J. exp. Zool. **177**, 119—128 (1971).
Yamada, T., Roesel, M. E., Beauchamp, J. J.: Cell cyle parameters in dedifferentiating iris epithelial cells. J. Embryol. exp. Morph. **34**, 497—510 (1975).
Yamada, T., Takata, C.: An autoradiographic study of protein synthesis in regenerative tissue transformation of iris into lens in the newt. Develop. Biol. **8**, 358—369 (1963).
Yamada, T., Reese, D. H., McDevitt, D. S.: Transformation of iris into lens in vitro and its dependency on neural retina. Differentiation **1**, 65—82 (1973).
Yoshida, K., Katoh, A.: Crystallin synthesis by chicken lens. I. Analysis of radioactive crystallins by cellulose acetate membrane electrophoresis. Exp. Eye Res. **11**, 122—131 (1971a).
Yoshida, K., Katoh, A.: Crystallin synthesis by chick lens. II. Changes in synthetic activities of epithelial and fiber cells during embryonic development. Exp. Eye Res. **11**, 184—194 (1971b).
Yoshida, K., Katoh, A.: Crystallin synthesis by chicken lens. III. mRNA stabilization under in vitro culture conditions. Exp. Cell Res. **71**, 361—371 (1972).
Young, R. W.: The renewal of photoreceptor cell outer segments. J. Cell. Biol. **33**, 61—72 (1967).
Young, R. W.: A difference between rods and cones in the renewal of outer segment protein. Invest. Ophthalmol. **8**, 222—231 (1969).
Young, R. W.: Visual cells. Sci. Amer. **223** (No. 4), 80—91 (1970).
Young, R. W.: An hypothesis to account for a basic distinction between rods and cones. Vision Res. **11**, 1—5 (1971).
Young, R. W., Bok, D.: Participation of the retinal pigment epithelium in the rod outer segment renewal process. J. Cell Biol. **42**, 392—403 (1969).
Young, R. W., Bok, D.: Autoradiographic studies on the metabolism of the retinal pigment epithelium. Invest. Ophthalmol. **9**, 524—536 (1970).
Zalik, S. E., Scott, V.: Cell surface changes during dedifferentiation in the metaplastic transformation of iris into lens. J. Cell Biol. **55**, 134—146 (1972).
Zalik, S. E., Scott, V.: Sequential disappearance of cell surface components during dedifferentiation in lens regeneration. Nature (New Biol.) **244**, 212—214 (1973).
Zalik, S. E., Yamada, T.: The cell cycle during lens regeneration. J. exp. Zool. **165**, 385—394 (1967).
Zalokar, M.: Contribution à l'étude de la régénération du cristallin chez le Triton. Rev. suisse Zool. **51**, 443—521 (1944).
Zapisek, W. F., Papaconstantinou, J.: Degradation of ribosomal RNA during chicken lens development. Biochim. Biophys. Acta **299**, 603—611 (1973).
Zelenka, P., Piatigorsky, J.: Isolation and in vitro translation of δ-crystallin mRNA from embryonic chick lens fibers. Proc. nat. Acad. Sci. (Wash.) **71**, 1896—1900 (1974).
Zwann, J.: Lens-specific antigens and cytodifferentiation in the developing lens. J. cell. Physiol. **72** (Suppl. 1), 47—71 (1968).
Zwann, J.: Mitotic activity in the lens rudiment of the chicken embryo before and after the onset of crystallin synthesis. II. Immunofluorescence studies. Wilhelm Roux' Arch. Entwickl.-Mech. Org. **175**, 13—25 (1974).
Zwaan, J., Hendrix, R. W.: Changes in cell and organ shape during early development of the ocular lens. Amer. Zool. **13**, 1039—1049 (1973).
Zwaan, J., Ikeda, A.: Macromolecular events during differentiation of the chicken lens. Exp. Eye Res. **7**, 301—311 (1968).

Chapter 7

The Visual Pigments of Geckos and other Vertebrates: an Essay in Comparative Biology

F. CRESCITELLI, Los Angeles, California, USA

With 37 Figures

Contents

A. Introduction

The exploration of a specific biological phenomenon is often facilitated by use of an organism appropriately suited to the investigation of the phenomenon. The squid with its giant axon for the study of excitation and conduction and the aglomerular fish for exploring the functions of the nephron are examples of this. In visual science the lateral eye of *Limulus* has similarly made possible significant understanding in the properties of visual systems. Nature is bountiful in providing investigators with organisms especially or uniquely suitable to their requirements. In the study of visual pigments I have found the gecko photopigments to have apparently unusual properties, and the study of these properties has given me new insights into the nature of these interesting and important biochemical substances. It may be that the gecko pigments represent some aberrant diversion of evolution and have little relevance to the fundamental behavior of vertebrate visual pigments. On the other hand, it may be that the gecko visual proteins possess in an exaggerated or available form properties shared by all retinal photopigments, and that a study of these proteins is worthwhile from a fundamental point of view. I have adopted the latter interpretation, and in this chapter I wish to present the results of these studies, many of which are still unpublished.

I. Historical

Geckos constitute a distincitive family of lizards inhabiting the warmer regions around the earth and utilizing a variety of habitats. They are quick, agile animals, mainly insectivorous as to diet, and have relatively large eyes. Nocturnal and crepuscular geckos are common, but diurnal species are also known to occur. The eyes of geckos reflect the adaptation to the photic environment. Some diurnal forms have circular pupils, yellow lenses, and densely packed visual cells with fine outer segments (TANSLEY, 1964). These have been called cones. Nocturnal and crepuscular geckos often have mobile, vertical pupils, colorless lenses, and visual cells with large, cylindrical outer segments. They have been designated as rods (TANSLEY, 1964; UNDERWOOD, 1970). Some investigators have employed geckos in visual studies because they considered the gecko retina to be of the pure rod class (CROZIER and WOLF, 1939).

In contrast to this sharp rod-cone duality, WALLS (1934) introduced the idea that the gecko visual cell is phylogenetically unique. According to him, the so-called rods of geckos are, in fact, phylogenetically transmuted cells having evolved from the so-called cones or photocytes of diurnal ancestors. The principal arguments adduced for this transmutation hypothesis are morphologic in nature: the presence of oil droplets, some colored, in cells with rod-shaped outer segments; the presence of foveae; and the identification in different species of homologous visual cells in apparently transitional stages of transmutation. In Figure 1, for example, are shown the three characteristic types of visual cells (type A singles, type B doubles, and type C doubles) as noted in several species of geckos, includ-

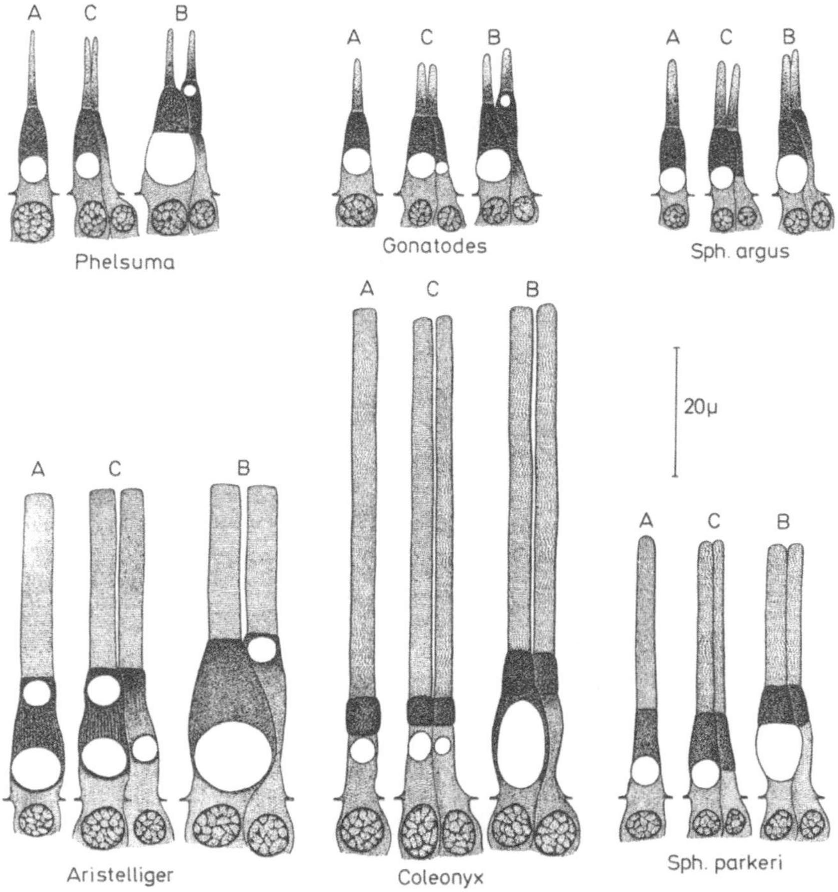

Fig. 1. The three main classes of visual photoreceptors in geckos: the type A singles, and two classes of doubles, type B and C. This sequence includes typically diurnal species like *Phelsuma* and *Gonatodes* that have cones and typical nocturnal forms like *Coleonyx* whose visual cells are rod-like. Note absence of oil droplets in all cells of some species and their presence, variably, in other species. Note presence of paraboloids even in the rods of *Coleonyx*. Both types of doubles are asymmetric in respect to the cytology of the two members, the type B doubles grossly so. This chart is compiled from the writings of WALLS, of UNDERWOOD and of TANSLEY

ing the diurnal *Phelsuma* whose cells have fine outer segments and the nocturnal *Coleonyx* whose cells have long cylindrical outer segments. Oil droplets are found in some visual cells but are missing in others. In addition to this transmutation of primary photocytes to secondary scotocytes, WALLS conceived of a tertiary phylogenetic change in geckos that were assumed to have reverted to a diurnal habit, examples being *Lygodactylus* and *Phelsuma*. The study by TANSLEY (1964) gave support to these ideas. She noted the eye of the diurnal *P. madagascariensis* to have a round pupil, a pale yellow lens, visual cells with fine outer segments, a colorless oil droplet in one type of cell, and a shallow lateral fovea. TANSLEY considered her results with the nocturnal *Tarentola*, the less nocturnal *Hemidacty-*

lus, and the diurnal *Phelsuma* to be in general support of the WALLS hypothesis. Here then is one unique feature of geckos for visual science: One has available in living forms several related species in transitional stages between the nocturnal and the diurnal retinal condition. A few investigators have already exploited the possibilities offered by this naturally occurring situation (CRESCITELLI, 1972).

From all we know concerning the biological distribution of visual pigments, rhodopsin, with an absorption peak in the region of 500 nm, should be the photosensitive substance of the rods of nocturnal geckos, and this, in fact, was suggested by earlier students of gecko ophthalmology (WALLS, 1942; UNDERWOOD, 1954) although Walls considered that rhodopsin has been "re-invented so many times". The truth of the matter is that though the extracted gecko retinal pigments are chemically based on retinal (the aldehyde of vitamin A_1), the spectral absorbance maximum is not at 500 nm but at spectral locations of significantly longer wavelengths than 500 nm. This suggests a unique property of the protein moiety (opsin) that is able to shift the spectral maximum of the chromophoric group beyond the position of the usual rhodopsin of terrestrial animals. Here then is the central biochemical question applied to a scotopic system of pigments: By what mechanism is the opsin able to adjust the spectral absorbance to a given region of the visible spectrum? If there is any hope of getting an answer to this question it will have to be worked out with a retinal system like that of the nocturnal geckos in which the amount of pigment, though still impracticably small, is greater than the amount of cone pigment and can be readily extracted from the visual cell.

II. Gekko gekko: The Protagonist of the Theme

Though this account concerns the gecko pigments, most of the specific information to be presented was obtained in experiments with the photopigments of the retina of the Tokay gecko. *G. gekko*. This rather large and common lizard is found in the Philippines, Ceylon, India, East and Southeast Asia, Indonesia, and in a number of islands in the Pacific area. This gecko is placed with the Gekkonidae by UNDERWOOD (1954). It has large eyes, colorless lenses, and vertical pupils that constrict down to form four characteristic pinholes (Fig. 2). *G. gekko* has aroused some interest among visual scientists, and a number of studies have been published that deal with the retina of this lizard. The photoreceptors have been examined by electron microscopy, and in all respects their types and patterns are typical for geckos. Numerically, the most significant cells are the type A singles, the type B doubles, and the less numerous type C doubles, all lacking an oil droplet (Fig. 3). As in all geckos, these are rather precisely arranged in a characteristic gecko pattern with rows of type B doubles alternating with rows each of which is made of both type A singles and type C doubles (UNDERWOOD, 1970). The visual cells of *G. gekko* have large cylindrical outer segments (PEDLER and TILLY, 1964), and although these cells have been called rods, COHEN (1972) called attention to the finding that some of the outer segment discs are open and continuous with the cell membrane, a characteristic usually associated with the ultrastructure of cones. The large size of the visual cells of this gecko has encouraged

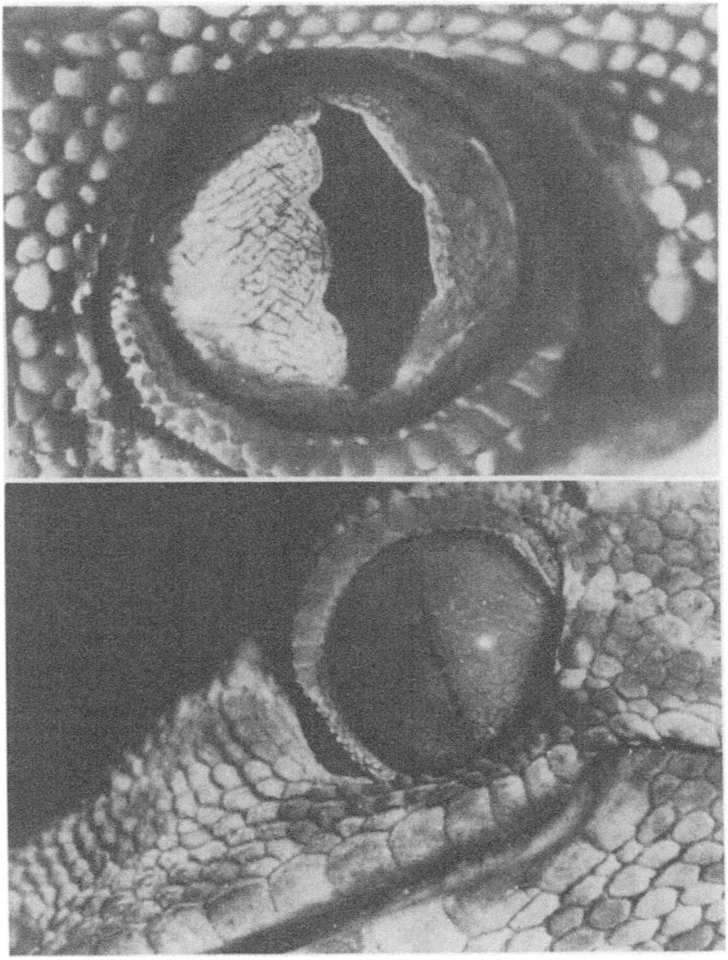

Fig. 2. The pupil of *G. gekko* partly dilated *(above)* and constricted *(below)*. Note the lobes at each of the iris edges; when the iris constricts, this leads to an aperture with four pinholes. UNDERWOOD (1954) designated this as a *Gekko*-type pupil and considered it to have arisen from the straight vertical pupil present in some genera

investigators to explore its properties by means of microelectrodes. A significant resting potential of 30-40 mV has been reported (KLEINSCHMIDT, 1973) and light has been shown to hyperpolarize the membrane and to increase the membrane resistance, i.e., to decrease the membrane permeability to ions (TOYODA et al., 1969). This finding accords with the results obtained with a number of other vertebrate photoreceptors. Additionally, intracellular recordings of the visual cells are being used to examine the process of local adaptation that occurs in photoreceptors and the interactions that these photoreceptors have with each other (KLEINSCHMIDT, 1973; PAK et al., 1973).

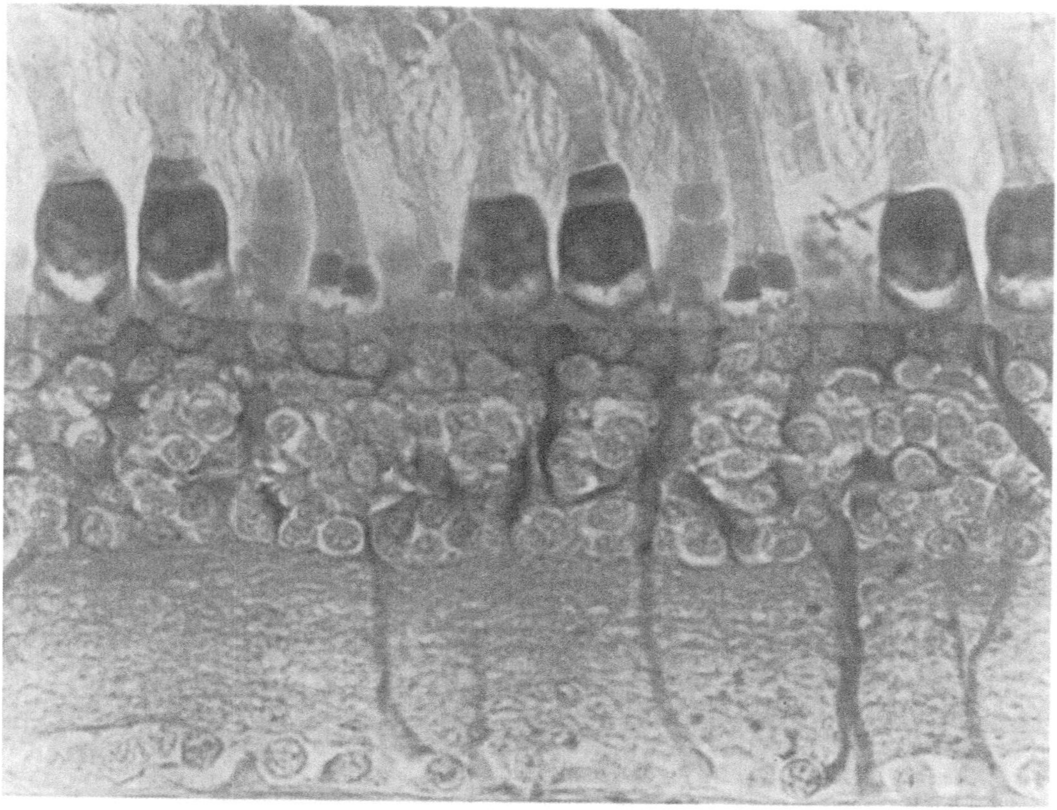

Fig. 3. Section of retina of *G. gekko* showing six type B doubles and two type C doubles. Type A singles are not clearly defined in this section. Note the more external positioning of the type B outer segments as compared with the type C segments. This photograph is published with the permission of Dr. D. I. HAMASAKI

The diverse morphologic types of visual cells might be expected to be associated with spectral diversity and this, in fact, has been demonstrated by means of retinal extracts and by microspectrophotometric examinations of individual gecko visual cells. A typical analysis of a retinal extract (Figs. 4 and 5) shows the spectral absorbances of two photopigments extracted into 2% digitonin made up in veronal-HCl buffer at pH 7.0 and measured at low temperature (3.5° C). In light of results to be described later in this chapter, the presence of chloride ions as well as the low temperature used in the measurements needs to be pointed out. The initial absorbance (curve 1, Fig. 4) has a well defined peak close to 521 nm. After exhaustive bleaching with light at 620 nm (curve 2), density was lost at wavelengths longer than 448 nm (the isosbestic wavelength) and was gained at shorter wavelengths. The difference spectrum for this first bleach (curve 1, Fig. 5) indicates pigment loss with maximum at 522 nm and a product with maximum at approximately 395 nm. Following this initial bleach, the extract was exposed to light at 520 nm until no further change occurred. The result (curve 3, Fig. 4) indicates the

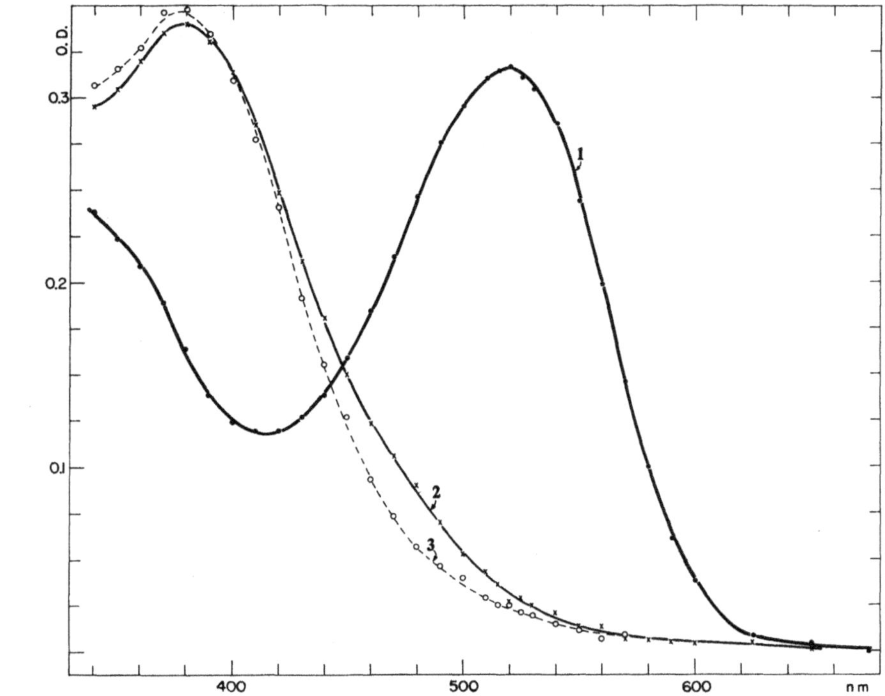

Fig. 4. Partial bleaching of a retinal extract of *G. gekko*. The solubilizer was 2% digitonin (Merck) made up in veronal-HCl buffer, pH 7.1. Curve 1: absorbance of unbleached extract at 2.4° C. Curve 2: after exhaustive bleaching with light at 620 nm. Curve 3: after terminal exhaustive bleach with light at 520 nm. Note absence of metarhodopsin III in the transition from curve 1 to curve 2. This is to be contrasted with the results of Fig. 14 with frog rhodopsin

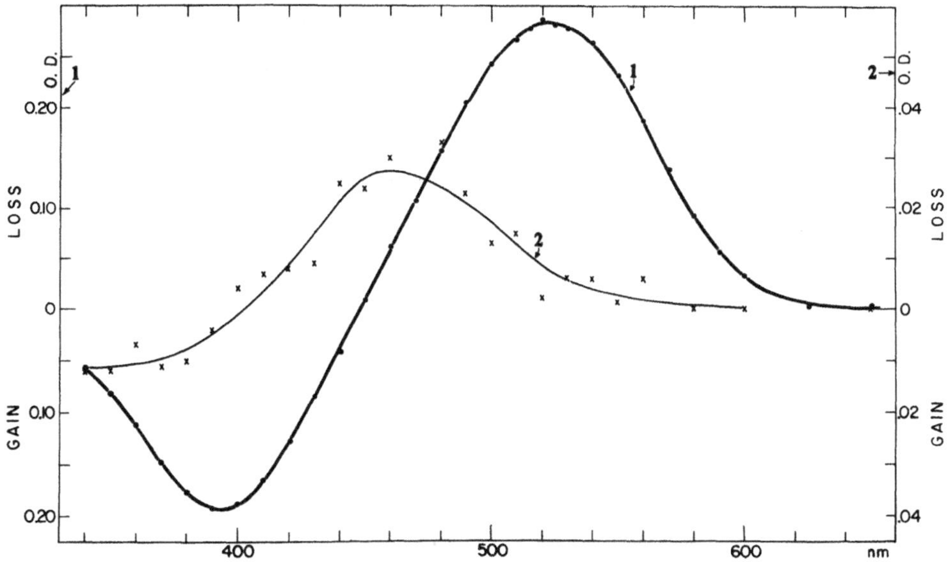

Fig. 5. Difference spectra obtained from data of Fig. 4. Curve 1: 1-2 difference. Curve 2: 2-3 difference. The results indicate the presence of two photopigments in the extract

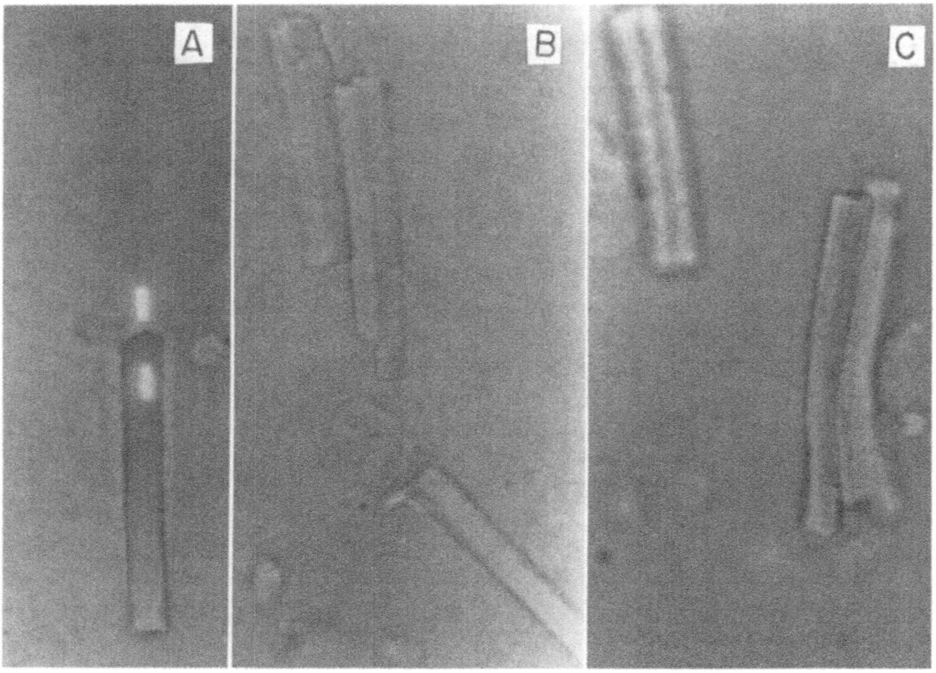

Fig. 6 A-C. Infrared photographs of outer segment types encountered in microspectrophotometric examination. All from *G. gekko*. (A) Single rod with two beams: the test beam within the rod, the reference beam in the medium outside. (B) A single rod *(lower oblique)*, a double with staggered members, and a double *(upper)* with both members aligned and in register. Magnification is indicated by size of beam which was 2.5 × 6.0 μm. (C) One of two triple rods that were encountered. Above the triple, and out of focus is a double

photic destruction of a second component absorbing at shorter wavelengths. The difference spectrum for this second bleach (curve 2, Fig. 5) suggests the presence in the extract of a pigment with maximum in the general region of 460 nm. On a density basis, this second component was present in this extract as 8.9% of the total photopigment density, a figure that is close to the values I originally reported for several extracts of the retinas of *G. gekko* (CRESCITELLI, 1963). The results summarized in Figures 4 and 5 are typical of many obtained with this species; all show a dual system of photopigments in the density proportion indicated.

Only two visual pigments have been recorded from extracts of the gecko retina, and there is no reason to doubt that each of these is located in separate visual cells. MARKS (1965a) mentioned finding a 520-pigment, and later LIEBMAN (1972) reported the recordings of two photopigments, one at 518 nm, the other at 467 nm. In collaboration with Dr. E. LOEW and Dr. H. J. A. DARTNALL (CRESCITELLI et al., 1976), I investigated, with the Sussex microspectrophotometer, the visual pigments of three species of geckos: *G. gekko*, *G. monarchus*, and *Homopholis wahlbergii*. A dual system of photopigments was recorded from the isolated outer segments of all three species. In *G. gekko*, the species studied most intensive-

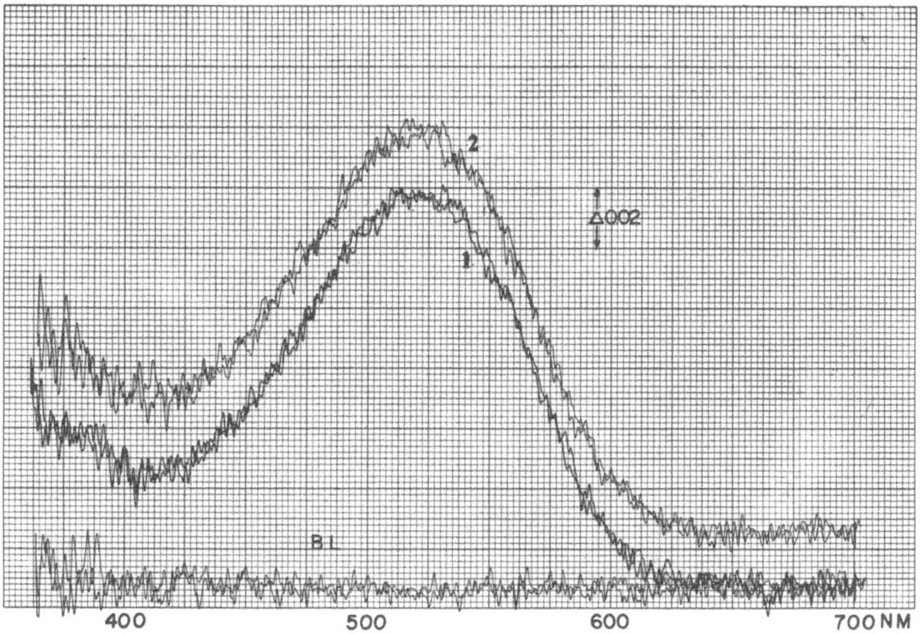

Fig. 7. Microspectrophotometric recordings from both ends of a single outer segment. Recordings from each end consisted of two spectral scans, the first from 700 nm to 350 nm the second in the reverse direction. As shown by duplication, no significant bleaching was caused by test beam. Lower trace *(BL)* is a double-scan recording of base line, with both beams in the medium. Temperature: 23° C

ly, the two pigments displayed spectral absorbance curves with maxima at 521 nm and 467 nm. Figure 6 pictures the types of isolated outer segments of *G. gekko* from which recordings were made. Similar outer segments were obtained in retinal squashes of the other two species. The photographs include a single outer segment (Fig. 6A) in which the 521-pigment was found. Some 53 such singles were examined and all contained the 521-pigment. None was found to be empty of photosensitive component. The 521-pigment was present equally in both ends of the outer segment (Fig. 7), and when bleached in stages the spectral maximum did not shift, indicative of but one photosensitive component (Fig. 8). In agreement with what is known about vertebrate visual pigments (LIEBMAN, 1972; HÁROSI and MACNICHOL, 1974) the chromophore was seen to be oriented in the outer segment in a typical manner, for by using light polarized transversely and parallely to the long axis of the cell, a dichroic ratio of 2–3:1 was obtained. In addition to single outer segments containing the 521-pigment, we have also found, though less frequently, single outer limbs with the 467-pigment.

For the double outer segments (Fig. 6B) in the retinal squashes, we encountered examples of doubles whose two members both contained the 521-pigment (Fig. 9) and examples whose two members both had the 467-pigment, the latter being of less frequent occurrence. In addition, we measured doubles in which the two members were spectrally different, with the 521- and 467-pigments occurring separately (Fig. 10). In such doubles both pigments were photolabile, and red light

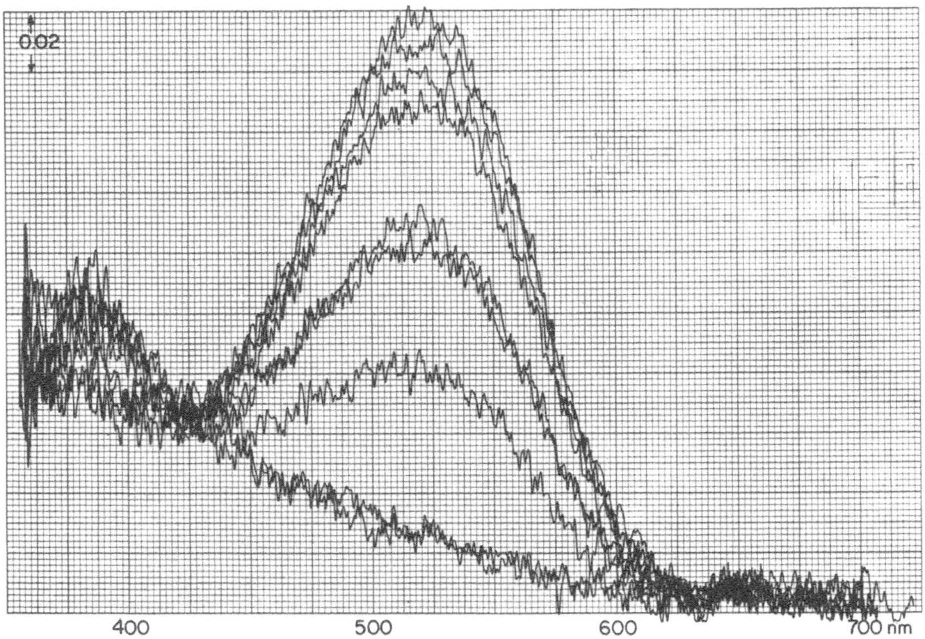

Fig. 8. Recordings from two single rods that were together. Test beam passed through both rods. They were then bleached in stages with light at 520 nm, except for final stage which was a total white light bleach. Maximum for each stage is fixed in position indicative of but a single pigment in both cells. Temperature: 25° C

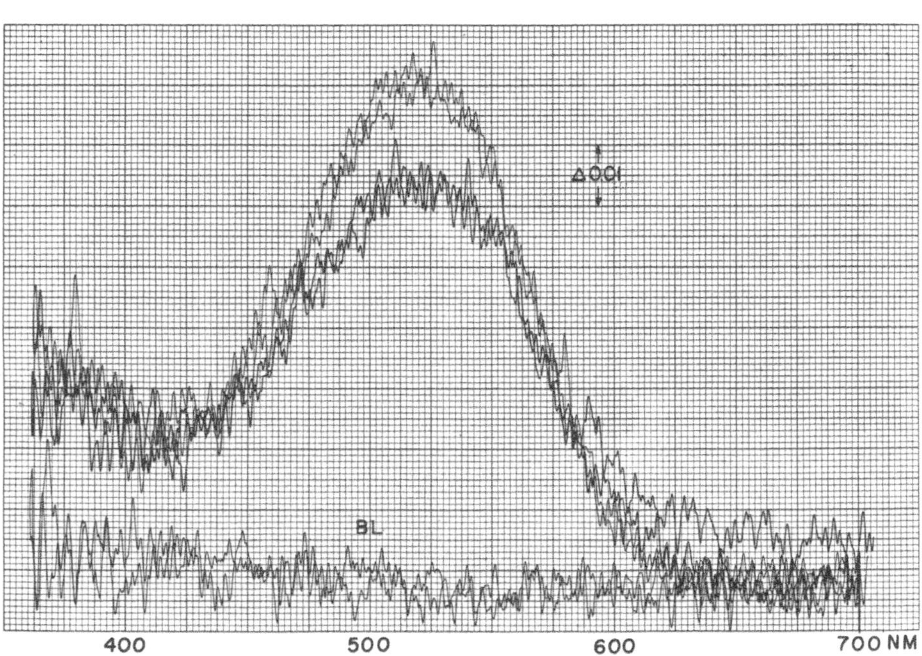

Fig. 9. Recordings from the two members of a double showing the 521-pigment in both members. *BL* is base line. Temperature: 23° C

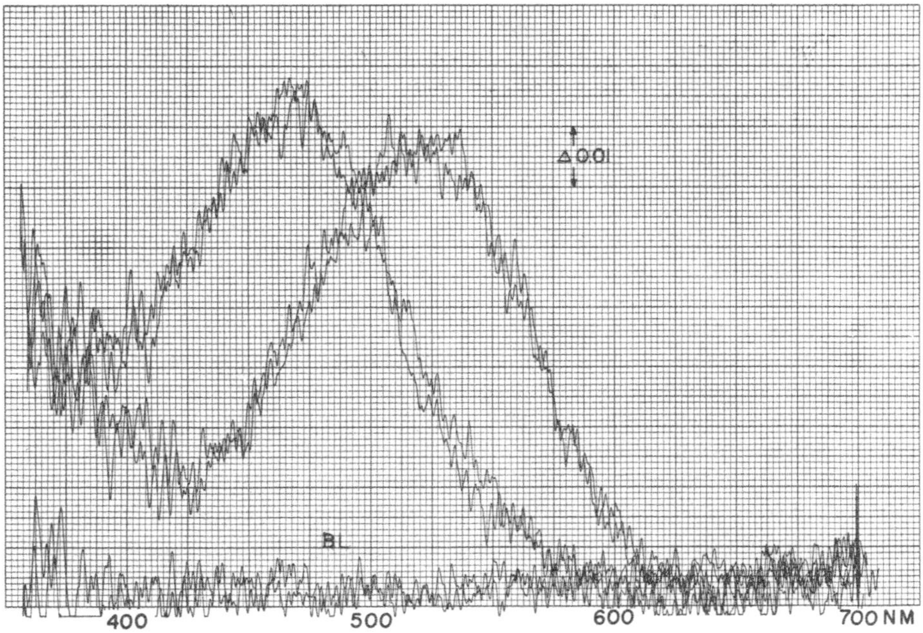

Fig. 10. Recordings from the two members of a double showing the 521-pigment in one member and the 467-pigment in the second member. *BL* is bas line. Temperature: 23° C

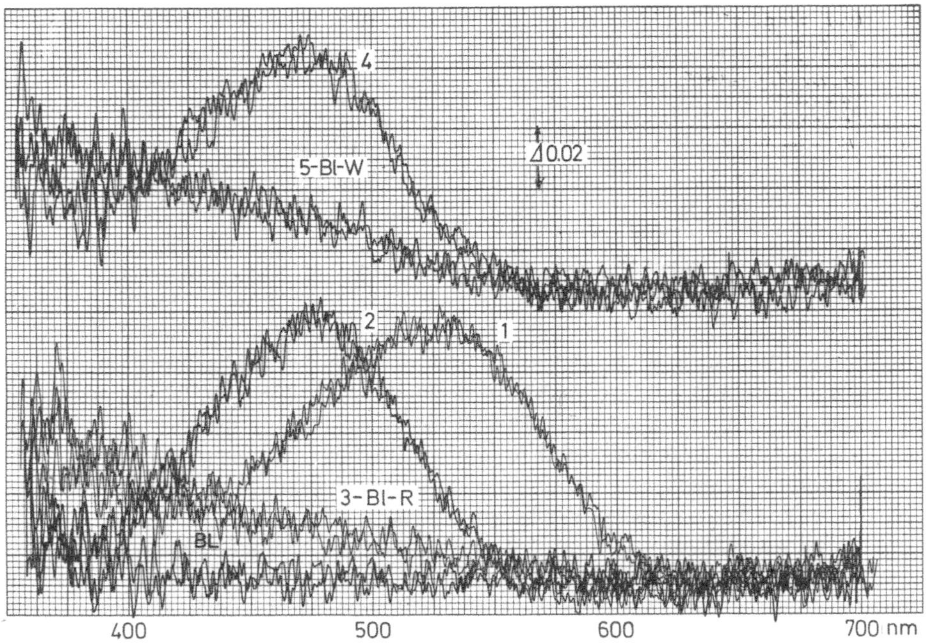

Fig. 11. Double-scan recordings from the two members of a double and partial bleaching of the two photopigments. Curve 1: scans of first member. Curve 2: scans of second member. Curve 3: scans of first member after a bleach of the entire field by red light *(Bl-R)*. Curve 4: scans of second member after the red bleach showing that the second pigment was still present. Curve 5: bleaching of the field with white light which resulted in loss of pigment from second member. *BL* is base line. These records were made with cells from Wahlberg's gecko. Temperature: 23° C

was able to remove the 521-pigment selectively (Fig. 11). As in the case of the single outer segments, each of these two photopigments in the 521-467 doubles displayed linear dichroism, and each was found at both ends of the outer segment. In summary then, the retina of *G. gekko* was found to contain two types of singles: S521 and S467; and three types of doubles: D521-521, D467-467, and D521-467. As yet we do not have a sufficient number of measurements to state, with any degree of certitude, the frequency of occurrence of these different spectral types. It is unnecessary to point out that these were the results with retinal squashes in which the integrity of the retinal mosaic is destroyed and in which the possibility exists of some doubles breaking up into single outer segments. It will eventually be necessary to identify, if possible, the spectral visual cell types in the intact retina. For this a photographic or cytochemical method may be required.

During our survey we encountered two triple cells (Fig. 6c) which occur as rare types in the gecko retina (DUNN, 1966a). Being careful to avoid the contact region between two neighboring members, we were able to record a 521-spectrum from each of two members and a 467-spectrum from the third member. This is insufficient to establish whether or not all triples are of this spectral designation.

It is not possible to identify these spectral types with the morphologic types A, B, and C of Fig. 1, because in preparing the retinal squashes all morphologic identifications are lost, and all we have are the isolated outer segments of Figure 6. Regardless of this, it is established that the same photopigment may occur in pairs that are not twins, employing the term twins to mean pairs of conjugate members that are cytologically identical and similar in size and retinal position. In this sense, neither the type B, nor type C doubles are twins (Figs. 1 and 3), and the gecko retina probably has few or no twins. Yet many double outer segments, dissimilar morphologically, house the same photopigment in both members. This bears on the general question, initiated with studies of teleost fishes, that different photopigments are associated with cytologically differing visual cells (SVAETICHIN et al., 1965; LIEBMAN, 1972; HÁROSI and MACNICHOL, 1974; STELL and LIGHTFOOT, 1975; SCHOLES, 1975; MUNZ and MACFARLAND, this volume). The finding that in geckos the same pigment may occur in morphologically different cells confirms the report of STELL and LIGHTFOOT (1975) that a blue-absorbing pigment occurs in both short single cones and in miniature single cones of the goldfish retina. It is an experimental fact that in *G. gekko* two spectral classes of double cells have evolved, one class having the same pigment (either 521 or 467), the second class having the two pigments segregated in the two members. The meaning of this is puzzling. Either this is an accident of development without any significance as to functional differences in these two classes, or else some subtle functional specialization exists which eludes speculation. We know nothing yet of the synaptic connections of the gecko visual cells.

Armed with this information, it is now possible to compare some properties of the gecko visual pigments while solubilized in aqueous digitonin and while in the natural environment of the outer segment. In the first place, it appears that a similar dual system of photopigments is demonstrated by both microspectrophotometry and the method of extraction. Attempts at discovering a red-absorbing photopigment have failed with both methods, and it now begins to appear that a visual pigment at longer wavelengths, so common in other vertebrates, may be lacking in geckos. The two pigments that have been definitely established are typical vertebrate visual pigments in conforming to the principle deduced by DARTNALL (1953) in constructing his nomogram (Fig. 12). Of special importance is the demonstration that the spectral absorbance of the 521-pigment *in situ* is in such good accord with the absorbance of the extract of this pigment, the latter absorbance being measured at low temperature (2.9° C) and in the presence of veronal-HCl buffer. This accord (Fig. 13), based not on difference spectra but on directly measured absorbances, appears to settle—for the gecko system at least—the question as to whether or not absorbance curves obtained with extracts are precisely applicable to the spectral absorbance of the pigment within the visual cell. The accord between the two absorbance curves in Fig. 13 depends on the extract being at low temperature. This is because the 521-pigment in digitonin displays a temperature-dependent spectral absorbance which is not the case for the pigment in the outer segment. In a later section, the role of temperature will be pointed out and this matter will be documented.

Another property of the 521-pigment that appears to be the same for the pigment *in situ* and in solution is related to the absence of a meta-III intermediate in the cycle of bleaching. This property, so unlike the behavior of other vertebrate visual pigments that have been studied, will be documented and discussed in a later section of this chapter.

It is one aim of visual science to relate the spectral absorbances of retinal photopigments to action spectra based on some visual response. Some success has been achieved in this direction, probably the most notable being the comparison of human rhodopsin with the human scotopic sensitivity function (CRESCITELLI and DARTNALL, 1953; WALD and BROWN, 1958). In the case of *G. gekko*, the action spectra that are available indeed suggest a role for both photopigments. In the interest of historical accuracy, it was, in fact, the finding by DENTON (1956) that the scotopic sensitivity function of the Tokay gecko does not accord with the classical position of rhodopsin that led me to the initial examination of the gecko system (CRESCITELLI, 1956). This examination quickly revealed that rhodopsin is not a component of the gecko retina, and that in *G. gekko* the spectral absorbance agrees roughly with Denton's sensitivity function (CRESCITELLI, 1963). Pursuing this matter further by means of the electroretinogram, I recorded two distinct b-waves: an initial, faster and more red-sensitive component (b_1) and a second, slower and more blue-sensitive wave (b_2) (CRESCITELLI, 1966). These could be partially isolated by selective bleaching, and in this way approximate spectral sensitivity curves for b_1 and b_2 were obtained. The sensitivity curve for b_1 has a maximum close to 520 nm and presumably reflects the activity of the 521-system, while the curve for b_2, probably contaminated by some b_1 contribution, is located at shorter wavelengths and involves the activity of the 467-photochemical system. Thus the dual-pigment system is correlated with a dual-response system in the living retina. Even if these are the only spectral mechanisms of the gecko retina, such a dual system poses the question as to its possible function in a nocturnal animal. One speculative answer is that it represents the vestige of a color vision system of the diurnal phylogenetic ancestor.

This, in summary, is the retinal system whose study will serve as the continuing theme of this chapter. In order to lend force to some of the newer findings, I

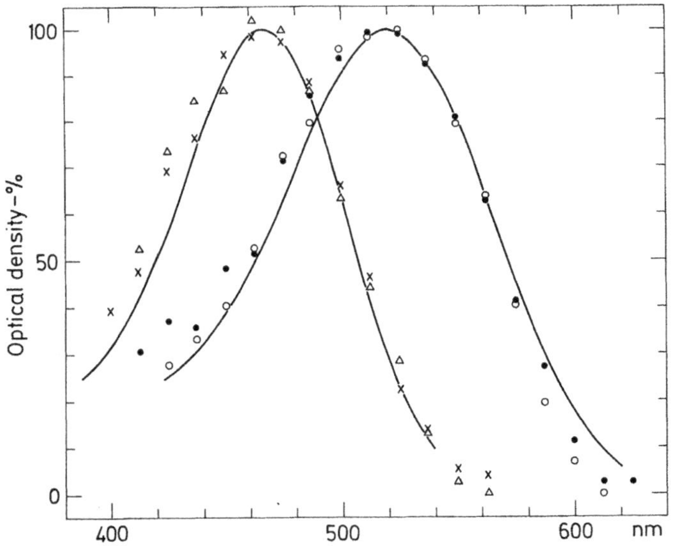

Fig. 12. Comparison of microspectrophotometric absorbances with nomogram curves for a 521- and a 467-pigment. Data points were from recordings made from the members of two separate 521/467 doubles. Full lines were derived from the DARTNALL nomogram (WYSZECKI and STILES, 1967). All curves normalized with maxima set a 100%. Open circles and crosses were from a double at 6° C; the solid circles and triangles were from the second double at 23° C

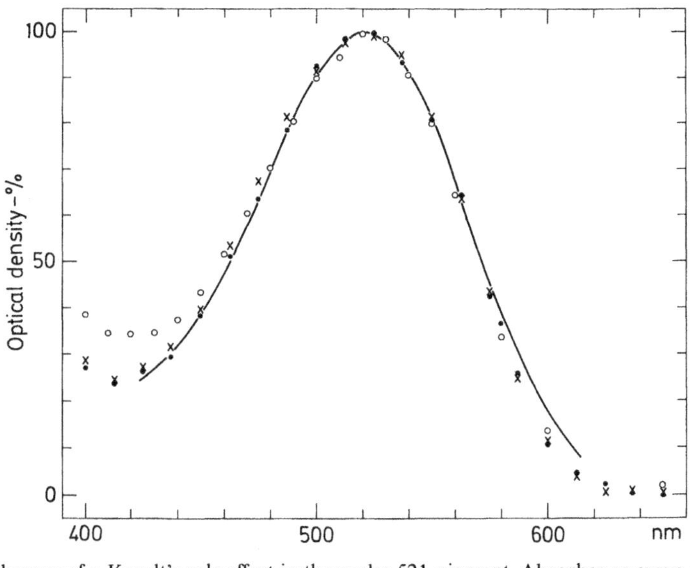

Fig. 13. Absence of a Kundt's rule effect in the gecko 521-pigment. Absorbance curve of extract *(open circles)* was obtained with pigment dissolved in 2% digitonin in veronal-HCl buffer (pH 7.2) and with temperature at 2.9° C. This is the directly measured spectrum corrected only by subtracting from it the 467-pigment absorbance which made up 8% of total photopigment density. Solid circles are means of three recordings from three single rods, measured at 6° C. Crosses are similar recordings from three other single rods but at 23° C. Solid line is a 521-nomogram curve. All data were normalized with maxima set at 100%

have, when possible, compared the gecko photochemical results with analysis of the systems in other vertebrates, especially that of the much-studied *Rana pipiens*. It will be recalled that the visual cells of this frog consist of single red rods, single green rods, single cones, and double cones (NILSSON, 1964). The red rods and the accessory member of the double cones, in the adult frog, contain a 502-pigment, the green rods a pigment at 432 nm, and the single cones, as well as the chief member of the double cones, a 575-pigment (LIEBMAN and ENTINE, 1968). These are the results of a microspectrophotometric inquiry. Extracts of the retina contain rhodopsin at 502 nm, and in some species *(R. crancrivora)* the green rod pigment at 433 nm (DARTNALL, 1967). The green rod pigment is more labile than is rhodopsin and will be destroyed by NH_2OH at concentrations of this reagent that do not affect the rhodopsin.

In the light of this background, I will now proceed to examine the properties of the photopigments extracted out of the retinas of geckos, especially *G. gekko*.

B. Gecko Photochemical System—a Comparative Examination

I. Metarhodopsin III in Frog Retinal Extracts

Extracts of the frog retina, when bleached at low temperatures, display as products of bleaching, not only metarhodopsin I (M-I) and metarhodopsin II (M-II) but a third colored product, metarhodopsin III (M-III), also called pararhodopsin. Historically, the knowledge of M-III goes back to 1938 when LYTHGOE and QUILLIAM (1938a) described a colored intermediate in the bleaching sequence of frog rhodopsin, to which they assigned the descriptive designation of transient orange. This was detected by rapidly bleaching rhodopsin at 3° C and then following the subsequent changes in extinction in the region between 470 and 490 nm. The transient orange, with peak absorbance near 470 nm, was noted seconds after the bleach and was seen to decay in the dark, over a period of some 40 min at 3° C, to form a product designated as indicator yellow by LYTHGOE and QUILLIAM. The spectral absorbance of transient orange was found to be independent of pH; its extinction was as large as that of the parent pigment; and it was considered to be a distinct substance in the pathway between rhodopsin and indicator yellow. It required 25 years before another report appeared mentioning such a transient intermediate in the bleaching of extracts of rhodopsin (MATTHEWS et al., 1963). This intermediate, absorbing at 465 nm, was shown to appear when M-II was kept in the dark at 3° C. The authors considered that the 465-pigment originates from M-II, is photochemically converted back to M-II, and is destroyed in the dark by NH_2OH. This 465-pigment is probably identical with transient orange and in agreement with this is the fact that the isosbestic wavelength of 412 nm for the conversion of transient orange to indicator yellow is identical with the isosbestic wavelength of 412.5 nm for the thermal conversion of

M-II to the 465-pigment. The 465-pigment was also considered to be a distinct substance, not on the direct pathway of decay of M-II, but a side product, bypassing the main sequence. This is the reason for pararhodopsin as the alternative name for M-III. MATTHEWS et al. (1963) speculated that the 465-pigment has 13-*cis* retinal as the prosthetic group. A somewhat different view was expressed by OSTROY et al. (1966) who used cattle rhodopsin. Based on several experimental observations, including one that M-III appears under conditions when M-II is not effectively irradiated, they concluded that M-III contains the all-*trans* retinal just as does M-II. Moreover, they proposed on the basis of experiments that involved the illumination of M-III that M-III yields in part rhodopsin and isorhodopsin, both *cis*-pigments, i.e. 11-*cis* and 9-*cis*. OSTROY et al. did not accept the idea that M-III is a side product of the main sequence of decay, but suggested on the basis of experimental findings that it is a direct intermediate between M-II and N-retinylidene-opsin (indicator yellow). These investigators also speculated on the possible nature and function of M-III, i.e. that M-III involves a reprotonation of the Schiff-base nitrogen of M-II, and that M-III has properties similar to the retinene isomerase of HUBBARD (1956).

II. Metarhodopsin III in the Visual Cell

It is generally accepted by most investigators that M-III is an intermediate in the sequence of products following the bleaching of rhodopsin in the visual cell. MATTHEWS et al. (1963) reported the occurrence of both the M-II and the 465-pigment in the frog retina. Several investigators have definitely identified M-III as a relatively long-lived intermediate occurring in the bleaching of rhodopsin in the frog retina (FRANK, 1969; BAUMANN, 1972). BAUMANN'S paper proposed a model for the thermal decay of M-II to retinal *via* two alternative pathways: one through M-III and the second directly to retinal. This model has received support from the work of BOWMAKER (1973) and of GYLLENBERG et al. (1974). Of these two pathways in the frog retina, the one leading through M-III appears to be an important one, for BAUMANN (1972) found the reaction rate constant of formation of M-III from M-II to be about double the constant for hydrolysis of M-II to retinal.

A number of recent studies have also supported the idea that M-III participates in the cycle of events of bleaching that occur in the mammalian eye, including the eye of man. RUSHTON (1956) noted that the difference spectrum due to the bleaching of human rhodopsin *in situ* varied according to the duration of the bleaching exposure. After a 10-min bleach, the difference spectrum agreed with the reconstructed human rhodopsin absorbance curve of CRESCITELLI and DARTNALL (1953). But after a bleach of less than 2 min, the resulting spectrum was shifted to significantly longer wavelengths. Of three possible explanations for this shift, RUSHTON favored the idea that an evanescent orange product, perhaps LYTHGOE and QUILLIAM'S transient orange, was formed in the rods and persisted long enough to distort the difference spectrum resulting from a relatively short bleach. The formation and decay of an orange product following a short bleaching exposure of the eye of several mammals has been noted in a number of investigations (HAGINS, 1957; WEALE, 1967; RIPPS and WEALE, 1969; ALPERN, 1971; BONDS and MACLEOD, 1974). In harmony with these studies is the finding of M-III in extracts of the mammalian retina. This intermediate has been noted in extracts of the cattle retina (OSTROY et al., 1966), the rabbit retina (CRESCITELLI, unpublished), and the human retina (CRESCITELLI and PEARLMAN, unpublished). It does appear that M-III is a significant intermediate in frogs and mammals and may have some, as yet undisclosed, physiologic role. Whatever this role may be, it will now be shown that it is not an essential one in all visual systems, for M-III is not apparent in the gecko system, as the following section shows.

III. The Gecko 521-Pigment: A System Without M-III

Whatever the physiologic significance of M-III may be, it is a fact that this transient product is not a component of bleaching either extracts of the gecko 521-pigment or of outer segments containing this pigment. In order to be certain of my results, I first established the point that I was able to demonstrate the occurrence of M-III in frog rhodopsin. An extract containing rhodopsin when bleached with light at 600 nm gave only the final bleached product with no evidence of M-III if the experiment was made at 25.8° C. Compared to this, another sample of the same extract when bleached quickly with light at 500 nm and at 2.8° C changed from curve 1, the rhodopsin spectrum, to curve 2 with a maximum at 468 nm (Fig. 14). At this low temperature the 468-product decayed slowly, requiring hours to disappear, but when the temperature was raised to 25.7° C it decayed quickly to yield curve 3. The difference spectrum for this thermal decay (curve 2 minus curve 3) gave a density loss maximum at 469 nm, close to the peak of curve 2. In agreement with MATTHEWS et al. (1963), the 469-product was destroyed by NH_2OH, for it failed to appear—even at 2.3° C—when bleaching was carried out in the presence of added NH_2OH. Alternatively, the addition of NH_2OH to an extract that contained M-III caused a rapid disappearance of this product.

In contrast to these positive results, similarly prepared extracts of the gecko retina failed to show any indications of an intermediate comparable to frog M-III (Fig. 4). This negative result was obtained with extracts made using phosphate, borate, and veronal buffers and at several pH values. In no case was evidence obtained for the presence of an M-III following bleaching at these low temperatures. In this laboratory and with the same equipment, BOWMAKER (1973) failed to see an M-III in the isolated gecko retina, altough it did appear in the bleaching of the frog retina. In the microspectrophotometric study of CRESCITELLI et al. (1976), the bleaching of the 521-pigment at 6° C revealed no evidence of a relatively long-lasting M-III (Fig. 15). Accordingly, the results are in accord for extracts, for the isolated retina, and for the isolated outer segments: M-III does not appear in the cycle of products of bleaching under conditions that lead to M-III for the frog rod system. It is of some interest here to point to the finding by DONNER et al. (1974) that a metaporphyropsin III apparently does not occur in the crucian carp pigment.

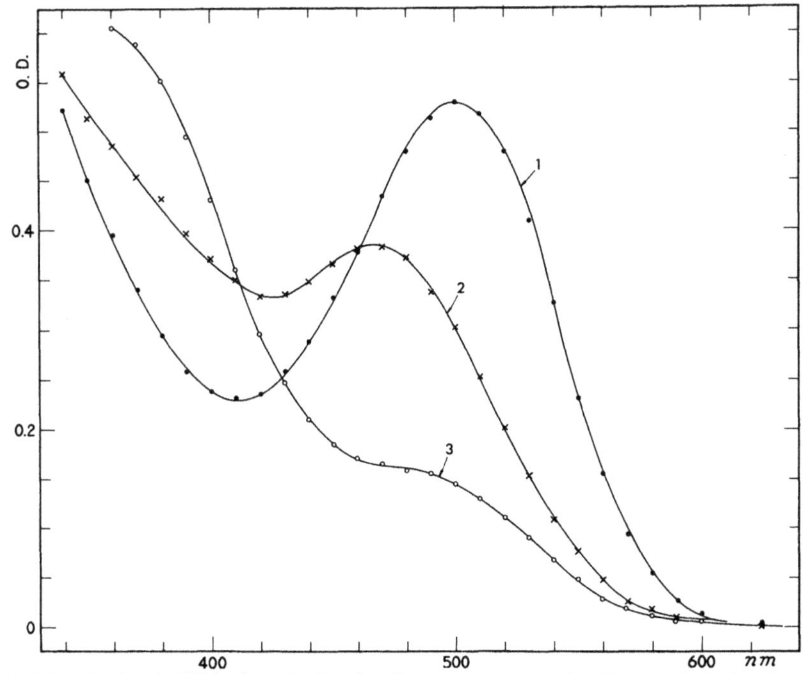

Fig. 14. Metarhodopsin III in frog rhodopsin. Curve 1: spectral absorbance of rhodopsin at 2.8° C. Curve 2: after an exhaustive bleach with light at 500 nm. Curve 3: after warming to 25.7° C and waiting for 5.5 hrs; the M-III decayed slowly in the dark

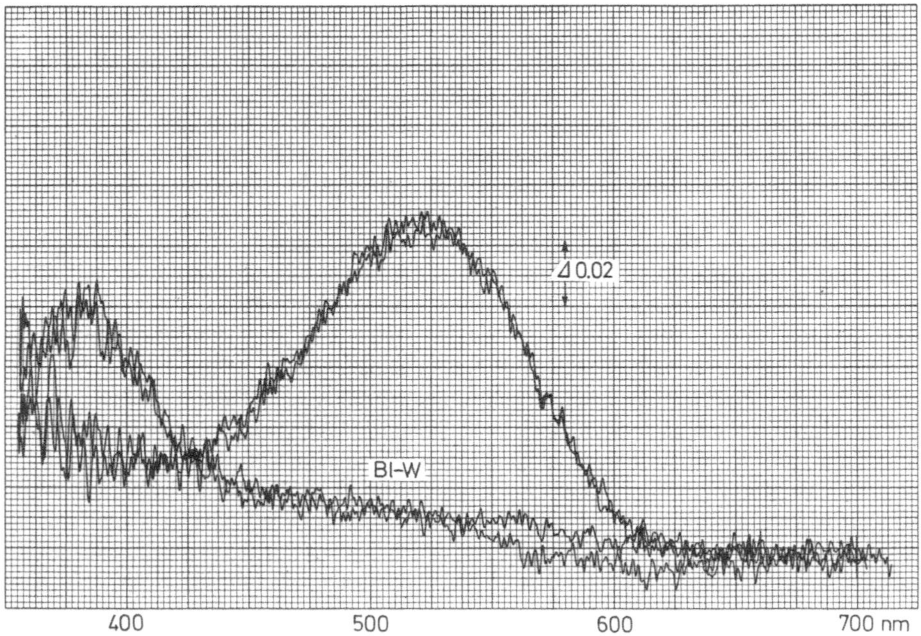

Fig. 15. Microspectrophotometric recordings from a single outer segment at 6° C before and after a total white light bleaching *(Bl-W)*. Note absence of metarhodopsin III

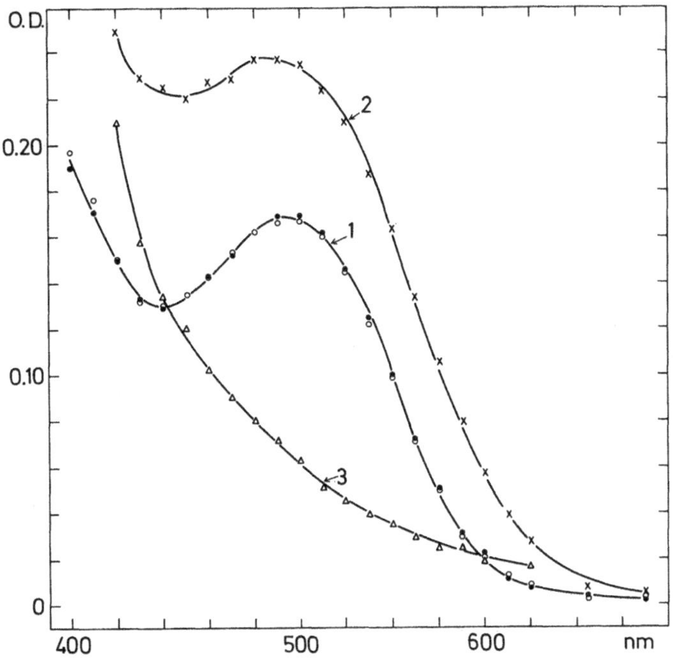

Fig. 16. Sensitivity of the gecko 521-pigment to NH_2OH in comparison with the inertness of frog rhodopsin. Curve 1: spectral absorbance of rhodopsin before *(filled circles)* and 51 hrs after adding NH_2OH to a concentration of 0.01 M *(open circles)*. Gecko pigment before (curve 2) and 2 h after (curve 3) adding the same NH_2OH to the same concentration, 25.6° C

IV. Sensitivity to Hydroxylamine

Hydroxylamine (NH_2OH) does not alter the spectral absorbance of many visual pigments whether these are solubilized in digitonin or present within the visual cells. Not all visual pigments are so inert to this reagent, and even different pigments extracted out of the same retina have differing sensitivities to NH_2OH. The green rod pigment of certain anuran amphibians, for example, is destroyed by concentrations of NH_2OH that leave the rhodopsin untouched (REUTER, 1966; DARTNALL, 1967). A similar situation exists for the two pigments of *G. gekko*, but in this case it is the 521-component that is more sensitive. A direct test has shown that this component is labile to concentrations of NH_2OH that do not affect frog rhodopsin. In one such experiment (Fig. 16), the two different sets of data points of curve 1 indicate that an exposure for 51 hrs to NH_2OH (0.01 M) resulted in no loss of frog visual purple. In contrast, the gecko pigment (curve 2) was almost completely lost in 2 hrs by this same concentration of NH_2OH (curve 3). This lability of the gecko pigment to this reagent has meant that it cannot be added to an extract before a bleaching analysis, but it must be added terminally after a total light bleach if one wishes to obtain difference spectra relatively undistorted by the usual products of bleaching produced in the absence of NH_2OH. This lability toward NH_2OH does not extend to the 467-pigment. This is illustrated by Fig. 17

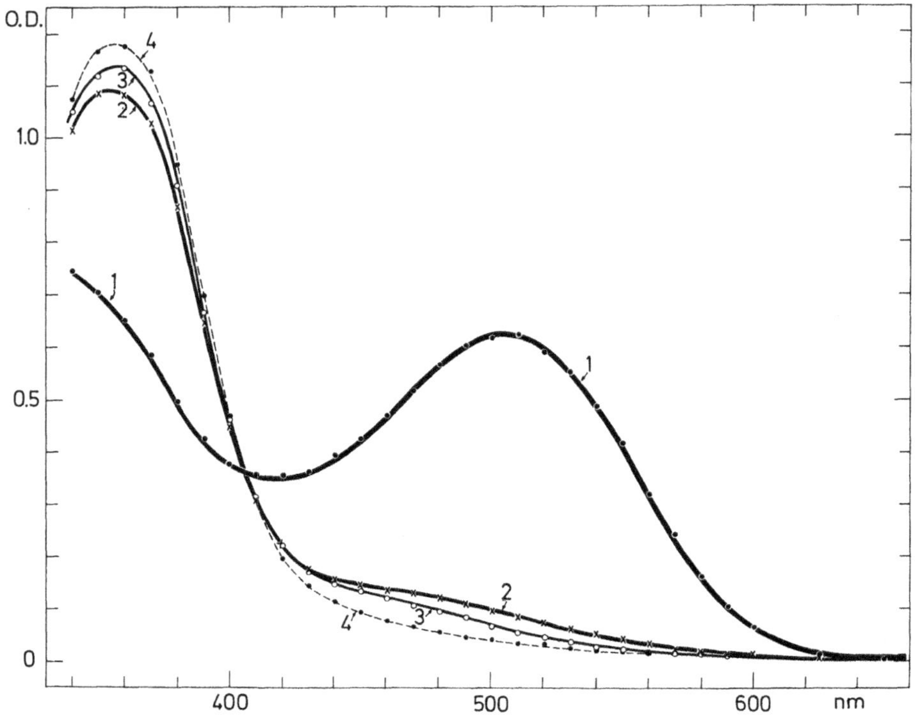

Fig. 17. Differential sensitivity of the two gecko pigments to NH_2OH. Curve 1: absorbance of an extract containing the two pigments. Curve 2: same extract 21 hrs in the dark after adding NH_2OH (0.01 M). Curve 3: a 620 nm bleach to remove the remnant 521-pigment. Curve 4: a 500 nm bleach to remove the 467-pigment. Temperature was 2.7° C

in which the extract containing the two pigments (curve 1) was treated with NH_2OH for 21 hrs in the dark during which most of the 521-pigment was destroyed (curve 2). The remainder of the 521-pigment was then removed by an exposure to light at 620 nm (curve 3). Following this, the extract was then bleached with light at 500 nm (curve 4) which removed the 467-pigment as indicated by the fact that the difference spectrum loss for the 500 nm bleach was maximal at 468 nm, and the product produced was maximal at about 364 nm, close to the maximum for the oxime of retinaldehyde. The density of the 467-pigment surviving the extended action of NH_2OH which destroyed nearly all of the 521-pigment was 7.5% of the total photopigment density. This means that the 467-component in this extract was virtually untouched, while the 521-pigment was lost. The basis for this differential sensitivity to NH_2OH on the part of the two gecko pigments is not known, but it could be a difference in the opsins of the two such that the chromophoric structure of P467 is protected from attack by the hydrophilic NH_2OH as in the case of frog rhodopsin. Whatever the reason, it is clear that this difference in sensitivity has no relation to those molecular features of the pigments that determine the spectral location. In the case of anurans, it is the blue-absorbing pigment that is the more sensitive; for the gecko system, it is the pigment at 521 nm.

V. Sensitivity to Borohydride

Over a wide range of pH values, rhodopsin is neither reduced, nor is its spectral absorbance altered by potassium or sodium borohydride ($NaBH_4$) as has been indicated by WALD and HUBBARD (1960), BOWNDS and WALD (1965), AKHTAR et al. (1965), HELLER (1969), and ZORN (1971). For my own satisfaction, I have tested this point with frog and rabbit rhodopsins and found that these are indeed stable toward added $NaBH_4$. If, however, rhodopsin is bleached and then treated with $NaBH_4$ the system becomes open to attack, apparently at the M-II product; the result of this reduction has been named dihydro-metarhodopsin II by AKHTAR et al. (1965) and N-retinyl opsin by BOWNDS (1967). Analysis of the reduced product revealed that the retinyl group is attached to the ε-amino group of a lysine residue, thus suggesting that this is also the primary site of attachment in native rhodopsin. The analysis by BOWNDS (1967) also provided some information about the amino acids adjacent to the lysine residue in N-retinyl opsin. A preponderance of these turned out to be amino acids with nonpolar side chains, and BOWNDS concluded that in native rhodopsin the retinyl group exists in an hydrophobic environment, thus accounting for some of its stability to aqueous reagents. BOWNDS was aware of the possibility of differences that might occur in visual pigments from other species in regard to the nature of the environment of the chromophoric group.

The 521-pigment of *G. gekko* is one such pigment, for it is destroyed by $NaBH_4$ in the dark and at low temperature. Moreover, the two gecko pigments are not equally sensitive to this reagent, P-467 persisting after P-521 has been destroyed by added borohydride. A complete experiment is summarized in Fig. 18 which includes the original spectrum (curve 1), read at 3.3° C, and the spectrum after adding a small grain of $NaBH_4$. The photopigment was quickly destroyed within 30 min, but readings could not be obtained sooner than this because of the bubbling due to the liberated hydrogen. The extract was then kept in the refrigerator at 5° C for 17 hrs, and the next morning the spectrum was reread, this time at 20° C. As curve 2 shows, the major portion of the pigment was lost but a definite elevation remained with a peak between 460 and 470 nm. This elevation was removed by exposing the extract to room light (curve 3) and the difference curve (2 minus 3), shown in the inset, reveals the loss of a pigment with maximum at about 465 nm, thus identifying it as P-467. Thus while P-521 was destroyed within 30 min by the $NaBH_4$, P-467 remained in its entirety after some 17 hrs.

VI. Sensitivity to Temperature

In my original studies of the gecko photopigments (CRESCITELLI, 1956, 1963), I failed to report any unusual behavior toward temperature. This was apparently due to the fact that I employed a constant temperature and did not investigate the effect of varying this experimental parameter. Recently, I returned to my old notebooks and discovered that in 1963 I did indeed observe some puzzling behavior in relation to changes in temperature. Consequently, I incorporated a low

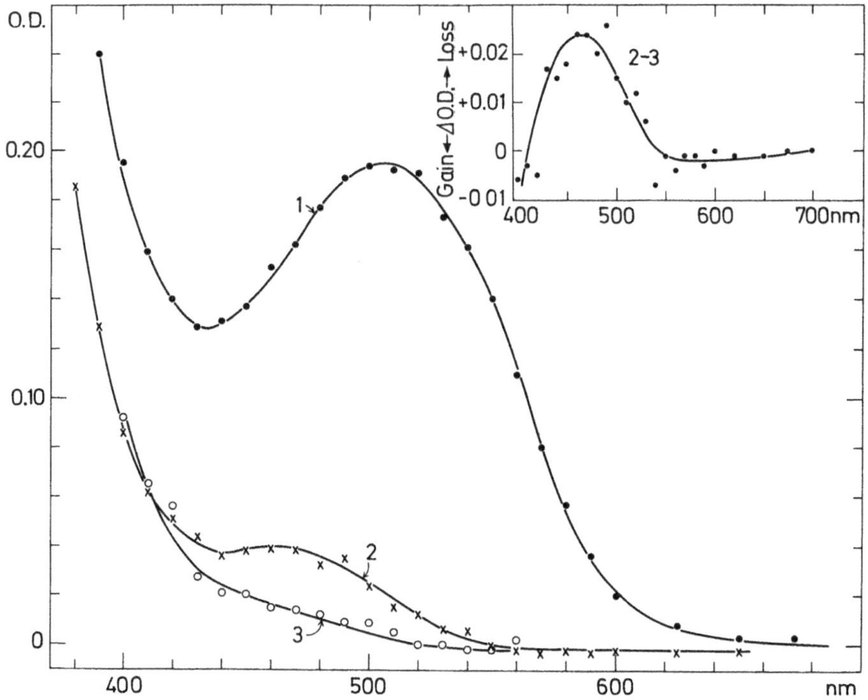

Fig. 18. Differential sensitivity of the two gecko pigments to $NaBH_4$. Curve 1: absorbance of an extract containing the two pigments. Curve 2: after destruction of all the 521-pigment by $NaBH_4$. Curve 3: photic bleaching of the remaining 467-pigment (*inset*) by white light. Temperature was 2.8° C

temperature control system into my spectrophotometer chamber and immediately observed that the spectral properties of gecko retinal extracts do in fact vary with temperature and in a manner that cannot be ignored. Two reports have already appeared on this matter (CRESCITELLI, 1974 a, b), and in this chapter I wish to put together additional information that I have gained since publishing the initial statements. I shall start by relating the results of experiments with the standard rhodopsin of *R. pipiens*. The contrast in behavior of rhodopsin and the gecko pigment will serve as an object lesson in comparative biology.

1. Frog Rhodopsin

Like most investigators of visual pigments, I have become accustomed to assume that the spectral properties of rhodopsin solutions are but *little* influenced by temperature in the range of 2° to 25° C. This is true, but it is not true that there is *no* effect. As pictured in Fig. 19, the spectral curves at 2.8° and 25.5° C, though generally similar, are not identical. The wavelengths for maximal absorbance at these two respective temperatures are at 502 and 498 nm, not a large difference but one that is repeatedly observed. At 25.5° C the extinction is seen to be less than for 2.8° C at wavelengths longer than about 470 nm, the greatest difference

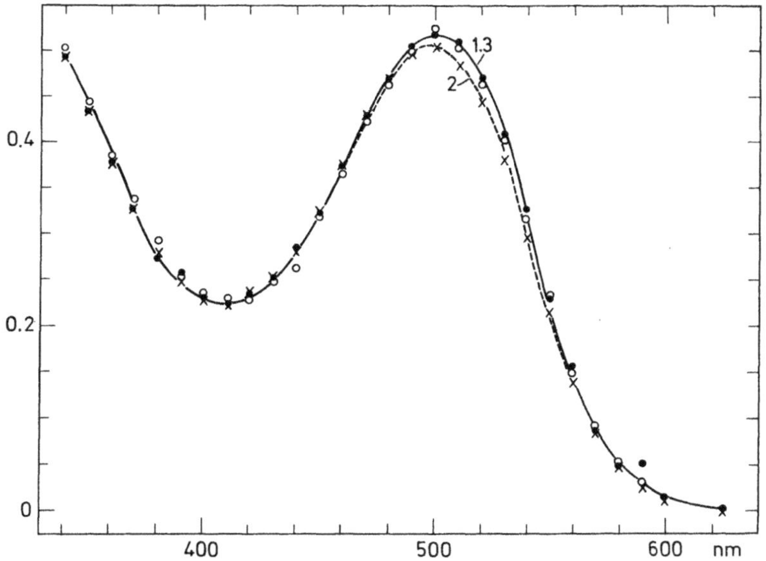

Fig. 19. Frog rhodopsin: spectral absorbance initially at 2.8° C (*filled circles*, curve 1), after raising temperature to 25.5° C (*crosses*, curve 2), and after return of temperature to 2.4° C (*open circles*, curve 3)

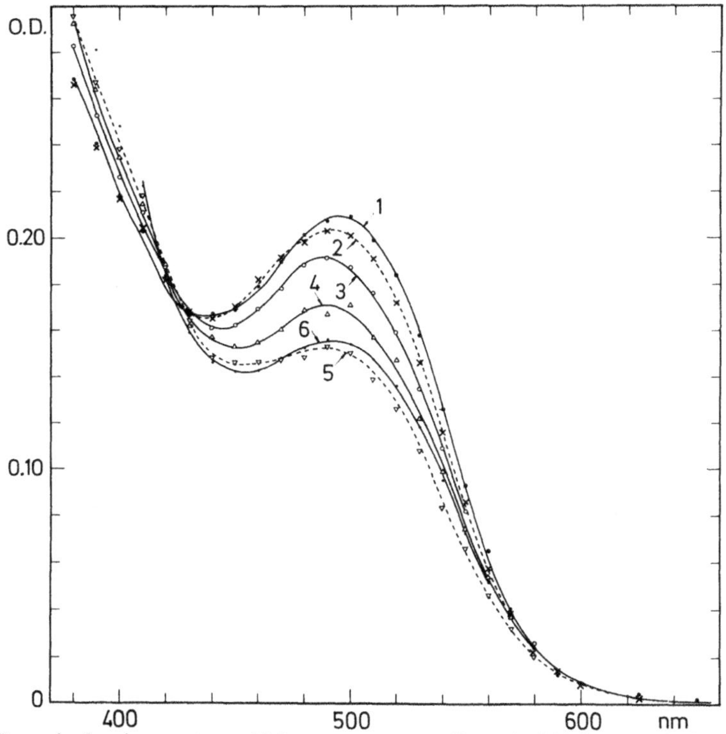

Fig. 20. Frog rhodopsin: spectra at higher temperatures. Curve 1: 3.1° C; curve 2: 30.1° C; curve 3: 45.1° C; curve 4: 50.1° C; curve 5: 50.1° C for 30 more min. Curve 6: after return of temperature to 3.1° C

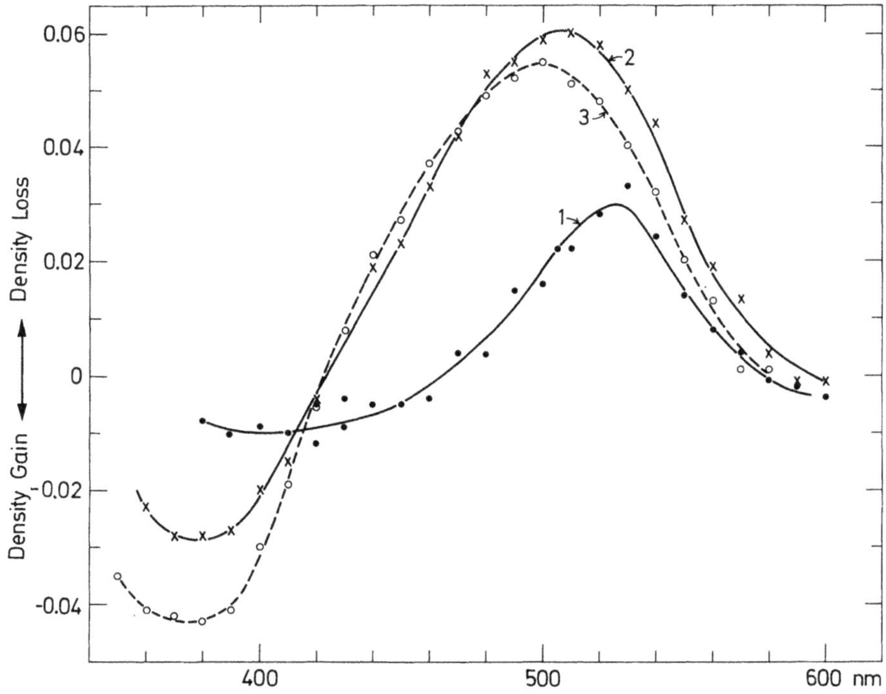

Fig. 21. Frog rhodopsin: thermal difference spectra. Curve 1: difference spectrum for range of reversible temperature effects; this is the result of a temperature transition from 3.2° to 30.4° C. Curve 2: the result of a transition to 50.1° C; initial curve being read at 3.1° C, final curve at 50.1° C. Curve 3: result of a transition to 50.1° C; initial curve being read at 3.1° C, final curve after temperature had been restored to 3.1° C. These difference spectra should be viewed in relation to the spectral absorbances of Fig. 20

being at 530 nm. This effect of going from 2.8° to 25.5° C is easily reversed by returning to the low temperature (Fig. 19, *open circles*). Such minor, reversible alterations in spectrum of frog rhodopsin have been obtained in temperature steps up to about 35° C. At temperatures above this, progressive and irreversible thermal damage occurs, the rate and extent of which is related to the temperature (Fig. 20). The irreversible thermal action appears to be of a qualitatively different character than the reversible thermal effects of temperatures lower than 35° C. This is suggested by the dissimilar form of the difference spectra obtained with the two types of action (Fig. 21). For the reversible transition (curve 1), the density loss is maximal at 526 nm, and the slight increase in density at lower wavelengths is poorly defined spectrally. In comparison, a temperature step from 3.1° to 50.1° C resulted in a loss with maximum at 506 nm, a well-defined gain at 380 nm and a ratio loss/gain of 2.0 (curve 2). Since the action of a temperature of 50.1° C is largely irreversible (curves 5, 6; Fig. 20), it is possible in such experiments to return to the low temperature (without recovering density) and to read the spectrum of the denatured extract at the initial temperature. In this way, a correctly expressed difference curve can be obtained and is shown as curve 3, Fig. 21. This curve has a maximum loss close to 500 nm and a density gain maximum at 375 nm; it indicates a simple irreversible destruction of rhodopsin by the higher temperature. Consequently, while the effect of the lower temperature (curve 1) is mainly the result of an hypsochromic shift of the pigment absorption, the effect of the higher temperature (curve 3) is a thermal destruction that cannot be repaired.

It is the thermal denaturation and not the thermal hypsochromic shift that has occupied the attention of previous workers in this area of study. Interestingly, this subject also attracted the interest of that great pioneer, Willy KÜHNE. Along with EWALD (EWALD and KÜHNE, 1878), KÜHNE tested the effect of various temperatures on frog rhodopsin while in solution in bile and also while intact within the rods. They noted a greater sensitivity to heat for the pigment color in solution as compared with the color in situ. The lowest temperature at which significant bleaching occurred was just above 50° C. EWALD and KÜHNE reported no regeneration of color on cooling for either the retina or for the extracted rhodopsin. Later, LYTHGOE and QUILLIAM (1938 b) examined the kinetics of thermal bleaching for frog rhodopsin solubilized in 1% digitonin. They concluded that this process follows first-order reaction kinetics, that it is affected by pH, and that NaCl added to the extract retards the rate of breakdown. According to LYTHGOE and QUILLIAM the product of bleaching, indicator yellow, is the same for the actions of heat and light. HUBBARD'S study of cattle rhodopsin (HUBBARD, 1958) revealed, however, that light and heat bleach by different mechanisms, light acting by isomerizing the 11-*cis* retinal, heat by denaturing the opsin. Moreover, the pigment was noted to be thermally more stable when in the rods than when solubilized in digitonin, and opsin was less stable than rhodopsin in both conditions. It is clear that none of these early investigators were concerned with the effects of temperatures in the range between 2° and 25° C, which is the temperature region where the thermal hypsochromic shift occurs without the complication of irreversible thermal damage. This mild temperature effect may be interpreted in general terms as a reversible change in structure of the opsin that affects the rigid interaction between the retinaldehyde and protein in a manner that leads to a small displacement of the absorption to shorter wavelengths. It is not related to the Keilin-Hartree effect (YOSHIZAWA and WALD, 1966), which apparently is the result of microcrystals in the solution forming under certain conditions when the extract is cooled to *very low* temperatures.

The reversible thermometric effect shown by retinal extracts of *R. pipiens* brings up a question as to the specific cellular source of the pigment giving this effect. As reported by LIEBMAN and ENTINE (1968), a pigment with an absorbance maximum of 502 nm is found in the red rods of this frog and also in the accessory member of the double cones. There is no way of knowing if, and how much, of the photopigment in digitonin extracts is the cone-502 pigment. Perhaps the small thermometric effect of these extracts is due to the behavior of the cone-502 and not of the rhodopsin of the red rods. This is an idea suggested by SILLMAN (1976) who reported a small, but a well-defined thermometric effect in extracts of the bullfrog retina. At the present state of knowledge, the merits of this idea cannot be evaluated. The difference, if any, between rod and cone rhodopsin is of some interest from several points of view, but nothing can be said about this until a method is devised to separate and study the separate pigments.

2. The Gecko Pigment

a) The Thermometric-Hypsochromic Effect: A Thermochromic Property

The extracted visual photopigments of geckos undergo a significant hypsochromic shift that is easily and repeatedly demonstrable (CRESCITELLI, 1974a, b). In the range from 2° to about 25° C the spectral absorbance shifts to progressively shorter wavelengths as the temperature is raised in stepwise fashion. In a typical experiment (Fig. 22) the absorbance maximum went from 506 nm at 3.2° C (curve 1) to 491 nm at 26.4° C (curve 6). This shift of 15 nm is to be compared with the 4 nm shift for frog rhodopsin when subjected to nearly the same temperature step (Fig. 19). Associated with the thermal absorbance shift of the gecko pigment (Fig. 22) is a loss of extinction at wavelengths greater than about 463 nm (the isosbestic wavelength) the magnitude of this loss being maximal at 543 nm. In warming to 26.4° from 3.2° C (curves 1 to 6), the loss at 543 nm is 35% of the initial density, a figure which is considerably greater than the 7% value for frog rhodopsin (Fig. 19). As in the case of frog rhodopsin, both the hypsochromic shift and the extinction loss can be reversed by recooling the warmed extract, which is illustrated in Fig. 22 by the solid circles of curve 1. The extinction loss and recovery for three successive temperature steps are shown in Fig. 23, which are the results of an experiment, not with *G. gekko* but with *G. stentor*. This figure illustrates three important characteristics in the behavior of these gecko pigments to mild temperatures: (1) the rapid loss of density to an increase in temperatures; (2) the rapid recovery of the extinction upon cooling; and (3) the approximate steady state level attained at each temperature, a level that is lower at each higher temperature.

The extinction loss occurring at the wavelength of maximum loss follows the temperature change faithfully as is indicated in experiments in which both temperature and density, as functions of time, were measured during a slow increase and decrease of temperature of the spectrophotometer chamber. The change in temperature was produced by circulating water from a constant temperature bath through the temperature blocks of the chamber. Temperature was recorded by means of a thermistor sealed into one of the blocks, the block system having been previously calibrated against a second thermistor placed inside a cuvette filled with digitonin as in the actual experiments. The progressive loss in extinction (plotted upward) in a slow temperature step from 15° to 25° C and the reversal of this loss (plotted downward) in going from 25° to 5° C are given in Fig. 24. The temperatures in the cell, as derived from the calibration curves, are given as solid lines, while the experimentally measured changes in density (at wavelength of maximum loss) are given as open circles. It is evident that the changes in density, whether losses or gains, are as good indicators of temperature as is the thermistor. In other words, the pigment behaves as a thermometer, and this behavior may be designated as a thermometric or thermochromic property, suggestive of a temperature-sensitive conformational transition of the opsin. These results do not suggest a qualitatively different behavior from frog rhodopsin; they suggest rather the same behavior but greatly magnified. As a matter of fact it was the discovery of this magnified effect in the gecko pigments that led me to search for it in frog rhodopsin.

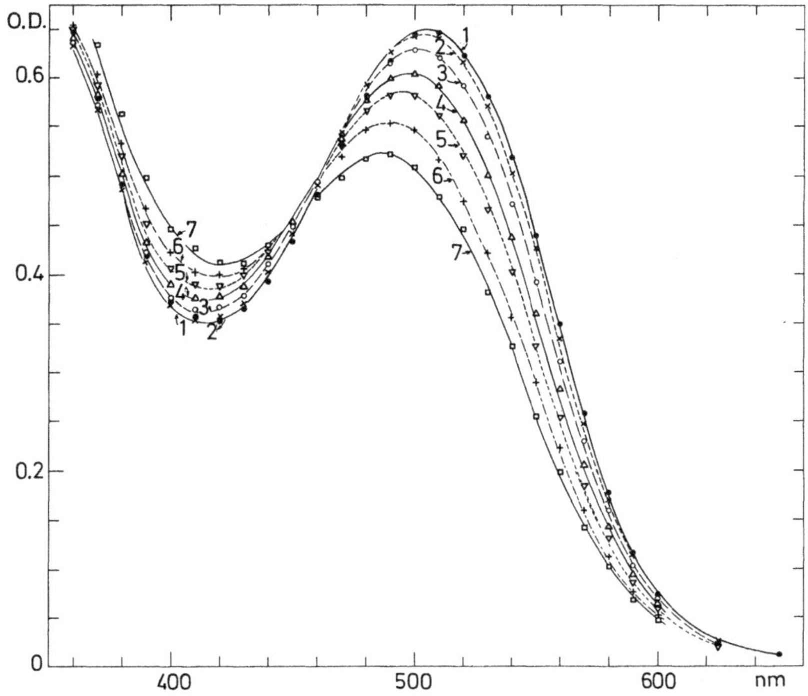

Fig. 22. The gecko pigment: thermochromic effect. Curve 1: solid line, 3.2° C; curve 2: 6.2° C; curve 3: 11.4° C; curve 4: 16.7° C; curve 5: 21.3° C; curve 6: 26.4° C. The temperature was then reduced to 2.3° C, and the data for curve 1 *(solid circles)* show the complete reversibility of the effect. Finally, the temperature was raised to 30.5° C and curve 7 was obtained

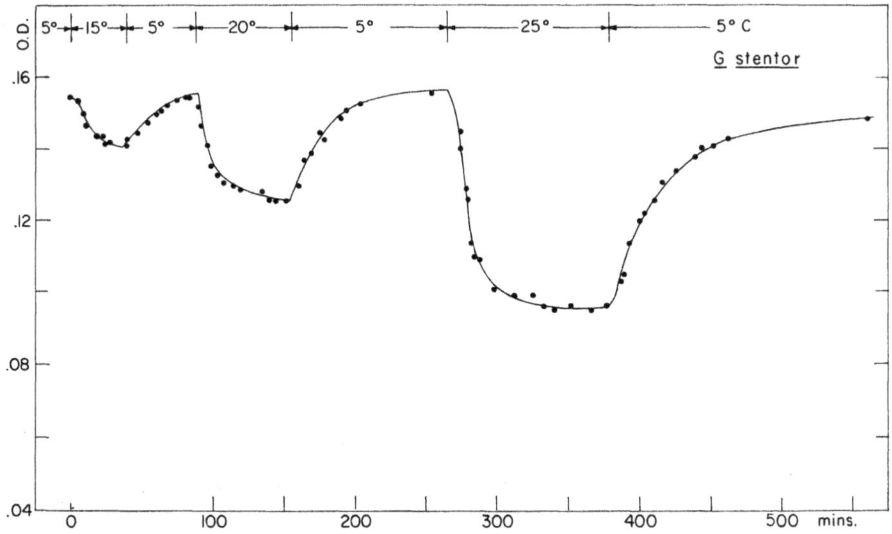

Fig. 23. Thermal effects for the pigment of *G. stentor*. Three successive temperature transitions are pictured: 5° to 15° to 5°, 5° to 20° to 5°, and 5° to 25° to 5° C. The relatively rapid loss and recovery of density (at maximum) are shown

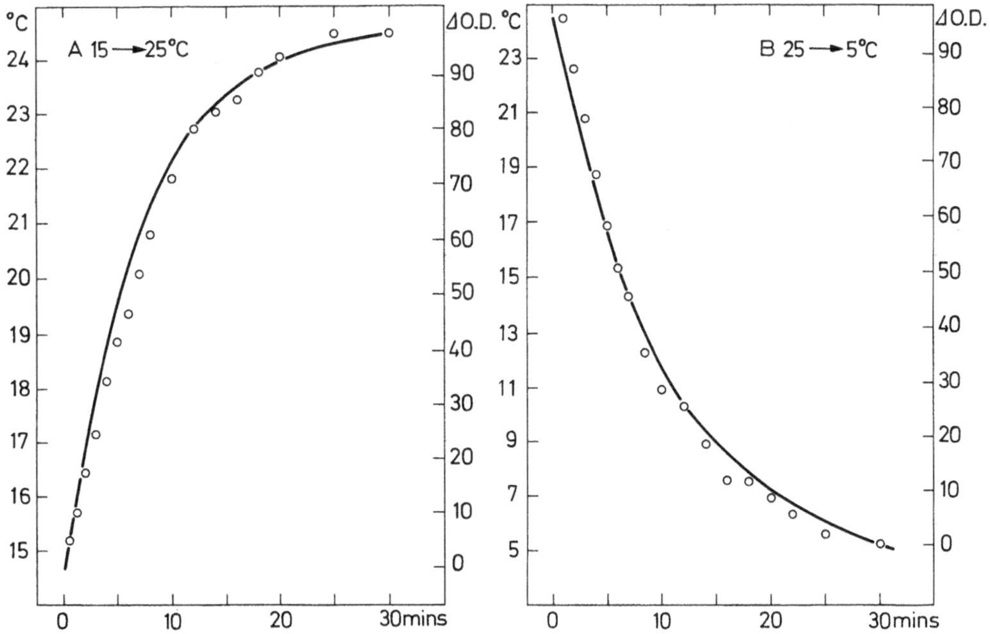

Fig. 24. Course of pigment change in transition from 15° to 25° C *(A)* and of recovery in transition from 25° to 5° C *(B)*. Pigment density change at wavelength of maximum change (525 nm) is shown as *open circles* with total change adjusted to same scale as temperature change. Pigment change is plotted with increasing loss upward *(A)* and increasing gain downward *(B)*. Pigment was from *G. gekko*. *Solid lines* represent temperature change as recorded by calibrated thermistor in block

In my original description of the thermometric effect (CRESCITELLI, 1974a), I conceived of this effect as an equilibrium between a thermal bleaching and a regenerative process, the equilibrium being shifted toward greater bleaching at the higher temperatures within the thermometric range. As the result of further analysis, I have abandoned this idea of an equilibrium, because I have been unable to detect the components of such an equilibrium at any given thermometric temperature. In the experiment of Fig. 25, for example, the sample of an extract was brought to 25.5° C and was then photically bleached in three separate, and approximately equivalent, exposures to light at 630 nm. Another sample of the same extract was bleached completely also at 25.5° C in one step by exposure to the same light. The three separate partial bleachings gave results indicated by the three different symbols, while the result of the single total bleach is plotted as a solid line. It appears that there is no significant difference in the results obtained by the three separate partial bleachings and that these are similar to each other and to the single total bleach. The maximum density loss for each of these bleaches is at 504 nm which, compared to the difference curve for a low temperature bleach (Fig. 5), is some 19 nm farther toward the blue end of the spectrum. In another and similar experiment to that of Fig. 25, the analysis was carried out at 14.6° C. In this case, the same difference spectra were obtained for the total and for the three partial bleaches, but the maximum density loss was at 515 nm. In

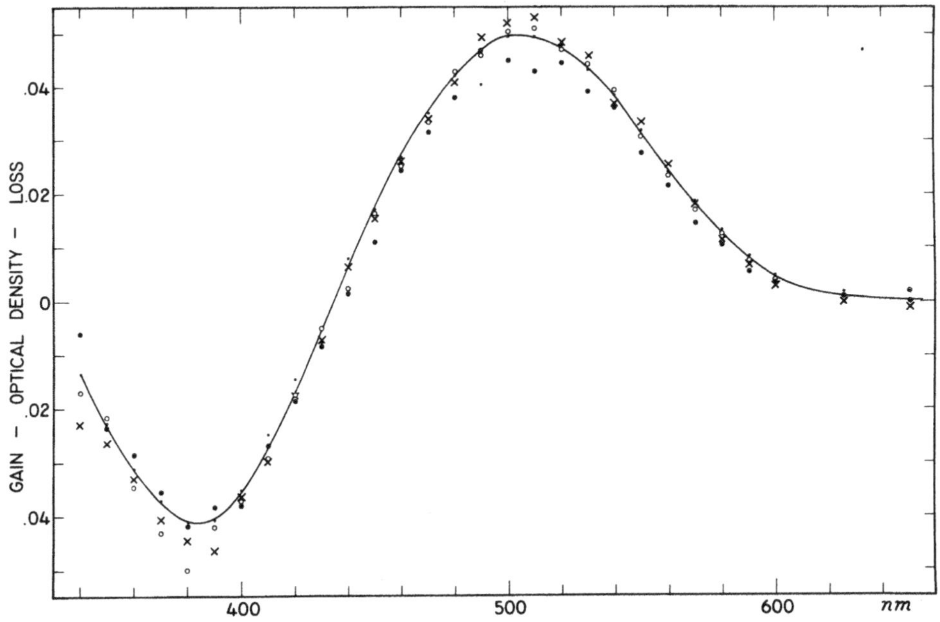

Fig. 25. Partial bleaching of an extract at 25.5° C. Solid line: Difference spectrum for a total bleach of an extract with light 630 nm. The three symbols represent three successive partial bleaches with light at 630 nm of another sample of the same extract at 25.5° C. The first, second, and third bleaches are given by solid circles, crosses and open circles, respectively. Each of the three bleaches was adjusted to give about the same density loss. Note the maximum loss at 504 nm, significantly different for the difference spectrum for a bleaching at 2.4° C (Fig. 5)

other words, at any given temperature in the thermochromic range, there is a single photosensitive spectral entity whose absorbance is shifted hypsochromically with higher temperature. There is no evidence of an equilibrium between bleached and unbleached pigment molecules. The hypsochromic shift and loss in extinction are related phenomena, the latter being the result of the shift of the absorbance curve to shorter wavelengths. This relationship is demonstrated by Fig. 26 which contains, plotted together on separate ordinate scales, the hypsochromic shift (Δ nm) and the extinction loss (Δ O.D.) at maximum. The two scales are adjusted so that the total change for both functions are graphically equal. Both are plotted as functions of temperature and both follow the identical function as they should, if the effect of temperature is, as I now postulate, the simple translation of the absorbance curve to shorter wavelengths as the temperature is raised. Apparently, the gecko visual pigment exists at various transitional spectral states at different temperatures. At least this is the situation for the pigment in 2% digitonin solution. In 1% digitonin the thermometric effect is present but is of lesser magnitude.

The thermochromic effect of extracts immediately poses the question as to whether or not such an effect occurs with the pigment in the natural environment of the cell. Accordingly, in the microspectrophotometric study of CRESCITELLI et al. (1976), we carried out some measurements with the pigment at 23° and 6° C.

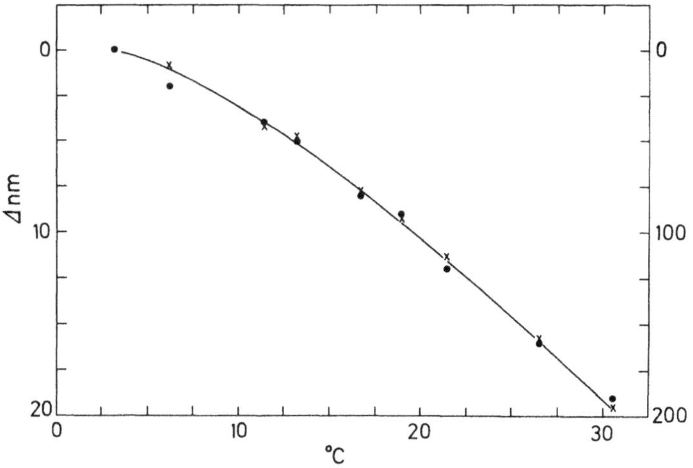

Fig. 26. The hypsochromic shift (*Δ* nm, *solid circles*) and the loss of density at maximum (*Δ* O.D., *crosses*) plotted as functions of temperature. The *Δ* nm scale *(left)* is plotted directly, while the *Δ* O.D. scale *(right)* was adjusted in range to equal the *Δ* nm range

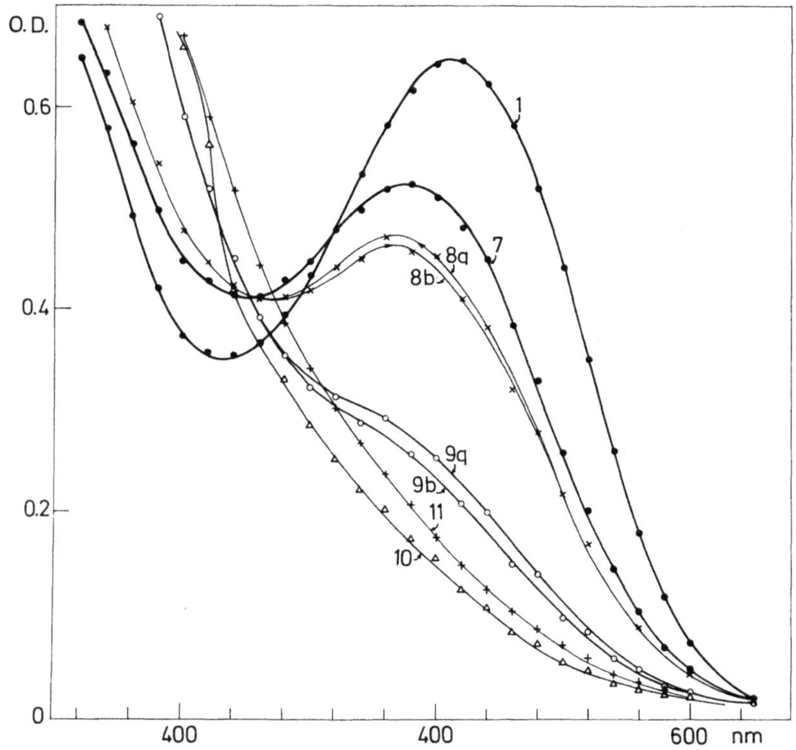

Fig. 27. The effect of higher temperatures. This is a continuation of the experiment of Fig. 22, curves 1 and 7 being duplicated from curves 1 and 7 of Fig. 22. Curve 1: 2.3° C; curve 7: 30.5° C; curve 8a: read after 45 min at 33.2° C; curve 8b: 33.2° C, read 10 min after 8a; curve 9a: read after 115 min at 36.9° C; curve 9b: 36.9° C, read 10 min after 9a; curve 10: after 100 additional min at 36.9° C; curve 11: temperature reduced to 2.7° C

Not unexpectedly, we found the spectral absorbance of the 521-pigment to be the same at these two temperatures (Figs. 12 and 13). It appears, therefore, that the thermochromic behavior is a property of the pigment when solubilized in 2% aqueous digitonin and not of the pigment in the natural state. This finding eliminates many questions that would have arisen if the spectral absorbance of the retina changed with temperature since, we are dealing here with a cold-blooded lizard. It also means that 521 nm is the true spectral maximum of the natural pigment, and any comparison of gecko spectral data obtained with extracts in digitonin cannot be biologically relevant unless the extract is at low temperature during the measurements. This difference in behavior on the part of the pigment *in situ* and in solution is yet another example of the stabilization conferred upon the visual protein by the environment of the cell (HUBBARD, 1958).

b) The Irreversible Thermal Effect

A difference in behavior of the extracted pigment occurs at temperatures above 25° C. The exact temperature at which such a different behavior appears cannot be stated precisely, because this varies according to certain chemical conditions of the extract. The beginning of this new behavior is suggested by the action of a temperature of 30.5° C in the analysis displayed in Fig. 22. The thermochromic behavior for this extract is shown in curves 1 through 6, which illustrate the effects of five successive temperature steps starting from 3.2° C (curve 1) to 26.4° C (curve 6). After the data points for curve 6 were read, the temperature was rapidly reduced to 2.3° C, and the data points for this temperature (curve 1, *solid circles*) indicate the reversibility of this action. The next change was to increase the temperature to 30.5° C (curve 7). As shown by the figure, the isosbestic wavelength for this temperature is at about 447 nm, some 16 nm lower than the crossover wavelength for the 26.4° curve. In addition, the action of higher temperatures involves: (1) a progressive loss of density that does not reach a steady state level and (2) the failure of recovery of density on subsequent cooling. These two features are shown in Fig. 27, the results of which are direct continuations of the sequence presented in Fig. 22, the numbering of the curves being kept in sequence in these two figures. Curve 7, which represents the spectral absorbance at 30.5° C for both figures, indicates that there was little progressive loss of pigment during the time required to obtain these readings. This follows from the fact that the forward readings, at even wavelengths, accord with the back readings, at odd wavelengths, which were taken 10 min later. Unlike this, the outcome of exposing the extract to 33.2° C for 45 min first led to curves 8a (forward readings) and 8b (back readings) taken 10 min later. The result indicates a significant loss (0.0010/min) between these two sets of readings with the extract at 33.2° C. Following this, the temperature was raised to 36.9° C for 115 min and the forward readings (curve 9a) and back readings (curve 9b) read as before. In this case the pigment loss between readings was at the rate of 0.0017/min. The extract was kept at 36.9° C for 100 additional min, after which the data points for curve 10 were taken and indicated a complete loss of pigment. The loss is not recoverable as indicated by curve 11, which was made after the temperature was reduced to 2.7° C. Although a change occurred (whose significance is discussed in the next section), the important fact illustrated by curve 11 is that no large quantity of photopigment was restored by lowering the temperature. This is in direct contrast to the results of Fig. 22, in which complete recovery of pigment density was obtained in lowering the temperature after a sequence to 26.4° C. It is a fact worth noting that a difference of about 10° C, i.e., from 26.4° to 36.9° C, makes the difference between a completely reversible and an irreversible change in color.

I have already pointed out the difference in thermal behavior on the part of frog rhodopsin and the gecko pigment. To summarize the contrast, I have plotted the percent loss of pigment density, at wavelength of maximum loss, as functions of temperature for both pigments (Fig. 28). The data points below 30° C for the gecko pigment and below 40° C for rhodopsin represent steady state levels reached after 30 min exposure to the temperature. Roughly, these two ranges are the reversible temperature domains for the two systems. Because of progressive pigment destruction, the data points for temperatures above these two respective ranges are not steady state values but readings taken arbitrarily at 30 min as before. Fig. 28 serves to show that in the thermochromic range from 5° to 25° C, where the reversible gecko pigment decrement is about 25%, the comparable rhodopsin loss is 4% at most. This low figure for rhodopsin places it out of the range of notice for the usual types of measurements and probably explains why it is that this effect of temperature has not been seen or reported in so many previous studies of frog rhodopsin. Figure 28 also reveals a discontinuity in the gecko pigment curve, and this is approximately at the temperature at which progressive and irreversible thermal damage begins. For rhodopsin it is the 40°–50° C range which represents the beginning of thermal denaturation examined by previous investigators (EWALD and KÜHNE, 1878; LYTHGOE and QULLIAM, 1938; HUBBARD, 1958).

If the thermal destruction of the gecko photopigment is incomplete, and this destruction is stopped quickly by lowering the temperature, the pigment remaining undergoes a bathochromic shift, i.e., in reverse of that which occurs in raising the temperature in the thermometric range and like that following the reduction of temperature in the same range. In other words, a thermochromic effect can be seen to occur buried below the thermal denaturation, which, in turn, suggests that any thermal denaturation is preceded by the pigment entering the reversible thermal transition state. Such an experimental result is included in Fig. 29 which first presents the outcome of raising the temperature from 3.7° C (curve 1) to 35.4° C (curve 2). Subsequently, the temperature was reduced 3.9° in 10 min, and 40 min later the data points for curve 3 were read. These data points show that while little, if any, photopigment recovery occurred, the pigment that remained was able to shift its absorbance bathochromically when the extract was cooled. The pigment remaining, not only responded in the usual way to cooling, but it was also able to respond in the usual way to light. This fact is documented in curve 4 of Fig. 29, in which the bleaching action of a light at 620 nm is given (curve 4). The points made by Figs. 27 and 29 may be summarized by the statement that temperatures above 30° C lead to irreversible denaturation, but that at any point during this process, the destruction may be stopped by cooling, and the pigment that remains behaves normally in respect to thermometric temperatures and to light.

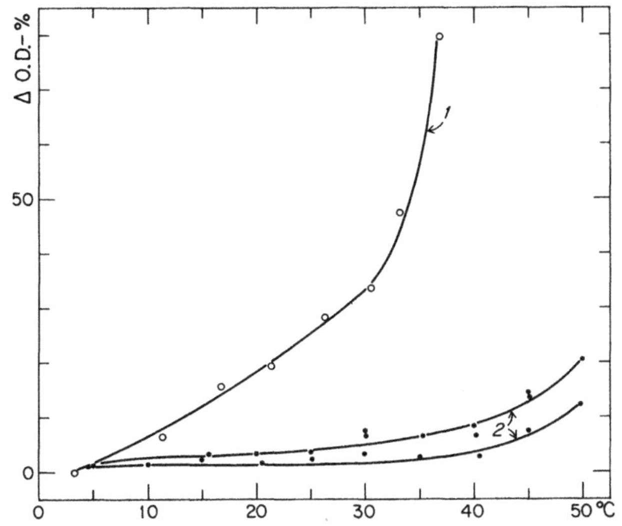

Fig. 28. Optical density loss at wavelength of maximal loss for gecko pigment (curve 1) and for frog rhodopsin (curve 2). The two segments of curve 2 represent the range of loss found in rhodopsin. Density loss is expressed as percentage of photopigment present, this being obtained by total bleaching of the 521-pigment and of rhodopsin

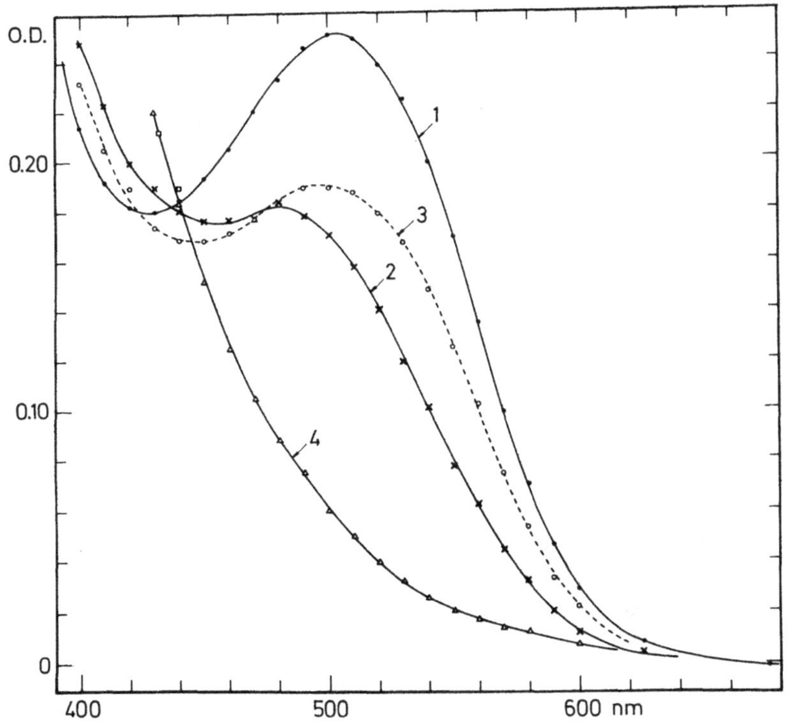

Fig. 29. Curve 1: 3.7° C; curve 2: 35.4° C; curve 3: 3.9° C; curve 4: 3.9° C, after bleaching with light at 620 nm

Just as the 521-pigment in the natural environment of the cell was found to be stabilized against the thermometric effect, so was it found to be similarly stabilized against the thermal denaturation of higher temperatures. It was not necessary to employ the microspectrophotometer to test this point. Rather, two tubes were taken, each holding two freshly removed retinas in contact with the vitreous that was removed along with the retinas. One tube, the experimental, was placed in a water bath at 30° C for 3.5 hrs in total darkness. The second tube, the control, was kept in the dark at 5° C for the same period. The retinas in each tube were then extracted with 0.3 ml of 2% digitonin adjusted to pH 7.7 with borate buffer, all treatments being the same for both batches of retina. The 521-pigment content was determined for each of the two extracts, bleaching and measurements being carried out at 4° C. The density of the 521-pigment in the control retinas was found to be 0.165; of the experimental retinas, 0.280. There is no evidence here of destruction of the photopigment while exposed to 30° C for 3.5 hrs. The comparable experiment with an *extract* that was exposed to 30° C for 3.5 hrs revealed a loss of 79% of the total 521-pigment. Stabilization of the gecko pigment while in the cellular environment applies to both the reversible thermochromic effect and the more irreversible denaturing action of higher temperatures.

3. The Temperature Behavior of P-467

The preceding experiments have dealt with the 521-pigment in the extracts and is the one more easily studied. The examination of the 467-pigment is not a straightforward problem, and I have been forced first to irreversibly destroy P-521 by heat and then to examine the extracts for the presence or absence of P-467. It will be recalled that P-467 is present in every retinal extract of *G. gekko* and that it is present in such extracts in the proportion of 7 to 10% of the total photopigment density (CRESCITELLI, 1963). The behavior of P-467 after P-521 was thermally destroyed is summarized in Fig. 30 which, in fact, is a continuation of Fig. 27. After a long-term exposure to 36.9° C the extract was recooled to 2.7° C, yielding the data points for curve 11 (Fig. 27), which is the same as curve 1 of Fig. 30. The content of the photopigment at this stage was then analyzed by the method of partial bleaching. With light at 620 nm a pigment loss of 0.043, with spectral maximum at 520 nm, was found (curve 1—2, Fig. 30). This was the remnant of P-521 that had survived the heat treatment and was photically bleached out by the red light. Next, the extract was exposed to light at 500 nm, which resulted in a density loss of 0.046 with maximum at 467 nm (curve 2–3, Fig. 30). This value of 0.046 represents 7.8% of the total initial photopigment density in this extract which means that while some 92% of the P-521 was being thermally degraded, all of the second pigment was able to survive the temperature treatment of 36.9° C. It has already been pointed out that P-467 is more stable to added NH_2OH and to borohydride than is P-521. What these differences in lability on the part of the two pigments suggest is that not all the photopigments in a given retina have the same properties which, in turn, lead to the conclusion that these properties have a basis in the intrinsic molecular characteristics and are not solely induced by the nature of the extrinsic environment of the analysis. It is already known that intrinsic

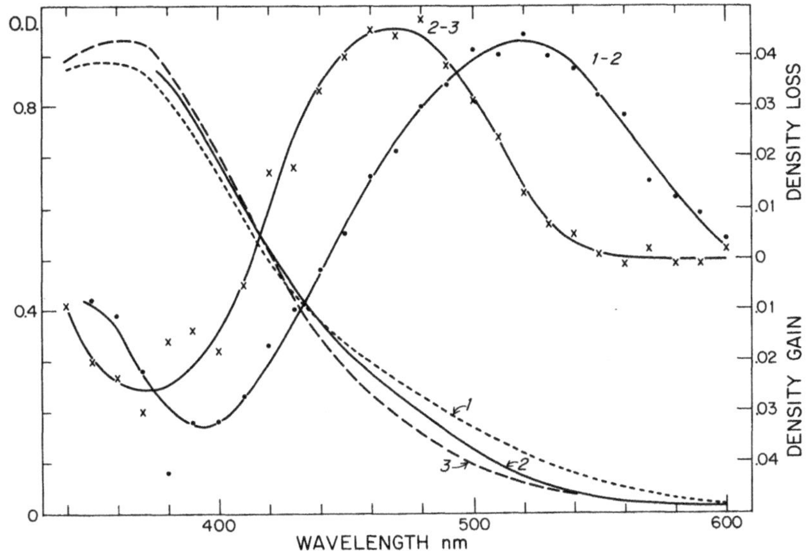

Fig. 30. The thermal behavior of P-467. This is a direct continuation of Fig. 27, curve 1 being a duplication of curve 11 of Fig. 27. Curve 1: 2.7° C, curve 2; after exhaustive bleaching with light at 620 nm, 2.7° C. Difference curve 1–2 is the effect of this bleach. Curve 3: bleaching with light at 500 nm at 2.7° C. Difference curve 2–3 is the effect of this bleach

differences occur between iodopsin and rhodopsin of the chicken and between frog rhodopsin and the green rod pigment. The existence of differences in lability to the chemical and physical environment of the extraction procedure could account for the failure to detect pigments by this method in cases in which the existence of the photopigment is known from other procedures. Such a case is the cone pigment P-575_1 of *R. pipiens* detected by LIEBMAN and ENTINE (1968) with the microspectrophotometry. Yet in all the hundreds of extracts prepared from the retinas of this and other frogs, no certain or confirmed trace of this photopigment has been reported.

VII. The Sensitivity to Inorganic Ions: The Ionochromic Effect

The discovery that certain inorganic ions have significant influence on the spectral absorbance of the solubilized gecko photopigment was made accidently as the outcome of attempts to study the role of various physical and chemical factors on the thermal behavior of the gecko pigment. Historically, few investigators have looked at the effects of ions, other than the hydrogen ion, on the properties of visual pigments. EWALD and KÜHNE (1878) mentioned some results obtained in the thermal bleaching of frog retinas immersed in distilled water and in solutions of NaCl of several different concentrations. This salt was found to have no significant influence on the lowest temperature at which the retina changed color. Later, LYTHGOE and QUILLIAM (1938b) reported the retarding and accelerating actions of low (0.01 M) and high (2.0 M) concentrations of NaCl, respectively, on the thermal bleaching of frog rhodopsin in solution. To my knowledge, however, no one has yet reported that the spectral absorbance of a visual pigment may be altered by inorganic ions.

In preliminary attempts to discover whether the thermochromic property of the gecko pigment is affected by such factors as pH, nature of buffer, nature of solubilizer, concentration of solubilizer, etc., I was soon puzzled by the finding that sometimes absorbance and difference spectra not only did not accord with the spectral position of 521 nm assigned to the gecko pigment but also did not accord with each other. In all cases, the spectral maxima were not at 521 nm but at various, significantly shorter, wavelengths. It is unnecessary to relate the numerous futile excursions and diversions that this led to, but eventually I found the explanation to be in the specific ionic composition of the extracting solution.

1. Nature of the Effect: A Chloride-Specific Action

The chloride effect is demonstrated by the spectral absorbance data of Fig. 31. The extract was prepared from retinas carefully washed before extraction with twice-distilled water. The pigment was solubilized in 2% digitonin (Merck) buffered to pH 7.7 with 0.0016 M sodium borate, all made up in the same water. The spectral absorbance of this extract, measured at 1.8° C, is shown as curve 1. Comparing this curve with curve 1 of Fig. 4, one notes the spectral maxima to be significantly different in position, that of Fig. 4 being at about 521 nm and that of Fig. 31 being at 483 nm. Anyone looking at these two curves could justifiably conclude that they were derived from pigments of two different species. Yet both were made using extracts of the retinas of *G. gekko*. The curve for Fig. 4 was obtained from an extract made with the 2% digitonin dissolved in a veronal-HCl buffer at pH 7.0. This comparison of Figs. 4 and 31 illustrates the dilemma that faced me when first I encountered this situation. The second curve of Fig. 31 serves to show how this dilemma was resolved. After the readings for curve 1 were taken, the cuvette was removed from the spectrophotometer, transferred to the dark room without exposure to light, and drop of NaCl solution was added to the extract to make a final concentration of 0.016 M NaCl. The dilution factor required to obtain this figure was computed from weighings of the cuvette plus extract before and after adding the drop. After correcting for dilution, the effect of the NaCl is shown as curve 2 of Fig. 31. Compared to curve 1, one can see a multiple effect of the NaCl: (1) the extinction at maximum increased by 4.6%, from 0.328 to 0.343; (2) the wavelength for maximal absorption shifted bathochromically from 483 nm to 505 nm; (3) the ratio density at minimum to density at maximum decreased from 0.79 (curve 1) to 0.66 (curve 2); and (4) the entire curve was translated toward the red. The effect is so striking that I had difficulty in accepting it until I had repeated it a number of times using varying concentrations of NaCl, employing several buffers, and varying the pH from 7.2 to 8.9. The result was qualitatively the same for all conditions—a bathochromic shift on adding the NaCl.

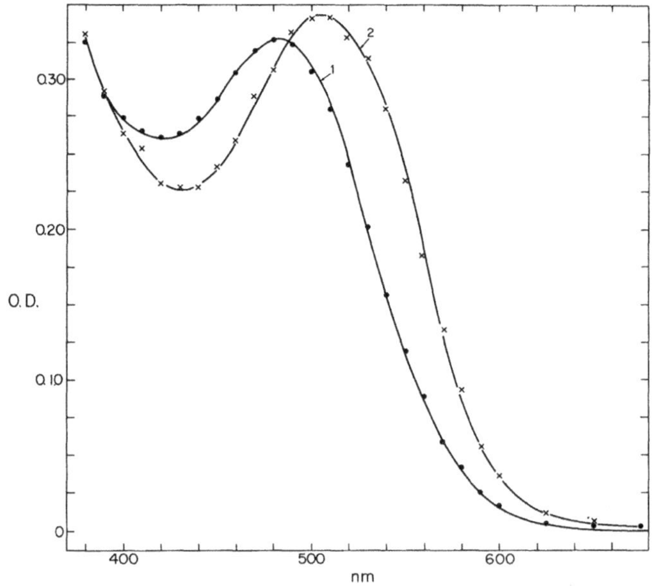

Fig. 31. The chloride ion effect. Curve 1: a gecko extract prepared so as to minimize the presence of inorganic ions; made up in 2% digitonin with borate buffer (0.0016 M), pH 7.7, 1.8° C. Curve 2: recorded a few minutes after adding a drop of NaCl in the borate buffer to make NaCl at 0.016 M, 2.5° C. Correction has been made for dilution with the NaCl solution

2. Action of Other Ions

A similar qualitative effect was obtained with the chlorides of other neutral salts: KCl, CsCl, RbCl, LiCl, $CaCl_2$, $MgCl_2$, $BeCl_2$, and even with the organic chloride, choline chloride. Evidently, the nature of the cation is not significant, and the results point to the anion as the critical factor. I already knew that the bathochromic shift was not produced by phosphate or borate ions, since these buffers were employed in producing the pigment in the hypsochromic state. I also knew that the veronal buffer containing HCl yielded the pigment in its normal position at 521 nm (Fig. 4). Testing various anions, I found that sulphate, fluoride, iodide, nitrate, and thiocyanate were without effect. Of all the ions tested only bromide was found to have an action comparable to that of chloride. It will be recalled that a number of physical properties, such as the ionic radius, the ionic hydration energy, etc., are more nearly alike for chloride and bromide ions than for chloride and the other two halogen ions. From the tests made to date, it does appear that the effect is specific, and since chloride is the ion of universal biological occurrence, it is this ion which becomes of primary interest. For the sake of economy, all of the numerous data that I accumulated in support of these statements will not be shown in this chapter. Hopefully, there will be an opportunity in a later publication to present this evidence. In this chapter I will employ the term "chloride effect or action" to signify the bathochromic shift produced by both chloride and bromide ions when they are added to an extract in the hypsochromic ion-deficient state.

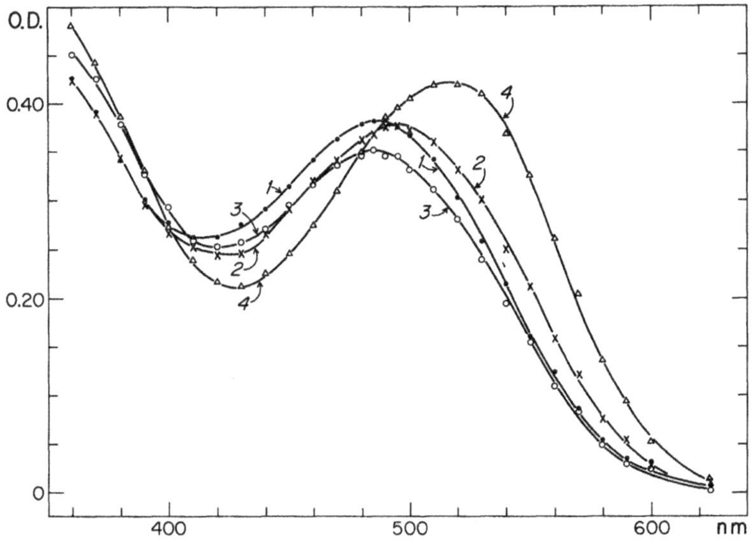

Fig. 32. The reversibility of the chloride effect. Curve 1: pigment in its ion deficient state; curve 2: after adding NaCl to make a concentration of 4.7×10^{-4} M; curve 3: borate buffer added to dilute the NaCl to 2.9×10^{-4} M; curve 4: NaCl added to yield a concentration of 2.48×10^{-1} M. Dilution effects have been corrected, 2.5° C

3. Reversibility of the Chloride Effect

In producing the bathochromic shift, the chloride ion is not bound irreversibly to the visual pigment molecule, because the shift can be quickly and easily reversed by diluting the extract with borate buffer. Such reversibility is demonstrated in Fig. 32 which portrays first of all the pigment in the chloride ion-deficient hypsochromic state (curve 1), having a spectral absorbance maximum at 486 nm. Sodium chloride in small amount was then added to make a concentration of 4.7×10^{-4} M. This caused a shift to curve 2 with a maximum at 495 nm, well below the maximal shift of some 25–30 nm that can be produced by higher concentrations of NaCl. Sodium borate buffer was next added to dilute the NaCl down to 2.9×10^{-4} M, which is closer to the threshold concentration for the chloride effect. The spectral absorbance quickly shifted back to 486 nm (curve 3), illustrating the easy reversibility of this ion effect. Finally, NaCl was added in amounts to give a maximal bathochromic shift, and this produced curve 4 whose maximum is at 517 nm, a shift of 31 nm from the initial spectrum. Whatever the mechanism of this chloride action, it is clear that this ion can quickly move into and out of the site of its action in the visual pigment molecule.

4. The Chloride Effect: Quantitative Aspects

The rather striking, and perhaps unusual, action of chloride in shifting the spectrum of the gecko pigment invites many questions, one of which concerns the quantitative aspects. An experiment that illustrates the mode of action of NaCl in varying concentrations is presented in Fig. 33. This experiment was made with a single sample of an extract in a cuvette, to which was successively added a small volume (about 0.01 ml) of successively increasing concentrations of NaCl dissolved in sodium borate buffer. The successive dilution factors were calculated by weighings before and after adding each drop. In such experiments it was not unusual to be able to add as many as six drops and still retain a concentration of photopigment sufficient to take meaningful readings. The seven curves of Fig. 33 represent the seven successive spectra, starting with curve 1 taken before any NaCl was added and proceeding in sequence following each addition. Except for curve 1, where it was unnecessary, the density readings have all been corrected for dilution. This series of curves illustrates several important features in the action of the chloride ion. The first is that a significant effect occurred with relatively low concentrations of this ion. Curve 2 was obtained with NaCl in the amount of 2.7×10^{-4} M. This concentration produced a shift of spectral maximum from 495 nm to 502 nm. Such a concentration argues for a rather special type of ion effect quite unlike that of the usual lyotropic concentrations of neutral salts acting on proteins (von HIPPEL and SCHLEICH, 1969). A second feature of these seven curves is the successive increase in maximum extinction with successive increases in chloride concentration. This increase is not accompanied by a parallel increase in the density of the minimum, so that the ratio density at minimum/density at maximum becomes smaller. In visual pigment extractions the magnitude of this ratio is usually considered to be an index of extract purity, and special procedures are often required to obtain small ratios. Here the "purification" was accomplished by adding table salt. A third feature that is worthy of note is that the increase in extinction and the bathochromic shift do not always change in parallel. A shift in wavelength can occur with little or no change in maximum extinction. Such behavior is always observed to occur at concentrations of NaCl just above the threshold values for a bathochromic shift. A summary of the quantitative aspects of the chloride action is given as Fig. 34 which lists three properties of the spectral curves as functions of the NaCl concentration. These three properties—the increase in maximum extinction, the wavelength shift of the peak, and the wavelength shift of the 50% density point of the long-wavelength limb of the curve—are all seen to change roughly in parallel with increasing chloride concentration and to reach maximum values at approximately the same concentration.

Two further points are worth mentioning in respect to this chloride bathochromic effect. One is that the maximum shift of 25 nm to 30 nm is as great or greater than many shifts that occur naturally from one visual pigment to another. In the case of the metamorphosing *R. pipiens*, for example, the red rods have a P-527_2 which changes to a P-502_1 in the adult frog (LIEBMAN and ENTINE, 1968). This involves the changeover from an A_2 to an A_1 system, which means the deletion of a double bond in the conjugated system. And yet the gecko pigment in solution is capable of the same or even greater change in absorbance, utilizing the same prosthetic group but by means of an interaction with chloride ions. The second point is that the maximum wavelength to which the pigment is shifted by chloride is the normal wavelength (521 nm) for the system *in situ*. I have never succeeded in moving the spectral absorbance to longer wavelengths than this, even with excessive concentrations of NaCl. This means that the chloride effect is,

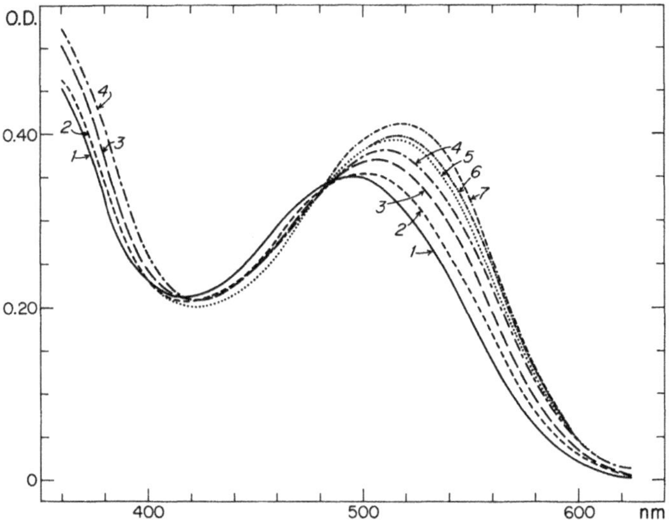

Fig. 33. Titration of gecko pigment with increasing concentrations of NaCl. Curve 1: pigment in ion-deficient state. Curve 2 to 7: increasing NaCl as follows: 2.7×10^{-4} M (curve 2), 8.25×10^{-4} M (curve 3), 1.81×10^{-3} M (curve 4), 4.65×10^{-3} M (curve 5), 1.04×10^{-2} M (curve 6), 3.75×10^{-1} M (curve 7). Temperature was 3.2° C throughout. Corrected for dilution

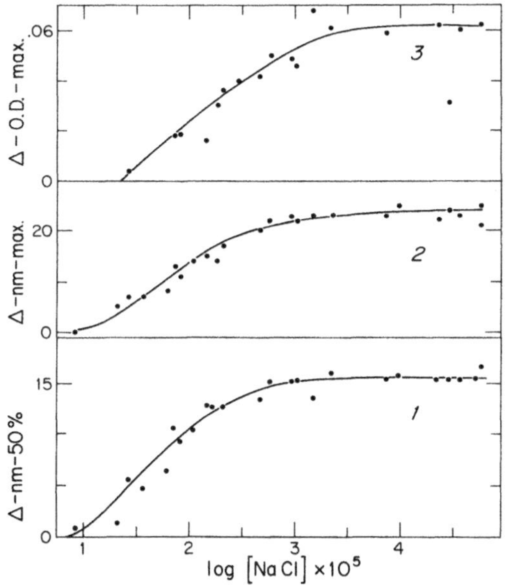

Fig. 34. Three characteristic of the chloride effect as functions of chloride concentration. Curve 1: bathochromic shift of the 50% ordinate density on long-wavelength portion of absorbance spectrum. Curve 2: bathochromic shift of spectral maximum. Curve 3: density increase at wavelength of maximum increase. NaCl concentration in terms of molarity. Data obtained from spectra that were corrected for the effects of dilution. All measurements made at low temperature

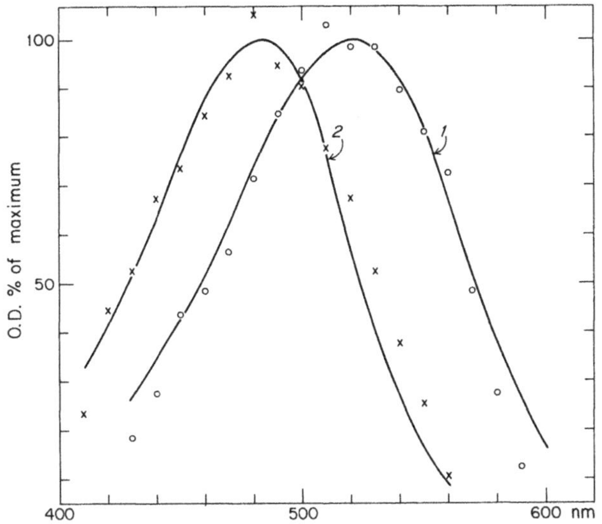

Fig. 35. Comparison of photic difference spectra for an extract with NaCl added (7.9×10^{-2} M) and without NaCl (curves 1 and 2, respectively). The solid lines represent DARTNALL nomogram curves for a pigment at 522 nm and at 483 nm. Analysis carried out at low temperature

in fact, a deficiency effect which is repaired by the chloride ion. The normal spectral absorbance cannot be moved to longer wavelengths than the limits set by the intrinsic properties of the molecule. I know nothing yet about the ionic situation within the outer segments. If chloride is also required to adjust the photopigment to its normal or physiologic absorbance, the chloride apparently can be easily removed from the gecko pigment, whereas this is not so for the photopigments of other vertebrates with which I have worked, including the rhodopsin of man.[1]

5. Difference Spectra

In its hypsochromic, chloride ion-deficient state, the gecko pigment is photosensitive and responds to light by bleaching. The difference spectrum obtained in such a photic bleach is typical in conforming to the nomogram of DARTNALL (1953) for vertebrate visual pigments. This point is brought out by the difference curves of Fig. 35. For this experiment, which is typical of many such experiments, an extract was divided into two samples. To one, NaCl was added to make a final NaCl concentration of 7.9×10^{-2} M. The other was left in the chloride ion-deficient state. Both samples were bleached with light at 600 nm until no further change was obtained. Hydroxylamine was then added to both samples to give a final NH_2OH concentration of 0.01 M. After reading the spectral absorbance of these two bleached samples, the difference curves were computed and are shown in Fig. 35. The sample with added NaCl is characterized by a typical absorbance loss with maximum close to 520 nm and conforms to a nomogram curve at this location (curve 1). The second sample is represented by an absorbance loss maximal at 483 nm and accords with a nomogram curve in this spectral position

[1] Since this statement was written I have established that a chloride effect does occur in a suspension of retinal rod particles.

(curve 2). Not shown in this figure are the product difference spectra. These are the same in having maxima at about 366 nm, characteristic of the oxime for retinaldehyde. The results in Fig. 35 confirm the interpretation already made that a deficiency of chloride ions simply translates the absorbance curve to shorter wavelengths without affecting the basic form of the curve. A deficiency of chloride results in an ionochromic action in the hypsochromic direction; the addition of chloride repairs this action.

VIII. Color and the Sulfhydryl Group: The Thiochromic Effect

It is a well-recognized assertion of biochemistry that disulphide and thiol groups play an important role in maintaining certain essential structures of proteins (CECIL, 1963). A corollary of this is that certain biological functions of proteins depend on these essential structures. The possible role of thiol groups in such functions has been tested by the use of reagents thought to specifically poison the sulfhydryl group. In this section, I propose to report the action of two thiol-inhibiting agents on the spectral absorbance of pigment-521 of *G. gekko*. Since the color of the visual pigment is the property related to function, this is probably the simplest system to examine this general problem.

1. Reversible Hypsochromic State

In its primary structure, cattle rhodopsin is stated to have sulfur-containing amino acids, and, presumably, the gecko visual pigments do also, although no data are available to support this. It is also stated that rhodopsin contains sulfhydryl groups, and that in the process of bleaching, titratable thiol groups appear (WALD and BROWN, 1952; OSTROY et al., 1966; HELLER, 1968). The original suggestion that the thiol group may be involved in the linkage of retinaldehyde to the opsin (WALD and BROWN, 1952) has been rejuvenated by HELLER (1968) who proposed a model in which the terminal carbon of retinyl forms a link, through sulfur, between lysine and cysteine residues of the opsin. To assay the possible role of thiols in the reactions of rhodopsin, WALD and BROWN (1952) employed the mercurial reagent *p*-chloromercuribenzoate (PMB) which reacts with thiol groups to form mercaptides. This reaction is useful for it may be reversed by adding the appropriate competing thiol. WALD and BROWN found that PMB was able to inhibit the regeneration of cattle rhodopsin, an inhibiton that was reversed in part by the addition of glutathione. In the case of cattle rhodopsin, it is apodictic that PMB does not alter the spectral location of the 498 nm peak (WALD and BROWN, 1952; OSTROY et al., 1966; ZORN, 1971). Chicken iodopsin was shown to be bleached in the presence of concentrations of PMB greater than 3×10^{-5} M (WALD et al., 1955), and a similar bleaching of cattle rhodopsin in the dark was shown in response to added *p*-mercuriphenylsulfonate (PMPS) by EARNSHAW and FUJIMORI (1973), but concentrations greater than 1×10^{-3} M were needed.

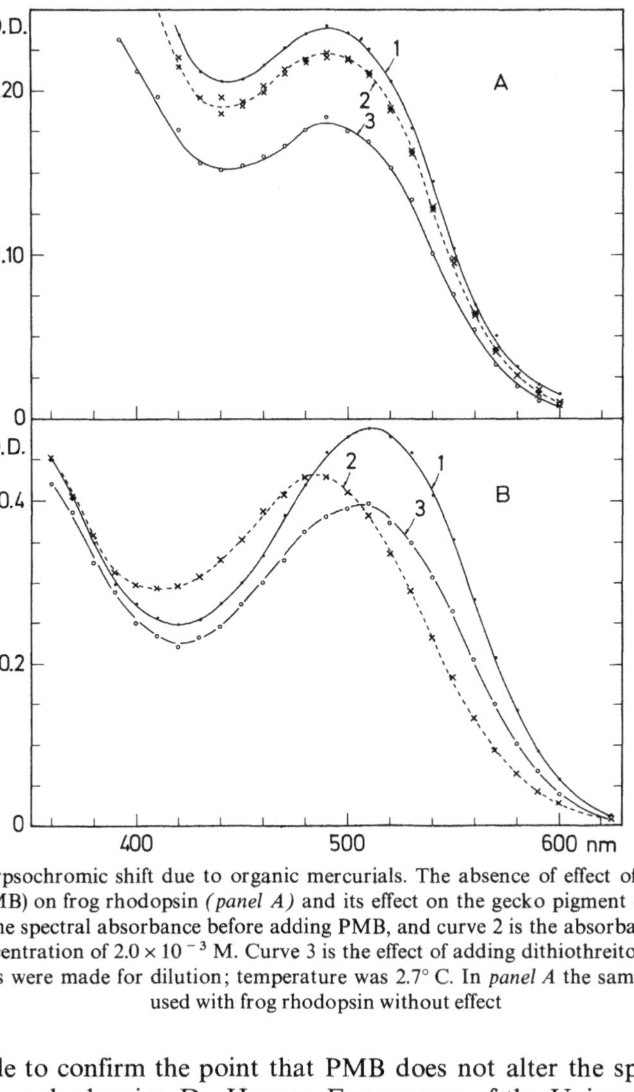

Fig. 36. The hypsochromic shift due to organic mercurials. The absence of effect of *p*-chloromercuribenzoate (PMB) on frog rhodopsin *(panel A)* and its effect on the gecko pigment *(panel B)*. *Panel B:* curve 1 is the spectral absorbance before adding PMB, and curve 2 is the absorbance after adding PMB to a concentration of 2.0×10^{-3} M. Curve 3 is the effect of adding dithiothreitol (2.8×10^{-2} M). No corrections were made for dilution; temperature was 2.7° C. In *panel A* the same procedure was used with frog rhodopsin without effect

I am able to confirm the point that PMB does not alter the spectral absorbance of many rhodopsins. Dr. HECTOR FERNANDEZ of the University of Southern California kindly gave me two live specimens of the Arctic sculpin, *Icelus spatula*, which he collected at Point Barrow, Alaska. The rhodopsin prepared from the retinas of these fishes has a spectral absorbance with maximum close to 507 nm. Adding PMB to make a concentration of 1×10^{-3} M and keeping the solution in the dark for 16 hrs led to no loss of pigment and no change in the absorbance curve. This rhodopsin is inert to this concentration of the mercurial compound. The same is true for frog rhodopsin, and, in additon, the photic bleaching sequence of frog rhodopsin through metarhodopsin III is unaffected by the presence of PMB. Taken along with previous statements in the literature, the results give no reason to suspect that the spectral absorbance of rhodopsins is altered by this sulfhydryl inhibitor.

Such is not the case for P-521 of *G. gekko*. I have already published a brief note (CRESCITELLI, 1975a) in which it was pointed out that PMB causes a rapid, hypsochromic shift of 25 nm to 30 nm, that this is rapidly and completely reversed by dithiothreitol (DTT), and that the shift is prevented by the prior addition of DTT to the extract. Dithiotreitol, introduced by CLELAND (1964) for the protection of SH groups, has been extensively employed in protecting monothiols from oxidation in numerous investigations involving the activity of the sulfhydryl group. Typical responses of the gecko pigment system to PMB and DTT are shown in Fig. 36 along with a parallel control experiment with frog rhodopsin to show the lack of response on the part of the latter pigment. In this figure, the curves numbered 1 in both frames represent the spectral absorbances of the extract from the frog (*A*) and *G. gekko* (*B*). Both extracts were prepared similarly using the same 2% digitonin solution made up in borate buffer at pH 8.5. The effect of adding PMB to a concentration of 2.5×10^{-3} M is shown as the curves numbered 2. Except for the dilution effect, there was no change for the frog extract, but the gecko pigment shifted in typical fashion from 510 nm to 485 nm. To be certain that there was no delayed effect of the PMB on the frog rhodopsin, I allowed the extract to remain for 2 more hrs in contact with the PMB and then repeated the spectral measurents. The X-symbols for curve 2A indicate no delayed effect of PMB on rhodopsin. The next step in the procedure was the addition of DTT to both extracts to a concentration of 2.8×10^{-2} M. The outcome of this addition (curves numbered 3) was the shift back to normal on the part of the gecko pigment, while the rhodopsin curve was unaffected. The data for this figure were not corrected for dilution but, if corrected, they would show no significant, progressive loss with time for any of the treatments. The results of Fig. 36 show the selective effect of these sulfhydryl agents on the gecko pigment in contrast with frog rhodopsin, both systems being treated identically.

The gecko pigment responds similarly to another sulfhydryl reagent, *p*-mercuriphenyl sulfonic acid (PMPS). This compound, more soluble in aqueous solutions than PMB and therefore more convenient, similarly shifted the gecko spectral absorbance to shorter wavelengths, and DTT was able to reverse this effect.

2. Difference Spectra

The photosensitivity is not lost when the visual pigment is treated with these SH-reagents and the spectrum is shifted to shorter wavelengths. This point is documented by Fig. 37 which gives the difference spectra obtained as the result of bleaching two samples of the same extract with light at 620 nm, so as to remove only P-521. Curve 1 relates to the sample treated with DTT, while curve 2 applies to the second sample exposed to PMB. These two difference curves show maximum density losses at 517 nm and 496 nm, respectively. Curve 1 accords with a nomogram spectrum for a 517-pigment (*open circles*), while the second curve agrees with a nomogram based on a 496-pigment (*triangles*). These photic bleachings were carried out (necessarily) without added NH_2OH which accounts for the fact that the agreement between experimental data and nomogram data is deficient for the short-wavelength limbs of the two curves. Altogether, the results

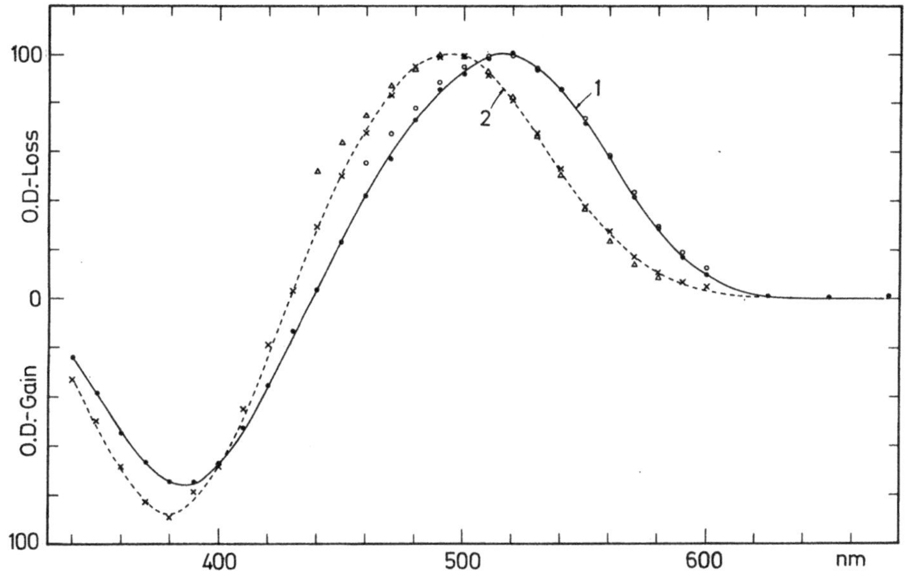

Fig. 37. Photic difference spectra. Two aliquots of same extract taken; aliquot 1 (curve 1) treated with DTT (3.3×10^{-2} M) and PMB (2.42×10^{-3} M); aliquot 2 (curve 2) treated only with PMB (2.42×10^{-3} M). Both aliquots were exhaustively bleached with light at 620 nm; temperature was 2.8° C. Curves 1 and 2 (*solid circles* and *crosses*) are the respective difference spectra. Open circles and triangles are respective nomogram curves for a pigment at 517 nm and one at 496 nm

point to the conclusion that P-521 of *G. gekko* can be shifted to a state that absorbs light at higher energy, but that in this state neither the photic responses to light nor the basic form of the absorbance curve is greatly altered. In this respect, the action of these sulfhydryl reagents resembles the effect of thermometric temperatures and of chloride deficiency. All three conditions lead to a qualitatively similar change in the photopigment absorbance.

C. Interpretations and Discussion

I. Biological Considerations

The choice of nocturnal geckos for the study of visual cells and pigments has proved to be a happy one for several reasons. For one, the large outer segments housing photopigments that are extracted by digitonin and that are analyzable with the microspectrophotometer, have made it possible to integrate knowledge gained by both approaches. Both methods have revealed the existence of but two pigments, one absorbing in the region 520–530 nm, the second at 460–470 nm. For *G. gekko* the spectral maxima are at 521 nm and 467 nm. Only one photopigment was found in every outer segment examined, and no pigment absorbing at longer wavelengths was detected by either extraction or microspectrophotometry. The presence of such a dual system prompts me to speculate regarding the phylogeny of gecko visual receptor systems. According to WALLS (1934), the ancestral forms of geckos were diurnal in habit and possessed photopic visual cells or photocytes. In the transition to a nocturnal habit, the visual cells transmuted in stages to scotopic receptors or scotocytes. According to this transmutation hypothesis, visual cells in intermediate or transitional stages of transmutation are to be found in modern genera of geckos. (UNDERWOOD (1968, 1970, 1971) contributed to this hypothesis and suggested that the retinal receptors of living diurnal lizards offer a possible model of the ancestral condition. Using *Anolis lineatopus* as one example, UNDERWOOD proposed the following transitions. The major (type A_1) singles of an *Anolis*-like ancestor transmuted to the type A singles of geckos, and in the process the outer segments were enlarged; the yellow oil droplets became colorless in some and were lost in other species (Fig. 1). We have no knowledge of the visual pigment within the type A_1 cells of *Anolis*, but in this hypothetical change to type A gecko cells, photopigments appeared that are easily extracted by digitonin. Continuing, according to UNDERWOOD, the gecko type C doubles originated by a merging of both singles (types A_1 and A_2) of an *Anolis*-like ancestor. Again, the outer segements became enlarged, and the oil droplets were discarded in some species. If in the ancestral animal the types A_1 and A_2 had different pigments, this would account for the dissimilar pigments in the two members of 521–467 doubles. Unfortunately, we do not know anything about the photopigments within the A_1 and A_2 cells of *Anolis* or of any other diurnal lizards. Finally, UNDERWOOD conceived the type B gecko doubles as originating from the standard type B doubles of diurnal lizards. Here it would be useful to know whether the type B doubles of diurnal lizards have the same or different photopigments in the two members, but this information is lacking. Neither I, nor anyone else, has succeeded in extracting visual pigments from diurnal lizards (CRESCITELLI, 1972), and the microspectrophotometric technique may be difficult because of the small size of the outer segments.

The apparent absence of a receptor absorbing in the red is worthy of brief mention. Such a receptor is found in turtles (LIEBMAN, 1973), but turtles are phylogenetically distinct from lizards (Chapter 1). There is some reason to believe, however, that diurnal lizards possess such "red" receptors. The spectral sensitivities of several species have been found to possess maxima or indications of maxima in the regions of 560 nm, 580 nm, and 600 nm (FORBES et al., 1960; HAMASAKI, 1968). According to the transmutation hypothesis, the absence of "red" receptors in geckos suggest that in the transition to the nocturnal habit these receptors were either lost or changed. The alternative is the abandonment of the transmutation hypothesis with the suggestion that geckos have had an independent phylogeny isolated from that of other lacertilian families.

The good agreement in spectral absorbance of the 521-pigment when solubilized and when in its natural environment (Fig. 13) has helped to resolve a problem of long-standing regarding the relevance of spectra of solubilized visual pigments to action spectra. The relevance of this comparison was questioned by HECHT and WILLIAMS (1922) who concluded that the human scotopic sensitivity curve is similar in form but bathochromically shifted by 7–8 nm as compared with the difference spectra of rabbit and monkey rhodopsins as given by KÖTTGEN and ABELSDORFF (1896). This bathochromic shift was interpreted to be the operation of Kundt's rule, whereby a pigment absorption is shifted to longer wavelengths when transferred from a medium of lower to one of higher refractive index. Kundt's rule was also invoked by WALD and BROWN (1958) in explaining the 7 nm discrepancy between the difference spectrum of human retinal rod particles and the spectrum of human rhodopsin in aqueous digitonin. Quite apart from the problem of the validity of Kundt's rule as well as of published measurements that appear to disagree with the interpretation of HECHT and WILLIAMS, the straightforward accord between the directly measured spectra of the extract and of the pigment *in situ* (Fig. 13) is evidence that for the gecko system a bathochromic displacement does not occur when the pigment is shifted from the aqueous environment to the environment of the cell. The accord seen in Fig. 13 was based on the stipulation that the extract absorbance was measured at low temperature and in the presence of chloride ions. Failure to follow these conditions would have meant that the extract absorbance would have shifted to shorter wavelengths and have displayed an apparent Kundt's rule effect. It is true of course that the temperature and the ion effects are particularly associated with the gecko photopigments but, as Fig. 19 shows, even frog rhodopsin reacts to higher temperature by a small but definite hypsochromic shift. In the future, therefore, investigators would do well to understand the influences of various physical and chemical conditions on their systems before comparing biochemical data with biological action spectra.

The absence of an M-III intermediate in the 521-pigment of *G. gekko* is another property that is the same for the pigment *in situ* and in the solubilized state (Figs. 14 and 15). This absence could be the result of a very rapid decay of M-III in the gecko system for which I have no information, but this qualification would alter the definition of this intermediate as we know it in rhodopsin systems. In any case, the absence of M-III at ordinary temperatures in the gecko retina offers the opportunity to test certain questions associated with the occurrence of M-III. One such question concerns the origin of the long-delayed electrical response of the frog retina known as the e-wave (CRESCITELLI and SICKEL, 1968) and of the delayed impulse discharge associated with the e-wave (SICKEL and CRESCITELLI, 1967; PICKERING and VARJÚ, 1967). These discharges are preceded by a silent period whose duration is longer as the stimulus strength is increased. On the basis of convincing evidence (PICKERING, 1968; PICKERING and VARJÚ, 1969), the long delay in appearance of discharges has been interpreted as the result of two long-persisting after-stimulus events: one, an inhibition which decays first; the second, an excitatory process which develops later and outlasts the inhibition. This would account for a silent period succeeded by a discharge. My studies of the developing frog tadpole eye (CRESCITELLI, 1970) agree with this. An e-wave appears in the early tadpole eye and here, too, a post-stimulus inhibition seems to account for the long delay in onset of the e-wave. A second flash inserted just before the expected appearance of the e-wave to an initial flash will prevent the appearance of this e-wave, although it will be followed, as expected, by its own e-wave after the usual delay. Additionally, a second flash inserted at the first appearance of the e-wave to an initial flash will dramatically cut off the completion of this e-wave but will give rise to its own e-wave. Recently, CHINO and STURR (1975) suggested that the cycle of inhibition and excitation is associated with a cycle of activity in cones and rods, respectively. Another suggestion is that inhibition and excitation are related to the long-persisting intermediates of bleaching rhodopsin, with M-II being the cause of inhibition and the decay of M-II and the persistence of M-III being the cause of excitation. There are several reports in the literature that relate the presence of M-II to the reduced sensitivity to light following a light flash and the rapid recovery of sensitivity after the flash to the decay of M-II (DONNER, 1973). Incidentally, as DONNER (1973) pointed out, the testing of the relationship between the recovery of sensitivity and the decay of M-II is more "straightforward" in a system lacking an M-III intermediate. This suggests that for a study of dark adaptation and photoproduct decay the gecko system, which lacks the M-III intermediate, is preferable to the frog retina in which this problem already has been examined (DONNER and REUTER, 1967, 1968). A photochemical basis for the anomalous early dark adaptation in man has been suggested by RUSHTON and POWELL (1972) who explained the unusually low sensitivity immediately after a flash and the unusually rapid rate of recovery for the first 8 min as being the result of a transient product of bleaching, an X-opsin. The sensitivity-reducing effect of M-II in the case of the frog appears to be localized to the receptors that contain the M-II (DONNER and REUTER, 1968). It is interesting to note, therefore, that a long silent period followed by a *negative* e-wave occurs in the tadpole eye at a stage when the outer segment discs have formed but before the synaptic bodies have vesicles (CRESCITELLI, 1970). This again suggests that local receptor events, possibly photochemical in origin, are responsible for the silent period and for the e-wave in the frog retina. Since M-III outlasts M-II, the hypothesis can be posed that M-III is excitatory and is responsible for the e-wave. This hypothesis can be tested by examining the gecko retina for an e-wave. The finding of this component would force me to abandon the hypothesis.

In one respect the 521-pigment behaves differently when in situ and when solubilized. The pigment *in situ* displays no thermochromic effect (Figs. 12 and 13), so there are no problems here regarding the possible effects of temperature in the behavior of the visual cells. The discovery of the thermochromic effect of the 521-pigment in solution prompted me to inquire into the thermal behavior of visual pigments of fishes inhabiting very cold waters or atypically hot waters. For the former fishes, I obtained from Dr. Hector Fernandez of the University of Southern California a living specimen of the Arctic cod, *Arctogadus borisovi*. This species inhabits the waters of the ice flow where temperatures may descend to 1.9° C. The fish was collected in summer at Pt. Barrow, Alaska and flown to the laboratory where it was kept alive at low temperature for about a year, its food being chopped beef. An extract prepared from the retinas of this fish proved to have a single photosensitive pigment with spectral maximum at 505 nm and behaved like ordinary rhodopsin except for its unusual thermal lability. Kept in the refrigerator at 5° C the extract lost 34% of its rhodopsin in 21 days and 52% in 32 days. The spectral behavior to a change in temperature from 3.3° to 25.5° C was quite unlike the gecko thermochromic action. This temperature step resulted in loss of pigment that was not restored by return to the low temperature. Even a change to 11.6° C produced a small but irreversible loss. This suggests that adaptation to the Arctic temperatures has resulted in a molecule such that when dissolved in digitonin it is irreversibly denatured by an unusually small increase in temperature. Apparently, the critical temperature for such denaturation is well below that for both frog rhodopsin and the gecko pigment as shown in Fig. 28. In the future, if more fishes become available, I hope to examine the thermal behavior of the cod pigment further, especially the pigment *in situ* within the cells.

I have also initiated a program intended to examine the question as to whether or not the critical temperature for irreversible denaturation is raised in the case of pigments from fishes inhabiting hot water. For this I have chosen the desert pupfishes, some species of which are acclimated to temperatures of 40–45° C. The pigment in these fishes is a vitamin A_2-based photopigment, which adds another complication, but in addition their small size, the fact that they are protected, and the fact that the extracts have relatively little photopigment, and have in them undersirable impurities, have made it impossible for me to reach a conclusion on the thermal properties of the pupfish pigments. It is known that temperature may influence the proportion of rhodopsin to porphyropsin in the golden shiner (ALLEN and MCFARLAND, 1973), but this is probably an indirect effect rather than an action on the properties of the visual pigments.

The comparative thermal properties of visual pigments have broad biological significance beyond the narrow field of visual science. Without reviewing the extensive literature, I may cite a few examples to illustrate the phenomenon of adaptation in the properties of proteins to the temperature habitat of the organism. In one example (KOMATSU and FEENEY, 1970), a comparison was made of the thermal behavior of the enzyme, muscle fructosediphosphate aldolase from two species of Antarctic fishes, from the temperature-zone rainbow trout, and from the rabbit. The temperatures for half-inactivation of the four enzymes were found to be 29°, 38°, 44°, and 55° C, respectively. SOMERO and HOCHACHKA (1968) determined the thermal optimum for muscle pyruvate kinase from an Antarctic fish and from the rainbow trout. This optimum was found to be at about 0.5° and 15° C for the two fishes. At the other end of the biological temperature scale, there is evidence of a relative increase in thermostability on the part of proteins of thermophilic microorganisms (SINGLETON and AMELUNXEN, 1973). FUJITA and IMAHORI (1974) examined the circular dichroism, at 219 nm, of D-glyceraldehyde-3-phosphate dehydrogenase from the thermophilic *Thermus thermophilus*. The dichroism was lost by warming the enzyme to 75° C but was regained on cooling. Warmed to 86° C, however, the secondary structure was irreversibly affected. The thermophilic enzyme was similar to its mesophilic counterpart in respect to molecular weight, secondary and quaternary structure, amino acid composition, and behavior toward sulfhydryl reagents. Further information has been contributed by OHTA (1967) and by OHTA et al. (1967). Using the purified thermostable protease from the thermophilic *B. thermoproteolyticus*, these investigators determined that this enzyme has a compact globular structure with an amino acid composition lacking in cysteine and cystine but high in hydrophobic amino acids. This structure and composition was considered by the authors to account for the unusual thermal stability of this enzyme.

II. Biochemical Considerations

In some respects, there is a difference in behavior of the 521-pigment when dissolved in digitonin and when in its natural environment in the cell. The increased thermal lability of the extracted pigment to temperatures in the thermometric range (5°–25° C) and in the temperatures for irreversible denaturation (above 30° C) is one such example. Apparently, digitonin opens up the molecule, thus rendering it more susceptible to temperature. The thermochromic effect, reversible as it is, is probably a minor structural change in the chromophore. The mechanism by which digitonin is able to influence some properties of the 521-pigment, such as the responses to temperature, while leaving other properties unaffected, such as the form and location of its spectral absorbance (Fig. 13), is an intriguing question.

The significant hypsochromic displacement of the 521-pigment brought on by the organic mercurials and its ready reversibility by dithiothreitol (Fig. 36) suggests that the solubilized photopigment contains reactive SH groups that are intimately associated with the retinal-protein interaction at the chromophoric site. It will be recalled that HELLER (1968) proposed a model in which one sulfhydryl group was viewed as linking retinal between lysine and cysteine residues, an idea reiterated by GIRSCH and RABINOVITCH (1971). Additionally, there is the model by GALINDO (1967) that invokes a charge transfer interaction between an electron of the retinal C_{11}–C_{12} double bond and an orbital of sulfur in cysteine. The possibility of a charge transfer interaction involving the -SH group is also suggested by FUJIMORI (1975). If we accept the idea that proteins have both reactive and unreactive sulfhydryls and that unreactive groups may be made reactive by denaturing the protein (CECIL, 1963), then we may suppose that digitonin acts in this way and exposes the SH group associated with the chromophore. This action is unique for the 521-pigment of the gecko, because the rhodopsins of frog, cattle, fish, etc., are not so affected by the digitonin. Nevertheless, there may be conditions in which even these rhodopsins may be made responsive to reagents which they normally resist. EARNSHAW and FUJIMORI (1973) and FUJIMORI (1975) demonstrated an effect of urea in accelerating the dark bleaching of cattle rhodopsin by *p*-mercuriphyenylsulfonate and by iodoacetamide. ZORN (1971) mentioned that exposure of rhodopsin solubilized in emulphogene to 35° C or to 2.0 M urea at 3° C caused the visual pigment to become reactive to *p*-hydroxymercuribenzoate and to become susceptible to reduction by $NaBH_4$. It has also been reported that cattle rhodopsin purified so as to remove the lipid moiety and dissolved in dodecyltrimethylammonium bromide is thermally sensitive and is rapidly destroyed at room temperature (HONG and HUBBELL, 1972). While these effects are not exactly identical to the mildly produced and readily reversible actions displayed by the gecko pigment, they do suggest that conditions may eventually be found in which the rhodopsins may reveal the thermochromic behavior, the ion effects, and the other activities mentioned in this chapter. If this should ever come to pass it will be a definite advance, for it will mean that a ready and cheap source of photopigment will become available in place of the expensive, meager, and sometimes unavailable gecko material.

The sensitivity of the unbleached 521-pigment to $NaBH_4$ is of interest in connection with the problem of the nature of the chemical moiety to which retinal is bound in the native visual pigment. Lysine is usually identified as the site of binding, but others have suggested that it is phosphatidylethanolamine. ABRAHAMSON and WIESENFELD (1972) have published a critical review of this problem and have pointed out one possible flaw in the experiments (BOWNDS, 1967; AKHTAR et al., 1967) identifying lysine as the binding site. These experiments involved a reduction by $NaBH_4$ of cattle rhodopsin *after illumination*. Under these circumstances, a product of bleaching, and not native rhodopsin, could have been reduced, and this could have allowed time for the chromophore to migrate from the primary site in native rhodopsin to a secondary site in the product, so that the identified moiety, lysine, may not have been the original site of attachment in the native pigment. My purpose in this paragraph is to point out, to those who believe this problem to be still unsettled, that the gecko pigment, reduced as it is by $NaBH_4$ *without illumination*, offers a system in which the flaw mentioned by ABRAHAMSON and WIESENFELD is avoided. Nature has partially performed the experiment for us.

The results reported in this chapter have demonstrated that in respect to temperature and to certain added reagents, the gecko 521-pigment responds in a manner that is different from the rhodopsins of several vertebrates. There is one photosensitive pigment, retinochrome from the eyes of cephalopods, that is similar to the gecko pigment in its responses to NH_2OH and $NaBH_4$ (HARA and HARA, 1972, 1973). Retinochrome is located in the visual cells in the structures internal to the rhabdomeres, which house the visual pigment. Unlike the visual pigment, retinochrome is a chromoprotein with the all-*trans* retinaldehyde as the prosthetic group. Light bleaches retinochrome and isomerizes the all-*trans* retinal to the 11-*cis* form. HARA and HARA (1972) suggested that cephalopod rhodopsin and retinochrome form a mutually linked system in which the all-*trans* retinal resulting from the bleaching of rhodopsin serves to supply the all-*trans* retinal required to regenerate retinochrome, which in turn supplies 11-*cis* retinal for the regeneration of rhodopsin. Whatever the merits of this suggestion, retinochrome and squid rhodopsin are apparently different in respect to the microenvironment of chromophore, for the retinochrome, but not the rhodopsin, is readily attacked in the dark by $NaBH_4$ and NH_2OH. In this respect, retinochrome and the gecko pigment behave similarly. Retinochrome also reacts to inorganic ions by a change in color (AZUMA et al., 1975), but this behavior is unlike the gecko chloride effect. In response to added NaCl, solubilized retinochrome shifts its spectral maximum hypsochromically by 2–13 nm, according to the concentration of NaCl. The chlorides of potassium, calcium, lithium, and magnesium produce a similar qualitative shift, although the quantitative behavior varies with the cation.

III. Speculations

1. The Thermochromic Shift

I have been unable to find in the biochemical literature an action of temperature on proteins quite like the thermochromic gecko effect. There are reports of reversible temperature-based spectral shifts, but most of these involve temperatures well above the 5°–25° C thermometric range of this chapter. As an example,

FOSS (1960) reported a denaturation "blue" shift in the ultraviolet region when ribonuclease was heated to above 70° C. There is of course the well-known "blue" shift in color of the lobster carapace on cooking the crustacean. From the shell, WALD et al. (1948) extracted a water-soluble astazanthin-protein complex named crustacyanin. This had an absorption band at 625 nm and when heated to 60° C became denatured to a product with absorption at 530 nm. On cooling, the blue color was restored. The gecko pigment is unusual in showing such a relatively large spectral shift at such low temperatures. Presumably, the microenvironment of the chromophore is a unique one. Since this environment is apparently hydrophobic in nature (BOWNDS, 1967; WU and STRYER, 1972), the action of temperature on hydrophobic bonds needs to be examined. It is known, however, that in the range below 58° C the strength of the hydrophobic bond increases with increasing temperature (SCHERAGA et al., 1962). Accordingly, the weakening of such bonds cannot be the explanation for the gecko thermochromic effect. Nevertheless, the hydrophobic environment may have a role in stabilizing hydrogen bonds which are known to become weaker at increasing temperature (SCHERAGA et al., 1962). It is known (CRESCITELLI et al., 1966) that the alpha-helical conformation is present in rhodopsin and that a portion of this helical structure is lost on photic bleaching. This suggests that this specific secondary structure is part of the environment of the chromophore. This in turn suggests that the thermochromic behavior may be due to disruption of hydrogen bonds, which are less protected in the gecko pigment than in the rhodopsins that lack the thermochromic behavior. In the future, if sufficient pigment can be obtained, an examination will be made of the circular dichroism as a function of temperature in the thermometric range.

2. The Ion Effects

The vertebrate visual pigments absorb light over a spectral range of some 200 nm and pigments exist, even in the same species, whose absorbance is adjusted to different specific spectral regions within this range. It is of interest from both chemical and biological viewpoints to understand the mechanisms by which this diversity and adjustment is accomplished. Many speculations have been published regarding these mechanisms (MANTIONE and PULLMAN, 1971; ABRAHAMSON and WIESENFELD, 1972), and some of these depend on the use of chemical models whose relevance to the real situation is doubtful. In geckos, however, we are provided with real visual pigments whose spectral absorbance is easily and reversibly adjustable over a range of some 20–30 nm by means of certain specific ions. No such behavior has been reported for the rhodopsins that have been so thoroughly studied.

Regarding the mechanism whereby chloride or bromide ions are able to maintain or to restore the normal location of the 521-pigment, it is easier to eliminate certain old ideas than to suggest new ones. The specificity of the action eliminates the ionic strength as a factor. The usual lyotropic series of anion effects (VON HIPPEL and SCHLEICH, 1969) do not apply to the specific actions of chloride and bromide on the gecko pigment. Moreover, the concentration of the lyotropic ions are normally one or two orders of magnitude higher than for the gecko ion effect.

In a number of studies, a case was made for the idea that certain anions influence hydrophobic proteins through their action on the structure of water. One group of anions, such as I^-, Br^-, and Cl^-, are thought of as water-structure breakers and are considered to increase the water solubility of hydrophobic groups and thus to destabilize protein complexes—a salting-in. In contrast to these chaotropic ions are the smaller univalent anions, like F^-, which are water-structure makers and which act antichaotropically by salting-out hydrophobic complexes (HANSTEIN et al., 1971). The specificity of the chloride gecko effect, along with the aberrant position of I^-, argues against a mechanism involving salting-in or salting-out. The speed and easy reversibility of the gecko effect and the fact that it occurs at temperatures approaching 0° C is against any hypothesis involving the making and breaking of covalent bonds. This leaves the suggestion that the chloride effect is an action at a critical site of the opsin, resulting in a change in the charge interaction between protein and retinal.

It has been stated many times that rhodopsin is a chromoprotein with the 11-*cis* isomer of retinal joined as a Schiff base to the ε-amino group of a lysine residue. Model Schiff bases of retinal with amines do not absorb at the wavelengths required for the case of many visual pigments. Consequently, two additional mechanisms have been adduced to explain the requisite bathochromic displacement. One is that the nitrogen of the chromophore is protonated (KROPF and HUBBARD, 1958), and the second is that a specific charge interaction exists between the retinal conjugated system and the opsin. There is good evidence now that the nitrogen is indeed protonated (LEWIS et al., 1973; OSEROFF and CALLENDER, 1974), and it is known from the circular dichroism in the visible region that an intimate stereospecific relationship between retinal and opsin does indeed exist (CRESCITELLI and SHAW, 1964). Up to this point the situation is quite reasonable, but beyond this the facts become fewer and speculations greater especially in regard to the role of the protein.

KROPF and HUBBARD (1958) conceptualized the interaction between opsin and prosthetic group as an influence of negatively charged groups of the opsin as stabilizing the positive charge of the nitrogen, thus lowering the energy of the first excited state and decreasing the energy of transition from the ground to the first excited state of the chromophore. This would result in a shift of absorbance to longer wavelengths. This hypothesis assigns to such protein anionic groups a critical function, and the nature, location, and behavior of such groups would determine the degree of interaction and the position of the spectral absorbance. A somewhat similar view was expressed by SUZUKI and KITO (1972) who emphasized the role of distance between the protonated nitrogen and the opsin anionic group. Nothing is known about the hypothetical anionic groups, but chemical models have been used to study the nature of this charge interaction. BLATZ et al. (1972) have examined the spectral absorbance of the N-retinylidene-n-butylammonium ion (NRBA) coupled with chloride, bromide, or iodide. The sizes of these halogen ions increase progressively from 1.36 for F^-, 1.81 for Cl^-, 1.95 for Br^-, and 2.16 Å for I^- (PAULING, 1960). BLATZ (1972) and BLATZ et al. (1972) found the spectrum of the NRBA to shift progressively in going from the chloride (441 nm), the bromide (451 nm), and the iodide (460 nm). ABRAHAMSON and WIESENFELD (1972) reported the results of quantum mechanical calculations in which the effect on the absorbance maximum of the 11-*cis* retinylidene-iminium ion was determined by placing negative point charges at various carbon atoms. The calculated absorbance maxima did not predict the full range of the bathochromic spectral displacement of visual pigments in nature, but the authors considered that more than 80% of the range was covered by the calculations. Similar calculations for the 11-*cis* 3-dehydro system gave figures which accounted for about 70% of the natural range. It is somewhat arbitrary, therefore, as to whether the theoretical predictions do indeed account for the distribution of visual pigments in nature. One interesting outcome of these quantum mechanical calculations was the result that point charges placed near the nitrogen cause no bathochromic shift. This is interesting, because it is known that vertebrate visual pigments exist that absorb at lower wavelengths and do not require an explanation in terms of a special interaction between retinal and protein. Such is the anuran green rod pigment which absorbs in the region of 432 nm for the adult and 438 nm for the tadpole (Liebman and Entine, 1968).

This is the context within which the chloride effect needs to be understood. In a general way, the chloride action appears to fit in with the idea of a secondary interaction between the polyene chain of retinal and the opsin. I must assume that chloride (or bromide) ions, because of their appropriate size, attach to a critical region of the opsin and specifically perturb the charge distribution and the interaction between opsin and prosthetic group. Perhaps chloride alters the conformation of the opsin in a way to influence the interactions at the chromophoric region. Whatever may be the mechanism of action of chloride, the problem remains as to why the gecko pigment, and not the rhodopsins, can be prepared in an ion-deficient hypsochromic state. The specific and easily reversible ion-binding properties of the gecko 521-pigment are also intriguing in light of the fact that visual pigments are components of excitable membranes that are involved in ion transfer processes within the living visual cell.

D. Appendix

After this chapter was written additional information appeared that is relevant to the story of the gecko visual pigments. Upon further experimentation, I have encountered two new chloride effects to be added to the account given above. Both these new effects involve the chloride and bromide ions specifically as in the case of the ionochromic effect. The first of these two is concerned with the finding that the 521-pigment, but not the 467-pigment, is bleached in the dark when the extract is made with 1% Triton-X-100 in Tris-maleate buffer at pH 7.2. This bleaching occurs very slowly with the extract at 5° C or below but is hastened when the temperature is increased to any higher level up to 25° C. This bleaching is irreversible but is significantly slowed in the presence of added chloride or bromide. The second of these new effects involves a similar protection of the 521-pigment against destruction, in the dark, by added NH_2OH. This means that all three chloride effects may be different manifestations of the same basic mechanism, possibly a conformational response of the opsin resulting from the binding of chloride to specific sites. The interest and significance of these chloride responses is enhanced by the finding that chicken iodopsin also displays a chloride color effect. This was first reported by Knowles (Biochem. and Biophys. Commun., [73], 56–62, 1976) and confirmed by Fager (Biophysic. Journ. [17], 16A, 1977). The latter report plus a private communication that I have received from Dr. R.S. Fager strongly suggest that the gecko pigment and the iodopsin behavior toward chloride are basically identical phenomena. If this is so, it is of significance in relation to Walls' interpretation of the gecko rods as transmuted cells having a phylogenetic origin in photopic receptors. It is also of interest that the 467-pigment, in addition to being resistant to temperature, to Triton-X-100, to NH_2OH and to $NaBH_4$, cannot be extracted in the chloride-deficient, hypsochromic state and so cannot be demonstrated to have a chloride effect. The two pigments of the Tokay gecko have inherently different molecular properties.

E. Acknowledgment

The research reported in this chapter was supported by Grant R01-EY00100-19 of the National Eye Institute, National Institutes of Health, U.S.P.H.S.

References

Abrahamson, E.W., Wiesenfeld, J.R.: The structure, spectra and reactivity of visual pigments. In: Photochemistry of Vision, Handbook of Sensory Physiology VII/1, Dartnall, H.J.A., ed. Heidelberg: Springer-Verlag 1972.

Akhtar, M., Blosse, P.T., Dewhurst, P.B.: The reduction of a rhodopsin derivative. Life Sci. **4**, 1221—1226 (1965).

Akhtar, M., Blosse, P.T., Dewhurst, P.B.: The active site of the visual protein, rhodopsin. Chem. Commun. 631—632 (1967).

Allen, D.M., McFarland, W.N.: The effect of temperature on rhodopsin-porphyropsin ratios in a fish. Vision Res. **13**, 1303—1309 (1973).

Alpern, M.: Rhodopsin kinetics in the human eye. J. Physiol. (Lond.) **217**, 447—471 (1971).

Azuma, K., Azuma, M., Sakaguchi, K., Kito, Y.: Reversible spectral change of squid retinochrome by salts. Nature (Lond.) **253**, 206—207 (1975).

Baumann, C.: Kinetics of slow thermal reactions during the bleaching of rhodopsin in the perfused frog retina. J. Physiol. (Lond.) **222**, 643—663 (1972).

Blatz, P.E.: The formation of long wavelength absorbing species from short wavelength absorbing linear conjugated polyenes. Photochem. Photobiol. **15**, 1—6 (1972).

Blatz, P.E., Mohler, J.H., Navangul, H.V.: Anion-induced wavelength regulation of absorption maxima of Schiff bases of retinal. Biochemistry **11**, 848—855 (1972).

Bonds, A.B., MacLeod, D.I.A.: The bleaching and regeneration of rhodopsin in the cat. J. Physiol. (Lond.) **242**, 237—253 (1974).

Bowmaker, J.F.: The photoproducts of retinal-based visual pigments in situ: a contrast between *Rana pipiens* and *Gekko gekko*. Vision Res. **13**, 1227—1240 (1973).

Bownds, D.: Site of attachment of retinal in rhodopsin. Nature (Lond.) **216**, 1178—1181 (1967).

Bownds, D., Wald, G.: Reaction of the rhodopsin chromophore with sodium borohydride. Nature (Lond.) **205**, 254—257 (1965).

Cecil, R.: Intramolecular bonds in proteins. I. The role of sulfur in proteins. In: The Proteins (Neurath, H. (ed.). New York: Academic Press 1963, Vol. I, pp. 379—476.

Chino, Y. M., Sturr, J. F.: Rod and cone contributions to the delayed response of the on-off ganglion cell in the frog. Vision Res. **15**, 193—202 (1975).

Cleland, W. W.: Dithiothreitol, a new protective reagent for SH groups. Biochemistry **3**, 480—482 (1964).

Cohen, A. I.: Rods and cones. In: Physiology of Photoreceptor Organs. Handbook of Sensory Physiology VII/2 Fuortes, M. G. F. (ed.). Heidelberg: Springer-Verlag 1972.

Crescitelli, F.: The nature of the gecko visual pigment. J. gen. Physiol. **40**, 217—231 (1956).

Crescitelli, F.: The photosensitive retinal pigment system of *Gekko gekko*. J. gen. Physiol. **47**, 33—52 (1963).

Crescitelli, F.: The spectral sensitivity of the gecko eye in relation to the state of adaptation. Vision Res. **6**, 129—142 (1966).

Crescitelli, F.: The e-wave and inhibition in the developing retina of the frog. Vision Res. **10**, 1077—1091 (1970).

Crescitelli, F.: The visual cells and visual pigments of the vertebrate eye. In: Photochemistry of Vision. Handbook of Sensory Physiology VII/1 (Ed. Dartnall, H. J. A.) Heidelberg: Springer-Verlag 1972.

Crescitelli, F.: The transmuted visual pigments of geckos. Exp. Eye Res. **18**, 307—322 (1974a).

Crescitelli, F.: The gecko visual pigments I. The thermosensitive property. Vision Res. **14**, 243—253 (1974b).

Crescitelli, F.: The gecko visual pigments — II. Color and the sulfhydryl group. Vision Res. **15**, 743—745 (1975a).

Crescitelli, F.: The gecko visual pigments; ionic effects on color. 26th Ann. Fall Meeting, Amer. Physiol. Soc., San Francisco, Calif. **34** (1975b).

Crescitelli, F., Dartnall, H. J. A.: Human visual purple. Nature (Lond.) **172**, 195—196 (1953).

Crescitelli, F., Dartnall, H. J. A., Loew, E.: The gecko visual pigments — III. A microspectrophotometric study. Abst., Fed. Proc. **35**, 562 (1976).

Crescitelli, F., Mommaerts, W. F. H. N., Shaw, T. I.: Circular dichroism of visual pigments in the visible and ultra-violet spectral regions. Proc. Nat. Acad. Sci. (Wash.) **56**, 1729—1734 (1966).

Crescitelli, F., Shaw, T. I.: The circular dichroism of some visual pigments. J. Physiol. (Lond.) **175**, 43—45P (1964).

Crescitelli, F., Sickel, E.: Delayed off-responses recorded from the isolated frog retina. Vision Res. **8**, 801—816 (1968).

Crozier, W. J., Wolf, E.: The flicker response contour for the gecko (rod retina). J. gen. Physiol. **22**, 555—566 (1939).

Dartnall, H. J. A.: The interpretation of spectral sensitivity curves. Brit. med. Bull. **9**, 24—30 (1953).

Dartnall, H. J. A.: The visual pigment of the green rods. Vision Res. **7**, 1—16 (1967).

Denton, E. J.: The responses of the pupil of *Gekko gekko* to external light stimulus. J. gen. Physiol. **40**, 201—215 (1956).

Donner, K. O.: Rod dark-adaptation and visual pigment photoproducts. In: Biochemistry and Physiology of Visual Pigments (Ed. Langer, H.) Berlin-Heidelberg-New York: Springer-Verlag 1973.

Donner, K. O., Hemilä, S., Hongell, K., Reuter, T.: Long-lived photoproducts of porphyropsin in the retina of the crucian carp (*Carassius carassius*). Vision Res. **14**, 1359—1370 (1974).

Donner, K. O., Reuter, T.: Dark-adaptation processes in the rhodopsin rods of the frog's retina. Vision Res. **7**, 17—41 (1967).

Donner, K. O., Reuter, T.: Visual adaptation of the rhodopsin rods in the frog's retina. J. Physiol. (Lond.), **199** 59—87 (1968).

Dunn, R. F.: Electron microscopy studies on the photoreceptor cells of the gecko, *Coleonyx variegatus*. Ph. D. Thesis, Univ. of. Calif., Los Angeles (1965).

Dunn, R. F.: Studies on the retina of the gecko *Coleonyx variegatus*. I. The visual cell classification. J. Ultrastruct. Res. **16**, 651—671 (1966a).

Dunn, R. F.: Studies on the retina of the gecko *Coleonyx variegatus*. II. The rectilinear visual cell mosaic. J. Ultrastruct. Res. **16**, 672—684 (1966b).

Dunn, R. F.: The dimensions of rod outer segments related to light absorption in the gecko retina. Vision Res. **9**, 603—609 (1969).

Dunn R. F.: The ultrastructure of the vertebrate retina. In: The Ultrastructure of Sensory Organs. Amsterdam: North-Holland 1973.

Earnshaw, W., Fujimori, E.: Dark bleaching of rhodopsin by organic mercurial. FEBS Lett. **34** 137—139 (1973).

Ewald, A., Kühne, W.: Untersuchungen über den Sehpurpur. III. Veränderungen des Sehpurpurs und der Retina im Leben. Untersuchungen aus dem Physiologischen Institut der Universität Heidelberg **1**, 370—455 (1878).

Forbes, A., Fox, S., Milburn, N., Deane, H. W.: Electroretinograms and spectral sensitivities of some diurnal lizards. J. Neurophysiol. **23** 62—73 (1960).

Foss, J. G.: Temperature difference spectra. Biochem. biophys. Res. Commun. **2**, 193—197 (1960).

Frank, R. N.: Photoproduces of rhodopsin bleaching in the isolated, perfused frog retina. Vision Res. **9**, 1415—1434 (1969).

Fujimori, E.: Charge-transfer interaction involving sulfhydryl groups of bovine serum albumin. FEBS Lett. **57** 37—41 (1975).

Fujita, S. C., Imahori, K.: The thermostability of secondary structure of D-glyceraldehyde-3-phosphate dehydrogenase from *Thermus thermophilus*. In: Peptides, Polypeptides and Proteins. Blout, E. R. (ed.) New York: Wiley, 1974.

Galindo, I. G.: A charge transfer process in the visual pigments. Bull. Math. Biophys. **29**, 677—690 (1967).

Girsch, S. J., Rabinovitch, B.: Bleaching of rhodopsin in the dark. Biochem. biophys. Res. Commun. **44**, 550—556 (1971).

Gyllenberg, G., Reuter, T., Sippel, H.: Long-lived photoproducts of rhodopsin in the retina of the frog. Vision Res. **14**, 1349—1357 (1974).

Hagins, W. A.: Rhodopsin in the mammalian retina. Thesis, University of Cambridge (1957).

Hamasaki, D. I.: The spectral sensitivity of the lateral eye of the Green Iguana. Vision Res. **8**, 1305—1314 (1968).

Hanstein, W. G., Davis, K. A., Hatefi, Y.: Water structure and the chaotropic properties of haloacetates. Arch. Biochem. **147**, 534—544 (1971).

Hara, T., Hara, R.: Cephalopod retinochrome. In: Photochemistry of Vision, Handbook of Sensory Physiology VII/l. Dartnall, H. J. A. (ed.). Heidelberg: Springer-Verlag 1972.

Hara, T., Hara, R.: Biochemical properties of retinochrome. In: Biochemistry and Physiology of Visual Pigments. Langer, H. (ed.). Heidelberg: Springer-Verlag 1973.

Hárosi, F. I., MacNichol, E. F.: Visual pigments of goldfish cones. Spectral properties and dichroism. J. gen. Physiol. **63**, 279—304 (1974).

Hecht, S., Williams, R. E.: The visibility of monochromatic radiation and the absorption spectrum of visual purple. J. gen. Physiol. **5**, 1—33 (1922).

Heller, J.: Structure of visual pigments. II. Binding of retinal and conformational changes on light exposure in bovine visual $pigment_{500}$. Biochemistry **7**, 2914—2920 (1968).

Heller, J.: Comparative study of a membrane protein. Characterization of bovine, rat and frog visual $pigment_{500}$. Biochemistry **8** 675—679 (1969).

Hong, K., Hubbell, W. L.: Preparation and properties of phospholipid bilayers containing rhodopsin. Proc. nat. Acad. Sci (Wash.) **69** 2617—2621 (1972).

Hubbard, R.: Retinene isomerase. J. gen. Physiol. **39**, 935—962 (1956).

Hubbard, R.: The thermal stability of rhodopsin and opsin. J. gen. Physiol. **42**, 259—280 (1958).

Kleinschmidt, J.: Adaptation properties of intracellularly recorded *Gekko* photoreceptor potentials. In: Biochemistry and Physiology of Visual Pigments. Langer, H. (ed.). Heidelberg: Springer-Verlag 1973.

Komatsu, S. K., Feeney, R. E.: A heat labile fructose diphosphate aldolase from cold-adapted antarctic fishes. Biochem. biophys. Acta **206**, 305—315 (1970).

Köttgen, E., Abelsdorff, G.: Absorption und Zersetzung des Sehpurpurs bei den Wirbeltieren. Z. Psychol. Physiol. Sinnesorg. **12**, 161—184 (1896).

Kropf, A., Hubbard, R.: The mechanism of bleaching rhodopsin. Ann. N.Y. Acad. Sci. **74**, 266—280 (1958).

Lewis, A., Fager, R. S., Abrahamson, E. W.: Tunable laser resonance Raman spectroscopy of the visual process. I. The spectrum of rhodopsin. J. Raman Spectrosc. **1**, 465—470 (1973).

Liebman, P. A.: Microspectrophotometry of photoreceptors. In: Photochemistry of Vision, Handbook of Sensory Physiology VII/1. Dartnall, H. J. A. (ed.). Heidelberg: Springer-Verlag 1972.

Liebman, P. A.: Microspectrophotometry of visual receptors. In: Biochemistry and Physiology of Visual Pigments (Ed. Langer, H.) Berlin-Heidelberg: Springer-Verlag 1973.
Liebman, P. A., Entine, G.: Visual pigments of frog and tadpole (*Rana pipiens*). Vision Res. **8**, 761—775 (1968).
Lythgoe, R. J., Quilliam, J. P.: The relation of transient orange to visual purple and indicator yellow. J. Physiol. (Lond.) **94**, 399—410 (1938a).
Lythgoe, R. J., Quilliam, J. P.: The thermal decomposition of visual purple. J. Physiol. (Lond.) **93**, 24—38 (1938b).
Mantione, M. J., Pullman, B.: A quantum mechanical investigation of the bathochromic shift in visual pigments. Int. J. quant. Chem. **5**, 349—360 (1971).
Marks, W. B.: Comment made (p. 323). In: Colour Vision, Physiology and Experimental Psychology. DeReuck, A. V. S., Knight, J. (eds.). Ciba Found. Symps. London: J. and A. Churchill 1965a.
Marks, W. B.: Visual pigments of single goldfish cones. J. Physiol. (Lond.) **178**, 14—32 (1965b).
Matthews, R. G., Hubbard, R., Brown, P. K., Wald. G.: Tautomeric forms of metarhodopsin. J. gen. Physiol. **47**, 215—240 (1963).
Nilsson, S. E. G.: An electron microscopic classification of the retinal receptors of the leopard frog (Rana pipiens). J. Ultrastruct. Res. **10**, 390—416 (1964).
Ohta, Y.: Thermostable protease from thermophilic bacteria. II. Studies on the stability of the protease. J. biol. Chem. **242**, 509—515 (1967).
Ohta, Y., Ogura, Y., Wada, A.: Thermostable protease from thermophilic bacteria. I. Thermostability, physicochemical properties, and amino acid composition. J. biol. Chem. **241**, 5919—5925 (1966).
Oseroff, A. R., Callender, R. H.: Resonance Ramam spectroscopy of rhodopsin in retinal disk membranes. Biochemistry **13**, 4243—4248 (1974).
Ostroy, S. E., Erhardt, F., Abrahamson, E. W.: Protein configuration changes in the photolysis of rhodopsin. II. The sequence of intermediates in the thermal decay of cattle rhodopsin in vitro. Biochim. biophys. Acta **112**, 265—277 (1966).
Ostroy, S. E., Rudney, H., Abrahamson, E. W.: The sulfhydryl groups of rhodopsin. Biochim. biophys. Acta. **126**, 409—412 (1966).
Pak, W. L., Grabowski, S. R., Pinto, L. H.: Receptor adaptation and receptor interactions: some results of intracellular recordings (preliminary note). In: Biochemistry and Physiology of Visual Pigments. Langer, H. (ed.). Heidelberg: Springer-Verlag 1973.
Pauling, L.: The Nature of the Chemical Bond. Ithaca, New Cork: Cornell Univ. Press 1960.
Pedler, C., Tilly, R.: The nature of the gecko visual cell. A light and electron microscopic study. Vision Res. **4**, 499—510 (1964).
Pickering, S. G.: The extremely long latency responce from on-off retinal ganglion cells: Relationship to dark adaptation. Vision Res. **8**, 383—387 (1968).
Pickering, S. G., Varjú, D.: Ganglion cells in the frog retina: Inhibitory receptive field and long-latency response. Nature (Lond.) **215**, 545—546 (1967).
Pickering, S. G., Varjú, D.: Delayed responses of ganglion cells in the frog retina: The influence of stimulus parameters upon the length of the delay time. Vision Res. **9**, 865—879 (1969).
Reuter, T.: The synthesis of photosensitive pigments in the rods of the frog's retina. Vision Res. **6**, 15—38 (1966).
Ripps, H., Weale, R. A.: Rhodopsin regeneration in man. Nature (Lond.) **222**, 775—777 (1969).
Rushton, W. A. H.: The difference spectrum and the photosensitivity of rhodopsin in the living human eye. J. Physiol. (Lond.) **134**, 11—29 (1956).
Rushton, W. A. H., Powell, D. S.: The early phase of dark adaptation. Vision Res. **12**, 1083—1093 (1972).
Scheraga, H. A., Némethy, G., Steinberg, I. Z.: The contribution of hydrophobic bonds to the thermal stability of protein conformations. J. biol. Chem. **237**, 2506—2508 (1962).
Scholes, J. H.: Colour receptors and their synaptic connexions in the retina of a cyprinid fish. Phil. Trans. B. **270**, 61—118 (1975).
Sickel, W., Crescitelli, F.: Delayed electrical responses from the isolated frog retina. Pflügers Arch. ges. Physiol. **297**, 266—269 (1967).
Sillman, A. J.: A thermosensitive property of bullfrog rhodopsin. Vision Res. **16**, 419—420 (1976).
Singleton, R., Amelunxen, R. E.: Proteins from thermophilic microorganisms. Bact. Rev. **37**, 320—342 (1973).

Somero, G. N., Hochachka, P. W.: Effect of temperature on catalytic and regulatory functions of pyruvate kinases of the rainbow trout and the antarctic fish *Trematomus bernacchii*. Biochem. J. **110**, 395—399 (1968).

Stell, W. K., Lightfoot, D. O.: Color-specific interconnections of cones and horizontal cells in the retina of the goldfish. J. comp. Neurol. **159**, 473—501 (1975).

Suzuki, T., Kito, Y.: Absorption spectra of TCA-denatured rhodopsin and of a Schiff base compound of retinal. Photochem. Photobiol. **15**, 275—288 (1972).

Svaetichin, G., Negishi, K., Fatehchand, R.: Cellular mechanisms of a Young-Hering visual system. In: Colour Vision, Physiology and Experimental Psychology, DeReuck, A. V. S., Knight, J. (eds.). Ciba Found. Sympos. London: J. and A. Churchill 1965.

Tansley, K.: The gecko retina. Vision Res. **4**, 33—37 (1964).

Toyoda, J., Nosaki, H., Tomita, T.: Light-induced resistance changes in single photoreceptors of *Necturus* and *Gekko*. Vision Res. **9**, 453—463 (1969).

Underwood, G.: Reptilian retinas. Nature (Lond.) **167**, 183—185 (1951).

Underwood, G.: On the classification and evolution of geckos. Proc. zool. Soc. London **124**, 469—492 (1954).

Underwood, G.: Some suggestions concerning vertebrate visual cells. Vision Res. **8**, 483—488 (1968).

Underwood, G.: The Eye. In: Biology of the Reptilia. Gans, C., Parsons, T. S. (eds.). New York: Academic Press 1970.

Underwood, G.: Vertebrate comparative anatomy at the cellular level: visual cells and phylogeny. Report of the NATO Advance Study Institute. Vertebrate evolution: mechanism and process, 64—69 (1971).

Von Hippel, P. H., Schleich, T.: The effects of neutral salts on the structure and conformational stability of macromolecules in solution. In: Structure and Stability of Biological Macromolecules. Timasheff, S. N., Fasman, G. D. (eds.). New York: Marcel Dekker, Inc. 1969.

Wald, G., Brown, P. K.: The role of sulfhydryl groups in the bleaching and synthesis of rhodopsin. J. gen. Physiol. **35**, 797—820 (1952).

Wald, G., Brown, P. K.: Human rhodopsin. Science (N. Y.) **127**, 222—226 (1958).

Wald, G., Brown, P. K., Smith, P. H.: Iodopsin. J. gen. Physiol. **38**, 623—681 (1955).

Wald, G., Hubbard, R.: Enzymic aspects of the visual processes. In: The Enzymes (Ed. Boyer, P., Lardy, H., Myrbäck, K.) New York: Academic Press 1960.

Wald, G., Nathanson, H., Jencks, W. P., Tarr, E.: Crustacyanin, the blue carotenoid of the lobster shell. Biol. Bull. **95**, 249—250 (1948).

Walls, G. L.: The reptilian retina. Amer. J. Ophthal. **17**, 892—915 (1934).

Walls, G. L.: The Vertebrate Eye and its Adaptive Radiation. Michigan: Cranbrook Institute of Science Bull. **19**, pp. 785 (1942).

Weale, R. A.: On an early stage of rhodopsin regeneration in man. Vision Res. **7**, 819—827 (1967).

Wu, C.-W., Stryer, L.: Proximity relationships in rhodopsin. Proc. nat. Acad. Sci. (Wash.) **69**, 1104—1108 (1972).

Wyszecki, G., Stiles, W. S.: Color Science. New York: John Wiley and Sons, Ind. 1967.

Yoshizawa, T., Wald, G.: Visual pigments and the Keilin-Hartree effect. Nature (Lond.) **212**, 483—485 (1966).

Zorn, M.: The reduction of rhodopsin with sodium borohydride under non-bleaching condition. Biochem. biophys. Acta **245**, 216—220 (1971).

Chapter 8

Vision in Turtles

A. M. Granda and C. A. Dvorak, Newark, Delaware, USA

With 33 Figures

Contents

A. Introduction

Perhaps more than in any other *class*, vision evolved to its greatest complexity in reptiles (Walls, 1942). Apart from the birds, to which reptiles form a natural connection, vision and its attendant structures in other animals are more constrained. Fishes and amphibians, for example, animals restricted largely or completely to aquatic environments, have difficulty in seeing for any great distance. Water is a poor medium for light transmission: It is easily set in motion by wave action and its transparency is complicated by turbidity. Because of these factors, the virtuous properties of light are very much degraded or practically non-existent. In many animals of these *classes*, visual structure is less elaborated. Some mammals, too, display a debilitation of visual structure and function that is especially noteworthy in view of their other biological advantages. Many mammals are nocturnal, or they emphasize other sensory attributes. Only in primates, perhaps in squirrels too with their cone-rich retinas, does vision predominate to the same extent as in reptiles and birds.

Reptiles were the first vertebrates to completely transfer their lives to a land habitat. To them air offered a clear medium for vision with great advantages secured by the rectilinear passage of light through it. Life on land, requiring higher standards of vision, allowed for new anatomy and functions to develop. Of the reptiles, the *order Chelonia*, alternatively known as *Testudinata*, the turtles, show perhaps the greatest elaboration of structure.

In the sections to follow, we will discuss this elaboration of the chelonian eye, its structure, and the anatomy and photopigments peculiar to it. There will follow a discussion of results obtained through electrophysiology, an especially fruitful progression over the past several years. We will also discuss results obtained through psychophysical procedures and endeavor finally to unite the relationships into a picture of Chelonian vision that is well worked out and remarkably advanced.

B. Structure

I. Anatomy and General Morphology

Turtles share the general vertebrate plan of an eyeball encased in three coats: sclera, choroid, and retina. The eyeball forms a dark chamber filled with clear fluids under pressure. At the anterior pole, there is a transparent cornea of varying refractive power, depending on habitat. Within the eye is a lens whose means of accommodation is unusual in terms of its evolution (DUKE-ELDER, 1958). The lens is pliable and transparent. Accommodation is obtained by striated ciliary muscles cantilevering from a ring of scleral ossicles to deform the lens (Fig. 1) rather than by a to-and-fro movement as it is in fishes. For binocular vision, there is assistance in convergence by a *transversalis muscle* which orients the lens nasally. The mechanisms for both accommodation and convergence are newly evolved for reptiles.

To this general reptilian plan, turtles add some features and differences. Specifically, there is a sphincter muscle that participates in the act of accommodation; there are ciliary processes that assist in this mechanism. There are, however, no elaborated structures to increase metabolic nourishment through increased circulation. Lizards have a *conus papillaris*, and snakes have a *vasculosa retinae* for greater efficiency, but the retina of the turtle is avascular; it is nourished only by the choroid layer, and the nourishment is indirect through diffusion.

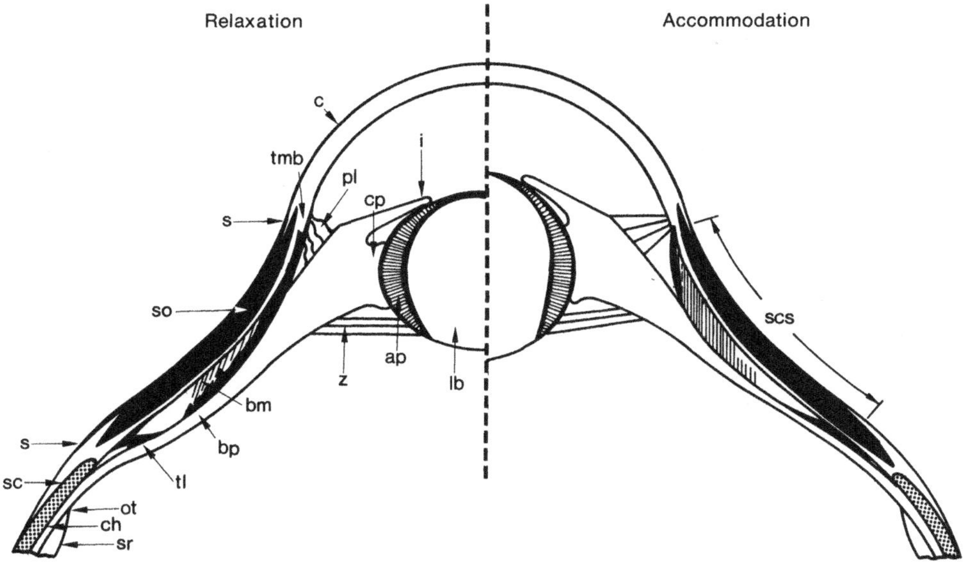

Fig. 1. Composite diagram of accommodation in reptiles. *ap*: annular pad of lens; *bm*: Brücke's muscle; *bp*: base plate of ciliary body; *c*: cornea; *ch*: choroid; *cp*: ciliary process; *i*: iris; *lb*: lens body; *ot*: ora terminalis; *pl*: pectinate ligament; *s*: sclera; *sc*: scleral cartilage; *scs*: sclerocorneal sulcus; *so*: scleral ossicle; *sr*: sensory retina; *tbm*: tendon of Brücke's muscle; *tl*: tenacular ligament; *z*: zonule. (From WALLS, 1942)

The turtle eye is fairly small compared to birds and primates, their nearest competitors for visual primacy. The cornea is thick and shows little change in curvature in the limbus region of the scleral junction. Birds show a sulcus here to a greater degree (DUKE-ELDER, 1958). In the corneal epithelium are the scleral ossicles—bony, thin plates that overlap (Fig. 2, p. 454). There are varying numbers of plates, depending on species (DUKE-ELDER, 1958; UNDERWOOD, 1970). Their value lies in the process of accommodation, where they act to stiffen the sclerocorneal sulcus, which otherwise would be evaginated by intraocular pressure during the process (BEER, 1898).

The pupil is round and despite claims to the contrary somewhat responsive to light, although accommodation confounds the effect. The pupil is fairly small in most species. The freshwater turtle *Pseudemys*, for example, has a pupillary opening of about 1.7 mm in diameter. The sea turtle *Chelonia* has a smaller opening, about 0.7 mm under natural conditions. The drug pilocarpine dilates the pupil of *Pseudemys* a little, but atropine has no effect (NORTHMORE, personal communication).

The lens is unusually soft. Freshwater turtles have a flatter lens that sea turtles, which is to be expected: The sea turtle lens must carry the entire accommodative burden, since the benefit of any corneal refraction would be cancelled largely by immersion in sea water (Fig. 3, p. 455). EHRENFELD and KOCH (1967) reported *Chelonia mydas*, the green turtle, to be emmetropic in water but extremely myopic in air. The land turtle, *Gopherus polyphemus*, is somewhat hyperopic in air, but in water it is unable to compensate for the loss of corneal refraction. The freshwater turtle *Clemmys insculpta* accommodates well in both media, as does *Pseudemys*.

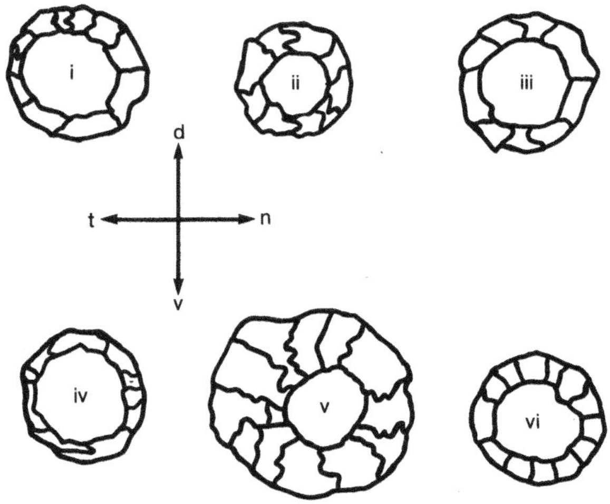

Fig. 2. Scleral ossicles of turtles. (i) *Dermatemys* (orientation uncertain); (ii) *Malacochersus;* (iii) *Platysternon;* (iv) *Carettochelys;* (v) *Eretmochelys;* (vi) *Emydura.* Compass directions refer to dorsal-ventral, temporal-nasal. (After UNDERWOOD, 1970).

NORTHMORE worked out the schematic eye for *Pseudemys.* The pertinent metrics are given in Table 1 (p. 456).

In several freshwater species the irises are beautifully colored: browns, reds, yellows, and greens. In *Pseudemys* there is, in addition, a black horizontal line across the iris which BROWN (1969) correlated with a high-density receptor area across the retina, the linear *area centralis.* The two operate in parallel. The black line is stabilized to the horizon by reflexive eye and head movement and so presumably is the retinal area of greatest resolution. In the land turtle *Terrapene carolina,* the iris of the male is red, while that of the female is brown, an interesting sexual dimorphism (DUKE-ELDER, 1958). On the other hand, the irises of sea and terrestrial turtles are normally quite dull and the coloration not marked.

The retinas are avascular, as they are in most reptiles. There is an embryologic attempt at a glial conus on the optic disc in some species, but none lasts into adulthood. The absence of nonvisual structures is advantageous in avoiding interferences in light reception, but absence of direct circulation sets limits on metabolic utilization, where nutrients must diffuse to their hosts over appreciable distances. The lizards which are visually dominated solve this problem by a *conus papillarus* (noted above), a vascularized structure that later evolved into the pecten of birds (WOOD, 1917). Turtles have no such structure.

II. Retina

1. Photoreceptors

There are single and double cones, and single rods in the eyes of most turtles. The single cones predominate and are of distinct sizes housing separate photopigments. In *Pseudemys scripta elegans,* where our knowledge is greatest, the cones

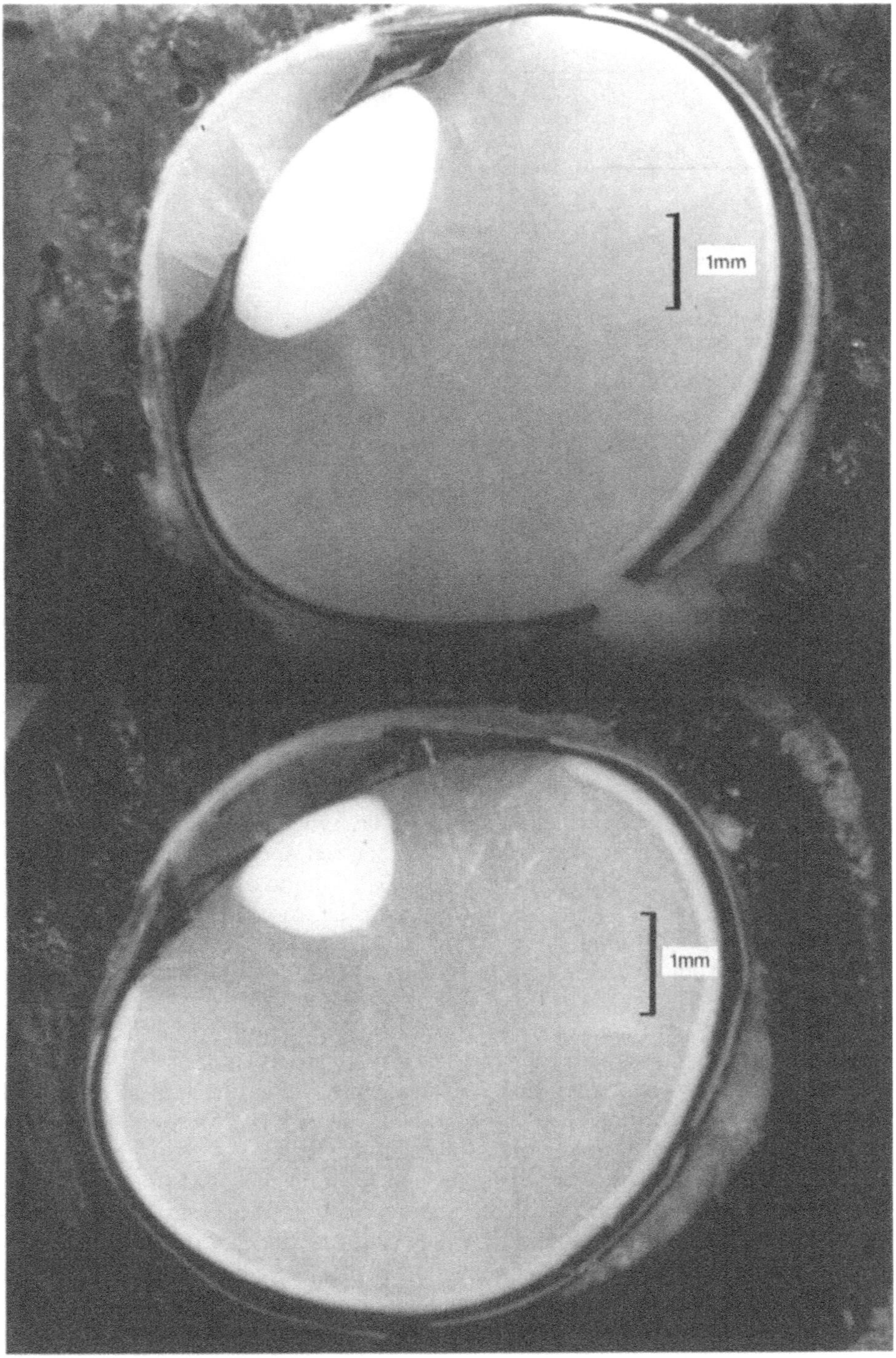

Fig. 3. Frozen hemisection of eyeball and lens. Top photograph: *Pseudemys*. Note shape of lens. Bottom photograph: *Chelonia*. The lens is more strongly curved. (The photographs were kindly provided by D. P. M. NORTHMORE)

Table 1. Schematic eye of *Pseudemys*

Axial positions	Lens—anterior vertex	1.25
	Lens—posterior vertex	3.12
	Retina—junction with pigment layer	7.56
	Pupil	1.25
Radii of curvature	Cornea	3.00
	Lens—anterior surface	2.84
	Lens—posterior surface	2.37
Cardinal points	Principal foci: anterior	−2.94
	posterior	7.56
	Principal points: first	0.93
	second	1.16
	Nodal points: first	2.53
	second	2.76
Refractive indices	Aqueous humour	1.335
	Lens	1.494
	Vitreous humour	1.335
	Pupillary diameter	1.74

All dimensions are given in millimeters relative to the vertex of the cornea. The eye is constructed to be emmetropic, i.e., distant objects are brought in focus at the junction of the retina and pigment layer by adjusting the bulk refractive index of the lens. Data are from 11 animals, 18 eyes (Northmore, personal communication).

outside the *area centralis* have outer segments up to 30–40 μm in length, with a 2 μm base diameter tapering to 1 μm at the tip on the outer segment [LIEBMAN and GRANDA (1971), but see BAYLOR and FETTIPLACE (1975) for a shorter outer segment measured length of 15 μm]. The single cones are easily identified by the presence of oil droplets. Brilliantly colored, they range, in *Pseudemys*, from rich reds to pale yellows. They are graded in size, the red ones being large, from 5.8–7.1 μm in diameter, the orange-pale yellows, 3–5.8 μm in diameter, and the small clear ones, 2.4–3.7 μm in diameter (GRANDA and HADEN, 1970). In the *area centralis*, BROWN (1969) found these diameters to be much smaller, with average diameters of 3.0 μm for red, 3.4 μm for orange, 3.0 μm for yellow, and 1.7 μm for clear droplets.

The relative frequencies of occurrence of the different droplets in *Pseudemys* are in fair agreement, as reported by several investigators (Table 2).

The photoreceptors containing red oil droplets have the largest inner and outer segments, according to BAYLOR and FETTIPLACE (1975), while those with yellow and clear droplets are progressively smaller (Table 3).

Most investigators have not distinguished separate classes of orange and yellow droplets, as they appear to shade imperceptibly into one another. But BAYLOR and FETTIPLACE (1975) found that they could reliably locate an orange droplet in one member of a double cone, the other member containing no droplet at all, by focusing the microscope down in different planes when the receptors were oriented vertically on a microscope slide. The orange droplets were found to lie in a plane slightly above the plane containing the red, yellow, and clear droplets. The latter three classes were always located in single cones; orange droplets were never seen there.

Table 2. Relative frequencies of oil droplets in *Pseudemys*

Color	PEIPONEN (1964)	BROWN[a] (1969)		GRANDA and HADEN (1970)	BAYLOR and FETTIPLACE (1975)	
Red	0.36		0.38	0.38		0.35
Orange		0.23			0.18	
	0.45		0.46	0.47		0.45
Yellow		0.23			0.27	
Clear	0.19		0.16	0.15		0.20

[a] Taken from the paracentral retina, 210 µm above the most central part of the *area centralis*. The measurements from other investigators are not localized but probably are outside the area of greatest receptor density.

Table 3. Measured diameters of single cones in *Pseudemys*

Color	Inner segment[a]	Outer segment[a]
Red	11.7	2.5
Yellow	9.1	2.2
Clear	7.1	1.5

[a] Inner segments measured at the level of the ellipsoid, outer segments close to the base. Measurements are in micrometers.

The double cones contain a principal member that is in apposition to a second member in the lower myoid region. The principal cone closely resembles the single cone, for it has an oil droplet. The second or accessory member is usually larger and, as noted, has no droplet.

Rods, too, are found in the turtle eye, although there is a long history almost to the present day that claims that they are absent. "Prior to 1877", WALLS wrote, "about everyone who described a chelonian retina saw rods in it, but since that time, owing to one of the few mistakes (and weighty authority) of Max Schultze, the turtles have been placed among the pure-cone reptiles. They do, however, possess droplet-free element with heavy, cylindrical outer segments ..." (WALLS, 1942). So they do. In *Pseudemys*, rods measure 3 µm in diameter by 40 µm long (LIEBMAN and GRANDA, 1971). The rod nuclei tend to be slightly below those in cones. The synaptic bases of rods are simple spherules. Cones, on the other hand, have complicated pedicles with recessed junctions (LASANSKY, 1971).

a) Photopigments

Turtles are distributed on land, in fresh waters, and in the sea. While several families, the *Emydidae*, *Chelydridae*, and *Dermatemyidae* are freshwater animals, the *Cheloniidae* and *Dermochelidae* are confined largely to warm seas and the *Testudinidae* to land. LIEBMAN and GRANDA (1971) microspectrophotometrically isolated sets of photopigments in two species, *Chelonia mydas mydas* belonging to the family *Cheloniidae*, and *Pseudemys scripta elegans* belonging to the family

Emydidae. The photopigments conformed to the vitamin A_1-vitamin A_2 environmental distribution. Creatures confined to land and in the sea share a broad assignment to vitamin A_1 or, as now named, retinol pigments. The pigments of freshwater animals, in contrast, have their peak absorbances shifted to longer wavelengths. WALD (1939a, b, 1941) showed these latter pigments in fishes to be based on vitamin A_2, or 3-dehydroretinol. He codified a simple system based on the combination of retinal (the aldehyde form of retinol or vitamin A_1) with rod or cone opsins to give pigments absorbing at 500 nm (rhodopsin) and 562 nm (iodopsin). Similarly for 3-dehydroretinal (the aldehyde form of 3-dehydroretinol or vitamin A_2), a system shifted to longer wavelengths, there are pigments absorbing at 522 nm (porphyropsin) and 620 nm (cyanopsin) (WALD, 1958, 1959; WALD et al., 1953).

The classification is far too simple. Visual pigments based on both retinal and 3-dehydroretinal have overlapping peak spectral regions. In some freshwater fishes both systems are present even in a single eye (BRIDGES, 1965, 1972). The classification, however, holds well enough for turtles. The *Chelonia* are sea creatures and possess *retinal* pigments, while *Pseudemys*, animals that live in freshwater ponds and streams, have *3-dehydroretinal* pigments.

The photopigments for *Chelonia* have peak absorbances at 440 nm, 502 nm, and 562 nm measured microspectrophotometrically (LIEBMAN and GRANDA, 1971). All these pigments were found in cone receptors identified by the geometry of the outer segments and by the presence of oil droplets. In *Pseudemys*, a similar identification was made. The 3-dehydroretinal pigments here have peak absorbances at 450 nm, 518 nm, and 620 nm. Smoothed density spectra for *Chelonia* and *Pseudemys* appear in Fig. 4 (after LIEBMAN and GRANDA, 1971). *Pseudemys* rods contain a 518_2 pigment. This pigment, or at least one that peaks in the same spectral region, is also found in the green-absorbing cones and in the accessory members of double cones. *Chelonia* rods, on the other hand, contain a 502_1 pigment. This pigment, too, is contained in single cones and in accessory members of double cones.

The specific density of measured turtle visual pigments recorded by microspectrophotometry in single cells can be obtained by knowing the measured extinction at peak, and the path length. Cone and rod pigments for both *Chelonia* and *Pseudemys* show a value of 0.013/μm similar to other vertebrates (LIEBMAN and GRANDA, 1971).

b) Oil Droplets

HANNOVER (1843) first reported colored oil droplets in the turtle. He described them as intensely colored and highly refractile. The oil droplets are judiciously placed in the optical pathway, where they intercept light destined for the outer segments. KRAUSE (1863) supposed that there might be a universal photopigment with critically placed, optically dense oil droplets acting as cut-off filters so as to permit only certain wavelengths to affect the outer segments. Some wavelengths, he reasoned, would stimulate only certain cones, and other wavelengths would affect certain other cones. But in turtles, LIEBMAN and GRANDA (1971) showed there to be several photopigments present, and KRAUSE's color-vision theory is without merit.

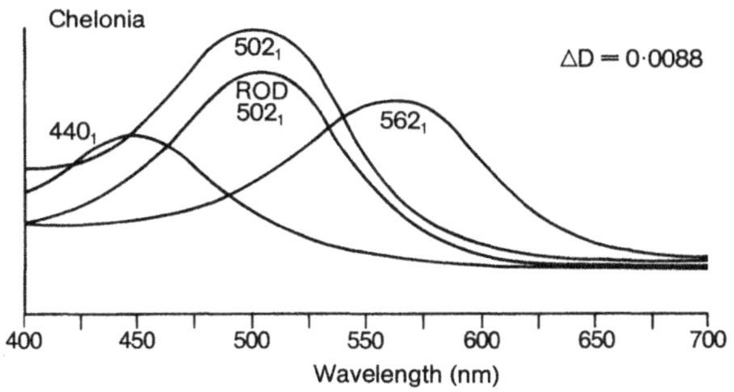

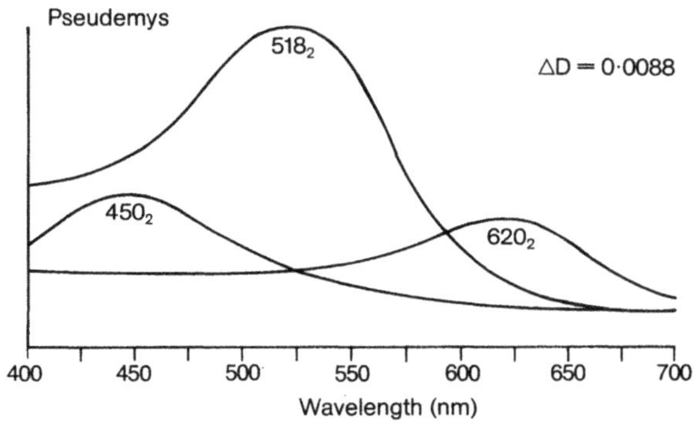

Fig. 4 A and B. Smoothed density spectra. (A) *Chelonia*; (B) *Pseudemys*. (After LIEBMAN and GRANDA, 1971)

The fact that they are optically dense and act as sharp cut-off filters suggested to WALLS and JUDD (1933) that the droplets protected the eye from glare and stray light. Since light scatters as a function of frequency—blue light more than red—it benefits a diurnal animal to enhance its resolution of form by eliminating extraneous rays. All known oil droplets cut off in the shorter wavelengths, passing light in the longer wavelengths to a greater or lesser degree. There are red, orange, yellow, and very pale yellow-green colored droplets, but there are no blue ones. There are clear ones, of course, but they affect the distribution of light very little in the visual region.

Measurement of the oil droplets by the direct method of microspectrophotometry shows transmittance in the long wavelengths for all colors with a plateau near one absorbance unit in the short wavelengths (Fig. 5a, b). These cut-off curves are featureless and show none of the spectral fine structure that should be obtained from the various carotenoids thought to be contained in the droplets (WALD and ZUSSMAN, 1937). The curves shown in Fig. 5a, b are not far different from those obtained in turtles by FUJIMOTO et al. (1957), STROTHER (1963), and LIEBMAN (1969), or in pigeons by STROTHER (1963), and KING-SMITH (1969).

It was LIPETZ who first suggested that the colored droplets might contain immeasurably dense pigments (LIEBMAN and GRANDA, 1975). With small and

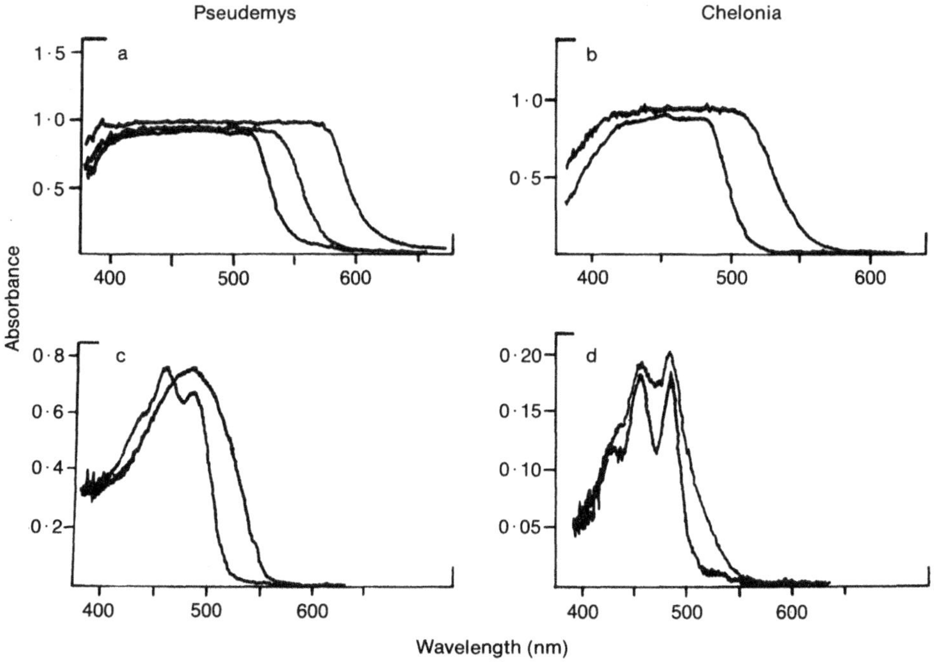

Fig. 5a–d. Spectra of individual cone oil droplets with the plateau of each occurring at one absorbance unit: (a) *Pseudemys*, yellow, orange, and red droplets; (b) *Chelonia*, yellow and orange droplets; from left to right. Resolved oil droplets diluted in mineral oil: (c) *Pseudemys*, three-fingered spectrum of yellow droplet (left trace), single-peaked spectrum of red and orange droplets (right trace); (d) *Chelonia*, sharply fingered spectrum of yellow droplet (lower trace), orange droplet (upper trace). (From LIEBMAN and GRANDA, 1975)

highly dense droplets it was very possible to exceed the physical limitations of the microspectrophotometer, either by the leakage of stray light around the droplet into the photomultiplier detector or by exceeding the range of absorbances the instrument was capable of measuring.

Absorbance, of course, depends on path-length, and LIEBMAN and GRANDA (1975) reduced that pathway in individual droplets by dissolving them in mineral oil. The new, diluted droplet was then smeared in a thin layer on a microscope slide and measured through its center in a microspectrophotometer. Using a conservation law (where volume and concentration preserve a constant), and the Beer-Lambert Law of $D = \varepsilon cl$ (where D is optical density, ε is extinction coefficient, c is concentration, and l is optical path length), it can be shown that the absorbance in the original droplet will be a function of the ratio of radii of the original and diluted droplets:

$$D_1 = D_2(R_2/R_1)^2 \tag{1}$$

where R_1 is the radius of the original droplet seen in planar view, and R_2 is the radius of the diluted droplet. Where the diluted droplets flatten to form odd shapes, the effects of these nonspherical forms are self-canceling and lead to the same relationship as the one for spheres.

Table 4. Photopigment-oil droplet pairings[a]

Color	*Pseudemys scripta*			*Chelonia mydas*		
	450_2	518_2	620_2	440_1	502_1	562_1
Red			2			
Orange			1			11
Yellow		1			5	10
Clear	1			3	2	2

[a] Entries refer to the number of confirmed pairings.

Results obtained in this fashion show resolved spectra, with fine structure giving calculated peak absorbances of 50 or more for the red droplets in *Pseudemys*, 6–9 for the orange droplets, and 12–18 for the yellow droplets. Results from unpublished work on *Emys blandings*, also a freshwater turtle, are essentially corroborative: Mean absorbances were 30 for the red droplets, 6–7 for the orange droplets, and 5–6 for the yellow droplets (LIPETZ, personal communication). In *Chelonia*, where there are no red droplets, the orange droplets measured 4–7 absorbance units and the yellow droplets 3–4.

Based on the resolved spectra for *Pseudemys* (Fig. 5c), the red droplet appears to contain the carotenoid pigment *astaxanthin*, a β-carotene derivative. It is also the carotenoid pigment found in orange droplets, although at lesser density. The yellow droplets possibly contain *zeaxanthin*, also a β-carotene, based on the peak profile of its absorbance curve, but the identification is still tentative.

In *Chelonia* (Fig. 5d), the orange and yellow diluted droplets are similar, but they do not quite match and are not identified. The droplets are not present in so great a density in their natural state as the droplet densities found in *Pseudemys*. The reason is not known.

Each cone photoreceptor contains a single photopigment, and in single cones and chief members of double cones, a single oil droplet. An obvious question is whether they are conjoined in specific combinations. GRANDA and LIEBMAN (unpublished observations) made positive identifications of photopigment-oil droplet pairings in *Pseudemys* and in *Chelonia*. In *Pseudemys*, although the pairings are few, the 3-dehydroretinol pigment peaking at 620 nm was found only with red or orange droplets, the pigment 518_2 only with yellow droplets, and a clear droplet with the pigment 450_2. In *Chelonia*, the sampling was larger. The clear droplets were found with the blue-absorbing pigment 440_1; yellow droplets mainly and a few clear ones were found with the pigment 502_1; and orange and yellow droplets were found with the pigment 562_1. A few clear droplets were also found here. The results are summarized in Table 4.

In *Pseudemys* the pairings are unique; they are largely so in *Chelonia*. One could very well expect the transmissions of the oil droplets to be in agreement with the maximum light-absorbing capabilities of the pigments to which they are conjoined. Correspondingly, one should not expect to find short-wavelength pigments, 440_1 (retinol) or 450_2 (3-dehydroretinol), combined with any of the colored oil droplets. Such combinations afford little advantage to these photopigments, for they block almost all of the light capable of being absorbed. The findings are largely in accord with these restrictions.

The fact that the orange droplet is paired with the red absorbing photopigment 620_2 is in accord with two independent pieces of evidence. One finding is that the orange droplet is a red one diluted to lower concentration; both droplets have the same absorbance spectrum (Fig. 5c). The second piece of evidence comes from RICHTER and SIMON (1974) in double cones, where the principal cone contains a photopigment similar to the one found in single red cones. That member always carries an orange droplet (BAYLOR and FETTIPLACE, 1975).

c) Cell Connections

Pseudemys cones make connections with other cones of like pigment (BAYLOR and HODGKIN, 1973), with rods (SCHWARTZ, 1975a; but see also COPENHAGEN and OWEN, 1976), with bipolars (LASANSKY, 1971; RICHTER and SIMON, 1975), and with horizontal cells (MILLER et al., 1973; SIMON, 1973; FUORTES and SIMON, 1974; SAITO et al., 1974). The intracellular records show that cones and horizontal cells influence each other reciprocally, but evidence of structural reciprocity, in terms of specific presynaptic horizontal cell processes penetrating cone synapses, has not been revealed by electron microscopy (LASANSKY, 1971, 1972) nor by freeze-fracturing (SCHAEFFER and RAVIOLA, 1976). The evidence that they must exist, however, is strong (see below).

The specialized contacts of the cone pedicle in turtle consists of basal contacts and several kinds of invaginated synapses (cf. Fig. 6). In some invaginations there are two processes belonging to horizontal cells that form a *dyad*. The two processes are on either side of the wedge-shaped synaptic ridge. Where there is an additional central process near the apex of the synaptic ridge, the formation of the two lateral processes and the central process in called a *triad*. These arrangements have been observed in other vertebrate retinas (LASANSKY, 1972). The pedicle is characterized by synaptic vesicles and ribbons which appear throughout the synaptic ridge. A specialized structure, the arciform density, separates the ribbon from the top of the synaptic ridge.

The specialized contacts of the lateral processes in the invaginated junctions are both proximal where the junction is at the sides of the synaptic ridge, and distal (occurring only in dyadic arrangements) where the junction is at the distal end of the medial gap. The nature of these contacts is obscure but LASANSKY (1972) entertains the possibility that proximal junctions may be presynaptic to horizontal cells with a specific polarized response, while distal junctions are opposite in direction of influence and polarity. This structural system, he suggests, may operate in the local feedback circuit for receptors and horizontal cells described by BAYLOR et al. (1971) (see below).

Each single cone pedicle in *Pseudemys* has roughly 200 junctions, about 30–40 lateral processes, and 12–15 central processes in triadic invaginated arrangements (LASANSKY, 1972).

In Golgi-stained sections, LASANSKY finds only horizontal cells as lateral processes. The basal contacts appear to be the site of contact between cone pedicles and bipolar cell processes. It is also probable that the third member of the triad is

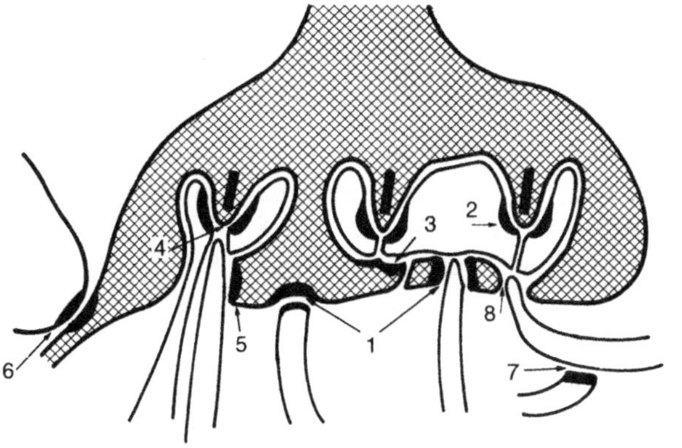

Fig. 6. Diagrammatic representation of a cone pedicle in turtle retina. Junctions indicated are (*1*) basal; (*2*) proximal with lateral process; (*3*) distal with lateral process; (*4*) apical in a triad; (*5*) distal in a triad; (*6*) adjacent cone pedicle; (*7*) *en passant* with basal process; (*8*) basal process ending. (From LASANSKY, 1972)

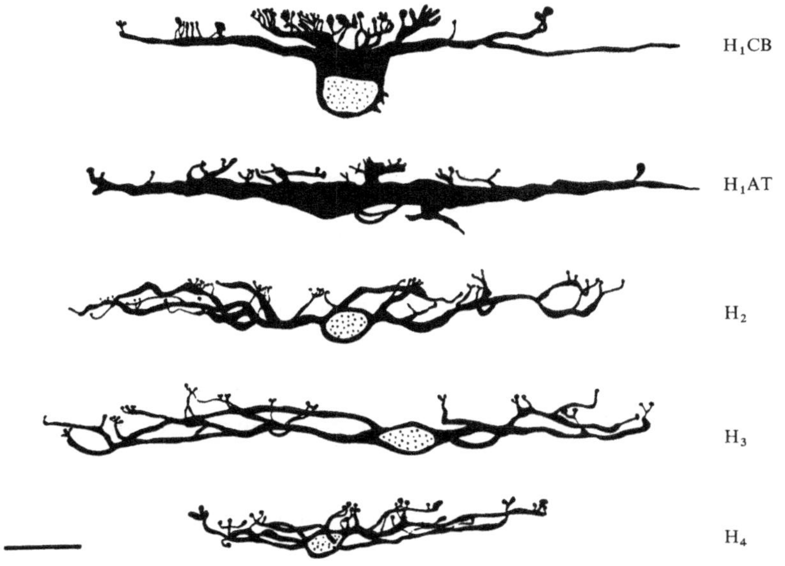

Fig. 7. Camera lucida drawings of five types of horizontal cells in *Pseudemys*. See text for details. Marker = 20 μm. (The drawings were kindly provided by H. F. LEEPER)

a bipolar process, the arrangement lending itself to the formation of center-surround relationships which are functionally present at this level.

Horizontal cells in the turtle retina are of two classes as in other vertebrate retinas. There are luminosity (L) cells and chromaticity (C) cells. The L-cells are made up of two subtypes according to SIMON (1973) and SAITO et al. (1974). L_1-cells have thick dendrites, no apparent somas, and large receptive fields in excess of 750 μm radius. L_2-cells have thin dendrites symmetrically arranged about cell bodies with smaller receptive fields of about 300 μm radius. Both L_1- and L_2-cells are now recognized as part of the same complex, the two parts being connected by a very long, high-resistance axon (Fig. 7). LEEPER (personal communication) differentiated five classes of horizontal cells in *Pseudemys* on the bases of specific contacts from separate classes of cones. H_1AT cells are similar to L_1 cells previously described. They are irregular in shape with tuberous processes and have no nuclei. These cells are the axon terminals of the L-cell complex noted above. They appear to contact rods, chief members of double cones, and probably single red-absorbing cones. H_1CB cells are similar to L_2 cells. They are oblong, stellate cells that contact rods, chief members of double cones, some accessory members of double cones, and probably both red- and green-absorbing single cones. H_2 cells contact green and blue single cones but do not contact the green-absorbing accessory members of double cones. H_3 cells contact only blue cones, which are small in size and oblique. H_4 cells contact only accessory members of double cones. The specificity of the anatomy is interesting. There is much work to do to confirm the functional properties involved here.

Two kinds of C-cells are differentiated on the basis of function. One type is depolarized by red light and hyperpolarized by green. A second type is triphasic in response, depolarizing to green light and hyperpolarizing to blue and red light (FUORTES and SIMON, 1974; YAZULLA, 1976a).

Horizontal cells are electrically coupled through gap junctions to cells of the same type. These connections may explain the differences in their receptive fields sizes and properties (SIMON, 1973).

2. Bipolars, Amacrines, and Ganglion Cells

The anatomy of bipolars, amacrines, and ganglion cells has not been extensively studied, although several investigations are underway.

On the basis of fluorescence microscopy (RICHTER and SIMON, 1975; YAZULLA, 1976b) and Golgi impregnations (LASANSKY, unpublished results cited in RICHTER and SIMON, 1975; DVORAK and GRANDA, unpublished observations, cf. Fig. 8), there appear to be at least two crude classifications of turtle bipolar cells based on morphology. The somas of one type are found in the inner nuclear layer, as expected; the other type has somas in the proximal outer nuclear layer and are identified as displaced bipolar cells. Most bipolar cells have Landolt clubs (LASANSKY, 1971) but some do not (YAZULLA, personal communication). VERRIEST et al. (1959, cited in UNDERWOOD, 1970) claimed that 50% of the outer nuclear layer consists of displaced bipolar cells in *Pseudemys*.

RICHTER and SIMON (1975) showed that the surround of center-hyperpolarizing, red-sensitive bipolar cells may be determined by L_1-cells, but this influence may be indirect, since well-defined chemical synapses have not been identified for horizontal-bipolar cell pathways (LASANSKY, 1972).

The morphology of amacrine and ganglion cells is even more obscure, although one would expect great complexity on the basis of the very wide inner plexiform layer with several layers of terminal arborizations that occur in diurnal reptiles (RAMÓN Y CAJAL, 1892). Recent electrophysiologic work on retinal gan-

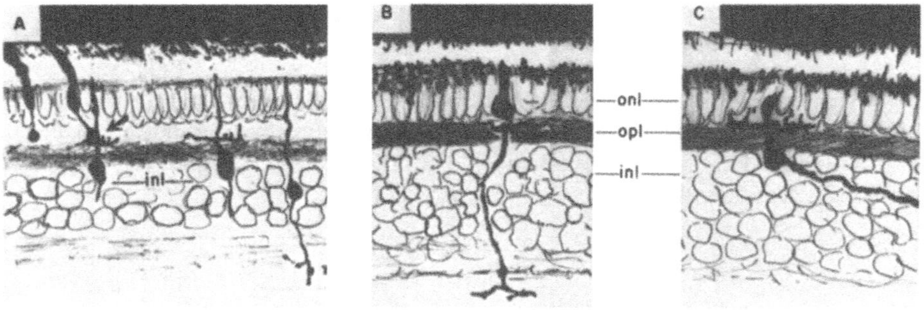

Fig. 8 A–C. Bipolar cells in *Pseudemys* retina. (A) Three cells of similar type, with somas in the inner nuclear layer (INL). All have Landolt clubs. The left-most one has its lateral dendritic processes directly below a cone *(arrow)*. (B) A displaced bipolar cell in the outer nuclear layer (ONL). The dendritic portion of this cell is *below* the cell body as it must be to contact terminal receptor processes. The inner plexiform layer (IPL) contains the terminal portion of this cell. (C) A bipolar cell whose axon transverses the retina obliquely. (DVORAK and GRANDA, unpublished observations)

glion cells in *Pseudemys* by BOWLING (1976) suggests a large distribution of various types. As more work is done on amacrine and ganglion cells in the turtle, evidence, both functional and structural, undoubtedly will support the statement made by RAMÓN Y CAJAL (1892): "In general, the reptiles can be considered to be the vertebrate in which the amacrine and ganglion cells have reached their highest degree of evolution and perfection."

Based on electronmicroscopic findings, there are about 500,000 ganglion cell axons in the optic nerve of *Pseudemys*, with 23% being unmyelinated, 0.3–0.4 μm in diameter; and 77% being myelinated, 1.0–4.0 μm in diameter (BOWLING, 1976).

The little we know about centrifugal effects is contained in two papers by MARCHIAFAVA (1976), and CERVETTO et al. (1976), where they show excitatory postsynaptic potentials appearing in amacrine and ganglion cells in response to optic nerve stimulation. The details of the anatomic loop and the contacts involved remain to be worked out.

3. Receptor Orientation

The receptors in the turtle eye are oriented toward the center of the pupil (BAYLOR and FETTIPLACE, (1975). At the posterior pole of the eye, the receptors are perpendicular to the retinal surface. Where the receptors are situated more anteriorly, there is a progressive deviation of up to 60° from the normal to the surface, so as to preserve their line of orientation to the pupil center.

Not all of the receptor is tilted. The outer segment, oil droplet, ellipsoid, and paraboloid are tilted for best orientation; the remainder of the cell remains fairly normal to the retinal surface at all locations.

The alignment of the receptors is a virtue shared with other vertebrates (LATIES et al., 1968). It allows for increased efficiency of quantal absorption in the outer segments.

C. Electrophysiology

This section surveys the recent electrophysiological literature of turtle vision. The approach has been especially fruitful in revealing new phenomena at the cellular level. There is remarkable consistency of results obtained in this way with the results from anatomical and psychophysical studies.

I. Receptor Mechanisms

There is general agreement about the electric responses of retinal neurons to light in vertebrates. Studied in mudpuppy (BORTOFF, 1964; WERBLIN and DOWLING, 1969), in carp (TOMITA, 1965), in gecko (TOYODA et al., 1969), and in turtle (BAYLOR and FUORTES, 1970), the experiments are consistent in showing that the membrane potential of the photoreceptor, recorded typically at the inner segment, increases with absorption of photons in the cell's outer segment. The recorded response is a slow hyperpolarization in vertebrates (cf. responses and time marker in Fig. 9a). The hyperpolarization contrasts to the depolarizing response observed in the photoreceptors of most invertebrates (for review, see FUORTES and O'BRYAN, 1972; TOMITA, 1972).

Turtle cones hyperpolarize during increases in membrane resistance (BAYLOR and FUORTES, 1970), a confirmation of the result obtained originally from photoreceptors in mudpuppy (BORTOFF and NORTON, 1967), and in gecko (TOYODA et al., 1969). This light-elicited increase in membrane resistance results in a reduction of the steady current that enters the outer segment in the dark (PENN and HAGINS, 1969, 1972; HAGINS et al., 1970). BAYLOR and FUORTES noted that turtle cone responses are increased or decreased as hyperpolarizing or depolarizing currents are passed into the cell.

The hyperpolarization with illumination is due to a decrease in sodium conductance across the outer segment membrane (BAYLOR and FUORTES, 1970). The membrane potential in turtle cones is determined in the dark largely by the sodium ion concentration; the peak hyperpolarization, on the other hand, is governed by potassium ion concentration. A change of sodium ion concentration affects the potential of the plateau of the response, but it does not affect the peak of the response (CERVETTO, 1973). These results confirm similar ones obtained extracellularly in the pigeon (ARDEN and ERNST, 1970, 1972), frog (SILLMANN et al., 1969a, b), and rat (YOSHIKAMI and HAGINS, 1973). In these cases, the size of the photoresponse is linearly related to the logarithm of the external sodium ion concentration.

Since the outer segments of turtle photoreceptors have a high permeability for sodium in the dark, a current flow is established that enters the photoreceptor at the outer segment and exits at the inner segment. With illumination, the current flow is shut off because of a decrease in sodium permeability, and the resultant membrane potential is determined largely by the concentration of potassium ions. A sodium-potassium ATPase transport exchange at the inner segment (HAGINS et

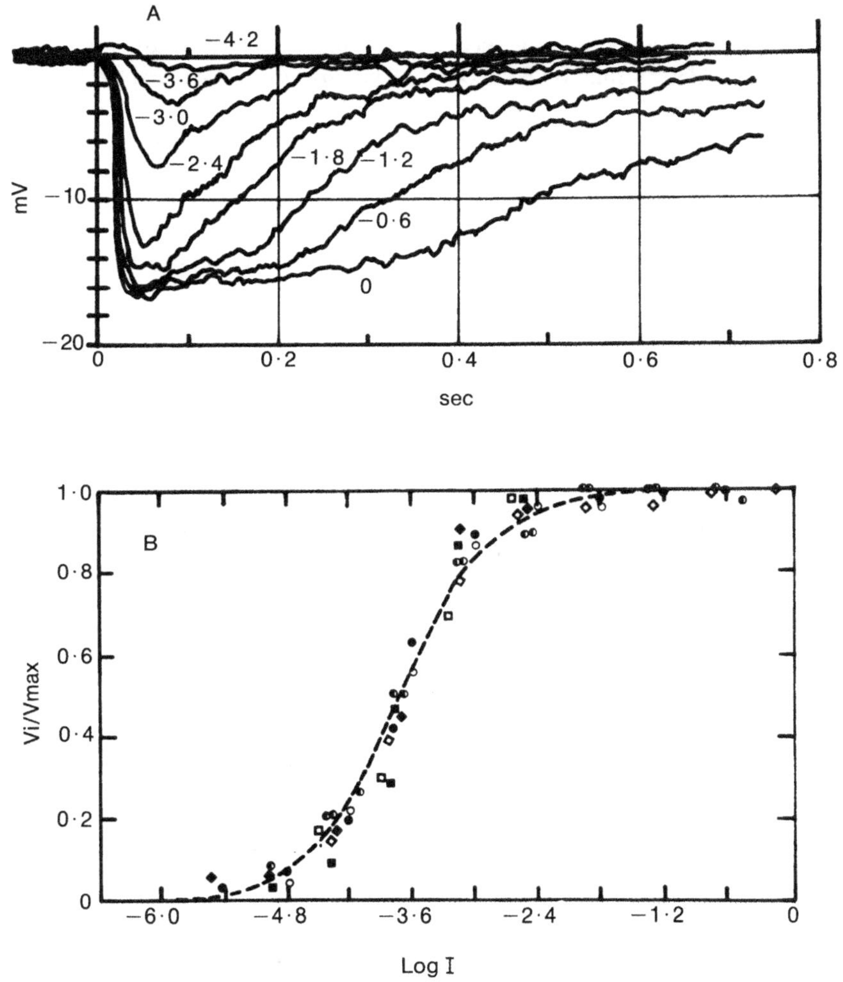

Fig. 9 A and B. (A) Superimposed cone responses to 10 msec flashes of light. Brightest flash delivered at about 8.5×10^6 absorbable photons. Values on each curve are density units. (B) Peak hyperpolarization as a function of intensity, normalized for different cells. The line plots Equation (2) in text. (From BAYLOR and FUORTES, 1970)

al., 1975) can provide the appropriate concentration gradients to maintain the dark currents from the inner to the outer segment (HODGKIN, 1972).

The kinetics of the response depend somewhat on the electric properties of the cell. BAYLOR et al. (1974a) measured resistances of 50–170 megohms in cones of *Pseudemys*. The effect of light is to increase these resistances by about 25%. With absorption of light, the membrane time constant increases typically from about 5–6.25 ms, a correspondence in membrane capacity of about 37 pF that is essentially independent of light absorption. This value yields a capacity value per unit area of 0.6 μF cm^{-2} for a turtle cone (BAYLOR et al., 1974a) and is comparable to that of most cellular and artificial membranes (JAIN, 1972). The resistance change accompanying light, then, accounts for the observed increase in the time constant. The range of values determined by BAYLOR et al. (1974a) do not exceed 8.4 ms, a value far less than the 50–150 ms required for an illuminated cone to develop its maximum hyperpolarization. The time course of the potential measured across a cone inner segment membrane, therefore, is due to some other process that interacts with the ionic currents.

With values determined for the resting potential, the maximum hyperpolarization, and the time constant for the cell, the proportion of blocked ionic channels at any time can be calculated (BAYLOR et al., 1974a). A theory to explain the sequence of these channel closings was suggested by FUORTES and co-workers. Absorption of photons results in a chain of chemical reactions whose end-product is an intermediary substance that interacts with the channels and affects their permeability (FUORTES and HODGKIN, 1964; BORSELLINO et al., 1965; BAYLOR and FUORTES, 1970). YOSHIKAMI and HAGINS (1971) found that absorption of light results in an increase in the concentration of intracellular calcium. According to them, calcium serves as the internal transmitter that blocks sodium channels in the plasma membrane. Raising the concentration of extracellular calcium, and thereby intracellular calcium, simulates the effect of light (YOSHIKAMI and HAGINS, 1973).

In support of this model CONE (1973) calculated that the absorption of a single photon results in the release of at least 25 transmitter particles. The equilibrium constant for the binding of a transmitter particle by a sodium channel is about 6×10^{-7}, the range where calcium ion is known to act as an internal transmitter in muscle. According to CONE, the absorption of a single photon closes about $^1/_{1000}$ of the open channels, based on the fractional decrease in dark current. This value is of the same order of magnitude as the value arrived at by BAYLOR et al. (1974a), who measured conductance changes. There is, then, for dim light a direct proportion between effectively absorbed photons and the fraction of channels blocked, and hence the amplitude of the observed hyperpolarization. This linearity for dim lights has been confirmed by superposition (BAYLOR and HODGKIN, 1973).

Since the amount of internal transmitter released must compete for open ionic channels to block, there is a nonlinear relation between the fraction of channels blocked and the transmitter concentration. The potential developed across the receptor membrane, V, that derives from this relationship relates to light intensity, I, as

$$\frac{V}{V_{max}} = \frac{I}{I+\sigma}, \tag{2}$$

where σ is the value of I for which $V = \frac{V_{max}}{2}$ (Fig. 9b). The sigmoid curve that results from V plotted as a function of $\log I$ is a rectangular hyperbola. The curve reflects the underlying Michaelis-Menten equation (MICHAELIS and MENTEN, 1913, cited in NEAME and RICHARDS, 1972). For dim I, where $I+\sigma \doteq \sigma$, the $V \doteq V_{max} \cdot \frac{I}{\sigma}$, and the responses are linear. As I increases, V approaches V_{max}, and absorbed photons have little effect. This saturation of receptor response

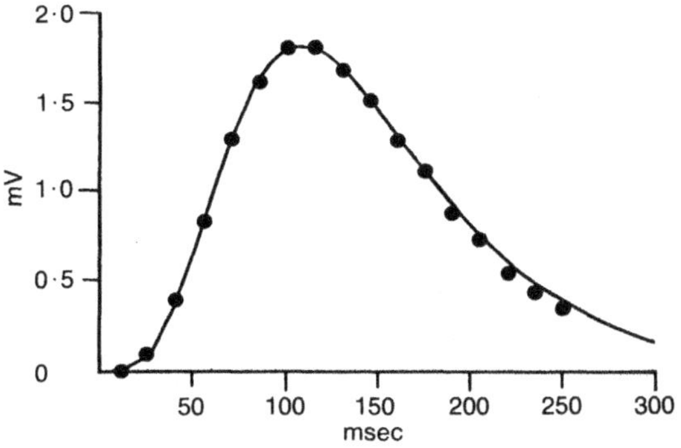

Fig. 10. Response of a red-sensitive cone to a dim flash of light 11 msec in duration. The curve was calculated from Eq. (4). (From BAYLOR et al., 1974a)

occurs at a level of illumination for which less than 1% of the pigment molecules are bleached (BAYLOR and FUORTES, 1970). It appears to be the blocking of ion channels that governs the cone membrane potential under conditions ranging from darkness to bright light. For large I, there are a limited number of channels for more and more released transmitter. The observed relationship is explained if the transmitter is assumed to bind reversibly to the channels, with bound and free transmitter molecules reaching equilibrium (CONE, 1973).

BAYLOR et al. (1974a, b) and BAYLOR and HODGKIN (1974) developed a model that not only accounts for the potentials developed in a receptor but also the kinetics of the response under differing states of adaptation. Their model is a detailed expansion of the afore-mentioned one proposed by FUORTES and HODGKIN (1964) in which light initiates a chain of reactions with characteristic time constants. This chain of events determines the waveform of the receptor response. For both turtle cones (BAYLOR et al., 1971) and rods (SCHWARTZ, 1973b), the Fuortes-Hodgkin equation provides a good fit to the rising phase of response $U(t)$:

$$U(t) = AQT^{n}e^{-T}, \tag{3}$$

where A is a constant, Q is proportional to the intensity, and $T = t/\tau$, where τ is a delay constant for each of n stages. In their kinetic analysis, BAYLOR et al. (1974a) determined that a better overall fit to the asymmetric waveform is given by

$$U(t) = AQe^{-T}(l - e^{-T})^{n-l}, \tag{4}$$

where $n = 6$ or 7 (Fig. 10). The two basic assumptions that govern this equation are that 1) light absorption initiates a chain of reactions that results in the release of an internal transmitter that blocks ionic channels, and 2) equilibrium is rapidly established between blocking molecules and the fraction of channels blocked. This scheme is shown in Fig. 11. It should be noted that Q_{10} for the rising phase of the response of a turtle cone is 1.8 (15°–25° C), corresponding to an activation energy of about 10 kcal/mol, too high for the chain of events producing the

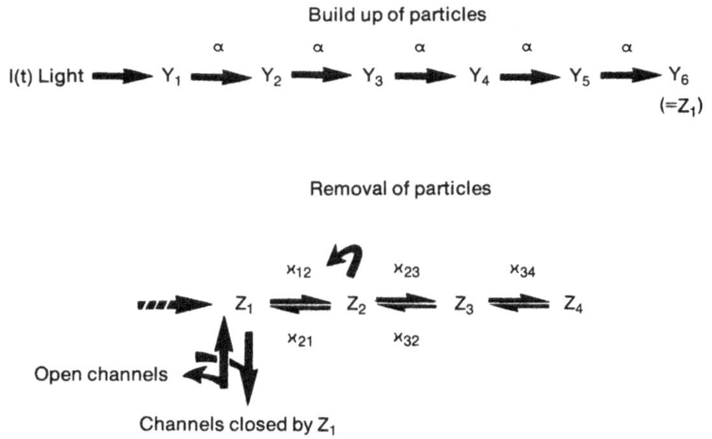

Fig. 11. Scheme of hypothetical reactions controlling concentrations of internal transmitter (blocking molecule). Y, Z are stages in the build-up or destruction of particles, respectively; α, are constants. (From BAYLOR et al., 1974b)

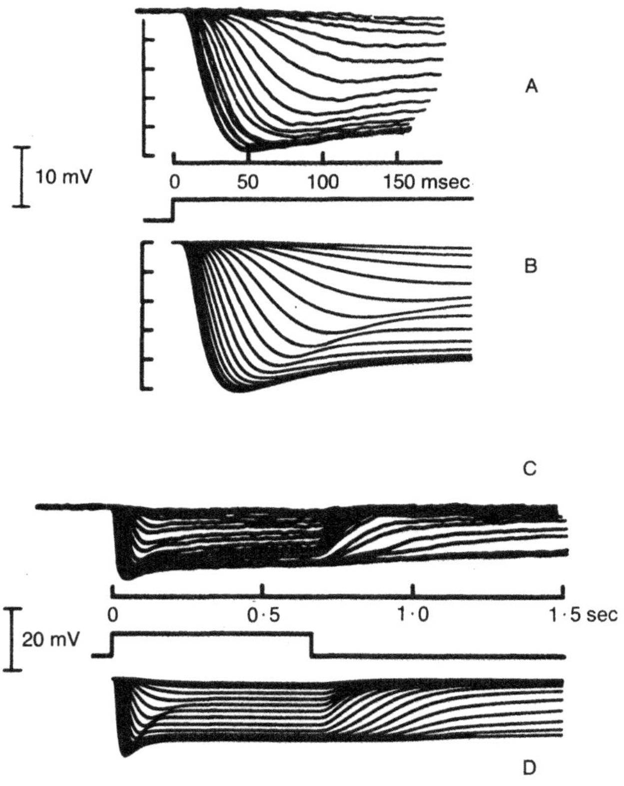

Fig. 12. Cone and model responses to a 0.67 sec step of light. A and C are cone responses; B and D are simulations. See text for description of the model. (Taken from BAYLOR et al., 1974b)

response solely to involve simple aqueous diffusion as the cause for delay of the particle to reach an ionic channel (BAYLOR et al., 1974a).

In summary, the time scale and sensitivity of the hyperpolarizations developed in turtle cone receptors can be accounted for to a large extent. The experiments referred to provide a quantitative anlaysis of the photoresponses based on ionic currents and their control by the blocking of channels of selective permeability. The time course requires several stages of delay by means of an internal transmitter that binds selectively and reversibly to the channels of passage. The success of present understanding is evident in the reconstruction of turtle cone responses to many conditions of illumination (Fig. 12).

II. Spatial Characteristics of Retinal Cells

In recording from turtle cones, BAYLOR and FUORTES (1970) noted that scattered light presented a problem in the precise determination of the effects of illumination. For small spots of light (< 100 µm diameter), retinal scattering causes cones near the spot boundary to absorb light. A correction for the effects of scattering can be made (MASTERS, 1955; BAYLOR et al., 1971). However, even when corrected, the response curves for spot radii of 4 and 70 µm do not compare well (Fig. 13), as they have differing slopes (BAYLOR et al., 1971). No adjustment of stimulus intensity to compensate for scattering can yield a matching of responses for spot radii of 4 and 70 µm.

If the spot size is increased to 600 µm, the waveform of the response is altered so that a delayed depolarizing wave is elicited. The presumed function of a turtle photoreceptor — to signal the rate of absorption of quanta in its own outer segment — is also dependent on the pattern of illumination over relatively large retinal areas that encompass hundreds of other cones. BAYLOR et al. (1971) found that cones less than 40 µm apart have mutually excitatory, i.e., summative, influences (Fig. 14). This conclusion was tested and confirmed by simultaneous recording from cones about 25 µm apart and observing the effects of light and polarizing currents on them (BAYLOR et al., 1971).

Cones much farther apart have inhibitory rather than excitatory influences on one another, and illumination of the "far" surround produces a depolarization in a cone (Fig. 14). This interaction should be chemical in nature and mediated by cells other than receptors, since the polarity of the interaction is reversed. BAYLOR et al. (1971) demonstrated that this effect was due to horizontal cell activity: Hyperpolarization of a horizontal cell with applied current caused a depolarization of nearby cones (Fig. 15). They noted that transmission from receptors to horizontal cells is probably chemical, since larger (> 50 mV) signals can be recorded from horizontal cells than from cones (25 mV maximum). Many cones feed into a single horizontal cell, which in turn hyperpolarizes with the cones in response to a flash of light. When the horizontal cell hyperpolarizes, it causes nearby cones to depolarize. Synaptic processes at the base of the receptors appear to provide the anatomic basis for both of the above interactions: Cones contact cones, and horizontal cell processes impinge on the terminals (see Section B).

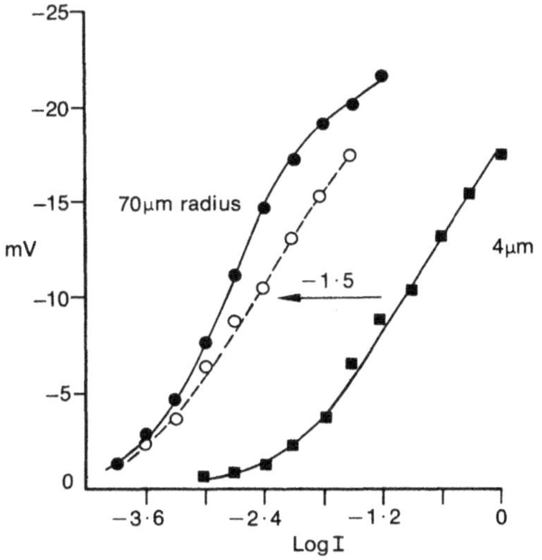

Fig. 13. Peak response amplitude of a turtle cone to spots of light 4 and 70 µm radius (see text). Compensating for scattered light shifts the 4 µm curve 1.5 log units to the left, but it does not match the response curve for 70 µm. (From BAYLOR et al., 1971)

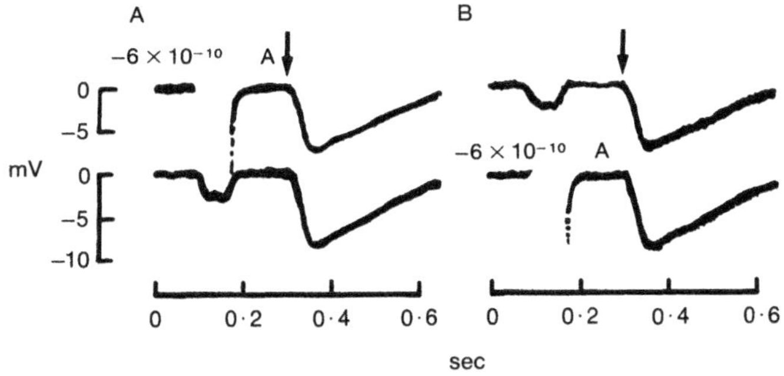

Fig. 14 A and B. Demonstration of reciprocity of receptor interactions between two cones 25 µm apart. Current pulses and light flashes *(arrows)* have the same effect on one cell as the other. In *A*, current is injected into one; in *B*, into the other. (From BAYLOR et al., 1971)

The possible functional role of the horizontal cell feedback onto turtle cones was considered by O'BRYAN (1973), who analyzed their depolarizing effects on cones and suggested that such an arrangement could subserve contrast enhancement. Another possibility involves adaptation, for a bright stimulus would initially hyperpolarize and possibly severely desensitize a cone. The negative feedback via horizontal cells would then depolarize the cone, hence increasing its possible response range (BYZOV, 1969).

Horizontal cell feedback has implications for color processing as well, for the cones themselves become wavelength dependent (FUORTES et al., 1973). Thus, a green cone that hyperpolarizes directly to 550 nm light might depolarize to

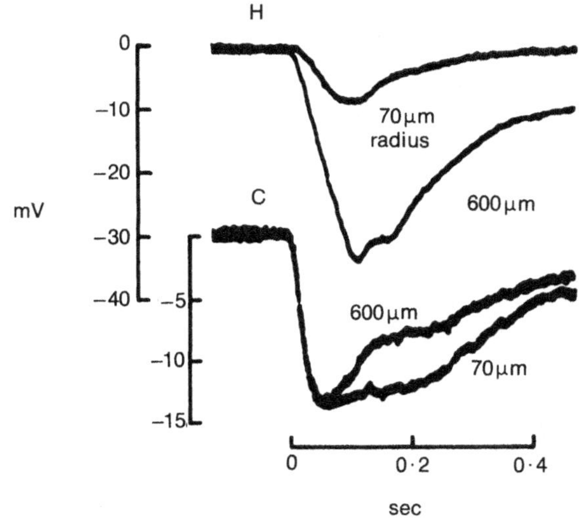

Fig. 15. Cone (*C*) and horizontal cell (*H*) records. Increasing the stimulus area to 600 μm increases response in H, but causes a depolarizing effect in C, after peak response of *C*. (From BAYLOR et al., 1971)

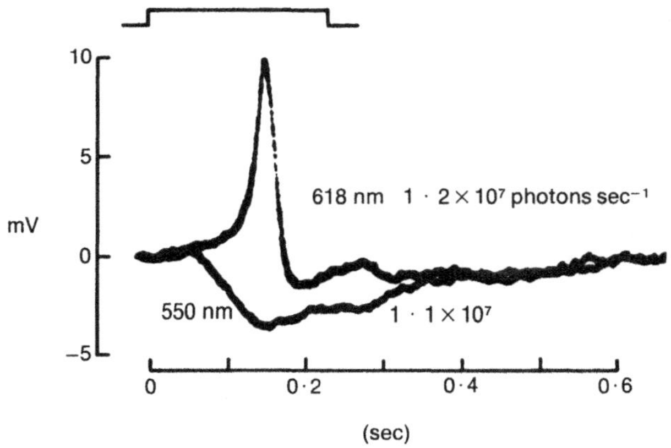

Fig. 16. Green-sensitive cone responses to red and green annuli (500–1250 μm radii). Red annulus depolarizes this cell; green does not. The noted photon flux is over 50 μm^2. (From FUORTES et al., 1973)

680 nm light due to the activity of red cones being fed back through horizontal cells (Fig. 16).

The enhancement that cones cause in nearby cones is organized quite differently, with respect to color, from that of the feedback network. Enhancement between cones occurs only for cones of the same class (BAYLOR and HODGKIN, 1973; O'BRYAN, unpublished results, cited in FUORTES et al., 1973). It would of course be confusing for cones of different classes to be coupled, for little or no wavelength discrimination would be possible. It appears, then, that to a large degree both spatial and chromatic antagonisms are established at the most peripheral level possible in the retina.

The receptive field of a single cone is much larger than the cone itself, which might lead one to suspect that spatial summation of the cones would be reflected in second-order neurons. Bipolar cells in the turtle retina exist in numbers sufficient to afford a 1:1 ratio with cones (BUSH, 1973), so their spatial resolution might be expected to be similar to that of cones. Recording from red-sensitive, center-hyperpolarizing bipolar cells which mainly receive direct, noninverting inputs from red cones, RICHTER and SIMON (1975) determined that area-sensitivity curves for bipolar cells reflect spatial summation of cones for spot diameters up to about 200 µm (Fig. 17). The bipolar-to-ganglion-cell ratio of about 4:1 (BUSH, 1973) suggests sufficient convergence in the latter stage of the receptor-bipolar-ganglion cell pathway to predict that many ganglion cell receptive field centers would differ from the area-sensitivity relations that exist at the receptors. LIPETZ and HILL (1970), BAYLOR and FETTIPLACE (1976), and BOWLING (1976) reported that most turtle ganglion cell receptive field centers are approximately circular and range in diameter from 75 to 450 µm, 200 to 500 µm, and 100 to over 1000 µm, respectively. Receptive field centers of up to 200 µm in diameter could possibly have their spatial properties determined by the spatial summation of the receptors, whereas the larger receptive field centers could not and would depend on the convergence of bipolar cell inputs to establish the spatial properties of centers over, say, 500 µm. Surface projections demonstrate the response profile and spatial localization of a receptive field for a ganglion cell in *Pseudemys* (Fig. 18).

Extensive spatial summation does occur in the distal retina, although not in the centers of bipolar cell receptive fields. First noted in the fish retina by TOMITA, et al. (1958), horizontal cells respond to light over extraordinarily large retinal areas. This summation has been confirmed by many investigators and studied analytically by NAKA and RUSHTON (1967), MARMARELIS and NAKA (1972), and DVORAK and TAYLOR (1977). The last paper indicates that the lower limit of spatial summation in horizontal cells is determined by the coupled interactions of the receptors. In the turtle, the spatial properties of the receptive field of horizontal cells have been studied by BYZOV (1966), FUORTES (1972), SIMON (1973), and SAITO et al. (1974). The work of BYZOV (1966) indicates that a distribution of horizontal cell types exists on the basis of receptive field size. Based on electrophysiology there are two types of luminosity horizontal cells with corresponding morphologies and summation areas (Fig. 19).

One fortunate result of such classification is that some of the wiring of retinal cells can be revealed on the basis of areal summation. For example, RICHTER and SIMON (1975) noted that L_1-cells are probably responsible for the surrounds of the receptive fields of certain bipolar cells. In ganglion cells it becomes difficult to determine the direct input, although MAKSIMOVA (1969) and NAKA and NYE (1971) demonstrated that the effects of current passed into a fish horizontal cell can be observed in the surround activity of ganglion cell receptive fields. SCHWARTZ (1973a), however, pointed out that *on-off* cells of the proximal turtle retina give responses to peripheral illumination that are not directly explained by horizontal cell responses. Instead, he suggested, the surrounds of ganglion cell receptive fields are more directly determined by bipolar cells. If true, then based on the spatial properties of turtle retinal cells, it appears that receptors feed onto bipolar

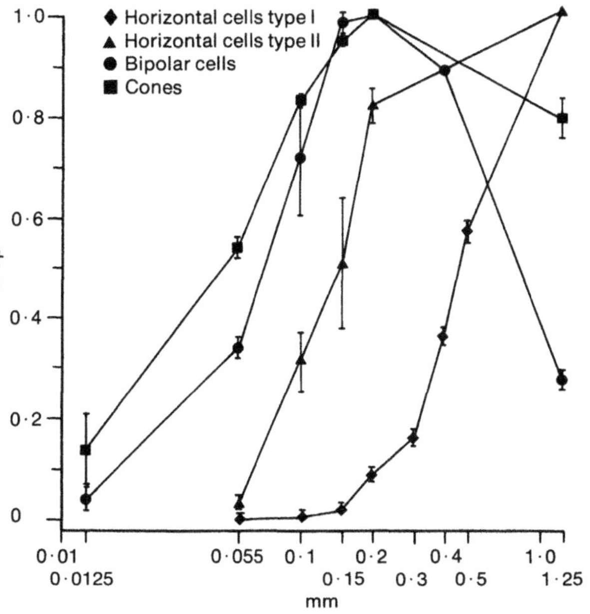

Fig. 17. Area-normalized sensitivity curves for red-sensitive cells in the outer retina. Abscissa is radius of the circular flash; *Sp* is the sensitivity based on peak response to 200 μm flash for cones and bipolars, 1250 μm flash for horizontal cells. (From RICHTER and SIMON, 1975)

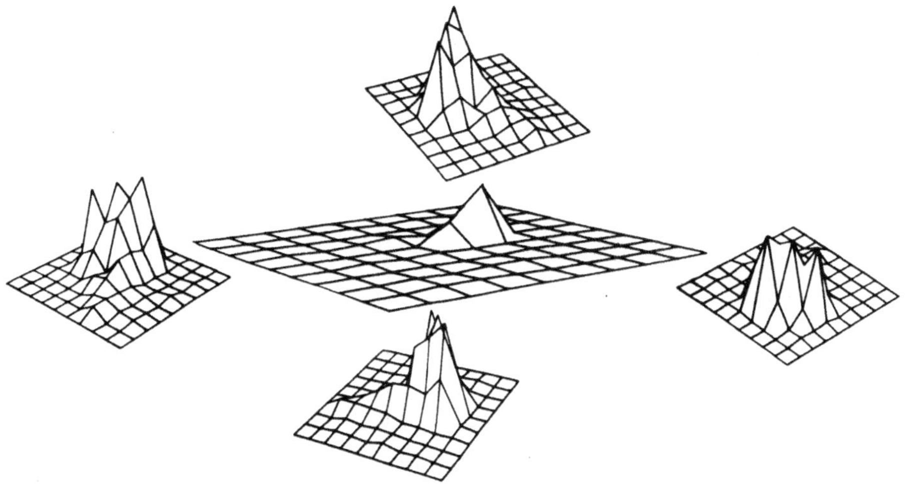

Fig. 18. Computer generated surface projection of a turtle ganglion cell field response profile. The center grid maps out a visual space 45° × 45° with a 15° × 10° receptive field. The other four projections are done in greater detail, each from a different perspective. In the center projection each square is 5° × 5°; in the others each square is 2° × 2°. (DVORAK and FULBROOK, unpublished observations)

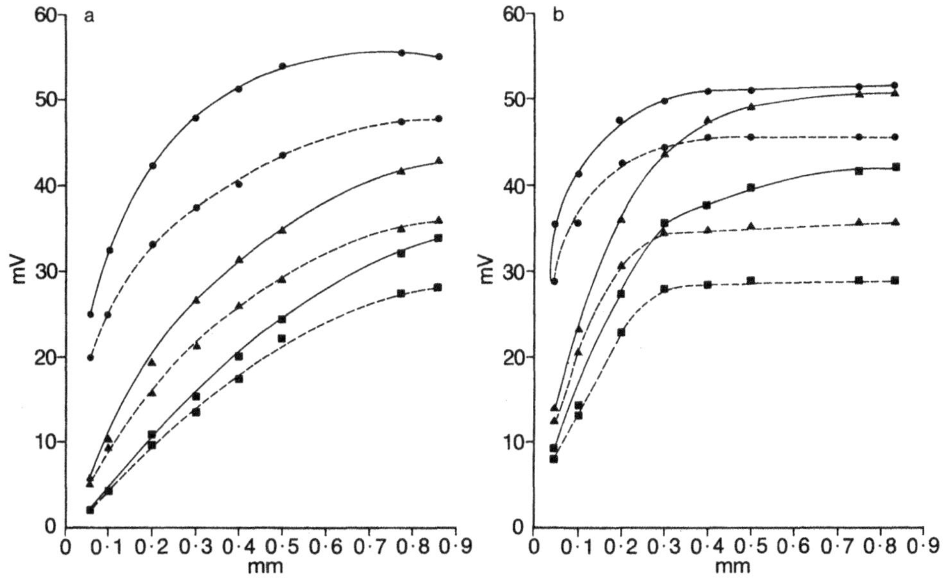

Fig. 19A and B. Graphs of L_1 and L_2 horizontal cell responses to increasing area of illumination. Different symbols represent different intensities of the 400 ms flash. (A) L_1 cell responses increase up to an 833 μm radius; (B) L_2 cell responses up to about 300 μm radius. (From SIMON, 1973)

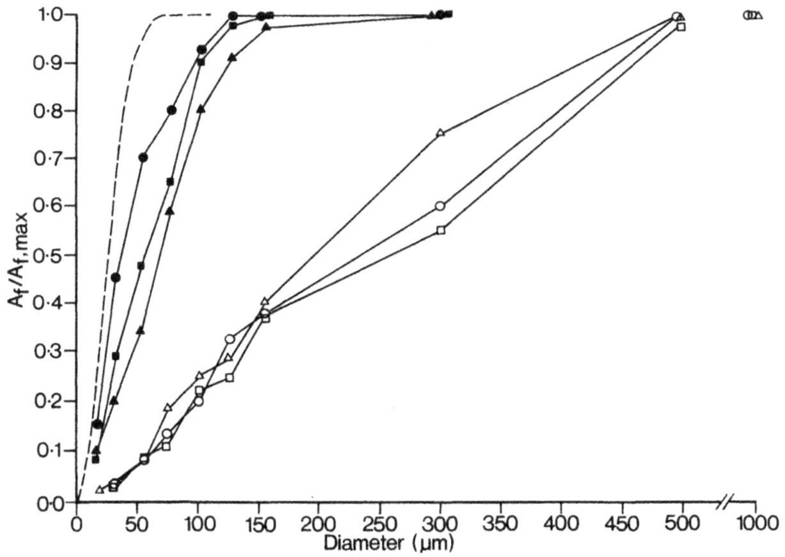

Fig. 20. Area-normalized amplitude graph for red cones *(filled)* and rods *(open)* of *Chelydra serpentina.* Rods interact over a distance 5× greater than cones do. The dotted line plots the effect that scattered light only would have were none of the cells coupled. (From SCHWARTZ, 1975a)

and horizontal cells, and horizontal cells feed forward to bipolar cells and back onto receptors. Bipolar cells now have a well-defined center-surround organization and feed forward onto ganglion, and probably amacrine, cells.

The retinas of *Chelydra* contain many more rods than the retinas of durnal turtles (UNDERWOOD, 1970). Rod properties have been studied in *Chelydra* by SCHWARTZ (1973b, 1975a, b, 1976), and COPENHAGEN and OWEN (1976).

SCHWARTZ (1973b, 1975a, 1976) reported that rods of the turtle retina show increased responses for spot diameter up to 500 μm; increasing the stimulus diameter from 100 to 500 μm produces a fivefold increase in response amplitude (SCHWARTZ, 1975a), as shown in Fig. 20. COPENHAGEN and OWEN (1976) found similar summative interactions, although only up to 300 μm stimulus diameter. They suggest that the discrepancy is due to either differences between rods in various retinal regions or imperfect centering of the stimulus by SCHWARTZ. In any event, turtle rods apparently interact with one another to a greater extent than do cones. The results of COPENHAGEN and OWEN show that for dim, diffuse stimuli, only about 4% of the response is the result of light absorbed in the impaled rod. Like cones, turtle rods appear capable of developing potentials in response to quanta absorbed by neighboring rods, even though they absorb no quanta themselves. It has been estimated that each turtle rod can interact in this fashion with up to 200 other rods (COPENHAGEN and OWEN, 1976).

Other interactions in the distal turtle retina occur. It appears that cones have an excitatory effect on the rods (SCHWARTZ, 1975b; COPENHAGEN and OWEN, 1976), although the extent to which this excitation affects the rod response is very sensitive to the level of adaptation (COPENHAGEN and OWEN, 1976). Also, turtle rods do not receive the recurrent inhibition from horizontal cells that cones do (SCHWARTZ, 1975b); the only influences on rods by cones and other rods are apparently excitatory in nature. This fact is in contrast to the aforementioned excitatory and inhibitory interactions that turtle cones are subject to.

III. Processing of Wavelength Information

1. Intracellular Recordings

a) Photoreceptors

The cones of the freshwater turtle *Pseudemys* contain pigments that absorb maximally at 620, 518, and 450 nm (LIEBMAN and GRANDA, 1971), and this absorption is reflected in the action spectra derived from single cones by BAYLOR and HODGKIN (1973) (Fig. 21). Light is filtered by the oil droplets in the inner segment of single and principal cones, and the action spectral peaks are shifted accordingly to 630, 550, and 460 nm.

For each absorbed photon, red- and green-sensitive cones yield mean peak hyperpolarizations of 25 μV; blue-sensitive cones yield 12 μV (BAYLOR and HODGKIN, 1973). These values are based on absolute flash sensitivities of 250, 250, and 120 $\mu V\ photon^{-1}\ \mu m^2$ for the red-, green- and blue-sensitive cones, respectively. Absolute flash sensitivity, S_F, is defined as the peak hyperpolarization

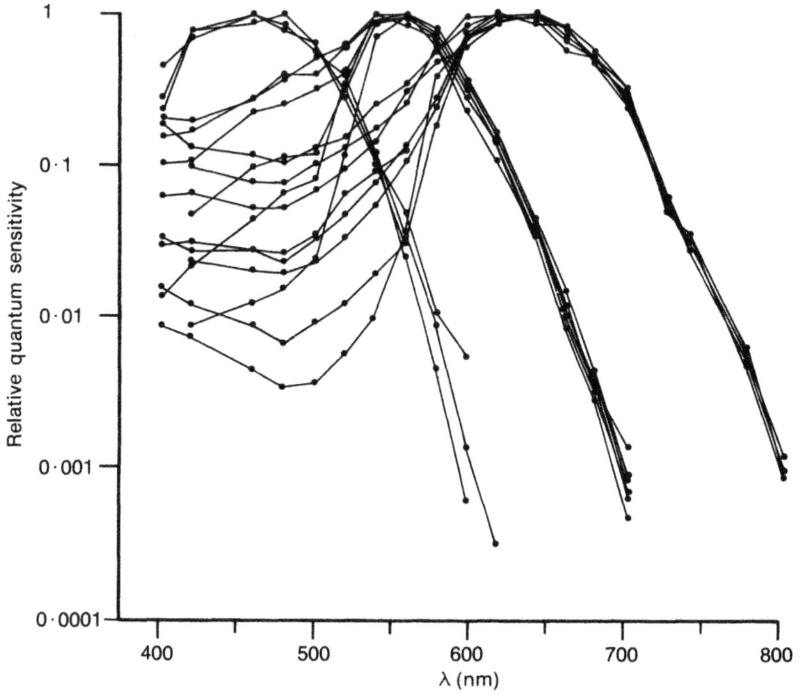

Fig. 21. Spectral sensitivity of turtle cones (see text). (From BAYLOR and HODGKIN, 1973)

divided by the product of the flash intensity and duration. To obtain the response per absorbed photon, an effective collecting area of 10 μm^2 was assumed by BAYLOR and HODGKIN as an estimate of the effective collecting area of a cone. In a more detailed treatment of the photon-catching capabilities of turtle cones, BAYLOR and FETTIPLACE (1975) indicated that the sensitivity of cones at the optimal wavelength was in the order: red receptors > green receptors > blue receptors. However, the low frequency of impaling blue-sensitive cones and the difficulty of obtaining stable records make these properties uncertain.

On the basis of action spectra derived from double cones, RICHTER and SIMON (1974) and BAYLOR and FETTIPLACE (1975) concluded that the chief member contains photopigment 620_2 similar to the one found in red-sensitive cones, and the accessory member contains photopigment 518_2, a pigment also contained in the rods (LIEBMAN and GRANDA, 1971). The spectral sensitivity is broad (Fig. 22) and shows inflections near 550 nm and 600–640 nm. The results are in keeping with a double-cone system that behaves as if the members were electrically coupled. The two signals generated by their respective pigments can then leak into one another and apparently do. Such color confusion is puzzling, yet a comparable but indirect mixing occurs in single-cone horizontal cell networks where the horizontal cells show inputs from more than one cell type.

While it is the case that single cones contain unique pigments, their responses are affected by distant cones via horizontal cell feedback. Cones, for example, that

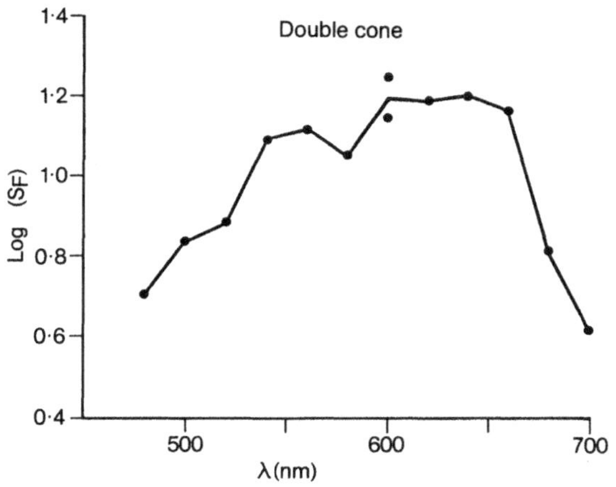

Fig. 22. Spectral sensitivity of a double cone in the turtle (see text). (From BAYLOR and FETTIPLACE, 1975)

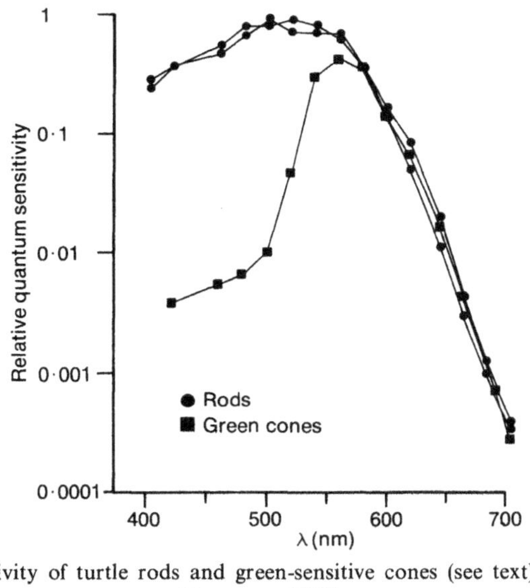

Fig. 23. Spectral sensitivity of turtle rods and green-sensitive cones (see text). (From BAYLOR and HODGKIN, 1973)

hyperpolarize to green light show marked depolarization to annuli of red light (Fig. 16) (FUORTES et al., 1973). The implications are important, for under large field illumination where horizontal cell feedback occurs, the principle of spectral univariance advanced by NAKA and RUSHTON (1966) does not hold: A turtle cone signals more than just the number of photons absorbed according to the pigment contained; it can indicate some further measure of the stimulus wavelength by the polarity of its response. For stimuli less than 100 μm diameter, however, spectral univariance largely holds, and a turtle cone can relay with fidelity information about the effective rate at which it catches photons (BAYLOR and HODGKIN, 1973).

Spectral sensitivities of turtle rods have been reported in *Pseudemys* (BAYLOR and HODGKIN, 1973), as well as in *Chelydra* (SCHWARTZ, 1975b; COPENHAGEN and OWEN, 1976). The action spectra of *Pseudemys* rods and green-sensitive cones are compared in Fig. 23, which shows the effects of the yellow droplet. Since the curves agree at longer wavelengths, BAYLOR and HODGKIN concluded that both cells contain the same pigment. One might wonder if the accessory cones have the same action spectrum as rods, since they both have the same 518_2 pigment (LIEBMAN and GRANDA, 1971), and neither has an oil droplet. The accessory cones, however, are coupled to their red-sensitive partners, and the direct determination has not been made. The spectra, however, almost certainly must match, and it would be surprising if they did not. While the pigments are the same or at least similar in green cones, accessory cones and rods, the kinetics of their responses are different. Integration times as defined by BAYLOR and HODGKIN are very much longer for rods than for any of the cones: Mean values are 550 ms for rods and 213 ms for green-sensitive cones. Their absolute flash sensitivities are also different: 1300 μV $photon^{-1}$ μm^2 for rods and 250 μV $photon^{-1}$ μm^2 for green-sensitive cones; red-sensitive cones are similar to green.

Spectral univariance applies to dark-adapted rods, as no horizontal cell feedback apparently occurs and cone input does not affect their responses (COPENHAGEN and OWEN, 1976). For rods not rigorously dark-adapted, it appears that the response is influenced by input from the red cones, and univariance fails.

b) Horizontal Cells

The transmission of wavelength information to turtle horizontal cells is similar to other retinas, where there are chromaticity (C) and luminosity (L) cell types. Both biphasic and triphasic C-cells have been reported (FUORTES and SIMON, 1974; YAZULLA, 1976a). Biphasic cells include an R/G cell that depolarizes to red light and hyperpolarizes to green light and a G/B cell that depolarizes to green light and hyperpolarizes to blue light. There is also a triphasic G/RB cell that depolarizes to green light but hyperpolarizes to red and blue light. The production of these responses can be accounted for by the feedback of horizontal cells onto receptors as discussed above (FUORTES et al., 1973; FUORTES and SIMON, 1974). For example, the response of an R/G cell can be explained by hyperpolarizing green cones feeding into it. Red light, which has little direct effect on green cones, is absorbed by nearby red-sensitive cones which cause the depolarization of the green cones via horizontal cells and the subsequent depolarization of the C-cells into which the green cones feed. The organization of the outer plexiform layer, then, with horizontal cells feeding back onto receptors, accounts for the color opponency which it is possible to demonstrate in both cones and horizontal cells. Triphasic cells can be explained in similar fashion.

c) Bipolar Cells

The spectral properties of turtle bipolar cells are also amenable to classification as L- and C-types (YAZULLA, 1976b). Red cones dominate both center and

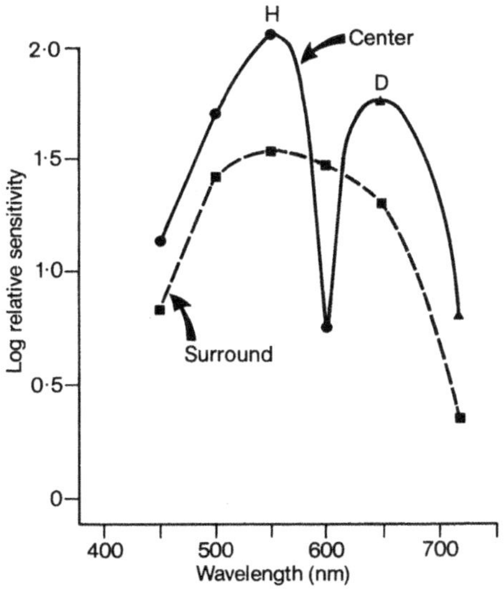

Fig. 24. Spectral sensitivity of the center (solid line) and surround (dashed line) of a bipolar cell whose center both hyperpolarizes (H) and depolarizes (D). The surround only hyperpolarizes. (From YAZULLA, 1976b)

surround inputs to L-type bipolars, whereas C-type bipolars have color-opponent centers but nonopponent surrounds (Fig. 24). In YAZULLA's analysis, a linear combination of L_1 horizontal cell input sensitivity and C-type (R/GB) horizontal cell input sensitivity can match the sensitivity of the C-type bipolar cell surround. Reciprocal connections for feedback described above in the horizontal cell-cone arrangement do not explain this organization for bipolar cells. Direct horizontal cell input to bipolar cells has not yet been demonstrated, but some way to transfer the necessary information must obviously exist.

d) Ganglion Cells

Ganglion cell responses also reflect the large numbers of red-sensitive cones. With the exception of rod-mediated responses, all of the optic nerve fibers recorded from by BOWLING (1976) have maximum sensitivity to red light, although the action spectra show different shapes that indicate additional input from cones other than red ones. Ganglion cells also show the same spatial center-surround antagonism that occurs in bipolar cells. The receptive fields are circular in shape, with a small percentage of the cells having differences in spectral sensitivity between the center and surround. Spectral curves are most sensitive to red light of 630 nm but are broad enough to include input from the green-sensitive cones (BOWLING, 1976). The ganglion-cell surrounds show similar spectral properties. About half the ganglion cells recorded from by BOWLING show the influence of rods, an interesting finding, since the number of rod receptors in *Pseudemys* is fairly low (estimates of 10% or less). Rod and cone receptor signals feed into the

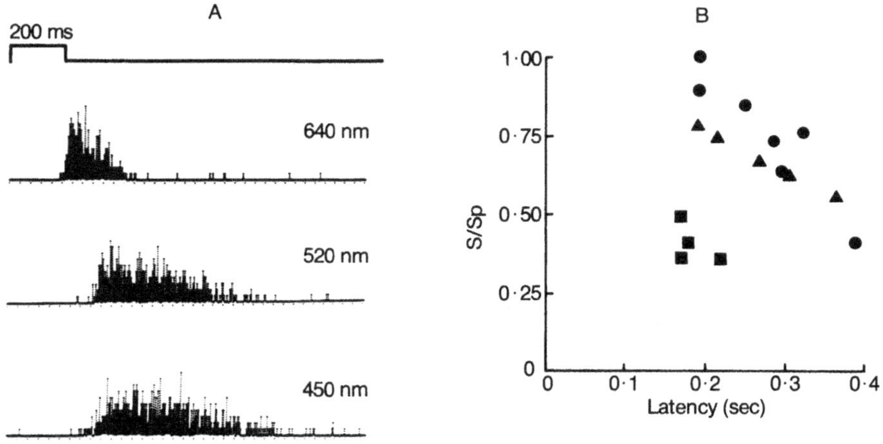

Fig. 25A and B. (A) Post-stimulus tune histograms of spike discharge in turtle ganglion cell axon to equal quantum input of 200 msec duration at the wavelengths indicated. The cell is blue-sensitive. Although the response to 450 nm light is largest, its latency and time-to-peak still remain longest. (B) S/S_p is the number of spikes expressed as a fraction of the peak spike output at 450 nm (plotted at unity). For an equivalent spike fraction, 640 nm light (squares) always produces shorter latencies than 450 nm light (circles) or 520 nm light (triangles). For any particular latency the light intensity required for 450 nm light is very much more than for 640 nm light, even though the cell is blue-sensitive and produces a greater number of spikes. (DVORAK and FULBROOK, unpublished data)

same ganglion cell, for BOWLING derived different sensitivity spectra under appropriate conditions of adaptation, but the significance of the interaction is unknown.

In the optic tectum, ROBBINS (1972) demonstrated an opponent color system for red and blue lights.

e) Temporal Characteristics of Color

Vertebrate rods have slower waveforms than cones. Furthermore, in the turtle *Pseudemys*, each cone mechanism has its own characteristic time course. Other than spectral sensitivity, a main difference between red- and green-sensitive cones is that response waveforms of the former have shorter times-to-peak, a mean value of 125 ms as opposed to 165 ms. Rods are very much longer: 315 ms under comparable conditions (BAYLOR and HODGKIN, 1973). Double cones, too, show similar dynamics, even when the two cones with different photopigments are electrically coupled. The response of the double cone peaks at 120 ms to a 680 nm flash, but a longer 143 ms to a flash at 550 nm (RICHTER and SIMON, 1974).

Higher order cells preserve these temporal differences as shown in the horizontal cell's response (FUORTES and SIMON, 1974) and in the discharge pattern of ganglion cells (Fig. 25). For comparable inputs, the discharge pattern to red light has a shorter latency and a smaller duration of discharge than to green light or blue light. Blue light produces the longest latency and discharge duration.

There are ample data collected from single retinal cells of the turtle to indicate the presence of sophisticated color vision solely on electrophysiological grounds.

The important points are that the responses of most cells are dominated by the activity of red cones, and where the inputs from other cone types can be demonstrated, their responses are distinguishable from those of the red cones on a temporal basis.

2. Electroretinography

Electrical measurements from the entire eye of *Pseudemys* show it to have major sensitivity in the red region of the visible spectrum (ARMINGTON, 1954; DEANE et al., 1958; GRANDA, 1962; GRANDA and STIRLING, 1965, 1966). That fact is not surprising, for the red-sensitive cones are the most numerous of the cone receptors. In the dark-adapted eye, the most sensitive region is near 640 nm, with a hump at 570–580 nm, and very little evidence of activity below 480 nm (Fig. 26). These measures of sensitivity are taken from mass potentials (electroretinograms) developed across an eye stimulated by light.

The lights required for well-defined electroretinograms are intense and very much above absolute threshold. Weaker stimulus lights require response-averaging (GRANDA and STIRLING, 1966). Spectral sensitivity curves obtained in this fashion show a blue-green process below 500 nm available in dim light, as well as the usual high red sensitivity.

In *Chelonia*, whose photopigments are based on vitamin A_1, the entire spectral sensitivity curve shifts to shorter wavelengths. GRANDA and O'SHEA (1972) obtained action spectra that showed a main peak at 520 nm with secondary peaks at 600 nm and 450–460 nm. Figure 26 also shows a relative spectral sensitivity function of *Chelonia*. Compared to *Pseudemys*, the red sensitivity for *Chelonia* is displaced about 40 nm to the shorter wavelengths. *Chelonia* are very sensitive to the blue spectral region; *Pseudemys* are not.

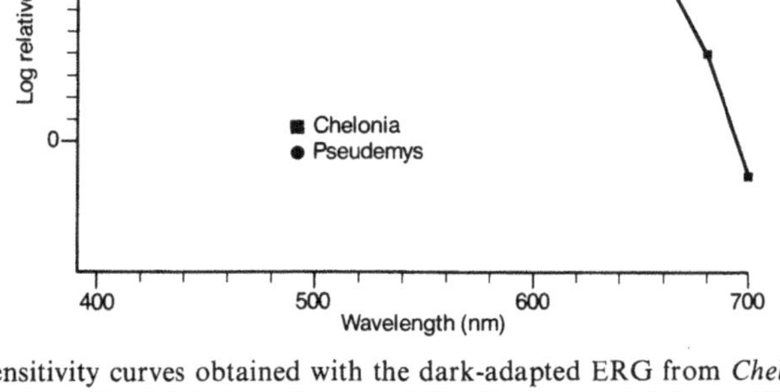

Fig. 26. Spectral sensitivity curves obtained with the dark-adapted ERG from *Chelonia* and *Pseudemys*. (Combined data from GRANDA, 1962; GRANDA and O'SHEA, 1972)

D. Psychophysics

Traditionally, psychophysics attempts to uncover sensory functioning by using the entire information-processing system of the organism, and in those cases where it can be successfully correlated with physiology, biochemistry, and neuroanatomy in the same animal, the picture that emerges is much more clearly drawn. But in animal work, experimental surgery and manipulation are relatively easy to pursue, and it is the difficulty in arranging a behavioral display for a particular function under controlled conditions that is problematic.

Turtles can learn to solve mazes and press keys for rewards of food and water. They can also be taught to make behavioral responses to avoid electric shock. Applied to visual problems under a variety of conditions, these techniques permit the determination of threshold signals that are sent to the brain for behavioral responses.

I. Color Sensitivity

Freshwater turtles show spectral functions that are sensitive to red light under light-adapted conditions (WOJTUSIAK, 1933; GRAF, 1967) and to blue light under conditions of dark adaptation (ARMINGTON, 1954; SOKOL and MUNTZ, 1966). This conclusion was corroborated by GRANDA et al. (1972), who investigated the temporal course of dark adaptation for colored lights in *Pseudemys*. Dark-adaptation threshold curves for red light reached stable levels in about 90 s, the same time required for a tungsten source of *white* light (ZWICK and GRANDA, 1968). The dark adaptation curves for shorter wavelength lights took longer to reach asymptote, with several discontinuities that indicated the influence of more than one receptor system (Fig. 27). Spectral sensitivity curves may be obtained across these curves at specified times in the dark, ranging from early dark adaptation to as long as 45 min. There is a progressive shift from a red-dominated sensitivity early in dark adaptation to an enhanced blue sensitivity after prolonged periods in the dark (Fig. 28). The increase in blue sensitivity is more than 1.6 logarithmic units. The main peak shifts from 633 nm to 466 nm.

It is possible to use chromatic adaptation to isolate chromatic mechanisms when sensitivity functions are determined against red, green, and blue backgrounds. Threshold values determined in this way showed large decreases in sensitivity appropriate to the spectral regions where they were adapted. Theoretical spectral curves were derived by assuming summing functions for the separate processes to equate the experimental dark-adapted spectral sensitivity curve, i.e., the dark curve is the sum of three underlying processes (Fig. 29). The peaks agree well with the functions derived from intracellular recordings from single cones (BAYLOR and HODGKIN, 1973).

The method developed by GRANDA and co-workers for psychophysical experiments in *Pseudemys* involves conditioned shock avoidance to light by head withdrawal (MAXWELL and GRANDA, 1975). The method is simple in principle and capitalizes on the limited behavioral repertoire of the animal. Head withdrawal is

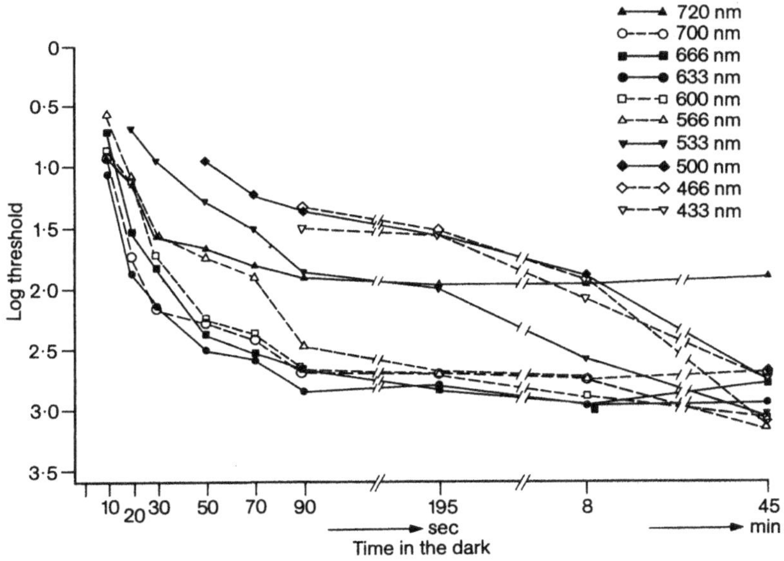

Fig. 27. Dark adaptation functions for *Pseudemys* measured with test lights ranging from 433 to 720 nm. The values plotted on the abscissa are continous to 90 s; thereafter, there are interruptions at indicated points which permit a compression of the time scale. (From GRANDA et al., 1972)

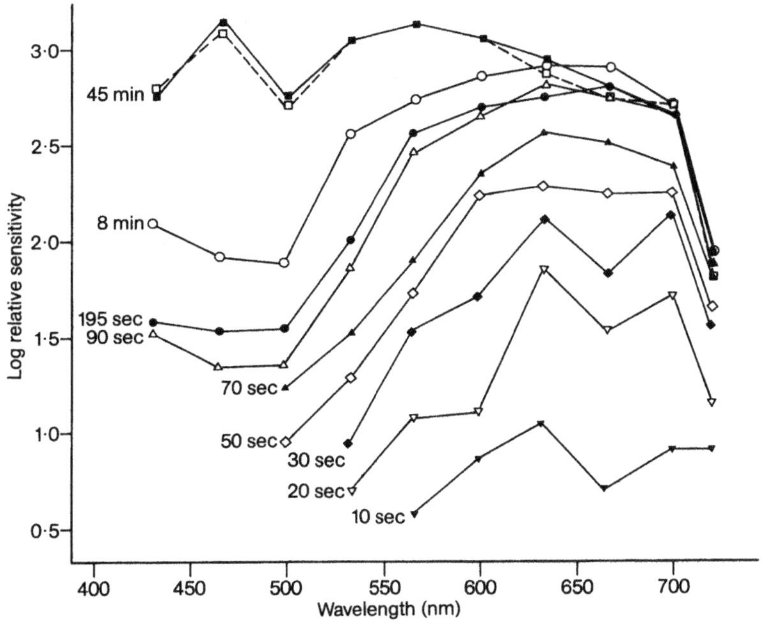

Fig. 28. Spectral sensitivity functions for *Pseudemys*. The number placed by each curve indicates the time in the dark at which the curve was defined. (From GRANDA et al., 1972)

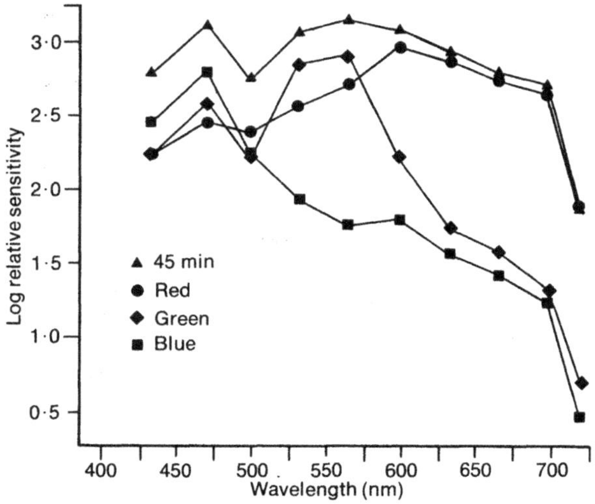

Fig. 29. Dark-adapted spectral sensitivity of *Pseudemys* obtained at 45 min in the dark is seen as the sum of three underlying processes. (From GRANDA et al., 1972)

part of the turtle's defense against danger, and discriminative shock applied to the jaw distinguishes stimuli requiring head withdrawal from those that do not. A line attached to the jaw permits the actuation of automatic programming equipment. The animal is then merely required to pull its head back a small distance in order to signal a threshold detection. Stimuli are tracked in a modified up-and-down procedure where the intensity of the next stimulus presentation is determined by the behavioral response to the last. The method concentrates the intensity of stimulation around the threshold, resulting in less variance and more reliability. Figure 30 shows the apparatus.

II. Spatiotemporal Characteristics

The behavioral procedures as described have been applied to problems of spatial-temporal integration with good success. The concern here is to describe the ability of *Pseudemys* to integrate quanta over time and areal extent to particular limits. Bloch's law is such a relationship that defines the domain where integration occurs up to some critical value of stimulus duration for a particular intensity, after which temporal integration stops and threshold is determined by intensity alone. The point of critical duration is not invariant but changes according to parametric changes in areal extent, color, and adaptational state of the animal.

The effect of target size on temporal summation is shown in Fig. 31, where the animal is fairly well light-adapted, and the cones may be presumed to be the major influence on visual behavior. Thresholds for each target size decrease in orderly fashion as stimulus duration increases. The slopes approximate -1.0, the

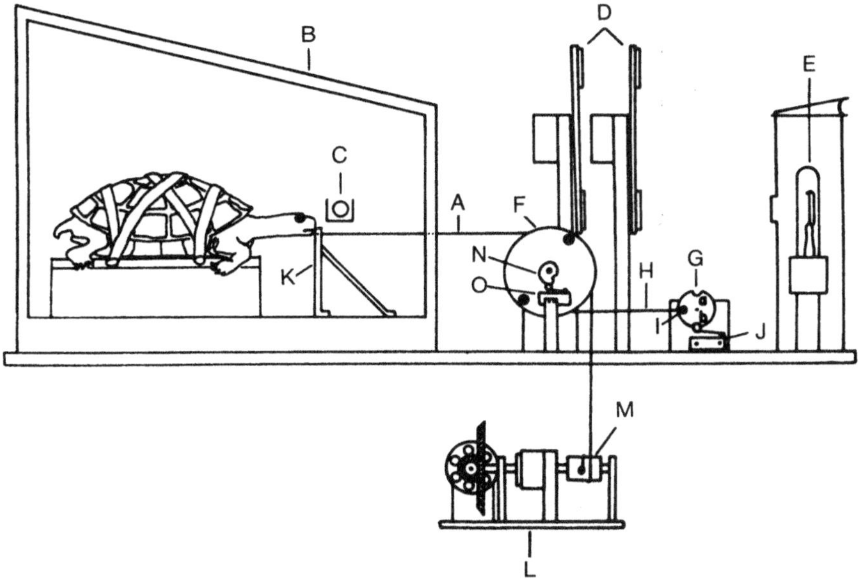

Fig. 30. Apparatus for measuring thresholds in turtles. *A*: leader line; *B*: Chamber; *C*: milk-glass; *D*: filter wheels; *E*: lamp; *F*: spring-loaded pulley; *G*: motorized disc; *H*: line; *I*: brass swivel; *J*, *O*: microswitch; *K*: plastic stop; *L*: microwinch; *M*: take-up spool; *N*: cam. (From MAXWELL and GRANDA, 1975)

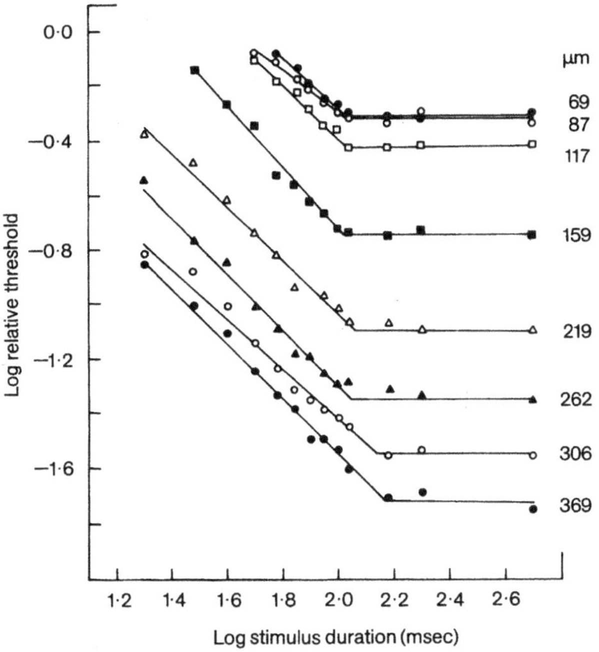

Fig. 31. Critical duration thresholds as a function of stimulus size: retinal image diameter ranged from 69 to 369 μm. For increasing stimulus size, the critical duration increases. (MAXWELL, unpublished data)

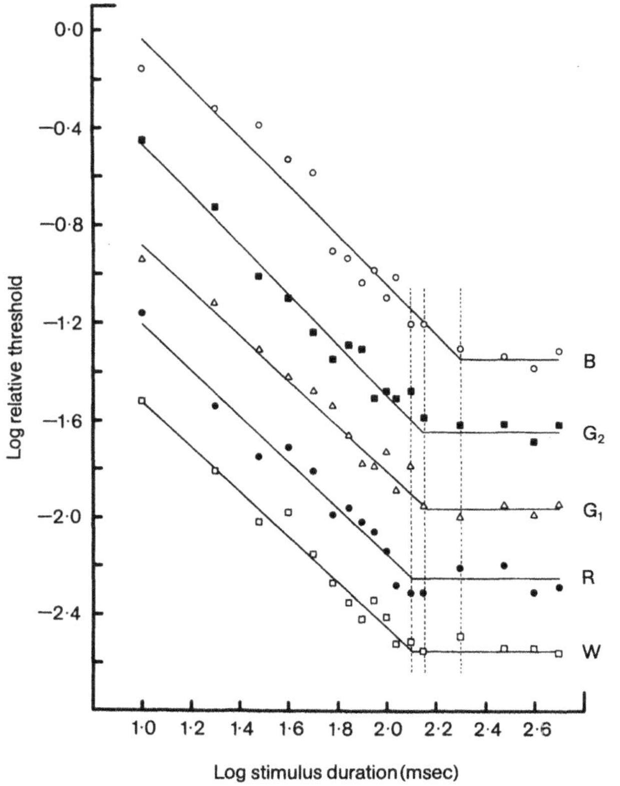

Fig. 32. Critical duration thresholds as a function of wavelength: 125 msec for 620 nm (*R*) and white (*W*) lights, 150 msec for 566 and 533 nm (G_1 and G_2) and 200 ms for 466 nm (*B*). Lines are arbitrarily displaced vertically for clarity. (MAXWELL, unpublished data)

value demanded by Bloch's law for complete reciprocity. With increases in stimulus durations, a time is reached where threshold is no longer affected. That point is known as the critical duration, beyond which the slope of the function is zero.

Increases in target size lower thresholds and increase critical duration. Those targets sampled in Fig. 31 subtend retinal diameters of 69 μm to 369 μm, with a range of critical durations from 107 ms for the smallest target to 148 ms for the largest, a substantial increase.

Wavelength, too, affects critical duration. The five curves shown in Fig. 32 span the spectrum, with tungsten *white* light at the bottom, blue at the top, and the two green curves separated by 30 nm in the midrange. The summation parts of each function show slopes little different from −1.0. Both red and tungsten *white* ligth curves have about the same duration of 125 ms. The predominant long-wavelength receptor system probably operates in both conditions. The two green curves have very similar critical durations of 145 and 148 ms. They probably reflect the activity of the green-sensitive mechanism. The critical duration for blue light is longer still, about 200 ms. That order of increase is reflected in the intracellular data of BAYLOR and HODGKIN (1973), where time to maximum hyperpolarization for the red-sensitive cones is about 100–140 ms, that for the green-sensitive cones about 140–180 ms. There are no values for the blue-sensitive cones.

Such functions as have been gathered to date indicate that critical durations in *Pseudemys* are rather long. None is shorter than 100–110 ms under any condition. Critical durations increase both with larger target areas and with dark adaptation. Lights of different wavelengths discriminate reliable critical function differences between red, green, and blue mechanisms that are of the same order and magnitude as the intracellular time-to-peak data collected from cones of different classes (BAYLOR and HODGKIN, 1973).

III. Increment Threshold

The data that define threshold conditions for particular target sizes of known quantal input can be used to estimate the signal in individual cones generated at behavioral threshold (FAIN et al., 1977). The task could be difficult, since receptors are coupled and the current developed on excitation spreads to other cells. Increment thresholds overcome this difficulty by assuring that the differences among receptors will be much less important in the number of photons absorbed than at absolute threshold. The receptor coupling will allow for an equalization of the voltages generated.

In *Pseudemys* the sensitivity of the red-sensitive cones varies with background intensity according to the relationship

$$\frac{1}{S_F} = \frac{1}{S_F^D} + \xi I_B, \tag{5}$$

where S_F is the flash sensitivity, S_F^D is the cone sensitivity in the dark, I_B is background intensity, and ξ is a nearly constant-value parameter of 20 s V^{-1} $flash^{-1}$ (BAYLOR and HODGKIN, 1974). At threshold, the receptor response peak amplitude equals the product of increment flash intensitiy and flash sensitivity. The voltage signal V_T at increment threshold ΔI_T is then given by

$$\Delta V_T = S_F \Delta I_T, \tag{6}$$

where ΔV_T is constant with background. The two equations can be combined to give the Weber relation at behavioral threshold

$$\Delta I_T = k(I_B + I_0), \tag{7}$$

where $I_0 = V_T/kS_F^D$ and the Weber fraction, k, is equal to $\Delta V_T \xi$. With a determination of k, the previously determined value of ξ can be combined with it to give ΔV_T.

The increment measurements are shown in Fig. 33 with a value for k of 10^{-4} $flash^{-1}$. From the values of k and ξ, ΔV_T is approximately 5–10 µV, allowing for errors in variability and experimentation. This means that receptor responses of no more than 10 µV are sufficient for visual threshold in red-sensitive cones. A very small voltage can therefore change the signalling properties between receptors and other cells.

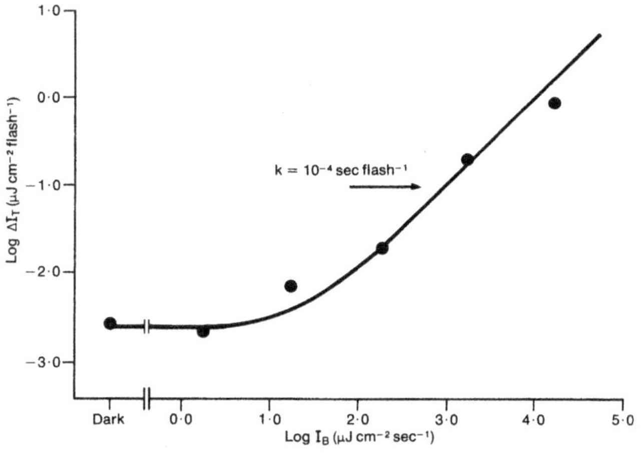

Fig. 33. Increment threshold curve for *Pseudemys* for white light (see text). (From FAIN et al., 1977)

E. Conclusions

Turtles are visual animals. They have elaborate and efficient mechanisms for the registration of images. Color especially is keenly appreciated, with much of the information processing and coding taking place in the retinal structure.

The working of the turtle eye has been described largely for the freshwater turtle, *Pseudemys*. In this animal the separate investigatory disciplines of photochemistry, anatomy, electrophysiology, and behavioral psychophysics have met with remarkable convergence.

Photopigments measured by microspectrophotometry are shown to reside uniquely in separate photoreceptors. Intracellular recordings from cones have action spectra that match the absorption spectra of the photopigments when suitably modified by the transmission spectra of the oil droplets they contain. The photopigments and oil droplets are uniquely paired.

The rich colors of the droplets have always attracted interest. They appear to play a role in color contrast and in protection from glare. The measured densities of the oil droplets are extremely high, many times more than required for the simple task of absorbing scattered blue light. Diurnal birds have the same arrangement of colored droplets in individual receptors, but in the higher primates the tasks relating to color contrast and glare use the more recently evolved arrangement of macular pigment and yellow lens. The exact role of the oil droplet is still not understood.

The eye is functionally sensitive to red light, a conclusion supported by the preponderance of red-sensitive cones, by electroretinographic action spectra from the entire eye, and by spectral sensitivity curves derived from threshold responses to colored light. The results are in keeping with a vitamin A_2 system and its absorptive emphasis in the longer wavelengths. Not all turtle pigments are based on vitamin A_2. Sea turtles have a system based on vitamin A_1, with photopigments and electroretinographic action spectra to corroborate that fact. The distribution of visual pigment absorption maxima in turtles is widespread. The pigments are varied in response to environmental demands.

The photoreceptors of turtles respond to light with slow hyperpolarizations. That message, uniquely interpreted by the photopigments contained in the outer segments, is communicated to other photoreceptors via cellular contacts that allow coupling to take place. What one photoreceptor does, affects its neighbors directly. With cells of the same type, there is the benefit of spatial enhancement through spread of photocurrent. There are also increases in cells' sensitivity with increases in areas of illumination. When photoreceptors of different types communicate, e.g., different cones, there is an elaborate coding via horizontal cells that feeds back the primary response to nearby cones of a different class. Both the primary and fedback responses in the same photoreceptor are separated by polarity and time. Here, then, is the first evidence of opponent coding that will prove so important and basic in more central stations of the visual system.

The spatial organization that accrues from the previously described arrangements of photoreceptor-horizontal cell communication provides for center-surround antagonisms in bipolar cells. That antagonism extends to color relationships in the center but apparently not in the surround. The arrangement set up here is then communicated to the ganglion cells for transmission centrally. The characteristics of these cells—their sensitivity to movement, the recoding into spike discharges—is presently being investigated.

Especially interesting has been the opportunity to compare threshold measures obtained in controlled behavioral situations with the known characteristics of retinal cells. This comparison is very exciting, for it affords a direct statement of the signals processed by individual cells during a behavioral act of sensory judgment. When an increment threshold is determined, it is possible to state the signals produced in individual photoreceptors that govern that discrimination.

Perhaps most surprising is the conclusion that much of the information processed by the visual system in turtles, information that results in visual behavior at threshold, is determined in the far periphery—in the photoreceptors themselves. How extensively that may hold is not known, but it appears that behavioral displays that rely on an appreciation of color or on certain spatial and temporal parameters find their limits largely set by the action of the cones and rods.

F. Acknowledgements

This work was supported by Grant EY 01540 from the National Eye Institute, National Institute of Health. We thank D.A. Baylor, L.E. Lipetz, J.H. Maxwell, D.P.M. Northmore and S. Yazulla for their critical reviews of the manuscript. We are especially grateful to Mrs. C. Groot for her careful editing and typing.

References

Arden, G. B., Ernst, W.: The effect of ions on the photoreceptors of pigeon cones. J. Physiol. (Lond.) **211**, 311—339 (1970).

Arden, G. B., Ernst, W.: A comparison of the behaviour to ions of the P III component of the pigeon cone and rat rod electroretinogram. J. Physiol. (Lond.) **220**, 479—497 (1972).

Armington, J. C.: Spectral sensitivity of the turtle, *Pseudemys*. J. Comp. Physiol. Psychol. **47**, 1—6 (1954).

Baylor, D. A., Fettiplace, R.: Light path and photon capture in turtle photoreceptors. J. Physiol. (Lond.) **248**, 433—464 (1975).

Baylor, D. A., Fettiplace, R.: Transmission of signals from photoreceptors to ganglion cells in the eye of the turtle. Cold Spring Harb. Symp. Quant. Biol. **40**, 529—536 (1976).

Baylor, D. A., Fuortes, M. G. F.: Electrical responses of single cones in the retina of the turtle. J. Physiol. (Lond.) **207**, 77—92 (1970).

Baylor, D. A., Hodgkin, A. L.: Detection and resolution of visual stimuli by turtle photoreceptors. J. Physiol. (Lond.) **234**, 163—198 (1973).

Baylor, D. A., Hodgkin, A. L.: Changes in time scale and sensitivity in turtle photoreceptors. J. Physiol. (Lond.) **242**, 729—758 (1974).

Baylor, D. A., Fuortes, M. G. F., O'Bryan, P. M.: Receptive fields of cones in the retina of the turtle. J. Physiol. (Lond.) **214**, 265—294 (1971).

Baylor, D. A., Hodgkin, A. L., Lamb, T. D.: The electrical response of turtle cones to flashes and steps of light. J. Physiol. (Lond.) **242**, 685—727 (1974a).

Baylor, D. A., Hodgkin, A. L., Lamb, T. D.: Reconstruction of the electrical responses of turtle cones of flashes and steps of light. J. Physiol. (Lond.) **242**, 759—791 (1974b).

Beer, T.: Die Akkomodation des Auges bei den Reptilien. Pflügers Arch. **69**, 507—568 (1898).

Borsellino, A., Fuortes, M. G. F., Smith, T. G.: Visual responses in *Limulus*. Cold Spring Harb. Symp. Quant. Biol. **30**, 429—443 (1965).

Bortoff, A.: Localization of slow potential responses in the *Necturus* retina. Vision Res. **4**, 627—636 (1964).

Bortoff, A., Norton, A. L.: An electrical model of the vertebrate photoreceptor cell. Vision Res. **7**, 252—263 (1967).

Bowling, D. B.: Properties of ganglion cells in the retina of the turtle. Unpublished doctoral dissertation, University of Colorado, 194 p. (1976). (Written under D. A. Baylor.)

Bridges, C. D. B.: Absorption properties, interconversions, and environmental adaptation of pigments from fish photoreceptors. Cold Spring Harb. Symp. Quant. Biol. **30**, 317—334 (1965).

Bridges, C. D. B.: The rhodopsin-porphyropsin visual system. In: Handbook of sensory physiology Vol. VII/I. New York: Springer 1972.

Brown, K. T.: A linear area centralis extending across the turtle retina and stabilized to the horizon by non-visual cues. Vision Res. **9**, 1053—1062 (1969).

Bush, W. G.: A qualitative and quantitative electron microscopic study of the retina of the turtle, *Pseudemys scripta elegans*. Unpublished doctoral dissertation submitted to the University of Delaware, 127 pp. (1973).

Byzov, A.: Investigation of the receptive field of cells: source of the S-potentials of the turtle and frog retina. Biofizica **11**, 861—870 (1966).

Byzov, A. L.: Role of horizontal cells in the mechanism of retinal adaptation. Neurofiziologiya **1**, 210—217 (1969). [In English translation, Neurosci. Transl. **14**, 63—70 (1970-1).]

Cervetto, L.: Influence of sodium potassium and chloride ions on the intracellular responses of turtle photoreceptors. Nature **241**, 401—403 (1973).

Cervetto, L., Marchiafava, P. L., Pasino, E.: Influence of efferent retinal fibres on responsiveness of ganglion cells to light. Nature **260**, 56—57 (1976).

Cone, R. A.: The internal transmitter model for visual excitation: some quantitative implications. In: Biochemistry and physiology of visual pigments. Langer, H. (ed.). Berlin: Springer 1973.

Copenhagen, D. R., Owen, W. G.: Functional characteristics of lateral interactions between rods in the retina of the snapping turtle. J. Physiol. (Lond.) **259**, 251—282 (1976).

Deane, H. W., Enroth-Cugell, C., Gongaware, M. A., Neyland, M., Forbes, A.: Electroretinogram of freshwater turtle. J. Neurophysiol. **21**, 45—61 (1958).

Duke-Elder, S.: System of ophthalmology, Vol. I. The eye in evolution. St. Louis: C. V. Mosby, 1958.

Dvorak, C. A., Taylor, T. M.: Computer analysis of spatial integration in vertebrate retinal horizontal cells. IEEE Trans. Biomed. Eng. **24**, 149—152 (1977).

Ehrenfeld, D. W., Koch, A. L.: Visual accommodation in the green turtle. Science **155**, 827—828 (1967).

Fain, G. L., Granda, A. M., Maxwell, J. H.: Voltage signal of photoreceptors at visual threshold. Nature **265**, 181—183 (1977).

Fujimoto, K., Yanase, T., Hanaoka, T.: Spectral transmittance of retinal coloured oil globules reexamined with microspectrophotometer. J. Physiol. Soc. Jpn. **7**, 339—346 (1957).

Fuortes, M. G. F.: Responses of cones and horizontal cells in the retina of the turtle. Invest. Ophthalmol. **11**, 275—284 (1972).

Fuortes, M. G. F., Hodgkin, A. L.: Changes in time scale and sensitivity in the ommatidia of *Limulus*. J. Physiol. (Lond.) **172**, 239—263 (1964).

Fuortes, M. G. F., O'Bryan, P. M.: Generator potentials in invertebrate photoreceptors. In: Handbook of sensory physiology, Vol. VII/2. Berlin: Springer 1972.

Fuortes, M. G. F., Schwartz, E. A., Simon, E. J.: Colour-dependence of cone responses in the turtle retina. J. Physiol. (Lond.) **234**, 199—216 (1973).

Fuortes, M. G. F., Simon, E. J.: Interactions leading to horizontal cell responses in the turtle retina. J. Physiol. (Lond.) **240**, 177—198 (1974).

Graf, V.: A spectral sensitivity curve and wavelength discrimination for the turtle *Chrysemys picta picta*. Vision Res. **7**, 915—928 (1967).

Granda, A. M.: Electrical responses of the light- and dark-adapted turtle eye. Vision Res. **2**, 343—356 (1962).

Granda, A. M., Haden, K. W.: Retinal oil globule counts and distributions in two species of turtles: *Pseudemys scripta elegans* (Wied) and *Chelonia mydas mydas* (Linnaeus). Vision Res. **10**, 79—84 (1970).

Granda, A. M., O'Shea, P. J.: Spectral sensitivity of the green turtle *(Chelonia mydas mydas)* determined by electrical responses to heterochromatic light. Brain Behav. Evol. **5**, 143—154 (1972).

Granda, A. M., Stirling, C. E.: Differential spectral sensitivity in the optic tectum and eye of the turtle. J. gen. Physiol. **48**, 901—917 (1965).

Granda, A. M., Stirling, C. E.: The spectral sensitivity of the turtle's eye to very dim lights. Vision Res. **6**, 143—152 (1966).

Granda, A. M., Maxwell, J. H., Zwick, H.: The temporal course of dark-adaptation in the turtle, *Pseudemys*, using a behavioral avoidance paradigm. Vision Res. **12**, 653—672 (1972).

Hagins, W. A., Penn, R. D., Yoshikami, S.: Dark current and photocurrent in retinal rods. Biophys. J. **10**, 380—412 (1970).

Hagins, W. A., Robinson, W. E., Yoshikami, S.: Ionic aspects of excitation in rod outer segments. In: Energy transformation in biological systems. Ciba Found. Symp. **31**, 169—189 (1975).

Hannover, A.: Über die Struktur der Netzhaut der Schildkröte. Arch. Anat., Physiol. Wissen. Med., 314—317 (1843).

Hodgkin, A. L.: Address of the President at the Anniversary Meeting, 30. November 1971. Proc. R. Soc. Lond. (Biol.), **B.180**, v-xx (1972).

Jain, M. K.: The bimolecular lipid membrance: a system. New York: Van Nostrand-Reinhold 1972.

King-Smith, P. E.: Absorption spectra and function of the colored oil drops in the pigeon retina. Vision Res. **9**, 1391—1399 (1969).

Krause, W.: Über die Endigung der Muskelnerven. Z. rat. Med. **20**, 1—18 (1863).

Lasansky, A.: Synaptic organization of cone cells in the turtle retina. Philos. Trans. R. Soc. Lond. (Biol.), **262**, 365—381 (1971).

Lasansky, A.: Cell junctions at the outer synaptic layer of the retina. Invest. Ophthalmol. **11**, 265—275 (1972).

Laties, A. M., Liebman, P. A., Campbell, C. E. M.: Photoreceptor orientation in the primate eye. Nature (Lond.), **218**, 172—173 (1968).

Liebman, P. A.: Microspectrophotometry of retinal cells. Ann. N.Y. Acad. Sci. **157**, 250—264 (1969).

Liebman, P. A.: Microspectrophotometry of photoreceptors. In: Handbook of sensory physiology, Vol. VII/I, New York: Springer 1972.

Liebman, P. A., Granda, M.: Microspectrophotometric measurements of visual pigments in two species of turtle, *Pseudemys scripta* and *Chelonia mydas*. Vision Res. **11**, 105—114 (1971).

Liebman, P. A., Granda, A. M.: Super dense carotenoid spectra resolved in single cone oil droplets. Nature **253**, 370—372 (1975).

Lipetz, L. E., Hill, R. M.: Discrimination characteristics of the turtle's retinal ganglion cells. Experientia **26**, 373—374 (1970).
Maksimova, Y. M.: Effect of intracellular polarization of horizontal cells on the activity of the ganglionic cells of the retina of the fish. Biofizika **14:3**, 537—544 (1969).
Marchiafava, P. L.: Centrifugal actions on amacrine and ganglion cells in the retina of the turtle. J. Physiol. (Lond.) **255**, 137—155 (1976).
Marmarelis, P. Z., Naka, K. I.: Spatial distribution of potential in a flat cell: application to the catfish horizontal cells layers. Biophys. J. **12**, 1515—1532 (1972).
Masters, J. I.: Some applications in physics of the P function. J. Chem. Phys. **23**, 1865—1874 (1955).
Maxwell, J. H., Granda, A. M.: An automated apparatus for the determination of visual thresholds in turtles. Physiol. Behav. **15**, 131—132 (1975).
Michaelis, L., Menten, M. L.: Die Kinetik der Invertinwirkung. Biochem. Z. **49**, 333—339 (1913).
Miller, W. H., Hashimoto, Y., Saito, T., Tomita, T.: Physiological and morphological identification of L- and C-type S-potentials in the turtle retina. Vision Res. **13**, 443—447 (1973).
Naka, K.—I., Nye, P. W.: Role of H-cells in the organization of the catfish retinal receptive field. J. Neurophysiol. **34**, 785—801 (1971).
Naka, K. I., Rushton, W. A. H.: An attempt to analyse colour reception by electrophysiology. J. Physiol. (Lond.) **185**, 556—586 (1966).
Naka, K. I., Rushton, W. A. H.: The generation and spread of S-potentials in fish *(Cyprinidae)*. J. Physiol. (Lond.) **192**, 437—461 (1967).
Neame, K. D., Richards, T. G.: Carrier transport. Elementary kinetics of membrane carrier transport. Oxford: Blackwell Scientific Publications 1972, pp. 16—40.
O'Bryan, P. M.: Properties of the depolarizing synaptic potential evoked by peripheral illumination in cones of the turtle retina. J. Physiol. (Lond.) **235**, 207—223 (1973).
Peiponen, V. A.: Zur Bedeutung der Ölkugeln im Farbensehen der Sauropsiden. Ann. Zool. Fenn. **1**, 281—302 (1964).
Penn, R. D., Hagins, W. A.: Signal transmission along retinal rods and the origin of the electroretinographic a-wave. Nature **223**, 201—205 (1969).
Penn, R. D., Hagins, W. A.: Kinetics of the photocurrent of retinal rods. Biophys. J. **12**, 1073—1094 (1972).
Ramon Y Cajal, S.: The structure of the retina. Compiled and translated by S. A. Thorpe and M. Glickstein from editions published in 1892, 1894, and 1933. Springfield, Illinois: Carles C. Thomas 1972.
Richter, A., Simon, E. J.: Electrical responses of double cones in the turtle retina. J. Physiol. (Lond.) **242**, 673—683 (1974).
Richter, A., Simon, E. J.: Properties of centre-hyperpolarizing red-sensitive bipolar cells in the turtle retina. J. Physiol. (Lond.) **248**, 317—334 (1975).
Robbins, D. O.: Coding of intensity and wavelength in optic tectal cells of the turtle. Brain, Behav., Evol. **5**, 124—142 (1972).
Saito, T., Miller, W. H., Tomita, T.: C- and L-type horizontal cells in the turtle retina. Vision Res. **14**, 119—123 (1974).
Schaeffer, S. F., Raviola, E.: Ultrastructural analysis of functional changes in the synaptic endings of turtle cone cells. Cold Spring Harb. Symp. Quant. Biol. **40**, 521—528 (1976).
Schwartz, E. A.: Organization of on-off cells in the retina of the turtle. J. Physiol. (Lond.) **230**, 1—14 (1973a).
Schwartz, E. A.: Responses of single rods in the retina of the turtle. J. Physiol. (Lond.) **232**, 503—514 (1973b).
Schwartz, E. A.: Responses of bipolar cells in the retina of the turtle. J. Physiol. (Lond.) **236**, 211—224 (1974).
Schwartz, E. A.: Rod-rod interaction in the retina of the turtle. J. Physiol. (Lond.) **246**, 617—638 (1975a).
Schwartz, E. A.: Cones excite rods in the retina of the turtle. J. Physiol. (Lond.) **246**, 639—651 (1975b).
Schwartz, E. A.: Electrical properties of the rod syncytium in the retina of the turtle. J. Physiol. (Lond.) **257**, 379—406 (1976).
Sillman, A. J., Ito, H., Tomita, T.: Studies on the mass receptor potential of the isolated frog retina. I. General properties of the response. Vision Res. **9**, 1435—1442 (1969a).
Sillman, A. J., Ito, H., Tomita, T.: Studies on the mass receptor potential of the isolated frog retina. II. On the basis of the ionic mechanism. Vision Res. **9**, 1443—1451 (1969b).

Simon, E. J.: Two types of luminosity horizontal cells in the retina of the turtle. J. Physiol. (Lond.) **230**, 199—211 (1973).
Sokol, S., Muntz, W. R. A.: The spectral sensitivity of the turtle, *Chrysemys picta picta*. Vision Res. **6**, 285—292 (1966).
Strother, G. K.: Absorption spectra of retinal oil globules in turkey, turtle and pigeon. Exp. Cell. Res. **29**, 349—355 (1963).
Tomita, T.: Electrophysiological study of the mechanisms subserving color coding in the fish retina. Cold Spring Harb. Symp. Quant. Biol. **30**, 559—566 (1965).
Tomita, T.: Light-induced potential and resistance changes in vertebrate photoreceptors. In: Handbook of sensory physiology, Vol. VII/2. Berlin: Springer 1972.
Tomita, T., Tosaka, T., Watanabe, K., Sato, Y.: The fish EIRG in response to different types of illumination. J. Physiol. Soc. Jpn. **8**, 41—50 (1958).
Toyoda, J., Nosaki, H., Tomita, T.: Light-induced resistance changes in single photoreceptors of *Necturus* and *Gekko*. Vision Res. **9**, 453—463 (1969).
Underwood, G.: The eye. In: Biology of the reptilia. New York: Academic Press 1970.
Wald, G.: On the distribution of vitamin A_1 and A_2. J. gen. Physiol. **22**, 391—415 (1939a).
Wald, G.: The porphyropsin visual system. J. gen. Physiol. **22**, 775—794 (1939b).
Wald, G.: The visual systems of euryphaline fishes. J. Gen. Physiol. **25**, 235—245 (1941).
Wald, G.: Retinal chemistry and the physiology of vision. In: Visual problems of colour. National Physical Laboratory Symposium No. 8. London: H.M. Stationary Office 1958
Wald, G.: The photoreceptor process in vision. In: Handbook of physiology. I. Neurophysiology, Vol. 1. Washington, D.C.: American Physiological Society, 1959.
Wald, G., Zussman, H.: Carotenoids of the chicken retina. Nature **140**, 197 (1937).
Wald, G., Brown, P. K., Smith, P. H.: Cyanopsin, a new pigment of cone vision. Science **118**, 505—508 (1953).
Walls, G. L.: The vertebrate eye and its adaptive radiation. Bloomfield Hills, Michigan: The Cranbrook Institute of Science 1942, p. 611.
Walls, G. L., Judd, H. D.: The intra-ocular colour-filters of vertebrates. Br. J. Ophthal. mol. **17**, 641—675, 705—725 (1933).
Werblin, F. S., Dowling, J. E.: Organization of the retina of the mudpuppy *Necturus maculosus*. II. Intracellular recording. J. Neurophysiol. **32**, 339—355 (1969).
Wojtusiak, R. J.: Über den Farbensinn der Schildkröten. Z. vergl. Physiol. **18**, 393—436 (1933).
Wood, C. A.: The Fundus Oculi of Birds. Chicago: H. A. Fox, 1917.
Yazulla, S.: Cone input to horizontal cells in the turtle retina. Vision Res. **16**, 727—735 (1976a).
Yazulla, S.: Cone input to bipolar cells in the turtle retina. Vision Res. **16**, 737—744 (1976b).
Yoshikami, S., Hagins, W. A.: Light, calcium and the photocurrent of rods and cones. Biophys. J. **11**, Abstract TPM-E 16 (1971).
Yoshikami, S., Hagins, W. A.: Control of the dark current in vertebrate rods and cones. In: Biochemistry and physiology of visual pigments. Heidelberg: Springer 1973.
Zwick, H., Granda, A. M.: Behaviorally determined dark adaptation functions in the turtle, *Pseudemys*. Psychon. Sci. **11**, 239—240 (1968).

Chapter 9

Adaptive Radiation of the Pineal System

D. I. HAMASAKI and D. J. EDER, Miami, Florida, USA

With 11 Figures

Contents

A. Introduction

I. General Considerations

In lower vertebrates an intracranial structure exists which contains typical photoreceptors and which responds to light falling on the surface of the head by an inhibition of neural activity. In higher vertebrates this structure lacks photoreceptors but its biochemical activity is influenced by light falling on the lateral eyes. This remarkable structure is the Pineal, and it is found in almost all vertebrates.

The pineal develops embryologically as an evagination of the diencephalon as do the lateral eyes and is of neuroepithelial origin. The morphologic appearance of the pineal is considerably different in the different classes of vertebrates. During the course of evolution there has been a systematic alteration of the pineal from a directly photosensitive structure to an indirectly photosensitive secretory organ. The functions attributed to the pineal have been as diversified as its morphologic appearance. There is evidence that the pineal influences body color changes, gonadotropic activity, locomotor activity, and other related activities. The pineal has been called a photoneuroendocrine organ which synthesizes a hormone important for the animal. Thus, for an organ which has long been considered vestigial and rudimentary, there is much experimental evidence to show that it is an important, functional organ.

In this chapter we shall examine the morphology, physiology, biochemistry, and function of the pineal complex in the different classes of vertebrates. This systematic survey will demonstrate the interesting changes undergone by the pineal during the course of evolution. We shall try to relate these adaptive changes to the function of the pineal.

In order to keep this chapter manageable, we shall concentrate mainly on the recent observations which pertain to the evolution of the pineal system. The reader is referred to STUDNIČKA (1905), TILNEY and WARREN (1919), and KITAY and ALTSCHULE (1954) for the earlier studies on the pineal. The proceedings of three excellent symposia on the pineal have been published (KAPPERS and SCHADÉ, 1965; WOLSTENHOLME and KNIGHT, 1971; REITER, 1970). In addition, three books, one dealing more with the superficial component of the pineal system (EAKIN, 1973), one dealing more with pineal system in the higher vertebrates (WURTMAN et al., 1968), and one dealing with mammalian pineal biochemistry (QUAY, 1974) have also appeared. Two review articles on the pineal have been published (AXELROD, 1974; REITER, 1973). For the physiologic findings on the pineal, the reader is referred to DODT's chapter in Volume VII/3B of The Handbook of Sensory Physiology.

II. Terminology

Because of the great diversity of the pineal, many names have been applied to it. We shall adopt the following terminology. When referring to the entire pineal complex or speaking in general terms, we shall refer to the *pineal system* or the

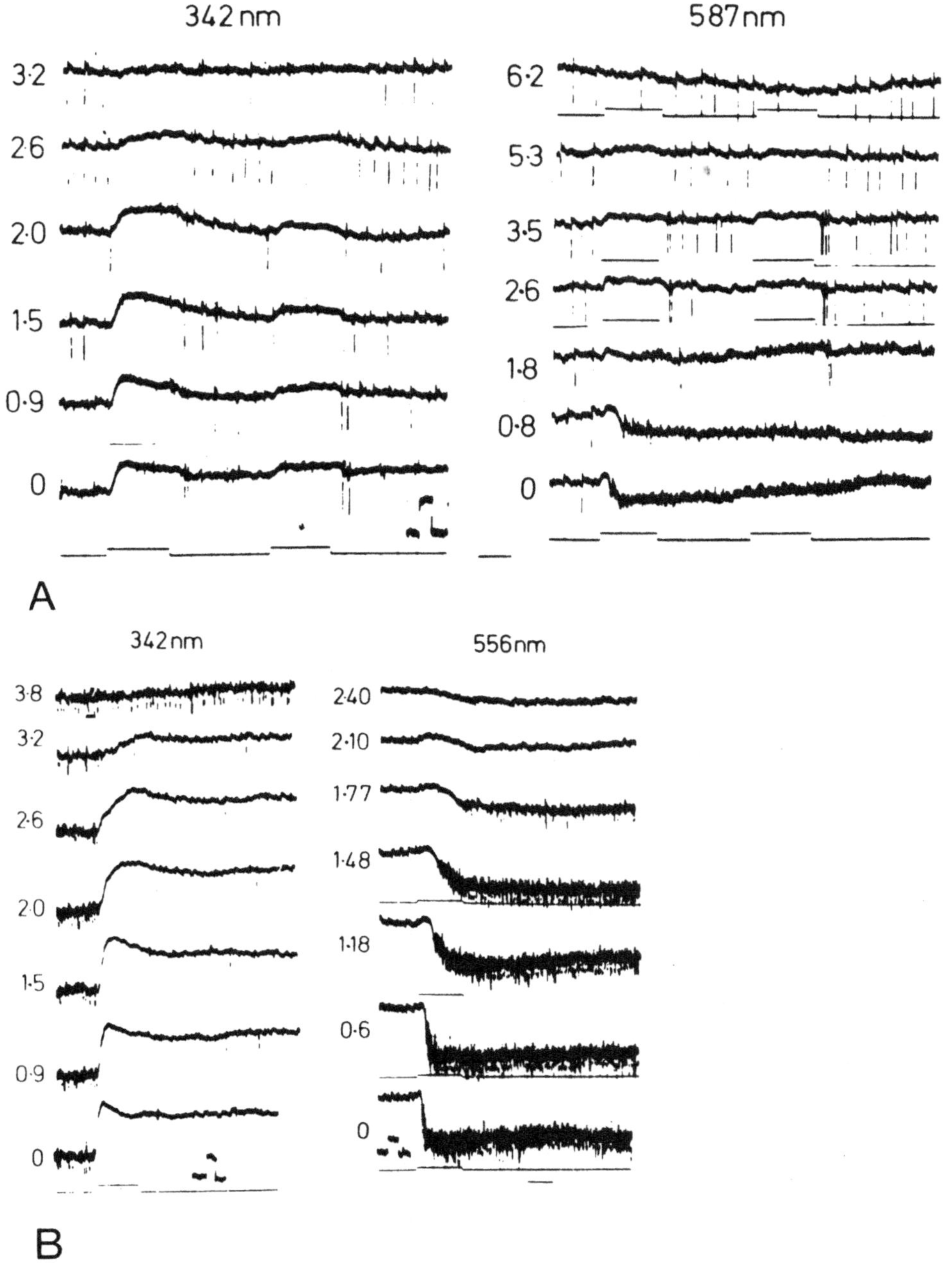

Fig. 1 A and B. Recordings from frontal organ of *R.esculenta* (from HAMASAKI, 1970). (A) Achromatic system—both short and long wavelength stimuli elicit negative-going (negativity = upward deflection) slow potentials and inhibition of spike activity. (B) Chromatic system—short wavelength stimuli elicit negative-going slow potential and inhibition of spike activity while long wavelength stimuli elicit positive-going slow potential and increase spike activity. Number to left of each recording represents density of neutral filter used to reduce stimulus intensity. Wavelength of stimulus shown above each set of recordings. Calibration 100 μV and 0.5 sec

pineal. The pineal system in some of the lower vertebrates consists of two parts. In frogs, the superficially situated element will be called the *frontal organ* (*Stirnorgan* in German). In lizards, the superficially situated structure will be called the *parietal eye*. In cyclostomes (lampreys), the homologue of the parietal eye and frontal organ is called the *parapineal organ*.

The intracranially situated component of the pineal system is present in almost all vertebrates, and will be called the *epiphysis* or *epiphysis cerebri*. The epiphysis has been referred to as the pineal gland, the pineal organ, or the pineal body.

The nerve fiber tract passing from the parietal eye to the brain will be called the *parietal nerve*. The nerve fiber bundle passing from the frontal organ to the brain has been called the pineal nerve in the literature, but we shall refer to it as the *frontal organ nerve*. The nerve fibers passing from the epiphysis to the brain will be referred to as the *epiphyseal tract*.

In electrophysiologic recordings from the pineal system slow potential changes and unitary spike activity can be obtained (Fig. 1). Slow potentials represent mass potential changes occurring in many neurons, while spike activity is usually recorded from individual units. We shall adopt Dodt's terminology of achromatic and chromatic units. *Achromatic units* respond to all wavelength stimuli by a decrease of spontaneous neural activity or inhibition (Fig. 1A). The *chromatic units* respond to short wavelength stimuli by inhibition and to medium wavelenght stimuli by an increase in neural activity or excitation (Fig. 1 B). When the neural activity of the pineal is affected by light falling on it, the pineal is said to be *directly photosensitive*. If the influence of light is mediated *via* the lateral eyes, the pineal then is *indirectly photosensitive*.

B. Historical Highlights

The pineal system has a long and interesting history. It begins with the ancient Indian philosophers who considered the pineal an organ of clairvoyance and the early Greek philosophers who suggested that the pineal acted as a sphincter which controlled the amount of "vital substance" passing between the third and forth ventricles, thereby regulating the "stream of thought". Descartes embellished this belief in the 17th century because he considered it the residence of the soul. An engaging résumé of the early history of pineal study has been presented by Kappers (1965a). By the beginning of the 1900's anatomists had extended light microscopy nearly to its limits in their studies on a variety of pineal systems. Studnička (1905) differentiated three cell types in the pineals of various animals and stated that one of these closely resembled the photoreceptor cells of the lateral eye. N. Holmgren (1918) not only identified distinct photosensory cells in the pineals of frogs and sharks but also identified the material in the lumen as parts of outer segments. He concluded from this latter observation that the photoreceptors were undergoing continual renewal.

During the next four decades two independent, slowly evolving lines of study emerged from a pair of significant observations. McCord and Allen (1917) artificially elicited the widely studied, photically-related skin lightening response in tadpoles by adding bovine pineal extracts to the media contraining the tadpoles. Rowan (1925) manipulated the avian breeding cycle by subjecting birds to varying light regimes. The former study was considered significant as a pharmacologic rather than a physiologic demonstration and the latter because it revealed that the endocrine system was open to environmental influences. These observations provided a framework within which the separate lines of study in pineal endocrinology and photosexual stimulation developed. In retrospect, these lines of study might have produced less confusing results and merged sooner had they had the benefit of two subsequent discoveries. Lerner et al.'s identification (Lerner et al., 1958) of melatonin as a biologically active product of pineal biosynthesis and Kappers' finding (Kappers, 1960) that the pineal in higher vertebrates is an innervated structure. With these discoveries a number of laboratories reopened the questions asked during the previous 40 years with the result that the pineal has now been identified as a light-sensitive neuroendocrine transducer which, especially in the mammal, secretes substances that modulate reproductive physiology and behavior. It is not merely a vestigial organ as had been supposed.

Credit for the first consolidated hypothesis of pineal neuroendocrine transduction is usually accorded to Bagnara (1960). Briefly, the hypothesis states that the pineal in the amphibian can be stimulated by darkness to release a melanophore-contracting agent (melatonin) which then acts on the dermal melanophores to produce body-blanching. The advancement of this hypothesis was unusual due to its emphasis on pineal stimulation by darkness, thus resolving at least one confusing point of those previous experiments which had concentrated on seeking effects of stimulation by light. The idea of the pineal as a photoendocrine transducer received considerably more attention following the first direct observation of the effect of light on the mammalian pineal by Fiske et al. (1960): increased light exposure reduced the pineal weight of otherwise normal rats. Wurtman et al. (1961) then suggested that an appropriate photoperiod could reduce the formation of pineal antigonadotrophic substance, thereby allowing the gonads to become stimulated. This suggestion, made with the benefit of knowledge that melatonin might serve as a mediator of mammalian pineal action, finally provided a mechanism linking studies of the pineal as a quasi-endocrine organ to those of the photosexual response. Shortly thereafter Quay (1963) discovered the circadian rhythm present in the melatonin precursor serotonin. Largely due to the efforts of Axelrod, Wurtman, Klein, and their coworkers (see Axelrod, 1974; Klein and Weller, 1970), the succeeding decade saw the discovery of a number of neurohumoral and enzymatic control mechanisms which are presently thought to cause the pineal's rhythmic and light-dependent activities, at least in the higher vertebrates (i.e., birds and mammals).

Kappers' (1960) work, tracing the innervation of the mammalian pineal, has already been mentioned. He found that the pineal receives adrenergic innervation from the superior cervical ganglion. By virtue of a functional requirement for this innervation to be intact, the pineal is not a typical endocrine organ (see Wurtman et al., 1968, pp.145-147). In higher animals photic modulation of the biochemical control mechanisms is due apparently to messages from the retina, which reach the superior cervical ganglion *via* a multisynaptic pathway involving the inferior accessory optic tracts and retinohypothalamic pathway (MOORE et al., 1968; MOORE and KLEIN, 1974).

Meanwhile EAKIN and WESTFALL (1959, 1960) and STEYN (1959, 1960) turned the electron microscope on the directly photosensitive parietal eyes of lizards and reported that the cells which projected into the lumen of the parietal eye were made up of stacked disks and had organelles which were similar to those found in the photoreceptors of the lateral eyes. Subsequent studies of the fine structure of the pineal have confirmed and extended these observations. The beautiful comparative studies of KAPPERS (1965b), OKSCHE (1971), and of COLLIN (1971) and their coworkers have provided much information of the phylogenetic evolution of the pineal system from the directly photosensitive system of lower vertebrates to the indirectly photosensitive system present in birds and mammals.

Finally, with respect to directly photosensitive pineals, DODT and HEERD (1962) reported that illumination of the frog frontal organ affected the neural activity recorded from the frontal organ nerve. Particularly significant was the observation that darkness rather than light stimulated the neural activity of a large number of the fibers in this nerve. Subsequently DODT and his coworkers demonstrated similar direct pineal photosensitivity in a variety of lower vertebrates. Such responses have yet to be recorded and related to known biochemical events in the indirectly photosensitive pineals of birds and mammals.

C. Biochemistry

I. Melatonin

The pineal, as a biosynthetically active organ, is unusual partly because the manufacture and release of its synthetic end product, melatonin, is regulated by light. Melatonin, an indole amine synthesized almost exclusively by pineal tissues (QUAY, 1965a), is capable of precipitating the body-blanching reaction in tadpoles at the remarkably low concentration of 10^{-13} g/ml aquarium water (LERNER and CASE, 1959). The low concentrations apparently required for its biological activity *in vivo* and its short half-life in plasma (minutes: KOPIN et al., 1961) have both hampered assessment of its function, although a number of effects related to growth, the thyroid, adrenals, gonads, dermal melanophores, brain activity, and photoperiodic timing have been proposed (for review, see WURTMAN et al., 1968; QUAY, 1970).

Melatonin synthesis occurs by the following steps (summarized by WURTMAN et al., 1968; AXELROD, 1974):

	TROH		AAD		NAT		HIOMT	
Tryptophan	----	5-HTP	---	5-HT	- - -	AcHT	-----	Melatonin
	1		2		3		4	

where TROH	=	tryptophan hydroxylase
5-HTP	=	5-hydroxytryptophan
AAD	=	aromatic amino acid decarboxylase
5-HT	=	serotonin (5-hydroxytryptamine)
NAT	=	serotonin N-acetyltransferase
AcHT	=	N-acetylserotonin
HIOMT	=	hydroxyindole O-methyltransferase

Step 1 requires NADPH and tetrahydrobiopterin
Step 2 requires pyridoxal phosphate
Step 3 requires acetyl-Co A
Step 4 requires S-adenosylmethionine.

The notion that the pineal may contain active substances besides melatonin to mediate the aforementioned effects has received attention from more than one laboratory (McCord and Allen, 1917; Pavel and Petrescu, 1966; Vaughan et al., 1974).

II. Innervation and Influence of Light

In certain stages of their life cycles some lower vertebrates (e.g., fish, amphibians) are able to lighten in color when placed in a dark environment (a phenomenon not to be confused with color adapting to light or dark backgrounds). The mechanism responsible for initiating body-blanching is direct pineal photostimulation leading presumably to melatonin release into the circulation. Evidence in favor of melatonin as the mediating agent is circumstantial but compelling (Bagnara and Hadley, 1970).

The avian and mammalian pineals also secrete melatonin into the blood (Kappers, 1971), an activity which technically makes the pineal an endocrine organ by definition. Yet it is also an innervated structure, a fact which has impeded assessment of its endocrine function by the classical endocrinologists' tests. Although melatonin in the higher vertebrates has, at present, poorly defined target organ effects, its synthesis, secretion, and target organ actions are known to be subject to photoperiodic control. In higher vertebrates (and some lower ones as well) this control arrives at the pineal *via* an adrenergic innervation, the details of which will be presented in the appropriate phylogenetic sections to follow. Generally, however, control is exerted by pineal Beta-adrenergic receptors which receive noradrenalin discharge from sympathetic fibers originating in the superior cervical ganglia (Axelrod, 1974). These ganglia themselves receive neural input from the retina and also from a biological clock located in the brain. Beta-adrenergic receptor stimulation promotes increased activity of one or more enzymes (Steps 3 and 4 above) which in turn enhances the synthesis of melatonin. Melatonin synthesis appears highest at night so it is not surprising that NAT and HIOMT enzyme activities are highest in the dark (Klein and Weller, 1970; Wurtman et al., 1963). Retinal stimulation by light depresses these enzyme activities by depressing sympathetic discharge.

Diurnal variations in some pineal enzymes and products continue to occur despite the absence of rhythmic photoperiodic stimulation, i.e., in constant darkness, but are abolished in constant light. Rhythms behaving in this manner satisfy the definition of true circadian rhythms and, as such, can be reset to a phase-shifted light cycle even though they are not totally driven by it. On the other hand, at least one substance, HIOMT does not seem to exhibit a consistent, truly circadian rhythm (AXELROD et al., 1965); the diurnal variations in its activity are apparently driven entirely by changes in the photic environment. Because investigations of pineal rhythms, their neural dependence, and biochemical mechanisms have been carried out primarily in mammals and some birds, a more complete description of these interactions is available in those sections of this chapter.

D. Phylogenetic Examination of the Pineal

I. Prehistoric Animals

Many of the skulls from the Paleozoic and Mesozoic eras show a parietal foramen (EDINGER, 1955). Because there is good correlation between the size of the foramen and the size of the parietal eyes in existing lizards, EDINGER concluded that the measurement of the foramen size gives a valid estimation of the size of the eye. Rather than considering the absolute size of the parietal foramen, she compared its size with that of the foramen magnum. While the ratio in present day lizards is close to 5:1 (foramen magnum: parietal foramen), the ratio is close to 1:1 in the terrestrial Permian and the semi-aquatic Triassic forms.

In addition to these measurements, EDINGER (1956) noted that the parietal foramen was paired in many of the extinct vertebrates. The paired foramina may be early indications of the dual aspect of the pineal system in some of the living vertebrates.

II. Agnatha: Cyclostomata

1. Morphology

In cyclostomes, the lowest form of existing vertebrates, the pineal system is made up of the epiphysis and the parapineal organ. Both structures are situated intracranially with the epiphysis lying dorsal to the parapineal to read then organ Fig. 2A).

The retina of the epiphysis is everted with the outer segments of the photoreceptors protruding into the lumen (see review by COLLIN, 1971). The disks of the outer segment are formed by an infolding of the plasma membrane as they are in the cones of the lateral eyes (Fig. 2B). Approximately 130 disks form the fan-shaped outer segment. The distal part of the inner segment contains the 9+0 centriole and is rich in mitochondria. Lysosomes and a well-developed Golgi

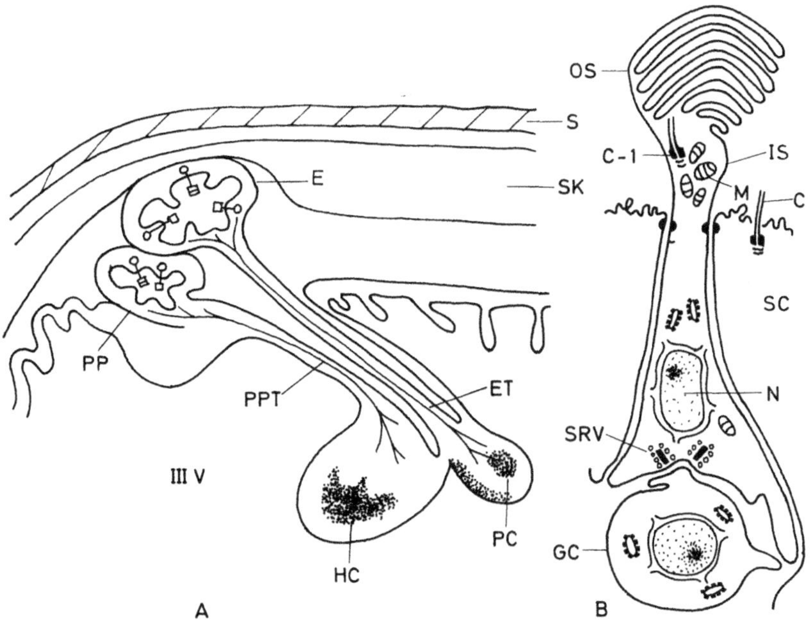

Fig. 2A and B. Pineal system of lamprey. (A) Schematic drawing of a midsagittal section through brain. (B) Schematic drawing of photoreceptor and supporting cell. Abbreviations: *C* cilium; *C-1* centriole; *D* dendrite of ganglion cell; *E* epiphysis; *ET* epiphyseal tract; *GC* ganglion cell; *HC* habenular commissure; *IS* inner segment; *M* mitochondria; *N* nucleus; *OS* outer segment; *PC* posterior commissure; *PP* parapineal organ; *PPT* parapineal tract; *S* skin; *SC* supporting cell; *SK* skull; *SRV* synaptic ribbon and vesicles; *III V* third ventricle

complex are present in the proximal portion. Numerous vesicles, some of which are dense-cored, surround the Golgi apparatus. Tubular and saccular endoplasmic reticulum and free ribosomes are also present in the proximal region of the photoreceptors.

The terminals of the photoreceptors are associated with one or more dendritic processes of the ganglion cells. The synaptic terminals are filled with clear vesicles (50 nm) which are accumulated around synaptic ribbons. About 1 to 7 ribbons are found in each terminal.

The ganglion cells lie near the ventral surface of the retina and contain masses of endoplasmic reticulum and ribosomes. Approximately 600 unmyelinated nerve fibers form the parapineal-epiphyseal tract which has been traced to the region of the posterior commissure. The supporting cells possess the 9 + 2 centriolar apparatus and many microvilli on the luminal surface. Crystalline structures which have been identified as guanine are also found in the supporting cells (MEINIEL, 1971).

The retina of the parapineal organ is also everted with the outer segments projecting into the lumen. The parapineal organ is oriented in such a way that the photoreceptors face ventrally, i.e., away from the surface of the skull. The photoreceptors of the parapineal organ are fewer and not as well organized as those of the epiphysis (MEINIEL, 1971).

2. Physiology

Slow potential changes and unitary spike activity have been recorded from the epiphysis of *Petromyzon fluviatilis* (MORITA and DODT, 1973). The units are active in the dark and light stimulation decreases the firing rate. The spectral sensitivity curve peaks at 525 nm and the absolute intensity threshold is approximately 10^{-5} lm/m^2. Most of the units are achromatic units, i.e., inhibited by all wavelength stimuli. The slow potential changes are of one polarity and are associated with inhibition of spike activity. With steady illumination, there is a linear relationship between the firing rate and the logarithm of the luminance of the steady light (DODT, 1971). The ERG-determined spectral sensitivity of the lateral eye of *P. fluviatilis* peaks between 525-530 nm (MORITA, 1973).

3. Biochemistry

We are not aware of any biochemical studies on the pineal system of lampreys.

4. Function

The diurnal body color changes in larval and adult lampreys (*Lampetra planeri*) were first shown by YOUNG (1935). This rhythm persists in blinded (removal of both lateral eyes) animals and also in normal animals kept in constant darkness. Pinealectomy abolishes the response in larval lampreys but in adults, both the pineal and the lateral eyes must be removed to abolish the rhythm. YOUNG concluded that the pineal system controlled the paling response by an inhibition of posterior pituitary secretion.

Recent studies (EDDY and STRAHAN, 1968; JOSS, 1973b) have confirmed YOUNG'S observation in *Lampetra* but the role of the pineal appears to be different in the different genera of lampreys. In *Geotria*, injection of pineal extracts or melatonin causes paling of the body. In *Lampetra*, the pineal contains a paling substance but injection of melatonin has no effect on the melanophores. In *Mordacia*, the melanophores react neither to pineal extracts nor to melatonin.

Pinealectomy at the start of the spawning run will retard egg development, ovulation, and development of secondary sex characteristics in female lampreys (JOSS, 1973a). In males, pinealectomy retards spermatogenesis and spermiation. These observations indicate that the pineal is progonadotrophic. Since these same changes are noted following hypophysectomy, the action of the pineal may be by way of the pituitary.

III. Chondrichthyes; Elasmobranchii

1. Morphology

In elasmobranchs, the epiphysis is situated on the dorsal surface of the brain. It extends from the roof of the diencephalon to its end vesicle which is attached to the roof of the skull at about the interocular line (Fig. 3A). The surface of the head does not show any specialization, although the absence of cartilage over the end

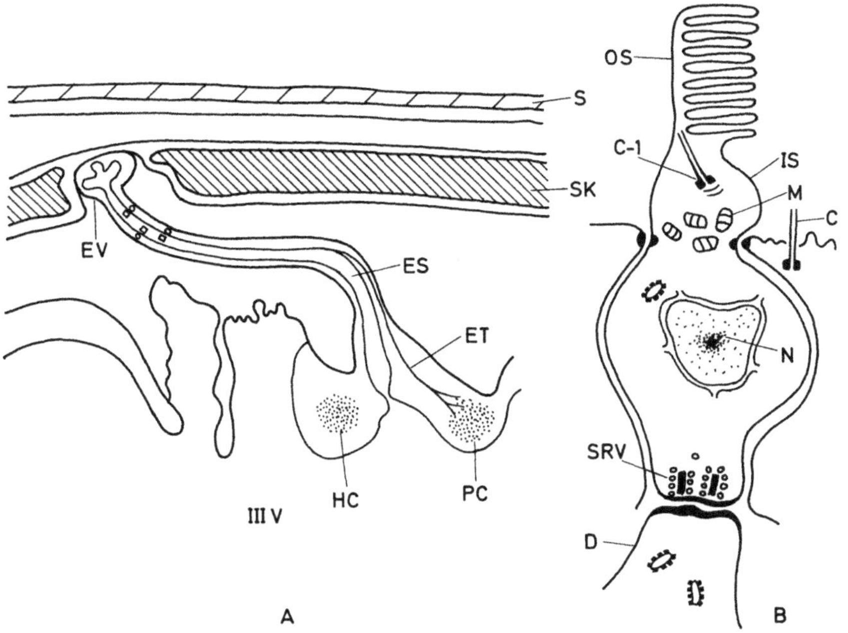

Fig. 3A and B. Pineal system of elasmobranch. (A) Schematic drawing of midsagittal section through brain. (B) Schematic drawing of photoreceptor and supporting cell. Abbreviations as in Fig. 1, and *EV* vesicle of epiphysis; *ES* epiphyseal stalk

vesicle, allows approximately 7x more light to enter this region of the skull (GRUBER et al., 1975).

Three types of cells are found in the epiphysis of *Scyliorhinus canicula* (RÜDEBERG, 1969). The photoreceptors line the lumen and have morphologic features of cones. The outer segments are about 5 µm long and contain up to 150 disks (Fig. 3B). The terminals of the photoreceptors contain many clear and occasional dense-cored vesicles. Synaptic ribbons surrounded by clear vesicles are only occasionally seen.

The ganglion cells are few in number and their axons run in the outer parenchymal wall of the epiphysis. The fibers are all unmyelinated and run to the habenular commissure. The exact termination of the fibers is unknown. The supporting cells extend from the basement membrane to the luminal border where they give rise to many microvilli and occasional cilia. A well-developed Golgi complex and smooth endoplasmic reticulum indicate active biosynthesis by the supporting cells.

2. Physiology

Both slow potential changes and spike activity have been recorded from the epiphysis of *S. canicula* (HAMASAKI and STRECK, 1971). Illumination of the skin overlying the end vesicle of the epiphysis results in a positive-going slow potential and a decrease in spike activity. The degree and duration of the inhibitory process is directly related to the stimulus intensity. The absolute intensity threshold is 4×10^{-4} lm/m^2 and the spectral sensitivity curve peaks at about 500 nm. These values are comparable to those obtained from the lateral eye of *S. canicula*.

3. Biochemistry

We are not aware of any studies dealing with the biochemical activity of the epiphysis of elasmobranchs.

4. Function

The earlier studies dealing with the control of body color of sharks have been summarized by WARING (1963). In a recent study, WILSON and DODD (1973) reported that blinding results in prompt darkening followed by gradual paling in *S. canicula*. The "background response," contraction of the melanophores when the animal is placed over a light background, is abolished in the blind shark. However, if the animal is placed in the dark, further paling of the body occurs indicating that the lateral eyes are not necessary for the paling response. Removal of the epiphysis does not affect the background response but abolishes the paling response. These observations indicate that the epiphysis controls the paling response while the lateral eyes control the background response.

IV. Osteichthyes; Teleostei

1. Morphology

The epiphysis forms the main component of the pineal system in bony fishes. The parapineal organ has been observed in only three out of some twenty species studied (RÜDEBERG, 1971). It is much smaller and has not been examined in detail.

The epiphysis is a relatively large, saccular structure situated on the dorsal surface of the diencephalon and telencephalon (Fig. 4 A). It lies immediately below the meninges and can usually be located by the pigment-free region of the meninges. In many species, the structure of the skull is specialized to allow more light to reach the pineal (BREDER and RASQUIN, 1950). The body of the pineal is connected by a stalk to the roof of the diencephalon between the habenular and the posterior commissures. The pineal lumen is continuous with the third ventricle.

The usual three types of cells are found in the teleost ephiphysis (see RÜDEBERG, 1971, for different species studied). The outer segment of the photoreceptors (Fig. 4 B) protrudes into the lumen and is made up of about 20-30 disks in the eel, *Anguilla anguilla* (RÜDEBERG, 1971). As in the lampreys, the outer segments are fan-shaped. Free ribosomes and a well-developed Golgi apparatus are found in the inner segment. Although synaptic vesicles are numerous in the terminal regions, the synaptic ribbons are few in number.

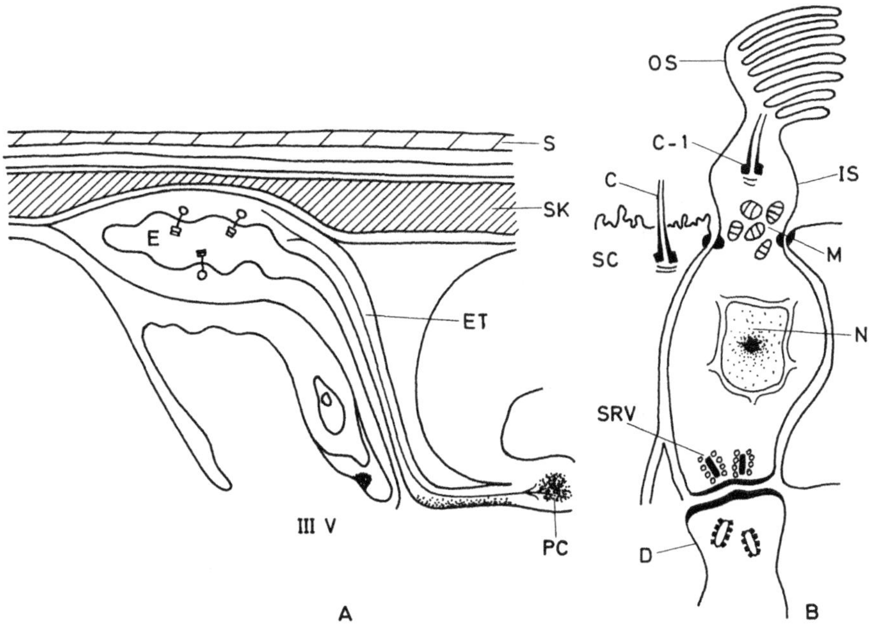

Fig. 4 A and B. Pineal system of teleost. (A) Schematic drawing of midsagittal section through brain. (B) Schematic drawing of photoreceptor and supporting cell. Abbreviations as in Fig. 1

The supporting cells have laminar cytoplasmic projections which form sheaths around receptor cells. The presence of smooth and rough endoplasmic reticulum, a well-developed Golgi apparatus, and free ribosomes indicate intense biosynthetic activity of the supporting cells.

Many ganglion cells are present in the pineal of the goldfish (TAKAHASHI, 1969) but the numbers are relatively few in *A. anguilla* (RÜDEBERG, 1971). The axons of the ganglion cells are unmyelinated and pass to the habenular region.

2. Physiology

As in lampreys and sharks, the ganglion cells are active in the dark and stimulation by light decreases the firing rate (MORITA, 1966b; MORITA and BERGMANN, 1971). In *Salmo irideus*, the threshold stimulus intensity is 3×10^{-5} lm/m^2, and the spectral sensitivity curve peaks between 500 and 533 nm (MORITA, 1966b). There is no Purkinje shift with light adaptation. The majority of the units are achromatic units but few chromatic units are present (excited by long wavelength stimuli).

With steady illumination of the epiphysis, there is a linear relationship between the firing rate of a ganglion cell and the logarithm of luminance of the steady light (MORITA, 1966b). The range of intensities over which this linearity holds is about 5-6 log units.

3. Biochemistry

Relatively high concentrations of HIOMT have been isolated from the pineal (and also retina) of the trout, *S. gairdneri irideus* (Quay, 1965a). *In vivo* uptake of ^{14}C-5HTP (Oguri et al., 1968) and synthesis of 5-HT from injected 5-HTP (Hafeez and Quay, 1969) by the epiphysis have been shown. Microspectrofluormetric studies of the pineal of the pike, *Esox lucius*, have shown the presence of 5-HT (Owman and Rüdeberg, 1970). The existence of the precursors and enzymes necessary to synthesize melatonin in the fish pineal is supported by the isolation of melatonin from the pineal of the salmon, *Oncorhynchus tshawytscha* (Fenwick, 1970a). The immature salmon was found to have 5 to 6 times more melatonin than the spawning salmon.

4. Function

Many species of fish show a diurnal variation in body color which persists in continuous darkness and in blinded animals (von Frisch, 1911; Waring, 1963; Schäfer, 1964). In blinded minnows, *Phoxinus phoxinus*, the response to ambient illumination has the same intensity threshold and spectral sensitivity (Schäfer, 1964) as the neural discharges recorded from the epiphysis of *S. irideus* (Morita, 1966b). Epiphysectomy alters the diurnal rhythm in trouts (Hafeez and Quay, 1970), and melatonin, either injected or placed in the aquarium water, causes body-blanching in trouts (Hafeez, 1970). This effect of melatonin, however, is not observed in all fish (Fain and Hadley, 1966; Reed et al., 1969).

The gonadotrophic effect of the pineal system has not been investigated extensively in fishes. There is, however, some evidence to indicate that the seasonal or cyclical reproductive behavior in fishes may be influenced by the epiphysis. Young guppies, *Lebistes*, fed dessicated bull pineal show a delayed appearance of the secondary sexual characteristics (Krockert, 1936). Epiphysectomy in the same species, on the other hand, leads to a slight acceleration of sexual development of males (Pflugfelder, 1954). These antigonadotrophic effects, however, have not been reported for all species (Schonherr, 1955; Rasquin, 1958; Pang, 1967; Peters, 1968). These contradictory findings may be related to other factors (temperature, age, season, photoperiod) which also influence gonadal development. For example, Fenwick showed that epiphysectomy in the goldfish had no effect on gonadal size unless the operation was performed just prior to the spawning period. Thus the negative findings of Peters (1968) in the same species may have been due to the time of the year that the experiments were conducted.

Intraperitoneal injection of melatonin (50 μg/fish for 50 days) inhibits gonadal growth in the goldfish exposed to long daylengths, but not those goldfish kept on short-day light regime (Fenwick, 1970b). Fenwick thus concluded that melatonin application is only effective at a time when the gonads are normally responsive to melatonin. Because the melatonin-treated fish had larger pituitary gonadotrophs, Fenwick suggested that melatonin acted on the hypothalamic center which controls the release of pituitary gonadotrophins.

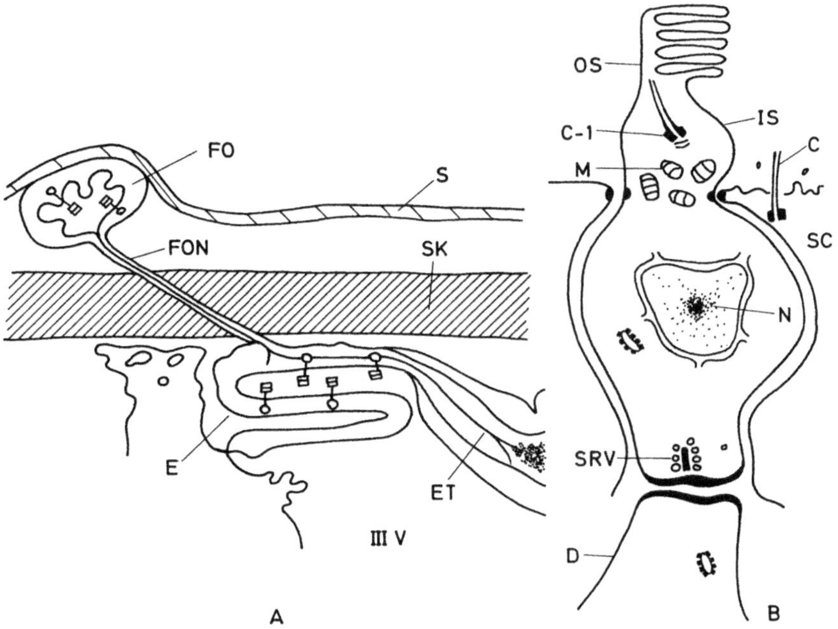

Fig. 5A and B. Pineal system of Anuran amphibian. (A) Schematic drawing of midsagittal section through brain. (B) Schematic drawing of photoreceptor and supporting cell. Abbreviations as in Fig. 1, and *FO* frontal organ; *FON* frontal organ nerve

V. Amphibia

1. Morphology

In urodeles, the tailed amphibians, the pineal system consists of only the intracranial epiphysis. In anurans, the tailless amphibians, the pineal system is made up of the epiphysis and the frontal organ. The frontal organ is homologous with the parapineal organ of lampreys but is situated extracranially. In most frogs, the frontal organ is a small (approx. 0.1 mm), round structure embedded in the skin slightly rostral to the interocular line (Fig. 5A). In *Hyla*, the frontal organ is present only in tadpoles. In *R. temporaria* and *R. esculenta*, the outer segments are 2.5–3.1 μm long (Fig. 5B) and consist of 60–110 disks (OKSCHE and von HARNACK, 1963). The disks are neatly stacked in some of the photoreceptors but in others the disks are fragmented and show signs of degeneration. Macrophages, with engulfed lamellar material, are found in the lumen of the frontal organ (KELLY and SMITH, 1964).

The inner segments of the photoreceptors contain numerous mitochondria, smooth and rough endoplasmic reticulum, and a well-developed Golgi complex (OKSCHE and VON HARNACK, 1963). Synaptic ribbons with clear vesicles aligned around them are found in the terminal region of the photoreceptors. Occasional dense-cored vesicles are also found in the terminal region (OKSCHE and VON HARNACK, 1963).

The supporting cells surround the photoreceptors and tight junctions bind them to the receptors. Cilia and microvilli are present on the luminal surface and the organelles point to active biosynthesis by the supporting cells. Lamellar material, similar to the myeloid bodies in the pigment epithelium of the lateral eyes, is also found in the supporting cells (KELLY and SMITH, 1964).

The ganglion cells are situated on the peripheral margins of the frontal organ and can be identified by their large nuclei and Nissl substance (EAKIN, 1973). The frontal organ nerve emerges from the ventral surface of the organ and passes between the skin and skull for a short distance before entering the cranium. It is made up of 1 to 22 myelinated and 35 to 146 unmeylinated nerve fibers (OKSCHE and VAUPEL-VON HARNACK, 1965). The fibers pass into the epiphysis and join the fibers which form the epiphyseal tract (DODT et al., 1971). After sectioning the frontal organ nerve, degenerating fibers have been found near the subcommisural organ (PAUL, 1972).

The intracranial epiphysis is a flat, saccular structure situated on the dorsal surface of the brain just rostral to the optic tectum. The lumen of the epiphysis is open to the third ventricle. The usual three types of cells are present in the epiphysis (OKSCHE and VAUPEL-VON HARNACK, 1965; KELLY and SMITH, 1964). A fourth type of cell, which may be an interneuron, has been identified in Golgi preparations (see OKSCHE, 1971). Morphologically, the cells in the epiphysis resemble those in the frontal organ.

The epiphyseal tract of *R. esculenta* is made up of 60 to 80 myelinated and many more unmyelinated fibers (OKSCHE and VAUPEL-VON HARNACK, 1965). In degeneration studies, fibers have been traced to the posterior commissure, to the nuclear areas of the periventricular grey, to the vicinity of the nucleus of Darkschewitsch, and to the pretectal region (PAUL et al., 1971).

2. Physiology

Slow potential changes and unitary spike activity have been recorded from the cut end of the frontal organ nerve (DODT and HEERD, 1962; BAUMANN, 1962). Two types of units are found. The achromatic units are inhibited by all wavelength stimuli and are associated with negative-going slow potentials (Fig. 1 A). The spectral sensitivity curve peaks at 575 nm, and the absolute intensity threshold is 3.6×10^{-6} lm/m^2. This threshold level is 4.0–5.0 log units lower than that of the chromatic system. After exposure to strong adapting lights (10^7 times threshold), the achromatic units require 15 min of dark adaptation before the sensitivity returns to the preadaptation level. BLOCH's law is valid for stimulus durations up to 0.1 sec which is comparable to the value for the lateral eyes. The achromatic responses are probably carried by the myelinated fibers (DODT, 1964).

The chromatic system of the frontal organ is made up of two components; an inhibitory component which is most sensitive to short wavelength stimuli (λmax = 355 nm), and an excitatory component which is most sensitive to medium wavelength stimuli (λmax = 515 nm). The inhibitory component is associated with negative-going and the excitatory component with positive-going slow potentials (see Fig. 1 B). The properties of the chromatic system are unusual in two respects. First, the chromatic system does not show significant adaptation, i.e., after exposure to strong adapting lights, the threshold recovers to the pre-adaptation level within 15 sec. And second, the chromatic system can integrate light flux over a period of 128 sec, i.e., BLOCH's law is valid for stimulus durations from 2.5 msec to 128 sec (DODT and HEERD, 1962; DONLEY, 1975).

Recordings have been made from the epiphysis of 12 species of anurans. In all cases, the units are active in the dark and respond to direct stimulation by a decrease in the firing rate (DODT and JACOBSON, 1963; MORITA, 1965). The threshold stimulus intensity is 3.6×10^{-6} lm/m^2 and the spectral sensitivity curve peaks at 500 nm in the dark adapted and at 560 nm in the light adapted preparations of *R. esculenta* (DODT and MORITA, 1964). All of the units are achromatic units except for a minority of units in *R. catesbiana*.

With steady illumination of the epiphysis, there is a linear relationship between the firing rate and the logarithm of the luminance of the steady light (MORITA and DODT, 1965). This linearity holds over a range of 5.0–6.0 log units.

3. Biochemistry

Microspectrophotometry of single photoreceptors from the frontal organ of *R. catesbiana* shows maximal photopigment density between 550–580 nm (HARTWIG and BAUMANN, 1974). These photoreceptors probably correspond to the achromatic units in the frontal organ. Similar studies on the photoreceptors isolated from the epiphysis of *R. temporaria* show maximum density at 502 nm.

There is good evidence that melatonin is synthesized by the epiphysis of amphibians. Serotonin, a precursor of melatonin, is preferentially taken up by the epiphysis of larval *Xenopus* (CHARLTON, 1966), and HIOMT, the enzyme necessary to convert AcHT to melatonin, has been isolated by chromatography from the epiphysis and retina of both urodelean (*Necturus maculosus*) and anuran (*R. pipiens* and *Bufo marinus*) amphibians (QUAY, 1965a; BAKER et al., 1965). In addition, melatonin has also been isolated from the diencephalon of *Xenopus* larvae (VAN DE VEERDONK, 1967).

4. Function

The control of body pigmentary change by the pineal system has been extensively studied in amphibians (see review by BAGNARA and HADLEY, 1970). Briefly, larval amphibians show the paling or body-blanching response, viz., lightening of the body color when the animal is placed in darkness. The threshold luminance necessary to elicit this response from blinded tadpoles (*R. esculenta*) is 2.0 log units higher than the threshold luminance necessary to alter the firing rate of epiphyseal units (BOGENSCHÜTZ, 1965). The difference may be partly due to the attenuation of light flux by the tissue overlying the epiphysis (see DODT, 1973).

As in other lower vertebrate classes, blinding does not affect the amphibian paling response but epiphysectomy abolishes it (BOGENSCHÜTZ, 1965). Removal of the frontal organ of larval or adult frogs (*Hyla*) also has no effect on the paling response but if both the frontal organ and lateral eyes are removed, the paling response is reduced or abolished (EAKIN, 1973). Hypophysectomy does not alter either the degree or time course of the paling response (BAGNARA, 1964).

The time course of the paling response can be accurately mimicked by placing an appropriate concentration of melatonin in the aquarium water (BAGNARA and HADLEY, 1970). Because the effect of melatonin application is not altered by hypophysectomy, BAGNARA (1964) concluded the melatonin was acting directly on the melanophores.

These observations (for additional data see BAGNARA and HADLEY, 1970) support the hypothesis that the paling response is controlled by the epiphysis and that the epiphysis is stimulated by darkness to release melatonin, which then acts directly on the melanophores (BAGNARA, 1960).

The influence of the pineal system on reproduction has not been studied in detail in amphibians. The spermatogenic response to chorionic gonadotrophins in *R. esculenta* is inhibited by bovine pineal homogenates (JUSZKIEWICZ and RAKALSKA, 1963). This antigonadotrophic effect, however, cannot be produced by injections of melatonin, indicating that there may be another active factor in the homogenate (JUSZKIEWICZ and RAKALSKA, 1965).

In addition to the pigmentary and gonadotropic effects, there is good evidence that the pineal system acts as a transducer to link environmental light conditions with certain behavioral patterns in amphibians (see ADLER, 1970). In terrestrial salamanders *(Plethodon glutinosus)* and green frogs *(R. clamitans)*, the circadian locomotor rhythm can be phase-shifted in the laboratory by shifting the light cycle, even in blinded animals. In the green frogs, the entrainment of the locomotor rhythm can be abolished by removal of the frontal organ (ADLER, 1971), while in blinded salamanders, the reaction time to a light stimulus is greatly increased by covering the skull over the forebrain and midbrain with opaque paint (ADLER, 1969).

There is also evidence that the pineal system may be one of the sensory systems used for compass orientation by amphibians (see ADLER, 1970). The ability of toads and salamanders to return to former breeding ponds is well established. While the more obvious visual and olfactory cues have been demonstrated to play an important role, it has been shown that toads with sectioned olfactory tracts can home as well as sham operates (OLDHAM, 1966). It has also been shown that toads return to the site of their breeding pond even after the pond and surrounding regions have been destroyed by building operations (HEUSSER, 1960). The evidence at present indicates that amphibians use various receptor systems to supply orientation information and that the pineal system may be one of these systems. It has been suggested that amphibians use the pineal system to determine the approximate position of the sun which then allows the animal to move at a fixed angle to the sun (see ADLER, 1970).

Another capability of amphibians which has been attributed to the pineal system is the ability to detect the e-vector of linearly polarized light (TAYLOR and ADLER, 1973). Salamanders trained to seek a goal (shore) in a training tank under linearly polarized light, will subsequently orient in a correspondingly appropriate direction in a uniformly circular test tank, where the only obvious cue is the orientation of the e-vector of the polarized light illuminating the experimental chamber. This is true for both sighted and blinded salamanders. Surgical insertion of an opaque plastic shield (but not a clear one) between the skin and the skull abolishes the specificity of the orienting response; orientation becomes random. Because the e-vector of linearly polarized light bears a fixed relationship to the sun's position in nature, it may be used as a cue for spatial orientation by amphibians.

VI. Reptilia

The pineal system reaches a pivotal point in its evolutionary development in the reptiles. The most eyelike structure is the parietal eye of lizards but not all lizards have a parietal eye. Well-developed photoreceptors are also found in the epiphysis of lizards but not of snakes where the chief cell lacks organized outer segments and more closely resembles the secretory pinealocytes of higher vertebrates. In crocodilians, the pineal is reportedly completely absent.

1. Morphology

The parietal eye of lizards develops as a second evagination from the dorsal surface of the diencephalon. It can be noted on the surface of the skull by the translucent scale covering the parietal eye (Fig. 6a), which on transillumination appears as a bright spot (Fig. 6b). The size of the parietal foramen in a 120 cm iguana *(I. iguana)* is approximately 2 mm.

The dorsal surface of the parietal eye is made up of tall columnar cells and is called the lens (Fig. 8A). The retina is everted with the outer segments of the photoreceptors protruding into the lumen (Fig. 6d). If the eye is sectioned in the horizontal plane, i.e. parallel to the top of the skull, photoreceptors are found to be aligned radially in the plane of the section (Fig. 6c). Thus light passing into the eye will pass perpendicular to the axis of these outer segments. The photoreceptors on the ventral portion of the eye are directed vertically and light will pass through them parallel to the axis.

The photoreceptors of the parietal eye are the best developed of all the pineal photoreceptors (COLLIN, 1971; EAKIN, 1973). The length of the outer segment is about 10 µm and contains 250 or more disks in *Sceloporus occidentalis* (EAKIN and WESTFALL, 1959, 1960) while in *Anguis fragilis*, the outer segment is 10–17 µm long and contains from 250 to 435 disks (PETIT, 1968). The disks are formed by an infolding of the plasma membrane (Fig. 7b) and do not show any sign of degeneration as in other pineal photoreceptors (Fig. 7a, 8B). Synaptic ribbons and clear vesicles are found in the terminal regions of the photoreceptors (Fig. 7d). In *Lacerta*, dense-cored vesicles are also found in the terminals (COLLIN, 1971).

The supporting cells surround the receptors and form tight junctions with them on the luminal border. In addition to the usual organelles, the supporting cells contain melanin pigment granules which migrate in response to ambient light levels (EAKIN, 1973).

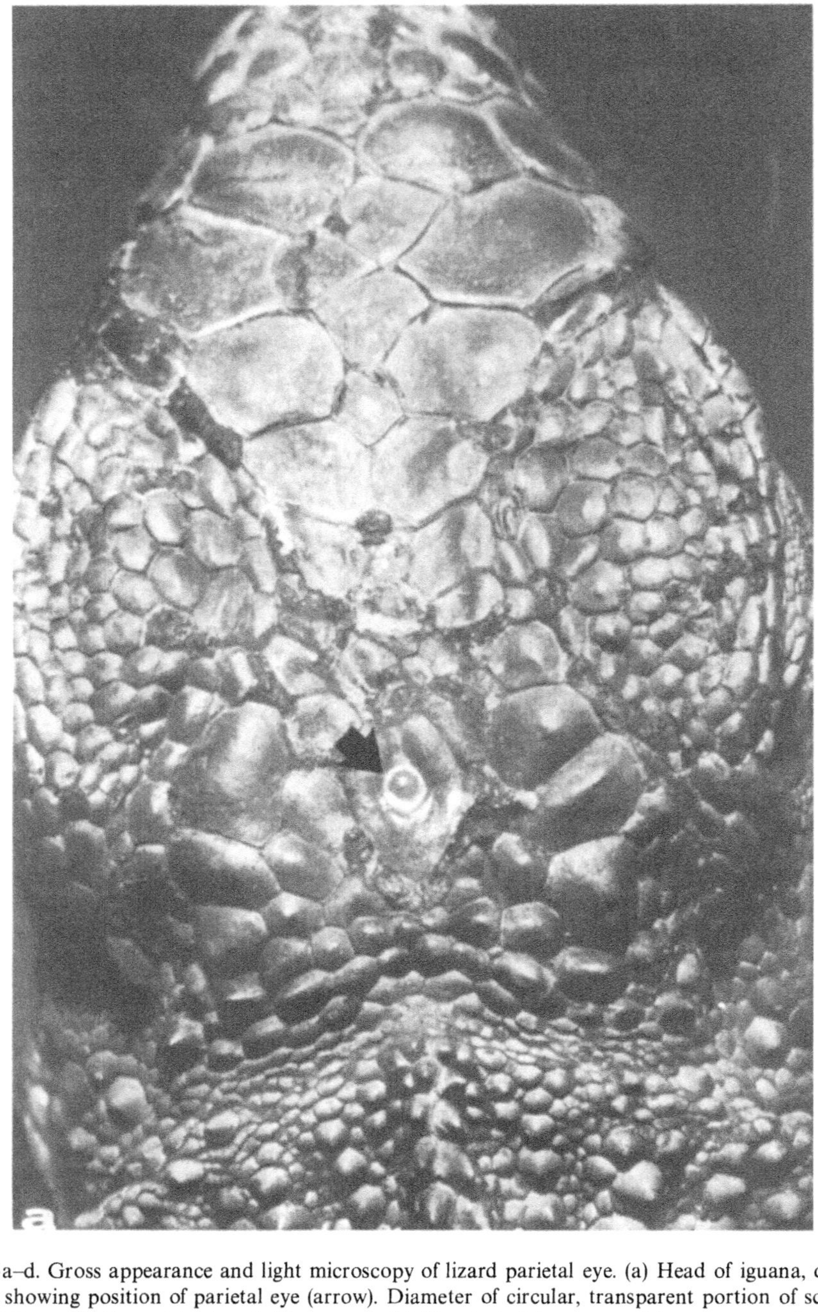

Fig. 6a–d. Gross appearance and light microscopy of lizard parietal eye. (a) Head of iguana, dorsal view, showing position of parietal eye (arrow). Diameter of circular, transparent portion of scale is about 2 mm. (b) Same intact parietal eye as in (a) illuminated by means of laryngoscope placed in animal's mouth to demonstrate eye location in skull foramen. (c and d) Horizontal section of iguana light-adapted parietal eye showing everted retina structure with outer segments (*OS*) protruding into lumen. Abbreviations: *L* cellular lens filling majority of lumen; *P* photoreceptor cell layer; *G* ganglion cell layer; *M* macrophage; *IS* inner segments of photoreceptor cells; *SC* supporting cells containing melanin granules. Diameter of circular eye itself is about 0.4 mm

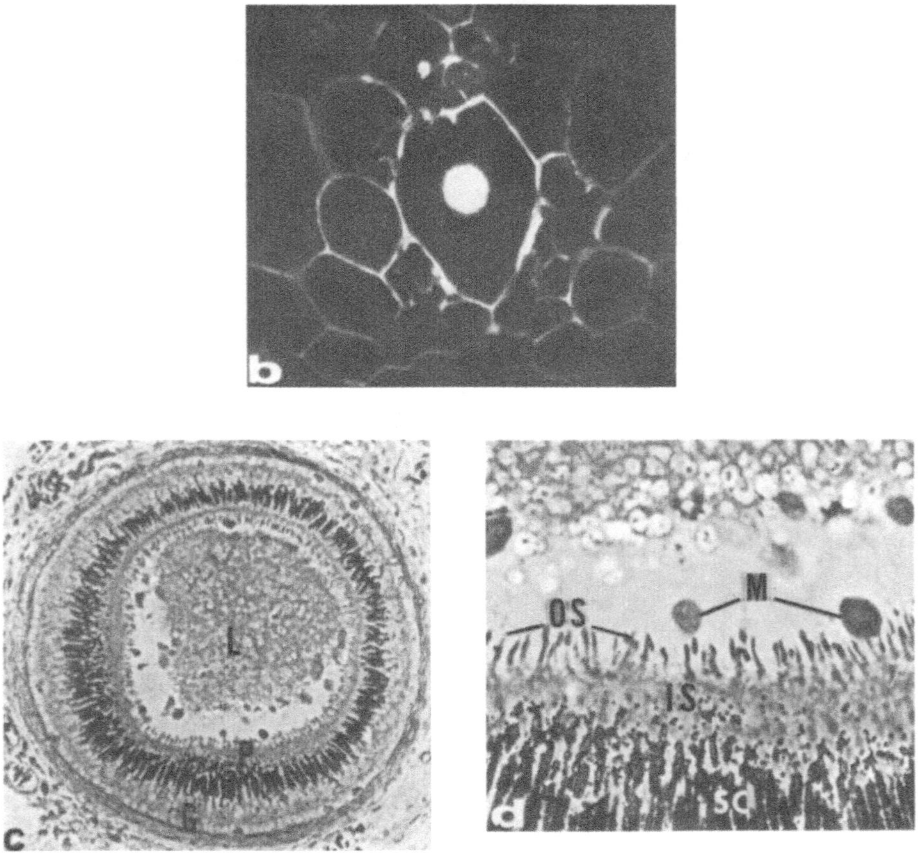

The ganglion cells are characterized by a large nucleus, many mitochondria, and the presence of Nissl substance (EAKIN, 1973). There are approximately 273 ganglion cells (from nerve fiber counts) in the parietal eye of *S. occidentalis* which is approximately 1/10 the number of photoreceptors (EAKIN, 1973).

The parietal nerve is made up almost exclusively of unmyelinated nerve fibers (EAKIN, 1973). It runs for a short distance on the ventral surface of the skull and then passes through the meninges into the epiphysis. It runs along the left side of the epiphysis and becomes part of the epiphyseal tract.

The morphology of the epiphysis varies considerably in the different orders of reptiles (see COLLIN, 1971). In lizards and turtles, the epiphysis is a tall saccular structure extending from the roof of the diencephalon to the surface of the brain. In snakes, it is more solid and parenchymous, and in crocodilians, the epiphysis is reportedly absent.

Differences can also be noted in the fine structure. In lizards, many of the main cells retain characteristics of photoreceptors (Fig. 8B) with well-organized outer segments (COLLIN, 1971; OKSCHE, 1971). However, the outer segments of other photoreceptors within the same epiphysis are vacuolated and disorganized (called rudimentary photoreceptor cells, COLLIN, 1971). In turtles *(Pseudemys scripta*

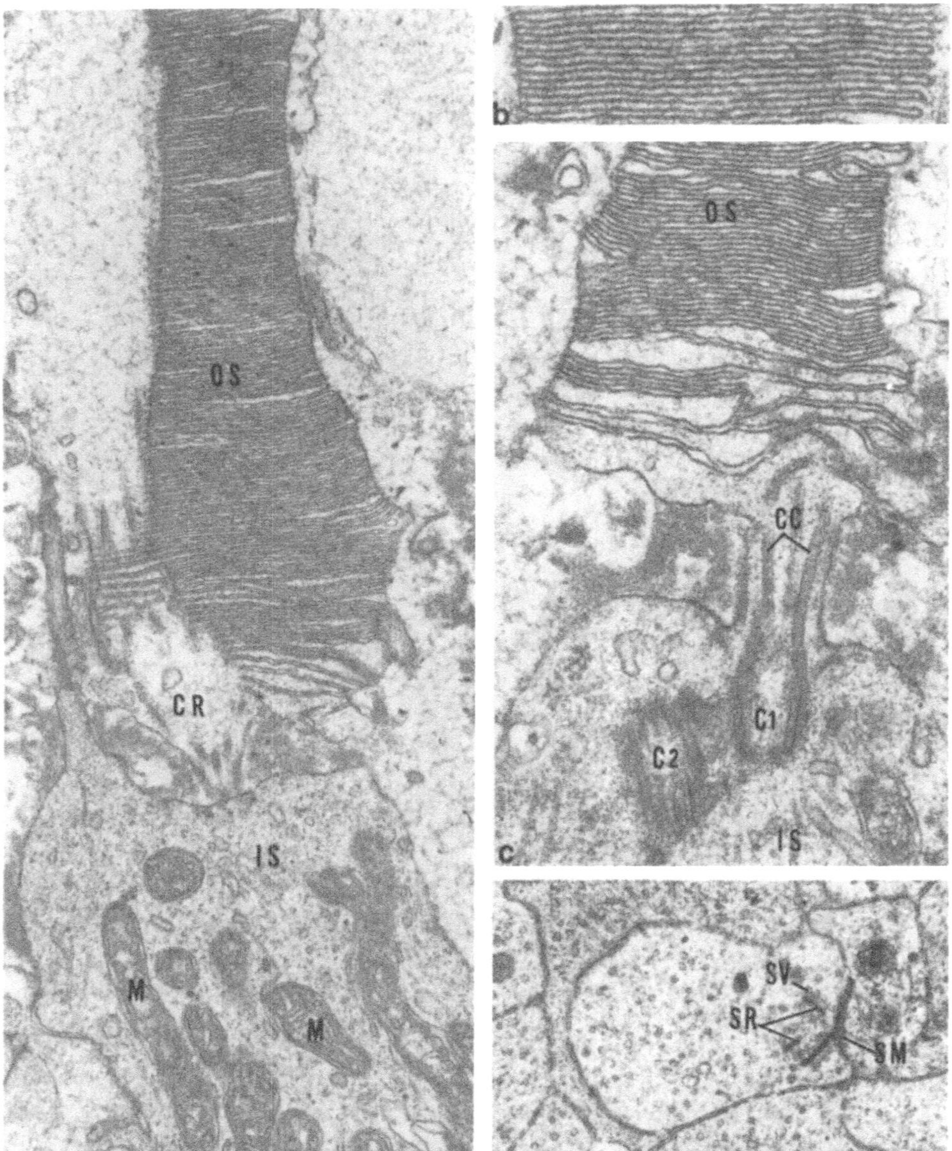

Fig. 7a–d. Electron microscopy of lizard parietal eye photoreceptor structures. *a*, *b*, *c* from *Iguana iguana*, *d* from *Anolis equestris*. Calibration mark = 1 micron. (a) Typical photoreceptor cell structures similar to those of lateral eye. *OS* outer segment; *CR* connecting region; *IS* inner segment; *M* mitochondria. (b) Higher magnification of outer segment material demonstrating conelike lamellar infolding of membrane. (c) Study of connecting region between outer (*OS*) and inner (*IS*) segments of photoreceptor cell at level centrioles (C_1 and C_2) and connecting cilia (*CC*). (d) Synaptic pedicle of photoreceptor cell. *SR* synaptic ribbon; *SV* synaptic vesicles; *SM* specialization of postsynaptic membrane

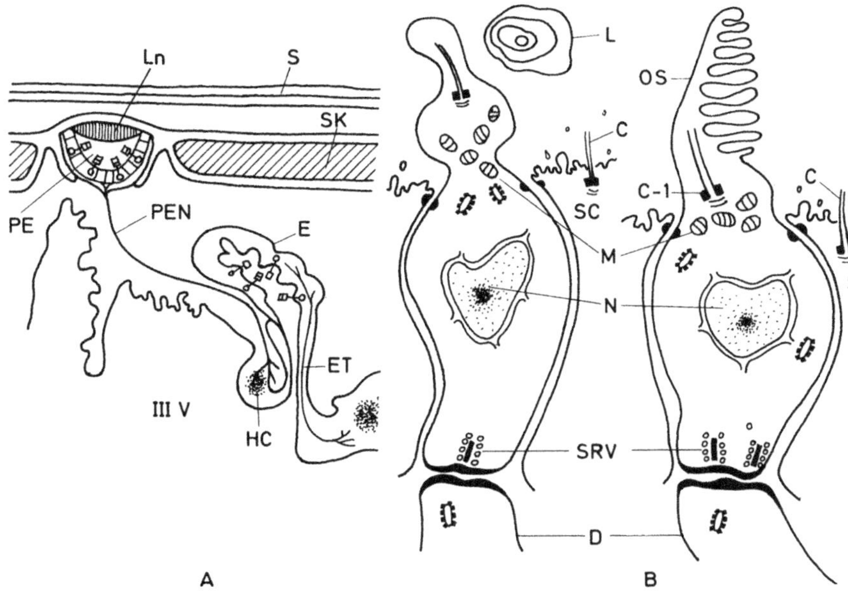

Fig. 8A and B. Pineal system of diurnal lizard. (A) Schematic drawing of midsagittal section through brain. (B) Schematic drawing of photoreceptor from epiphysis (left) and from parietal eye (right). Abbreviations as in Fig. 1, and *L* lamellar material; *Ln* lens; *PE* parietal eye; *PEN* parietal eye nerve

elegans), typical photoreceptors are found only on the posterior wall of the epiphysis, and in other parts, they appear rudimentary (VIVIEN-ROELS, 1970). In snakes, the outer segments of the chief cells are empty sacs and resemble more the mammalian secretory pinealocytes (COLLIN, 1971). These cells, however, retain the same organelles as the photoreceptors including the 9+0 centriolar apparatus (COLLIN, 1971).

In lizards *(L. vivipara)*, the photoreceptors make typical ribbon type synaptic contacts with ganglion cells (COLLIN and KAPPERS, 1971). In turtles, the basal process of the "photoreceptors" terminate on the basement membrane of the epithelium and in these regions clear and dense-cored vesicles are found (VIVIEN-ROELS, 1970). It is suggested that the contents of these vesicles are emptied into the pericapillary spaces. In snakes, the clear and dense-cored vesicles are found in the cytoplasm as well as in the terminal regions (COLLIN, 1971).

The epiphyseal tract in lizards and turtles is made up mainly of unmyelinated fibers which run to the habenular region. Its termination is unknown. In snakes, the tract has been traced to the region of the posterior commissure (QUAY et al., 1968).

Accompanying the transition of the epiphysis from a photosensitive to a secretory organ is an increase in the sympathetic innervation (see KAPPERS, 1971). In lizards, the sympathetic fibers run in the perivascular space and may enter the parenchyma of the epiphysis. In turtles and snakes, the sympathetic fibers are found, not only in the perivascular space, but also invade the parenchymatous tissue to form simple appositional contacts with the chief cells.

2. Physiology

Both slow potentials and unitary spike activity have been recorded from the parietal eye of three species (*Lacerta*, *Iguana*, and *Anolis*) of lizards (MILLER and WOLBARSHT, 1962; DODT and SCHERER, 1968; HAMASAKI, 1968a). All of the units are the chromatic type: they are inhibited by short wavelength (λmax = 460 nm) and excited by medium wavelength (λmax = 520 nm) stimuli. The absolute intensity threshold for the parietal eye of *Lacerta* and *Iguana* is about 20 lm/m^2. The slow potentials are of opposite polarity: the negative-going slow potential is associated with an increase in the firing rate, and the positive-going slow potential with a decrease in the firing rate (DODT and SCHERER, 1968). Unlike the homologous frontal organ of frogs, the physiologic properties of the parietal eye are not unusual. The parietal eye shows significant changes in threshold after light adaptation (DODT and SCHERER, 1968; HAMASAKI, 1968a), and the critical stimulus duration for a constant response is about 0.1 sec as in the lateral eyes (DODT and SCHERER, 1968).

The units in the epiphysis of lizards are active in the dark and are inhibited by light (HAMASAKI and DODT, 1969). The absolute intensity threshold for the exposed epiphysis is approximately 4 lm/m^2 for *Lacerta*. All of the units are achromatic with peak sensitivity at about 600 nm. The spectral sensitivity curve is markedly distorted by the presence of a rich blood supply over the epiphysis. The ERG-determined spectral sensitivity of the lateral eyes is made up ot three components with peaks at 460 nm, 520 nm, and 600 nm (HAMASAKI, 1968b). In the pineal system of lizards, the blue and green systems are found in the parietal eye, while the red system is found in the epiphysis.

There have been no reports of recordings from the epiphysis of snakes or turtles.

3. Biochemistry

CRESCITELLI and EAKIN (see EAKIN, 1973) were not able to extract any photopigment from the parietal eye of *S. occidentalis* which is not unexpected inasmuch as photopigments have not been extracted from the all-cone retinas of the lizard's lateral eyes.

Serotonin and possibly 5-HTP have been identified by microspectrofluorometric analysis of the epiphysis of *L. vivipara* (MEINIEL et al., 1973). The fluorescent material is found in the granules as well as throughout the cytoplasm. HIOMT has been isolated by chromatographic methods from the epiphysis of lizards (QUAY, 1965a) and snakes (QUAY et al., 1968). The presence of these precursors and the enzyme HIOMT indicates that melatonin is probably synthesized by the epiphysis of lizards and snakes.

4. Function

Although lizards show a broad range of dermal color changes, the physiologic basis for these alterations has not been established conclusively (see WALLS, 1942). The body-blanching response is present in normal and blinded lizards, but parietalectomy does not lead to any obvious dermal color change (EAKIN, 1973).

Whether the epiphysis has any influence on the dermal color changes has not been examined.

While it is well established that the photoperiod (and temperature) is important for gonadal recrudesence in lizards (LICHT, 1967), the experiments designed to determine the receptor have given conflicting results. LICHT and PEARSON (1970) showed that parietalectomy does not accelerate testicular recrudesence in *A. carolinensis* as had been reported earlier by CLAUSEN and PORIS (1937). However, in the night lizards, *Xanthusia vigilis*, parietalectomy results in an acceleration of reproductive behavior (STEBBINS, 1970), and in *S. occidentalis*, parietalectomized females had heavier ovaries (by 40%) than the sham operates (STEBBINS and COHEN, 1973). The males, however, did not show any significant differences in testicular weight.

Lizards show circadian locomotor activity cycles which can be entrained by light (EAKIN, 1973). In a field study, STEBBINS and EAKIN (1958) reported that parietalectomized lizards (*S. occidentalis*) are exposed for longer periods in sunlight and are found at greater distances from the release point than are the sham operates. In the night lizard, *X. vigilis*, locomotor activity is higher in animals which had their parietal eye occluded (GLASER, 1958). These observations point to the parietal eye as the photoreceptive organ which synchronizes the activity pattern with the light level. GLASER and also STEBBINS and EAKIN have thus suggested that the parietal eye acted as a dosimeter (measuring light flux) which then influences the locomotor activity of the animal.

Experiments designed to determine what role the epiphysis plays in the other order of reptiles have not been carried out.

VII. Aves

1. Morphology

In birds, the epiphysis is the only component of the pineal system. It is saccular in some birds (e.g., *Passer domesticus*) and more parenchymal in others (e.g., *Gallus gallus*). It extends dorsally from the roof of the diencephalon between the cerebrum and the cerebellum to end immediately beneath the meninges (Fig. 9A). The chief cells or pinealocytes retain some features of photoreceptors but most investigators consider them to be secretory, rather than sensory cells. The outer segments are empty sacs and, if lamellar structures are present (Fig. 9B), they are usually ectopic to the sacs (OKSCHE and KIRSCHSTEIN, 1969; COLLIN, 1971). The distal portion of the inner segment retains the 9+0 centriolar apparatus and is rich in mitochondria. A well-developed Golgi apparatus is also found in the inner segment and many dense-cored vesicles are found in the cytoplasm. The pinealocytes terminate either on the basement membrane or may penetrate the membrane to pass into the pericapillary space (COLLIN, 1971). "Synaptic ribbons" surrounded by clear vesicles are seen in the terminals as well as throughout the cytoplasm (COLLIN, 1971). The cell membranes near the "synaptic ribbons" do not show any specialization, and there is question whether these are true synaptic

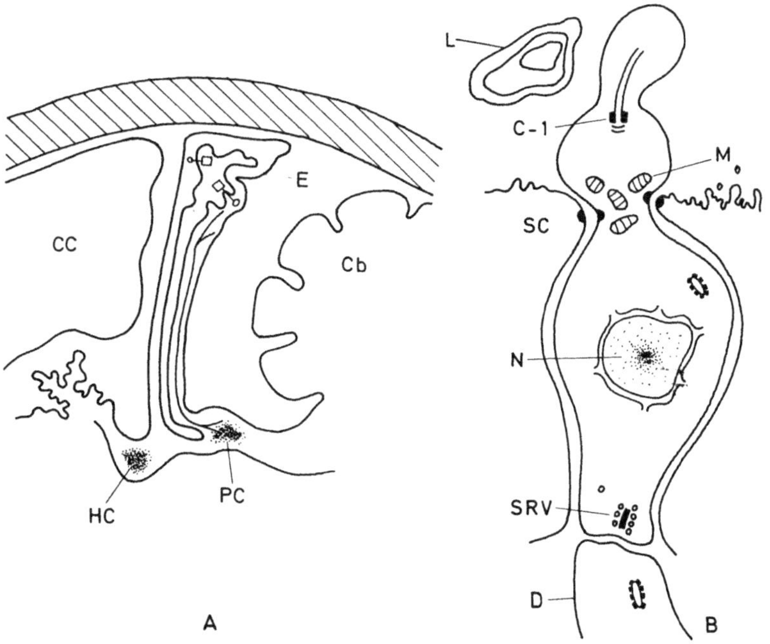

Fig. 9 A and B. Pineal system of bird. (A) Schematic drawing of midsagittal section through brain. (B) Schematic drawing of a chief cell. Abbreviations as in Fig. 1, and *CC* cerebral cortex; *Cm* cerebellum

junctions (COLLINS, 1971). Typical synaptic junctions with pre- and postsynaptic membrane specialization are seen in the white-crowned sparrow *(Z.l. gambelii)* and in the house sparrow *(P. domesticus)* (OKSCHE et al., 1972; OKSCHE and KIRSCHSTEIN, 1969). Ganglion cells have been observed but they are very rare (UECK, 1970). Unmyelinated nerve fibers are found in the stalk of the epiphysis of *Passer* (OKSCHE and KIRSCHSTEIN, 1969). The termination of these fibers is unknown.

The avian epiphysis is well innervated by noradrenergic fibers (KAPPERS, 1971). The fibers run with the blood vessels in the perivascular space and in some species penetrate into the parenchymal region. Bilateral superior cervical sympathetectomy results in complete degeneration of the fibers.

2. Physiology

Electrophysiologic experiments have been conducted on the epiphysis of pigeons (MORITA, 1966) and on sparrows (RALPH and DAWSON, 1968; HAMASAKI, in OKSCHE and KIRSCHSTEIN, 1969). In all cases, responses could not be elicited by direct stimulation or stimulation of the lateral eyes. Very recently, HERBUTÉ and BAYLÉ (1974) reported that in the quail *(Coturnix coturnix japonica)*, light stimulation of the lateral eyes results in a 30–40% reduction in multiunit activity of the epiphysis. Bilateral section of the optic nerves abolished the response.

3. Biochemistry

Melatonin and its synthetic enzymes are highly localized in the avian pineal although the enzyme HIOMT, for instance, has also been reported in some avian retinas (QUAY, 1965a). Generally the pineal synthetic enzyme activities and product concentrations appear higher in birds than in other animal groups but large species variations exist (AXELROD and LAUBER, 1968; RALPH, 1970; BINKLEY et al., 1973). Serotonin, melatonin, NAT, and HIOMT all vary with a diurnal rhythmicity in birds as they do in mammals but the species, age, photic environment, and even photic history apparently exert considerable influence over the existence and phase of the rhythms. Over a normal 24-h period avian melatonin and NAT reach their highest levels in the dark while serotinin reaches its highest level in the light (QUAY, 1966a; HEDLUND and RALPH, 1967; RALPH et al., 1967; BINKLEY et al., 1973); these variations conform to those usually observed in mammals. In contrast, avian HIOMT activity is highest in constant light and lowest in constant darkness; corresponding variations in its activity are found in a normal 24-hrs light-dark cycle (AXELROD et al., 1964; WINGET et al., 1967; LAUBER et al., 1968; ALEXANDER et al., 1970). This is the reverse of that typically seen in mammals (WURTMAN et al., 1964) and continues to be questioned on the basis of the different species and analytical methods used by various investigators (SAYLER and WOLFSON, 1969; PELHAM and RALPH, 1972; BINKLEY et al., 1973). Efforts to relate the rise and fall of product concentrations and enzyme activities systematically to locomotor activity rhythms have met with mixed results. Similary, previous suggestions of a lack of substrate specificity in avian vs. mammalian HIOMT have been questioned on the basis of assay procedure (see PELHAM and RALPH, 1972). The integrated nature of avian pineal biochemical rhythms remains to be settled.

HIOMT differences between light and dark are abolished in the rat by blinding or bilateral superior cervical ganglionectomy (WURTMAN et al., 1964; MOORE et al., 1967), an effect which neither operation accomplishes in chicks (LAUBER et al., 1968). This difference points to less dependent relationship between the avian pineal and its neural innervation despite the fact that sympathetic innervation does reach the avian pineal from the superior cervical ganglion (KAPPERS, 1965; HEDLUND and NALBANDOV, 1969). It also suggests the existence of an extraocular photoreceptive organ activated by light passing through the thin avian skull, which influences the pineal in much the same manner as in the newborn rat (ZWEIG et al., 1966). On the other hand, superior cervical ganglionectomy does abolish the diurnal serotonin rhythm observed in the intact mature quail (HEDLUND and RALPH, 1968), implying perhaps age or species variations in this extraocular photoreceptive ability.

Duck pineal explants in organ culture respond directly to light with increased *in vitro* HIOMT activity and melatonin synthesis (ROSNER et al., 1971), yet direct pineal photosensitivity has not been convincingly demonstrated either by histologic or electrophysiologic methods (for review, see RALPH, 1970; however, cf. HERBUTÉ and BAYLÉ, 1974).

Although much investigative effort has been directed at examining the role of HIOMT as the regulatory enzyme in avian melatonin synthesis, it presently appears that NAT is a more attractive candidate for this function, at least on a daily basis in chickens (BINKLEY et al., 1973). NAT has been observed to increase by a factor of 27, with peak activity occurring at night in phase with pineal melatonin content; HIOMT values were seen to vary by only 20% with an inconsistent phase relation to melatonin content. NAT, therefore, seems more likely to control the melatonin rhythm. If NAT serves to regulate melatonin synthesis in a phasic, daily manner, then perhaps HIOMT reacts to environmental lighting in a more tonic, seasonal manner, thereby providing a seasonally-adjusted diurnally rhythmic melatonin output (BARFUSS and ELLIS, 1971; BINKLEY et al., 1973).

4. Function

A number of functions, most of them relating to reproduction and daily activity rhythms, have been attributed to the avian pineal and to melatonin. RALPH (1970) has reviewed the extensive and confusing literature on whether the avian pineal is in fact pro-or antigonadotrophic. With respect to both pinealectomy and administration of pineal extracts, there is an array of evidence to support both conclusions but more recent evidence supports a progonadotrophic role (SAYLER and WOLFSON, 1969). In any case, the action of melatonin may vary with species, age, sex, and photic environment.

Male testicular maturation and function (FOA, 1912; SHELLABARGER, 1952, 1953; SAYLER and WOLFSON, 1968 a, b), and circadian locomotor activity (GASTON and MENAKER, 1968) can be manipulated by both pinealectomy and changing photoperiods. These functions, therefore, have often been used as assays of pineal activity under various regimes of photoperiodic stimulation. The effects of such manipulations as painting the scalp with radioluminous paint (KATO et al., 1967) or injecting India ink under the scalp (MENAKER, 1968; GWINNER et al., 1971), when assayed by the aformentioned methods, are best attributed to a yet-to-be-identified, extraretinal, nonpineal photoreceptor (see BENOIT, 1964), one element of which may reside in the median eminence of the hypothalamus (STETSON, 1971). Depending on the species, the eyes may or may not participate in mediating these responses (MENAKER et al., 1970; GWINNER et al., 1971; HOMMA and SAKAKIBARA, 1971; HOMMA et al., 1972).

Most convincing is the demonstration that the pineal gland functions, at least in some species, as a pacemaker for circadian rhythms (GASTON and MENAKER, 1968). While not directly photosensitive (MENAKER et al., 1970), it is required for the maintenance of circadian locomotor activity functions once the phasing of the cycles has been entrained.

Biochemically, a strong case for direct pineal regulation of avian seasonal reproductive rhythms has not been made. Yet most observations are consistent with the notion that perhaps through its activity as a neuroendocrine transducer at the diurnal level it mediates the effects of light on the gonads (BARFUSS and ELLIS, 1971). In this manner, it may synchronize the seasonal migratory and reproductive rhythms to conditions of changing daylength, which were observed by ROWAN (1925) so many years ago.

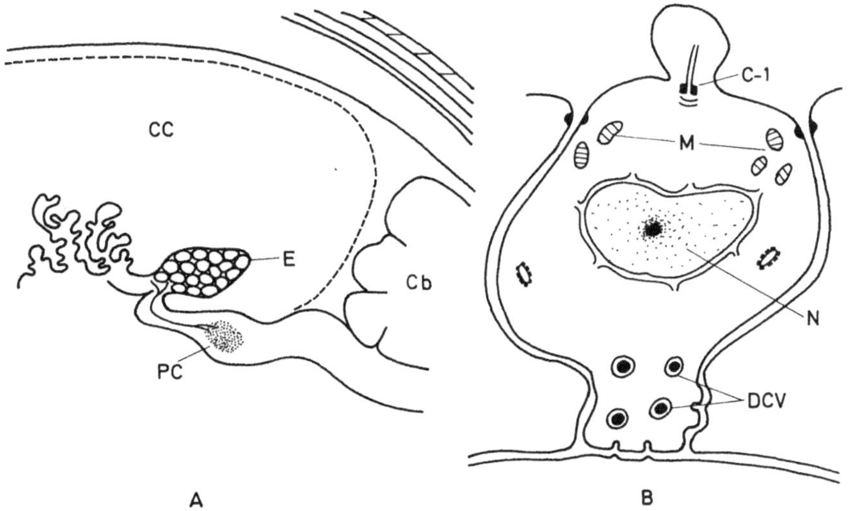

Fig. 10A and b. Pineal system of mammal. (A) Schematic drawing of midsagittal section through brain (B) Schematic drawing of pinealocyte. Abbreviations as in Fig. 1, and *DCV* dense-cored vesicles

VIII. Mammalia

1. Morphology

The epiphysis, the only component of the pineal system in mammals, is situated on the dorsal surface of the diencephalon immediately rostral to the superior colliculus (Fig. 10A). Although it is part of the brain and is connected to the habenular region by a stalk, there is no obvious neural connection with the brain in most mammals (KAPPERS, 1971). The epiphysis, however, does receive rich autonomic innervation via the *nervi conarii (vide infra)*, and the epiphyseal output is exclusively humoral. The ferret, a seasonally breeding carnivore, is an exception since its epiphysis contains a collection of ganglion cells which send axons to the brain through the epiphyseal stalk (HERBERT, 1971).

The evolution of the chief cell from a photosensitive element into a completely secretory structure is culminated in the mammalian epiphysis. Adult pinealocytes (Fig. 10B) contain no evidence of an outer segment or lamellar material although they still retain the 9+0 centriolar apparatus (WOLFE, 1965; COLLIN, 1971). In embryonic material, however, a vestige of an outer segment is seen as a bulbous outpouching from the ciliary region (CLABOUGH, 1970, in rat and hamster; COLLIN, personal communication in mole and hedgehog). The presence of outer segments may be related to the direct photosensitivity of the epiphysis reported in newborn rats (ZWEIG et al., 1966).

The organelles in the pinealocytes are similar to those found in the chief cells of lower vertebrates, viz., abundant mitochondria, smooth and rough endoplasmic reticulum and a prominent Golgi complex (WURTMAN et al., 1968). The processes of the pinealocytes travel for some distances from the cell body and end next to other pinealocytes or near perivascular areas or near interstitial trabeculae (WURTMAN et al., 1968). In some species, a glial basal lamina separates the terminals of the pinealocytes from the vascular compartments while in other species, the basal lamina is absent.

Both clear and dense-cored vesicles are found throughout the cytoplasm of the pinealocytes. Some of these vesicles are accumulated around electron-dense rods which then resemble the true synaptic ribbons of photoreceptors. These structures are found throughout the cytoplasm but when situated near a membrane, the usual differentiation of the pre- and postsynaptic membranes is not seen. Thus there is strong doubt whether these represent synaptic elements. However, there is recent evidence that indicates that these structures are not mere vestiges of an early photoreceptor heritage but are dynamic organelles of the pinealocytes. In guinea pigs on a 12:12 light regime there is a diurnal variation in the numbers of "synaptic ribbons" such that the numbers of ribbons is highest (25x) during the dark phase (VOLLRATH, 1973). It should be noted that the number of ribbons begins to decrease an hour before the lights are turned on and begins to rise 4 h before the lights are turned off. It has not been determined whether this rhythm can be phase-shifted. Animals kept in continuous darkness for 2 months give variable results but animals kept in continuous light for the same period, unexpectedly show significantly fewer "synaptic ribbons" than the controls (VOLLRATH and HUSS, 1973). This discrepancy might be resolved if the long light period damaged the photoreceptors. Unfortunately, the condition of the retinas of the lateral eyes was not examined.

In most animals, the input to the epiphysis is formed by the sympathetic fibers arising from the superior cervical ganglion (KAPPERS, 1971). Section of these fibers results in complete degeneration of the fibers in the epiphysis and abolition of the influence of light on its biochemical activity. In the macaque monkey, preganglionic parasympathetic fibers enter the pineal and synapse with intrapineal nerve cells (KENNY, 1961). Information about the photic environment is transmitted to the epiphysis by a pathway which includes the anterior accessory optic tract, medial forebrain bundle, pre- and postganglionic sympathetic fibers (MOORE et al., 1968).

2. Physiology

Electrophysiologic studies have been carried out on the epiphysis of the rat (TAYLOR and WILSON, 1970; SHAPIRO and SALAS, 1971). Both groups report that the units in the pineal are spontaneously active and are affected by stimulation of the lateral eyes but not by direct illumination. Other observations, however, gave contradictory results: TAYLOR and WILSON report that illumination inhibits spontaneous activity while SHAPIRO and SALAS report excitation; TAYLOR and WILSON present evidence that the pathway to the pineal is via the sympathetic, while SHAPIRO and SALAS report that superior cervical ganglionectomy does not abolish the response.

It is thus apparent that further electrophysiologic studies of the mammalian pineal system are required.

3. Biochemistry

This section reviews only indole amine synthesis and its regulation in the mammalian pineal primarily because more attention has been given to this group of compounds than to any other. There are, however, a number of other substances found in the pineal whose significance remains incompletely disclosed (QUAY, 1974).

The mammalian pineal in particular has been viewed as a neuroendocrine transducer because the biosynthesis of its end product, melatonin, is a rhythmic function under photic control (AXELROD, 1974). Biochemically, the mammalian pineal contains melatonin and all the enzymes necessary for its manufacture from tryptophan. The terminal enzyme in this pathway, HIOMT, has for several years been considered the "melatonin-forming enzyme" and unique to the pineal in mammals (AXELROD et al., 1961). Recently, however, it has been found along with other pineal substances in small concentrations in other tissues such as the retina, Harderian gland, and brain (WURTMAN et al., 1968; CARDINALI and ROSNER, 1971; VLAHKES and WURTMAN, 1972; see discussion in SHEIN, 1971).

The rhythmic nature of pineal melatonin synthesis has been most heavily studied in mammals, especially in the rat. The pineal content of melatonin, serotonin, and AcHT and the enzyme activity of NAT all vary with true circadian rhythms (RALPH et al., 1971; LYNCH, 1971; QUAY, 1963; BROWNSTEIN et al., 1973; KLEIN and WELLER, 1970). These rhythms persist in blinded animals and in intact animals deprived of environmental light cues. Furthermore, the rhythms can be slowly reset in intact animals to a phase-shifted light cycle. They are abolished, however, in sighted animals exposed to constant light (SNYDER et al., 1965; RALPH et al., 1971). In contrast, the rhythm exhibited by mammalian HIOMT appears entirely exogenous in character, that is, its variations in activity are apparently driven entirely by the imposed light cycle. Variations in HIOMT activity disappear when the animal is blinded or deprived of environmental light cues (WURTMAN et al., 1964; AXELROD et al., 1965; however, cf. RALPH et al., 1971; NAGLE et al., 1972).

In rats under a normal 24-hrs light cycle, melatonin, AcHT, NAT, and HIOMT reach their highest levels in darkness, during which time serotonin reaches its lowest. Rhythmic changes in pineal constituents that are truly circadian ordinarily show some anticipation of light or dark onset. However, an unscheduled light onset can rapidly depress NAT activity and raise the serotonin level (ILLNEROVA, 1971; KLEIN and WELLER, 1972). In constant darkness the true circadian rhythms persist with the result that the pineal melatonin content rises and falls in phase with the rat's normally nocturnal locomotor activity cycle. In diurnally active mammals such as monkeys the same pineal rhythms seem to exist with two exceptions. First, in constant darkness the melatonin rise occurs out of phase with the normally diurnal locomotor activity cycle. Second, light stimulation has been related to both elevated and depressed HIOMT activities (QUAY, 1966b; MOORE, 1969). The latter exception is significant insofar as one result was found under stimulation during a normal 24-hrs light cycle and the other under stimulation by constant light.

Part of the problem in evaluating the controlling effect of light on pineal rhythms may lie in the nature of its presentation as a stimulus. The health of the retina should be considered during the assessment of pineal function under various light regimes. Retinal damage can occur during exposure to a continuously lighted environment (NOELL et al., 1966) yet this problem and its possible effect on pineal responses to various light regimes has been considered overtly in only a single pineal study (CARDINALI et al., 1972).

The manner of neural control over pineal biochemical rhythms has been examined more extensively in mammals than in any other group. The lateral eyes provide the only known pathway for photic input to the mammalian pineal (exception: neonatal rats, ZWEIG et al., 1966; MACHADO et al., 1969) both for driving the HIOMT cycle and for phasing the endogenous rhythms in NAT and other substances (CARDINALI et al., 1973; see also AXELROD, 1971; MOORE and KLEIN, 1974). The eye-to-pineal neural pathway, as traced by KAPPERS (1960) and by MOORE et al. (1968) is torturous and passes through the superior cervical ganglia. As presently understood this sympathetic innervation represents the major and perhaps the entire neural pathway to the pineal. Neural inputs to the superior cervical ganglia apparently contain at least two components, one from the retina via the inferior accessory optic tract and another from a biological clock located perhaps in the suprachiasmatic nucleus of the hypothalamus (AXELROD, 1974; MOORE and KLEIN, 1974). Photic information necessary for setting the phase of the biological clock to the environmental light cycle appears to pass from the retina to the suprachiasmatic nucleus along a retinohypothalamic pathway (MOORE and KLEIN, 1974). What sort of neural activity or information passes along these pathways for the purpose of modulating pineal biochemical activity remains an enigma despite several efforts.

A model of neural regulation of mammalian pineal biochemical activity that is consistent with most biochemical evidence is possible. As mentioned previously, control of pineal biochemical rhythmicity and its sensitivity to light is exerted through pineal Beta-adrenergic receptors. Noradrenalin discharge reaches the receptor sites from nerve terminals of sympathetic fibers originating in the superior cervical ganglia (AXELROD, 1974). The amount of noradrenalin present in the sympathetic nerve terminals varies, according to the most recent evidence, in a true circadian manner. This particular finding and the method of noradrenalin assay used to obtain it differ from those of previous studies (WURTMAN et al., 1967; MOORE et al., 1968). Stimulation of the Beta-adrenergic receptors by noradrenalin promotes greatly increased NAT (but not HIOMT) synthesis. This stimulatory effect is blocked by inhibitors of protein synthesis (KLEIN and WELLER, 1970; KLEIN et al., 1970; see discussion in AXELROD, 1974). Interestingly, while noradrenalin itself promotes production of melatonin (SHEIN and WURTMAN, 1969), direct electrical stimulation of the sympathetic ganglia enhances NAT activity but depresses HIOMT activity (BROWNSTEIN and HELLER, 1968; VOLKMAN and HELLER, 1971). The mediating agent between the Beta-receptors and increased NAT activity is presumably cyclic AMP. Its intracellular levels are raised by increased adenyl cyclase activity under the influence of continuous occupation of the Beta-receptors by noradrenalin (DEGUCHI and AXELROD, 1972; DEGUCHI, 1973).

In this scheme NAT rather than HIOMT seems more likely to act as the regulator of circadian rhythms in melatonin synthesis (see discussion in SHEIN, 1971). A circadian rhythm in noradrenalin release drives a similar rhythm in NAT activity. Increased NAT activity increases the amount of AcHT synthesized and, by mass action, the amount of melatonin. This whole process is usually most active at night. Whether NAT or HIOMT acts as the rate controlling agent in melatonin synthesis is conjectural. This is partly because, as noted earlier, the terminal enzyme in the melatonin pathway, HIOMT, does not vary in a truly circadian manner but melatonin itself does. After NAT was found to vary with a circadian rhythm sufficient to explain the observed variations in melatonin, the manner of noradrenalin release from the sympathetic nerve terminals in the pineal underwent re-examination. One presently accepted interpretation puts noradrenalin in control of the NAT rhythm via a cyclic AMP mechanism as mentioned above. The neurotransmitters or neurotransmitter receptors on the postsynaptic membrane responsible for the totally light-dependent HIOMT cycle are presently undetermined. Certainly an intact adrenergic innervation seems required (reviewed by QUAY, 1974) although a cholinergic mechanism may also exist (WURTMAN et al., 1969; see discussions in and following KAPPERS, 1971; WURTMAN, 1971). There is evidence that HIOMT shows additional variations over seasonal and estrous cycles and could thus serve to impose a rhythm of that nature on the rhythm of melatonin synthesis (WURTMAN et al., 1965; see discussion following HERBERT, 1971).

4. Function

In an effort to demonstrate that the mammalian pineal functions as an endocrine organ, many manipulations have instead demonstrated it to be more or less expendable with only transient or marginal effects accompanying its departure. As an innervated structure it has not responded in the classical manner to standard endocrinologic procedures such as the removal-and-replacement-with-extract and the transplantation experiments (see WURTMAN et al., 1968, CHAPTER 5). Furthermore, REITER (1973) has persistently argued that any demonstration of physiologic function for the light-responsive pineal requires consideration of the lighting conditions appropriate for that animal's normal behavior. The arbitrary act of imposing standard laboratory lighting can mask the very functions being sought especially in the burrowing and nocturnal animals such as the hamster and rat that are used for much pineal experimentation. Nevertheless, considerable evidence implicates pineal involvement in mammalian gonadal growth and function and particularly in the timing of reproduction to seasonal cycles (FRANSCHINI et al., 1971; HERBERT, 1971; see REITER, 1973; QUAY, 1974). Possible functions of the pineal, which were mentioned in a previous section, have been reviewed extensively (WURTMAN et al., 1968).

In mechanistic terms, the pineal is thought to exert its influence by one of two alternative possibilities. First, pineal products may interact directly with receptors in the brain, presumably including those located in the hypothalamic neuroendocrine centers. Observed endocrine effects are then secondary results of this primary sphere of influence. The second possibility is that pineal products act directly on individual endocrine organs such as the gonads, thyroid, and adrenal cortex, as well as other organs such as the liver. These two alternatives are not necessarily mutually exclusive.

In the discussion that follows, melatonin remains a central consideration. However, two laboratories have suggested that arginine vasotocin, a polypeptide identified in the mammalian pineal (PAVEL and PETRESCU, 1968) might mediate some of the gonadal effects of pineal activity previously attributed to melatonin (VAUGHAN et al., 1974).

Very few statements regarding pineal function in the mammal can be made with certainty. The mass of experimental evidence does favor certain conclusions, however, particularly with regard to a gonad-regulating function. Curtailing the amount of light to which an adult golden hamster is normally exposed causes severe atrophic gonadal changes (HOFFMAN and REITER, 1965) which can be related to changes in pineal biochemical activity already described. Pinealectomized and superior cervical ganglionectomized animals fail to show this atrophic response (REITER and HESTER, 1965; REITER, 1967). Immature animals also show pineal-related gonadal responses to light inasmuch as pinealectomized prepubertal rats deprived of light mature earlier than similarly handled, intact controls. Finally, anestrous ferrets placed in a stimulatory light regime can be brought into estrus earlier in the year than is normal, an effect which pinealectomy or superior cervical ganglionectomy abolishes. Pinealectomized ferrets still enter estrus at the normal time but subsequent cycles with a longer than usual periodicity (HERBERT, 1971). The results of these studies highlight two difficulties in drawing conclusions about the pineal-gonad relationship. First, the pineal may seem pro- or antigonadotrophic depending on the species, seasons, or other factors and second, the pro- and antigonadal effects may appear opposite in immature and adult animals.

The nonadditive nature of light stimulation and pinealectomy in the immature rat has implied that both manipulations operate by reducing the availability of a single inhibitory substance, namely melatonin (WURTMAN et al., 1961). Subsequent studies with numerous other animals under varying conditions have tended to confirm this idea. Nevertheless, one observation seems particularly relevant here (REITER, 1973): the seasonally changing photoperiod is one but by no means the only cue used to time sexual competence in wild animal populations. Because the pineal operates as an environmental sensor and can be "sensitized" by any number of seemingly anomolous factors (i.e., olfactory bulb transection), it should not be surprising that conflicting results are often obtained under the arbitrary conditions imposed by the laboratory environment.

Pituitary luteinizing hormone (LH) has been suggested to be under pineal influence and therefore a mediator of pineal-gonadal effects. Melatonin administered at the time of the normal proestrus surge in serum LH prevents the rise in LH which ordinarily causes ovulation (REITER and SORRENTINO, 1971). Apparently melatonin works at the level of the pituitary or brain rather than at the level of the ovary. Indeed, melatonin deposited into the median eminence and midbrain reticular formation prevents the accumulation of LH in the pituitary (FRASCHINI et al., 1968). Furthermore, intraventricular but not subcutaneous presentation of melatonin restricts cyclic ovulation in normal rats (COLLU et al., 1971).

Similar experiments in which follicle stimulating hormone (FSH) was measured following intraventricular administration of pineal substances suggests a multichannel, reciprocally operated regulatory mechanism of both FSH and LH release existing at the level of the median eminence (see FRASCHINI et al., 1971). Therefore, despite the notion that melatonin is a true endocrine agent, secreted perhaps into a portal vascular system that carries it from the pineal to its target area (KAPPERS, 1971), its rather direct action in this regard on sensitive reticular formation receptors via the cerebrospinal fluid leaves the question largely open (ANTON-TAY, 1971).

Studies initiated by SASCHIERE et al. (1963) and amplified subsequently by many workers have assigned an antithyrotrophic role to melatonin. Other studies (reviewed by QUAY, 1974) have suggested a pineal role in adrenal cortex regulation. The mechanisms involved remain obscure; indeed the very existence of an adrenocorticotrophic role for the pineal is in serious doubt.

The pineal may exert some influence over salt (particulary K^+) and water balance, especially in the brain (QUAY, 1965b). Such influence could considerably alter cortical neural activity (NIR et al., 1969) and perhaps provide another tool for understanding the mechanisms behind pathologic seizure or provide a physiologic correlate to the observed phenomenon of cortical spreading depression. Despite the fundamental implications, this work has not been seriously pursued. Finally, a large area of conjecture exists between pineal biochemical findings and presumed physiologic effects of that biochemical activity. Quite simply, no mechanism has yet linked in a consistent manner the periodic variations in melatonin synthesis, which are so evident on a daily basis, to the estrous and seasonal cycles in mammalian physiology. It is these longer duration cycles that are presumed to be at least partially under pineal control. Therein remains yet another pineal enigma.

E. Adaptive Radiation of the Pineal System

1. Morphology

The morphologic changes undergone by the pineal system and its chief cell during the course of evolution are shown in Fig. 11. In all vertebrates, the pineal system develops as an evagination of the diencephalon as do the lateral eyes. The intracranial epiphysis is present in all classes of vertebrates while the second component is found intracranially in lampreys and some teleosts and extracranially in frogs and lizards. The pineal photoreceptors resemble those of the lateral eyes in the following ways. First, they are derived from ependymal cells and have the 9+0 ciliary process. Second, the outer segments consist of stacked disks formed by an infolding of the plasma membranes as in the cones. Third, the organelles in the inner segment are similar to those found in the photoreceptors of the lateral eyes. And fourth, synaptic ribbons surrounded by clear vesicles are found in the terminal regions.

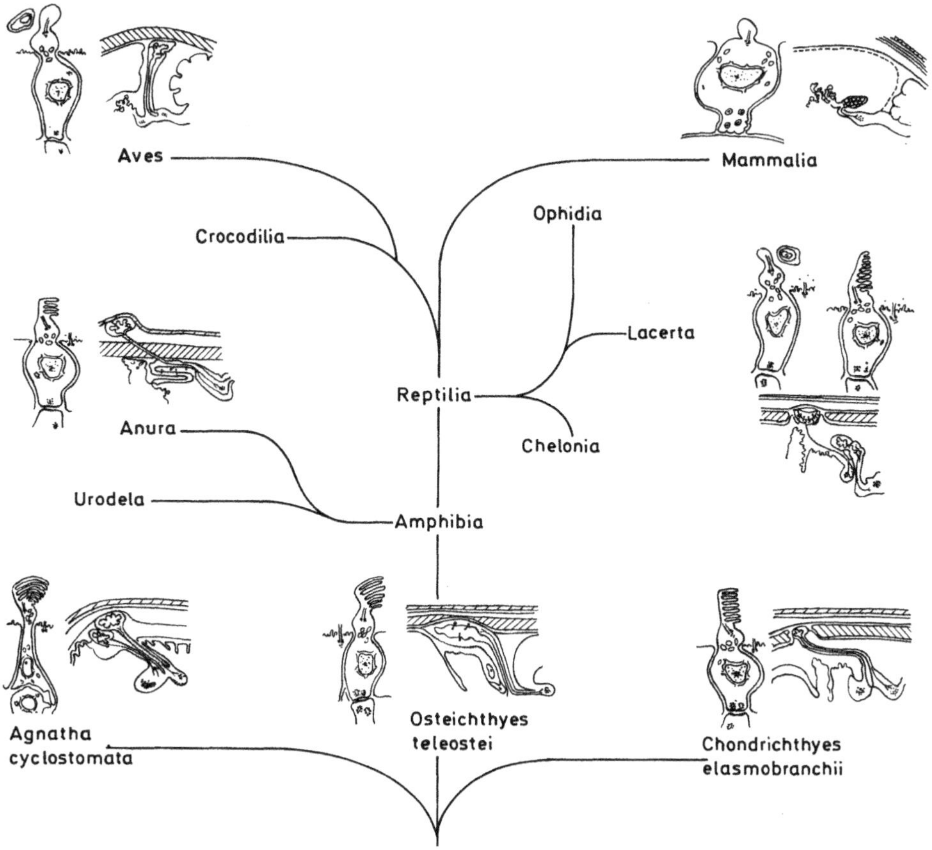

Fig. 11. Schematic representation of morphologic changes undergone by pineal system during course of evolution

The pineal photoreceptors differ from those of the lateral eyes in the following ways. First, the outer segments are shorter (maximum 450 disks) and may contain vesiculations and other signs normally associated with degeneration. Second, dense-cored vesicles are found in the cytoplasm as well as in the terminals. Third, the dendrites of the ganglion cells do not interdigitate deeply into the photoreceptors as they do in the lateral eyes. And fourth, oil droplets are not found in the pineal photoreceptors.

During the course of evolution, there has been a progressive alteration of the chief cell of the epiphysis from a photoreceptor to a secretory cell. Because of the marked difference in the morphologic appearance of the photoreceptors and the pinealocytes, there has been debate whether these are homologous structures. The recent demonstration of the 9+0 centriolar apparatus, the presence of outer segments in embryonic mammalian tissues and the presence of "synaptic ribbons" in avian and mammalian pinealocytes point a common origin of the chief cells of the pineal. The isolation of melatonin and its precursors from the pineal of different classes of vertebrates supports this conclusion.

The ability of the pineal system of lower vertebrates to synthesize melatonin indicates that the secretory property of the pineal has always been present, and it is only the *direct* photosensitivity which is lost in the higher vertebrates. Thus the main difference between the lower and higher classes of vertebrates is whether the pineal system is directly or indirectly photosensitive. This idea, however, does not include the role played by the neural impulses which are modulated by ambient illumination and are carried to the brain by the axons of the ganglion cells. A discussion of the function of this pathway must await the determination of the endings of these fibers.

2. Physiology

The electrophysiologic observations on the epiphysis are remarkably similar in those animals (lampreys, sharks, teleosts, frogs, and lizards) which show direct photosensitivity. All of the units are spontaneously active in the dark, and direct illumination of the epiphysis results in a decrease in the firing rate. With some minor exceptions, all of the units are achromatic units with peak sensitivity at 500 nm in sharks and frogs, 525 nm in lampreys and teleosts, and 600 nm in lizards. These values are comparable to those obtained from the lateral eyes of the same species.

The sensitivity of the epiphysis is also comparable to that of the lateral eyes of the same species (DODT, 1973). This brings up an interesting point regarding the relationship between the morphology and physiology of the epiphyseal retinas. As described, the epiphyseal photoreceptors have morphologic features of cones and have considerably fewer disks (10x fewer than in the frog's rods). In addition, the epiphyseal retinas are two-neuron systems lacking the intermediate neurons (horizontal, bipolar, and amacrine cells). In spite of these morphologic features, the sensitivity of the epiphyseal retinas is as high as that of the lateral eyes.

With steady illumination of the epiphysis, there is a linear relationship between the firing rate and the logarithm of the luminance of the steady light in frogs, lampreys, and teleosts. Whether this relationship also exists in sharks and lizards has not been .examined. This linearity holds over a range of 5.0–6.0 log units. Thus the frequency of nerve impulses passing to the brain is related to the ambient light level. The significance of this will be discussed in the function section.

The physiologic properties of the superficial component of the pineal system are very different from those of the epiphysis. The evolutionary movement of the frontal organ and parietal eye out of the cranial cavity is accompanied by the acquisition of a short wavelength sensitive system. The strong absorption of short wavelength energies by the skin and skull may account for this movement. There is, in addition, another significant change in the physiologic properties. In both superficial components, a color-coded antagonistic system is present. Thus short wavelength stimuli decrease the firing rate while medium wavelength stimuli excite the system. At first hand, this system is reminiscent of the S-potentials of fish and also the opponent color ganglion cells of fish and monkeys although it is probably not spatially arranged as in the receptive fields of color-coded units in fish and monkeys.

The question then arises regarding the significance of this opponent color system. Because the lateral eyes of lizards are well suited for color vision, it is difficult to believe that the parietal eye has evolved to mediate color vision. There is, however, another function to consider which takes advantage of a system which is antagonistically stimulated by short and medium wavelengths lights. This will be considered in the function section.

3. Biochemistry and Function

The biochemical and functional evolution of the pineal are considered together in this section mainly because all functions yet attributed to the pineal are inevitably entwined with the molecule melatonin. We approach the task of evaluating functional evolution with a special caution due in part to the apparent large scale discontinuity existing between lower and higher vertebrates, the latter group consisting of birds and mammals. In this regard one notes immediately that melatonin's function as a body-blanching hormone, which is particularly noticeable in juvenile amphibians, is totally absent in the higher vertebrates. It has been replaced there by widely-studied rhythmic and gonad-related functions, some of which may also be found in lower groups. Accompanying this apparent change in function is a change in structure. The directly photosensitive pineal structures disappear and are replaced by an increasingly well-defined sympathetic innervation which endows the pineal with an indirect photosensitivity dependent upon the lateral eyes. In short, it is sometimes difficult to imagine that the single substance melatonin, which has been found so far in every pineal when sought, is capable of mediating the diverse effects attributed to it throughout the vertebrate classes. Except for the disappearance of the body-blanching reaction in birds and mammals, one then wonders whether the changing pineal role generated by an evolutionary view such as this one is perhaps more artifactual than real. For example, biochemists and behaviorists have been notable in their efforts to explain the link between pineal mechanisms and certain circadian and reproduction-related rhythms, in mammals especially. This same interest, with few exceptions, has not motivated similar studies in the lower vertebrates. Is one justified, therefore, in suggesting that a pineal-related evolutionary trend exists coincident with the trend away from the largely poikilothermic temperature control mechanisms of lower vertebrates and toward the homeothermic mechanisms of higher vertebrates? Namely, does the role of the pineal actually evolve away from one of mediating the body-blanching response and toward one of regulating rhythmic and reproductive responses? Clearly gaps exist (also cf. BAKKER, 1975).

Within this limitation, it is still possible to investigate trends. Biochemically, the molecule melatonin and its method of synthesis appear identical in all animal groups. The enzymes of the pathway have been found to vary in their substrate specificity between classes with the most specific substrate requirements found in mammals (AXELROD and LAUBER, 1968; AXELROD and VESEL, 1970). This view has been challenged on the basis of an imbalance in cofactors required during enzyme analysis (PELHAM and RALPH, 1972). In addition, the "melatonin enzyme", HIOMT, was thought for several years to be uniquely localized in the mammalian pineal even though it had been found in several extrapineal locations in lower vertebrates. This view is changing as more mammalian tissues are examined for HIOMT. Nevertheless, the idea that melatonin formation becomes more and more the exclusive function of pineal tissues as one progresses up the vertebrate evolutionary ladder presently prevails (QUAY, 1974).

Certainly the most obvious effect of pineal activity in the lower vertebrates is body-blanching in a darkened environment. Ordinarily this is a rhythmic process inasmuch as it occurs with day-night regularity. In contrast to the 24-hrs body-blanching rhythms of some invertebrate species, this vertebrate function does not apparently exhibit true circadian components. Without much doubt the pineal and not the lateral eye is the photoreceptive organ for the body-blanching response. In contrast to the color-adaptive responses, which are mediated by the lateral eyes and are apparently protective in nature, the body-blanching response (darkening of the skin in the light, blanching in the dark) may serve to protect the organism from the deleterious effects of light (for a discussion of this topic see PROSSER and BROWN, 1961). Because some evidence implies a reproduction-related function for the pineal in lower vertebrate groups it is hard to ignore the possibility of a dual role for pineal activity. The presence of indications of autonomic innervation and possible associated secretory structures in and around the pineal, which increase in complexity through the fish-amphibian-reptile sequence, can only enhance this possibility (see KAPPERS, 1965, 1971). Which role represents a major and which a minor function is an impossible decision at this time due especially to the limitations outlined at the start of this section. Endogenous biochemical rhythms have not, to our knowledge, been reported in the lower vertebrate groups. Thus, one possible evolutionary trend in function suggests the following: pineal activity in those animals with the least autonomic dependence is more closely constrained and driven by the environmental light cycle than it is in the higher groups, in which pineal autonomic innervation is extensively developed.

Despite this constraint, there seem to be some endogenous, nonbiochemical rhythms that are dependent on the pineal such as circadian locomotor rhythmicity in the salamander (ADLER, 1969). Such a finding is intriguing in light of the well-documented ability of many amphibians to use a "sun compass" in spatial orientation (FERGUSON, 1971). Management of direction finding through solar cues, especially by using the e-vector of linearly polarized sky light, which has a fixed relationship to the sun's position, has been linked to an extraoptic (pineal?) photoreceptor in the salamander (TAYLOR and ADLER, 1973). The successful use of such a cue implies the use of a clock. If the pineal indeed maintains or monitors a true circadian rhythm in these animals and is capable of discriminating polarized light, it would become a most attractive candidate for a practically self-contained organ of orientation in these animals.

Some species of fish also maintain an ability to orient by means of a sun compass (GOODYEAR and FERGUSON, 1969) or polarized light detector (WATERMAN and FORWARD, 1972; FORWARD and WATERMAN, 1973). Experimental evidence does not exclude the pineal but strongly suggests the involvement of the lateral eyes presumably because in these animals light can enter the rod outer segments from directions considerably beyond their normal acceptance angles

(SNYDER, 1973). Moreover, due to the structure of the pineal itself as a photoreceptive organ, it is just such a mechanism that makes that structure an attractive polarized light detector in amphibians (ADLER and TAYLOR, 1973; see also following discussion of reptiles).

Because the members of the reptile class have undergone diverse evolutionary histories with a resulting diversity in form and behavior, it is probably unwise to seek unifying relationships among them by means of the pineal. The extracranial photoreceptive portion of the pineal complex is extraordinarily well developed among a number of reptile groups, especially in the New Zealand tuatara and in several lizard families. The extensive reptilian fossil record suggests a remarkably large, sometimes paired parietal eye structure. The function of this extraordinary photoreceptor in these ancient reptiles can only be inferred from present day species. Most such inferences invoke a dosimetric or orienting role *(vide infra)*.

Almost all nocturnal lizards lack parietal eyes as also do snakes. Observations of this nature have permitted speculation that burrowing (such as that experienced by snakes during their evolution) and nocturnal habits represent lifestyles for which a median eye is ill suited. As such the parietal eye does not contribute to the successful functioning of the organisms; hence its evolutionary demise in these groups (WALLS, 1942; EAKIN, 1973; GUNDY, 1974). Conclusions such as the preceding one must be interpreted cautiously, however, since inconsistencies exist. Positive correlations relating the presence or absence of the parietal eye to burrowing or tree-climbing lifestyles or high lattitude environments seem to hold only in severely limited taxonomic groups.

Turtles also lack the extracranial portion of the pineal and crocodilians, the group which is supposedly most closely related in an evolutionary sense to the birds, reportedly lack the pineal complex altogether. Interestingly, both these groups have largely evolved in the direction of aquatic living. Nevertheless, with respect to the entire reptile class, few solid clues for the wide diversity in pineal appearance have been offered.

Re-emphasis of an important point is offered here. The pineal complex is comprised of two portions, both of which have an evolutionary history of photosensitivity. One portion is the intracranially located epiphysis. Throughout vertebrate evolution this portion has tended to remain ventral to the other portion (except in lampreys), to favor the right side of the midline somewhat in its position and neural relationships, and to associate to a degree with autonomic nerve fibers and closely neighboring secretory structures. Only hints of its direct photosensitivity have so far been discovered in the higher vertebrates; it responds to indirect photostimulation by modifying its secretory activity. The other portion of the pineal is the parietal eye and its homologues (frontal organ, parapineal organ). This portion has tended to move extracranially or at least dorsally (except in lampreys), to favor the left side of the midline slightly in its position and neural relationships, and to appear as a photosensory, not a secretory, organ. This portion is often absent (e.g., sharks, salamanders, geckos, turtles, birds, mammals), but when present may influence the secretory activity of the epiphysis. The two portions differ physiologically in ways which may reflect their respective functional contributions to the pineal complex as a whole. This becomes very apparent in the reptiles.

Among those reptilian species that do possess a parietal eye, its dorsomedial location especially in the diurnal lizards would seem to give these animals an efficient means to monitor exposure to sunlight. The functioning of the reptilian parietal eye as a dosimeter for solar radiation in order to aid thermal regulation and prevent dessication has been proposed before (STEBBINS and EAKIN, 1958; see EAKIN, 1973). This function seems to complement many physiologic properties of the pineal complex, the epiphysis in particular. For one thing, although not directly tested in reptiles, the epiphysis of all other animals examined to date exhibits an inverse, linear relationship between the log of the ambient illumination and neural firing rate. That is, regardless of its previous exposure to adapting illumination, the epiphysis regulates its neural firing rate in a linear fashion to signal the intensity of incident steady light. This holds true over a very wide intensity range and contrasts with the lateral and parietal eye responses, which signal intensity changes over more limited dynamic ranges that depend heavily on the adaptive state of the eye. The presence of a response that is linear with log intensity and independent of adaptation appears to conform well to a proposed dosimetric function. In addition, the reptilian epiphysis responds with greatest sensitivity to long wavelengths in the visible spectrum, a feature that seems of certain benefit to animals that must regularly modify their exposure to the sun in order to thermoregulate properly. In short, the epiphysis has certain physiologic properties that appear to enhance the performance of the pineal complex as a dosimeter of solar radiation.

Ironically, it is the parietal eye that first became the focus of the presumed pineal dosimetric function, primarily on the basis of behavioral evidence. Certain physiologic and structural properties of the parietal eye may permit it to function in this regard as well but one at those properties is not an ability to signal the intensity of ambient light without regard to its adaptive state. However, parietal eye placement above light-absorbing layers of skin and bone does permit it to respond to the potentially life-threatening ultraviolet end of the spectrum, an ability which has been confirmed by electrophysiologic methods. Furthermore, the color discriminating ability of this eye can enable it to monitor the quality of daylight, which is known to vary regularly throughout the day (GATES, 1966). It may be this ability that brings at least a degree of photoperiodic control to these animals' activity rhythms. A measure of this control apparently disappears upon surgical removal of the parietal eye (STEBBINS and EAKINS, 1958; EAKINS, 1973), that is, parietalectomized animals tend to remain active and exposed to daylight longer than do sham-operated controls.

Some of the other presumed pineal functions, such as an antithyrotropic role, have been regarded as secondary effects of this increased daylight exposure and its resulting increase in body temperature and locomotor activity. Because the parietal eye is capable of assessing the regularly changing color quality of daylight, it might well contribute to the phasing of an internal biological clock. Furthermore, the everted structure and radial arrangement of the photoreceptor outer segments may enable it to make use of this clock as an aid in orientation *(vide infra)*. In the earlier behavioral studies, removal of perhaps the orienting instead of the dosimetric ability was what accompanied parietalectomy and thus that may have become the primary (but not necessarily exclusive) cause of the increased activity and exposure. It can be suggested that because the parietalectomized animals might have more difficulty orienting, they spend more time than usual in travel and in accomplishing their daily chores, tend to wander farther from home, and additionally are unable to set the phase of their internal clocks as accurately as their sham-operated counterparts.

It was hinted earlier that light entering photoreceptor cell outer segments from beyond their usual acceptance angles might allow them to respond differentially to polarized light. Photoreceptor outer segments are highly dichroic at right angles to their long axes (DENTON, 1959; see BROWN, 1972). A significant number of parietal eye outer segments protrude into the lumen on a plane that is perpendicular to the incident light; photon absorption will therefore occur from the side in these receptors (Fig. 6C). The centrally projecting ring of photoreceptors could thereby perceive the polarization of the sky, which varies regularly throughout the dawn, daylight, and twilight hours (see TAYLOR and ADLER, 1973). Such information, if evaluated against an internal clock, could provide the organism with a means of directional orientation using this environmental cue. Behavioral experiments with directionally orienting amphibians have already raised this mechanism as a possible one (TAYLOR and ADLER, 1973).

The homeothermic birds and mammals do not maintain the extracranial, photoreceptive portion of the pineal complex. Furthermore, any direct photosensitivity of the intracranial, epiphyseal portion is largely, if not totally, replaced by an indirect photosensitivity which depends on the lateral eyes and is mediated by a well-defined sympathetic innervation. Any such innervation detected in lower vertebrate classes has been described as present only in a much less extensive, perhaps rudimentary form. Neural signals required for driving the pineal metabolic rhythms, which are so clearly evident in the birds and mammals, derive a photic component from the eyes and an endogenously rhythmic component most likely from somewhere in the hypothalamus. The photic component in some way entrains the rhythmic component to maintain a specified phase relationship with the environmental light cycle. Avian and mammalian pineals differ somewhat in their functional dependence on neural information delivered by the sympathetic route, with the avian pineal appearing more aloof in this regard.

Most pineal biochemical rhythms occur with a 24-h periodicity yet the functions to which the pineal was early related are reproductive or seasonal in nature and, furthermore, are largely under photoperiodic control. Despite the fact that the search for mechanisms in this area has produced results that contain disturbing inconsistencies when viewed across the many species studied, certain overall impressions are inescapable.

First, the pineal in higher vertebrates is a neuroendocrine transducer with biochemical properties that vary on a circadian time scale. This is not to imply that the pineal in lower vertebrates is not also a neuroendocrine transducer but simply to indicate that this property exists in a form that is to date more evident and has drawn considerably more attention in the higher than in the lower vertebrates. Evidence for pineal functional participation on a circadian basis, especially in maintaining feeding and locomotor activity patterns, is stronger in birds than in mammals, due largely to the interest of MENAKER and his colleagues (BINKLEY et al., 1972; MENAKER, 1972). Some level of pineal participation in this regard, far from being marginal or transitory, seems lifelong (QUAY, 1972).

Second, even if the pineal is not absolutely required for reproductive activity, it does exert some influence over the timing of reproductive cycles. Seasonal phasing of mating in order to insure subsequent survival of a maximum number of offspring is certainly a function for which a number of environmental cues are available, including photoperiodic ones. Despite the historical connection between the pineal and photoperiodic modulation of reproduction, discrete mechanisms for this effect are obscure. The activity of the enzyme HIOMT does change in such a way as possibly to impose a rhythm on melatonin synthesis that is phased to the estrous cycle in both monestrous and polyestrous mammalian species (QUAY, 1963, 1964; see HERBERT, 1971 and discussion following). In addition, a feedback relationship has been demonstrated between the ovary and the pineal in the rat in which estradiol inhibits the ability of noradrenalin to increase pineal adenyl cyclase activity (WEISS and CRAYTON, 1970). Finally, evidence exists which indicates a close relationship between some circadian and seasonal activities, especially in the birds. The seasonal urge to migrate and the direction of migration are known to be subject to photoperiodic manipulation and are thought to be functions of physiologic state (EMLEN, 1969). The circadian rhythm in seasonal migratory restlessness *(Zugunruhe)* is in many species greatest at night, totally out of phase with circadian activity rhythms normally present throughout the rest of the year. The pineal, known to be required for maintenance of other avian rhythms, is also required for this one (MCMILLAN, 1972). Ample species exist in all vertebrate classes from cyclostomes to mammals in which migration is a highly organized, regular, reproductively related activity. Nevertheless no unified model has yet materialized to explain the influence which the photically modulated circadian rhythms in pineal biochemistry must have over seasonal and estrous cycle physiology and behavior.

In summary, if the exact nature of the cause-effect relationship between the pineal and its target organs is unclear it is because so many experimental variables remain to be defined (WURTMAN, 1971). The evolutionary experiment with directly photosensitive pineals, which persist throughout the largely poikilothermic lower vertebrates, enters a denouement in the homeothermic birds and mammals. Yet in its place another experiment continues, this one with indirectly photosensitive pineals which require the presence of lateral eyes and special nerve pathways in order to maintain contact with the photic environment (CHASE et al., 1969). During the change from direct to indirect photosensitivity a number of rhythmic processes have become more evident. This suggests a far different evolutionary role for the pineal than the vestigial existence to which it seemed doomed for so many years.

F. Acknowledgements

We wish to thank E.B. DAVIS for assistance with the electron micrographs, Mel JOHNSON for photographic assistance, Dianne MORGAN for her drawings and secretarial help, and O. NAVARRO for general assistance. We also thank Drs. A. OKSCHE and J.-P. COLLIN for their generous help.

DJE gratefully acknowledges postdoctoral fellowship support successively from Fight for Sight, Inc. (N.Y.) F-266 and the National Institutes of Health (Washington, D.C.) 1-F22-EY04084-01.

DIH was supported by a grant EY00376-07 from the National Eye Institute, National Institutes of Health, Bethesda, Maryland.

References

Adler, K.: Extraoptic phase shifting of circadian locomotor rhythm in salamanders. Science **164**, 1290—1292 (1969).

Adler, K.: The role of extraoptic photoreceptors in amphibian rhythms and orientation: A review. J. Herpetol. **4**, 99—112 (1970).

Adler, K.: Pineal end organ: role in extraoptic entrainment of circadian locomotor rhythm in frogs. In: Menaker, M. (ed.): Biochronometry. pp. 342—350. Washington, D.C.: Nation. Acad. Sci. (1971).

Adler, K., Taylor, D. H.: Extraocular perception of polarized light by orienting salamanders. J. comp. Physiol. **87**, 203—212 (1973).

Alexander, B., Dowd, A. J., Wolfson, A.: Pineal hydroxyindole-O-methyltransferase (HIOMT) activity in female Japanese quail (Coturnix coturnix japonica). Neuroendocrinology **6**, 236—246 (1970).

Anton-Tay, F.: Pineal-brain relationships. In: Wolstenholme, G. E. W., Knight, J. (eds.): The pineal gland, pp. 213—227. Edinburgh and London: Churchill Livingstone 1971.

Axelrod, J.: Neural control of indoleamine metabolism in the pineal. In: Wolstenholme, G. E. W., Knight, J. (eds.): The pineal gland. pp. 35—52. Edinburgh and London: Churchill Livingstone 1971.

Axelrod, J.: The pineal gland: A neurochemical transducer. Science **184**, 1341—1348 (1974).

Axelrod, J., Lauber, J.: Hydroxyindole-O-methyltransferase in several avian species. Biochem. Pharmacol. **17**, 828—830 (1968).

Axelrod, J., McLean, P. D., Albers, R. W., Weissbach, H.: Regional distribution of methyltransferase enzymes in the nervous system and glandular tissues. In: Kety, S. S., Elkes. J. (eds.): Regional neurochemistry. pp. 307—311. Oxford: Pergamon Press 1961.

Axelrod, J., Vesell, E. S.: Heterogeneity of N- and O-methyltransferases. Mol. Pharmacol. **6**, 78—84 (1970).

Axelrod, J., Wurtman, R. J., Snyder, S. H.: Control of hydroxyindole O-methyltransferase activity in the rat pineal gland by environmental lighting. J. Biol. Chem. **240**, 949—955 (1965).

Axelrod, J., Wurtman, R. J., Winget, C. M.: Melatonin synthesis in the hen pineal gland and its control by light. Nature (Lond). **201**, 1134 (1964).

Bagnara, J. T.: Pineal regulation of the body lightening reaction in amphibian larvae. Science **132**, 1481—1483 (1960).

Bagnara, J. T.: Independent actions of pineal and hypophysis in the regulation of chromatophores of anuran larvae. Gen. comp. Endocrin. **4**, 299—306 (1964).

Bagnara, J. T., Hadley, M. E.: Endocrinology of the amphibian pineal. Am. Zool. **10**, 201—216 (1970).

Baker, P. C., Quay, W. B., Axelrod, J.: Development of hydroxyindole-O-methyltransferase activity in the eye and brain of amphibian, Xenopus laevis. Life Sci. **4**, 1981—1987 (1965).

Bakker, R. T.: Dinosaur renaissance. Sci. Am. **232**, 58—78 (1975).

Barfuss, D. W., Ellis, L. C.: Seasonal cycles in melatonin synthesis by the pineal gland as related to testicular function in the house sparrow (Passer domesticus). Gen. comp. Endocrin. **17**, 183—193 (1971).

Baschieri, L., DeLuca, F., Cramarossa, L., DeMartino, C., Oliverio, A., Negri, M.: Modifications of thyroid activity by melatonin. Experientia **19**, 15—17 (1963).

Baumann, Ch.: Lichtabhängige langsame Potentiale aus dem Stirnorgan des Frosches. Pflüg. Arch. ges. Physiol. **276**, 56—65 (1962).

Benoit, J.: The role of the eye and the hypothalamus in the photostimulation of the gonads in the duck. Ann. N.Y. Acad. Sci. **117**, 204—216 (1964).

Binkley, S., Kluth, E., Menaker, M.: Pineal function in sparrows: circadian rhythms and body temperature. Science **174**, 311—314 (1971).

Binkley, S., Kluth, E., Menaker, M.: Pineal locomotor activity: level and arrythmia in sparrows. J. comp. Physiol. **77**, 163—169 (1972).

Binkley, S., MacBride, S. E., Klein, D. C., Ralph, C. L.: Pineal enzymes: Regulation of avian melatonin synthesis. Science **181**, 273—275 (1973).

Bogenschütz, H.: Extraokulare Steuerung des Farbwechsels bei Kaulquappen. Experientia **21**, 451 (1965).

Breder, C. M., Rasquin, P.: A preliminary report on the role of the pineal organ in the control of pigment cells and light reactions in recent teleost fishes. Science **111**, 10—12 (1950).

Brown, P. K.: Rhodopsin rotates in the visual receptor membrane. Nature New Biol. **236**, 35—38 (1972).

Brownstein, M. J., Heller, A.: Hydroxyindole-O-methyltransferase activity: Effect of sympathetic nerve stimulation. Science **162**, 367—369 (1968).

Brownstein, M., Saavedra, J. M., Axelrod, J.: Control of pineal N-acetylserotonin by a beta adrenergic receptor. Mol. Pharmacol. **9**, 605—611 (1973).

Cardinali, D. P., Larin, F., Wurtman, R. J.: Action spectra for effects of light on hydroxyindole-O-methyltransferase in rat pineal, retina and Harderian gland. Endocrinology **91**, 877—886 (1972).

Cardinali, D. P., Nagle, C. A., Denari, J. H., DePerez Bedes, G., Rosner, J. M.: Lack of effects of light on the rat pineal in organ culture. Gen. comp. Endocrin. **21**, 573—577 (1973).

Charlton, H. M.: The uptake of ^{14}C 5-hydroxytryptamine creatinine sulfate and ^{14}C methylmethionine by the epiphysis of Xenopus laevis Daudin. Comp. Biochem. Physiol. **17**, 777—785 (1966).

Chase, P. A., Seiden, L. S., Moore, R. Y.: Behavioral and neuroendocrine responses to light mediated by separate visual pathways in the rat. Physiol. Behavior **4**, 949—952 (1969).

Clabough, J. W.: Ultrastructural study of pineal cytogenesis in fetal rats and hamsters. Anat. Rec. **166**, 291 (1970).

Clausen, H. J., Poris, E. G.: The effect of light upon sexual activity in the lizard, Anolis carolinensis, with especial reference to the pineal body. Anat. Rec. **69**, 39—53 (1937).

Collin, J.-P.: Differentiation and regression of the cells of the sensory line in the epiphysis cerebri. In: Wolstenholme, G. E. W., Knight, J. (eds.): The pineal gland, pp. 79—125. Edinburgh and London: Churchill Livingstone 1971.

Collin, J.-P., Kappers, J. A.: Synapses of the ribbon type in the pineal organ of Lacerta vivipara (Reptiles, Lacertilians). Experientia **27**, 1456—1457 (1971).

Collu, R., Fraschini, F., Martini, L.: Blockade of ovulation by melatonin. Experientia **27**, 844—845 (1971).

Deguchi, T.: Role of the beta adrenergic receptor in the elevation of adenosine cyclic 3′, 5′-monophosphate and induction of serotonin N-acetyltransferase in rat pineal glands. Mol. Pharmacol. **9**, 184—190 (1973).

Deguchi, T., Axelrod, J.: Control of circadian change of serotonin N-acetyltransferase activity in the pineal organ by the β-adrenergic receptor. Proc. Nat. Acad. Sci. USA **69**, 2547—2550 (1972).

Denton, E. J.: The contributions of the oriented photosensitive and other molecules to the absorption of whole retina. Proc. Roy. Soc. B, **150**, 78—94 (1959).

Dodt, E.: Aktivierung markhaltiger und markloser Fasern im Pinealnerven bei Belichtung des Stirnorgans. Prog. Brain Res. **5**, 201—205 (1964).

Dodt, E.: Differences in data processing upon light stimulation between retina and pineal organ. Nova Acta Leopoldina N.F. **37**, 137—144 (1971).

Dodt, E.: The parietal eye (Pineal and parietal organs) of lower vertebrates. In: Jung, R. (ed.): Handbook of Sensory Physiology. vol. VII/3B. pp. 113—140. Berlin and New York: Springer-Verlag 1973.

Dodt, E., Heerd, E.: Mode of action of pineal nerve fibers in frogs. J. Neurophysiol. **25**, 405—429 (1962).

Dodt, E., Jacobson, M.: Photosensitivity of a localized region of the frog diencephalon. J. Neurophysiol. **26**, 752—758 (1963).

Dodt, E., Morita, Y.: Purkinje-Verschiebung, absolute Schwelle und adaptives Verhalten einzelner Elemente der intrakranialen Anuren-Epiphyse. Vision Res. **4**, 413—421 (1964).

Dodt, E., Scherer, E.: Photic responses from the parietal eye of the lizard, Lacerta sicula campestris (De Betta). Vision Res. **8**, 61—72 (1968).

Dodt, E., Ueck, M., Oksche, A.: Relations of structure and function: The pineal organ of lower vertebrates. From: Proc. J. E. Purkyne Centenary Symposium. Prague (1971).

Donley, C. S.: Color opponent slow potential interactions in the frontal organ of the frog: Rana pipiens. Vision Res. **15**, 245—251 (1975).

Eakin, R. M.: The third eye. Univ. California Press: Berkeley, Calif. 1973.

Eakin, R. M., Westfall, J. A.: Fine structure of the retina in the reptilian third eye. J. Biophys. Biochem. Cytol. **6**, 133—134 (1959).

Eakin, R. M., Westfall, J. A.: Further observations on the fine structure of the parietal eye of lizards. J. Biophys. Biochem. Cytol. **8**, 483—499 (1960).

Eddy, J. M. P., Strahan, R.: The role of the pineal complex in the pigmentary effector system of the lampreys, Mordacia mordax (Richardson) and Geotria australis Gray. Gen. comp. Endocrin. **11**, 528—534 (1968).

Edinger, T.: The size of parietal foramen and organ in reptiles. Bull. Museum comp. Zool., Harvard Coll. **114**, 3—34 (1955).

Edinger, T., Paired pineal organs. In: Kappers, J. A. (ed.): Progress in Neurobiology. Amsterdam: Elsevier 1956.

Emlen, S. T.: Bird migration: Influence of physiological state upon celestial orientation. Science **165**, 716—718 (1969).

Fain, W. B., Hadley, M. E.: In vitro response of melanophores of Fundulus heteroclitus to melatonin, adrenalin and noradrenalin. Am. Zool. **6**, 596 (1966).

Fenwick, J. C.: Demonstration and effect of melatonin in fish. Gen. comp. Endocrin. **14**, 86—97 (1970a).

Fenwick, J. C.: The pineal organ: photoperiod and reproductive cycles in the goldfish, Carassius auratus, L. J. Endocrin. **46**, 101—111 (1970b).

Ferguson, D. E.: The sensory basis of orientation in amphibians. Ann. N.Y. Acad. Sci. **188**, 30—36 (1971).

Fiske, V. M., Bryant, K., Putnam, J.: Effect of light on the weight of the pineal in the rat. Endocrinology **66**, 489—491 (1960).

Foa, C.: Hypertrophie des testicules et de la crête après l'extirpation de la gland pineale chez le coq. Arch. Ital. Biol. **57**, 233—252 (1912).

Forward, R. B., Jr., Waterman, T. H.: Evidence for e-vector and light intensity pattern discrimination by the teleost Dermogenys. J. comp. Physiol. **87**, 189—202 (1973).

Franschini, F., Collu, R., Martini, L.: Mechanisms of inhibitory action of pineal principles on gonadotropin secretion. In: Wolstenholme, G. E. W., Knight, J. (eds.): The pineal gland, pp. 259—278. Edinburgh and London: Churchill Livingstone 1971.

Fraschini, F., Mess, B., Piva, F., Martini, L.: Brain receptors sensitive to indole compounds: Function in control of luteinizing hormone secretion. Science **159**, 1104—1105 (1968).

Gaston, S., Menaker, M.: Pineal function: The biological clock in the sparrow? Science **160**, 1125—1127 (1968).

Gates, D. M.: Spectral distribution of solar radiation at the earth's surface. Science **151**, 523—529 (1966).

Glaser, R.: Increase locomotor activity following shielding of the parietal eye in night lizards. Science **128**, 1577—1578 (1958).

Goodyear, C. P., Ferguson, D. E.: Sun-compass orientation in the mosquitofish, Gambusia affinis. Anim. Behav. **17**, 636—640 (1969).

Gruber, S. H., Hamasaki, D. I., Davis, E. B.: Window to the epiphysis in sharks. Copeia **1975**, 379—381 (1975).

Gundy, G. C.: The evolutionary history and comparative morphology of the pineal complex in Lacertilia. Doct. Diss. Univ. Pittsburgh, Pa. (1974).

Gwinner, E. G., Turek, F. W., Smith, S. D.: Extraocular light perception in photoperiodic responses of the white-crowned sparrow (Zonotrichia leucophrys) and of the golden-crowned sparrow (Z. atricapilla). Z. vergl. Physiol. **75**, 323—331 (1971).

Hafeez, M. A.: Effect of melatonin on body coloration and spontaneous swimming activity in rainbow trout, Salmo gairdneri. Comp. Biochem. Physiol. **36**, 639—656 (1970).

Hafeez, M. A., Quay, W. B.: Histochemical and experimental studies of 5-hydroxytryptamine in pineal organs of teleosts (Salmo gairdneri and Atherinopsis californiensis). Gen. comp. Endocrin. **13**, 211-217 (1969).

Hafeez, M. A., Quay, W. B.: The role of the pineal organ in the control of phototaxis and body coloration in rainbow trout (Salmo gairdneri Richardson). Z. vergl. Physiol. **68**, 402—416 (1970).

Hamasaki, D. I.: Properties of the parietal eye of the green iguana. Vision Res. **8**, 591—599 (1968a).
Hamasaki, D. I.: Spectral sensitivity of the lateral eye of the green iguana. Vision Res. **8**, 1305—1314 (1968b).
Hamasaki, D. I.: Interaction of excitation and inhibition in the stirnorgan of the frog. Vision Res. **10**, 307—316 (1970).
Hamasaki, D. I., Dodt, E.: Light sensitivity of the lizard's epiphysis cerebri. Pflüg. Arch. ges. Physiol. **313**, 19—29 (1969).
Hamasaki, D. I., Streck, P.: Properties of the epiphysis cerebri of the small-spotted dogfish shark, Scyliorhinus caniculus, L. Vision Res. **11**, 189—198 (1971).
Hartwig, H.-G., Baumann, Ch.: Evidence for photosensitive pigments in the pineal complex of the frog. Vision Res. **14**, 597—598 (1974).
Hedlung, L., Nalbandov, A. V.: Innervation of the avian pineal body. Am. Zool. **9**, 1090 (1969).
Hedlund, L., Ralph, C. L.: Daily variation of pineal serotonin in Japanese quail and Sprague-Dawley rats. Am. Zool. **7**, 712 (1967).
Hedlund, L., Ralph, C. L.: Effect of photoperiod phase and ganglionectomy on pineal serotonin in Japanese quail. Am. Zool. **8**, 756 (1968).
Herbert, J.: The role of the pineal gland in the control by light of the reproductive cycle of the ferret. In: Wolstenholme, G. E. W., Knight, J. (eds.): The pineal gland, pp. 303—327. Edinburgh and London: Churchill Livingstone 1971.
Herbuté, S., Baylé, J. D.: Multiple-unit activity in the pineal gland of the Japanese quail: Spontaneous firing and responses to photic stimulations. Neuroendocrinology **15**, 52—64 (1974).
Heusser, H.: Über die Beziehungen der Erdkröte, Bufo bufo L., zu ihrem Laichplatz II. Behav. **16**, 94—109 (1960).
Hoffman, R. A., Reiter, R. J.: Pineal gland: Influence on gonads of male hamsters. Science **148**, 1609—1611 (1965).
Holmgren, N.: Zur Frage der Epiphysen Innervation bei Teleostiern. Folia Neurobiol. **10**, 1—35 (1917—1918).
Homma, K., McFarland, L. Z., Wilson, W. O.: Response of the reproductive organs of the Japanese quail to pinealectomy and melatonin injections. Poultry Sci. **46**, 314—319 (1967).
Homma, K., Sakakibara, Y.: Encephalic photoreceptors and their significance in photoperiodic control of sexual activity in Japanese quail. In: Menaker, M. (ed.): Biochronometry. pp. 333—350. Washington, D. C.: National Acad. Sci. 1971.
Homma, K., Wilson, W. O., Siopes, T. D.: Eyes have a role in photoperiodic control of sexual activity of Coturnix. Science **178**, 421—423 (1972).
Illnerova, H.: Effect of light on the serotonin content of the pineal gland. Life Sci. **10**(I), 955—960 (1971).
Joss, J. M.: Pineal-gonad relationships in the lamprey Lampetra fluviatilis. Gen. comp. Endocrin. **21**, 118—122 (1973a).
Joss, J. M.: The pineal complex, melatonin, and color change in the lamprey Lampetra. Gen. comp. Endocrin. **21**, 188—195 (1973b).
Juszkiewicz, T., Rakalska, Z.: Antioestrogenic effects of bovine pineal glands. Nature (Lond.) **200**, 1329—1330 (1963).
Juszkiewicz, T., Rakalska, Z.: Lack of the effect of melatonin on the frog spermatogenic reaction. J. Pharm. Pharmacol. **17**, 189—190 (1965).
Kappers, J. A.: The development, topographic relations and innervation of the Epiphysis cerebrii in the albino rat. Z. Zellforsch. **52**, 163—215 (1960).
Kappers, J. A.: Preface to Structure and Function of the Epiphysis cerebri. Prog. Brain Res. **10**, ix-xii (1965a).
Kappers, J. A.: Survey of the innervation of the epiphysis cerebri and the accessory pineal organs of vertebrates. Prog. Brain Res. **10**, 87—153 (1965b).
Kappers, J. A.: The pineal organ: An introduction. In: Wolstenholme, G. E. W., Knight, J. (eds.): The pineal gland. pp. 3—34. Edinburgh and London: Churchill Livingstone 1971.
Kappers, J. A., Schadé, J. P., eds.: Structure and function of the epiphysis cerebri. Prog. Brain Res. vol. 10. Amsterdam: Elsevier 1965.
Kato, M., Kato, Y., Oishi, T.: Radioluminous paints as activator of photoreceptor systems studied with swallow-tail butterfly and quail. Proc. Jap. Acad. Sci. **43**, 220—223 (1967).
Kelly, D. E., Smith, S. W.: Fine structure of the pineal organs of the adult frog, Rana pipiens. J. Cell Biol. **22**, 653—674 (1964).

Kenny, G. C. T.: The "nervus conarii" of the monkey (an experimental study). J. Neuropathol. exptl. Neurol. **20**, 563—570 (1961).
Kitay, J. I., Altschule, M. D.: The pineal gland—A review of the physiologic literature. Cambridge, Mass.: Harvard Univ. Press 1954.
Klein, D. C., Bert, C. R., Weller, J.: Melatonin synthesis: Adenosine 3′,5′-monophosphate and norepinephrine stimulate N-acetyltransferase. Science **168**, 979—980 (1970).
Klein, D. C., Weller, J. L.: Indole metabolism in the pineal gland: A circadian rhythm in N-acetyltransferase. Science **169**, 1093—1095 (1970).
Klein, D. C., Weller, J. L.: Rapid light-induced decrease in pineal serotonin N-acetyltransferase activity. Science **177**, 532—533 (1972).
Kopin I. J., Pare, C. M. B., Axelrod, J., Weissbach, H.: The fate of melatonin in animals. J. biol. Chem. **236**, 3072—3075 (1961).
Krockert, G.: Die Wirkung der Verfütterung von Schilddrüsen- und Zirbeldrüsen-Substanz an Lebistes reticulatus (Zahnkarpfen). Zeit. ges. exp. Med. **98**, 214—220 (1936).
Lauber, J. E., Boyd, J. E., Axelrod, J.: Enzymatic synthesis of melatonin in avian pineal body: Extraretinal response to light. Science **161**, 489—490 (1968).
Lerner, A. B., Case, J. D.: Pigment cell regulatory factors. J. invest. Dermatol. **32**, 211—221 (1959).
Lerner, A. B., Case, J. D., Takahashi, Y., Lee, T. H., Mori, W.: Isolation of melatonin, the pineal gland factor that lightens melanocytes. J. Am. Chem. Soc. **80**, 2587 (1958).
Licht, P.: Environmental control of annual testicular cycles in the lizard, Anolis carolinensis. I. Interaction of light and temperature in the initiation of testicular recrudescence. J. exp. Zool. **165**, 505—516 (1967).
Licht, P., Pearson, A. K.: Failure of parietalectomy to affect the testes in the lizard Anolis carolinensis. Copeia **1970**, 172—173 (1970).
Lynch, J. J.: Diurnal oscillations in pineal melatonin content. Life Sci. (I) **10**, 791—795 (1971).
Machado, C. R. S., Machado, A. B. M., Wragg, L. E.: Circadian serotonin rhythm control: Sympathetic and nonsympathetic pathways in rat pineals of different ages. Endocrinology **85**, 846—848 (1969).
McCord, C. P., Allen, F. P.: Evidences associating pineal gland function with alteration in pigmentation. J. exp. Zool. **23**, 207—224 (1917).
McMillan, J. P.: Pinealectomy abolishes the circadian rhythm of migratory restlessness. J. comp. Physiol. **79**, 105—112 (1972).
Meiniel, A.: Étude cytophysiologique de l'orange parapinéal de Lamperta planeri. J. Neuro-Visc. Rel. **32**, 157—199 (1971).
Meiniel, A., Collin J.-P., Hartwig, H. G.: Pinéale et troiséme oeil de Lacerta vivipara (J.), au cours de la vie embryonnaire et postnatale. Z. Zellforsch. **114**, 89—115 (1973).
Menaker, M.: Light perception by extra-retinal receptors in the brain of the sparrow. Proc. Am. Psychol. Assoc. 76th Ann. Conv., pp. 299—300 (1968).
Menaker, M.: The pineal organ and circadian rhythmicity. Gen comp. Endocrin. **18**, 608 (1972).
Menaker, M., Roberts, R., Elliot, J., Underwood, H.: Extraretinal light perception in the sparrow, III: The eyes do not participate in photoperiodic photoreception. Proc. Nat. Acad. Sci. USA **67**, 320—325 (1970).
Miller, W. H., Wolbarsht, M. L.: Neural activity in the parietal eye of a lizard. Science **135**, 316—317 (1962).
Moore, R. Y.: Pineal response to light: Mediation by the accessory optic system in the monkey. Nature (Lond.) **222**, 781—782 (1969).
Moore, R. Y., Heller, A., Bhatnagar, R. K., Wurtman, R. J., Axelrod, J.: Central control of the pineal gland: Visual pathways. Arch. Neurol. **18**, 208—218 (1968).
Moore, R. Y., Heller, A., Wurtman, R. J., Axelrod, J.: Visual pathways mediating pineal response to environmental light. Science **155**, 220—223 (1967).
Moore, R. Y., Klein, D. C.: Visual pathways and the central neural control of a circadian rhythm in pineal serotonin N-acetyltransferase activity. Brain Res. **71**, 17—33 (1974).
Morita, Y.: Extra- und intracelluläre Abteilungen einzelner Elemente des lichtempfindlichen Zwischenhirns anurer Amphibien. Pflüg. Arch. ges. Physiol. **286**, 97—108 (1965).
Morita, Y.: Absence of electrical activity of the pigeon's pineal organ in response to light. Experientia **22**, 402 (1966a).
Morita, Y.: Entladungsmuster pinealer Neurone der Regenbogenforelle (Salmo irideus) bei Belichtung des Zwischenhirns. Pflüg. Arch. ges. Physiol. **289**, 155—167 (1966b).

Morita, Y.: Action spectra of ERG from lampreys, Petromyzon fluviatilis and Lampetra japonica. Physiol. Soc. Japan, April 1973.

Morita, Y., Bergmann, G.: Physiologische Untersuchungen und weitere Bemerkungen zur Struktur des lichtempfindlichen Pinealorgans von Pterophyllum scalare Cuv. et Val. (Cichlidae, Teleostei). Z. Zellforsch. **119**, 289—294 (1971).

Morita, Y., Dodt, E.: Nervous activity of the frog's epiphysis cerebri in relation to illumination. Experientia **21**, 221—222 (1965).

Morita, Y., Dodt, E.: Slow photic responses of the isolated pineal organ of lamprey. Nova Acta Leopoldina. N.F. **38**, 331—339 (1973).

Nagle, C. A., Cardinali, D. P., Rosner, J. M.: Light regulation of rat retinal hydroxyindole-O-methyltransferase (HIOMT) activity. Endocrinology **91**, 423—426 (1972).

Nir, I., Behroozi, K., Assael, M., Ivriani, I., Sulman, F. G.: Changes in the electrical activity of the brain following pinealectomy. Neuroendocrinology **4**, 122—127 (1969).

Noell, W. K., Walker, V. S., Kang, B. S., Herman, S.: Retinal damage by light in rats. Invest. Ophthal. **5**, 450—473 (1966).

Oguri, M., Omura, Y., Hibiya, T.: Uptake of ^{14}C-labelled 5-hydroxytryptophane into the pineal organ of rainbow trout. Bull. Japan Soc. Sci. Fisheries **34**, 687—690 (1968).

Oksche, A.: Survey of the development and comparative morphology of the pineal organ. Prog. Brain Res. **10**, 3—29 (1965).

Oksche, A.: Sensory and glandular elements of the pineal organ. In: Wolstenholme, G. E. W., Knight, J. (eds.): The pineal gland. pp. 127—146. Edinburgh and London: Churchill Livingstone 1971.

Oksche, A., Kirschstein, H.: Elektronenmikroskopische Untersuchungen am Pinealorgan von Passer domesticus. Z. Zellforsch. **102**, 214—241 (1969).

Oksche, A., Kirschstein, H., Kobayashi, H., Farner, D. S.: Electron microscopic and experimental studies of the pineal organ in the white-crowned sparrow, Zonotrichia leucophrys gambelli. Z. Zellforsch. **124**, 247—274 (1972).

Oksche, A., Vaupel-von Harnack, M.: Elektronenmikroskopische Untersuchungen an den Nervenbahnen des Pinealkomplexes von Rana esculenta L. Z. Zellforsch. **68**, 389—426 (1965).

Oksche, A., von Harnack, M.: Elektronenmikroskopische Untersuchungen am Stirnorgan von Anuren (zur Frage der Lichtrezeptoren). Z. Zellforsch. **59**, 239—288 (1963).

Oldham, R. S.: Spring movements in the American toad, Bufo americanus. Cand. J. Zool. **44**, 63—100 (1966).

Owman, C., Rüdeberg, C.: Light, fluorescence, and electron microscope studies on the pineal organ of the pike, Esox lucius L., with special regard to 5-Hydroxytryptamine. Z. Zellforsch. **107**, 522—550 (1970).

Pang, P. K. T.: The effect of pinealectomy on the adult killifish, Fundulus heteroclitus. Am Zool. **7**, 715 (1967).

Paul, E.: Innervation und zentralnervöse Verbindungen des Frontalorgans von Rana temporaria und Rana esculenta. Z. Zellforsch. **128**, 504—511 (1972).

Paul, E., Hartwig, H.-G., Oksche, A.: Neurone und zentralnervöse Verbindungen des Pinealorgans der Anuren. Z. Zellforsch. **112**, 466—493 (1971).

Pavel, S., Petrescu, S.: Inhibition of gondatropin by a highly purified pineal peptide and by synthetic arginine vasotocin. Nature (Lond.) **212**, 1054 (1966).

Pelham, R. W., Ralph, C. L.: Pineal hydroxyindole-O-methyltransferase (HIMOT) in the chicken: Effect of diurnal lighting and substrate concentration. Life Sci. (II) **11**, 51—59 (1972).

Peters, R. E.: Failure to detect an effect of pinealectomy in goldfish. Gen. comp. Endocrin. **10**, 443—445 (1968).

Petit, A.: Ultrastructure, innervation et fonction de l'épiphyse de l'orvet (Anguis fragilis L). Z. Zellforsch. **96**, 437—465 (1969).

Pflugfelder, O.: Wirkungen partieller Zerstörungen der Parietalregion von Lebistes reticulatus. Arch. Entwicklung mech. Organ. **147**, 42—60 (1954).

Prosser, C. L., Brown, F. A., Jr.: Comparative animal physiology. Philadelphia: W. B. Saunders 1961.

Quay, W. B.: Circadian rhythm in pineal serotonin and its modifications by estrous cycle and photoperiod. Gen. comp. Endocrin. **3**, 473—479 (1963).

Quay, W. B.: Circadian and estrous rhythms in pineal melatonin and 5-hydroxyindole-3-acetic acid. Proc. Soc. exptl. Biol. Med. **115**, 710—713 (1964).

Quay, W. B.: Retinal and pineal hydroxyindole-O-methyltransferase activity in vertebrates. Life Sci. **4**, 983—991 (1965a).

Quay, W. B.: Experimental evidence for pineal participation in homeostasis of brain composition. Prog. Brain Res. **10**, 646—653 (1965 b).

Quay, W. B.: Rhythmic and light-induced changes in levels of pineal 5-hydroxyindoles in the pigeon (Columba livia). Gen. comp. Endocrin. **6**, 371—377 (1966 a).

Quay, W. B.: 24-hour rhythms in pineal 5-hydroxyindole-0-methyltransferase activity in the macaque. Proc. Soc. exptl. Biol. Med. **121**, 946—948 (1966 b).

Quay, W. B.: Endocrine effects of the mammalian pineal. Am Zool. **10**, 237—246 (1970).

Quay, W. B.: Pineal homeostatic regulation of shifts in the circadian activity rhythm during maturation and aging. Trans. N.Y. Acad. Sci. **34**, 239—254 (1972).

Quay, W. B.: Pineal chemistry in cellular and physiological mechanisms. Springfield. Ill.: C. C. Thomas 1974.

Quay, W. B., Kappers, J. A., Jongkind, J. F.: Innervation and fluorescence histochemistry of monoamines in the pineal organ of a snake (Natrix natrix). J. Neuro-Visc. Rel. **31**, 11—25 (1968).

Ralph, C. L.: Structure and alleged functions of avian pineals. Am. Zool. **10**, 217—235 (1970).

Ralph, C. L., Dawson, D. C.: Failure of the pineal body of two species of birds (Coturnix coturnix japonica and Passer demesticus) to show electrical responses to illumination. Experientia **24**, 147—148 (1968).

Ralph, C. L., Hedlund, L., Murphy, W. A.: Diurnal cycles of melatonin in bird pineal bodies. Comp. Biochem. Physiol. **22**, 591—599 (1967).

Ralph, C. L., Mull, D., Lynch, H. J., Hedlund, L.: A melatonin rhythm persists in rat pineals in darkness. Endocrinology **89**, 1361—1366 (1971).

Rasquin, P.: Studies in the control of pigment cells and light reactions in recent teleost fishes. Bull. Am. Mus. Natur. Hist. **115**, 1—68 (1958).

Reed, B. L., Flinnin, B. C., Ruffin, N. E.: The effects of melatonin and epinephrine on the melanophores of fresh water teleosts. Life Sci. **8**, 113—120 (1969).

Reiter, R. J.: The effect of pineal grafts, pinealectomy and denervation of the pineal gland on the reproductive organs of male hamsters. Neuroendocrinology **2**, 138—146 (1967).

Reiter, R. J.: Comparative endocrinology of the pineal (A symposium). Am. Zool. **10**, 187—267 (1970).

Reiter, R. J.: Comparative physiology: pineal gland. Ann. Rev. Physiol. **35**, 305—328 (1973).

Reiter, R. J., Hester, R. J.: Interrelationships of the pineal gland, the superior cervical ganglia and the photoperiod in the regulation of the endocrine systems of hamsters. Endocrinology **79**, 1168—1170 (1965).

Reiter, R. J., Sorrentino, S., Jr.: Factors influential in determining the gonad-inhibiting activity of the pineal gland. In: Wolstenholme, G. E. W., Knight, J. (eds.): The pineal gland. pp. 329—344. Edinburgh and London: Churchill Livingstone 1971.

Rosner, J. M., de Perez Bedes, G. D., Cardinali, D. P.: Direct effect of light on duck pineal explants. Life Sci. (11) **10**, 1065—1069 (1971).

Rowan, W. C.: Reaction of light and bird migration and developmental changes. Nature (Lond.) **115**, 494—495 (1925).

Rüdeberg, C.: Light and electron microscopic studies on the pineal organ of the dogfish, Scyliorhinus canicula, L.Z. Zellforsch. **96**, 548—581 (1969).

Rüdeberg, C.: Structure of the pineal organs of Anguilla L. and Lebistes reticulatus Peters (Teleostei). Z. Zellforsch. **122**, 227—243 (1971).

Sayler, A., Wolfson, A.: Influence of the pineal gland on gonadal maturation in the Japanese quail. Endocrinology **83**, 1237—1246 (1968 a).

Sayler, A., Wolfson, A.: Role of the eyes and superior cervical ganglia on the effects of light on the pineal and gonads of the Japanese quail. Arch. Anat. Histol. Embryol. **51**, 615—626 (1968 b).

Sayler, A., Wolfson, A.: Hydroxyindole-O-methyltransferase (HIOMT) activity in the Japanese quail in relation to sexual maturation and light. Neuroendocrinology **5**, 322—332 (1969).

Schäfer, O.: Spektrale Empfindlichkeit und absolute Schwelle des Farbwechsels geblendeter Elritzen (Phoxinus phoxinus L.). Biol. Zbl. **83**, 47—66 (1964).

Schonherr, J.: Über die Abhängigkeit der Instinkthandlungen vom Vorderhirn und Zwischenhirn (Epiphyse) bei Gasterosteus aculeatus L. Zool. Jahrb. Abt. Allgem. Zool. Physiol. **65**, 357—386 (1955).

Shapiro, S., Salas, M.: Effects of age, light and sympathetic innervation on electrical activity of the rat pineal gland. Brain Res. **28**, 47—55 (1971).

Shein, H. M.: Control of melatonin synthesis by noradrenaline in rat pineal organ cultures. In: Wolstenholme, G. E. W., Knight, J. (eds.): The pineal gland, pp. 197—212. Edinburgh and London: Churchill Livingstone 1971.

Shein, H. M., Wurtman, R. J.: Cyclic adenosine monophosphate: Stimulation of melatonin and serotonin synthesis in cultured rat pineals. Science **166**, 519—520 (1969).

Shellabarger, C. J.: Pinealectomy vs. pineal injection in young cockerel. Endocrinology **51**, 152—154 (1952).

Shellabarger, C. J.: Observations of the pineal in the white leghorn capon and cockerel. Poultry Sci. **32**, 189—197 (1953).

Snyder, A. W.: How fish detect polarized light. Invest. Ophthal. **12**, 78—79 (1973).

Snyder, S. H., Zweig, M., Axelrod, J., Fischer, J. E.: Control of the circadian rhythm in serotonin content of the rat pineal gland. Proc. Nat. Acad. Sci. USA **53**, 301—305 (1965).

Stebbins, R. C.: The effect of parietalectomy on testicular activity and exposure to light in the Desert Night Lizard (Xanthusia vigilis). Copeia **1970**, 261—270 (1970).

Stebbins, R. C., Cohen, N. W.: The effect of parietalectomy on the thyroid and gonads in free-living western fence lizards (Sceloporus occidentalis). Copeia **1973**, 662—668 (1973).

Stebbins, R. C., Eakin, R. M.: The role of the "third eye" in reptilian behavior. Am. Mus. Nov. **1870**, 1—40 (1958).

Stetson, M. H.: Neuroendocrine control of photoperiodically induced fat deposition in white-crowned sparrows. J. exper. Zool. **176**, 409—413 (1971).

Steyn, W.: Ultrastructure of pineal eye sensory cells. Nature (Lond.) **183**, 764—765 (1959).

Steyn, W.: Observations on the ultrastructure of the pineal eye. J. Roy. Micr. Soc. **79**, 47—58 (1960).

Studnička, F. K.: Die Parietalorgane. In: Oppel, A. (ed.): Lehrbuch der vergleichenden mikroskopischen Anatomie der Wirbeltiere, part 5. Jena: Gustav Fischer 1905.

Takahashi, H.: Light and electron microscopic studies on the pineal organ of the goldfish, Carassius auratus L. Bull. Fac. Fish., Hokkaido Univ. **20**, 143—158 (1969).

Taylor, A. N., Wilson, R. W.: Electrophysiological evidence for the action of light on the pineal gland in the rat. Experientia **26**, 267—269 (1970).

Taylor, D. H., Adler, K.: Spatial orientation by salamanders using plane-polarized light. Science **181**, 285—287 (1973).

Tilney, F., Warren, L. F.: The morphology and evolutional significance of the pineal body. The American Anatomical Memoirs, No. 9. Philadelphia: The Wistar Institute 1919.

Ueck, M.: Weitere Untersuchungen zur Feinstruktur und Innervation des Pinealorgans von Passer domesticus L. Z. Zellforsch. **105**, 276—302 (1970).

Van de Veerdonk, F. C. G.: Demonstration of melatonin in Amphibia. Curr. Mod. Biol. **1**, 175—177 (1967).

Vaughan, M. K., Vaughan, G. M., Klein, D. C.: Arginine vasotocin: Effects on development of reproductive organs. Science **186**, 938—939 (1974).

Vivien-Roels, B.: Ultrastructure, innervation et fonction de l'épiphyse chez les Chéloniens. Z. Zellforsch. **104**, 429—448 (1970).

Vlahakes, G. J., Wurtman, R. J.: A Mg^{2+} dependent hydroxyindole-O-methyltransferase in rat Harderian gland. Biochem. Biophys. Acta **261**, 194—197 (1972).

Volkman, P. H., Heller, A.: Pineal N-acetyltransferase activity: Effect of sympathetic stimulation. Science **173**, 839—840 (1971).

Vollrath, L.: Synaptic ribbons of a mammalian pineal gland circadian changes. Z. Zellforsch. **145**, 171—183 (1973).

Vollrath, L., Huss, H.: The synaptic ribbons of the guinea-pig pineal gland under normal and experimental conditions. Z. Zellforsch. **139**, 417—429 (1973).

Von Frisch, K.: Beiträge zur Physiologie der Pigmentzellen in der Fischhaut. Pflüg. Arch. ges. Physiol. **138**, 319—387 (1911).

Walls, G. L.: The vertebrate eye and its adaptive radiation. Bloomfield Hills, Mich.: Cranbrook Institute of Science 1942.

Waring, H.: Colour change mechanisms of cold-blooded vertebrates. London and New York: Academic Press 1963.

Wartman, S. A., Branch, B. J., George, R., Tayler, A. N.: Evidence for a cholinergic influence on pineal hydroxyindole-O-methyltransferase activity with changes in environmental lighting. Life Sci. (I) **8**, 1263—1270 (1969).

Waterman, T. H., Forward, R. B., Jr.: Field demonstration of polarotaxis in the fish Zenarchopterus. J. exp. Zool. **180**, 33—54 (1972).

Weiss, B., Crayton, J.: Gonadal hormone as regulators of pineal adenyl cyclase activity. Endocrinology **87**, 527—533 (1970).

Wilson, J. F., Dodd, J. M.: The role of the pineal complex and lateral eyes in the colour change response of the dogfish, Scyliorhinus canicula L. J. Endocrin. **58**, 591—598 (1973).

Winget, C. M., Warren, C. A., DeRoshia, C. W.: Interrelationships of the pineal gland, the diencephalon and the pituitary (Gallus domesticus). Am. Zool. **7**, 732 (1967).

Wolfe, D. E.: The epiphyseal cell: an electron-microscopic study of its intercellular relationships and intracellular morphology in the pineal body of the albino rat. Prog. Brain Res. **10**, 332—376 (1965).

Wolstenholme, G. E. W., Knight, J., eds.: The pineal gland. Edinburgh and London: Churchill Livingstone 1971.

Wurtman, R. J.: Summary of symposium. In: Wolstenholme G. E. W., Knight, J. (eds.): The pineal gland. pp. 379—389. Edinburgh and London: Churchill Livingstone 1971.

Wurtman, R. J., Axelrod, J., Fischer, J. E.: Melatonin synthesis in the pineal gland: Effect of light mediated by the sympathetic nervous system. Science **143**, 1328—1330 (1964).

Wurtman, R. J., Axelrod, J., Kelly, D. E.: The pineal. New York and London: Academic Press 1968.

Wurtman, R. J., Axelrod, J., Phillips, L. S.: Melatonin synthesis in the pineal gland: Control by light. Science **142**, 1071—1073 (1963).

Wurtman, R. J., Axelrod, J., Sedvall, G., Moore, R. Y.: Photic and neural control of the 24-hour norepinephrine rhythm in the rat pineal gland. J. Pharmacol. exptl. Ther. **157**, 487—492 (1967).

Wurtman, R. J., Axelrod, J., Snyder, S. H., Chu, E. W.: Changes in the enzymatic synthesis of melatonin in the pineal during the estrous cycle. Endocrinology **76**, 798—800 (1965).

Wurtman, R. J., Roth, W., Altschule, M. D., Wurtman, J. J.: Interactions of the pineal and exposure to continuous light on organ weights of female rats. Acta Endocrin. **36**, 617—624 (1961).

Young, J. Z.: The photoreceptors of lampreys. II. The function of the pineal complex. J. exp. Biol. **12**, 254—270 (1935).

Zweig, M., Snyder, S. H., Axelrod, J.: Evidence for a nonretinal pathway of light in the pineal gland of newborn rats. Proc. Nat. Acad. Sci. USA **56**, 515—520 (1966).

Chapter 10

The Avian Eye and its Adaptations

DAVID B. MEYER, Detroit, Michigan, USA

With 22 Figures

Contents

L'Oiseau c'est un aile guidée par un oeil, ce que exige la précision et la vitesse des fonctions rétiniennes ...
ROCHON-DUVIGNEAUD (1943).

A. Introduction

The avian eye is considered to be "supreme amongst all living creatures" and capable of attaining "an order of excellence unmatched in any other species not excepting man" (DUKE-ELDER, 1958). Many structural and functional adaptations have become involved in the perfection of visual processes in birds which, according to PUMPHREY (1948a) represent the culmination of phylogenetic development toward diurnal vision. This chapter is concerned with several of these adaptations, the most obvious of which are the oil droplet inclusions and glycogen deposits within their photoreceptors, the well-developed areae and foveae within their retina and the highly vascularized pecten within their vitreous body. A detailed description of the avian eye is not intended; excellent sources are available on this subject: WOOD (1914), SLONAKER (1918), FRANZ (1934), WALLS (1942), ROCHON-DUVIGNEAUD (1943), POLYAK (1957), DUKE-ELDER (1958), and HODGES (1974).

I. Characteristics of the Avian Eye

In general, the avian eye (Fig. 1) combines a remarkably large size with a superb accommodative mechanism and a relatively thick, neatly stratified retina devoid of blood vessels and capable of high visual acuity, color discrimination, and movement detection. Characteristic features of such a retina include a large retinal image, a fascinating array of visual cells (rods, single, and double cones), an efferent centrifugal system, elaborate and diverse areas for acute vision (with one or two foveae), and an additional nutritive source, the pecten (Fig 1). Pure rod retinae found in some nocturnal vertebrates do not exist in birds, but another feature, a tapetum lucidum, retinal in nature, has been recently described in several goatsuckers *(Caprimulgidae)* (NICOL and ARNOTT, 1974).

The large size (absolute, not relative) of the avian eye is indicative of the importance of vision to birds. It permits the formation of a large retinal image, a desirable feature for improving visual acuity. The two eyes often weigh more than that of the brain, leaving little room in the head for anything else (Fig. 2). The ostrich eye, for example, is 50 mm in diameter, the largest of any land vertebrate (WALLS, 1942) and the eyes of the tawny owl constitute one-third of the entire weight of the head (LEUCKART, 1876). In general, nocturnal predators possess the largest eyes, water and swamp birds the smallest. WALLS (1942) classifies the avian eye into three groups based upon their morphology: flat, globose, and tubular (Fig. 3). The flat shape (Fig. 3a) is characteristic of most birds; it possesses a short anteroposterior diameter. The globose eye (Fig. 3b) is found in diurnal birds which require high resolution for great distances, e.g., crow. In tubular eyes (Fig. 3c) the ciliary region represents a broad concave zone created and supported

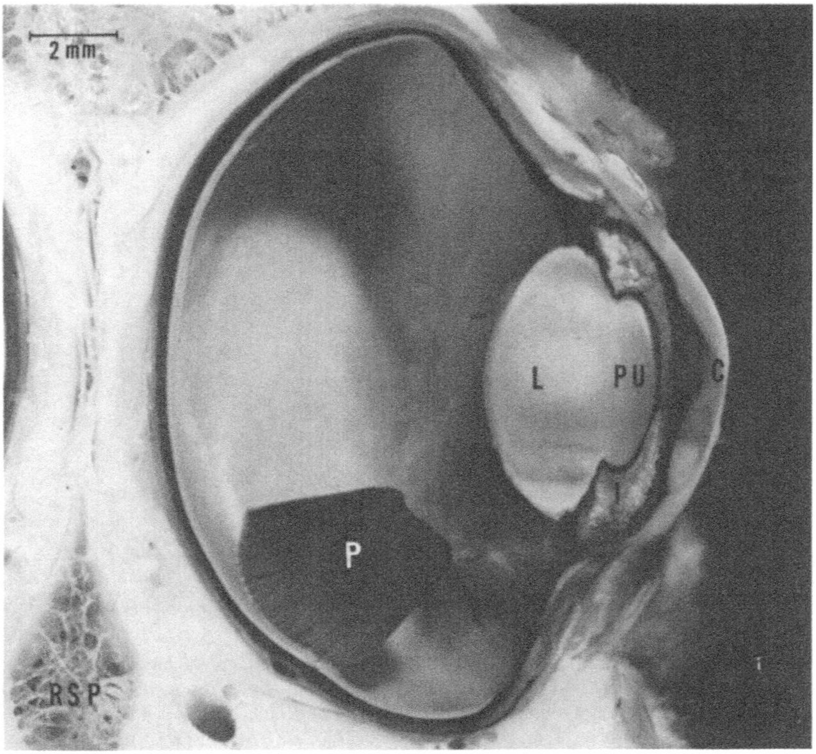

Fig. 1. Transverse section of eye in pigeon *(Columba livia)* prepared by WINGSTRAND and MUNK (1965). Pecten (*P*) lies in rostro-ventral portion of bulbus and projects toward lens (*L*). *PU*, pupil. *C*, cornea. *I*, iris. *RSP*, rostrum sphenoidalum. This figure is reproduced through courtesy of Drs. K. G. WINGSTRAND and O. MUNK

by a well-developed ring of bones, the scleral ossicles, rendering the anteroposterior axis often longer than the other diameters. Characteristic of most owls and some eagles, this phylogenetic increase in axial length has been implicated as an adaptation toward both accommodation and greater visual acuity, the increased "throw" of the image being the desired effect. The avian retina lies almost entirely in the image plane permitting a sharp focus of all distant objects within the visual angle, and the elaboration of a so-called panoramic field of view.

B. Accommodative Adaptations

The power and range of accommodation in birds is probably greater than in any other vertebrate group. PUMPHREY (1948a) reports that most birds have a range of 20 diopters, twice the maximum accommodation of a 20-year-old man. The greatest range is exhibited by aquatic birds, particularly the cormorant

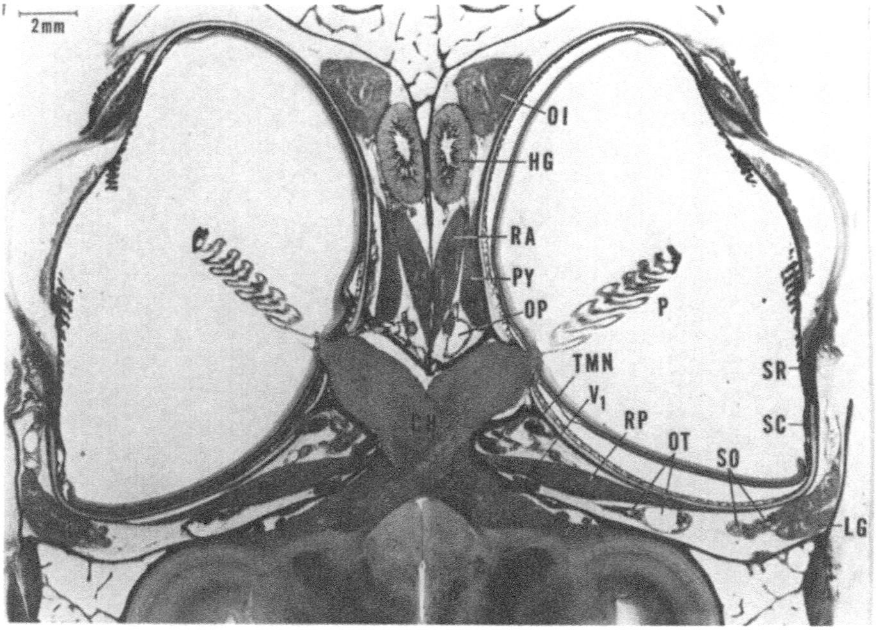

Fig. 2. Light photomicrograph of transverse section through the head of pigeon (*Columba livia*) at level of optic chiasma (*CH*) and attachment of pecten (*P*) to linear optic disc (WINGSTRAND and MUNK, 1965). Other structures include Harderian (*HG*) and lacrimal (*LG*) glands, inferior oblique (*OI*), pyriformis (*PY*), anterior rectus (*RA*), posterior rectus (*RP*) muscles, ophthalmic vein (*OP*), ophthalmotemporal (*OT*) and supraorbital (*SO*) arteries and veins, scleral ossicle (*SR*), tendon of nictitating membrane (*TMN*) and deep trigeminal ramus (V_1). This figure is reproduced through courtesy of Drs. K. G. WINGSTRAND and O. MUNK

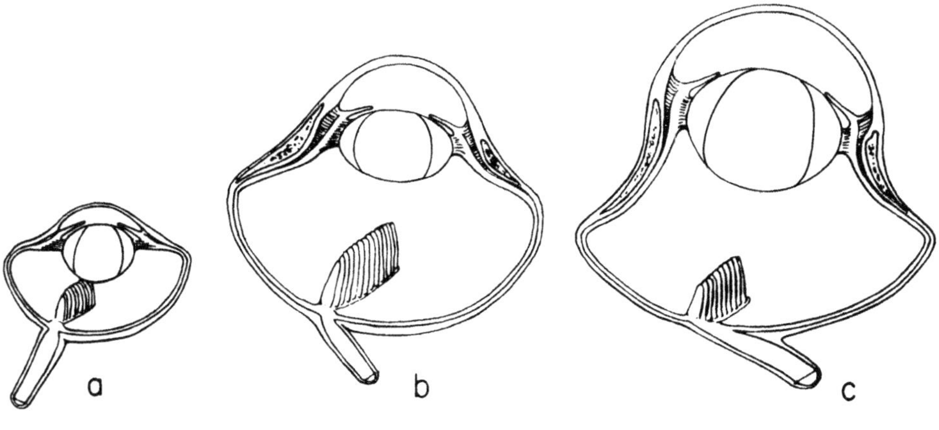

Fig. 3a–c. Three morphologic appearances of avian eye as revealed by the ventral half of left eyeball (After SOEMMERRING, 1818). (*a*) flat. (*b*) globose. (*c*) tubular

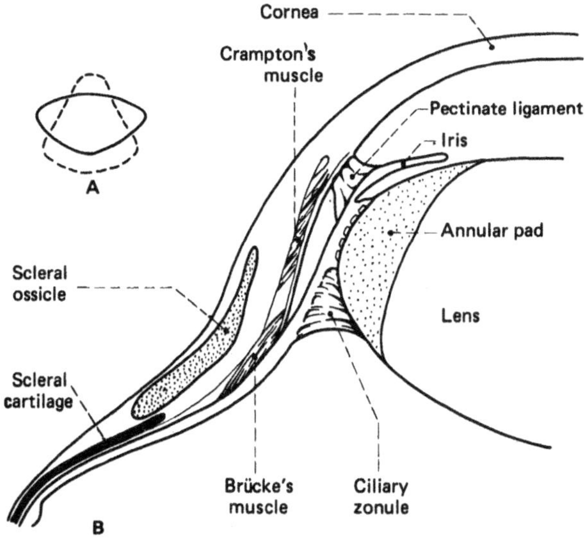

Fig. 4a and b. Avian accommodation. (a) Lens of cormorant at rest (solid line) and fully accommodated (dotted line) for near vision. Anterior lenticonus is evident. (From PUMPHREY, 1961, after FRANZ and HESS.) (b) Accommodative apparatus in hawk. (Modified from WALLS, 1942)

(HESS, 1909–1910) and the dipper (GOODGE, 1960), both of which have a potential accommodation of almost 50 diopters. Nocturnal birds accommodate the least (2–4 diopters), whereas chickens and pigeons have a range of 8–12 diopters (HESS, 1909–1910). Avian accommodation is always for near vision and incorporates a variety of mechanisms which are not the same in all species of birds. Several anatomical and physiological adaptations, unique to birds (and reptiles, except snakes), are believed to be responsible for them, some involving a change in the curvature of the cornea, others a deformation of the lens. Retinal movement, particularly in the region of the fovea, has also been suggested as a supplemental mechanism (WALLS, 1942). Such movement which resembles the fine adjustment of a microscope is implied from the presence of muscular tissue (smooth or striated) in the fundal part of the choroid. Found exclusively in birds but not all of them, this tissue courses "rivet-fashion" through the thickness of the choroid so that upon contraction the choroid is thinned and the retina is drawn backward for sharp focussing. Such muscular activity may also serve to regulate the flow of blood and thus the blood volume within the choroid.

The muscles of the avian ciliary body are believed to be responsible for both corneal and lenticular accommodation, but considerable variation appears to exist in the arrangement and thus precise function of these muscles. Best developed in birds with powerful accommodation, the ciliary muscle mass has been described most often to consist of 1–3 separate skeletal muscles all arranged in a meridional direction. Four well-developed ciliary muscles have been described in one species, the red-tailed hawk *(Buteo borealis)* (LORD, 1956). Whether the ciliary muscle mass in this bird or any other species represents a group of separate and independent muscles or merely artefactual separations (resulting from microtechnical manipulations) of one or two larger muscles was evident to early investigators, e.g., CANFIELD (1886), and still remains unanswered in many species. When two are present they most often appear as an anterior Crampton's and a posterior Brücke's muscle (Fig. 4b). In turn, Brücke's muscle is sometimes divided into two separate units, the anterior muscle being designated as Müller's, the posterior portion remaining as Brücke's muscle. According to WALLS (1942), Müller's muscle represents the posterior portion of Crampton's muscle.

I. Corneal Accommodation

The ability of the avian cornea to undergo alterations in its radius of curvature and thus to function in accommodation has been advocated by most investigators, e.g., HESS (1909–1910), SLONAKER (1918), WALLS (1942), GUNDLACH et al. (1945), LORD (1956), and PUMPHREY (1961). On the basis of ophthalmometric measurements of the pigeon cornea during drug-induced accommodation, GUNDLACH et al. (1945) maintain that the cornea represents the chief accommodating organ in this species. It provides the pigeon with such a great range of accommodation (17 diopters) that these authors find little need for any other device.

Crampton's muscle has been described to possess a direct tendinous attachment to the cornea along its entire margin and thus has been regarded by most investigators as the most important muscle of corneal accommodation. Arising from the inner aspect of the sclera at the posterior region of the ciliary body, this meridional band of striated muscle, when contracted, merely pulls the corneal margin inward affecting a change in its curvature. Crampton's muscle is reported to be small or absent in diving birds (perhaps because the cornea is of little optical use under water) and best developed in hawks and owls (WALLS, 1942; TRIPATHI, 1974). According to FRANZ (1913), however, it is only weakly developed in owls, an observation recently supported by STEINBACH and MONEY'S (1973) failure to detect (via mirrors) any changes in the curvature of the cornea in the Great Horned Owl *(Bubo virginianus)* during accommodation. This muscle is also said to be lacking in the cormorant (WALLS, 1942; DUKE-ELDER, 1958; TRIPATHI, 1974), a diving bird with the greatest breadth of accommodation (40–50 diopters), but according to HESS (1909–1910) it is present and functions in lenticular accommodation rather than corneal. In the English sparrow *(Passer)* SLONAKER (1918) does not recognize Crampton's muscle whereas LORD (1956) describes it as well-developed in this species. SLONAKER (1918) believes that corneal accommodation is accomplished by the contraction of the entire ciliary muscle which operates as a single individual muscular sheet attached at both ends to the sclera. Contraction of this muscle, he says, reduces the equatorial diameter of the sparrow eye producing both an increase in intraocular pressure and axial diameter and a concomitant forward displacement of the cornea. GUNDLACH et al. (1945) report that it is the tension produced by contraction of the ciliary muscles, particularly Müller's muscle which brings this about. All agree, however, that the especially deep corneoscleral sulcus (supported by a ring of 10–18 scleral ossicles) facilitates this process, perhaps by placing the ciliary body in a more advantageous functional position and at the same time serving as an essential rigid support for the rapid and strong muscular activity characteristic of striated muscle.

II. Lenticular Accommodation

The principal means of accommodation in higher vertebrates (Sauropsidans, mammals) is by deformation of the lens. In most birds this is brought about by direct squeezing of the lens by the ciliary body and iris. The avian lens is always in firm contact with the ciliary body (via ciliary processes) and the iris, both of which contain well-developed striated muscle. The avian lens is also unusually soft and malleable in birds with strong accommodation and develops an extensive equatorial thickening, the annular pad (Ringwulst) for contact with the ciliary processes (Fig. 4 b).

Brücke's muscle (Fig. 4 b) is regarded by most investigators as the principal muscle of lenticular accommodation. Arising from the inner aspect of the sclera and inserting into the posterior portion of the ciliary body, the contraction of this meridional muscle thrusts the ciliary body inward (axially) causing the ciliary processes to push the lens evenly along the entire extent of its equatorial annular pad. The lens becomes deformed by this maneuver, its anterior surface being pushed forward against the anterior chamber (pupil) and iris. Forward movement of the lens itself is prevented by the ciliary zonule and pectinate ligament (Fig. 4 b), whereas posterior displacement is prevented by a relatively unyielding vitreous body. Because the peripheral portion of the anterior surface of the lens abuts against the relatively firm iris, it is particularly the central portion of the anterior lens surface which undergoes the greatest change in curvature. Contraction of the striated iridial sphincter increases the convexity of this surface even more and produces an anterior lenticonus (Fig. 4 a) which further increases the refractive power of the lens. The role of the iridial sphincter in this case is an accommodative one. It is best developed (of all vertebrates) in amphibious birds where it causes the most extreme anterior deformation of the lens and helps to account for the great range of accommodation in these species.

According to Walls (1942), the annular pad acts as an architectural column transmitting the stress of the ciliary body directly to the soft lenticular center. Apparently this function coincides with the maximum development of the annular pad in birds with a high degree of accommodation, less in nocturnal species, still more poorly formed in diving and flightless birds, and smallest or vestigial in running birds. It comprises half the width of the lens in the martin, one-third to one-fourth in the eagle and only one-twentieth in the cormorant (Rochon-Duvigneaud, 1920 b). According to Slonaker (1918), however, the annular pad plays no role in accommodation but serves only a nutritive function. He describes it merely as a glandular epithelium which secretes a nutritive material into the lenticular chamber (remnant of the embryonic lens vesicle) for subsequent use by the lens. Slonaker (1918) maintains that the lens in the English sparrow *(Pas-*

ser) is very firm (a characteristic of eyes with poor accommodation) and does not change shape during accommodation. Instead, he believes that the lens (and cornea as well) is moved forward by the intraocular pressure produced by the inward movement of the sclera (in the region of the scleral ossicles) during contraction of the ciliary musculature.

An interesting and unique adaptation that has been associated with avian accommodation is the presence of a conspicuous quantity of glycogen in the central part of the lens (RABAEY, 1963; KASKEL et al., 1975). Comprising generally about 10% of the dry weight of the lens, this polysaccharide is believed by RABAEY (1963) to increase the refractive index of the lens' center (nucleus). Because lenticular glycogen is absent in birds considered to be the most primitive, he also suggests that it is "a characteristic in the evolution of birds only and unique phenomenon in the vertebrate-range." Glycogen has been detected in lenses of several mammals (rat, guinea pig, calf, ox), however, and is believed to represent an energy source for lens metabolism (KASKEL et al., 1975). Birds utilizing both aerial and aquatic vision have other devices for achieving accommodation, e.g., nictitating membrane with highly refractive window (diving ducks, loons, auks); duplex optical system with optically asymmetrical lens (kingfisher).

C. Retinal Adaptations

The avian retina possesses many structural features which have been implicated as favorable adaptations for a variety of visual functions, e.g., acuity, color discrimination, movement detection, sensitivity. The relative thickness of the retina itself has been associated with improved visual acuity and birds, in particular, are characterized by noticeably thicker retinae than most other vertebrates, including man. An abundance of retinal neurons, including thinner and more densely packed cones, and a low degree of summation at the ganglion cell level accounts for this appearance which together with other adaptations, e.g., oil droplets, glycogen deposits in rods and accessory cones, specialized areae and foveae, pecten, may contribute to the relatively greater visual acuity of birds compared with other animals. Data on the visual acuity of different birds have been tabulated by DONNER (1951) and updated most recently by FISCHER (1969). They reveal that birds exhibit a wide range of resolving power, the visual acuity being much better in diurnal birds than nocturnal species and best developed in diurnal birds of prey. Old world vultures *(Gyps, Neophron percnopterus ginginianus)* have twice the resolving power of man (FISCHER, 1969). The maximal acuity of the buzzard *(Buteo buteo)* corresponds close to the human, that of the kestrel *(Falco tinnunculus)* close to the rhesus monkey (OEHME, 1964). On the basis of foveal cone dimensions, however, ROCHON-DUVIGNEAUD (1919b) believes the buzzard has four times the visual acuity of man.

It has been repeatedly shown that the avian retina mediates color vision. Initially demonstrated in the chicken (WATSON, 1915; LASHLEY, 1916) and pigeon (WATSON, 1915; HAMILTON and COLEMAN, 1933) color discriminatory abilities have more recently been reported to exist in other diurnal birds, e.g., emus (NEUMANN, 1962), hummingbirds (POLEY, 1968), honeycreepers (WINKEL, 1969), gulls (THOMPSON, 1971), weaverbirds (MAYR, 1972a), daws (WESSELS, 1974), and even some nocturnal species, e.g., *Athene noctua* (MEYKNECHT, 1941, *Strix aluco* (FERENS, 1947; MARTIN, 1974). The spectral sensitivities of these birds are not the same, however, and several visual systems have been described: trichromatic in the pigeon (HAMILTON and COLEMAN, 1933; RIGGS et al., 1972), tetrachromatic in the daw (WESSELS, 1974) and either dichromatic or an anomalous trichromatic system in the tawny owl (MARTIN, 1974). Involved in an understanding of these systems is the presence in avian cones of brightly colored oil droplet inclusions and photosensitive visual pigments.

The avian retina is also capable of a great range of movement detection. Pigeons, for example, can discriminate the rate of movement of an object as slow as 15°/h (MEYER, 1964) and yet have the highest flicker fusion frequency recorded in vertebrates (150/S) (DODT and WIRTH, 1953). Such exceptional discriminatory ability is undoubtedly needed for controlling rapid flight movement and developing navigational skills for migration, e.g., detection of movements of the sun, changes in altitude, and may involve special functional adaptations. The pecten and foveae, for example, have been implicated as possible mechanisms in movement detection and more recent studies have provided evidence that the retinal ganglion cells are highly specialized in birds for the detection of different types of movements. Perhaps involved directly or indirectly with these cells are centrifugal nerve fibers emanating from the brain and representing a unique efferent innervation of the avian retina.

I. Visual Cells

1. Structure

The avian retina features three types of visual elements (rods, single cones, and double cones) (Fig. 5a, b) which are characterized by their structural and chemical compartmentation. Their outer segments resemble those of other vertebrate visual cells and consist of stacks of membranous discs housing the photosensitive visual pigments. Within the inner segments the mitochondria are concentrated distally to form the so-called ellipsoid, whereas most of the other organelles, e.g., Golgi apparatus, endoplasmic reticulum, reside in the proximal portion or myoid. The myoid represents the principal site of protein synthesis in visual cells; most of the ribonucleic acid is localized here. Further compartmentation may involve the presence of brightly colored or colorless oil droplet inclusions (lipids and carotenoid pigments) at the distal extremities of the ellipsoids and characteristically shaped glycogen deposits within the myoids (RABINOVITCH et al., 1954; YAMADA, 1962; O'RAHILLY and MEYER, 1963) (Fig. 5c).

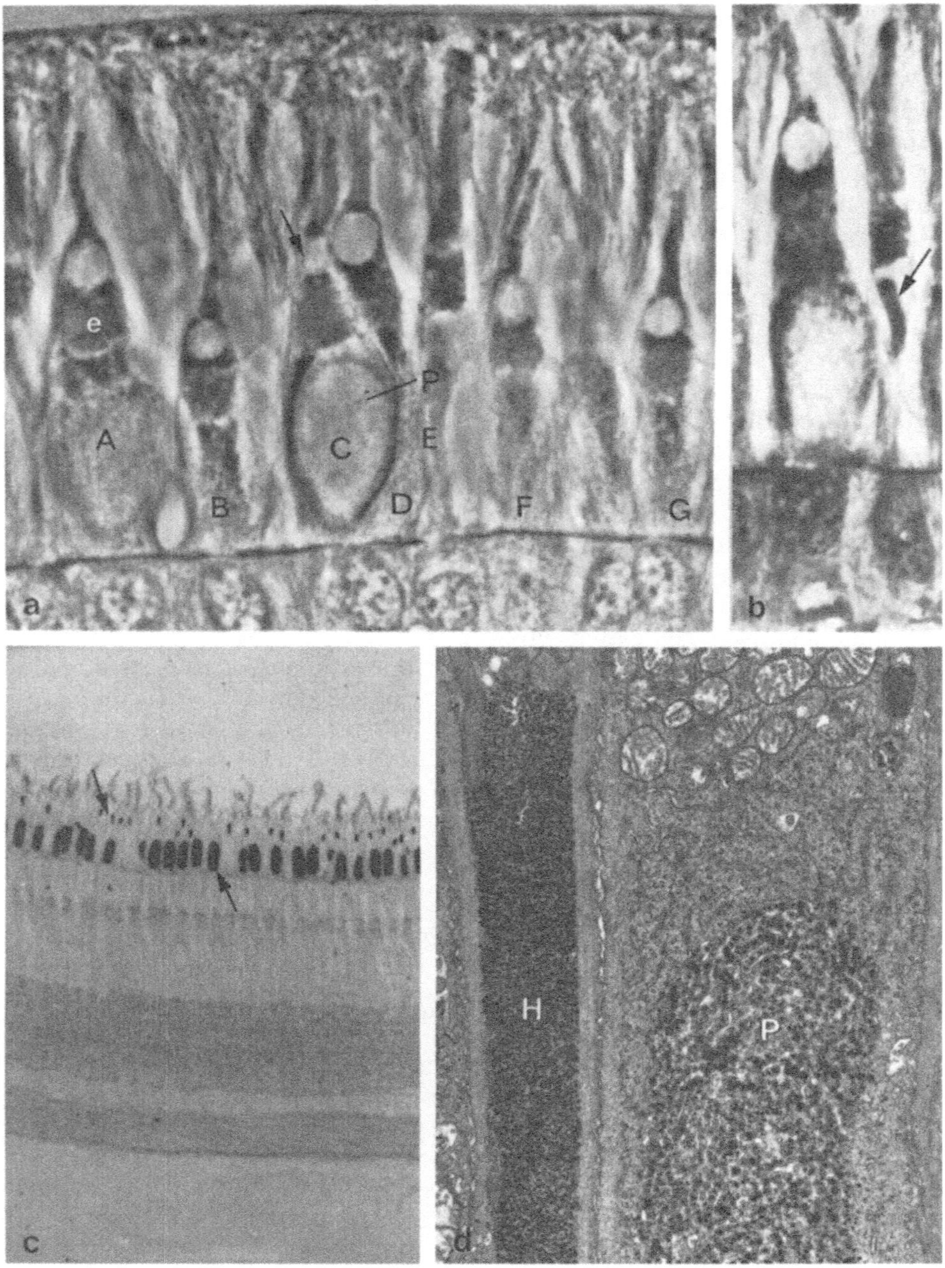
e
P
A
C
E
B
D
F
G
a
b
c
H
P
d

2. Classification

a) Rods

Avian rods are characterized by relatively large, cylindrical outer segments (Fig. 5a), the visual pigment-containing discs of which undergo continual formation and destruction. Like other vertebrate rods, new discs are being continually formed at the base of the outer segment while older ones are being intermittently removed at the tip and phagocytized by the pigment epithelium (YOUNG, 1967; YOUNG and BOK, 1969) where they constitute the lamellar inclusion bodies (phagosomes). The precise mechanism of rod detachment varies, however. In the human (SPITZNAS and HOGAN, 1970) the pigment epithelium possesses broad, rampart-like villous sheets which encompass the tips of the rod outer segments and pinch off 30–40 discs at a time. In the rhesus monkey (YOUNG, 1971) the rods themselves pinch off the distal fragments (8–30 discs at a time). In birds, e.g., the Japanese quail *(Coturnix coturnix japonica)*, on the other hand, both the pigment epithelial processes and the rod itself participate in the process (MEYER et al., 1973).

Avian rod inner segments lack oil droplets but can be easily recognized in light microscopic preparations by applying the periodic acid-Schiff (PAS) reaction (Fig. 5c) which selectively stains their glycogen-containing hyperboloids residing in the myoids (MEYER and MAY, 1973). As the name implies, the myoid has the ability to shorten (contract) or elongate (relax) in response to illumination and in coordination with the movement of pigment into and out of the processes of the pigment epithelium surrounding the visual cells. In the light, the rod myoids elongate and the cone myoids (except the accessory member of the double cone) contract concomitant with the migration of pigment toward them. Opposite reactions occur in the dark. These photomechanical changes, which represent adaptations for both diurnal and nocturnal vision, are quite rapid and extensive in birds.

b) Cones

Avian cones are distinguished by the presence of oil droplets (most often brightly colored, but also colorless) and relatively short, conical outer segments (Fig. 5a). Disc removal in vertebrate cones is a controversial issue at present. Unlike vertebrate rods, protein renewal in cones has repeatedly been shown

Fig. 5. (a) Phase contrast photomicrograph of visual cells in *Gallus domesticus* (MEYER and COOPER, 1966). A, double cone seen from side. Large oil droplet vacuole resides in chief cone immediately external to ellipsoid (e). *B*, single cone with oil droplet vacuole. *C*, Accessory and *D*, chief components of double cone each with oil droplet vacuole: oval in accessory cone *(arrow)*, round in chief cone. *E*, rod. *F* and *G*, single cones. Note large, oval, unstained paraboloid (*P*)occupying most of inner segment of accessory cone. Rod hyperboloid is not in focus. Unstained, Kolmer's fixative, bleached × 2000. (b) Phase contrast photomicrograph of double cone and rod in *Gallus*. Arrow indicates rod hyperboloid. Same preparation as above. (c) Periodic acid-Schiff preparation of retina in *Gallus* showing positive reactions *(glycogen)* in rod hyperboloid *(upper arrow)* and accessory cone paraboloid (lower arrow). X 82. (d) Electron micrograph of hyperboloid (*H*) and paraboloid (*P*) in rod and accessory cone, respectively of Japanese quail *(Coturnix coturnix japonica)*. Note smooth endoplasmic reticulum in paraboloid and absence of membrane system in hyperboloid. × 12000 autoradiographically to exhibit a diffuse pattern instead of the localized concentrations of radioactivity (representing new protein or membrane components) indicative of the continual production of new discs at the base of the outer segments (YOUNG, 1969). Nevertheless, electron microscopy has revealed the presence of phagosomes within the pigment epithelium of the rod-free human fovea (HOGAN, 1972) and has provided fine structural evidence that the cones in the human (HOGAN et al., 1974) and tree squirrels (ANDERSON and FISHER, 1975) undergo disc removal in a similar fashion as rods. Unfortunately, very little information is available on avian cones. YOUNG (1974) cites an unpublished study by GODFREY (1973) that during development of the chicken *(Gallus domesticus)* there is no discrete concentration of new protein in the outer segments of developing cones (type not indicated). Apparently disc removal does not occur during development of this species since phagosomes are not observed until after hatching (YEW, 1974).

Avian single cones (Fig. 5a) contain a relatively large, circular oil droplet of variable coloration. It is red or yellow in weaverbirds (MAYR, 1972b) and red or yellow-green in *Hieratus fasciatus* and *Milvus milvus* (GALLEGO et al., 1975), whereas in the chicken it has been reported to be red or yellow (SCHULTZE, 1867; MORRIS and SHOREY, 1967), red or yellow-green (GALLEGO et al., 1975), predominantly red (MEYER and COOPER, 1966), or a multitude of colors (HOFFMANN, 1877). On the basis of their oil droplet and mitochondrial densities in electron micrographs MORRIS (1970) has distinguished three types of single cones in the chicken.

Double cones are present in all vertebrate groups except placental mammals and represent one of the most obscure elements of the vertebrate retina (WALLS, 1942). Each consists of two intimately associated, but independent cones, dissimilar in size, shape, and structure: a tall, thin chief cone and a broad short accessory cone (Fig. 5a). They are best visualized when both components lie side by side in the plane of section (Fig. 5a, C and D) rather than perpendicular to it (Fig. 5a, A). The significance of double cones is not known although many theories have been advanced including the obvious economical advantage for both metabolism and space (KALBERER and PEDLER, 1963). They do respond differently to light, the chief cones contracting, whereas the accessory cone is unresponsive. WALLS (1942) associates a high number of double cones with animals possessing strong diurnal habits, although he considers an equal number of double and single cones as the maximum proportion. In the chicken they are twice as numerous as single cones throughout the retina and equal in number to rods in all regions except a circumscribed area posterior to the pecten (MEYER and MAY, 1973). According to MATSUSAKA (1963a) these double cones occupy 4 times the area of a single cone and 15 times that of a rod. At the level of their inner segments he calculates that 82% of the retinal area consists of double cones. In the great Tit *(Parus major)* the distribution of single and double cones assumes a definite pattern, each single cone being surrounded by four double cones (ENGSTRÖM, 1958). Between each group are rods and other single cones.

The chief cone is the tallest and thinnest of the two components of the double cone. It always possesses a large circular oil droplet which may assume a variety of colors. In the chicken, for example, it has been described as yellow (SCHULTZE, 1867; HOFFMANN, 1877; MEYER and COOPER, 1966; GALLEGO et al., 1975) or green

(MORRIS and SHOREY, 1967). In weaverbirds (Ploceidae) it is green (MAYR, 1972b); in two nocturnal species (*Carina noctua* and *Asio flammeus*) it appears yellow-orange.

The accessory component of the double cone is readily identified in PAS preparations owing to the conspicuous barrel-shaped deposit of glycogen (paraboloid) within its myoid (Fig. 5c). The presence of an oil droplet, however, has been very controversial. It has not been observed in most birds, including the pigeon (COHEN, 1963), weaverbirds (MAYR, 1972b), *C. noctua* and *A. flammeus* (GALLEGO et al., 1975), but in the chicken it has been pictorially documented from light microscopy as a small, oval yellowish-green droplet (MEYER and COOPER, 1966; MEYER and MAY, 1973), although in this species it apparently also appears in many other forms, e.g., colorless (GALLEGO et al., 1975), a yellow lentiform body (SCHULTZE, 1872), a granular vesicle (MORRIS and SHOREY, 1967), yellow pigment granules (KRAUSE, 1894) or a diffusion of oil droplet pigments (WALLS and JUDD, 1933). In the Japanese quail (*C. coturnix japonica*) from one to five small droplets constitute the accessory cone oil droplet (HAZLETT et al., 1974); in *H. fasciatus* and *M. milvus* it is small and colorless (GALLEGO et al., 1975).

3. Oil Droplets

Oil droplets of various colorations, sizes and shapes can be easily observed in situ by the microscopic examination of fresh, flat retinal preparations which readily permits quantitative analyses of their relative proportions and topographical distribution throughout the entire retina (MEYER and MAY, 1973). The colored mosaic thus revealed has aroused considerable interest, particularly regarding the possible role it plays in the visual process. PEIPONEN (1964) studied the distribution of oil droplet coloration in dorsal and ventral halves of the retinae from 44 birds of different feeding habits and found that the coloration varied with the region of the retina and the living habits of the birds. In diurnal birds, for example, the red and yellowish droplets appear to predominate although both KRAUSE (1894) and PEIPONEN (1964) have reported that the diurnal swift has only 6% red-orange compared to the nocturnal owl's (*Hirundo rustica*) 10%; PEIPONEN (1964) also found about 10% of these droplets in a crepuscular bird (*Caprimulgus europaeus*). According to ROCHON-DUVIGNEAUD (1920), birds frequently move their head so as to aim images on different parts of the retina in order to examine them with these different color filters and diagnose their nutritional value. In this regard SCHWARTZKOPFF (1960) has pointed out that diurnal birds, in contrast to man, can discriminate colors in every portion of their visual field. Such exceptional discrimination is independent of the relative brightness and size of the stimulus and that learned in one part of the visual field can be immediately generalized to another part (BLOCH and MATURANA, 1971). The coloration, as well as anatomical location of the droplets directly in the path of incoming light (Fig. 5) destined for the photosensitive pigment in the cone outer segments in which they reside has implicated them as color discriminatory devices, or more simply as bandpass filters for absorbing light rays. Their specific coloration is not believed to be influenced by sex, iridial, or feather pigmentation (HAHN, 1916), although the relative proportions of their colorations have been reported to be affected by castration, maturity, season of the year, as well as sexual and spermatogenic activity (PÉZARD, 1931; 1957a, b; 1964). Theories of oil droplet function are discussed by HAHN (1916), UCHIYAMA (1930), WALLS and JUDD (1933), WALLS (1942), ROCHON-DUVIGNEAUD (1943), DONNER (1953, 1960), DUKE-ELDER (1958), MEYER et al. (1965), ALPERN (1968), KING-SMITH (1969, 1971), THOMPSON (1971) and most recently by MUNTZ (1972), RODIECK (1973), and SILLMAN (1973).

Oil droplets are not confined to birds but have also been described in amphibians, reptiles and non-placental mammals. In diurnal birds they are generally colored red, orange, yellow, or green, whereas in nocturnal species they appear colorless or pale yellow [1]. All of these colors are apparently unaffected by light, and their relative proportions and distribution are characteristic for a given species. In *G. domesticus*, for example, three oil droplet colors have been recognized and each has been associated with a particular cone type: red in single cones, yellow in chief, and yellowish green in the accessory member of the double cone (MEYER and COOPER, 1966). Such cone-oil droplet associations, however, have not been confirmed in this or any avian species and recent studies utilizing the electron microscope have still provided controversial data, particularly with respect to the presence or absence of an oil droplet in the accessory cone (MORRIS and SHOREY, 1967; MORRIS, 1970).

Each colored oil droplet in birds consists of one or more stable carotenoid pigments dissolved in a mixture of lipids (KÜHNE and AYRES, 1878; WAELCHLI, 1881; LOEVENICH et al., 1943), the majority of which are nonpolar. Astaxanthin has been substantiated as the carotenoid in the red oil droplets (KUHN et al., 1939; V. STUDNITZ et al., 1943; LOEVENICH et al., 1943; WALD, 1948; GOODWIN, 1954; STROTHER and WOLKEN, 1960), whereas considerable controversy still exists with respect to the identification of the carotenoids involved in the other avian colored droplets (MEYER et al., 1965). LIEBMAN and GRANDA (1975) have recently shown that these carotenoid pigments are extremely dense in the turtle oil droplets. Astaxanthin, for example, was found to cut out virtually all light at the peak of its absorbance band. Preliminary studies on avian red oil droplets (species not indicated) also report extremely high optical density. Further insight into the visual function of the colored droplets perhaps can be obtained by a careful examination (visual behavior, electroretinography) of birds which have been experimentally deprived of any carotenoid pigmentation. It is well-documented, for example, that animals are unable to synthesize carotenoids de novo, but must obtain them from ingested food (PALMER and KEMPSTER, 1919) (see FOX and VEVERS, 1960, for the most up-to-date review of this subject). Japanese quail with "colorless" oil droplets have been obtained by dietary exclusion of carotenoid pigments from the diet of their parents and colonies of them have been successfully maintained on carotenoid-free rations (vitamin A added) with no apparent

[1] Almost every color of the visible spectrum has been reported in the cone oil droplets of birds, but it is now generally accepted that many of these colors, particularly those in the blue-violet range are merely the result of optical phenomena owing to the crude lenses used in the past (see WALLS and JUDD, 1933). Data on oil droplet coloration and distribution in birds are provided by DOBROWOLSKY (1871), WAELCHLI (1883), KRAUSE (1894), HAHN (1916), ROCHON-DUVIGNEAUD (1920b), ERHARD (1924), COULOMBRE (1955), DÜCKER (1963), PEIPONEN (1964), MAYR (1972b), and MEYER and MAY (1973).

interference in their normal growth and reproductive capacities (MEYER, 1971). The retinae of such birds possess no demonstrable alteration in visual cell morphology or ultrastructure (MEYER et al., 1971). Oil droplets are still present in all cones, although they have been completely depleted of carotenoids and now contain only lipids. Preliminary visual behavioral studies suggest that the absence of these pigments has little, if any, qualitative effect on the color discriminatory abilities of the Japanese quail (BOWMAN, unpublished observations).

The remainder of the oil droplet appears to consist entirely of neutral lipids which are mainly mono-, di- and triacylglycerides (HUDSON et al., 1971). Further data are now available on the lipid components of these oil droplets as well as their fatty acid composition (JOHNSTON and HUDSON, 1976). They indicate a marked similarity between the fatty acids linked to the acylglycerides of the oil droplets and those found in some of the phospholipid components of the cone outer segment membranes (JOHNSTON and HUDSON, 1974). These lipids are apparently the only component in the colorless droplets.

WALLS and JUDD (1933) suggest that cone oil droplets, regardless of their coloration, also serve as an aid for better visual acuity. All of them filter the light selectively and in so doing they enhance contrast, reduce glare, and lessen chromatic aberration.

4. Glycogen Bodies

The rod hyperboloids and accessory cone paraboloids (Fig. 5) have been implicated as possible adaptations for the improvement of visual acuity. Referred to by early investigators as refractile bodies, it is now well established that they are composed primarily of glycogen particles, although the ultrastructural organization and the chemical reactivity of these bodies are not the same. In chicken rods the hyperboloid measures, on the average, about 1–2 μm in thickness and 7–10 μm in length (MEYER and COOPER, 1966). Its slightly thicker distal or basal end lies close to the ellipsoid; its proximal tapered apex terminates about half way between the ellipsoid and the external limiting membrane. In unstained, phase contrast preparations in which the glycogen has been removed by solvents, the contour of the hyperboloid can still be recognized. It is smooth in texture and relatively dark in contrast to the clear, pale cytoplasm of the inner segment which surrounds it (Fig. 5b). In electron microscopic preparations the particles of hyperboloid glycogen appear to lie free; a membrane system characteristic of the paraboloid glycogen is lacking (Fig. 5d). In the latter, the glycogen particles lie between sacs of smooth endoplasmic reticulum (COHEN, 1963; EICHNER and THEMANN, 1963; MATSUSAKA, 1963a, b). In phase contrast, unstained preparations, the paraboloid appears relatively clear; its oval contour is recognized because it represents almost the entire bulk of the accessory cone inner segment. Regardless of these differences, the hyperboloid and paraboloid both possess a high index of refraction (KING-SMITH, 1971), as do the oil droplets, which perhaps enables them to concentrate the light onto the region of the photosensitive visual pigments in the outer segments.

Morphological changes have been observed in these bodies during both light and dark adaptation (VAN GENDEREN STORT, 1887). In the light, the accessory cone inner segments contract causing a swelling of the paraboloids which then permits a greater amount of light to be concentrated upon and absorbed by the cone outer segments. During dark adaptation similar changes take place in the rod inner segment and hyperboloid allowing for their greater efficiency (KING-SMITH, 1971). It must be emphasized, however, that most authorities regard these glycogen concentrations as merely energy sources for visual cell metabolism, perhaps substituting for the Müller cell glycogen which is much reduced in these avascular retinae. Both glycogen synthetase and phosphorylase activities have recently been demonstrated within the cytoplasmic matrices of the paraboloid (AMEMIYA, 1975) and their glycogen content has been shown to decrease in the light (EICHNER and THEMANN, 1963). In turn, electron microscopy has revealed that light adaptation produces an increase in the density of the paraboloid membrane system and a concomitant obscuring of the glycogen granules, whereas dark adaptation brings about denser glycogen granules and an obscuring of the membrane system (MATSUSAKA, 1963a). For some unknown reason, fluorescent light has been shown to bring about an increase in myoid glycogen in the pigeon (MARSHALL et al., 1972).

5. Visual Pigments

The outer segments of the photoreceptors contain the visual pigments which are responsible for the absorption of light incident upon the retina. These visual pigments are conjugated proteins that are classically pictured as consisting of an opsin prosthetic group joined by means of a Schiff base linkage to a vitamin A chromophore. The chromophore, which may be based either on the vitamin A_1 aldehyde (retinal) or on the vitamin A_2 aldehyde (dehydroretinal) is always present in the native pigment in the 11-*cis* configuration. The action of light is to isomerize the chromophore from the 11-*cis* to the all-*trans* form, thereby initiating a complex sequence of conformational changes that eventually results in excitation of the photoreceptor (see DARTNALL, 1972, for detailed information).

The visual pigments of the avian retina have long interested investigators concerned with the physiology and biochemistry of vision. (See LYTHGOE, 1972; and SILLMAN, 1973, for lists of avian visual pigments.) To a large degree, this interest stems from the fact that the avian retina is unique in that most birds possess a retina with a very high percentage of cones, those photoreceptors utilized under photopic conditions for acute vision and color discrimination. The visual pigments of cones have proven extremely elusive to standard methods of extraction and, therefore, the cone—rich avian retina has always been appealing as a good source of material offering a fair chance of success. To a degree, albeit a small one, the attention given to the avian retina has been rewarded, for the only visual pigment ever extracted which is generally accepted as originating in cones is the avian visual pigment, iodopsin. Iodopsin was first detected by WALD (1937) in digitonin extracts prepared from chicken retinas and its peak absorbance (λ_{max}) was later firmly established at 562 nm (BLISS, 1946; WALD et al., 1955). A visual pigment with λ_{max} identical to that of chicken iodopsin has also been extracted from the retina of the domestic turkey (CRESCITELLI et al., 1964), but it is still far from clear whether or not all birds possess iodopsin in their cones. Analyses of extracts have yielded inconsistent and sometimes confusing results. Thus, although WALD (1958) briefly mentions extracting iodopsin from the retina of the pigeon, BRIDGES(1962) reports the extraction of a visual pigment with λ_{max} not at 562 nm but at 544 nm, and SILLMAN (1969) was unable to detect either VP562 or VP544 in extracts prepared from the pigeon retina. In fact, extracts have been prepared from the retinas of 19 diurnal species of birds other than the chicken, turkey, or pigeon, but none have contained any visual pigment save those which could be interpreted as originating in the rods (CRESCITELLI, 1958a, b; BRIDGES, 1962; SILLMAN, 1969). However, there are many reasons why one pigment might lend itself to extraction but another might not and, therefore, the failure to extract cone pigments from these many species of birds is certainly no reason to conclude that they are absent. Indeed, as a result of their microspectrophotometric observations LIEBMAN (1964)[2] and LIEBMAN and HAILMAN (1967)[2] conclude that cones containing iodopsin are present not only in the retina of the chicken and of the pigeon, but also in the retina of the laughing gull *(Larus atricilla)*. Unfortunately, the optical problems inherent in measuring spectra in an outer segment as small as that of an avian cone prevents a very accurate determination of the λ_{max} and, therefore, LIEBMAN (1964) and LIEBMAN and HAILMAN (1967) could only place the peak absorbance of the cone pigment in the general region of 560–575 nm. Nevertheless, their findings constitute some support for the idea that iodopsin is a rather widely occurring avian visual pigment. If one accepts that idea, the question then arises as to whether or not iodopsin is the only cone pigment in the avian retina. The microspectrophotometric observations yielded no evidence of BRIDGES' 544 nm pigment nor of any other cone pigment aside from iodopsin, yet there are data from electrophysiologic experiments performed on the pigeon, the chicken, and the daw *(Corvus monedula)* which suggest the presence of photopic visual pigments that absorb at shorter wave lengths than iodopsin (IKEDA, 1965; GRAF and NORREN, 1974; NORREN, 1975). Several morphologically distinct cone types are known to be present in the avian retina and it is quite conceivable that these different cone types contain different visual pigments. LIEBMAN (1964) and LIEBMAN and HAILMAN (1967) studied only a few cones and it is possible, therefore, that they sampled only those cells which happen to contain iodopsin. The presence of more than one cone pigment in the avian retina would certainly be consistent with the condition in all the other vertebrate classes where at least some species possess retinas with multiple cone pigment systems (BROWN and WALD, 1963; MARKS et al., 1964; MARKS, 1965; LIEBMAN and ENTINE, 1968; LIEBMAN, 1972; RUSHTON, 1972).

The question of whether or not iodopsin is the only cone pigment in the avian retina is of great importance with respect to the manner in which birds discriminate colors. That birds do indeed perceive color is indicated not only by their cone rich retinas but also by a fair amount of behavioral and electrophysiologic evidence (e.g., PORTER, 1904, 1906; WOOD, 1925; HAMILTON and COLEMAN, 1933; DONNER, 1953; IKEDA, 1965; HAILMAN, 1966; KLOPFER, 1968; HODOS, 1969; GRANDA and YAZULLA, 1971; RIGGS et al., 1972; GRAF and NORREN, 1974; WESSELS, 1974; NORREN, 1975) but the mechanism responsible for avian color discrimination is not yet clear. One popular suggestion is that color discrimination in birds involves the use of but one cone pigment in conjunction with three

[2] Unpublished observations first reported by LIEBMAN (1972).

different types of colored oil droplets, thus providing three cone types, each with a unique spectral absorbance (SCHULTZE, 1866; WALD and ZUSSMAN, 1938; HAILMAN, 1964; KING-SMITH, 1969). The failure of LIEBMAN (1964) and of LIEBMAN and HAILMAN (1967) to detect any cone pigment other than iodopsin is consistent with this idea. Without arguing the case for or against the suggestion, it is important to note that such a mechanism of avian color discrimination would be basically different from the multiple cone pigment mechanism thought to be employed by other vertebrates which are known to perceive color. Certainly the mere presence of colored oil droplets is not a compelling argument against the presence of several cone pigments since some species of turtles have retinas with a well-developed array of colored oil droplets as well as multiple cone pigments (LIEBMAN, 1972).

In contrast to cone pigments, rod pigments are readily extracted and, therefore, substantially more information is available with respect to the visual pigments of the avian rods. All of the prevalent avian rod visual pigments examined either in solution or microspectrophotometrically absorb maximally at 500 nm–506 nm. In addition, in some species of Passerines a very small amount of a second, presumably rod, visual pigment has been detected with λ_{max} 480 nm–490 nm. Thus, the rod pigments of the avian retina are relatively constant in their spectral location and do not seem to vary with the behavior or environment of the species. This is similar to the case in the mammals (CRESCITELLI, 1958a, b) but is strikingly different from the situation in the amphibians or in the fishes where it has been possible to make some correlations between the animal's environment and the visual pigment content of the rods (e.g., WALD et al., 1957; MUNZ, 1958a, b; MUNTZ, 1964; SCHWANZARA, 1967; REUTER et al., 1971). Since the major variation in absorbance among visual pigments is attributed to structural differences in the opsin portion of the different visual pigment molecules, one can conclude that the relative constancy of the spectral absorbance of the avian rod pigments is due, for the most part, to the fact that their protein structures are very similar. However, visual pigments based upon the dehydroretinal chromophore do exhibit absorbance maxima about 20 nm higher than comparable visual pigments which incorporate the retinal chromophore. Thus, to some degree the constancy of the λ_{max} of the avian rod pigments is explained by the fact that all avian visual pigments, including iodopsin and VP544, are based on the retinal chromophore. This is much the same as what is seen in mammals and reptiles but is unlike the situation in amphibians and fishes where variations in the nature of the chromophore are in part responsible for differences in visual pigment absorbance (e.g., CRESCITELLI, 1958a, b; DARTNALL et al., 1961; SCHWANZARA, 1967; REUTER et al., 1971).

II. Ganglion Cells

MATURANA (1962) describes five classes of retinal ganglion cells in the pigeon, each of which differs in its response to a particular movement, e.g., vertical edge, horizontal edge, general edge, directional moving edge, convex edge. He describes these responses as being highly specialized and unambiguous (in contrast to mammalian responses). MATURANA believes that each is part of a deterministic type of visual response, which is fundamentally different in its organization from that observed in mammals. He points out that a neocortex is absent in birds with the result that an interpretation of the visual image perceived by the avian retina must be done at least at the ganglion cell level of the retina itself. As asserted in a subsequent publication, each selective response "arises from the selective sensitivity of each neuron to a particular spatiotemporal configuration in its afferent influences, and is independent of specific pathways" (MATURANA and FRENK, 1963). An electronic model of the pigeon retina capable of eliciting the specific responses of the ganglion cells has been constructed by RUNGE et al. (1968).

In the chicken, MILES (1972a) could not find the ganglion cell orientation detectors described in the pigeon by MATURANA (1962). Ganglion cells in the chicken were found to be capable of only "a rudimentary analysis of visual information at the retinal level."

III. Centrifugal Fibers

Centrifugal fibers constitute an efferent retinal pathway unique to birds. Its anatomical existence has been conclusively demonstrated in birds; studies on other vertebrates have been inconclusive (OGDEN, 1968). Arising from cell bodies in the midbrain (isthmo-optic nucleus), the axons of the centrifugal fibers form a special tract (isthmo-optic tract) to the optic chiasma where they completely cross to join the optic nerve from the contralateral retina (COWAN, 1970). Within the retina, the centrifugal fibers traverse the nerve fiber and ganglion cell layers in order to terminate among cells (primarily amacrines) in the inner nuclear and inner plexiform layers. MATURANA and FRENK (1965) describe two patterns of centrifugal fiber terminals in the pigeon retina: divergent and convergent, both of which contact amacrine and displaced ganglion cells. The significance of this centrifugal system has not been established. MATURANA and FRENK (1965) believe it provides some sort of localized control of retinal functions. They may modulate the output of the amacrine cells (DOWLING and COWAN, 1966), perhaps by the release of an oscillatory mechanism (OGDEN, 1968). From a series of experimental studies (MILES, 1970; 1972a, b, c), MILES (1972d) concludes that the centrifugal system in birds enhances retinal excitability by suppressing the inhibitory mechanisms acting upon the retinal ganglion cells. Greater visual sensitivity is acquired by this activity, which may thus function to improve visual search in shadowed areas and early detection of targets (ROGERS and MILES, 1972).

IV. Areae and Foveae

1. Characteristics

Improved visual acuity in birds may be correlated with the presence within the cone-rich avian retina of well-developed areae and foveae (often one or two), capable of housing an extremely dense population of cones and rods. These so-called areas of acute vision represent circumscribed thickenings of the retina brought about by an increase in the number of visual cells (predominantly cones) per unit of retinal area, combined with a concomitant increase in the number of bipolar and ganglion cells involved with them. The visual cells are characteristically thinner and longer here than elsewhere in the retina, thus improving the resolving power, and the nuclear layers are noticeably thicker, accounting for most of the increased thickness of the area. In the pigeon, for example, the ganglion cell density in the area is about 4 times that in the periphery (CHIEVITZ, 1889), in the owl about 3 times (OEHME, 1961). Areae such as these are present in some representatives of all vertebrate classes but not all, and attain their greatest perfection in diurnal birds where almost always one, sometimes two, and even three distinct areae may be developed. Occasionally requiring microscopic confirmation, these areae occupy different positions in the fundus (an important feature in connection with eye movements and space perception) but are always found superior to the upper end of the pecten (CHIEVITZ, 1891; MANN, 1924a). The most common area is located in the central region of the fundus slightly above and at the nasal side of the optic nerve (and thus pecten) and is termed the central[3] (nasal, principal, posterior) area; a second area occupies a more lateral, or eccentric position, temporal to and slightly above the optic nerve and is designated as the lateral (temporal, eccentric, anterior) area. The shape of these two areae may be oval, or circular, whereas a third area has the form of a long band or ribbon extending horizontally for a variable distance across the central part of the retina.

An additional region of high cellular density has been identified by GALIFRET (1968) in the posterodorsal portion of the pigeon retina in the so-called red field. Characterized by a relatively high concentration of cones bearing red and orange oil droplets and a most complex inner plexiform layer as revealed by electron microscopy (YAZULLA, 1974), this field corresponds to a dorsal area which receives the image of the pigeon's "pecking field". It serves as an acuity area for near vision when the pigeon is pecking for food on the ground. A mediolateral "yellow field" has also been described in the pigeon; it is dominated by yellow oil droplet-bearing cones and is believed to be specialized for far vision (RODIECK, 1973).

A depression or fovea of variable depth and size is generally observed within either a central or lateral area (thus central or lateral fovea), or within both, and represents a further adaptation for diurnal vision and greater visual acuity. An area may, and indeed often does exist without a fovea, but a fovea is only found within an area. Nevertheless, a single and double foveae have been described to be independent of any area, e.g., swallows, kingfisher (KOLMER, 1936).

Foveae are caused by a radial displacement of the more internal layers of the retinal area resulting in a shallow, saucer-shaped (Fig. 6a) or deep funnel-shaped (convexiclivate) (Fig. 6b) cavity filled with vitreous humor. Foveal depths vary; see OEHME (1961) for a tabulation of such data in different birds. Visual cell density is greater in the fovea than elsewhere in an area. Cell density counts of

[3] In primates (including man), the term macula lutea, is synonymous with the central area (WALLS, 1942), being located in the central region of the retina, but it lies lateral to the optic nerve and could likewise by rightfully called a lateral area (with a lateral fovea). Cognizance should be made of this ambiguity, as well as to the fact that the macula is pigmented and should not be used to designate nonpigmented areae in other vertebrates (WALLS, 1937).

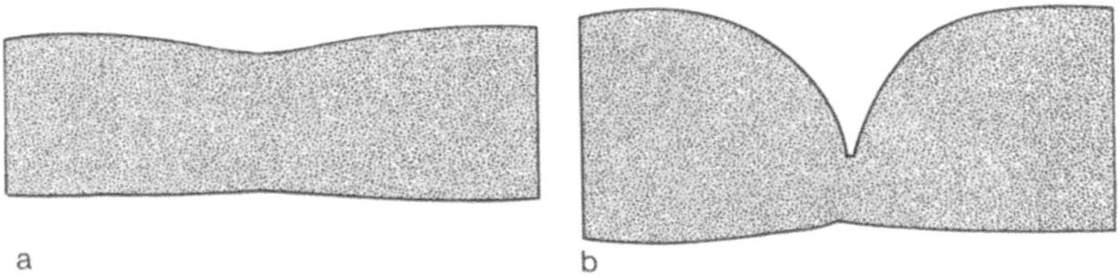

Fig. 6a and b. Comparison of foveae in American golden eagle *(Aquila chrisaëtos)* (after POLYAK, 1957). (a) Shallow saucer-shaped lateral fovea. (b) Convexiclivate central fovea

photoreceptor and ganglion cells in the central fovea of several birds, e.g., hawks, owls, tern, blue jay, and sunbird, have revealed that the convergence ratio (receptors per ganglion cell) clearly attains its lowest value at this site (FITE, 1974). According to many authors, including PUMPHREY (1948a) a one to one relationship between cones and ganglion cells exists at the center of the fovea, each one being represented by a single optic nerve fiber. The maximum perfection of the fovea is exemplified by the central fovea of swift-flying diurnal birds of prey which is characterized by a remarkably deep and well-excavated pit (Fig. 6b) containing extremely thin cones with yellow oil droplets. In the buzzard, ROCHON-DUVIGNEAUD (1919b) calculates that these cones are four times thinner than those in the human fovea, thus producing a resolving power four times greater than in man. According to WALLS (1942), the fovea of some hawks and eagles are supreme providing a visual acuity at least eight times that of man.

2. Functions

Several functions have been attributed to the fovea. Early investigators believed that the depression merely represented a structural feature designed to permit incident light to reach the foveal receptors with a minimum of obstruction and in so doing to improve visual acuity. WALLS (1937, 1940, 1942) does not accept this explanation but contends that the margins of the fovea, particularly the convexiclivate type (Fig. 6b) act like a convex lens and magnify the retinal image (13% linearly, 30% in area) (Fig. 7) owing to the substantially higher index of refraction of the retina than the vitreous, e.g., 0.0097 in the chicken (VALENTIN, 1879a), 0.0084 in the parrot (VALENTIN, 1879b). Thus, according to WALLS, the deeper the fovea, the greater its magnifying power. PUMPHREY (1948b) does not accept this interpretation but regards it as erroneous, as well as incomplete because it neglects spherical aberration which would interfere with the sharpness of the retinal image. Depicting POLYAK's (1941) profile of the convexiclivate, central fovea of the golden eagle *(Aquila chrisaëtus)* (Fig. 6b) as a geometric model consisting of quadrants of two identical circles tangent to each other, PUMPHREY (1948b) calculates that foveal refraction would be unable to bring about any improvement in visual acuity.

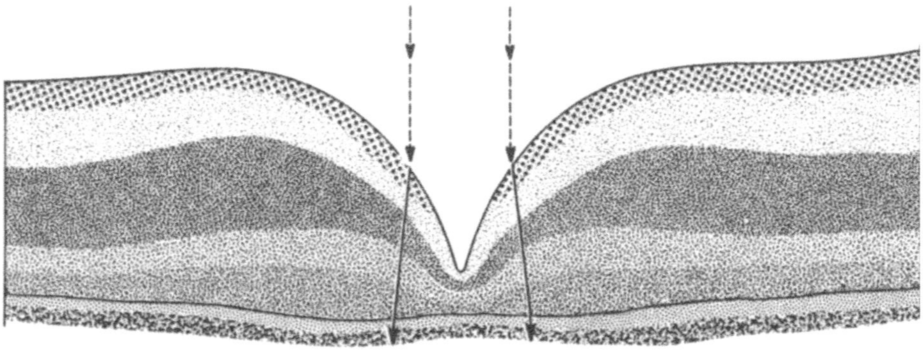

Fig. 7. Magnifying effect of avian fovea as depicted by WALLS (1942). Arrows represent incident light which, when falling upon foveal clivus, becomes refracted laterally resulting in enlargement of image at visual cell level. Foveal depression and retinal stratification as illustrated by ROCHON-DUVIGNEAUD (1943) in *Astur palumbarius*

He finds instead that this type of fovea has the ability to emphasize angular displacements by being able to transform a radially symmetrical image into an asymmetrical one (unless there is exact coincidence between the axes of symmetry of the fovea and the object). He postulates, therefore, that such a fovea is capable of carrying out two cognate functions, "the maintenance of accurate fixation and the sensitive appreciation of angular movements of a fixed object." High visual acuity in birds is associated always with their ability to be sensitive to movement or to be able to fix an object once sighted. The convexiclivate shape of the eagle fovea, PUMPHREY (1948 b) states, reflects the highest development of a fovea as a fixation device without sacrificing almost maximal acuity and movement sensitivity. A more pronounced convexiclivate fovea, as found in the kingfisher *(Alcedo)*, sacrifices some acuity and movement sensitivity for a better fixation device. The shallow, saucer-like profile of the lateral fovea (Fig. 6a), on the other hand, involves minimal refraction and functions in binocular vision only (PUMPHREY, 1948b); it cannot tolerate aberration and loss of definition (PUMPHREY, 1948a).

KREITHEN and KEETON (1974) present evidence that the fovea (in *Columba livia*) functions in the detection of polarized light and is therefore indirectly involved in navigation. They believe that the fovea acts as a radial analyzer and responds differently to planes of polarization. The sloping walls of the fovea may reflect light so that it strikes the opposite side of the fovea at an angle different from the axis of the visual elements. Evidence is available, they maintain, that such right angle illumination evokes responses from the photoreceptors which differ depending upon the varying planes of polarized light.

Data on the location and shape of these areae and the depth of their foveae (shallow or deep) have been collated from 101 avian species by SLONAKER (1897); they include observations by MÜLLER (1863), ENGELMANN (1884), CHIEVITZ (1891) and himself, but overlook considerable data by KRAUSE (1894). More recent findings have also been compiled by KAJIKAWA (1923) and ROCHON-DUVIGNEAUD (1943); KAJIKAWA's data include other pertinent information, e.g., ocular diameter, pecten mass, number of pecten pleats and ciliary folds.

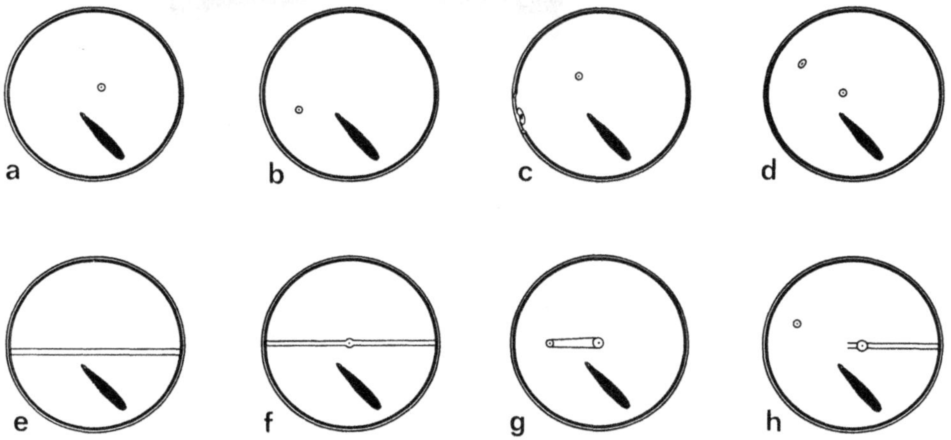

Fig. 8a–h. Different arrangements of retinal areae and foveae in right fundus of birds. Black, elongated, oval structure directed diagonally in lower half of each fundus represents pecten. It is included merely as landmark; its shape and size should be ignored. (a) Central fovea only, characteristic of most avian fundi. Steller jay *(Cyanocitta stelleri)* (after WOOD, 1917). (b) Lateral fovea only, found in owls and swifts. European barn owl *(Strix flammea)* (after WOOD, 1917). (c) Bifoveate fundus of kingfisher *(Alcedo ispida)* with extremely eccentric lateral fovea (after WOOD, 1917). (d) Bifoveate fundus of bittern *(Botaurus lentigenosus)* with superiorly located lateral fovea (after WOOD, 1917). (e) Linear area only, afoveate. Manx shearwater *(Puffinus puffinus)* and fulmar petrel *(Fulmaris glacialis)* (after LOCKIE, 1952). (f) Linear area with enclosed central area and fovea. American coot *(Fulica americana)*. (g) Linear area connecting lateral and central areae, both with fovea. Sparrow hawk *(Falco sparverius)* (after SLONAKER, 1897). (h) Linear area with enclosed central area (and fovea) and separate lateral area (and fovea). Tern *(Sterna hirundo)* (after SLONAKER, 1897)

3. Classification

The avian fundus can be classified into several groups (WOOD, 1917; WALLS, 1942; DUKE-ELDER, 1958) (Fig. 8): (1) complete absence of an area (thus afoveate), (2) Central area only, almost always with a fovea (Fig. 8a), (3) Lateral area only, always with a fovea (Fig. 8b), (4) Two areae, central and lateral, each with a fovea (bifoveate) (Fig. 8c,d), (5) Linear area only, without a fovea (Fig. 8e), (6) Two areae, central and linear, with a central fovea (Fig. 8f), (7) Three continuous areae, central, lateral, and linear with a central and lateral fovea (bifoveate), (Fig. 8g), and (8) Three areae, central, lateral, and linear with a central and lateral fovea (bifoveate). The lateral area (with its fovea) is independent of the other two (Fig. 8h).

The complete absence of an area is rare; WOOD (1917) acknowledges it in only one species (*Lophortyx californicus vallicola*, the California Valley quail). WALLS (1942), however, states that areae are often absent in domestic birds, although CHIEVITZ (1891) and SLONAKER (1897) both report a central area in the most domesticated species, *G. domesticus*, apparently on the basis of a slight retinal thickening as determined microscopically. In this species, however, the greatest concentration of cones occurs in a topographical segment which corresponds to the site of the lateral area of some birds (MEYER and MAY, 1973).

a) Central Area and Fovea

A central area is characteristic of most birds. SLONAKER (1897), for example, noted it in 95% of his specimens where it was usually found alone (54%) and most often with a fovea. The central fovea (Fig. 8a) therefore, represents the most common type of avian fovea; it has been recorded in all birds except owls, *Apus apus* (OEHME, 1962), the California Valley quail (WOOD, 1917) and the domestic chicken *(G. domesticus)* (SCHULTZE, 1866; WALLS and JUDD, 1933; WALLS, 1942; ROCHON-DUVIGNEAUD, 1943).

The central fovea is generally conceded to be involved in monocular vision (MÜLLER, 1863). As such, it functions independently of the central fovea of the other eye and is apparently of use to flying birds only for seeing and avoiding obstacles. In the ring plover, finch, and magpie, the central fovea contains a greater proportion of cones bearing yellow oil droplets, but other colored (red, orange) and colorless droplets are also found there (ROCHON-DUVIGNEAUD, 1919b). On the other hand, green oil droplets residing in double cones have been reported to predominate over red and yellow (single cones) in the fovea of several members of the weaverbird family (MAYR, 1972b). Blackbirds and starlings completely lack double cones in their foveae (OEHME, 1962). In diurnal birds foveal rods have been reported by some investigators (FRITSCH, 1911), denied by others (SLONAKER, 1918).

b) Lateral Area and Fovea

Solitary lateral areae with foveae have been repeatedly reported in owls (Fig. 8b) including the nocturnal flightless parrot owl, *Strigops habroptilus* (WALLS, 1942), where they are generally conceded to be involved in binocular vision, although several anatomical features, e.g., geometric relationship to optic axis, total decussation of optic nerves, have rendered this interpretation objectionable to some authors, e.g., ROCHON-DUVIGNEAUD (1920a). Their fovea is unusual among vertebrates in that it possesses both rods and cones (OEHME, 1961; FITE, 1973), the rods being the predominate element, as in the rest of the retina. Double cones have not been seen in these lateral foveae; only single cones with colorless or pale yellow oil droplets have been described there (ROCHON-DUVIGNEAUD, 1919b; OEHME, 1961). FITE (1973) suggests that either of these foveal photoreceptors (rods or cones) may be involved in the discrimination of fine spatial detail in the horned owl *(Bubo virginianus)*, whereas MARTIN and GORDON (1974) consider the foveal cones to be either vestigial, or if operational, of insufficient number to contribute to visual performance. Since color discrimination has been demonstrated in *Strix aluco* (MARTIN, 1974), however, they infer that the cones of this owl are indeed functional, but only at photopic levels of illumination.

A well-developed convexiclivate lateral fovea has been described in the swift by ROCHON-DUVIGNEAUD (1943) and recently confirmed by OEHME (1962) who noted only the presence of single cones there. This is the only fovea observed in the swift retina by these authors in contradiction to KAJIKAWA (1923) and WALLS (1942), who report only a central fovea. To add to the confusion a linear area may also be present, although it requires microscopic identification.

c) Central and Lateral Areae and Foveae

When both central and lateral areae are present each contains a fovea. Such bifoveate retinae are characteristic of many birds which obtain their food "on the wing," e.g., passerines, hummingbirds, bitterns (Fig. 8d), and kingfishers (Fig. 8c) and thus require accurate perception of distance and speed. The depths and relative positions of the two foveae exhibit considerable variation depending upon the species. In general, the central area and its foveal depression is better developed than its lateral counterpart. According to SLONAKER (1897), the central fovea (being concerned, he believes, with monocular vision) varies little in retinal position, whereas the location of the lateral fovea (being related to binocular vision) is variable depending upon the position of the eye in the head. The more lateral the eyes, the more lateral becomes the location of the lateral fovea and thus the greater becomes the interfoveal distance. On the other hand, the more the eyes face forward and the more binocular vision comes into play, the closer become the two foveae culminating in a less distinct central fovea or its disappearance by fusion. In birds of prey, the lateral fovea provides direct forward vision for its rapid descent toward its prey (DELAGE, 1920). In the kingfisher *(Alcedo)*, which relies upon both aerial and aquatic vision, the lateral fovea is positioned very eccentrically (merely 1.5 mm from the ora serrata) (Fig. 8c) and, in coordination with a uniquely shaped lens, seems to function in binocular vision under water (KOLMER, 1924).

d) Linear Area

A ribbon-like or linear area is characteristic of many avian species, including entire families, but is rarely observed in other vertebrates (MUNK, 1970). CHIEVITZ (1891), SLONAKER (1897), WOOD (1917), ROCHON-DUVIGNEAUD (1943), LOCKIE (1952), and DUIJM (1958) have recorded its occurrence in a variety of birds and a collation of their findings is presented by DUIJM (1958) and more recently reproduced by PEARSON (1972). To these should be added several other pertinent studies, e.g., KRAUSE (1894), KAJIKAWA (1923), FRANZ (1934), which have been overlooked. In general, linear areae appear to be associated with birds which inhabit open spaces (FRANZ, 1934; DUIJM, 1958), seek their nourishment from the ground (KAJIKAWA, 1923), swim (FRANZ, 1934), and undergo backward head bobbing during walking (WHITESIDE, 1966). According to DUIJM (1958), this area is generally maintained in a horizontal position similar to that occupied by the lateral semicircular canal.

Several specialized functions have been attributed to the linear area. These include the minimization of eye movement (WALLS, 1942), movement detection (PUMPHREY, 1961; MUNK, 1970), spatial orientation (by determining the normal position of the eye with respect to the horizontal plane) (DUIJM, 1958), and the localization of the clear visual horizon (PENNYCUICK, 1960). CRESCITELLI (1972) regards this area as an adaptation for higher visual acuity along a horizontal field and suggests that it represents a retinal system for movement detection which birds have retained and improved from their reptilian ancestors.

Linear areae are generally associated with a central area (and fovea) and sometimes with both central and lateral foveae. Only very rarely (e.g., in the swift) has a linear area been found with only a lateral area (and fovea) (OEHME, 1962). The

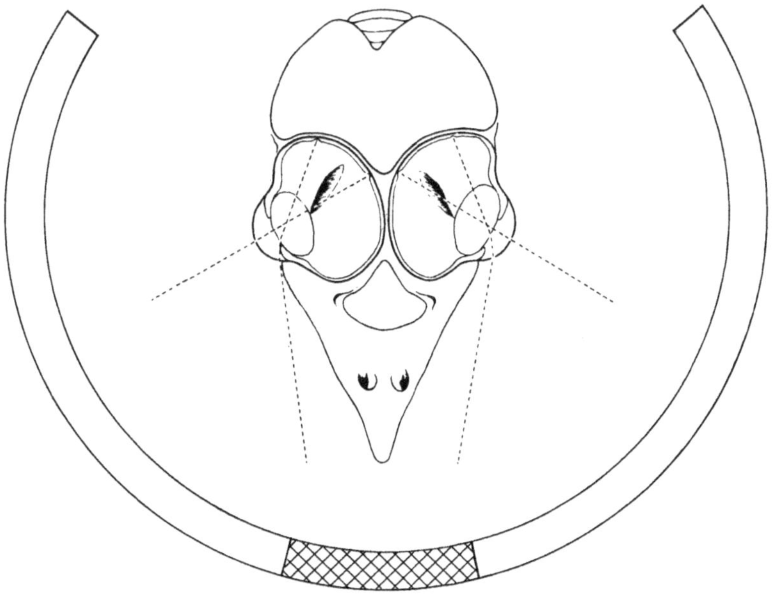

Fig. 9. Projection of visual fields of bifoveate avian eye indicating visual trident (after WALLS, 1942). Head and eyes are those of barn swallow showing central and lateral foveae with their visual axes (modified from POLYAK, 1957) and visual fields: binocular (cross-hatched) for two lateral foveae, monocular for each of central foveae

solitary presence of a linear area is likewise infrequent being limited to some aquatic birds, e.g., some gulls, shearwaters (Fig. 8e), and flamingos. According to CHIEVITZ (1891), a trough-like fovea is present macroscopically within some of these linear areae, but SLONAKER (1897) points out that these depressions cannot be recognized microscopically. LOCKIE (1952) has failed to discern a fovea in the Manx shearwater [*Puffinus puffinus* (Brünn)] and the Fulmar petrel [*Fulmaris glacialis* (L.)]; O'DAY (1940), on the other hand, acknowledges the existence of a central fovea and linear area in the giant petrel (*Macromectes giganteus giganteus*, Gmelin).

When the linear area is associated with only a central area (and its fovea), the latter generally lies within the middle of the ribbon (Fig. 8f). Such an arrangement has been found in many ground feeders, swimmers, divers, and waders.

The presence of three distinct retinal areae (central, lateral, and linear), two of which (central and lateral) possess a fovea (Figs. 8g, h) is a unique avian adaptation aimed at providing a superb range of monocular and binocular vision. The bifoveate condition permits the formation of three separate and distinct visual fields (visual tridents) (Fig. 9), two lateral monocular fields (one for each eye) and a central binocular field; the linear area no doubt enhances the ability to follow moving objects. Such an arrangement is characteristic of diurnal birds of prey which require a very high degree of visual acuity and stereoscopic vision, e.g., eagles, hawks, falcons, vultures, buzzards.

Two fundic patterns have been recognized involving these three areae. In the most common pattern, which is characteristic of the diurnal birds of prey, the linear area connects the lateral and central areae (with foveae) which lie at each end of it (Fig. 8 g). The other pattern described only in the tern *(Sterna)* (KRAUSE, 1894) has a linear area enclosing a central area (and fovea) either in the middle or at one end and a completely separate, foveate lateral area (Fig. 8 h).

ROCHON-DUVIGNEAUD (1920a) theorizes that in these bifoveate conditions both foveae function independently, constituting a sort of compass by giving separate points of distinct vision: a marvelous apparatus for the estimation of distances. This view is supported by POITIER (1923) from his observations on the flying habits of bifoveate birds. He believes that the lateral fovea serves to fix the prey while the central fovea functions as an "alarm" indicator for the many obstacles which may easily get into the way of its flight pattern.

Data on foveal thickness and interfoveal distance in the sparrow hawk, goshawk, and buzzard have been obtained by ROCHON-DUVIGNEAUD (1919a) in relation to the length of the optic axis. They reveal that the central fovea is slightly larger and deeper than the temporal one and possesses slightly longer, and thus more numerous cones. More significant, however, are the findings in *F. buteo* (SCHULTZE, 1866), *B. vulgaris* (KRAUSE, 1894) and *F. sparverius* (WALLS and JUDD, 1933) that in contrast to single foveae, both foveae of a bifoveate retina consist entirely of cones with yellow oil droplets, although red droplet-bearing cones are found in the rest of the retina. This is not the case in the kestrel *(F. tinnunculus)*, however, where some cones with red oil droplets are present among the majority of extremely small and pale yellow droplets of other cones (ROCHON-DUVIGNEAUD, 1919b).

D. Pecten

The complete avascularity of the avian retina appears to be associated with the presence of a bizarre supplemental nutritional device (WALLS, 1942), the pecten oculi (Fig. 1). Arising from the site of exit of the optic nerve (linear in birds) and projecting a variable distance into the vitreous body in the ventral and temporal quadrant of the fundus (Fig. 2), the pecten consists almost exclusively of capillaries and extravascular pigmented stromal cells. Lacking both muscular and nervous tissues, it is unique to birds and has been implicated in many functions (some visual), the most plausible of which is the nutrition of the inner layers of the avascular avian retina.

I. Vascular Supply

The primary blood supply to all vertebrate retinae is the choroid, which in birds is relatively thick, its external vascular layer (venous lakes) being traversed in some species by muscular cords (smooth or striated), particularly near the fovea (KAJIKAWA, 1923). According to DUKE-ELDER (1958) such an arrangement may regulate the amount of blood present in the choroid and serve for the fine positioning of the fovea during accommodation.

The vascular system of the avian pecten has been described by many investigators utilizing serial sections and vascular injection techniques. The most extensive anatomical studies have been conducted on the sparrow *(Passer)* (SLONAKER, 1918; WINGSTRAND and MUNK, 1965), the pigeon *(Columba)* and blackbird *(Turdus)* (ROHEN, 1955) and several other species (BEAUREGARD, 1876; KIJOWSKA, 1925; TANAKA, 1938 b).

The arterial supply of the pecten has been repeatedly shown to be completely independent from that of the choroid. Although blood reaches both organs from the external ophthalmic artery via the ophthalmotemporal artery (temporal ophthalmic artery) or its rete mirabile, the arteries to the pecten are always found to be separate branches without connection to choroidal vessels. According to WINGSTRAND and MUNK (1965), the external ophthalmic artery in *Columba* forms a large capillary network (rete mirabile ophthalmicum) in the posteroventral wall of the orbit which gives rise, among others, to the ophthalmotemporal artery (Fig. 2). A second capillary network (rete mirabile pectinis) has been reported by early investigators (BEAUREGARD, 1876; VIRCHOW, 1900; v. HUSEN, 1913) to be formed by this artery on its way toward the pecten, but WINGSTRAND and MUNK (1965) could not find it. Indeed, ROHEN (1955) failed to detect either network in *Columba* or *Turdus*, and SLONAKER (1918) did not see them in *Passer*. In this species the ophthalmotemporal artery together with anastomotic branches from the internal ophthalmic artery form a complete arterial circle around the optic nerve. In the vicinity of the posterior aspect of the nerve, the ophthalmotemporal artery gives off usually one (SLONAKER, 1918) or several (2–4) (WINGSTRAND and MUNK, 1965) posterior pecten arteries which enter the bulbus. In addition, WINGSTRAND and MUNK (1965) describe a small anterior pecten artery originating from either the short ciliary artery or artery of the pyriformis (both branches of the ophthalmotemporal artery) entering the bulbus rostral to the optic nerve (see their Fig. 8). Upon arrival at the base of the pecten, each pecten artery divides to form a basilar artery which courses along the longitudinal extent of the base of the pecten and gives rise, in ladder-like fashion, to an arteriole for each pleat (Fig. 10). Within the pecten these arterioles ascend toward the bridge in a rather straight manner giving off on their way the extensive and unique capillaries with which they merge before reaching the bridge. The capillaries empty into venules, all of which course parallel to the arterioles and drain into the basilar vein, a companion of the basilar artery, at the root of the pecten. According to SLONAKER (1918; *Passer*) and ROHEN (1955; *Columba* and *Turdus*) the arterioles occupy the middle of the pleats and the venules are found at their angles. TANAKA (1938 b) and WINGSTRAND and MUNK (1965), however, find a more variable situation, the arterioles and venules residing at either location. TANAKA (1938 b) examined 27 avian species and observed that no two pectens possess the same arrangement of blood vessels.

Of particular significance to the vascular supply of the pecten is the presence of a venous ring encircling the optic nerve, communicating with the cavernous sinus (WINGSTRAND and MUNK, 1965), and receiving among others, the basilar vein. Calling it the venous plexus of the optic nerve (Plexus venosus fasc. optici),

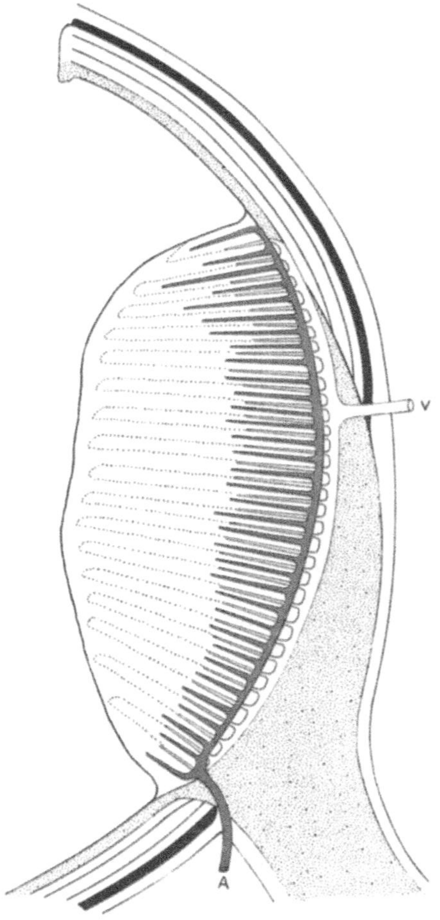

Fig. 10. Diagrammatic representation of arterial (*A*) supply and venous (*V*) drainage of pecten in sparrow *(Passer domesticus)* (after WOOD, 1917 and SLONAKER, 1918)

ROHEN (1955) describes this ring in *Columba* and *Turdus*, as several sinus-like veins anastomosing with one another many times to form a dense circular plexus around the entire optic nerve at the optic chiasma. He finds typical epithelioid, arteriovenous anastomoses connecting this venous ring with arteries supplying the bulbus, but does not detect an arterial ring in these species. This venous arrangement, he believes, may serve to regulate the flow of blood into the eye and no doubt into the capillary system of the pecten and choroid.

II. Classification

The ophthalmoscopic appearance of the avian fundus in many birds has been beautifully reproduced by WOOD (1917) who has provided much comparative data on the size, position, and morphology of the pecten, as well as the relative location of the fovea. On the basis of its morphology, the pecten can be grouped into three types: conical, vaned, and pleated (Fig. 11). The pleated pecten has been extensively studied by both light and electron microscopy; histological studies on the other two types could not be found.

1. Conical (Conus-like) Pecten

The conical pecten has been described only in the kiwi *(Apteryx)* (LINDSAY-JOHNSON, 1901; WOOD, 1907), the only genus in the family Apterygidae. Resembling the conus papillaris of reptiles, it arises trumpet-shaped from the center of a circular optic disc and extends into the vitreous as a heavily pigmented (brown-black) cone, without pleats or vanes, nearly as far as the lens (Fig. 11a). WALLS (1942) believes it is degenerative; LINDSAY-JOHNSON (1901) puts it in an intermediate position between the vestigial pecten-like formations of various mammals and the fully developed and pleated avian pecten. According to LINDSAY-JOHNSON (1909), the conical pecten of *Apteryx* is identical in size and position to that observed in agouti, a rodent. BACSICH et al. (1966b) likewise report a similar conus-like structure in the golden hamster *(Cricetus auratus)*, but in some colonies and not in others. They associate its presence with a strong hypermetropia.

The conus papillaris of reptiles, considered by many investigators to be homologous to the avian pecten (e.g. VIRCHOW, 1900; MENNER, 1938) assumes three morphologic appearances (with concomitant fine structural differences) which could be construed to represent stages in its evolution toward the avian pecten. The most primitive type of conus appears in snakes as a vestigial pseudo-cone consisting of blood vessels (hyaloid) of classic ultrastructure accompanied by nonmyelinated nerves and a mass of pigmented connective tissue (NGUYEN H. ANH, 1971). Because it shows no ultrastructural feature comparable to the other types of conus, nor to the pecten for that matter, NGUYEN H. ANH (1971), does not consider the pseudo-cone as a homologous structure to the pecten. In *Anguis fragilis* (slow-worm), *Zonosaurus ornatus*, *Lacerta viridis*, and *Lacerta muralis* it is in the form of a thin black stalk, the histologic organization of which varies with the species. It is most primitive in *Anguis* (NGUYEN H. ANH, 1969) where it displays several features unrelated to the avian pecten, e.g., nonmyelinated nerves in close relation to stromal cell processes (suggestive of Schwann cell relationships), flattened vascular endothelia, pericyte-like cells containing myofibrils. Relatively smooth endothelia with no surface foldings or fenestrations are also found in the gecko *(Eublepharidae*, DUNN, 1968). In *Zonosaurus*, *Lacerta* (NGUYEN H. ANH, 1974), and *Iguana iguana* (DIETERICH and DIETERICH, 1975b), however, some of the capillary endothelia in the conus have developed microfolds and a perivascular membrane, both characteristic features of the pecten, although some capillaries with smooth surfaced endothelia, as well as nonmyelinated nerve fibers scattered throughout the perivascular spaces are still retained. On the basis of its fine structure NGUYEN H. ANH (1974) considers this type of conus to be an intermediate stage between that developed in *Anguis* and that observed in *Oplurus*. In the latter, an Iguanid, and *Chamaeleon lateralis* (NGUYEN H. ANH, 1970) the conus papillaris apparently reaches its greatest development. Here it is in the form of a thin transparent plate consisting of a regularly folded membrane about a central axis emerging from the optic disc. Morphologically resembling the pecten of the ostrich, its fine structure possesses features intermediate between those of *Anguis* and the avian pecten. For example, relatively long microfolds are now consistently found on both surfaces of its capillary endothelia, although not as well developed as those in the pecten, and nerve fibers are no longer present throughout the body of the organ but are now confined to the base of the conus in the vicinity of the large blood vessels located there. In several birds DIETERICH and DIETERICH (1975a) find more highly differentiated endothelial microfolds (apically and basally) in diurnal species than nocturnal ones.

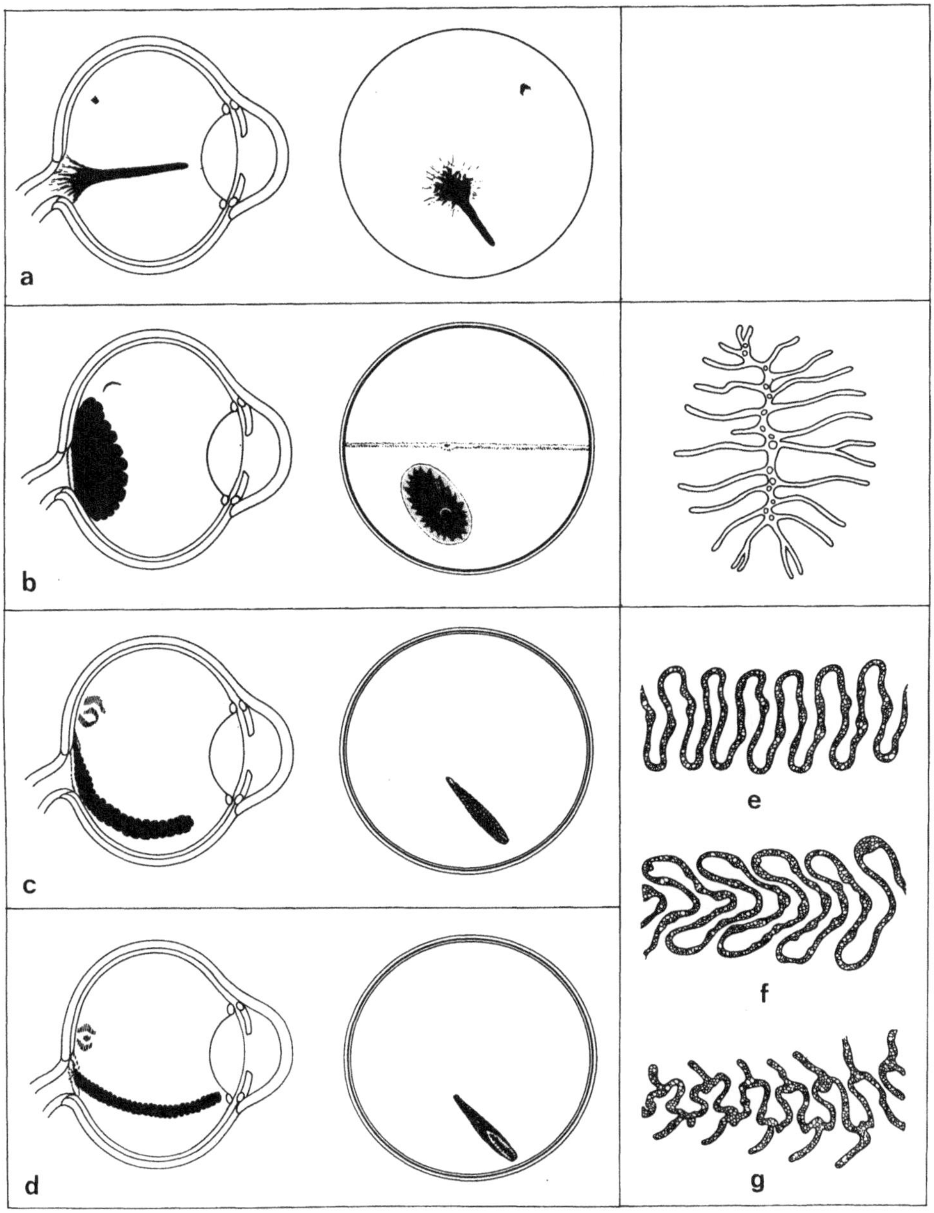

Fig. 11a–g. Morphologic types of avian pectens as depicted from lateral views of eyeball *(left column)*, from ophthalmoscopic examination of living fundus *(middle column)* and from transverse sections of pecten itself *(right column)*. (a) Conical pecten of kiwi *(Apteryx mantelli)* (after WOOD, 1917). Data on histology of this type of pecten have not been found. (b) Vaned pecten of ostrich *(Struthio camellus)* (after WOOD, 1917). Right column shows arrangement of vanes as illustrated by FRANZ (1934). (c) Pleated pecten of common wood pigeon *(Columba palumbus)* (after WOOD, 1917). This type of pecten arises from entire surface of linear optic disc and immediately slopes away from visual axis. (d) Pleated pecten of Wonga Wonga dove *(Leucosarcia picata)* (after WOOD, 1917). This type of pecten arises from entire linear optic disc and projects along concave wall of bulbus until it reaches or almost touches lens, generally near equator. Right column opposite pleated pectens (c and d) illustrates three types of pleat arrangement described by TANAKA (1938a). (e) and (f) most common patterns. (g) Rare pattern; each pleat has 1–3 membranous projections on its crest *(Alcedo atthis japonica)*

2. Vaned Pecten

The vaned pecten is characteristic of two families (Rheidae and Struthionidae), namely the rheas (VRABEC, 1958) and the ostrich, and apparently the tinamous (Tinamidae) (WALLS, 1942). It consists of a central vertical lamina which takes origin from a linear optic disc and a series of lateral vanes (25–30) arising from it (Fig. 11 b).

3. Pleated Pecten

The pleated pecten (Fig. 11 c, d) is by far the most common type and is found in all other avian families including Casuariidae (Cassowarys) (KAJIKAWA, 1923) and Dromiceidae (emus) (VRABEC, 1958). It arises from a linear optic disc as a single accordion-pleated lamina held in place at its free (apical) end (except in a few species) by a more heavily pigmented crest or bridge of tissue (Fig. 12); a bridge is lacking in owls, *Podargus* and *Haliaetus albicilla* (FRANZ, 1909).

Although the location of the pecten remains constant, i.e., its base is always directed along the former course of the embryonic choroid (retinal) fissure, its mass (length, width, height), shape, number and arrangement of pleats (Fig. 11 e, f, g), extent of pigmentation, and relationship to the ventral ciliary body varies considerably in different avian species. Within a particular species, however, these features remain relatively constant, so much so that individual birds can often be identified merely by the appearance of their pecten and its association with other ocular components.

In order to gain some insight into the significance of the pecten in the everyday operation of the avian eye, much comparative data have been compiled on the dimensions of the pecten in different birds, particularly with respect to establishing possible correlations with visual abilities and living habits (SOEMMERRING, 1818; WAGNER, 1837; BEAUREGARD, 1876; CARRIÉRE, 1885; Franz, 1909; KAJIKAWA, 1923; BACSICH and GELLÉRT, 1935; TANAKA, 1938a; BERTOLINI, 1958). These data have also been compared with the number of scleral ossicles (GIEBEL, 1857; GADOW and SELENKA, 1891), the development of the annular pad (FRANZ, 1909), the diameters of the eyeball (KAJIKAWA, 1923; BERTOLINI, 1958) and the cornea (BACSICH and GELLÉRT, 1935), as well as the type of area or fovea and the number of ciliary folds (KAJIKAWA, 1923).

In general, the information at hand indicates that the size of the pecten and the number of its pleats do not necessarily coincide with the size of the bulbus, but appear to be directly related to the behavior of the bird toward illumination and its general level of activity. Active, diurnal birds with high visual acuity and monocular rather than binocular perception tend to possess a larger and more pleated pecten, nocturnal species with poor vision a smaller one of simpler morphology. Nevertheless, no relationship, has been found to exist between the size of the pecten and the number of retinal cones (LEINER, 1950), although a rather strong correlation has been reported between corneal diameter and the length of the pecten (BACSICH and GELLÉRT, 1935). The size of the pecten in the living bird, for example, can be ascertained quite accurately from the diameter of the cornea.

With respect to eating habits, the predacious birds (hawks and eagles) tend to have the largest pectens followed by insectivorous and then seminivorous birds.

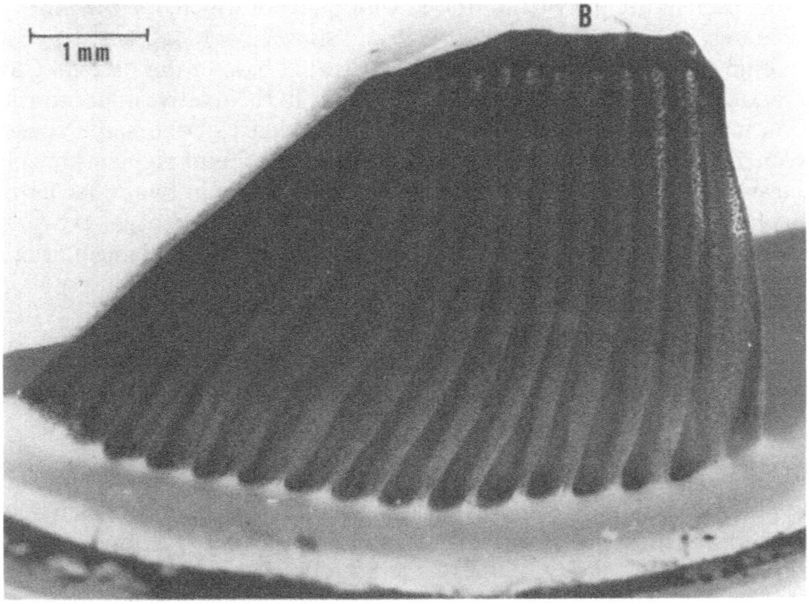

Fig. 12. Lateral view of pecten in pigeon (*Columba livia*) showing pleats (18–19 in number) and bridge (*B*). This figure is reproduced through courtesy of Drs. K. G. WINGSTRAND and O. MUNK

Most of the available data concern the number of pleats and although these values are found to be independent of the mass of the pecten (BACSICH and GELLÉRT, 1935) they do provide a good indication of the developmental strength with respect to ocular size (FRANZ, 1934). When the pecten is unfolded (by cutting it from the optic disc and bridge) its length varies between one-half and one-third of the greatest diameter of the eyeball (GIEBEL, 1857). The highest number of pleats has been recorded in *Corvus* (30). As a group the singing birds have the most pleats (20–30), nocturnal predators e.g., *Caprimulgus*, *Podargus*, the least (3–4). Asymmetrical variations in the number of pleats have been reported, particularly in *Corvus corone* (WAGNER, 1837) and *Aquila* (FRANZ, 1909).

In his morphologic classification of the pecten, WOOD (1917) groups them into three varieties based upon their appearance in the ophthalmoscope and not upon their structural organization e.g., vaned or pleated. His type I, for example, is characterized by the fact that it never extends into the vitreous any farther than the length of its widest segment. Such a pecten is found in owls and eagles, which have a pleated type of pecten, as well as in the Nubian ostrich (Fig. 11 b), which is of the vaned type. It equally covers the optic disc and has the advantage of not obstructing the incoming light rays. Types II and III are all of the pleated variety. A type II pecten does not project far into the vitreous, but immediately slopes away from the visual axis (Fig. 11 c). A type III pecten, on the other hand, projects deep into the vitreous almost up to the lens by following a straight or curved course along the concave wall of the bulbus (Fig. 11 d). Some birds possess pectens which occupy an intermediate position between types II and III.

Most pectens are heavily infiltrated with pigment which, for the most part, is more densely concentrated in the bridge and shows a gradual, but slight diminution within the pleats (from the bridge toward the base of the pecten). Considerable variation has been shown to exist, however, in the relative concentration and distribution of this pigment both within individuals of the same species and between different species. FRANZ (1909), for example, found pigment-free pleats in both eyes of one specimen of *Haliaetus albicilla*, but not in two other individuals of the same species. According to BEAUREGARD (1876), pigment is very rare in the pectens of the penguin and magpie; he contends that the pigmentation of the pecten bears no relation to its plumage.

III. Histologic Organization (Pleated)

Our understanding of the cytologic organization of the pecten has been gleaned, for the most part, from electron microscopic investigations of the pleated type of pecten. Few, if any, studies have been carried out on the other types of pecten. The following discussion, therefore, applies to the pleated pecten which is composed of three basic components: capillaries (with infrequent pericytes), pigmented stromal cells, and a superficial covering membrane. Considerable differences in the cellular arrangement of these constituents are observable in the pleats and bridge.

1. Pleats

The histologic organization of a typical pleat (*G. domesticus*) as seen with the light microscope is shown in Figure 13a. Only capillaries and pigmented stromal cells can be identified. The fine structural relationships of these components in the pigeon pleat has been diagrammatically illustrated by WINGSTRAND and MUNK (1965) and is reproduced in Fig. 13b.

a) Capillaries

The capillaries in the pleats have aroused particular interest because of the unique morphology of their endothelia. Noticeably thick when examined by light microscopy (Fig. 13a), each endothelial cell is relegated to a thin cytoplasmic plate which gives off extensive plasmalemmal-enclosed cytoplasmic processes radially from both their internal (apical, luminal) and external (basal, abluminal) surfaces (Fig. 14). In the pigeon (RAVIOLA and RAVIOLA, 1967), chicken (SEMBA, 1962; DIETERICH et al., 1973), and sparrow (JASIŃSKI, 1973) the cytoplasmic plate varies in thickness generally 0.2–0.4 μm, except in the nuclear region, and has been observed to thin down to less than a micrometer (500 Å). The usual organelles have been described in this cytoplasmic plate, including fairly abundant free ribosomes, electron dense bodies and scarce rough endoplasmic reticulum.

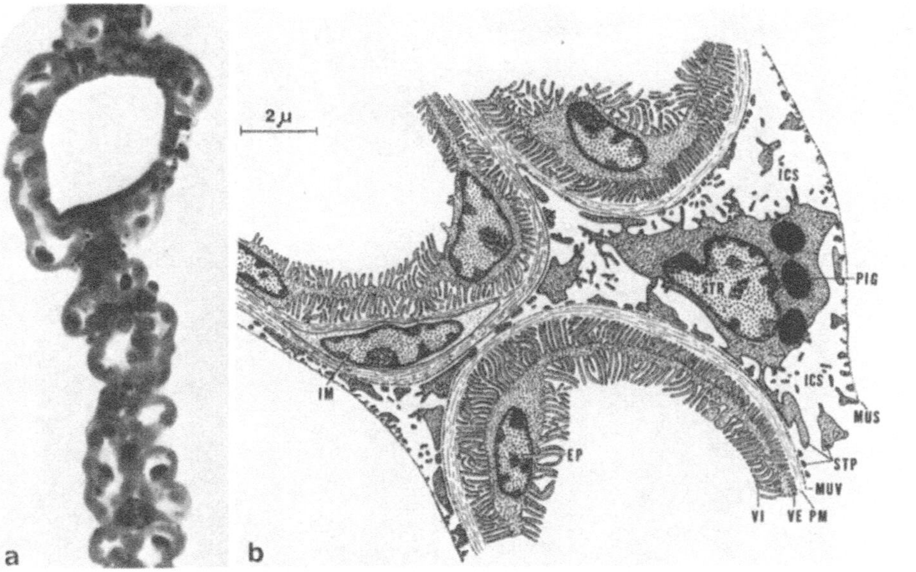

Fig. 13. (a) Light photomicrograph of portion of pleat in *Gallus domesticus*. Only capillaries and pigmented stromal cells are evident. Larger vessel possesses flattened endothelium, whereas smaller vessels present bulky lining cells. Masson, × 400. (b) Diagram of ultrastructural organization of pleated portion of pecten in pigeon (*Columba livia*) as seen in cross-section by electron microscope. Capillary endothelial cells (*EP*) are seen as thin cytoplasmic plates with microvillous-like folds projecting from their internal (luminal) (VI) and external (abluminal) (*VE*) surfaces. Each capillary is surrounded by fibrous, PAS-positive, perivascular membrane (*PM*) which also encloses pericytes or intramural cells (*IM*). Stromal cells (*STR*) containing pigment granules (*PIG*) occupy intercellular spaces (*ICS*) between capillaries and give off numerous processes (*STP*), many of which end on mucopolysaccharide membrane covering perivascular membranes (*MUV*) or surface of pecten pleats (*MUS*). Note that intercellular space (*ICS*) forms continuous cavity between these membranes occupied only by stromal cells and their processes; superficial membrane (*MUS*) contacts either these processes or intercellular space directly. This figure is reproduced through courtesy of Drs. K. G. WINGSTRAND and O. MUNK

The processes emanating from the endothelial plate likewise vary in size, shape and orientation; DIETERICH and DIETERICH (1974) found them to be relatively sparse in the owl (*Tyto alba*), more abundant and uniform in the kestrel (*F. tinnunculus*), and highly developed in the sparrow (*P. domesticus*). Originally termed apical microvilli and basal infoldings (TANAKA, 1960) they have been repeatedly shown in serial sections to be primarily broad, tongue-like microfolds at least 1 μm in length and depth and 400–800 Å in thickness.

The internal microfolds are more numerous than the external folds and tend to be oriented with their long axes parallel to the long axis of the capillary. Those arising from the cytoplasmic plate are fairly erect, their distal ends ending unbranched and freely in the lumen (Fig. 15). In the nuclear region, however, they become shorter, more scarce and disordered and may be replaced by cytoplasmic swellings (FISCHLSCHWEIGER and O'RAHILLY, 1966). The internal microfolds represent the innermost lining of the capillaries. As such, they are coated in *Gallus*

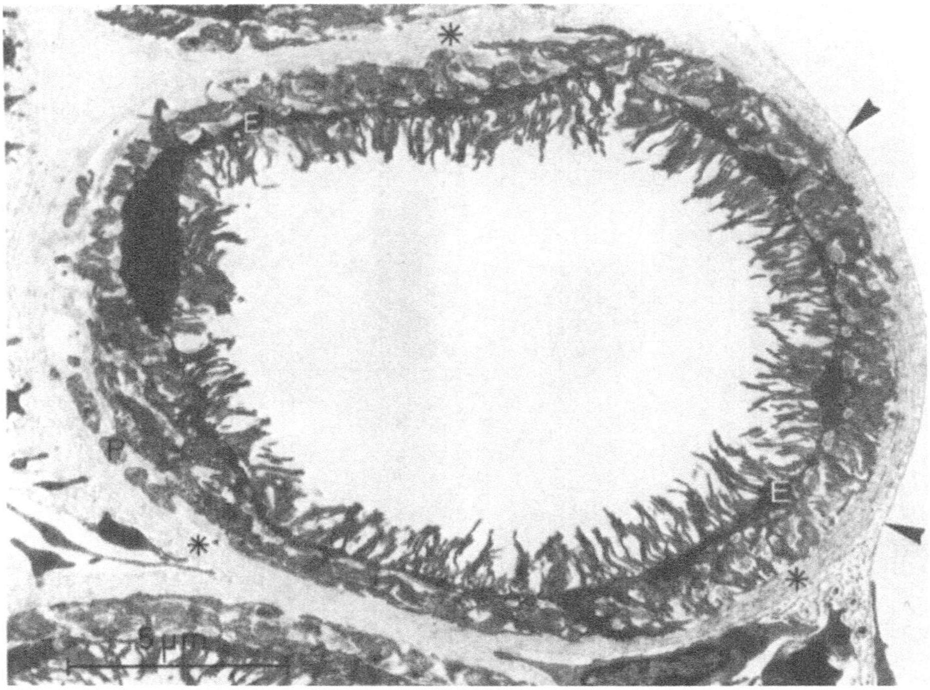

Fig. 14. Transmission electron micrograph of capillary in pleated portion of pecten in *Gallus domesticus* showing thin cytoplasmic plate of endothelium (*E*), as well as its internal and external processes. Both perivascular membrane (*) and processes of pericytes (*P*) are seen in close contact with external processes of endothelium. Superficial covering membrane surrounding pecten is indicated by arrows. At lower right is apparently process of hyalocyte adhering to this covering membrane. This figure is reproduced through courtesy of Drs. C.E. DIETERICH, H.J. DIETERICH, M.A. SPYCHER, and M. PFAUTSCH (1973)

(SEAMAN, 1966) and *Larus* (WELSCH, 1972), but not in the pigeon (RAVIOLA and RAVIOLA, 1967), by an extracellular mucopolysaccharide-protein complex and possess alkaline phosphatase activity (WELSCH, 1972; v. BASSEWITZ et al., 1975) (Fig. 16). Aminopeptidase activity has also been localized in the microfolds (WELSCH, 1972).

The external microfolds, on the other hand, tend to be shorter, and more irregular than the internal folds and to be arranged with their long axes perpendicular to the long axis of the capillary. They branch frequently and their distal ends bend hook-like before terminating in the perivascular membrane which surrounds all capillaries in the pleats (Fig. 15).

Both systems of microfolds serve to increase significantly the surface area of the capillary endothelia. According to JASIŃSKI (1973), this amounts to 20% increase in *Passer*. Further increase in surface area has been accomplished experimentally in *Gallus* by SEAMAN and HIMELFARB (1963) who reported the transformation of the microfolds into a complex membranous system after intravenous injections of diamox.

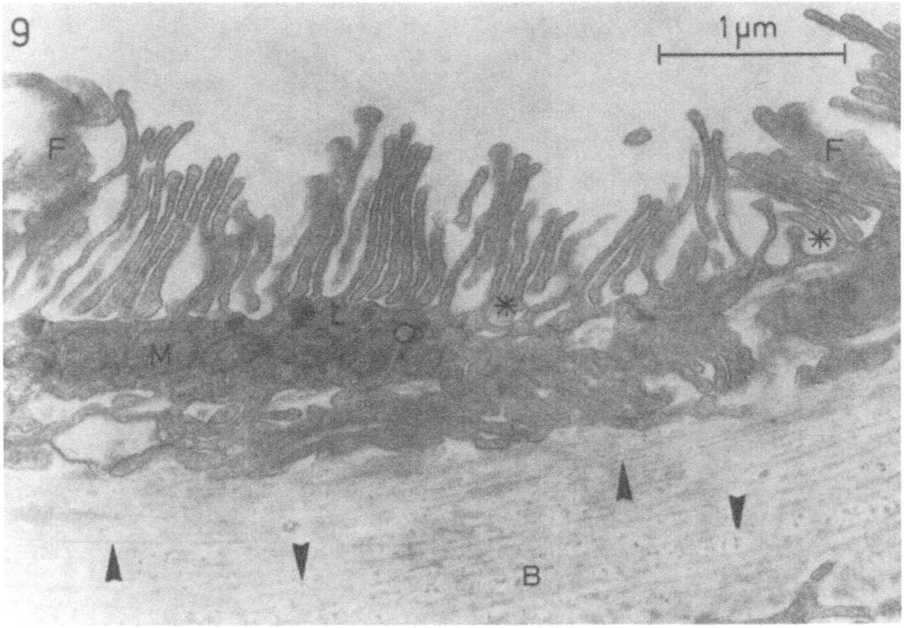

Fig. 15. Transmission electron micrograph of capillary endothelium in pleated portion of pecten of *Gallus domesticus*. Internal processes (*F*) are relatively thin and straight compared to bulky and twisted configurations of outer processes. Thin cytoplasmic plate of endothelial cell, particularly narrow in some places (*) contains mitochondria (*M*), cytosomes (*L*), and filaments (*O*). Perivascular sheath (*B*) contains numerous filaments (arrows) often showing cross striations. Figure is reproduced through courtesy of Drs. C.E. DIETERICH, H.J. DIETERICH, M.A. SPYCHER, and M. PFAUTSCH (1973)

Freeze-fracture studies of the microfold membrane substructure (DIETERICH et al., 1973) reveal a similar uniform distribution of particles in both internal and external folds.

The endothelial cells of the pecten are attached to one another on their lateral surfaces by several types of intercellular junctions which appear to exhibit considerable variation as reported by different investigators. In *G. domesticus*, for example, terminal bar complexes are described by SEMBA (1962) and FISCHLSCHWEIGER and O'RAHILLY (1964, 1966), whereas only the zonula adhaerens or desmosome component of this complex is reported by DIETERICH et al. (1973) and FIELDING (1972) respectively. In *P. domesticus* only zonulae adherentes are evident (Fig. 17) (JASIŃSKI, 1973); in *Larus ridibundus* both zonulae occludentes and desmosomes are present (WELSCH, 1972). Desmosomes are also observed in *Columba palumbus*, *Phasianus colchicus*, *Corvus corax*, and *Surnia ulula* (NGUYEN H. ANH et al., 1967).

In contrast to ordinary capillaries which are enclosed by a thin basement membrane, the pecten capillaries are surrounded by a thick, filamentous and trilaminar membrane, rich in alkaline phosphatase activity (WELSCH, 1972). Termed the perivascular membrane (endothelial support lamina of FIELDING, 1972; vessel sheath of DIETERICH et al., 1973) it appears in the light microscope as

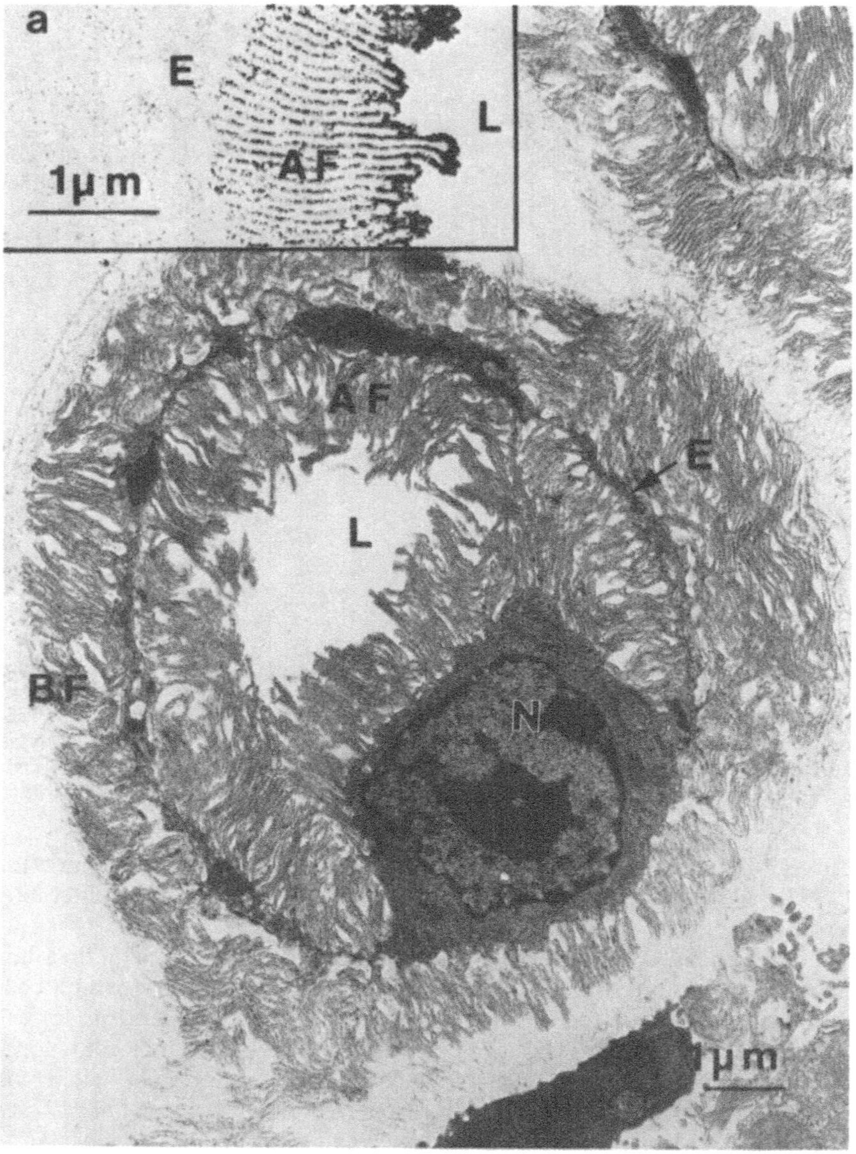

Fig. 16. Transmission electron micrograph of capillary in pecten of sparrow *(Passer domesticus)* showing thin cytoplasmic plate comprising endothelial cell (*E*) and extensive apical (*AF*) and basal (*BF*) microfolds given off by it. *L*, lumen; *N*, nucleus. × 12000. Insert a, localization of alkaline phosphatase activity along margins of apical microfolds (*AF*) of endothelium (*E*). X 19000. Figure is reproduced through courtesy of Drs. D. B. VON BASSEWITZ, H. J. DIETERICH, and C. E. DIETERICH (1975)

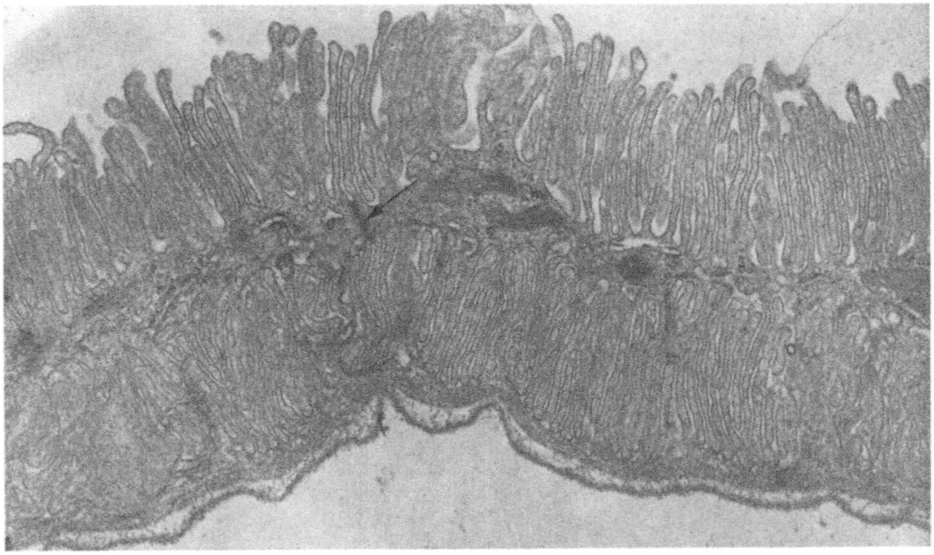

Fig. 17. Transmission electron micrograph of capillary wall in pleated portion of pecten of sparrow (*Passer domesticus*) (JASIŃSKI, 1973). Arrow indicates junction (zonula adherens) of two endothelial cells. Contrary to general trend, microfolds of this capillary are organized in same plane. × 27040. Figure is reproduced through courtesy of Dr. JASIŃSKI

a highly refractile and homogenous intermediate layer surrounded internally and externally by typical periodic acid-Schiff–positive basement membranes (O'RAHILLY and MEYER, 1961). In the pigeon these basement membranes consist ultrastructurally of a thin condensation of amorphous mucopolysaccharide material, the outer being more distinct than the inner. The thick intermediate hyaline layer, in turn, contains lamellae of numerous collagenous filaments (with distinct periodicities) embedded in a homogeneous ground substance of low electron density; no contractile elements are observed (WINGSTRAND and MUNK, 1965). In the chick this perivascular membrane consists of as many as 15 discontinuous layers of basement membrane-like material each separated by 100–250-Å-thick filaments, often displaying cross-striations (Fig. 15) (DIETERICH et al., 1973).

Fusiform cells termed pericytes (periendothelial, subendothelial, or intramural cells) are infrequently found enclosed within the perivascular membrane (Figs. 13b, 14). Their cell body contains the typical organelles, some individual filaments, and various inclusion bodies. Fine processes from these cells extend between the lamellae and may directly contact the external microfolds of the endothelium (SEAMAN and STORM, 1963; RAVIOLA and RAVIOLA, 1966; DIETERICH et al., 1973).

b) *Pigmented Stromal Cells*

The extravascular portion of the pecten pleat consists of only one morphologically-discernible cell, the pigmented stromal cell (Fig. 18). These cells are scattered throughout an extensive intercellular space, a continuous cavity between the capillaries and the superficial covering membrane, into which they send many

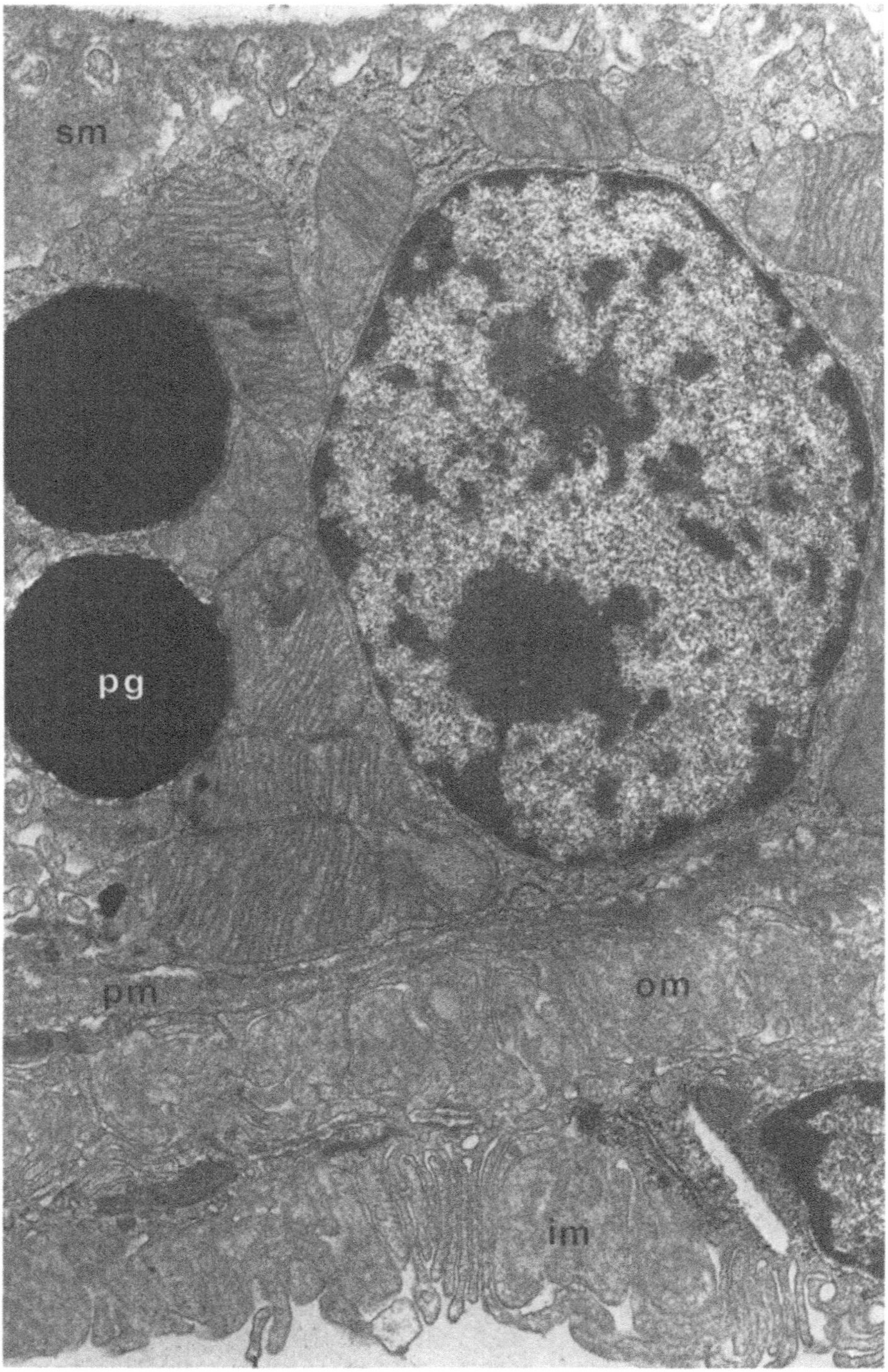
sm
pg
pm
om
im

interdigitating cytoplasmic processes (Fig. 13b). Perfusion studies have shown that this space is readily accessible to electron dense markers (e.g., lanthanum particles) which become uniformly distributed between the capillaries and stromal cells with no apparent barrier to their passage into the vitreous humor (FIELDING, 1972).

The cell bodies of the stromal cells generally contact other stromal cells and one or two capillaries. Their thin processes extend to the superficial covering membrane where they terminate usually with round feet, and to the perivascular membranes of the capillaries where they often end as flattened feet (WINGSTRAND and MUNK, 1965). The rounded feet perhaps account for the raspberry-textured appearance of the surface of these cells when observed with the scanning electron microscope (Fig. 19) (DIETERICH et al., 1973). In *Larus* electron-dense material resembling hemidesmosomes are present in the cytoplasmic side of these feet (WELSCH, 1972). Dispersed throughout the cytoplasm of the stromal cells are granules of melanin (BACSICH and GELLÉRT, 1935), sometimes over 2 µm in diameter (DIETERICH, 1973), numerous free ribosomes, large mitochondria (as large as 2.8 µm in the pigeon) (RAVIOLA and RAVIOLA, 1967), bundles of filaments and the other usual organelles. According to PORTE et al. (1968) the stromal cells at the base of the pecten are continuous with the retinal Müller cells from which they arise.

2. Bridge

In contrast to the pleats, the bridge of the pecten is characterized by the irregular distribution and compact arrangement of the stromal cells resulting in a concomitant reduction in the intercellular spaces and in the number of capillaries.

a) Capillaries

The capillaries in the bridge resemble typical endothelial tubes lacking microfolds (WINGSTRAND and MUNK, 1965; PORTE et al., 1968), although some still retain these surface modifications (RAVIOLA and RAVIOLA, 1967). A thick bilaminar adventitial coat consisting of an inner PAS-positive basement membrane and an outer fibrillar layer surrounds these vessels (RAVIOLA and RAVIOLA, 1967).

b) Pigmented Stromal Cells

As demonstrated by FISCHLSCHWEIGER and O'RAHILLY (1968) the stromal cells of the chick bridge are more heavily concentrated in its distal portion where they constitute the only cells present. Here they form several continuous, epithelioid layers of tall, slender, heavily pigmented cells with basal nuclei and numer-

Fig. 18. Transmission electron micrograph of portion of stromal cell in pleated portion of pecten of sparrow *(Passer domesticus)* (JASIŃSKI, 1973). Stromal cell is lodged between superficial covering membrane (*sm*) and perivascular membrane (*pm*) of capillary. It possesses pigment granules (*pg*) and many large mitochondria. Irregular inner (*im*) and outer (*om*) microfolds characterize the capillary endothelium. This figure is reproduced through courtesy of Dr. JASIŃSKI

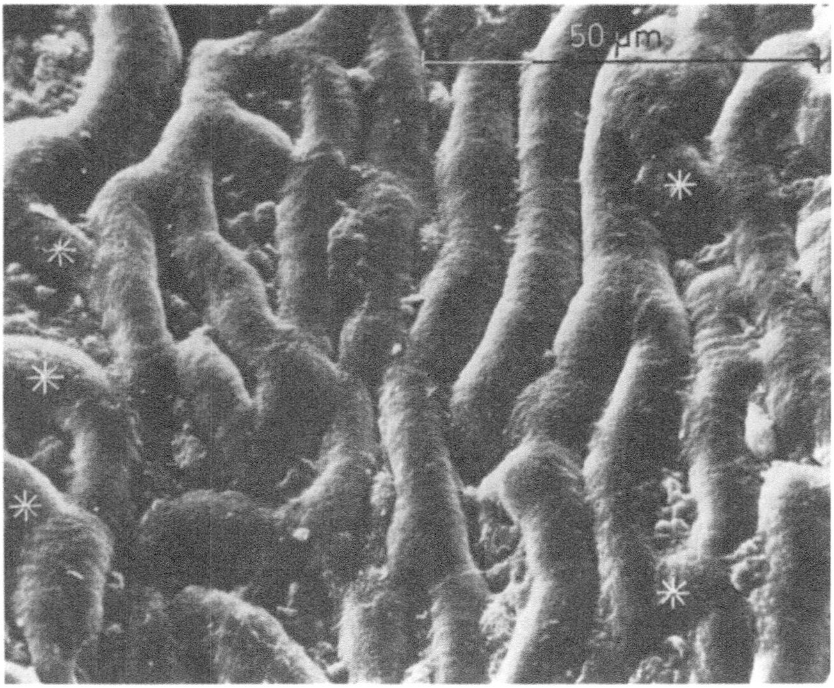

a

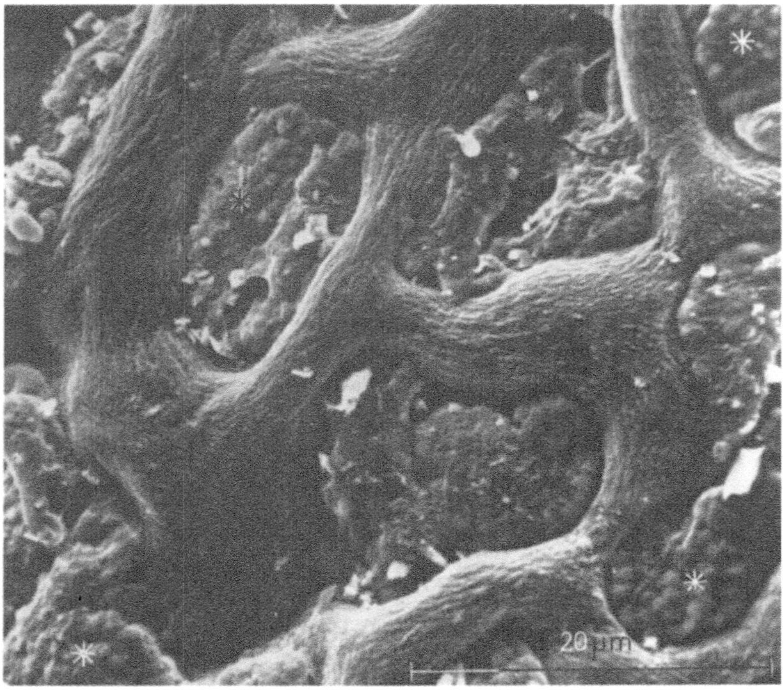

b

ous apical (distal) processes directed toward the surface. This distal portion of the bridge undoubtedly corresponds to the broad zone of alkaline phosphatase activity detected histochemically by WELSCH (1972). Although precise localizations were not recognizable in his light microscopic preparations, the absence of enzymatic activity in the blood vessels of the bridge strongly implicated the pigmented stromal cells as the custodians of this enzyme. In the chicken, BAWA and YASHROY (1972) find approximately 3.7 times more alkaline phosphatase activity in the pecten (per gram of tissue) than in the retina and note a marked reduction in this activity in the pecten (contrary to that in the retina) following dark adaptation. According to FISCHLSCHWEIGER and O'RAHILLY (1968), pigment granules, mitochondria, and rough endoplasmic reticulum are found randomly distributed throughout the stromal cells whereas some cytoplasmic constituents, e.g., Golgi complexes and secretory vesicles, are concentrated in the supranuclear and apical regions, respectively, a polarization which they point out is suggestive of a secretory activity. RAVIOLA and RAVIOLA (1967) describe round or oval, PAS-negative, nonmembrane bound bodies (2–6 μm in diameter) of flocculent material in the stromal cells of some of the pigeons they examined but do not indicate the intracytoplasmic distribution of this unknown substance.

The intercellular spaces in the bridge are small, but fairly numerous and are characterized in the electron microscope by the presence of a meshwork of connective tissue elements, e.g., fibroblasts and collagenous fibrils (PORTE et al., 1968). These collagenous, as well as reticular fibers, have been selectively stained and observed in the light microscope by BACSICH and GELLÉRT (1935).

3. Superficial Covering Membrane

The entire pecten, pleats and bridge, is enveloped by a PAS-positive membrane which is continuous at the base of the pecten with the internal limiting membrane of the retina (O'RAHILLY and MEYER, 1961). Termed the superficial covering membrane (vitreopectineal limiting membrane), it separates the stromal cells and capillaries from the vitreous (Figs. 13, 14) and represents vitreal fibrils (PAS-positive component) plus the basal lamina portion of the internal limiting membrane which becomes invaginated by the developing pecten (GÄRTNER, 1966). Because the PAS-positive fibrillar component is not very evident in electron micrographs, the majority of investigators have designated this basal lamina alone as the superficial covering membrane which in the pigeon, for example, measures 300Å (WINGSTRAND and MUNK, 1965). On the pleats it presents a

Fig. 19a and b. Scanning electron micrograph of outer (vitreal) surface of pleated portion of pecten in *Gallus domesticus* (DIETERICH et al., 1973). (a) Low power view revealing strongly anastomosing (*) network of irregular capillaries. × 950. (b) Higher magnification showing furrowed appearance of capillary walls and raspberry-textured surface of the pigmented stromal cells (*) situated between capillaries. Large intercellular pores described here by DIETERICH et al. (1973) may represent cellular-free surface areas between stromal cell processes described by WINGSTRAND and MUNK (1965) and depicted in Fig. 13b. × 2600. Figure is reproduced through courtesy of Drs. C. E. DIETERICH, H. J. DIETERICH, M. A. SPYCHER, and M. PFAUTSCH smooth contour adhering closely to the superficial surfaces of the capillary walls and stromal cells. When viewed with the scanning electron microscope (DIETERICH et al., 1973), the capillary walls appear furrowed. The stromal cells possess a raspberry-textured surface (Fig. 19) and pores are sometimes observed between them. According to RAVIOLA and RAVIOLA (1967), the covering membrane represents one of two continuous membranes (basement laminae) which, with an intervening discontinuous layer of stromal cells, is interposed between the vitreous body and the pecten capillaries.

On the bridge the superficial covering membrane covers the epithelial-like layer of stromal cells which is very irregular in contour on its distal portion owing to the presence of various depressions. Collagenous filaments from the vitreous penetrate into these depressions and serve for the attachment of the bridge to the vitreous body. Detected by early histologists, these filaments are stained blue in Azan preparations and are readily seen by light microscopy penetrating from the vitreous deep into the substance of the bridge (WINGSTRAND and MUNK, 1965).

4. Cortical Tissue Layer and Hyalocytes

According to BALAZS et al. (1964), a thin gelatinous "cortical tissue layer" of the vitreous body adheres to the external surface of the pecten and retina in the chicken. Formerly termed the hyaloid membrane, this layer contains a uniform population of cells which, on the basis of their staining and morphologic characteristics, are considered by BALAZS et al. (1964) to be "connective tissue cells of the macrophage type." Designated by BALAZS et al. (1959) as hyalocytes, these cells have also been called endothelial-like cells (SEMBA, 1962), peri-pectinate cells (FISCHLSCHWEIGER and O'RAHILLY, 1966), and free vitreal cells (PORTE et al., 1968). They have not been observed in the pigeon (RAVIOLA and RAVIOLA, 1967).

On the basis of their fine structure, e.g., well-developed rough endoplasmic reticulum, extensive Golgi complexes, voluminous dense bodies, and granulomembranous vacuoles PORTE et al. (1968) agree that hyalocytes possess fibroblastic properties, as well as some characteristics reminiscent of phagocytic and resorptive activities. They suggest that these cells may actively intervene in certain metabolic processes concerned with hemovitreal exchange.

IV. Functions

Published accounts of the pecten first appeared about 300 years ago (1673–1675). Regarded as "one of the greatest puzzles in comparative ophthalmology" (WALLS, 1942), there have been a great many functions attributed to it, most of which lack critical experimental substantiation. [For interesting historical notes see WINGSTRAND and MUNK (1965).] Much indirect evidence has been obtained, however, from microscopic (light and electron), histochemical, and experimental investigations, but more conclusive proof is still required to determine the essential functions of the pecten. It is quite likely, for example, that it serves more than one role in the normal operation of the avian eye (TURCHINI, 1933). Since the end of the century periodic reviews have appeared on this subject (almost every 15 years), a true indication of the interest this organ has evoked in the field of comparative ophthalmology (BEAUREGARD, 1876; VIRCHOW, 1900; FRANZ, 1909; WOOD, 1914; V.SZILY, 1922; KAJIKAWA, 1923; MANN, 1924a; FRANZ, 1934; MENNER, 1938; WALLS, 1942; DUKE-ELDER, 1958; WINGSTRAND and MUNK, 1965). In the most recent survey WINGSTRAND and MUNK (1965) recognized 30 different functional theories which they organized into a table of nine major groups provided with the most pertinent references (both advocators and contenders). For the most part, these functions involve the vascular system, the pigmented stromal cells and both the morphology and intraocular position of the pecten itself. Still unanswered is WALLS' (1942) plea for more information concerning "the *true* size of the pectens—the area over which they expose the blood circulating in them, the rate at which the blood is changed for fresh, the permeability of their vessels, and so on," all essential data for a more reliable interpretation of the pecten's function.

1. Functional Theories Involving the Vascular System

According to the most popular and long-standing theory, the pecten acts as a supplemental nutritive organ, providing oxygen and nutrients for the inner portion of the retina. Its richness in blood vessels, particularly capillaries, its pleated arrangement to increase its surface area, and its advantageous intraocular location in close proximity to a retina devoid of blood vessels are grounds enough to persuade early investigators of this important function. More convincing evidence, however, was initially provided by ABELSDORFF and WESSELY (1909), who observed ophthalmoscopically in several birds, including the pigeon and an owl, that fluids (in this case an intravenously injected fluorescent marker, potassium fluorescein) readily pass into the vitreous from the pecten. TURCHINI (1933) observed this phenomenon using acriflavine, but RODRIGUEZ-PERALTA (1968) failed to substantiate it using a different fluorescent label (diaminoacridines). More recently, however, BELLHORN and BELLHORN (1975), employing sodium fluorescein have confirmed and pictorially documented ABELSDORFF and WESSELY's and TURCHINI's findings by means of rapid serial fundus angiography (Fig. 20). In their experiments with *C. domestica*, *Rhea americana albescens*, *Otus asio naevus*, and *Alectoris grecachukar*, BELLHORN and BELLHORN (1975) recorded an initial choroidal flush (Fig. 20a) followed by the filling of the pecten (Fig. 20b) and finally an outward spreading of fluorescein from the pecten into the vitreous (Fig. 20c). Subsequent transvitreal passage of such fluids toward the retina is undoubtedly facilitated by the liquid nature of the vitreous in the vicinity of the retina and pecten in contrast to the gelatinous state of the vitreous anteriorly (Fig. 21) (BALAZS et al., 1959, 1960). The presence of a completely gelatinous vitreous body in owls (BALAZS et al., 1959) can perhaps be correlated with the relative weak development of the pecten in these species.

Experimental evidence that nutrients derived from the pecten are important for the maintenance of the retina has been controversial. WINGSTRAND and

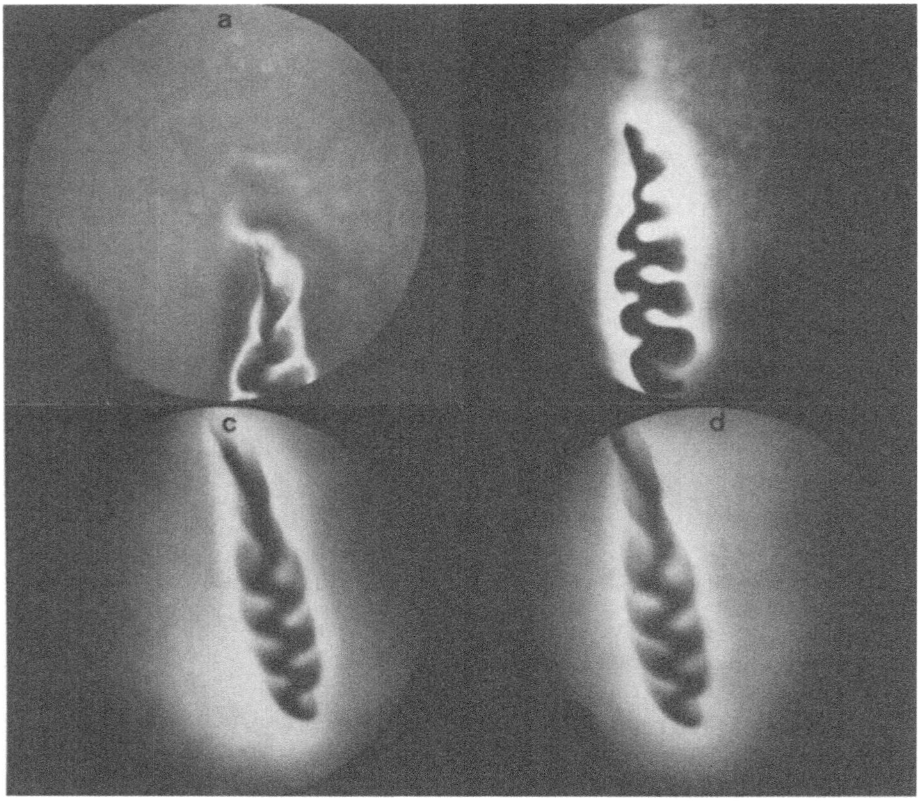

Fig. 20a–d. Fluorescein fundus photographs of pecten of screech owl *(Otus asio naevius)* following intravenous injection of sodium fluorescein (BELLHORN and BELLHORN, 1975). (a) Choroidal flush and faint appearance of dye along margin of pecten, (b) Complete filling of pecten. (c) Fluorescein begins to diffuse into vitreous. (d) Relatively diffuse fluorescence. Figure is reproduced through courtesy of S. KARGER AG, Basel and Drs. R. W. BELLHORN, and M. S. BELLHORN

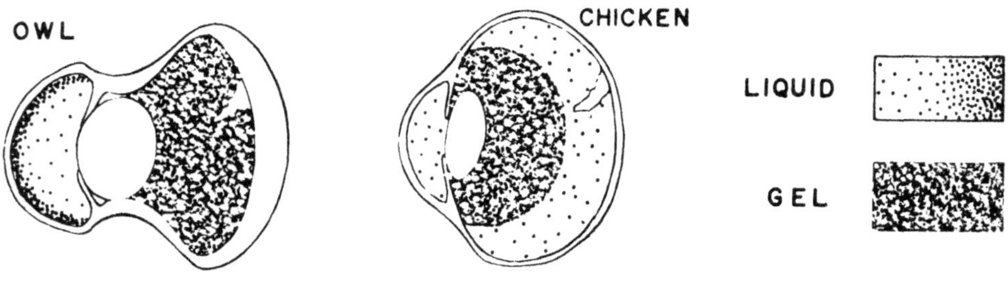

Fig. 21. Physical state of vitreous body in owls *(Bubo virginianus virginianus, Strix varia varia)* and chicken *(Gallus gallus)* (after BALAZS, 1960). Entire vitreous body of owls is gel whereas in chicken (and pigeon and turkey as well) it is solid gel behind lens and liquid next to retina and pecten

MUNK (1965) have shown that operative elimination of the arterial supply to the pecten not only results in a complete degeneration of this organ, but also produces well-developed degenerative changes in the retina beginning with, and sometimes confined to, its innermost layers (nerve fiber and ganglion cell layers). These results indicate, they maintain, that the "intact pecten is necessary for the maintenance of a normal retina." Since oxygen is involved in this process, the authors then measured its pressure in different parts of the vitreous and calculated its rate of diffusion in order to establish the possible existence as well as the physiologic significance, if any, of diffusion gradients from the pecten to the retina. They find that such a gradient does indeed exist and that the diffusion of oxygen is of sufficient rate to be metabolically significant for the maintenance of the inner layers of the retina. Experimental destruction of the pecten by photocoagulation (FRANCOIS and NEETENS, 1974) and intraocular electrocautery (BRACH, 1975), on the other hand, has had no visible effect on the histology of the retina. Laser energy has also been used to destroy the pecten but with inconsistent results (BACSICH et al., 1966a; BACSICH and GELLÉRT, 1968). BRACH (1975), who claims to have produced total ablation of the chick pecten contends that the retinal atrophy observed by WINGSTRAND and MUNK (1965) is a result of operational damage to the optic nerve axons owing to accidental destruction of the blood supply to the optic nerve head. He believes, therefore, that the pecten is not concerned primarily with the nutritional supply of the retina. In support of this supposition he points out that the avian retina undergoes anaerobic metabolism (KREBS, 1927) and lacks both mitochondria and succinic dehydrogenase activity in its innermost portion (HUGHES et al., 1972). Although BRACH'S (1975) pictorial evidence for total ablation of the chick pecten is not very convincing, he does find alterations in the pH of the vitreous caused by the ablation of the pecten and suggests that perhaps this organ is involved in the regulation of intraocular pH. Nevertheless, the presence of a uniformly electron lucent and structureless intercellular space within the pecten undoubtedly facilitates gaseous exchange between the capillaries of the pecten and the vitreous (JASIŃSKI, 1973).

Likewise, the development of both internal and external microfolds, surface modifications which greatly increase the surface area of the capillaries, promote fluid transport to and from these vessels and lend further support to the nutritional role of this organ. In addition, high concentrations of carbonic anhydrase (LEINER, 1950; KAUTH and SOMMER, 1953) and alkaline phosphatase (FRIESS, 1969; WELSCH, 1972) activities both of which are known to be involved in active transport processes have been detected in these endothelial cells. According to BAWA and YASHROY (1972), this alkaline phosphatase activity varies with the intensity of light; it is greatest during light adaptation (3.7 times greater than in the retina), when the increased energy demands from the avascular retina require the release of additional nutrients from the pecten. In an ultracytochemical study V. BASSEWITZ et al. (1975) localizes this enzyme at the internal (luminal) microfolds of the endothelium (Fig. 16), which may therefore represent the precise site of active transport. Likewise, on the basis of intravitreal and intravenous injections of a fluorescent marker (diaminoacridines), the endothelial membrane on this side is disclosed as a hematic barrier (RODRIGUEZ-PERALTA and KAIRYS, 1968). The pigmented stromal cells of the pecten have been implicated as the energy source for this transport because they exhibit a high metabolism (SEAMAN and HIMELFARB, 1963).

On the basis of ultrastructural studies it is now well-established that the anatomical blood-vitreal barrier in the pecten comprises the following structures: endothelial cytoplasm, two basement membranes, a discontinuous layer of pigmented stromal cells, and a superficial covering membrane, all of which may be influential in limiting the exchanges of materials. Likewise, the possibility of intercellular transport between the endothelial cells should not be ruled out, although in some species junctional complexes believed to prevent the passage of substances, e.g., zonulae occludentes, have been recognized there. Nevertheless, little is known of the permeability characteristics of these and other intercellular junctions, but some interesting data have been obtained on the possible route and mechanism of intracellular transport in the capillary walls of the pecten utilizing intravenously injected, electron dense markers. Ferritin, for example, enters this endothelium by pinocytosis and may be transported through its cytoplasm within a system of vesicles (SEAMAN and STORM, 1966), although these are few and some of the ferritin is found free in the cytoplasm (SEAMAN, 1966). Lanthanum, on the other hand, is believed to traverse the blood-vitreal barrier mainly by diffusion (FIELDING, 1972). It becomes uniformly distributed in the extracellular spaces surrounding the capillaries and stromal cells and apparently experiences no further obstacle in its passage toward the vitreous.

The significance of the vascular system in the functioning of the pecten is further supported by the existence of two extrabulbar vascular mechanisms believed to regulate the flow of blood into this organ: rete mirabile (pectinis), arteriovenous anastomoses. According to KAJIKAWA (1923), in many birds the arteries destined for the pecten course through two separate and distinct capillary networks or rete mirabile (ophthalmicum and pectinis) of unknown significance before giving rise to the true arteries of the pecten. He surmises, however, that they may be concerned with maintaining a uniform arterial pressure for the pecten vessels. ROHEN (1955) notes that blood flow into the pecten and choroid, although independent of each other, is regulated by a system of extrabulbar arteriovenous anastomoses capable of diverting the blood into a prominent venous plexus. ROHEN describes such a system in *Columba* and *Turdus* and believes it is understandable for an organ subjected to fluctuating demands, e.g., providing nutrients for diurnal or nocturnal retinae, regulating intra-ocular pressure.

WALLS (1942) contends that the increased vascularity contributed by the pecten is associated with the high level of metabolism exhibited by the avian retina. He feels that the choroidal vessels cannot adequately supply the entire retina, indeed even the photoreceptors themselves. According to MANN (1924a, b), the intraocular location of the pecten increases the efficiency of the eye as an optical system without interfering with its metabolism by depriving it of any of its blood supply. She regards the pecten as a phylogenetic adaptation pointing to specialization in the direction of improved vision rather than any other sense in these animals. She points out that the visual acuity of animals possessing a vascular retina (e.g., man, most mammals, some amphibians) can never equal that of a bird because the presence of blood vessels in front of the sensory elements of the retina interferes appreciably with the clearness of the visual image.

In supplying the retina with nutrients, WALLS (1942) admits that heat is thereby given off, but claims that this is not needed and is only incidental to its main nutritional function. KAJIKAWA (1923) and BACSICH and GELLÉRT (1935) however, contend that this heat loss is sufficient enough to maintain the avian eye at a constant temperature. They believe that this represents a primary function of the pecten; it protects the eye against cooling, particularly during flight. At the same time excess heat from the sun may also be removed by these capillaries (GRIFFIN, 1953).

2. Functional Theories Involving Stromal Cells

The stromal cells represent the major extravascular elements in the avian pecten. As such they are most suspect of playing a role in secretion and light absorption, two functions attributed to the pecten. LINDSAY-JOHNSON (1901) postulates that the stromal cells secrete a fluid for the nourishment of the vitreous and neighboring structures. ABELSDORFF and WESSELY (1909) and v. HUSEN (1913) believe that it is the vitreous humor itself that is produced by the pecten. Although ABELSDORF and WESSELY point out that this secretory role is minor to that of the choroid, they do not indicate the cells responsible for this secretion. They base their conclusion upon experiments involving the operative removal of the ciliary processes which results not only in an almost complete preservation of this humor, but likewise in the development of a strongly hypertrophied pecten. KAJIKAWA (1923) supports this functional interpretation and implicates the vascular endothelia as the secretory elements. Moreover, on the basis of histochemical studies, AGARWAL et al. (1966) believe that mucopolysaccharides are synthesized by the pecten; they find similarly stained (alcian blue and PAS-positive) carbohydrates in both the pecten and the vitreous.

Recent electron microscopic studies have provided more convincing evidence for the secretory abilities of some stromal cells, although the precise nature of the secretory products and their ultimate disposition have still not been clearly established. Nevertheless, owing to the almost complete investment of the pecten by the vitreous, it is assumed that most secretions are destined for this body.

In an electron microscopic study of the chick pecten, FISCHLSCHWEIGER and O'RAHILLY (1968) characterize the pigmented stromal cell in the distal part of the bridge as secretory in nature. Here they form several epithelial-like layers of tall, slender cells possessing an intracellular polarization of cytoplasmic organelles (secretory vesicles, Golgi apparatus) and occupying a zone within the pecten of high alkaline phosphatase activity (WELSCH, 1972). According to JASIŃSKI (1973), the stromal cells in the pleats are characterized by large mitochondria and abundant ribosomes indicative of a high metabolic activity, perhaps, he conjectures, the production of the superficial covering membrane. Indeed, the high metabolic activity of these cells was previously advanced by SEAMAN and HIMELFARB (1963) from experimental studies in which intravenous injections of diamox (a known specific inhibitor of carbonic anhydrase and thus an inducer of decreased intraocular pressure) cause striking increases in the number of cytoplasmic filaments and in the size and number of mitochondria. They conclude that these cells may also provide the energy needed for the transport of fluids occurring in this organ.

The fact that fluid may be given off into the vitreous body by the pecten, whether this be secreted by stromal cells or merely pass actively or passively from them or the capillaries, has led many investigators, e.g., RABL (1900), LEPLAT (1912), KAJIKAWA (1923), and MANN (1924a, b), to believe that the pecten is involved in the regulation of intraocular pressure, a function normally attributed to the ciliary processes owing to its continual production of aqueous humor. Evidence that the pecten may operate as a marginal apparatus for the maintenance of this pressure has also been provided by SEAMAN and HIMELFARB (1963) who find that both the pecten and ciliary body undergo similar ultrastructural changes when the intraocular pressure is experimentally reduced with intravenous injections of diamox.

The light absorptive properties of the pecten are due to the densely packed melanin pigment granules dispersed throughout the cytoplasm of the stromal cells. Melanin granules, themselves, represent a functionally important constituent of the pecten because by absorbing the diffuse light rays entering the eye they prevent any internal reflections which may interfere with the production of a clear image (BEAUREGARD, 1876; MANN, 1924a). BAWA and YASHROY (1974) contend that the absorption of light by the pigment granules, particularly in the bridge, occurs at the same time as the physiologic activity of the retina is increased by the exposure to light (BAWA and YASHROY, 1972). They propose that in the pecten, the absorption of light results in a rise in temperature which, in turn, increases the rate of metabolism and stimulates the secretory cells (e.g., those described by FISCHLSCHWEIGER and O'RAHILLY, 1968) to secrete nutrients required to meet the increased demands of the metabolically-active, avascular, avian retina.

3. Functional Theories Involving Pecten as a Whole

The fairly constant anatomical position of the pecten within the vitreous body has motivated many functional theories, the most plausible of which is concerned with its influence on vision, particularly the perception of movement. The pecten lies almost in the plane of one of the secondary optic axes (MANN, 1924a, b) and consequently casts a shadow upon the retina. MANN (1924a, b) maintains that this shadow falls mainly along its own base (linear optic disc) and very slightly to each side, the net result being a slight enlargement of the blind spot (upper part of the visual field) and no interference with the area or fovea, which in most cases lies above the upper end of the pecten. Not being convinced of this, MENNER (1938) conducted an extensive ophthalmoscopic examination of 56 different species of living birds and recorded the fundic shadows produced by a variety of pectens of different sizes and shapes. In contrast to MANN (1924a), he believes that these shadows, although never completely symmetrical in both eyes, are of significant dimensions to influence retinal responses. The shadow figures he illustrates vary between a large triangle as seen in the diurnal predators which possess the largest pectens (Fig. 22A) and a narrow elongated oval appearance exemplified by the small pectens of nocturnal birds (Fig. 22C). The large triangular shadows occupy a relatively large surface area in the central regions of the fundus, sometimes including the fovea (e.g., *Circus auruginosus*), whereas the oval shadows cover only a small portion of the inner nasal retina. Between these two extremes MENNER (1938) found a continuous series of shadows (from triangular to oval) with those of the insectivorous birds residing in the middle (Fig. 22B). By arranging all of the species examined by him in a continuous transitional series, he finds a striking correlation between the size of the pecten shadow and both the eating habits and the mode of life. Conjecturing that the pecten and its shadow must play some essential role in the vision of birds, he suggests that they are concerned with increasing the retina's sensitivity for the perception of movement. Although WALLS (1942) fails to be convinced by these data, maintaining, like MANN (1924a), that the pecten's shadow is confined primarily to its base, CROZIER and

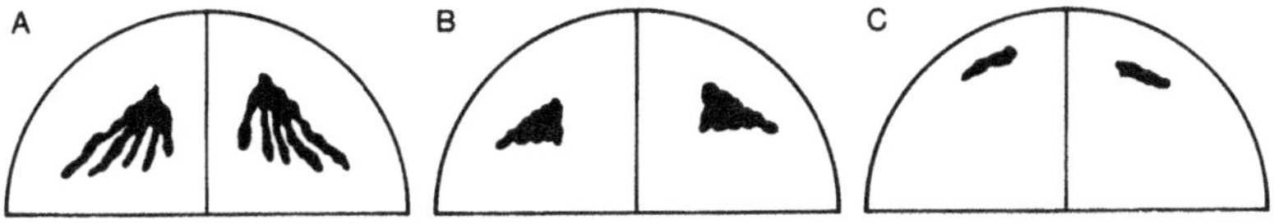

Fig. 22A–C. Pecten shadows as illustrated by MENNER (1938). (A) *Falco subbuteo*, northern hobby, diurnal predator. (B) *Emberiza citrinella*, insectivorous yellowhammer. (C) *Syrnium aluco*, nocturnal owl

WOLF (1944a, b) provide experimental evidence in support of MENNER's theory. They found that the pecten shadows in the zebra finch (*Taeniopygia castinotis*, Gould) with an all-cone retina (1944a) and in the house sparrow *(P. domesticus)* with a duplex retina (1944b) cause a considerable increase of critical fusion frequency (when the dark spaces are brief with respect to the illumination). No evidence was found, however, that the pecten has any influence on rod excitation (CROZIER and WOLF, 1944b). Going a bit further THOMSON (1928, 1929) speculates that light rays coming from the upper and anterior part of the visual field and striking the bridge of the pecten are not only absorbed, but, more significantly, are reflected onto the retina after being deprived of much of their "dazzle and glare." Owing to its dense pigmentation and adherent vitreal material, THOMSON thus compares the bridge to a dark mirror, which he theorizes greatly improves the visual acuity of birds in strong light. Light falling upon the bridge theoretically would produce two images (a direct and a reflected) which would help in estimating distances and spaces, a sort of modified stereoscopic vision adapted to the unique requirements of birds.

"In a spirit of pure speculation," GRIFFIN (1952, 1953) suggests another visual function for the pecten and its shadow. He conjectures that the pecten could serve to orient birds in space by acting as a sextant, casting a shadow upon the retina and permitting the latter to estimate the angular position and movement of the sun. As pointed out by DETWILER (1956), however, this theory pertains only to diurnal migratory birds and does not explain the migratory abilities of birds which travel at night when the sun is not available as a celestial beacon. Nevertheless, more recent studies on bird navigation have both supported (MATHEWS, 1968) and denied (KEETON, 1969) that birds can utilize the position of the sun for navigational purposes.

Scanning electron microscopic studies of the entire pigeon pecten has suggested to TUCKER (1975) that the overall construction of the pecten serves to stabilize the vitreous humor and thus to protect the retina from excessive movements of the vitreous body.

Still another theory considers the pecten as an intraocular shade against the glare of the sun (BARLOW and OSTWALD, 1972). Its shadow covers one part of the retina (including the central area) from light scattered from an image on another part of the retina.

The pecten has been implicated in many other functions for which there is no apparent anatomical basis. The complete absence of nervous and muscular tissues in this organ, for example, renders untenable such properties as sensory perception (FRANZ, 1908), registration of intraocular pressure and/or radiating heat and controlled movement, as in accommodation (HOME, 1796; WOOD, 1907). Early experiments by BEAUREGARD (1875, 1876) involving elimination of the extrinsic eye muscles reveal that the pecten does not move voluntarily, but that the oscillations observed ophthalmoscopically are involuntary movements resulting from muscular activities of the eyeball and nictitating membrane. Electrical stimulation of the pecten in enucleated eyes has also substantiated this inactivity (BEAUREGARD, 1876). Nevertheless, WOOD (1907), among others, (e.g., OWEN, 1866; ZIEM, 1891; RABL, 1900) considers the pecten as an erectile organ capable of controlled movement by being alternately filled with, and emptied of, blood at short notice. He ascribes an accommodative function to this activity, the pecten pushing the lens forward "by sort of hydraulic pressure" when filled (erect) and restoring the lens to its original shape when emptied. ABELSDORFF (1910) refuted this theory experimentally noting that the pecten never moves when accommodation is electrically stimulated in curarized birds.

E. Summary

This chapter is concerned with the structural and functional adaptations which have evolved in the avian eye in its demand for visual perfection in a diurnal, aerial environment. Such adaptations involve mechanisms designed to produce a greater amplitude of accommodation, as well as retinal features favorable for improved visual acuity, color discrimination, and movement detection.

Visual acuity is enhanced by several means. The large size of the avian eye permits the formation of a large image upon a relatively thick retina packed with visual cells and associated neurons. The visual cells are longer and thinner in such retinae and contain specific inclusion bodies (oil droplets and glycogen bodies) characteristic of Sauropsidan photoreceptors. Since incident light must traverse these inclusions before striking the photosensitive outer segments, they may influence the light in such a way as to reduce glare and minimize chromatic aberration (oil droplets) and concentrate the light rays upon the outer segments (glycogen bodies).

Specialized retinal areae with foveal depressions represent one of the most conspicuous adaptations for visual acuity. They are extremely well-developed and diverse in birds where they may assume several shapes (circular, oval, linear) and occupy different fundic positions. A classification of the arrangement of these areae (and foveae) in avian fundi is presented in this chapter and discussed with respect to distribution and possible functional implications. Generally, one, two, or three separate areae may be present in the same fundus and the depths of their foveae may vary. The central fovea is concerned with monocular vision and functions independently from the central fovea of the other eye. The lateral fovea, on the other hand, is involved in binocular vision (with the lateral fovea of the other eye). Foveal depressions are believed to enhance visual acuity by magnifying the retinal image (WALLS). Monofoveate retinae are the most common, whereas bifoveate (lateral and central) retinae appear to be specializations for seeking food "on the wing." Many functions have been attributed to the linear areae, the most plausible of which appears to be its association with improved visual acuity along a horizontal plane. The epitomy of areal and foveal development throughout the vertebrates is attained in birds which require great visual acuity and stereosopic vision, e.g., diurnal birds of prey. They possess three distinct retinal areae (central, lateral, linear) with two foveae (central and lateral), a bifoveate condition which permits the formation of three visual fields (visual trident): two independent, lateral, monocular fields and a single, central binocular one and the linear area which perhaps aids in following moving prey along a horizontal plane.

A possible retinal adaptation for color discrimination in birds is the presence of brightly colored oil droplets in their cones. These carotenoid-lipid inclusions are discussed with respect to their location, coloration, distribution, and chemical content. Conclusive evidence of their involvement in avian color vision is still lacking.

Several adaptations have been discussed with respect to the bird's ability to detect movement. Retinal ganglion cells, for example, have been shown to possess highly specialized and unambiguous responses to particular movements and the convexiclivate fovea is believed to function in maintaining an accurate fixation during movement perception and causing an improvement in sensitivity to angular movements of a fixed object. An efferent centrifugal system (conclusively demonstrated only in birds) is believed to act upon these ganglion cells, but this is believed to represent a suppression of their inhibitory mechanisms resulting in an improvement in visual sensitivity. The pecten has also been implicated as a possible adaptation for movement detection. Its anatomical location within the vitreous serves to cast a shadow upon the retina, thus increasing the retina's sensitivity for movement detection.

The major portion of this chapter has been devoted to a detailed discussion of the pecten. Unique to the avian eye, this highly vascular, intravitreal organ has been analyzed with respect to its vascular supply, fine structure, and function. A direct relationship appears to exist between the size of the pecten and the bird's habits toward food and light: larger and more pleated in diurnal predators. The pleats of the pecten consist of only two elements: capillaries and pigmented stromal cells surrounded by a PAS-positive membrane. The capillaries of the pleat are unique; their endothelia are characterized by an extensive number of the apical and basal microfolds which greatly increase the surface area of these cells. In the bridge of the pecten the capillaries lack microfolds and resemble typical endothelial tubes. Their pigmented stromal cells, however, become epithelial-like, contain alkaline phosphatase activity, and appear to be secretory in functions.

The possible functions of the pecten are discussed in the light of the available histologic, histochemical, and experimental data on the major constituents of the pecten (capillaries, pigmented stromal cells), as well as the anatomical position and morphology of the pecten itself within the vitreous body. The most plausible function still remains that the pecten acts as a supplemental nutritive organ providing the inner portion of the avascular avian retina with oxygen and nutrients. Other possible functions, e.g., movement perception, secretion, regulation of intra-ocular pressure, intra-ocular shade, and navigational aid are also discussed.

References

Abelsdorff, G.: Über das Verhalten des Pekten bei der Akkommodation des Vogelauges. Arch. vergleich. Ophthalmol. **1**, 290—292 (1910).

Abelsdorff, G., Wessely, K.: Vergleichend physiologische Untersuchungen über den Flüssigkeitswechsel des Auges in der Wirbeltierreihe. Arch. Augenheilk. **64**, 65—125 (1909).

Agarwal, P., Singh, K., Dayal, Y.: Retinal vascular patterns. I. Lower Animals. Orient. Arch. Ophthalmol. **4**, 35—46 (1966).

Alpern, M.: Distal mechanisms of vertebrate color vision. Ann. Rev. Physiol. **30**, 279—318 (1968).

Amemiya, T.: Electron microscopic and cytochemical study on paraboloid glycogen of the accessory cone of the chick retina. Histochemistry **43**, 185—192 (1975).

Anderson, D. H., Fisher, S. K.: Disc shedding in rodlike and conelike photoreceptors of tree squirrels. Science **187**, 953—955 (1975).

Bacsich, P., Chisholm, I. A.: Further observations on the effect of the ophthalmic laser on the pecten of the bird's eye. J. Anat. (Lond.) **102**, 576—578 (1968).

Bacsich, P., Chisholm, I. A., Gellért, A.: The riddle of the pecten in the avian eye and the probable usefulness of the laser in the analysis of its function. In: Aspects of comparative ophthalmology, Graham-Jones, O. (ed.). Oxford: Pergamon 1966a.

Bacsich, P., Chisholm, I. A., Gellért, A., Savay, J.: A reptilian type conus papillaris in the eye of the golden hamster. In: Aspects of comparative ophthalmology, Graham-Jones, O. (ed.). Oxford: Pergamon Press 1966b.

Bacsich, P., Gellért, A.: Beiträge zur Kenntnis der Struktur und Funktion des Pectens im Vogelauge. Albrecht v. Graefes Arch. Ophthal. **133**, 448—460 (1935).

Balazs, E. A.: Physiology of the vitreous body. In: Importance of the vitreous body in retina surgery, with special emphasis on reoperations, Schepens, C. L., (ed.). St. Louis: C. V. Mosby 1960.

Balazs, E. A., Laurent, T. C., Laurent, U. B. G., Deroche, M. H., Bunney, D. M.: Studies on the structure of the vitreous body. VIII. Comparative biochemistry. Arch. Biochem. **81**, 464—479 (1959).

Balazs, E. A., Toth, L. Z. F., Eckl, E. A., Mitchell, A. P.: Studies on the structure of the vitreous body. XII. Cytological and histochemical studies on the cortical tissue layer. Exp. Eye Res. **3**, 57—71 (1964).

Barlow, H. B., Ostwald, T. J.: Pecten of the pigeon's eye as an inter-ocular shade. Nature (Lond.) **236**, 88—90 (1972).

Bassewitz, D. B. von, Dieterich, H. J., Dieterich, C. E.: Ultrahistochemische Untersuchungen am Kapillarendothel des Pecten oculi. Verh. anat. Ges. **69**, 631—633 (1975).

Bawa, S. R., Yashroy, R. C.: Effect of dark and light adaptation on the retina and pecten of the chicken. Exp. Eye Res. **13**, 92—97 (1972).

Bawa, S. R., Yashroy, R. C.: Structure and function of the vulture pecten. Acta. anat. (Basel) **89**, 473—480 (1974).

Beauregard, H.: Note sur l'etude ophthalmoscopique du peigne de l'oeil des oiseaux. C.R. Soc. Biol. (Paris) **27**, 132—135 (1875).
Beauregard, H.: Recherches sur les réseaux vasculaires de la chambre postérieure de l'oeil des vertébrés. Ann. Sci. Naturelles (Zoologie) **4**, 1—158 (1876).
Bellhorn, R. W, Bellhorn, M. S.: The avian pecten. 1. Fluorescein permeability. Ophthal. Res. **7**, 1—7 (1975).
Bertolini, M.: Contributo alla conoscenze del petinne dell'occhio degli uccelli. Biol. lat. (Milano) **11**, 625—642 (1958).
Bliss, A. F.: The chemistry of daylight vision. J. gen. Physiol. **29**, 277—297 (1946).
Bloch, S., Maturana, H. R.: Oil droplet distribution and colour discrimination in the pigeon. Nature (New Biol.) **234**, 284—285 (1971).
Brach, V.: The effect of intraocular ablation of the pecten oculi of the chicken. Invest. Ophthalmol. **14**, 166—168 (1975).
Bridges, C. D. B.: Visual pigments of the pigeon (Columba livia). Vision Res. **2**, 125—137 (1962).
Brown, P. K., Wald, G.: Visual pigments in human and monkey retinas. Nature (Lond.) **200**, 37—43 (1963).
Canfield, W. B.: Vergleichend anatomische Studien über den Accommodationsapparat des Vogelauges. Arch. mikr. Anat. **28**, 121—170 (1886).
Carrière, J.: Die Sehorgane der Tiere. Vergleichend-anatomisch dargestellt. München-Leipzig: K. Oldenburgh 1885.
Chievitz, J. H.: Untersuchungen über die Area centralis retinae. Arch. Anat. Physiol. (Anat. Abt.) **13**, (Suppl.) 139—196 (1889).
Chievitz, J. H.: Über das Vorkommen der Area centralis retinae in den vier höheren Wirbeltierklassen. Arch. Anat. Physiol. **15**, 311—334 (1891).
Cohen, A. I.: The fine structure of the visual receptors of the pigeon. Exp. Eye Res. **2**, 88—97 (1963).
Coulombre, A. J.: Correlations of structural and biochemical changes in the devoloping retina of the chick. Amer. J. Anat. **96**, 153—189 (1955).
Cowan, W. M.: Centrifugal fibres to the avian retina. Brit. med. Bull. **26**, 112—118 (1970).
Crescitelli, F.: The natural history of visual pigments. Ann. N.Y. Acad. Sci. **74**, 230—255 (1958a).
Crescitelli, F.: The natural history of visual pigments. Proc. Ann. Biol. Colloq. (Oregon State University) **19**, 30—51 (1958b).
Crescitelli, F.: The visual cells and visual pigments of the vertebrate eye. In: Handbook of Sensory Physiology, Photochemistry of Vision, Dartnall, H. J. A., (ed.). Berlin: Springer, Vol. VII/1, 1972.
Crescitelli, F., Wilson, B. W., Lilyblade, A. L.: The visual pigments of birds. I. The turkey. Vision Res. **4**, 275—280 (1964).
Crozier, W. J., Wolf, E.: Theory and measurement of visual mechanisms. X. Modifications of the flicker response contour, and the significance of the avian pecten. J. gen. Physiol. **27**, 287—313 (1944a).
Crozier, W. J., Wolf, E.: Flicker response contours for the sparrow, and the theory of the avian pecten. J. gen. Physiol. **27**, 315—324 (1944b).
Dartnall, H. J. A.: Photochemistry of vision. Handbook of Sensory Physiology, Vol. VII/1. Berlin: Springer 1972.
Dartnall, H. J. A., Lander, M. R., Munz, F. W.: Periodic changes in the visual pigment of a fish. Progr. Photobiol. Proc. Inst. Congr. 3rd, pp. 203—213 (1960).
Delage, Y.: Suggestion sur la raison d'être de la double fovea des rapaces diurnes. C.R. Acad. Sci. (Paris) **170**, 425—431 (1920).
Detwiler, S. R.: The eye and its structural adaptations. Amer. Scient. **44**, 45—72 (1956).
Dieterich, C. E.: Zur Frage der Differenzierung melaninhaltiger Pigmentzellen im Wirbeltierorganismus. Dargestellt am Beispiel der Feinstruktur einiger Melanozytenformen des Sehorgans. Med. Welt. (N.F.) **24**, 1364—1371 (1973).
Dieterich, C. E., Dieterich, H. J.: Vergleichend-elektronenmikroskopische Untersuchungen am Kapillarendothel des Conus papillaris von Echsen (Sauria). Verh. anat. Ges. **69**, 643—650 (1975a).
Dieterich, C. E., Dieterich, H. J., Spycher, M. A., Pfautsch, M.: Fine structural observations of the pecten oculi capillaries of the chicken. Z. Zellforsch. **146**, 473—489 (1973).
Dieterich, H. J., Dieterich, C. E.: Vergleichende elektronenmikroskopische Untersuchungen am Pecten oculi der Vögel und Conus papillaris der Reptilien. Verh. anat. Ges. **69**, 635—642 (1975b).
Dieterich, H. J., Dieterich, C. E.: Licht- und elektronenmikroskopische Untersuchungen am Pecten oculi verschiedener Vogelarten. Verh. anat. Ges. **68**, 467—478 (1974).

Dobrowolsky, W.: Zur Anatomie der Retina. Arch. Anat. Physiol. wiss. Med. **38**, 221—236 (1871).
Dodt, E., Wirth, A.: Differentiation between rods and cones by flicker electroretinography in pigeon and guinea pig. Acta physiol. scand. **30**, 80—89 (1953).
Donner, K. O.: The visual acuity of some passerine birds. Acta zool. fenn. **66**, 2—40 (1951).
Donner, K. O.: The spectral sensitivity of the pigeon's retinal elements. J. Physiol. **122**, 524—537 (1953).
Donner, K. O.: On the effect of the coloured oil droplets on the spectral sensitivity of the avian retina. Proc. XIIth Int. Ornithol. Congr. Helsinki. **1**, 167—172 (1960).
Dowling, J. E., Cowan, W. M.: An electron microscope study of normal and degenerating centrifugal fibre terminals in the pigeon retina. Z. Zellforsch. **71**, 14—28 (1966).
Dücker, G.: Spontane Bevorzugung arteigener Farben bei Vögeln. Z. Tierpsychol. **20**, 43—65 (1963).
Duijm, M.: On the position of a ribbon-like central area in the eyes of some birds. Arch. Néerl. Zool. **13**, (Suppl. 1) 128—145 (1958).
Duke-Elder, S.: System of ophthalmology, Vol. 1. The eye in evolution. St. Louis: C.V. Mosby, 1958.
Dunn, R. F.: The morphology of the pecten oculi and the conus papillaris. 26th Ann. Meeting Electron Microscopic Soc. Amer., 1968.
Eichner, C., Themann, H.: Zur Frage der Glykogenlokalisation in der Netzhaut der Ratte, des Huhnes und des Goldfisches. Z. mikr.-anat. Forsch. **70**, 330—343 (1963).
Engelmann: Über Bewegung der Zapfen und Pigmentzellen. Copenhagen: Congres Internat. med. **8**, 1884 (Cited by Chievitz, 1891).
Engström, K.: On the cone mosaic in the retina of Parus major. Acta zool. (Stockh.) **39**, 65—69 (1958).
Erhard, H.: Messende Untersuchungen über den Farbensinn der Vögel. Zugleich ein Beitrag zur Lehre von den Schmuckfarben. Zool. Jahrb. (Abt. Allg. Zool. Physiol.) **41**, 489—552 (1924).
Ferens, N. B.: On the ability of colour-discrimination of the tawny owl (Strix aluco aluco L.). Bull. Int. Acad. Sci. Cracovie **B2**, 309—336 (1947).
Fielding, M.: Ultrastructure of the pecten oculi in the domestic fowl. J. Anat. (Lond.) **113**, 295—297 (1972).
Fischer, A. B.: Laboruntersuchungen und Freilandbeobachtungen zum Sehvermögen und Verhalten von Altweltgeiern. Zool. Jb. Abt. System. **96**, 81—132 (1969).
Fischlschweiger, W., O'Rahilly, R.: The ultrastructure of the pecten oculi. Anat. Rec. **148**, 282 (1964).
Fischlschweiger, W., O'Rahilly, R.: The ultrastructure of the pecten oculi in the chick. Acta anat. **65**, 561—578 (1966).
Fischlschweiger, W., O'Rahilly, R.: The ultrastructure of the pecten oculi in the chick. II. Observations on the bridge and its relation to the vitreous body. Z. Zellforsch. **92**, 313—324 (1968).
Fite, K. V.: Anatomical and behavioral correlates of visual acuity in the Great Horned Owl. Vision Res. **13**, 219—230 (1973).
Fite, K. V.: The vertebrate fovea: a comparative study. Anat. Rec. **178**, 355 (1974).
Fox, H. M., Vevers, G.: The nature of animal colours. New York: MacMillan, 1960.
Francois, J., Neetens, A.: Comparative anatomy of the vascular supply of the eye in vertebrates. In: The Eye, Vol. 5, Comparative Physiology, eds. H. Davson, L. T. Graham, Jr. New York: Academic Press 1974.
Franz, V.: Das Pecten, der Fächer, im Auge der Vögel. Biol. Zbl. **28**, 449—468 (1908).
Franz, V.: Das Vogelauge. Zool. Jahrb. **28**, 73—282 (1909).
Franz, V.: Sehorgan. In: Lehrbuch der vergleichenden mikroskopischen Anatomie, 7th part, ed. A. Oppel. Jena: G. Fischer 1913.
Franz, V.: Höhere Sinnesorgane. 1. Vergleichende Anatomie des Wirbeltierauges. In: Handbuch der vergleichenden Anatomie der Wirbeltiere, 2(2), eds. L. Bolk, E. Göppert, E. Kallius, W. Lubosch. Berlin-Wien: Urban-Schwarzenberg 1934.
Friess, A. E.: Zur Topochemie der Vogelretina. Verh. anat. Ges. **64**, 89—91 (1969).
Fritsch, G.: Der Ort des deutlichen Sehens in der Netzhaut der Vögel. Arch. mikr. Anat. **78**, 245—270 (1911).
Gadow, H., Selenka, E.: Vögel. In: Bronn's Klassen und Ordnungen Tier-Reichs. **6(4)**, I. Anat., 423—442 (1891).
Galifret, Y.: Les diverses aires fonctionnelles de la rétine du Pigeon. Z. Zellforsch. **86**, 535—545 (1968).
Gallego, A., Baron, M., Gayoso, M.: Organization of the outer plexiform layer of the diurnal and nocturnal bird retinae. Vision Res. **15**, 1027—1028 (1975).
Gärtner, J.: Electron microscopic observations of the relationship between vitreous body and retina. Mod. Probl. Ophthal. (Basel) **4**, 67—75 (1966).

Genderen Stort, A. G. H. van: Über Form- und Ortsveränderungen der Netzhautelemente unter Einfluß von Licht und Dunkel. Albrecht v. Graefes Arch. Ophthal. **33**, 229—292 (1887).
Giebel, C.: Über den Sklerotikalring, den Fächer und die Hardersche Drüse im Auge der Vögel. Z. ges. Naturw. **9**, 388—419 (1857).
Godfrey, A.: Autoradiographic studies on the development of visual cells in the chicken (Gallus domesticus). Unpublished (1973).
Goodge, W. R.: Adaptations for amphibious vision in the dipper (Cinclus mexicanus). J. Morph. **107**, 79—91 (1960).
Goodwin, T. W.: Carotenoids: Their comparative biochemistry. New York: Chemical Publ. Co. 1954.
Graf, V., Norren, D. V.: A blue sensitive mechanism in the pigeon retina: λ_{max}400 nm. Vision Res. **14**, 1203—1209 (1974).
Granda, A. M., Yazulla, S.: The spectral sensitivity of single units in the nucleus rotundus of a pigeon, *Columba livia*. J. gen. Physiol. **57**, 363—384 (1971).
Griffin, D. R.: Bird navigation. Biol. Rev. **27**, 359—393 (1952).
Griffin, D. R.: Sensory physiology and the orientation of animals. Amer. Scient. **41**, 209—244 (1953).
Gundlach, R. H., Chard, R. D., Skahen, J. R.: The mechanism of accommodation in pigeons. J. comp. Psychol. **38**, 27—42 (1945).
Hahn, E.: Über den Farbensinn der Tagvögel und die Zapfenölkugeln. Z. wiss. Zool. **116**, 1—42 (1916).
Hailman, J. P.: Coding of the colour preference of the gull chick. Nature (Lond.) **204**, 710 (1964).
Hailman, J. P.: Mirror-image preferences for background and stimulus-objekt in the gull chick (Larus atricilla). Experientia (Basel) **22**, 257—258 (1966).
Hamilton, W. F., Coleman, T. B.: Trichromatic vision in the pigeon as illustrated by the spectral hue discrimination curve. J. comp. Psychol. **15**, 183—191 (1933).
Hazlett, L. D., Meyer, D. B., Susan, S. R.: Visual cell ultrastructure in the Japanese quail (Coturnix coturnix japonica). Anat. Rec. **178**, 371—372 (1974).
Hess, C.: Die Akkommodation bei Tauchervögeln. Arch. vergleich. Ophthal. **1**, 153—164 (1909—1910).
Hodges, R. D.: The histology of the fowl. New York: Academic Press 1974.
Hodos, W.: Color discrimination deficits after lesions of the nucleus rotundus in pigeons. Brain Behav. Evol. **2**, 185—200 (1969).
Hoffmann, C. K.: Zur Anatomie der Retina. III. Über den Bau der Retina bei den Vögeln. Niederl. Arch. Zool. **3**, 217—231 (1877).
Hogan, M. J.: Role of the retinal pigment epithelium in macular disease. Trans. Amer. Acad. Ophthal. Otolaryng **76**, 64—80 (1972).
Hogan, M. J., Wood, I., Steinberg, R. H.: Phagocytosis by pigment epithelium of human retinal cones. Nature (Lond.) **252**, 305—307 (1974).
Home, E.: The Croonian lecture on muscular motion. Phil. Trans. B. **85**, 1—26 (1796).
Hudson, R. A., Johnston, D., Meyer, D. B.: The Chemical composition of retinal oil droplets. I. Ophthal. Res. **2**, 217—222 (1971).
Hughes, J. T., Jerrome, D., Krebs, H. A.: Ultrastructure of the avian retina. An anatomical study of the retina of the domestic pigeon (Columba liva) with particular reference to the distribution of mitochondria. Exp. Eye Res. **14**, 189—195 (1972).
Husen, E. von: Zur Kenntnis des Pectens im Vogelauge. Zool. Jb. (Abt. 2) **36**, 215—266 (1913).
Ikeda, H.: The spectral sensitivity of the pigeon (Columba livia). Vision Res. **5**, 19—36 (1965).
Jasiński, A.: Fine structure of capillaries in the pecten oculi of the sparrow, Passer domesticus. Z. Zellforsch. **146**, 281—292 (1973).
Johnston, D., Hudson, R. A.: Phospholipids of the cone-rich chicken retina and its photoreceptor outer segment membranes. Biochim. biophys. Acta. **369**, 269—277 (1974).
Johnston, D., Hudson, R. A.: Isolation and composition of the carotenoid-containing oil droplets from cone photoreceptors. Biochim. Biophys. Acta **424**, 235—245 (1976).
Kajikawa, J.: Beiträge zur Anatomie und Physiologie des Vogelauges. Albrecht v. Graefes Arch. Ophthal. **112**, 260—346 (1923).
Kalberer, M., Pedler, C.: The visual cells of the alligator: an electron microscopic study. Vision Res. **3**, 323—329 (1963).
Kaskel, D., Schedtler, C. M., Hockwin, O., Ciolo, S.: Localization of glycogen in lenses of different species, demonstrated by histochemical methods. Ophthal. Res. **7**, 65—72 (1975).
Kauth, H., Sommer, H.: Das Ferment Kohlensäureanhydratase im Tierkörper. IV. Über die Funktion des Pekten im Vogelauge. Biol. Zbl. **72**, 196—209 (1953).

Keeton, W.T.: Orientation by pigeons: Is the sun necessary? Science **165**, 922—928 (1969).

Kijowska, W.: The development and distribution of blood vessels in bird's eyes (in Polish). Bull. Acad. pol. Sci. Lettres, Cl. Sci. Nat. Ser. **3B**, 993—999 (1925).

King-Smith, P.E.: Absorption spectra and function of the coloured oil drops in the pigeon retina. Vision Res. **9**, 1391—1399 (1969).

King-Smith, P.E.: Special senses. In: Physiology and biochemistry of the domestic fowl, eds. D.J. Bell, B.M. Freeman. New York: Academic Press 1971.

Klopfer, P.H.: Stimulus preferences and discrimination in neonatal ducklings. Behaviour **32**, 309—314 (1968).

Kolmer, W.: Über das Auge des Eisvogels (Alcedo attis attis). Pflüger's Arch. ges. Physiol. **204**, 266—274 (1924).

Kolmer, W.: Haut und Sinnesorgane. 2nd part. Auge In: v. Möllendorf's Handbuch mikroskop. Anat. **3/2**, eds. W. Kolmer, H. Lauber. Berlin: Springer 1936.

Krause, W.: Die Retina. V. Die Retina der Vögel. Int. Mschr. Anat. Physiol. **11**, 1—66 (1894).

Krebs, H.A.: Über den Stoffwechsel der Netzhaut. Biochem. Z. **189**, 57—99 (1927).

Kreithen, M.L., Keeton, W.T.: Detection of polarized light by the homing pigeon, Columba livia. J. comp. Physiol. **89**, 83—92 (1974).

Kuhn, R., Stene, J., Sörensen, N.A.: Über die Verbreitung des Astaxanthins im Tier- und Pflanzenreich. Ber. dsch. chem. Ges. **72**, 1688—1701 (1939).

Kühne, W., Ayres, W.C.: On the stable colours of the retina. J. Physiol. **1**, 109—130 (1878).

Lashley, K.S.: The color vision of birds. I. The spectrum of the domestic fowl. J. Animal Behav. **6**, 1—26 (1916).

Leiner, M.: Über die Bedeutung des Pekten im Vogelauge. Verh. Dtsch. zool. Ges. **15**, 117—123 (1950).

Leplat, G.: Recherches sur le développement et la structure de la membrane vasculaire de l'oeil des Oiseaux. Arch. Biol. (Liège) **27**, 403—523 (1912).

Leuckart, R.: Organologie des Auges. Vergleichende Anatomie. Hdb. ges. Augenheilk. **2**, 145—266 (1876).

Liebman, P.A.: Microspectrophotometry of photoreceptors. In: Handbook of Sensory Physiology, vol. VII/1, ed. H.J.A. Dartnall. Berlin: Springer 1972.

Liebman, P.A., Entine, G.: Visual pigments of frog and tadpole (Rana pipiens). Vision Res. **8**, 761—775 (1968).

Liebman, P.A., Granda, A.M.: Super dense carotenoid spectra resolved in single cone oil droplets. Nature (Lond.) **253**, 370—372 (1975).

Lindsay-Johnson, G.: Contributions to the comparative anatomy of the mammalian eye, chiefly based on ophthalmoscopic examination. Phil. Trans. B **194**, 1—82 (1901).

Lindsay-Johnson, G.: Ein Versuch zur Klassifizierung der Säugetiere, Reptilien und Amphibien in Familien und Ordnungen nach den ophthalmoskopischen Erscheinungen des Augenhintergrundes und dem während des Lebens auftretenden Grade der Exophorie. Sitz. Ges. naturf. Freunde, Berlin. **5**, 249—264 (1909).

Lockie, J.D.: A comparison of some aspects of the retinae of the Manx Shearwater, Fulmar Petrel, and House Sparrow. Quart. J. micr. Sci. **93**, 347—356 (1952).

Loevenich, H.K., Studnitz, J. von, Wigger, H.: Zur Charakteristik der Zapfenölkugeln. Naturwissenschaften **31**, 568—569 (1943).

Lord, R.D. Jr.: A comparative study of the eyes of some falconiform and passeriform birds. Amer. Midl. Natur. **56**, 325—344 (1956).

Lythgoe, J.N.: List of vertrebrate visual pigments. In: Handbook of Sensory Physiology, Vol. VII/1, Photochemistry of vision, ed. H.J.A. Dartnall, Berlin-Heidelberg-New York: Springer 1972.

Mann, I.C.: The function of the pecten. Brit. J. Ophthal. **8**, 209—226 (1924a).

Mann, I.C.: The pecten of Gallus domesticus. Quart. J. Micr. Sci. **68**, 413—442 (1924b).

Marks, W.B.: Visual pigments of single goldfish cones. J. Physiol. (Lond). **178**, 14—32 (1965).

Marks, W.B., Dobelle, W.H., MacNichol, E.F.: Visual pigments of single primate cones. Science **143**, 1181—1183 (1964).

Marshall, J., Mellerio, J., Palmer, D.A.: Damage to pigeon retinae by moderate illumination from fluorescent lamps. Exp. Eye Res. **14**, 164—169 (1972).

Martin, G.R.: Colour vision in the tawny owl *(Strix aluco)*. J. comp. physiol. Psychol. **86**, 133—141 (1974).

Martin, G. R., Gordon, I. E.: Increment-threshold spectral sensitivity in the tawny owl *(Strix aluco)*. Vision Res. **14**, 615—621 (1974).

Mathews, L. H.: Bird navigation. 2nd ed. London: Cambridge Univ. Press 1968.

Matsusaka, T. F.: Electron-microscopic observations on cytology and cytochemistry of the paraboloid glycogen of chick retina. Jap. J. Ophthal. **7**, 238—253 (1963a).

Matsusaka, T. F.: Observations on the cytology and histochemistry of paraboloid glycogen of the chick retina. Acta Soc. Ophthal. Jap. **67**, 1151—1167 (1963b).

Maturana, H. R.: Functional organization of the pigeon retina. In: Information processing in the nervous system. Amsterdam: Excerpta Med. Found. 1962.

Maturana, H. R., Frenk, S.: Directional movement and horizontal edge detectors in the pigeon retina. Science **142**, 977—979 (1963).

Maturana, H. R., Frenk, S.: Synaptic connections of the centrifugal fibers in the pigeon retina. Science **150**, 359—361 (1965).

Mayr, I.: Das Farbunterscheidungsvermögen einiger Ploceidae (Passeriformes, Aves) im kurzwelligen Bereich des Spektrums. Vision Res. **12**, 509—517 (1972a).

Mayr, I.: Verteilung, Lokalisation und Absorption der Zapfenölkugeln bei Vögeln (Ploceidae). Vision Res. **12**, 1477—1484 (1972b).

Menner, E.: Die Bedeutung des Pecten im Auge des Vogels für die Wahrnehmung von Bewegungen, nebst Bemerkungen über seine Ontogenie und Histologie. Zool. Jahrb., Abt. Allg. Zool. Physiol. Tiere **58**, 481—538 (1938).

Meyer, D. B.: The effect of dietary carotenoid deprivation on avian retinal oil droplets. Ophthal. Res. **2**, 104—109 (1971).

Meyer, D. B., Cooper, T. G.: The visual cells of the chicken as revealed by phase contrast microscopy. Amer. J. Anat. **118**, 723—734 (1966).

Meyer, D. B., Cooper, T. G., Gernez, C.: Retinal oil droplets. In: The Structure of the Eye, II Symposium, ed. J. W. Rohen. Stuttgart: Schattauer 1965.

Meyer, D. B., Hazlett, L. D., Susan, S. R.: The fine structure of the retina in the Japanese quail (Coturnix coturnix japonica). I. Pigment epithelium and its vascular barrier. Tissue Cell **5**, 489—500 (1973).

Meyer, D. B., May, H. C. Jr.: The topographical distribution of rods and cones in the adult chicken retina. Exp. Eye Res. **17**, 347—355 (1973).

Meyer, D. B., Stuckey, S. R., Hudson, R. A.: Oil droplet carotenoids of avian cones. I. Dietary exclusion: Models for biochemical and physiological studies. Comp. Biochem. Physiol. **40B**, 61—70 (1971).

Meyer, M. E.: Discriminative basis for astronavigation in birds. J. comp. physiol. Psychol. **58**, 403—406 (1964).

Meyknecht, J.: Farbensehen und Helligkeitsunterscheidung beim Steinkauz (Athene noctua vidalii A. E. Brehm). Ardea **30**, 129—174 (1941).

Miles, F. A.: Centrifugal effects in the avian retina. Science **170**, 992—995 (1970).

Miles, F. A.: Centrifugal control of the avian retina. II. Receptivé field properties of cells in the isthmo-cells. Brain Res. **48**, 65—92 (1972a).

Miles, F. A.: Centrifugal control of the avian retina. II. Receptive field propertößs of cells in the isthmo-optic nucleus. Brain Res. **48**, 93—113 (1972b).

Miles, F. A.: Centrifugal control of the avian retina. III. Effects of electrical stimulation of the isthmo-optic tract on the receptive field properties of retinal ganglion cells. Brain Res. **48**, 115—129 (1972c).

Miles, F. A.: Centrifugal control of the avian retina. IV. Effects of reversible cold block of the isthmo-optic tract on the receptive field properties of cells in the retina and isthmo-optic nucleus. Brain Res. **48**, 131—145 (1972d).

Morris, V. B.: Symmetry in a receptor mosaic demonstrated in the chick from frequencies, spacing and arrangement of the types of retinal receptor. J. comp. Neurol. **140**, 359—398 (1970).

Morris, V. B., Shorey, C. D.: An electron microscope study of types of receptor in the chick retina. J. comp. Neurol. **129**, 313—340 (1967).

Müller, H.: Über das Vorhandensein zweier Foveae in der Netzhaut vieler Vogelaugen. Klin. Monatsbl. **1**, 438—440 (1863).

Munk, O.: On the occurrence and significance of horizontal band-shaped retinal areae in teleosts. Vidensk. Meddr. dansk. naturh. Foren. **133**, 85—120 (1970).

Muntz, W. R. A.: Vision in frogs. Sci. Amer. **210**, 111—119 (1964).

Muntz, W. R. A.: Inert absorbing and reflecting pigments. In: Handbook of Sensory Physiology VII/1, Photochemistry of Vision, ed. H. J. A. Dartnall. Berlin: Springer 1972.
Munz, F. W.: Photosensitive pigments from the retinae of certain deep-sea fishes. J. Physiol. (London) **140**, 220—235 (1958a).
Munz, F. W.: The photosensitive retinal pigments of fishes from relatively turbid coastal waters. J. gen. Physiol. **42**, 445—459 (1958b).
Neumann, G. H.: Das visuelle Lernvermögen eines Emus. J. Ornithol. **103**, 153—165 (1962).
Nguyen H. Anh, J.: Le cône papillaire des reptiles. I. Ultrastructure chez l'Orvet (Anguis fragilis, Anguides). Z. mikr.—anat. Forsch. **81**, 96—110 (1969).
Nguyen H. Anh, J.: Le cône papillaire des reptiles. II. Ultrastructure chez l'Oplure, Oplurus cyclurus (Iguanides). Z. mikr.—anat. Forsch. **82**, 17—28 (1970).
Nguyen H. Anh, J.: Le pseudo-cone papillaire des ophidiens. Ultrastructure chez la Vipère, Vipera aspis (Viperides). Z. mikr.—anat. Forsch. **83**, 289—298 (1971).
Nguyen H. Anh, J.: L'ultrastructure du cone papillaire chez Zonosaurus ornatus (Gerrhosauridae). Acta anat. **88**, 44—55 (1974).
Nguyen H. Anh, J., Delaville, A., Coujard, R.: L'endothélium capillaire du peigne de l'oeil des oiseaux. Z. mikr.—anat. Forsch. **77**, 432—441 (1967).
Nicol, J. A. C., Arnott, H. J.: Tapeta lucida in the eyes of goatsuckers (Caprimulgidae). Proc. roy. Soc. B **187**, 349—352 (1974).
Norren, D. V.: Two short wavelength sensitive cone systems in pigeon, chicken and daw. Vision Res. **15**, 1164—1166 (1975).
O'Day, K.: The fundus and fovea centralis of the albatross (Diomedea cauta cauta Gould). Brit. J. Ophthal. **24**, 201—207 (1940).
Oehme, H.: Vergleichend histologische Untersuchungen an der Retina von Eulen. Zool. Jahrb. (Anat. Abt.) **79**, 439—478 (1961).
Oehme, H.: Das Auge von Mauersegler, Star und Amsel. J. Ornithol. **103**, 187—212 (1962).
Oehme, H.: Vergleichende Untersuchungen an Greifvogelaugen. Z. Morph. Ökol. Tiere **53**, 618—635 (1964).
Ogden, T. E.: On the function of efferent retinal fibres. In: Structure and function of inhibitory neuronal mechanisms, eds. C. von Euler, S. Skoglund, U. Soderberg. New York: Pergamon Press 1968.
O'Rahilly, R., Meyer, D. B.: The development and histochemistry of the pecten oculi. In: The Structure of the Eye, ed. G. K. Smelser. New York: Academic Press 1961.
O'Rahilly, R., Meyer, D. B.: Etude Histologique et Histochimique des Cellules visuelles de la rétine du poulet. Ann. Histochim. **8**, 281—282 (1963).
Owen, R.: On the anatomy of vertebrates, Vol. 2, Birds and mammals. London: Longmans, Green 1866.
Palmer, L. S., Kempster, H. L.: Relation of plant carotenoids to growth, fecundity, and reproduction in fowls. J. biol. Chem. **39**, 299—312 (1919).
Peiponen, V. A.: Farbensehen bei den Vögeln und die Zapfenölkugeln. Ornis Fenn. **36**, 88—94 (1959).
Peiponen, V. A.: Zur Bedeutung der Ölkugeln im Farbensehen der Sauropsiden. Ann. Zool. Fenn. **1**, 281—302 (1964).
Pearson, R.: The avian brain. New York: Academic Press 1972.
Pennycuick, C. J.: The physical basis of astro-navigation in birds: theoretical considerations. J. exp. Biol. **37**, 573—593 (1960).
Pézard, A. G.: Influence du sexe, de la maturité et de la castration sur la rétine des Oiseaux. C. R. Soc. Biol. (Paris) **108**, 722—723 (1931).
Pézard, A. G.: Influence de la lumière et de l'activité spermatogénétique sur les boules colorées de la rétine des Oiseaux et des Sauriens. C. R. Soc. Biol. (Paris) **161**, 840—842 (1957a).
Pézard, A. G.: Influence de la spermatogenèse sur les boules colorées de la rétine des Sauriens. Congrès de Physiologistes de Langue française (Genève). **49**, 352 (1957b).
Pézard, A. G.: Les boules colorées des cellules visuelles de certains Vertébrés et la relation de leur couleur avec les fonctions sexuelles. Arch. Anat. micr. Morph. exp. **53**, 45—56 (1964).
Poitier, P.: Interpretation physiologique de la double fovea des Rapaces diurnes. C. R. Soc. Biol. (Paris) **88**, 330—331 (1923).
Poley, D.: Experimentelle Untersuchungen zur Nahrungssuche und Nahrungsaufnahme der Kolibris. Bonner Zool. Beitr. **1**, 111—156 (1968).
Polyak, S. L.: The retina. Chicago: University of Chicago Press 1941.
Polyak, S.: The vertebrate visual system. Chicago: The University of Chicago Press 1957.

Porte, A., Stoeckel, M. E., Brini, A.: Structure du pecten oculi chez le poulet. Arch. Ophthal. **28**, 7—26 (1968).
Porter, J. P.: A preliminary study of the psychology of the English sparrow. Amer. J. Psychol. **5**, 313—346 (1904).
Porter, J. P.: Further study of the English sparrow and other birds. Amer. J. Psychol. **17**, 248—271 (1906).
Pumphrey, R. J.: The sense organs of birds. Ibis. **90**, 171—199 (1948a).
Pumphrey, R. J.: The theory of the fovea. J. exp. Biol. **25**, 299—312 (1948b).
Pumphrey, R. J.: Sensory Organs: Vision. In: Biology and Comparative Physiology of Birds, Vol. 2 ed., A. J. Marshall. New York: Academic Press 1961.
Rabaey, M.: Glycogen in the lens of bird's eyes. Nature (Lond.) **198**, 206—207 (1963).
Rabinovitch, M., Mota, J., Yoneda, S.: Note on the histochemical localization of glycogen and pentose-polynucleotides in the visual cells of the chick (Gallus gallus). Quart. J. micr. Sci. **95**, 5—10 (1954).
Rabl, C.: Über den Bau und die Entwicklung der Linse. III. Die Linse der Säugetiere. Z. wissen. Zool. **67**, 1—138 (1900).
Raviola, E., Raviola, G: Electron microscopy of pigeon's pecten oculi. Anat. Rec. **154**, 407 (1966).
Raviola, E., Raviola, G.: A light electron microscopic study of the pecten of the pigeon eye. Amer. J. Anat. **120**, 427—462 (1967).
Reuter, T. E., White, R. H., Wald, G.: Rhodopsin and porphyropsin fields in the adult bullfrog retina. J. gen. Physiol. **58**, 351—371 (1971).
Riggs, L. A., Blough, P. M., Schafer, K. L: Electrical responses of the pigeon eye to changes in wavelength of the stimulating light. Vision Res. **12**, 981—991 (1972).
Rochon-Duvigneaud, A.: La double fovea rétinienne des Repaces diurnes. C. R. Acad. Sci. (Paris) **169**, 43—46 (1919a).
Rochon-Duvigneaud, A.: Quelques données sur la fovea des oiseaux. Ann. Oculist. (Paris) **156**, 717—732 (1919b).
Rochon-Duvigneaud, A.: La situation de foveae simples et doubles dans la rétine des oiseaux et la problème de leurs relations fonctionelles. Ann. Oculist. (Paris) **157**, 673—680 (1920a).
Rochon-Duvigneaud, A.: La Vision et l'oeil des Oiseaux. Bull. Biol. de la France et le da Belgique **54**, 109—167 (1920b).
Rochon-Duvigneaud, A.: Les Yeux et la Vision des Vertébrés. Paris. Masson 1943.
Rodieck, R. W.: The vertebrate retina. Principles of structure and function. San Francisco: W. J. Freeman 1973.
Rodriguez-Peralta, L. A.: Hematic and fluid barriers of the retina and vitreous body. J. comp. Neurol. **132**, 109—124 (1968).
Rodriguez-Peralta, L. A., Kairys, S.: Role of the pecten in the hematic and fluid barriers of the vitreal chamber of birds. Anat. Rec. **160**, 492 (1968).
Rogers, L. J., Miles, F. A.: Centrifugal control of the avian retina. V. Effects of lesions of the isthmo-optic nucleus on visual behaviour. Brain Res. **48**, 147—156 (1972).
Rohen, J.: Arterio-Venose-Anastomosen in der Orbita bei Vögeln. Morph. Jb. **95**, 364—383 (1955).
Runge, R. G., Uemura, A., Viglione, S. S.: Electronic synthesis of the avian retina. I.E.E.E. Trans. Biomed. Eng. **BME-15**, 138—151 (1968).
Rushton, W. A. H.: Visual pigments in man. In: Handbook of Sensory Physiology, Vol. VII/1, Photochemistry of Vision, ed. H. J. A. Dartnall. Berlin: Springer 1972.
Schultze, M.: Zur Anatomie und Physiologie der Retina. Arch. mikr. Anat. **2**, 175—286 (1866).
Schultze, M.: Über Stäbchen und Zapfen der Retina. Arch. mikr. Anat. **3**, 215—245 (1867).
Schultze, M.: Die Retina. In: Handbook der Lehre von den Gewebe des Menschen und der Thiere, ed. S. Stricker. Leipzig: Engelmann 1872.
Schwanzara, S. A.: The visual pigments of freshwater fishes. Vision Res. **7**, 121—148 (1967).
Schwartzkopff, J.: Physiologie der höheren Sinne bei Säugern und Vögeln. J. Ornith. **101**, 61—91 (1960).
Seaman, A. R.: Ultrafine architecture of the avian pecten oculi. Albrecht v. Graefes Arch. Ophthal. **107**, 1—15 (1966).
Seaman, A. R., Himelfarb, T. M.: Correlated ultrafine structural changes of the avian pecten oculi and ciliary body of Gallus domesticus: Preliminary observations on the physiology: I. Effects of decreased intraocular pressure induced by intravenous injection of acetazolamide (Diamox). Amer. J. Ophthal. **56**, 278—296 (1963).

Seaman, A. R., Storm, H.: A correlated light and electron microscope study on the pecten oculi of the domestic fowl (Gallus domesticus). Exp. Eye Res. **2**, 163—172 (1963).

Seaman, A. R., Storm, H.: The uptake and transport of colloidal particles in the capillaries of the "pecten oculi" of the domestic fowl (Gallus domesticus). Anat. Rec. **154**, 488 (1966).

Semba, T.: The fine structure of the pecten studied with the electron microscope. I. Chick pecten. Kyushu J. med. Sci. **13**, 217—232 (1962).

Sillman, A. J.: The visual pigments of several species of birds. Vision Res. **9**, 1063—1077 (1969).

Sillman, A. J.: Avian vision. In: Avian biology, Vol. III, eds. D. S. Farner, J. R. King, K. C. Parkes. New York: Academic Press 1973.

Slonaker, J. R.: A comparative study of the area of acute vision in vertebrates. J. Morph. **13**, 445—493 (1897).

Slonaker, J. R.: A physiological study of the anatomy of the eye and its accessory parts of the English Sparrow (Passer domesticus). J. Morph. **31**, 351—459 (1918).

Soemmerring, D. W: De oculorum hominis animaliumque sectione horizontale commentatio. Göttingen: Vandenhoeck and Ruprecht 1818.

Spitznas, M., Hogan, M. J.: Outer segments of photoreceptors and the retinal pigment epithelium. Arch. Ophthal. **84**, 810—819 (1970).

Steinbach, M. J., Money, K. E.: Eye movements of the owl. Vision Res. **13**, 889—891 (1973).

Strother, G. K.: Absorption spectra of retinal oil globules in turkey, turtle and pigeon. Exp. Cell. Res. **29**, 349—355 (1963).

Strother, G. K., Wolken, J. J.: Microspectrophotometry. I. Absorption spectra of colored oil globules in the chicken retina. Exp. Cell Res. **21**, 504—512 (1960).

Studnitz, G. von, Neumann, H. J., Loevenich, H. K.: Die Natur der Ölkugeln und Sehstoffe. Pflügers Arch. ges. Physiol. **264**, 652—663 (1943).

Szily, A. von: Vergleichende Entwicklungsgeschichte der Papilla nervi optici und der sog. axialen Gebilde. I. Morphogenese des Sehnerveintrittes und des „Fächers" beim Hühnchen als Beispiel für den Typ „Vogel". Albrecht v. Graefes Arch. Ophthal. **107**, 317—431 (1922).

Tanaka, A.: Electron microscopic study on the avian pectines. I. (Japanese) Dobitsugasku Zasshi. **69**, 314—317 (1960).

Tanaka, H.: On the Pecten of Japanese Birds. Jap. J. med. Sci. I. Anatomy **7**, 95—132 (1938a).

Tanaka, H.: The blood-vessel distribution and pigmentation in Pectens of Japanese birds. Jap. J. med. Sci. I. Anatomy **7**, 133—151 (1938b).

Tansley, K.: Vision in vertebrates. London: Chapman and Hill, 1965.

Thompson, G.: The photopic spectral sensitivity of gulls measured by electroretinographic and pupillometric methods. Vision Res. **11**, 719—731 (1971).

Thomson, A.: The riddle of the pecten with suggestions as to its use. Trans. ophthal. Soc. U. K. **48**, 293—331 (1928).

Thomson, A.: The pecten, considered from an environmental point of view. Ibis **12**, 608—639 (1929).

Tripathi, R. C.: Comparative physiology and anatomy of the outflow pathway. In: The Eye, Comparative physiology. vol. 5, eds. Davson, H., Graham, L. T. Jr. New York: Academic Press 1974.

Tucker, R.: The surface of the pecten oculi in the pigeon. Cell Tiss. Res. **157**, 457—465 (1975).

Turchini, J.: Contribution a l'etude du peigne des oiseaux. C.R. Ass. Anat. **28**, 677—680 (1933).

Uchiyama, T.: Beiträge zur Morphologie des Lipoidstoffwechsels III. Über die Öltropfen und Glykogengebilde in den Sehzellen der Netzhaut des Haushuhns. Virchows Arch. path. Anat. **27**, 631—641 (1930).

Valentin, G.: Ein Beitrag zur Kenntnis der Brechungsverhältnisse der Thiergewebe. Arch. ges. Physiol. **19**, 78—105 (1879a).

Valentin, G.: Fortgesetzte Untersuchungen über die Brechungsverhältnisse der Thiergewebe. Arch. ges. Physiol. **20**, 283—314 (1879b).

Virchow, H.: Fächer, Zapfen, Leiste, Polster, Gefäße im Glaskörperraum von Wirbeltieren sowie damit in Verbindung stehende Fragen. Anat. Hefte **10**, 720—837 (1900).

Vrabec, F.: Anatomical study of the pecten in Dromaeus novaehollandiae. Acta. Soc. Zool. Bohemoslov. **22**, 31—33 (1958).

Waelchli, G.: Mikrospektroskopische Untersuchungen der gefärbten Kugeln in der Retina von Vögeln. Albrecht v. Graefes Arch. Ophthal. **27**, 303—319 (1881).

Waelchli, G.: Zur Topographie der gefärbten Kugeln der Vogelnetzhaut. Albrecht v. Graefes Arch. Ophthal. **29**, 205—223 (1883).

Wagner, R.: Beiträge zur Anatomie der Vögel. Abh. math.—phys. Classe Akad. Wissen **2**, 271—308 (1837).
Wald, G.: Photo-labile pigments of the chicken retina. Nature (Lond.) **140**, 545—546 (1937).
Wald, G.: Galloxanthin, a carotenoid from the chicken retina. J. gen. Physiol. **31**, 377—383 (1948).
Wald, G.: Retinal chemistry and the physiology of vision. In: Visual Problems of Colour, Symp. Nr. 8, Nat. Phys. Lab. H. M. S. O., London, pp. 7—61 (1958).
Wald, G., Brown, P. K., Brown, P. S.: Visual pigments and depths of habitat of marine fishes. Nature (Lond.) **180**, 969—971 (1957).
Wald, G., Brown, P. K., Smith, P. H.: Cyanopsin, a new pigment of cone vision. Science **118**, 505—508 (1953).
Wald, G., Brown, P. K., Smith, P. H.: Iodopsin. J. gen. Physiol. **38**, 623—681 (1955).
Wald, G., Zussman, H.: Carotenoids of the chicken retina. J. Biol. Chem. **122**, 449—460 (1938).
Walls, G. L.: Significance of the foveal depression. Arch. Ophthal. **18**, 912—919 (1937).
Walls, G. L.: Postscript on image expansion by the foveal clivus. Arch. Ophthal. **23**, 831—832 (1940).
Walls, G. L.: The vertebrate eye and its adaptive radiation. Bloomfield Hills, Mich.: Cranbrook Inst. Sci., Bull. No. **19**, (1942).
Walls, G. L., Judd, H. D.: The intraocular filters of vertebrates. Brit. J. Ophthal. **17**, 641—675; 705—725 (1933).
Watson, J. B.: Studies on the spectral sensitivity of birds. Papers Dept. Marine Biol. Carnegie Inst. Wash. **7**, 87—104 (1915).
Welsch, U.: Enzymhistochemische und feinstrukturelle Beobachtungen am Pecten oculi von Taube (Columba livia) und Lachmöve (Larus ridibundus). Z. Zellforsch. **132**, 231—244 (1972).
Wessels, R. H. A.: Tetrachromatic vision in the daw (Corvus monedula L.). Thesis, Utrecht (1974).
Whiteside, T. C. D.: The head movement of walking birds. J. Physiol. (Lond.) **188**, 31P—32P (1966).
Wingstrand, K. G., Munk, O.: The pecten oculi of the pigeon with particular regard to its function. Biol. Skr. Dan. Vid. Selsk. **14**, 1—64 (1965).
Winkel, W.: Experimentelle Untersuchungen an Zuckervögeln (Coerebidae) im Funktionskreis der Nahrungssuche: Über die Bedeutung von Farben, Formen und Zuckerkonzentrationen. Z. Tierpsychol. **26**, 573—608 (1969).
Wood, C. A.: The eyes and eyesight of birds, with especial reference to the appearances of the fundus oculi. Ophthalmology **3**, 377—398 (1907).
Wood, C. A.: Comparative ophthalmology. In: American Encyclopedia and Dictionary of Ophthalmology, Vol. IV. Chicago: Cleveland Press 1914.
Wood, C. A.: The fundus oculi of birds especially as viewed by the ophthalmoscope. Chicago: Lakeside Press 1917.
Wood, C. A.: Color sense of the bower bird. Amer. J. Ophthal. **8**, 120—122 (1925).
Yamada, E.: Some observations on the fine structure of the vertebrate retina. 19th Int. Congr. Ophthal. New Delhi (1962).
Yazulla, S.: Intraretinal differentiation in the synaptic organization of the inner plexiform layer of the pigeon retina. J. comp. Neurol. **153**, 309—324 (1974).
Yew, D. T.: The development of the retina in staged chicken embryos (Gallus domesticus) (26-hatching): A histological, histochemical and electron microscopic study. Doctoral Dissertation, Wayne State University (1974).
Young, R. W.: The renewal of photoreceptor cell outer segments. J. Cell Biol. **33**, 61—72 (1967).
Young, R. W.: A difference between rods and cones in the renewal of outer segment protein. Invest. Ophthal. **8**, 222 (1969).
Young, R. W.: Shedding of discs from rod outer segments in the rhesus monkey. J. Ultrastruct. Res. **34**, 190—203 (1971).
Young, R. W.: Biogenesis and renewal of visual cell outer segment membranes. Exp. Eye Res. **18**, 215—223 (1974).
Young, R. W., Bok, D.: Participation of the retinal pigment epithelium in the rod outer segment renewal process. J. Cell. Biol. **42**, 392—403 (1969).
Ziem, C.: Geschichtliche Notiz über den Fächer im Auge der Vögel. Z. Psychol. Sinnesorg. **6**, 474 (1891).

[4] The section on Visual Pigments was written by Dr. A. J. Sillman, University of California, Davis.

Chapter 11

The Topography of Vision in Mammals of Contrasting Life Style: Comparative Optics and Retinal Organisation

AUSTIN HUGHES, Canberra City (Australia)

With 26 Figures

Contents

Glossary of Abbreviations

CMF	Cortical magnification factor, mm/°
CNS	Central nervous system
CSF	Contrast sensitivity function
LSF	Line spread function
MTF	Modulation transfer function
PND	Posterior nodal distance, mm
PSF	Point spread function
RMF	Retinal magnification factor, mm/°

The physiologist, who is conversant with natural history in general, is fortified against uncertain opinions, and the showy but flimsy textures of verbal sophistry. An hypothesis, which to others appears perfectly adequate to the object in view, is not convincing to him. He rises above the particular object to which it is accommodated, in order to appreciate its value; as we ascend an eminence to gain a commanding view of a district ...

The basis of our physiological principles is rendered broader and deeper, in proportion as our survey of living beings is more extensive ... These enlarged views, which unfold to us the natural play of the animal mechanism, are our surest guide in the study of its deranged motions, an essential criterion for estimating the nature and degree of the deviation, and an important indication of the means by which it may be corrected

> ... On such researches, and such studies, on a foundation no less extensive than the whole empire of living nature, the science of medicine must be established; if indeed, it is destined to occupy the rank of a science.
>
> LAWRENCE: Lectures on Comparative Anatomy, Physiology, Zoology and the Natural History of Man, 1844, London.

A. Introduction

> The eye to this day gives me a cold shudder

wrote DARWIN (1860),

> to suppose that the eye with all its inimitable contrivances ... could have been formed by natural selection, seems, I freely confess, absurd in the highest degree.

but

> when I think of the fine known gradations, my reason tells me I ought to conquer the cold shudder.

Our knowledge of these gradations and the great variety of ocular organisation has been summarised in several large works in the last four decades (FRANZ, 1934; WALLS, 1942; ROCHON-DUVIGNEAUD, 1943; POLYAK, 1957; DUKE-ELDER, 1958). Of these authors perhaps WALLS (1942) has done most to emphasize, and stimulate interest in, the extreme evolutionary adaptive lability of the eye. His book has served for some 30 years as a compendium of visual anatomy and physiology; although an essential reference, because unequalled in scope, many of its sections contain much that is misleading according to the current consensus. What appear in isolation to be relatively unimportant errors, such as

> where the eyes are placed laterally as in the rabbit, there is usually no area centralis ... and there are no spontaneous movements at all ...
> no non-simian mammal can converge its eyes ...
> among the carnivores, the domestic cat is pre-eminent with ... a binocular field of 130° and a total field of 287° ...

are quite common and may suggest to the reader outmoded qualitative distinctions between species, and further the erroneous generalisations which are so common in comparative literature.

In spite of the vast increase in comparative visual information since the appearance of WALLS' book, it remains difficult to make a quantitative comparative survey of even relatively straightforward problems, such as the extent of the visual field, because of the dearth of reliable data for not only *ferae naturae* but also the common laboratory animals. The situation is improving, but much systematic "bread-and-butter" work remains which might readily be undertaken with the armamentarium of the most underpriviliged laboratories of these stringent times. However, such work can be tedious and certainly lacks the charisma of the in-vogue, front-line topics; the information backlog must be reduced in a process of attrition by a very limited number of comparatively orientated physiologists and anatomists.

Revision of WALLS' book in its entirety would now be a formidable task in which much of its idiosyncratic flavor and breadth of insight could be lost; the aim of this chapter is much more modest. It deals with some of the topics of WALLS' Chapter 10, "Adaptations to Space and Motion," in a modern context and is primarily concerned with the mammals. The topics subsequently dealt with may appear to be selected rather arbitrarily, but their common thread is that they are factors which set limits on the extent and quality of the spatial contrast information which a given species may acquire from its environment; little attention is given to the mechanisms for its analysis.

I. The Theme: Some General Observations on the Relationship Between Object, Image, and Visual Spaces

Light may arrive from all directions at any point in an illuminated environment so that global information about the visual organisation of the surroundings is potentially available to an eye wherever its anterior nodal point is situated. No eye can accept light from much more than a hemispherical field, and two are only just capable of providing complete coverage of the surroundings if they are appropriately positioned relative to one another. Even then, only that portion of the total optic field which is served by retina can be significant in the determination of the behavioral capacity of a species.

The topographic organisation of the visual environment is retained in its projective transformation into the retinal image and remains embodied in several subsequent central stages of neural processing as well as providing a substrate for the organisation of the motor system in terms of external space (e.g., APTER, 1946). Oculocentric distance along visual lines of direction in object space (HERING, 1879) transforms into a continuum of conjugate image space shells which may be selectively brought to the effective receptor plane by an accommodatory mechanism. Relative and absolute depth information is potentially available from the slight differences between the retinal images of the common field of the two eyes. In some species, at least, this binocular parallax is detected by the neural mechanism of receptive field disparity (BISHOP, 1973) which appears to underlie the faculty of stereopsis. The intrinsic limitations on the maximum size of the uniocular retinal field mean that there is a tendency for an inverse relationship to exist between the extent of the binocular and lateral visual fields in a given species. The balance between the extent of the two fields, and thus the proportion of the global array to which the species has parallel access, must be subject to considerable selective pressure and does not correlate well with phylogenetic status; what ecologic and behavioral factors actually determine the width of the binocular field remains to be established.

GIBSON (1950) emphasised that visual information about environmental organisation is primarily contained in the intensity variations of light arriving from different directions (but see BOYNTON, 1975). Thus, at a *point of observation*, he distinguished the *ambient optic array* from ambient light as:

> A nest of solid angles, not rays, each solid angle corresponding to one of the large faces or small facets of the environment. The solid angles are separated by contours or contrasts.

The flow of structural information between object and image may be treated directly (GABOR, 1961; WINTHROP, 1971) or, more conventionally, in terms of the quantal nature of light which sets the ultimate limit on the information capacity of the ambient optic array. Of course, the eye can never attain this capacity; the mamalian eye is effectively insensitive to polarisation and thus the accessible potential information of the ambient array is reduced by 50%; again, losses in the ocular media may amount to some 50% of the light incident on the cornea and of this perhaps only 20% is absorbed by photopigment (HECHT et al., 1942). Thus, under the conditions of a typical calculation (HISDAL, 1967), the *foveally* detectable information in a photopic photon beam amounted to only $1/10^5$ of that potentially available. Although the information content of the ambient optic array is substantially lost in the formation of the image at high levels of illumination, it is found that noise arising from the statistical fluctuation in photon arrival at low intensities may so limit the spatial information content of the array that the optical apparatus of the eye is equal to its efficient transfer to the image plane and the neural equipment to its detection; under these conditions the match between the information capacity of the ambient array and of the eye is much better. Within the eye, however, at even low photopic levels the quantal nature of light must be taken into account during high resolution tasks (PIRENNE, 1962; COHN, 1976).

Scatter, diffraction and aberrations of the optical system are more significant factors than quantum efficiency in setting a limit on the transfer of information from the ambient optic array to the image space; the efficacy of this process is summarised by its three dimensional modulation transfer function (FRIEDEN, 1967). The neural apparatus only samples that shell of the image space, the effective image shell, which is intersected by the photoreceptor apertures and whose information capacity is described by a surface modulation transfer function. The potential information of the effective image shell is even further reduced by the effects of retinal sampling grain and noise introduced during retinal transduction and transmission. For an idealised retina, SNYDER et al. (1977) have considered the influence of the above factors on information capacity defined as a function of the number of discriminable intensity levels and resolvable points, ‚the number of discriminable pictures.‘ For a specific sampling grain this measure of information capacity may be shown to be optimal at a given level of illumination.

A 2.4 mm diameter human pupil strikes an optimal balance between the deleterious effects of diffraction and aberration on the retinal image quality and this is expressed in the corresponding maximal image spatial cut-off frequency, *f*, of some 60 cy/°. In accordance with SHANNON'S (1949) sampling theorem for optimal systems the foveal cone separation is found to be matched to this cut-off frequency at a value of $(1/2f)°$. One-to-one connectivity between foveal photoreceptors and ganglion cells (MISSOTTEN, 1974) ensures that this potential for optimal resolution is relayed to the CNS with the result that maximum behavioral acuity almost attains the value of optimal resolution in the retinal image.

Convergence of connectivity from photoreceptors to ganglion cells ensures that, outside the foveal centre, the latter set the upper limit on *neural* resolution but whether neural or optical factors predominate in limiting overall peripheral acuity remains an equivocal question. No satisfactory human peripheral MTF measurements have been made to establish optical quality and very little peripheral acuity data is available from which optical influences have been excluded. In conjunction with the unavailability of satisfactory ganglion cell distributions for man it is not surprising that the question of whether neural sampling in peripheral retina is optimally matched to the image remains undecided at present. WEALE (1976) has suggested that the total peripheral photoreceptor density of man is well matched to the number of degrees of freedom in the local retinal image but his computations are based on geometric considerations and do not incorporate the influence of oblique aberrations on the local MTF.

The possibility of locally matching neural and optical resolution is a simple example of many possible strategies which the CNS may employ to avoid redundancy and achieve an optimal system (ATTNEAVE, 1954; BARLOW, 1961). A comparative survey indicates that the optic quality and retinal provision for different regions of the optic array are rarely uniform and vary considerably between species. The great anisotropy of ganglion cell distribution in some species indicates that the overall neural sampling distribution of the retina is definitely not matched to the more symmetrical variation of image quality. Such optic and retinal heterogeneity could reflect adaptive minimisation of redundancy by matching of the sampling apparatus to a required subset of the information distributed in those invariant features of the retinal image, the overt image of the visual horizon or the less obvious flow patterns which occur during locomotion, which arise out of its projective relationship with the environment. Alternatively, like the balance between the overall extents of the binocular and total fields of view or the local specialisation of a retinal region for photopic colour vision, it may arise as an adaptive solution to the conflicting requirements of the species life style and not from anisotropy of the image, optic array, or environment. In spite of such differentiation of the optics and retina, it remains possible for the isotropic potential of the optic array to be exploited in full by serial sampling which involves eye, head and body movements whose extent and frequency may, therefore, be expected to correlate with the various specialisations in retinal topography.

It is generally established for vertebrate species that the visual apparatus does not simply relay the variations in light intensity across the retinal image. Early mammalian single unit recordings, first by KUFFLER (1953) in the retina and HUBEL and WIESEL (1959) in the visual cortex, revealed that many receptive fields at these levels are organised in the cat for the detection of contrast and contour, respectively. The visual system is thus appropriately equipped in this animal, and man, with the neural apparatus for the extraction of the optic array from ambient light in accordance with GIBSON'S (1950) psychophysical theories. It appears possible, but not certain however, that the manner in which visual information is processed both peripherally and centrally may be different in other mammals depending upon whether they are organised for frontal or lateral vision.

At the distances encountered in terrestial, but not astronomical, observation, the abstraction of the everyday physical world exemplified in the infinite rectangular coordinate system of Euclidean space appears to be very accurate. The projective transformation of the external world into a spherical image space results in general distortion of relative object separations, dimensional changes when objects recede or approach and other non-Euclidean phenomena. The subjective *visual space* which is centrally derived from the retinal image is further dependent upon the nature of the transformations effected by the retinal and subsequent neural processes and might not be expected to be Euclidean. Although many reports have been made of anisotropies in subjective visual space (HELMHOLTZ, 1924; LE GRAND, 1967; PIRENNE, 1970), and it has been suggested that the metric of binocular space is hyperbolic (LUNEBERG, 1947), the majority of such anomalous perceptions are difficult to demonstrate and not universally accepted as occurring. The mature human observer appears capable only of a projective, rather than a Euclidean, interpretation when faced with dynamic figural transformations, and thus assumes a rigid Euclidean organisation of the stimuli (GUNNAR JOHANSSEN, 1975). Higher-level processing effectively inverts the projective transformation in man, thus giving rise to phenomena like size constancy (LEGRAND, 1967) and ensures a subjective visual space which corresponds quite closely in its properties to a Euclidean abstraction rather than to a retinal image space.

The gross correspondence between the properties of human visual space, whether subjectively or behaviorally established, and those of Euclidean space indicate that the various stages of visual processing are effectively "transparent." If there is advantage for man in this arrangement, it seems probable that it will occur in other species. In spite of the obvious differences between the optic arrays of various environments, it remains that the underlying spatial, metric, and optical principles are common to all species and, in spite of gross differences in neural substrate, their subjective visual spaces might therefore be expected to be similar to that of man in qualitative, if not quantitative, properties. The visual space of other species is, of course, only accessible through behavioral techniques and has not yet been subject to serious investigation. An underlying uniformity of perception across a wide range of classes of animal, however, is implicit in the universal effectiveness of the few principles of camouflage employed throughout the animal kingdom (COTT, 1966). If the MÜLLER-LYER illusion does indeed have its origin in perspective effects (GREGORY, 1963; GREGORY and HARRIS, 1975) then perhaps it should not be surprising that both fly and man are subject to it (GEIGER and POGGIO, 1975).

B. The Field of View of the Animal

I. The Uniocular Optic and Retinal Fields

The uniocular field of vision behaviorally demonstrable in an animal has in the past been treated as equivalent to the field of view of the eye's dioptric apparatus subject to its restriction by the optic adnexa. The long-standing assumption that the extent of the retina is not a factor which limits the extent of the behavioural fields of view was first made explicit by VAKKUR and BISHOP (1963) in their cat schematic eye, but has since been demonstrated to be invalid in that species (HUGHES, 1976a). It is consequently necessary to define two distinct *uniocular* fields: (1) the *optic field of an eye*—the solid angle subtended at its anterior nodal point by that region of visual space from which light may be refracted through the lens; and (2) the *retinal field of an eye*—that portion of its optical field which is encompassed by the external projections of its retinal margins.

The absolute uniocular fields are measured on the freshly enucleated eye or in situ if the field is free of obstruction by the nose and optic adnexa. The relative, or obscured, field includes the influence of the adnexa and must be measured in situ.

Most published measurements of the extent of the uniocular field have been obtained with the technique of transcleral illumination. TSCHERMAK-SEYSENEGG (1902), ROCHON-DUVIGNEAUD (1922), and other workers assessed the horizontal angular range over which light from a small source was refracted into the eye by viewing its image transclerally. The technique was extended by THIEULIN (1927) and PISA (1939) to other planes. Although commonly accepted as satisfactory (WALLS, 1942; DUKE-ELDER, 1958), this method establishes the extent of neither the optical nor the retinal fields; in the cat it indicates (VAKKUR and BISHOP, 1963) a field of greater extent than the uniocular retinal field yet smaller than the optical field. The perimetric determination of the ophthalmoscopically observed limits of the retinal uniocular field reveals that it serves only 143° of the 181° cat uniocular optic field and some 179° of 200° in man (unpublished observations). By contrast, the 192° rabbit (PISA, 1939; HUGHES, 1971, 1972) and 205° rat (LASHLEY, 1932; HUGHES, 1977c) uniocular optic fields are entirely served by the retina. These and other unpublished observations substantiate the belief that the uniocular optic field is of relatively constant size in most species but WALLS' (1942) view that

> the cone of space subtended by the retina is surprisingly uniform throughout the vertebrates. It is rarely much greater or less than 170°

appears to be ill founded; the uniocular retinal fields of the cat and rat alone range from 143° to 205°, respectively. Obviously the uniocular retinal field must replace the optical field in consideration of the behavioural range of vision in a given species; calculation from the dioptric apparatus (VAKKUR and BISHOP, 1963) is not adequate.

II. The Cyclopean Fields

1. Nomenclature

The horizontal separation of the two eyes, advantageous because the resulting disparate retinal images enable binocular depth perception, introduces a difficulty into the specification of the fields of view *of the animal*. For this purpose a single reference point is required relative to which the angular projections of the ophthalmoscopically observed delimiting landmarks of the fundus may be subtended. The cyclopean projection centre situated half-way between the anterior nodal points of the two eyes (HELMHOLTZ, 1924) has been adopted for this purpose. In the limit, at measuring distances so great that the interocular separation is relatively negligible, the cyclopean projection point and the anterior nodal points are effectively coincident. We thus define two *cyclopean* static visual fields for the animal with its eyes in the primary position, i.e., centred in the head with their visual axes parallel:

The *cyclopean optic field of an animal* is the solid angle subtended at the cyclopean projection centre by those regions of space from which light may be refracted through the lens of either eye.

The *cyclopean retinal field of an animal* is that portion of its cyclopean optical field which is encompassed by the external projection of the retinal margins of *either* eye.

So defined, the angular limits of the optical and retinal cyclopean fields are the same as for the corresponding uniocular fields when referred to the visual axes. This would not be true for closer measurement distances. The absolute or relative fields must be established from the measurements of the appropriate uniocular field.

The *nasal limits* of either the retinal or optical uniocular fields determine the corresponding binocular fields.

The *binocular field of an animal* is that common portion of either cyclopean retinal or optic fields from which light may be simultaneously refracted through the lenses of both eyes. Both relative and absolute fields may again be derived.

2. Measurement and Construction

The extents of the binocular and panoramic fields of view were at first indirectly assessed. MÜLLER (1826) employed the angle between the orbital margins to make estimates of the binocular field, but the relationship varies extensively between species. A rough idea of the degree of laterality of the eyes in different forms may be gained from the divergence of the corneal axes (GROSSMAN and MAYERHAUSEN, 1877) or the optic axes (Fig. 1) (JOHNSON, 1901). The subsequent, more direct, estimates of fields based upon transcleral illumination (ROCHON-DUVIGNEAUD, 1922) suffer from similar defects to those of the uniocular fields and require, with the possible exception of estimates of the relative binocular fields, replacement by measurements based upon direct observation of the fundal landmarks.

The elimination of eye movements and assurance of the primary eye position are difficult to achieve in the conscious animal so that the relative cyclopean fields are not determined directly. The rabbit is an exception to this generalisation because of the "freeze" posture elicited by a loud noise. In this condition the eyes return to the primary position for a sufficient length of time to permit the cyclopean fields to be directly measured (HUGHES, 1971).

The cyclopean fields of species whose fixation area is ophthalmoscopically well defined, such as cat, monkey, and man, are instead constructed from the appropriate relative or absolute uniocular fields by superposition of the visual axes. In such animals the ability to define the visual pole and projection of the visual axis enables the anatomic position of rest of the eyes in the paralysed or dead preparation to be compared with the primary position. In man (DUKE-ELDER, 1932) and cat (BISHOP et al., 1962) the difference is small. If this is true of

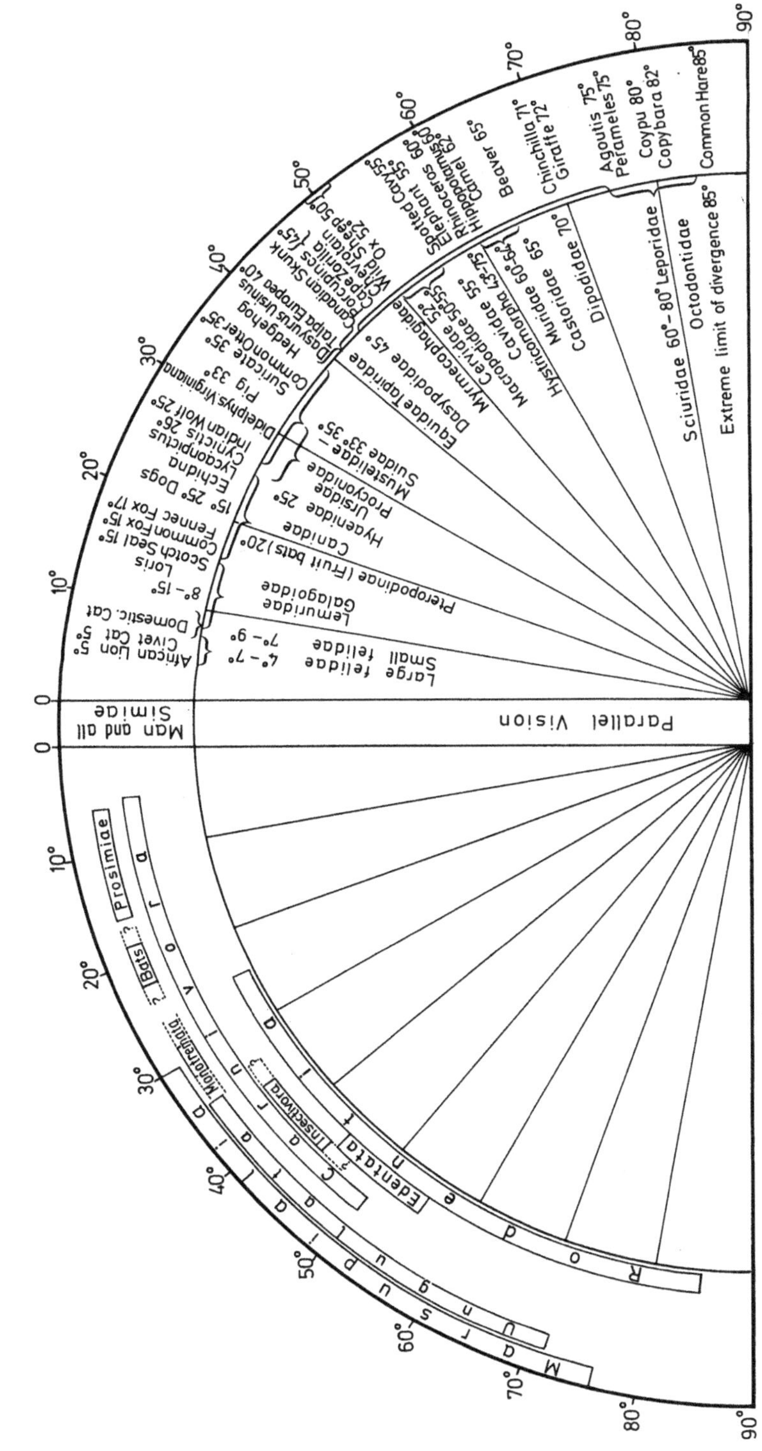

Fig. 1. Divergence of the optic axes in the mammalia; a crude index of binocular field width (after JOHNSON, 1901)

other forms, the results of PISA (1939) and other workers may acceptably define the binocular fields of the conscious animal.

In rabbit, rat, and other species the fixation area is not ophthalmoscopically visible, and examination of the retinal ganglion cell distribution does not enable a visual pole to be specified accurately with respect to the projection of some other visible landmark. The error entailed in acceptance of the fields defined for the anatomic position of rest as a substitute for those in the primary position cannot be assessed nor the cyclopean be constructed from the uniocular fields when the projection of the visual axes is unknown. Indeed, according to DUKE-ELDER (1958),

> unless there is an area centralis of acute vision ... the whole concept of fixation along a visual axis is meaningless; only in those animals provided with an area of acute vision is such a concept possible ...

In spite of the absence of an obvious visual pole both rat (LASHLEY, 1938) and rabbit (VAN HOF and LARGERS VAN HASELEN, 1973) behave as if they possess a fixation axis projecting dead ahead which suggests, contrary to DUKE-ELDER'S generalisation, that some part of the retina may be functionally differentiated.

In man (VAN BUREN, 1963), monkey (STONE et al., 1973), and cat (STONE, 1966) the gross, histologically defined decussation line passes through the fixation area. Although the decussation line of neither rat nor rabbit has been defined anatomically, its cortical representation as the boundary between the electrophysiologically mapped areas VI and VII has been specified in terms of its projection into the visual field relative to that of the optic nerve head [rat (ADAMS and FORRESTER, 1968; MONTERO et al., 1968; MONTERO et al., 1973), and rabbit (THOMPSON et al., 1950; HUGHES, 1971)]. In both species the plane of projection of the decussation line contains the behavioural fixation axis; by analogy with cat and monkey, it is presumed that the functional visual axis also lies in this plane so that the cyclopean may be constructed from the uniocular fields.

3. Extent of the Cyclopean Panoramic Field

The absolute cyclopean fields of the rabbit (HUGHES, 1971), rat (HUGHES, 1977c), and cat (HUGHES, 1976a) are presented in Fig. 2 as constructed according to the previously outlined principles from ophthalmoscopically defined uniocular fields. The remarkable differences between the visual fields of frontal and lateral eyed species are well illustrated.

The whole field of the rabbit encompasses nearly an entire sphere, but contrary to earlier reports (DUKE-ELDER, 1958), there is no sign of binocular overlap to the rear of the head in the horizontal plane (HUGHES, 1971), although such a field exists at greater elevations (PISA, 1939; HUGHES, 1971). Again, the alleged 287° horizontal field of the cat (THIEULIN, 1927; DUKE-ELDER, 1958) has not been confirmed (HUGHES, 1976a); the cyclopean optical field of the cat extends for only 210° and the retinal field for 187°. The blind field to the rear of the cat in the horizontal plane is thus 173° compared with 1° or 2° in the rabbit. The extent of these estimated cyclopean fields is substantially confirmed for the cat by the behavioral measurements of SHERMAN (1973).

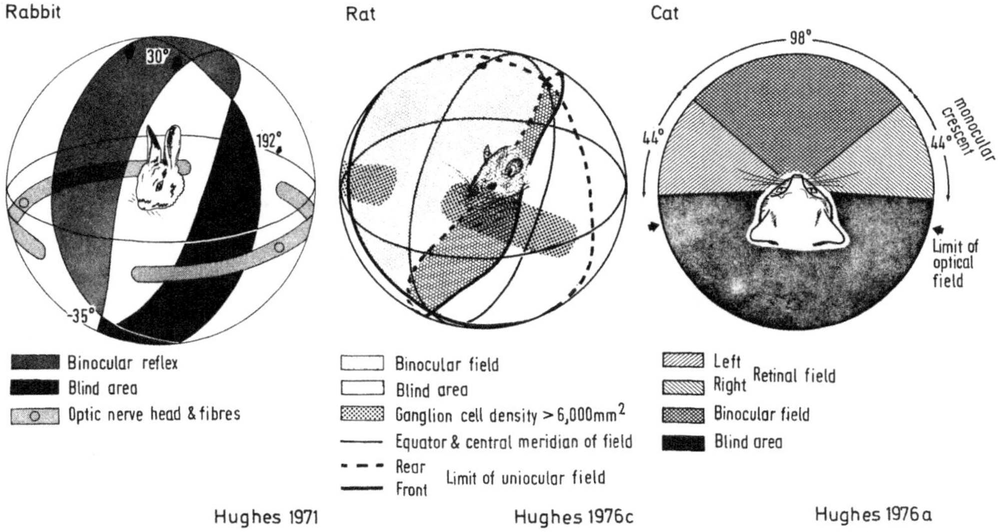

Fig. 2. Ophthalmoscopically defined absolute ocular fields of the rabbit, rat, and cat. The equatorial cyclopean fields of rabbit and rat are almost equally extensive, but, in spite of this, rat binocular field is almost three times wider than that of the rabbit; it is obvious that the extent of binocular and cyclopean fields need not vary in strict inverse ratio. Cat binocular field is only slightly wider than that of the rat, but its cyclopean field is much more restricted in extent; cyclopean field width is thus not a simple passive consequence of the binocular field width, and its limited extent in the cat is an adaptive feature of the frontal syndrome (HUGHES, 1972, 1976a, 1977c)

The extensive horizontal panoramic fields of the rabbit, 360°, and rat, 320°, are attained by retinal exploitation of the entire uniocular optic field. The less extensive panoramic field of the cat is compatible with a 38° difference between the retinal and uniocular optic fields (HUGHES, 1976a). The discrepancy between optic and retinal uniocular fields of view may thus generally be greater for frontal than for lateral eyed species.

4. Extent of the Binocular Field

The rat may be described as possessing lateral eyes, its 320° panoramic field exceeds that of the cat by some 135° and approximates the 360° of the rabbit, yet it possesses an 80° binocular field (HUGHES, 1977c) which approximates the 99° of the cat (HUGHES, 1976a) and considerably exceeds the 24° of the rabbit (PISA, 1939; HUGHES, 1971). The extensive binocular and panoramic fields of the rat arise from the retinal exploitation of its large uniocular optic field.

It is very noticeable that the spherical panoramic fields of view of the rat and rabbit are accompanied by extensive upper binocular fields. It has not been established whether these regions are binocularly represented at the cortical level or whether the animals attain stereoscopic vision in them. In man, monkey, and cat, the upper field of view is very limited.

The striking contrast between the elevated optic axis of ground dwelling species such as the rabbit and rat and its depression in a taller species with little to

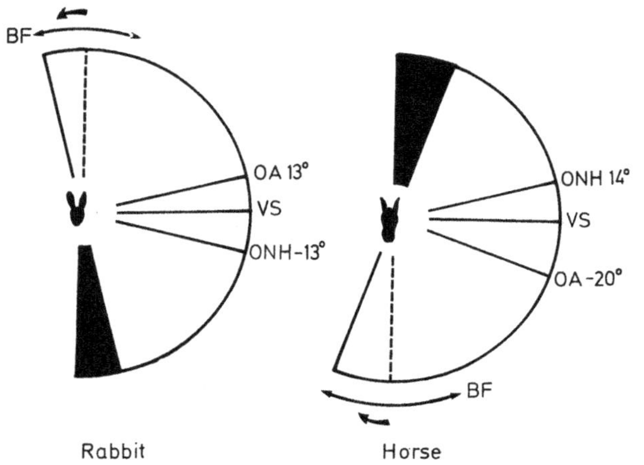

Fig. 3. Coronal sections through absolute visual fields of rabbit and horse. The rabbit, which is vulnerable to aerial predators, shows upper-field binocular overlap, *BF*, but the larger, cursorial horse has binocular overlap in region of inferior field occupied by forefeet. Lateral projection of the visual streak, *VS*, optic axis, *OA*, and optic nerve head, *ONH*, are also indicated (HUGHES, 1971, and unpublished observations)

fear from elevated predators is revealed by comparison of coronal sections of the uniocular field of rabbit and horse. The horse possesses an effective binocular field below its head (PISA, 1939), the inverse of the arrangement in the rabbit (Fig. 3).

The recent quantitative revisions of the visual fields in species so familiar as cat and rabbit suggest that the literature dealing with more exotic forms requires confirmation before its detailed consideration. The qualitative description of the fields in other species which is provided by WALLS (1942), ROCHON-DUVIGNEAUD (1943) and DUKE-ELDER (1958) is probably sufficiently accurate for general pu. poses.

III. The Functional Basis of Species Differences in the Extent of Panoramic and Binocular Fields

1. Panoramic Field

XENOPHON (1905) describes the hare as lacking

> keen sight, for many reasons; for it has its eyes prominent and the lids small, affording no protection for the eye balls; and their sight is on this account dim and dispersed.

According to the Greek centrifugal theory of vision, the wide field of sight is disadvantageous, but the modern centripetal theory indicates that for such hunted species

> there can be no doubt that the divergence of the optic axes combined with a large cornea, is a great source of protection to the animals by the increased range of vision it gives them

(JOHNSON, 1901). The hare and rabbit, both with the habit of "freezing", can spot enemies coming in any direction without a revealing head or eye movement. In spite of this advantage, the cyclopean retinal, or panoramic field, is effectively absent to the rear of the head in frontal eyed species such as primates and carnivores. Why is this so?

According to WALLS (1942):

> the visual field of the single eye is roughly constant at 170° or so (thus), we may consider the angular width of the binocular field to be quite directly related to the angle between the two optic axes, which in itself will depend upon the position of the eyes in the head.

The binocular and panoramic fields would thus vary inversely in extent. Given WALLS's (1942) belief in a universal

> urge to binocularity

then a trend towards the disappearance of the rear panoramic field would appear to be inevitable.

The previous demonstration that the uniocular retinal field may vary over a considerable range between species means, however, that the magnitudes of the binocular and panoramic fields may, in general, attain some degree of independence of each other. It certainly remains true that the limitations on the extent of the uniocular optic field ensure that only a small binocular field is possible in those lateral-eyed species for which, because of defencelessness or lack of cover in their environment (WALLS, 1942), an almost complete panoramic field of view is vital. But the suggestion that

> binocular animals have made a sacrifice; they have abandoned the enormous biological advantage of panoramic vision in order to have their eyes pointing forward

(BLAKEMORE et al., 1972) it not the primary explanation. The rat attains a large binocular field, only 20% smaller than that of the cat, but its large uniocular fields generate a panoramic field 72% more extensive than in the cat (HUGHES, 1977c) and only slightly smaller than that of the rabbit. The absence of a panoramic field to the rear of the head in frontal species is not universally restricted by the limited extent of the uniocular field alone. Some 38° of the cat, and 30° of the human uniocular fields remain unserved by retina and would be thought to be potentially available to extend the field to the rear of the head without reduction of the binocular field (HUGHES, 1976a).

The loss of the rear panoramic field in frontal species possibly arises out of a need for parallel or only slightly divergent *optic axes*, the adaptive significance of which is discussed later, when the extent of the temporal uniocular optic field cannot be increased. That loss of the rear panoramic field is acceptable as secondary to some other requirement in these species is perhaps more indicative of a false attribution of importance to it in their life styles than of a "sacrifice" on their part.

> The predaceous animal can afford not to have such "eyes in the back of his head," because his offensive weapons, teeth and claws, give him immunity from stealthy attack

(Walls, 1942). This observation is in keeping with the presence of more lateral eyes in those small predators, such as weasels, which are themselves preyed upon but does not account for the extensive blind regions to the rear of the head in the poorly armed and nonpredaceous primates. However, only

> a few specialised arboreal mammal-eating predators have been developed, including snakes, monkey-eating eagles, and several carnivores. But on the whole arboreal animals are less subject to predation than terrestial forms. Many are relatively unarmed, and depend upon inconspicuousness (sloths), unpalatibility (koalas) or rapid escape (squirrels and monkeys)

(Haines, 1958).

Consideration of the static cyclopean fields is a very limited approach. The disappearance of the rear panoramic field in frontal species is much less significant when the mobility of the eyes and head are taken into account. The *static* cyclopean retinal field of man extends for only 208° (Hartridge, 1919) compared with the horizontal limits for the *dynamic* field of 288°. Even the 72° dynamic blind area to the rear of the head is totally eliminated if the head is permitted to rotate.

2. Role of the Binocular Field

Stereopsis. Among mammals only the larger whales are said to lack a binocular field, but what advantage is obtained by its possession? The human binocular absolute luminance threshold for 50% seeing is 0.1 $\log_{10}$ units lower than the uniocular threshold (Pirenne, 1962). Visual acuity (Bárány, 1946) and the threshold contrast for sinusoidal gratings (Campbell and Green, 1965) are improved to a similar extent in binocular relative to uniocular viewing. Such threshold benefits have been suggested to be the reason for the development of a binocular field in abyssal-fishes and owls (Walls, 1942), but among the mammals, at least, the frontal binocular field is generally regarded as the substrate for stereoscopic depth perception (Bishop, 1973), in which the relative separation of objects in depth is determined from the disparity between the retinal images of the common field of the two eyes. Johnson (1901) rejected Berlin's (1893) assumption that this ability should be attributed to all species evincing depth perception, because monocular individuals perform equivalently in many tests; this conclusion has been supported by many later experiments (Walk, 1965). The isolation of binocular disparity (see Bishop, 1973) from monocular cues for relative depth has, however, become relatively straightforward since the advent of the anaglyphic random dot stereogram (Julesz, 1971) which has already been employed to demonstrate stereopsis in the monkey (Bough, 1970). The cat alone among the remaining mammals has been tested for stereopsis, by another method, and was concluded to possess it (Fox and Blake, 1971); more recently, the range of fusion and quantitative stereopsis have been established in the same animal (Packwood and Gordon, 1975).

The low ganglion-cell density at the fixation area and small interocular distance of many small mammals has been regarded as an inadequate basis for useful stereopsis. It is true that the theoretical maximum stereoacuity of these species appears poor when compared with that of man, if expressed as an angle, but its value is of greater significance in terms of a linear displacement *over the range of interest to the animal.* High angular thresholds may be quite adequate for small animals which are concerned with nearby objects.

The neural mechanism of stereopsis in mammals (BISHOP, 1973) is believed to depend upon *partial* decussation at the chiasma and subsequent convergence of the projections of the left and right hemiretinas to cells selectively sensitive to image disparity. Such cells have been recorded in monkey (HUBEL and WIESEL, 1970), cat (BARLOW et al., 1967; NIKARA et al., 1968) and goat (CLARKE and WHITTERIDGE, 1976). The established presence of the whole or part of this apparatus in a mammalian species would be presumptive evidence for stereopsis.

Stereopsis in Lateral-Eyed Species. The phrase "binocular vision" is rarely defined, yet clearly has not the same implications to all authors; an unfortunate situation, because its use is often accompanied by judgments of its relative quality in different species. Common usage tends to equate frontality with binocular vision

> among the mammals only the carnivores, lower primates and man have developed binocular vision to any extent ...

(BISHOP, 1973). For HARTRIDGE (1952) the binocular fields of

> animals in which the two eye axes are parallel have the greatest overlap, and therefore possess the most perfect binocular vision

with "a single mental impression" over the widest field. But these usages fail to separate consideration of the binocular field width from that of stereoscopic vision quality. The inevitable outcome is that it is still common to find the corollary of the above judgements, that binocular vision is poorest in lateral eyed species, is used to imply that they also lack stereopsis (ELLIOT SMITH, 1930; CARTMILL, 1974)

> Laterally placed eyes, of course, preclude stereoscopic vision

(VINCENT, 1912). If binocular vision is primarily concerned with stereoscopic depth judgement, as previously argued, then comparison of its quality between species would best be made in terms of their differences in eye separation, visual acuity or behaviorally determined stereoacuity. Discussion of the variations in binocular field width is taken up again subsequently.

The continued implicit belief that in lateral-eyed species

> this great range of lateral vision makes, of necessity, vision in the direct line ahead correspondingly bad

(HARRIS, 1904) may provide supplementary reasons for the neglect of the limited binocular field of many such animals. It has, however, not received experimental justification and usually appears ill-founded in either confusion of the optic and visual axes or the well-entrenched, but not necessarily valid, belief that peripheral optics must be so poor as to inevitably preclude adequate forward vision.

The 360° panoramic field of the rabbit (PISA, 1939; HUGHES, 1971), 320° of the goat (HUGHES and WHITTERIDGE, 1973), and 300° of the grey squirrel (HALL et al., 1971; LANE et al., 1971) qualify these animals as lateral eyed. Many workers would thus not describe them as binocular species in spite of the potential for stereopsis in their respective 24°, 60°, and 60° binocular fields. In the goat the temporal fixation area some 50° divergent from the optic axis is yet of potential neural resolution better than that of the cat (HUGHES and WHITTERIDGE, 1973), thus suggesting that its peripheral optics are good. The adequacy of the peripheral optics of rabbit and squirrel is implied in their use of the binocular field as their behaviorally favoured fixation area (VAN HOF et al., 1973; POLYAK, 1957). Presumptive evidence for the presence of stereopsis is also available. The electrophysiologically mapped cortical representation of the binocular field receives a bilateral input in rabbit (THOMPSON et al., 1950; HUGHES, 1971), goat (CLARKE and WHITTERIDGE, 1976), and squirrel (HALL et al., 1971; LANE et al., 1971). This binocular area contains disparity-sensitive binocular neurones in both rabbit (VAN SLUYTERS and STEWART, 1974a) and goat (CLARKE and WHITTERIDGE, 1976), but such cells need not unequivocally establish a substrate for stereopsis. JOSHUA and BISHOP (1970) suggest that the equal vertical and horizontal disparity sensitivity of cat visual area I neurones rather implicates them in the mechanism of binocular fusion. However, in the sheep at least, horizontal disparity sensitivity appears to predominate in the binocular neurones of visual area II (CLARKE et al., 1976), and they are described as organised into "depth" columns similar to those reported for the cat by BLAKEMORE (1970). The good depth judgement of sheep (WALK and GIBSON, 1961), the well-known agility of goats.

> In North Africa domestic goats are becoming pests of fully grown fruit trees, for they have learned to climb on high branches and browse off the foliage.

(HAINES, 1958)—and of squirrels (CARTMILL, 1974; POLYAK, 1957), also strongly imply the possession of stereopsis in these species.

A recent extensive study of single units in the rabbit binocular area (VAN SLUYTERS and STEWART, 1974a, 1974b) places great emphasis on differences between their organisation and that of the corresponding units of the frontal-eyed cat. Disparities for many units were so large as to appear nonfunctional; binocular units often had as inputs two different classes of receptive field, and developmental plasticity was markedly lacking (VAN SLUYTERS and STEWART, 1974b; CHOW and SPEAR, 1974). But while awaiting confirmation of these features, it is worth emphasising the *correspondences* between rabbit and cat binocular cortex organisation; some 70% of the rabbit units *were* binocular; 50% *did* have similar inputs from each eye and *some* disparities were in a range that could be functionally useful. These components of the cortical apparatus could alone suffice to provide a substrate for stereopsis in spite of the puzzling anomalies.

At present there is a great deal of evidence to support the view that many features of organisation in the central provision for the binocular field are common to forms with frontal or lateral eyes, and it would be wise not to dismiss the possibility of stereopsis in any species of mammal which possesses a binocular field.

Stereopsis Among Nonmammalian Vertebrates. Since WALLS (1942) said of stereopsis.

> There is no more vexed question in all of comparative ophthalmology than the one whether this binocular factor in depth perception exists for vertebrates below the mammals,

there has been little progress towards a solution. A binocular field is present in nearly all vertebrates, but in the opinion of many workers (HARRIS, 1904; COLLINS, 1921), although not all (WALLS, 1942), the alleged absence of a partial retinal decussation precludes the possibility of stereopsis in the nonmammalian classes. Thus INGLE (1970) regarded the frog as lacking stereopsis and attributed its depth perception to motion parallax or information from an accommodatory center (INGLE, 1968). However, from the functional point of view it is not important whether decussation occurs at the chiasm or at some subsequent stage in the visual pathway. The tectal representation of the binocular field in the submammalian species has been most thoroughly examined for *Rana* and *Xenopus* and found to obtain an input from both eyes (GAZE and JACOBSEN, 1962; KEATING and GAZE, 1970; GAZE et al., 1970); the ipsilateral retinotectal projection arises from the contralateral tectum and passes to the ipsilateral side via the postoptic commissures (KEATING and GAZE, 1970). Cells in the binocular area respond to inputs from either eye (FITE, 1969; SKARF, 1971) between which predominantly horizontal disparities have been demonstrated in unparalyzed preparations (RAYBOURN, 1975), thus providing a mechanism for the stereopsis previously suggested only by the presence of an ipsilateral tectal representation. A similar ipsilateral projection has been reported for the axolotl (BRÄNDLE and STIRLING, 1975). A recent examination of the influence of lenses and prisms on snapping behaviour in toads (COLLET, 1977) indicates that, under monocular conditions, the distance of prey is judged by means of accommodatory cues in agreement with INGLE'S hypothesis (1970; 1971; 1972; 1973) but that disparity cues were employed during binocular judgements.

The presence of superposed ipsi- and contralateral projections in the tectal representation of the binocular field has not been established for other nonmammalian vertebrates, such as fish (JACOBSEN and GAZE, 1964; SCHWASSMAN and KRUGER, 1965; SCHWASSMAN, 1968), alligator (HERIC and KRUGER, 1965), or pigeon (HAMDI and WHITTERIDGE, 1954). A detailed search for an ipsilateral representation in the tectum of foveate marine teleosts was not successful and revealed binocular activation only in the nucleus praetectalis (SCHWASSMANN, 1968). Four binocular units were recorded in the pigeon cerebellum (CLARKE, 1974). However, other regions of the brain in these species remain unexplored. A very recent report (PETTIGREW and KONISHI, 1976) described the visual "Wulst" of the owl as containing many binocular neurones, in spite of an allegedly monocular thalamus, the characteristics of which are similar to those of cat and monkey visual cortex. Tests for stereopsis were described as "promising." It thus appears probable that investigation of the telencephalic regions of other submammalians will reveal binocular organisation and that, in common with WALLS (1942), we might feel justified in regarding the presence of the binocular field in these classes as presumptive evidence of stereopsis. Whether or not we would *be* justified remains to be seen, at least for species with independent eye movements. HARKNESS (1977) has made a convincing demonstration, although not employing pure disparity cues, that chameleons base their judgement of prey distance on accommodatory cues under both monocular and binocular conditions. However, the animal remains untested for ability to sense pure disparity cues.

It is somewhat ironic, now the reality of a supraoptic, chiasm has been demonstrated in the owl, that early suggestions (see PRINCE, 1956) of a decussation in reptiles and snakes should be gaining support from work on another nonmammalian species. A genuine, primary, ipsilateral projection from the retina is now established as present in the frog (LAZAR and SZEKELY, 1969; CURRIE and COWAN, 1974), but it does not supply the tectum. Its physiologic significance remains to be established.

3. Teleology of Binocular Vision

The Relationship Between Binocular Field Width, Phylogeny, and Life Style. Binocular vision is almost universal amongst the vertebrate species in the majority of whom it presumably provides a basis for stereopsis. Any explanation of the variability of binocular field *width* between species would therefore be expected to be generally applicable amongst the vertebrate classes.

The absolute horizontal extent of the binocular field in both fishes and reptiles is said to rarely exceed 40° (KAHMANN, 1930, 1935), while the upper limit for birds, with the exception of some raptorial species, is about 50° (WALLS, 1942). Although these measurements still require confirmation, their contrast with the mammalian range of 25°–100° for the subprimates and up to 130° for the primates is obvious to direct inspection. Among the nonmammalian vertebrate classes, the anurans contain the only known species which attain binocular fields as wide as those to be found in the higher mammals; in *Rana* and *Xenopus* the binocular field is from 80° to 100° wide (GAZE and JACOBSEN, 1962; GAZE et al., 1970). Similar observations led JOHNSON (1901) to the conclusion that the width of the binocular field, and in his opinion the inversely related extent of the panoramic field is determined by the phylogenetic status of the species,

> the higher the order, the nearer the axes approach parallel vision, although the range is considerable ...

The possession of an extensive panoramic field was even regarded by some as a "primitive" feature (PARSONS, 1927).

Although adopted by a few later workers (PARSONS, 1927; POLYAK, 1957), the hypothesised phylogenetic correlation was contemporarily challenged by HARRIS (1904),

> the presence of binocular vision in animals seems to depend upon their feeding habits rather than upon their social position, if I may use the term, in the animal scale ...
>
> lateral placed eyes being found for the most part in the herbivores and rodents, that is to say, in animals whose food does not run away from them, whilst the most perfect binocular vision is found in the carnivora ... and in the primates.

HARRIS (1904) realised the difficulty of applying his generalization to the submammalian classes when couched in terms of absolute binocular field widths and was suitably cautious. Subsequent results have shown, however, that within one class it is often true that predators possess wider binocular fields than prey. The binocular fields of predatory fish, such as trout or pike, 30°–40° (KAHMANN, 1935), may be of twice the width of those in herbivorous or bottom-feeding fish, 10°–20° (WALLS, 1942). Among reptiles the same trend is to be found; the herbivorous *Testudo* binocular field is 18° wide, while that of the predatory snapping turtle, *Chelydra*, extends for 38° (WALLS, 1942). The largest binocular fields amongst the birds are to be found in the raptores; the hawks, kestrels, and eagles have binocular fields alleged to be from 35°–50° in width (PRINCE, 1956) and the owl some 60° (WALLS, 1942), although the tubularity of the owl eye makes its effective retinal binocular field only some 40° wide (PETTIGREW, personal communication). The generality of this correlation provides prima facie evidence for the possibility of dealing with the variation of binocular field width on a common basis throughout the vertebrate kingdom. The theory of HARRIS, however, fails to account for the observations which support JOHNSON'S (1901) phylogenetic correlation, and upon cross-class comparison such puzzles arise as to why the width of the binocular field in the predatory crocodile is as narrow (25°) as that of the rabbit and smaller than in the ungulates.

A further difficulty with HARRIS' theory (1904) is that the majority of the primates, although possessing the widest binocular fields, are not predators. HARRIS was content to accept this apparent anomaly as related to the interconnected development of sight and the use of the hand as a prehensile organ:

> Binocular vision is clearly of great assistance in the accurate use of the hand for fine movements, which, in proportion to its specialisation as a fine and delicate instrument, has been accompanied by an increasing development of brain power

(HARRIS, 1904).

The evolution of binocular vision among the primates has subsequently tended to be treated in isolation from that of the other vertebrates. The importance of the eye-hand relationship was again taken up by ELLIOT SMITH (1912, 1928, 1930) and his school (WOOD JONES, 1926; LE GROS CLARK, 1934, 1962, 1970; HILL, 1972) whose general thesis, the *arboreal theory* (HOWELLS, 1947), is that the frontal-eyed, prehensile, extant primates have evolved from nonprehensile lateral-eyed precursors (LE GROS CLARK, 1970) under the influence of selection pressures arising in arboreal locomotion per se.

The frontal eye position and wide binocular field of the primates was initially regarded as a result of the change of emphasis from smell to vision during primate evolution, of general benefit in providing a large field of stereoscopic vision (WOOD JONES, 1926), but simply a consequence of the recession of the snout (ELLIOT SMITH, 1912; WOOD JONES, 1926). The view that the wide binocular field and stereoscopic vision are adaptations for the accurate judgement of distance during arboreal locomotion (COLLINS, 1921) was adopted subsequently (ELLIOT SMITH, 1924).

Nevertheless, HAINES (1958) has pointed out that although the primates show structural specialization for the arboreal habitat, there are few such gross changes amongst the *arboreal* carnivores:

> ... the Indian mongoose *Herpestes mungo* is an accomplished climber, and takes readily to trees, but is structurally very similar to the terrestial *H. ichneumon*. The pine marten, *Martes martes*, can run down a squirrel leaping through the canopy and walk down a tree head first, yet its hands and feet show little difference from a terrestial polecat *Mustela putorius* ...

In a similar vein CARTMILL (1972, 1974) argues that the grey squirrel is the equal of a saltatory lemuroid in its arboreal agility, and yet its orbital convergence, optic axis divergence and manus evince no sign of primatelike change relative to the more terrestial ground squirrel. In fact, the majority of tree-dwelling species do not converge with the primates in anatomic specialisation; *there are many different arboreal locomotory niches*, and the contrast between the primates and squirrels is alone sufficient contraindication of a common selection pressure arising in arboreal locomotion (CARTMILL, 1974).

In his *visual predation hypothesis* CARTMILL (1972, 1974) adopts WOOD JONES' (1926) view that grasping extremities arose not as adaptations to arboreal locomotion per se but for movement amongst the finest branches. In contrast to his expectation from the arboreal theory, he notes that orbital convergence is greatest in the slow, visually directed insect predators, such as *Loris* and *Nycticebus* which employ their grasping hand to take prey, rather than in the swift acrobatic species. Similar orbital convergence, after allometric correction, is reported for the marsupial manual insectivores *Cercatetus* and *Burramys* (CARTMILL, 1974). The increased orbital convergence of these manual predators is regarded (CARTMILL, 1972, 1974) as an adaptation of similar significance to that in other hunting species, such as cats, and taken to imply an increased field of stereoscopic vision. By analogy, therefore, CARTMILL suggests that the prehensile hands, frontal eyes, and, implicitly, the large binocular field of extant primates arose from a common ancestral form which inhabited the fine branches of the canopy and undergrowth of the tropical forests and employed its grasping hands for visually directed manual predation on insects and other prey.

Such primate precursors would be consistent with HARRIS' (1904) theory of predation as the "fount" of binocularity, but the problem of the nonpredatory frugivorous primates is simply evaded. CARTMILL (1972, 1974) has made no attempt to demonstrate the correlation, which is necessary to his hypothesis, between the binocular field width and the degree of orbital convergence; the binocular fields of nonpredatory primates appear to be as wide as those of predatory forms, and it is this, not the degree of orbital convergence, which is of functional significance. Indeed, in criticising the limited evidence for CARTMILL'S hypothesis SZALAY (1975) has pointed out that the most convergent orbits amongst the Malagasy strepsirhines are to be found in the phytophagous, forearm-dominated *Palaeopropithecus*. The retention of convergent orbits in this form and in extant phytophagous primates clearly indicates that visually directed manipulatory habits could be argued to provide teleologic justification for their evolution without any intermediate stage of visual predation.

An uncommon question: What functional advantage is gained by a wide binocular field? It would not in general be accepted that the advantage of the wide binocular field is that it simply

> affords a double assurance of information pick-up in a wider angle of the field ahead

(GIBSON, 1966), although this is to some extent true under quantum-limited scotopic conditions. Neither does it appear that the binocular field width of a given species correlates with the speed attained in normal locomotion. The 25° binocular field of the racing pigeon, which is typical of a granivorous bird, is thus markedly smaller than the 40°–60° of the hawk, although both forms are recorded as attaining air speeds in excess of 60 m.p.h. and even higher ground speeds; the binocular field of the slow-moving owl is similar in size to that of the swifter hawks. The binocular field of the slow loris is double the magnitude common in ungulates which may attain speeds of up to 60 m.p.h. During smooth forward locomotion, adequate stereoscopic coverage of a much wider range of space than that to be entered by the body is provided by a relatively narrow binocular field; progressive change of course is accompanied by a gradual realignment of the binocular field towards the destination. It has been suggested that the irregular paths of swift predators engaged in the chase may put a premium on a wide binocular field in order that knowledge of the spatial organisation of the *environment* is immediately available during massive readjustments of the direction of movement. However, momentary consideration indicates that the prey lead the chase without apparent inconvenience, although commonly possessed of only narrow binocular fields. CARTMILL (1972) has suggested more specifically that cats and prosimians are better able to compensate for the evasive movements of their prey because of their wide field of stereoscopic vision. There is undoubtedly some truth in this argument even when applied to the *relatively* wider binocular fields of fish and reptilian predators rather than to extremely frontal species, but it is unsatisfactory as a fundamental explanation. In spite of wide binocular fields in the prosimians, the ballistic "smash and grab" technique of insect capture in *Galago* takes little account of prey movement (BISHOP, 1964), and stalking by neither *Loris* nor *Nycticebus* disturbs the prey until the hand touches it; compensation for flight reaction is not made in any of these prosimians, and the unitary prehensive pattern is readily disrupted (BISHOP, 1964). The *Tupaioidae* are able to grab flies in a similar fashion (LE GROS CLARK, 1927; POLYAK, 1957), in spite of a much narrower binocular field. Again, this explanation fails to account for the retention of the wide field in nonpredatory primates.

Such ad hoc explanations for the extent of the binocular field in a given species hinder the isolation of a common underlying function. According to DUKE-ELDER (1958), the wide binocular field simply provides stereopsis over a more extensive range of visual space than a narrow field; its application must be varied and very general:

> the extent of binocularity, on the other hand, is determined by the greater value of the fine judgement rendered possible by binocular vision in pursuit and attack by the predator, in its ordinary activities by the arboreal animal, or by the Primate the eyes of which have become accurately correlated with the use of its hands.

In the classic analysis, as summarised by WALLS (1942), the quality of "solid" vision in stereopsis is seen to be of some benefit in all species, so that there is a "*universal* urge to binocularity." The width of the binocular field attained by any species follows passively and inversely upon its need for panoramic vision. No specific adaptive significance is seen in the correlation between predatory or arboreal life styles and the presence of the wide binocular field; the wide field is expressed because of the low pressure for panoramic vision in these niches.

LINDSAY JOHNSON'S (1901) observation that there is a phylogenetic trend towards wider binocular fields is true in the sense that the mean width of the binocular field for a given class appears to increase from the fish to the mammals; within a class, however, the range is wide, and there are both fish and primates which possess binocular fields more than 100° wide. In the context of the classic analysis this would require that the importance of the panoramic field decrease throughout the phylogenetic series rather than that the importance of the binocu-

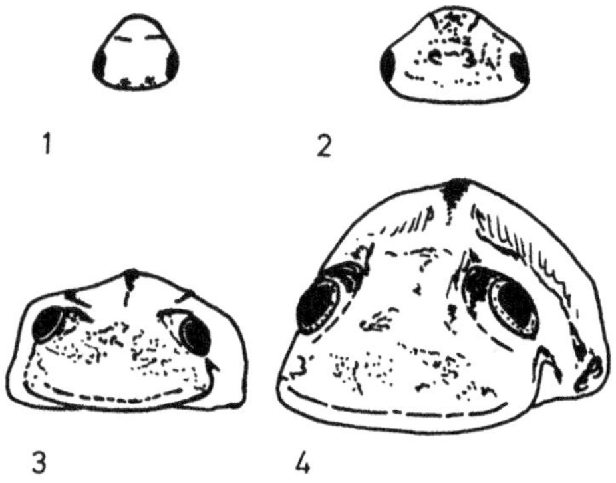

Fig. 4. Shift of eyes from their lateral position in *Xenopus laevis*, stage-46 larva, *1*, stage 48,2, to their more frontal location in the 20-mm-long juvenile, 3, and 34-mm-long juvenile, 4. (after R. M. GAZE, 1970)

lar field increase. An appropriate functional basis for the absolutely narrow binocular fields of fish and reptilian *predators* would thus be their common lack of effective brood care, independent lives when small (MITCHELL, 1912), and vulnerability to secondary carnivores when adult. The premium on panoramic vision would be greater for such forms than for bird and mammalian predators, which tend to lead a sheltered life when young.

The crocodile is a good case in point. It has been estimated that the survival rate of crocodiles to maturity is only 1% (STEVENSON-HAMILTON, 1912), compared to about 30% for the lion, in spite of an agressive nature at hatching and considerably more parental supervision than received by the majority of poikilotherm young; the majority are eaten by birds, cats, and other crocodiles. The surprisingly small 25° binocular field of the voracious and nearly invulnerable adult might thus be determined by its needs when immature.

The generality of this explanation is made suspect by the fact that the young of most *predatory* fish or reptiles are subject to the same degree of predation and cannibalism as their prey, or the young crocodile, and yet their binocular field may substantially exceed that of the adult crocodile in width, for instance, it is 40° wide in the pike (KAHMANN, 1935). It seems rather that the life style of the crocodile sets a low premium on a wide binocular field; stereoscopic vision necessary for its rapid forward lunges towards prey would be adequately provided by the narrow binocular field, and the animal appears to exploit its wide panoramic field to detect adjacent prey which are taken with a side sweep of the head and jaws which requires little judgment of depth.

Again, it is hard to believe that the adult frog or toad is less vulnerable than a young crocodile, and yet both forms surrender the advantages of panoramic vision at metamorphosis when the eyes rotate forward to give rise to a wide binocular field with a complementary blind area to the rear of the head [Fig. 4 (GAZE, 1970) and Fig. 5a].

Such instances make it impossible to accept WALLS' (1942) concept of a uniform universal urge to the development of a wide binocular field. Among the mammals it is obvious that the width of the binocular field may vary to some extent independently of panoramic field width; the cat and primate binocular fields are significantly different, at 100° and 130°–140°, respectively, in spite of almost identical optic panoramic fields. It is thus concluded that *the relative width of the panoramic and binocular fields is determined by the interaction of separate selective pressures which are not universal in intensity but predominate only in certain lifes styles.* The wide binocular field of the frog thus fulfills some specific need which is not shared by the crocodile. It remains to specify in more detail just what aspects of a species life style determine the pressure for a wide binocular field; it is clear from the differing widths of the frog and crocodile binocular fields that it is not simply predation per se.

Praxic space and binocular field width: a systematic survey.

> There is the Yak, as great as a cloud shadowing the sky. Big as it is, it cannot catch a mouse.
>
> CHUANG TSU: Inner Chapters

The class of *visually* controlled activities involved in directed locomotion, prey capture, nest building, and other manipulations of the environment have been described as *praxic* behaviour (TREVARTHEN, 1968). Almost any part of the body may be employed for praxic activity; the horse kicks accurately with the rear legs or the crocodile sweeps prey into the water with a blow from its tail, but such instances are exceptions in the behavioral repertoire. In vertebrates such visually controlled actions are confined to a limited anterior segment of the egocentric behavioral space which, with the head in the primary position, is symmetrically disposed about the midsagittal plane and may be described as the *central praxic field.* The extent of this field is determined by the nature of the individual species' manipulatory organs and by the manner in which they are employed; its limits would be delineated by the locus beyond which reorientation of the body preceeds manipulation, directed activity, or a strike.

This section investigates the possibility that the width of the binocular field increases pari passu with that of the central praxic field. It is obvious that much of praxic behavior requires judgment of relative distances such as is provided by binocular stereopsis; during complex activities, depth information may be simultaneously required at several points in the field of view so that the more extensive the praxic range, the more a wide binocular field might be of assistance. This attempt at a general synthesis of the systematic data does not presume the width of the binocular field to be determined by one factor. Quite apart from the use of space by a given species, it is obvious that reliance upon scotopic vision also leads to a wide binocular field.

The recent revision of long-accepted measurements of the binocular and panoramic fields in common forms such as the cat suggests that the comparative literature contains many errors; the discussion of comparative aspects of binocular field width should therefore currently be based upon gross differences alone. This requirement is sustained by the fact that functional interpretation of binocular field width is implicitly couched in terms of the behavioral binocular field but at present is necessarily, although improperly, supported by reference to measurements of the optical rather than the retinal field.

Fish do not possess prehensile organs and

> on the whole are rather conspicuous for the lack of ability to manipulate inanimate objects

(THORPE, 1963). Tasks such as nest building are carried out by mouth so that the praxic field is little more than a strip of space running forward from the head. The binocular fields of this class are correspondingly narrow, rarely exceeding 40° in width; the limited extent of the 14° binocular field in the archer fish should not be surprising, as it was to KAHMANN (1935), because the animal always orientates so that its shot is directed along the midsagittal plane and is thus served by the narrow field. The extreme parallelism of the optic axes and frontal situation on the eyes in some deep sea fishes (WALLS, 1942; DUKE-ELDER, 1958) is misleading if interpreted as signifying a wide behavioral binocular field because of the restrictions set upon the retinal extent by the tubular form of the eye. A very wide 120° overhead binocular field and substantial frontal binocular field have been described for the bullhead (Polyak, 1957), but this is a predominately scotopic, nonvisual form, and the extensive binocular overlap is most likely an adaptation for increased sensitivity such as is found in certain mammals.

The development of forelimbs among the reptiles is not accompanied by an increase in the extent of their praxic space. *Chelonians* such as the snapping turtle may hold down food with the forelimbs while eating but,

> generally speaking, however, the limbs of existing reptiles play little if any part in the capture of food. The head is the main instrument of predation ...

(Bellairs, 1969) just as in the fish. Usually the prey is not stalked but waited for by the motionless reptile.

> Capture of agile prey is finally effected by a lunge or strike, or sudden rush culminating in a snap

(Bellairs, 1969). The central praxic space of these forms, like their binocular fields, is very narrow. Even in the anole and chameleon, which have wide-ranging eye movements and good depth judgment, the binocular field is relatively narrow. As among the fishes, it is the predatory species which have the widest binocular fields, especially when given to the sessile expectation of prey and attack by rapid lunges such as are employed by tree snakes and long necked turtles. The need for judgment of angled strikes without a revealing reorientation of the body and eyes, as well as for the guidance of head and jaw position during the impressively swift attack (Frazetta, 1966), could be understood to set a premium on a wider field of steroscopic vision in these forms, but its magnitude rarely exceeds 40°.

The 100° binocular field of the frog or toad is wide in comparison with those of the majority of vertebrate species and singular amongst the poikilotherms. This greater width is not accompanied by the use of the forelimbs for prey capture, although food may be held in the mouth or ejected by manipulation, nor is it associated with a more pronounced noctural component in the habits of these amphibians than evinced by more lateral eyed poikilotherms. The frog and toad are distinguished, however, by an extensive praxic space which arises from their leaping to prey over a 100° wide "jump-and-snap zone" without reorientation before attack (Fig. 5a) (Ingle, 1970); the width of the binocular field approximates this zone. The initial judgment of prey position must be especially accurate in the frog because its eyes are partially withdrawn during the leap, and visual feedback is eliminated (Fig. 5c). It remains puzzling why the binocular field is not better expressed in some fish, such as the mud-skipper, whose sessile predatory activies resemble those of the frog (Munk, 1970) but whose binocular field is not more than 40° across (Walls, 1942).

Although measurements are not plentiful, it appears that insectivorous birds which pursue prey with rapid darting adjustment of flight, such as swifts and swallows (Polyak, 1957), possess binocular fields which are wide relative to the granivorous forms (Walls, 1942). In spite of their subjection to greater brood care than is common amongst the poikilotherms, the birds do not reveal a tendency towards substantially increased binocular field widths. It is noteworthy that the largest fields are encountered amongst these raptorial forms which take their prey with two talon-equipped feet. The victim is fixated during approach, and the relative position of the feet with respect to it judged within the peripheral binocular field (Fig. 5m). The increased angular range of nearby praxic space which results from the use of the feet as well as the mouth during predation is thus again correlated with a wider binocular field, but it apparently extends for only some 40°–60°, even amongst the owls and hawks (Walls, 1942; Prince, 1956; Duke-Elder, 1958).

The release of the forelimbs from purely load-bearing functions among a great variety of mammalian forms, including rodents, marsupials, carnivores, and primates, gives rise to the potential for more complex manipulative activity and increases the volume of local space in which praxic behaviour is possible. In contrast to the submammalian classes, the simple manipulation of food is carried out by many nongrazing mammalian species; rats will handle grain and other objects (Barnett, 1963) and have in common with the *sciurids* the bimanual holding of food (Polyak, 1957). A great variety of skill is revealed in the use of visually directed manipulation during killing and feeding among the dasyurid marsupials, insectivores, and carnivores (Eisenberg and Leyhausen, 1972). In general, the frequency and complexity of such behaviour is related to the extent of grasping ability in the forepaws as well as pronative and supinative ability in the forelimbs. Such attributes are commonly correlated with arboreal skill so that among the *Viverridae* those species which employ the forelimb to pin down prey

Fig. 5. Instances of praxic activity which are discussed in the text

for the kill and feeding are highly arboreal, while those which rarely pin their prey and never hold food are highly digitigrade with little forelimb flexibility (EISENBERG and LEYHAUSEN, 1972). But even among the cursorial carnivores, the forelimbs always function as more than simple props (EWER, 1973). In parallel with this increase in the extent of praxic space, the binocular field widths encountered in this group are much greater than among herbivores and extend from 60° to 130°.

Only among species equipped for extensive abduction and adduction of the forelimb, and thus commonly demonstrating some degree of arboreal skill, is access to the widest range of manipulative space made available. The use of this range of space under visual control would alone be expected to lead to the development of a wide binocular field. According to the praxic space hypothesis, therefore, neither arboreal nor predatory habits lead per se to the widening of the binocular field, but this occurs because of the increased range of praxic behaviour common to *some species* pursuing these and other life styles. Unfortunately we have not sufficient information to examine the validity of the hypothesis in a wide variety of species and must deal only with certain obvious instances.

Consider dogs and cats. The dog has, as an endowment of its cursorial origin, only limited ability to rotate the forelimbs or grasp with the paws (EDEY, 1968). Such space to which the forelimbs have access is simply not employed for praxic or even manipulatory activities. The *Canidae* kill small prey with a relatively undifferentiated thorax bite, shaking and tossing (Figs. 5d, e) (FOX, 1969); large species are dealt with by repeated crushing and tearing bites. The only use of the paws is to restrain, and occasionally strike, small prey before the kill and to hold the carcass when eating. In association with the limited praxic behaviour, we find the binocular field of the dog to be markedly smaller than that of the cat and to range in general from 70° to 80°: dingo, 70°, cocker spaniel, 80° (personal observation), and greyhound, 80° (PISA, 1939). Larger fields have been reported for some domestic forms (PISA, 1939), but these are usually "sports" of brachycephalic appearance subject to human fancy breeding.

By contrast, however, the *Felidae* combine visually directed behaviour over a wide range of local space with the most extensive binocular fields outside the primates. A few species of *Canidae* may approach the cat in the use of cover during visual stalking, but the *visually* directed killing bite is employed with unique precision by the *Felidae* upon even large prey. Visual direction of the bite towards the skull base is common to some marsupial carnivores and viverrids, but the characteristic felid kill (Fig. 5h), severs the spinal cord after two canine teeth have passed between the neck vertebrae (LEYHAUSEN, 1965). The precise location of this bite is uniquely aided by manipulation;

> it is only within the family Felidae that we find a definite selection favouring the use of the forepaws in restraining prey with a specialised clasping movement prior to the administration of a killing bite

(EISENBERG and LEYHAUSEN, 1972). The visual domination of the extensive manipulatory repertoire of the domestic cat is well known but is common to other Felidae; the leap by which birds or butterflies may be "fielded" with one or both paws (Fig. 5a) is a widespread example of such praxic behaviour (EDEY, 1968) and characteristically differentiates this family from the Canidae which employ the mouth to catch insects (ROMANES, 1882).

The singularly wide 100° binocular field of the domestic cat is thus seen as accommodating a variety and range of praxic behavior remarkable among subprimates. The combined use of the mouth and paws during the initial stage of an attack requires the simultaneous visual fixation of the neck of the prey and accurate but independent direction of the left and right forepaws. The extensive field of feline steroscopic vision is ideally suited to such a task.

A similar argument may be applied to primates which must accurately and independently locate their hands in the peripheral visual field while the eyes simultaneously fixate a goal during brachiation, leaping, clinging and an enormous variety of other focal praxic activities (Fig. 5i, j, k, l). The range of manipulatory and locomotory praxic space employed by carnivorous and frugivorous primates alike is greater than that of any other species of vertebrate and, in accordance with the praxic space hypothesis, is correspondingly found to be associated with the widest binocular fields of any species of animal.

It is curious that in spite of their quite narrow 60° binocular fields (LANE et al., 1971), the squirrel and tree shrew demonstrate considerable visually directed manipulative activity (see POLYAK, 1957); it is probable that the range of space accessible to the forepaws of these forms is relatively as great as in the cat. It is conspicuous, however, that both species, although essentially arboreal, spend a great deal of time on the ground surface, where they are particularly vulnerable to predators, and the need for extensive panoramic vision may result in restricted expression of the binocular field.

In order to differentiate the praxic-space hypothesis from the theory of the binocular field held by ELLIOT SMITH (1924), it is to be emphasized that the presence of the wide binocular field is now suggested to be correlated with an extensive range of praxic space and not with skilled manipulation per se. Thus BISHOP (1964) has suggested that racoons may be superior in manipulatory skill to prosimians, and the otter is equally dexterous (POLLOCK, 1921; MAXWELL, 1960, 1963), yet, from measurements made on photographs, neither of these animals appears to possess a binocular field greater than 70° wide; their optic axes diverge more than in dogs (JOHNSON, 1901). Both species have a wide range of manipulatory space, but the skilled use of the hand appears to have developed for rooting crustacea out of crevices by means of tactile cues in the absence of visual control (EDEY, 1968). Thus BLANQUART DES SALINES (BUFFON, 1812) described the racoon as dexterously opening oysters

> as well as all other of his operations, by feeling alone, seldom making use of either his eye or his nose.

The animal is nocturnal and fish are taken by touch (ERRINGTON, 1967).

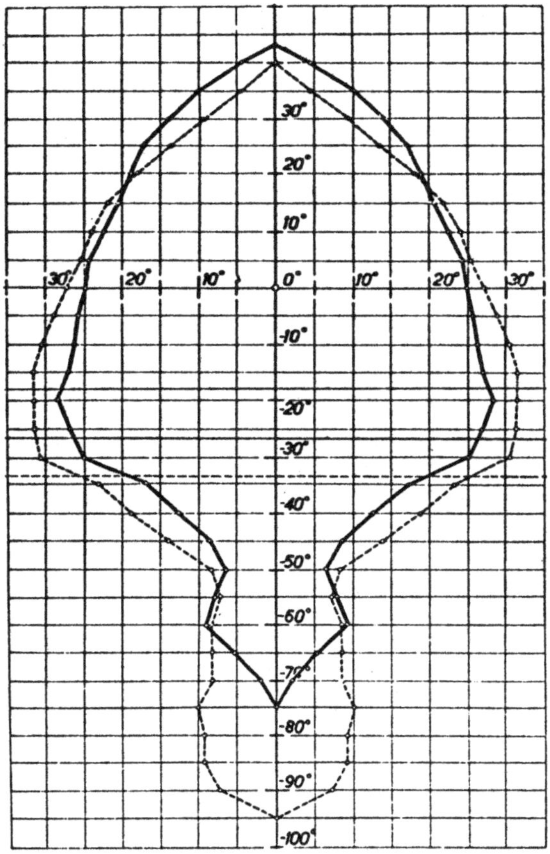

Fig. 6. The absolute *(hatched line)* and restricted *(bold line)* binocular fields of horse. Maximum width is about 55° at the equator; it attains a frontal elevation of 40° and declination of 75° to deal with ground near feet. This extensive inferior binocular field is reflected in the retinal ganglion-cell topography by presence of an anakatabatic region (Fig. 20) common to retina of other grazing species (PISA, 1939)

Although the otter usually hunts fish by sight, its praxic field is small during this activity, because they are captured by mouth (Fig. 5u); the forepaws are not employed until eating on land (EDEY, 1968; MAXWELL, 1960) (Fig. 5t); food is readily discriminated by touch in turbid water (EDEY, 1968). Such narrow binocular fields in species with well-developed tactile and manipulatory skills strongly supports emphasis upon praxic acitivity rather than manipulation as a primary factor in the determination of binocular field width. It might be suggested that these forms require panoramic vision to such an extent as to prevent expression of a wide binocular field, but the otter, at least, is not subject to significant natural predation (HARRIS, 1968), and such an explanation is unconvincing.

A more ambitious formulation of the praxic-space hypothesis might attempt to account for the shape of the binocular field, but that would be premature. It is worthy of note, however, that the binocular field of artiodactyls and perissodactyls is extended *not in width but downwards* along a narrow strip in the midsagittal plane, for more than 90° in the horse (Fig. 6), to deal with the limited region of space accessible to the nonabductable forefeet (PISA, 1939).

The assumption that a wide binocular field *necessarily* implies great emphasis on stereopsis is implicit in the surprise often encountered upon drawing attention to the 80°–100° wide fields of rat, mouse, and hamster. These preyed-upon species can be no less subject to pressure for an extensive panoramic field than the tree shrew or squirrel, and yet, in spite of a much inferior range of praxic behaviour, their binocular fields are distinctly wider. It is notable, however, that these rodents also possess *overhead* binocular fields which are as wide as their frontal binocular fields. The continuous overhead coverage required to give warning of aerial predators could, however, be equally provided by the abutment of the upper margins of the two uniocular fields. It has been suggested (HUGHES, 1971) that uniocular field overlap in the upper binocular field of the rabbit, which is about 35° wide, ensures continuous coverage of the sky when the head is rotated and blocks off the field of one eye, but this explanation hardly suffices for the 100° wide overhead field of the rat or mouse. The small interocular distances and the low innervation densities of the retinal regions dealing with the upper field ensure that stereopsis within such an overhead binocular field would be very crude and appropriate to such short distances as to be of little use in the avoidance of aerial attacks. The most acceptable explanation of these very extensive fields of binocular overlap characteristic of nocturnal or, more generally, scotopic forms is that they simply increase sensitivity in conditions of low illumination. Such a conclusion does not, of course, preclude stereopsis within the frontal binocular field of these species.

Predation, arboreality, and manipulative skill per se have been rejected above as determinants of binocular field width. The range of visually directed praxic activity, not necessarily involving forelimb manipulation, and the need to exploit over a wide field the lowering of the luminance threshold obtained by binocular viewing have been suggested as alternate primary influences. Although the concept of a *universal* urge to binocularity modulated by the need for panoramic vision has been set aside, it is apparent that interaction between the pressures selective for the two fields must exist in certain life styles. However, even with only these three effective influences on the relative extents of the binocular and panoramic fields, it is found that the variety of possible explanations for the condition in a given species is an *embarras de richesse* which tends to make indulgence in them appear facile. But such attempts are justifiable if based on sound behavioral data.

Nevertheless, many hoary problems remain unsolved. Why do the nonsaltatory, cursorial, herbivorous forms such as the ox so greatly exceed even the nonmammalian predators in binocular field width when their limited manipulative skills are equally dependent upon specializations of mouth, teeth, and tongue, rather than on their nonabductable forelimbs, and they appear to be in similar need of protective panoramic vision? Why is the binocular field of the dog so large when it is obviously a species relatively deficient in visually directed manipulative skill and possessed of a restricted praxic field? Perhaps the greater extent of the cat binocular field should be accounted for in terms of specialisation of increased scotopic sensitivity rather than by emphasis on its skills! The best hope for a satisfactory future exposition of this subject clearly lies in the combination of accurate optic measurements, extensive behavioral studies in the natural habitat, and the techniques of numerical taxonomy.

The observations of CANELLA (1936a, b) indicate that monocular vision in all classes of vertebrates lacks any obvious qualitative deficit relative to binocular vision during everyday locomotion, jumping, or dealing with food. Monocular frogs (see INGLE, 1972, 1973), kestrels, and cats appear to be as capable as binocular animals in capturing their aerial prey. If the total loss of stereopsis and other binocular cues to depth, such as image size and binocular perspective difference (OGLE, 1950), is attended by so little apparent deficit in normal life, it is not surprising that the peculiar adaptive significance of the wide peripheral binocular field proves elusive.

An implication of the praxic-space hypothesis is that the peripheral binocular field is primarily organised to deal with nearby space. At present, stereoscopic vision has been studied in primates alone and then only at small eccentricities; little is known of the quality of fine judgment in the peripheral binocular field. Recent work by BLAKEMORE (1970) shows that over a 3° range of disparity bridging the fixation point, there obtains a sensation of depth proportional to the displacement of an object from either the fixation point or another object within this range; the region, although more extensive than was originally conceived

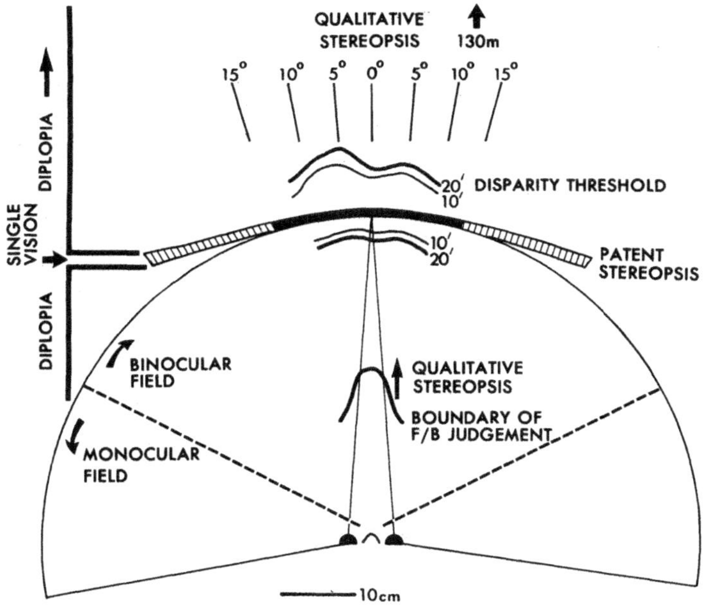

Fig. 7. A diagrammatic horizontal section through human binocular field which summarizes spatial distribution of the various qualities of stereopsis for a given degree of vergence. It is emphasised that depth judgement is not confined to the region of patent stereopsis. The lack of information about peripheral stereopsis is clear. [Compounded from the data of OGLE (1950), FLOM (1957), and BLAKEMORE (1970)]

(OGLE, 1950), is described as providing *patent stereopsis*. As shown in Fig. 7, vision within this range may be single or diplopic; outside it extends the area of *qualitative diplopic stereopsis* in which fore/aft judgments of depth relative to the fixation point may be made without relative depth sensation. The distribution of stereoacuity and patent and relative stereoscopy at eccentricities greater than 16° from fixation has been little studied, but Fig. 7 combines the results of three workers (OGLE, 1950; FLOM, 1957; BLAKEMORE, 1970). It is generally agreed that, although stereoacuity decreases in the peripheral binocular field, the range of patent stereopsis increases. If peripheral stereoacuity continues to be proportional to visual acuity (OGLE, 1950), then from WERTHEIM (1894) and BLAKEMORE (1970) we would estimate that from 0.4′ at the fixation point it will drop to some 15′ on the horopter at an eccentricity of 50°. The spatial displacement represented by this threshold would vary with the angle of vergence of the eyes, but during fixation at 50 cm it would represent about 5 mm at an eccentricity of 50°. Although obviously much cruder than central stereopsis, the peripheral binocular field of man thus has the potential for making the quite fine spatial judgments attributed to it by DUKE-ELDER (1958).

For other species nothing is known concerning the quality of peripheral stereopsis, but neurophysiologic results from the cat are compatible with the suggestion that the animal possesses a PANUM's fusional shell similar to that in man (JOSHUA and BISHOP, 1970). A recent investigation has shown PANUM's area to be some 50′ wide for crossed disparity in the cat, compared with 10′ in man, but the range of qualitative stereopsis, unlike that of man, is not much greater than that of fusion (PACKWOOD and GORDON, 1975).

4. Frontal Vision

The eyes of the Primates and Felidae are described as frontally situated; this is a loosely defined tripartite syndrome involving a small divergence of the optic axes, the absence of a panoramic field to the rear of the head, and, as commonly understood, a wide binocular field. A satisfactory explanation of frontality would encompass all of these features, but for many writers the implicitly associated wide binocular field simply,

> provides an essential basis for stereoscopic vision

(LE GROS CLARKE, 1962), and the frontal condition requires no further explanation.

This view may be traced back to ELLIOT SMITH (1928, 1930) whose still-current hypotheses (LE GROS CLARKE, 1962; HILL, 1972; SPATZ, 1968) have dominated primatologic thought on the evolution of binocular vision for the last half-century in spite of their inadequate experimental basis. For ELLIOT SMITH, binocular vision is acquired only at the level of the cat or monkey; the presence of the frontal eyes enables "macular" vision of unique quality, and stereoscopic vision is possible only after the attainment of macular vision (ELLIOT SMITH, 1928, 1930). With these features the vision of man and the higher primates, the "New Vision," is unique amongst all the animals,

> a vision of the world was thus revealed to man, with an appreciation of form, colour, size and space and a fuller understanding of distance and movement

(ELLIOT SMITH, 1930). But the argument of this chapter is the antithesis of his view. Binocular vision is assumed to be universal among mammals, the common mammalian area centralis is argued to have the potential for all the attributes of macular vision, except its high resolution, and stereoscopy is suggested to not justifiably be regarded as absent from any mammal possessing a binocular field; the vision of primates is not qualitatively different from that of many subprimates. If therefore, all of ELLIOT SMITH's envisaged advantages of frontality are already in the possession of nonfrontal precursors then the evolution of frontal vision requires a justificatory basis different to that generally proposed.

It cannot be accepted that frontality may be equated with the acquisition of a large binocular field; the owl and some deep sea fish are undoubtedly frontal species, but their binocular fields are quite narrow because of the tubular structure of their eyes. It is the parallelism of the optic axes and, equally important, the absence of the rear panoramic field which are diagnostic. It has been argued that the inherent limitation on the extent of the uniocular optic field makes the loss of the rear panoramic field inevitable in a species which acquires a large binocular field. This might be valid for species with tubular eyes but is quite inappropriate for man or cat, because both forms fail to exploit some 30° of their uniocular optic fields. Any explanation of their frontality should encompass this fact.

The convergence of Primates and Felids in their development of frontal vision was regarded by HARRIS (1904) as not so much directed towards increase in the extent of the binocular field as arising from the common need for predators and manipulators to ensure that

> the two maculae are brought to bear on the same object.

This erroneous belief is founded on the still-encountered equation of the optic and visual axes. JOHNSON (1901) writes of man and monkey *alone* as possessed of "parallel vision" when the eyes are in the position of rest. Divergent optic axes are understood as implying divergent areas of good vision. Of course, it is now generally realized that the optic axes have not been found to be parallel in any species. Universally among investigated mammals there is substantial evidence that a subregion of the area retinae, the visual pole (BISHOP et al., 1962), receives the image of a distant point in both eyes simultaneously subject to the error introduced by small, dissociated, involuntary eye movements. Effectively parallel fixation axes are thus compatible with widely divergent optic axes. The situation during the observation of nearby objects is dependent upon whether the species is capable of vergence, but this has so far been unequivocally demonstrated only for primate and cat (HUGHES, 1972).

It remains possible, however, that frontality arises in man from a need to bring both *optic* axes to bear on the same object! The results of CAMPBELL and GREEN (1965) indicate that optic attenuation significantly limits the human contrast sensitivity for high spatial frequencies imaged at the fovea. Because the human eye resembles a Gaussian system to the extent that its image quality deteriorates with increasing eccentricity from the optic axis (see WEALE, 1956), it follows that the neural potential of the fovea to achieve low-contrast sensitivity thresholds at high-spatial frequencies would ideally be optimal upon fusion of the visual and optic axis; in man they diverge by only 6° (HELMHOLTZ, 1924). If the achievement of such fusion is the primary adaptation in frontality, the fabric of the optical and neural apparatus remaining unchanged, then restriction of the panoramic field to the rear of the head follows as a necessary consequence. The temporal optic hemifield of the human eye, defined by the 98° angle between the limit of the uniocular optic field and optic axis in the horizontal plane (HELMHOLTZ, 1924; HARTRIDGE, 1919), is as large as encountered in any mammalian species. Increase of the temporal hemifield to encompass the rear field could only occur with lateral rotation of the optic away from the visual axis and consequent loss of image quality at the fovea.

This explanation of frontality is probably as applicable to the remaining foveate Simians as it is to man, but its general validity might be questioned in relation to the Prosimians and cat, because their low ganglion-cell density, high receptor-to-ganglion-cell-convergence ratio, and lack of a fovea (ROHEN and CASTENHOLTZ, 1967) suggests that optic attenuation may be much less significant to central resolution than in Simians. However, measurements of optic quality available for the cat show that two necessary requirements for the applicability of the above hypothesis may be met. Not only is the optic quality of the cat eye reported to be maximal close to the optic axis (BONDS, 1974), but also the resolution of high spatial frequencies appears to be significantly limited by optic factors—although the situation remains open to discussion. Sufficient information is not available to discuss the relevance of this explanation of frontality to the owl and deep sea fishes. It should not be too readily discarded for these species; the owl (FITE and ROSENFIELD-WESSELS, 1975) and at least one deep sea fish (WALLS, 1942) possess a fovea, a structure usually associated with high-resolution vision of such quality that oblique imagery may be undesirable.

C. Comparative Optics

I. The Schematic Eye

1. Introduction

Numerous optic, physiologic, and ophthalmologic problems involving the passage of light through the eye may only be solved by means of a quantitative model of the average dioptric system. Such a *schematic* eye enables the compound coaxial optic apparatus of the lens and cornea to be treated as a whole. A model human eye based on ray tracing was presented by YOUNG (1801), but an analytic treatment of the eye only became possible with GAUSS' exposition (1841) of the assumptions under which a compound system may be specified in terms of three cardinal points. For the human eye the early work of LISTING (1845) and HELMHOLTZ (1856, 1866, see 1924) culminated in GULLSTRAND'S schematic eye (1908). The descriptive mensuration of the animal eye began perhaps with NEWTON'S treatment of the sheep globe (ca. 1680; BREWSTER, 1860), in which the ellipticity of the cornea was first recorded (Fig. 8A); but attempts at an analytic optic treatment of species other than man awaited MATTHIESSEN (1879–1893) and HIRSCHBERG (1882); the provision and use of schematic eyes has, however, flourished only recently as a basic tool for neuro-ophthalmologic studies. In spite of an extensive qualitative literature on species differences in ocular organisation, the *comparative* quantitative study of dioptric systems remains relatively neglected.

The Gaussian theory contains certain assumptions—that light travels rectilinearly, is homogeneous, is incident close to the axis of spherical refracting surfaces arranged coaxially and separating homogeneous media—which are untrue and set varying limitations on the adequacy of a schematic eye as a model of the real dioptric system. The visual axis of man is sufficiently close to the optic axis for rays which reach the fovea to be regarded as paraxial and their course determined from a Gaussian schematic eye. The greater divergence of the optic and visual axes in most other species means, however, that the optic path described by an axial schematic eye is not appropriate to the retinal area of best vision, and its usefulness is therefore much reduced. The improvement of schematic eyes to embrace marginal and oblique rays poses formidable theoretical and computational difficulties, yet it is probably true that some of the most interesting contrasts in optic organisation between species will be encountered in the mechanisms for imaging from marginal and oblique rays. It may be that the inherent limitations of the Gaussian approach has discouraged its more general application, but it is surprising that schematic eyes are available for so few species in spite of the fact that interest in such models was well developed by the middle of the last century.

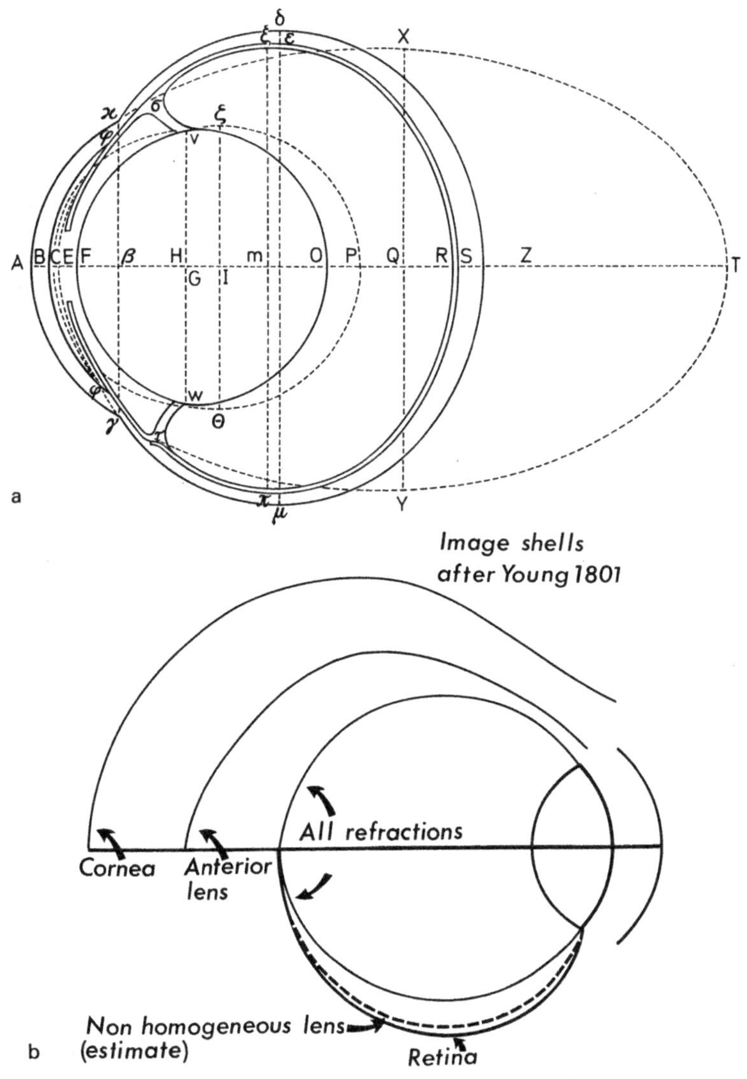

Fig. 8(a). Newton's schematic diagram of the sheep eye in which the ellipticity of a cornea was first described. Earliest known quantitative description of the optics of an eye (BREWSTER, 1860). (b) THOMAS YOUNG's schematic cross section of human eye, illustrating image shells obtained by calculation of net refraction achieved by successive surfaces in the dioptric apparatus. He recognized the presence of oblique astigmatism and, in assessing the influence of nonhomogeneity of lens refractive index, did not bring the final peripheral image shell into conjunction with the retina. (After YOUNG, 1801)

2. Methods

The essential purpose of a schematic eye calculation is to locate the cardinal points of a complex optic system. These comprise the primary and secondary focal points, principal points and nodal points which, between them, uniquely determine the behaviour of paraxial rays. For a *thin lens* the optical centre acts as principal and nodal points, and its separation from the focal points defines the focal length of the lens. A *thick lens* or *reduced equivalent system* behaves for incident rays like an equivalent thin lens placed at the first encountered principal point and for emergent rays like the same thin lens placed at the subsequent principal point; focal lengths are thus defined for such a system in terms of the principal points rather than an optical center. If the refractive index of the first and last media of the system is not the same then, as in the eye, a ray which enters the first principal point does not leave the second principal point at the same angle to the optic axis; under these conditions that property belongs to the nodal points. The separation of the posterior nodal point from the posterior focal point, the posterior nodal distance or PND, is therefore a very important parameter in visual neurophysiology, because it defines both image scale and surface curvature.

It is possible to straightforwardly determine the three posterior cardinal points after measurement of the refractive error, vitreous refractive index, and PND of an eye. The PND may be obtained from measurements of the transclerally viewed image of an object of known angular dimension (VAKKUR et al., 1963) or somewhat more satisfactorily from the angular dimensions of fundus features of known size (HUGHES, 1976a). So far, however, the cardinal points derived this way have been employed only to confirm those based on calculation from the position and curvature of the refractive surfaces of the eye (VAKKUR et al., 1963; HUGHES, 1976a).

The computation of a complete schematic eye begins with the use of thick-lens theory to develop equivalent lenses for the cornea and crystalline lens. It is finished when a further equivalent lens is derived by combination of the equivalent corneal and crystalline lenses to represent the behaviour of the entire optic system. The theoretical basis of the computation is presented in this handbook by WESTHEIMER (1972b) and more extensively by GULLSTRAND (1924) and DUKE-ELDER (1970); SOUTHALL (1964) provides a convenient computational paradigm.

The calculation of a schematic eye requires knowledge of the position and radius of curvature of each refracting surface and the refractive index of each of the media in the modeled globe. These parameters should ideally be obtained from the living eye with an intact circulation; however, in spite of the availability of suitable techniques (HELMHOLTZ, 1924; SORSBY et al., 1957; DUKE-ELDER, 1970), current human and other schematic eyes are based upon measurements of enucleated and dissected globes. Changes in the dioptric components during growth have, however, been followed by such in vivo methods in rabbit (LUDLAM and TWAROWSKI, 1973) and children (LUDLAM et al., 1972). For optimal results the necessary parameters should be measured and a separate schematic eye computed for each globe studied (VAKKUR and BISHOP, 1963), averaging being only subsequently carried out on the parameters of the equivalent optic systems. The majority of schematic eyes are, however, computed from averaged data and are thus somewhat less satisfactory. The availability of the dimensions and positions of the dioptric components of the eye enables a great variety of optical calculations including estimation of the field of view, entrance and exit pupils, contact lens form, etc. The full schematic eye with extensive measurements of the globe components is thus more useful than a simple knowledge of the cardinal points of its equivalent dioptric system.

A common method for obtaining the necessary measurements has been to photograph rostral views of the globe and lens immersed in saline after enucleation (VAKKUR and BISHOP, 1963), but in other instances (HERMANN, 1958; SUTHERS and WALLIS, 1970), measurements were made on photographs of the globe hemisected after rapid enucleation and freezing. In both techniques the excised eye should be subject to an internal hydrostatic pressure similar to that in life (25 cm H_2O), but there is a great risk of distortion or displacement of the lens if a cannula is introduced into small eyes. Freezing, on the other hand, may disturb the relative positions and dimensions of the optic components of the eye upon volume expansion of the vitreous by about 9% (FISCHER, 1907; HUGHES, 1975b).

When measuring a small globe for the computation of its schematic eye, it is worth taking note of the extraordinary accuracy that is required if the original refractive state is to be reconstructed. In a 6 mm diameter eye such as that of a rat, an uncertainty of 0.05 mm in the separation of the focal and receptor planes is equivalent to an indeterminacy of the refractive state of about 1 D. Even with great care some dimensional uncertainty remains, e.g., the thickness of the choroid when full of blood may not be known so that the receptor plane cannot be accurately sited in the model (VAKKUR and BISHOP, 1963).

The measurement of the refractive index of the fluid media is straightforward when an Abbé refractometer is employed, and the results of VALENTIN (1879a, b) and FREYTAG (1910) remain representative. The disruption necessary when taking samples of the lens and cornea for use in such an instrument might be thought to be a source of considerable error, but the refractive index of any specific region of the lens appears to be determined entirely by the local protein concentration (PHILLIPSON, 1969). The examination of local samples in the Abbé refractometer gives similar values to more sophisticated techniques. In most schematic eyes the need to measure the lens refractive index is avoided as described below, but in more ambitious models it is necessary to obtain the description of the refractive-index variation across the plane containing the anteroposterior axis of the lens—the isoindicial equation. With assumptions, such a curve may be fitted to local samples (HELMHOLTZ, 1924; it may be obtained by a scanning refractometer (rabbit, NAKAO et al., 1968; cat, monkey, man, NAKAO et al., 1969) or computed from the protein-concentration profile established by microradiography (rat, PHILLIPSON, 1969).

The use of the dimensions of the excised lens generates an unaccommodated schematic eye for the rabbit, cat and rat because its shape is invariant during accommodation in these species. Frozen sections through the equator of the monkey eye would give lens dimensions for the unaccommodated state, but those obtained from the excised lens lead to an accommodated schematic eye (VAKKUR, 1967).

3. Choice of the Schematic Eye Model

An equivalent lens for the cornea is straightforward to derive, and its combination with the equivalent crystalline lens to form the schematic eye is readily achieved (SOUTHALL, 1964); it is the derivation of the equivalent crystalline lens which necessitates careful consideration and compromise. The difficulty arises, because the refractive index of the lens is not uniform throughout.

That the refractive index of the human and other crystalline lenses varies both radially and axially is a long-established fact. As early as 1801, THOMAS YOUNG extensively discussed the influence of this gradient in reducing the separation of the image shell and the retina as determined by ray tracing through the refracting surfaces (Fig. 8B). BREWSTER (1816) employed polarized light to demonstrate the variation in optic density of the lenses of various species. Both YOUNG and BREWSTER were aware of the potentiality of the gradient for reducing aberrations, but quantitative treatment was not possible in the absence of its analytic description. Stimulated by the examination of the fish lens, MAXWELL (1854) demonstrated that perfect imagery would be possible in a medium in which the refractive index varied inversely as the distance from a given point, but only if the object and image lay within the medium (MAXWELL, 1858). If a similarly organised lens were to be placed in a homogeneous medium, then, as demonstrated earlier by YOUNG (1801), perfect imagery was not possible, although some advantage in minimizing the aberrations would accrue from the gradient of index (FLETCHER et al., 1954) The effect of the gradient was clearly understood by YOUNG (1801) to make the lens more powerful than if it were uniformly constructed of material of the same index as the core (HELMHOLTZ, 1924).

One obvious solution has been to dispose of the refractive index measurements on the crystalline lens and to represent the lens by an equivalent organ of identical shape but of uniform and higher refractive index; this is the Gaussian homogeneous lens model which has been extensively employed for man (LISTING, 1845; HELMHOLTZ, 1924), cat (VAKKUR and BISHOP, 1963), rabbit (HUGHES, 1972), and rat (BLOCK, 1969; HUGHES, 1977c). It is to be emphasized that the effective, or *total* (SENFF, 1846) refractive index employed for the homogeneous lens considerably exceeds the actual values observed at the center of the crystalline lens core (HELMHOLTZ, 1924). Gross errors are introduced if it is replaced, as in MASSOF and CHANG'S "revised" rat schematic eye (1972), by a mean index arbitrarily derived from actual measurements of the crystalline lens index.

The homogeneous lens is computed by a relaxation technique whereby its effective refractive index is chosen so that the back vertex power of the model and the crystalline lens are identical (VAKKUR and BISHOP, 1963; BLOCK, 1969). The principal points of such a lens model remain necessarily separated by a greater interval than those of the corresponding crystalline lens (HELMHOLTZ, 1924; GULLSTRAND, 1924), and the power of the homogeneous lens is greater. The misrepresentation of the principal points resulting from the use of the homogeneous lens in the cat schematic eye is small and introduces a refractive error of only 0.5 D (VAKKUR and BISHOP, 1963). In small eyes, such as those of goldfish or rat, the error arising from the use of a homogeneous lens may amount to 4 D, so that a more sophisticated model lens is required.

The refractive contribution of the lens to the schematic eye is made up of the sum of its normal surface contributions, additional surface contributions resulting from inhomogeneity, and the transfer contribution due to refraction by the intervening nonhomogeneous medium. It is the additional surface contribution and the transfer component, described as the core of the lens by MATTHIESSEN, which is problematic to represent in the schematic eye. Many authors attempted to use approximate laws for the variation in index across the lens (GULLSTRAND, 1924), but these were inadequate for a variety of reasons. MATTHIESSEN (1887) stated that the indicial equation for the lens of many species was such that the effective total index was as much greater than the refractive index of the crystalline core as the index of the core is greater than that of the cortex surface. Comparison of the total indices for cat, rabbit, and rat in Table 1 with the core and cortex values indicates the inadequacy of this model.

A substitute for the inaccurate homogeneous model of the human lens was presented by GULLSTRAND (1908; 1924) in the form of the *equivalent core lens*. To derive this model, GULLSTRAND approximated the human indicial equation along the optic axis by means of a series expansion. After indirectly determining the various constants, the power and principal points of the core lens were derived by integration. For representative purposes these properties of the lens core were then employed to define a unique *equivalent core lens*. This consists of a pair of refracting surfaces forming a homogeneous lens of refractive index equal to the crystalline lens center and regarded as suspended in a homogeneous medium of the same refractive index as that of the vertices of the crystalline lens cortex. The advantage of the equivalent core lens is that

> so far as the imagery laws of the first order are concerned the entire lens system then has precisely the optical characteristics of the actual crystalline lens

(GULLSTRAND, 1924). The principal points occupy positions exactly corresponding to those of the real core lens. A more recent seven-shell lens with aspheric vertices has been employed to simulate spherical aberration of the organ (LOTMAR, 1971).

In so far as all adult vertebrate crystalline lenses contain a central core of higher refractive index and density than the surrounding cortex, the equivalent core-lens model might be regarded as less arbitrary than the homogeneous lens. It must be realized, however, that the equivalent core lens is determined by calculation and not by anatomic demonstration—it does not have any exact physical counterpart and is simply a model of the real lens core or transfer component.

The in vivo state of the lens could not be determined for man by measurements on the excised organ because of the changes which occur in its form upon cutting the zonular fibres; GULLSTRAND (1924) thus derived the power of the lens from consideration of the aphakic eye. The stiffness of the cat, rabbit (FISHER, 1971), and rat lenses is too great for such a change of shape to occur upon their excision, so that their power may be directly determined by measurement of their vertex powers when suspended in saline. However, in spite of the availability of actual isoindicial, rather than approximated, curves for the meridional sections of lenses in a variety of species, there have been no derivations of core lens schematic eyes for forms other than man.

The availability of the isoindicial curves for the rat lens (PHILLIPSON, 1969) suggested the computation of a *two-shell* model lens in an attempt to overcome the shortcomings of the homogeneous lens for this small eye (HUGHES, 1976b). The central meridional isoindicial contours of the rat lens are circular in section and correspond closely to a real spherical core which may be extracted from the crystalline lens and is found to have a quite uniform and high refractive index. A two-shell model is thus well supported anatomically but is an approximation in that the cortex is treated as homogeneous. After specification of the core and cortex refractive indices in terms of the limiting values of the crystalline lens, then its spherical form means that only the equivalent core (or central shell) radius remains as a variable for matching the power of the two-shell lens to that of the crystalline lens. This model does not, of course, correspond to that derived by GULLSTRAND in that no integration of the indicial equation is employed, but it is an improvement on the homogeneous lens model and unique for a circular central shell. The two-shell rat lens model is less powerful than the homogeneous lens by 3.5 D, but in the complete schematic eye, the influence of the greater separation of the homogeneous lens principal points predominated over its greater power, and the refractive state is some 4.3 D more *hypermetropic* than that of the two lens shell schematic eye.

The isoindicial curves computed from the refractive-index gradient across the lens of various species have enabled NAKAO and his colleagues to attempt computer-ray trace models of the lens for the assessment of aberrations resulting from marginal incidence. Some 2000 shells were employed to approximate the continuous gradient of refractive index in the rabbit lens (NAKAO, 1968). Although an important improvement in the methodology of this field, it is clear that problems remain, because the computed focal length of this lens is not in agreement with either that of the rabbit schematic eye (HUGHES, 1972) or the confirmatory results of LUDLAM and TWAROWSKI (1973); the flaw may lie in the freezing of the lens or in the computation itself. Two human schematic eyes based upon multiple shell lens models have been recently reported—one in brief (NAKAO et al., 1970), the other—with 98 shells—in more detail (POMERANTZEFF et al., 1971).

Simple geometric ray tracing is not adequate for the rigorous analysis of media with continuously varying refractive index. WOOD (1911) illustrated this by the fact that light initially travelling parallel to linear isoindicial contours undergoes a refraction towards the denser medium, which ray tracing would not lead one to expect. Numerical solutions of the differential equation which describes the ray path in inhomogeneous media have been developed (MARCHAND, 1970) but do not permit the contributions of the system components to be independently explored.

The problems encountered in the attempt to derive an accurate paraxial schematic eye thus result from the structure of the lens, which is also the obstacle to the extension of the schematic treatment to give a wide-angle schematic eye dealing with marginal rays and oblique incidence. YOUNG (1801) began the exploration of the peripheral image shell form in man, and several investigators have followed up his work. GULLSTRAND (1924) regarded the requisite information for this investigation as unavailable. The treatment of cornea form has been sophisticated in wide-angle schematic eyes (STINE, 1934; DRASDO and FOWLER, 1974) but relatively crude lens models are employed to consider obliquities of incidence up to 100° without verification from experimental measurement or from continuous media treatment of the lens.

The mathematical apparatus required for the general treatment of optic systems is provided by Hamiltonian optics (SYNGE, 1937; BUCHDAHL, 1970) which enables the whole system to be discussed in terms of its characteristic function. The difference between the characteristic function ideal for some specific purpose and its realisable form define the aberrations of the system. An early treatment of aberration in continuous nonhomogeneous media by BUCHDAHL has been extended by SANDS (see SANDS, 1973) to encompass media with discontinuities. SANDS's results have been put into a form suitable for computer analysis by MOORE (1971), and the implications of various radial and axial index gradients investigated. Current interest in such techniques is directed towards the exploitation of newly developing procedures for the production of inhomogeneous glass. The ability to treat inhomogeneous media, marginal rays, and oblique incidence make Hamiltonian optics a potentially powerful, although formidable, tool for the investigation of the isoindicial contours of the lenses of different species. MOORE (1971) has demonstrated that the axial gradient of index is similar in its influence to an aspheric surface, but that radial gradients enable the very ready control of the Seidel aberration coefficients. If the acceptable levels of peripheral aberrations differ in the effective image shells of frontal- and lateral-eyed species, then adaptations for their control might be revealed by comparison of the radial-index gradients.

4. Schematic Eye Parameters for Different Mammalian Species

GULLSTRAND's human schematic eye (1909) has not so far been replaced; for most purposes this would be unnecessary, because its dimensions and parameters are within 2% of the means for a population of 1000 eyes measured by STENSTRÖM in vivo (1946). Complete schematic eyes are now available for a variety of vertebrates, including the cat (VAKKUR and BISHOP, 1963; VAKKUR et al., 1963), rabbit (HUGHES, 1972), rat (HERMANN, 1958; BLOCK, 1969; HUGHES, 1977c), monkey (VAKKUR, 1967), gecko and iguana (CITRON and PINTO, 1973), pigeon (MARSHALL et al., 1973), frog (DU PONT and DE GROOT, 1976), and goldfish (CHARMAN and TUCKER, 1973); extensive measurements of the ocular dimensions of echolocating bats have also been provided by SUTHERS and WALLIS (1970). A detailed comparison of the results of these reports would be enervating, but in view of their value as a tool for comparative optical work, they have been summarised in Table 1.

These results are based upon measurements of populations of different size and techniques of varying reliability. The pooling of dimensions to give mean values before computation of the average eye also eliminates the influence of their covariation and will inevitably widen the confidence limits that can be put on the mean parameters for the whole eye. These effects are particularly important in small eyes and can make the schematic eye useless for the estimation of the mean refraction of the animal at the effective image plane (HUGHES, 1977c). It should be noted that the final refractive state of the eye is calculated from the difference between the length of the dioptric path to the plane of focus and the length of the eye to the effective image shell, i.e., from the out-of-focus distance. Certain authors have adjusted the calculated refraction by manipulation of the position of the effective image shell to match the retinoscopic refraction of the eye. In view of the results outlined in Section CIII, this practice (VAKKUR and BISHOP, 1963) is not advised, and the eye should, if adjusted, be brought to emmetropia. In other instances the schematic eye refraction is taken as that computed (HUGHES, 1972; HUGHES, 1977c) for comparison with the results of optometric refraction. A supplementary emmetropic eye is easily computed.

5. Comparative Ocular Structure

The system of equations which define the Gaussian schematic eye shows that neither its refractive state nor light-gathering efficiency are influenced by overall changes of size. W are led, therefore, to consider what factors may account for the 68-fold range of axial length to be found between the 0.8-mm-long eye of the shrew and the 54-mm eye of the baleen whale.

The size of the eye is related to that of the body according to HALLER (1769):

> Magnitudo ocularum est fere in ratione inversa animalium. Balaenae, Rhinoceroti, Elephanto parvi sunt oculi
> (The size of the eye varies inversely as the size of the body. The eyes are small in whales, rhinoceros and elephant)

Table 1

P = position R = radius relative lo corneal vertex (mm)	Man unaccom-modated		Man accom-modated		Monkey unaccom-modated		Monkey accom-modated	
	P	R	P	R	P	R	P	R
Anterior cornea	0.0	7.7	0.0	7.7	0.0	6.36	0.0	6.36
Posterior cornea	0.5	6.8	0.5	6.8	0.61	5.14	0.61	5.14
Anterior lenticular	3.6	10.0	3.2	5.33	2.25	?	1.75	3.58
Core anterior	4.146	7.911	3.873	2.655	—	—	—	—
Core posterior	6.565	−5.76	6.528	−2.655	—	—	—	—
Posterior lenticular	7.2	−6.0	7.2	−5.33	6.46	?	6.46	−4.55
Retinal surface	—	—	—	—	—	—	—	—
Inner limiting membrane	—	—	—	—	—	—	—	—
Choroid/retina	24.0	—	24.0	—	17.97	—	17.97	—
Posterior scleral	—	—	—	—	18.97	−11.0	18.97	−11.0
Cornea refractive index	1.376		1.376		1.376		1.376	
Aqueous refractive index	1.336		1.336		1.336		1.336	
Vitreous refractive index	1.336		1.336		1.336		1.336	
Lens capsule refractive index	1.386		1.386		—		—	
Lens core refractive index	1.406 eq.		1.406 eq.		—		—	
Effective index homogeneous lens	—		—		?		1.463	
Cornea, power Diopters	43.05		43.05		51.541		51.541	
Lens, power Diopters	19.11		33.06		38.168		60.109	
Eye, power Diopters	*58.64*		*70.57*		*82.916*		*102.807*	
Anterior Focal length; eye. mm	−17.055		−14.169		−12.0604		− 9.727	
Posterior focal length; eye. mm	22.785		18.930		16.113		12.995	
Anterior principal point; eye. mm	1.348		1.772		1.523		1.602	
Posterior principal point; eye. mm	1.602		2.086		1.857		2.013	
Anterior focal point; eye. mm	−15.707		−12.397		−10.538		− 8.125	
Post focal point; eye. mm	24.387		21.016		17.970		15.009	
Anterior nodal point; eye. mm	7.088		6.533		5.575		4.870	
Posterior nodal point; eye. mm	7.332		6.847		5.910		5.282	
Out of focus distance. mm	1.105		5.07		0.0		− 2.962	
Post nodal distance. mm	*17.055*		*14.169*		*12.061*		*12.689*	
Refractive state schematic. Diopters	1.0		− 9.6		0.0		−19.080	
Retinoscopic refractive state. Diopters	1.0		—		—		—	
Retinal magnification. mm/°	0.29 mm/°		—		0.25 mm/°		—	

Sources of data: man, HELMHOLTZ (1924); monkey, VAKKUR (1967); cat, VAKKUR and BISHOP (1963), HUGHES, (1976a); rabbit, HUGHES (1972); rat, HUGHES (1977c); goldfish, CHARMAN and TUCKER (1973); pigeon, MARSHALL et al. (1973); gecko and iguana, CITRON and PINTO (1973); frog, DU PONT and DE GROOT (1976). Refractive indices, FREYTAG (1910).

(FRANZ, 1934; WALLS, 1942; ROCHON-DUVIGNEAUD, 1943), but the eyes of these species are *large* in their absolute dimensions; the rule is obviously misstated and should refer to relative eye size. The universal application of the law is complicated by other generalizations, such as LEUCKART's law (1876) which states that swifter species have larger eyes.

The axial length of the eye varies almost linearly with the logarithm of body weight over a wide variety of species and great range of body weight. The allometrically more desirable plot of the logarithm of the axial length of the eye against the logarithm of body weight thus takes the form of Fig. 9A. At a given weight there is a three- to fourfold range of eye length, but this is reduced if the birds are treated as a separate distribution; in accordance with LEUCKART's law, their eyes are consistently larger than those of mammals of the same weight. It is

Cat		Rabbit		Rat		Bat	Pigeon		Frog		Goldfish		Gecko (Nocturnal)		Iguana (diurnal)	
P	R	P	R	P	R		P	R	P	R	P	R	P	R	P	R
0.0	8.57	0.0	7.5	0.0	2.965	See SUTHERS and WALLIS (1970) for incomplete schematic data	0.0	3.44	0.0	3.08	0.0	3.0	0.0	4.48	0.0	2.6
0.68	7.89	0.4	7.1	0.26	2.705		—	—	0.15	2.93	0.036	2.964	0.44	4.48	0.15	2.6
5.2	7.2	3.3	6.9	0.881	2.34		1.77	4.17	1.7	2.9	2.036	1.1	1.7	3.26	1.07	4.83
—	—	—	—	1.778	0.958		—	—	—	—	—	—	—	—	—	—
—	—	—	—	3.695	0.958		—	—	—	—	—	—	—	—	—	—
13.7	−8.05	11.2	5.8	4.591	2.34		4.7	3.08	5.91	2.5	4.236	1.1	7.4	3.05	4.22	2.17
—	—	—	—	5.981	—		11.27	—	—	—	5.446	—	—	—	—	—
—	—	—	—	—	—		—	—	7.7	4.28	5.546	—	11.0	—	8.7	—
21.83	—	17.9	—	6.154	—		11.62	8.41	—	—	—	—	—	—	—	—
22.3	−12.5	18.1	−9.8	6.292	3.23		—	—	8.0	4.6	4.2	—	—	6.5	—	4.9
1.376		1.376		1.38			—		1.39			—	1.36		1.35	
1.336		1.337		1.337			1.334		1.338			1.335	1.336		1.336	
1.336		1.337		1.337			1.341		1.338			1.337	1.336		1.336	
1.40		1.407		1.39			1.383		1.39			—	—		—	
1.448		1.457		1.5			1.409		1.47			—	—		—	
1.554		1.6		1.683			1.408		1.65			1.69	1.415		1.4	
											Water	Air				
38.914		44.6		112.649			—		109.1		0.8	129	—		—	
52.964		75.0		243.991			—		198.3		—	—	—		—	
77.983		*100.6*		*300.705*			*126*		*244.8*		*467*	*532*	*111.33*		*155.71*	
−12.823		−9.9		−3.326			−7.91		−4.08		−2.86	−1.88	−8.9		−6.45	
17.132		13.3		4.446			10.61		5.47		2.87	2.51	12.0		8.58	
4.514		4.0		1.651			0.76		2.32		1.39	0.92	1.44		1.87	
5.022		4.7		1.732			0.85		2.34		1.39	1.06	1.47		1.67	
−8.309		−5.9		−1.675			−7.15		−1.76		−1.46	−0.96	−7.5		−5.64	
22.153		18.0		6.178			11.46		—		4.26	3.57	3.6		10.3	
8.823		7.3		2.771			3.46		3.71		1.4	1.55	4.47		2.9	
9.330		8.0		2.852			3.55		3.73		1.4	1.69	4.51		3.79	
0.323		0.1		0.026			0.0		—		0.0	—	—		—	
12.5		*9.9*		*3.326*			*7.91*		*4.08*		*2.86*	*1.88*	*6.5*		*4.9*	
1.5		0.7		1.755			—		—		0.0	−130	—		—	
1.5		2.0		9.4			2.4		4.0		6.9	—	3.0		1.5	
0.21 mm/°		0.173 mm/°		0.06 mm/°			0.14 mm/°		0.08 mm/°		0.05 mm/°	—	0.11 mm/°		0.086 mm/°	

clear from the graph that only the central region of the distribution approximates the slope of 0.3 which would be expected (GUNTHER, 1975) for a linear dimension in such a plot; in small animals the eye length increases with body weight more rapidly than expected, and for the larger species its size is relatively independent of body weight. These results suggest that, given the nature of the elements common to its construction and the physical world to which it is adapted, a globe from 35 mm to 50 mm in diameter suffices to meet the most stringent requirements of species over a 10^4-fold range of body weight. This point was put by D'ARCY THOMPSON (1961):

> The eye, in short, can never be very small and need never be very big; it has its own conditions and limitations apart from the size of the animal. But the insect's eyes tell another story. If a fly had an eye like ours, the pupil would be so small that diffraction would render a clear image impossible. The only alternative is to unite a number of small and optically isolated simple eyes into a compound eye, and in the insect Nature adopts this alternative possibility.

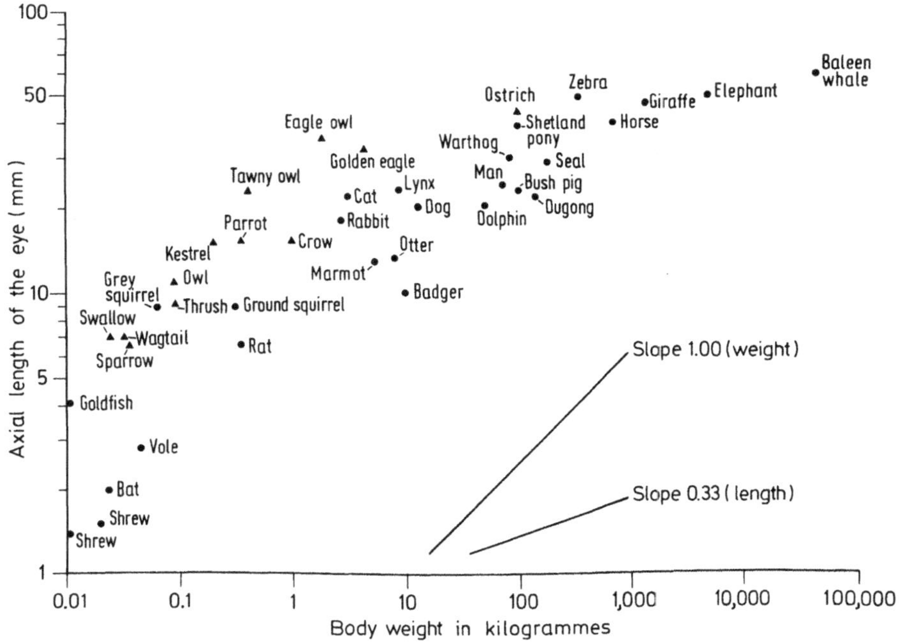

Fig. 9(a). Relationship between the axial length of the eye and body weight in birds *(triangles)* and other forms *(discs)*. Length of mammalian eye increases more rapidly with increase in weight than allometric proportion would suggest for the small species and more slowly for the large; in the intermediate range alone is the expected slope of 0.33 attained. An eye of 35–50 mm diameter meets the requirements of a variety of larger species over a 10^5-fold range of body weight. Sources various: schematic eyes if available, measurements from published photographs (e.g., ROCHON-DUVIGNEAUD, 1943), and unpublished data. Data has diverse origins but demonstrates a trend; too much reliance should not be put on individual points. (b). Retinal magnification factor (RMF) in mm/° as a function, of the axial length of the eye, L. The range of RMF at any given axial length is small considering the diverse sources of the data. The ratio of the posterior nodal distance (PND) to the axial length of the eye is thus relatively fixed and varies little between nocturnal and diurnal forms. The scatter diagram for this assortment of nocturnal and diurnal forms is well fitted by a regression line of slope of 0.011 mm/° per mm L and thus PND = 0.6 L. The RMF was derived from schematic eye data if available (circled points; birds triangles and others, discs). In other instances the PND was estimated from the mean radius of curvature of the eye cup over an arc extending 45° each side of the optic axis in published photographs of eye cross sections

We have already seen that the refractive state of an eye remains constant with isomorphic scaling of the dioptric apparatus, however, the wave nature of light ensures that in an ideal optical system the highest spatial frequency, f, in the image shell is proportional to the pupil diameter, D, thus $f = D/\lambda$ (WESTHEIMER, 1972a). The illumination of a given area of image surface varies as the pupil area, D^2, and inversely as PND^2; illumination per unit retinal area thus remains constant with isomorphic scaling of an eye at a relative value indicated by $(D/PND)^2 = (1/F_{no.}2)^2$. It was on the basis of this relationship that WALLS, (1942) wrote,

> an eye which is simply larger will not, then, have brighter images and greater overall sensitivity in dim light,'

but the last statement is not true. The absolute light gathering power of a scaled eye increases with pupil area, D^2, so that the *potential* absolute threshold

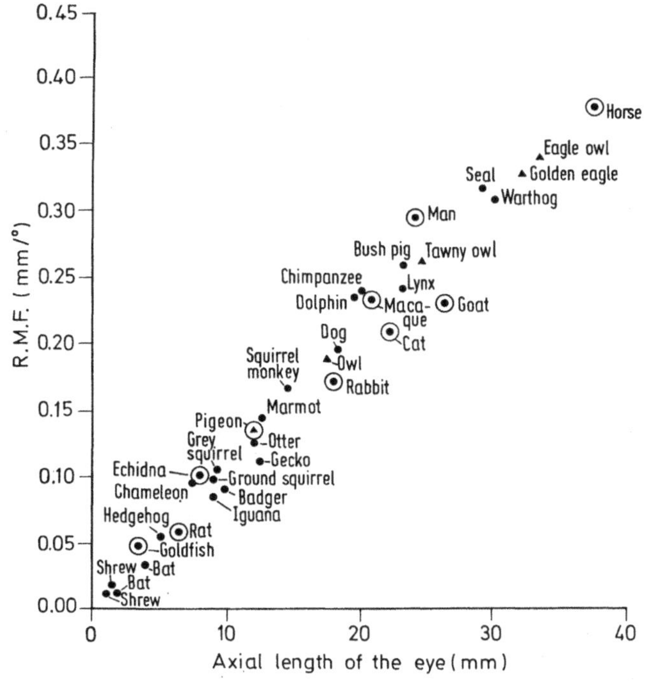

b

and increment sensitivity are improved because of the increased amount of light per unit solid angle of the image space (BARLOW, 1964); whether such potential is attained depends upon the strategy employed in sampling the image space. However, from the point of view of improving both resolution and absolute threshold it may be argued that nocturnal and diurnal species gain advantage from increased eye size which alone would not, therefore, enable the distinction of the two lifestyles. Whether or not a relatively large eye *is* developed depends upon the pressures encouraging reliance upon vision in a given species' niche; the eye of the mole (QUILLAM, 1966) is relatively small for an animal of its size.

Of course, the pupillary enlargement necessary to achieve an improved quantum capture and better resolution in the image plane need not be obtained by isomorphic scaling. An alterative is to increase the $F_{no.}$ of the eye by enlarging the pupil and thus transferring higher spatial frequencies to the brighter image plane. The quantum content per unit solid angle of the image space is the same for such an eye as for an isomorphically scaled eye of identical aperture and both forms require the same degree of pooling per unit solid angle in the photoreceptor system if the potential increase in sensitivity or absolute threshold is to be realised. However, to take advantage of the potential for increased resolution in the instance of increasing aperture alone it is necessary to reduce the sampling grain, assuming the original eye to be of optimal design, and this requires smaller photoreceptors. It turns out that the lower limit on eye size, in fact the whole design strategy for diffraction limited eyes organised for optimal resolution pivots around the minimum possible photoreceptor size. Isomorphic scaling is the optimal strategy for such eyes.

It has been demonstrated (SYNDER and MENZEL, 1975; PASK and SNYDER, 1975) that the angular acceptance of a photoreceptor, and thus its independence as a sampling element, is minimised when its waveguide parameter, V, has a value of about 2.5. Receptor diameter is then defined in terms of the wavelength of light, λ, and the difference between the refractive index of the receptors, n_1, and their surrounding medium, n_2,

$$D = V\lambda/\pi \sqrt{n_1^2 - n_2^2}\,.$$

The materials available in the eye set a minimum photoreceptor diameter of about 1–2 μ; if V were reduced to obtain a smaller photoreceptor then crosstalk with adjacent receptors would occur and the acceptance angle increase.

Now, we have that the highest spatial frequency in the image plane is $f = D/\lambda$ cy./°; from sampling theory the minimum number of sampling elements per unit solid angle of retina which could resolve this frequency, f cy/°, is given by $(2f)^2$ and their angular separation, $\Delta\phi$, therefore by $\Delta\phi = 1/2f$; thus $\Delta\phi = \lambda/2D$ if they are arranged in a square matrix. The smallest possible eye capable of a given resolution with a square sampling array would have cones of the minimum diameter packed shoulder to shoulder and a PND sufficient to make the angle d/PND equal to the SHANNON-NYQUIST angular sampling rate $\Delta\phi$ = d/PND = 1/2f = $\lambda/2D$ (KIRSCHFELD, 1976; SNYDER and MILLER, 1977). The optimal ratio of PND and D for a diffraction limited eye, PND/D = $2d/\lambda$ = $F_{no.}$ is thus defined and for typical values of d = 2μ and λ = 0.5μ would be $F_{no.}$ = 8. SNYDER and MILLER (1977b) have derived from signal to noise considerations that the angular photoreceptor diameter, $\Delta\varrho$, which is optimal for a diffraction limited eye is given by $\Delta\varrho = 0.586\ \lambda/D$ which very closely approximates the maximum space available at the optimal sampling rate.

Would there be advantage in employing photoreceptors larger than the minimum possible size for an eye of optimal diffraction limited design? From above we see that for a given cut-off frequency, f, or pupil size, D, a larger photoreceptor of diameter, d, can only be accomodated by increase of PND and thus a decrease in the F_{no} and 'speed' of the eye. A further disadvantage of increase in PND whether under isomorphic of anisomorphic scaling results from a proportional increase in retinal noise if the tissues remain similar in properties (BARLOW, 1964). If the PND is alone increased then the eye is in this respect at an immediate disadvantage; by contrast, in isomorphic scaling the increased light capture and reduction in photon noise by the enlarged pupil would more than offset the increased retinal noise. The presence of retinal noise, however, reduces the expected reduction of absolute threshold and increment sensitivity upon isomorphic scaling of the eye (BARLOW, 1964). In real eyes, both uncorrected spherical and chromatic aberration also increase with the pupil diameter, although the latter varies inversely as PND so that isomorphic scaling leaves it unaltered. This suffices, however (BARLOW, 1964), to again reduce the improvement in image quality to be expected from an enlarged pupil, but the influence of aberration may be reduced by anisomorphic scaling. Quite apart from competition by other organs for space in the head, it is apparent that, for a diurnal diffraction limited eye with an adequate light supply, the minimum photoreceptor diameter and associated minimum PND have considerable inherent advantages, an increase in photoreceptor size would be positively disadvantageous. Whether a diffraction limited eye achieves its potential resolution is dependent upon the luminance level of the environment but SNYDER and MILLER (1977b) point out that large and small eyes of optimal design, which are diffraction limited to high and low spatial frequencies respectively, achieve their maximum resolution at the same external luminance level.

Given such a theoretical background, the limited range of cone density encountered *in the areas of best vision* of vertebrates becomes comprehensible. Characteristic values would range between the 160000 cones/mm² in the human fovea (ÖSTERBERG, 1935) and the minimum densities of some 400000 cones/mm² encountered in the central areas of the hawk and eight other species of bird (FITE and ROSENFELD-WESSELS, 1975); early reports of foveal receptor densities in the order of 1000000 cones/mm² in eagle (POLYAK, 1957) and hawk (ROCHON-DUVIGNEAUD, 1919) have not been confirmed in recent studies. Away from the central area quite different considerations apply. Thus peripheral cones may be very large and have been argued to play a non-optimal resolution task as re-radiators of light to the rod system (MILLER and SNYDER, 1973).

Can these principles be discerned in the structure of real eyes? They are certainly helpful to qualitative understanding but in most species the eye is a compromise between conflicting requirements and a general analytic treatment of its design is not yet possible. Few species have a pure cone, diffraction limited, central area in which photopic resolution is determined by the photoreceptor matrix; it is known amongst the homeotherms in primates and birds and may occur in some small, so-called, pure cone eyes such as those of tree shrew and ground squirrel. The high resolution of birds is legendary and is accompanied by the equally well known large eyes which would be expected from the above analysis. The pressure for high diffraction limited acuity in birds obviously leads to considerable difficulty in housing the necessarily large eyes; MILLER and SNYDER (1977) and SNYDER and MILLER (1977a) have suggested that the convexiclivate foveal pit of the bird is employed as the negative element of a telephoto lens system which, by magnifying the local image at the foveola, enables the eye to take advantage of an increased aperture to achieve resolution usually appropriate to a much longer eye.

As the environmental illumination is decreased, the diffraction limited eye suffers a reduction in its resolving power determined by the increasing photon noise; this may, over a limited range of intensity, be offset by increase in the pupil size until aberrations intervene to limit resolution once more. At the largest pupil diameter some further advantage may be gained by increased pooling in the sampling matrix. Under these conditions we are considering quantum, rather than diffraction, limited resolution so that design considerations change and are more difficult to discuss in detail.

The body of the globe may generally be approximated by a sphere and thus, if a limited area of the retina attains a diffraction limited sampling rate then its associated PND sets a radius, to a first approximation, for the remainder of the retina. Similarly, the evolutionary pressure on arhythmic and nocturnal forms to attain a low absolute threshold and high increment sensitivity may lead to a large diameter globe; one which is much larger than a diffraction limited globe would needs be in order to achieve the possibly modest photopic resolution of that species. On this basis it is perhaps comprehensible that eyes of quantum limited design may not possess regions for maximal photopic resolution which are of optimal diffraction limited design; the mere presence of both rods and cones in the central region of the area centralis of most mammalian species is sufficient to exclude optimal design. In the cat area centralis there are some 25000 cones/mm^2 and 360000 rods/mm^2 (STEINBERG et al., 1973). Under photopic conditions the pupil may be optimally adapted to the photopic receptor sampling density but the cone size and associated PND cannot be optimally matched. The maximum possible resolution of the cat cone population could be achieved by an eye of only one quarter of the size, but of course, such an eye would have a higher absolute threshold under conditions of low illumination. Eyes predominantly designed for crepuscular and nocturnal conditions would thus be expected to be larger than diffraction limited considerations would predict from their behavioural resolution. Clearly the whole strategy of eye design requires consideration of the relative importance of vision at different intensity levels with a resultant optimisation beyond the capacity of current models. No account has been taken here of the retinal reorganisation which occurs during dark adaptation (PIRENNE, 1962) or of the contending requirements of the peripheral retina for photopic and scotopic vision.

It does not follow, a priori, that the size of a diffraction limited eye should be calculable from its behavioural photopic resolution. Optimisation in the design is between the sampling matrix and image; only if there is 1:1 connectivity between the sampling elements and the brain, as in the human fovea, would behavioural resolution represent the potential of the sampling matrix. In the central area of most eyes, and the peripheral retina of all, the output of the photopic receptors converges onto ganglion cells with possible loss of independence and reduction of the behavioural resolution. However, the fine receptor matrix is employed in generating outer retinal subunits which are sensitive to movement thus justifying analysis in terms of diffraction limited organisation.

A similar argument has been put forward in relation to rod diameters. Central rod densities appear to cover a similar range to those of cones, with characteristic values of 400000 rods/mm^2 for rat (CONE, 1963) and about 400000 rods/mm^2 in cat (STEINBERG et al., 1973) in spite of a 3.5 fold difference in eye size. Anatomical convergence is so obvious in the first stages of rod connectivity that larger rods might be thought adequate. As RUSHTON (1962) asks,

'Would we not do just as well with giant rods a hundred times as big as they are?'

Although suggesting that speculation would outstrip propriety, he hazards that the answer might lie in the need for a fine scotopic grain for movement detection. The possibility that the rod matrix is employed in some form of discrete sampling is inherent in the recent observation of WEALE (1976) that the information content of the image plane is in reasonably close agreement with that attainable by sampling with the entire receptor matrix, not just the cone matrix, in both central and peripheral retina. Whether observations of the real optical system support this theoretical treatment remains to be seen.

Examination of Fig. 9A clearly shows that the dimensions of small mammalian eyes increase more rapidly in proportion to body weight than would be expected from allometric considerations; the requirements of these species thus seem to be constrained by the limited space available for housing the eyes in the head. In contrast, the larger animals have eyes of relatively uniform size over a considerable range of body weight which, unless there is some inherent limitation on, say, the size which a crystalline lens may achieve, indicates their needs to be met by, relatively, very small globes. Such a situation is very plausible. The need of large diffraction limited eyes for high resolution vision is predominantly centred on the detection and identification of either prey or predator at a distance adequate to respectively enable certain escape or stealthy approach. The inherent properties of an environment set limits on these requirements which are fairly independent of the size of its denizens e.g., distance vision is massively degraded by quantum limitations on moonless nights, no diurnal forest species can see further than the herbage density permits. Only a few environments, such as the open plain or that occupied by the raptorial birds, set a premium on high resolution vision. Thus the eagle, a relatively small animal, has perhaps the most acute vision of any species; its resolution would be quite unneccessary to the herbivorous elephant with its freedom from carnivorous predators. The possibility of very large quantum limited eyes arising in large animals which are active at very low light intensities seems to have been excluded by the development of alternate sensory systems; quantal limitations in such environments make adequate resolution impossible within a reasonable integration time.

These observations are clearly at variance with KIRSCHFELD'S (1976) concept of 'subjective resolution;' over a range of species from insects to mammals it is argued that resolution normalised to body height is constant within an order of magnitude. KIRSCHFELD suggests that small animals deal with the world at shorter distances and can thus resolve an object into the same number of points as does a larger from at a greater distance. There may be some truth in this hypothesis when applied to the whole gamut of species but within a group like the homeotherms it is a very crude approximation as indicated by comparison of the acuity of man and squirrel monkey (MERIGAN, 1976).

A given axial length may be achieved for an emmetropic eye by the choice of a great variety of corneal powers, lens powers, and separations of these components. Such differences are actually found to occur in the eyes of different species and may be attributed to specializations for nocturnal and diurnal vision. WALLS (1942) describes the lens of nocturnal species as relatively more powerful and situated further back than in a diurnal form of the same size. The more rearward situation of the posterior nodal point in such an eye results in a smaller PND and RMF (RMF = retinal magnification factor = $(\mathrm{PND}\,\pi/180)\,\mathrm{mm}/^\circ$); the brightness of the retinal image is improved, because it varies inversely as the area of retina illuminated, i.e., proportional to $(1/\mathrm{PND})^2$ but at the expense of a smaller image. A plot of the RMF against the axial length of the eye shows the extent to which it varies with nocturnal and diurnal life styles at a given eye size. It is clear from the *estimated* RMF's plotted in Fig. 9B that the diurnal eyes tend to greater magnification than the nocturnal, but the range is limited and rarely exceeds 20% of the mean value. The trend to strong proportionality between RMF, or PND, and

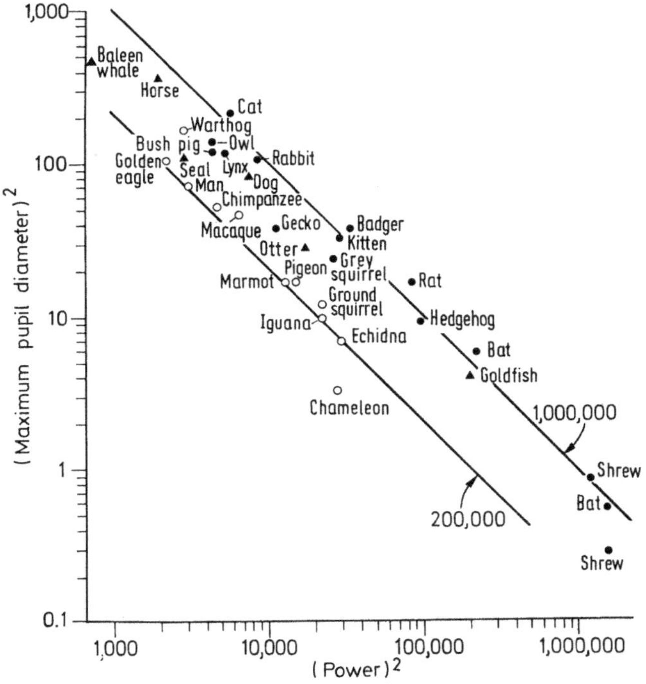

Fig. 10. Light-collection factor; square of maximum entrance pupil diameter, D_m (estimated as 90% of lens equatorial diameter in mm corrected for corneal magnification if not known from direct measurement) plotted against retinal distance factor, the square of the power of the whole optic system of the eye, estimated as $(1/\mathrm{PND})^2$ determined from the radius of curvature of the eye cup, in meters, if not available from a schematic eye. Animals whose points fall on a given isoillumination line have the same retinal illumination under given conditions of ambient lighting when the pupil is fully dilated. Of the available species it is obvious that the nocturnal forms lie close to the 10^6 illumination line, for which the f No. is given by $\mathrm{PND}/D_m = 0.95$, and the diurnal forms lie close to the 2×10^5 line corresponding to an f No. of 2.1. The range of maximum f Nos. is thus small for a wide variety of forms and the diurnal and nocturnal animals differ only fivefold at most, in their retinal illumination with fully dilated pupils. From the legend of Fig. 9 B we may conclude for extreme nocturnal forms that $D_m = 0.63L$ and for extreme diurnal species that $D_m = 0.29L$. Note that kitten, cat, and rat have similar light collecting power in spite of the difference in size of their eyes

axial length is apparent, but quite small variations in PND have a significant influence on the retinal image brightness because of the inverse square relationship between the two factors.

The amount of light which enters the eye is determined by the area of the entrance pupil; for a given axial length its maximal value shows a good correlation with nocturnal and diurnal habits. A useful index of the light-gathering power of the eye may thus be obtained as $p^2/(1/\mathrm{PND})^2 = (1/\mathrm{F}_{NO}^2)$, where p is the maximal entrance pupil diameter and units arbitrary. Sufficient complete schematic eyes are not available for an informative graph to be made from their parameters, but the necessary measurements may be approximated as described in the legend of Figure 10, in which a function of the maximum entrance pupil area has been plotted against the image area factor.

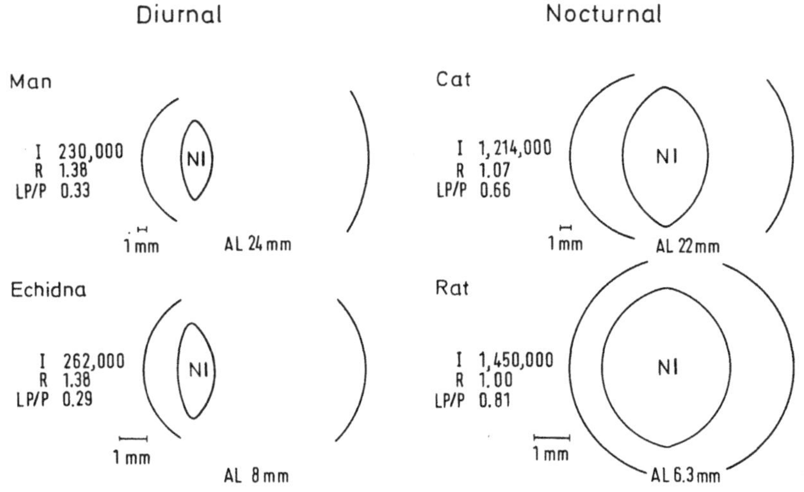

Fig. 11. Human, echidna, cat, and rat eyes scaled to the axial length of the rat eye. Optic design is obviously related to the nocturnal/diurnal division rather than to absolute size. Retinal illumination index, I = pupil area/(1/posterior nodal distance)^2, is independent of scaling and clearly differentiates diurnal man and echidna from nocturnal cat, and rat. As discussed in text, the retinal image in the scaled diurnal eyes is about one-third larger than in nocturnal forms (R = scaled image size for animal/scaled image size for rat) thus enabling increased resolution with loss of brightness. Smaller, brighter image of the scaled nocturnal eyes is achieved, because the ratio of lens power to total power of eye, LP/P, is greater than in the diurnal species

Species located on the same $-45°$ isoillumination line in Figure 10 will have retinal images of the same illumination when their pupils are fully dilated under common conditions of illumination. It is striking to note the manner in which the indices for the eyes of obviously nocturnal species fall on, or close to, the 10^6 illumination index line. The cat and, potentially, the 1-day kitten eyes are thus of similar light-gathering power to those of the rat. From a given nocturnal eye, such as that of the rabbit, we may pass on the graph to a diurnal eye of either similar PND but smaller maximal entrance pupil, e.g., gecko, or to one of larger PND and similar maximal entrance pupil, e.g., eagle. The range in $(1/\mathrm{PND})^2$ at a given pupil size between nocturnal and diurnal species is somewhat less than 1 log unit, and a similar range of pupil area is found at a given PND. The illumination index thus differs fivefold between truly diurnal species, such as the marmot, and nocturnal types, such as the hedgehog and rat. Among the smaller eyes it is the reduced PND and RMF which predominate in establishing the retinal illumination. Among the larger it is the relatively increased pupil area which plays the major part.

Schematic cross sections of the eyes of man, echidna, and cat have been normalized to the rat eye on the basis of axial length (Fig. 11) in order to illustrate the influence of the factors discussed in this section on optic organisation. Various parameter values for each eye in its original and scaled-down form accompany the figures.

The cat and rat are arrhythmic species primarily adapted to nocturnal conditions but evincing diurnal activity. The eyes of man and echidna represent those of typical diurnal adaptation, although the two species are again flexible in behavior. Although reportedly a nocturnal animal, the echidna in fact spends a great deal of its time abroad during the day (COLEMAN, 1934, 1935) and has been found to possess cones (unpublished observations) and is thus not a pure-rod retina as previously described (O'DAY, 1939; WALLS, 1942). The fivefold greater illumination index of cat and rat over the human and echidna eyes reflects these differences in behavior. It should be noted that the relative proportions of the index attributable to the PND and pupil area change with scaling of the eye, but the overall index remains independent of size for a given optic organisation. The position of the lens and its power in relation to that of the cornea and maximum pupil diameter are very similar in the normalized sections of each nocturnal and diurnal pair in spite of the 300% difference in the absolute size of their globes.

The more posterior position of the lens in the scaled nocturnal relative to the diurnal eye has been shown to increase the image brightness. The consequent increased separation of the lens from the cornea necessitates a greater lens power in order to maintain that of the eye, and an additional increment of power is also required to return the eye to emmetropia. To effect this, it is necessary that the lens increase in surface curvature; this must be accompanied by an increase in thickness if the equatorial diameter and effective pupil aperture are to be maintained; a further increase in lens thickness is necessary if the eye is to obtain a larger pupil. The rearward displacement of the lens facilitates the housing of its greater equatorial diameter. The corneal base may also be relatively enlarged in order to enable the larger pupil aperture to be employed, but its extent does not play an important part in defining the total field of the eye.

The optic organisation of cat and rat, in spite of almost identical retinal illumination indices and total normalized power, differ much more than that of the human and echidna. The equatorial plane of the rat lens is situated in a similar relative position to that of the cat, but the lens itself is relatively thicker and more powerful. Because of the similarity of total normalized power and refractive state in the eyes of these two species, it follows that the rat cornea must provide a smaller proportion of the total power than does that of the cat, some 43% compared with 55%. The consequent relatively reduced curvature of the rat cornea is accompanied by an increased height which exposes more of the peripheral lens to oblique rays than in the cat eye. No obvious explanation of this is available in terms of the previous discussion, but it may be pertinent that the rat visual axis is some 60° anterior to its optic axis; an increase in the effective entrance pupil for oblique rays when the pupil is fully dilated could be of great benefit for forward vision.

II. Nonaxial Optics

1. Marginal Rays

a) Spherical Aberration

Marginal rays incident upon a spherical refracting surface are brought to a nearer focus than paraxial rays—spherical undercorrection. This *spherical aberration* may be calculated for the human (LE GRAND, 1942) or animal cornea upon the assumption that its surface *is* spherical in form, but, in fact, it is elliptical in man and flattens perceptibly towards the periphery (DUKE-ELDER, 1958; HAGE and BERNY, 1973). Such flattening is characteristic of an *aplanantic* surface in which the peripheral power is lowered to eliminate or minimise spherical aberration. The positive aberration of the human cornea may have been reduced by peripheral flattening but is not eliminated and is especially prominent in the aphakic eye when the pupillary diameter reaches 6 mm with the result that scotopic vision is poor (HAGE and BERNY, 1973). Peripheral flattening of the cat cornea has been reported (VAKKUR et al., 1963) and is apparent in the rabbit (HUGHES, 1972). The rat cornea is spherical (HUGHES, 1977c).

A homogeneous lens of conventional schematic eyes also demonstrates strong spherical undercorrection. By contrast, the actual crystalline lenses of cat (VAKKUR and BISHOP, 1963) and rat (HUGHES, 1977c) are found to display negative spherical aberration in the "cone of confusion" of their emergent ray bundles. A detailed analysis of the refractive-index gradient in the rabbit crystalline lens has shown it to offset the spherical surface-induced undercorrection to the extent that the system is overcorrected; marginal rays are thus brought to a focus further away than the paraxial rays.

The negative aberration of the lens may obviously play a part in offsetting the effects of the positive spherical aberration of the cornea. In spite of corneal aberration, at pupil diameters of less than 2 mm the human eye is almost free of overall aberration (HAGE and BERNY, 1973). Most commonly, the positive aberration of the cornea predominates (STINE, 1930) up to a maximum of some 7 D extra power, which makes the refraction myopic through the pupil periphery. Considerable variation of power occurs from one quadrant of the pupil to another (VAN DEN BRINK, 1962), and its pattern may change during accommodation.

The spherical cornea and homogeneous lens of the GULLSTRAND (1924) schematic eye lead to pronounced undercorrection along the optic axis; the introduction of an accurately aspheric cornea halves this aberration (LOTMAR, 1971). In the real eye, the remaining aberration is offset by the lens. The negative aberration of the lens has been simulated in recent extensions of the Gullstrand schematic eye either by the introduction of an arbitrary aspheric rear surface or, more satisfactorily, by the use of an approximate seven-shell model (LOTMAR, 1971). An accurate model lens has not been so far employed in a human schematic eye, although the extent of the crystalline lens axial aberration has been computed from the refractive-index gradient of five human lenses by NAKAO et al. (1969).

Little information is available about the spherical aberrations of lens and eye in species other than man. The obvious negative aberration of the cat and rat lens show that compensation for corneal undercorrection would be possible as in man. The overall aberration of the cat eye is not known, but its analysis in the rat (HUGHES, 1977b) provides an interesting contrast to man. The optometric refraction of the rat eye through the peripheral pupil by means of marginal rays indicated some 8 D greater hypermetropia than for central axial rays. It was also found that neurophysiologic refraction indicates as much as 10 D of hypermetropia when the pupil is dilated, although the eye appears to be emmetropic when the pupil is constricted (HUGHES, 1977b). It is interesting to note that the overall aberration of the rat eye is negative, like that of the lens, but that in man it is usually positive, like of the cornea. Examination of Figure 11 reveals that the rat lens takes precedence over the cornea in determining the power of the eye, whereas for the human the reverse is true. These gross aberrations predominate in the rat, as in man, at pupil diameters such as would be encountered only under scotopic conditions when quantum limitations restrict their influence on resolution.

b) Entrance and Exit Pupils

The need to consider marginal rays arises when dealing with pupillary optics. The entrance pupil of the eye is the image of the real pupil formed by the cornea and determines the quantity of light which enters the eye. Calculation of its diameter when small is relatively straightforward (BENNETT and FRANCIS, 1962), but ray tracing, either calculated or geometric, is strictly required to specify its parameters when large. By this means both marginal rays and deviation of the cornea from sphericity may be taken into account. The average *calculated* entrance pupil diameter for an 80% dilated real pupil in the cat (VAKKUR and BISHOP, 1963) was, however, only 3% larger than the measured entrance pupil when derived according to Gaussian theory.

For the geometric specification of blur-circle diameters in the defocused eye, it is necessary to compute the size and position of the *exit pupil*, which is the image of the real pupil in the crystalline lens. Invariably only approximations of these parameters are obtained. If the homogeneous lens model is regarded as acceptable, it is possible to determine the exit pupil size and position for a small real pupil either by vergence calculation according to the Gaussian approximation or by ray tracing, but only the latter method may be used for larger pupils which accept marginal rays. The significance of such calculations is debatable, because they should be performed for a non-homogeneous thin-shell lens. No experimental determinations appear to have been made of the exit-pupil dimensions in the mammalian eye, so the error inherent in the common approximate computations is not known. For critical work the exit-pupil positions and sizes reported for the cat (VAKKUR and BISHOP, 1963) and rat (HUGHES, 1977c) should be regarded with scepticism until confirmed experimentally.

2. Oblique Rays

a) Oblique Axial Aberrations

The difficulties entailed in tracing rays which enter the eye oblique to the optic axis have already been attributed to the gradient of refractive index through the lens. YOUNG (1801) computed oblique-ray image shells for the human eye on the assumption of a homogeneous lens, and since that time there has been little improvement in the methods employed. The intrinsic complications of the subject are great; the shell of focus for horizontal and vertical sections through an oblique ray bundle differ in position, the oblique astigmatic difference or STURM'S interval, by an amount varying in man between 2 and 10 D at 60° eccentricity. In some people the retina lies between these two shells, in others to one side (FERREE and RAND, 1932). Other aberrations such as coma are potentially significant but may be minimized by the refractive-index gradient of the lens (NAKAO et al., 1968) so that their consideration requires a schematic eye with at least a thin-shell lens model. The PND of the optic system decreases for rays of increasing obliquity, and thus the overall refractive error of the eye depends upon the extent to which the retinal shell is deformed from a sphere to accommodate this. The retinal magnification factor may thus decline in the periphery, but its magnitude is determined not only by the PND but also by the angle subtended between the local retinal surface and the eccentric chief rays. A full treatment of peripheral imagery has not been attempted for any species. Some wide-angle schematic eyes have been developed for man (LOTMAR, 1971; DRASDO and FOWLER, 1974) but are

Table 2

	Horizontal field			Vertical field
	Calculated		Measured	Measured
	Minimum	Maximum		
	/°	/°	/°	/°
Cat	158	214	181	180
Rabbit	168	212	192	180
Rat	147	238	205	198
Goat	—	—	190	186
Horse	—	—	193	178

Sources of data: cat, VAKKUR and BISHOP (1963), HUGHES (1976a); rabbit, HUGHES (1972); rat, HUGHES, (1977c); goat, HUGHES and WHITTERIDGE (1973); horse, HUGHES (unpublished observations).

based on homogeneous lenses and must therefore be accepted with caution; agreement between their performance and that of the eye may have been obtained simply by arbitrary modification of their parameters.

b) Uniocular Optical Field of the Eye

A primary application of wide-angle human schematic eyes has been the investigation of the nonlinear projection of the visual field onto the retina. The peripheral retinal magnification factor is said to be halved relative to its central value in man (DRASDO and FOWLER, 1974). The computed intraocular angle subtended between an image and the optic axis was some 90% of that subtended by the object over most of the uniocular field, although in the extreme periphery this value dropped to 70%. In the cat (HUGHES, 1976a) the measured variation of RMF was found to be much smaller and the retinal image to be a rather better central projection than calculated for man.

The theoretical extent of the uniocular field of an eye unobstructed by the ocular adnexa should strictly be assessed by means of a schematic eye equipped with a nonhomogeneous lens. For some purposes this may be avoided by employing the observation that a ray which subtends a straight line of the equatorial margin of the opposite side of the lens, when refracted by the cornea, will pass into the vitreous humour, while one which grazes the lens vertex will not (VAKKUR and BISHOP, 1963). Back tracing of these rays sets an upper and lower limit on the extent of the visual field in the plane considered. Such limited tracing has been made for several species of mammal, the results of which are included in Table 2. It is to be emphasised that, contrary to frequently encountered statements in the literature (DUKE-ELDER, 1958), the extent of the cornea does not necessarily determine that of the optic field of view; the relative position of the lens and cornea is more important.

Criticism has already been made of the methods commonly employed for the determination of the extent of the uniocular optical field. Thus the transcleral illumination technique reveals the cat optic field to be 150° (VAKKUR and BISHOP, 1963), whereas the use of perimetric ophthalmoscopy indicates 181° (HUGHES,

1976a). In contrast to the transcleral technique, in which the dense pigment and thickening at the ciliary body may restrict observation of the image, the results obtained by ophthalmoscopy usually agree substantially with the mean of the calculated limits; rabbit (HUGHES, 1972), cat (HUGHES, 1976a), rat (HUGHES, 1977c), goat (HUGHES and WHITTERIDGE, 1973). The range of horizontal uniocular optical field is from about 180° in cat (HUGHES, 1976a) to 205°–207° in rat (LASHLEY, 1932; HUGHES, 1977c); a horse eye with dilated pupil was found to possess a horizontal field of only 193° (HUGHES, unpublished observations) in contrast to the 215°–228° commonly attributed. The vertical field is much more uniform between species and extends for about 180°.

The elongated pupillary aperture of the horse, goat, and other ungulates has been suggested to extend the uniocular field, but the validity of this assumption remains untested. It doubtlessly does increase the effective entrance pupil for oblique rays passing to the forward fixation area.

III. Refractive State of the Vertebrate Eye

1. Axial Refraction

a) The Retinoscopic Refractive State

Modern visual dioptrics began with KEPLER'S (1604) hypothesis that the vertebrate eye is effectively a *camera obscura* in which the objects of the external world are imaged upon the superficial retina by means of the spherical refracting surfaces of cornea and lens. SCHEINER (1619) was able to confirm this theory to the extent of demonstrating the presence of an image in the retina, but it remained for MÜLLER (1853) to conclude that the visual process is initiated, not on the vitread retinal surface, but at specialized photoreceptors in the sclerad border.

We have previously seen how the extent of the retina, the uniocular field, and the relative position of the two eyes determine the angular extent of the external object space encompassed by the visual field. It is characteristic of an optic system of power, D, however, that image quality is optimal in terms of resolution and contrast at only one distance, v, which is conjugate with a unique object distance, u, so that for a simple thin lens,

$$v = \frac{1}{1/u + D}$$

The photoreceptors define the effective image plane of the eye, and the conjugate object plane is thus established by the power of the optic system, D. Many species employ an accommodatory mechanism to vary their optic power and thus obtain good imagery over a wider range of object distance. In spite of this dynamic range of accommodation, only two static refractive states are commonly considered, the anatomic and the physiologic conditions of rest. The former is established with the lens in its state of maximum flatness and accommodation relaxed, either voluntarily or by means of drugs, while the latter incorporates some degree of accommodatory tone under defined field conditions. If parallel light from distant objects is brought to a focus in the photoreceptor plane then, for rough ophthalmologic purposes, the eye is described as emmetropic, and the optimal images of nearer objects may be displaced from the photoreceptor plane—although other factors may permit some depth of field to exist. There is no a priori reason to expect the anatomical or physiologic resting refractive condition to be emmetropic in all species; it is arguable that animals dealing with close objects (WALLS, 1942) might find a myopic refraction, perhaps focussed at the hyperfocal distance, more suited to their requirements. More extreme myopia, or even hypermetropia, need not be disadvantageous if the accommodatory mechanism is such as to permit reflex emmetropia.

The objective assessment of the refractive state of an eye is almost invariably attempted with retinoscopy. By this means the vergence of the emergent rays of the fundus reflex is measured and the extent of the separation of their surface of origin from the posterior focus of the eye thus established. If the emergent rays are parallel, then the principle of optic-path reversibility establishes that incoming rays from distant objects will be brought to a focus in the plane of origin of the fundus reflex; this condition is commonly described as emmetropic upon the implicit assumption that the reflex originates at the photoreceptor plane. There is no evidence that is does; such "objectively" measured emmetropia may represent subjective ametropia; however, the majority of the literature is based upon this procedure. The classic *conception* of the static mammalian refractive state with relaxed accommodation (e.g., LANG and BARRETT, 1887) is well summarized by JOHNSON (1901) on the basis of his ophthalmoscopic and retinoscopic observations of 182 species:

> A slight degree of hypermetropia, i.e. under 1 Diopter, may be said to be the rule throughout the higher Mammalia, whilst higher hypermetropia, i.e. 2 D to 5 D, is found in the wild species of the Rodents, the Edentata, and the Marsupials. *Simple emmetropia is rarely met with; indeed I very much doubt whether it is compatible with sight of a high degree of perfection.* From examination of the vision among a large number of native youths from the Upper and Lower Congo, the Nile Valley, and the Niger Territory, I found that every one with vision beyond the standard of 6/6 or 6/5 was hypermetropic to at least 0.5 to 0.75 Diopter (manifest), an amount closely corresponding with that found in the great majority of Primates and Carnivores.

Not all authors have obtained results in agreement with JOHNSON (1901). For instance, the rat eye has been described as from −13 D to −3 D myopic [scleral image focusing (LASHLEY, 1932); retinoscopy, vessel ophthalmoscopy, single-unit receptive field size (BROWN and ROJAS, 1965); retinoscopy (MONTERO et al., 1968)] as well as from 9 D to 20 D hypermetropic [retinoscopy (GLICKSTEIN and MILLODOT, 1970); retinoscopy (HUGHES, 1977a); retinoscopy (BLOCK, 1969)]. Again, dogs have been described as considerably myopic (UBERREITER, 1959), slightly myopic (WORFOLD, 1965), emmetropic (VOGT, 1936), or occasionally as hypermetropic (SMYTHE, 1961).

It is curious that the rhinoceros is so consistently described as myopic that such ametropia is sine qua non for the animal even in children's literature (Fig. 12). This prejudice—because it is apparently not based upon refraction—appears to have been well established by 1801 (LEIGH THOMAS, 1801), but to date no additional information about the rhinoceros eye is available which might engender doubt about Thomas's conclusion that,

> 'if we should ever become aquainted with the natural habits of this animal, his vision will be found to be as perfect as that of any other of the same class.'

Fig. 12. Acceptance of nonemmetropic refractive states in the animal eye is so prevalent that even the children's literature describes the rhinoceros as short-sighted! (After "Rupert the Rhinoceros" Little Golden Books, USA)

Such variability of results is understandable but necessitates careful consideration of published animal refractions. Ophthalmoscopy is not an acceptably accurate technique for refraction, and retinoscopy requires a modicum of skill. The state of accommodation of the animal or observer is rarely monitored during either procedure, and retinoscopy eliminates only the influence of the observer if a cycloplegic is not employed. Mydriasis concurrent with cycloplegic retinoscopy may, however, introduce peripheral aberrations so that the refraction does not represent the state of the eye when the pupil is small under normal photopic conditions. Additional discrepancies will arise if the results do not refer to a common spectacle lens plane.

It is essential that studies of the refractive state in a given species be based upon a reasonably large sample in order to allow for a range of ametropia within the population. The distribution of refractive error in large populations has only been established for macaque, chimpanzee (Young, 1963; Young and Farrer, 1964), and man (see Duke-Elder, 1970). In man the distribution is not normal (Stromberg, 1936).

The great excess of human hypermetropes of low degree over the statistical expectation suggests covariation between the various optic components of the eye during growth to produce a much smaller spread of refractive error than would occur if their values were uncorrelated; these compensatory processes have also been demonstrated in the rabbit (Ludlam and Twarowski, 1973). There is no information about the population distribution of refractive errors or refraction during growth in other species, and it cannot be said whether "emmetropization" is universal. If the small population refractions available for other mammals are assumed to be normal distributions, then their standard deviations are very similar for a variety of eye sizes and range from 1 D to 2 D (Fig. 13), as in the human population.

JOHNSON (1901) considers myopic refraction common only in domestic animals, such as horses (LANG and BARRETT, 1887), cattle (DUBAR and THIEULIN, 1927), and especially hutched animals. Support for the long-debated influence of the visual environment on the refractive state of the eye is available from studies on monkeys (YOUNG, 1961, 1963; YOUNG and LEARY, 1973) and cats (ROSE et al., 1974). Street cats were found to show a mean refraction of 1.15 D $\pm$0.18 D whereas caged animals were overtly myopic at -0.62 ± 0.24 D with insignificant difference of mean axial length between the groups.

The 3 D myopic refraction of the rat eye (LASHLEY, 1932) and -15 D for teleosts (FRANZ, 1931) were accepted by both WALLS (1942) and DUKE-ELDER (1958), although conflicting results were available, as adaptations for near viewing. More recently, however, both rat (BLOCK, 1969; HUGHES, 1977b) and the teleosts (BARON and VERRIER, 1951; BAYLOR and SHAW, 1962; CHARMAN and TUCKER, 1973) have been shown to be of hypermetropic refraction. The consensus of modern workers is in agreement with the view of JOHNSON (1901) that mean retinoscopic hypermetropia is universal among wild mammalian species, which is puzzling; emmetropia or myopia is teleologically more comprehensible.

b) *The Effective Image Shell*

It is to be emphasized that retinoscopy simply indicates the separation of the reflective layer from the posterior focal plane of the eye and does not specifically determine the refraction of the photoreceptor plane. The surprising universal mean hypermetropia revealed by retinoscopy may thus be of little relevance in the consideration of subjective or behavioral refraction at the photoreceptor plane.

It is important for the ensuing discussion to identify the exact position along the photoreceptor at which light is brought to a focus during normal vision. Light has commonly been regarded as equally effective if in focus at any point along the axis of the photoreceptor outer segments. WALLS (1942) thus suggested that the elongation of the outer segment might be employed in certain species to minimize accommodation by the production of a greater tolerance for the plane of focus. The more recent concept of the photoreceptor as a waveguide (ENOCH, 1961) and/or "funneling" device (RUSHTON, 1956, 1963; DONNER and REUTER, 1965) implicates the inner segments of both rods and cones (BRINDLEY, 1966) as collectors of light for the outer segment, the basal diameter of the inner segment determining the receptor's entrance aperture (SNYDER and MENZEL, 1975). Under these conditions the directional properties of the receptors (STILES and CRAWFORD, 1933; BAYLOR and FETTIPLACE, 1975) require that the receptor axis project to the center of the exit pupil of the eye for optimal light capture. The placement of the exit pupil forward of the centre of curvature of the retina thus ensures that the more peripheral receptors would be set at an angle to the radial axis of the retina. Such oblique orientation of the peripheral receptors has indeed been reported in monkey, rabbit (ENOCH, 1972), fish, frog, and gecko (LATIES and ENOCH, 1971) as well as psychophysically demonstrated in man (ENOCH and HOPE, 1972). In all of these species the receptor orientates about the *beginning* of the inner segment, not about the junction of the inner and outer segments, in a manner consistent with its presumed role as the entrance aperture. Optimal resolution would thus be achieved when the circles of least confusion of an image for which the eye is focused are coplanar with the receptor entrance apertures; these are aligned in the eye along the outer limiting membrane. The outer limiting membrane may therefore be taken to represent the presumed effective posterior focus of the eye and defines a shell in the light flux through the eye which is subsequently referred to as the *effective image shell*. It is to this shell that light will be brought to a focus during accommodation and subjective refraction.

c) Glickstein and Millodot's Explanation of Mean Retinoscopic Hypermetropia

In 1894 BEER noted that the true refraction of the photoreceptor surface is not obtained by focussing on the retinal blood vessels during *ophthalmoscopy*. In man the separation of the effective image shell from the vessels is necessarily small, relative to the posterior focal distance of the eye. The error in assessing the refractive error introduced by the separation of the plane of reflection from the effective image shell would thus make a subjectively emmetropic eye appear spuriously myopic if the plane of reflection lay behind the receptors and hypermetropic if in front. The almost universal mean hypermetropia described by JOHNSON (1901) and KAHMANN (1930) thus suggests a common reflecting surface forward of the receptors in the majority of species. GLICKSTEIN and MILLODOT (1970) found the magnitude of the retinoscopic error to be inversely proportional to the square of the posterior focal length of the eye, thus explaining WALLS' (1942) observation that

> As the size of the eye diminishes from that of a cat to that of a mouse... an increase of the hypermetropia from half a diopter or so to five, seven or even ten diopters (occurs).

Such an explanation would also account for the reduction of the mean hypermetropia of 2–3 D in human infants (GRAHAM and GRAY, 1963; SEBA, 1974) to some 0.5 D in the adult population as the eye grows from 17.9 $\pm$ 0.1 mm (SORSBY et al., 1961) to 24.0 $\pm$ 0.03 mm (STENSTRÖM, 1948). A similar change occurs in the refractive state of the growing rabbit eye (LUDLAM and TWAROWSKI, 1973) and during the growth of the eel eye just before breeding (WALLS, 1942).

Recent, careful retinoscopic investigations of a wide variety of species have been combined in Figure 13, with those of GLICKSTEIN and MILLODOT (1970) to display the relationship between the refractive state and axial length of the eye. These data are well fitted by a curve derived from the theoretical equation relating axial length and refractive state. The plane of reflection was estimated by GLICKSTEIN and MILLODOT (1970) to be displaced on average by 135 μm vitread of the posterior focus of the eye, a position which is consistent with reflection from the boundary between the vitreous and retina in the functionally emmetropic globe. It was thus concluded that *the almost universal hypermetropia established by retinoscopy results from a systematic error of method, and that the true mean refractive state of most, if not all, wild vertebrate species is close to emmetropia in the effective image plane.*

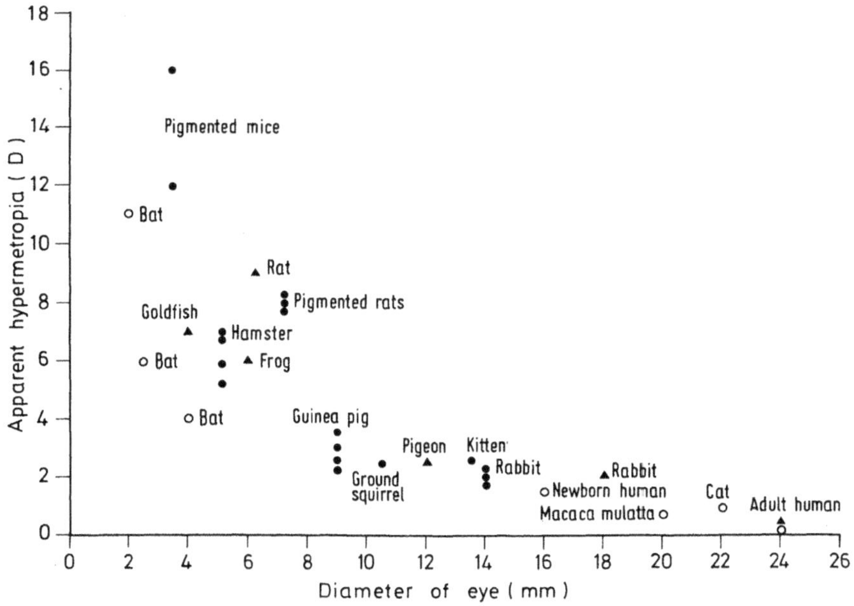

Fig. 13. Retinoscopic refraction as a function of the axial length of the eye. The graph well illustrates the marked hypermetropia of small eyes. (After GLICKSTEIN and MILLODOT, 1970, with a few additions)

These results clearly suggest the abandonment of WALLS' (1942) view that their hypermetropia shows

> the indifference of mice and mouse sized mammals in general to any refinements of vision relating to resolving power

and that

> many mud-grubbing, small eyed fishes are hypermetropic, indicating a loss of importance of vision to them.

d) Experimental Evidence for Glickstein and Millodot's Theory

The hypothesis of GLICKSTEIN and MILLODOT (1970) accounts well for the relationship between eye size and apparent retinoscopic hypermetropia but still requires proof by the demonstration that mean retinoscopic hypermetropia is accompanied by mean emmetropia at the effective image shell, and that the predominant fundus reflex arises at the border between the retina and vitreous humor. In keeping with the hypothesis, it is well known that a mean hypermetropia of about 0.5 D is detected in humans by retinoscopy when the eye is subjectively emmetropic (BALDWIN, 1964; FRY, 1967; SOUTHALL, 1926).

Evidence for the emmetropic state of the effective image shell is rare in other species of mammals which evince overt retinoscopic hypermetropia. Reports of the retinoscopic refraction of the cat indicate a mean hypermetropia from +0.8 D to 1.5 D (HESS and HEINE, 1898; VAKKUR et al., 1963; ROSE et al., 1974). The refractive state at the effective image shell of the cat eye is in principle readily determined by electrophysiologic refraction during which the receptive field map or, for certain classes of retinal single unit, the response to light spots (HILL and IKEDA, 1971; IKEDA and WRIGHT, 1972) and to grating patterns (BARLOW and LEVICK, 1965) are used as a direct criterion for an optimal image during differential focusing. Deviation from ideal focus within ± 0.5 D is demonstrable by these means or by the use of cerebral-evoked potentials (BERKLEY and WATKINS, 1973). Unfortunately, none of these investigations provide information about the refractive state of the cats before contact lenses were fitted; the spectacle power required for best focus need not therefore represent the refractive state of the effective image plane in the normal animal. Single-unit refraction (HUGHES, 1977b) suggest the effective image plane to be emmetropic in the rat, but the great depth of field of small eyes when the pupil is constricted limits the sensitivity of the method, and aberrations preclude its use with a large pupil.

Emmetropia at the effective image plane of overtly hypermetropic eyes has, however, been demonstrated in nonmammalian species, such as pigeon, by means of an ERG criterion (MILLODOT and BLOUGH, 1971), for the frog by the refraction of single units stimulated with moving stimuli (MOSER and KRUEGER, 1972; MOSER, 1973), and by the determination of single-unit acuity thresholds in the fish (MEYER and SCHWASSMANN, 1970).

The necessary conclusion from GLICKSTEIN and MILLODOT'S (1970) hypothesis, that the fundus reflex originates at the vitreous/retina junction, receives little support in the literature. Alternative suggested sites have been the choroid (WALLS, 1942), BRUCH'S membrane (WEALE, 1966), or the pigment epithelium (FRY, 1967), but all of these lie behind the effective image shell and would lead to an apparent universal myopia, not hypermetropia. According to CAMPBELL and GUBISCH (1966), if polarised light is used to illuminate the fundus, then some 20% of the reflex is depolarised with a red cast, which suggests it to have penetrated the choroid. WEALE (1966) claims that bleaching of the photoreceptor pigments reduces the reflection of the polarized and depolarized components alike so that they must both arise beyond the photoreceptors. In contrast, however, RÖHLER et al. (1969) describe the depolarized reflex component as arising forward of the receptor layer, in the manner required by the above hypothesis, but it should be noted that this component forms a relatively small proportion of the reflex. Recent confirmations of VALENTIN'S (1879) report that the retinal refractive index is larger than that of the vitreous (NORDENSEN, 1934; AJO, 1949) established that a specular reflection must occur at the vitreous/retina interface. MILLODOT (1972) has computed this to be 0.06% of the incident beam, about 50% of the proportion actually recorded in man, so that the calculated front surface reflection explains 70% of the total human specular reflex. In a detailed discussion of this problem, MILLODOT (1972) fails to account for WEALE'S (1966) claim that pigment bleaching influences both specular and diffuse components of the reflex; this must be done before the subject can be regarded as even partially resolved.

The fundus reflex employed for retinoscopy in the cat is commonly regarded as originating at the highly reflective tapetum (WEALE, 1953; BONDS et al., 1972). This layer is separated from the outer limiting membrane, or presumed effective image shell, by a distance of some 50 μm which corresponds to a refraction of

0.35 D, but, because the tapetum lies behind the posterior focus, overt myopia would be expected rather than hypermetropia. It remains to be determined whether the dynamic refraction of the anesthetized cat is sufficiently hypermetropic to mask this apparent myopia or whether a significant component of the reflex arises from a region in front of the posterior focus.

Evidence that the retinoscopic reflex does arise at the retina surface has only been put forward for the rat (HUGHES, 1977b). An objective optometer was employed to show that the refraction of the surface nerve fibers and blood vessels is at 8.9 D, in good agreement with the standard retinoscopic refraction of 9.4 D, when the pupil is small. The reflex employed in retinoscopy is thus concluded to arise close to the retina/vitreous junction. The photoreceptor layer could not be directly refracted, but a mean of −2.1 D was determined for the orange choidal vessels which lie in the pigment cell layer. The presumed effective image plane lies some 0.036 mm vitread of the pigment layer, which is equivalent to a 2.5 D change in refraction (HUGHES, 1977b). The refraction of the image plane is thus about +0.5 D and indicates the eye to be close to emmetropia, in contrast to claims that the rat is functionally an extreme hypermetrope (MASSOF and CHANG, 1972).

The implication of GLICKSTEIN and MILLODOT'S theory that retinoscopically hypermetropic species are behaviorally emmetropic is thus substantiated to some extent by the technique of refracting the effective image plane, but evidence for the retina/vitreous junction as the site of origin of the fundus reflex is limited. A few other reports conflict with the theory. SIVAK (1974) describes the mean refractive state of 6.5 D in the goldfish as being in agreement with the refraction of the photoreceptor plane, rather than that of the retinal surface as computed from measurements made on frozen sections of the eye. In this instance, however, a simple calculation reveals that a 0.07 mm change in length of the globe is equivalent to a 1 D change in refraction. Loss of ocular pressure or vitreal expansion upon freezing are just two possible sources of substantial dimensional change capable of introducing considerable error into the reconstructed refractions. The computation of the refractive state of a small eye with any useful reliability requires extraordinary accuracy of measurement. Again, SUTHERS and WALLIS (1970) describe the 1.68 mm-long eye of the bat *Myotis* to be only 10 D hypermetropic as determined by retinoscopy, rather than some 65 D calculated from the great retinal thickness relative to the posterior focal length of the eye. No explanation of the discrepancy is available, but it may arise in the difficulty of refracting such small eyes.

2. Peripheral Refraction

It may no longer be true to describe the hemispheric field of view of the eye as

> wider than that of any optical instrument

(HELMHOLTZ, 1924), but its extent remains impressive. Yet even in man there has been little attention paid to the refraction of the uniocular field peripheral to the optic and visual axes.

It is not satisfactory to treat peripheral rays which pass through the dioptric system at considerable obliquity to the optic axis as if there is a single focal image shell which may coincide with the effective image shell. Because of oblique astigmatism, such rays give rise to tangential and sagittal image shells (BENNETT and FRANCIS, 1962; SOUTHALL, 1964) which bridge the shell of least confusion and also, near the visual axis, the retina. More complex phenomena which appear in systems with finite aperture, such as asymmetrical coma, may be treated analytically by Seidel's formulae (SOUTHALL, 1964), but recourse must be taken to ray tracing for the consideration of even moderate physiologic obliquities, especially in view of the nonhomogeneous structure of the lens. In the absence of any analytic treatment of the oblique peripheral rays of the vertebrate eye, it is unjustifiable to hold a priori expectations of the peripheral refraction. The retinal image of the human eye has, however, been reported to be stigmatic only within 10° of the visual axis (FEREE and RAND, 1932); beyond this eccentricity even axially emmetropic eyes demonstrate astigmatic errors of great variety. At an eccentricity of 55°–70°, the unpractised eye has been described as benefiting little from the correction of these refractive errors (DOBREE and WEALE, 1954; MILLODOT et al., 1975); however, LEIBOWITZ et al. (1972) obtained improvement of peripheral motion sensitivity after the correction of peripheral optics. Although up to 8 D of spherical and 2.25 D of cylindrical correction could be necessary, certain subjects required less than 1 D correction some 60° eccentric which indicated that peripheral vertical or horizontal ametropia is not a *necessary* outcome of the dioptric structure of the eye.

A recent investigation of peripheral refraction (MILLODOT and LAMONT, 1974) which employed retinoscopy, refractometry, and subjective refraction confirmed the observations of FEREE and RAND (1932) and reported marked peripheral astigmatism. The ametropia was less than theoretically predicted (LE GRAND, 1967), not because of corneal flattening, so it would appear that peripheral aberrations are indeed to some degree minimized by adjustment of the SIEDEL coefficients of the nonhomogeneous lens.

It is essential to note that off-axis retinoscopy is more complex to interpret than on-axis retinoscopy, and caution is required in accepting published results (KEATING, 1975).

The possible importance of peripheral optics in species other than man was recognised by JOHNSON (1901):

> In our eye, possessing only a small angle of acute vision, a blurred image produced at a point near the periphery is of no consequence, but in the case of those animals which have a widely extended sensitive area and need to see all around them, a sharp focus for peripheral rays is of much greater importance.

Not only species with an extensive visual streak and lateral eyes like the rabbit ($\lambda = 77°$) but also those with more limited areas set at a considerable angle to the optic axis, rat ($\lambda = 60°$) and goat ($\lambda = 47°$) may require good peripheral optics. The goat area centralis is obviously organised as a region of very acute vision with its peak ganglion-cell density of 15000 cells/mm^2 (HUGHES and WHITTERIDGE, 1973) in spite of the great obliquity of its visual axis relative to the optic axis. The problems of oblique imagery may have been solved in such eyes, because there is *no* analytic demonstration that this is impossible. Alternatively, it may suffice for the majority of mammals with lower acuity than primates that central on-axis imagery is of such quality, in spite of a deterioration with increasing eccentricity that the most peripheral images remain adequate to the needs of the neural apparatus. This is a problem in comparative optics which remains to be investigated, but such results as are available have been obtained by the modulation transfer function technique and will be dealt with subsequently.

At present there have been no systematic investigations of peripheral image quality and refraction in any mammalian species. Small changes of the refractive state across the pigeon retina have been reported by MILLODOT and BLOUGH (1971), but we have seen that these need not represent the refractive state at the effective image shell. Ray traces based upon oversimplified schematic eye assumptions (NYE, 1973) cannot provide support for these observations. Care must be taken before attributing teleologic significance to such gradients of refractive state. In the rat it is found that optometric refraction of the temporal retina along the visual axis reveals some 5 D of hypermetropia rather than the 9 D at the retinal surface observed along the optic axis. Gross examination of the eye indicates, however, that the retina is only two-thirds of its central thickness in the corresponding temporal region, so that the systematic retinoscopic error is smaller, and the observed refractions are compatible with emmetropia at the effective image shell even in the peripheral retina (HUGHES, 1977b). In the rabbit, DEGRAAUW and VAN HOF (1977) claim the anterior field to be myopic in contrast to hypermetropia on the optic axis. However, HUGHES and VANEY (1977) find both wild and laboratory rabbits to a evince a slight hypermetropic refraction in the pigment layer from the binocular into the temporal field at the level of the visual streak. This finding is in keeping with the electrophysiological refractions of MEYER et al. (1972) which indicated near emmetropia over the whole field.

An asymmetry of the lens position in the eye, or a difference in the distance from the center of the lens to the peripheral and central retina, should not, of itself, be assumed to necessarily imply a difference in the refractive state of these regions. Such an assumption is implicit in discussion of the ramp retina of the horse (NICOLAS, 1930; DUKE-ELDER, 1958) and would be true if the focal length of the dioptric system were constant with varying eccentricity as is implied by MILLODOT and BLOUGH (1971) for the pigeon. However, in cat (VAKKUR et al., 1963; HUGHES, 1976a) and rabbit (HUGHES, 1972) the distance between lens and retina is reduced by about 10% towards the periphery but is accompanied by a change of PND, and thus no substantial alteration of refraction need be expected.

A further complication in assessing variations of refraction with retinal eccentricity is possible in species with lateral eyes and accommodation along the visual axis. MEYER and SCHWASSMANN (1970) describe the refraction of the bass *Paralabrax* as emmetropic in nasal, central, and temporal retina when the fish is paralysed with Flaxedil. Under Tricaine anesthesia, the lens moves forward, leaving the central retina emmetropic but bringing the temporal retina to -13 D of myopia and the nasal retina to a complementary refraction of 10 D of hypermetropia.

3. Epilogue

The great variation in the degree of retinoscopic hypermetropia which has been established among the mammalian species thus appears to represent a remarkable uniformity of functional mean emmetropia. Among the vertebrates in general, the current acceptance of teleost refraction as hypermetropic rather than myopic leaves the lamprey eye as the last alleged instance of static myopic refraction with active accommodation for distance vision (FRANZ, 1934; WALLS, 1942). There is no evidence for specialization of the refractive state with relaxed accommodation for a specific life style or habitat. These studies are, however, based upon the refractive state under a cycloplegic, deep sleep, or death, and the physiologic accommodatory condition of rest (WESTHEIMER and BLAIR, 1973) may be quite different; in contrast to the static refraction, it may evince correlations with life style or size of a given species. The little information available on this topic is outlined in Section IV.

IV. Accommodation

1. Range of Accommodation

The ability of the young human eye to increase the refractive power of its dioptric apparatus by some 15 D during accommodation for nearby objects has attracted considerable interest over several centuries, so that an enormous body of literature has accumulated which incorporates much comparative material. Confirmation of a great deal of this information by modern techniques is badly required. The comparative study of a mammalian accommodation has been relatively neglected, and our knowledge of its range has changed little since the studies of HESS and HEINE (1898) and KAHMANN (1930).

Apart from man, the monkey possesses the greatest range of accommodation so far established among the mammals. HESS and HEINE'S (1898) report of some 10–12 D of accommodation is substantiated by 12 D obtained upon brain stimulation (JAMPEL and MINDEL, 1967), 11 D from parasympathetic stimulation, and a claimed 28 D from drug action (CHIN et al., 1968).

The accommodation of the cat eye remains the most extensively studied after that of man. HESS and HEINE (1898) found the mean refractive state of the anesthetised cat to change by some 2 D upon electrical stimulation. A 3.5 D range was described by HARTRIDGE and YAMADA (1922) after examination of the influence of atropine and eserine on the refractive state—more recently the latter agent has been reported to induce some 8 D of accommodation (IKEDA and WRIGHT, 1972). Autonomic stimulation has revealed a wide variety of accommodatory ranges in the cat, 4–5 D (OLMSTED and MORGAN, 1941), 6 D (ARMALY, 1959), and even 11.5 D (MORGAN et al., 1943; OLMSTED, 1944); behavioral testing suggests some 3–4 D (BLOOM and BERKELEY, 1976). By contrast, the dog is reported to evince only about 1 D of accommodation (HESS and HEINE, 1898; ROBERTS, 1956).

Neither BARRETT (1898) nor HESS and HEINE (1898) were able to demonstrate accommodation in the rabbit, although some 2–4 D were claimed to be detectable in movements of the Purkinje images induced by the electrical stimulation of the freshly enucleated globe (SMITH, 1898) or, more recently, upon the stimulation of the sympathetic supply to the eye (OLMSTED and MORGAN, 1941). MEYER et al. (1972) were unable to obtain evidence of accommodation when employing single-unit refraction in conscious rabbits; the minimum visibile was constant for distances greater than 45 cm at an unknown pupil size.

Accommodation has not been demonstrated in the rat. LASHLEY (1932) concluded that it is not present because of the absence of a well-developed ciliary muscle in this species. The very small pupil of the rat can endow it with great depth of field under photopic conditions (BLOCK, 1969), and this has been demonstrated to be functionally significant by single-unit refraction and consideration of behavioral performance (HUGHES, 1977b).

Away from the common laboratory species, our information is even more limited. Rodents and ungulates in general are described as lacking accommodation (KAHMANN, 1930; WALLS, 1942; DUKE-ELDER, 1958) although apparently with little experimental basis. PRINCE et al. (1960) draw attention to the conventional ciliary apparatus of the ungulates. In a comparative examination of the ciliary muscle, WOOLF (1956) agrees with ROCHON-DUVIGNEAUD (1943) in describing it as well-developed in the ungulates and primates although effectively absent from the majority of rodents. WALLS (1942) resolves the paradoxical presence of an extensive ciliary muscle in ungulates which allegedly lack accommodation by concluding that the lens is too stiff to alter its shape significantly, but, as we shall see, this does not preclude the possibility of an alternative arrangement in these mammals. It will be obvious that there is a need for a fresh examination of this topic.

It is possible that the description of the "ramp retina" in the horse (WALLS, 1942; DUKE-ELDER, 1958) has reduced interest in the dynamic accommodation of herbivorous species because of its seeming redundancy. The "discovery" of the ramp retina is commonly attributed to NICOLAS (1930), but, in fact, it appears to have been derived from his data by WALLS (SIVAK and ALLEN, 1975). He pointed out that the ground surface at decreasing distances projects systematically at increasing elevations in the effective image shell and suggested that the lens-to-retina distance varies in such a fashion as to ensure the degree of myopia at a given elevation appropriate for focussing the region of the object plane, which gives rise to the local image.

The asymmetry of the horse eye has recently been confirmed (SIVAK and ALLEN, 1975), but the lens-to-retina distance is relatively uniform in the upper eye and the ramp unverified. However, these authors assume that the refractive state of the eye along a given axis is related to the corresponding PND, or its approximation, when in fact they may vary independently. The retinoscopic assessment of the horse eye (SIVAK and ALLEN, 1975) indicated no more than 1 D variation of refraction over the vertical field, but, again, the refractive state of the effective image shell may vary more or less than this. A serious difficulty is that an asymmetry sufficient to produce a gradient of only 1 D, or even a depth of field equivalent to this, would suffice to provide satisfactory imaging of objects from 1 m to infinity. Much more sophisticated testing would be required to establish whether a ramp retina is functionally present. There has never been any suggestion that the ramp retina operated for frontal vision along the fixation axis of the horse; it is possible that the moderate development of the ciliary muscle (PRINCE et al., 1960) suffices for the adjustment of focus in the anterior field.

The continuous assessment of the accommodatory state has recently been achieved for the human eye by the use of the infrared optometer (CAMPBELL and ROBSON, 1959) and subsequently, by the same technique, for the conscious pretrigeminal cat (ELUL and MARCHIAFAVA, 1964). The static refraction of the sleeping or anesthetized cat eye was found to be 1.5 D, but upon waking it changed by −0.5 D to +1.0 D. Some 3–4 D of accommodatory power then remains to deal with objects up to 20 cm from the eye. It is claimed that the conscious cat evinces a decrease in its refractive power from the condition of rest when looking to the distance, 6 m. The physiologic tonic refractive state of the cat eye thus appears to be myopic in contrast to the hypermetropic anatomic condition of rest (WESTHEIMER and BLAIR, 1973) which can be determined under a cycloplegic or after death. The interpretation of ELUL and MARCHIAFAVA'S (1964) results in terms of the state of focus at the effective image shell is complicated by the factors discussed previously in relation to the static refraction of the cat.

WESTHEIMER and BLAIR (1973) find the refractive state of the monkey to be myopic when the visual field is empty, but, curiously, during sleep and anesthesia this changes to more pronounced *myopia* rather than to hypermetropia as in the cat under the same conditions. In man, the physiologic accommodatory resting state is similarly about −1.0 to −1.5 D under empty field conditions (CAMPBELL, 1953; TOATES, 1972) and when the pupil is small (HENNESSY et al., 1976) the accommodatory system rests in focus close to the hyperfocal distance.

2. The Mechanism of Mammalian Accommodation

The foundation of modern understanding of the human accommodatory process lies in LANGENBECK'S (1849) use of the Purkinje images to demonstrate its origin in the change of shape of the lens. In HELMHOLTZ'S (1924) view, the contraction of the ciliary muscle relaxes the zonular fibres and reduces tension on the lens capsule so that its inherent elasticity acting upon the matrix induces a more spherical shape; the excised lens thus assumes its form of greatest power. FINCHAM'S (1937) observation that the *decapsulated* human lens takes up the unaccommodated form led WEALE (1962) to put forward the current theory in which the lens matrix is regarded as possessed of its own elasticity, overcome by the capsule during accommodation, which tends to return the lens to an unaccommodated form. The work of FISHER (1969a, 1969b, 1971, 1973) has outlined in most elegant fashion the respective roles of the capsule and matrix in defining the properties of the human lens.

The monkey lens has not been investigated by FISHER, but the range of accommodation of this animal, its endowment with ciliary muscle, and the lens shape indicate the presence of an accommodatory mechanism similar to that of man. Support for this view is available from VAKKUR (1967), who finds the monkey schematic eye to be some 19 D myopic when based upon the dimensions of the excised lens—a refractive state consistent with the excised lens having taken up its accommodated form.

Fortunately, FISHER (1969a, 1971) has also applied his techniques to the examination of the rabbit and cat lenses. The value of Young's modulus of polar elasticity for the adult cat lens is 21 times and that of the young rabbit 5 times that for the human lens (FISHER, 1971). This lack of compressibility in conjunction with the low capsular energy of the cat and rabbit lenses (FISHER, 1969b) makes an accommodatory change of lens shape in these animals impossible on mechanical grounds alone. The same conclusion almost certainly applies to the lens of the rat, which is even more spherical and contains a solid core. In contrast to the monkey schematic eye, those of the cat (VAKKUR and BISHOP, 1963), rabbit (HUGHES, 1972) and rat are all close to emmetropic when constructed from the dimensions of the excised lens, a result consistent with Fisher's observations.

The evidence for accommodation in the rabbit is poor, but the cat is undoubtedly capable of changing its refractive state in spite of its stiff lens. From their cat schematic eye, VAKKUR and BISHOP (1963) concluded that a change in lens shape and power would be a relatively inefficient means of altering the refractive state of the eye. The forward movement of the principal planes subsequent to such a change of shape would be more effective but is, of course, precluded by FISHER'S results (1971). Their suggestion that the lens may translate in toto during accommodation appears to be the true foundation of the process in the cat.

The forward movement of the lens anterior pole during parasympathetic stimulation was measured by O'NEILL and BRODKEY (1969) in the cat and found to amount to some 0.6 mm at most; in a later paper (O'NEILL and BRODKEY, 1970) this estimate is reduced without explanation to 0.3 mm. The larger of the two estimates would amount to some 2 D of accommodation (VAKKUR and BISHOP, 1963) if it entirely represented translation of the lens rather than change of shape. Apparently larger displacements of the lens were observed by ARMALY (1959) during parasympathetic stimulation. The bulging forward of the pupil during accommodation in the *conscious* cat has been reported by HUGHES (1973), and its magnitude is consistent with the known range of cat accommodation if achieved by translation. Support for VAKKUR and BISHOP'S hypothesis thus arises at present from the argument that change in the shape of the cat lens is mechanically impossible, rather than a direct demonstration that axial translation occurs. A recent investigation has now shown only very small changes in the power of the cat lens subject to circumferential forces thus confirming the above arguments (SUNDERLAND and O'NEILL, 1976).

KEPLER (1611) suggested that the human eye achieves accommodation by axial movements of the lens. Calculation from GULLSTRAND'S schematic eye readily reveals that a 2 mm forward movement of the lens changes the refractive state of the human eye by only 1.24 D, so that the contribution of this mechanism to the total accommodatory power could only be small. Although the front surface of the human lens is known to move forward by some 0.3 mm during accommodation (FINCHAM, 1937), this is usually regarded as a result of change of thickness and curvature. Recent slit-lamp results of PATNAIK (1967) suggest that in fact the core centre may also shift forward by some 0.17 mm under these conditions, although it is still not certain that this arises from equatorial translation rather than lens-matrix redistribution. The classic experiment of HENSEN and VÖLCKER (1868) demonstrated the forward movement of the ciliary body in cat, ape, and man during accommodation, but this translatory motion would appear to be more effectively conveyed to the lens in the cat than in the primate eye. It is consequently of great interest to note that the anatomy of the ciliary regions is very different in cat and man (Fig. 14) (KAHMANN, 1930; WALLS, 1942; ROHEN, 1952, 1964), but in the absence of a comparative mechanical analysis, no functional interpretation of the dissimilarities can be made. It is doubtless pertinent,

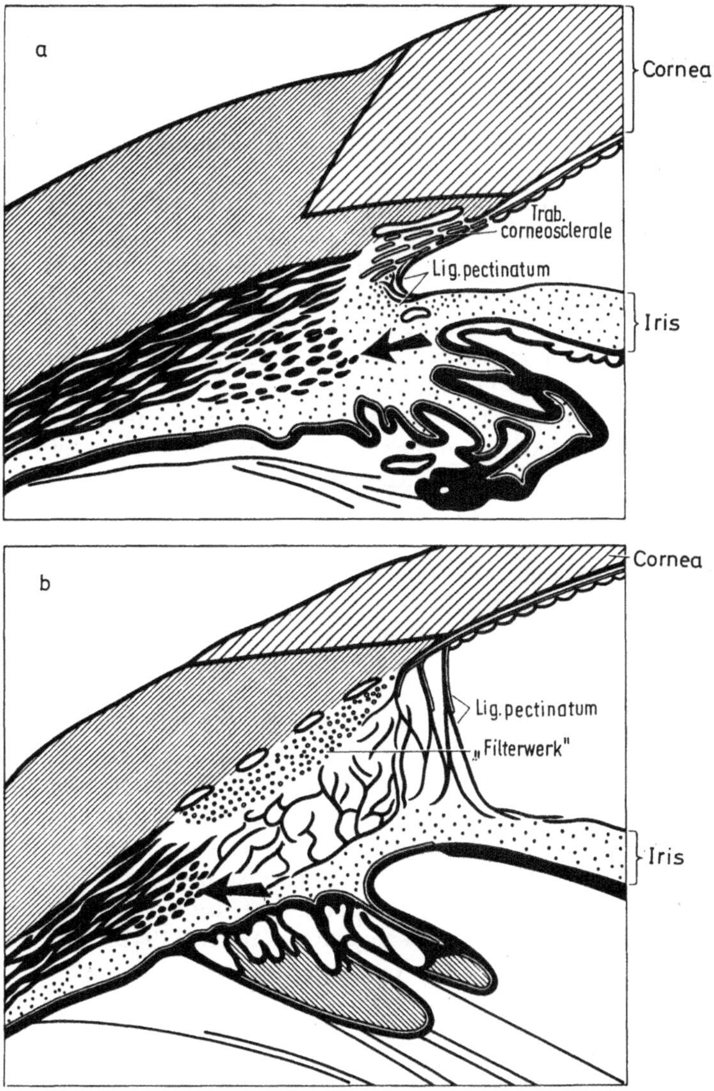

Fig. 14. Ciliary region of primate (*a*) and carnivore (*b*, cat). The transversely cut, circular muscle fibers are abundant in the former but weakly developed in the latter whose accommodation appears to be achieved by axial translation of the lens alone. (Courtesy ROHEN, 1964)

however, that the cat ciliary body is predominantly equipped with the meridional or longitudinal fibers of BRÜCKE (PRINCE et al., 1960; ROHEN, 1964), which are so appropriately oriented for the anteroposterior translation of the lens. In man it is the circular MÜLLER'S fibres which predominate; these are very sparse in the cat.

Unless the cat ciliary muscles deteriorate with age, it would appear that the difference between its accommodatory mechanism and that of man must make us doubt WALL'S (1942) claim that the cat

> Has but half the accommodation of a thirty year old man and loses even this in old age.

3. Epilogue

The elimination of corneal power when the fish eye is under water makes the position of the lens vertex especially easy to monitor and convenient for the study of accommodation under relatively natural conditions. In both elasmobranchs (NICOL, 1963) and teleosts (BREIT, 1957; PUMPHREY, 1961), accommodation appears to occur by displacement of the lens. The lens of the rock bass moves in a rostrocaudal direction during accommodation so as to alter the refraction of the frontal field (SIVAK and HOWLAND, 1973). Behavioral testing can elicit the entire 28 D range of accommodation which may be pharmacologically induced in this species (SIVAK, 1973). The alewife and silverside have been described as possessing up to 18 D of accommodation (BAYLOR and SHAW, 1962)—a range again very substantial in comparison with that of most mammals. CHARMAN and TUCKER (1973) have argued, however, that the goldfish and some other teleosts may not require an accommodatory mechanism because of the depth of focus of the eye.

Little reliable evidence of accommodation is available for birds, but it is claimed to be well developed among the diving species in which it is employed to offset the loss of corneal power when submerged. The dipper is described as possessing a cornea of some 40 D power and 48 D of accommodation (GOODGE, 1960); the robin has only 7.5 D of accommodation, but this is respectable in comparison with the range in mammals.

Among the fishes (TAMURA, 1957; SIVAK, 1973) and birds, it has been suggested that the magnitude and axis of accommodation vary according to feeding habits and life style, but our knowledge of the accommodatory powers of mammals is so limited that little can be said of such correlations. The otter is reputed to possess the greatest accommodatory power of the mammalian subprimates for the same purpose as the diving birds (WALLS, 1942; ROCHON-DUVIGNEAUD, 1943; DUKE-ELDER, 1958), so that it is able to maintain a similar acuity in air and water (SCHUSTERMAN and BARRETT, 1973). By contrast, the more aquatic sea lion eye is emmetropic in water and depends upon a flattened cornea and stenopaeic aperture for the formation of a good image in air (WALLS, 1942; SCHUSTERMAN, 1972). The range of accommodation among the seriously investigated mammals appears to follow the extent of their praxic skill. It is greatest in monkey, less evident in the paw-dominated cat, feeble in the dog, and apparently absent from the nonmanipulative rabbit.

WALLS (1942) lists several mechanisms whereby image quality can be made relatively independent of object distance in species lacking a dynamic accommodatory system. Of these, doubt has already been cast in previous sections on the ramp retina and the usefulness of elongated receptors. The stenopoeic aperture is another matter and is a definite asset under photopic conditions. The rat possesses no discernible ciliary muscles (LASHLEY, 1932; WOOLF, 1956) but is yet concerned with viewing at short distances and handles objects. The absence of accommodation is only behaviorally significant if the retinal image quality for the range of distance significant to that species is substantially worse than the *neural*, not the photoreceptor, "grain" employed in the specific acuity task considered. Under photopic conditions, the tiny pupil of the rat ensures such small blur circles that this does not occur, and its eye possesses an enormous depth of field

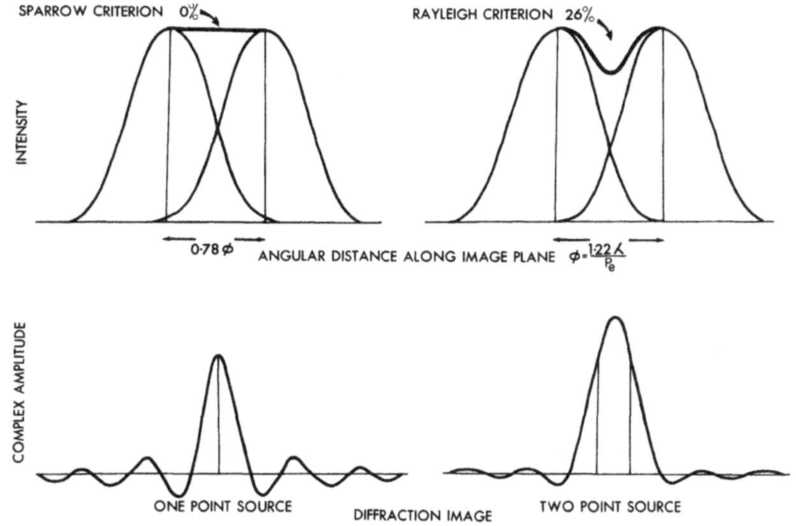

Fig. 15. Summed intensity distribution for the PSF of two points separated by distances appropriate to the SPARROW criterion and the RAYLEIGH criterion. It is to be noted that shape of PSF for one point remains quite different from that for two, even when they are closer together than suggested by the SPARROW criterion, and the central trough is absent

(HUGHES, 1977b). A curious but significant outcome of the dynamic accommodatory process in man (MOSES, 1970; ENOCH, 1975) and cat (VAN ALPHEN, 1961) has recently been demonstrated to be stretching of the retina with consequent change in the spatial direction associated with a given retinal locus.

V. Image Quality

The contents of the object space within the field of view of an optic system may be regarded as if formed by arrays of point sources of given luminance; their conjugate image may correspondingly be regarded as compounded from the elementary images of these point sources. Even in an ideal optic system, however, the image of a point source is not stigmatic because the inevitable presence of a finite aperture ensures a diffraction pattern consisting of a central Airy's disc concentric with a series of light and dark rings. The intensity distribution across the central disc is Gaussian and represents the pointspread function (PSF) of the exit pupil of the optic system (see WESTHEIMER, 1972, 1973, and FRY, 1970, for theoretical treatment). The dimensions of the PSF thus limit the amount of object-space detail which may be represented in the image space; the images of two points separated by less than 0.78 of the PSF radius, show only a single intensity peak (SPARROW, 1916) (Fig. 15). The angular diameter of the first zero of the PSF, as subtended from the exit pupil, is given in radians by $2.44\lambda/p_e$, where λ is the wavelength of the light when passing through the optic media in nm and p_e is the exit pupil diameter in mm. In real optic systems, however, the pointspread function width varies inversely as the pupil diameter only over limited range; above this it increases in size with the pupil. Focus error, aberrations arising with the larger pupils, and light scatter in the optic media compound with the effects of diffraction to generate a pointspread function with a Gaussian core, which closely matches the ideal diffraction image in man when the pupil is small (CAMPBELL and GUBISCH, 1966), and an exponential skirt which represents the influence of all other factors. The pointspread function is thus a useful end product by means of which the overall quality of an optic system may be judged in terms of object-space resolution under given conditions.

Just like any other two-dimensional signal, the PSF may be represented as compounded from sinusoidal Fourier components at different orientations. The finite width of the PSF ensures that, unlike the Dirac delta function employed in systems analysis, its Fourier spectrum covers only a finite spatial-frequency range. Widening of the PSF along a given diameter thus corresponds to the failure of the optic system to transmit high spatial frequencies of the corresponding orientation; the Fourier transform of the PSF specifies the bandpass range, or modulation transfer function (MTF), of the system. This function displays the ratio of contrast in the image to that in the object for the spatial frequencies below the cut-off frequency of the system. It is a convenient approximation related to the Rayleigh criterion to assume the half-height width of the PSF as equal to one period of the Fourier component at the cut-off frequency.

The PSF and MTF are thus inverse transforms and may be obtained from one another by calculation if the PSF is rotationally invariant; otherwise the MTF is a complex function, the phase factors of which must be taken into account if gross errors in the PSF and subsequently calculated image distributions are to be avoided (RÖHLER and HILZ, 1965). Only for grating images can rotational invariance be safely ignored in such calculations.

The image quality of the human eye was first adequately established by FLAMANT (1955), who projected a line of light onto the retina and measured the intensity distribution across its aerial image formed in front of the eye by rays reflected from the fundus. The measurement of the linespread function (LSF) is technically easier than that of the PSF because of the greater amount of light available. The square root of the double-pass LSF transform is required to specify the system performance as its single passage MTF. The aerial image technique has been technically improved by various workers (PSF determination, RÖHLER, 1959, 1962; LSF, WESTHEIMER and CAMPBELL, 1962; CAMPBELL and GREEN, 1965) to culminate in the work of CAMPBELL and GUBISCH (1966), who regard the 1′ half-height width of the LSF core as determined by the diffraction image width alone at the optimal 2.4 mm pupil diameter in man. The MTF has been directly established for man by projection of grating patterns onto the retina and measurement of their contrast loss (WESTHEIMER, 1963; RÖHLER et al., 1969) with results similar to those obtained by calculation from the LSF.

The aerial-image, double-pass technique was first employed for the cat eye by MARRIOT and MORRIS (1961) and WESTHEIMER (1962), but the pupils were dilated and a relatively poor image reported. The MTF was assessed from photographs of the retinal image of projected grating patterns by WÄSSLE (1971), and the calculated PSF found to be of 4′–5′ half-height width for a 3 mm pupil; the image

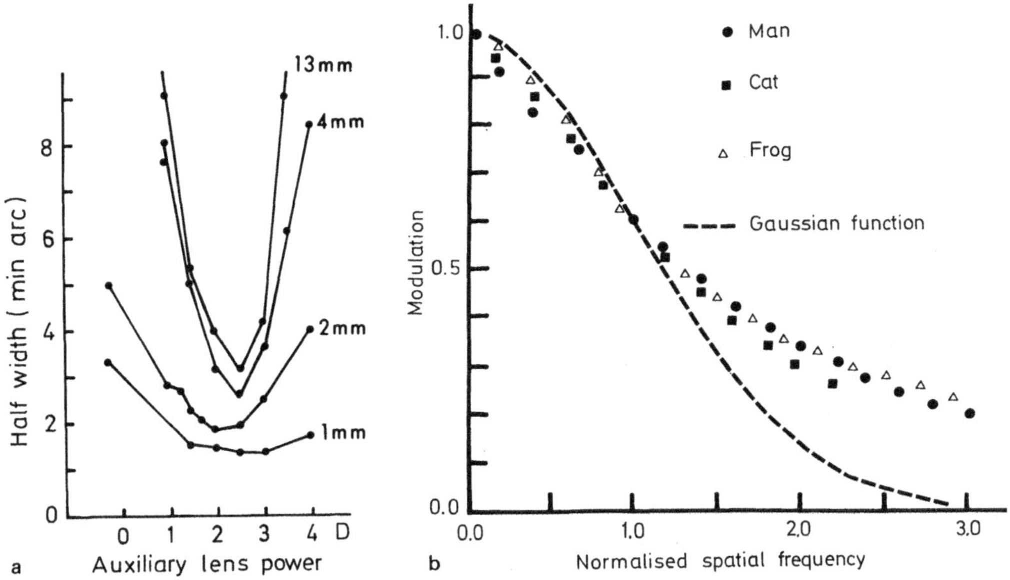

Fig. 16(a). Cat PSF half-height half-width for various degrees of defocusing and pupil size. (After ENROTH-CUGELL and ROBSON, 1975). (b) The normalized modulation transfer functions of man, cat, and frog are very similar in form but quite different from a Gaussian distribution. (After KRUEGER and MOSER, 1973)

quality of the cat eye is thus substantially poorer than that of man. Subsequent assessment of the cat LSF (BONDS et al., 1972; BONDS, 1974) confirm the findings of WÄSSLE (1971).

The results of these double-pass methods have been criticized by IKEDA and WRIGHT (1973) and ENROTH-CUGELL and ROBSON (1974) as leading to a spurious underestimation of resolving power. IKEDA and WRIGHT (1972) consequently substituted fluorescein angiography to make the smallest vessels of the retina a source from which to photographically obtain the single-pass LSF. For a 1.5 mm pupil, they observed a width at half amplitude of 4′—no better than WÄSSLE'S double-pass result, yet obtained with a smaller pupil. IKEDA and WRIGHT (1972) claim, however, that the result is an improvement on previous double-pass findings, because they compare the half-width at half-height with previous values for the half-height width. WÄSSLE'S (1971) results were again confirmed by the single-pass technique of ENROTH-CUGELL and ROBSON (1974), who employed a light guide which extends just beyond the retina to deliver light falling on the probe tip to a photocell when the image of a line source is scanned across it. The half-height width for a 3 mm pupil was 4.3′. Their results for a range of defocusing and pupil sizes are shown in Fig. 16A, where the 1 mm pupil is accompanied by an LSF half-height width of 3′, the narrowest so far demonstrated for the cat.

Although the half-height width of the cat LSF is similarly described by various observers for equivalent conditions, the extent of the tail of the distribution has been subject to discussion. The LSF of WÄSSLE (1974) falls to zero at about 7′ radius, whereas BONDS (1974) describes a wider "skirt"; the agreement between the results is very good when the animals with the best optics are compared, in that the LSF of BONDS (1974) is only some 3% of its peak amplitude at 7′ radius. That an extensive skirt does exist around the PSF and LSF has been demonstrated by means of an intraocular fiber optic probe by ROBSON and ENROTH-CUGELL (1975), who suggest that 20% of light entering the retina is scattered more than 15′ from its geometrically determined position. None of these studies take into account the standing zero frequency illumination which may result from the eye operating like an integrating sphere—possibly an important consideration when a tapetum is present as in the cat.

Both *single*-pass methods mentioned above establish the MTF of the optic system without incorporating the effects of transmission through the retina. In one instance, the retinal surface vessels act as a source, and in the other the probe extends beyond the retina. The problems of specifying the plane of origin of the fundus reflex, especially in the cat eye, make it difficult to assess the extent to which the retinal MTF is involved in the *double*-pass methods. In man, as we have seen, there have been substantial arguments that the fundal reflex arises behind the receptors (WEALE, 1966), so that the double-pass method may give rise to spuriously large values of LSF width because of scatter in the post-receptor layers. In the cat, however, the similarity of the single- and double-pass LSF half-height widths at equivalent pupil sizes suggests that such scatter may be of significance to resolution only in so far as it possibly broadens the PSF "tail." The same argument indicates that the retinal MTF plays little part in the determination of the overall optic quality of the living cat eye. This conclusion is supported by direct measurements of the retinal MTF of human (OHZU and ENOCH, 1972), squirrel monkey, and rat (OHZU et al., 1972). The human foveal MTF is superior to that for the axial optic system as determined by CAMPBELL and GREEN (1965) in spite of the use of material which had been deprived of a blood supply for 10 min and thus probably did not represent the optimal in vivo performance. The MTF of the squirrel monkey and rat retina was not grossly different from that of man and thus much superior to that of the optics in the latter species. The rat retinal MTF undergoes a precipitous decline in high spatial-frequency transmission between 2 and 11 min after the cessation of the blood supply (OHZU et al., 1972); the increased scatter appears to arise in the neurones of the retinal layers overlying the photoreceptors.

Of the remaining mammals a detailed MTF, measured shortly after death, is available only for the enucleated eyes of cow and pig (RÖHLER, 1962). The frog PSF is some 11′ wide at half amplitude (KRUEGER and MOSER, 1971, 1972) at an unknown pupil diameter, and the eagle LSF is about 1′ wide for a 6 mm pupil (SHLAER, 1972), which suggests that its optics are superior to those of man in whom the same LSF width is obtained only at smaller pupil diameters. This suggestion is substantiated by the observation (FOX et al., 1976) that at high luminances the falcon resolves a 160 cycle/° grating, whereas the human limit is 60 cycles/°. Such a performance by the bird requires an LSF width of some 22″ which is quite outside the human range. The shape of the frog, cat, and human MTF is very similar (KRUEGER and MOSER, 1973) but not fitted by a Gaussian curve (Fig. 16B). The similarity of the MTF of different species was assumed by HUGHES and WÄSSLE (1977) in order to obtain the rat PSF width at half ampli-

Table 3

	Man	Cat	Rat
PSF. half-height width for 3 mm pupil	1′	4–5′	30′
Width on retina	0.005 mm	0.018 mm	0.03 mm
Optimal PSF. half-height width for species	1′	3′	13′
Width of pupil	2.4 mm	1.0 mm	0.3 mm
Approx. ambient illumination for this pupil size (Fig. 17)	2.7 lg(cd/m^2)	2.5 lg(cd/m^2)	2 lg(cd/m^2)

Sources of data: man, CAMPBELL and GUBISCH (1966), WOODHOUSE (1975); cat, WÄSSLE (1971); optimal PSF. is smallest value from ENROTH-CUGELL and ROBSON (1974); rat, smallest size assumed optimal, HUGHES and WÄSSLE (1977).

tude from the cut-off frequency observed for a grating image projected onto the retina. The half-height width for a 0.33 mm pupil under photopic conditions was some 13′, which is about double the calculated diffraction limited value.

The power of the Fourier method for the overall description of optic systems (WARD and DAVIS, 1972) has not yet been extensively applied to comparative investigations. The systematic investigation of the influence of pupil size and defocus has begun for man (CAMPBELL and GREEN, 1965), cat (ROBSON and ENROTH-CUGELL, 1974; BONDS, 1974), and rat (HUGHES and WÄSSLE, 1977). The influence of oblique incidence has barely been touched upon; the results of RÖHLER (1962) enable estimation of the half-height width of the LSF for eccentricities up to 50° in man, and BONDS (1974) has presented LSF measurements ranging 30° either side of the visual axis in the cat. Both sets of results indicate a substantial peripheral deterioration in image quality; in the cat this finding is challenged by WÄSSLE (personal communication) from results obtained by the grating-projection technique. The interpretation of the peripheral observations on image quality is at present difficult, because the pointspread function is almost certainly not rotationally symmetrical for oblique rays, and the MTF should take on complex components (RÖHLER and HILZ, 1965). In addition, the studies of cats have been made with contact lenses in position, and no examination of their influence on oblique rays has been reported.

The variety of factors which influence the PSF and MTF make essential the standardisation of at least pupil diameter and focus when comparing the optic image quality of different species. If the resolving power of the optic apparatus is to be compared with behavioral acuity for a range of species, it is convenient to have the PSF specified in terms of its angular extent in the object space common to them all. From Table 3 it is apparent that between man and rat there is a 30-fold range of optic resolution, or PSF half-height width, when the pupil is 3 mm in diameter which alone indicates the difference in quality of the optical systems. Under these conditions, however, the differing PNDs' of the eyes ensure that they would not be operating at the same F_{no}. Discussions involving relative acuity might more suitably be served by comparison of the PSFs' when the pupil diameters are in the ratio of the PNDs' of the eyes concerned. In this instance, the rat with a 0.5 mm diameter pupil is to be compared to a man with a 3.0 mm pupil when their PSFs' differ only 13-fold in half-height-width. There is little evidence,

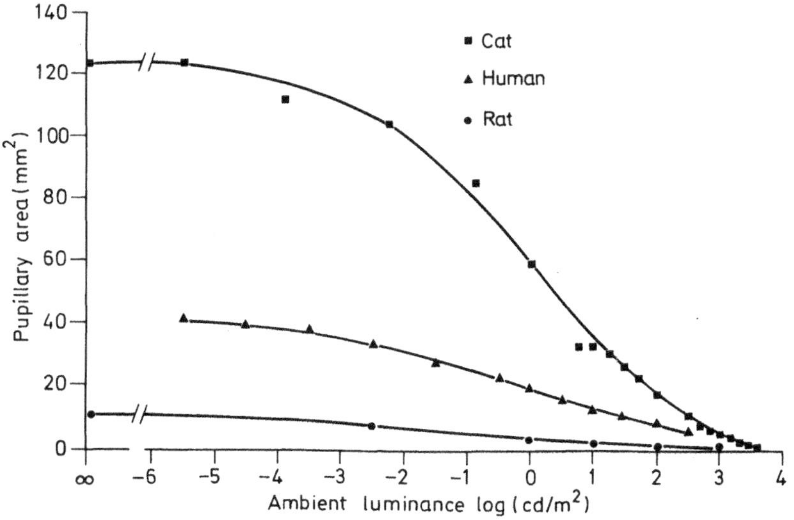

Fig. 17. Pupillary area as a function of light intensity for man, cat, and rat. Curves follow a very similar course with changing intensity. (After DE GROOT and GEBHARD, 1952; WILCOX and BARLOW, 1975; HUGHES, 1977c)

however, that different species operate at the same F_{no} over a range of conditions of illumination.

The opposing effects on the PSF half-height width of diffraction and aberration ensure that in man its value is minimal for a 2.4-mm-diameter pupil (CAMPBELL and GUBISCH, 1966). A similar minimum PSF width for a pupil diameter between 2 and 3 mm has been claimed for the cat (BONDS, 1974), but the results of ENROTH-CUGELL and ROBSON (1974) indicate a monotonic decrease down to 3′ for a 1-mm pupil; the smallest rat pupil tested, 0.33 mm, gives a minimum 13′ wide PSF at half-height (HUGHES and WÄSSLE, 1977). Resolution is thus optimal at quite different pupil diameters in these three species.

Under natural conditions, the pupil diameter is set by the pupillary light reflex, and an alternative means of comparing the optic performance of different species is to consider image quality at a given light intensity in a common environment. The optimal pupil sizes of man, cat, and rat are then revealed to be attained at similar photopic luminances characteristic of an external daytime environment of about 1000 cd/m^2 (man, DE GROOT and GEBHARD, 1952; cat, WILCOX and BARLOW, 1975; rat, HUGHES, 1977c).

The relationship between pupillary diameter and light intensity is of similar form for the investigated mammals (Fig. 17), but the pupillary area of cat and rat can change over a 135-fold range (BARLOW and WILCOX, 1975; HUGHES, 1977c), in contrast to the 10-fold range of man. At diameters other than that optimal for geometric optical resolution, it does not follow that the *performance of the system* is not optimal; quantum limitations on light flux may prevent attainment of the apparent geometric resolution unless the pupil size increases. The optimal diameter of the pupil in terms of physically achievable resolution thus changes with light intensity, even when the limitations of the neural apparatus are not taken into account. The introduction of an artificial pupil larger or smaller than the *natural* pupil of man reduces resolution (CAMPBELL and GREGORY, 1960; WOODHOUSE, 1975) to some extent at all diameters. WOODHOUSE and CAMPBELL (1975) conclude from the small magnitude of this effect that the primary role of the pupil is not simply to establish optimal resolution.

The resolving power of the cat and rat is sufficiently low for their pupils to be able to contract to a much smaller size than in man before diffraction limits behavioral acuity. For such arrhythmic nocturnal species, this range permits rapid adaptation and greater contrast sensitivity in the early stages of dark adaptation when passing from light to dark. As WOODHOUSE and CAMPBELL (1975) point out, WALLS (1942) suggested that

> the slit pupil is thence in a sense paradoxical, for though it is an adaptation to nocturnality it has nothing whatever to do with seeing in a dim light

but now the paradox is resolved. The more limited pupillary diameter range of man confers similar advantages which have been argued to be of greater functional importance than optimisation of resolution (WOODHOUSE and CAMPBELL, 1975).

From a systems point of view, it thus appears that comparison of the image quality in different species requires full knowledge of its behaviorally encountered light intensity range under normal conditons. The striking influence of the ambient light intensity on behaviour patterns in forest and bush has been outlined for the nocturnal Malagasy *Lepilemur* (PARIENTE, 1974) in a model study representative of what is required for other species. It is obviously possible that the peripheral aberrations which come into play as the pupil dilates may be so adapted to the habitat and life style of each of a given group of species as to result in ecologically correlated variations of rank order of their image quality when the ambient light intensity is changed.

D. Retinal Topography

I. The Photoreceptor Layer

The discriminatory capacity of an animal at a given eccentricity from its fixation axis is potentially limited as much by the neural apparatus of the corresponding retinal region and subsequent stages of the central visual system as by the quality of the local retinal image. MÜLLER'S (1853) demonstration that the photoreceptors are the first stage of neural processing and potentially set the ultimate limitations on its performance was not long followed by SCHULTZE'S (1866) extensive comparative study of the retina of nocturnal and diurnal species, which established the duplex nature of the receptor layer—the rods serving scotopic and the cones photopic vision. The distribution of rods and cones within one retina may follow topographically similar or quite different patterns, but, so far, only a few mammalian species have been quantitatively examined.

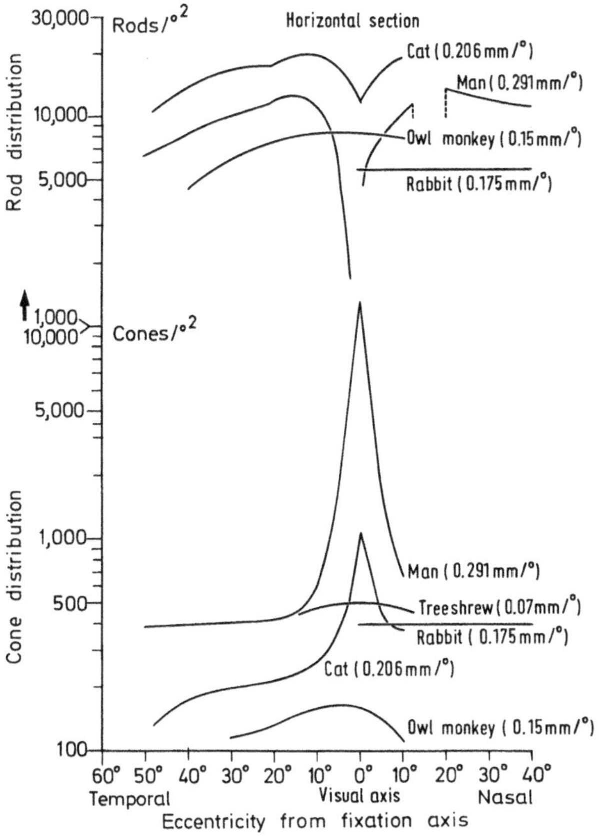

Fig. 18. Rod and cone densities/°² as measured in the horizontal plane about the visual axis of various mammalian species. Values were converted from published densities/mm² by means of the retinal magnification factors shown. Sources: man, ÖSTERBERG, 1935; tree shrew, ROHEN and CASTENHOLZ; rabbit, HUGHES, 1971; cat, STEINBERG et al., 1973; monkey, OGDEN, 1975

The maps of human receptor distribution provided by ÖSTERBERG (1935) clearly demonstrate the anatomic basis of HERSCHEL'S statement (BREWSTER, 1830) that

> a faint star ... will yet totally disappear, as if suddenly blotted out, when the eye is turned full upon it...

in the absence of scotopic receptors from the central fovea (Fig. 18). An increased cone density and correlated reduction in rod density occurs at the cat area centralis, as in the fovea of man, but the rod population is much less reduced as befits this nocturnal species (Fig. 18); such infiltration of the cones in the area is without detriment in the cat because of its lower photopic acuity than man. In the nocturnal owl monkey (Fig. 18) (OGDEN, 1975), not only are the rod and cone populations intermingled near to the fixation axis, but their distributions follow a similar course, and the relative proportions remain constant; absolute densities, however, differ considerably between the two populations. When a visual streak is present in the retinal ganglion-cell layer, then the underlying photoreceptor distribution also shows increased density in a band-shaped region as in rabbit (Fig. 18) (HUGHES, 1971), cat (HUGHES, 1978), and turtle (BROWN, 1969).

Such duplex retinas form the majority of those encountered among mammalian species. Pure rod or cone retinas have been reported, but it is currently debatable whether such a description is strictly true for any mammal. The rat has been alleged to possess only rods (LASHLEY, 1932; MESSING, 1972), but several investigators identify cones (WALLS, 1934; SIDMAN, 1958; SOKOL, 1970) in the functionally useful concentration of 1:100 rods (HUGHES, 1977c). The tarsier retina is said to be of the pure-rod type (POLYAK, 1957; ROHEN and CASTENHOLTZ, 1967) but so small a proportion as an easily overlooked single cone per 1000 rods (KOLMER, 1930), a ratio found in the rod dominated retina of the lemuriform *Cheirogaleus* (ROHEN and CASTENHOLTZ, 1967), may be functionally important, because it can make a substantial contribution to diurnal acuity. The exclusively cone populations reported for the tree shrew (ROHEN and CASTENHOLTZ, 1967; WOOLLARD, 1926) and ground squirrel (DOWLING, 1964) have also been questioned; rods have been alleged in the former species upon the basis of light microscopy (POLYAK, 1957), and peripheral rodlike structures have been described in electronmicroscopy studies (SAMORAJSKI et al., 1966). The limited summation available from these reputedly small rod populations would bring their functional value into question, except that recent electronmicroscopy enables the differentiation of substantial numbers of the receptor population in some allegedly pure-cone retinas as a new rodlike class which might commonly be confused with cones under light microscopy (WEST and DOWLING, 1975). Such receptors form 40%, 10%, and 5% of the grey squirrel, prairie dog, and ground squirrel retinas, respectively, and show little centroperipheral change in proportion. The ERG b-wave threshold differs among these species in accordance with the "rodlike receptor" proportions. Reexamination of other rod- or cone-dominated retinas with the new methods appears to be in order.

The physiologic properties of the great majority of ganglion cells indicate that they receive both rod and cone inputs when they overlay the two receptor types in a given retinal region; cat (BARLOW et al., 1957a, 1957b; DAW and PEARLMAN, 1969; ANDREWS and HAMMOND, 1970a, 1970b), monkey (GOURAS and LINK 1966). Apart from the mesopic state (ANDREWS and HAMMOND, 1970a), a ganglion cell responds to just one receptor type, to rods under scotopic and cones under photopic conditions (GOURAS and LINK, 1966), with its receptive field size changing little. A small number of pure-rod receptive fields have been described in cat retina (ANDREWS and HAMMOND, 1970a), and pure-cone midget fields may be recorded from the perifoveal region of the macaque (GOURAS, 1971), but in general the double photoreceptor distribution maps into a single retinal ganglion-cell distribution. The topographic implications of this arrangement in terms of differences in photopic and scotopic resolving power, light sensitivity, etc., across the retina are dependent upon the unknown quantitative features of connectivity between the photoreceptors and ganglion cells.

It is often stated that, other than near a foveal pit, the retinal ganglion cell and underlying cone distributions follow each other closely. This is certainly true in the commonly considered duplex retinas which possess a diurnal cone dominated area—man (OPPEL, 1965), macaque (ROLLS and COWEY, 1970), and cat (STEINBERG, 1973); in such species the rod distribution is inversely correlated with the ganglion-cell density, and the total receptor density is relatively uniform. Similarly, the retinal ganglion cell and receptor distribution across the supposedly pure-cone retina of the tree shrew parallel each other (ROHEN and CASTENHOLTZ, 1967). However, the area of the predominantly nocturnal duplex retina of *Aotes*, the owl monkey (OGDEN, 1975; ROHEN and CASTENHOLTZ, 1967), is accompanied by little change in rod, cone, or total receptor density, and yet the ganglion-cell distribution independently reveals a marked peak density (JONES, 1965); even in the supposedly pure-rod retina of the tarsier the retinal ganglion-cell density is substantially elevated at the fovea (ROHEN and CASTENHOLTZ, 1967) in spite of only a small receptor density increase. The alleged relationship between cone and ganglion-cell distribution is thus perhaps true for diurnal forms but is quite without general applicability.

Concentration on human foveal vision has led to an excessive emphasis in the literature on the limitations set by the photoreceptor matrix for visual resolution. The work of ÖSTERBERG (1935) has shown for the whole retina of man that 120 million rods and 6.5 million cones project to 1.2 million optic nerve fibers (POTTS et al., 1972), giving ratios of 95:5.4:1 for rod:cone:optic nerve fibers, respectively. Information available for the cat retina (STEINBERG et al., 1973; HUGHES, 1975a; HUGHES and WÄSSLE, 1976) indicates even greater overall convergence, with 204 million rods and 3 million cones projecting to 193000 optic nerve fibers and thus leading to 1000:15:1 for the rod:cone:optic nerve fiber ratio. Obviously gross overall ratios may mask local regions of lesser convergence, such as the 1:1 coupling of cones to ganglion cells at the primate fovea (BOYCOTT and DOWLING, 1969), but in general the potential upper limit set on resolution by the photoreceptor matrix cannot be approached, because there are usually not enough ganglion cells in any local region to relay the output of the corresponding receptor cells by "private" channels.

Several workers have sought quantitative details of local variation in receptor-to-ganglion cell ratios by means of gross local measurements of the thickness of the three nuclear layers of the retina (CHIEVITZ, 1889; ZURN, 1902; ROHEN and CASTENHOLTZ, 1967). Counts of the relative nuclear densities in central and peripheral retina have also been made for man, simians, prosimians, guinea pig, ground squirrel and rabbit (VILTER, 1949, 1954; OPPEL, 1965; ROHEN and CASTENHOLTZ, 1967; HUGHES, 1971). These ratios may be misleading if not corrected for thin-section sampling errors (ABERCROMBIE, 1946) but do indicate different classes of organisation of the central and peripheral retina which correlate with the balance of nocturnal and diurnal activity in the life style of given species. Such methods do not, however, permit the actual connectivity of the cells to be deduced, so that it would be possible for each ganglion cell to be connected to all overlying receptors in spite of a low convergence ratio.

Recent investigations of monkey (DOWLING and BOYCOTT, 1966; BOYCOTT and DOWLING, 1969; KOLB, 1970), and cat (BOYCOTT and KOLB, 1973) outer plexiform layers have begun the provide specific information about receptor-to-bipolar connectivity. Some 15–45 rods supply each rod bipolar in monkey, and a given rod may supply more than one such bipolar, whereas over much of the retina a cone may supply a single midget invaginating bipolar, possibly a midget flat bipolar, and a flat bipolar which receives an input from about 7 other cones. The cat invaginating-bipolar contacts no fewer than 4 cones and up to 10, the flat bipolar from 8 to 14 cones, so that the potential for a fine cone-sampling-matrix appears to have been lost before the inner plexiform layer. The area of the rod and cone bipolar fields, the ratio and numbers in which the various bipolar classes project to the ganglion cells, and the variety of ganglion-cell classes at different eccentricities remain unknown until the details of the inner plexiform layer connections are worked out. Until then, our knowledge of relative scotopic and photopic resolving powers in any given region of retina must be arrived at from other considerations than the photoreceptor distribution.

Although receptor-to-ganglion cell convergence prevents the achievement of the receptor matrix potential for the accurate location of stimuli in visual space, it does not follow that the matrix resolution is totally lost. The discrete elements of the photoreceptor matrix may be employed with little convergence in the generation of functional subunits which precede the ganglion cells. Thus the rabbit has a behavioral acuity amounting to some 20′ at best, yet its directional units show differential response to stimuli separated by only 6′ (BARLOW and LEVICK, 1964), and the eye can track OKN stimuli moving at a rate of less than 1.8′/s (COLLEWEIJN, 1971; OYSTER et al., 1972). The technique of assessing visual acuity by means of the optokinetic response (CATFORD and OLIVER, 1973) thus gives values similar to those obtained by standard methods for central vision in man when convergence from receptor to ganglion cell is 1:1. In other species the acuity so determined is spuriously high compared with that determined behaviorally.

II. Retinal Organisation

1. The Area Retinae

That human oblique vision is less acute than central vision was well known by the time of KEPLER (1604); he attributed the difference to the combined effects of variation in retinal constitution and poor peripheral optics. The significance of the specialized region of retina bearing yellow pigment, the *macula lutea*, and the thinned retina at its center had not, however, been correctly interpreted as late as 1830 (BREWSTER, 1830), in spite of its location on the visual axis. Indeed, at that time, the region was thought to contain an aperture, the *foramen centrale*, which gave rise to the blind spot. Not until MÜLLER (1856) was the organisation of the *fovea centralis*, its significance in central vision and the possibilities for variation of organisation across the retina and between species fully understood. Subsequent investigations emphasized fine foveal receptors, rods in some species but usually cones, clearing of the bipolar and ganglion cells from the foveal pit and the absence of overlying blood vessels as specializations for high resolution vision in a wide variety of vertebrates (POLYAK, 1957).

The absence of a foveal pit from the localized concentration of specialized nerve cells and receptors encountered at the intersection of the fixation axis with the retinas of investigated mammalian subsimians did not prevent MÜLLER (1861) from recognising their functional equivalence to the foveal elements as a means of improving local resolution; he described such regions as *central areas*. Among the mammals and most other vertebrates with nonfrontal eyes, such areas are not central but situated in temporal retina so as to form the visual pole of the forward fixation axis. For these regions, therefore, it is best to employ the term *area retinae temporalis*.

The macroscopic examination of the eye cup and microscopic study of radial sections from the retinas of a wide variety of species (CHIEVITZ, 1889, 1891; SLONAKER, 1897; ZURN, 1902; WOOD, 1917; BRÜCKNER, 1961a, 1961b) has since revealed two common forms of retinal ganglion-cell distribution which appear to be specialised for increased resolution, the usually round *area retinae* of MÜLLER and a "bandlike area" or *visual streak*, which extends horizontally across the retina. Similar classes of sensitive area were described upon the basis of ophthalmoscopic appearance of the fundus blood vessels in mammals (JOHNSON, 1901).

In the subsequent discussion, an *area retinae* or visual streak is assumed to be established by the presence of a locally increased ganglion-cell density; increased retinal thickness or photoreceptor density, the absence of overlying major blood vessels, or reduced convergence from receptors to ganglion cell may indicate or substantiate the presence of an area but do not define it. So far no mammal has been found to lack both an area and a visual streak.

This approach conflicts to some extent with that of ROHEN and CASTENHOLTZ (1967), who employ the extent of the change in primate local receptor to ganglion-cell convergence ratio in passing into the *area retinae* from the periphery to define a degree of "centralisation" (ROHEN, 1958). The foveate arrhythmic form, homo, or nocturnal species, tarsius, show obviously increased convergence in passing towards the fovea with ratios of 16:2:1 and 40:8:1, respectively (peripheral receptor: central receptor: ganglion cell) just as do the afoveate arrhythmic forms such as galago, 30:10:1. Among these species the centralization is even more obvious when the receptor data is divided into rod and cone components, and all are regarded as possessed of differentiated area retinae. Centralization in the tree shrew cone retina, 1.6:0.8:1, and cheirogaleus rod retina, 60:30:1, is less obvious and described as „*ohne differenzierte Netzhautmitte*" (ROHEN and CASTENHOLTZ, 1967), yet both cheirogaleus (WOLIN and MASSOPUST, 1967) and tupaia (Fig. 20c) possess area retinae which are clearly differentiated by an increase in ganglion-cell density. A parallel increase in receptor density masks the absolute increase in ganglion-cell density at the area when differentiation is judged only in terms of the local convergence ratio. The absence of centralization cannot therefore be equated with the absence of an area.

In some species, such as guinea pig and squirrel, the retinal ganglion-cell distribution is relatively uniform; depending upon its density, or preferably its angular resolution, such retinas are regarded as "universally peripheral" or "universally macular." It may be wondered why such organisation is not more common; why should an area centralis or visual streak be differentiated in other retinae? It will be appreciated, however, that all regions of the visual field are not necessarily of equal significance to a given species; a need for economy in usage of neural apparatus would therefore lead to differential organization of the retina.

Retinal uniformity is most conspicuously lacking in those species which manifest high visual acuity and whose area retinae set a disproportionately high demand on central tissues for the processing of their output. The sharply circumscribed extent of such retinal regions may thus be explained in terms of the incapacity of the CNS to make central provision for the output of an entire uniform high-resolution retina with mechanisms equivalent to those it can provide for a circumscribed area centralis. If the ganglion-cell density of the entire cat retina were the same as that at the peak of its present area centralis and cortical provision were to be proportionately increased, then the feline area 17 would need to be increased from 340 mm^2 in area to 50000 mm^2—5 times larger than the visual cortex of man! Given that a gradient of density is necessary in some species, then many different distributions are possible, depending upon the means by which sampling redundancy is minimised; some of these are discussed subsequently.

It does not follow, however, that the total sampling distribution across the retina is determined simply by considerations of resolution; the variety of receptive field classes present in the retina indicate that separate sampling distributions exist for other classes of information. However, the theoretical apparatus for treating such distributions is not available, and thus we confine the discussion to consideration of a system for spatial resolution such as is represented by the cat brisk-sustained class. Even in such a system it is necessary to consider sensitivity as well as spatial resolution when arriving at conditions for optimal sampling. Limitations on the independence of light sensitivity and resolution in an optimally organized system are set by the sampling theorem. Treatment of resolution in terms of the density of image sampling points neglects the finite area of the sampling apertures.

Consider an eye in which diffraction and aberrations limit the maximum spatial frequency at some local region of the image to a cut-off value of f cy/°. A system which optimally samples this region of the image is required by SHANNON's sampling theorem (SHANNON and WEAVER, 1949) to have a minimum angular sampling interval $\Delta\phi = 1/2f$ if its matrix is square, although a hexagonal matrix, the Quincuncial array for packing orchard trees (Sir THOMAS BROWNE, 1912), is more economical (PETERSON and MIDDLETON, 1962; SNYDER and MILLER, 1977b). If finite sampling apertures are employed rather than points then light capture and sensitivity are increased without detriment to the cut-off frequency up to a limiting angular diameter of 1/f if photon noise is ignored. It follows, in such a system whose aperture area $\pi(1/2f)^2$ is associated with a sampling density $(2f)^2$, that coverage, the average number of apertures—or receptive field centres in the case of ganglion cells—which overlap at a retinal point (CLELAND et al., 1975), should be uniform across the retina at the value of π given by $\pi(1/2f)^2 \cdot (2f)^2$ regardless of sampling density. However, the modulation of the output signal from a sampling aperture decreases as its diameter increases and may become lost in photon noise; SNYDER and MILLER (1977b) thus find for an aperture of uniform sensitivity in square array that the diameter, $\Delta\varrho$, is optimal at $\Delta\varrho = 0.586/f$ for a given cut-off frequency. The aperture sensitivity profile for a ganglion cell receptive field appears to be defined by a Gaussian profile (ENROTH-CUGELL and ROBSON, 1966; WÄSSLE and CREUZFELDT, 1973) for which (SNYDER, 1977) the diameter, $\Delta\varrho_G$ defined as 2× the diameter at 50% sensitivity, is given by $\Delta\varrho_G =$

0.75/f; whence/coverage may be derived as 0.56 π regardless of sampling density but dependant upon the form of the sensitivity profile of the aperture and the nature of the sampling array. Application of the result is not straightforward. Is the peripheral sampling frequency matched to the cut-off frequency of the image as the theory requires? Does the ganglion cell density of the class involved in resolution relate to acuity according to the requirements of the sampling theorem? Are *on* and *off* cells to be treated separately or as one system? Coverage values are available for the brisk transient class of the cat but the properties of these cells do not appear to be suited to resolution tasks and are not explicitly treated by the theory in the above form. The uniform coverage across the cat retina of 4–6 for the brisk transient class (CLELAND et al., 1975) thus remains in tantalising agreement with both the prediction of constancy and the value required if *on* and *off* systems were separately treated. WÄSSLE and PEICHL (personal communication) have obtained a similar value of 3–4 for cat horizontal cell coverage. The high coverage value of 30 obtained by FISCHER (1973) for the ganglion cell population of the cat retina treated as a whole is large because that population encompasses several classes of receptive field. Most of these classes cannot be incorporated into the above theoretical treatment at present because of the complex nature of their aperture sensitivity profiles. The uniformity of the whole population coverage, however, indicates the general applicability of an inverse relationship between receptive field diameter and sampling density and thus suggests that a theoretical treatment similar to that above may be applicable.

An interesting formalization of these factors for a preliminary model retina has been put forward by SNYDER et al. (1977), who describe the information capacity of a local retinal region as the product of the number of discriminable points with the natural logarithm of the number of discriminable intensity levels. Consideration of sampling aperture, sampling density, quantum capture, and noise within the system establishes that the "number of pictures" is optimal at only one intensity of illumination for a given sampling density with optimal sampling apertures. It follows that all regions of a uniform retina, whether nocturnal or diurnal, have their optimal information capacity at the same level of illumination. If the habits of the species confine it to purely nocturnal or diurnal conditions of illumination, then the illumination levels it encounters would vary little more than 1 log unit, and such a retina would be very efficient. The gradient of sampling density to be found in the retinas of arrhythmic, crepuscular, and auroral species conversely provides retinal regions operating optimally at the differing illumination intensities at which these species are active. It could be argued that the circumscribed area in such forms results from the need for a gradient of sampling density to generate retinal areas optimised for differing levels of illumination. This seems unlikely when the form of the gradients is so obviously adapted to features other than the distribution of light intensity across the visual field. The range of optimal illuminations for such a retina thus appears more probably an outcome of the presence of the ganglion-cell density gradient rather than its primary cause. Similar theoretical treatment of the duplex retina with its *superposed* nocturnal and diurnal systems is a more complex matter for which the necessary information is not yet available. The model has provided some notable insights into the design of compound eyes (SNYDER et al., 1977).

2. The Fovea

The fovea is present at the center of the area retinae of many species of fish, reptile, bird, and all simians (WALLS, 1942; ROCHON-DUVIGNEAUD, 1943; DUKE-ELDER, 1958). In the majority of forms the cones—in tarsius the rods—of the fovea have more elongated outer segments than in the remaining retina, so that the outer limiting membrane may be displaced vitread to form what is described as a *fovea externa*. The long, narrow shape of the central cones is concluded by MILLER and SNYDER (1973) to be more effective in light retention than that of peripheral cones.

The *fovea interna* consists of a localized thinning of the retina accompanied by displacement of ganglion cells to the surrounding perifoveal region as well as reduction in, or complete elimination of, the inner nuclear layer. The fovea may be a concave depression with a relatively flat bottom, concaviclivate, or a narrow but deep pit with convex walls, convexiclivate. In the most extreme convexiclivate fovea the outer nuclear layer may be thinned but more commonly its thickness increases.

CHIEVITZ (1889, 1891) and ZURN (1902) failed detect a fovea interna in a wide variety of subprimate mammals, including mouse, squirrel, hare, shrew, hedgehog, sheep, goat, ox, deer, pig, polecat, dog, and cat. ZURN (1902) described a fovea externa in the cat because of retinal thinning. STONE (1965) reported a 10% reduction in the outer nuclear layer thickness near the area and two layers of ganglion cells instead of the one characteristic at the periphery. BRÜCKNER (1961a) recently claimed a fovea externa to be present in several feline and canid retinas including tiger, puma, panther, lion, leopard, cheetah, wolf and one cat but not in the jaguar, lynx, two cats, and four dogs. However, examination of the alleged cat fovea reveals a depression accompanied by massive reduction of the outer nuclear layer, which is very characteristic of feline central retinal degeneration as described and illustrated by BELLHORN and FISCHER (1970), and BELLHORN et al. (1974); the opposite of what is commonly found in a concaviclivate retina. In all of the other species, except the tiger, the outer nuclear layer is also somewhat reduced, which suggests a pathologic process rather than a fovea.

The function of the fovea pit is a long-standing problem. WALLS (1937) argued on the basis of the refractive-index difference between retina and vitreous, which was reported by VALENTINE (1879b), that the convexiclivate pit of birds and reptiles magnifies the central image and thus aids resolution. Quantitative analysis by PUMPHREY (1948a) revealed that such a fovea would reduce acuity rather than increase it at the foveal center, but at the same time it would induce asymmetry into an off-axis symmetrical image and aid in alignment of the visual axis during tracking. Recently SNYDER and MILLER (1977) have argued that the photoreceptor matrix of some birds is too coarse to account for their claimed behavioural acuity. They suggest that the concave bottom of the convexiclivate fovea, rather than its sides (WALLS, 1937) acts as the negative component of a telephoto lens and provides magnification of a limited part of the local image. The flatter temporal foveae of birds and simians could not function in the manner of the convexiclivate fovea and, according to PUMPHREY (1948a), cannot have much refractive significance. WALLS (1942) regarded the concaviclivate as poorly developed or degenerate relative to the convexiclivate fovea, but it would now appear to be functionally different.

WOLFF (1940) argued that the retinal layers are displaced at the fovea in order to allow direct access of light to the receptors without reduction of image quality by scattering in overlying tissue; the disadvantage of an inverted retina is thus offset. WEALE(1966) rejected this explanation because the depolarized nonspecularly reflected component of the fundus reflex, which arises during polarized illumination, is similar in proportion from foveal and perifoveal retina, thus indicating that scatter is not introduced by the thickening of the inner nuclear and ganglion cell layers. Instead, WEALE (1966) suggests that blood vessels cannot be tolerated in a high-resolution *area* because of a deleterious influence on image quality, so that the majority of the retinal tissue must be shifted to a region of more adequate nutrient supply. The outer nuclear and bacillary layers are adequately supplied by the choroidal circulation (POLYAK, 1957). An apparent conflict with this theory occurred in *Aotes* retina, where a vessel-free central area was reported to be present in the absence of a fovea, but the problem was resolved when DE OLIVERA and RIPPS (1968) were able to demonstrate the presence of central vessels in this species in contrast to the situation in the rhesus monkey fovea.

It is only necessary, however, to consider the extreme convexiclivate foveae, which are found in the bird retina in spite of the total absence of retinal vasculature, to realise that WEALE'S emphasis on overlying blood vessels rather than neural tissue as the prime factor leading to the formation of the fovea is not adequate in its generality. It is possible that interpretation of the origin of the reflected depolarized light component is in some way inadequate; little is known of its site of origin, and difficulty with it has already been encountered in the interpretation of retinoscopy.

A direct attempt to assess the influence of the inner retinal layers on image quality at the photoreceptor plane has been made by OHZU et al. (1972). In human retinas already showing some signs of deterioration, although very swiftly prepared, the foveal MTF is superior to that of the perifoveal region. The deterioration in quality of image after isolation appears to arise at the inner layers of the retina. The fresh human peripheral retina was not examined, but the MTF of 11-minute-old fovea was superior to fresh rat retina, so that OHZU et al. (1972) suspected that the inner retinal layers were to some extent detrimental to image quality in even fresh human retina. If this is confirmed, their reduction in thickness at the fovea would, contrary to the conclusion of WEALE (1966), be an asset.

If the foveal ganglion cells of man overlay their photoreceptors of origin, then they would require a local density of some $160000/mm^2$ and some 30 layers, compared with the maximum of six actually found in the perifoveal region. If retinal scattering is such as to require the foveal tissue to be *thinner* than elsewhere in the retina, then the displacement of this *potential* ganglion-cell accumulation to the perifoveal region must effect an enormous improvement in the local MTF of the area. In a few species, *Lemur catta* and *Indri indri* (ROHEN and CASTENHOLTZ, 1967), the area is actually dome-shaped, as this hypothetical structure would be in man, and is suggested by WALLS (1942) as inevitably effecting some reduction of image size by refraction.

3. Macular, Central and Peripheral Vision

The phrase "macular vision" is often employed to describe central vision in man, but its common extension to vision mediated by the area retinae in the great preponderance of species which lack a macula lutea is unjustified. In fact, the term is probably best avoided altogether, because its use by some authors, e.g., ELLIOT SMITH (1928), has extensive theoretical implications which survive in the backwaters of the literature although insupportable on the basis of current comparative visual physiology. For ELLIOT SMITH, macular vision was the climax of evolutionary development, *the New Vision*, qualitatively different from anything developed in the subprimates. He regarded frontal eyes, conjugate eye movements, and optic fiber decussation as necessary precursors to the development of true macular vision. Once present, the macula leads during evolution to the

> progressive transformation of the brain and eventually to the making of the human mind

ELLIOT SMITH,, 1930). High resolution vision, colour vision, and stereoscopic vision combined with a sophisticated hand lead to the mastery of space. Even the enlargement of the brain arises from the presence of a macula.

DETWILER (1943) criticized these views and pointed out that the anatomic basis of ELLIOT SMITH's"true macular vision" appears simply to be the fovea and its associated area. However, in many vertebrate species the fovea has developed in the absence of frontal eyes or hemidecussation, and in some birds it is even unrelated to binocular vision, because it deals with the lateral field (WALLS, 1942; DUKE-ELDER, 1958). In such forms there is no trend towards the development of primate-like features, and encephalisation is no greater than in related species with nonfoveate retinas. The evolutionary sequence envisaged by ELLIOT SMITH (1930) as leading to the development of macular vision is thus not a *necessary* chain of events.

ELLIOT SMITH (1930) and many recent workers wrote of the fovea and its area as if it represented an evolutionary functional discontinuity relative to an area retinae lacking the fovea. There is evidence which suggests that the foveal outflow of the macaque retina, unlike the projection of an area in other species, does not reach the superior colliculus (HENDRICKSON et al., 1970); WILSON and TOYNE, 1970; BUNT et al., 1975), but, in general, the known properties of a concaviclivate fovea with an area differ only qualitatively from an nonfoveate area retinae; the convexiclivate fovea may have quite different properties. Features such as high resolution, colour, and stereoscopic vision, which are attributed by ELLIOT SMITH (1928, 1930) to foveate macular vision, are certainly well developed when a fovea is present but are demonstrable in species with only an area retinae.

The pure-cone, rod-free, central zone of the human macula is the basis of high photopic resolution, but vision in this region is qualitatively distinguished from that of the peripheral retina by its scotopic blindness. The high acuity of the area retinae, whether equipped with a fovea or not, again ensures its qualitative differentiation from the periphery because of its use as the image target for the fixation reflex of man and cat (DREHER and ZERNICKI, 1969) and as the region about which directed saccades maintain image centering during fixation (DITCHBORN, 1973). These features, in conjunction with the differentiated appearance of the human macular region, have led to a general tendency to treat central vision as if it is qualitatively, rather than quantitatively, different from peripheral vision in all species. In the majority of mammals, however, the central area is duplex (Fig. 18), so that the discontinuity of the human pure-cone macular region is absent and the common organization of central and peripheral retina more apparent; it is possible that the photopic and scotopic acuities in the area of such eyes could be very similar, provided that quantum limitations are not involved, and there is no question of a diurnal or nocturnal physiologic scotoma. Even for man the emphasis on the differences between central and peripheral vision is difficult to justify, because most retinal properties such as rod/cone and receptor/ganglion cell ratios, visual acuity, and colour vision, as well as higher level functions like stereoscopic vision and pattern recognition, change in uniform fashion along a centroperipheral gradient over much of which no discontinuity is apparent.

The attention-attracting power of peripheral moving stimuli has often been interpreted as signifying a lower threshold in that region of field and employed to differentiate central and peripheral vision on a qualitative basis:

> Laterally there are two monocular visual fields, each selectively sensitive to change or motion.

(TREVARTHEN, 1968). Measurement of the velocity at which slowly displaced stimuli are just perceived as moving does not support this view, because the *central* movement threshold is found to be much lower than the peripheral (AUBERT, 1886, 1887; RUPPERT, 1908; LICHTENSTEIN, 1963), and the threshold gradient follows a smooth curve as a function of eccentricity. However, the *highest velocity* at which movement is detectable *is* greater for the peripheral than for central field (BHATIA, 1975) but, again, no qualitative distinction is apparent. It remains possible that the subjective impression of qualitatively distinct salience for peripheral, rather than central, moving stimuli reflects some underlying organizational discontinuity between the two regions. The motion sensitivity of the cat Y-cell, or brisk-transient receptive fields, has led to the suggestion that similar cells underly the peripheral motion salience of man (e.g., NELSON, 1974). The implied and ill-defined qualitative difference between central and peripheral motion sensitivity is suggested to be paralleled by the alleged absence of such brisk-transient receptive fields, α-cells, from the cat area centralis (FUKUDA and STONE, 1974), and by their high peripheral percentage. Such suggestions are premature; a more exhaustive investigation of α-cell distribution (WÄSSLE et al., 1975) has indicated, by contrast, that the ratio of α-cells to the total ganglion-cell population is uniform throughout the retina and provides no support for the qualitative differentiation of central and peripheral vision.

III. The Ganglion Cell Layer

The radial organisation of the retina from the photoreceptors to the ganglion cell layer ensures that, away from a fovea, the perikarya of the neural elements overlay the region of the retinal image dealt with by their receptive fields. The local ganglion-cell density thus establishes the maximum number of independent optic nerve fiber pathways passing to the brain from unit area of retina, or solid angle of field, so that the form of the retinal ganglion-cell distribution determines the retinal sampling distribution over the uniocular field and the available input for more central regions of the nervous system.

At subsequent stages in the visual pathways it is found that the topographic mapping of the visual field onto various regions of the CNS is such that retinal regions of high ganglion-cell density obtain proportionately greater representation than those of low density. The topography of central stations on the visual pathway is thus *usually* determined by the effective peripheral innervation density as for other sensory modalities (WHITTERIDGE, 1973).

The retinal ganglion-cell density map therefore provides not only a rough guide to the variation in quality of vision with retinal eccentricity (although with caveats detailed later) but is also representative of the central neural provision for each region of the uniocular field. A knowledge of the relative positioning of the two eyes and of the retinal ganglion-cell distribution thus provides an easily accessible basis for investigating adaptations in the use of visual space which are peculiar to a given life style and characteristic of a species entire visual system.

Just as the total photoreceptor population determines the potential maximum number of independent data available to the retina, so does the retinal ganglion-cell population set a limit on the proportion of this information which is passed on to the central nervous system via that "bottleneck" on information flow in the visual system, the optic nerve (RUSHTON, 1962). Before dealing with the topography of the ganglion cell layer, it is worth briefly considering its total channel capacity as indicated by the optic nerve fiber population.

1. Channel Capacity of the Retinal Ganglion Cell Layer as Determined from the Optic Nerve Count

The optic nerve contains the axons of the retinal ganglion cells which are passing to the brain and a population of efferent fibers on their way to the retina. In the pigeon such efferent fibers amount to only about 1% of the total fiber population of the nerve (COWAN and POWELL, 1963), and similar proportions have been reported by BROOKE et al. (1965) in monkey and cat. The rabbit is also said to possess efferent optic nerve fibers (CRAGG, 1962), but in no mammalian species has the proportion been described as over 1% of the total nerve fiber population and in fact the presence of efferent fibers has been questioned for the cat by BRINDLEY and HAMASAKI (1966). It is acceptable therefore to employ the optic nerve fiber total count as an index of the total retinal ganglion-cell population of a given species.

A great range in the magnitude of the total optic nerve fiber population was documented among vertebrate species by the classic light microscopic studies of BRUESCH and AREY (1942), but so far no teleologic basis has been put forward to explain the distribution of high and low counts; the optic nerve fiber population did not appear to correlate with eye size, laterality of eye position, or primary adaptation to photopic or scotopic vision.

At present, however, it appears that all of the earlier work requires reexamination; MATURANA (1959) has demonstrated that light microscopy underestimates the anuran optic nerve fiber population by a factor of thirty because of its failure to resolve a large population of unmyelinated nerve fibers. Subsequently it was noted that even in the rat optic nerve, where all the fibers are myelinated, the light microscope fails to reveal 60% of the fiber population identifiable by means of the electron microscope (FORRESTER and PETERS, 1967). Similarly, some 50% of pigeon myelinated fibers, and unmyelinated fibers amounting to 30% of the total count, remained unresolved by light microscopy (BINGGELI and PAULE, 1969).

It is surprising to find that, in spite of these results, several recent investigators of duck (O'FLAHERTY, 1971) and teleost (TAPP, 1973) have employed the electron microscope to determine fiber-diameter spectra but depend upon light microscopy for their total counts. It is commonly assumed that light microscopy will resolve fibers of at least 0.6 μm diameter (FORRESTER and PETERS, 1967) or less under phase contrast, 0.25 μm (POTTS et al., 1972), but resolution may be drastically reduced by slight deviations from optimal conditions. A recent reexamination of the cat optic nerve by electron microscopy (HUGHES and WÄSSLE, 1975) indicates the light microscopic study of DONOVAN (1967) to have recorded only 6% of fibers 1 μm diameter and 75% of 2 μm fibers; it is to be noted, however, that BISHOP and CLARE (1955) employed light microscopy and obtained a fiber-diameter spectrum identical to that of the electron microscope study. Whatever the reasons for the poor performance of light microscopy under certain conditions, it obviously must not be employed without electron microscopic controls. The absence of such controls means that reservations must be entertained about the recent, and otherwise model, attempt to completely enumerate the axons of the human optic nerve (POTTS et al., 1972). by phase microscopy of thin sections. The resulting total of 1.1–1.3 million fibers was almost identical to that of BRUESCH and AREY (1942) instead of showing an increase of up to 50% such as has occurred upon electron microscopic reexamination of other species with larger modal-axon diameters.

Total fibre counts based upon electron microscope examination are available for only a few mammals: rat (FORRESTER and PETERS, 1967; HUGHES, 1977), 118000; cat (HUGHES and WÄSSLE, 1976), 193000; rabbit (VANEY and HUGHES, 1976), 395000; hamster (TIAO and BLAKEMORE, 1976) 119000; the consideration of so few species can be of little value; there is more profit in comparative examination of the local speculations for resolution in the area retinae and fovea and the overall retinal ganglion-cell distribution.

2. Quantitative Mammalian Retinal Ganglion-Cell Topography

Quantitative maps of ganglion-cell density distributions have been constructed from counts along radial retinal sections of several mammalian species including rat (LASHLEY, 1932), rabbit (SENEVIRATNE, 1963), man (VAN BUREN, 1963), suiformes (LUCK, 1965), baboon (WHITTERIDGE, 1965), and pinnipeds (LANDAU and DAWSON, 1970), but the method is subject to error (e.g., ABERCROMBIE, 1946), and more satisfactory results are obtained by counting ganglion cells in flat mounts of the entire retina (STONE, 1965).

The definitive criterion for the identification of a neurone of the retinal ganglion-cell layer as a retinal ganglion cell is that it sends an axon into the optic

Table 4

	Cat	Rabbit	Pigeon
A. Optic nerve count	193000	397000	4200000
B. Ganglion cell-layer neurone count	190000–260000	450000–520000	2380000
Ratio of A/B	0.98–0.75	0.88–0.75	0.57

Sources of data; cat, HUGHES and WÄSSLE (1976), WÄSSLE et al. (1975), HUGHES (1975a); rabbit, VANEY and HUGHES (1976); pigeon, BINGGELI and PAULE (1969).

nerve. Unfortunately, neither the dendritic morphology nor the axon are visible in Nissl-stained flat mounts, and the cells of the retinal ganglion-cell layer must be classified by the appearance of their perikarya alone.

The presence of Nissl substance enables the differentiation of neurones of the ganglion-cell layer from the glia and vascular elements (STONE, 1965; HUGHES, 1975a). It is, however, difficult to asses the influence of displaced amacrine cells (CAJAL, 1892; GALLEGO, 1971) on the count. If such cells show Nissl substance, they will be miscounted as neurones, but if stained like cells of the inner nuclear layer they could be classified as glia. The total fiber count of the optic nerve may be compared with the integrated ganglion-cell count from the map as a test of the criteria (HUGHES, 1975a), but for most species the necessary information is not available for such a control. The only results based upon integrated retinal whole-mount density maps and electron microscope nerve fiber total counts are contained in Table 4. Recently, TIAO and BLAKEMORE (1976) reported some 119000 optic nerve fibres and 114000 ganglion cells in hamster retina, a ratio of 1.04. VANEY (1977) has employed the advantageous arrangement of the rabbit retinal fibre bundles to count the fibre efflux from a specific retinal area and thus establish the local ganglion cell count. His streak ganglion cell densities are within 80% and peripheral densities within 60–70% of previous presumed-ganglion-cell maps (VANEY and HUGHES, 1976). A substantial proportion of the difference is accounted for by neurones which do not project to the optic nerve.

Earlier results like those of AREY and GORE (1942), who obtained a ratio of 0.94 of optic nerve fiber to ganglion cells in the dog, require reexamination with modern methods before acceptance. The above ratios, ignoring miscounted glia, clearly indicate the possibility of a much smaller proportion of internuncial retinal neurones in the mammals than in the birds, which is in keeping with known retinal histology (CAJAL, 1955). That cells with the appearance of neurones remain after optic-nerve section has been reported for the cat (LEINFELDER, 1938) and rat (EAYRS, 1952) but denied for rabbit (JAMES, 1933) and kitten (STONE, 1965). Again, modern criteria must be applied for the reconsideration of these results. Although FUKUDA (1977) has claimed agreement between published rat optic nerve counts (118000, FORRESTER and PETERS, 1967) and his integrated ganglion cell density map (115000), it has been pointed out by HUGHES (1977d) that a more satisfactory integration of his map indicates the presence of 145000 ganglion cells. HUGHES' unpublished neurone count for the rat indicates between 125000 and 190000 cells in contrast to 120000 optic nerve fibre in an animal of the same strain (HUGHES, 1977d). It remains to be seen whether the excess cells are really neurones or mis-classified glia. For the present, the maps subsequently

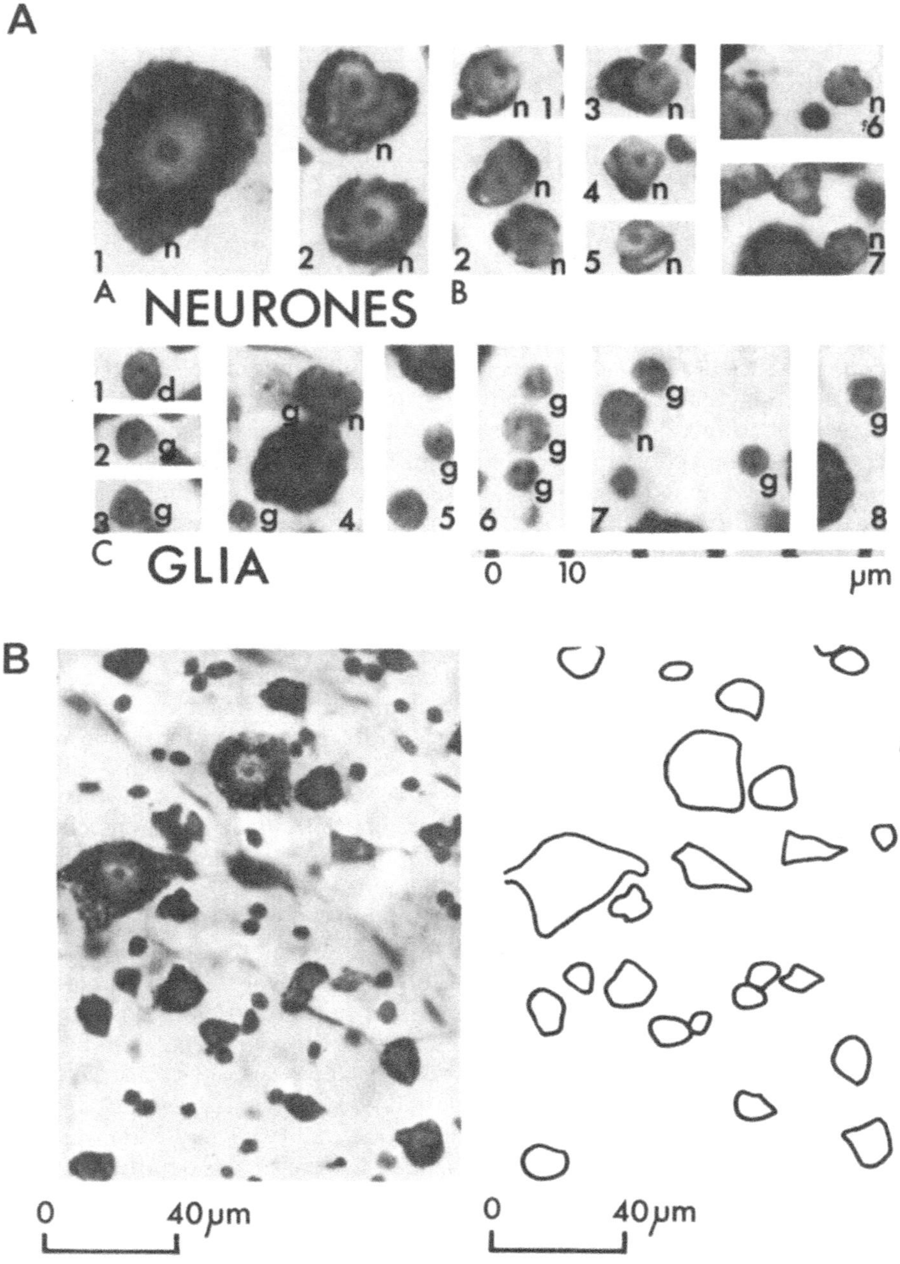

Fig. 19. Criterion examples of ganglion cells in retina of cat: *A*, neurones of retinal ganglion-cell layer, *n*, identified by their Nissl staining. Doubtful cell, *d*, and glial cells of the same layer, *g*. The neurones are presumed ganglion cells; *B*, a photograph of an area of cat visual streak accompanied by the outlines of presumed ganglion cells which have been identified under oil immersion. (After HUGHES, 1975a)

presented must be regarded as provisional and as showing the distribution of *presumed ganglion cells.*

Some criterion examples of retinal ganglion cells and glia from the cat retina (HUGHES, 1975a) are presented in Fig. 19a, and a region of the cat visual streak shown in Fig. 19b with a key to the identified ganglion cells. The general appearance of the cell types is similar to those of the cat in the majority of mammalian species examined.

The recent development of methods which enable the retrograde transport of various identifiable materials, such as cobaltous chloride (LINCOLN and MASON, 1974) or horseradish peroxidase (HRP; LA VAIL and LA VAIL, 1974), from the axon terminal up to its perikaryon promises more certain identification of the retinal ganglion cells, but difficulties remain. BLUNT et al. (1974) have reported on the passage of HRP along the optic nerve of the rat but find that the enzyme does not enter a substantial population of small cells in the retinal ganglion-cell layer. These cells look like small neurones but may be interneurones, nontransporting ganglion cells, or ganglion cells which project to sites other than those injected.

Quantitative ganglion-cell distribution maps based upon flat-mount preparations are now available in the literature for several mammalian species: cat (STONE, 1965; HUGHES, 1975a), rabbit (HUGHES, 1971), goat (HUGHES and WHITTERIDGE, 1973), and plains and tree kangaroos (HUGHES, 1975a), owl monkey (WEBB and KAAS, 1976), rat (FUKUDA, 1977), hamster (TIAO and BLAKEMORE, 1976). From these and the author's unpublished material it is possible to select a representative series of mammalian retinas such as is illustrated in Fig. 20.

No mammal so far investigated has been found to lack both streak and an area. Among the species which lack a visual streak, the characteristic feature of the isodensity contours is the relatively high degree of radial symmetry. Three classes of such distribution may be distinguished:

a) Low Ganglion-Cell Density with Little Centroperipheral Density Gradient. This is typical of small nocturnal inhabitants of the undergrowth such as the guinea pig (Fig. 20a), mouse, weasel, etc. In a more arrhythmic species, such as the rat, the area may be better defined (Fig. 20b). It is not, however, true that

> The mouse and rat have nothing that can be recognised as indicating an area where vision is anything but indifferent!

(PRINCE, 1956). All encountered nocturnal forms do have a region of higher density about the fixation axis. In the rat (HUGHES, 1977c), and probably other nocturnal forms, this region is accompanied by an increased cone density in the subadjacent receptor layer.

b) High Ganglion-Cell Density with Little Centroperipheral Density Gradient. Such a distribution is characteristic of the predominantly diurnal retina such as is found in the tree shrew or squirrel (Fig. 20c). The ganglion-cell density changes no more than three- or fourfold from the area to the periphery, the condition described as "universal macularity" by ROCHON-DUVIGNEAUD (1943).

c) Steep Centroperipheral Density Gradient. Such a distribution reveals a precisely localised visual pole and is to be found in man (Fig. 20e) (VAN BUREN, 1963), baboon (WHITTERIDGE, (1965), and other primates; although the area retinae is associated with a fovea in both of these species, this is not inevitable.

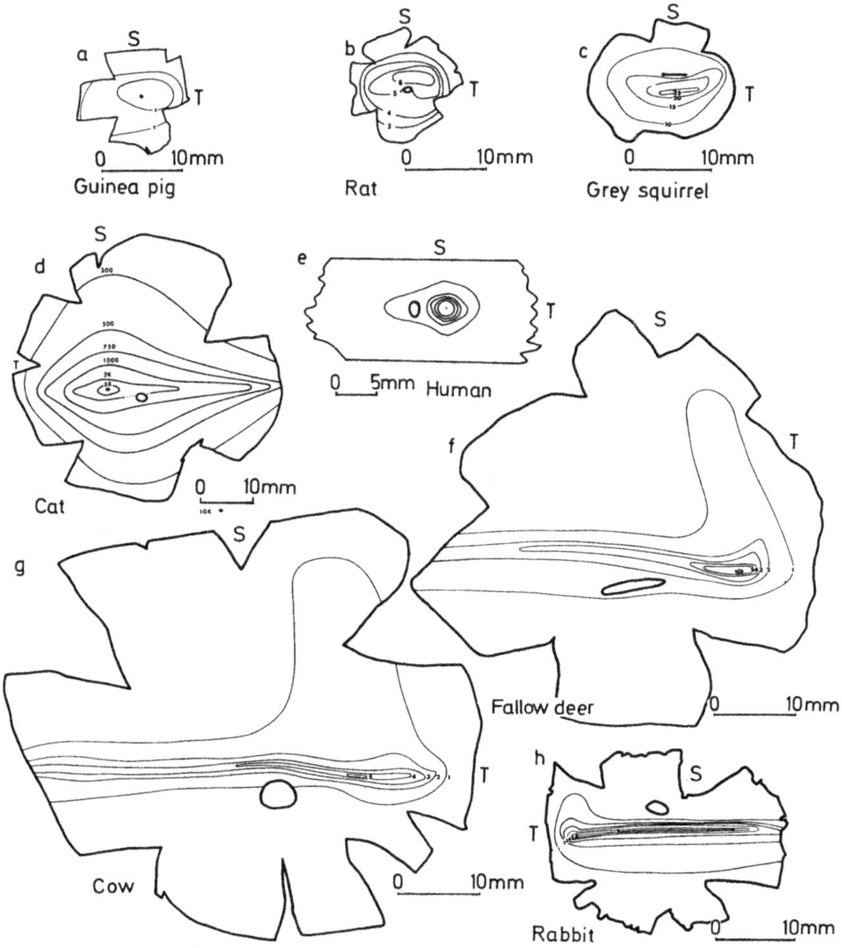

Fig. 20. Presumed ganglion-cell density distribution maps derived from counts made on the Nissl-stained retinas of various mammalian species. The teleologic basis for the differing forms of these maps is discussed in the text. The isocount lines indicate changes of density in steps of 1000 ganglion cell/ mm^2 or multiples thereof; the human map, after VAN BUREN (1963), is an exception in that it indicates the number of layers of ganglion cells between one and five. All species so far examined have revealed some degree of area differentiation

The construction of ganglion-cell distribution maps has revealed that a large number of mammals possess mixed retinas in which a visual streak is combined with an area of the type (c) above.

d) Central or Temporal Area Superposed on a Visual Streak. The best investigated retina containing a relatively central area in association with a visual streak is that of the cat (Fig. 20d), but a similar arrangement is to be found in most carnivores, including the dog, mongoose, and polecat. No different in principle is the temporal area found associated with the visual streak in the more lateral-eyed herbivorous species, such as the fallow deer (Fig. 20f), sheep, goat, horse, or red kangaroo. The more temporal situation of the area than in carnivores simply enables the visual axis to remain in the parasagittal plane in spite of the lateral position of the eyes. The relative quality of the area and streak differs between species; the cow is an extreme member of this group, with a very poorly defined area (Fig. 20g) and almost qualifies for the final class of distribution.

e) Visual Streak Unassociated with an Area. This distribution is the most extreme deviation from that found in man but is rarely encountered in pure form among the mammals. The rabbit (Fig. 20h) shows it to best effect, but it is also present in the ground squirrel and mongolian gerbil.

Among many of the herbivores there is a conspicuous upward bulge of the lower density isocount contours in a region bridging the decussation line above the visual pole; its upward extension is much greater in the tall species (Fig. 20g) than the short (Fig. 20h). This feature is as distinctive as the area or visual streak in the isocount maps and is subsequently referred to as "the anakatabatic area".

3. Retinal Ganglion-Cell Topography in Nonmammalian Classes

Extensive macroscopic examinations of the nonmammalian vertebrate fundus have been published (CHIEVITZ, 1889, 1891; SLONAKER, 1897; WOOD, 1917; FRANZ, 1934; WALLS, 1942; ROCHON-DUVIGNEAUD, 1943; DUKE-ELDER, 1958), although there are very few quantitative ganglion-cell distribution maps available for this group: frog (JACOBSEN, 1962), pigeon (WHITTERIDGE, 1965; BINGGELI and PAULE, 1969), and the kelp bass (SCHWASSMANN, 1968). No vertebrate species whose retina has been microscopically examined has been found to lack at least the rudiments of an area retinae; the term "amacular fundus" coined by WOOD (1917) is applicable only to the macroscopic appearance of the retina. The fovea is singularly lacking among the mammalian subprimates but common in fish, reptiles, and birds. With the exception of the birds, the circular or oval area retinae is located for forward vision so that its nasal or temporal retinal position is determined by the laterality of the eye position in the head. Among birds, however, macroscopic examination indicates examples of a central area for lateral viewing, two areas (bimacular fundus), or even two areas superposed on a visual streak, the *infulabimacular fundus* (WOOD, 1917), but these organizations require quantitative substantiation. Macroscopic examination also reveals the visual streak to be common among birds (CHIEVITZ, 1889, 1891; SLONAKER, 1897; WOOD, 1917; O'DAY, 1940; DUIJM, 1959); in addition, it is present in certain sharks (FRANZ, 1931) and teleosts (BUTCHER, 1938; ENGSTRÖM, 1963; MUNK, 1970) but not in many other fish (KAHMANN, 1934; TAMURA, 1957; SCHWASSMANN, 1958), anurans (CHIEVITZ, 1889, 1891; SLONAKER, 1897; GAUPP, 1904; JACOBSEN, 1962), chelonians (HESS, 1913; BROWN, 1969), lacertilia (SLONAKER, 1897; KAHMANN, 1936), and crocodilia (CHIEVITZ, 1889, 1891). Unique among the studied vertebrate retinae are those of two cyprinodontid fish (MUNK, 1970) which have a *bi-infular fundus* containing two parallel visual streaks about which more will be said later. In that the common belief that the visual streak is rare among the mammals has been dispelled by quantitative examination of ganglion-cell distribution, so must the results of macroscopic examination on other vertebrates await detailed reexamination before its absence in any species is regarded as established.

IV. Function of the Visual Streak

1. Current Theories

It was clear to the earliest students of the vertebrate area retinae that there is no correlation between phylogenetic status and retinal ganglion-cell topography. The great variety of area shapes also suggests that their form is not simply an outcome of some convenient morphogenic process which is without functional significance. The possibility of a relationship between the life style of a species and the form of its area was raised by CHIEVITZ (1889), but neither he, SLONAKER (1897), nor ZURN (1902) attempted functional interpretation of ganglion-cell topography, and neglect of this topic continues in the classic accounts of retinal reorganisation (FRANZ, 1934; WALLS, 1942; ROCHON-DUVIGNEAUD, 1943; DUKE-ELDER, 1958).

The elongated visual streak was obviously understood by JOHNSON (1901) to be of protective value; thus he refers to the ungulates as species which

> graze in the proper sense of the word. They poke their heads down into the long sharp grass, while they are notoriously those which have most need of being on the alert for Carnivores, their natural enemies,

and couples the need for panoramic vision with an extended area:

> those animals which have a widely extended sensitive area and need to see all round them.

The idea is more explicit in VINCENT (1912), who suggests that the central area

> seems to be larger in those animals where the divergence of the optic axes is great and when combined with nearly spherical lenses and large corneas it gives the animal a wide range of vision.
> The ability to see on all sides gives the animal greater security from enemies ...

A comparative study of retinal organisation among the Suidae led LUCK (1965) to suggest that the visual streak provides more than panoramic vision. The prominent streak of the wart hog implies that its organisation

> accords well with its habitat and habits. It is the eye of an animal living by day in the open country and in one plane. Food and foe alike come to it from the ground and this horizontal plane of interest is reflected by the retinal horizontal streak area.

The absence of the streak in the bush pig retina is suggested by LUCK (1965) to result from the equal importance of the vertical and horizontal planes as a source of danger in its natural habitat. The linear area centralis of the turtle led BROWN (1969) to a similar conclusion, that in an animal

> which never rises far above water or ground, most visual stimuli which are important must come from the region of the horizon. Hence the linear and horizontal area centralis appears to be a highly functional specialisation for an animal living close to the earth's surface!

This suggestion is in striking contrast to the description of the bird visual streak as employed in navigation during flight at considerable height above the ground (PENNYCUICK, 1960; MATHEWS, 1968)! But the ad hoc nature of this hypothesis is indicated by the absence of a streak in the pigeon (BINGGELI and PAULE, 1969), which is a skilled navigator, and other migrating birds (DUIJM,

1959). Other theories are equally unsubstantiated. Thus, for PUMPHREY (1948b) the visual streak in birds of the sea, desert, and open spaces

> suggests a device for accurately fixating the horizon as a datum to which other objects in the visual field can be referred.

Why such a reference should be needed when absent from forest-dwelling species, which appear upon macroscopic examination to possess only circular areas' remains unexplained. In any event, it remains to be demonstrated that the bird streak is specifically adapted for use during flight. Its presence in the albatross (O'DAY, 1940) suggests, in view of the long diurnal periods spent on the wing, that the streak plays some role in flight. According to WOOD (1917) the streak is common in ground feeding birds and

> probably enables the bird to widen the limits of his field of distinct vision (especially while feeding) without being obliged to move the whole head.

At least such a hypothesis is applicable to the other vertebrate classes containing members with visual streaks.

2. Counter-Examples to the Current Hypotheses: The "Terrain" Theory

The availability to the author of retinal ganglion-cell distributions for nearly 100 mammalian species makes it possible to assess the general applicability of these hypotheses more rigorously. The common extension of JOHNSON'S (1901) theory that the streak is correlated with lateral eyes, a wide panoramic field, grazing habits, and subjection to predation is well substantiated by its presence in the rabbit, hare, and ox as well as deer, sheep, horse, and plains kangaroo (HUGHES, 1975b) (Fig. 21a) in association with an area temporalis.

Unfortunately for the attractive simplicity of this view, there are several counter-examples available. The presence of the visual streak in the crocodile (CHIEVITZ, 1889) reveals that this distribution may occur in predatory as well as preyed-upon species. Of course, the crocodile is lateral eyed, but that the streak is not confined to *lateral eyed* predators has been recorded in the literature since Slonaker's report of a bandlike area in the relatively frontal-eyed fox and cat (1897). An "arm of high ganglion-cell density" in the cat distribution was confirmed by STONE (1965) and subsequently shown to be prominent enough to merit description as a visual streak (HUGHES, 1975a). Various stages of streak development are present in the more frontal-eyed carnivores, such as the dog (Fig. 21b) (collie, greyhound, beagle, dingo) as well as the fox, polecat, and mongoose. The observation (WHITTERIDGE, personal communication) that the blood vessels of the cheetah fundus (JOHNSON, 1901) suggest the presence of a streak led to the discovery of a well-developed example of such a distribution (Fig. 21c) in conjunction with a circular area. In this instance, we find a frontal-eyed predator with nearly parallel optic axes to be in possession of a very pronounced visual streak. It thus does not follow that a bandlike retinal area

> invariably accompanies a particular position of the eyes.

(PRINCE, 1956), nor is it uniquely associated with either prey or predatory species.

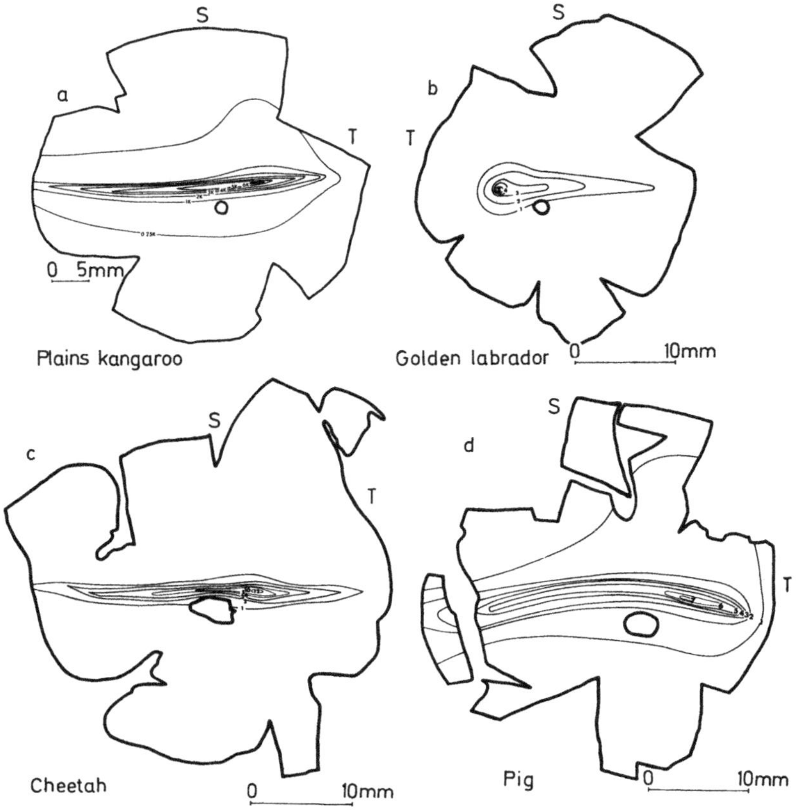

Fig. 21. More retinal ganglion-cell distribution maps which are discussed in text. Conventions same as for Figure 20. Note well-developed visual streak and area centralis in retina of the frontal-eyed cheetah

Of the original conglomerate hypothesis only the association betweeen the presence of a visual streak and open-country habitat (LUCK, 1965; WHITTERIDGE, 1973) has not so far been contested. It is striking that the best-developed visual streaks among the investigated carnivores are those of the cheetah, which depends upon open country for its hunting, and the greyhound, which is a specialized desert hunting dog (FIENNES and FIENNES, 1968). Numerous other forms substantiate this correlation, including the sea and desert birds (PUMPHREY, 1948), open plains and desert-living mammals, such as the wart hog (LUCK, 1965), ground squirrel, mongolian gerbil, antechinomys and its placental counterpart, notomys (HUGHES, unpublished). In keeping with the association between the visual streak and an open-country habitat is the presence of this organ in bottom-feeding sharks (MUNK, 1970) and the mudskipper of the intertidal flats (MUNK, 1970). However, even this generalizations has counter-examples. Thus, the visual streak of the pig (Fig. 21 d) is as well developed as that of the ox, and yet the former species has been a forest dweller since its origin in the oligocene (COLBERT, 1967). Again, the grey kangaroo visual streak is as prominent as that of the plains kangaroo (HUGHES, 1975 a), and yet the former is characteristic of wooded country and never enters open grass or shrub-covered plains (FRITH, 1973).

The visual streak is, however, by no means universal even among terrestial animals. It is entirely absent among small nocturnal species of the thick undergrowth, such as the hedgehog, mouse, guinea pig, and to a lesser extent, the rat, whose retinal ganglion-cell population is fairly uniformly distributed across the retina.

Amongst arboreal species, such as the tree shrew, tree kangaroo (HUGHES, 1975a), grey squirrel, squirrel glider, and macaque, it is found that the ganglion-cell distribution is quite radially symmetrical with the projection of its center of symmetry aligned for forward vision.

The conclusion to be reached from these results is that the visual streak is common to terrestial species whose field of view is not completely obscured by nearby vegetation and not simply to those which occupy open country, as suggested by LUCK (1965); the terrain surface is overt in all but the most overgrown woods, above the stream bed to a fish, or above the sea to a flying or floating bird.

3. The Retinal Image of a Stationary Eye Above an Extensive Terrain

The advantage of the visual streak is readily understood from the projective transformation of an extensive terrain into the retinal image. Consider an object as it moves longitudinally along the surface of an extensive plain away from a stationary eye at height, h, whose effective image shell is bisected by the image of the horizon. As the object distance, d, increases, its image passes down the superior retina towards the projection of the horizon line at the equator of the globe. The angular elevation, θ, of the object's image above that of the horizon line may be regarded with little error ($<7\%$ when d is in excess of $2.5\,h$) as varying inversely with the object distance:

$$\theta = \arctan h/d \approx h/d \text{ radians}.$$

The image of the terrain from $d = 2.5\,h$ to infinity is thus packed into a strip of the image shell subtending only 21° of the 90° superior quadrant. Beyond $100\,h$ the entire terrain is imaged in a strip only 0.6° wide.

The invariant features of this image were first noted by GIBSON (1950):

> An out-of-doors world is one in which the lower portion of the visual field ... is invariably filled by a projection of the terrain. The upper portion of the visual field is usually filled with a projection of the sky. Between the upper and lower portions is the skyline ... always cutting the normal visual field in a horizontal section ... It is little noticed, but on the average and over the ages it must have determined the fundamental pattern of retinal images for all or most terrestial animals.

The majority of terrestial objects and events are thus represented in a horizontal strip of the retinal image, and their detection is most economically ensured by the concentration of ganglion cells in the corresponding region of the retina to form a visual streak. The effective evolution of such a streak requires, however, that the receptors which supply the input to the streak be kept in fixed relationship with the retinal image of the terrain and horizon. The validity of the "terrain" theory thus necessitates that the streak be horizontal under normal conditions and resist displacement when the orientation of the head is changed.

The specific examination of the visual streak's projection into the visual field of the rabbit (HUGHES, 1971), goat (HUGHES and WHITTERIDGE, 1973), and cat (HUGHES, 1976a) indicates that it is indeed horizontal and aligned close to the visual horizon of these animals when they are consicious. Although some authors (SLONAKER, 1897; WOOD, 1917) appear to accept that the streak is oblique in certain birds, it is in fact found to be horizontal (DUIJM, 1959) if the normal head position is taken into account, according to the suggestion of PUMPHREY (1948b). The necessary dynamic reflex maintenance of the horizontal alignment of the visual streak over a wide range of passive head rotation has been demonstrated to exist in turtle (BROWN, 1969), rabbit (HUGHES, 1971), and goat (FORRESTER, 1975) according to the requirements of the above hypothesis.

For arboreal, as for terrestrial, forest species, it is unlikely that obscuration of the terrain is total. The absence of the visual streak among arboreal forms is rather accounted for by the lack of a fixed relationship between the image of the visible terrain and any part of the retina. The situation for a climbing animal which swings or jumps through the branches while viewing the world from different orientations is quite different to that of many birds which fly in relatively constant orientation and with an overt horizon.

So little light penetrates the undergrowth and hedge-rows from dusk to dawn that for the small nocturnal animals

Vision at a distance is therefore altogether impossible in a dim light

(WALLS, 1942). The distant terrain would thus be invisible even if not obscured and a visual streak of no benefit. It is more puzzling to find that the secondarily terrestrial baboon (WHITTERIDGE, 1965) and man (VAN BUREN, 1963) show only slight horizontal ellipticity of their ganglion-cell isodensity maps in lieu of a visual streak. However, in spite of their terrestrial habits, these species retain a much greater variety of eye and head positions than is encountered in streak-bearing quadrupeds, and it is subsequently suggested that their sampling distribution is adapted to the analysis of the dynamic retinal image.

4. Quantitative Models of the Retinal Ganglion-Cell Topography

It is a commonplace of information theory that a finite-state information channel like the optic nerve attains its maximal information-transmission rate only when its states are maximally uncertain—a condition achieved if they are equiprobable or when the probability of activation of each element is equal. Any statistical nonuniformity or invariance in the information source which is not matched by a corresponding modification of the sampling distribution ensures nonoptimal encoding. It is thus arguable that the sampling distribution should match the distribution of source information when encoding by sampling over an information source like the retinal image.

The accepted concept of the evolutionary adaptation of the retina to the needs of a species implicitly accepts some approach to optimal encoding with respect to both *habitat* and *life style*. The information source to which the sampling distribution is adapted cannot be regarded as the retinal image of the habitat but only as those components, the *information array*, which are important to the animal in its specific lifestyle. The visual world extends about an animal in all directions, but vulnerability to rapacious birds makes a high-resolution survey of the entire sky necessary for small species although redundant for large forms. Again, although the terrain may be statistically invariant around an animal, the extent and quality of its visual analysis is weighted according to life style by the degree of laterality of the eyes and the retinal provision for the analysis of the lateral field. It thus follows that a comparison between a sampling distribution and the invariant features of the typical retinal image cannot *necessarily* be expected to give immediate insight into its functional significance but, at least, the attempt is worthwhile.

The open-country (LUCK, 1965) and terrain theories of the streak employ the high probability, assessed on the basis of ecologic knowledge, that the retinal image of streak-bearing species will contain an invariant and foreshortened image of an extensive terrain to hypothesise that the distribution is adapted to deal with the ground surface and its contents. A variety of quantitative distributions are compatible with this hypotheses, depending upon the feature to which the sampling array is optimally matched.

Certain cautions are necessary before theoretical treatment. The sampling distribution is discussed as if involved in a *generalized resolution task;* its elements are treated as if functionally equipotential, and the sampling theorem is regarded as determining the resolution obtainable from a region of given sampling density (SHANNON and WEAVER, 1949). It is obvious that the presentation is based upon the assumption of an idealized environment which may be realized in life only as stochastic features of the retinal image. The inadequacies of these assumptions are taken up subsequently, and many of the terms discussed in more detail.

a) Constant Angular Resolution

First consider a visual streak or entire retina across which the sampling rate is uniform. From the sampling theorem we have that the angular resolution, α, achieved by an ideal system varies inversely as the effective sampling rate, $\alpha = 2/r$. Now the angular magnitude, ϕ, of the image of an object with a frontal dimension, w, varies inversely as its distance, d, from the eye according to the law of visual angle,

$$\phi = \arctan w/d \approx w/d \text{ radians } (e < 7\% \text{ if } w/d < 0.4)\,.$$

Setting w equal to the absolute frontal dimension, s, of the minimum separabile, we obtain $\phi = \alpha$ and may write

$$s = 2 \cdot d/r\,.$$

Within such a region of uniform sampling rate, the size of resolvable detail, the minimum separabile, is uniquely correlated with stimulus distance and increases as the stimulus approaches the eye regardless of its path. A visual streak of uniformly elevated sampling rate relative to the peripheral retina, such as is envisaged by MUNK (1970), gains the advantage of an increased resolution for the terrain and its contents. A uniform sampling rate over the entire retina would be suited to a static eye whose homogeneous environment contains information uniformly distributed over all directions or whose distance vision is so limited in extent as to minimize the effects of perspective in the retinal image and render ineffective the subsequently described distribution.

b) Constant Size Resolution in the Local Frontal Plane

It is clear from above that if $r \propto d$, the absolute size of the minimum separabile would remain constant with increasing stimulus distance. This could *not* be easily achieved when images from objects at different distances have no differential distribution across the retina. But consider an eye of which the information array consists of the objects and features of an extensive terrain above which it is set at height, h, with the image of the horizon line coincident with the equator of the globe. Because the terrain and its content at a given distance project to a unique elevation in the image, θ, it is possible for the local sampling rate to be arranged to vary as the distance of the corresponding terrain and thus inversely as its elevation on the superior retina. We may write

$$r = \frac{2 \cdot h}{s} \cdot \frac{1}{\theta}$$

so that the dimension of the minimum separabile remains constant with increasing distance until $r \neq k/\theta$. The possibility of only a finite maximal sampling rate ensures that this must occur for some limiting distance, d_r. A plot of r against θ is thus a truncated rectangular hyperbola for such a retina; the horizontal sampling rate is uniform along the retina at any elevation.

The reduction in angular subtense and decreasing elevation of the images of a uniformly distributed and homogeneous population of objects with increase of their distance from the eye along a flat terrain is described by GIBSON (1950) as a gradient of visual texture in the retinal image. It has been suggested (HUGHES, 1975a) that the sampling profile across the streak is matched to the gradient of visual texture in the image, but difficulty arises with this terminology. Obviously such a gradient exists for any *specific* population of objects, but in the natural image the nearer objects reveal, as do the distant, grades of texture up to the limit of optical resolution, so that the gradient of visual texture is not objectively demonstrable without reference to higher level feature-recognition processes. The more fundamental description of this model is that it eliminates redundancy introduced into the image by the law of visual angle and simultaneously achieves constant size resolution in the local *frontal* plane along an extended flat terrain.

The terrain of the real world is populated with objects of vertical extension which is substantial in relation to the height of the eye. When close to the eye the vertical projection of such stimuli in the retinal image extends over regions of ganglion-cell density not matched to their distance from the eye. However, because the vertical angular subtense of such stimuli in the retinal image decreases in proportion to their distance from the eye, it follows that their images become progressively better confined to a region of ganglion-cell density matched to their distance if they move away along the terrain surface. The sampling profile which achieves constant size resolution is thus most appropriate under circumstances in which the information array consists of the terrain and its contents very distant in terms of h.

Examination of the projection of the visual streak in several species (e.g., rabbit, HUGHES, 1971) indicates that the vertical ganglion-cell profile is fairly symmetrical about the region close to the image of the horizon line. *The high-density region of the streak is thus adapted to the image of the upper field as well as to that of the terrain*, a situation not encompassed in the open-country or terrain theories of the streak but compatible with the constant size resolution or "inversion of the law of visual angle" model. Although less clearly defined, an aerial horizon similar to that of the ground may be regarded as implicit in the inferior retinal image of the sky. If, for instance, the distribution of rapacious birds forms the information array for the eye, then the relatively limited range of their altitude would suffice to stochastically determine its parameters. Qualitative consideration suggests that variation in the vertical sampling rate of the inferior retina according to a rectangular hyperbola or an approximate pragmatic distribution would be again appropriate and lead to some symmetry in the sampling distribution about the projection of the horizon line.

c) Constant Size Resolution Along the Longitudinal Plane

A further theoretically possible distribution exists if the information array for the eye is the terrain surface and contents treated as

> a complex symmetrical, two dimensional pattern viewed from an angle such that its retinal or photographic image is not symmetrical. It will be economical to transmit a description of the pattern as if it were in the frontal plane and symmetrical (eliminating the redundancy of symmetry ...)

(ATTNEAVE, 1954).

To effect such optimal sampling in the longitudinal plane the sampling provision must be the same for unit distance along the terrain regardless of its distance from the eye. The local sampling rate must therefore vary inversely as the square of its angular elevation above the image of the horizon line [the square relation arises from oblique viewing of the surface (BAIRD, 1970)]. Such a distribution would convert the longitudinal into a pseudofrontal projection of the terrain and compensate for perspective distortion.

d) The Quantitative Description of the Real Visual Streak

The above analysis suggests acuity profiles matched to several invariant features of the retinal image. Under optimal conditions the sampling density profile should be matched to the predicted acuity profile by the sampling theorem. If all retinal ganglion cells are of classes to which the above analysis applies then the ganglion cell density profile in some species may have one of the predicted forms; if not, the models will be applicable only to the distribution of the appropriate subclasses. If the system is subject to other influences, such as the lattice noise discussed in Section E III, then the number of independent samples predicted by the model will be less than the number required by the systems' degraded sampling efficiency. A direct comparison of the predicted and actual acuity distributions is currently possible for man alone; comparison of the above models with ganglion cell distributions is the only possibility for other species.

The constant angular-resolution model described above is approximated by the relatively uniform ganglion-cell distributions to be found in a few diurnal and nocturnal retinas, but no species has been encountered with a visual streak whose vertical-sampling profile falls off as the predicted $1/\theta$ or $1/\theta^2$. However, the above analysis may have been in vain; in both monkey and man the behaviorally determined acuity varies as the ganglion-cell density rather than as its square root. *If* such apparent "oversampling" occurs in other species then matching of the ganglion-cell distribution to the information array would require the streak *density*, rather than linear sampling, profile to fall off as $1/\theta$ or $1/\theta^2$ for the two models above. Of these, the former distribution has been observed in the rabbit retina (Fig. 22b).

The vertical-density profile of the visual streak in the majority of examined species does not match the models outlined above, and only slight compensation for the influence of the law of visual angle could be effected by them. On the basis of the assumptions made earlier, the visual streak may be regarded as transforming the spherical isoacuity shells of the space about an eye with a uniform retina into a set of oblate spheroid shells which are more appropriate to objects distributed over an extended terrain (Fig. 22a). However, even for the rabbit, in spite of the form of the density profile, it is not obvious whether the streak may be understood to provide strictly quantitative constant size resolution over a limited range of space or partial compensation over a greater region. The finite peak density of a real visual streak means that it can provide constant-size resolution only to the distance, d_t, which is uniquely determined by the angle subtended by the half value of the density profile. In the rabbit this angle is such that the peak must project at a declination of 2.3°; strict compensation would then occur from the animal to a distance of $28h$; beyond this, resolution would deteriorate. As pointed out previously, this distance is too short for the streak to be effective in dealing with the normal objects of an animal's environment, and it seems more probable that the peak count is directed toward the horizon with only partial offsetting of the influence of the law of visual angle.

It is clear from the proportional relation between θ and h that the terrain from infinity to some distance d will subtend a smaller angle when h is small. If the streak is organised to play a role in offsetting the influence of the law of visual angle on the absolute size of the minimum separabile, then it is a necessary consequence that the half-height, half-width of the streak profile will be narrowest in species with the eyes just above the terrain surface, intermediate in low species such as rabbit, and widest in tall forms such as sheep and horse. The actual presence of such a difference between distributions, with the sheep half-height width at 20° and that of the rabbit at 5.6°, is substantiation of the role suggested for the streak in the model above. The anakatabatic region in the temporal retina of grazing species, which provides increased resolution of the terrain from which food is taken and the feet placed during locomotion, is more elevated in tall than short species.

The above discussion is centered on the role of the streak in the resolution of static features of the environment, but it does not follow that it plays no part in movement detection. Both resolution and the displacement threshold vary in man according to the local ganglion-cell density. MUNK's (1970) emphasis on the streak as providing high sensitivity of movement for the detection of predator or prey rather than as simply increasing resolution is thus apparently unnecessary, because the former may be consequent upon the latter.

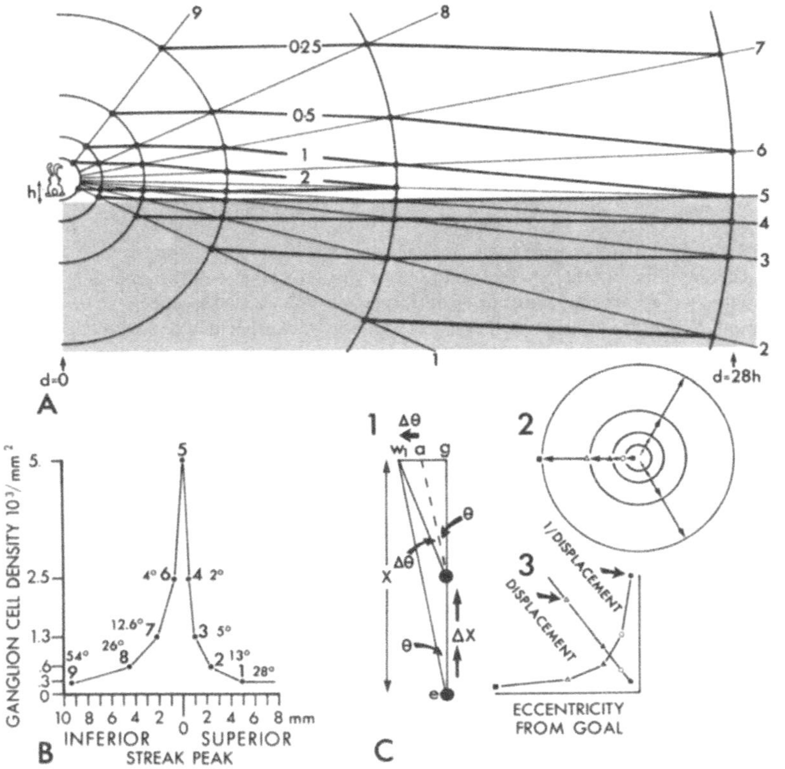

Fig. 22a–c. Vertical cross section of rabbit visual streak ganglion-cell density distribution is shown in (b). Projection into the visual field of points at which ganglion-cell density is successively halved, 5, 6, 7 etc., and 5, 4, 3, etc., is shown in (a) as a series of correspondingly numbered lines which radiate from rabbit eye at height, *h*, above ground surface. Series of arcs centered on the eye are situated at successively doubled distances. Bold lines thus join angular points in space at which the projected ganglion-cell density is the same. *If ganglion-cell density determines acuity in the rabbit*, as appears to be true for man, then bold lines define profiles along which *absolute size* of minimum separabile is constant. Decline of ganglion-cell density along a rectangular hyperbola in superior retina ensures that if the peak density projects to the ground at a distance $28h$ away from eye, then the inferior borders of absolute-size isoresolution profiles lie parallel to the surface. Thus, local absolute size of the minimum separabile along surface would remain invariant with distance, and the influence of the law of visual angle on resolution is compensated for up to the distance, $d = 28h$. Benefit of this arrangement is very limited at such short distances, and it seems more probable that complete compensation is not employed but the streak rather projects to horizon. Arcs represent constant-size, minimum-separabile profiles for a retina with uniform ganglion-cell density. The more appropriate form of rabbit profiles for a species confined to a terrestial habitat is immediately apparent. [c (1)] Plan view of an eye, *e*, moving towards a goal, *g*, in a frontal plane which contains a point w_1. [c (2)]. A representation of image shell in eye, *e*, which contains the frontal-plane image during such movement. [c (3)] Angular displacement of a series of points in the frontal-plane image shell, *c (2)*, as a function of their initial eccentricity when distance of eye from frontal plane changes by an amount, ΔX. If eye approaches frontal plane containing its goal, *g*, then for a displacement, ΔX, image of point w_1 will appear to be displaced radially, *c* (2), from image of goal by an amount, $\Delta\theta$, which is proportional to its initial eccentricity from goal, [c (2)] and [c (3)]. Angular resolution required to detect flow, $\Delta\theta$, of a given image point upon a given displacement, ΔX, of the eye decreases proportionally to eccentricity from image of goal. The equivalent acuity thus falls off along a rectangular hyperbola, $k\ 1/\Delta\theta$, [c (3)], just as the human acuity profile (WERTHEIM, 1894). If the human eye fixates its goal, then a displacement of the eye, ΔX, which produces a liminal image flow, $\Delta\theta$, near to fovea would also produce liminal flow in all parts of the retina served by such an area

That retinal topography is adapted to the information array of the visual image is strongly supported by the singular retinal organization of two species of cyprinodontid fish (MUNK, 1970). These animals possess a lateral streak which projects to the visual horizon and presumably deals with the stream bed. In addition, there is a second, parallel streak which projects its peak density at an elevation of 40° above the horizontal to deal with a 97° diameter circular field above the fishes' head. The retinal region between the two streaks is relatively disorganised and appears almost free of retinal ganglion cells.

The explanation of this arrangement appears to be that these fish are surface feeders for whom events above the water surface form an important part of the information array. Light rays entering the water are refracted at the surface according to SNELL'S law, which means that a 180° range of incident rays is compressed by refraction into some 97°. Compression is most marked for the region of field close to the water surface where insects abound and least for that overhead. The ganglion-cell distribution thus appears organised to provide high resolution in the region of greatest image compression and to invert the influence of SNELL'S Law on the absolute size of the minimum separabile. The disorganized region of retina between the two visual streaks accepts rays from the stream bed which are subject to total internal reflection at the stream bed which are subject to total internal reflection at the surface, a possible source of confusion to the fish if analyzed in detail.

V. Teleology of the Circumscribed Area Centralis

Certain species, such as squirrels, have long been described as possessed of "universal macularity" because of the relatively uniform organization and high ganglion-cell packing densities of their central and peripheral retina. However, when the small size of these eyes is taken into account, their high ganglion-cell densities are seen to be necessary for even moderate resolution; in all species which attain high *angular* resolution the retina appears to develop a circumscribed area centralis. Such a region is commonly but a few degrees in diameter, and it has been suggested that more extensive distributions are precluded by their concomitant requirement for a greater area of visual cortex for the analysis of their output.

If present in *mammals*, the area centralis is situated centrally in frontal eyes and temporally in lateral eyes; when the eyes are close to the anatomic position of rest, it invariably projects into the binocular field and is so placed as to obtain the image formed by parallel rays from the intersection of the horizon with the median plane. This region of the image is unique in that it contains the goal during forward linear motion, and the presence of a retinal area capable of high resolution could be of obvious benefit.

The circumscribed area centralis, unlike the visual streak, has attracted little discussion concerning the teleologic basis of its form. In man (OPPEL, 1965), goat (HUGHES and WHITTERIDGE, 1973), and cat (HUGHES, 1975b) the ganglion-cell density profile of the area centralis is well approximated by a rectangular hyperbola. Although the isocount lines of these areas are not usually circular, their symmetry about the peak count is very pronounced. It is improbable that these features are without functional significance; the great variety of ganglion-cell distributions so far described argues against their arising simply as a by-product of the morphogenic process.

During forward locomotion, the images of points which make up the bounding surfaces of objects describe well-defined trajectories which are a function of their position relative to the image of the goal. Such *optical flow* (GIBSON, 1966) is radially symmetrical about the image of the goal, and the image flow pattern is thus centered on the area centralis when the goal is fixated. It is suggested that the high degree of radial symmetry in the density profile of the area centralis is an adaptation to this relatively common dynamic information array; some quantitative support for this thesis is available.

Let the goal of a rectilinearly moving animal be contained within a frontal plane at distance x. Two points in the plane which are separated from the goal by distances w_1 and w_2 will subtend angles θ_1 and θ_2 according to the ratio defined by $\theta_1/\theta_2 = w_1/w_2$. If the distance of the eye from the frontal plane changes by an amount Δx (Fig. 22c1), then the image of a given point moves across the retina along an axis with its origin at the image of the goal (Fig. 22c2). The incremental magnitude of this image displacement, $\Delta\theta$, is proportional to the initial separation, θ, of the point from the image of the goal (Fig. 22c3); optical-flow velocity is thus zero at the goal image and increases proportional to eccentricity under these conditions.

Now in the eye of man under photopic conditions not only acuity (WERTHEIM, 1894) but also sensitivity for movement (BOURDON, 1902; AUBERT, 1886) decline along a rectangular hyperbola with increasing eccentricity from the fovea. When expressed in terms of threshold rather than sensitivity (WEYMOUTH, 1958), this simply means that the angular subtense of both the minimum separabile and limen for movement detection increase proportional to eccentricity. It therefore follows that if a man fixates his goal during rectilinear motion then the radially symmetrical image flow pattern overlays a retinal distribution of movement sensitivity (MC'COLGIN, 1960) which closely matches its form. Thus, for man, *a displacement of the eye perpendicular to the frontal plane, Δx, which produces a liminal image flow near to the fovea would also produce a liminal flow in all parts of the retina served by such an area.*

It is well established for man that optic flow patterns are important cues to the organisation of the visual environment (GIBSON, 1950, 1966). It is obvious, however, that under normal conditions the locomotion-induced flow patterns of the retinal image are much more complex than those analyzed above (GORDON, 1965; KOENDERINK and VAN DOORN, 1975; NAKAYAM and LOOMIS, 1974); objects at varying distances, the longitudinal extension of the terrain, and failure to fixate the goal may eliminate the symmetry of the flow pattern or its matching to the retinal limen distribution. Nevertheless, if the above flow pattern is a stochastically significant feature of the retinal image about the fixation point, then it may be economical to eliminate its inherent redundancy by distributing the available ganglion cells, which are concentrated in the region for the provision of increased resolution, according to the above profile.

The lack of behaviourally determined acuity profiles for species other than man, and the difficulties encountered in attempting their derivation from the ganglion-cell distribution, make it impossible to decide whether the above hypothesis is applicable to other forms whose area centralis ganglion-cell density profile approximates a rectangular hyperbola. That an area may evolve which is not adapted to the radial flow pattern discussed above is indicated by the presence of lateral areas in some birds. Such areas may be primarily concerned with the provision of increased resolution and would not be expected, under the above hypothesis, to show the same distribution of acuity as a frontally projecting area. At present, nothing is known of the behavioral potential of lateral areas in birds, and their quantitative anatomy has only been subjected to preliminary examination (FITE and ROSENFIELD-WESSELS, 1975).

The high ganglion-cell density of the area centralis is thus suggested to have the role of increasing both static and dynamic resolution, but the form of the density profile is hypothesised as matched to the *differential* of the image during goal-fixated forward locomotion. It might be suggested that the density profile of the visual streak is not for the compensation of the effects of the law of visual angle in the eye of the static animal but is rather matched to the flow pattern in lateral-eyed species during forward locomotion. In this instance, however, the lowest density in the streak would be that projecting laterally and the highest to front and rear in a distribution not encountered thus far.

E. Visual Acuity

I. Theoretical Resolving Power

Most visual behavior, especially that encompassed under the term pattern recognition, is beyond the scope of this chapter because its understanding mainly involves parameters of the CNS. The optic and neural factors discussed here rather define the bounds of possible performance in visual behaviour. Traditionally, visual acuity has been a popular topic for interpretation in morphologic terms; many problems remain, however, as will become obvious in due course.

PIRENNE (1962) defines visual acuity as

> the reciprocal of the visual angle, in minutes, subtended by the smallest detail which can be seen under given conditions.

Both he and OGLE (1969) give good general discussions of parameters influencing its magnitude. Many kinds of acuity may be established according to the nature of this detail: alignment acuity (SALOMON, 1947), parallelism acuity (ANDREWS, 1967), curvature acuity (OGILVIE and DAICAR, 1967), vernier acuity (BERRY, 1948), and stereoscopic acuity (BERRY, 1948), among others. Acuity is thus an operationally defined feature of an animal's behaviour which may vary with conditions of measurement, such as the level of illumination, and is determined by optic, retinal, and central factors. Most of the above types of acuity are predominantly determined by central factors (ANDREWS et al., 1973), have unknown mechanisms (WESTHEIMER, 1976, 1977) and thus, although dependant on retinal organisation, reveal little about it and have not been systematically investigated in other species. These *hyperacuities* (WESTHEIMER, 1975) enable discrimination of differential spatial localisation in the order of a few seconds of arc, much finer than the minimum photoreceptor separation, and at first consideration appear to be beyond the capabilities of the optical or neural apparatus. Hyperacuity is unrelated to the super-resolution mentioned below, in which the diffraction limit in the image may be locally exceeded, and is not to be confused with 'aliasing' in which the presence of a periodic stimulus can be detected by a receptor array spaced at some sub-harmonic interval, although it cannot be resolved in the usual sense. Such a phenomenon is only possible when the optical quality of the image is better than that of the sampling matrix and thus might be demonstrable in peripheral retina or when viewing interference fringes which enable optical limitations to be bypassed. Reports of anomalous resolution in insects (BURT and CATTON, 1962) were suggested by BARLOW (1965) to arise as edge effects when stimulus gratings were moved past a frame; since then no technically unchallengable evidence in support of anomalous resolution has been put forward (see HORRIDGE, 1975). Aliasing or anomalous resolution may account for phenomena observed with high frequency interreference gratings of period below that detectable via the optical apparatus of the eye (BYRAM, 1944). In fact, some of the improvement in resolution encountered with such fringes at the fovea may result from inability to see the transition from cut-off frequency resolution to 'presence' detection in the tiny visible region of grating (CAMPBELL and GREEN, 1965). Certain other 'so-called' acuities which establish a *minimum visibile* are neglected here because they reduce to a brightness discrimination task (HECHT and MINTZ, 1939; PIRENNE, 1962) rather than one of identification of the spatial location of stimuli. This discussion is thus limited to relating the *minimum separabile* to morphologic features of the eye and is centered on acuity established from static grating-pattern stimuli for which a few comparative results are available. Results obtained with the Landolt C pattern cannot be compared, because this amounts to a minimum visibile task and gives quite different results to grating acuity in man (SHLAER, 1937; MILLODOT et al., 1975). For a retinal illumination of 3.2×10^3 trolands, a Landolt C gap of 28″ is resolvable compared with a grating-bar width of 35″. In man, at least, it has been shown (GILBERT and FENDER, 1969; WESTHEIMER, 1972a) that visual acuity with a stabilised retinal image is only very slightly lower than for that measured with a freely moving eye. No consideration of dynamic acuity is therefore presented in the following discussion but it has been treated in some detail by DITCHBURN (1973).

It has already been pointed out that the image of even a diffraction-limited optic system has an inherent grain, because any point in object space gives rise in the image plane to a light distribution of finite and irreducible dimensions, the *Airy Disc;* for a human eye with an optimal 2.0 mm pupil (WOODEHOUSE, 1975), $\lambda = 555$ nm and $\mu_v = 1.336$ the Airy disc is 52′ in radius. Under other conditions of illumination, it has been demonstrated that the natural pupil diameter is optimized to give the best possible behavioural resolution (CAMPBELL and GREGORY, 1960; WOODEHOUSE, 1975).

The optic resolving power of the eye is often defined in terms of the RAYLEIGH (1879) criterion, according to which the images of two point sources cease to be resolvable when their centers are separated by the Airy disc radius, and the summed intensity distribution evinces a 26% trough between its peaks. The assumption of such a degree of limiting contrast is arbitrary, and the true value for any system depends upon the nature of the receptor which is employed to examine the intensity distribution. The SPARROW (1916) criterion, achieved when the Airy discs are separated by 0.78 or less of their radius, is sometimes presented as a limiting condition in which resolution is impossible, because there is no trough in the summed intensity distribution (Fig. 15). For a 2-mm pupil the optic resolving power is thus 52′ according to the Rayleigh criterion and 40′ according to the Sparrow criterion. Two-point discrimination is, however, more complex than implied by such an approach, and its inadequacy as a basis for specifying resolution has been emphasized by TORALDO DI FRANCIA (1955) who writes

> one cannot escape the discouraging conclusion that a very common sentence like: The resolving power of such an instrument has such a value has no meaning. Resolving power is not a well-defined physical quantity.

The summed intensity distribution for two points, no matter how close they are to one another, is always theoretically different from that for one point, and is thus subject only to practical limitations on its discrimination. The consideration of limitations on resolution thus requires a more general approach. The fundamental *invariant* which limits the information capacity of an optic system is its number of degrees of freedom defined in terms of the two-dimensional spatial band width, object area, polarization states, temporal band width, and observation time. An increase in the performance of one of these components may be attained upon the surrender of capacity from another. Super-resolving systems which exceed the classic resolution limit are thus possible if the object field is restricted. A theoretical and experimental discussion of such systems is available (BACHL and LUKOSZ, 1967).

However, when information is *uniformly* distributed throughout the image space, a straightforward and theoretically sound basis for the definition of an optic minimum separabile is possible. According to the Fourier approach discussed above, we know that the image cannot contain a spatial frequency component greater than the MTF cut-off frequency defined by the optic system's PSF width. The minimum separabile thus cannot be smaller than the period of an extended sinusoidal grating of cut-off frequency, because the contrast between the peaks and troughs of the intensity distribution is then zero.

The *attainment* of even this degree of resolution is, however, dependent upon the receptor distribution. HELMHOLTZ (1924) first pointed out that between two

Table 5

	Man		Cat		Rat	
MTF. cut-off frequency	60	cycles/°	20	cycles/°	4.6	cycles/°
RMF	0.291	mm/°	0.222	mm/°	0.06	mm/°
Theoretical necessary sampling density for optic information	179000	mm^2	32000	mm^2	24000	mm^2
Cone density	150000	mm^2	25000	mm^2	4000	mm^2
Theoretical cone cut-off frequency	56	cycles/°	17.5	cycles/°	1.86	cycles/°
Ganglion-cell density	150000	mm^2	10000	mm^2	6000	mm^2
Theoretical ganglion-cell cut-off frequency	56	cycles/°	11	cycles/°	2.3	cycles/°
Behavioral cut-off frequency (acuity)	64	cycles/°	6	cycles/°	1.5	cycles/°

Sources (man, cat, rat, optimal values, respectively): MTF.: CAMPBELL and GREEN (1965); ENROTH-CUGELL and ROBSON (1974); 1-mm pupil: HUGHES (1976); RMF.: LE GRAND (1967); VAKKUR and BISHOP (1963); HUGHES (1977c); Cone density: ÖSTERBERG (1935); STEINBERG et al. (1973); HUGHES (1977c); Ganglion-cell density: 1:1 cone to ganglion cell assumed, see BOYCOTT and DOWLING (1969), MISSOTTEN (1975), HUGHES (1975b), and HUGHES (1977c); Behavioural acuity: SHLAER (1937), BLAKE et al. (1974), and HERMANN (1958).

potentially resolvable image points a further receptor must be interdigitated to detect the reduction of intensity between them. This result is essentially that pronounced in the sampling theorem of CAUCHY and SHANNON (YEN, 1956; CAUCHY, 1841; SHANNON and WEAVER, 1949); a one-dimensional distribution is completely specified by its amplitudes at successive intervals of $1/2f$ if it has no Fourier components of higher spatial frequency than f cycles per degree. The resolution of the 52′ two-point separation of the 2 mm pupil RAYLEIGH criterion thus requires receptor spacing of 26′. The sampling theorem is readily extended to deal with a two-dimensional distribution, such as the retinal image, for which it may be shown (REZA, 1961; BRILLOUIN, 1962) that all of the available information may be recovered from a *square* sampling matrix with receptor spacing of $1/2f$ degrees as above. A more realistic hexagonal array may be shown to be optimal with a somewhat greater spacing of $1.15/2f$ (PETERSON and MIDDLETON, 1962; SNYDER and MILLER, 1977b).

It is thus possible, given the MTF or PSF of an optic system, to estimate its cut-off spatial frequency, f cycles/degree, by means of extended gratings and obtain the minimum retinal sampling density adequate to recover all the spatial information present in the image. Correction is required if the test fields are of limited extent (SIMON and DENIEUL, 1973). A comparison of the theoretical and actual sampling density of the cone or ganglion-cell distributions indicates whether the potential resolution of the optic system could be approached by the neural apparatus—but, of course, does not indicate whether it is. Sufficient information is available to permit this exercise to be carried out for only three species of mammals (Table 5).

Persons of analytic disposition, and I must confess to have one with "too strict an incredulity" in mind, are commonly unsatisfied with attempts to apply the sampling theorem to real systems because they find the conditions of its mathematical derivation immediately unsatisfied. Certain problems may arise simply from incomplete statements of the theorem such as that above, which omits reference to signals at the cut-off frequency, f_c. In an excellent treatment of the subject BRACEWELL (1967) presents a more complete version of the theorem,

A function whose Fourier transform is zero for $|f|>f_c$ is fully specified by values spaced at equal intervals not exceeding $1/2f_c$ save for any harmonic terms with zeros at the sampling points.
but equal intervals are not obligatory; irregularity in the sampling points is not necessarily, but may be (BRACEWELL, 1965; FRENCH et al., 1977), detrimental to performance and does not preclude theoretical treatment (YEN, 1956). Inability to meet the requirements of ideal bandlimiting is more often seen as a barrier to application of the theory. However, the whole conception of bandwidth limitation as applied to real systems is essentially paradoxical (SLEPIAN, 1976); a bandwidth limited signal must be infinite in duration and a finite duration signal must be infinite in bandwidth, yet we commonly deal with signals as if of both finite duration and limited bandwidth. Clearly the practical application of these concepts requires consideration of the *effective bandwidth* of a signal specified by some upper frequency beyond which its energy is small enough to be ignored with a resultant error acceptable in the system under consideration (CHIRLIAN, 1967). It is noteworthy that the diffraction limited image is one in which there really is an upper cut-off-frequency above which there is zero modulation.

Photoreceptor densities and optical image quality in homeotherm eyes are such that undersampling at this layer is improbable. Available data does suggest, however, that optical quality may be superior to the sampling capability of the retinal ganglion cell array in the central areas of most species, and the periphery of all. If the ganglion cell had direct access to the image plane then the effect of this would be to cause the periodic upper sidebands in the frequency domain (BRACEWELL, 1965) to overlap the distribution centred on zero frequency with consequent inability to recover the spectrum of the original signal and resultant distortion. However, such a result may be quite acceptable in some systems and, in any case, it is apparent that the ganglion cell layer may well be matched to the optimal image of the overlying retina by a neural low pass filter compounded in the prior stages of retinal processing. It remains to be seen whether the cut off frequency of retinal ganglion cells involved in resolution tasks is matched to their local density or to the image quality, when it is superior in resolution.

Other limitations such as the inevitable restriction of the system to a finite array introduce further sources of distortion but these can be small and are discussed by BRACEWELL (1965) and by DITCHBURN (1976) in a lucid and brief account. More important from the point of view of practical application of sampling theory is consideration of retinal noise and photon noise which, at all realistic light intensities, may swamp the tiny modulation of the sampling aperture resulting from signals above cut-off-frequency. A compact discussion of the design requirements for maximising signal to noise ratio in a diffraction limited retina has been provided by SNYDER and MILLER (1977b).

Tentatively accepting the data upon which Table 5 is based, we may conclude that the foveal photoreceptor and ganglion-cell matrix of man are closely matched to the quality of the retinal image (see also BURTON, 1973); the match is also good for the cat photoreceptor matrix, although not for its ganglion-cell distribution, and is poorer still for both rat photoreceptor and ganglion-cell matrices. As suggested earlier, the behavioral cut-off frequency is most closely related to the theoretical ganglion cell cut-off frequency but, of course, this is demonstrated above only for the retinal elements near the fixation axis and not those of the peripheral retina.

II. Optical Limitations on Visual Acuity

The direct formation of sinusoidal interference grating patterns in the human retina (LE GRAND, 1937; CAMPBELL and GREEN, 1965; WESTHEIMER, 1960) is relatively uninfluenced by the refractive state of the eye and may thus be employed to determine the resolution of the neural visual apparatus with the optics "short-circuited." Comparison of a contrast sensitivity function determined this way with one obtained by direct viewing of a sinusoidal grating enabled CAMPBELL and GREEN (1965) to establish that nearly one-third of the loss in contrast of high-frequency components of the retinal image arose in the human *foveal* region from the limitations of the optic rather than the neural apparatus. A subsequent examination of more eccentric temporal perifoveal images by GREEN (1970) indicated that up to 4° from the fixation point the image quality was poorer for normally viewed than for interference gratings (Fig. 23–4 + 5). At greater eccentricities he concluded that optic quality was of little significance in limiting the performance of the subjects employed.

MILLODOT et al. (1975) report that the correction of refractive errors has no influence on peripheral grating, or Landolt C, acuity. By contrast, LEIBOWITZ et al. (1972) describe considerable improvement in peripheral-movement detection thresholds after vertical and horizontal cylindrical refractive corrections, but such corrections were not necessary in all subjects. Oblique astigmatism was not compensated and could play an important part in limiting performance under natural conditions. It is important to realise, however, that image quality is not simply determined by focus and that diffraction, scatter, etc., may play important roles. An investigation of the peripheral retina at 20° to 70° eccentricity by means of the interference technique (FRISEN and GLANSHOLM, 1975) contrasted with the findings of GREEN (1970) in attributing a reduction of peripheral acuity of up to 40% to optic factors. Of course the absolute quality of the retinal image is not established by this method, and it must have undergone considerable deterioration relative to that on the optic axis in order to have such influence on the much-reduced peripheral neural acuity. This view is substantiated by the preliminary human peripheral LSF measurements of RÖHLER (1962).

For the remaining mammals there is little direct experimental information about optic limitations on resolution. According to BONDS (1974), the sensitivity profile of some cat retinal ganglion-cell receptive fields is so narrow as to be necessarily defined by the PSF; the theoretical treatment of WÄSSLE and CREUTZFELDT (1973) shows that the receptive-field diameter measurements require significant correction for optic blurring if less than 15′ in extent even in the central field. The 135-fold range of pupil area of the cat (WILCOX and BARLOW, 1975) means that the optic limitations on neural resolution will vary significantly with the natural pupil size. The LSF determined by ENROTH-CUGELL and ROBSON (1974) appears narrowest for a 1 mm pupil at 2.8′ half-height width in contrast to BONDS' claim that image quality reduces again for pupils less than 3 mm wide. It is improbable, for the theoretical reasons outlined previously, that the cat should suffer in behavioral resolution because of optic factors when the pupils are 1–3 mm in width. However, BONDS (1974) points out that for larger pupils the best retinal ganglion cell MTF obtained by ENROTH-CUGELL and ROBSON (1965) is of inferior quality to the optic MTF with a 4.5 mm pupil but superior to that with a 13.5 mm pupil. CAMPBELL et al. (1973) report that the evoked potentials at the cortical projection of the area centralis establish a contrast-sensitivity function which drops to 4% at 5 cycles/°, a frequency at which BONDS' optic MTF for the cat is only reduced to 60%, and conclude that the optic quality of the cat retinal image is much better than the resolving power of its visual system. However, the validity of this conclusion is doubtful, because CAMPBELL et al. (1973) employed a 6 mm pupil, yet compare their results with those BONDS obtained with a 3 mm pupil. The difference in pupil size can have substantial implications; the MTF for one of BONDS' cats dropped to 5% (BONDS, 1974) at 5 cycles/° for a 3.0 mm-wide pupil but only attained 2.5 cycles/° with a 4.8 mm pupil. A 6-mm pupil would thus impose a significantly greater limitation on the overall cat contrast-sensitivity function than a 3-mm pupil. The behavioral contrast-sensitivity function obtained for the cat by BLAKE et al. (1974) would be even more subject to influence by optics than that of CAMPBELL et al. (1973).

According to BONDS (1974), the peripheral optic quality of the cat eye is poorer than the central; the LSF increases twofold in width over 15°–30° eccentricity. However, he calculates that the 0.6° LSF

10% width at 30° eccentricity would lead to only a 3% increase in the mean local receptive field diameter of 2.5° (FISCHER and MAY, 1970), and thus peripheral optic factors have little influence on resolving power.

HUGHES and WÄSSLE (1977) have estimated the rat MTF to have a cut-off frequency of some 13′ with a 0.3-mm pupil and between 25′ and 36′ for a 2.5–3.0-mm pupil. The behavioral acuity of 40′ (HERMANN, 1958) clearly indicates that resolution in this species is unlikely to be substantially limited by the quality of its optics at photopic levels of illumination.

III. Neural Limitations on Visual Acuity

The decline of visual acuity with increasing eccentricity from the human fixation axis has been subject to numerous attempts at explanation in morphologic terms since its first detailed description by AUBERT and FÖRSTER (1857). However, current work makes it impossible to agree with the recent statement of LE GRAND (1967) that for man

> it is obvious that the dioptrics of the eye, which are of fairly good quality up to 30° eccentricity, have no effect whatsoever on peripheral acuity,

and the neural component of the acuity gradient requires separate specification. Neither peripheral acuity nor optics has been investigated in other mammalian species, and the topic must be considered almost solely in terms of the organization of the human eye.

The various methods of measuring visual acuity reveal a qualitatively similar gradient across the retina (Fig. 41, LE GRAND, 1967), but under the same conditions the use of the Landolt C consistently indicates higher peripheral values (BERKLEY et al., 1975) than grating acuity. The latter is, however, a more satisfactory measure and amenable to both optic and anatomic interpretation. In spite of this, few measurements of human grating acuity are available, and of these the majority deal only with the temporal horizontal meridian. The descriptions provided by DOR (1873), WERTHEIM (1894), FICK (1898), and BERKLEY et al. (1975) agree quite closely with each other, but the results of WERTHEIM (1894) alone provide complete isoacuity contours for the whole field of vision. These still stand as exemplary work but suffer the disadvantage that they are expressed in relative terms, and the conditions of illumination at which they were obtained is not known. The human isoacuity contours are elliptical, with the minor vertical axis in the ratio of 0.7 to the horizontal axis. Lower field acuity is somewhat greater than upper (WERTHEIM, 1894; LOW, 1943; MILLODOT and LAMONT, 1974). The decline in human acuity along the horizontal meridian in Wertheim's results was shown by WEYMOUTH (1958) to be well fitted by a rectangular hyperbola out to an eccentricity of some 50°. It is thus of interest to see if this form can be explained on an anatomic basis.

Certain difficulties first require outlining. The acuity curve of WERTHEIM (1894) for the temporal retina is shown in Fig. 23—3, in terms of its inverse—the minimal angle of resolution; the bar width or half period in minutes of a just resolvable grating—, where it is seen to be similar in form to the Landolt C acuity curve determined by MANDLEBAUM and SLOANE (1947) at their highest light

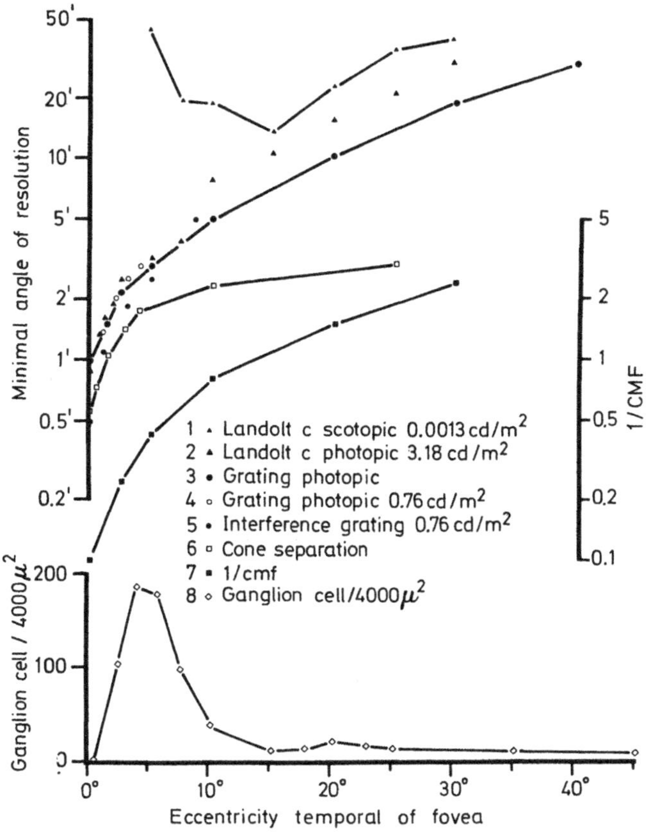

Fig. 23. Minimal angle of resolution—the bar width of a just-resolvable grating—as a function of retinal eccentricity, illumination, and stimulus form. Greater resolution obtained by the use of interference fringes as grating stimuli (GREEN, 1973) indicated that optic quality limits performance of foveal neural apparatus during normal viewing of a grating. Other sources: photopic grating normalized to 1′ acuity, WERTHEIM (1894); photopic and scotopic Landolt C, MANDELBAUM and SLOANE (1947). Cone separation (ÖSTERBERG, 1935) is plotted over the the same range of eccentricity and may be seen to account for neural resolution within some 2° of visual axis. At greater eccentricities, potential resolution of cone matrix is not achieved because of convergence in receptor to ganglion-cell connections. Displacement of foveal ganglion cells to the perifoveal region in man (OPPEL, 1965) makes it difficult to demonstrate a direct relationship between ganglion-cell separation and resolution. However, in other primate species, the cortical magnification factor has been shown to vary as the ganglion-cell *density* COWEY and ROLLS, 1968). The inference that this is also true in man is reasonable; the above demonstration (COWEY and ROLLS, 1974) that the reciprocal of the CMF varies as the minimal angle of resolution thus indirectly supports the thesis that human acuity varies as ganglion-cell *density*

intensity (Fig. 23—2). It is to be emphasised that both of these curves change as the light intensity is reduced (MANDLEBAUM and SLOANE, 1947) into the scotopic range where, as in Fig. 23—1, the resolution becomes fairly constant across the entire hemifield. A morphologic explantation attempts only to deal with extreme photopic conditions and cannot encompass the functional reorganisation of the retina. The illumination sensitivity of the acuity curves also introduces the possibility of obtaining low peripheral acuity readings because of the reduction in illumination intensity arising with oblique-ray incidence and the reduced apparent pupil size (WEALE, 1956).

It is characteristic of attempts to account for acuity in anatomic terms that only *retinal* structure is considered, but it is obviously possible that subsequent stages in the visual pathway set their own limitations, and this has been demonstrated to occur. Thus, acuity for vertical or horizontal gratings is the same at a given region of the visual field over a range of some 25° from the fixation axis (BERKLEY et al., 1975), but acuity for oblique gratings is markedly reduced (HIGGINS and STULTZ, 1950; APPELLE, 1972) up to an eccentricity of some 8°–18° (BERKLEY et al., 1975). This "oblique effect" has been denied in infants (TELLER et al., 1974) but recently was claimed at 6 weeks and more convincingly demonstrated at over 13 weeks of age (LEEHEY et al., 1975), so that it may not be environmentally determined. The phenomenon is regarded as neural rather than optic in origin, because it remains when interference fringes are employed as stimuli (CAMPBELL et al., 1966; MITCHELL et al., 1973). Cortical-evoked potentials are found to be greater for vertical and horizontal than for oblique grating stimuli in man (MAFFEI and CAMPBELL, 1970; FREEMAN and THIBOS, 1973), but such a difference was not found in the electroretinogram. Explanation of this aspect of grating acuity is thus sought at the cortical rather than retinal level in terms of the proportions of single units sensitive of specific stimulus orientations (BERKLEY et al., 1975).

Central photopic acuity was first explained in terms of foveal-cone separation by HELMHOLTZ (1924), but extension of the quantitative comparison to the peripheral retina awaited ÖSTERBERG'S (1935) account of the human retinal-cone distribution. Several authors [POLYAK, 1941; LUDVIGH, 1941; DOESSCHATE, 1946; (Fig. 43), LE GRAND, 1967] then found satisfactory agreement to extend for only a degree or two of eccentricity, acuity subsequently falling off more rapidly than cone separation (Fig. 23—6). Even when the possibility of optic impairment of resolution is eliminated by the use of interference fringes as stimuli, it is found (GREEN, 1970) that cone separation and the sampling theorem account for the attained neural resolution over no more than 2° from the fixation point.

The accepted conclusion that acuity at greater eccentricities is primarily determined by the local ganglion-cell provision (DOESSCHATE, 1946; WEYMOUTH, 1958) is exceedingly difficult to establish quantitatively in man. WEYMOUTH (1957) employed odd results of POLYAK (1941) to obtain a crude quantitative human ganglion-cell distribution which he related to the acuity gradient. The ganglion-cell distribution map of VAN BUREN (1963) is of little use in quantitative discussion, because it does not provide densities but only the local number of cell layers. A relationship between ganglion cell provision and acuity is, however, apparent in these maps, because their contours are of similar elliptical form to the isoacuity lines of WERTHEIM (1894). No ganglion-cell density map based on a human retinal flat mount is available, but VILTER (1949, 1954) and OPPEL (1965, 1967) provide uncorrected ganglion-cell distributions for a horizontal thin section of human retina (Fig. 23—8). It is apparent that the contamination of ganglion cells connected to perifoveal receptors with those displaced from the fovea makes the comparison between ganglion cell distribution and acuity difficult. Correction for the ganglion-cell connectivity pattern may become feasible (MISSOTTEN, 1974). At present, it is possible to relate acuity to the ganglion-cell distribution via the cortical magnification factor.

In many species of animal, it has been found that the distance in millimeters devoted to the representation of 1° of the visual field at its projection on some central station of the visual pathway, the magnification factor (MF), is proportional to the ganglion-cell density of the corresponding retinal region: frog tectum (JACOBSEN, 1962), cat LGN, rabbit colliculus (SENEVIRATNE, 1963), cat and baboon colliculus (VEJBAESYA, 1968), cat LGN (SANDERSON, 1971), rabbit colliculus, cortex, and LGN (HUGHES, 1971), and monkey cortex (ROLLS and COWEY, 1970), owl monkey (WEBB and KAAS, 1976). This relationship cannot be regarded as inevitable, because it does not apply to the projection of the visual field parallels onto the rabbit visual cortex. However, ROLLS and COWEY (1970) describe both squirrel monkey and macaque cortical magnification factors (CMF) as proportional to the topographically corresponding ganglion-cell density from 50° to 7° eccentricity and extrapolate from the cone separation to conclude that this relationship applies throughout the perioveal and foveal regions. At present, however, the gradient of monkey visual acuity has not been determined. but DANIEL and WHITTERIDGE (1961) did show that *human* visual acuity varies as the *monkey* CMF and thus (ROLLS and COWEY, 1970) as the *monkey* retinal ganglion-cell density.

More recently, COWEY and ROLLS (1974) have employed the results of BRINDLEY and LEWIN (1968), who plotted the projection into the visual field of phosphenes induced during cortical stimulation at known locations, to determine CMF's for man and to demonstrate their proportionality (Fig. 23—7) to acuity measured at the same eccentricity by WERTHEIM (1894). No satisfactory human ganglion-cell distribution map is available, but the uncorrected results of VILTER (1954) and, to a lesser degree, of OPPEL (1965, 1967) suggest that from 50° to 10° eccentricity the cortical magnification factor varies roughly as ganglion-cell density. The form of the ganglion-cell density/CMF plot is so similar to those for *saimiri* and *macaque* (ROLLS and COWEY, 1970) as to stand as presumptive evidence that proportionality between the two factors may be extrapolable into the foveal region in similar fashion as for the two monkeys (ROLLS and COWEY, 1970).

FRISEN and FRISEN (1976) have recently claimed that the angular frequency of human retinal ganglion cells, corrected for peripheral reduction in RMF, is simply proportional to the corresponding local acuity established by an interference fringe technique (FRISEN and GLANSHOLM, 1975). A similar claim has been made by DRASDO (1977) who reports proportionality between acuity as established by WERTHEIM (1894) and the corresponding ganglion cell frequency, obtained as the mean of data from several orientations in the publications of three authors (VILTER, 1954; OPPEL, 1967; VAN BUREN, 1963) and also corrected for change in RMF with eccentricity. FRISEN and FRISEN suggest that their result should be no surprise; indeed, it is what might be expected a priori from sampling theory considerations! However, it would necessarily follow from their findings, and those of DRASDO (1977), that CMF, because it is proportional to acuity, must also be proportional to the ganglion cell frequency, i.e. the square root of the ganglion cell density. DRASDO (1977) clearly misinterprets ROLLS and COWEY'S results (1970) as indicating this to be true for the monkey and thus in accordance with his conclusions for man; in fact, ROLLS and COWEY report the macaque CMF to be proportional to ganglion cell density and not its square root. If FRISEN and FRISEN (1976) and DRASDO (1977) are correct, it would mean that man is unique amongst the investigated mammals, assuming the validity of earlier data, in failing to have his CMF proportional to ganglion cell density and, perhaps more important, quite different in organisation to the macaque. However, the ganglion cell distribution data employed by FRISEN and FRISEN, and DRASDO, are not of such quality to support confidence in their conclusions, especially in view of the conflicting indications from other sources. All the data was obtained from thin sections which give unreliable counts (see STONE, 1965); their shrinkages are unknown and probably different; the counts have not been subjected to an ABERCROMBIE correction (ABERCROMBIE, 1946), whose magnitude changes as a function of ganglion cell size with eccentricity; the ganglion cell criteria are unspecified and probably vary between the sources (the values of VAN BUREN differ by at least one order of magnitude from those of VILTER with which DRASDO pools them) and so on. A definitive count is required.

It thus remains almost certain, as suggested above, that the peripheral visual acuity of man and monkey is proportional to the corresponding ganglion cell density and not, as might be expected from elementary application of the sampling theorem, to its square root. Resolution would thus decrease more rapidly with increasing eccentricity than ganglion cell proximity and clearly indicates that *total* ganglion cell density cannot be a limiting factor for human peripheral acuity. It is curious therefore, that the absolute foveal acuity of man is well accounted for in terms of the square root of the sampling density at the receptor and ganglion cell levels (Table 5). The apparent surplus of ganglion cells over the theoretical requirements thus seems to increase with eccentricity from the fovea and suggests the possibility of a progressive change in retinal organisation which follows a simple law.

To avoid the problem of the displaced ganglion cells of the fovea, consider the ganglion cell densities, *D*, to be expressed in terms of their receptive field densities per square degree instead of perikaryon density on the retina. The theoretical and actual spatial frequency of the image above which the ganglion cell matrix is incapable of resolution, *fp*, for the peak density of the fovea is then given as $f_p = \sqrt{D_p}/2$ in cy/°. If the cut-off frequency elsewhere on the retina is accepted as falling off proportional to the local ganglion cell density then at the point *i* it is given as $f_i = (\mathrm{Di}/\mathrm{Dp}).(\sqrt{\mathrm{Dp}})/2$ in cy/° whence it may readily be derived that the ratio of the actual ganglion cell density at the point i to the effective density, De (i.e. that required by an optimal system to obtain the same cut-off frequency, fi), is given by Di/De = Dp/Di. It is possible that various factors interact to give an approximation to this relationship but, if not, we seek a single influence to explain why the local surplus of ganglion cells increases as the ratio of the peak to local density and thus inversely as acuity. Given the previous approximation of the human acuity curve to a rectangular hyperbola it may also be concluded that the surplus ratio increases proportional to eccentricity.

Behavioural acuity on the fixation axis and the corresponding central ganglion cell density are available for five species of mammal thus enabling an examination of the general applicability of the sampling theorem at the visual pole (Fig. 24). Agreement between theoretical and behavioural acuities is fair but the behavioural values for cat and rabbit a significantly lower than the ganglion cell density would suggest. A degree of uncertainty must, however, by inherent in such comparisons. The acuity data of the cat was obtained from the results of BLAKE et al. (1974) which suggest a cut-off frequency of some 6 cy/°; new techniques, BISTI and MAFFEI (1974), MITCHELL et al. (1977), have brought reports of a 9 cy/° cut-off frequency and even higher values have been mooted, JACOBSEN et al. (1976), although accompanied by some curious anomalies. In tests of stripes versus uniform fields, or even of horizontal versus vertical stripes, the animal may be able to detect the presence of the stimulus by means of an 'aliasing' phenomenon because its eye is not diffraction limited. Acuities in the order of 9 cy/° in the cat are close to the predictions based on the total ganglion cell density of the central area being representative for cells involved in resolution tasks. The applicability of the sampling theorem to the *total* ganglion cell density in the cat central

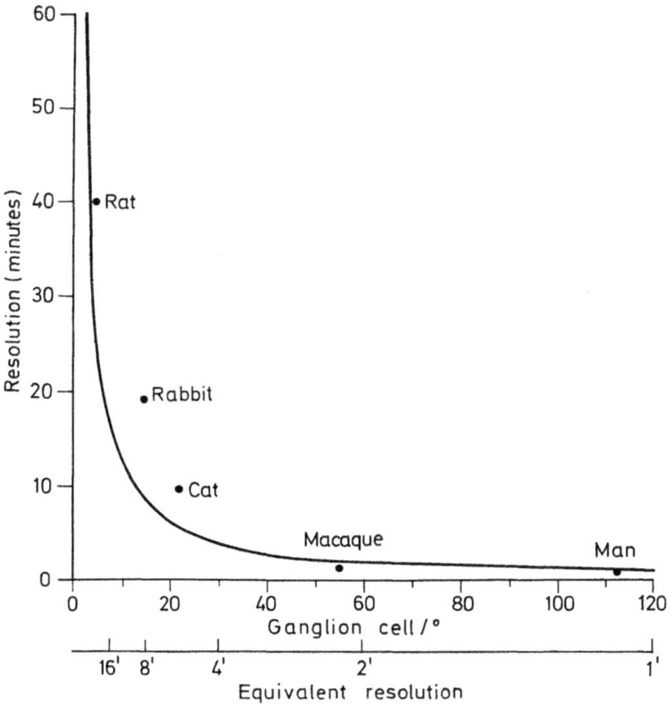

Fig. 24. The points indicate the behaviourally determined grating resolution in min/cycle, plotted against the square root of the ganglion cell density, ganglion cells/°, for the region of retina bearing the highest ganglion cell density on the visual axis. A second scale, equivalent resolution, indicates the width in minutes of the theoretical minimum resolvable grating period determined as [60/(ganglion cells/°)]/2 in accordance with SHANNON's sampling theorem. The continuous line indicates the relationship theoretically applicable if the potential resolution of the ganglion cell density were achieved behaviourally. The deviation from the line is significant for rabbit and cat. Sources: rabbit; VAN HOF (1967), HUGHES (1970); macaque (ROLLS and COWEY (1970); remainder, see Table 5

area thus remains in doubt; the situation in the less well investigated rabbit remains more obscure.

We are thus left with the puzzle as to why there is an apparent *systematic* failure of the sampling theorem away from the human fovea, some doubt as to its applicability in the central area of cat and complete uncertainty as to its relevance in the central areas of other non-primate mammalian species. The peripheral retina of species other than man cannot be discussed in this context as peripheral acuity measurements are not available. The interference fringe method of retinal stimulation indicates that the majority of the peripheral reduction in acuity in man is attributable to neural factors and that a simple explanation of the peripheral deviations from the sampling theorem is not probable in terms of either refractive error or the greater obliquity of the pupil leading to increased PSF width. The oblique angle of illumination of the pupil by peripheral rays reduces the peripheral retinal illumination and might thus lower the measured acuity relative to that obtainable with the same area on axis but this effect cannot explain the marked deviations from the expected relationship within the central 30° of the retina. Possible sources of this discrepancy are worth discussing in detail because of their more general significance.

1. Nonhomogeneity in the Retinal Ganglion Cell Receptive Field Population

It has long been known that the retinal ganglion cell population of the cat is made up of cells with both *on*- and *off*-center concentric receptive fields (KUFFLER, 1953). Although these two classes have been regarded as involved in resolution tasks, it is not known whether they play independent roles. The description of the rabbit retina contrasted markedly with that of the cat in that at least 40% of the encountered cells possessed nonconcentric receptive fields which could be grouped into about six classes (BARLOW et al., 1964; LEVICK, 1967). The early hints that the cat also possesses nonconcentric receptive fields which could play no part in acuity tasks (RODIECK and STONE, 1965; STONE and FABIAN, 1966) have subsequently been validated. Cat retinal units are now grouped into three major categories which are based upon a variety of classificatory techniques and termed X, Y, and W (STONE and FUKUDA, 1974), or brisk-sustained, brisk-transient, and sluggish (CLELAND et al., 1971; CLELAND and LEVICK, 1974a); within the last group some six subclasses have been described (STONE and FUKUDA, 1974; CLELAND and LEVICK, 1974b) which contain receptive fields similar to all except the orientation selective retinal units of the rabbit. Nonconcentric units have recently been reported in the macaque retina (DE MONASTERIO and GOURAS, 1975), in addition to some 20 varieties of colour and broad-band receptive fields; the extent to which these units are pooled for resolution tasks remains to be determined. Among other mammalian species, the presence of nonconcentric units has been reported in the ground squirrel (MICHAEL, 1968), sheep (HUGHES and WHITTERIDGE, 1973), and rat (HUGHES, 1976).

Of all these receptive field classes, only the category corresponding to the brisk-sustained cells, X cells, of the cat is generally regarded as capable of subserving resolution (CLELAND et al., 1971). The discrepancy between the acuity achieved by a given area of nonfoveal human retina, or of non-primate mammalian central area, and that calculated from its corresponding ganglion-cell density might thus arise simply because such cells form only a fraction of the local ganglion-cell population.

The identification of the proportion of the local ganglion-cell population suited to resolution tasks is a formidable problem. The physiologic rate of encounter of receptive field types does not appear to correspond with the absolute proportion of the population which makes up a given class. This is known for certain in one instance. Morphologic criteria have enabled the retinal ganglion cells of the cat retina to be grouped into three classes termed α, β, and γ, in order of decreasing perikaryal size and axon diameter (BOYCOTT and WÄSSLE, 1974). Various techniques have enabled the identification of these with the three major receptive field classes (STONE and FUKUDA, 1974; CLELAND and LEVICK, 1974a, b). In a detailed study, some 60% of the peripheral receptive fields examined by CLELAND and LEVICK (1974b) are brisk-transient, whereas histologic evidence indicates that only 3% (WÄSSLE et al., 1975) of the total ganglion-cell population is made up of the corresponding (CLELAND et al., 1975) α cell bodies. Attempts to relate the two sets of proportions by corrections based on theoretical considerations (FUKUDA and STONE, 1974) have not proved satisfactory, and the degree to which electrode impedance is involved in selective sampling is disputed (STONE, 1973; LEVICK and CLELAND, 1974).

The three major cat retinal ganglion-cell classes may be distinguished in peripheral retina by means of perikaryal features alone, but in regions of higher ganglion-cell density the ganglion-cell diameter spectrum loses its distinct modes. In these regions the use of an arbitrary transition size to divide the spectrum (FUKUDA and STONE, 1974) of all areas of retina is not justified. However, by this means FUKUDA and STONE (1974) suggest that the proportions of the various classes, especially the α cells, undergo significant change across the temporal retina. By contrast, in a wider ranging study, WÄSSLE et al. (1975) conclude the α cell proportion to remain relatively constant over the entire retina and publish a map of α-cell distribution (Fig. 25) which is not greatly different in form from that of the total ganglion-cell distribution (Fig. 20). ROWE and STONE (1976) have attempted to estimate the proportions of α, β and γ cells in the periphery and visual streak of the cat retina by dividing the soma diameter spectrum of both regions into large, medium and small modes by means of the criterion

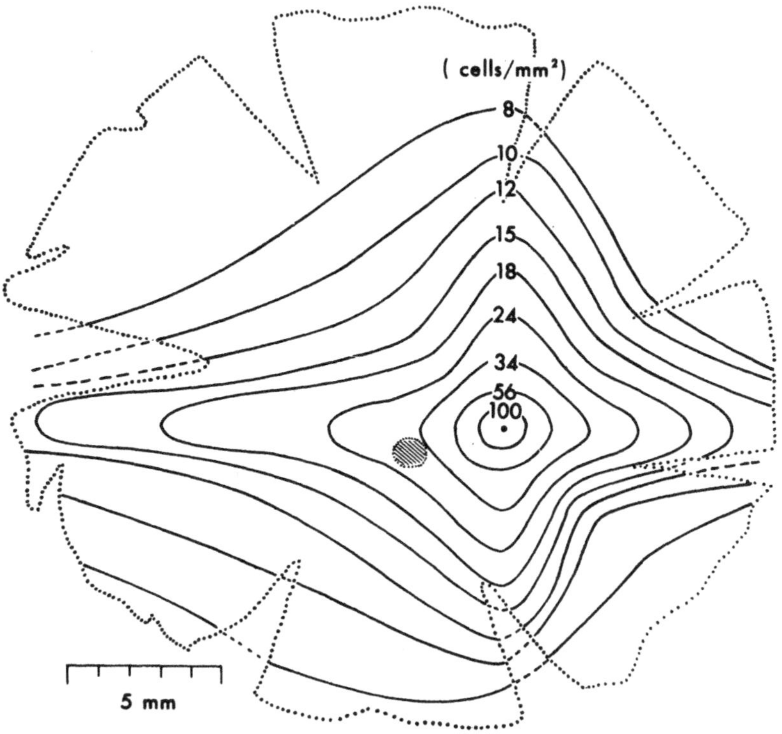

Fig. 25. α-type ganglion-cell density map of a Nissl-stained cat retina showing cruciform distribution which contains area centralis and visual streak. Compare with total ganglion-cell density distribution map of Fig. 20 (Courtesy of WÄSSLE et al. and J. comp. Neurol., 1975)

diameters of 14 μ and 22 μ. They report an increase in the percentage of small cells when passing from the periphery to streak retina in accordance with expectations based upon a similar increase in their electrophysiological encounter rates for units of the corresponding W class. The possibility that the increase in the percentage of small cells indicates, not an increase in the λ cell proportion but rather, a decrease in the β cell mean diameter on the high density streak is dismissed by the unsatisfactory argument that such an increase in the β cell proportion is not to be observed in a region of similar density but adjacent to the area centralis. HUGHES (1978) regards as unjustified the use of peripheral soma criterion diameters in the streak retina and contests the conclusion of ROWE and STONE (1976) that the γ cell proportion increases in the visual streak. The natural criterion for division of the soma diameter spectrum into modes is the dip to be seen between the γ and β modes in all regions, except those of highest density in the streak and area centralis. This method is admittedly subject to error because of the overlap in most regions between the γ and β modes but it is a more satisfactory approximation to the required information than that of ROWE and STONE (1976). Examination of diameter spectra for a sequence of retinal regions which approach either area or streak reveals that the β mode narrows and that its mean moves towards the more constant γ mode mean in agreement with the results of WÄSSLE and BOYCOTT (1975) and quite in keeping with a relatively constant proportion of the classes in spite of a substantial increase in the small cell population. Now ROWE and STONE (1976) report the small cell proportion, which they equate with W units or γ cells, to fall off from nearly 70% on the streak to some 40% below it. By contrast HUGHES (1977e) reports the peripheral γ cell mode to represent up to 65% of the population, a trend suggested by the optic nerve fibre diameter spectra (HUGHES and WÄSSLE, 1976), and finds the proportion to fall in passing to the streak. It is apparent that the peripheral cell densities of ROWE and STONE (1976) are substantially lower than those of WÄSSLE et al. (1975) or HUGHES (1976b) and it appears possible that they do not identify a proportion of the peripheral γ cell class as ganglion cells. In the central area HUGHES (1977c) reports the γ cell proportion to fall off relative to the streak and WÄSSLE (personal communication), employing techniques for direct identification of cell class, finds that the β proportion amounts to some 70% of the total. Such a value would be necessary from sampling theory if cat acuity is confirmed to be greater than 9.0 cy/° (MITCHELL et al., 1977). The range of density variation of β or, in the inverse sense, of γ cells is thus at least 2:1 across the retina. That encounter rates for given receptive-field classes do differ, in the rabbit between central and peripheral retina (OYSTER et al., 1971) and in the cat between central and peripheral retina (CLELAND and LEVICK, 1974a), and central and streak retina (STONE et al., 1975) is by no means conclusive because of the sampling problem when recording.

It is to be emphasized, however, that the deviation of the behavioral acuity profile by other than a constant factor from that computed on the basis of the total ganglion-cell distribution would not be accounted for by a nonhomogeneous human retinal ganglion-cell population if its composition were uniform throughout the retina. Only a progressive decrease, with increasing eccentricity, in the proportion of the cell class subserving resolution would suffice to explain the observed phenomena without the introduction of other factors.

2. Differential Projection of the Retinal Ganglion Cells

Even if the ganglion-cell population were homogeneous, the theoretical potential resolution could not be attained if the ganglion-cell axons project to different destinations, so that only a fraction of the total population is involved in a resolution task at a given central region. In the cat it appears that the cells most likely to subserve resolution, the brisk-sustained population, project predominantly to the LGN with only minor contributions to other visual centers (CLELAND and LEVICK, 1974; FUKUDA and STONE, 1974). Studies on the retrograde transport of horseradish peroxidase from the cat LGN and superior colliculus to the retina substantiate this differential projection (KELLY and GILBERT, 1975). Similar work on the macaque indicates that the great majority of ganglion cells project to the LGN (BUNT et al., 1975). In these two species and man, at least, the retinogeniculate pathway appears to be the major projection involved in resolution (BRINDLEY, 1970), and the output of the majority of suitable ganglion cells is thus available to the cortex, so that their effective density is not reduced by differential projection.

3. Variation in Receptive Field Size

Not only does sampling theory specify a minimum rate of sampling (2f/°) which is necessary to achieve a given spatial frequency resolution (f cycles/°), but it also indicates an optimum sampling aperture width; apertures smaller than this unnecessarily reduce sensitivity without improving resolution at a given sampling density. Several studies of receptive-field-centre diameter have described a range of sizes within a given region of cat retina (FISHER, 1973) which implies that the system operates non-optimally. However, recent work (CLELAND and LEVICK, 1974) suggests that the range of centre size *for a given class* is more restricted and amounts to only ±10% for the brisk sustained units at the area centralis; such a range would influence the cut-off-frequency, or sensitivity, very little. In the peripheral retina, however, the diameter range increases so that for brisk sustained units it is ±50% at 15° eccentricity. Such a range could limit resolution to an extent dependent upon the relation between its mean and the local optimum diameter: if similarly organized in man it might contribute to the increase with eccentricity in the discrepancy between theoretical and attained resolution but is unlikely to be a major factor.

4. Change in Peripheral Magnification Factor

In most mammalian eyes the PND, and consequently the corresponding local RMF, decreases with eccentricity for chief rays. This phenomenon must be taken into account when estimating the potential maximum resolution to be obtained from a given local ganglion cell density. Measurements of the reduction in RMF have been provided only for the cat eye (HUGHES, 1976a) although theoretical estimates have been made for man which are based upon unverified homogeneous lens schematic eyes (DRASDO and FOWLER, 1974; FRISEN and FRISEN, 1976; LOTMAR, 1971). The peripheral reduction of RMF is unlikely to be a major factor underlying the discrepancy between the theoretical and measured sampling densities in the human eye because the divergence between the two is well developed within the central 30° of retina where the RMF is relatively constant.

5. Reduction of Sampling Efficiency by a Non-uniform Retinal Mosaic

Irregularity in the sampling matrix need not result in a reduced cut-off frequency relative to an optimal, regular matrix if an appropriate reconstitution function is employed (YEN, 1956). A substantial reduction in the cut-off-frequency will, however, occur if the signals from an irregular lattice are treated as though from a regular lattice; a situation which seems more probable for the eye. A matrix noise such that the standard deviation of sampling position is equal to the mean sampling interval would require a fourfold increase in sampling density if the cut-off frequency of an undisturbed matrix is to be achieved (FRENCH, SNYDER and STAVENGA, 1977). In man, the sampling matrix at the foveal centre is determined by the optimal cone packing lattice; away from this region the constraints on ganglion cell packing are reduced and increased noise in the sampling point distribution might be expected. Under the conditions outlined above this would result in an increasing divergence with eccentricity between the theoretical and attained resolution for a region of given ganglion cell density, just as is found. The matrix of ganglion cell distribution in mammals is certainly not regular for any identifiable class and the degrading influence of such an irregular matrix on the performance of the system at a given ganglion cell density must at least play a part in limiting acuity. Increasing irregularity in the sampling matrix with eccentricity is a strong contender for the explanation of the progressively greater divergence of theoretical and actual sampling density with eccentricity.

A curious relationship arises from the above discussion. Consider two regions of the human visual field with acuities a_1 and a_2, total ganglion cell densities D_1 and D_2, effective densities D_{1e} and D_{2e} and cortical magnification factors in the corresponding regions of area 17 of CMF_1 and CMF_2; then we may write,

$$a_1/a_2 = D_1/D_2 = \sqrt{D_{1e}}/\sqrt{D_{2e}} = \mathrm{CMF}_1/\mathrm{CMF}_2 \quad \text{and thus}$$
$$(\mathrm{CMF}_1)^2/(\mathrm{CMF}_2)^2 = D_{1e}/D_{2e}\,.$$

The cortex being of relatively constant thickness and cell density, it follows from the ratio of cortical area for regions 1 and 2 that the ratio of the number of cortical cells in regions dealing with equal elements of the visual field is in the same ratio as the minimum sampling densities theoretically required to attain the acuities achieved by the areas! The implication is thus that there is a constant provision of some k cortical cells for each independent sampling element at the retinal level. At the human fovea, where the sampling theorem appears to apply directly, it is possible to estimate the value of k as 600 from published cone and cortical cell densities. It seems improbable that several interacting factors, each varying with eccentricity, should result in the properties outlined above; one predominant process would be more likely to lead to such simple arithmetical relationships. A straightforward progressive decrease in the number of sampling points, resulting from a systematic decrease with eccentricity in the proportion of cells involved in resolution tasks, or a loss of independence in the sampling points, as a result of increasing lattice irregularity, would appear to be the major contenders as explanations for the results. Morphological identification of the cells involved in resolution tasks would readily distinguish between the two hypotheses.

IV. Neural Compensation for Optical Impairment of Image Quality

It is a suggestion of long standing that, rather than adding to the optic degradation of image quality, the neural apparatus of the visual system actually compensates for the effects of optic blurring. Originally considered by MACH (1865) as an explanation of subjective border enhancement, this hypothesis has more recently been presented with detailed theoretical substantiation (RATLIFF, 1965). CAMPBELL and GUBISCH (1966), however, find little direct evidence for its general validity *at threshold.* The neural apparatus actually *overcompensates* for optical limitations at low spatial frequencies but in the range above 10 cycles/°, which is important for the definition of spots and edges; the system performance is substantially impaired by the central processing. In agreement with MACH (1865) and BARLOW (1961), CAMPBELL and GUBISCH (1966) conclude that border enhancement is simply a by-product of the mechanism which makes visual sensation independent of average illumination.

The possibility of neural "deblurring" has recently been reintroduced for suprathreshold conditions (HAY and CHESTERS, 1972; GEORGESON and SULLIVAN, 1975). Although a theoretically plausible phenomenon, the evidence in its favor requires cautious appraisal in view of the recent improved estimates of human retinal image quality.

V. The Contrast Sensitivity Function of the Entire Visual System

The form of the MTF for the dioptric apparatus of the eye shows that it transmits spatial frequencies ranging from just below cut-off to zero with equal facility; comparison of such optic systems may thus be readily made by means of their cut-off frequencies. It is, however, a very limited approach to compare the quality of vision of different species in terms of their behavioral grating acuity, the cut-off frequency for the whole visual system, because the sensitivity of the CNS to spatial frequencies varies throughout the range transmitted. This sensitivity may be expressed as the reciprocal of the threshold contrast which is required for the detection of a given frequency; the whole system is thus best described by the contrast-sensitivity function, CSF, which specifies this value for all transmitted spatial frequencies and was first determined for man by CAMPBELL and ROBSON (1968). More complete specification of the system capability is given when the function is determined over a range of light intensities (VAN NES and BOUMAN, 1967) and orientations (CAMPBELL et al., 1966). Although of interest, the function has not thus far been specified for regions away from the fixation axis.

More recently, CAMPBELL and MAFFEI (1970) have shown that the CSF may be obtained from averaged cortical-evoked potentials in man. This objective technique was subsequently employed to obtain a CSF for the cat (CAMPBELL et al.,

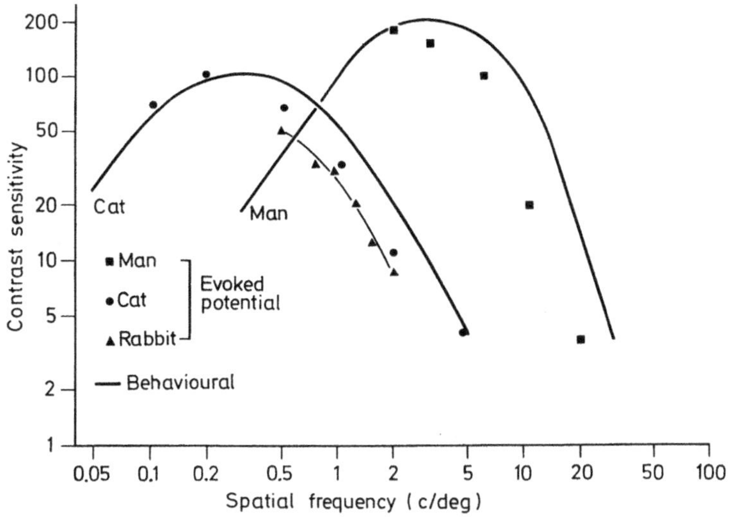

Fig. 26. Behavioral contrast sensitivity functions for man and cat *(bold lines)* compared with functions derived from averaged evoked potentials of man, cat, and rabbit (*square*, *disc*, and *triangle*, respectively). Sources: BISTI and MAFFEI (1974), CAMPBELL et al. (1973), and KULIKOWSKI (1977)

1973) which has now been confirmed by direct behavioral observations (BLAKE et al., 1974; BISTI and MAFFEI, 1974). Behavioral CSF's are also available for macaque (DE VALOIS et al., 1974), squirrel monkey (MERIGAN, 1976), and rabbit (KULIKOWSKI, 1977). The satisfactory cross-species comparison of the CSF ideally requires the function to have been established not only at similar light intensities but also under ecologically characteristic levels of illumination for each form. In the absence of such wide-ranging information, the shape of the available CSF's is similar for all examined species (Fig. 26); it differs from the optical MTF in the possession of a peak sensitivity and a low frequency cut-off. Identified species differences are thus confined to variation in the position of the curve on the frequency axis; the high frequency cut-off of the cat visual system is three octaves lower than that of man and monkey but accompanied by similarly displaced peak sensitivity and low-frequency cut-off. The loss of high-frequency resolution by the cat, relative to the primates, is thus complemented by an extended low-frequency range which is unavailable to man under conditions of continuous photopic viewing. This shift in the position of the CSF on the frequency axis is not determined by the size of the animal, because the squirrel monkey high-frequency cut-off is similar to that of man rather than cat (CAMPBELL et al., 1969; MERIGAN, 1976). CAMPBELL et al. (1973) suggest that the enhanced low-frequency sensitivity of the cat arises because it deals with a closer visual world than man or primates; this seems unlikely in view of the close observation employed by primates during manipulation and their greater range of accommodation than the cat. KIRSCHFELD (1976) has recently argued for relative constancy of the minimum separabile in terms of object space measurement when normalised to body height for a wide range of forms from insects to large mammals. As a gross approximation this appears to be valid but within the homeotherms alone it is unsatisfactory as suggested above.

The low-frequency reduction in contrast sensitivity which distinguishes the CSF from the MTF of the optic system does not occur under low-illumination intensities (ROBSON and CAMPBELL, 1965) or for short exposure times (SCHOBER and HILZ, 1965; RÖHLER and HILZ, 1966) and is usually accounted for in terms of lateral inhibitory processes (WESTHEIMER, 1972a).

Attempts at explaining visual acuity in terms of retinal morphology and physiology have been of great benefit to the understanding of visual function. The attainment of a similar neurophysiological basis for more complex aspects of visual perception has until recently appeared remote. Now the CSF provides a much more comprehensive description of the visual capabilities of the CNS than does resolution alone, and yet it retains the appearance of tractability to analysis. Its parameters may be established by means of either behavioral studies or objective examination of cortical-evoked potentials. Certain features of the CSF clearly arise during central processing (MAFFEI and CAMPBELL, 1970) and, like its orientation selectivity (CAMPBELL and KULIKOWSKI, 1966), could be explicable in terms of known features of cortical physiology, although the relationship is not as clear-cut as is sometimes made out (BISTI and MAFFEI, 1974).

In principle, a knowledge of the CSF is much more fundamental than it may appear. At threshold, this function defines the spatial-frequency information available to the CNS. Within a limited range, the visual system does appear to operate linearly (CAMPBELL and ROBSON, 1968) so as to enable the prediction of its response to summed spatial-frequency components. Thus, close to threshold, it is possible to describe patterns in terms of their Fourier components and to employ the CSF to set bounds upon the range of patterns the visual system might discriminate. The extent to which a suprathreshold treatment of this nature is justified remains to be established; in this region, however, the system must evince some nonlinearity. The more complex question of whether the visual system actually analyzes its input in terms of spatial-frequency components is beyond the scope of this work but has been discussed in some detail (CAMPBELL, 1974; ROBSON, 1975). Recent results (HENNING, HERTZ and BROADBENT, 1975) are inconsistent with the hypothesis that the visual system analyses spatial patterns in independent, narrowly tuned bands of spatial frequency.

F. Conclusion

Taxonomists are generally characterized as "splitters" or "lumpers" according to the readiness with which they subdivide the living kingdom; the history of comparative visual physiology suggests it to be more inbued with the spirit of the former. This "set" for the discovery of contrast between species has often led to the establishment of artificial dichotomies on the basis of preliminary results and a consequent premature emphasis on explanatory theorizing. No species of mammal has so often been set apart on the basis of visual physiology as the rabbit, which is normally taken as representative of lateral-eyed mammals in general. Undoubtedly the most obvious organizational feature of a given visual system is its relative exploitation of the frontal and lateral fields of view, and it is natural to anticipate the possibility of equally overt contrasts in the organisation of the CNS which are consequent upon the disposition of the eyes. The extreme laterality of the rabbit eye thus appears to predispose to the acceptance of results indicating its qualitative difference from the cat, monkey, and man, whereas distinctions between the latter species tend to be underemphasised.

Several early investigators reported that the rabbit, in contrast to other mammalian species, did not show optokinetic nystagmus (URBANTSCHITSCH, 1910; BARTELS, 1920), but subsequent technical improvements (TER BRAAK, 1963; BRECHER, 1936) enabled its ready demonstration, and now it is regarded to play a greater role in rabbit eye movements than in most other mammals (HUGHES, 1971). Associated with the extreme stability of the rabbit eye, which arises from its powerful optokinetic mechanism (COLLEWIJN, 1971), is the often-claimed absence of spontaneous eye movements. BRECHER (1936) regarded the role of the rabbit's eyes as *Warnorgane* to be so preeminent as to be consistent with the absence of voluntary eye movements. This view, supported by the great extent of the animal's visual streak, has been long accepted and regarded as qualitatively distinguishing the rabbit from other mammalian species. Nevertheless, spontaneous eye movements are readily visible in the rabbit (HUGHES, 1971; COLLEWIJN, 1977), even vergence is now reported (COLLEWIJN and ZUIDAM, 1977); it remains true that these movements are much less frequent than in man or cat (PRITCHARD and HERON, 1960; WINTERSON and ROBINSON, 1975; STRYKER and BLAKEMORE, 1972), and monkey (FUCHS, 1967). It appears, however, that the rabbit does not follow small objects with smooth tracking as do cat, dog, and monkey (RADEMAKER and TER BRAAK, 1948). The *Willkürliche Fixation* described for the rabbit by FLEISCH (1922) was shown to be a special instance of optokinetic nystagmus (BRECHER, 1936). Certainly the extent of the visual field and area centralis would be expected to quantitatively influence the patterns of eye and head movements in different species, but the available evidence does not support extensive qualitative distinctions between mammalian species.

Of course it remains possible that there are substantial differences in the neuronal organization of mammalian forms, but again caution is recommended in classification. Some years ago, when it was believed that mammalian retinal ganglion cells possessed only concentric receptive fields, MATURANA (1964) discriminated between an indeterminate mammalian visual system, characteristic of animals possessing neocortex to which nonspecific information is supplied and deterministic systems in amphibians, reptiles, and birds which lack neocortex and whose retinal output is specialized and unambiguous. The report of nonconcentric units in the rabbit (BARLOW et al., 1964) caused this concept to give way to that of a spectrum of animals in which the rabbit fell between the highly encephalized species, such as monkey and cat, whose retinas were equipped only with concentric units and the forms, such as reptiles and birds, which lack neocortex and possess complex retinas. It has been suggested (HUGHES, 1971) that peripheral preprocessing by nonconcentric units is required in the rabbit in order to reduce the enormous demands on cortical area which would occur if its visual-streak projection were analysed in similar fashion to those of the smaller cat area centralis.

The subsequent discovery of substantial nonconcentric retinal unit populations in cat and other mammalian species has substituted quantitative for these qualitative distinctions. Reference to current observations of higher levels of the visual system might again enable qualitative distinctions to be established. Although the cat LGN (WILSON and STONE, 1975; CLELAND et al., 1975) is now known to receive the projections of some nonconcentric units, as does that of the rabbit (OYSTER et al., 1971), the rabbit visual cortex is described by many authors to be very different from that of cat and monkey (OGAWA et al., 1968; HUGHES, 1971; VAN SLUYTERS and STEWART, 1974a). The situation outlined above in relation to the retina should, however, serve as a warning against too readily emphasizing distinctions based upon encounter rates in single-unit populations.

Nevertheless, we continue to find the size of the lateral and binocular fields of the rabbit implicated in accounting for all manner of observations discrepant with findings in the cat. Thus, rabbit monocular and binocular cortical units (CHOW and SPEAR, 1974; VAN SLUYTERS and STEWART, 1974b) are described as unhindered by deprivation during development or by selective experience (MIZE and MURPHY, 1973); a great contrast with the situation in the cat. BLAKEMORE and VAN SLUYTERS (1975) accept this negative evidence and set out to account for the discrepancy between the two species. In view of the substantial discrepancies which exist in this field between the findings of those working on *one* species (e.g., HUBEL and WIESEL, 1963; BARLOW and PETTIGREW, 1971; BUISSERT and IMBERT, 1976), it might be argued that such an attempt is very premature. BLAKEMORE and VAN SLUYTERS (1975) suggest that plasticity such as found in the cat enables matching of binocular receptive field properties in the two eyes, a condition perhaps necessary only for highly binocular species. They emphasise the panoramic nature of rabbit vision and the bizarre features of some of its binocular units. The presence of a proportion of binocular fields which are capable of subserving stereopsis in the rabbit is ignored, in spite of the fact that these units might equally require plasticity for "matching" of their left and right eye contributions, regardless of the animal's binocular field width. The traditional predisposition to differentiate the lateral-eyed species thus tends to override the available evidence, but even such distinctions as appear well founded should not be generalized before demonstration of their applicability to lateral-eyed species other than the rabbit.

The tendency to turn quantitative distinctions between species into qualitative distinctions is, perhaps, no more evident than in discussions of visual acuity. The investigator only mildly afflicted with anthropomorphism describes the low acuity of the rat as 'poor acuity' but a more serious condition manifests itself in compound ignorance by the suggestion that 'rats are as blind as bats.' We would do well to remember Sir THOMAS BROWNE's dictum (1658; 1912),

'if the Eagle were judge, we might be blind ourselves.'

G. Acknowledgments

My debt to regular and transient devotees of Professor P. O. BISHOP's lunch table for their stimulating, critical, and sometimes tedious observations on topics dealt with above has already been pointed out to me by some of them, but I here give formal recognition of my gratitude.

I am greatly indebted to LYN SPEIGHT for her extensive assistance during the preparation of this chapter.

References

Abercrombie, M.: Estimation of nuclear population from microtome sections. Anat. Rec. **94**, 239—247 (1946).

Adams, A. D., Forrester, J. M.: The projection of the rat's visual field on the cerebral cortex. Quart. J. exp. Physiol. **53**, 327—336 (1968).

Ajo, A.: On the refractive index of the retina. Acta. physiol. scand. **13**, 130—149 (1949).

Andrews, D. P.: Perception of contour orientation in the central fovea. Part II. Spatial integration. Vision Res. **7**, 999—1013 (1967).

Andrews, D. P., Butcher, A. K., Buckley, B. R.: Acuities for spatial arrangement in line figures: Human and ideal observers compared. Vision Res. **13**, 599—620 (1973).

Andrews, D. P., Hammond, P.: Mesopic increment threshold spectral sensitivity of single optic tract fibres in the cat: Cone rod interaction. J. Physiol. (Lond.) **209**, 65—81 (1970a).

Andrews, D. P., Hammond, P.: Suprathreshold spectral properties of single optic tract fibres in cat under mesopic adaptation: cone: Rod interaction. J. Physiol. (Lond.) **209**, 83—103 (1970b).

Appelle, S.: Perception and discrimination as a function of stimulus orientation; the "oblique effect" in man and animals. Psychol. Bull. **87**, 266—278 (1972).

Apter, J. T.: Eye movements following strychninization of the superior colliculus of cats. J. Neurophysiol. **9**, 73—86 (1946).

Arey, L. B., Gore, M.: The numerical relationship between the ganglion cells of the retina and the fibres in the optic nerve of the dog. J. comp. Neurol. **77**, 609—617 (1942).

Armaly, M. F.: Studies on intraocular effects of the orbital parasympathetic pathway. Arch. Ophthal. (Chicago) **61**, 14—29 (1959).

Attneave, F.: Informational aspects of visual perception. Psychol. Rev. **61**, 183—193 (1954).

Aubert, H.: Die Bewegungsempfindung. Arch. ges. Physiol. **39**, 347—370 (1886).

Aubert, H.: Die Bewegungsempfindung. Arch. ges. Physiol. **40**, 459—480 (1887).

Aubert, H., Förster, R.: Beiträge zur Kenntniss des indirecten Sehens. (1) Untersuchungen über den Raumsinn der Retina. Arch. Ophthal. **3**, 1—37 (1857).

Bachl, A., Lukosz, W.: Experiments on super-resolution imaging of a reduced object field. J. opt. Soc. Amer. **57**, 163—169 (1967).

Baird, J. C.: Psychophysical Analysis of Visual Space. London: Pergamon Press 1970.

Baldwin, W. R.: Some relationships between ocular, anthropometric and refractive variables in myopia. Doctoral Thesis, Indiana University 1964.

Bárány, E. H.: A theory of visual acuity and an analysis of the variability of visual acuity. Acta ophthal. (Kbh.) **24**, 63—92 (1946).

Barlow, H. B.: Possible principles underlying the transformation of sensory messages. In: Sensory Communication, Rosenblith, W. A. (ed.). New York: M.I.T. & Wiley 1961.

Barlow, H. B.: Three points about lateral inhibition. In: Sensory Communication, Rosenblith, W. A. (ed.). New York: M.I.T. & Wiley 1961.

Barlow, H. B.: The physical limits of visual discrimination. In: Photophysiology. vol. ii, Giese, A. C. (ed.). New York: Academic Press 1964.

Barlow, H. B.: Visual resolution and the diffraction limit. Science **149**, 553—555 (1965).

Barlow, H. B., Blakemore, C. B., Pettigrew, J. D.: The neural mechanisms of binocular depth discrimination. J. Physiol. (Lond.) **193**, 327—342 (1967).

Barlow, H. B., Fitzhugh, R., Kuffler, S. W.: Change of organization in the receptive fields of the cat's retina during dark adaptation. J. Physiol. (Lond.) **137**, 338—354 (1957a).

Barlow, H. B., Fitzhugh, R., Kuffler, S. W.: Dark-adaptation, absolute threshold and Purkinje shift in single units of the cat's retina. J. Physiol. (Lond.) **137**, 327—337 (1957b).

Barlow, H. B., Hill, R. M., Levick, W. R.: Retinal ganglion cells responding selectively to direction and speed of image motion in the rabbit. J. Physiol. (Lond.) **173**, 377—407 (1964).

Barlow, H. B., Levick, W. R.: The mechanism of directionally selective units in the rabbit's retina. J. Physiol. (Lond.) **178**, 477—504 (1965).

Barlow, H. B., Pettigrew, J. D.: Lack of specificity of neurones in the visual cortex of young kittens. J. Physiol. (Lond.) **218**, 98P—100P (1971).

Barnett, S. A.: A Study in Behaviour. London: Methuen 1963.

Baron, J., Verrier, M. L.: Réfraction et cerveau des poissons à fovea. Contribution à l'etude des corrélations organiques. Bull. Biol. Fr. Belg. **85**, 105—111 (1951).

Barrett, J. W.: Do animals accommodate? Ophthal. Rev. **7**, 255—270 (1898).

Bartels, M.: Vergleichendes über Augenbewegungen. In: Handbuch der normalen und pathologischen Physiologie. Berlin: Springer **12**, 1920, pp. 1113—1165.

Baylor, D. A., Fettiplace, R.: Light path and photon capture in turtle photoreceptors. J. Physiol. (Lond.) **248**, 433—464 (1975).

Baylor, E. R., Shaw, E.: Refractive error and vision in fishes. Science, **136**, 157—158 (1962).

Beer, T.: Die Accommodation des Fischauges. Pflügers Arch. ges. Physiol. **58**, 523—650 (1894).

Bellairs, A.: The Life of Reptiles. London: Weidenfeld & Nicolson 1969.

Bellhorn, R. W., Aguirre, G. D., Bellhorn, M. B.: Feline central retinal degeneration. Invest. Ophthal. **13**, 608—616 (1974).

Bellhorn, R. W., Fischer, C. A.: Feline central retinal degeneration. J. anat. vet. med. Ass. **157**, 842—849 (1970).

Bennett, A. G., Francis, J. L.: Visual optics and the optical space sense. In: The Eye. Davson, H. (ed.). 2nd ed. London: Academic Press 1962.

Berkley, M. A., Kitterle, F., Watkins, D. W.: Grating visibility as a function of orientation and retinal eccentricity. Vision Res. **15**, 239—244 (1975).

Berkley, M. A., Watkins, D. W.: Grating resolution and refraction in the cat estimated from evoked cerebral potentials. Vision Res. **13**, 403—415 (1973).

Berlin, R.: Über die Schätzung der Entfernung bei Thieren. Z. vergl. Augenheilk. **7**, 1—97 (1893).

Berry, R. N.: Quantitative relations among vernier, real depth and stereoscopic depth acuities. J. exp. Psychol. **38**, 708 (1948).

Bhatia, B.: Minimum separabile as a function of speed of moving object. Vision Res. **15**, 23—33 (1975).

Bickerdyke, J.: The Book of the All-Round Angler. London: Upcott Gill 1889.

Binggeli, R. L., Paule, W. J.: The pigeon retina: Quantitative aspects of the optic nerve and ganglion cell layer. J. comp. Neurol. **137**, 1—18 (1969).

Bishop, A.: Use of the hand in lower primates. In: Evolutionary and Genetic Biology of Primates. Buettner-Janusch, J. (ed.). New York: Academic Press 1964.

Bishop, P. O.: Neurophysiology of binocular single vision and stereopsis. In: Handbook of Sensory Physiology, vii/3A, Jung, R. (ed.). Berlin: Springer-Verlag 1973, pp. 255—305.

Bishop, P. O., Kozak, W., Vakkur, G. J.: Some quantitative aspects of the cat's eye: Axis and plane of reference, visual field coordinates and optics. J. Physiol. (Lond.) **163**, 466—502 (1962).

Bishop, G. H., Clare, M. H.: Organisation and distribution of fibers in the optic tract of the cat. J. comp. Neurol. **103**, 269—304 (1955).

Bisti, S., Maffei, L.: Behavioural contrast sensitivity of the cat in various visual meridians. J. Physiol. (Lond.) **241**, 201—210 (1974).

Blake, R., Cool, S. J., Crawford, M. L. J.: Visual resolution in the cat. Vision Res. **14**, 1211—1217 (1974).

Blakemore, C.: The representation of three-dimensional visual space in the cat's striate cortex. J. Physiol. (Lond.) **209**, 155—179 (1970).

Blakemore, C., Fiorentini, A., Maffei, L.: A second neural mechanism of binocular depth discrimination. J. Physiol. (Lond.) **226**, 725—739 (1972).

Blakemore, C., Van Sluyters, R. C.: Innate and environmental factors in the development of the kitten's visual cortex. J. Physiol. (Lond.) **248**, 663—716 (1975).

Block, M. T.: A note on the refraction and image formation of the rat's eye. Vision Res. **9**, 705—711 (1969).

Bloom, M., Berkley, M. A.: Behavioural determination of the cat's near point of accommodation. ARVO Proc., Sarasota, 1976.

Bonds, A. B.: Optical quality of the living cat eye. J. Physiol. (Lond.) **243**, 777—795 (1974).
Bonds, A. B., Enroth-Cugell, C., Pinto, L. H.: Image quality of the cat eye measured during retinal ganglion cell experiments. J. Physiol. (Lond.) **220**, 383—401 (1972).
Bough, E. W.: Stereoscopic vision in the macaque monkey: A behavioural demonstration. Nature (Lond.) **225**, 42—44 (1970).
Bourdon, B.: La Perception Visuelle de l'Espace. Paris, 1902.
Boycott, B. B., Dowling, J. E.: Organisation of the primate retina: Light microscopy. Phil. Trans. B. **255**, 109—184 (1969).
Boycott, B. B., Kolb. H.: The connections between bipolar cells and photoreceptors in the retina of the domestic cat. J. comp. Neurol. **148**, 91—114 (1973).
Boycott, B. B., Wässle, H.: The morphological types of ganglion cells of the domestic cat's retina. J. Physiol. (Lond.) **240**, 397—419 (1974).
Boynton, R. M.: The visual system: environmental information. In: Handbook of Perception: Seeing. Vol. 1. Carterette, E. C. and Friedman, M. P. (eds.). Berlin: Springer 1975, pp. 285—306.
Bracewell, R.: The Fourier Transform and its Applications. New York: McGraw Hill (1965).
Brändle, K., Stirling, R. V.: Development of the ipsilateral visual projection in axolotls treated with thyroxine. J. Physiol. (Lond.) **250**, 30—31 P (1975).
Brecher, G. A.: Optisch ausgelöste Augen- und Körperreflexe am Kaninchen. Z. vergl. Physiol. **23** 374—390 (1936).
Brett, J. R.: The sense organs: The eye. In: The Physiology of Fishes. Brown, M. E. (ed.). New York: Academic Press 1957.
Brewster, D.: On the structure of the crystalline lens in fishes and quadrupeds as ascertained by its action on polarised light. Phil. Trans. 311—317 (1816).
Brewster, Sir David: A Treatise on Optics. London: Longman, Brown, Green & Longman, 1830.
Brewster, Sir David: Memoirs of the Life, Writings and Discoveries of Sir Isaac Newton. Edinburgh: Edmonston & Douglas 1860.
Brillouin, L.: Science and Information Theory. London: Academic Press, 1962.
Brindley, G. S.: The deformation phosphene and the funnelling of light into rods and cones. J. Physiol. (Lond.) **188**, 24—25 P (1966).
Brindley, G. S.: Physiology of the Retina and Visual Pathway. London: Edward Arnold 1970.
Brindley, G. S., Hamasaki, D. I.: Histological evidence against the view that the cat's optic nerve contains centrifugal fibres. J. Physiol. (Lond.) **184**, 444—449 (1966).
Brindley, G. S., Lewin, W. S.: The sensations produced by electrical stimulation of the visual cortex. J. Physiol. (Lond.) **196**, 479—493 (1968).
Brooke, R. N. L., Downer, J. de C., Powell, T. P. S.: Centrifugal fibres to the retina in the monkey and cat. Nature (Lond.) **207**, 1365—1367 (1965).
Brown, K. T.: A linear area centralis extending across the turtle retina and stabilized to the horizon by nonvisual cues. Vision Res. **9**, 1053—1062 (1969).
Brown, J. E., Rojas, J. A.: Rat retinal ganglion cells; receptive field organisation and maintained activity. J. Neurophysiol. **28**, 1073—1090 (1965).
Browne, T.: The Works of Sir Thomas Browne. Edinburgh; Grant 1912.
Brückner, R.: Beiträge zur Biologie des Auges. I. Mitteilung: Über die Netzhaut von Feliden und Caniden. Biol. Zbl. **80**, 37—66 (1961 a).
Brückner, R.: Beiträge zur Biologie des Auges. 2. Mitteilung: Über die Netzhaut von Huftieren. Biol. Zbl. **80**, 129—136 (1961 b).
Bruesch, S. R., Arey, L. B.: The number of myelinated and unmyelinated fibres in the optic nerve of vertebrates. J. comp. Neurol. **77**, 631—665 (1942).
Buffon, Count of.: Buffon's Natural History. Vol. 6. London: Symonds 1812.
Buchdahl, H. A.: An Introduction to Hamiltonian Optics. New York: Cambridge University Press 1970.
Buissert, P., Imbert, M.: Visual cortical cells: their developmental properties in normal and dark reared kittens. J. Physiol. (Lond.) **255**, 511—525 (1976).
Bunt, A. H., Hendrickson, A. E., Lund, J. S., Lund, R. D., Fuchs, A. F.: Monkey retinal ganglion cells: Morphometric analysis and tracing of axonal projections, with a consideration of the peroxidase technique. J. comp. Neurol. **164**, 265—286 (1975).
Bunt, A. H., Lund, R. D., Lund, J. S.: Retrograde axonal transport of horseradish peroxidase by ganglion cells of the albino rat retina. Brain Res. **73**, 215—228 (1974).
Burton, G. J.: Evidence for nonlinear response processes in the human visual system from measurements on the thresholds of spatial beat frequencies. Vision Res. **13**, 1211—1225 (1973).

Burt, E. T., Catton, W. T.: A diffraction theory of insect vision. I. An experimental study of visual acuity in certain insects. Proc. roy. Soc. B **157**, 53—67 (1962).

Butcher, E. O.: The structure of the retina of Fundulus heteroclitus and the regions of the retina associated with the different chromatophoric responses. J. exp. Zool. **79**, 275—293 (1938).

Byram, G. M.: The physical and photochemical basis of visual resolving power. Pt. II Visual acuity and the photochemistry of the retina. J. opt. Soc. Amer. **34**, 718—738 (1944).

Cajal, Ramón y.S.: La Rétine des Vértebres. Cellule **9**, 119 (1892).

Cajal, Ramón y.S.: Histologie du Systéme Nerveaux. Madrid: Conseijo Superior de Investigaciones Cientificas 1955.

Campbell, F. W.: Twilight myopia. J. opt. Soc. Amer. **43**, 925—926 (1953).

Campbell, F. W.: The transmission of spatial information through the visual system. In: The Neurosciences, Third Study Program. Cambridge, Mass.: M.I.T. Press 1974.

Campbell, F. W., Cooper, G. F., Robson, J. G., Sachs, M. B.: The spatial selectivity of visual cells of the cat and the squirrel monkey. J. Physiol. (Lond.) **204**, 120P—121P (1969).

Campbell, F. W., Green, D. G.: Optical and retinal factors affecting visual resolution. J. Physiol. (Lond.) **181**, 576—593 (1965).

Campbell, F. W., Gregory, A. H.: Effect of size of pupil on visual acuity. Nature (Lond.) **187**, 1121—1123 (1960).

Campbell, F. W., Gubisch, R. W.: Optical quality of the human eye. J. Physiol. (Lond.) **186**, 558—578 (1966).

Campbell, F. W., Kulikowski, J. J., Levinson, J.: The effect of orientation on the visual resolution of gratings. J. Physiol. (Lond.) **187**, 427—436 (1966).

Campbell, F. W., Maffei, L.: Electrophysiological evidence for the existence of orientation and size detectors in the human visual system. J. Physiol. (Lond.) **207**, 635—652 (1970).

Campbell, F. W., Maffei, L., Piccolino, M.: The contrast sensitivity of the cat. J. Physiol. (Lond.) **229**, 719—731 (1973).

Campbell, F. W., Primrose, J. A. E.: The state of accommodation of the human eye in darkness. Trans. ophthal. Soc. U.K. **73**, 353—361 (1953).

Campbell, F. W., Robson, J. G.: High speed infra-red optometer. J. opt. Soc. Amer. **49**, 268—272 (1959).

Campbell, F. W., Robson, J. G.: Application of Fourier analysis to the visibility of gratings. J. Physiol. (Lond.) **197**, 551—566 (1968).

Canella, F.: Quelques recherches sur la vision monoculaire. C.R. Soc. Biol. (Paris) **122**, 1221—1224 (1936a).

Canella, F.: Les problèmes du chiasma et de la vision binoculaire. Quelques recherches sur la vision monoculaire. J. Psychol. norm. path. **33**, 696—711 (1936b).

Cartmill, M.: Arboreal adaptations and the origin of the order Primates. In: The Functional and Evolutionary Biology of Primates. Tuttle, R. (ed.). Chicago: Aldine/Alherton 1972.

Cartmill, M.: Rethinking primate origins. Science, **184**, 436—443 (1974).

Catford, G. V., Oliver, A.: Development of visual acuity. Arch. Dis. Childh. **48**, 47—50 (1973).

Cauchy, A. D.: Memoire sur diverses formulés d'analyse. C.R. Acad. Sci. (Paris) Paris **12**, 283—298 (1841).

Charman, W. N., Tucker, J.: The optical system of the goldfish eye. Vision Res. **13**, 1—8 (1973).

Chievitz, J. H.: Untersuchungen über die Area centralis retinae. Arch. Anat. Physiol. Lpz. Anat. Abteil. Suppl., 139—396 (1889).

Chievitz, J. H.: Über das Vorkommen der area centralis retinae in den vier höheren Wirbelthierklassen. Arch. J. Anat. Physiol. Suppl. **15**, 311—334 (1891).

Chin, N. B., Ishikawa, S., Lappin, H., Davidowitz, J., Breinin, G. M.: Accommodation in monkeys induced by midbrain stimulation. Invest. Ophthal. **7**, 386—396 (1968).

Chirlian, P. M.: The effective bandwidth of a system. Quart. J. Applied Math. **25**, 311—312 (1967).

Chow, K. L., Spear, P. D.: Morphological and functional effects of visual deprivation on the rabbit visual system. Exp. Neurol. **42**, 429—447 (1974).

Citron, M. C., Pinto, L. H.: Retinal image: Larger and more illuminous for a nocturnal than for a diurnal lizard. Vision Res. **13**, 873—876 (1973).

Clarke, P. G. H.: The organization of visual processing in pigeon cerebellum. J. Physiol. (Lond.) **243**, 267—285 (1974).

Clarke, P. G. H., Whitteridge, D.: The cortical visual areas of the sheep. J. Physiol. (Lond.) **256**, 497—508 (1976).

Clarke, P. G. H., Donaldson, I. M. L., Whitteridge, D.: Binocular visual mechanisms in cortical areas I and II of the sheep. J. Physiol. (Lond.) **256**, 509—526 (1976).

Cleland, B. G., Dubin, W. M., Levick, W. R.: Sustained and transient neurones in the cat's retina and lateral geniculate nucleus. J. Physiol. (Lond.) **217**, 475—496 (1971).

Cleland, B. G., Levick, W. R.: Brisk and sluggish concentrically organised cells in the cat's retina. J. Physiol. (Lond.) **240**, 421—456 (1974a).

Cleland, B. G., Levick, W. R.: Properties of rarely encountered types of ganglion cells in the cat's retina and an overall classification. J. Physiol. (Lond.) **240**, 457—492 (1974b).

Cleland, B. G., Levick, W. R., Wässle, H.: Physiological identification of a morphological class of cat retinal ganglion cells. J. Physiol. (Lond.) **248**, 151—171 (1975).

Cleland, B. G., Morstyn, R., Wagner, H. G., Levick, W. R.: Long-latency retinal input to lateral geniculate neurones of the cat. Brain Res. **91**, 306—310 (1975).

Cohn, T. E.: Quantum fluctuation limit in foveal vision. Vision Res. **16**, 573—579 (1976).

Colbert, E. H.: Evolution of the Vertebrates. New York: Wiley 1967.

Coleman, E.: The Echidna under domestication. Vict. Nat. **51**, 12—21 (1934).

Coleman, E.: The Echidna under domestication. Vict. Nat. **52**, 151—154 (1935).

Collet, T.: Stereopsis in toads. Nature (Lond.) **267**, 349—351 (1977).

Collewijn, H.: The optokinetic system of the rabbit. Docum. ophthal. (Den Haag) **30**, 205—226 (1971).

Collewijn, H.: Eye and head movements in freely moving rabbits. J. Physiol. (Lond.) **266**, 471—498 (1977).

Collewijn, H., Zuidam, I.: Eye and head movements in the freely moving rabbit. Brain Res. **127**, 360—361 (1977).

Collins, E. T.: The Bowman Lecture: Changes in the visual organs correlated with the adoption of arboreal life and with the assumption of the erect posture. Trans. ophthal. Soc. U.K. **41**, 10—90 (1921).

Cone, R. A.: Quantitative relations of the rat electroretinogram. J. gen. Physiol. **46**, 1267—1286 (1963).

Cott, H. B.: Adaptive coloration in Animals. London: Methuen 1966.

Cowan, W. M., Powell, J. P. S.: Centrifugal fibres in the avian visual system. Proc. roy. Soc. **B 158**, 232—252 (1963).

Cowey, A., Rolls, E. T.: Human cortical magnification factor and its relation to visual acuity. Exp. Brain Res. **21**, 447—454 (1974).

Cragg, B. G.: Centrifugal fibres to the retina and olfactory bulb, and composition of the supraoptic commissures in the rabbit. Exp. Neurol. **5**, 406—427 (1962).

Currie, J., Cowan, W. M.: Evidence for the late development of the uncrossed retino-thalamic projections in the frog *Rana pipiens*. Brain Res. **71**, 133—139 (1974).

Daniel, P. M., Whitteridge, D.: The representation of the visual field on the cerebral cortex in monkeys. J. Physiol. (Lond.) **159**, 203—221 (1961).

Darwin, C. R.: In: The Autobiography of Charles Darwin and Selected Letters. Darwin, F. (ed.). London: Murray 1860.

Daw, N. W., Pearlman, A. L.: Cat colour vision: One cone process or several? J. Physiol. (Lond.) **201**, 745—764 (1969).

De Graauw, J. G., Van Hof, M. W.: The relation between behaviour and eye refraction in the rabbit. Brain Res. **127**, 360 (1977).

De Groot, S. G., Gebhard, J. W.: Pupil size as determined by adapting luminances. J. opt. Soc. Amer. **42**, 492—495 (1952).

De Monasterio, F. M., Gouras, P.: Functional properties of ganglion cells of the rhesus monkey retina. J. Physiol. (Lond.) **251**, 167—195 (1975).

De Oliveria, L. F., Ripps, H.: The "area centralis" of the owl monkey *(Aotes trivirgatus)*. Vision Res. **8**, 223—228 (1968).

Detwiler, S. R.: Vertebrate Photoreceptors. New York: Macmillan 1943.

De Valois, R. L., Morgan, H., Snodderly, D. M.: Psychophysical studies of monkey vision. III. Spatial luminance contrast sensitivity tests of macaque and human observers. Vision Res. **14**, 75—81 (1974).

Ditchburn, R. W.: Eye movements in relation to retinal action. Optica Acta. **1**, 171—176 (1955).

Ditchburn, R. W.: Eye Movements and Visual Perception. Clarendon Press: Oxford 1973.

Ditchburn, R. W.: Light. 3rd ed. Academic Press: London 1976.

Dobree, J. H., Weale, R. A.: (In: Weale, R. A.: Problems of Peripheral Vision, 1956) Brit J. Ophthal. **40**, 392—414 (1954).

Doesschate, J. ten.: Visual acuity and distribution of percipient elements on retina. Ophthalmologica (Basel) **112**, 1—18 (1946).
Donner, K. O., Reuter, T.: The dark adaptation of single units in the frog's retina and its relation to the regeneration of rhodopsin. Vision Res. **5**, 615—632 (1965).
Donovan, A.: The nerve fibre composition of the cat optic nerve. J. Anat. **101**, 1—11 (1967).
Dor, H.: Beiträge zur Elektrotherapie der Augenkrankheiten. Arch. Ophthal. **19**, 316—321 (1873).
Dowling, J. E.: Structure and function in the all-cone retina of the ground squirrel. In: The Physiological Basis for Form recognition. 17—23, N.I.H. sponsored at Brown University, Providence R.I. 1964.
Dowling, J. E., Boycott, B. B.: Organization of the primate retina: Electron microscopy. Proc. roy. Soc. **B 166**, 80—111 (1966).
Drasdo, N.: The neural representation of visual space. Nature (Lond.) **266**, 554—556 (1977).
Drasdo, N., Fowler, C. W.: Nonlinear projection of the retinal image in a wide angle schematic eye. Brit. J. Ophthal. **58**, 709—714 (1974).
Dreher, B., Zernicki, B.: Visual fixation reflex: behavioural properties and neural mechanism. Acta biol. exp. **29**, 359—383 (1969).
Dubar, J., Thieulin, G.: L'état de refraction des yeux des Mammifères domestiques. Rev. gén. Méd. vet. **36**, 361—565 (1927).
Dubin, M. W.: Anatomy of the vertebrate retina. In: The Eye: Comparative physiology. Davson, H. and Graham, L. T. (eds.). Vol. 6. London: Academic Press 1974, pp. 227—256.
Duijm, M.: On the position of a ribbon-like central area in the eyes of some birds. Arch. néerl. Zool. **13**, Suppl., 128—145 (1959).
Duke-Elder, W. S.: Text-book of Ophthalmology, Vol. 1. London: Henry Kimpton 1932.
Duke-Elder, W. S.: System of Ophthalmology. Vol. 1. In: The Eye in Evolution. London: Henry Kimpton 1958.
Duke-Elder, W. S.: System of Ophthalmology. Vol. 5. In: Ophthalmic Optics and Refraction. London: Henry Kimpton 1970.
Du Pont, J., De Groot, P. J.: A schematic dioptric apparatus for the frog's eye. Vision Res. **16**, 803—810 (1976).
Eayrs, J. T.: Relationship between the ganglion cell layer of the retina and the optic nerve in the rat. Brit. J. Ophthal. **36**, 453 (1952).
Edey, M.: The Cats of Africa. New York: Time-Life 1968.
Eisenberg, J. F., Leyhausen, P.: The phylogenesis of predatory behaviour in mammals. Z. Tierpsychol. **30**, 59—93 (1972).
Elliot Smith, G.: Presidential address to the Anthropological Section. In: Report of the British Association, Dundee, 575—598 (1912).
Elliot Smith, G.: Evolution of Man. London: Humphrey Milford 1924.
Elliot Smith, G.: The new vision. Nature (Lond.) **121**, 680—681 (1928).
Elliot Smith, G.: New light on vision. Nature (Lond.) **125**, 820—824 (1930).
Elul, R., Marchiafava, P. L.: Accommodation of the eye as related to behaviour in the cat. Arch. ital. Biol. **102**, 616—644 (1964).
Engström, K.: Cone types and cone arrangements in teleost retinae. Acta. zool. (Stockh.) **44**, 179—243 (1963).
Enoch, J. M.: Waveguide modes in retinal receptors. Science. **133**, 1353—1354 (1961).
Enoch, J. M.: Marked accommodation, retinal stretch, monocular space perception and retinal receptor orientation. Amer. J. Optom. Physiol. Opt. **52**, 376—392 (1975).
Enoch, J. M.: Retinal receptor orientation and the role of fiber optics in vision. Amer. J. Optom. **49**, 455—471 (1972).
Enoch, J. M., Hope. G. M.: An analysis of retinal receptor orientation. III. Results of initial psychophysical tests. Invest. Ophthal. **11**, 765—782 (1972).
Enroth-Cugell, C., Pinto, L. H.: Properties of the surround response mechanism of cat retinal ganglion cells and centre-surround interaction. J. Physiol. (Lond.) **220**, 403—439 (1972).
Enroth-Cugell, C., Robson, J. G.: The contrast sensitivity of retinal ganglion cells of the cat. J. Physiol. (Lond.) **187**, 517—552 (1966).
Enroth-Cugell, C., Robson, J. G.: Direct measurement of image quality in the cat eye. J. Physiol. (Lond.) **239**, 31P—33P (1974).
Errington, P. L.: Of Predation and Life. Iowa State Univ. Press. Ohio 1967.
Ewer. R. F.: The Carnivores. London: Wiedenfeld & Nicolson 1973.
Ferree, C. E., Rand, G.: Report of a Joint Discussion on Vision by the Physical and Optical Societies. Cambridge, Mass.: University Press 1932, p. 244.

Fick, A. E.: Über Stäbchensehschärfe und Zapfenschärfe. Arch. Ophthal. **45**, 336—356 (1898).
Fiennes, R., Fiennes, A.: The Natural History of the Dog. London: Weidenfeld & Nicholson 1968.
Fincham, E. F.: The mechanism of accommodation. Brit. J. Ophthal. **2** Monograph Suppl. 8 (1937).
Fischer, B.: The overlap of receptive field centres and representation of the visual field in the cat's optic tract. Vision Res. **13**, 2113—2120 (1973).
Fisher, B., May, H. O.: Invariances in the cat retina: Principles in the relations between sensitivity size and position of receptive fields of ganglion cells. Exp. Brain. Res. **11**, 448—464 (1970).
Fischer, F.: Über Fixierung der Linsenform mittels der Gefriermethode. Arch. Augenheilk. **56**, 342 (1907).
Fisher, R. F.: The elastic constants of the human lens capsule. J. Physiol. (Lond.) **201**, 1—19 (1969a).
Fisher, R. F.: The significance of the shape of the lens and capsular energy changes in accommodation. J. Physiol. (Lond.) **201**, 21—47 (1969b).
Fisher, R. F.: The elastic constants of the human lens. J. Physiol. (Lond.) **212**, 147—180 (1971).
Fisher, R. F.: Presbyopia and the changes with ageing in the human crystalline lens. J. Physiol. (Lond.) **228**, 765—779 (1973).
Fite, K. V.: Single unit analysis of binocular neurons in the frog optic tectum. Exp. Neurol. **24**, 475—486 (1969).
Fite, K. V., Rosenfield-Wessels, S.: A comparative study of deep avian foveas. Brain Behav. Evol. **12**, 97—115 (1975).
Flamant, F.: Etude de la répartition de lumière dans l'image rétinienne d'une fente. Rev. Opt. **34**, 433—459 (1955).
Fleisch, A.: Tonische Labyrinthreflexe auf die Augenstellung. **194**, 554—573 (1922).
Fletcher, A., Murphy, T., Young, A.: Solutions of two optical problems. Proc. roy. Soc. **A 223**, 216—225 (1954).
Flom, M. C.: The empirical longitudinal horopter in anomalous correspondence. Ph. D. Thesis, Univ. of California, Berkeley 1957.
Forrester, J. M.: Rolling movements of the sheep's eye. J. Physiol. (Lond.) **244**, 72 P (1975).
Forrester, J. M., Peters, A.: Nerve fibres in optic nerve of rat. Nature (Lond.) **214**, 245—247 (1967).
Fox, M. W.: Ontogeny of prey-killing behaviour in canidae. Behaviour **35**, 259—272 (1969).
Fox, R., Blake, R. R.: Stereoscopic vision in the cat. Nature (Lond.) **255**, 55—56 (1971).
Fox, R., Lehmkuhle, S., Westendorf, D. H.: Falcon visual acuity. Science **192**, 263—265 (1976).
Franz, V.: Die Akkommodation des Selachierauges und seine Abblendungsapparate, nebst Befunden und der Retina. Zool. Jb. Abt. allg. Zool. Physiol. Tiere **49**, 323—462 (1931).
Franz, V.: Vergleichende Anatomie des Wirbeltierauges. In: Handbuch der vergleichenden Anatomie der Wirbeltiere. Bolk, Göppert, Kallius, und Lubosch (eds.). **2**, Berlin: Urban & Schwarzenberg 1934, pp. 989—1292.
Frazetta, T. H.: Studies on the morphology and function of the skull in the Boidae Serpentes. II. Morphology and function of the jaw apparatus in *Python sebae* and *Python Molurus*. J. Morph. **118**, 217—296 (1966).
Freeman, R., Thibos, L.: Electrophysiological evidence that abnormal early visual experience can modify the human brain. Science, N.Y. **180**, 876—878 (1973).
French, A. S., Snyder, A. W., Stavenga, D. G.: Image degradation by a non-uniform retinal mosaic. Biol. Cybern. in press (1977).
Freytag, G.: Die Brechungsindices der Linse und der flüssigen Augenmedien bei der Katze und beim Kaninchen. Arch. vergl. Ophthal. **1**, 61—72 (1910).
Frieden, B. R.: Optical transfer of the three-dimensional object. J. opt. Soc. Amer. **57**, 56—66 (1967).
Frisen, L., Frisen, M.: A simple relationship between the probability distribution of visual acuity and the density of retinal output channels. Acta Ophthalmol. **54**, 437—443 (1976).
Frisen, L., Glansholm, A.: Optical and neural resolution in peripheral vision. Invest. Ophthal. **14**, 528—536 (1975).
Frith, H. J.: Wildlife Conservation. Sydney: Angus & Robertson 1973.
Fry, G. A.: Factors contributing to the discrepancy between subjective and stereoscopic determinations of the refraction of the eye. The O-Eye-O, **33**, 16—25 (1967).
Fry, G. A.: The optical performance of the human eye. Prog. in Optics **8**, 23—131 (1970).
Fuchs, A. F.: Saccadic and smooth pursuit eye movements in the monkey. J. Physiol. (Lond.) **191**, 609—631 (1967).
Fukuda, Y.: A three group classification of rat retinal ganglion cells; histological and physiological studies. Brain Res. **19**, 327—344 (1977).

Fukuda, Y., Stone, J.: The retinal distribution and central projection of Y, X and W cells of the cut's retina. J. Neurophysiol. **37**, 749—772 (1974).

Gabor, D.: Progress in Optics, vol. I. Wolf, E. (ed.) Amsterdam: North-Holland (1961).

Gallego, A.: Horizontal and amacrine cells in the mammal's retina. Vision Res. Suppl. **3**, 33—50 (1971).

Gaupp, E.: Lehre von den Eingeweiden, dem Integument und den Sinnesorganen. Ecker's u. Wiederheim's Anatomie des Frosches. Braunschweig **3** (1904).

Gauss, J. K. F.: Dioptrische Untersuchungen. Göttingen, 1841.

Gaze, R. M.: The Formation of Nerve Connections. London: Medical Books Ltd. 1970.

Gaze, R. M., Jacobson, M.: The projection of the binocular visual field on the optic tecta of the frog. Quart. J. exp. Physiol. **47**, 273—280 (1962).

Gaze, R. M., Jacobson, M.: The path from the retina to the ipsilateral optic tectum of the frog. J. Physiol. (Lond.) **165**, 73—74 (1963).

Gaze, M., Keating, M. J.: Further studies on the restoration of the contralateral retina tectal projection following regeneration of the optic nerve in the frog. Brain Res. **21**, 207—216 (1970).

Gaze, R. M., Keating, M. J., Szekely, G., Beazley, L.: Binocular interaction in the formation of specific intertectal neuronal connections. Proc. roy. Soc. B.**175**, 107—147 (1970).

Geiger, G., Poggio, T.: The Müller-Lyer-Figure and the fly. Science **190**, 479—480 (1975).

Georgeson, M. A., Sullivan, G. D.: Contrast constancy: Deblurring in human vision by spatial frequency channels. J. Physiol. (Lond.) **252**, 627—656 (1975).

Gibson, J. J.: The Perception of the Visual World. Boston: Houghton-Mifflin 1950.

Gibson, J. J.: The Senses Considered as Perceptual Systems. Boston: Houghton-Mifflin 1966.

Gilbert, D. S., Fender, D. H.: Contrast thresholds measured with stabilised and non-stabilised sine wave gratings. Optica acta **16**, 191—204 (1969).

Glickstein, M., Millodot, M.: Retinoscopy and eye size. Science, **168**, 605—606 (1970).

Goodge, W. R.: Adaptations for amphibious vision in the dipper *(Cinclus mexicanus)*. J. Morph. **107**, 79—91 (1960).

Gordon, D. A.: Static and dynamic fields in human space perception. J. opt. Soc. Amer. **55**, 1296—1303 (1965).

Gouras, P.: The effects of light adaptation on rod and cone receptive field organization of monkey ganglion cells. J. Physiol. (Lond.) **192**, 747—760 (1967).

Gouras, P.: The function of the midget cell system in primate color vision. Vision Res. Suppl. **3**, 397—410 (1971).

Gouras, P., Link, K.: Rod and cone interaction in dark-adapted monkey ganglion cells. J. Physiol. (Lond.) **184**, 499—510 (1966).

Graham, M. V., Gray, O. P.: Refraction of premature babies eyes. Brit. med. J. **1**, 1452—1454 (1963).

Green, D. G.: Regional variations in the visual acuity for interference fringes on the retina. J. Physiol. (Lond.) **207**, 351—356 (1970).

Gregory, R. L.: Distortion of visual space as inappropriate constancy scaling. Nature (Lond.) **199**, 678—680 (1963).

Gregory, R. L., Harris, J. P.: Illusion destruction by appropriate scaling. Perception **4**, 203—220 (1975).

Grossman, K., Meyerhausen, M.: Beitrag zur Lehre vom Gesichtsfeld bei Säugethieren. Arch. Ophthal. **23**, 217 (1877).

Gullstrand, A.: Die optische Abbildung in heterogenen Medien und die Dioptrik der Kristallinse des Menschen. K. sv. vet. Handl. **43**, 1—58 (1908).

Gullstrand, A.: Appendix in Helmholtz' Physiologische Optik, 3rd ed. 1909 (Rep. Dover, New York 1962 of trans. by J.P.C. Soothall for Am. Opt. Soc.) 1924.

Hage, S. G. el, Berny, F.: Contribution of the crystalline lens to the spherical aberration of the eye. J. opt. Soc. Amer. **63**, 205—211 (1973).

Haines, R. W.: Arboreal or terrestrial ancestry of placental animals. Quart. Rev. Biol. **33**, 1—23 (1958).

Hall, W. C., Kaas, J. H., Killackey, H., Diamond, I. T.: Cortical visual areas in the grey squirrel *(Sciurus carolinesis)*: A correlation between cortical evoked potential maps and architectonic subdivisions. J. Neurophysiol. **34**, 437—451 (1971).

Haller, A.: Elementa Physiologae Corporis Humani. Lausannae: Fransisci Grasset & Sociorum 1769.

Hamdi, F. A., Whitteridge, D.: The representation of the retina on the optic tectum of the pigeon. Quart. J. exp. Physiol. **39**, 111—119 (1954).

Harkness, L.: Chameleons use accommodation cues to judge distance. Nature (Lond.) **267**, 346—349 (1977).

Harris, C. J.: Otters. London: World Naturalist Series, Weidenfeld Nicholson 1968.

Harris, W.: Binocular and stereoscopic vision in man and other vertebrates with its relation to the decussation of the optic nerves, the ocular movements, and the pupil light reflex. Brain **27**, 107—147 (1904).
Hartridge, H.: The limit to peripheral vision. J. Physiol. (Lond.) **53**, xvii—xviii (1919).
Hartridge, H.: The special senses. In: Principles of Human Physiology. Evans, C. L. (ed.). London: Churchill 1952, pp. 377—455.
Hartridge, H., Yamada, K.: Accommodation and other optical properties of the eye of the cat. Brit. J. Ophthal. **6**, 481—492 (1922).
Hay, G. A., Chesters, M. S.: Signal-transfer functions in threshold and suprathreshold vision. J. opt. Soc. Amer. **62**, 990—998 (1972).
Hecht, S., Mintz, E. V.: The visibility of single lines at various illuminations and the retinal basis of visual resolution. J. gen. Physiol. **22**, 593—612 (1939).
Hecht, S., Shlaer, S., Pirenne, M. H.: Energy quanta and vision. J. gen. Physiol. **25**, 819—840 (1942).
Helmholtz, H. von: Über die Accommodation des Auges. Arch. Ophthal. **1** (1855).
Helmholtz, H. von: Handbuch der Physiologischen Optik (1856—1866), Gullstrand, A., Kries, J., Nagel, W. (eds.). 3rd ed. 1909. Rep. New York: Dover 1962, of trans. by J.P.C. Southall for Amer. opt. Soc., 1924.
Hendrickson, A. C., Wilson, M. E., Toyne, M. J.: The distribution of optic nerve fibres in *Macaca mulatta*. Brain Res. **23**, 425—427 (1970).
Hennessy, R. T., Iida, T., Shiina, K., Leibowitz, H. W.: The effect of pupil size on accommodation. Vision Res. **16**, 587—589 (1976).
Henning, G. B., Hertz, B. G., Broadbent, D. E.: Some experiments bearing on the hypothesis that the visual system analyses spatial patterns in independent bands of spatial frequency. Vision Res. **15**, 887—897 (1975).
Hensen, V., Völcker, C.: Experimental Untersuchungen über dem Mechanismus Akkommodation. Kiel 1868.
Heric, T. M., Kruger, L.: Organization of the visual projection upon the optic tectum of a reptile (*Alligator mississippiensis*). J. comp. Neurol. **124**, 101—111 (1965).
Hering, E.: Der Raumsinn und die Bewegungen der Augen. In: Handbuch der Physiologie, **3**. Hermann, L. (ed.) Leipzig: Vogel 1879, pp. 343—601.
Hermann, G.: Beiträge zur Physiologie des Rattenauges. Z. Tierpsychol. **15**, 462—518 (1958).
Hess, C.: Gesichtsinn.: Handbuch der vergleichenden Physiologie. Winterstein, H. (ed.). Jena: Fischer 1913.
Hess, C., Heine, L.: Arbeiten aus dem Gebiete der Accommodationslehre. Arch. Ophthal. **46**, 243—276 (1898).
Higgins, G. E., Stultz, K.: Variation of visual acuity with various test objects orientations and viewing conditions. J. opt. Soc. Amer. **40**, 135—137(1950).
Hill, R. M., Ikeda, H.: "Refracting" a single retinal ganglion cell. Arch. Ophthal. **85**, 592—596 (1971).
Hill, W. C. O.: Evolutionary biology of the primates. London: Academic Press 1972.
Hirschberg, J.: Zur Dioptrik und Ophthalmoskopie der Fisch- und Amphibienaugen. Arch. Anat. Physiol. Lpz. **6**, 493—526 (1882).
Hisdal, E.: Detectable information in a photon beam. J. opt. Soc. Amer. **57**, 35—43 (1967).
Horridge, G. A. (ed.): The Compound Eye and Vision in Insects. Oxford: Clarendon Press 1975.
Howells, W. W.: Mankind so far. New York: Doubleday 1947.
Hubel, D. H., Wiesel, T. N.: Receptive fields of single neurones in the cat's striate cortex. J. Physiol. (Lond.) **148**, 574—591 (1959).
Hubel, D. H., Wiesel, T. N.: Receptive fields of cells in striate cortex of very young, visually inexperienced kittens. J. Neurophysiol. **26**, 994—1002 (1963).
Hubel, D. H., Wiesel, T. N.: The period of susceptibility to the physiological effects of unilateral eye closure in kittens. J. Physiol. (Lond.) **206**, 419—436 (1970).
Hubel, D. H., Wiesel, T. N.: Cells sensitive to binocular depth in area 18 of the Macaque monkey cortex. Nature (Lond.) **225**, 41—42 (1970).
Hughes, A.: Topographical relationships between the anatomy and physiology of the rabbit visual system. Docum. ophthal. (Den Haag) **30**, 33—159 (1971).
Hughes, A.: A schematic eye for the rabbit. Vision Res. **12**, 123—138 (1972).
Hughes, A.: Vergence in the cat. Vision Res. **12**, 1961—1994 (1972).
Hughes, A.: Observing accommodation in the cat. Vision Res. **13**, 481—482 (1973).
Hughes, A.: A comparison of retinal ganglion cell topography in the plains and tree kangaroo. J. Physiol. (Lond.) **244**, 61—63 P (1975a).

Hughes, A.: A quantitative analysis of cat retinal ganglion cell topography. J. comp. Neurol. **163**, 107—128 (1975b).
Hughes, A.: A supplement to the cat schematic eye. Vision Res. **16**, 149—154 (1976a).
Hughes, A.: The refractive state of the rat eye. Vision Res. **17**, 927–939 (1977b).
Hughes, A.: A schematic eye for the rat. Vision Res. in press (1977).
Hughes, A.: Directional units in rat optic nerve. Brain Res. submitted (1977d).
Hughes, A.: The pigmented rat optic nerve: fibre count and diameter spectrum. J. comp. Neurol. **176**, 263—268 (1977).
Hughes, A.: A comparison of the retinal ganglion cell diameter spectrum in different regions of the cat retina. J. comp. Neurol., submitted (1977).
Hughes, A., Vaney, D. I.: Optometric refraction of the rabbit at various eccentricities. Vision Res. submitted (1977).
Hughes, A., Wässle, H.: The cat optic nerve: Fibre total count and diameter spectrum. J. comp. Neurol. **169**, 171—184 (1976).
Hughes, A., Wässle, H.: Optical image quality of the rat. (Appendix to Hughes, 1977c) in press, (1977b).
Hughes, A., Whitteridge, D.: The receptive fields and topographical organization of goat retinal ganglion cells. Vision Res. **13**, 1101—1114 (1973).
Ikeda, H., Wright, M. J.: Differential effects of refractive errors and receptive field organization of central and peripheral ganglion cells. Vision Res. **12**, 1465—1476 (1972).
Ikeda, H., Wright, M.: Optical quality of the cat's eye and human eye. J. Physiol. (Lond.) **232**, 34—35P (1973).
Ingle, D. (ed.): The Central Nervous System and Fish Behavior. Chicago: University of Chicago Press 1968.
Ingle, D.: Visuomotor functions of the frog optic tectum. Brain Behav. Evol. **3**, 57—71 (1970).
Ingle, D.: Prey-catching behaviour of anurans toward moving and stationary objects. Vision Res. Suppl. **3**, 447—456 (1971).
Ingle, D.: Depth vision in monocular frogs. Psychon. Sci. **29**, 37—38 (1972).
Ingle, D.: Evolutionary perspectives on the function of the optic tectum. Brain Behav. Evol. **8**, 211—237 (1973).
Jacobson, M.: The representation of the retina on the optic tectum of the frog. Quart. J. exp. Physiol. **47**, 170—178 (1962).
Jacobson, M., Gaze, R. M.: Types of visual response from single units in the optic tectum and optic nerve of the goldfish. Quart. J. exp. Physiol. **49**, 199—209 (1964).
Jacobsen, S. G., Franklin, K. B. J., McDonald, W. I. M.: Visual acuity of the cat. Vision Res. **16**, 1141—1143 (1976).
James, G. R.: Degeneration of ganglion cells following axonal injury; an experimental study. Arch. Ophthal. **9**, 338—343 (1933).
Jampel, R. S., Mindel, J.: The nucleus for accommodation in the midbrain of the Macaque. Invest. Ophthal. **6**, 40—50 (1967).
Johansson, G.: Visual motion perception. Sci. Am. **233**, 76—88 (1975).
Johnson, G. L.: Contributions to the comparative anatomy of the mammalian eye, chiefly based on ophthalmoscopic examination. Phil. Trans. **B194**, 1—82 (1901).
Jones, A. E.: The retinal structure of *(Aotes trivirgatus)* the owl monkey. J. comp. Neurol. **125**, 19—27 (1965).
Joshua, D. E., Bishop, P. O.: Binocular single vision and depth discrimination. Receptive field disparities for central and peripheral vision and binocular interaction on peripheral single units in cat striate cortex. Exp. Brain Res. **10**, 389—416 (1970).
Julesz, B.: Foundations of Cyclopean Perception. Chicago: University of Chicago Press 1971.
Kahmann, H.: Untersuchungen über die Linse, die Zonula ciliaris, Refraktion und Accommodation von Säugetieren. Zool. Jb. Abt. allg. Zool. Physiol. **48**, 509—588 (1930).
Kahmann, H.: Über das Vorkommen einer Fovea centralis im Knochenfischauge. Zool. Anz. **106**, 49—55 (1934).
Kahmann, H.: Über das foveale Sehen der Wirbeltiere. II. Gesichtfeld und Fovea centralis. Sitz. Ges. naturf. Freunde **8**, 361—376 (1935).
Kahmann, H.: Über das foveale Sehen der Wirbeltiere. I. Über die Fovea centralis und die Fovea lateralis bei einigen Wirbeltieren. Arch. Ophthal. **135**, 265—276 (1936).
Keating, M. J., Gaze, R. M.: The ipsilateral retinotectal pathway in the frog. Quart. J. exp. Physiol. **55**, 284—292 (1970).

Keating, M. P.: A theoretical analysis of off-axis streak retinoscopy. Amer. J. Optom. physiol. Optics. **52**, 750—757 (1975).

Kelly, J. P., Gilbert, C. D.: The projections of different morphological types of ganglion cells in the cat retina. J. comp. Neurol. **163**, 65—80 (1975).

Kepler, J.: Ad Vitellionem Paralipomena, quibus Astronomiae pars optica traditur. Francofurti, 1604. (Trans. F. Plehn; ed. M. von Rohr). Grundlagen d. geom. Optik. Ostwald's Klassiker d. exakt Wissensch. **198**. Leipzig: Akad. Verlagsgesellschaft 1922.

Kepler, J.: Dioptrice. Augsburg, 1611. (Trans. F. Plehn). Ostwald's Klassiker d. exakt. Wissensch. **144**. Leipzig: Engelmann 1904.

Kirschfeld, K.: The resolution of lens and compound eyes. In: Neural Processing in Visual Systems. Zettler, F., Weiler, R. (eds.). Berlin: Springer-Verlag 1976.

Koenderink, J. J., Van Doorn, A. J.: Invariant properties of the motion parallax field due to the movement of rigid bodies relative to an observer. Optica Acta, **22**, 773—791 (1975).

Kolb, H.: Organization of the outer plexiform layer of the primate retina: Electron microscopy of Golgi-impregnated cells. Phil. Trans. **B 258**, 261—283 (1970).

Krueger, H., Moser, E. A.: Refraktion und Abbildungsgüte des Froschauges. Pflügers Arch. ges. Physiol. **326**, 334—340 (1971).

Krueger, H., Moser, E. A.: The influence of the modulation transfer function of the dioptric apparatus on the acuity and contrast of the retinal image in *Rana esculenta*. Vision Res. **12**, 1281—1289 (1972).

Krueger, H., Moser, E. A.: On the approximation of the optical modulation transfer function (MTF) by analytical functions. Vision Res. **13**, 493—494 (1973).

Kuffler, S. W.: Discharge patterns and functional organisation of mammalian retina. J. Neurophysiol. **16**, 37—68 (1953).

Kulikowski, J. J.: Pattern and movement detection: A comparison between man and rabbit. (in press) 1977.

Landau, D., Dawson, W. W.: The histology of retinas from the Pinnipedia. Vision Res. **10**, 691—702 (1970).

Lane, R. H., Allman, J. M., Kaas, J. H.: Representation of the visual field in the superior colliculus of the grey squirrel *(Sciurus carolinensis)* and the tree shrew *(Tupaia glis)*. Brain Res. **26**, 277—292 (1971).

Lang, W., Barrett, J. W.: The refractive character of the eyes of mammalia. Arch. Augenheilk. **17**, 103—138 (1887).

Langenbeck, M.: Klinische Beiträge aus dem Gebiete der Chirurgie und Ophthalmologie. Göttingen, 1849.

Lashley, K. S.: The Mechanism of Vision. V. The structure and image-forming power of the rat's eye. J. comp. Psychol. **13**, 173—200 (1932).

Lashley, K. S.: The mechanism of vision. XV. Preliminary studies of the rat's capacity for detail vision. J. gen. Psychol. **18**, 123—193 (1938).

Laties, A. M., Enoch, J. M.: An analysis of retinal receptor orientation. I. Angular relationship of neighbouring photoreceptors. Invest. Ophthal. **10**, 69—77 (1971).

La Vail, J. H., La Vail, M. M.: The retrograde intraaxonal transport of horseradish peroxidase in the chick visual system. J. comp. Neurol. **157**, 303—357 (1974).

Lazar, G. V., Szekely, G. Y.: Distribution of optic terminals in the different optic centers of the frog. Brain Res. **16**, 1—14 (1969).

Leehey, S. C., Moskowitz-Cook, A., Brill, S., Held, R.: Orientational anisotropy in infant vision. Science **190**, 900—901 (1975).

LeGrand, Y.: Recherches sur la diffusion de la lumière dans l'oeil humain. Rev. Opt. **16**, 201—241 (1937).

Le Grand, Y.: Sur l'aberration sphérique de l'oeil. C. R. Acad. Sci. (Paris) **215**, 547 (1942).

Le Grand, Y.: Form and Space Vision. (Trans. by Millodot & Heath), London: Indiana Univ. Press 1967.

Le Gros Clarke, W. E.: Remarks on the tree shrew, *Tupaia minor*, with photographs. Proc. zool. Soc. Lond. **97**, 254—256 (1927).

Le Gros Clarke, W. E.: The Early Forerunners of man. London: Tindall & Cox 1934.

Le Gros Clarke, W. E.: The Antecedents of Man. Edinburgh: Edinburgh Univ. Press 1962.

Le Gros Clarke, W. E.: History of the Primates. London: British Museum 1970.

Leibowitz, H., Johnson, C., Isabelle, E.: Peripheral motion detection and refractive error. Science **177**, 1207—1208 (1972).

Leigh Thomas, H.: An anatomical description of a male rhinoceros. Phil. Trans. **91**, 145—152 (1801).

Leinfelder, P. J.: Retrograde degeneration in the optic nerves and retinal ganglion cells. Trans. Amer. ophthal. Soc. **36**, 307 (1938).

Leuckart, R. K. G. F.: Organologie des Auges. Vergleichende Anatomie. I. Handbuch der gesamten Augenheilkunde. Graefe-Saemisch. (ed.). **2**, 145—301. Leipzig: Engelmann 1876.

Levick, W. R.: Receptive fields and trigger features of ganglion cells in the visual streak of the rabbit's retina. J. Physiol. (Lond.) 188, 285—307 (1967).

Levick, W. R., Cleland, B. G.: Selectivity of microelectrodes in recordings from cat retinal ganglion cells. J. Neurophysiol. **37**, 1387—1393 (1974).

Leyhausen, P.: Über die Funktion der relativen Stimmungshierarchie. Z. Tierpsychol. **22**, 412—494 (1965).

Lichtenstein, M.: Spatiotemporal factors in cessation of smooth apparent movement. J. opt. Soc. Amer. **53**, 302—306 (1963).

Lincoln, D. W., Mason, C. A.: The use of cobalt sulphide precipitation technique to delineate neuronal projections in the rat brain. J. Physiol. (Lond.) **245**, 40—41 P (1974).

Listing, J. B.: Beitrag zur physiologischen Optik. Göttingen, 1845.

Lotmar, W.: Theoretical eye model with aspherics. J. opt. Soc. Amer. **61**, 1522—1529 (1971).

Low, F. N.: The peripheral visual acuity of 100 subjects. Amer. J. Physiol. **140**, 83—88 (1943).

Luck, C. P.: The comparative morphology of the eyes of certain African suiformes. Vision Res. **5**, 283—297 (1965).

Ludlam, W. M., Twarowski, C. J.: Ocular-dioptric-component changes in the growing rabbit. J. opt. Soc. Amer. **63**, 95—98 (1973).

Ludlam, W. M., Weinberg, S. S., Twarowski, C. J., Ludlam, D. P.: Comparison of cycloplegic and noncycloplegic ocular component measurement in children. Amer. J. Optom. **49**, 805—818 (1972).

Ludvigh, E.: Extrafoveal acuity as measured with Snellen test letters. Amer. J. Ophthal. **24**, 303—310 (1941).

Luneberg, R. K.: Mathematical Analysis of Binocular Vision. Hannover: Dartmouth Eye Institute 1947.

Mach, E.: Über die Wirkung der räumlichen Vertheilung des Lichtreizes auf die Netzhaut, I. In: Sitzungberichte der mathematisch-naturwissenschaftlichen Classe der kaiserlichen Akademie der Wissenschaften. **52**, 303—322 (1865).

Maffei, L., Campbell, F. W.: Neurophysiological localization of the vertical and horizontal visual coordinates in man. Science **167**, 386—387 (1970).

Mandelbaum, J., Sloan, L. L.: Peripheral visual acuity. Amer. J. Ophthal. **30**, 581—588 (1947).

Marchand, E. W.: Ray tracing in gradient index media. J. opt. Soc. Amer. **60**, 1—7 (1970).

Marriott, F. H. C., Morris, V. B.: The distribution of light in an image formed in the cat's eye. Nature (Lond.) **90**, 176—177 (1961).

Marshall, J., Mellerio, J., Palmer, D. A.: A schematic eye for the pigeon. Vision Res. **13**, 2449—2453 (1973).

Massof, R. W., Chang, F. W.: A revision of the rat schematic eye. Vision Res. **12**, 793—796 (1972).

Matthiessen, L.: Die Differentialgleichungen der Dioptrik der geschichteten Krystallinse. Pflügers Arch. ges. Physiol. **19**, 480—562 (1879).

Matthiessen, L.: Über die Beziehungen welche zwischen dem Brechungsindex des Kerncentrums der Krystallinse und der Dimensionen des Auges bestehen. Pflügers Arch. ges. Physiol. **27**, 510—523 (1882).

Matthiessen, L.: Beiträge zur Dioptrik der Krystallinse I. Z. vergl. Augenheilk. **4**, 1—39 (1886).

Matthiessen, L.: Über den physikalisch-optischen Bau des Auges der Cetaceen und der Fische. Pflügers Arch. ges. Physiol. **38**, 521—528 (1886).

Matthiesen, L.: Über den physikalisch-optischen Bau des Auges der Vögel. Pflügers Arch. ges. Physiol. **38**, 104—112 (1886).

Matthiesen, L.: Beiträge zur Dioptrik der Krystallinse, II & III. Z. vergl. Augenheilk. **5**, 21—44, 97—126 (1887).

Matthiessen, L.: Über den physikalisch-optischen Bau des Auges von Cervus alcesmos. Pflügers Arch. ges. Physiol. **40**, 314—323 (1887).

Matthiessen, L.: Über den physikalisch-optischen Bau des Auges von Knölwal (Megaptera boops, Fabr.) und Finwal (Balaenoptera musculus Comp.), Z. vergl. Augenheilk. **7**, 77—101 (1893).

Matthews, G. V. T.: Bird navigation. Cambridge Monographs in Expl. Biology No. 3. Cambridge: Cambridge Univ. Press 1968.
Maturana, H. R.: Number of fibers in the optic nerve and the number of ganglion cells in the retina of anurans. Nature (Lond.) **183**, 1406—1407 (1959).
Maturana, H. R.: Functional organization of the pigeon retina. Proc. IUPS 22nd Int. Cong. (Leiden) **3**, 170—178. Amsterdam: Excerpta Medica 1964.
Maxwell, G.: Ring of Bright Water. London: Longmans, Green & Co. 1960.
Maxwell, G.: The Rocks Remain. London: Longmans, Green & Co. 1963.
Maxwell, J. C.: Solutions of problems. Cambridge & Dublin Mathematical Journal **8**, 188 (1854). In: The Scientific Papers of J.C. Maxwell. I. Niven, W.D., London: Cambridge University Press 1890.
Maxwell, J. C.: On the general laws of optical instruments. Quart. J. Pure & Appl. Mathematics **VII** (1858). In: The Scientific Papers of J.C. Maxwell. I. London: Cambridge University Press 1890.
McColgin, F. H.: Movement thresholds in peripheral vision. J. opt. Soc. Amer. **50**, 774—779 (1960).
Merigan, W. H.: The contrast sensitivity of the squirrel monkey. Vision Res. **16**, 375—379 (1976).
Messing, R. B.: The sensitivity of albino rats to lights of different wavelengths: a behavioural assessment. Vision Res. **12**, 753—761 (1972).
Meyer, D. L., Meyer-Haame, S., Schaeffer, K.-P.: Electrophysiological investigation of refractive state and accommodation in the rabbit's eye. Pflügers Arch. ges. Physiol. **332**, 80—86 (1972).
Meyer, D. L., Schwassmann, H. O.: Electrophysiological method for determination of refractive state in fish eyes. Vision Res. **10**, 1301—1307 (1970).
Michael, C. R.: Receptive fields of single optic nerve fibres in a mammal with an all-cone retina. II. Directionally selective units. J. Neurophysiol. **31**, 257—267 (1968).
Miller, W. H., Snyder, A. W.: Optical function of human peripheral cones. Vision Res. **13**, 2185—2194 (1973).
Miller, W. H., Snyder, A. W.: Deep fovea of birds functions as telephoto lens. A.R.V.O. Symp. Rec. May, p. 26 (1977).
Millodot, M.: Reflection from the fundus of the eye and its relevance to retinoscopy. Fondazione Giorgio Ronchi **27**, 31—50 (1972).
Millodot, M., Blough, P.: The refractive state of the pigeon eye. Vision Res. **11**, 1019—1022 (1971).
Millodot, M., Johnson, C. A., Lamont, A., Leibowitz, H. W.: Effect of dioptrics on peripheral visual acuity. Vision Res. **15**, 1357—1362 (1975).
Millodot, M., Lamont, A.: Refraction of the periphery of the eye. J. opt. Soc. Amer. **64**, 110—111 (1971).
Millodot, M., Lamont, A.: Peripheral visual acuity in the vertical plane. Vision Res. **14**, 1497—1498 (1974).
Missotten, L.: Estimation of the ratio of cones to neurons in the fovea of the human retina. Invest. Ophthal. **13**, 1045—1049 (1974).
Mitchell, D. E., Freeman, R. D., Millodot, M., Haegerstrom, G.: Meridional amblyopia: Evidence for modification of the human visual system by early visual experience. Vision Res. **13**, 535—558 (1973).
Mitchell, D. E., Griffin, F., Tinney, B.: A behavioural technique for the rapid assessment of visual capabilities of kittens. Perception **6**, 181—193 (1977).
Mitchell, P. C.: The Childhood of Animals. London 1912.
Mize, R. R., Murphy, E. H.: Selective visual experience fails to modify receptive field properties of rabbit striate cortex neurons. Science **180**, 320—323 (1973).
Montero, V. M., Brugge, J. F., Beitel, R. E.: Relation of the visual field to the lateral geniculate body in the albino rat. J. Neurophysiol. **31**, 221—236 (1968).
Montero, V. M., Rojas, A., Torrealba, F.: Retinotopic organisation of striate and peristriate visual cortex in the albino rat. Brain Res. **53**, 197—201 (1973).
Moore, D. T.: Design of singlets with continuously varying indices of refraction. J. opt. Soc. Amer. **61**, 886—894 (1971).
Morgan, M. W., Mohney, J., Olmsted, J. M. D.: Astigmatic accommodation. Arch. Ophthal. **30**, 247—249 (1943).
Moser, E. A.: Retinoskopische und Neurophysiologische Refraktion beim Frosch. Inaugural Dissertation zur Doktorwürde, Munich University 1973.
Moser, E. A., Krueger, H.: Retinoscopic and neurophysiological refractometry in *Rana tempororia*. Pflügers Arch. ges. Physiol. **335**, 83 (1972).
Moses, R. A.: Adler's Physiology of the Eye: Clinical Application. Moses, R. A. (ed.). 5th ed. St. Louis: C. V. Mosby 1970, pp. 350—371.

Müller, H.: Zur vergleichenden Physiologie des Gesichtsinnes des Menschen und der Thiere. Leipzig: Cnobloch 1826 (cited by E. G. Boring, 1942).

Müller, H.: Über einige Verhältnisse der Netzhaut bei Menschen und Thieren. Verh. phys. med. Ges. Würzburg **3**, 336 (1853).

Müller, H.: Anatomisch-physiologische Untersuchungen über die Retina des Menschen und der Wirbelthiere. Leipzig 1856.

Müller, H.: Über das ausgedehnte Vorkommen einer dem gelben Fleck der Retina entsprechenden Stelle bei Thieren. Würzburg. Naturwiss. Z. **2**, 139—140 (1861).

Munk, O.: On the occurrence and significance of horizontal band-shaped retinal areae in teleosts. Vidensk. Meddr. dansk. naturh. Foren. **133**, 85—120 (1970).

Nakao, S. N., Fujimoto, S., Nagata, R., Iwata, K.: Model of refractive index distribution in the rabbit crystalline lens. J. opt. Soc. Amer. **58**, 1125—1130 (1968).

Nakao, S., Mine, K., Nishioka, K., Kamiya, S.: New schematic eye and its clinical applications. Abstracts 21st Int. Cong. of Ophthal., Mexico **E 102** (1970).

Nakao, S., Ono, T., Nagata, R., Iwata, K.: The distribution of refractive indices in the human crystalline lens. Jap. J. clin. Ophthal. **23**, 41—44 (1969).

Nakayama, K., Loomis, J. M.: Optical velocity patterns, velocity-sensitive neurons, and space perception: A hypothesis. Perception **3**, 63—80 (1974).

Nelson, J. I.: Motion sensitivity in peripheral retina. Perception **4**, 151—152 (1974).

Nicol, J. A. C.: Some aspects of photoreception and vision in fishes. Advanc. Marine Biol. **1**, 171—208 (1963).

Nicolas, E.: Veterinary and Comparative Ophthalmology. London: H. & W. Brown 1930.

Nikara, T., Bishop, P. O., Pettigrew, J. D.: Analysis of retinal correspondence by studying receptive fields of binocular single units in cat striate cortex. Exp. Brain Res. **6**, 353—372 (1968).

Nordenson, J. W.: Über den Brechungsindex der Netzhaut. Acta ophthal. (Kbh.) **12**, 171—175 (1934).

Nye, P. W.: On the functional differences between frontal and lateral visual fields of the pigeon. Vision Res. **13**, 559—574 (1973).

O'Day, K.: The visual cells of the Australian reptiles and mammals. Trans. ophthal. Soc. Aust. **1**, 12 (1939).

O'Day, K.: The fundus and fovea centralis of the albatross (*Diomedea cauta cauta*-Gould). Brit. J. Ophthal. **24**, 201—207 (1940).

O'Flaherty, J. J.: The optic nerve of the mallard duck: Fibre diameter frequency distribution and physiological properties. J. comp. Neurol. **143**, 17—24 (1971).

Ogawa, T., Karita, K., Tsuchiya, I.: Response characteristics of single neurons in the rabbit visual cortex. Tohoku J. exp. Med. **96**, 349—364 (1968).

Ogden, T. E.: The receptor mosaic of Aotes trivirgalus: distribution of rods and cones. J. comp. Neurol. **163**, 193—202 (1975).

Ogilvie, J., Daicar, E.: The perception of curvature. Canad. J. Psychol. **21**, 521—525 (1967).

Ogle, K. N.: Researches in Binocular Vision. Philadelphia: Saunders 1950.

Ogle, K. W.: Visual acuity. In. The Retina. U.C.L.A. Forum in Medical Sciences No. 8 Straatsma, B. R. et al. Berkley and Los Angeles: University of California Press 1969.

Ohzu, H., Enoch, J. M.: Optical modulation by the isolated human fovea. Vision Res. **12**, 245—251 (1972).

Ohzu, H., Enoch, J. M., O'Hair, J. C.: Optical modulation by the isolated retina and retinal receptors. Vision Res. **12**, 231—244 (1972).

Olmsted, J. M. D.: The role of the autonomic nervous system in accommodation for far and near vision. J. nerv. ment. Dis. **99**, 794—798 (1944).

Olmsted, J. M. D., Morgan, M. W.: The influence of the cervical sympathetic nerve on the lens of the eye. Amer. J. Physiol. **133**, 720—723 (1941).

O'Neill, W. D., Brodkey, J. S.: Linear regression of lens movement with refractive state. Arch. Ophthal. **82**, 795—799 (1969).

O'Neill, W. C., Brodkey, J. S.: A nonlinear analysis of the mechanics of accommodation. Vision Res. **10**, 375—391 (1970).

Oppel, O.: Quantitative Untersuchungen über die Retinaganglien und Optikusfusern. In: Eye Structure, II. Symp. Rohen, S. W. (ed.). Stuttgart: Schattauer-Verlag 1965, pp. 97—108.

Oppel, O.: Untersuchung über die Verteilung und Zahl der retinalen Ganglienzellen beim Menschen. Albrecht v. Graefes Arch. Klin. exp. Ophthalmol. **172**, 1—22 (1967).

Österberg, G.: Topography of the layer of rods and cones in the human retina. Acta. ophthal. (Kbh.) Suppl. **6** (1935).

Oyster, C.W., Takahashi, E., Collewijn, H.: Direction selective retinal ganglion cells and control of optokinetic nystagmus in the rabbit. Vision Res. **12**, 183—193 (1972).

Oyster, C.W., Takahashi, E., Levick, W.R.: Information processing in the rabbit visual system. In: Vision in the Rabbit. Van Hof, M.W. and Collewijn, H. (eds.). Docum. ophthal. (Den Haag) **30**, 161—204 (1971).

Packwood, J., Gordon, B.: Stereopsis in normal domestic cat, Siamese cat and cat raised with alternating monocular occlusion. J. Neurophysiol. **38**, 1485—1499 (1975).

Pariente, G.: Influence of light on the activity rhythms of two Malagasy lemurs: *Phaner furcifer* and *Lepilemur mustelinus leucopus*. In: Prosimian Biology. Martin, R.D., and Doyle, G.A., and Walker, A.C. (eds.). Part. 1. London: Duckworth 1974, pp. 183—198.

Parsons, J.H.: An Introduction to the Theory of Perception. London: Cambridge Univ. Press 1927.

Pask, C., Snyder, A.W.: Angular sensitivity of lens photoreceptor systems. In: Photoreceptor Optics. Snyder, A.W. & Menzel, R. (ed.). Berlin: Springer-Verlag 1975.

Patnaik, B.: A photographic study of accommodative mechanisms: changes in the lens nucleus during accommodation. Invest. Ophthal. **6**, 601—611 (1967).

Pennycuick, C.J.: The physical basis of astro-navigation in birds: Theoretical considerations. J. exp. Biol. **37**, 573—593 (1960).

Peterson, D.R., Middleton, D.: Sampling and reconstruction of wave number limited functions in N-dimensional Euclidean spaces. Inf. Control **5**, 279—323 (1962).

Pettigrew, J.D., Konishi, M.: Binocular neurones sensitive to orientation and disparity in the visual wulst of the owl. Science **193**, 675—678 (1976).

Phillipson, B.: Distribution of protein within the normal rat lens. Invest. Ophthal. **8**, 258—270 (1969).

Pirenne, M.H.: The visual process. In: The Eye, Davson, H. (ed.). London: Academic Press 1962.

Pirenne, M.H.: Optics, Painting and Photography. Cambridge: Cambridge Univ. Press 1970.

Pisa, A.: Über den binokularen Gesichtsraum bei Haustieren. Arch. Ophthal. **140**, 1—54 (1939).

Pollock, R.I.: On the external characters of some species of Lutrinae. Proc. zool. Soc. Lond. **91**, 535—546 (1921).

Polyak, S.: The Retina. Chicago: University of Chicago Press 1941.

Polyak, S.: The Vertebrate Visual System. Chicago: University of Chicago Press 1957.

Pomerantzeff, O., Govignon, J., Schepens, C.L.: Wide-angle optical model of the human eye. Ann. Ophthal. **3**, 815—819 (1971).

Potts, A.M., Hodges, D., Shelman, C.B., Fritz, K.J., Levy, N.S., Mangall, Y.: Morphology of the primate optic nerve. I. Method and total fibre count. Invest. Ophthal. **11**, 980—988 (1972).

Prince, J.H.: Comparative Anatomy of the Eye. Illinois: Thomas 1956.

Prince, J.H., Diesem, C.D., Eglitis, I., Ruskell, G.L.: Anatomy and Histology of the Eye and Orbit in Domestic Animals. Illinois: Thomas 1960.

Pritchard, R.M., Heron, W.: Small eye movements of the cat. Canad. J. Psychol. **14**, 131—137 (1960).

Pumphrey, R.J.: The theory of the fovea. J. exp. Biol. **25**, 299—312 (1948a).

Pumphrey, R.J.: The sense organs of birds. Ibis. **90**, 171—199 (1948b).

Pumphrey, R.J.: Concerning Vision. In: The Cell and the Organism. Ramsay & Wigglesworth (eds.). London: Cambridge Univ. Press 1961, pp. 193—208.

Pumphrey, R.J.: Sensory organs: Vision II. Biology & Comparative Physiology of Birds. Marshall, A.J. (ed.). New York: Academic Press 1961.

Quilliam, T.A.: The problems of vision in the ecology of *Talpa Europea*. Exp. Eye. Res. **5**, 63—78 (1966).

Rademaker, G.G., Ter Braak, J.W.G.: On the central mechanism of some optic reactions. Brain **71**, 48—76 (1948).

Ratliff, F.: Mach Bands: Quantitative Studies on Neural Networks in the Retina. London: Holden-Day 1965.

Raybourn, M.S.: Spatial and temporal organization of the binocular input to frog optic tectum. Brain Behav. Evol. **11**, 161—178 (1975).

Rayleigh, Lord: Investigations in optics with special reference to the spectroscope. Phil. Mag., 261—274 (1879).

Reza, F.M.: An introduction to information theory. London: McGraw-Hill 1961.

Roberts, S.R.: A system of testing vision in animals. J. Amer. vet. med. Ass. **128**, 544—546 (1956).

Robson, J. G.: Receptive fields. In: Handbook of Perception. Vol. 5. Carterette, E. C., and Friedman, M. P. (eds.). New York: Academic Press 1975, pp. 81—112.
Robson, J. G., Enroth-Cugell, C.: Direct measurement of image quality in the cat eye. J. Physiol. (Lond.) **239**, 30 P (1974).
Robson, J. G., Enroth-Cugell, C.: Scattered light and the retinal image in the cat eye. Proc. Aust. physiol. pharmacol. Soc., **6**, 202 (1975).
Rochon-Duvigneaud, A.: Quelques données sur la fovéa des oiseaux. Annls. Oculist **156** (1919).
Rochon-Duvigneaud, A.: Une méthode de détermination du champ visuel chez le vértébres. Ann. Oculist **159**, 561 (1922).
Rochon-Duvigneaud, A.: Les Yeux et la Vision des Vértébres. Paris: Masson 1943.
Rodieck, R. W., Stone, J.: Response of cat retinal ganglion cells to moving visual patterns. J. Neurophysiol. **28**, 819—831 (1965).
Rohen, J. W.: Der Ziliarkörper als funktionelles System. Morph. Jb. **92**, 415—440 (1952).
Rohen, J. W.: Anatomie des Auges. In: Der Augenarzt, Bd. I, Velhagen, K. (ed.). Leipzig: Thieme 1958, pp. 1—123.
Rohen, J. W.: Das Auge und seine Hilfsorgane, V. 3. In: Handbuch der Mikroskopischen Anatomie des Menschen. Möllendorff, W., Bargmann, W. (eds.). Berlin: Springer-Verlag 1964.
Rohen, J. W., Castenholz, A.: Über die Zentralisation der Retina bei Primaten. Folia primatol. **5**, 92—147 (1967).
Röhler, R.: Untersuchungen der Übertragungseigenschaften des Auges mit Strichgittern. Phys. Verh. Mosbach. **10**, 89 (1959).
Röhler, R.: Die Abbildungseigenschaften der Augenmedien. Vision Res. **2**, 391—492 (1962).
Röhler, R., Hilz, R.: Physical and physiological factors in visual modulation transfer. In: Performance of the Eye at Low Luminances. Bouman, M. A., and Vos, J. J. (eds.) Amsterdam: Excerpta Medica 1965.
Röhler, R., Miller, U., Aberl, M.: Zur Messung der Modulationsübertragungsfunktion des lebenden menschlichen Auges im reflektierten Licht. Vision Res. **9**, 407—428 (1969).
Rolls, E. T., Cowey, A.: Topography of the retina and striate cortex and its relationships to visual acuity in rhesus monkeys and squirrel monkeys. Exp. Brain Res. **10**, 298—310 (1970).
Romanes, G. J.: Animal Intelligence. London: Kegan, Paul, Trench & Co. 1882.
Rose, L., Yinon, U., Belkin, M.: Myopia induced in cats deprived of distance vision during development. Vision Res. **14**, 1029—1032 (1974).
Rowe, M. H., Stone, J.: Properties of ganglion cells in the visual streak of the cat's retina. J. comp. Neurol. **169**, 99—126 (1976).
Ruppert, L.: Ein Vergleich zwischen dem Distinktionsvermögen und der Bewegungsempfindlichkeit der Netzhautperipherie. Z. Sinnesphysiol. **42**, 409—423 (1908).
Rushton, W. A. H.: The difference spectrum and the photosensitivity of rhodopsin in the living human eye. J. Physiol. (Lond.) **134**, 11—29 (1956).
Rushton, W. A. H.: The retinal organisation of vision in vertebrates. In: Biological Receptor Mechanisms. Symp. Soc. exp. Biol. XVI. 1962, pp. 12—31.
Rushton, W. A. H.: The density of chlorolabe in the foveal cones of the protanope. J. Physiol. (Lond.) **168**, 360—373 (1963).
Salomon, A. D.: Visual field factors in the perception of direction. Amer. J. Psychol. **60**, 68—88 (1947).
Samorajski, T., Ordy, J. M., Keefe, J. R.: Structural organization of the retina in the tree shrew *(Tupaia glis)*. J. Cell Biol. **28**, 489—504 (1966).
Sands, P. J.: Inhomogeneous lenses, VI Derivatives of paraxial coefficients. J. opt. Soc. Amer. **63**, 1210—1216 (1973).
Sanderson, K. J.: Visual field projection columns and magnification factors in the lateral geniculate nucleus of the cat. Exp. Br. Res. **13**, 159—177 (1971).
Scheiner, C.: Oculus Hoc Est: Fundamentum Opticum. Innsbruck: Oenoponti 1619.
Schober, H. A. W., Hilz, R.: Contrast sensitivity of the human eye for square wave gratings. J. opt. Soc. Amer. **55**, 1086—1091 (1965).
Schultze, M.: Zur Anatomie und Physiologie der Retina. Arch. mikr. Anat. **2**, 175—286 (1866).
Schusterman, R. J.: In: Behaviour of Marine Animals: Vertebrates. Winn, H. E., and Olla, B. L. (eds.). New York: Plenum Press 1972.
Schusterman, R. J., Barrett, B.: Amphibious nature of visual acuity in the Asian "clawless" otter. Nature (Lond.) **244**, 518—519 (1973).
Schwassmann, H. O.: Visual projection upon the optic tectum in foveate marine teleosts. Vision Res. **8**, 1337—1348 (1968).

Schwassmann, H. O., Kruger, L.: Organization of the visual projection upon the optic tectum of some freshwater fish. J. comp. Neurol. **124**, 113—126 (1965).
Seba, J.: Ophthalmological findings in newborn children, Part I. Cs. Oftal. **30**, 42—47 (1974).
Seneviratne, K. N.: The representation of the subcortical centers of cat and rabbit. Edinburgh: Doctoral Thesis 1963.
Senff, R.: Sehen. In: Handwörterbuch der Physiologie. Wagners, R. (ed.) Bd III. 1846, p. 296.
Sherman, S. M.: Visual field defects in monocularly and binocularly deprived cats. Brain Res. **49**, 25—45 (1973).
Shlaer, R.: An eagles eye; Quality of the retinal image. Science **176**, 920—922 (1972).
Shlaer, S.: The relation between visual acuity and illumination. J. gen. Physiol. **21**, 165—188 (1937).
Shannon, C. E., Weaver, W.: The Mathematical Theory of Communication. University of Illinois Press 1949.
Sidman, R. L.: Histochemical studies on photoreceptor cells. Ann. N.Y. Acad. Sci. **74**, 182—195 (1958).
Simon, J. F., Denieul, P. M.: Influence of the size of test field employed in measurements of modulation transfer function of the eye. J. opt. Soc. Amer. **63**, 894—896 (1973).
Sivak, J. G.: Interrelation of feeding behaviour and accommodation lens movements in some species of North American freshwater fishes. J. Fish Res. Bd. Con. **30**, 1141—1146 (1973).
Sivak, J. G.: The refractive error of the fish eye. Vision Res. **14**, 209—213 (1974).
Sivak, J. G., Allen, D. B.: An evaluation of the "ramp" retina of horse eye. Vision Res. **15**, 1353—1356 (1975).
Sivak, J. G., Howland, H. C.: Accommodation in the northern rock bass in response to natural stimuli. Vision Res. **13**, 2059—2064 (1973).
Skarf, B.: Development of binocular single units in the optic tectum of frogs raised with disparate stimulation to the eyes. Brain Res. **51**, 352—357 (1971).
Slepian, D.: On bandwidth. Proc. I.E.E.E. **64**, 292—300 (1976).
Slonaker, J. R.: A comparative study of the area of acute vision in vertebrates. J. Morph. **13**, 445—500 (1897).
Smith, P.: On accommodation in the rabbit. Ophthal. Rev. **17**, 287—296 (1898).
Snyder, A. W.: Acuity of compound eyes: Physical luneleeleons and design. J. comp. Physiol. **116**, 161—182 (1977).
Snyder, A. W., Miller, W. H.: Bird resolving power; telephoto lens system of the avian eye. Nature (Lond.) in press (1977a).
Snyder, A. W., Miller, W. H.: Photoreceptor diameter and spacing for highest resolving power. J. opt. Soc. Amer. **67**, 696—698 (1977b).
Snyder, A. W., Menzel, R. (eds.): Photoreceptor Optics. Berlin; Springer-Verlag 1975.
Snyder, A. W., Stavenga, D. G., Laughlin, S. B.: Spatial information capacity of compound eyes. J. comp. Physiol. **116**, 183—207 (1977).
Snyder, A. W., Laughlin, S. B., Stavenga, D. G.: Information capacity of eyes. (Vision Res., in press) 1977.
Sokol, S.: Cortical and retinal spectral sensitivity of the hooded rat. Vision Res. **10**, 253—262 (1970).
Sorsby, A., Benjamin, B., Davey, J. B., Sheridan, M., Tanner, J. M.: Emmetropia and its aberrations. MRC Report 293. London: HMSO 1957.
Sorsby, A., Benjamin, B., Sheridan, M., Stowe, J., Leary, G. A.: Refraction and its components during the growth of the eye from the age of three. MRC. Spec. Rep. Series No. 301. London: HMSO. 1961.
Southall, J. P. C.: The optical theory of skiascopy. J. opt. Soc. Amer. **13**, 245—266 (1926).
Southall, J. P. C.: Mirrors, Prisms and Lenses. New York: Dover 1964.
Sparrow, C. M.: Spectroscopic resolving power. Astrophys. J. **44**, 76—86 (1916).
Spatz, W. B.: Die Bedeutung der Augen für die sagittale Gestaltung des Schädels von Tarsius. Folia primatol. **9**, 22—40 (1968).
Steinberg, R. H., Reid, M., Lacy, P. L.: The distribution of rods and cones in the retina of the cat. J. comp. Neurol. **148**, 229—248 (1973).
Stenström, S.: Investigation of the variation and the correlation of the optical elements of human eyes. Amer. J. Optom. Monog. **58** (1948).
Stevenson-Hamilton, J.: Animal Life in Africa. London: Alligator Books 1912.
Stiles, W. S., Crawford, B. H.: The luminous efficiency of rays entering the pupil at different points. Proc. roy. Soc. Lond. **B 112**, 428 (1933).
Stine, G. H.: Variations in refraction of visual and extravisual pupillary zones: skiascopic study. Amer. J. Ophthal. **13**, 101—112 (1930).

Stine, G. H.: Tables for accurate retinal localization. Amer. J. Ophthal. **17**, 314—324 (1934).
Stone, J.: A quantitative analysis of the distribution of ganglion cells in the cat retina. J. comp. Neurol. **12**, 337—352 (1965).
Stone, J.: The naso-temporal division of the cat's retina. J. comp. Neurol. **126**, 585—600 (1966).
Stone, J.: Sampling problems of microelectrodes assessed in the cat's retina. J. Neurophysiol. **36**, 1071—1079 (1973).
Stone, J., Fabian, M.: Specialised receptive fields of the cat's retina. Science **152**, 1277—1279 (1966).
Stone, J., Fukuda, Y.: Properties of cat retinal ganglion cells: A comparison of W-cells with X- and Y-cells. J. Neurophysiol. **37**, 722—748 (1974).
Stone, J., Leicester, J., Sherman, S. M.: The naso-temporal division of the monkey's retina. J. comp. Neurol. **150**, 333—348 (1973).
Stone, J., Rowe, M. H., Campion, J., Holländer, H.: Properties of ganglion cells in the visual streak of the cat's retina. Proc. Aust. physiol. pharmacol. Soc. **6**, 106—107 (1975).
Strömberg, E.: Über Refraktion und Achsenlänge des menschlichen Auges. Berlin: Karger 1936.
Stryker, M., Blakemore, C.: Saccadic and disjunctive eye movements in cats. Vision Res. **12**, 2005—2013 (1972).
Sunderland, H. R., O'Neill, W. D.: Functional dependence of optical parameters on circumferential forces in the cat lens. Vision Res. **16**, 1151—1158 (1976).
Suthers, R. A., Wallis, N. E.: Optics of the eyes of echolocating bats. Vision Res. **10**, 1165—1173 (1970).
Synge, J. L.: Geometrical Optics. Cambridge Tracts in Mathematics & Mathematical Physics. London: Cambridge University Press 1937.
Szalay, F. S.: Where to draw the nonprimate-primate taxonomic boundary. Folia primatol. **23**, 158—163 (1975).
Tamura, T.: A study of visual perception in fish especially on resolving power and accommodation. Bull. Jap. Soc. Sci. Fish. **22**, 536—557 (1957).
Tapp, E. L.: Axon numbers and distribution, myelin thickness and the reconstruction of the compound action potential in the optic nerve of the teleost, *Eugerres plumieri*. J. comp. Neurol. **153**, 267—274 (1973).
Teller, D., Morse, R., Borton, R., Regal, D.: Visual acuity for vertical and diagonal gratings in human infants. Vision Res. **14**, 1433—1439 (1974).
Ter Braak, J. W. G.: Untersuchungen über optokinetischen Nystagmus. Arch. néerl. Physiol. **21**, 309—376 (1936).
Thieulin, G.: Récherches sur le globe oculaire et sur la vision du chien et du chat. Paris: Thèse 1927.
Thompson, Sir D'Arcy: On Growth and Form. Cambridge Univ. Press 1961.
Thompson, J. M., Woolsey, C. N., Talbot, S. A.: Visual areas I and II of cerebral cortex of rabbit. J. Neurophysiol. **13**, 277—287 (1950).
Thorpe, W. H.: Learning and Instinct in Animals. London: Methuen 1963.
Tiao, Y. C., Blakemore, C.: Regional specialisation in the golden hamster's retina. J. comp. Neurol. **168**, 439—458 (1976).
Toates, F. M.: Accommodation function of the human eye. Physiol. Rev. **52**, 828—863 (1972).
Toraldo di Francia, G.: Resolving power and information. J. opt. Soc. Amer. **45**, 497—501 (1955).
Trevarthen, C. B.: Two mechanisms of vision in primates. Psychol. Forsch. **31**, 299—337 (1968).
Tschermak-Seysenegg, A.: Studien über das Binocularsehen der Wirbelthiere. Pflügers Arch. ges. Physiol. **91**, 1—20 (1902).
Uberreiter, O.: Examination of the eye and eye operations in animals. Advanc. vet. Sci. **5**, 2—73 (1959).
Urbantschitsch, E.: Kopfnystagmus. Mschr. Ohrenheilk. **44**, 1—14 (1910).
Vakkur, G. J.: Studies on optics and neurophysiology of vision. M.D. Thesis, University of Sydney 1967.
Vakkur, G. J., Bishop, P. O.: The schematic eye in the cat. Vision Res. **3**, 357—381 (1963).
Vakkur, G. J., Bishop, P. O., Kozak, W.: Visual optics in the cat, including posterior nodal distance and retinal landmarks. Vision Res. **3**, 289—314 (1963).
Valentin, G.: Ein Beitrag zur Kenntniss der Brechungsverhältnisse der Thiergewebe. Pflügers Arch. ges. Physiol. **19**, 78—105 (1879a).
Valentin, G.: Fortgesetzte Untersuchungen über die Brechungsverhältnisse der Thiergewebe. Pflügers Arch. ges. Physiol. **20**, 283—314 (1879b).
Van Alphen, G. W. H. M.: On emmetropia and ametropia. Ophthalmologica Suppl. (Basel) **142** (1961).
Van Buren, J. M.: The Retinal Ganglion Cell Layer. Thomas: Springfield Ill. 1963.

Van den Brink, G.: Measurements of the geometrical aberrations of the eye. Vision Res. **2**, 233—244 (1962).
Vaney, D. I.: A quantitative comparison between the ganglion cell population and the axonal outflow in the visual streak and periphery of the rabbit retina. Submitted (1977).
Vaney, D. I., Hughes, A.: The rabbit optic nerve: Fibre diameter spectrum, fibre count, and comparison with a retinal ganglion cell count. J. comp. Neurol. **170**, 241—251 (1976).
Van Hof, M. W.: Visual acuity in the rabbit. Vision Res. **7**, 749—751 (1967).
Van Hof, M. W., Lagers-Van Haselen, G. C.: The retinal fixation area in the rabbit. Exp. Neurol. **41**, 218—221 (1973).
Van Nes, F. L., Bouman, M. A.: Spatial modulation transfer in the human eye. J. opt. Soc. Amer. **57**, 401—406 (1967).
Van Sluyters, R. C., Stewart, D. L.: Binocular neurons of the rabbit's visual cortex: receptive field characteristics. Exp. Brain Res. **19**, 196—204 (1974b).
Van Sluyters, R. C., Stewart, D. L.: Binocular neurons of the rabbit's visual cortex: effects of monocular sensory deprivation. Exp. Brain Res. **19**, 196—204 (1974b).
Vejbaesya, C.: Studies on the connections of the visual system. Doctoral thesis, Edinburgh 1968.
Vilter, V.: Recherches biométriques sur l'organisation synaptique de la rétine humaine. C.R. Soc. Biol. (Paris) **143**, 830—832 (1949).
Vilter, V.: Asymmetrie cyto-architectonique de la fovea retinienne de l'homme. C.R. Soc. Biol. **148**, 220—223 (1954a).
Vilter, V.: Histologie et activité électrique de la rétine d'un mammifére strictement diurne, le Spermophile *(Citellus citellus)*. C.R. Soc. Biol. (Paris) **148**, 1768—1771 (1954b).
Vincent, S. B.: The Mammalian Eye. J. Anim. Behav. **2**, 249—255 (1912).
Vogt, A.: Die Kurzsichtigkeit. Zürich 1936.
Walk, R. D.: The study of visual depth and distance perception in animals. Advanc. Study Behav. **1**, 99—154 (1965).
Walk, R. D., Gibson, E. J.: A comparative and analytical study of visual depth perception. Psychol. Monogr. **75**, 1—44 (1961).
Walls, G. L.: The visual cells of the white rat. J. comp. Psychol. **18**, 363—366 (1934).
Walls, G. L.: Significance of foveal depression. Arch. Ophthal. **18**, 912—919 (1937).
Walls, G. L.: The vertebrate eye and its adaptive radiation. New York: Hafner 1942.
Ward, B., Davis, J. K.: The modulation transfer function as a performance specification for ophthalmic lens and protective devices. Amer. J. Optom. **49**, 234—259 (1972).
Wässle, H.: Optical qualitiy of the cat eye. Vision Res. **11**, 995—1006 (1971).
Wässle, H., Creutzfeldt, O. D.: Spatial resolution in the visual system: a theoretical and experimental study on single units in the cat's lateral geniculate nucleus. J. Neurophysiol. **36**, 13—27 (1973).
Wässle, H., Levick, W. R., Cleland, B. G.: The distribution of the alpha type of ganglion cells in the cat retina. J. comp. Neurol. **159** 419—438 (1975).
Weale, R. A.: The spectral reflectivity of the cat's tapetum measured in situ. J. Physiol. (Lond.) **119**, 30—42 (1953).
Weale, R. A.: Problems of peripheral vision. Brit. J. Ophthal. **40**, 392—415 (1956).
Weale, R. A.: Presbyopia. Brit. J. Ophthal. **46**, 660—668 (1962).
Weale, R. A.: Why does the human retina possess a fovea? Nature (Lond.) **212**, 255—256 (1966).
Weale, R. A.: Polarised light and the human fundus oculi. J. Physiol. (Lond.) **186**, 175—186 (1966).
Weale, R. A.: Ocular optics and evolution. J. opt. Soc. Amer. **66**, 1053—1054 (1976).
Webb, S. V., Kaas, J. H.: The size and distribution of ganglion cells in the retina of the owl monkey *Aotus Trivirgatus*. Vision res. **16**, 1247—1254 (1976).
Wertheim, M.: Über die indirekte Sehschärfe. Z. Psychol. Physiol. Sinnesorg. **7**, 172—189 (1894).
West, R. W., Dowling, J. E.: Anatomical evidence for cone and rod-like receptors in the grey squirrel, ground squirrel, and prairie dog retinas. J. comp. Neurol. **159**, 439—460 (1975).
Westheimer, G.: Modulation thresholds for sinusoidal light distributions on the retina. J. Physiol. (Lond.) **152**, 67—74 (1960).
Westheimer, G.: Line spread function of the living cat eye. J. opt. Soc. Amer. **52**, 1326 (1962).
Westheimer, G.: Optical and motor factors in the formation of the retinal image. J. opt. Soc. Amer. **53**, 86—93 (1963).
Westheimer, G.: Visual acuity and spatial modulation thresholds. In: Handbook of Sensory Physiology, VII/4, Jameson, D., and Hurvich, L. M. (eds.). Berlin: Springer-Verlag 1972a.

Westheimer, G.: Optical properties of vertebrate eyes. In: Handbook of Sensory Physiology VII/2. Fuortes, M. G. F. (ed.) Berlin: Springer-Verlag 1972b.
Westheimer, G.: Visual acuity and hyperacuity. Invest. Ophth. **14**, 570—572 (1975).
Westheimer, G.: Diffraction theory and visual hyperacuity. Am. J. Optom. Phys. Opt. **53**, 362—364 (1976).
Westheimer, G.: Spatial frequency and light-spread descriptions of visual acuity and hyperacuity. J. opt. Soc. Amer. **67**, 207—212 (1977).
Westheimer, G., Blair, S. M.: Accommodation of the eye during sleep and anaesthesia. Vision Res. **13**, 1035—1040 (1973).
Westheimer, G., Campbell, F. W.: Light distribution in the image formed by the living human eye. J. opt. Soc. Amer. **52**, 1040—1044 (1962).
Weymouth, F. W.: Visual sensory units and the minimal angle of resolution. Amer. J. Ophthal. **46**, 102—113 (1958).
Whitteridge, D.: Geometrical relations between the retina and the visual cortex. In: Mathematics and Computer Science in Biology and Medicine. London: M.R.C. 1965, pp. 269—276.
Whitteridge, D.: Projection of optic pathways to the visual cortex. In: Handbook of Sensory Physiology, VII/3 B, Jung, R. (ed.). Berlin: Springer-Verlag 1973, pp. 247—268.
Wilcox, J. G., Barlow, H. B.: The size and shape of the pupil in lightly anaesthetized cats as a function of luminance. Vision Res. **15**, 1363—1365 (1975).
Wilson, M. E., Toyne, M. J.: Retino-tectal and cortico-tectal projections in *Macaca mulatta*. Brain Res. **24**, 395—406 (1970).
Wilson, P. D., Stone, J.: Evidence of W-cell input to that cat's visual cortex via the C laminae of the lateral geniculate nucleus. Brain Res. **92**, 472—478 (1975).
Winthrop, J. T.: Propagation of structural information in optical wave fields. J. opt. Soc. Amer. **61**, 15—30 (1971).
Winterson, B. J., Robsinson, D. A.: Fixation by the alert but solitary cat. Vision Res. **15**, 1349—1352 (1975).
Wolff, E.: The Anatomy of the Eye and its Orbit. London: Lewis & Co. 1940.
Wolin, L. R., Massopust, L. C.: Characteristics of the ocular fundus in primates. J. Anat. **101**, 693—699 (1967).
Wood, C. A.: The Fundus Oculi in Birds. Chicago 1917.
Wood, R. W.: Physical Optics. New York: Macmillan 1911.
Woodhouse, J. M.: The effect of pupil size on grating detection at various contrast levels. Vision Res. **15**, 645—648 (1975).
Woodhouse, J. M., Campbell, F. W.: The role of the pupil light reflex in aiding adaptation to the dark. Vision Res. **15**, 649—653 (1975).
Wood Jones, F.: Arboreal Man. London: Edward Arnold 1926.
Woolf, D.: A comparative cytological study of the ciliary muscle. Anat. Rec. **124**, 145—163 (1956).
Woollard, H. H.: Notes on the retina and lateral geniculate body in *Tupaia*, *Tarsius*, *Nycticebus* and *Hapale*. Brain, **49**, 77—104 (1926).
Worfold, R. L.: Canine optics. Aust. J. Optom. **48**, 164—174 (1965).
Xenophon. Xenophon's Minor Works: Cynegeticus. (Trans. J.S. Watson). London: G. Bell & Sons 1905.
Yen, J. L.: On nonuniform sampling of band-width limited signals. I.R.E. Trans. on Circuit Theory, **3**, 251—257 (1956).
Young, F. A.: The effect of restricted visual space on the primate eye. Amer. J. Ophthal. **52**, 799—806 (1961).
Young, F. A.: The effect of restricted visual space on the refractive error of the young monkey. Invest. Ophthal. **2**, 571—577 (1963).
Young, F. A., Farrer, D. N.: Refractive characteristics of chimpanzees. Amer. J. Optom. **41**, 81—91 (1964).
Young, F. A., Leary, G. A.: Visual-optical characteristics of caged and semi-free ranging monkeys. Amer. J. Phys. Anthrop. **38**, 377—382 (1973).
Young, T.: On the mechanism of the eye. Phil. Trans. **91**, 23—88 (1801).
Zurn, J.: Vergleichend histologische Untersuchungen über die Retina und die Area centralis retinae der Haussäugethiere. Arch. Anat. Physiol. Anat. Abtheil. **102**, 99—146 (1902).

Author Index

Page numbers in *italics* refer to the bibliography

Subject Index

GPSR Compliance

The European Union's (EU) General Product Safety Regulation (GPSR) is a set of rules that requires consumer products to be safe and our obligations to ensure this.

If you have any concerns about our products, you can contact us on ProductSafety@springernature.com

In case Publisher is established outside the EU, the EU authorized representative is:

Springer Nature Customer Service Center GmbH
Europaplatz 3
69115 Heidelberg, Germany

Batch number: 10761693

Printed by Printforce, the Netherlands